FIRST MULTICOLOUR ILLUSTRATIVE EDITION

BASIC ELECTRONICS
SOLID STATE

FIRST MULTICOLOUR ILLUSTRATIVE EDITION

BASIC ELECTRONICS

SOLID STATE

[A Textbook for the Students of B.E./B. Tech. (Electronics, Communication, Electrical), B.Sc. (Electronics) Diploma in Electronics & Communication]

B.L. THERAJA

S. CHAND
PUBLISHING

S Chand And Company Limited
(ISO 9001 Certified Company)

S Chand And Company Limited
(ISO 9001 Certified Company)
Head Office: D-92, Sector–2, Noida – 201301, U.P. (India), Ph. 91-120-4682700
Registered Office: A-27, 2nd Floor, Mohan Co-operative Industrial Estate, New Delhi – 110 044, Phone: 011-49731800
www.**schandpublishing.com;** e-mail: **info@schandpublishing.com**

S. CHAND PUBLISHING

Marketing Offices:

Chennai	:	Ph: 23632120; chennai@schandpublishing.com
Guwahati	:	Ph: 2738811, 2735640; guwahati@schandpublishing.com
Hyderabad	:	Ph: 40186018; hyderabad@schandpublishing.com
Jalandhar	:	Ph: 4645630; jalandhar@schandpublishing.com
Kolkata	:	Ph: 23357458, 23353914; kolkata@schandpublishing.com
Lucknow	:	Ph: 4003633; lucknow@schandpublishing.com
Mumbai	:	Ph: 25000297; mumbai@schandpublishing.com
Patna	:	Ph: 4011400; patna@schandpublishing.com

First Edition 1985

Subsequent Editions and reprints 1987, 88, 89, 91 (Twice), 93, 95, 96, 97, 98, 99, 2000, 2001, 2002, 2003, 2004, 2005, First Multicolour Edition 2005; Reprints 2007, 2008 (Thrice), 2009 (Twice), 2010 (Twice), 2012, 2014 (Twice), 2015, 2016, 2017, 2018, 2019 (Twice), 2022, 2023 (Twice), 2024, 2025

Reprint 2026

ISBN: 978-81-219-2555-6 **Product Code:** H3BEL53BEEL10ENAE0XO

PRINTED IN INDIA

By Shri Shyamlal Printing Press Private Limited, Plot 20/4, Site-IV, Industrial Area Sahibabad, Ghaziabad – 201 010 and Published by S Chand And Company Limited, A-27, 2nd Floor, Mohan Co-operative Industrial Estate, New Delhi – 110 044.

Preface to the Fifth Edition

The author feels great pleasure in presenting to his esteemed readers a thoroughly revised fifth edition of this popular book **(IN MULTICOLOUR ILLUSTRATIVE EDITION)** which caters to the needs of students studying Electronics and Communication at the diploma and degree levels. The technical level and writing style used are intended to make the text student-oriented so that it is easily readable and readily understandable.

Most of the elementary and redundant material has been eliminated and replaced with fresh material. Similarly, quite a number of elementary Solved Examples have been deleted and supplemented with new examples. A new chapter entitled **Fibre Optics** has been added to this edition for the first time. Its purpose is to introduce **Fibre Optic Technology** which has become very popular in the last few years. In addition to their use in communication, **Optical Fibres** are widely used in medical applications, industry, local area communication and computer communication etc.

I am sure that with this thorough revision and additions, my numerous readers — both here and abroad — will find this edition more useful than the earlier ones.

In the end, the author takes this opportunity to thank his publishers particularly Sh. Ravindra Kumar Gupta CMD for bringing out the book in this **Attractive and Colourful Edition.** The Multicoloured edition of the textbook of Basic Electronics is the outcome of sincere and dedicated efforts of Mr. Shishir Bhatnagar, Mr. Riyaz Baqar, and all the Editorial Staff of S. Chand & Company Ltd.

B.L. THERAJA

Preface to the First Edition

During the last one decade or so, the author had been almost flooded with numerous requests from both teachers and students of **Electronics and Communication** to write a simple but comprehensive textbook on the above two subjects similar to the author's highly-popular text entitled 'Electrical Technology'. It is in deference to their repeated wishes that the present volume **Basic Electronics** is being presented (with all humility) to my esteemed readers both here and abroad. Needless to say, it is based on author's 35 years experience in teaching Electricity, Electronics and related subjects to a variety of engineering students.

Aims of the Book

The foremost and primary aim of the book is to meet the requirements of students pursuing following courses of study:

1. Diploma in Electronics and Communication Engineering (ECE) — 3-year course offered by various Indian and foreign Polytechnics and technical institutes like City and Guilds of London Institute (CGLI).
   It fully covers their syllabuses for the three-year course.
2. B.E. (Electronics & Commn.) — 4-year course offered by various Engineering Colleges. Efforts have been made to cover the papers : **Electronics -I & II** and **Pulse and Digital Circuits.**
3. B.Sc. (Electronics) — 3-year vocationalised course recently introduced by many Indian Universities.

Approach

The material has been presented in a very simple and logical manner using step-by-step development of the subject matter. Clarity and practicality are the keynotes of the text which has been written with shorter sentences, shorter paragraphs and more subheads. Successful efforts have been made to provide very concise and clear explanations of these circuits and devices which most of the students find difficult to comprehend. Examples are : waveshaping circuits, solid-state theory, concepts of transistor saturation and cut-off, various sinusoidal and non-sinusoidal oscillators, flip-flops, Boolean algebra, logic families and digital computers.

As will be found by the readers themselves, all these have been presented in a manner which apart from being easy to understand, is refreshingly original. All along, *author's intention has been to express, not to impress.*

Organisation

The book has been sectionalized into 39 chapters for step-by-step development.

Chapters 1—12 introduce basic electronic fundamentals including electronic components, network theorems and simple ac/dc circuits, tuning circuits and filters.

Chapters 13—18 present broad coverage of solid-state diodes like P-N junction, Zener, tunnel, PIN, Schottky, varactors, optoelectronic devices and regulated dc power supplies.

Chapters 19—31 discuss transistor theory, its use as a circuit element in different configurations, in single-stage and multistage amplifiers, feedback amplifiers, JFETs, MOSFETs and breakdown devices like UJT, SCR, Triac, Diac and SCS, various types of oscillators and multivibrators.

Chapters 32—39 introduce essential concepts of digital electronics, linear and digital ICs including different number systems, logic gates, Boolean algebra, logic families, transducers, electronic instruments and digital computer fundamentals.

Aids to Students

As a help to the students, many novel techniques, aids-to-memory, practical tips and thumb-rules have been provided throughout the text for quick and easy understanding:

1. all important points have been printed in **boldface** so that they may stand out clearly from the rest of the text.
2. a large number of unsolved numerical problems (with answers) have been provided at the end of each chapter. Their correct solution should help students to build up self-confidence.
3. about 500 solved examples have been beautifully, interwoven in the text which not only serve to elucidate the text but also demonstrate the application of theory to practical problems.
4. easy-to-draw and exceptionally lucid diagrams make it easy for the students to grasp the facts at a glance.

Objective Tests

Since such self-testing questions are becoming increasingly popular with the examiners, about 3,000 of them have been provided at the end of all chapters for the students to familiarise themselves with this question form.

It contains 'fill-in-the-blanks', 'true/false' and 'multiple choice' items.

Credits

Amongst the wide panorama of people who provided me help and motivation to write this book, I must thank

—my numerous students without whose inquisitive minds and interchange of ideas, I would not have been able to write this book.

—my publishers particularly Padam Shree L. Shyam Lal Gupta and his multitalented and dynamic progeny Sh. Ravinder Kumar Gupta for the personal interest they took in expeditious printing of the book through their thoroughly matured, meticulous and methodical Printing Manager Sh. H.C. Saxena.

—Mrs. Nirmal Gupta M.A. and Sh. R.C. Kumar MBA for the simple yet highly imaginative cover design.

—Sh. P.S. Bhatti with the fond hope that, as hitherto, he will leave no proverbial stone unturned to popularise the book from Borneo to Botswana during his frequent global jaunts.

—Ms. Vandana Dhingra of Women Polytechnic, Chandigarh for help in collecting text material.

—Ms. Philomina Perepatt : the quick-witted Kerala-born mini-computer who came all the way from that ever-green fairy land of Kerala (5000 km away as per her exaggerated estimate !) to render secretarial services in the preparation of this book.

Request

In the end, I would request my esteemed readers both here and abroad to kindly send me their valuable suggestions for the improvement of the book and to notify me of any errors they might come across while going through it. Both will be gratefully acknowledged.

B.L. THERAJA

Thanks

Author's sincere thanks are due to the following manufacturers for their kind co-operation in supplying the photographs, line drawings, half-tone blocks and technical data about the electronic components and equipment manufactured by them.

1. Advance Components and Instruments Pvt. Ltd., Belavadi Industrial Area, Mysore—571 186.
2. Agronic Instruments Pvt. Ltd., Ghatkopar (W) Bombay—400 086.
3. Al-Ameen Commercial and Industrial Co. Ltd., Crescent Road, Bangalore—560 001.
4. Applied Electronics Ltd., Wagle Industrial Estate, Thane, Bombay—400 604.
5. Arun Electronics Pvt. Ltd., S.V. Road, Bombay—400 072.
6. Bharat Capacitors Ltd., Jeedimetta, Hyderabad—500 855.
7. Bharat Electronics Ltd. (BEL), Bangalore.
8. Calcutta Resistance Co., BB Ganguly Road, Calcutta—700 012.
9. Data Control Engineering Corporation, Goregaon (East), Bombay—400 063.
10. Electromek (India), Roorkee—247 667.
11. Elsonic Santo Corpn., Palmgrove Road, Bangalore—560 047.
12. Hyderabad Batteries Ltd., Banjara Hills, Hyderabad—500 034.
13. Keltron, Trivandram.
14. Kothari Electronics & Industries Ltd., Nungambakkam High Road, Madras—600 034.
15. Meltron Semiconductors Ltd., Nariman Point, Bombay—400 021.
16. Omega Electronics, 36, Hathi Babu Ka Bagh, Jaipur—302 006.
17. Polytronics, Dattawadi, Pune—411 030.
18. Punjab Semiconductor Devices, Chandigarh.
19. Solid State Electronics Co. Pvt. Ltd., Sir M.V. Road Bombay—400 059.
20. The Motwane Mfg. Co Pvt. Ltd., Gyan Baug, Nasik Road—422 101.
21. Uptron Capacitors Ltd., Aishbagh, Lucknow—226 004.
22. Usha Electronics India Pvt. Ltd., Nehru Place, New Delhi —110 019.
23. Universal Semiconductors, Chinchwad, Pune—411 019.
24. Vasavi Electronics, 162, Vasavinagar, Secunderabad—500 003.

CONTENTS

1. CIRCUIT FUNDAMENTALS **1—10**

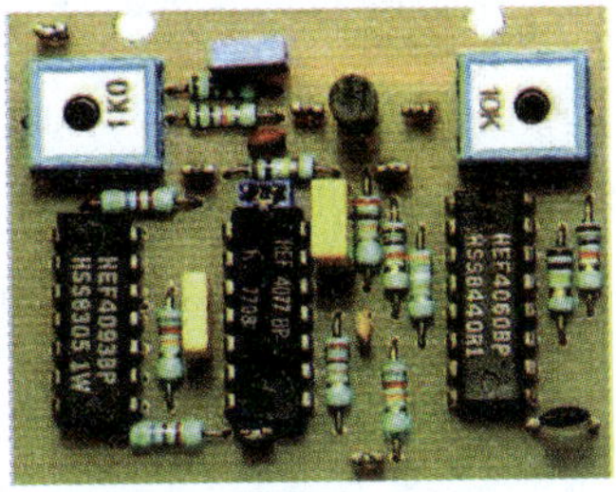

Zero Reference Level — Chassis Ground — Ohm's Law — Formula Variations of Ohm's Law — Graphical Representation of Ohm's Law — Linear Resistor — Non-linear Resistor — Cells in Series and Parallel — Conventional Problems — Self Examination Questions.

2. RESISTIVE CIRCUITS **11—29**

Series Circuit — Characteristics of a Series Circuit — The Case of Zero IR Drop — Polarity of IR Drops — Total Power — Series Aiding and Series Opposing

Voltages — Proportional Voltage Formula in a Series Circuit — Series Voltage Dividers — 'Opens' in a Series Circuit — 'Shorts' in a Series Circuit — Parallel Circuits — Laws of Parallel Circuits — Special Case of Equal Resistances in all Branches — Special Case of Only Two Branches — Any Branch Resistance —Proportional Current Formula — 'Opens' in a Parallel Circuit — 'Shorts' in a Parallel Circuit — Series-Parallel Circuits — Analysing Series Parallel Circuits — 'Opens' in Series-Parallel Circuits — 'Shorts' in Series-Parallel Circuits — Voltage Division in a Complex Series-Parallel Circuit — Conventional Problems — Self Examination Questions.

3. KIRCHHOFF'S LAWS **30—34**

General — Kirchhoff's Current Law — Kirchhoff's Voltage Law — Determination of Algebraic Sign — Assumed Direction of Current Flow — Conventional Problems — Self Examination Questions.

4. NETWORK THEOREMS **35—46**

General — Superposition Theorem — Ideal Constant-Voltage Source — Ideal Constant—current Source — Thevenin's Theorem — How to Thevenize a Circuit? — Norton's Theorem — How to Nortonise a Given Circuit — Maximum Power Transfer Theorem — Conventional Problems — Self Examination Questions.

5. PASSIVE CIRCUIT ELEMENTS **47—75**

General — Resistors — Resistor Types — Wire-wound Resistors — Carbon Composition Resistors — Carbon Film Resistors — Cermet Film Resistors — Metal Film Resistors — Power Rating — Value Tolerance — Variable Resistors — Potentiometers and Rheostats — Fusible Resistors — Resistor Colour Code —Resistance Colour Bands — Resistors under Ten Ohm — Resistor Troubles —

Checking Resistors with an Ohmmeter — Inductor — Comparison of Different Cores — Inductance of an Inductor — Another Definition of Inductance — Mutual Inductance — Coefficient of Coupling — Variable Inductors — Inductors in Series or Parallel without M — Series Combination with M — Stray Inductance — Energy Inductance — Energy Stored in a Magnetic Field — DC Resistance of a Coil — Troubles in Coils — Reactance Offered by a Coil — Impedance Offered by a Coil — Q-Factor of a Coil — Capacitors — Capacitor Connected to a Battery — Capacitance —Factors Controlling Capacitance — Types of Capacitors — Fixed Capacitors — Variable Capacitors — Voltage Rating of Capacitors — Stray Circuit Capacitance — Leakage Resistance — Capacitors in Series — Two Capacitors in Series — Capacitors in Parallel — Two Capacitors in Parallel — Energy stored in a Capacitor — Troubles in Capacitors — Checking Capacitors with Ohmmeter — Charging of a Capacitor — Capacitor Connected Across an AC Source — Capacitive Reactance — Self Examination Questions.

6. ENERGY SOURCES **76—84**

Primary and Secondary Cells — Cell and Battery — Voltage and Current of a Cells — Cell life — Different Types of Dry Cells — Carbon Zinc Cell — Alkaline Cell — Manganese Alkaline Cell — Nicke Cadmium Cell — Mercury Cell — Silver Oxide Cell — Lead Cells — Battery Rating — Testing Dry Cells — Photoelectric Devices — Photovoltaic Cell — Solar Cell — Conventional Problems — Self Examination Questions.

7. MAGNETISM AND ELECTROMAGNETISM **85—97**

Magnetic Materials — Ferrites — Types of Magnets — Demagnetising or Degaussing —Magnetic Shielding — Magnetic Terms and Units — Ohm's Law for Magnetic Circuit — Transformer — Transformer Working — Transformer Impedance — Can a Transformer Operate on DC ? — RF Shielding — Autotransformer — Impedance Matching — Conventional Problems — Self Examination Questions.

8. A.C. FUNDAMENTALS **98—113**

Introduction — Types of Alternating Waveforms — The Basic AC Generator — Some Definitions — Characteristics of a Sine Wave — Audio and Radio Frequencies — Different Values of Sinusoidal Voltage and Current — Phase of an AC — Phase Difference — Vector Representation of an Alternating Quantity — AC Through Pure Resistance Only — AC Through Pure Inductance Only — AC Through Pure Capacitance Only — Non-sinusoidal Waveforms — Harmonics —Conventional Problems — Self Examination Questions.

9. SERIES A.C. CIRCUITS **114—126**

R-L Circuit — Q Factor of a Coil — Skin Effect — R-C Circuit — Coupling Capacitor — R-L-C Circuit — Resonance in an R-L-C Circuit — Resonance Curve — Main Characteristics of Series Resonance — Bandwidth of a Tuned Circuit — Sharpness of Resonance — Tuning — Tuning Ratio — Radio Tuning Dial — Parallel Resonance —Conventional Problems — Self Examination Questions.

10. TIME CONSTANT **127—138**

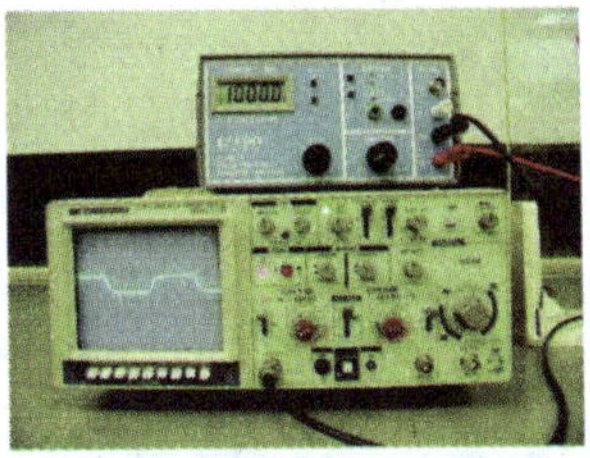

Rise and Fall of Current in Pure Resistance — Time Constant of an R-L Circuit — Circuit Conditions — Inductive Kick — Time Constant of an RC Circuit — Charging and Discharging of a Capacitor — Decreasing Time Constant — Flasher — Pulse Response of an RC Circuit — Effect of Long and Short Time Constants — Square Voltage Wave Applied to Short λ RC Circuit — Square Voltage Wave Applied to Long λ RC Circuit — Conventional Problems — Self Examination Questions.

11. TUNING CIRCUITS AND FILTERS **139—152**

What is a Tuning Circuit ? — Tuned Circuit — Operating Characteristics of a Tuning Circuit — Resonance — Actual Series Resonance — Is it Series or Parallel Resonance ? — Tuned Transformers — Double Tuned Transformers — Parallel Circuit — Coupled Circuits — Simple Coupled Circuits — Coefficient of Coupling — Filters — Filter Definitions — Types of Filter Circuits — Low-pass Filter — Highpass Filter — Bandpass Filter — Bandstop Filter — Multisection Filter Circuits — Uses of Filters — Conventional Problems — Self Examination Questions.

12. SOLID STATE PHYSICS **153—176**

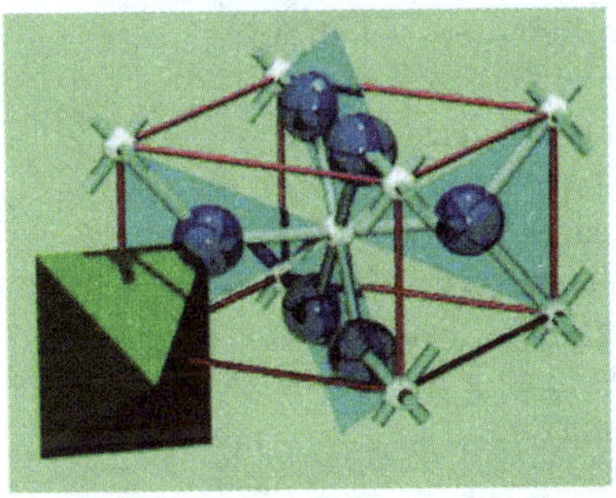

Definition of Matter — Crystalline Solids — Unit Cell — Forms of Matter — Atom and Molecule — Atomic Structure — Atomic Number (Z) — Atomic Mass Number (A) — Electron Orbits or Shells — Electron Distribution of Different Atoms — Electron Suborbits or Subshells — Valence Electrons — Orbital Energy — Normal, Excited and Ionised Atom — Orbital Energies in Hyrogen Atom — Energy Levels in an Isolated Atom — Energy Bands in Solids — Bonds in Solids — Valence and Conduction Bands — Conduction in Solids — Hole Formation and its Movement — Conductors, Semiconductors and Insulators — Types of Semiconductors — Intrinsic Semiconductors — Extrinsic Semiconductors — Majority and Minority Charge Carriers — Mobile Charge Carriers and Immobile Ions — Drift Current in Good Conductors — Drift Current in Intrinsic Semiconductors — Intrinsic Conduction — Conventional Problems — Self Examination Questions.

13. THE P-N JUNCTION **177—187**

The P-N Junction — Formation of Depletion Layer — Junction or Barrier Voltage (V_B) — Effect of Temperature on Barrier Voltage — Forward Biased P-N Junction — Forward V/I Characteristics — Reverse Biased P-N Junction — Reverse Saturation Current (I_s) or I_o) — Reverse V/I Characteristic — Combined Forward and Reverse V/I Characteristics — Junction Breakdown — Junction Capacitance — Equivalent Circuit of a P-N Junction — Self Examination Questions.

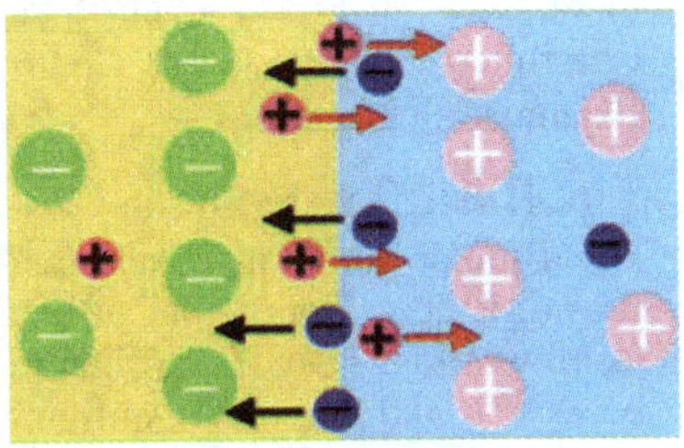

14. P-N JUNCTION DIODE 188—211

P-N Junction Diode — Diode Ratings or Specifications — Diode Testing — The Ideal Diode — The Real Diode — Diode Circuits with DC and AC Voltage Sources — Diode Fabrication — Grown Junction — Alloy Junction — Diffused Junction — Epitaxial Junction — Point Contact Junction — Clippers and Clampers — Clippers — Some Clipping Circuits — Clampers — Summary of Clamping Circuits — Conventional Problems — Self Examination Questions.

15. SPECIAL DIODES 212—231

Zener Diode — Voltage Regulation — Zener Diode as Peak Clipper — Meter Protection — Tunneling Effect — Tunnel Diode — Tunnel Diode Oscillator — Varactor — PIN Diode — Schottky Diode — Step Recovery Diode — Thermistors —Conventional Problems — Self Examination Questions.

16. OPTOELECTRONIC DEVICES 232—241

Introduction — Spectral Response of Human Eye — Light Emitting Diode (LED) — Photoemissive Devices — Photomultiplier Tube — Photovoltaic Devices — Bulk Type Photoconductive Cells — Photodiodes —P-N Junction Photodiode — PIN Photodiode — Avalanche Photodiode — Self Examination Questions.

17. DC POWER SUPPLIES 242—273

Introduction — Unregulated Power Supply — Regulated Power Supply — Steady and Pulsating DC Voltages — Rectifiers — Half-wave Rectifier — Full-wave Rectifier — Full-wave Bridge Rectifier — Filters — Series Inductor Filter — Shunt Capacitor Filter — Effect of Increasing Filter Capacitance —LC Filter — The CLC or Pi Filter — Bleeder Resistor — Voltage Regulation — Zener Diode Shunt Regulator — Transistor Series Voltage Regulator — Controlled Transistor Series Regulator — Transistor Shunt Voltage Regulator — Transistor Current Regulator — Voltage Dividers — Complete Power Supply — Voltage Multipliers — Half-wave Voltage Doubler — Full-wave Voltage Doubler — Voltage Tripler and Quadrupler Circuits — Troubleshooting Power Supplies — Controlled Rectification — Output Waveforms for Different Firing Angles — Output Voltage and Current Values in Controlled Rectifiers — Average Values for FW Controlled Rectifier — Silicon Controlled Rectifier (SCR) — Pulse Control of SCR — 90°Phase Control of SCR — 180° Phase Control of SCR — SCR Controlled Circuit — UJT Controlled Circuit — Conventional Problems — Self Examination Questions.

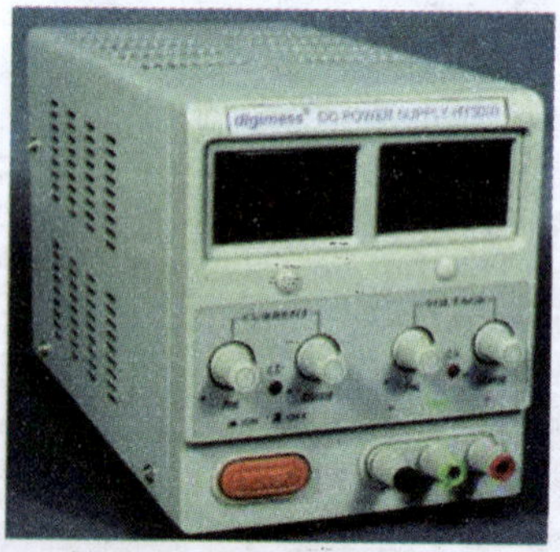

18. THE BASIC TRANSISTOR 274—287

The Bipolar Junction Transistor — Transistor Biasing — Important Biasing Rule — Transistor Currents — Summing Up — Transistor Circuit Configurations — CB Configuration — CE Configuration —Relations between α and β — CC Configuration — Relations between Transistor Currents — Leakage Currents in a Transistor — Thermal Runaway — Conventional Problems — Self Examination Questions.

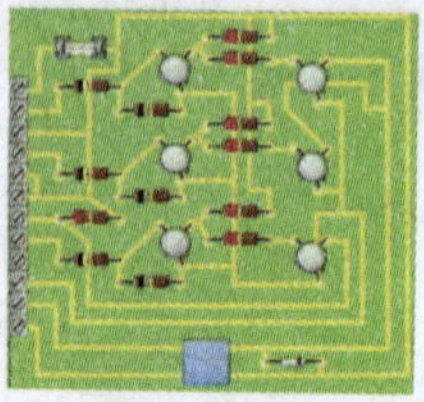

19. Transistor Characteristics and Approximations 288—314

Transistor Static Characteristics — Common Base Test Circuit — Common Base Static Characteristics — Common Emitter Test Circuit — Common Emitter Static Characteristics — Common Collector Static Characteristics — Different Ways of Drawing Transistor Circuits — Common Base Formulas — Common Emitter Formulas — Common Collector Formulas — The Beta Rule — Importance of V_{CE} Cut-off and Saturation Points — Normal DC Voltage Transistor Indications —Transistor Fault Location — Solving Universal Stabilization Circuit — Notation for Voltages and Currents — Increase / Decrease Notation — Applying AC to a DC Biased Transistor — Transistor AC/DC Analysis —Conventional Problems — Self Examination Questions.

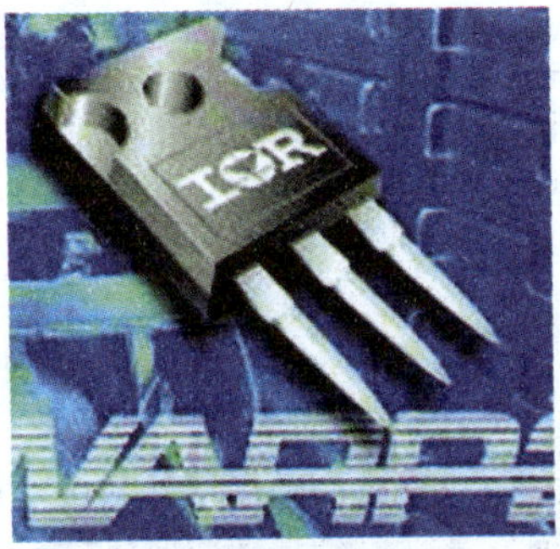

20. Load Lines and DC Bias Circuits 315—333

DC Load Line — Q-point and Maximum Undistorted Output — Need for Biasing a Transistor — Factors Affecting Bias Variations — Stability Factor — Beta Sensitivity — Stability Factor for CB and CE Circuits — Different Methods for Transistor Biasing — Base Bias — Base Bias with Emitter Feedback — Base Bias with Collector Feedback — Base Bias with Collector and Emitter Feedbacks — Voltage Divider Bias — Load Line and Output Characteristics — AC Load Line — Conventional Problems — Self Examination Questions.

21. Transistor Equivalent Circuits and Models 334—369

General — DC Equivalent Circuit — AC Equivalent Circuit — Equivalent Circuit of a CB Amplifier — Effect of Source Resistance RS on Voltage Gain — Equivalent Circuit of a CE Amplifier — Effect of Source Resistance RS — Equivalent Circuit of a CC Amplifier — Small-Signal Low-frequency Model or Representation — General — T-Model — Formulas for T-Equivalent of a CB Circuit — T-Equivalent of a CE Circuit — T-Equivalent of a CC Circuit — What are *h*-parameters ? — The *h*-parameter Formulas for Notation for Transistors — The *h*-parameters of an Ideal Transistor — The *h*-parameters of an Ideal CB Transistor — The *h*-parameters of an Ideal CE Transistor — Approximate Hybrid Equivalent Circuits — Typical Values of Transistor *h*-parameters — Hybrid Formulas for Transistor Amplifier — Approximate Hybrid Formulas —Conventional Problems — Self Examination Questions.

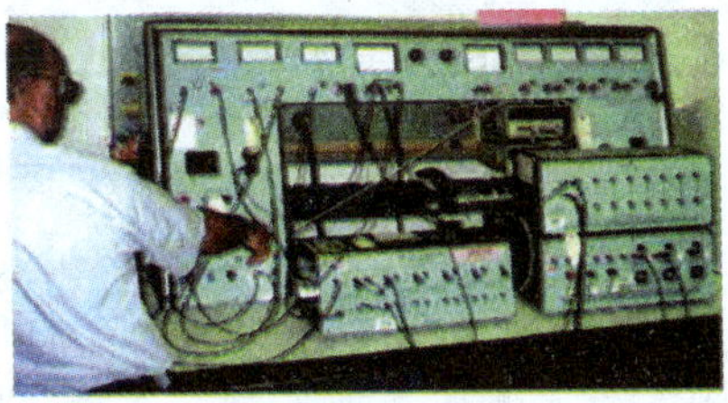

22. Single-Stage Transistor Amplifiers 370—401

Classification of Amplifiers — Common Base (CB) Amplifier — Various Gains of a CB Amplifier — Characteristics of a CB Amplifier — Common Emitter (CE) Amplifier — Various Gains of a CE Amplifier — Characteristics of a CE Amplifier — Common Collector (CC) Amplifier — Various Gains of a CC Amplifier —Characteristics of a CC Amplifier — Uses — Comparison of

Amplifier Configurations — Amplifier Classification Based on Biasing Condition — Graphic Representation — Class-A Amplifiers — Power Distribution in a Class-A Amplifier — Power Rectangle — Power Efficiency — Maximum AC Power in Load — Transformer-coupled Class-A Amplifier — Class-B Amplifier — Power Relations for Class-B Operation — Maximum Values — Class-B Push-Pull Amplifier — Crossover Distortion — Power Efficiency of Push-Pull Amplifiers — Complementary Symmetry Push-Pull Class-B Amplifier — Class-C Amplifier — Tuned Amplifier — Distortion in Amplifier — Non-linear Distortion — Intermodulation Distortion — Frequency Distortion — Phase or Delay Distortion — Noise — Self Examination Questions.

23. MULTISTAGE AMPLIFIERS **402—429**

General — Amplifier Coupling — RC-coupled Two-stage Amplifier — Advantages of RC Coupling — Impedance-coupled Two-stage Amplifier —Advantages of Impedance Coupling — .Transformer-coupled Two Stage Amplifier — Advantages of Transformer Coupling — Frequency Response — Applications — Direct-coupled Two-stage Amplifier Using Similar Transistors — Direct-coupled Amplifier Using Complementary Symmetry of Two Transistors — Darlington Pair — Advantages of Darlington Pair — Comparison between Darlington Pair and Emitter Follower — Special Features of a Differential Amplifier — Common Model Input — Differential Amplifier — Conventional Problems — Self Examination Questions.

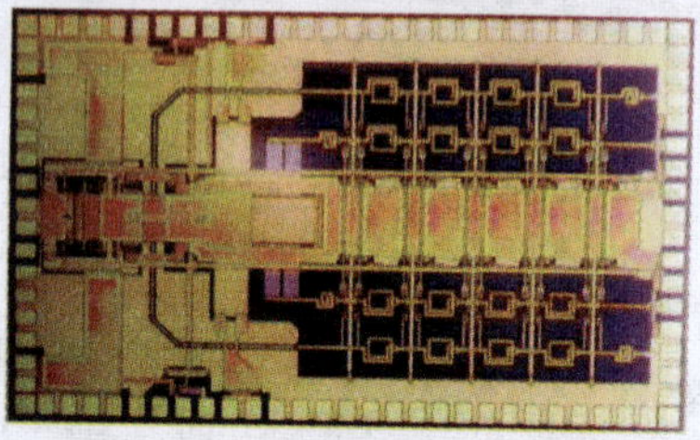

24. DECIBELS AND FREQUENCY RESPONSE **430—441**

The Decibel System — Other Expressions for Power Gain — Voltage and Current Levels — Characteristics of the Decibel System — Value of 1 dB — Zero Decibel Reference Level — Variations in Amplifier Gain with Frequency — Changes in Voltage and Power Levels — Causes of Gain Variations — Miller Effect — Cut-off Frequencies of Cascaded Amplifiers — Transistor Cut-off Frequencies — Alpha Cut-off Frequency — Beta Cut-off Frequency — The f_T of a Transistor — Relation Between f_α, f_β and f_T — Gain-Bandwidth Product — Conventional Problems — Self Examination Questions.

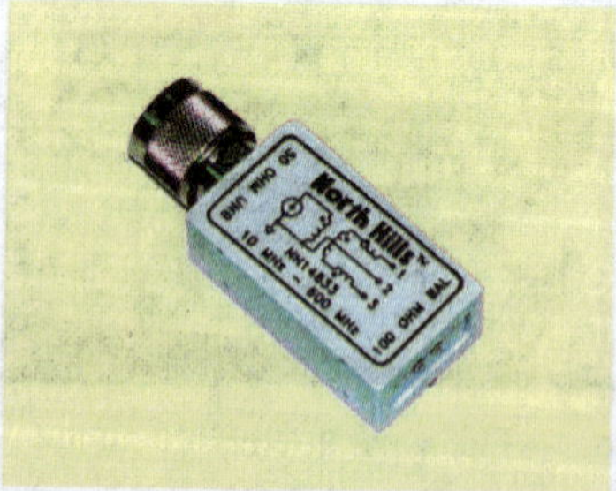

25. FEEDBACK AMPLIFIERS **442—452**

Feedback Amplifiers — Principle of Feedback Amplifiers — Advantages of Negative Feedback — Gain Stability — Decreased Distortion — Increased Bandwidth — Forms of Negative Feedback — Shunt-derived Series-fed Voltage Feedback — Current-series Feedback Amplifier — Voltage-shunt Negative Feedback Amplifier — Current-shunt Negative Feedback Amplifier —Conventional Problems — Self Examination Questions.

26. Field Effect Transistors 453—471

What is a FET ? — Junction FET (JFET) — Static Characteristics of a JFET — JFET Drain Characteristic with $V_{GS} = 0$ — JFET Characteristic with External Bias Transfer Characteristic — Small Signal JFET Parameters — DC Biasing of a JFET — DC Load Line — Common Source JFET Amplifier — JEFT on an IC Chip — Advantages of FETs — MOSFET or IGFET — DE MOSFET — Schematic Symbols for a DE MOSFET — Static Characteristics of a DE MOSFET — Enhancement-only N-channel MOSFET — Transfer Characteristic — FETs as Switches — FET Applications — MOSFET Handling — Self Examination Questions.

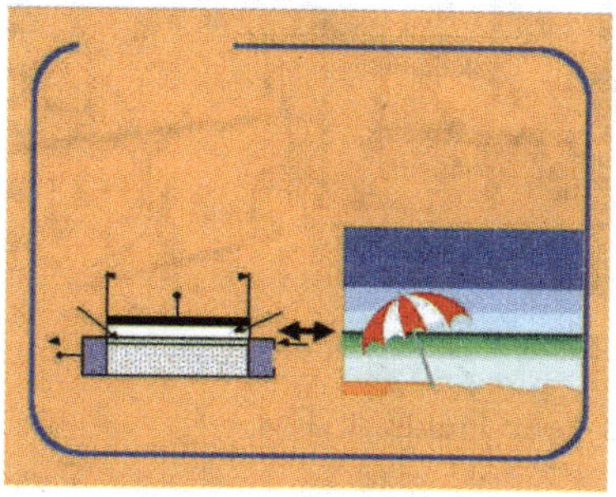

27. Breakdown Devices 472—484

What are Breakdown Devices ? — Unijunction Transistor — UJT Relaxation Oscillator — Silicon Controlled Rectifier — 90° Phase Control — Theft Alarm — Triac —Diac — Silicon Controlled Switch (SCS) — Self Examination Questions.

28. Sinusoidal Oscillators 485—503

What is an Oscillator ? — Comparison between an Amplifier and an Oscillator — Classification of Oscillators — Damped and Undamped Oscillations — The Oscillatory Circuit — Frequency of Oscillatory Current — Frequency Stability of an Oscillator — Essentials of a Feedback LC Oscillator — Tuned Base Oscillator — Tuned Collector Oscillator — Tuned Drain Oscillator (FET) — Hartley Oscillator — FET Hartley Oscillator — Colpitts Oscillator — Clapp Oscillator — FET Colpitts Oscillator — Crystals — Crystal Controlled Oscillator — Transistor Pierce Crystal Oscillator — FET Pierce Oscillator — Phase Shift Principle — Phase Shift Oscillator — Wien Bridge Oscillator — Self Examination Questions.

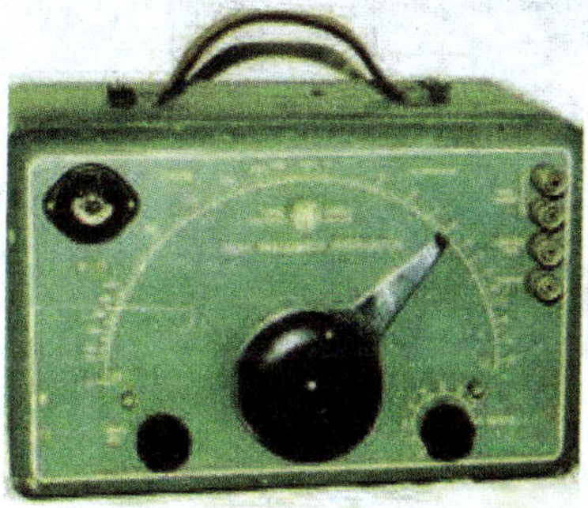

29. Nonsinusoidal Oscillators 504—518

Nonsinusoidal Waveforms —Classification of Nonsinusoidal Oscillators — Pulse Definitions — Basic Requirements of a Sawtooth Generator —UJT Sawtooth Generator — Multivibrators (MV) — Uses of Multivibrators — Astable Multivibrator — Monostable Multivibrator (MMV) — Bistable Multivibrator (BMV) — Schmitt Trigger—Transistor Blocking Oscillator—Self Examination Questions.

30 Modulation and Demodulation 519—548

Introduction — What is a Carrier Wave? — Radio Frequency Spectrum — Sound — Need for Modulation — Radio Broadcasting — Modulation — Methods of Modulation — Amplitude Modulation — Per cent Modulation — Upper and Lower Side Frequencies — Upper and Lower Sidebands — Mathematical Analysis of a Modulated Carrier Wave — Power Relations in an AM Wave — Forms of Amplitude

Modulation — Generation of SSB — Methods of Amplitude Modulation
— Block Diagram of an AM Transmitter — Modulating Amplifier Circuit — Frequency Modulation — Frequency Deviation and Carrier Swing — Modulation Index —Deviation Ratio — Per cent Modulation — FM Sidebands — Modulation Index and Number of Sidebands — Mathematical Expression for FM Wave — Demodulation or Detection — Essentials of AM Detection — Diode Detector for AM Signals — Transistor Detectors for AM Signals — FM Detection — Quadrature Detector — Frequency Conversion — Superheterodyne AM Receiver — FM Receiver — Comparison between AM and FM — The Four Fields of FM — Conventional Problems — Self Examination Questions.

31. INTEGRATED CIRCUITS 549—571

Introduction — What is an Integrated Circuit ? — Advantages of ICs — Drawbacks of ICs — Scale of Integration — Classification of ICs by Structure — Comparison between Different ICs — Classification of ICs by Function — Linear Integrated Circuits (LICs) — Digital Integrated Circuits — IC Terminology — How Monolithic ICs are Made ? — IC Symbols — Fabrication of IC Components — Complete Monolithic Integrated Circuits — Popular Applications of ICs — MOS Integrated Circuits — What is an OP-AMP ? — OP-AMP Symbol — Polarity Conventions — Ideal Operational Amplifier — Virtual Ground and Summing Point — Why V_i is Reduced to almost Zero ? — OP-AMP Applications — Linear Amplifier — Unity Follower — Adder or Summer — Subtractor — Integrator — Differentiator — Comparator — Self Examination Questions.

32. NUMBER SYSTEMS 572—593

Number of Systems —The Decimal Number System — Binary System — Binary to Decimal Conversion — Binary Fractions — Double-Dadd Method — Decimal to Binary Conversion — Shifting the Place Point — Binary Operations — Binary Addition — Binary Subtraction — Complement of a Number — 1's Complemental Subtraction — 2's Complemental Subtraction — Binary Multiplication — Binary Division — Shifting a Number to Left or Right — Representation of Binary Numbers as Electrical Signals — Octal Number System — Octal to Decimal Conversion — Decimal to Octal Conversion — Binary to Octal Conversion —Octal to Binary Conversion — Advantages of Octal Number System — Hexadecimal Number System — How to Count beyond F in Hex Number System? — Binary to Hexadecimal Conversion — Hexadecimal to Binary Conversion — Conventional Problems — Self Examination Questions.

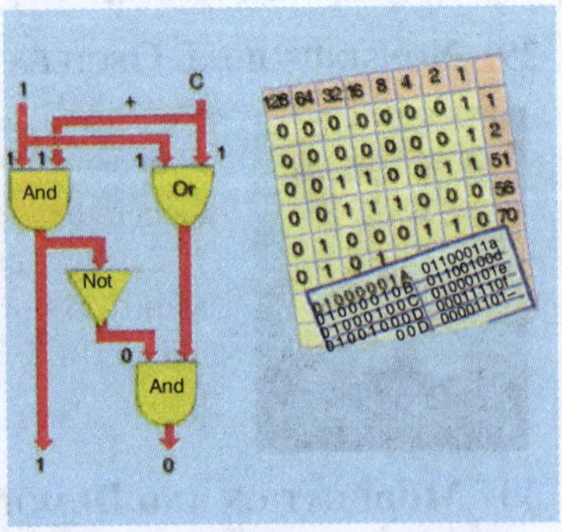

33. LOGIC GATES **528—547**

Definition — Positive and Negative Logic — The OR Gate — Equivalent Relay Circuit of an OR Gate — Diode OR Gate — Transistor OR Gate — OR Gate Symbolizes Logic Addition — Three Input OR Gate — Exclusive OR Gate — The AND Gate — Equivalent Relay Circuit of an AND Gate. — Diode AND Gate — Transistor AND Circuit — AND Gate Symbolizes Logic Multiplication — The NOT Gate — Equivalent Circuits for a NOT Gate — The— OT Operation — Bubbled Gates — The NOR Gate — NOR Gate is a Universal Gate — The NAND Gate —NAND Gate is a Universal Gate — The XNOR Gate — Logic Gates at a Glance — Adders and Subtractors — Half Adder — Full Adder — Parallel
Binary Adder — Half Subtractor — Full Subtractor —Conventional Problems — Self Examination Questions.

34. BOOLEAN ALGEBRA **616—624**

Introduction — Unique Feature of Boolean Algebra — Laws of Boolean Algebra — Equivalent Switching Circuits — De Morgan's Theorems — Duals — Conventional Problems — Self Examination Questions.

35. LOGIC FAMILIES **625—634**

Main Logic Families — Saturated and Non-saturated Logic Circuits — Characteristics of Logic Families — RTL Circuit — DTL Circuit — TTL Circuits — TTL Subfamilies —ECL Circuit Circuit — I^2 MOS Family — PMOS Circuit — NMOS Circuit — CMOS Circuit — Self Examination Questions.

36. TRANSDUCERS **635—653**

What is a Transducer ? — Classification of Transducers — Classification based on Electrical Principle Involved — Resistive Position Transducer — Resistive Pressure Transducer — Inductive Pressure Transducer — Capacitive Pressure Transducer — Self-generating Inductive Transducers — Linear Variable Differential Transformer (LVDT) — Piezoelectric Transducer — Strain Gauge — Temperature Transducers — Resistance Temperature Detectors — Thermistor — Thermocouples — Ultrasonic Temperature Transducers — Photoelectric Transducers — Various Types of Microphones — Carbon Microphone — Ribbon Microphone — Moving-Coil (MC) Microphone — Crystal Microphone — Ceramic Microphone — Capacitor Microphone — The Electret Microphone — The Loudspeaker — Self Examination Questions.

37. Electronic Instruments **654—684**

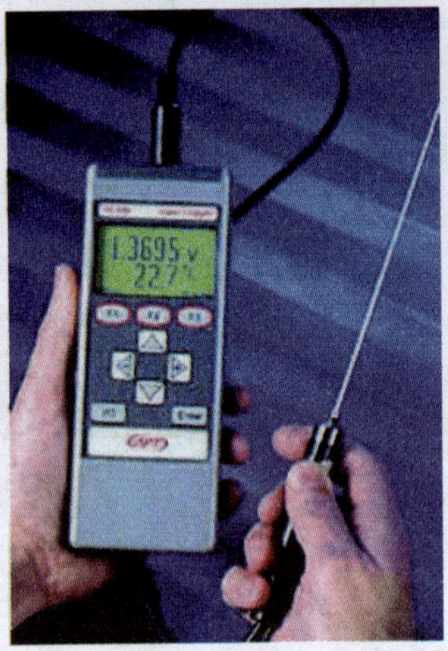

Introduction — Analog and Digital Instruments — Functions of Instruments — Electronic versus Electrical Instruments — Essentials of an Electronic Instrument — Measurement Standards — The Basic Meter Movement — Characteristics of Moving Coil Meter Movement — Variations of Basic Meter Movement — Converting Basic Meter to DC Ammeter — Multirange Meter — Measurement of Current — Converting Basic Meter to DC Voltmeter — Multirange DC Voltmeter —Loading Effect of a Voltmeter — Ohmmeter — The Multimeter — Rectifier Type AC Meter — Electronic Voltmeters — The Direct Current VTVM — Comparison of VOM and VTVM — Direct Current FET VM — Electronic Voltmeter for Alternating Currents — The Digital Voltmeter (DVM) —Cathode Ray Oscilloscope (CRO) — Cathode Ray Tube (CRT) — Deflection Sensitivity of a CRT — Normal Operation of a CRO — Triggered and Non-triggered Scopes — Dual Trace CRO — Dual Beam CRO — Storage Oscilloscope — Sampling CRO — Digital Readout CRO — Lissajous Figures — Frequency Determination with Lissajous Figures — Applications of a CRO — The Q Meter — Self Examination Questions.

38. Fibre Optics **685—703**

Fibre Optics — Structure of Optical Fibres — Classification of Optical Fibres — Plastic Fibres — Propagation of Light — Refraction and Snell's Law — Total Internal Reflection — Light Propagation through an Optical Fibre — Acceptance Angle and Numerical Aperture — Dispersion — Intermodal Dispersion — Intramodal Dispersion — Fibre Characteristics — Fibre Losses — Calculation of Losses — Choice of Wavelength — Optical Fibre Cable — Multifibre Cable — Splicing and Connectors — Splicing — Fusion Splices — Mechanical Splices — Connectors — Connection Losses — Fabrication of Optical Fibres — Fibre Optic Communications — Advantages of Optic Fibres — Disadvantages — Application of Fibre Optic Communication — Self Examination Questions.

Index **705 —715**

CHAPTER 1

Circuit Fundamentals

1. Zero Reference Level
2. Chassis Ground
3. Ohm's Law
4. Formula Variations of Ohm's Law
5. Graphical Representation of Ohm's Law
6. Linear Resistor
7. Non-linear Resistor
8. Work and Power
9. Cells in Series and Parallel

1.1. Zero Reference Level

In order to avoid errors in the measurement of various voltages in an electronic circuit, it is essential to select some common point which is *considered* to be at zero potential. All circuit voltages whether positive or negative, are measured with respect to this point. It can be any point in the circuit and need not be necessarily at 0 V. In Fig. 1.1 (*a*), negative battery terminal *i.e.*, point *D* is taken as reference point and voltages of other points in the circuit are stated with reference to this point.

For example, voltage of point *A* with respect to negative battery terminal (zero reference level) is +12 V, that of point *B* is + 6 V and that of point *C* is 0 V since it is directly connected to point *D*. In fact, negative battery terminal can be grounded as shown by 3 short lines in Fig. 1.1 (*b*).

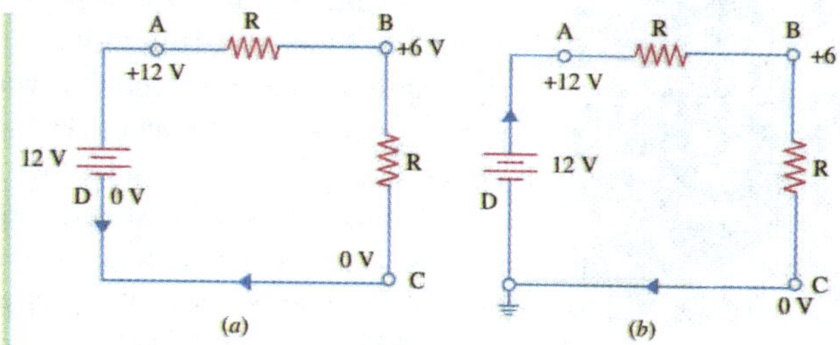

Fig. 1.1

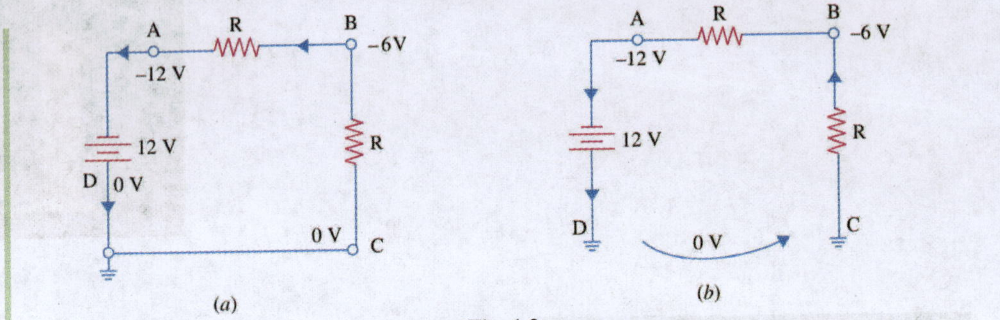

Fig. 1.2

It would make no difference to voltage measurements even if positive battery terminal is grounded as shown in Fig. 1.2 (*a*). The only difference is that now all circuit voltages would be negative with respect to the ground as shown in the figure. But their magnitude (which is our primary concern) remains unchanged. Of course, direction of current flow is reversed. Fig. 1.2 (*b*) shows another way of showing ground in circuit schematic diagrams.

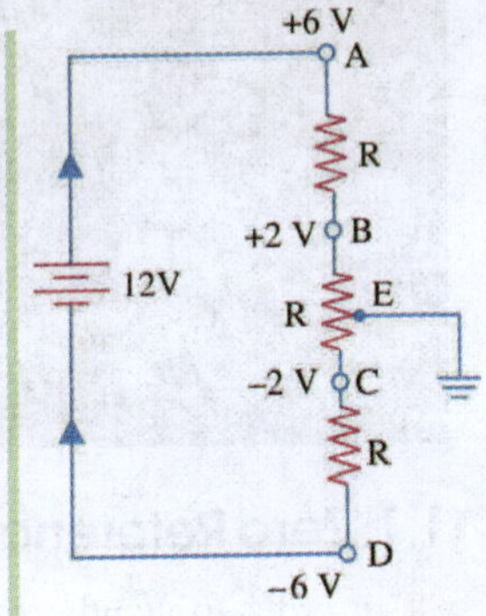

Fig. 1.3

In Fig. 1.3, three equal resistors are connected in series across a 12 V battery. The centre point *E* of the middle resistor has been grounded *i.e.*, fixed as zero reference level for voltage measurements. It is seen that point *B* is 2 V *above* point *E i.e.*, it is positive with respect to *E* whereas point *C* is 2 V *below* point *E i.e.*, it is negative with respect to *E*. Similarly, point *A* is 6 V *above E* and point *D* is 6 V *below E*. Another point worth noting is that conventional current always flows from a point at higher potential to one at lower potential. Since *E* is at a higher potential as compared to point *D*, current flows from *E* to *D*. For similar reasons, current flow is from *A* to *E*.

1.2. Chassis Ground

Generally, electronic components are mounted either on a conducting metal sheet called *chassis* or on a non-conducting plastic board with printed wiring. When using chassis, it is common practice to treat the body of the chassis itself as the common ground. Being a good conductor, it provides a return path for different currents as shown in Fig. 1.4 (*a*). All circuit voltages are measured with respect to chassis ground which is supposed to be at zero potential. It would be appreciated that selection of chassis as a common ground greatly simplifies the circuit wiring. In drawing simple circuit diagrams, it is common practice to show various electronic components connected to chassis ground with the help of ground symbols depicted in Fig. 1.4 (*b*).

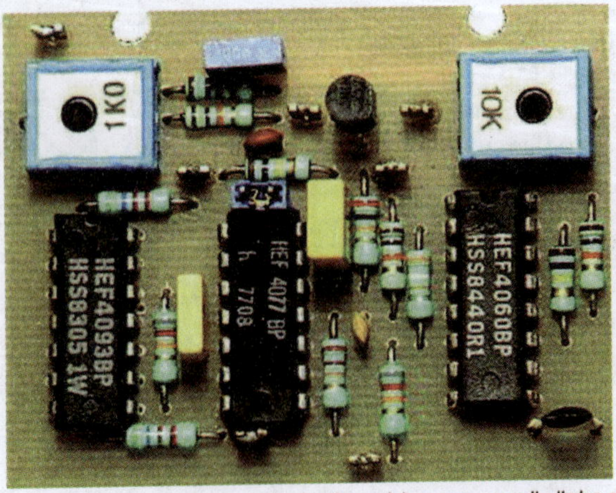

Electronic components including chips are usually linked together by fixing them to a printed circuit board, sometimes called a card.

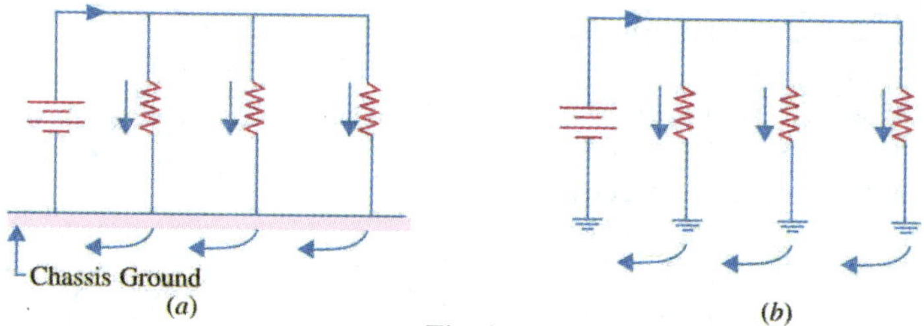

Fig. 1.4

In the case of printed-circuit boards (PCB) made of plastic, a rim of solder around the edge serves as the chassis ground as shown in Fig. 1.5. When a voltage source is connected between points A and B, its one end is grounded through B and the other is connected to the resistors R_1 and R_2.

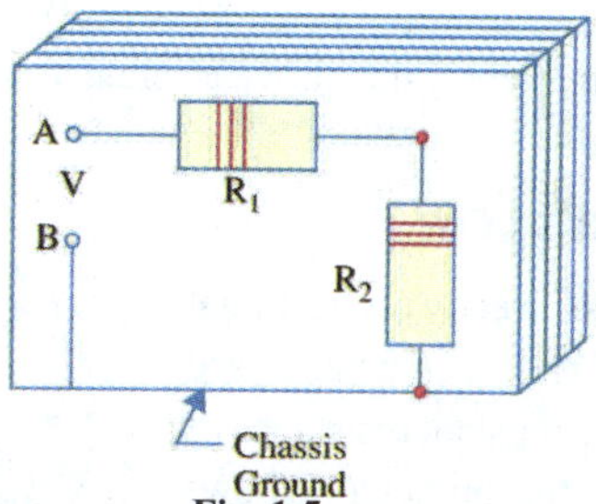

Fig. 1.5

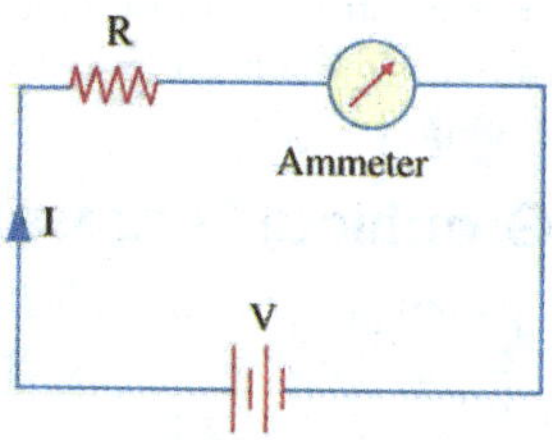

Fig. 1.6

1.3. Ohm's Law

In circuits energised by dc voltage sources, there exists a definite relationship between the current (I) that flows through a resistance (R) and the voltage (V) applied across the resistance (Fig. 1.6). This relationship is called Ohm's law and may be expressed by the equation.

German physicist
George Ohm
(1789-1854)

$$I = \frac{V}{R}$$

where I = current in amperes

V = applied voltage in volts

R = resistance in ohms

It is seen from the above formula that current is

1. directly proportional to applied voltage and
2. inversely proportional to resistance.

1.4. Formula Variations of Ohm's Law

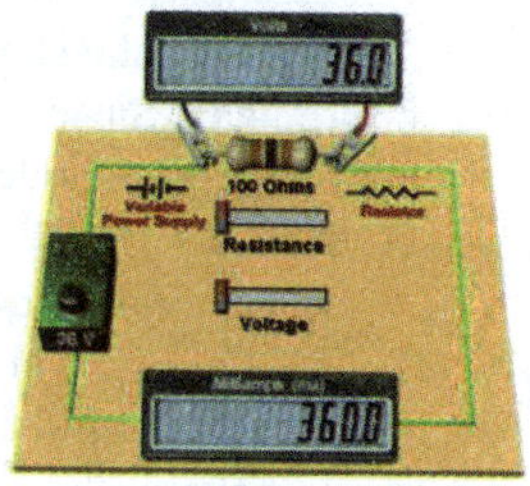

Ohm's law.

The three formula variations of Ohm's law are as under :

1. $I = V/R$ — for finding current
2. $R = V/I$ — for finding resistance
3. $V = I/R$ — for finding voltage

These formulae are an important aid to the understanding of circuit behaviour. They enable us to determine the value of any of the three quantities involved if the values of the other two are known. It makes it possible to design electronic circuits and to determine the values of various components mathematically thereby avoiding unnecessary measurements or experimentation and wastage of time.

The above formulae may be memorised by using the circle divided as in Fig. 1.7. To use the circle, all you need to do is to cover the factor you want and read the remaining formula. Remember that the above three quantities are given in the units of volt, ampere and ohm.

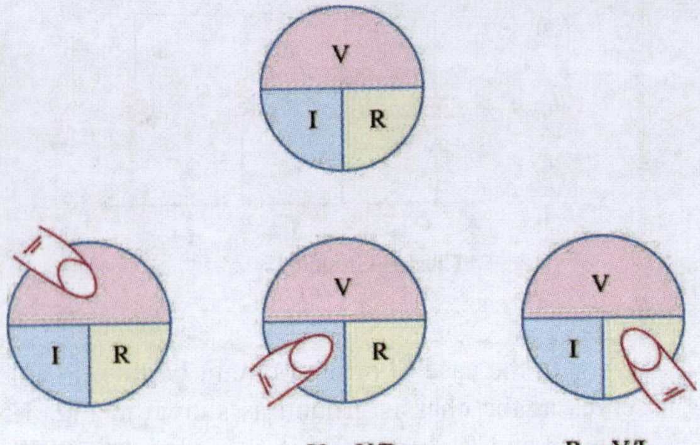

V = IR   V = V/R   R = V/I

Fig. 1.7

For electronic circuit calculations, the resistances are often in kilohms (kΩ) and currents in milliamperes (though voltage is usually in volts). In that case, the above circle can be modified as shown in Fig. 1.8.

1.5. Graphical Representation of Ohm's Law

The Ohm's formula $I = V/R$ states that V and I are directly proportional for any given value of R. This relationship between V and I can be analysed with the help of the circuit shown in Fig. 1.9 (*a*) where a constant resistance of 2 Ω has been taken. When V is varied from 0 V to 12 V, the ammeter A shows I values directly proportional to V as tabulated in Fig. 1.9 (*b*). For example, with 12 V, I equals 6 A, for 10 V, $I = 5$ A and so on.

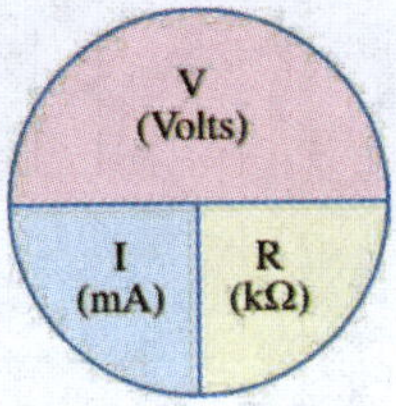

Fig. 1.8

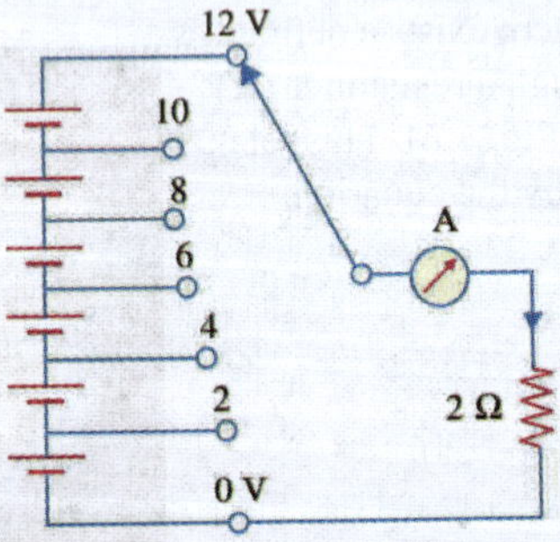

(*a*)

VOLTS	OHMS	CURRENT
0	2	0
2	2	1
4	2	2
6	2	3
8	2	4
10	2	5
12	2	6

(*b*)

Fig. 1.9

The above relationship can be shown graphically. The voltage values are marked along the X-axis (also called horizontal axis or abscissa) and current values along Y-axis (also called vertical axis or ordinate). Since values of V and I depend on each other, they are called variable factors. Between the two, V is the independent variable because we assign different values to it and note the resulting current I.

Generally, independent variable is plotted along the X-axis and dependent variable along Y-axis. The two scales need not be the same.

The graph shown in Fig. 1.10 is also known as volt-ampere (V-I) characteristic of R because it shows how much current the resistor allows for different voltages.

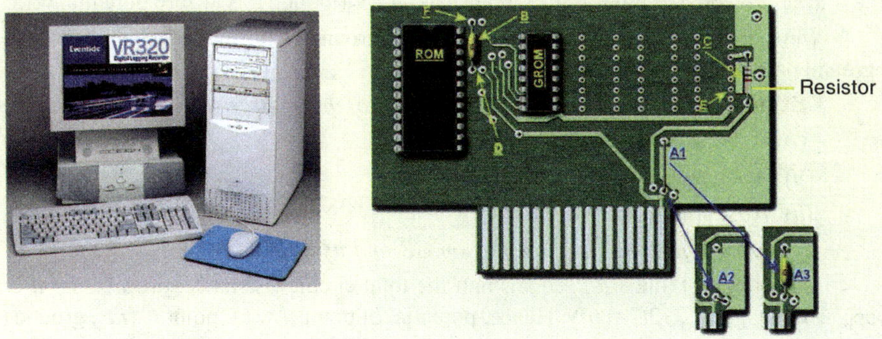

Resistors are found on circuit boards of computers.

1.6. Linear Resistor

A linear resistor is one whose value remains constant *i.e.*, it does not depend on applied voltage. The *V-I* characteristic of such a resistor is a straight line similar to the one shown in Fig. 1.10. Obviously, *V* and *I* are directly proportional.

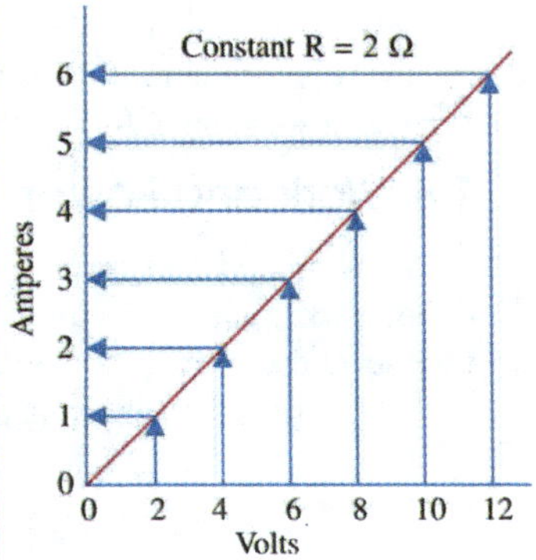

Fig. 1.10

1.7. Non-linear Resistor

It is that resistor in which *V* and *I* are not directly proportional to each other. If applied voltage is doubled, the resultant current is not exactly double of its previous value. Such a resistor has non-linear *V-I* characteristic. An example is tungsten filament in an electric bulb. Here, *R* increases with more current as the filament becomes hotter. Increasing the applied voltage does produce more current but it does not increase in the same proportion as *V*.

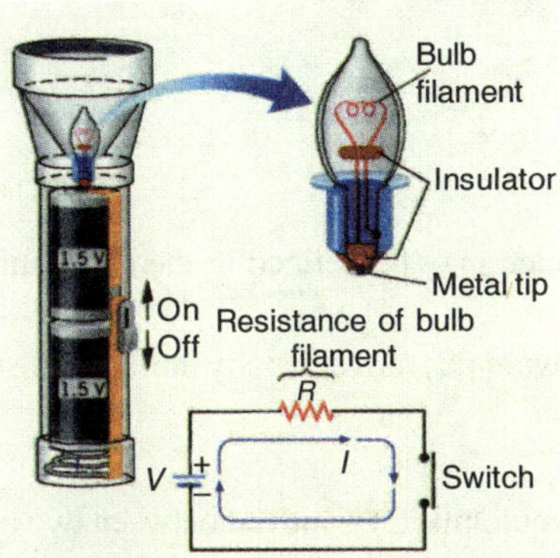

The circuit in this flashlight consists of a resistor (the filament of the light bulb) connected to a 3 V battery.

Example 1.1. *In the circuit of Fig. 1.11 (a), find*

(*i*) *circuit current, I*

(*ii*) *voltage of point A with respect to ground*

(*iii*) *voltage of point B with respect to A*

(*iv*) *voltage of point B with respect to ground.*

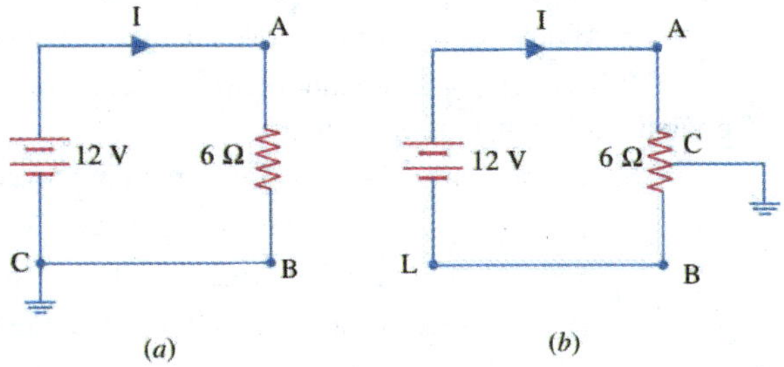

Fig. 1.11

Solution. (*i*) As per Ohm's law,

$$I = V/R = 12/6 = 2\text{ A}$$

(*ii*) Voltage drop across 6 Ω resistor

$$= IR = 2 \times 6 = 12\text{ V}$$

Since lower end of the resistor is grounded via point *B*, potential of point *A* w.r.t. ground is **+12 V.**

(*iii*) Potential of point *B* w.r.t. point *A* is –12 V (though it is at zero potential w.r.t. ground).

(*iv*) Since it is electrically connected to the ground by means of the conductor *BC*, point *B* is at ground potential *i.e.*, it is at **0 V.**

Example 1.2. *In Fig. 1.11 (b), mid-point C of the 6 Ω resistor has been grounded. Calculate*

(*i*) *potential of point A w.r.t. ground*

(*ii*) *potential of point B w.r.t. ground*

(*iii*) *p.d. between A and B*

(*iv*) *current flowing through portion CB and its direction.*

Solution. (*i*) Since R_{AC} equals half the total circuit resistance, drop across it is also half the applied voltage *i.e.*, 12/2 = 6 V. Hence, potential of point *A* w.r.t. point *C* (*i.e.*, ground) is **+6 V**

(*ii*) For similar reasons, potential of point *B* w.r.t. point *C* (*i.e.*, ground) is **–6V**

(*iii*) P.D. between *A* and *B* = 6 V – (– 6 V) = **+12 V**

(*iv*) Since, $R_{CB} = 3\Omega$ and $V_{CB} = 6$ V, $I_{CB} = 6/3 =$ **2 A**

This current must flow from higher to lower potential. Since point *C* is at 0 V and *B* is at – 6 V, current flows along *CB*.

1.8. Work and Power

Work and energy are the same thing but power is different because it is defined as the rate of doing work. Suppose, a battery of *V* volts drives a current of *I* amperes through a resistance of *R* ohms for *t* seconds. Then, total work done by the battery to maintain this current is

$$\begin{aligned} \text{W.D.} &= VIt \text{ joules} \\ &= I^2Rt \text{ joules} && \text{— eliminating } V \\ &= V^2 t/R \text{ joules} && \text{— eliminating } I \\ &= Wt \text{ joules} && \text{— putting } W = VI \end{aligned}$$

Unit of Work

1. The commonly-used unit of work is joule (*J*) which may be defined in the following two ways :

(*a*) It is equal to the work done when a force of 1 newton (N) moves a body through a distance of 1 metre (m) in its direction of application.

or

(*b*) It is equal to the work done when a charge of 1 coulomb (C) is moved between two points having a potential difference of 1 V.

$$\begin{aligned} 1 \text{ joule} &= 1 \text{ metre-newton} \\ &= 1 \text{ volt-coulomb} \end{aligned}$$

2. Another unit often employed in Semiconductor Physics is electron-volt (eV).

It is equal to the amount of work needed to move an electron between two points having a potential difference of one volt.

Since, there are 6.24×10^{18} electrons in one coulomb

$$\therefore \quad 1 \text{ J} = 6.24 \times 10^{10} \text{ eV}$$

$$1 \text{ eV} = 1/6.24 \times 10^{18} = 1.6 \times 10\times^{19} \text{ J}$$

Power. The electric power required to maintain this current is

$$P = \frac{\text{W.D.}}{t} \text{ watts} = \frac{VI\,t}{t} = VI \text{ watts}$$

$$= \frac{V^2 t}{Rt} = \frac{V^2}{R} \text{ watts} = \frac{I^2 \, Rt}{t} = I^2 R \text{ watts}$$

The bigger units are :

1 kilowatt (kW) = 1000 W = 10^3 W

1 megawatt (MW) = 1,000,000 W = 10^6 W

Example 1.3. *A 100 Ω resistor is required to be used in a circuit carrying a current of 0.15 A. What should be the power rating of the resistor?*

Solution. $P = I^2R = 0.15^2 \times 100 = 2.25$ W

In order to prevent overheating of such a resistor, its power or wattage rating should be nearly twice of that calculated above. Hence, a resistor of **5 W** power rating would be most suitable.

1.9. Cells in Series and Parallel

Electric cells may be connected either in series or in parallel to form *batteries*. Each of these combinations has a different value of the total voltage and current-delivering capacity.

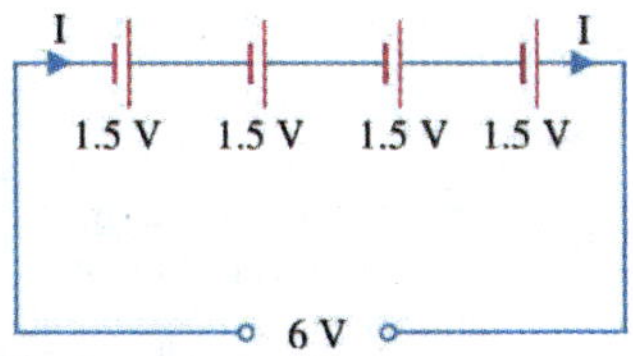

Fig. 1.12

1. Series Connection

In Fig. 1.12, four dry cells each of 1.5 V have been connected in series *i.e.,* from end-to-end.

The total voltage is 4 times the voltage of a single cell *i.e.,* 4 × 1.5 = 6 V. However, the current-delivering capacity of the series combination *does not exceed that of the single cell.* In case, cells of different emf's are connected in series, the current delivering capacity of such a combination is equal to that of the single cell which has the lowest current-delivering capacity.

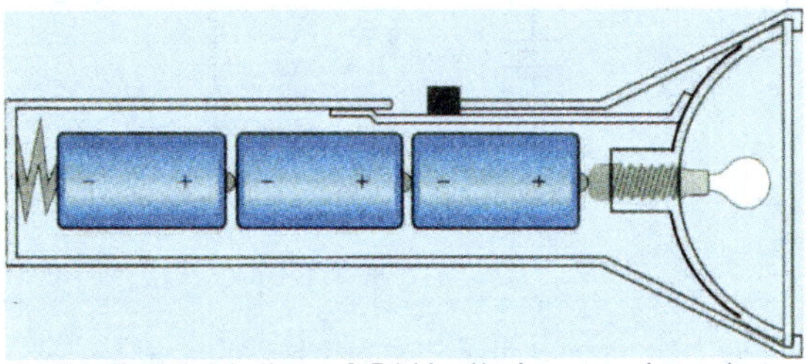

In this flashlight, three 1.5 V batteries are placed in series to produce a larger voltage.

Hence, series combination of cells is employed when higher *voltages* (non-currents) are required.

2. Parallel Connection

Such a combination is used when the purpose is to obtain more current than is available from a single unit. As shown in Fig. 1.13, total voltage available across output terminals *A* and *B* is equal to the voltage of a single cell. However, output current *I* is equal to the sum of four cell currents *i.e.,* $I = i_1 + i_2 + i_3 + i_4$. Normally, only those cells having identical emf's are connected in parallel.

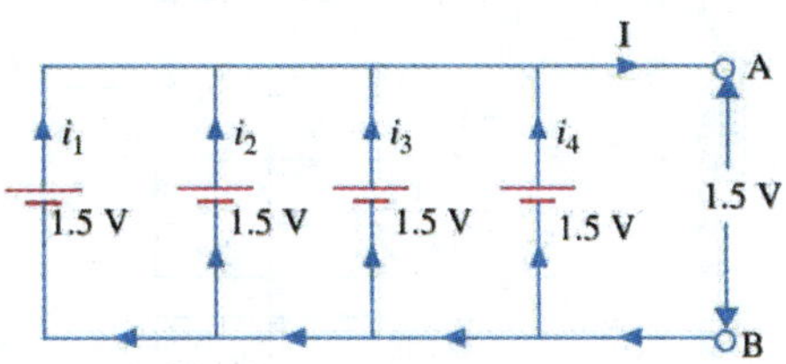

Fig. 1.13

Hence, parallel combination of cells is employed, where increased *current* (rather than voltage) is the main requirement.

However, series-parallel combination (Fig. 1.14) is employed where both higher voltage and increased current are required *i.e.,* greater power is required. Such connections are frequently found in many circuits including those in radio and television receivers.

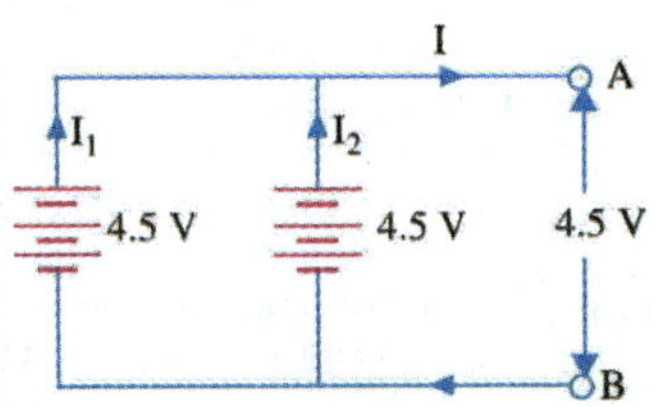

Fig. 1.14

CONVENTIONAL PROBLEMS

1. Find the potential of points (*i*) *C* and (*ii*) *D* in Fig. 1.15 if *C* is located at mid-point of the resistor *R*. What is the direction of current flow through *CD* ? **[–8 V; – 16 V; C to D]**

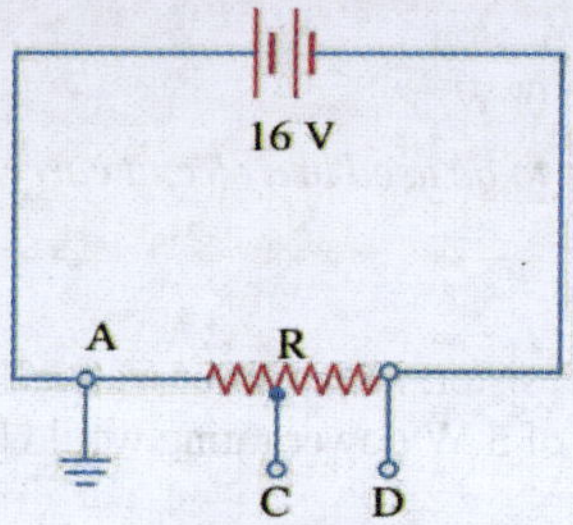

Fig. 1.15

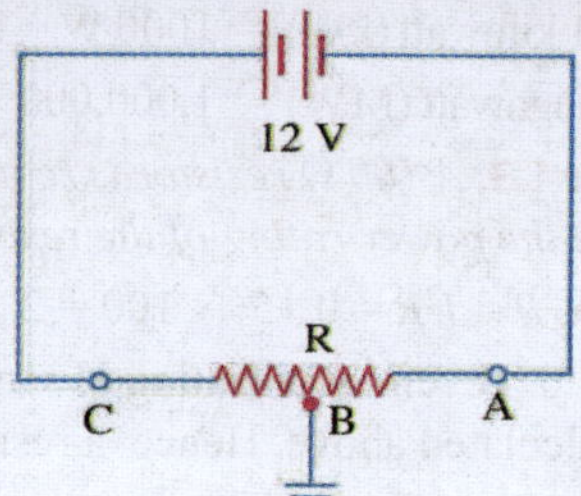

Fig. 1.16

2. In Fig. 1.16, the grounding point *B* is located at one-third of resistance *R* from *A*. Compute the potential of point (*i*) *A* and (*ii*) *C*. What is the p.d. between *A* and *C* ? What is the direction of flow of conventional current between points *A* and *C* ? **[(*i*) – 4 V (*ii*) +8 V; 12 V; C to A]**

3. Find the reading of the milliammeter connected in the circuit of Fig. 1.17. **[0.6 mA]**

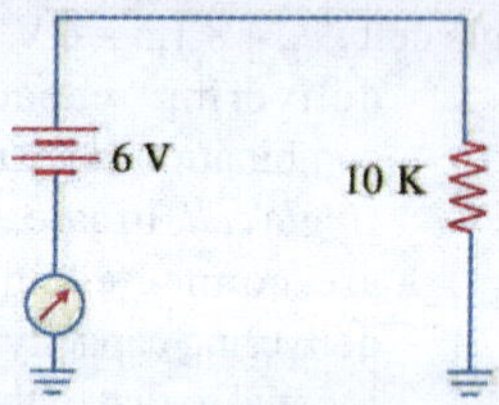

Fig. 1.17

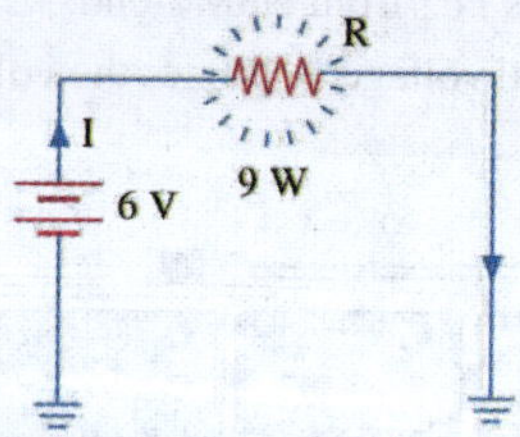

Fig. 1.18

4. Compute the value of resistor *R* in Fig. 1.18 if power dissipated by it is 9 W. **[4 Ω]**

5. What would be the voltage across points *A* and *B* in series-parallel combination of Fig. 1.19 if each cell has a voltage of 1.5V? **[3V]**

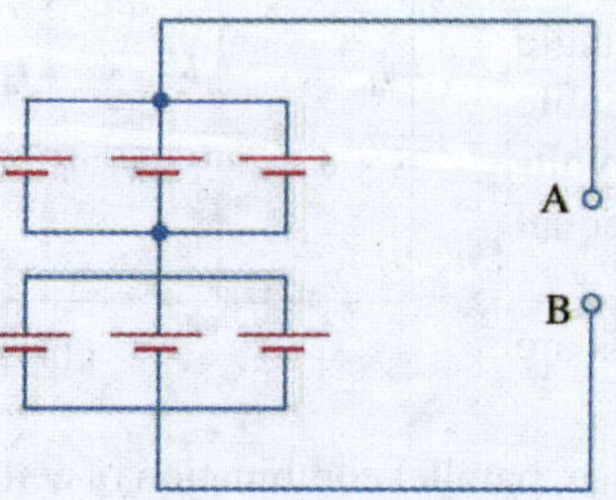

Fig. 1.19

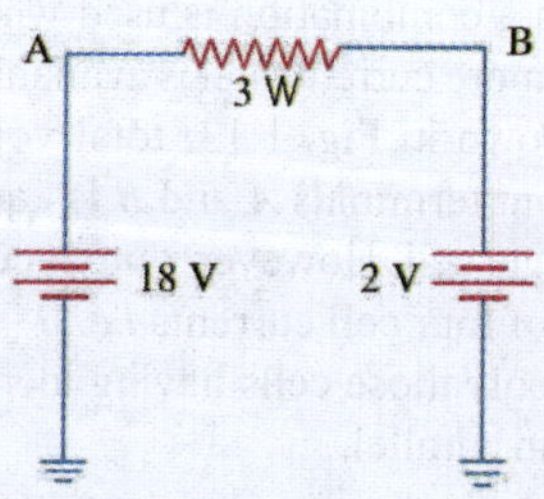

Fig. 1.20

6. Find the magnitude and direction of current in Fig. 1.20. **[2A; from A to B]**

7. In the series circuit of Fig. 1.21, compute the voltage to chassis ground of points *A*, *C* and *D*. What is the p.d. between *A* and *C* ? **[+10 V; –10 V; – 20 V; 20 V]**

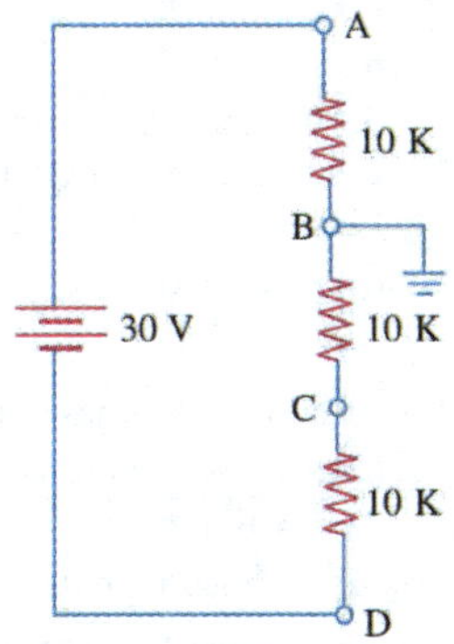

Fig. 1.21

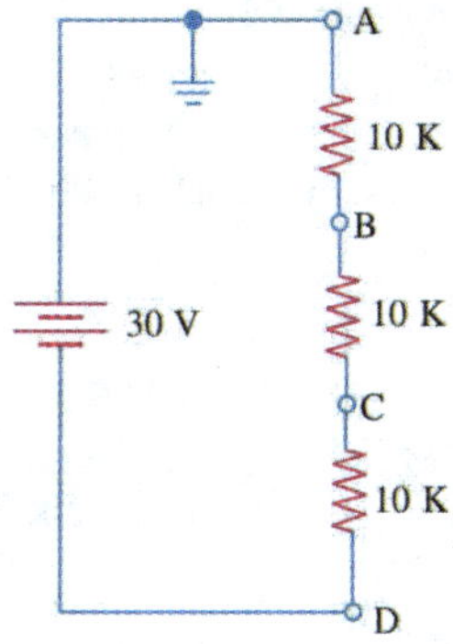

Fig. 1.22

8. In the network of Fig. 1.22, compute the potential of points *A*, *B*, *C* and *D*. Which point is at a higher potential : *B* or *C* ? **[0 V; – 0 V; – 20 V; – 30 V; B]**

SELF EXAMINATION QUESTIONS

A. Fill in the blanks by most appropriate word (s) or numerical values(s).

1. Chassis ground is assumed to have .... potential.
2. Ohm's law gives relationship between applied voltage, current and ....
3. The resistance of a circuit is equal to voltage divided by ....
4. Linear resistor is one whose .... remains constant.
5. Tungsten filament of an electric bulb represents an examples of .... resistance.
6. Basic unit of work is .... and that of power is ....
7. Wattage of a device is given by the product of .... and ....
8. Electron-volt is the unit of ....
9. For getting more current, cells should be connected in ....

B. Answer True or False

1. In an electronic circuit, it is immaterial whether positive or negative battery terminal is grounded.
2. The ends of a mid-earthed resistor have opposite polarity.
3. Ground is always at **0 V.**
4. In electronic circuits, generally, chassis is treated as the common ground.
5. A long and straight wire is called a linear resistor.
6. A non-linear resistor is one whose current does not very linearly with its voltage.
7. If voltage across a resistor is doubled, its power dissipation is quadrupled.
8. Electron-volt is the unit of potential.
9. When a few similar cells are connected in series, current is increased proportionately.
10. Parallel combination of cells is used when more current is needed.
11. Series-Parallel grouping of cells provided more power.

C. Multiple Choice Items

1. In an electronic circuit, common reference point
   (*a*) is always the negative battery terminal
   (*b*) is always the most positive point
   (*c*) is always the most negative point
   (*d*) may be any point
2. The term 'ground' spoken in connection with an electronic circuit means
   (*a*) a direct connection to earth through a wire
   (*b*) a common connection for all components
   (*c*) a short circuit
   (*d*) negative battery terminal
3. For doubling the current in a circuit of constant resistance, the applied voltage must be
   (*a*) kept the same (*b*) doubled
   (*c*) halved (*d*) quadrupled
4. Total current drawn from four 1.5 V cells connected in series is 1 ampere. Each cell supplies.........ampere(s).
   (*a*) 1 (*b*) 0.25
   (*c*) 1.5 (*d*) 4

5. Electron-volt is the unit of

(*a*) voltage (*b*) energy

(*c*) current (*d*) power

6. The current flowing through the PNP transistor of Fig. 1.23 is

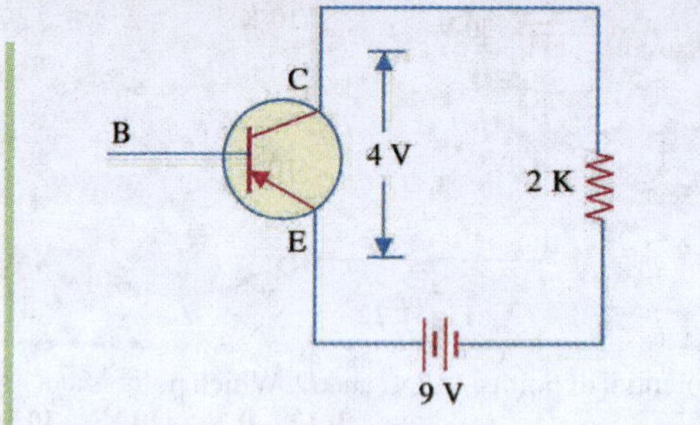

Fig. 1.23

(*a*) 2.5 mA (*b*) 2 mA

(*c*) 4.5 A (*d*) 6.5 mA.

7. A 100 Ω resistor is to be used in a circuit carrying a current of 0.5 A. Its power rating should be....watt.

(*a*) 50 (*b*) 25

(*c*) 200 (*d*) 500

8. A linear resistor is defined as a resistor

(*a*) whose value is independent of applied voltage

(*b*) whose *V-I* characteristic is a straight line

(*c*) whose current varies inversely with the applied voltage

(*d*) both (*a*) and (*b*)

ANSWERS

A. Fill in the blanks

1. zero 2. resistance 3. voltage, resistance 4. resistance 5. non-linear
6. joule, watt 7. voltage, current 8. energy 9. parallel

B. True or False

1. T 2. T 3. F 4. T 5. F 6. T 7. T 8. F 9. F 10. T 11. T

C. Multiple Choice Items

1. *d* 2. *b* 3. *b* 4. *a* 5. *b* 6. *a* 7. *b* 8. *d*

CHAPTER 2

Resistive Circuits

1. Series Circuit
2. The Case of Zero IR Drop
3. Total Power
4. Series-Aiding and Series-Opposing Voltages
5. Series Voltage Dividers
6. 'Opens' in a Series Circuit
7. 'Shorts' in a Series Circuit
8. Parallel Circuits
9. Laws of Parallel Circuits
10. Special Case of Only Two Branches
11. Any Branch Resistance
12. Proportional Current Formula
13. 'Opens' in a Parallel Circuit
14. 'Shorts' in a Parallel Circuit
15. Series-Parallel Circuits
16. 'Opens' in Series-Parallel Circuits
17. 'Shorts' in Series-Parallel Circuits

2.1. Series Circuit

When components in a circuit are connected end-to-end (Fig. 2.1) so that all the circuit current passes through each component, they form a series circuit. The three resistors R_1, R_2 and R_3 are in series with each other and the battery. The result is only one path for current flow. Hence, current I is the *same in all the three resistors.* Due to this current flow, voltage drops V_1, V_2, and V_3 occur across R_1, R_2 and R_3 respectively. Obviously,

$$V_1 = I R_1,\ V_2 = I R_2 \text{ and } V_3 = I R_3$$

The sum of these three voltage drops must equal the applied voltage.

$$\therefore\ V = V_1 + V_2 + V_3$$

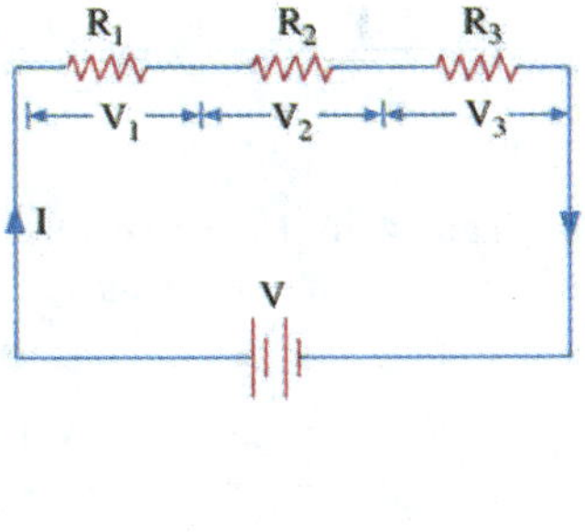

Fig. 2.1

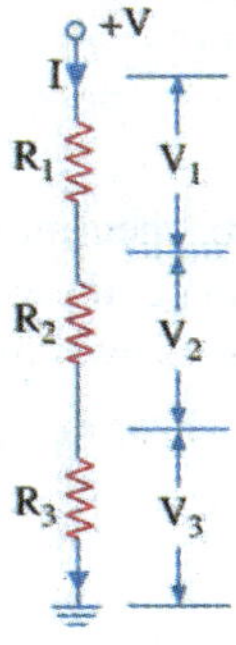

Fig. 2.2

Incidentally, circuit of Fig. 2.1 may be redrawn as shown in Fig. 2.2. In this diagram, the negative battery terminal has been grounded (even though we could ground the positive terminal as well).

2.2. Characteristics of a Series Circuit

A series resistive network has the following characteristics :

1. Total resistance equals the sum of all series resistances.

In the Fig. 2.3 are shown a few resistors connected in series across a voltage source.

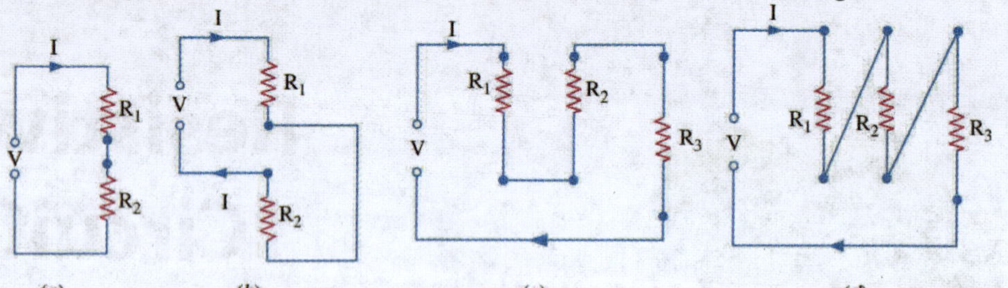

Fig. 2.3

In Fig. 2.3 (*a*)	$R = R_1 + R_2$
In Fig. 2.3 (*b*)	$R = R_1 + R_2$
In Fig. 2.3 (*c*)	$R = R_1 + R_2 + R_3$
In Fig. 2.3 (*d*)	$R = R_1 + R_2 + R_3$

Talking in terms of conductances, we have in Fig. 2.3 (*c*)

$$\frac{1}{G} = \frac{1}{G_1} + \frac{1}{G_2} + \frac{1}{G_3}$$

2. Current through all resistors is the same.

The value of circuit current is given by

$$I = \frac{\text{applied voltage}}{\text{total resistance}} = \frac{V}{R}$$

3. The sum of individual IR drops equals the applied voltage.

As seen from Fig. 2.1,

$$V = V_1 + V_2 + V_3$$

4. There is a stepped fall in voltage as we go from one end the battery to the other as shown in Fig. 2.4.

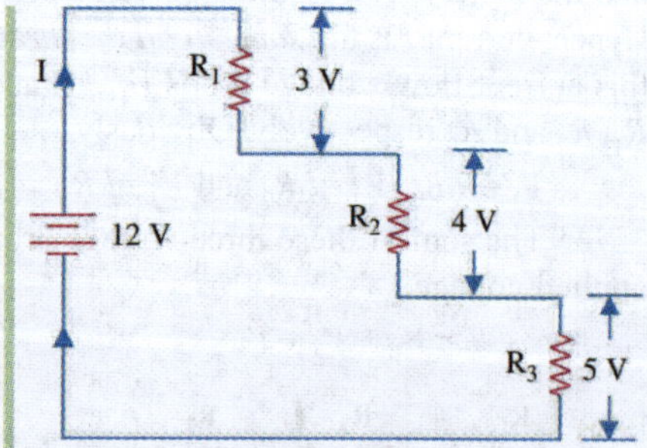

Fig. 2.4

2.3. The Case of Zero IR Drop

It is obvious that drop '*IR*' will be zero when either *I* is zero or *R* is zero. Now, for copper connecting wires, *R* is practically zero. Hence, there is no voltage drop across such interconnecting wires even though they may carry their normal current.

Similarly, there is no *IR* drop when *I* is zero *i.e.*, when applied voltage has been disconnected or there is an open in the circuit.

2.4. Polarity of IR Drops

The study of voltage polarities, whether positive or negative, is of extreme importance in transistor and semiconductor circuits. When voltage drop exists across a resistor, its one end must be more positive than the other. Otherwise, without a potential difference, no current could flow through the resistance to produce the IR drop. The polarity of this drop can be associated with the direction of current flow. If current enters a resistor at point A and goes out from point B, then A must be at a higher potential than B. In other words, A must be positive with respect to B.

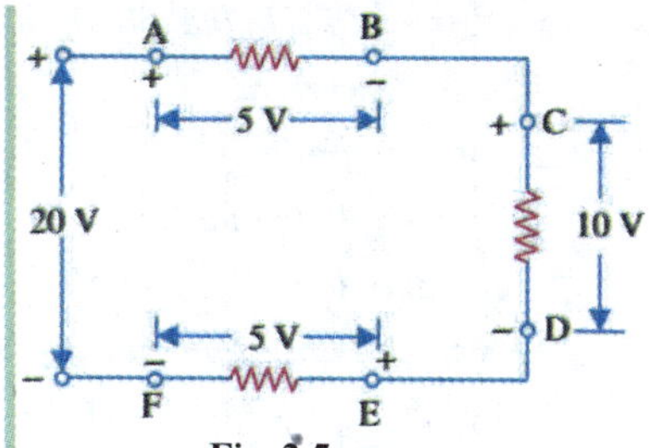

Fig. 2.5

It should be clearly understood that '+' and '–' polarities marked in Fig. 2.5 *relate to voltage drops across resistors only.* Otherwise, points B and C and, similarly, points D and E cannot be at different potentials because they are connected by a piece of conductor wire of zero resistance.

2.5. Total Power

The power needed to drive current through different resistors appears in the form of heat. Hence, total power supplied by the energy source must be equal to the sum of individual powers dissipated in different resistors.

$$\therefore \quad P = P_1 + P_2 + P_3 \text{......etc.}$$

2.6. Series-Aiding and Series-Opposing Voltages

In series-aiding combination, the voltage sources (cells or batteries) are connected in series such that positive terminal of one is joined to the negative terminal of the next. In this case, the total voltage across the circuit is the sum of all voltages or battery emf's as shown in Fig. 2.6 (*a*). Here, voltage applied across $AB = 6 + 6 = 12$ V and $I = \dfrac{12}{6} = 2$ A.

In series-opposing combination, positive terminal of one voltage source is connected to the positive terminal of the other source as shown in Fig. 2.6 (*b*). In this case, the net voltage is the difference of the two voltages and has the same polarity as the larger of the two voltages.

For example, in Fig. 2.6 (*b*), net voltage across AB is $12–6 = –6$ V. Hence, $I = 6/6 = 1$ A.

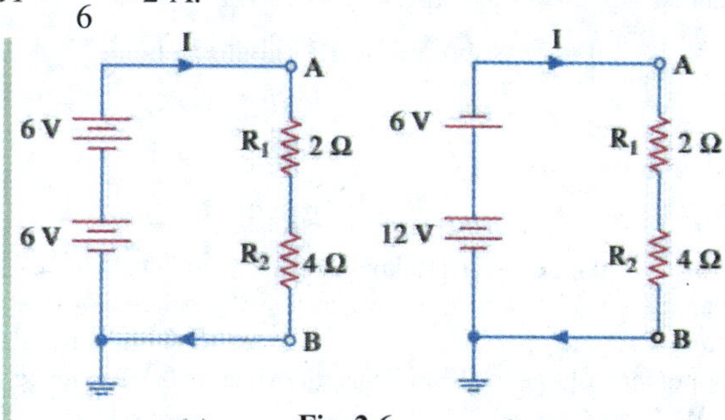

(*a*) Fig. 2.6 (*b*)

Example. 2.1. *In Fig. 2.7, compute*

1. *total circuit resistance*
2. *circuit current*
3. *p.d. between A and E*
4. *potential of point E*
5. *power supplied by the battery.*

Solution. 1. $R = R_1 + R_2 + R_3 = 2 + 3 + 1 = 6\ \Omega$

2. $I = V / R = 12/6 =$ **2 A**

3. $R_{AE} = 2 + 3 = 5\ \Omega$

   $V_{AE} = IR_{AE} = 2 \times 5$

   $=$ **10 V**

4. $V_E = 1 \times 2 =$ **+2 V** — above ground

5. $P = VI = 12 \times 2 =$ **24 W**

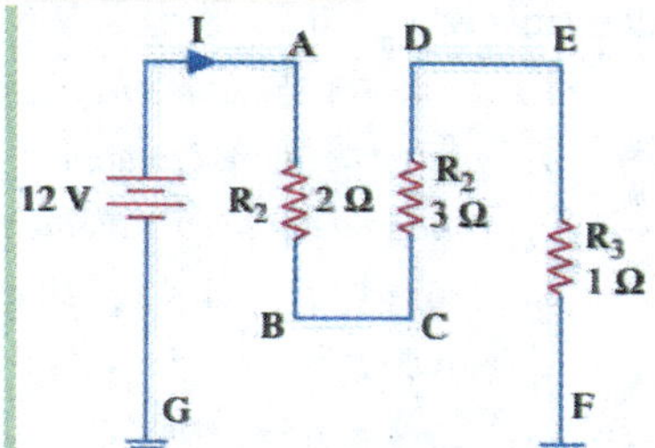

Fig. 2.7

Example. 2.2. *In the circuit of Fig. 2.8, calculate*

1. *circuit current*
2. *potential of point C*
3. *potential of point B*
4. *point of lowest potential*
5. *value of lowest potential.*

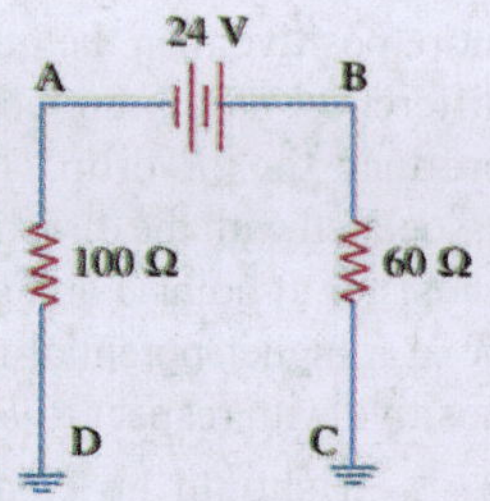

Fig. 2.8

Solution. 1. $I = 24/160 =$ **0.15 A**

2. being grounded, point C is at **0 V**

3. potential of point $B = 0.15 \times 60 =$ **+9 V**

4. point A has lowest potential *i.e.*, it has maximum negative potential,

5. potential of point $A = -0.15 \times 100 =$ **– 15 V** *i.e.*, 15 V below ground potential.

Example 2.3. *With reference to the circuit of Fig. 2.9, compute*

1. *circuit current*
2. *power supplied by the two batteries*
3. *power dissipated in the two resistors*
4. *potential of point A and its polarity*
5. *potential of point B and its polarity.*

Solution. 1.

$$\text{Net driving voltage} = 12 - 6 = 6 \text{ V}$$

$$\text{Total resistance} = 4 + 8 = 12\ \Omega$$

$$I = 6/12 = \mathbf{0.5\ A} \qquad \text{— from } A \text{ to } B$$

2. Power is supplied by 12 V battery only and none by 6 V battery. In fact, 6 V battery consumes power as explained below.

Power supplied by 12 V battery is

$$= 12 \times 0.5 = \mathbf{6\ W}$$

3.

$$P_1 = I^2R_1 = 0.5^2 \times 4 = \mathbf{1\ W}$$

$$P_2 = I^2R_2 = 0.5^2 \times 8 = \mathbf{2\ W}$$

Note. It is seen that total power dissipated in the two resistors is 1 + 2 = 3 W. But power supplied by the 12V battery = 6 W. Question is : where has the balance of (6 – 3) = 3 W gone? It has been *consumed* by the 6V battery. Remember that power is needed to drive current against an opposing voltage. In the present case, opposing voltage is 6V and current driven is 0.5 A. Hence, power consumed by 6 V battery is $P_3 = 6 \times 0.5 =$ 3 W.

So, out of the 6 W input power, 3 W are dissipated by the two resistors and 3 W are consumed by the 6 V battery.

4. Potential of point A is 12 – 6 = + 6 V *i.e.*, positive with respect to the ground.

5. Potential of point $B = V_A - 4 \times 0.5 = 6 - 2 = 4$ V positive w.r.t. ground.

Alternatively, $V_B = 0.5 \times 8 = +4$ V w.r.t. point C *i.e.*, ground.

Example 2.4. *For the circuit of Fig. 2.10, find*

1. *power delivered to element A*
2. *power delivered to element B*
3. *voltage drop across element B.*

Solution. 1. $P_A = 4 \times 5 =$ **20 mW**

2. Now, $P = 12 \times 5 =$ **60 mW**

Also $\qquad P = P_A + P_B$

or $60 = 20 + P_B$

$\therefore \quad P_B = \mathbf{40\ mW}$

3. Now, $V = V_A + V_B$

$\therefore\ 12 = 4 + V_B \quad \therefore\ V_B = \mathbf{8\ V}$

Alternatively, $P_B = V_B \times I$

or $40 = V_B \times 5$

$\therefore \quad V_B = \mathbf{8\ V}$

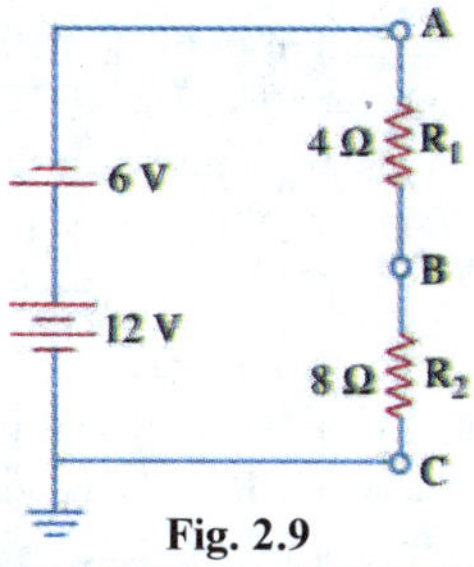

Fig. 2.9

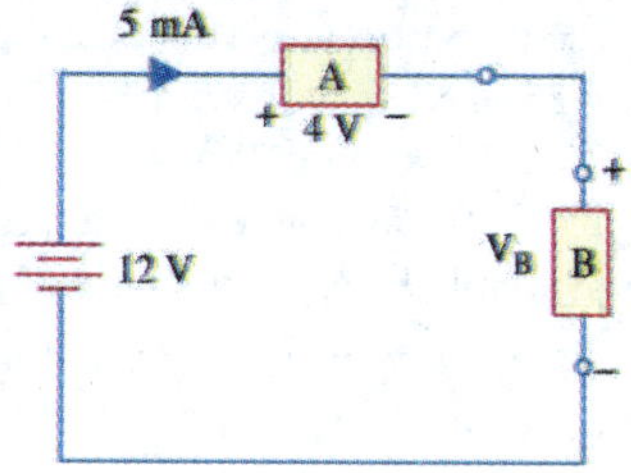

Fig. 2.10

2.7. Proportional Voltage Formula in a Series Circuit

In a series circuit, voltage drop varies directly with resistance. Hence, a simple relation can be found to calculate individual voltage drops without first finding the circuit current.

In Fig. 2.11 (*a*) is shown a 24 V battery connected across a series combination of three resistors, R_1, R_2 and R_3. Fig. 2.11 (*b*) shows a more popular way of drawing the same circuit.

Now, total resistance $R = R_1 + R_2 + R_3 = 12$ K.

According to Proportional Voltage Formula, various drops are

$$V_1 = V \times \frac{R_1}{R} = 24 \times \frac{2}{12} = 4\ \text{V}$$

$$V_2 = V \times \frac{R_2}{R} = 24 \times \frac{4}{12} = 8\ \text{V}$$

$$V_3 = V \times \frac{R_3}{R} = 24 \times \frac{6}{12} = 12\ \text{V}$$

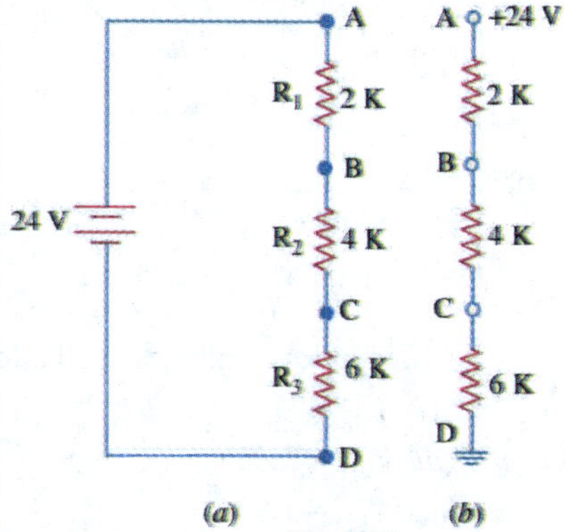

Fig. 2.11

2.8. Series Voltage Dividers

A simple series voltage divider consists of two or more resistors connected in series across a voltage source, say, a battery, as shown in Fig. 2.12. Current flowing through the resistors gives rise to voltage drops proportional to their resistances. These voltages can be used for loads needing voltages less than the battery voltage. In fact, such voltage-divider circuits are used when it is necessary to obtain different values of voltage from a single energy source. A typical example is when we use a single power supply V_{CC} to provide collector voltage and bias voltage for transistor bias circuit as shown in Fig. 2.13.

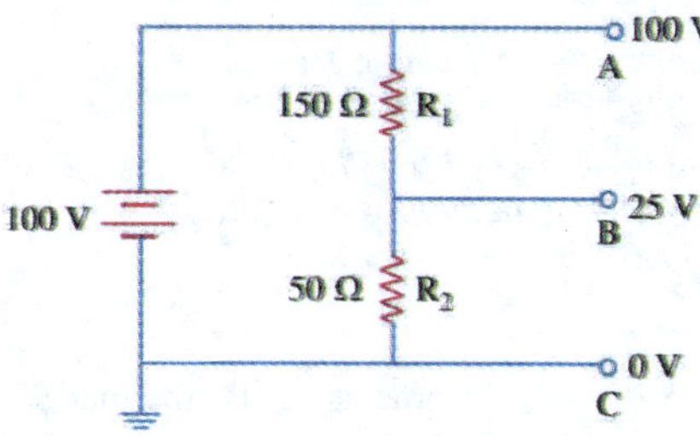

Fig. 2.12

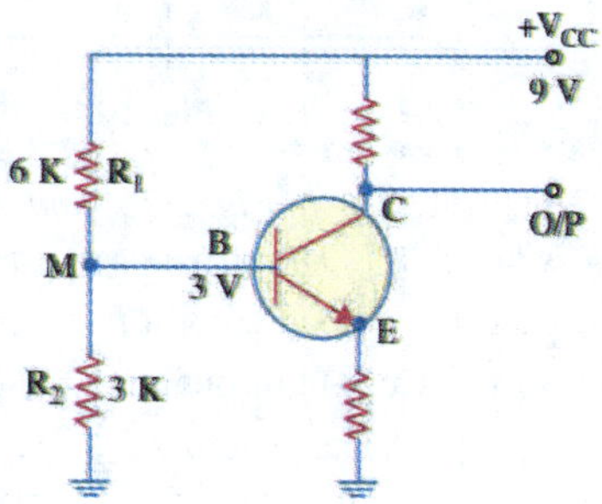

Fig. 2.13

In Fig. 2.12, two resistors of 150 Ω and 50 Ω are connected in series across a 100 V source. Voltage drop across $R_1 = 100 \times 150/200 = 75$ V. Similarly, drop across $R_2 = 100 \times 50/200 = 25$ V. As seen, point A is at 100 V whereas potential of point $B = 100 - 75 = 25$ V. Now, we have a choice of three voltages : 100 V, 75 V and 25 V. If we want 75 V, the pick off points are *A* and *B*. If we need 25 *V*, it is available between points *B* and *C*. Of course, 100 V is available between points *A* and *C*.

Now, take the transistor biasing circuit of Fig. 2.13. Here, collector voltage required is 9 V but base bias voltage required is 3 V. A simple series voltage divider network $R_1 - R_2$ is added to supply the two required voltages from a single source. As seen, total voltage across $R_1 - R_2$ is 9 V. Drop across $R_2 = 9 \times 3/9 = 3$ V. Hence, base *B* of the transistor is at + 3 V.

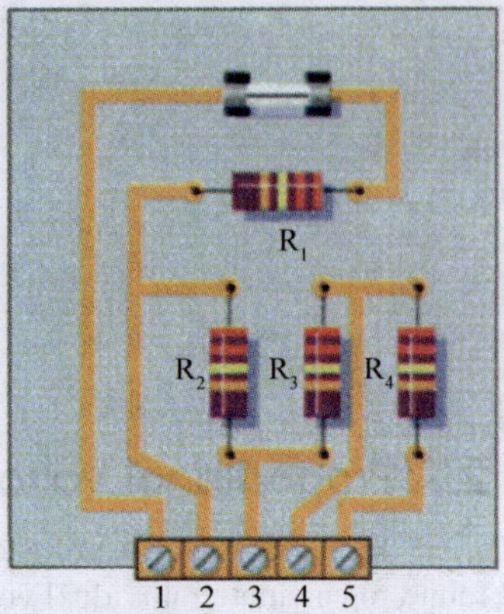

Voltage divider circuit board.

Example 2.5. *Find the values of different voltages that can be obtained from a 12 V battery with the help of voltage divider circuit of Fig. 2.14.*

Solution. $R = R_1 + R_2 + R_3 = 4 + 3 + 1 = 8\ \Omega$

drop across $R_1 = 12 \times 4/8 = 6$ V

$\therefore \quad V_B = 12 - 6 = 6$ V

drop across $R_2 = 12 \times 3/8 = 4.5$ V

$\therefore \quad V_C = V_B - 4.5 = 6 - 4.5 = 1.5$ V

drop across $R_3 = 12 \times 1/8 = 1.5$ V

Different load voltages available are :

(*i*) $V_{AB} = V_A - V_B =$ **6 V** (*ii*) $V_{AC} = 12 - 1.5 =$ **10.5 V**

(*iii*) $V_{AD} = 12$ V (*iv*) $V_{BC} = 6 - 1.5 =$ **4.5 V**

(*v*) $V_{CD} =$ **1.5 V**

Hence, there is a choice of five different voltages.

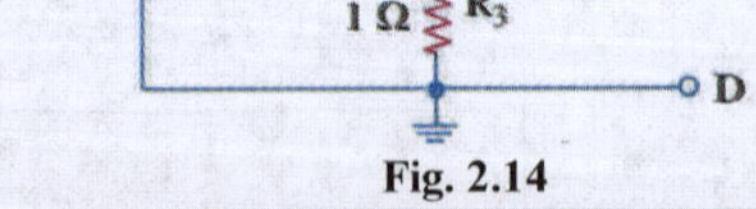

Fig. 2.14

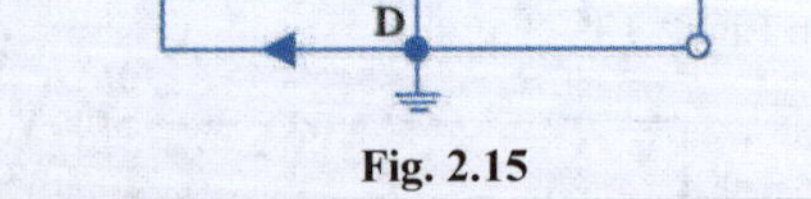

Fig. 2.15

Example 2.6. *In Fig. 2.15, find values of R_1 and R_2 if the voltage applied across load resistor of 1 K is to be 20 V. The maximum current which the battery can supply is 60 mA.*

Solution. Load current $I_L = 20/1\text{ K} = 20$ mA

Current through $R_2 = 60 - 20 = 40$ mA

Voltage across $R_2 = 20$ V — same as across the load

$R_2 = 20/40\text{ mA} =$ **500 Ω**

Drop through $R_1 = 100 - 20 = 80$ V

Current through $R_1 = 60$ mA

$R_1 = 80 / 60\text{ mA} =$ **1333.3 Ω**

2.9. 'Opens' in a Series Circuit

In a normal series circuit like the one shown in Fig. 2.16 (*a*), there is a current flow and voltage drops across different resistors are proportional to their resistances. If the circuit becomes open anywhere as shown in Fig. 2.16 (*b*), following two effects would be produced :

1. First, the 'open' will offer an infinite resistance. Hence, circuit current will become zero. Consequently, there would be no voltage drops across R_1 and R_2.

2. Second, *whole of the applied voltage would be felt across the 'open'*. The reason for this is that resistances R_1 and R_2 become negligible as compared to the infinite resistance of the 'open' which has practically whole of the applied voltage dropped across it (as per Proportional Voltage Formula).

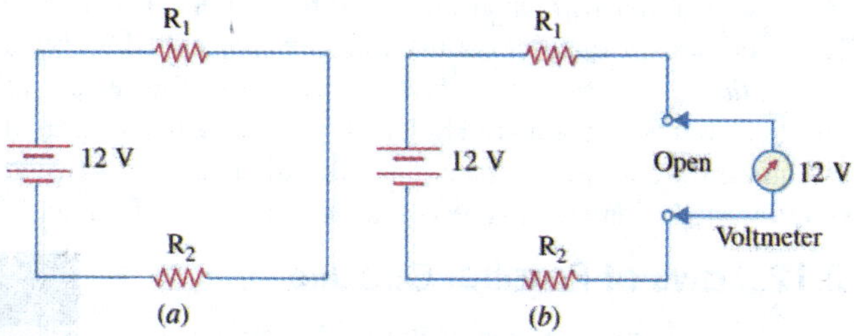

Fig. 2.16

2.10. 'Shorts' in a Series Circuit

A 'short' has practically zero resistance. Hence, it causes the problem of excessive current which, in turn, causes power to increase many times and circuit components to burn out.

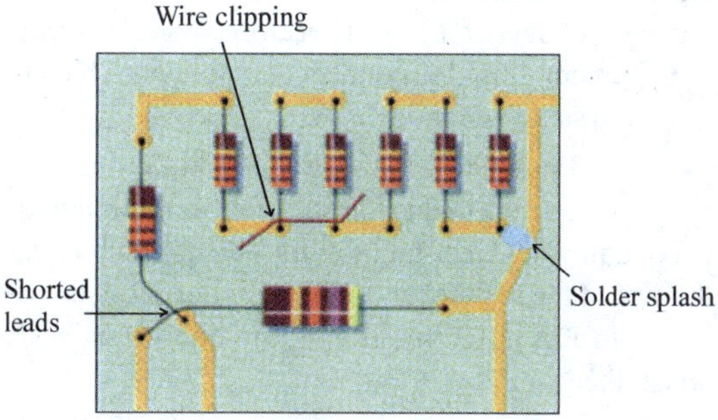

Examples of shorts on a PC board.

In Fig. 2.17 (*a*) is presented the normal series circuit where $V = 12$ V, $R = 6\ \Omega$, $I = 12/6 = 2$ A, $P = I^2R = 2^2 \times 6 = 24$ W.

In Fig. 2.17 (*b*), the 3 Ω resistance has been shorted out by a resistanceless copper wire. Now, total circuit resistance is $R = 1 + 2 + 0 = 3\ \Omega$. Hence, $I = 12/3 = 4$ A and power increases to $4^2 \times 3 = 48$ W.

Fig. 2.17 (*c*) shows the situation where both 2 Ω and 3 Ω resistances have been shorted out of the circuit. In this case, $R = 1\ \Omega$. $I = 12/1 = 12$ A, $P = 12^2 \times 1 = 144$ W.

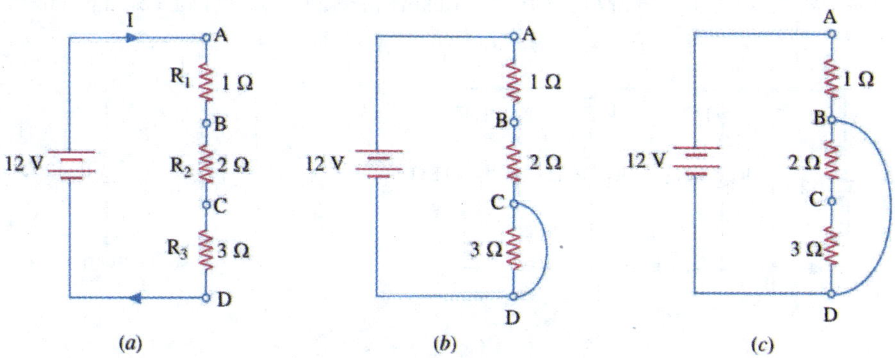

Fig. 2.17

Because of this excessive current (6 times the normal value) wires and other circuit components can become hot enough to ignite and burn. Hence, there should be a fuse which should open if there is too much current in the circuit.

2.11. Parallel Circuits

A parallel circuit is a *branched* arrangement in which two or more resistors are connected side by side across a single voltage source as shown in Fig. 2.18. Here, two resistors R_1 and R_2 are in parallel with each other and the battery. It is a two-branched circuit. Parallel connections are also called *multiple connections or shunt connections.* A very important feature of parallel circuits, as compared to series circuits, is that in parallel circuits different branch loads operate independently of each other. Hence, if any load is disconnected or turned off, other branch loads continue to operate.

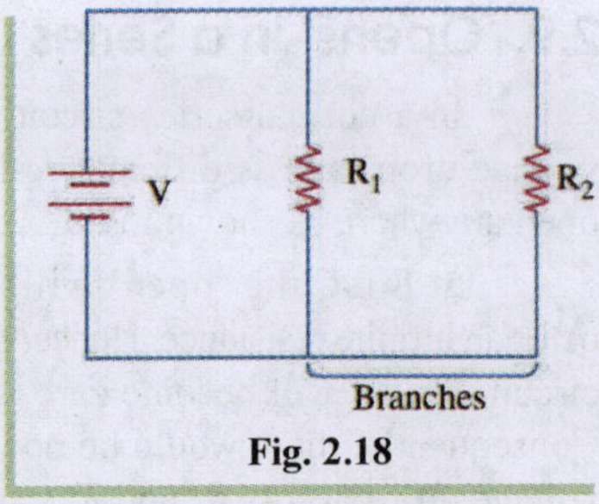

Fig. 2.18

2.12. Laws of Parallel Circuits

Some of the important features of a parallel circuit are as under :

1. Voltage across each branch is the same

For example, in Fig. 2.19 (*a*), voltage across all the three resistors R_1, R_2 and R_3 is the same *i.e.*, battery voltage of 12 V. Therefore, parallel circuit arrangement is used to connect components which require the same voltage.

A car's headlights are connected in parallel. Hence each headlight is exposed to the full potential difference supplied by the car's electrical system, giving maximum brightness. One headlight burns out, the other one will keep shining.

2. Reciprocal resistance formula

According to this formula, the *reciprocal* of the net or equivalent resistance of the entire circuit equals the sum of the reciprocals of the individual resistances.

In Fig. 2.19 (*a*), if R is the equivalent resistance, then

$$\frac{1}{R} = \frac{1}{R_1} + \frac{1}{R_2} + \frac{1}{R_3}$$

$$= \frac{1}{4} + \frac{1}{6} + \frac{1}{12} = \frac{1}{2} \qquad \therefore \; R = 2\,\Omega$$

It means that the three parallel-connected resistors of 4 Ω, 6 Ω and 12 Ω can be replaced by a single resistor of 2 Ω as shown in Fig. 2.19 (*b*). Talking in terms of conductances, it means that

$$G = G_1 + G_2 + G_3$$

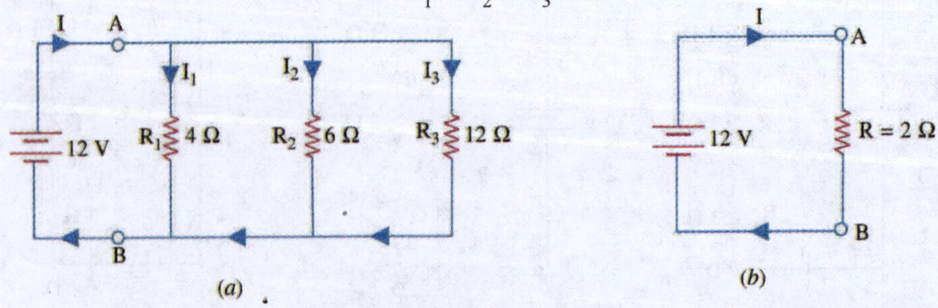

Fig. 2.19

3. Each branch current is given by V/R

In Fig. 2.19 (*a*) above, each branch current is given by

$$I_1 = \frac{V}{R_1} = \frac{12}{4} = 3\text{ A}$$

$$I_2 = \frac{V}{R_2} = \frac{12}{6} = 2 \text{ A}$$

$$I_3 = \frac{V}{R_3} = \frac{12}{12} = 1 \text{ A}$$

4. The sum of branch currents is equal to the total current supplied by the battery

As seen from Fig. 2.19 (*b*) on previous page,

Battery current = 12/2 = 6 A

Sum of branch currents = 3 + 2 + 1 = 6 A

2.13. Special Case of Equal Resistances in all Branches

If resistances in all branches of a parallel circuit are equal, then combined resistance equals the value of one branch resistance divided by the number of branches.

In Fig. 2.20 (*a*), three equal 60 Ω resistances are connected in parallel across terminals *A* and *B*. The equivalent resistance of the entire circuit = 60/3 = 20 Ω as shown in Fig. 2.20 (*b*). If there were four such resistances, then *R* would have been = 60/4 = 15 Ω. If five, then *R* = 60/5 = 12 Ω. In fact, as we keep adding more branch resistances, the total resistance keeps on decreasing.

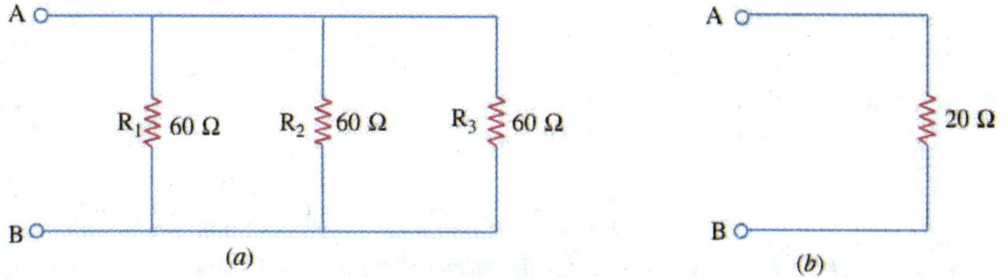

Fig. 2.20

It may seem somewhat unusual at first that putting more resistances into a parallel circuit lowers the total circuit resistance. However, it should become quite clear if one realizes that each resistor provides an additional current path and thus increases the total current. Now, with a *constant* voltage, an *increase* in current can only mean a *decrease* in circuit resistance.

2.14. Special Case of Only Two Branches

The combined resistance of two unequal resistances connected in parallel (Fig. 2.18) is given by

$$\frac{1}{R} = \frac{1}{R_1} + \frac{1}{R_2} = \frac{R_1 + R_2}{R_1 R_2}$$

or $$R = \frac{R_1 R_2}{R_1 + R_2}$$

It is easier to find combined resistance by the above relation directly instead of using the reciprocal formula.

It is seen that combination of two parallel resistances is equal to their product divided by their sum. Often, the combined resistance is written in the form

$$R = R_1 \parallel R_2$$

2.15. Any Branch Resistance

It can be verified that combined or equivalent resistance of a parallel circuit is less than the least amongst them. From example, in Fig. 2.21, $R = 2 \parallel 500 \parallel 1000 \parallel 10{,}000 = 1.98$ Ω.

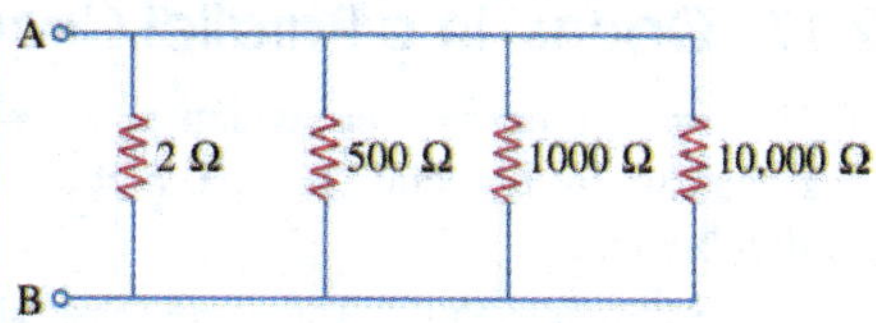

Fig. 2.21

It is less than the least of the three resistances *i.e.*, 2 Ω. In Fig. 2.21,

$$\frac{1}{R} = \frac{1}{2} = \frac{1}{500} + \frac{1}{1000} + \frac{1}{10,000}$$

or $$R = 1.98\ \Omega$$

It is less than the least branch resistance of 2 Ω.

2.16. Proportional Current Formula

Current is inversely proportional to resistance. Hence, current through each branch of a parallel circuit [Fig. 2.22 (*a*)] can be found by setting up an inverse formula like the one given below :

$$I_1 = I \cdot \frac{R_2}{R_1 + R_2}$$

$$I_2 = I \cdot \frac{R_1}{R_1 + R_2}$$

where I is the total current entering the parallel circuit.

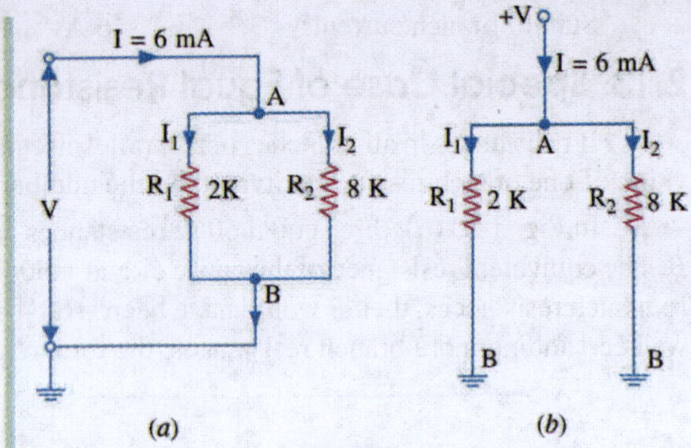

Fig. 2.22

Consider the parallel combination of two resistors shown connected in two different ways in Fig. 2.22.

$$I_1 = 6 \times \frac{8}{2+8} = 4.8\ \text{mA}; \qquad I_2 = 6 \times \frac{2}{2+8} = 1.2\ \text{mA}$$

Example 2.7. *A 12 V battery of negligible internal resistance is connected across a parallel combination of 4 K, 6 K and 12 K resistors as shown in Fig. 2.23. Compute*

1. *combined circuit resistance*
2. *current supplied by the battery*
3. *power supplied by the battery*
4. *power developed by each resistor*

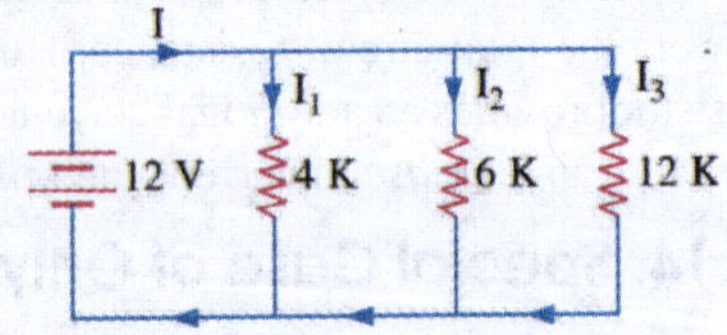

Fig. 2.23

Solution. 1. $\frac{1}{R} = \frac{1}{4} + \frac{1}{6} + \frac{1}{12} = \frac{1}{2}$

$R = \mathbf{2\ K}$

2. $I_1 = 12/4\ \text{K} = 3\ \text{mA}$

$I_2 = 12/6\ \text{K} = 2\ \text{mA}$ $\qquad I_3 = 12/12\ \text{K} = 1\ \text{mA}$

$I = I_1 + I_2 + I_3 = \mathbf{6\ mA}$ $\quad$ or $\quad I = \text{V/R} = 12/2 = 6\ \text{mA}$

3. Power supplied by the battery is

$P = VI = 12 \times 6 = \mathbf{72\ mW}$

4. $P_1 = I_1^2 R_1 = 3^2 \times 4 = \mathbf{36\ mW}$ $\quad P_2 = 2^2 \times 6 = \mathbf{24\ mW}$

$P_3 = 1^2 \times 12 = \mathbf{12\ mW}$

2.17. 'Opens' in a Parallel Circuit

Since an 'open' is equivalent to an infinite resistance, there would be no current in the part of the circuit where it occurs. In a parallel circuit, an 'open' can occur either in the main line or in any parallel branch.

As shown in Fig. 2.24 (*a*), an 'open' in the main line prevents flow of current *to all branches.*

Hence, none of the three bulbs glows. But full *applied voltage is available across the 'open'*.

In Fig. 2.24 (*b*), 'open' has occurred in branch circuit of bulb B_1. Since, there is no current in this branch, B_1 would not glow. However, as the other two bulbs remain connected across the voltage supply, they would keep operating normally.

It may be noted that if a voltmeter is connected across the open bulb, it will read full supply voltage of 220 V.

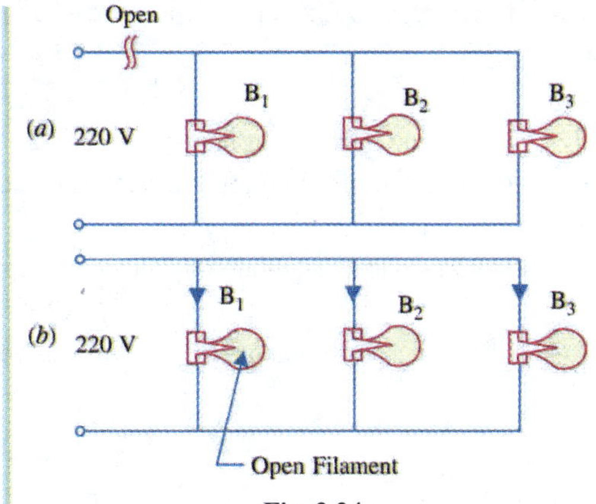

Fig. 2.24

2.18. 'Shorts' in a Parallel Circuit

Suppose a 'short' is placed across resistor R_3 as shown in Fig. 2.25. It becomes directly connected across the battery and draws almost infinite current because not only its own resistance but that of the connecting wires *AC* and *BD* is negligible. Due to this excessive current, the wires may get hot enough to burn out unless the circuit is protected by a fuse. Following points are worth noting :

1. not only is R_3 short-circuited but both R_1 and R_2 are also shorted out *i.e.*, *short across one branch means short across all branches.*

2. there is no current in the shorted resistors. If there were three bulbs, they will not glow.

3. the shorted components are not damaged.

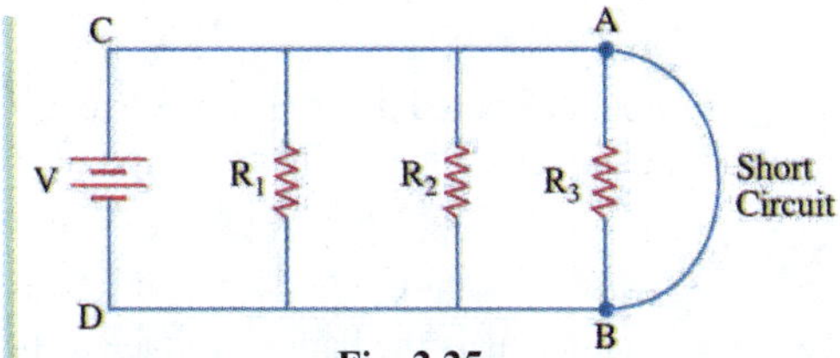

Fig. 2.25

For example, if we had three bulbs in Fig. 2.25, they would glow again when circuit is restored to normal condition by removing the short-circuit.

2.19. Series-Parallel Circuits

Series-parallel circuits are used where it is necessary to provide various amounts of current and voltages with single power supply. Electronic circuits are usually of this type because they generally use only one voltage source. In such circuits, some resistors are connected in parallel and then this parallel combination is connected in series with other resistors as shown in Fig. 2.26.

2.20. Analysing Series-Parallel Circuits

In most of the situations, it is helpful to reduce the given circuit to an equivalent series circuit. This can be done by first handling the parallel part of the circuit and then reducing it to an equivalent single resistance as shown in Fig. 2.26 (*c*).

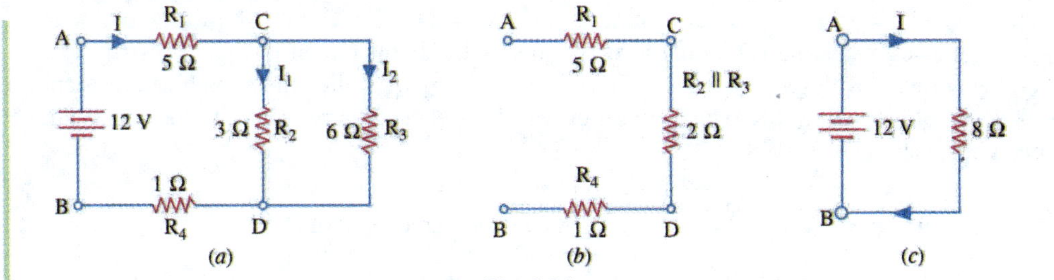

Fig. 2.26

First step is to find equivalent resistance of R_2 and R_3 which is

$$R_2 \parallel R_3 = \frac{R_2 \times R_3}{R_2 + R_3} = \frac{3 \times 6}{6 + 3} = 2\ \Omega$$

The circuit becomes as shown in Fig. 2.26 (*b*).

Second step is to find total resistance by addition.

$$R = 5 + 2 + 1 = 8\ \Omega$$

Third step is to calculate total current supplied by the battery by using Ohm's law.

$$I = 12/8 = 1.5\ \text{A}$$

The branch current I_1 and I_2 and various voltage drops can be found as under :

$$I_1 = I \frac{R_3}{R_2 + R_3} = 1.5 \times \frac{6}{9} = 1\ \text{A}$$

$$I_2 = I \frac{R_2}{R_2 + R_3} = 1.5 \times \frac{3}{9} = 0.5\ \text{A}$$

(or $\quad I_2 = I - I_1 = 1.5 - 1.0 = 0.5$ A)

Drop across 5 Ω resistor = 1.5 × 5 = 7.5 V

Drop across $R_2 = 3 \times 1 = 3$ V

Drop across $R_3 = 6 \times 0.5 = 3$ V

It should be noted that the above two 3 V drops are not two different voltages but actually the one and the same voltage of 3 V across points *C* and *D* is driving both currents I_1 and I_2.

Drop across $R_1 = 1.5 \times 1 = 1.5$ V

Sum of various drops = 7.5 + 3 + 1.5 = 12 V

This sum equals the battery voltage as it, in fact, should.

2.21. 'Opens' in Series-Parallel Circuits

As indicated earlier, an 'open' has the effect of introducing infinite resistance in the branch or leg of the circuit, where it occurs. The effect of an 'open' would be explained with the help of Fig. 2.27.

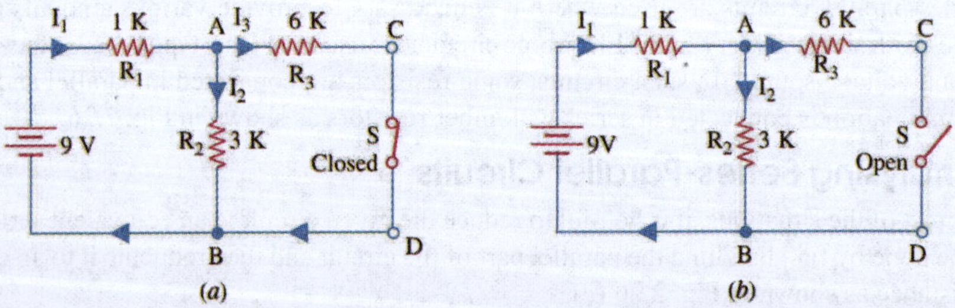

Fig. 2.27

Let us calculate the values of currents and voltages in the circuit of Fig. 2.27 (*a*) when it is working normally with switch *S* closed. Since R_2 and R_3 are in parallel, their combined resistance is $R_2 \parallel R_3 = 6 \times 3/(6 + 3) = 2$ K. This is in series with R_1. Hence, total resistance of the series-parallel circuit becomes = 1 + 2 = 3 K.

$$\therefore \qquad I_1 = 9/3\ \text{K} = 3\ \text{mA}$$

Branch currents can be found by using the relation given in Art. 2.16.

$$I_2 = 3 \times \frac{6}{9} = 2\ \text{mA}$$

$$I_3 = 3 \times \frac{3}{9} = 1 \text{ mA}$$

Drop across $R_1 = 3 \times 1 = 3$ V

Drop across $R_2 = 2 \times 3 = 6$ V

(same is the drop across R_3)

Of course, total voltage drop is equal to the battery voltage.

Now, let us consider the effect of opening the switch as in Fig. 2.27 (*b*). First thing to note is that because of 'open' in path *CD*, R_3 cannot be considered in parallel with R_2. In fact, now R_1 and R_2 become in series with the battery. Treated as a voltage divider,

$$\text{Drop across} \quad R_1 = 9 \times \frac{R_1}{R_1 + R_2} = 9 \times \frac{1}{4} = 2.25 \text{ V}$$

$$\text{Drop across} \quad R_2 = 9 \times \frac{3}{4} = 6.75 \text{ V}$$

Since there is no current through R_3, its voltage drop is zero. Moreover, point *C* in Fig. 2.27(*b*) is at the same potential as point *A*. Consequently, *p.d.* across the open terminals *C* and *D* is the same as across points *A* and *B i.e.*, 6.75 V.

Another point worth noting is that battery current has decreased from previous value of 3 mA to the present value of 9/4 K = 2.25 mA. Similarly, power supplied by the battery decreases from (9 × 3) = 27 mW to (9 × 2.25) = 20.25 mW.

2.22. 'Shorts' in Series-Parallel Circuits

As explained earlier, a short-circuit has practically zero resistance. Depending on where it occurs, it can change all the current and voltage values in the circuit. Let us calculate the normal current and voltage values etc., for the circuit shown in Fig. 2.28 (*a*).

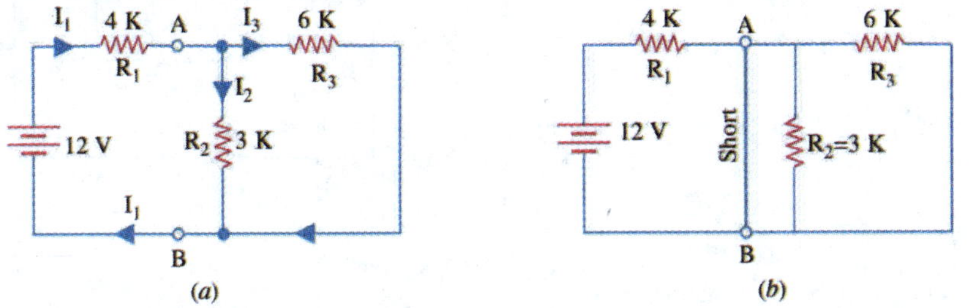

Fig. 2.28

Resistance to the right of points *A* and *B* is

$$R_2 \parallel R_3 = \frac{6 \times 3}{9} = 2 \text{ K}$$

∴ total circuit resistance = 4 + 2 = 6 K

∴ I_1 = 12/6 K = 2 mA $\qquad I_2 = 2 \times 6/9 = 1.33$ mA

$I_3 = 2 \times 3/9 = 0.67$ mA

Drop across $R_1 = 4 \times 2 = 8$ V $\qquad$ Drop across $R_2 = 1.33 \times 3 = 4$ V

(same is the drop across R_3)

Power supplied by the battery = 12 × 2 = **24 mW.**

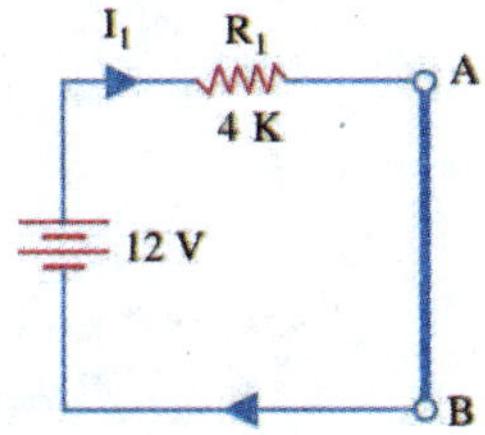

Fig. 2.29

Suppose, now, a short occurs across terminals *A* and *B* as shown in Fig. 2.28 (*b*). It not only shorts out R_2 but R_3 as well thereby reducing the given circuit to the series circuit shown in Fig. 2.29.

Battery current I_1 = 12/4 K = 3 mA

Power supplied by battery = 12 × 3 = 36 mW

As seen, the effect of 'short' is to increase the value of both these quantities.

2.23. Voltage Division in a Complex Series-Parallel Circuit

In Fig. 2.30 is shown a circuit which has many parallel paths. One can analyse it from any point by looking at the path or paths to the ground.

It is seen that

1. $V_6 = V_7$
2. $V_4 = V_5 + V_6 = V_5 + V_7$
3. $V_2 = V_3 + V_4 = V_3 + (V_5 + V_6) = V_3 + (V_5 + V_7)$
4. $V = V_1 + V_2 = V_1 + (V_3 + V_4)$
   $= V_1 + (V_3 + V_5 + V_6) = V_1 + (V_3 + V_5 + V_7)$

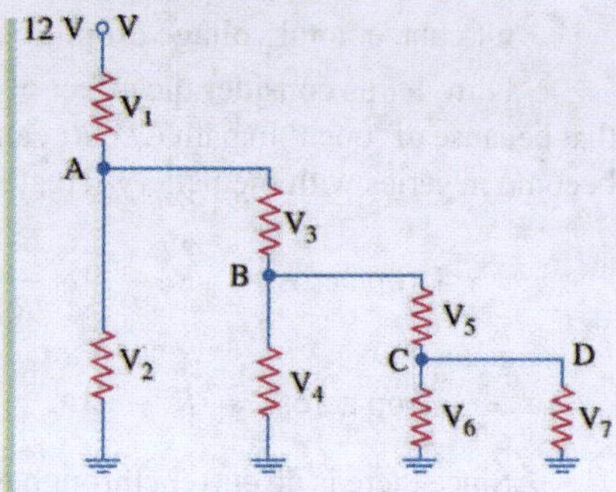

Fig. 2.30

Example 2.8. *For the circuit shown in Fig. 2.31, compute*

1. resistance of the entire circuit

2. current in each resistor

3. voltage drop across each resistor

All resistances are in kilohm.

Solution. 1. First, let us simplify parallel circuit *CD*.

R_{CD} = 8/2 = 4 K

R_{BCD} = 6 + 4 = 10 K

∴ R_{BE} = 10/2 = 5 K

$R = R_{AB} + R_{BE} + R_{EG}$

= 20 + 5 + 5 = 30 K

2. I = 15/30 K = **0.5 mA**

Since $R_{BCE} = R_{BFE}$

∴ $I_1 = I_2$ = 0.5/2 = **0.25 mA**

Similarly, $I_3 = I_4$ = 0.25/2

= **0.125 mA**

Current through 5 K resistor

= I = **0.5 mA**

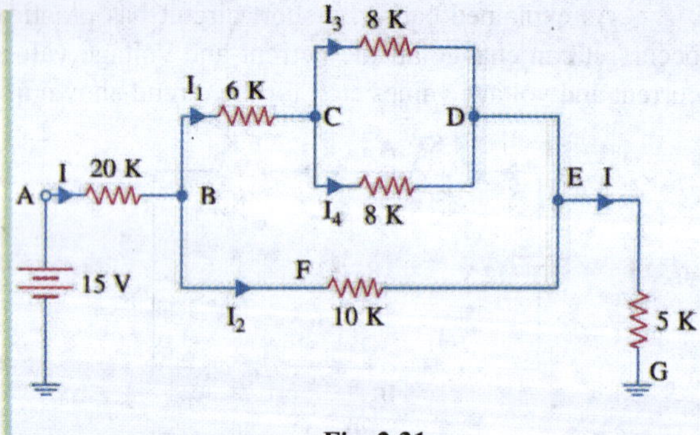

Fig. 2.31

3. V_{AB} = 20 × 0.5 = **10 V**

V_{BC} = 6 × 0.25 = **1.5 V**

V_{CD} = 8 × 0.125 = **1V** (or = 4 × 0.25 = 1 V)

V_{BFE} = 10 × 0.25 = **2.5 V** (same as V_{BCE})

V_{EG} = 5 × 0.5 = **2.5 V**

Example 2.9. (*a*). *Find the current and voltage drop across each resistor in Fig. 2.32.*

(*b*) *How will these value change if point D is shorted to ground accidentally ?*

Solution. (*a*) We will first find the combined resistance of the circuit shown in Fig. 2.32. We will use laws of series and parallel combination and start from point *D* upwards to point *A*. As seen, there are two 4 K resistors in parallel across point *C* (or *D*) and ground *G*.

R_{CG} = 4 / 2 = 2 K

resistance of *leg* $EG = 6 + 2 = 8$ K

Again, there are two 8 K resistors in parallel between points B (or E) and ground G.

$\therefore \quad R_{BG} = 8/2 = 4$ K

$\therefore$ resistance of *leg* $FG = 4 + 2 = 6$ K

Now, there are two 6 K resistors in parallel across point A (or F) and G.

$\therefore \quad R_{AG} = 6/2 = 3$ K

This 3 K resistor is in series with the topmost 3 K resistor. Hence, combined or equivalent resistance of the whole circuit is

$$R = 3 + 3 = 6 \text{ K}$$

battery current, $I = 12/6$ K $= 2$ mA

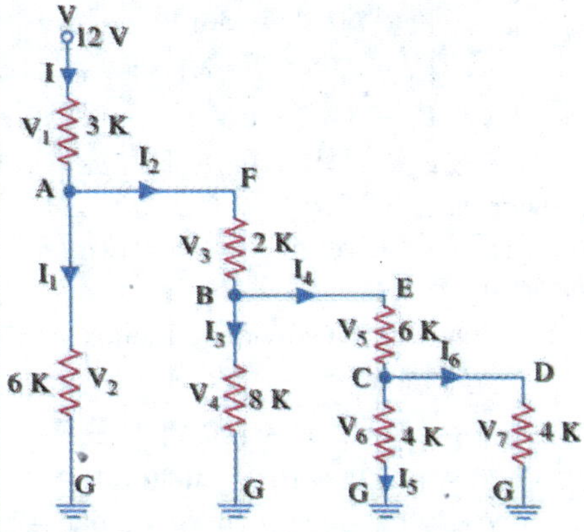

Fig. 2.32

This current gets divided into two equal parts at points A because $R_{FG} = R_{FG} = 6$ K.

$I_1 = 2/2$ K $=$ **1 mA** ; $I_2 =$ **1 mA**

Next, I_2 gets divided into two equal parts at point B for similar reasons as above.

$I_3 = 1/2 =$ **0.5 mA** ; $I_4 =$ **0.5 mA**

Finally, I_4 lets divided into two equal parts at point C.

$I_5 = 0.5/2 =$ **0.25 mA**

$I_6 = 0.25$ mA

$V_1 = 3 \times 2 =$ **6 V**; $\quad V_2 = 6 \times 1 =$ **6 V**

$V_3 = 2 \times 1 =$ **2 V**

$V_4 = 8 \times 0.5 =$ **4 V**

$V_5 = 6 \times 0.5 =$ **3 V**, $\quad V_6 = 4 \times 0.25 =$ **1 V**, $\quad V_7 =$ **1 V**

(*b*) When D is shorted to ground (Fig. 2.33), C also gets shorted. Hence, R_{CG} becomes zero as shown in Fig. 2.33. Now, $R_{BG} = 8 \times 6/14 = 24/7$ K.

Resistance of *leg* $FG = 2 + (24/7) = 38/7$ K

$$\therefore \quad R_{AG} = \frac{6 \times 38/7}{6 + (38/7)} = 2.85 \text{ K}$$

$\therefore$ total $R = 3 + 2.85 = 5.85$ K

$I = 12 \div 5.85 = 2$ mA (approximately)

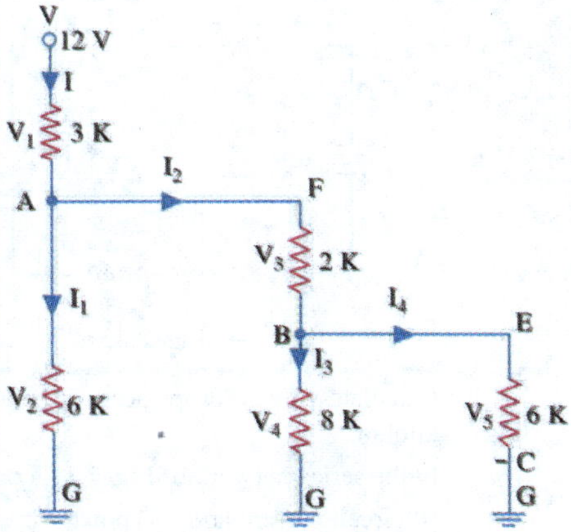

Fig. 2.33

Currents I_1 and I_2 can be found by using Proportional Current Formula (Art. 2.16)

$$I_1 = 2 \times \frac{38/7}{6 + 38/7} = \mathbf{0.95\ mA};$$

$$I_2 = 2 - 0.95 = \mathbf{1.05\ mA}$$

Again, there is division of current at point B

$$\therefore \quad I_3 = 1.05 \times \frac{6}{6 + 8} = \mathbf{0.45\ mA}$$

$$I_4 = 1.05 - 0.45 = \mathbf{0.6\ mA}$$

$V_5 = 6 \times 0.6 = \mathbf{3.6\ V}$; $V_4 = 8 \times 0.45 = \mathbf{3.6\ V}$

$V_3 = 1.05 \times 2 = \mathbf{2.1\ V}$; $V_2 = 6 \times 0.95 = \mathbf{5.7\ V}$

$V_1 = 3 \times 2 = \mathbf{6\ V}$

Example 2.10. *A tapped voltage divider is to be connected across a 220 V dc supply to provide outputs of (a) 15 mA at 100 V and (b) 20 mA at 160 V. Find the resistance of each section of the divider, its power dissipation and total resistance if the bleeder current passed by the divider is to be 60 mA.*

Solution. The divider circuit is shown in Fig. 2.34.

Section R_1

$I_2 = 25$ mA, $V_1 = 100$ V,

$R_1 = 100/25 \times 10^{-3} = \mathbf{4000\ \Omega}$

$P_1 = I^2_1 R_1 = (25 \times 10^{-3})^2 \times 4{,}000 = \mathbf{2.5\ W}$

or $P = V_1 I_1 = 100 \times 25 \times 10^{-3} = \mathbf{2.5\ W}$

Section R_2

$I_2 = 25 + 15 = 40$ mA; $V_2 = 160 - 100 = \mathbf{60\ V}$

$\therefore \quad R_2 = V_2 / I_2 = 60/40 \times 10^{-3} = \mathbf{1500\ \Omega}$

$P_2 = V_2 I_2 = 60 \times 40 \times 10^{-3} = \mathbf{2.4\ W}$

Section R_3

$I_3 = 40 + 20 = 60$ mA, $V_3 = 220 - 160 = \mathbf{60\ V}$

$R_3 = V_3 / I_3 = 60 / 60 \times 10^{-3} = \mathbf{1000\ \Omega}$

$P_3 = V_3 I_3 = 60 \times (60 \times 10^{-3}) = \mathbf{3.6\ W}$

Total resistance of the divider

$= 4{,}000 + 1{,}500 + 1{,}000 = \mathbf{6{,}500\ \Omega}$

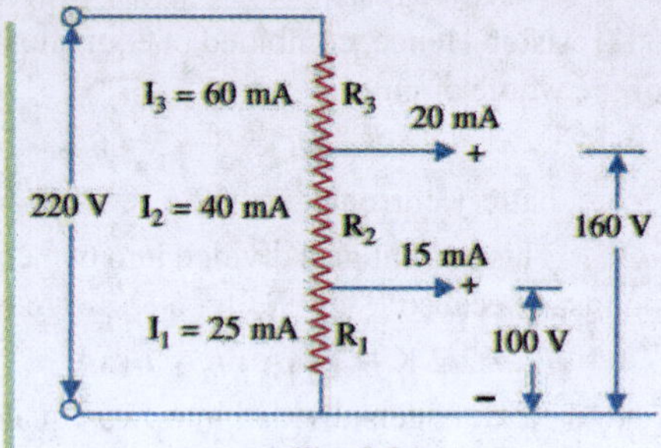

Fig. 2.34

CONVENTIONAL PROBLEMS

1. In the circuit of Fig. 2.35, compute
   (*i*) circuit current (*ii*) value of V_1
   All resistances are in kilohm. **[(*i*) 0.5 mA; (*ii*) 3.5 V]**

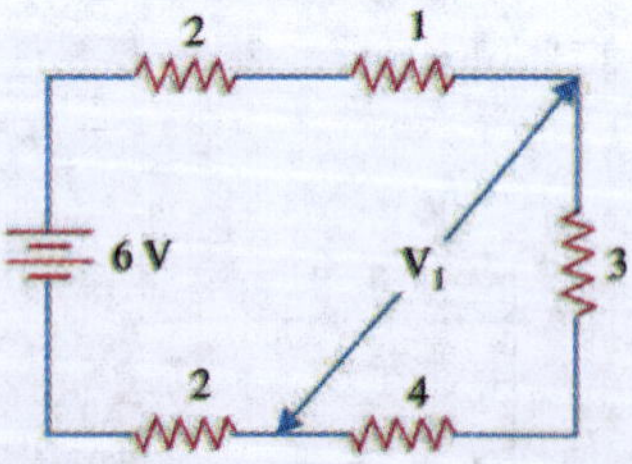

Fig. 2.35

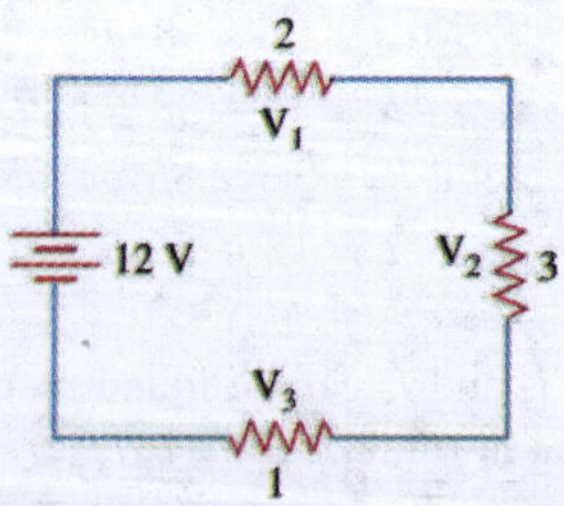

Fig. 2.36

2. Calculate voltage drops across the three series-connected resistors in Fig. 2.36. All values are in kilohm. **[4 V; 6 V; 2V]**
3. In the series network of Fig. 2.37, compute
   (*i*) circuit current and (*ii*) power supplied by the battery.
   How will these values change if a short is placed, between points *A* and *B* ? All resistances are in kilohm. **[(*i*) 2 mA; (*ii*) 60 mW; 5 mA; 150 mW]**

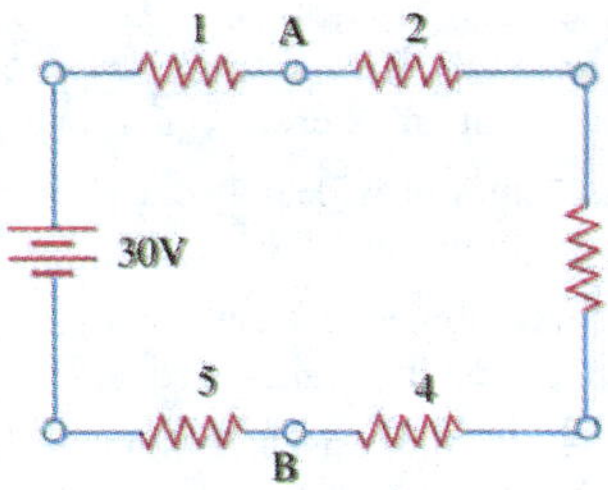

Fig. 2.37

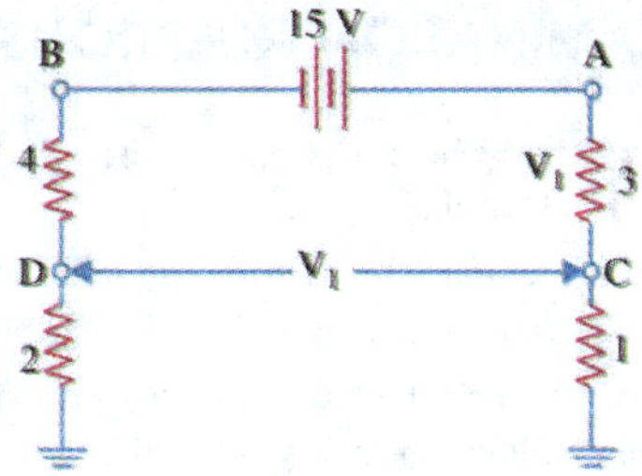

Fig. 2.38

4. In the circuit of Fig. 2.28, find

   (*i*) potentials of points *A* and *B* and (*ii*) value of V_1 and its polarity.

   All resistance values are in kilohm. **[(*i*) V_A = +6 V; V_B = – 9 V; (*ii*) 4.5 V with point C positive]**

5. In the series voltage-divider circuit of Fig. 2.39, rated load current is 2 mA and maximum current the battery can supply is 10 mA. Calculate the values of resistances R_1 and R_2 and their power ratings. **[500 Ω; 800 Ω ; 3 mW; 80 mW]**

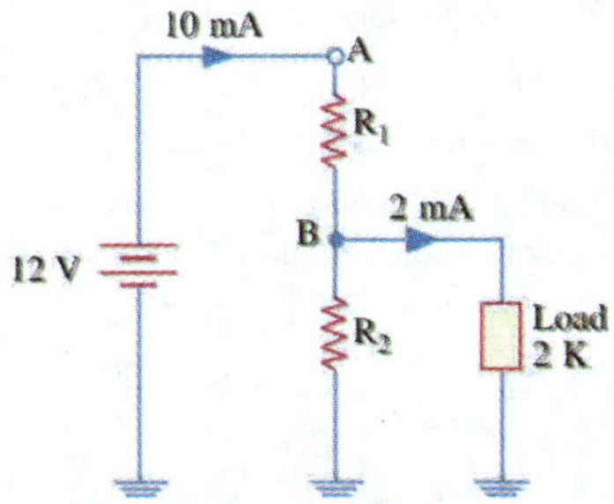

Fig. 2.39

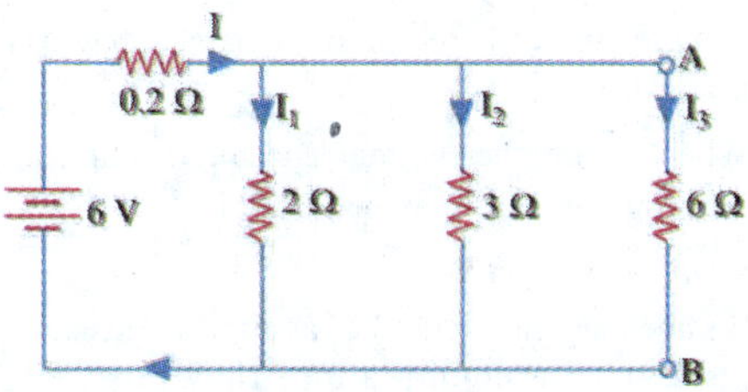

Fig. 2.40

6. In the network of Fig. 2.40, compute

   (*i*) equivalent resistance of the network

   (*ii*) values of branch currents

   (*iii*) current and power supplied by the battery

   (*iv*) current and power supplied by the battery if an accidental short occurs across points *A* and *B*.

   [(*i*) 1.2 Ω ; (*ii*) I_1 = 2.5 A, I_2 = 5/3 A, I_3 = 5/6 A ; (*iii*) 5 A ; 30 W; (*iv*) 30 A, 180 W]

7. In the circuit of Fig. 3.41, compute the value of circuit current. What is the value of the potential of point A with respect to the negative battery terminal ? How do you account for the fact that power supplied by 12 V battery is more than the sum of powers dissipated by the three resistors. **[1 mA, 8 V]**

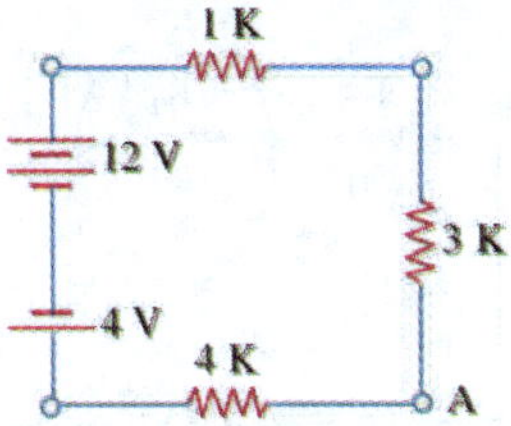

Fig. 2.41

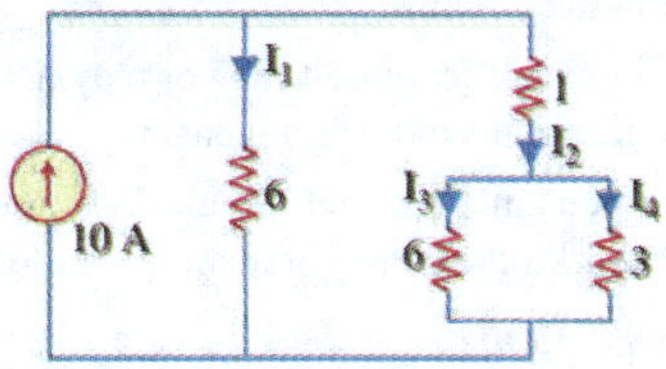

Fig. 2.42

8. Calculate the various branch currents of the circuit shown in Fig. 2.42. All resistance values are in ohms. **[I_1 = 10/3 A, I_2 = 20/3 A; I_3 = 20/9 A; I_4 = 40/9 A]**

SELF EXAMINATION QUESTIONS

A. Fill in the blanks by most appropriate word (s) or numerical value(s).

1. Resistors are said to be connected in series when .... current passes through each resistor.
2. A series network of resistors used for providing different voltages from a single voltage source is called ....
3. The same .... is present across all branches of a parallel circuit.
4. The total resistance of a parallel circuit is always .... than the lowest resistance connected in any of its branches.
5. Cells are connected in series when higher .... is required.
6. If more current is required, then cells may be joined in ....
7. When both higher voltage and higher current are required, cells may be joined in ....

B. Answer True or False

1. If same amount of current passes through each resistor of a combination, they must be connected in series.
2. In a series circuit, the sum of voltage drops across different components is equal to the applied battery voltage.
3. A 'short' in a series circuit always results in grounding of the voltage source.
4. 'Open' in a series circuit always results in damage to the connected apparatus.
5. A 'short' anywhere in parallel circuit leads to excessive power demand on the source of energy supply.
6. The combined resistance of a parallel circuit is sometimes less than the least resistance connected in the circuit.
7. An 'open' in a parallel circuit may or may not disable all the connected components.
8. A branch 'open' in a parallel circuit does not disrupt the working of other branches.

C. Multiple Choice Items

1. Two resistors are said to be connected in series when
   (*a*) both carry the same amount of current
   (*b*) total current equals the sum of branch currents
   (*c*) sum of *IR* drops equals battery emf
   (*d*) they provide only one path for the current flow
2. The combined resistance of two equal resistors connected in parallel is equal to
   (*a*) 1/2 the resistance of one resistor
   (*b*) twice the resistance of one resistor
   (*c*) the resistance of one resistor divided by the other
   (*d*) 1/4 the resistance of one resistor
3. When two resistances are connected in series, they have
   (*a*) same resistance value
   (*b*) same voltage across them
   (*c*) same current passing through them
   (*d*) different resistance values
4. If, in Fig. 2.43, voltmeter *V* measures 20 V, then value of circuit voltage *V* is about ....volt.
   (*a*) 40 (*b*) 70
   (*c*) 60 (*d*) 20

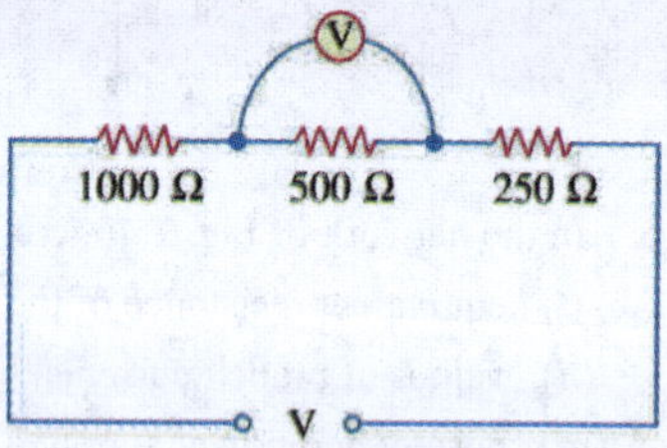

Fig. 2.43

5. If, in Fig. 2.44, R_2 becomes open-circuited, the reading of the voltmeter will
   (*a*) fall to zero (*b*) increase slightly
   (*c*) decrease slightly (*d*) become 200 V

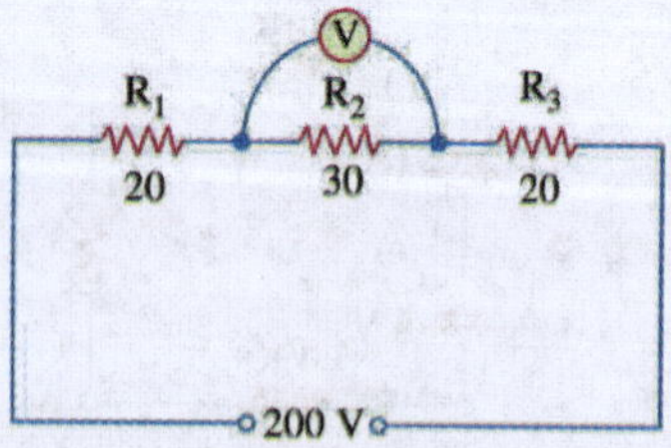

Fig. 2.44

6. In Fig. 2.45, V is an ideal voltmeter having infinite resistance. It will read ....volt.
   (*a*) 40 (*b*) 60
   (*c*) 0 (*d*) 100

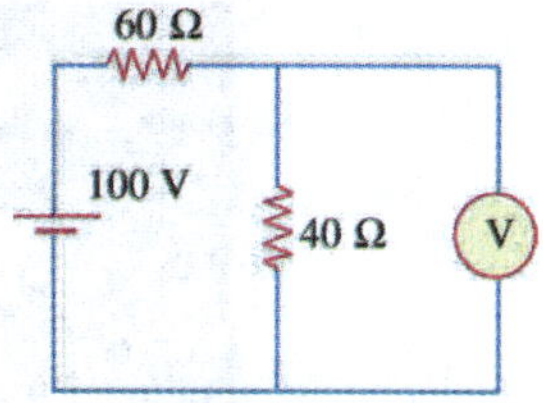

Fig. 2.45

7. The ideal voltmeter V of Fig. 2.46 will read............volt.

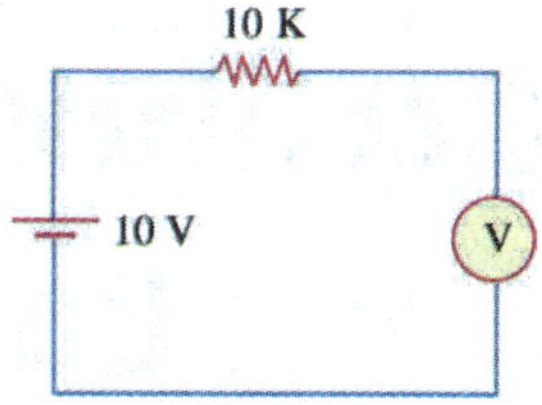

Fig. 2.46

(*a*) 0 (*b*) 10
(*c*) 5 (*d*) 100

8. In a parallel circuit, all components must
   (*a*) have same p.d. across them
   (*b*) have the same value
   (*c*) carry the same current
   (*d*) be switched ON or OFF simultaneously

A parallel circuit of light bulbs is shown in above Fig.

ANSWERS

A. Fill in the blanks

1. same **2.** voltage divider **3.** voltage **4.** less **5.** voltage
6. parallel **7.** series, parallel

B. True or False

1. F **2.** T **3.** F **4.** F **5.** T **6.** F **7.** T **8.** T

C. Multiple Choice Items

1. *d* **2.** *a* **3.** *c* **4.** *b* **5.** *d* **6.** *a* **7.** *b* **8.** *a*

CHAPTER 3

Kirchhoff's Laws

1. General
2. Kirchhoff's Current Law
3. Kirchhoff's Voltage Law
4. Determination of Algebraic Sign
5. Assumed Direction of Current Flow

3.1. General

We come across many circuits in which various components are neither in series nor in parallel nor in series-parallel. One example is a circuit with two or more batteries connected in its different branches. Another is an unbalanced bridge circuit. Here, rules of series and parallel circuits are inapplicable. Such circuits can be easily solved with the help of Kirchhoff's laws which are two in number :

1. Kirchhoff's Current Law (KCL)
2. Kirchhoff's Voltage Law (KVL)

3.2. Kirchhoff's Current Law

It states that **in any network of conductors, the algebraic sum of currents meeting at a point (or junction) is zero.**

Put in another way, it simply means that the total current **leaving** a junction is equal to the total current **entering** that junction.

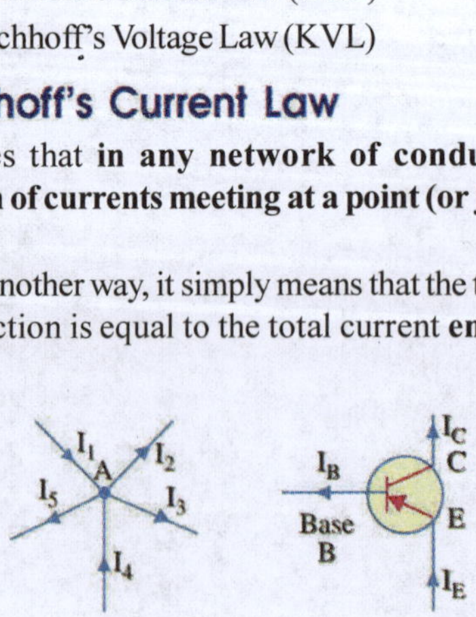

(a) (b)

Fig. 3.1

Explanation

Consider the case of five currents meeting at junction A of the network shown in Fig. 3.1(a). Let us adopt the following sign convention for determining the algebraic sign of different currents.

All currents *entering* the junction would be taken as **positive** whereas those *leaving* it would be taken as **negative**.

According to the above convention, I_1 and I_4 would be taken as positive whereas I_2, I_3 and I_5 would be taken as negative. Using KCL, we have

$$I_1 + (-I_2) + (-I_3) + I_4 + (-I_5) = 0 \quad \text{or} \quad I_1 - I_2 - I_3 + I_4 - I_5 = 0$$

$$\therefore \qquad \Sigma I = 0 \qquad \text{— at a junction}$$

Also, transposing the negative terms to the right-hand side, we get

$$I_1 + I_4 = I_2 + I_3 + I_5$$

incoming currents = outgoing currents

$$\text{or} \qquad I_{in} = I_{out} \qquad \text{— at a junction}$$

Similarly, in Fig. 3.1 (b), I_E is the emitter current directed *towards* the base. The two currents *i.e.,* base current I_B and collector current I_C are directed *away* from the base. Hence, according to KCL

$$I_E + (-I_B) + (-I_C) = 0 \quad \text{or} \quad I_E - I_B - I_C = 0 \quad \text{or} \quad I_E = I_B + I_C$$

3.3. Kirchhoff's Voltage Law

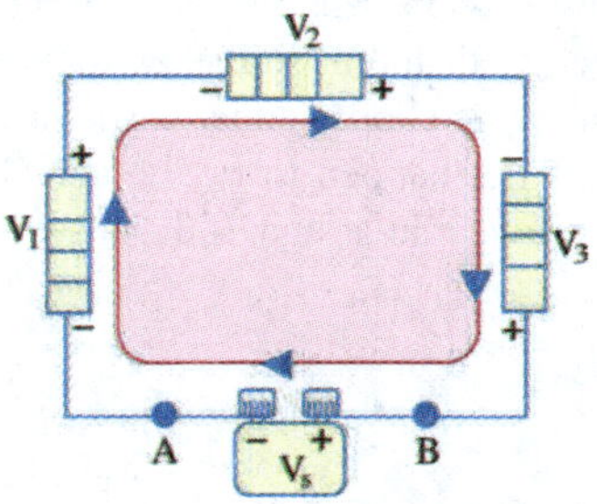

The sum of the voltage drops equals the source voltage.

It states that **the algebraic sum of all IR drops and EMFs in any closed loop (or mesh) of a network is zero.**

$$\text{or} \qquad \Sigma IR = \Sigma \text{ EMF} = 0 \qquad \text{— round a loop}$$

While applying the above two laws for circuit calculations, plenty of errors can occur unless proper algebraic signs are given both to IR drops and battery EMFs.

3.4. Determination of Algebraic Sign

We will follow a very simple sign convention which would apply equally to IR drops and battery EMFs.

Gustav R. Kirchhoff (1824-1887)

A *rise* (or increase) in voltage would be considered *positive* and given a +ve sign and a ***fall*** (or decrease) in voltage would be considered negative and hence given a –ve sign. Let us see how we will apply this convention (i) firstly, to battery EMFs or voltage sources and (ii) secondly, to IR drops across various branch resistors.

(*i*) Battery EMF

While going round a loop (in a direction of our own choice) if we go from the –ve terminal of a battery to its +ve terminal, there is a *rise* in potential, hence this EMF should be given +ve sign. If, on the other hand, we go from its +ve terminal to its –ve terminal, there is fall in potential, hence this battery EMF should be given –ve sign.

It is important to note that algebraic sign of battery EMF is independent of the direction of current flow (whether clockwise or anticlockwise) through the branch in which the battery is connected.

(*ii*) IR drops on resistors

If we go through a resistor in the *same* direction as its current, then there is a *fall* or *decrease* in potential for the simple reason that current always flows from a higher to a lower potential. Hence, this IR drop should be taken –ve. However, if we go around the loop in a direction *opposite* to that of the current, *i.e.,* if we go upstream, there is a rise in voltage. Hence, this IR drop should be a taken as positive.

It is clear from above that algebraic sign of IR drop across a resistor depends on the direction of current that resistor.

Consider the closed loop *ABCDA* of Fig. 3.2. Starting from point *A* if we go around this mesh in clockwise direction,* then different EMFs and IR drops will have following values and signs :

I_1R_1 is –ve	(fall in potential)
I_2R_2 is +ve	(rise ,,)
E_2 is –ve	(fall ,,)
I_3R_3 is –ve	(fall ,,)
E_1 is +ve	(rise ,,)
I_4R_4 is –ve	(fall ,,)

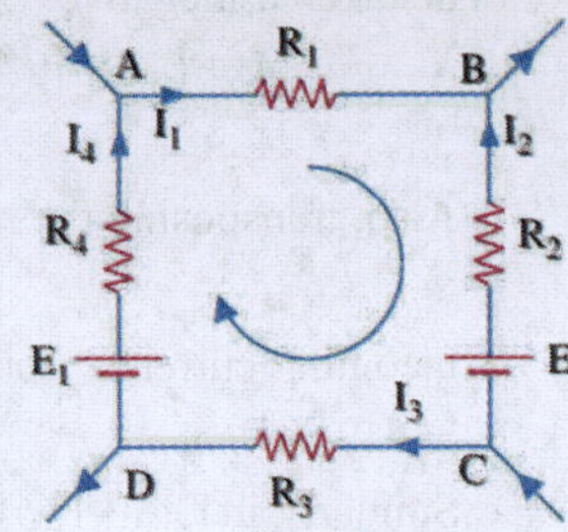

Fig. 3.2

Hence, according to KVL,

$$-I_1R_1 + I_2R_2 - E_2 - I_3R_3 + E_1 - I_4R_4 = 0$$

or $\quad -I_1R_1 + I_2R_2 - I_3R_3 - I_4R_4 = E_2 - E_1$ or $I_1R_1 - I_2R_2 + I_3R_3 + I_4R_4 = E_1 - E_2$

3.5. Assumed Direction of Current Flow

In applying Kirchhoff's laws to electrical or electronic networks, one is often faced with the tricky question of assuming proper direction of current flow through different branches of the network. However, the matter is quite simple because one is at full liberty to choose either clockwise or anticlockwise direction. If the assumed direction of current does not happen to be the actual direction, then on solving the question, this current would be found to have minus sign. If the answer is positive, then assumed direction is the correct direction of current flow.

Example 3.1. *Using Kirchhoff's laws, calculate the branch currents in the network of Fig. 3.3.*

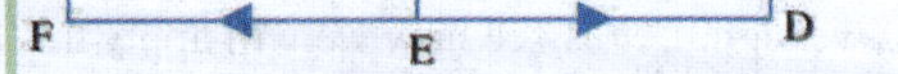

Fig. 3.3

Solution. As shown in the figure, we have taken currents I_1 and I_2 as *originating* from the *positive* terminals of the batteries. However, it is not always essential to do so because a stronger battery can drive current *into* one with lower emf *i.e.,* it can charge it. We will consider the following two closed loops :

(i) Loop ABEFA

Starting from point *A* and going clockwise round the loop, we have

$$-4I_1 - 8(I_1 + I_2) + 12 = 0 \quad \text{or} \quad 3I_1 + 2I_2 = 3 \quad \text{....(i)}$$

(ii) Loop BCDEB

Starting from point *B* and again going clockwise round the loop, we get

$$+2I_2 - 10 + 8(I_1 + I_2) = 0 \quad \text{or} \quad 4I_1 + 5I_2 = 5 \quad \text{...(ii)}$$

Multiplying Eq. (*i*) by 5 and Eq. (*ii*) by 2 and substracting one from the other, we get

$$15I_1 - 8I_1 = 15 - 10 \text{ or } \quad 7I_1 = 5 \quad \therefore I_1 = \mathbf{5/7\ A}$$

Substituting this value either in Eq. (*i*) or (*ii*), say, Eq. (*i*), we have

$$3 \times (5/7) + 2I_2 = 3 \quad \text{or} \quad I_2 = \mathbf{3/7\ A}$$

* We could as well choose to go along anticlockwise direction.

Since these have come out to be positive, it means that their assumed directions are the actual directions.

$$\text{Current through branch } BE = I_1 + I_2 = \frac{5}{7} + \frac{3}{7} = \mathbf{1\frac{1}{7}\ A}$$

Example 3.2. *In the grounded-emitter transistor amplifier circuit of Fig. 3.4, calculate the base current I_B and collector current I_C.*

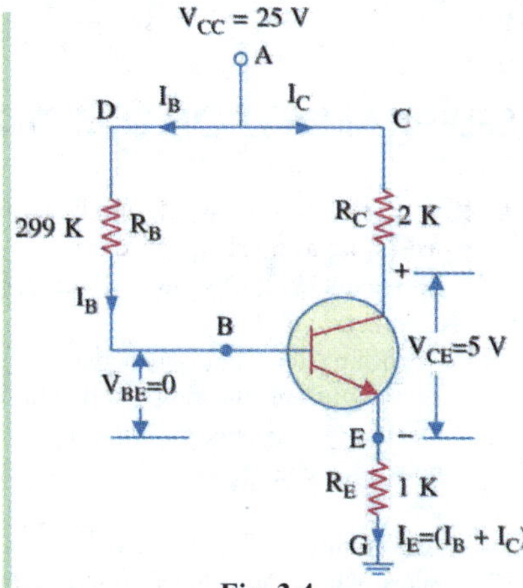

Fig. 3.4

Solution. We will consider two loops : base-current loop ADBEGA and collector-current loop ACBEGA.

Loop ADBEGA

As we move down from the positive terminal A of collector supply battery V_{CC}*,

(*i*) we first get $I_B R_B$. It would be taken negative because we are moving *alongwith* the curent *i.e.,* downstream.

(*ii*) then we meet V_{BE} which also would be taken –ve because we are going from its positive end to negative end.

(*iii*) next, we meet $I_E R_E = (I_B + I_C) R_E$ drop which will be taken as negative since we again are going downstream.

(*iv*) finally, we go from ground G (where –ve battery terminal is supposed to be connected) to positive battery terminal A. Hence, battery EMF of $V_{CC} = 25$ V would be taken positive.

Applying KVL to the loop, we get

$$-I_B R_B - V_{BE} - I_E R_E + V_{CC} = 0$$

or $$-299\, I_B - 0 - 1\,(I_B + I_C) + 25 = 0$$

or $$300\, I_B + I_C = 25 \qquad ...(i)$$

Since resistances are in kilohms, we are taking I_B and I_C to be in milliamperes so that their product gives volts (Art. 2.4).

Loop ACBEGA

$$-2I_C - 5 - 1(I_B + I_C) + 25 = 0 \quad \text{or} \quad I_B + 3\,I_C = 20 \qquad ...(ii)$$

Solving for I_B and I_C from Eq. (*i*) and (*ii*), we get,

$$I_C = 5975/899 = \mathbf{6.64\ mA}; \qquad I_B = 0.08\ \text{mA} = 80\ \mu\text{A}$$

CONVENTIONAL PROBLEMS

1. Use Kirchhoff's laws to find the magnitude and direction of current flow through the 10 Ω resistor of Fig. 3.5. **[1/23 A from A to B]**

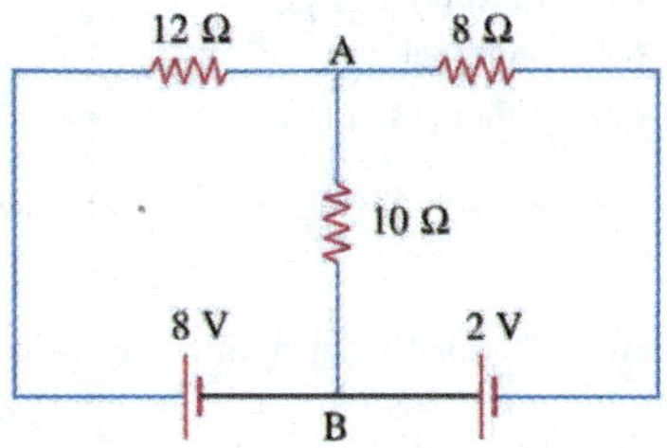

Fig. 3.5

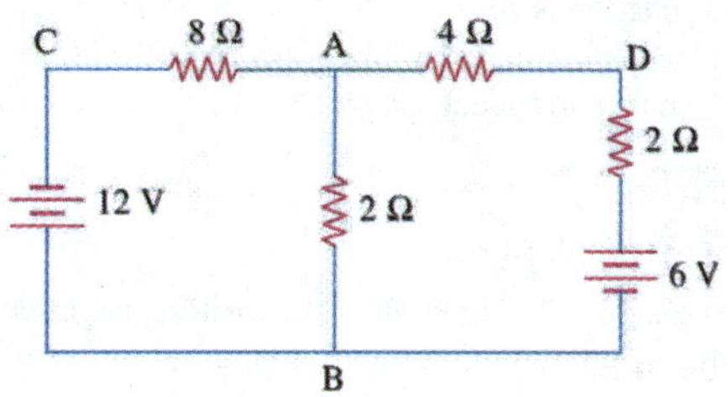

Fig. 3.6

* Please remember that in an NPN transistor, base is positive w.r.t. emitter (Art. 18.2).

2. With the help of Kirchhoff's laws, calculate the magnitude and direction of current flowing through branches *AC*, *AB* and *AD* of the circuit shown in Fig. 3.6.
What would be the potentials of points *A, C, D* and *E* if point *B* is grounded?

**[I_{AC} = 27/19 A; I_{BA} = 6/19 A; I_{DA} = 21/19 A;
V_A = – 12/19 V; V_C = –12 V; V_E = +6 V; V_D = +72 /19 V]**

SELF EXAMINATION QUESTIONS

A. Fill in the blanks with most appropriate word (s) or numerical value (s).

1. The two Kirchhoff's laws are : current law and .... law.
2. According to KCL, in any network, the .... sum of currents meeting at a junction is zero.
3. Generally, currents entering a junction in a network are taken as .... a those leaving it as ....
4. According to KVL, algebraic sum of all *IR* drops and EMFs in any closed loop in a network is ....
5. Generally, rise in voltage is considered .... whereas fall is taken as ....
6. The algebraic sign of *IR* drop depends on the .... of current flow through the connected resistor.

B. Answer True or False

1. In a circuit, the sum of all EMFs around a mesh or a closed loop must always equal zero.
2. Kirchhoff's laws are true for all electronic circuits.
3. Input currents at a junction in a network of conductors equal the output currents at that junction.
4. While applying Kirchhoff's laws to networks, one is at full liberty to assume the directions of flow of various branch currents.

C. Multiple Choice Items

1. According to KCL as applied to a junction in a network of conductors
   (*a*) total sum of currents meeting at the junction is zero
   (*b*) no current can leave the junction without some current entering it
   (*c*) net current flow at the junction is positive
   (*d*) algebraic sum of the currents meeting at the junction is zero
2. Kirchhoff's voltage law is concerned with
   (*a*) *IR* drops
   (*b*) battery EMFs
   (*c*) junction voltages
   (*d*) both (*a*) and (*b*)
3. According to KVL, the algebraic sum of all *IR* drops and EMFs in any closed loop of a network is always
   (*a*) zero
   (*b*) positive
   (*c*) negative
   (*d*) greater than unity
4. According to the commonly-used sign convention for voltages
   (*a*) a fall in voltage is considered positive
   (*b*) a rise in voltage is considered positive
   (*c*) *IR* drop is taken as negative
   (*d*) battery EMFs are taken as positive
5. The algebraic sign of an *IR* drop is primarily dependent upon the
   (*a*) amount of current flowing through it
   (*b*) value of *R*
   (*c*) direction of current flowing through it
   (*d*) battery connection
6. While applying Kirchhoff's laws to electronic circuits, assumed direction of current flow must be
   (*a*) clockwise
   (*b*) from left to right
   (*c*) anticlockwise
   (*d*) either (*a*) or (*b*)

ANSWERS

A. Fill in the blanks

1. voltage **2.** algebraic **3.** positive, negative **4.** zero **5.** positive, negative **6.** direction

B. True or False

1. *F* **2.** *T* **3.** *T* **4.** *T*

C. Multiple Choice Items

1. *d* **2.** *d* **3.** *a* **4.** *b* **5.** *c* **6.** *d*

4 CHAPTER

Network Theorems

1. General
2. Superposition Theorem
3. Ideal Constant-Voltage Source
4. Ideal Constant-Current Source
5. Thevenin's Theorem
6. How to Thevenize a Circuit ?
7. Norton's Theorem
8. How to Nortonise a Given Circuit
9. Maximum Power Transfer Theorem

4.1. General

As we know, a network is just a combination of various components such as resistors etc., interconnected in all sorts of manner. Most of these networks cannot be solved merely by applying laws of series and parallel circuits. Of course, Kirchhoff's laws can always be used. But often it makes the solution quite long and laborious. Hence, various network theorems have been developed which provide very short and time-saving methods to solve these complicated circuits. The reason is that such theorems enable us to convert the given complicated network into a much simpler one which can then be easily solved by only applying the rules of series and parallel circuits. We will discuss the following network theorems which find wide application in electronic and transmission circuits :

1. Superposition Theorem
2. Thevenin's Theorem
3. Norton's Theorem
4. Maximum Power Transfer Theorem

Though we will consider only dc networks in this chapter, these theorems are applicable to ac networks as well.

4.2. Superposition Theorem

According to this theorem, if there are a number of voltage or current sources acting simultaneously in a network, then each source can be treated as if it acts independently of the others.

Hence, we can calculate the effect of one source at a time and then superimpose *i.e.*, algebraically add the results of all the other sources.

Following steps are taken while applying this theorem to the solution of networks which contain more than one voltage or current source :

1. First, all sources except the one under consideration are removed. While removing these voltage sources, their internal resistances (if any) are left behind. While removing current sources, they are replaced by an **open** circuit since their internal resistance (by definition) is infinite (Art. 4.4).

2. Next, currents in various resistors and their voltage drops due to this single source are calculated.

3. This process is repeated for other sources taken one at a time.

4. Finally, *algebraic* sum of currents and voltage drops over a resistor due to different sources is taken. It gives the actual value of current and voltage drop in that resistor or component.

Example 4.1. *Using Superposition theorem, calculate current in each branch of the network shown in Fig. 4.1 (a).*

Solution. We will find two sets of branch currents : one when 6 V battery is not there and the other, when 12 V battery is not there. Let the different branch currents be I_1, I_2 and I as shown in Fig. 4.1 (*a*). In Fig. 4.1 (*b*), 6 V battery has been removed and then replaced by short- circuit (since its internal resistance is zero). Various branch currents are as under :

$R_{BD} = 4 \parallel 4 = 2\ \Omega$. Hence, total circuit resistance is = 6 + 2 = 8 Ω. Therefore, $I_1' = 12/8\ 1.5$A. At point *B*, this current divides equally into two parts.

$$I_1' = 0.75 \text{ A and } I_2' = 0.75 \text{ A}$$

In Fig. 4.2 (*a*) 12 V battery has been removed and replaced by short circuit (since its internal resistance is zero).

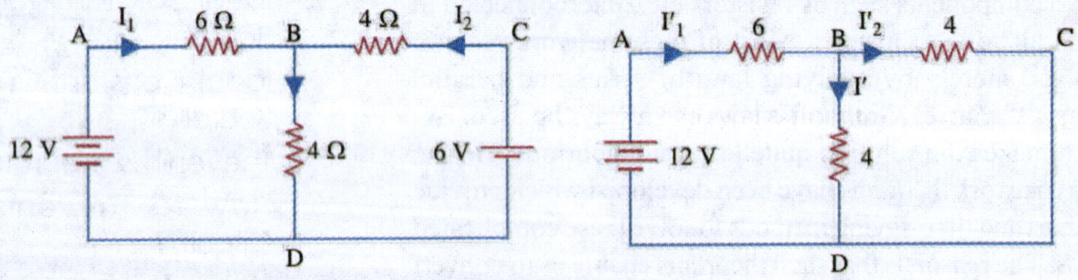

(*a*) (*b*)

Fig. 4.1

Here, $R_{BD} = 6 \parallel 4 = 2.4\ \Omega$, total resistance, $R = 4 + 2.4 = 6.4\ \Omega$. Hence, $I_2'' = 6/6.4 = 0.94$ A. This current divides at point *B* in the inverse ratio of the resistances of the two parallel paths.

$I_1'' = 0.94 \times 4/10 = 0.38$ A; $I'' = 0.94 \times 6/10 = 0.56$ A

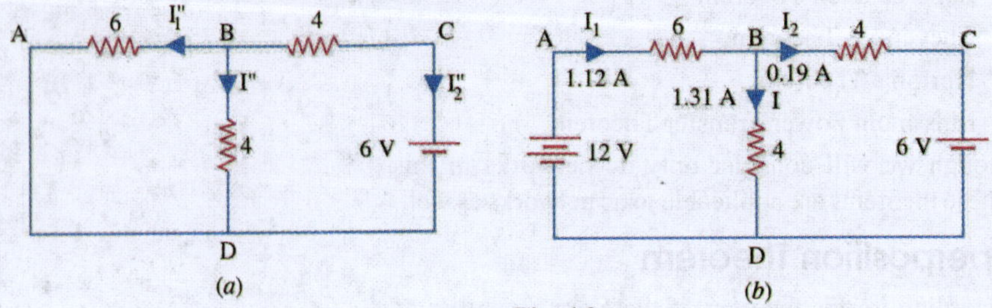

(*a*) (*b*)

Fig. 4.2

If we combine the results of Fig. 4.1 (b) and 4.2 (a), we get,

$$I_1 = I''_1 - I'_1 = 1.5 - 0.38 = \mathbf{1.12\ A}$$
$$I_2 = I''_2 - I'_2 = 0.94 - 0.75 = \mathbf{0.19\ A}$$
$$I = I' + I'' = 0.75 + 0.56 = \mathbf{1.31\ A}$$

Actual currents are shown in Fig. 4.2 (b).

Example 4.2. *The circuit of Fig. 4.3 (a) is excited by two voltage sources of zero internal resistances. Use Superposition Principle to find current flowing through the common resistance R_3 and voltage drop across it.*

Solution. Of course, it is understood in Fig. 4.3 (a) that negative terminal of each voltage source is grounded.

We will first replace V_2 by short-circuit and then V_1.

V_2 short-circuited

When V_2 is shorted, circuit becomes as shown in Fig. 4.3 (b) which further simplifies to that shown in Fig. 4.3 (c).

$$I_1 = 20/(6 + 4) = 2\text{ A}$$
$$V_{R4} = 6 \times 2 = 12\text{ V}$$

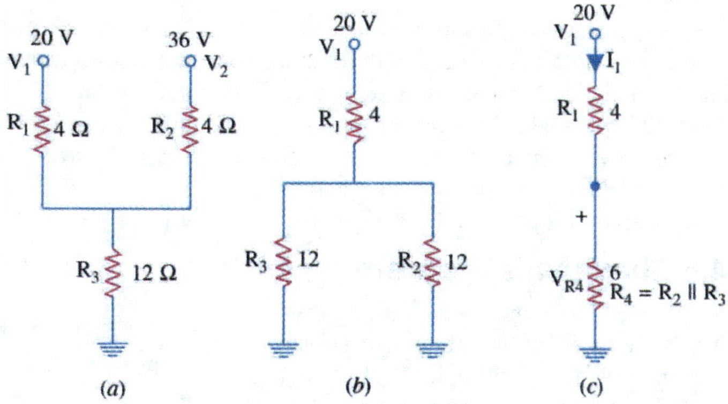

Fig. 4.3

V_1 short-circuited

Now, when source V_1 is shorted out leaving behind only V_2, the circuit becomes as shown in Fig. 4.4 (a) which further reduces to that shown in Fig. 4.4 (b).

$$I_2 = 36/15 = 2.4\text{ A}\ ;\ V_{R5} = 3 \times 2.4 = 7.2\text{ V}$$

Drop across R_3 due to both sources V_1 and V_2

$$= 12 + 7.2 = \mathbf{19.2\ V}$$

$$\text{Current through } R_3 = \frac{19.2}{12} = \mathbf{1.6\ A}$$

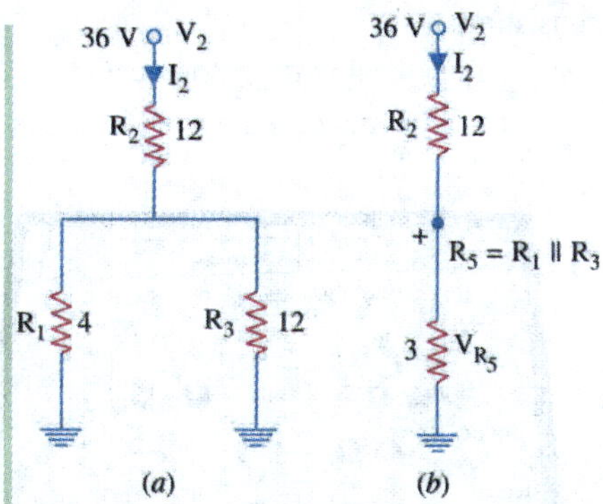

Fig. 4.4

4.3. Ideal Constant-Voltage Source

It is that voltage source or generator whose output voltage remains absolutely constant whatever the change in load current. Such a voltage source must possess *zero internal resistance so that internal voltage drop in the source is zero*. In that case, output voltage

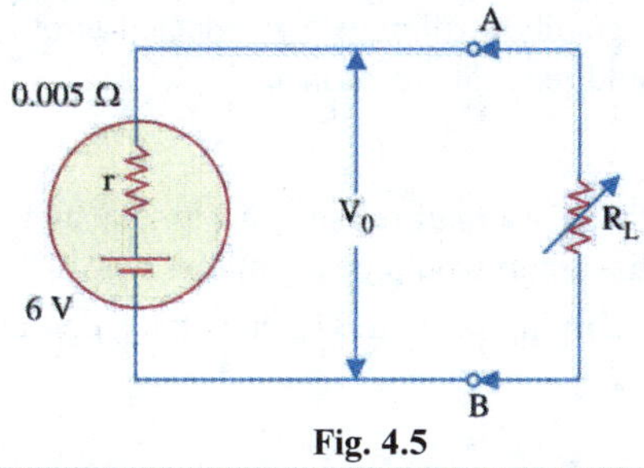

Fig. 4.5

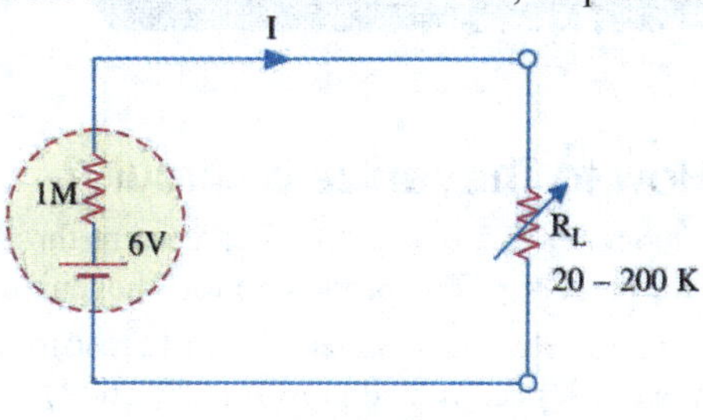

Fig. 4.6

provided by the source would remain *constant irrespective of the amount of current drawn from it*. In practice, none such ideal constant-voltage source can be obtained. However, smaller the internal resistance *r* of a voltage source, closer it comes to an ideal source described above.

Suppose, a 6 V battery has an internal resistance of 0.005 Ω (Fig. 4.5). When it supplies no current *i.e.*, it is on no-load, $V_0 = 6$ V *i.e.*, output voltage provided by it at its output terminals *A* and *B* is 6 V. If load current increases to 100 A, internal drop = 100 × 0.005 = 0.5 V. Hence, $V_0 = 6 - 0.5 = 5.5$ V.

Obviously, an output voltage of 5.5 – 6 V can be considered constant as compared to wide variation in load current from 0 A to 100 A.

4.4. Ideal Constant-Current Source

It is that voltage source whose internal resistance is infinity. In practice, it is approached by a source which possesses very high resistance as compared to that of the external load resistance. As shown in Fig. 4.6, let the 6 V battery have an internal resistance of 1 MΩ and let load resistance vary from 20 K to 200 K. The current supplied by the source varies from 6/1.02 = 5.9 μA to 6/1.2 = 5 μA. As seen, even when load resistance increases 10 times, current decreases by 0.9 μA. Hence, the source can be considered, for all practical purposes, to be a constant-current source.

4.5. Thevenin's Theorem

This theorem is very useful when we desire to know the amount of power, current or voltage drop in a particular component of a given circuit. With the help of this theorem, a normally complex circuit can be simplified to a *series* circuit* consisting of

(*i*) an ideal voltage source and

(*ii*) a resistance connected in series with it.

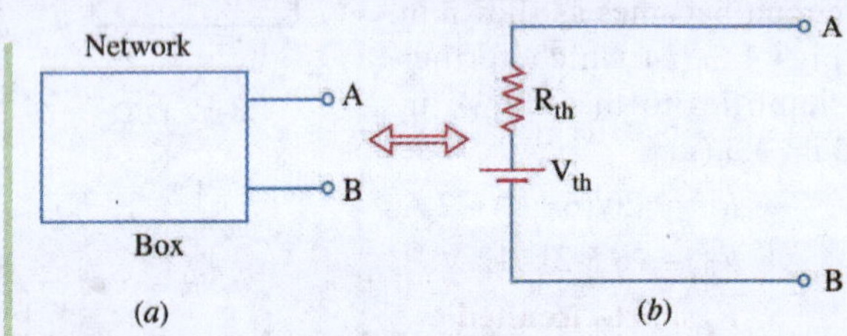

Fig. 4.7

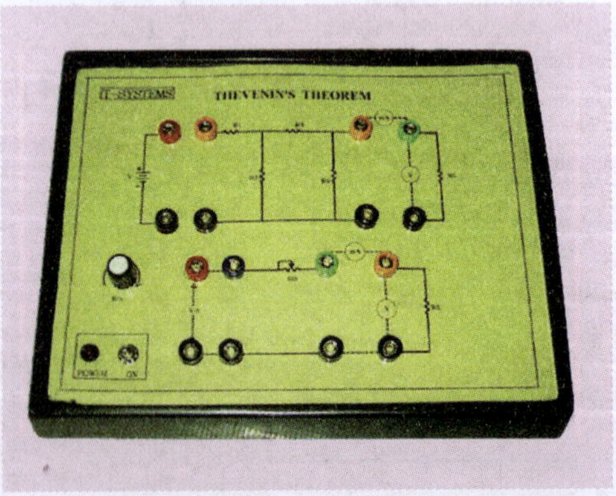

In Fig. 4.7 (*a*), imagine that the block contains a network connected to its terminals *A* and *B*. According to this theorem, the *entire* network connected to *A* and *B* can be replaced by a single voltage source V_{th} connected in series with a single resistance R_{th} across the same terminals.

V_{th} is actually the open-circuit voltage (V_{oc}) existing across terminals *A* and *B* and R_{th} is the resistance of the network as looked into or viewed back from the same terminals with all sources removed leaving behind their internal resistance if any. Actually, an ohmmeter connected across *A* and *B* would read this resistance.

4.6. How to Thevenize a Circuit ?

Suppose we are asked to find current through the 15 Ω load resistor in Fig. 5.8 by a using Thevenin's theorem. The procedure for Thevenizing this circuit would be as follows :

1. First Step. Disconnect the 15 Ω resistor from terminals *A* and *B* (incidentally, don't throw it away, we will need it again towards the end !).

* Norton's theorem reduces it to a parallel circuit (Art. 5.7).

2. Second Step. With load terminals *A* and *B* open, calculate the open-circuit voltage (V_{oc}) between them by any convenient method.

In the present case, V_{oc} is equal to the voltage drop across 12 Ω resistor because point *A* is at the same potential as point *C*.

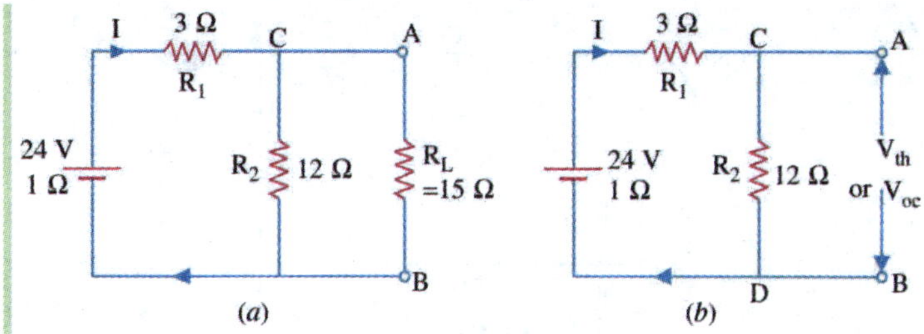

Fig. 4.8

$$V_{CD} = V_{AB} = V_{OC} = IR_2$$

Now, $$I = 24/(1 + 3 + 12) = 1.5 \text{ A}$$

∴ $$V_{CC} = 1.5 \times 12 = 18 \text{ V}$$

It is also called Thevenin voltage V_{th}.

3. Third Step. Now, remove the 24 V battery leaving behind its internal resistance of 1 Ω as shown in Fig. 4.9 (*a*).

When looked into from open-circuit terminals *A* and *B*, the circuit consists of two parallel paths : one having a resistance of 12 Ω and the other of (3 + 1) = 4 Ω. Hence, equivalent resistance of the network when viewed back from these two terminals is

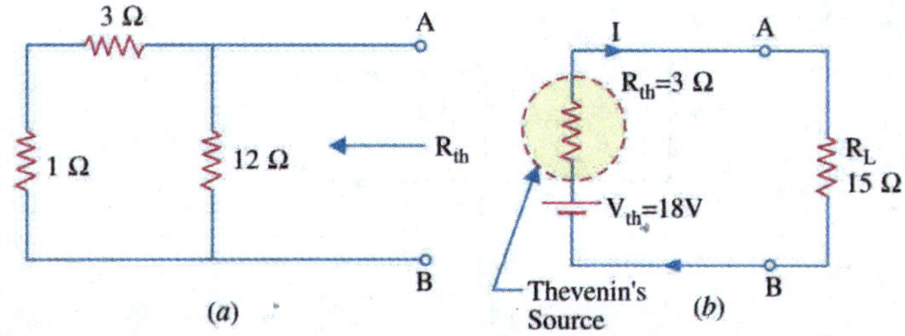

Fig. 4.9

$$= \frac{12 \times 4}{(12 + 4)} = 3 \ \Omega$$

It is also called Thevenin resistance R_{th}.

Consequently, as seen from points *A* and *B*, the given network can be reduced to a single voltage source (called Thevenin's source) of 18 V whose internal resistance R_{th} equals 3 Ω [Fig. 4.9 (*b*)].

4. Fourth step. Finally, connect back the load resistor to the terminals *A* and *B* thereby giving a simple series circuit shown in Fig. 4.9 (*b*). Finding current I through R_L should be no problem now.

$$I = 18/(3 + 15) = 1 \text{ A}$$

Example 4.3. *Apply Thevenin's theorem to find current through the 12 Ω resistor of the circuit shown in Fig. 4.10 (a).*

Solution. In Fig. 4.10 (*b*), point *A* is at the same potential as point *C* since there is no current through the 4 Ω resistor and hence no drop across it.

$$V_{th} = V_{CD} = I \times 6$$

Now $$I = 36/(3 + 6) = 4 \text{ A}$$

∴ $$V_{th} = 4 \times 6 = 24 \text{ V}$$

Since, 36 V battery has no internal resistance, it has been replaced by a short-circuit in Fig. 4.11 (*a*).

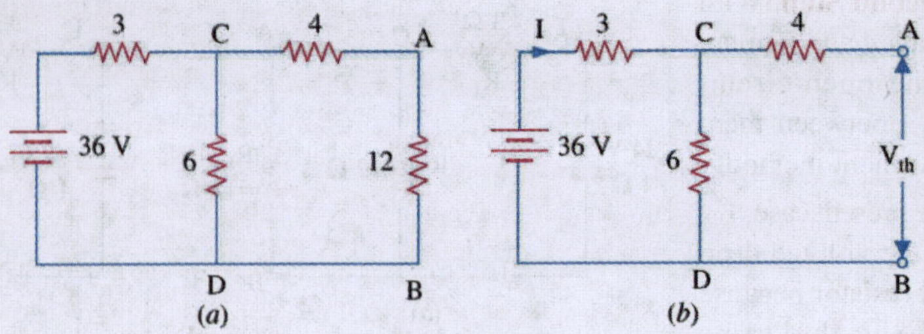

Fig. 4.10

$$R_{th} = 4 + \frac{6 \times 3}{(6+3)} = 6\ \Omega$$

Having found the Thevenin's source, the 12 Ω resistor is connected back in series with this source as shown in Fig. 4.11 (*b*). Current flowing through it is

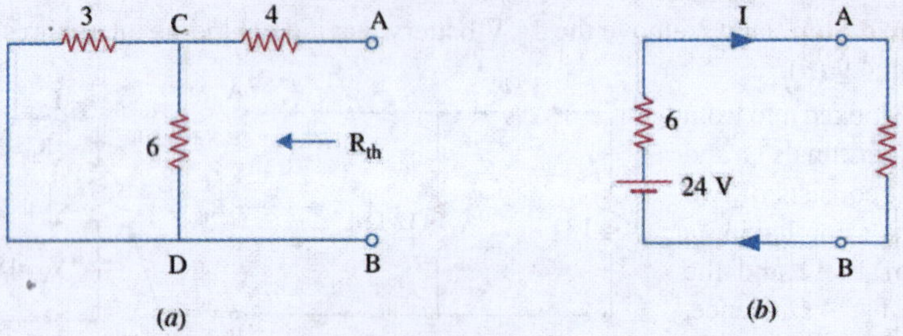

Fig. 4.11

$$I = 24/(12+6) = \mathbf{1.33\ A} \qquad \text{— from A to B}$$

Example 4.4. *Using Thevenin's theorem, calculate the current through the 4 K resistor of Fig. 4.12 (a).*

Solution. If we remove 4 K resistor, circuit becomes as shown in Fig. 4.12 (*b*). As seen, full current of 10 mA passes through 2 K resistor producing a drop of 10 × 2 = 20 V. Hence, V_B = 20 V. w.r.t. ground. Now, 12 V battery is connected in series with two resistors 3 K and 6 K which form a voltage divider circuit.

V_A = drop across 6 K resistor*

= 12 × 6/(6 + 3) = + 8 V — w.r.t. ground

p.d. between points *A* and *B* is

V_{th} = 20 – 8 = 12 V — with *B* at a higher potential

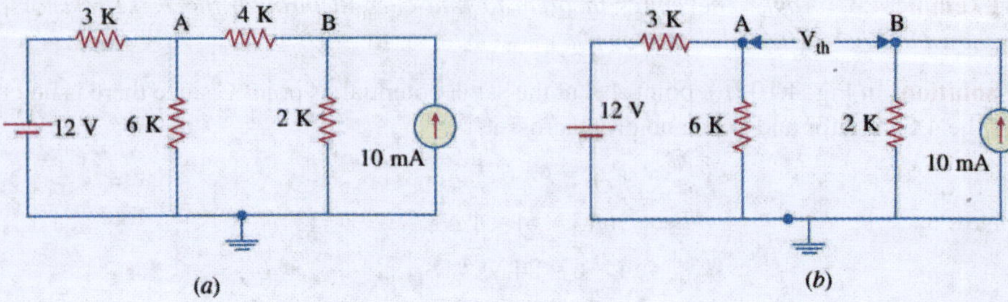

Fig. 4.12

* Or V_A = 12 — drop across 3 K resistor.

Now, we will find R_{th} *i.e.*, resistance of the network as looked back into the open-circuited terminals *A* and *B*. For this purpose, we will replace both the voltage and current sources. Since voltage source has no internal resistance, it would be replaced by a short-circuit *i.e.*, zero resistance. However, current source would be removed and replaced by an infinite resistance *i.e.*, an 'open' (Art. 4.4). In that case, the circuit becomes as shown in Fig. 4.13 (*a*).

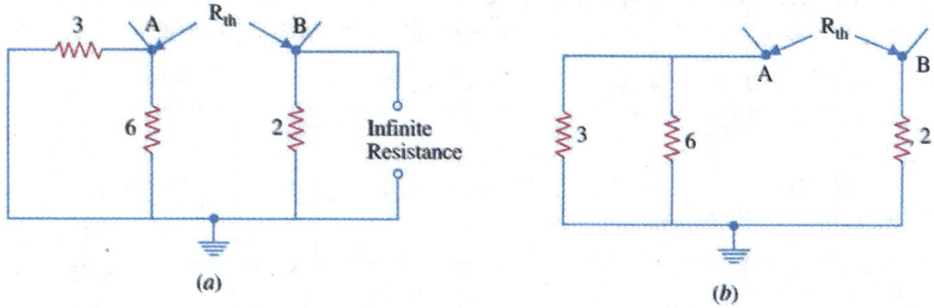

Fig. 4.13

As seen from Fig. 4.13 and Fig. 4.14 (*a*), the value of $R_{th} = 2 + 2 = 4\ \Omega$.

Hence, Thevenin source has a voltage of 12 V and an internal resistance of 4 Ω as shown in Fig. 4.14 (*b*).

∴ $I = 12/(4 + 4) = \mathbf{1.5\ A}$

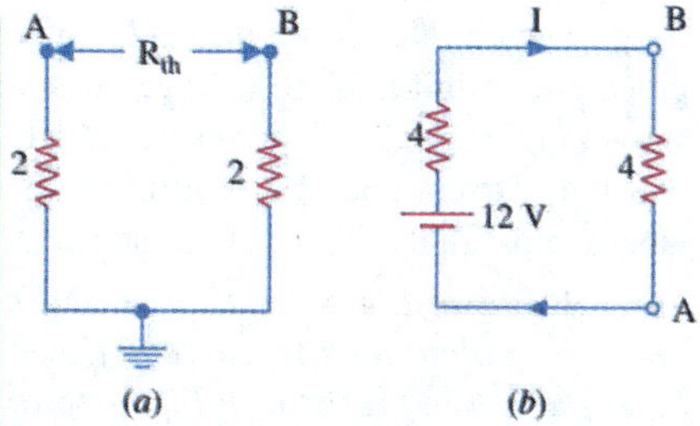

Fig. 4.14

4.7. Norton's Theorem

This theorem is used where it is easier to simplify a network in terms of currents instead of voltage. This theorem reduces a normally complicated network to a simple *parallel* circuit consisting of

(*a*) an ideal current source I_N of infinite internal resistance and

(*b*) a resistance R_N (or conductance $G_N = I/R_N$) in parallel with it as shown in Fig. 4.15.

Here, I_N is the current which would flow through a short circuit placed across terminals *A* and *B*. R_N is the circuit resistance looking back from the open *A-B* terminals. These terminals are not short-circuited for finding R_N but are 'open' as for calculating R_{th} for Thevenin's Theorem*.

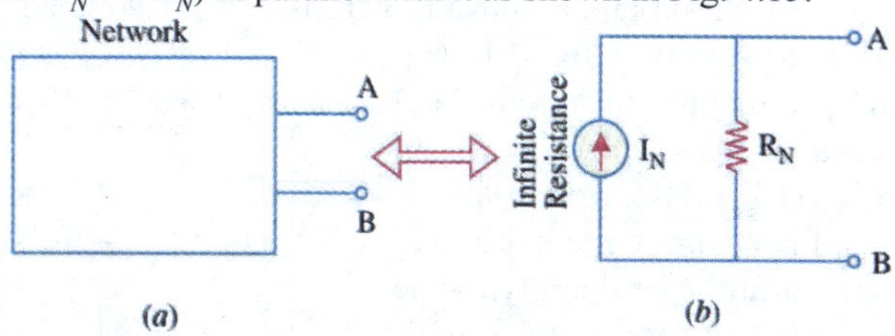

Fig. 4.15

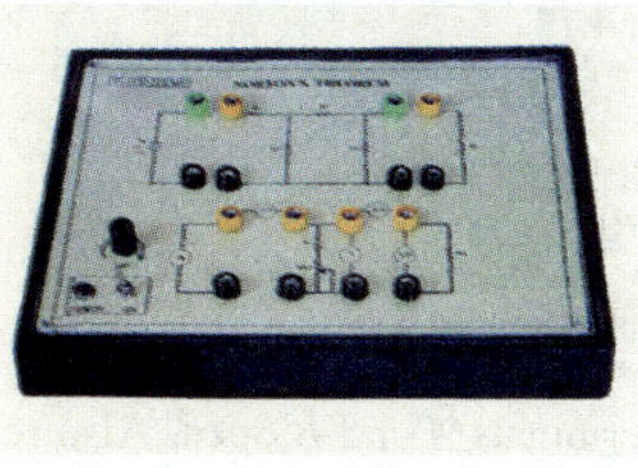

4.8. How to Nortonise a Given Circuit

Suppose we want to Nortonise the circuit shown in Fig. 4.16 (*a*) *i.e.*, we want to find Norton's equivalent of this circuit between terminals *A* and *B*.

The different steps are as under :

1. First Step. Put a short across terminals *A* and *B* [Fig. 4.16 (*b*)]. As seen, it results in shorting out 12 Ω resistor as

* In fact, this resistance is the same both for Thevenin and Norton equivalent circuits. In Norton's case, this resistor is in parallel with the current source whereas in Thevenin's case, it is in series with V_{th} (Art. 4.6).

shown separately in Fig. 4.16 (*c*).

$$\therefore \quad I_{SC} = 24/4 = 6 \text{ A}$$

This current is usually called Norton current I_N.

2. Second Step. Remove the short from terminals *A* and *B* so that they are again open.

3. Third Step. Remove the battery and replace it by its internal resistance which, in the present case, is zero. The resistance R_N of the circuit as viewed back or looked into from open terminals *A* and *B* is

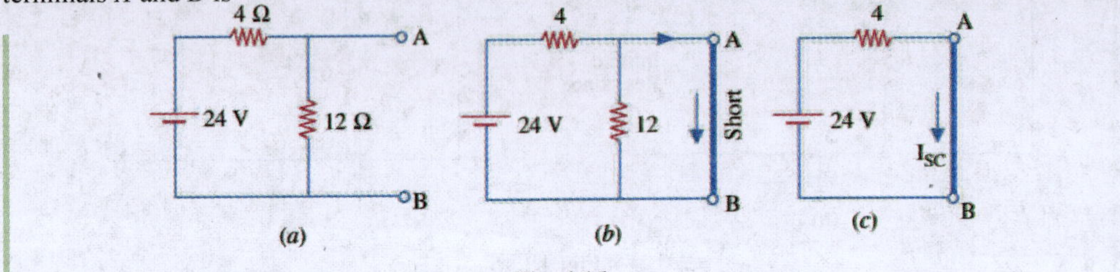

Fig. 4.16

$$R_N = 12 \parallel 4 = 3 \ \Omega$$

Hence, Norton's equivalent of the given circuit in Fig. 4.16 (*a*) becomes that shown in Fig. 4.17 (*b*). It consists of a 6 A constant-current source (of infinite resistance) in parallel with a 3 Ω resistance.

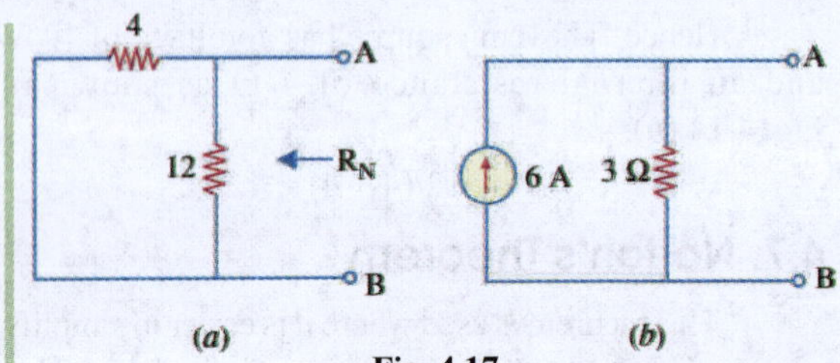

Fig. 4.17

Example 4.5. *Using Norton's theorem, calculate the current flowing through the 12 Ω resistor in Fig. 4.18 (a).*

Solution. It is the same circuit as shown in Fig. 4.10 (*a*) and solved earlier with the help of Thevenin's theorem (Example 4.3). When applying Norton's theorem, the procedure would be as under :

1. First Step. Remove 12 Ω resistor from terminals *A* and *B* and then put a short-circuit across them as shown in Fig. 4.18 (*b*).

The current passing through 4 Ω resistor is also the short-circuit current I_{SC} (also written as I_N). For finding this current, we have first to find *I* by simplifying the circuit. As seen, total circuit resistance

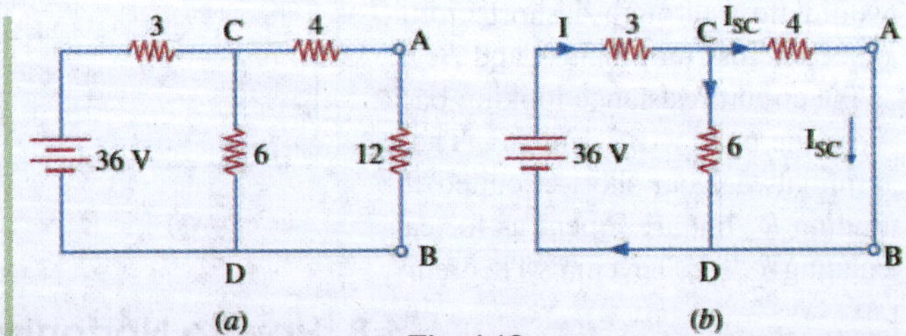

Fig. 4.18

$$= 3 + 6 \parallel 4 = 3 + 2.4$$

$$= 5.4 \ \Omega$$

$$I = 36/5.4 = 20/3 \text{ A}$$

This current divides at point *C* into two unequal parts.

$$I_{SC} = \frac{20}{3} \times \frac{6}{10} = 4 \text{ A}$$

2. Second Step. Remove the short circuit, thereby leaving terminals *A* and *B* open. Also remove the battery. Since its internal resistance is zero, it is replaced by a resistanceless piece of connecting wire thereby closing the circuit [Fig. 4.19 (*a*)]. The resistance R_N of the circuit as viewed from terminals *A* and *B* is

$= 4 + 6 \parallel 3 = 6\ \Omega$

Hence, the Norton's equivalent of the given circuit is as shown in Fig. 4.19 (*b*). Load current I_L through 12 Ω resistor can be found by using the Proportional Current Formula (Art. 2.16).

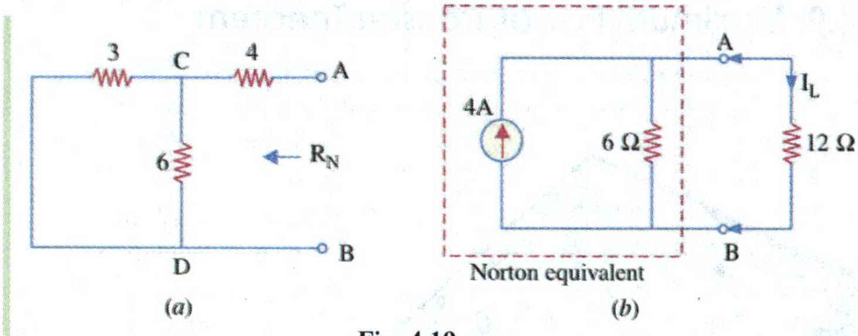

Fig. 4.19

$$\therefore \qquad I_L = 4 \times \frac{6}{(6+12)} = \mathbf{1.33\ A} \qquad \text{— from A and B.}$$

It is the same value as found in Example 4.3 earlier.

Example 4.6. *The circuit of Fig. 4.20 (a) is excited by a voltage source and a current source. Using Norton's theorem, calculate current through the 6 Ω resistor.*

Solution. In Fig. 4.20 (*b*), 6 Ω resistor has been removed and a short placed across terminals *A* and *B*. Short-circuit current I_{SC} (or I_N) is

$$= \text{current from voltage source + current of current-source.}$$

$$= \frac{60}{20} + 12 = 15\ \text{A}$$

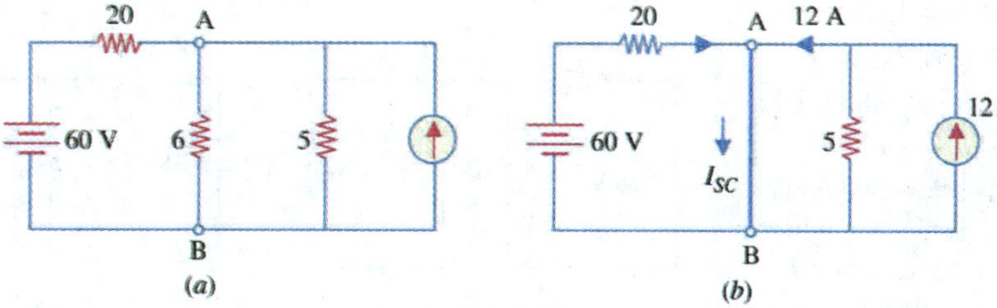

Fig. 4.20

It should be noted that short across *AB* also shorts out 5 Ω resistor. Hence, all the 12 A current passes through short-circuit and none through 4 Ω resistor.

In Fig. 4.21 (*a*), 'short' has been removed leaving terminals *A* and *B* open. Voltage source has been replaced by a connecting wire of zero resistance (since its internal resistance is zero). Current source has been replaced by an 'open' since it has infinite resistance.

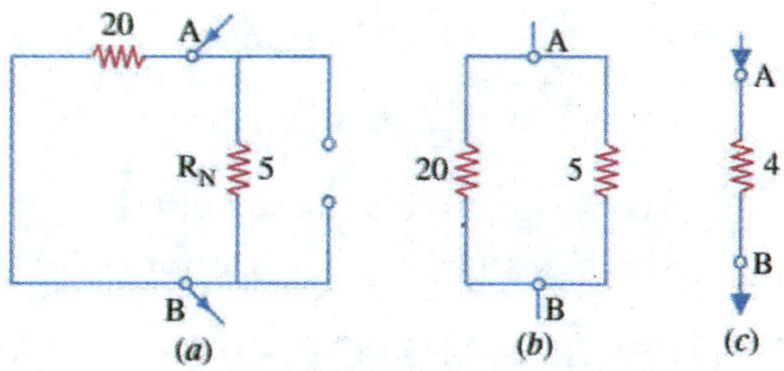

Fig. 4.21

When looked into the circuit from terminals *A* and *B*, there are two parallel paths between points *A* and *B* having resistances of 20 Ω and 5 Ω. Their combined resistance, as seen from Fig. 4.21 (*c*) is $= 20 \parallel 5 = 4\ \Omega$. Hence, $R_N = 4\ \Omega$.

The Norton's equivalent of the original circuit with respect to terminals *A* and *B* is shown in Fig. 4.22. The 6 Ω resistance has been connected back to the terminals *A* and *B* (from where it was removed earlier).

$$I_L = 15 \times 4/(6+4) = \mathbf{6\ A}$$

4.9. Maximum Power Transfer Theorem

This theorem is very useful for analysing electronic and communication networks where main consideration is to transfer maximum power to the load irrespective of the efficiency. Its application to power transmission and distribution networks is limited because in their case, the goal is high efficiency and not maximum power transfer.

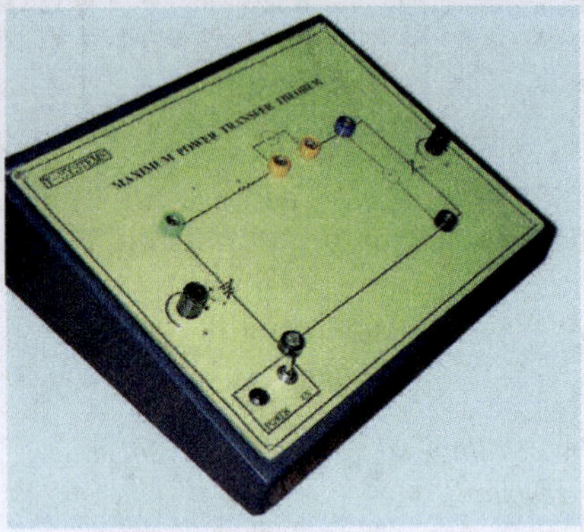

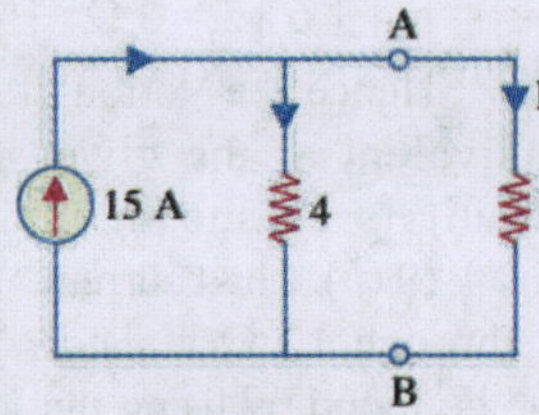

Fig. 4.22

When applied to dc networks, this theorem states that *a resistive load will abstract maximum power from a network when its resistance equals the resistance of the network as viewed from the output terminals with all voltage and current sources removed leaving behind their internal resistances if any.*

Example 4.7. *In the circuit of Fig. 4.23, find the value of load resistance R_L to be connected across terminals A and B which would abstract maximum power from the circuit. Also find the value of this maximum power.*

Solution. As seen, resistance of the network as viewed back from terminals A and B (with battery removed) is

$$= 4 + 6 \parallel 3 = 6\ \Omega$$

Hence, R_L should be equal to **6 Ω.**

Let us now find power developed in R_L for which purpose we have to find I_L.

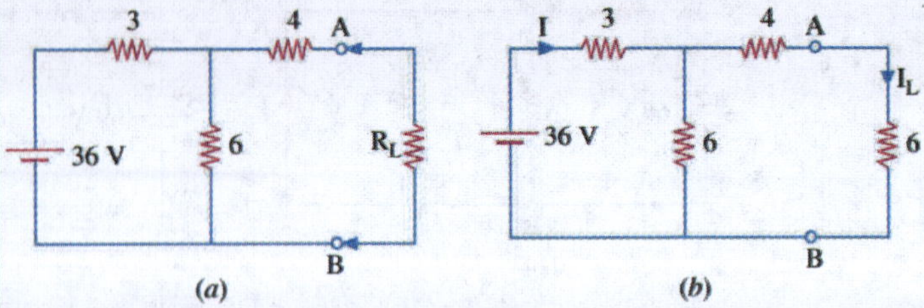

Fig. 4.23

In Fig. 4.23 (*b*),

Total circuit resistance $= 3 + 6 \parallel 10$

$$= 27/4\ \Omega$$

$$\therefore \quad I = 3 \div 27/4 = 16/3\ \text{A}$$

$$\therefore \quad I_L = \frac{16}{3} \times \frac{6}{(6+10)} = 2\ \text{A}$$

Max. power possible in load resistance $R_L = 2^2 \times 6 =$ **24 W**

It can be verified that if we have any other value of R_L, power drawn will be less than 24W.

CONVENTIONAL PROBLEMS

1. Use Superposition Principle to find current through 12 Ω resistor of Fig. 4.24. All resistance values are in ohms. **[34/21 A]**

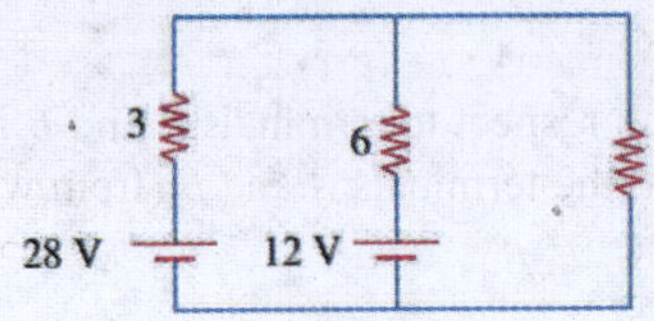

Fig. 4.24

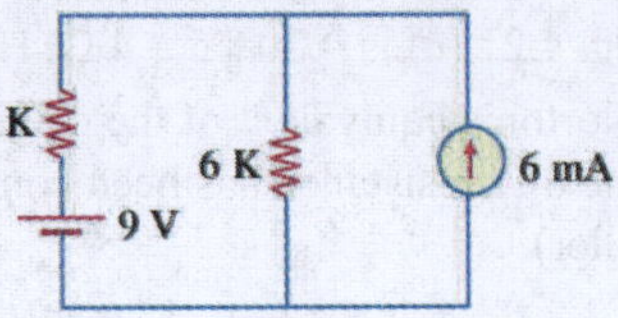

Fig. 4.25

2. Using Superposition Theorem, calculate the current flowing through 6 K resistor of the circuit shown in Fig. 4.25. **[3 mA]**

3. Make use of Thevenin's theorem to find current in the 12 Ω resistor of Fig. 4.26. **[0.6 A]**

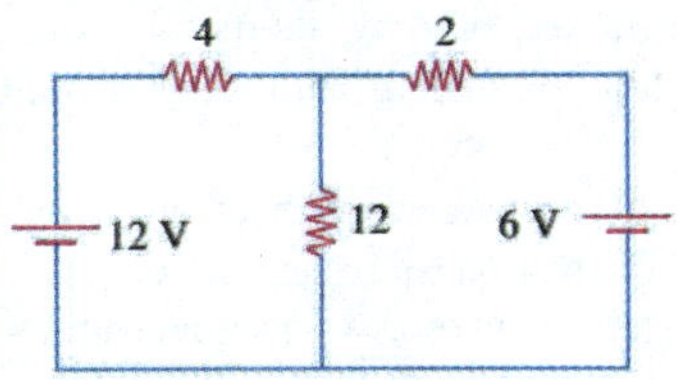

Fig. 4.26

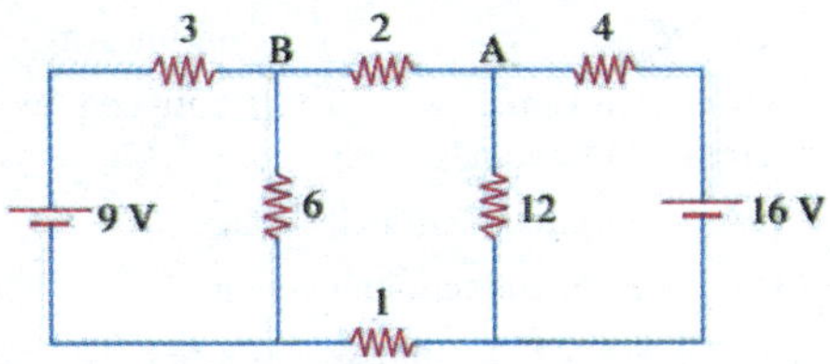

Fig. 4.27

4. In the circuit of Fig. 4.27, use Thevenin's theorem to find the magnitude and direction of flow of current through 2 Ω resistor. **[0.75 A from A to B]**

5. For the circuit of Fig. 4.25, find Norton's equivalent and hence find circuit through the 6 K resistor. **[I_N = 9 mA, R_N = 3 Ω, 3 mA]**

6. What is the Norton equivalent for the circuit shown in Fig. 4.26 ? Use it to find current flowing through 12 Ω resistor. **[I_N = 6 mA, R_N = 4/3 Ω; 0.6 A]**

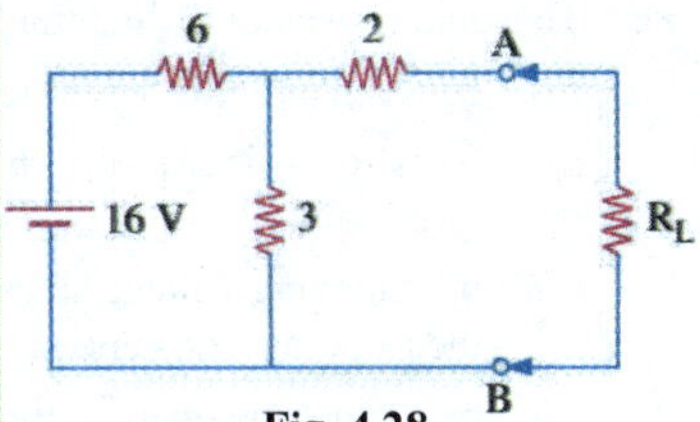

Fig. 4.28

7. According to Maximum Power Transfer Theorem, what should be the value of load resistance R_L to abstract maximum power from the 16 V battery shown in Fig. 4.28 ? What is the value of this power ? **[4 Ω, 16/9 W]**

SELF EXAMINATION QUESTIONS

A. Fill in the blanks with most appropriate word(s) or numerical value (s).

1. While using Superposition Theorem, all voltage sources except one are removed leaving behind their .... resistances if any.
2. An ideal constant-voltage source has .... resistance.
3. An ideal constant-current source has .... resistance.
4. According to Thevenin's theorem, any network with two open terminals can be replaced by a single voltage source V_{th} in .... with a single resistance R_{th}.
5. While finding R_{th}, all voltage sources are removed but not their .... resistances.
6. Norton's equivalent of a circuit consists of a constant-current source and a resistance in .... with it.
7. A load draws maximum power from a circuit when its resistance .... the circuit resistance when viewed back from output terminals.

B. Answer True or False

1. V_{th} is an open-circuit voltage.
2. I_N is an open-circuit current.
3. Superposition principle treats each voltage source as if it acts independently of others in a circuit.
4. A constant-voltage source has infinite resistance.
5. A constant-current source has zero resistance.
6. R_{th} is the same as R_N.

C. Multiple Choice Items

1. According to Superposition theorem as applied to circuits excited by more than one voltage source
   (*a*) current is contributed by each source
   (*b*) all voltage sources sends their currents in the same direction
   (*c*) current in any resistor equals the algebraic sum of currents which each source would send if acting alone
   (*d*) algebraic sum of voltage drops equals the sum of total applied voltage
2. Superposition theorem can be applied only to circuit having .... elements.
   (*a*) non-linear (*b*) passive
   (*c*) linear bilateral (*d*) resistive

3. The Superposition theorem is essentially based on the concept of
(a) duality (b) linearity
(c) reciprocity (d) non-linearity

4. While Thevenizing a circuit between two terminals, V_{th} equals
(a) short-circuit terminal voltage
(b) open-circuit terminal voltage
(c) EMF of the battery nearest to the terminals
(d) net voltage available in the circuit

5. Thevenin resistance R_{th} is found
(a) between any two 'open' terminals
(b) by short-circuiting the given two terminals
(c) by removing voltage sources along with their internal resistances
(d) between same open terminals as for V_{th}

6. Norton's equivalent of a circuit consists of a
(a) constant-current source with a conductance in parallel
(b) constant-current source in series with an infinite resistance
(c) constant-voltage source in parallel with a high resistance
(d) single current source and a single voltage source

7. While calculating R_{th}, constant-current sources in the circuit are
(a) replaced by 'opens'
(b) replaced by 'shorts'
(c) treated in parallel with other voltage sources
(d) converted into equivalent voltage sources

8. The Norton equivalent of a circuit consists of a 2 A current source in parallel with a 4 Ω resistor. Thevenin equivalent of this circuit is a .... volt source in series with a 4 Ω resistor.
(a) 2 (b) 0.5
(c) 6 (d) 8

9. If two identical 3 A, 4 Ω Norton equivalent circuits are connected in parallel with like polarity to like, the combined Norton equivalent circuit is
(a) 6 A, 4 Ω (b) 6 A, 2 Ω
(c) 3 A, 2 Ω (d) 6 A, 8 Ω

10. For abstracting maximum power from any two given terminals of a circuit, the load resistance across the terminals should be
(a) four times the internal resistance of the network
(b) equal to the circuit resistance when viewed back from the two terminals
(c) less than the circuit resistance
(d) greater than the circuit resistance.

ANSWERS

A. Fill in the blanks

1. internal **2.** zero **3.** infinite **4.** series **5.** internal
6. parallel **7.** equals

B. True or False

1. T **2.** F **3.** T **4.** F **5.** F **6.** T

C. Multiple Choice Items

1. *c* **2.** *c* **3.** *b* **4.** *b* **5.** *d* **6.** *a* **7.** *a* **8.** *d* **9.** *d* **10.** *b*.

CHAPTER 5

Passive Circuit Elements

1. Resistors
2. Wire-Wound Resistors
3. Carbon Composition Resistors
4. Carbon Film Resistors
5. Cermet Film Resistors
6. Metal Film Resistors
7. Variable Resistors
8. Fusible Resistors
9. Resistor Colour Code
10. Inductor
11. Variable Inductors
12. Inductors in Series
13. Energy Stored in a Magnetic Field
14. Capacitors
15. Capacitor Connected to a Battery
16. Capacitance
17. Variable Capacitors
18. Leakage Resistance
19. Capacitors in Series
20. Two Capacitors in Series
21. Capacitors in Parallel
22. Energy Stored in a Capacitor

5.1. General

Individual components which make up an electronic circuit are called *elements* or *parameters*. Most commonly-used elements in such circuits are :

1. resistors 2. inductors 3. capacitors

In resistors, current is *directly* proportional to the *applied voltage*. In inductors, voltage required is *directly* proportional to the rate of change of *current* whereas capacitors require current which is *directly* proportional to the rate of change of *voltage*.

5.2. Resistors

A resistor is an electrical component with a known specified value of resistance. It is probably the most common component in all kinds of electronic equipment ranging from a small radio to a colour television receiver. As its name suggests, a resistor *resists* or *opposes* the flow of current through it. Resistance is necessary for any circuit to do useful work. In

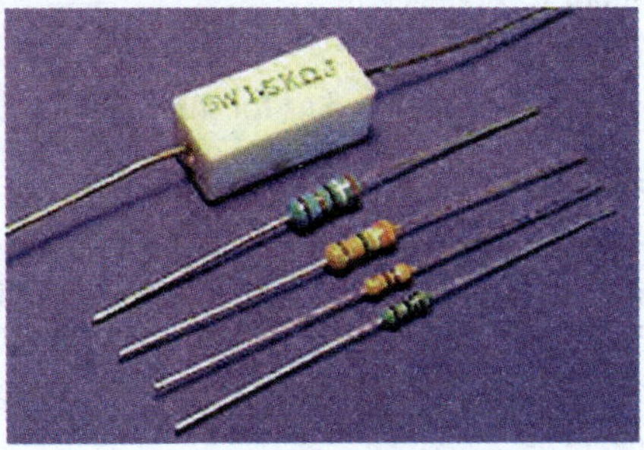

Resistors.

fact, without resistance, *every circuit would be a short circuit* !

Some of the common uses of resistors are :

1. to establish proper values of circuit voltages due to *IR* drops
2. to limit current and
3. to provide load

The two main characteristics of a resistor are its resistance and power rating. Resistors can be connected in the circuit in either direction because they have no 'polarity'.

5.3. Resistor Types

Resistors are mainly of two types and can be either of fixed or variable value.

1. wire-wound resistors
2. carbon resistors
   (*a*) carbon-composition type
   (*b*) carbon-film type
   (*c*) cermet-film type

Another type is called metal thin-film resistor.

5.4. Wire-Wound Resistors

They are constructed from a long fine wire (usually nickel-chromium wire) wound on a ceramic core. The length of the wire used and its resistivity determine the resistance of the unit. The wire is bare but the entire assembly is covered or coated with a ceramic material or special vitreous enamel (Fig. 5.1).

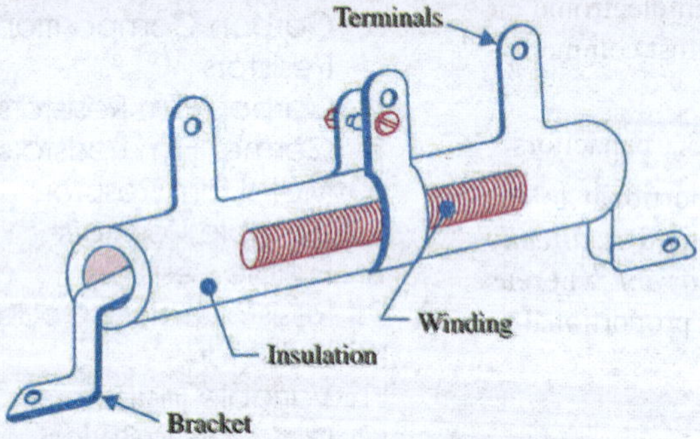

Fig. 5.1

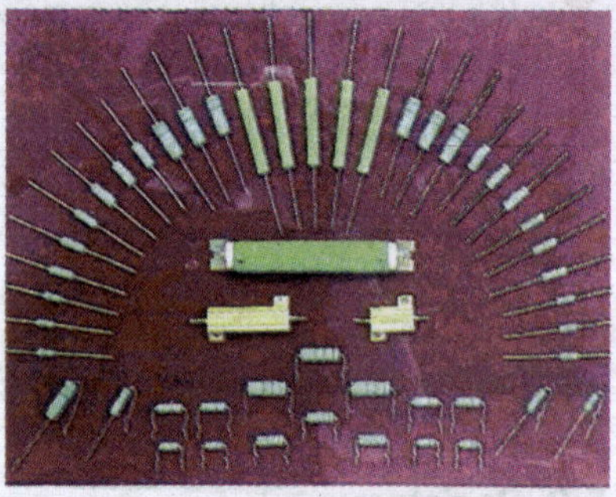

Wire wound resistor.

Such resistors are generally available in power ratings from 5 W to several hundred watts and resistance values from 1 Ω to 100 K. These can be of either fixed value or variable type.

Wire-wound resistors are used where

(*a*) large power dissipation is necessary
(*b*) precise and stable resistance values are required as for meter shunts and multipliers.

5.5. Carbon Composition Resistors

They are made of finely-divided carbon mixed with a powdered insulating material in suitable proportion. Often, the resistance element is a simple rod of pressed carbon granules which is usually enclosed in a plastic case for insulation and mechanical strength [Fig. 5.2 (*a*)]. The two ends of the carbon resistance element are joined to metal caps with leads of tinned wire for soldering its connections into a circuit.

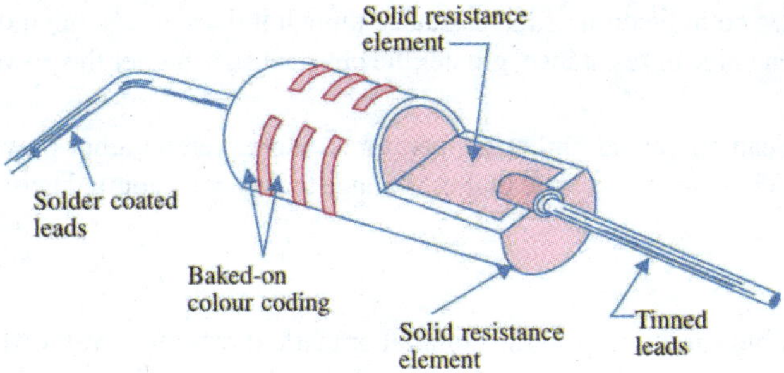

Fig. 5.2

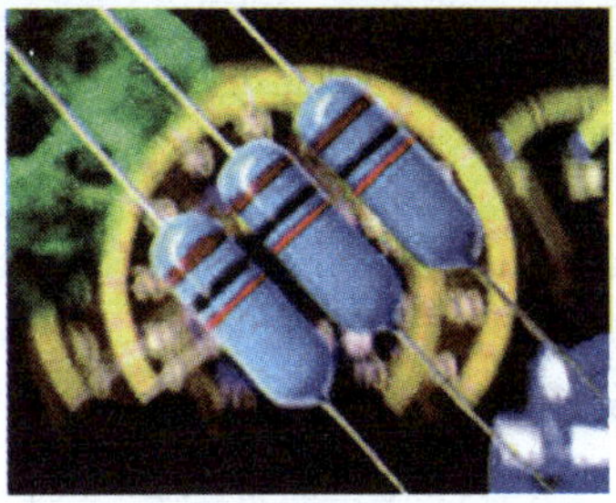
Carbon composition resistors.

Such resistors are available in power ratings of 1/10, 1/8, 1/4, 1/2, 1, 2 watt and in resistance values ranging from 1 Ω to 20 MΩ. Where power dissipation is 2 W or less, such resistors are preferred because they are smaller and cost less. Carbon resistors with power rating of 1 W or less are most common in electronic equipment.

5.6. Carbon Film Resistors

They consist of a high-grade ceramic rod or core (called the substrate) on which is deposited a thin resistive film of carbon. They are cheaper than composition resistors.

5.7. Cermet Film Resistors

They consist of thin carbon coating fired on to a solid ceramic substrate. The main purpose is to have more precise resistance values and greater stability with heat. Very often, they are made in a small square with leads to fit into a printed circuit board (PCB).

5.8. Metal Film Resistors

They are also referred to as thin-film resistors. They consist of a thin metal coating deposited on a cylindrical insulating support. The high resistance values are due to thinness of the film. Because it is difficult to produce films of uniform thickness, it is not possible to control their resistance values as accurately as in the case of wire-wound resistors. However, such resistors are free of trouble-some inductance effects so common in wire-wound resistors particularly at high frequencies.

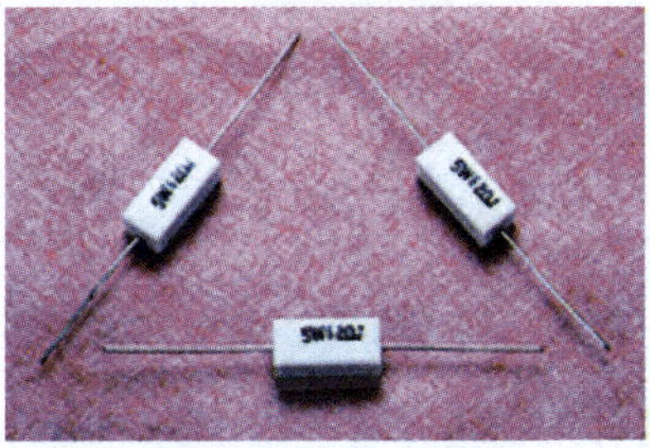
Metal oxide film resistors.

5.9. Power Rating

The power rating of a resistor is given by the maximum wattage it can dissipate without excessive heat. Since it is current which produces heat, power rating also gives some indication of the maximum current a resistor can safely carry. If the current exceeds this value, more heat will be produced than can be carried safely and the resistor will burn out. A 1/2 watt resistor, for example, can dissipate 1/2 watt of heat without damage whereas a 1W resistor can throw off twice as much heat. In a circuit, you may substitute 1 watt resistor of same resistance value for a 1/2 watt resistor but not *vice-versa*.

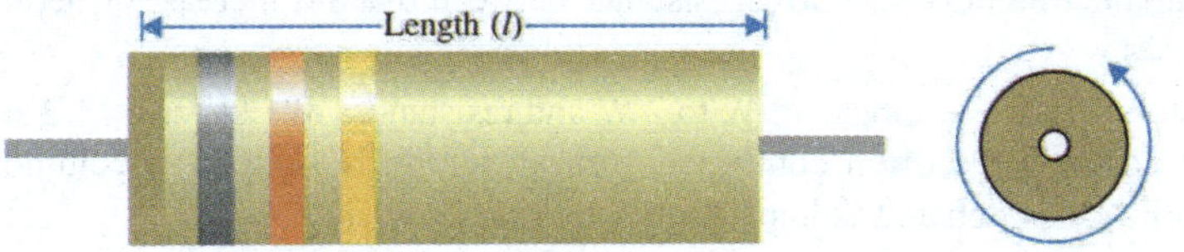

The larger the surface area of a resistor, the more power it can dissipate. The power rating of a resistor is directly related to its surface area.

The physical size of a resistor is no indication of its resistance though it does give some indication of its wattage rating. For a given value of resistance, greater the physical size, higher the power rating.

Also, higher-wattage resistors can operate at higher temperatures. Moreover, a higher power rating allows a higher voltage rating. This rating gives the highest voltage that may be applied across the resistor without internal arcing.

5.10. Value Tolerance

By tolerance is meant the possible variation from the nominal or marked resistance value of a resistor. It means that actual resistance of a resistor may be greater or lesser than its indicated value. All resistors are manufactured and sold with a specified tolerance. For example, a 1000 Ω resistor with a tolerance of 10% will have an actual resistance anywhere between 900 Ω and 1100 Ω *i.e.,* 100 Ω more or less than the rated value.

Carbon-composition resistors have tolerances of ± 5%; ± 10% and ± 20% whereas general-purpose wire-wound resistors usually have a tolerance of ± 5%.

5.11. Variable Resistors

These are the resistors whose resistance can be changed between zero and a certain maximum value. They can be wire-wound or carbon type. As shown in Fig. 5.3, the sliding arm has been attached to a shaft which can be rotated in almost a complete circle. As the shaft rotates, the point of

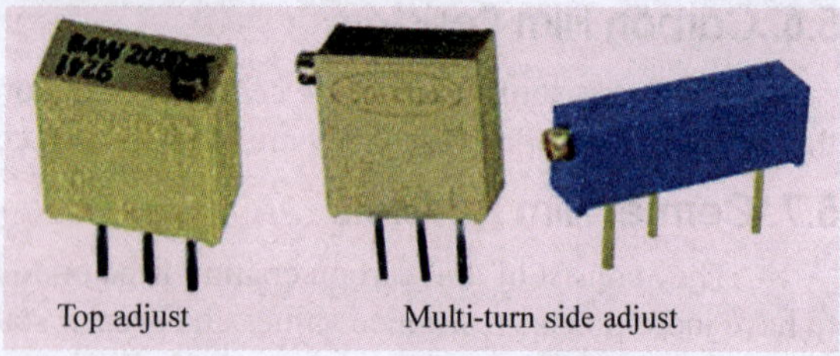

Variable resistors.

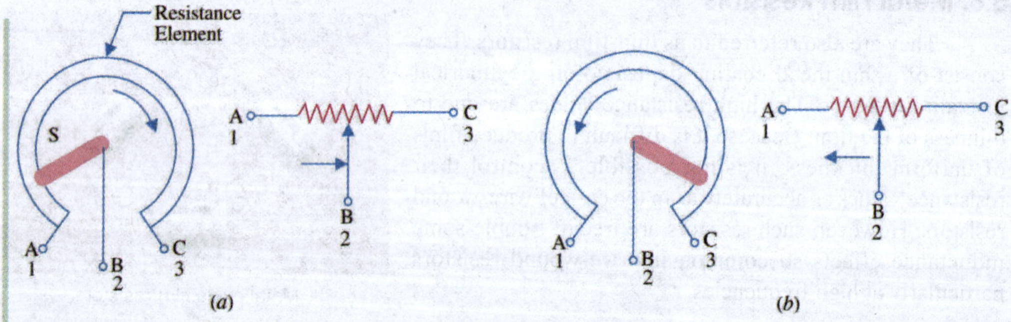

Fig. 5.3

contact of the sliding arm on the circular carbon-composition resistance element changes thus changing the resistance between arm terminal *B* and terminals of the stationary resistance *A C*. In Fig. 5.3 (*a*) as we move the sliding arm, the resistance between *B* and *A* increases whereas that between *B* and *C* decreases. In Fig. 5.3 (*b*), with the rotating of the arm, resistance between *B* and *C* increases whereas that between *B* and *A* decreases.

Carbon variable resistors of power ratings 1/2 W to 2 W and resistances of 1 kΩ to 5 MΩ are commonly available. Such controls are often combined with an OFF-ON switch—a common example being the power OFF-ON switch and volume control of a radio receiver.

5.12. Potentiometers and Rheostats

These are variable resistors either of carbon or wire-wound type often used for controlling voltage and current in a circuit.

(a) Potentiometers

They generally have carbon composition resistance element and are connected across a voltage source. They have *three* terminals, the centre one being connected to the variable arm which is used for varying voltage division in the circuit as shown in Fig. 5.4. By moving the variable arm *B* over the fixed resistance *R* between points *A* and *C*, any part of the input voltage can be tapped off. Since in Fig. 5.4, *B* happens to be at the middle value of *R*, output voltage is half the input voltage.

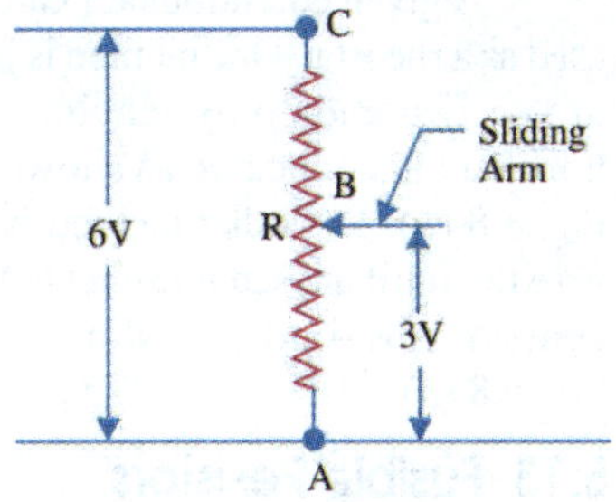

Fig. 5.4

Most variable resistors used in radios are potentiometers meant for controlling volume or tone. When used as a volume control, it picks off a voltage between zero and the full available voltage as shown in Fig. 5.5. By moving *B* up and down, any desired amount of voltage (signal) can be picked up between maximum signal value at point *A* and zero signal value at *C*.

The tone control circuit shown in Fig. 5.6 uses only two terminals on the potentiometer. The resistor allows the capacitor to by-pass to ground either more or less of high frequencies in an audio circuit. When *B* is at point *A*, the capacitor becomes a direct by-pass to ground for higher frequencies in the audio signal. Consequently, radio sound becomes 'bassy'. When *B* is at terminal *C*, it increases the amount of resistance in series with the capacitor. Hence, less amount of high frequencies is by-passed to ground and consequently, there is more trebble in the sound.

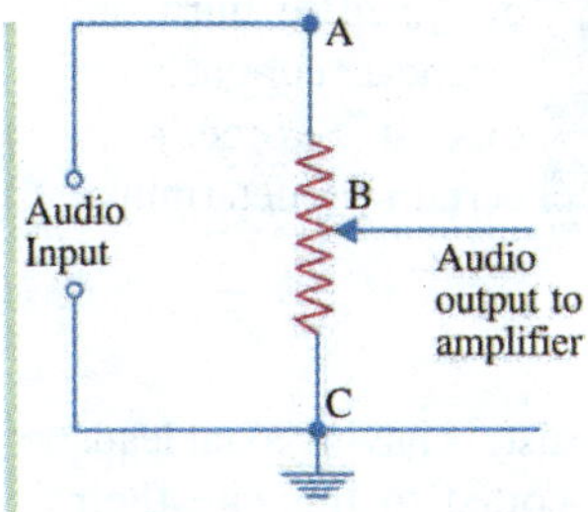

Fig. 5.5

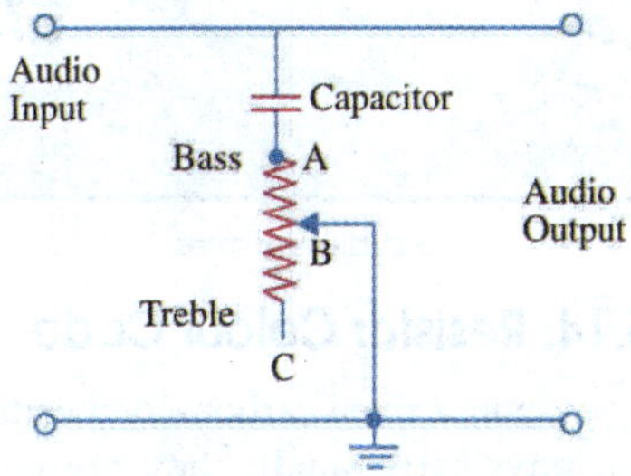

Fig. 5.6

As illustrated above, potentiometers are commonly used as control devices in amplifiers, TV sets and various types of meters. Typical applications include volume and tone controls, balance controls, linearity and brightness control in TV receivers etc.

(b) Rheostats

The resistance element of rheostats is made of high-resistance wire. It has two terminals and is connected in *series* with a circuit for adjusting the amount of current flowing through it. Rheostats are commonly used to control relatively high currents such as those found in motor and lamp loads. Fig. 5.7 shows how a rheostat can be connected into a lamp circuit for controlling its current. As seen, only resistance *BC* is connected into the circuit.

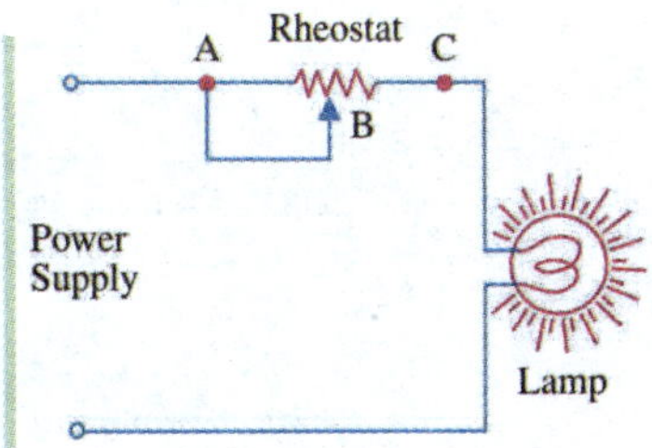

Fig. 5.7

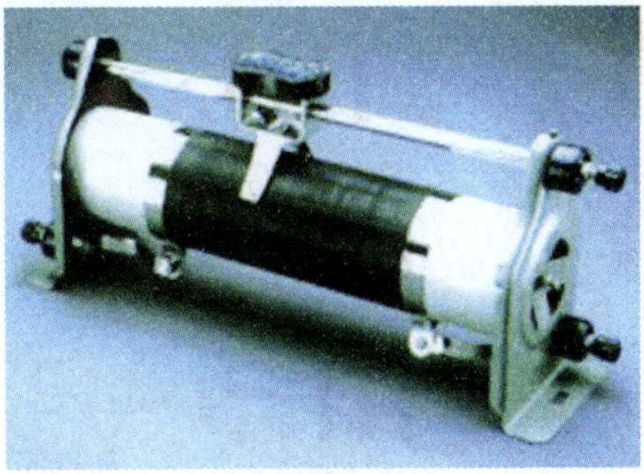
A rheostat.

Though similar in construction to potentiometers, they are usually larger in size because they possess much higher power rating.

A given potentiometer can be used as a rheostat. One method is just to use two ends only leaving the third end unconnected as shown in Fig. 5.8 (*a*). The other method is to wire the third unused terminal to the central terminal as shown in Fig. 5.8 (*b*).

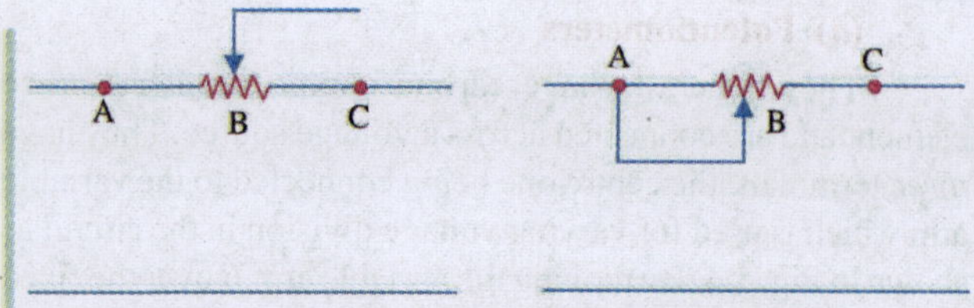

(*a*) (*b*)

Fig. 5.8

5.13. Fusible Resistors

These are wire-wound devices similar in appearance to ordinary wire-wound resistors. They are sometimes used in amplifiers and TV sets to protect certain circuits. They have resistance of less than 15 Ω. Their resistance element is quite similar to the fuse link in a cartridge fuse and is designed to burn out whenever current in the circuit exceeds a certain predetermined value.

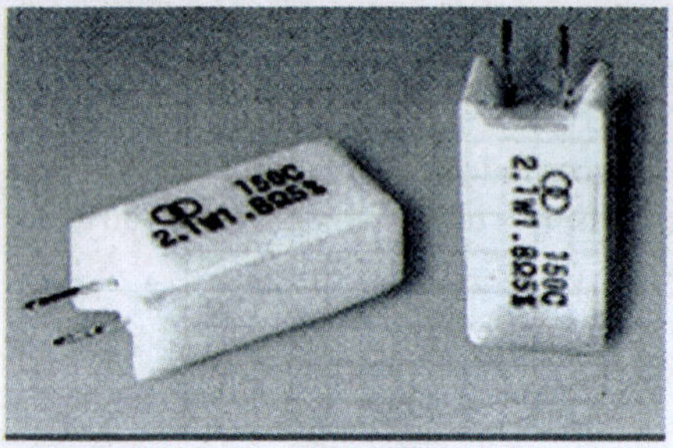

Fusible resistors.

5.14. Resistor Colour Code

Since fixed carbon-composition resistors having axial leads are physically small, they are colour-coded to indicate their resistance in ohms. The system is based on the use of colours (painted on the body of the resistor) as numerical values. In general, it should be remembered that dark colours like black and brown correspond to lowest numbers of zero and one respectively, the light colours to next higher numbers and lastly, white colour to nine. The colours used with the code and the numbers they represent are given in Table 5.1.

Table 5.1

Colour	*Value*
Black	0
Brown	1
Red	2
Orange	3
Yellow	4
Green	5
Blue	6
Violet	7
Grey	8
White	9

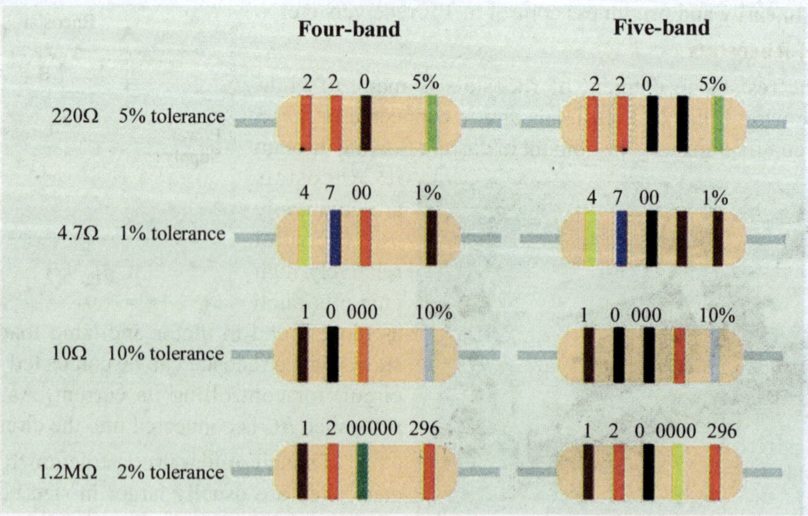

Some examples of the resistor colour code.

5.15. Resistance Colour Bands

These bands are printed around the body of the resistor near its one end (Fig. 5.9). Each colour stands for a digit. Often, there are four bands though sometime there may be five. In each case, *the first three bands give the resistance value.*

(*a*) Three Bands Only

They represent the resistance value as per the colour code. Absence of fourth band means a resistance tolerance of ± 20%.

(*b*) Four Bands Only

As before, the first three bands give resistance value and the fourth one gives tolerance.

Fourth gold ring means a tolerance of ± 5% whereas a fourth silver ring means a resistance tolerance of ± 10%.

(*c*) Five Bands

The first three bands, as usual, give resistance, fourth one gives tolerance and the fifth one indicates reliability level or failure rate for which colour code is,

Brown = 1% ; Red = 0.1% ; Orange = 0.01% ; Yellow = 0.001%

Starting from left to right, the colour bands (Fig. 5.9) are interpreted as follows :

1. The first band close to the edge indicates the first digit in the numerical value of the resistance.
2. The second band gives the second digit.
3. The third band is *decimal multiplier i.e.,* it gives the number of zeros after the two digits. It is important to note that if the third band is black, it means "*do not add zeros to the first two digits*". The resulting number is the resistance in ohms.
4. The fourth band gives resistance tolerance. If there is no fourth band, tolerance is understood to be ± 20%.

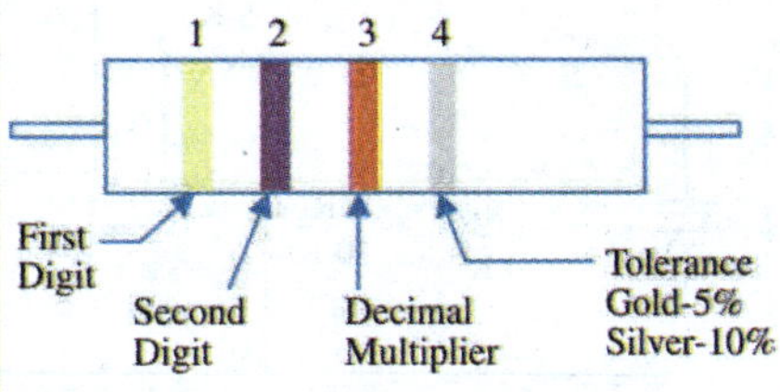

Fig. 5.9

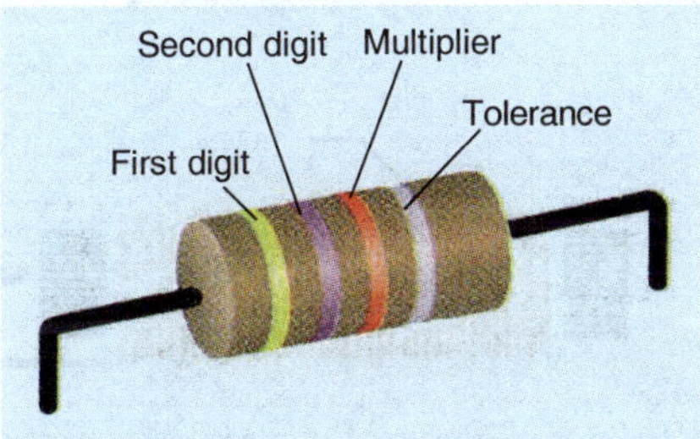

This Resistor has a value of 47,000Ω , ± 10%

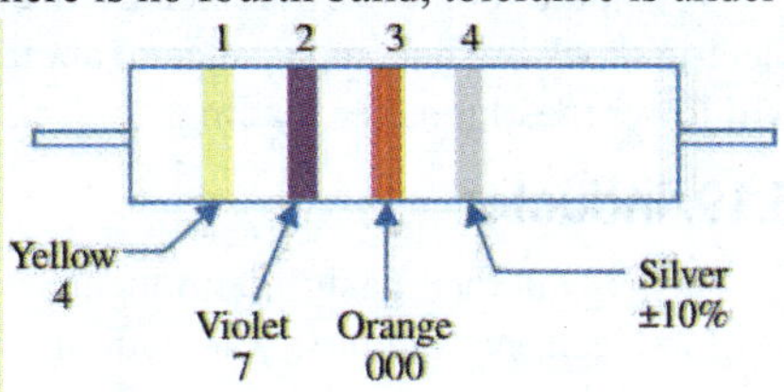

Fig. 5.10

The resistance value of the carbon-composition resistor shown in Fig. 5.10 is 47,000Ω ± 10%. It means that its value is 47,000 ± 4,700 Ω *i.e.,* it can lie anywhere between 42,300 Ω and 51,700 Ω.

5.16. Resistors under Ten Ohm

In their case, the third band is either gold or silver which serves as *fractional* multiplier. If third band is gold, multiply the first two digits by 0.1. If it is silver, then multiply by 0.01. However, fourth band, as before, gives tolerance.

As seen from Fig. 5.11, the value of resistor is 6.9 ± 10% ohm.

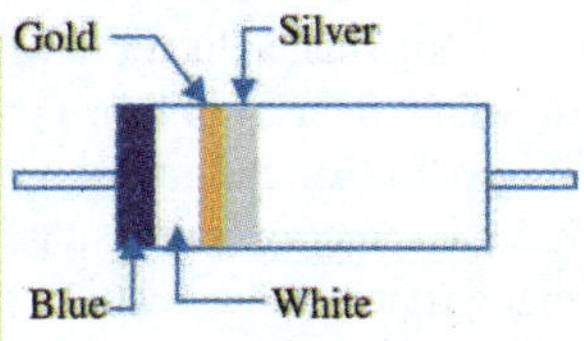

Fig. 5.11

5.17. Resistor Troubles

The most common trouble with resistors is 'open' which happens due to excessive current and heat. A charred or discoloured resistor should be discarded straight away *though it will usually check good with an ohmmeter.*

5.18. Checking Resistors with an Ohmmeter

Since an ohmmeter has it own voltage source, it is always used without any external power being supplied to the resistance under measurement.

The ohmmeter must have an ohm scale appropriate to the value of resistance being measured, otherwise it will not give correct value. In checking a 10 MΩ resistor, if the highest reading of the meter scale used is 1 MΩ resistor, the instrument will read infinity even if the resistor has its normal value of 10 MΩ. It is essential to use a scale of 100 MΩ for checking such high resistances.

Similarly, for checking resistances of value 10 Ω or less, a low ohm scale of 100 Ω or less should be used. Otherwise, the ohmmeter will read a low resistance such as zero thus indicating a short.

Another very important precaution one should take while checking a resistance is to make sure that there are no parallel paths across the resistance being measured. Otherwise, the measured resistance can be much lower than the actual resistance. As shown in Fig. 5.12 (*a*), the ohmmeter reads $R_1 \parallel R_2$ *i.e.,* 10/2 = 5 K. To check R_2 alone, its one end should be disconnected as shown in Fig. 5.12 (*b*).

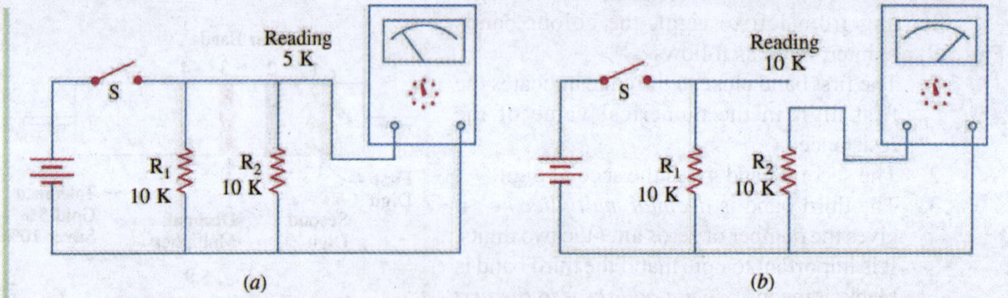

Fig. 5.12

Another point worth remembering is : not to touch the ohmmeter leads. There is no danger of electric shock but body resistance of about 50 K acting in parallel with the resistance being measured will lower the ohmmeter reading.

5.19. Inductor

It is another basic component commonly used in electronic circuits. It is nothing else but a coil wound on a core or former of some suitable material.

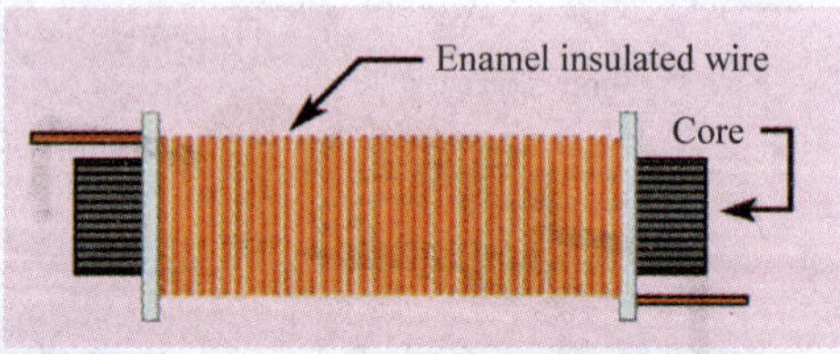

This one example of what an inductor might look like.

(*a*) Air-core Inductor

It consists of number of turns of wire wound on a former made of ordinary cardboard [Fig. 5.13 (*a*)]. Since there is nothing but air inside of the coil, an air-core inductor has the least inductance for a given number of turns and core length.

(*b*) Iron-core Inductor

It is that inductor in which a coil of wire is wound over a solid or laminated iron core [Fig. 5.13 (*b*)]. Putting iron inside an inductor has the effect of increasing its inductance as many times as the relative permeability (μ_r) of iron. In order to avoid eddy current loss, iron core is laminated *i.e.,* it is made up of thin iron laminations pressed together but

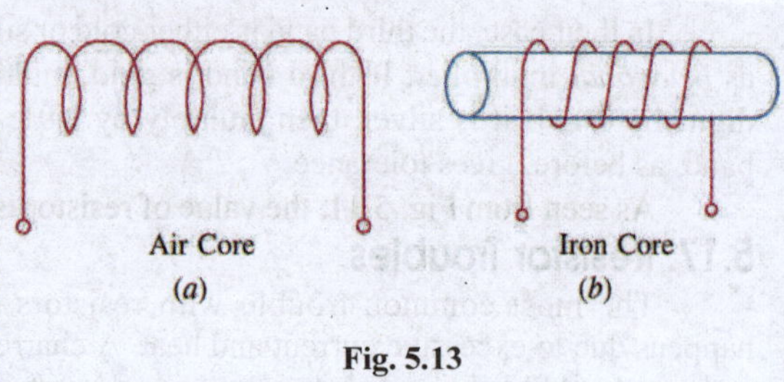

Fig. 5.13

insulated from each other. Sometimes, such an inductor is also called a choke.

The iron core has been found to work more efficiently particularly at low frequencies if it is in the form of a closed core *i.e.,* if the core not only goes through the centre of the coil but also surrounds it on its two sides as shown in Fig. 5.14.

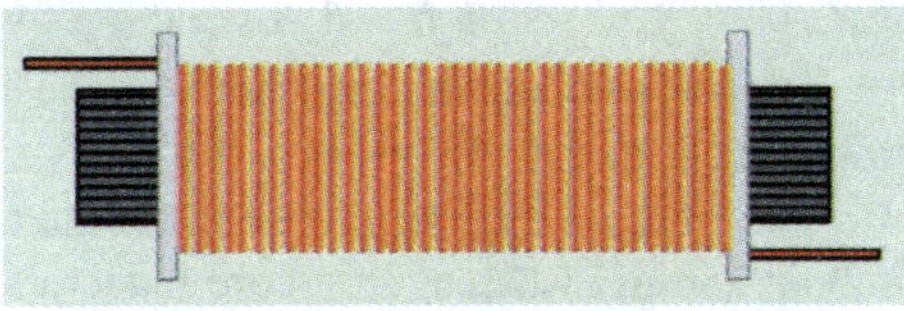

Laminated iron cores are generally used in low cost, low power inductors.

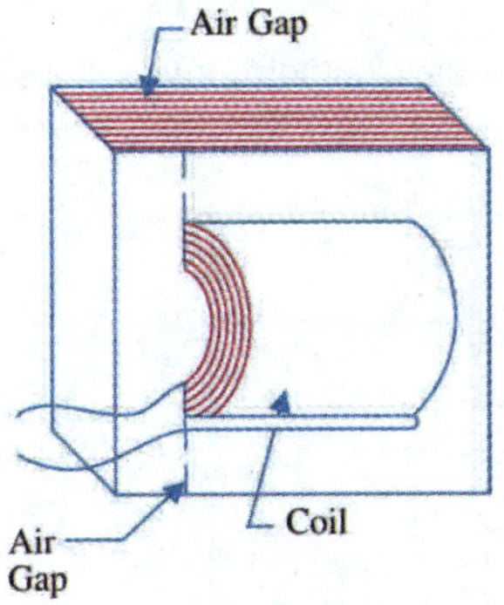

Fig. 5.14

(c) Ferrite-core Inductor

In this case, coil of wire is wound on a solid core made of highly ferromagnetic substance called ferrite. Ferrite is a solid material consisting of fine particles of iron powder embedded in an insulating binder. A ferrite core has minimum eddy current loss.

Ferrite core inductors are used in moderately high power systems.

The symbols for different types of inductors discussed below are shown in Fig. 5.15.

5.20. Comparison of Different Cores

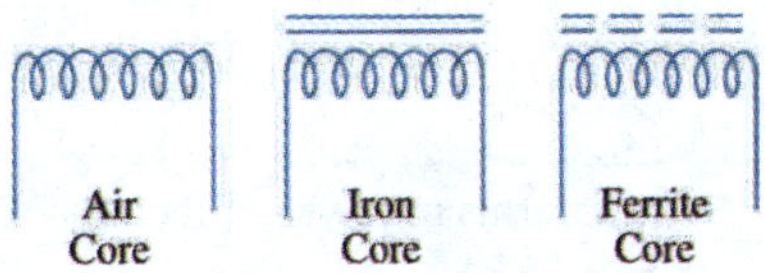

Fig. 5.15

In air-core coils, there are no core losses even at high frequencies but their inductances are limited to low values in the μH or mH range.

In iron-core coils, losses are minimal at low *i.e.,* audio frequencies but become considerable at high frequencies even when iron core is laminated. They possess comparatively much larger inductance as compared to air-core coils.

Ferrite-core coils have high inductance value with minimum eddy current and hysteresis losses even at very high frequencies. The built-in antennas used in transistor radios have ferrite core.

5.21. Inductance of an Inductor

It is found that whenever current through an inductor changes (*i.e.*, increases or decreases), a counter emf is induced in it which tends to oppose this change. This property of the coil due to which it opposes any change of current through it is called inductance (L). Its unit is henry (H). The inductance of a coil is given by

$$L = \frac{\mu_0 \mu_r A N^2}{l} \text{ henrys}$$

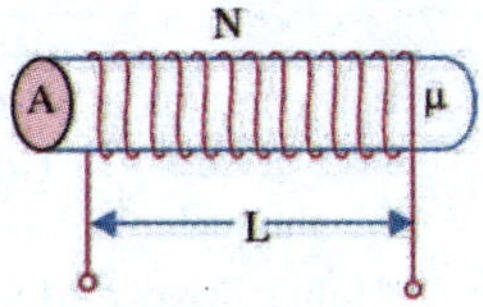

Fig. 5.16

It is seen that L varies

1. directly as relative permeability of the core material,
2. directly as core cross-sectional area,
3. directly as square of the number of turns of the coil,
4. inversely as core length.

5.22. Another Definition of Inductance

Suppose, current through an inductor is changed at the rate of di/dt because of which a counter emf of 'c' is induced in it. Then, it is found that

$$e = L\frac{di}{dt} \quad \text{or} \quad L = \frac{e}{di/dt}$$

If $di/dt = 1$ A/s and $c = 1$V, then $L = 1$H

Hence, a coil has an inductance of one henry if an emf of one volt is induced in it when current through it changes at the rate of 1 A/s.

The radio-frequency (RF) coil for the radio broadcast band 550-1650 kHz has an inductance of about 250 μH whereas iron-core inductors used for audio frequency have inductance values of about 1 to 25 H.

5.23. Mutual Inductance

When two coils are placed so close to each other (Fig. 5.17) that the expanding and collapsing magnetic flux of one coil links with the other, an induced emf is produced in the other coil. These two coils are then said to have mutual inductance (M).

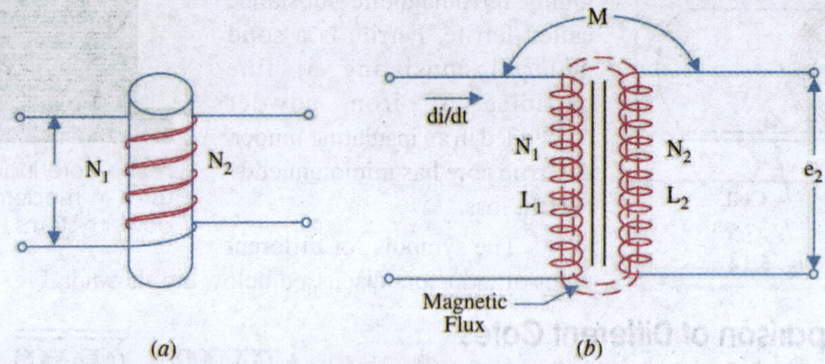

Fig. 5.17

In terms of physical factors

$$M = \frac{\mu_0 \mu_r A N_1 N_2}{l}$$

where l is the length of the magnetic path.

As shown in Fig. 5.17 (b), let the rate of current change through the first coil be di/dt. This changing current will produce a changing magnetic flux through it which will link partly or fully with the second coil. Hence, an induced emf 'e_2' (called mutually-induced emf) will be produced in the second coil. Its value is given by

$$e_2 = M\frac{di}{dt}$$

if $di/dt = 1$ A/s and $e_2 = 1$ V, then $M = 1$ H

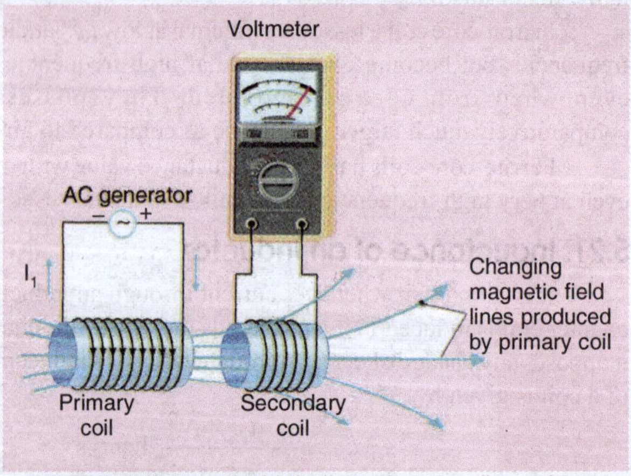

An alternating current I_1 in the primary coil creates an alternating magnetic field. This changing field induces an emf in the secondary coil.

Hence, two coils have a mutual inductance of one henry if a current change of one ampere/ second in one coil induces one volt in the other.

5.24. Coefficient of Coupling

Two coils are said to be magnetically coupled if full or part of the flux produced by one coil links with the other. Closer they are, tighter the coupling between them. If, on the other hand, no flux from one coil links with the other, then coupling between the two is zero.

This coupling effect is measured in terms of coefficient of coupling given by

$$k = \frac{M}{\sqrt{L_1 L_2}}$$

where L_1 = inductance of the first coil

L_2 = inductance of the second coil

M = mutual inductance between the two coils

When magnetic flux produced by one coil does not link with the other coil, then $k = 0$. If all the flux produced by one coil links with the other, $k = 1$. Air-core coils wound on the same former have k value of 0.05 to 0.3. Coils on iron core have k almost equal to unity.

5.25. Variable Inductors

The inductance of a coil can be varied by the three different methods shown in Fig. 5.18.

1. By using a tapped coil as shown in Fig. 5.18 (*a*). Here, either more or fewer turns of the coil can be used by connection to one of the taps on the coil.
2. By using a slider contact to vary the number of turn used as in Fig. 5.18 (*b*).

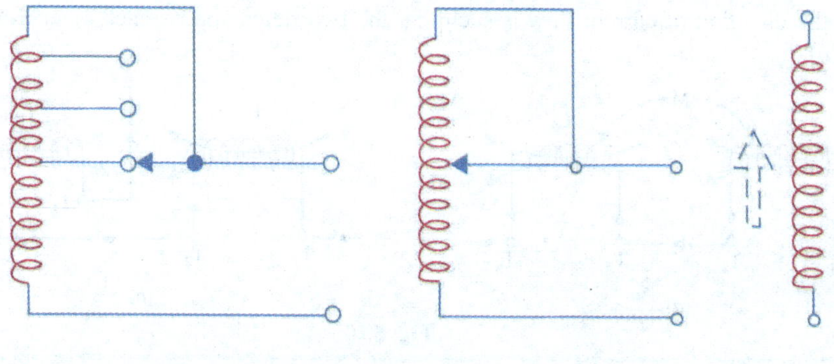

(*a*) (*b*) (*c*)

Fig. 5.18

These methods are used for large coils. It will be noted that the unused turns have been short-circuited to prevent the tapped coil from acting as an autotransformer otherwise the stepped up voltage could cause arcing across the turns.

3. Fig. 5.18 (*c*) shows the symbol for a coil with a ferrite slug which can be screwed in or out of the coil to vary its inductance.

5.26. Inductors in Series or Parallel without M

We will assume that the different coils connected in series or parallel have no mutual inductance *M*.

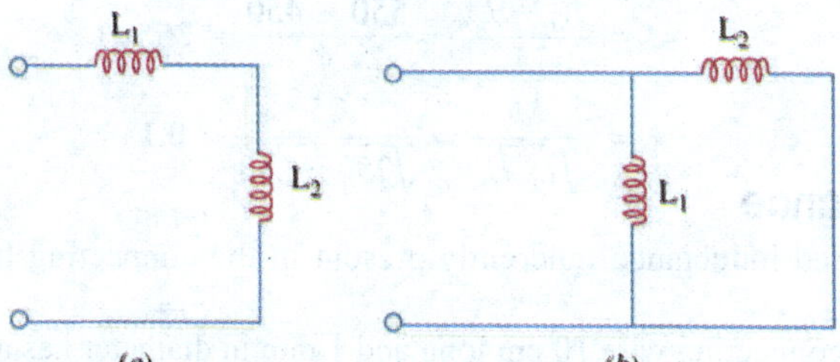

(*a*) (*b*)

Fig. 5.19

(*a*) **Series Combination**

Here, $L = L_1 + L_2$ — Fig. 5.19 (*a*)

In general, $L = L_1 + L_2 + L_3 + ...$

(*b*) **Parallel Combination**

Here, $L = L_1 \parallel L_2 = \dfrac{L_1 L_2}{L_1 + L_2}$ — Fig. 5.19 (*b*)

In general, $\dfrac{1}{L} = \dfrac{1}{L_1} + \dfrac{1}{L_2} + \dfrac{1}{L_3} + ...$

5.27. Series Combination with M

In this case, value of total inductance will depend upon

1. amount of mutual inductance present, and
2. whether the combination is series-aiding or series-opposing.

In series-aiding, the direction of the common current through the two coils is such that their magnetic fields are in the same direction *i.e.,* they aid each other as shown in Fig. 5.20 (*a*). In series-opposing, the direction of current flow is such that the two fields oppose each other as shown in Fig. 5.20 (*b*).

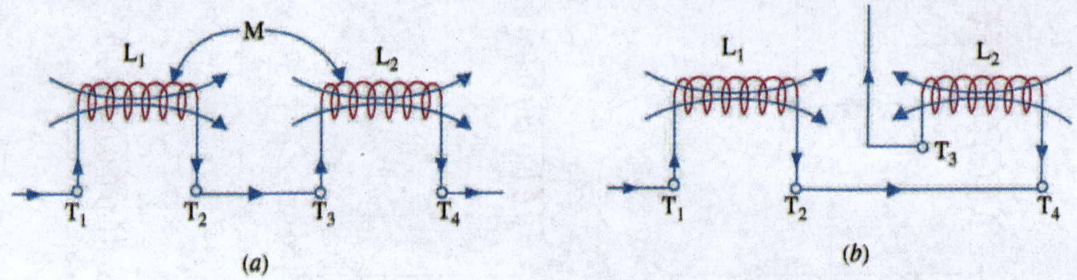

Fig. 5.20

For the circuit of Fig. 5.20 (*a*)

$$L_a = L_1 + L_2 + 2M$$

For the circuit of Fig. 5.20 (*b*)

$$L_o = L_1 + L_2 - 2M$$

Incidentally, it can be deduced from the above two equations that

$$M = \frac{L_a - L_o}{4}$$

Example 5.1. *Two coils each having an inductance of 250 μH have combined inductance of 550 μH when connected series-aiding and 450 μH when connected series-opposing. Calculate (i) their mutual inductance, and (ii) coefficient of coupling.*

Solution. (*i*) $M = \dfrac{L_a - L_o}{4} = \dfrac{\mathbf{550 - 450}}{4} = \mathbf{25\ \mu H}$

(*ii*) $k = \dfrac{M}{\sqrt{L_1 L_2}} = \dfrac{25}{\sqrt{250 \times 250}} = \mathbf{0.1}$

5.28. Stray Inductance

It is the unwanted inductance inherently present in the connecting leads and wire-wound resistors.

For example, a connecting wire 10 cm long and 1 mm in diameter has an inductance of about 0.1 μH. It is negligible at low frequencies but can be troublesome at MHz frequency range. That is why connecting leads are kept very short in high-frequency circuits.

Since ordinary wire-wound resistors have considerable stray inductance, non-inductive wire-wound resistors are used instead. These are wound in such a way that adjacent turns have currents in the opposite direction so that opposing magnetic fields cancel out the inductance.

5.29. Energy Stored in a Magnetic Field

When current passes through an inductor, magnetic flux is produced for which energy is supplied by the voltage source. This energy is safely stored in the field itself and is recoverable. The amount of energy stored is

$$E = \frac{1}{2} L I^2 \text{ joules}$$

where L is inductance in henrys and I is the steady current in amperes.

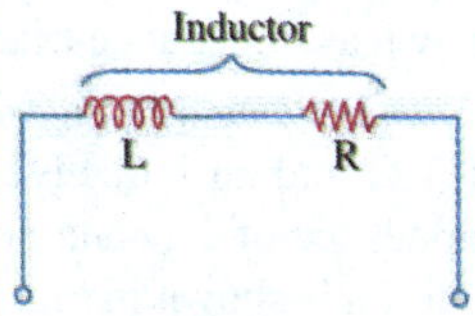

Fig. 5.21

5.30. DC Resistance of a Coil

It is the ohmic resistance of the wire used in making the coil. This resistance will depend on the thickness of the wire and its length *i.e.,* total number of turns. For radio-frequency coils with inductance values upto several mH requiring 10 to 100 turns of fine wire, the dc resistance is from 1 to 20 Ω approximately.

5.31. Troubles in Coils

The most common trouble is an 'open' in the winding. An ohmmeter connected across the coil will read infinity for the open circuit. As usual, while checking its resistance, the coil should be disconnected from the external circuit to eliminate any parallel paths which could affect the resistance reading.

5.32. Reactance Offered by a Coil

An inductor offers opposition to the passage of any *changing* or *alternating* current through it. This opposition is given the name of inductive reactance, X_L.

$$X_L = 2\pi f L = \omega L \text{ ohm}$$

where L = coil inductance in henrys

f = frequency of alternating current in Hz

ω = angular frequency in radian/second.

Like resistance, unit of inductive reactance is also ohm.

Obviously, $X_L = 0$ if $f = 0$, *i.e.,* a coil offers no reactance to the passage of direct current through it since frequency of such a nonchanging-current is zero. Of course, it does offer dc resistance possessed by it.

It may be noted that, unlike resistance, inductive reactance offered by a coil is not constant but depends on frequency of the alternating current passing through it. Higher the frequency, greater the reactance. Moreover, for a given frequency, X_L depends directly on coil inductance L.

Example 5.2. *Calculate the inductive reactance offered by a coil of inductance 250 μH to radio-frequency currents of frequencies (i) 1 MHz and (ii) 10 MHz.*

Solution. (*i*) $X_L = 2\pi f L$

$= 2\pi \times (1\times 10^6) \times (250 \times 10^{-6}) = \mathbf{1570\ \Omega}$

(*ii*) $X_L = 2\pi \times (10 \times 10^6) \times (250 \times 10^{-6}) = \mathbf{15{,}700\ \Omega}$

As seen, X_L increase proportionally with increase in frequency.

5.33. Impedance Offered by a Coil

A coil having both inductance (L) and resistance offers opposition in the form of both X_L and R. The combined opposition of X_L and R is known as *impedance* (Z).

However, it should be noted that X_L and R are not added arithmetically but *vectorially* as shown in Fig. 5.22 (*a*).

Here, $Z = \sqrt{R^2 + X_L^2}$ (and $\neq R + X_L$)

The right-angled triangle of Fig. 5.22 (*a*) is known as impedance triangle.

For example, an inductor coil having a resistance of 3 Ω and an inductive reactance of 4 Ω offers an impedance of 5 Ω and not (4 + 3) = 7 Ω. The vector addition is shown in Fig. 5.22 (*b*).

Incidentally, the angle φ in Fig. 5.22 represents the phase difference between the applied alternating voltage and circuit current.

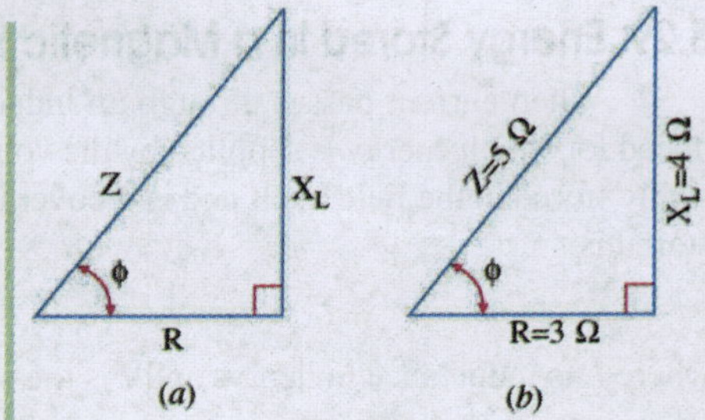

Fig. 5.22

Example 5.3. *A coil has a resistance of 30 Ω and an inductance of 127.3 mH. It is connected across a 200 V, 5 Hz ac supply. Find*

1. impedance *2. circuit current* *3. phase angle φ*

Solution.

$$X_L = 2\pi f L$$
$$= 2\pi \times 50 \times (127.3 \times 10^{-3}) = 40\ \Omega$$

1. $$Z = \sqrt{R^2 + X_L^2} = \sqrt{30^2 + 40^2} = \mathbf{50\Omega}$$

2. $$I = V/Z = 200/50 = \mathbf{4\ A}$$

3. $$\tan\phi = \frac{X_L}{R} = \frac{40}{30} = 1.3333 \qquad \therefore \phi = \mathbf{53^\circ}$$

5.34. Q-Factor of a Coil

The quality or merit of a coil is measured in terms of its *Q*-value given by

$$Q = \frac{X_L}{R} = \frac{2\pi f L}{R}$$

As seen, smaller the d.c. resistance of a coil as compared to its inductance, higher its *Q*-factor. In tuned radio receiver circuits, a high *Q*-coil is preferred because

1. it increases sharpness of tuning *i.e.*, makes the tuned circuit more selective,
2. it additionally increases its sensitivity.

5.35. Capacitors

Apart from resistors and inductors, a capacitor is the other basic component commonly used in electronic circuits. It is a device which

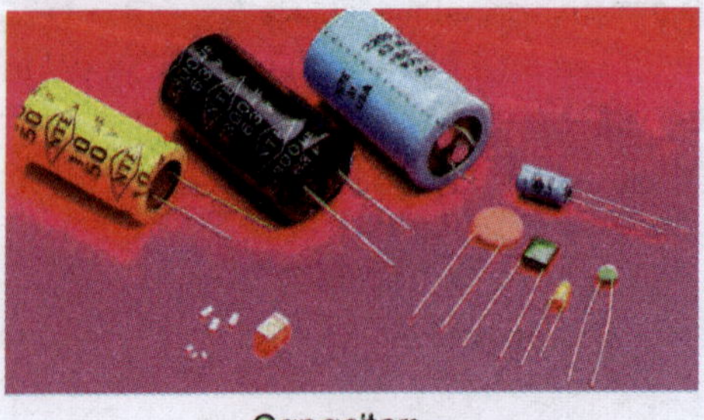
Capacitors.

1. has the ability to store charge which neither a resistor nor an inductor can do;
2. opposes any change of *voltage* in the circuit in which it is connected*;
3. blocks the passage of direct current through it.

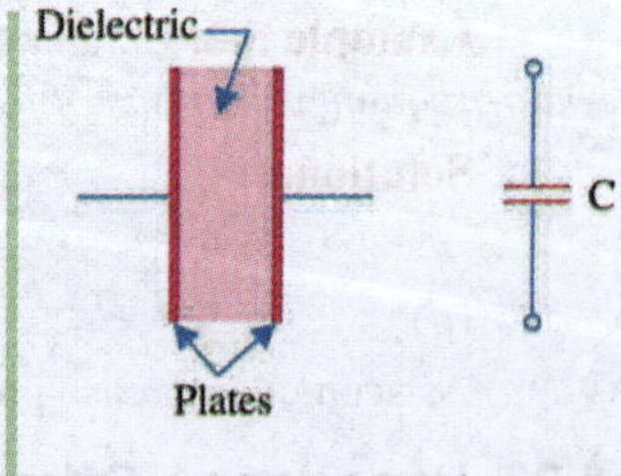

Fig. 5.23

* An inductor opposes change of current.

Capacitors are manufactured in various sizes, shapes, types and values and are used for hundreds of purposes.

Essentially, a capacitor consists of two conducting plates separated by an insulating medium called *dielectric* as shown in Fig. 5.23. The dielectric could be air, mica, ceramic, paper, polyester, polystyrene or polycarbonate plastics etc.

5.36. Capacitor Connected to a Battery

When switch S in Fig. 5.24 (a) is closed, the capacitor gets connected across the battery. There is momentary flow of electrons from plate M to plate N. The positive terminal of the battery attracts and pulls away negatively-charged electrons from plate M which consequently becomes positive. Similarly, as these electrons collect on plate N, it becomes negative. Hence, a potential difference is established between the two plates M and N. This *transient* flow of electrons from one plate to another through the connecting wires gives rise to charging current which establishes positive charge of $+Q$ coulomb on the plate M. The strength of this charging current is maximum in the beginning when the two plates are uncharged but it then decreases and finally ceases when p.d. across the two plates becomes slowly and slowly *equal and opposite* to the battery emf *i.e.,* V volts. The capacitor then becomes fully charged as shown in Fig. 5.24 (b).

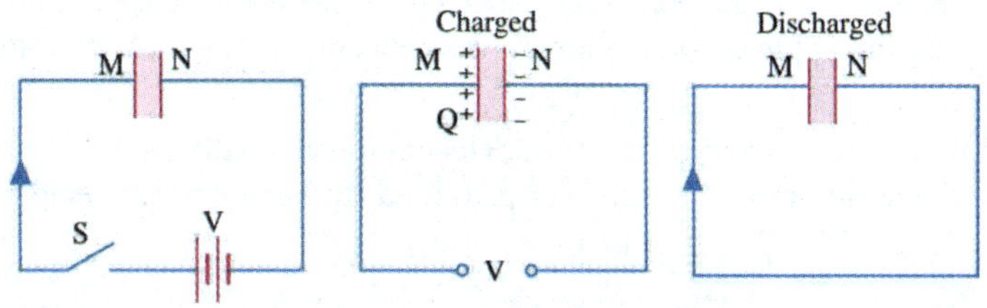

(a) (b) (c)

Fig. 5.24

Following two points should be noted :

1. no current can flow '*through*' the capacitor because of the presence of dielectric in the circuit which offers infinite resistance. The electric charge is momentarily displaced from one plate to another through the *external circuit only*;
2. as p.d. between the plates is increased, the dielectric medium comes under increasing stress. If this p.d. is increased, the stress in the dielectric increases till it can no longer bear it. At that stage, electrical breakdown occurs accompanied by a spark between the two capacitor plates. The maximum voltage per metre thickness which a medium can withstand without a rupture or breakdown is called its *dielectric strength*. Usually, it is given in kV/mm instead of the basic unit V/m. If the two leads of a charged capacitor are connected together as shown in Fig. 5.24 (c), the p.d. between the plates is equalized and the capacitor becomes discharged;
3. since there exists a p.d. between the two plates, an electric field is set up between them whose strength is given by

$$E = \text{V/d volt/metre}$$

5.37. Capacitance

It measures the ability of a capacitor to store charge. It may be defined as the amount of charge required to create a unit potential difference between its plates.

Suppose, we give $+Q$ coulomb of charge to one of the two plates of a capacitor and if a p.d. of V volts is established between them, then its capacitance is

$$C = \frac{Q}{V} \text{ farad}$$

If $Q = 1$ C and $V = 1$ volt, then $C = 1$ farad (F).

Hence, one farad is defined as the capacitance of a capacitor which requires a charge of one coulomb to establish a p.d. of one volt between its plates.

Capacitance of a capacitor may also be defined in terms of its property to oppose the change of voltage in the circuit. In that case,

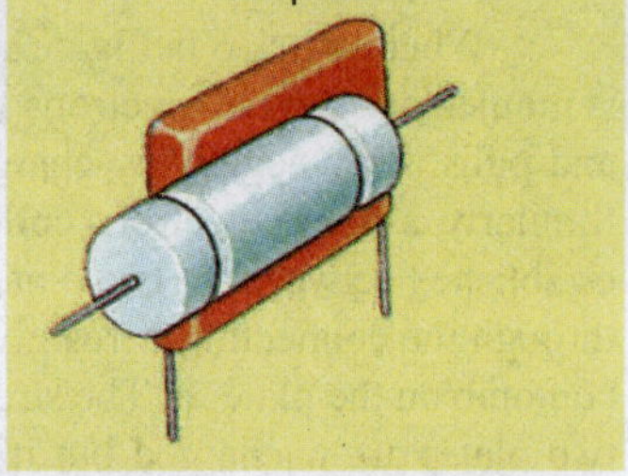

The amount of charge the device can store for a given voltage difference is called the capacitance.

$$C = \frac{i}{dv/dt}$$

where i = charging current

dv/dt = rate of change of voltage

i = 1 ampere, dv/dt = 1 volt/second

then C = 1 farad

Hence, one farad may be defined as the capacitance which will cause one ampere of charging current to flow when the applied voltage across the capacitor changes at the rate of one volt per second.

Farad is too large for practical purposes. Hence, much smaller units like microfarad (μF), nanofarad (nF) and micro-micro-farad (μμF) or picofarad (pF) are generally employed.

$$1 \text{ μF} = 10^{-6} \text{ F}$$

$$1 \text{ nF} = 10^{-9} \text{ F}$$

$$1 \text{ pF or } 1 \text{ μμF} = 10^{-12} \text{ F}$$

5.38. Factors Controlling Capacitance

The capacitance of a capacitor depends on the following factors :

1. Plate Area

Capacitance increases directly with increase in plate area (A).

2. Plate Separation

As plate separation (d) decreases, capacitance increases and *vice-versa*. Since plate separation often equals the thickness of the dielectric used, we may say that thinner the dielectric slab, greater the capacitance and *vice-versa.*

3. Type of Dielectric

It depends on the relative permittivity ε_r (previously called dielectric constant) of the dielectric medium used. Higher the value of ε_r, greater the value of capacitance.

Combining the above three factors, we get

$$C = \frac{\varepsilon_0 \ \varepsilon_r \ A}{d} \text{ farad}$$

where ε_0 is the absolute permittivity of vacuum = 8.854×10^{-12} F/m.

The relative permittivities of some dielectric media are listed in Table 5.2 on next page.

Table 5.2

Material	*Dielectric Constant or relative permittivity* (ε_r)
Air or vacuum	1
Bakelite	4.5—5.5
Ceramic	50—3000
Fibre	5—8
Glass	7—8
Mica	3—6
Paper (waxed)	3—5
Polysterene	2.6
Porcelain	5—6

Example 5.4. *What is the capacitance of a capacitor if a charging current of 100 mA flows when the applied voltage changes 20 V at a frequency of 50 Hz?*

Solution. $C = \dfrac{i}{dv/dt}$

Now, $i = 100 \text{ mA} = 100 \times 10^{-3} \text{ A} = 0.1 \text{ A}$

$$dv = 20 \text{ V}, dt = 1/50 = 0.02 \text{ s}$$

$$C = \frac{0.1}{20/0.02} = 0.1 \times 10^{-3} \text{ F}$$

$$= 100 \times 10^{-6} \text{ F} = \mathbf{100\ \mu F}$$

Example 5.5. *What is the capacitance of a parallel-plate capacitor of plate area 0.01 m² and air dielectric of thickness 0.01 m ?*

If the capacitor is given a charge of 500 μμC, what will be the p.d. between its plates?

How will this be affected if space between the two plates is filled with wax which has a relative permittivity of 4 ?

Solution. $C = \dfrac{\varepsilon_0\ \varepsilon_r\ A}{d}$

Remembering that for air, $\varepsilon_r = 1$, we get

$$C = \frac{8.854 \times 10^{-12} \times 1 \times 0.01}{0.01} = 8.854 \times 10^{-12} \text{ F} = \mathbf{8.854\ \mu\mu F}$$

Now, $C = \dfrac{Q}{V}$ or $V = \dfrac{Q}{C}$

$$\therefore \quad V = \frac{500\ \mu\mu C}{8.854\ \mu\mu F} = \mathbf{56.5\ V}$$

When wax is introduced, capacitance will increase four times.

$$\therefore \quad C = 4 \times 8.854 = \mathbf{35.4\ \mu\mu F}$$

Since charge remains constant, p.d. will *decrease* to one-fourth of its previous value because $V \propto 1/C$

$$\therefore \quad V = 56.5/4 = \mathbf{14.1\ V}$$

5.39. Types of Capacitors

All capacitors commonly used in electronic circuits may be divided into two general classes (*a*) fixed capacitors and (*b*) variable capacitors.

Fixed capacitors may be further sub-divided into (*i*) electrolytic and (*ii*) non-electrolytic capacitors.

5.40. Fixed Capacitors

These can be grouped into two classes as detailed below :

(a) Non-electrolytic type

It includes paper, mica and ceramic capacitors. Such capacitors have no polarity requirement *i.e.,* they can be connected in either direction in a circuit.

(*i*) Paper Capacitor

It consists of two tinfoil sheets which are separated by thin tissue paper or waxed paper. The sandwich of foil and paper is then rolled into a cylindrical shape and enclosed in a paper tube or encased in a plastic capsule. The lead at each end of the capacitor is internally attached to the metal foil. Fig. 5.25 (*a*) shows a 2.0 μF tubular paper capacitor of maximum ac voltage rating of 2000 V whereas Fig. 5.25 (*b*), shows polyester capacitor having moulded plastic box encapsulation suitable for use in printed circuit boards.

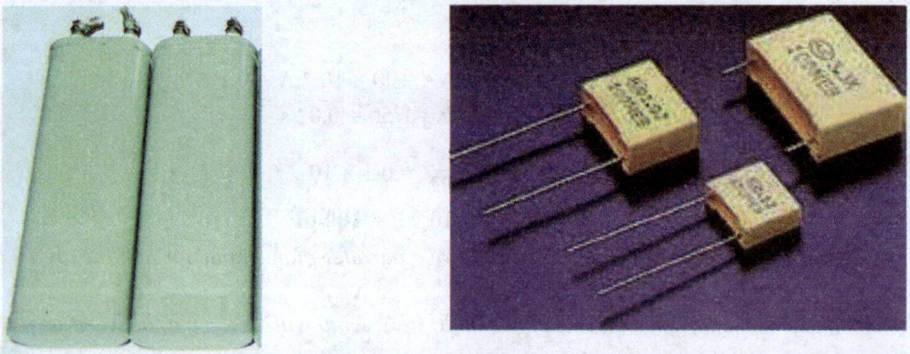

(*a*) (*b*)

Fig. 5.25

Paper capacitors have a capacitance range of 0.001 to 2.0 μF and working-voltage rating as high as 2000 V. These specifications are usually printed on the capacitor case. Paper capacitors have large physical size as compared to their capacitance and also become inefficient as the frequency of applied ac voltage exceeds a few megahertz. These facts prevent their use in most FM TV circuits except in low-frequency portions of the circuits *i.e.,* in audio stages.

These days such foil construction capacitors use thin plastic film instead of paper as the dielectric medium. Two most commonly-used plastic films have trademark names of Teflon and Mylar. Such capacitors have

1. high insulation resistance of 1000 M and above,
2. low losses, and
3. longer shelf life without breakdown as compared to paper capacitors.

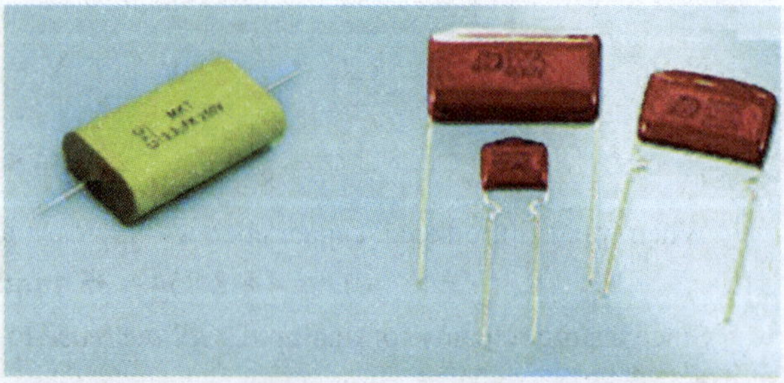

(*c*) (*d*)

Fig. 5.25

Fig. 5.25 (*c*) and (*d*) show tubular and flat metallised polyester film capacitors which are ideal for coupling and by-pass applications in radio, TV, hi-fi equipment and instrumentation etc.

(*ii*) Mica Capacitor

It is a sandwhich of several thin metal plates separated by thin sheets of mica. Alternate plates are connected together and leads attached for outside connections. The total assembly is encased in a plastic capsule or bakelite case as shown in Fig. 5.26. Such capacitors have small capacitance values (50 to 500 pF) yet high working voltage ratings (500 V and above).

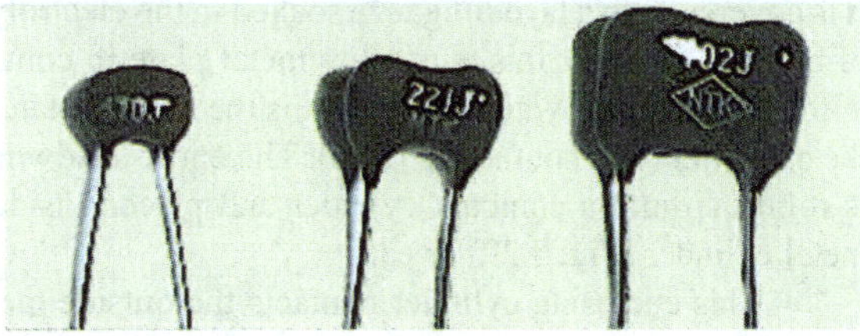

Fig. 5.26

Once such capacitors were used extensively in radio circuits but, of late, they have been superseded by ceramic capacitors because of excellent properties of ceramics and their economy.

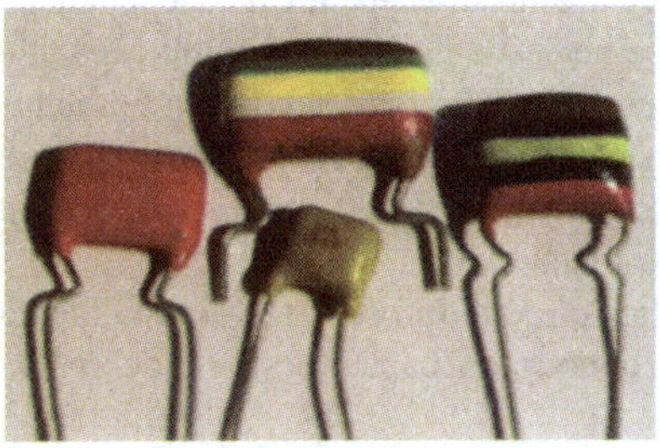

Fig. 5.27. The small value ceramic capacitors.

(*iii*) Ceramic Capacitors

Such capacitors have disc- or hollow tubular-shaped dielectric made of ceramic material such as titanium dioxide and barium titanate. Thin coatings of silver compound are deposited on both sides of the dielectric disc which act as capacitor plates. Leads are attached to each side of the disc and the whole unit is encapsulated in a moisture-proof coating (Fig. 5.27).

Because of the very high value of the dielectric constant of ceramics ($\varepsilon_r = 1200$), disc type capacitors have very large capacitances (upto 0.01 μF) compared to their size.

In the case of tubular ceramics, the hollow ceramic tube has a silver coating on the inside and outside surfaces. The capacitance range varies from 1 to 500 pF with working voltage rating exceeding 10 kV. Ceramic capacitors have many advantages as compared to mica and paper capacitors. These capacitors

1. are economical,
2. have very small size but large capacitance. Hence, they occupy less space.
3. have very high working-voltage rating,
4. have very low power factor (*i.e.*, loss) which further *decreases with increase in the frequency*. Hence, they are very useful for short-wave work in radio.

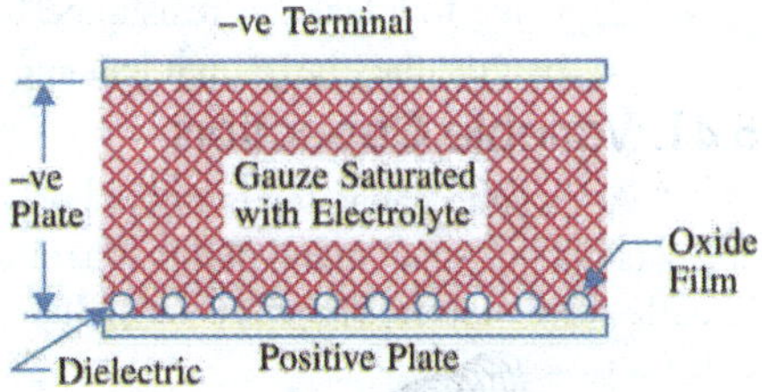

Fig. 5.28

(*b*) Electrolytic Capacitors

These capacitors are called *electrolytic* because they use an electrolyte (borax or a carbon salt) as *negative* plate. The capacitor consists of

1. a positive plate of aluminium;
2. an extremely thin (*i.e.,* molecular thin) insulating film of aluminium oxide (Al_2O_3) as dielectric medium. It is electrochemically deposited on the surface of anode itself (Fig. 5.28);
3. an electrolyte of borax (phosphorus or carbonate). As shown in Fig. 5.28, an absorbent gauze saturated with the electrolyte is kept in contact with the dielectric. The second aluminium plate serves merely as contact to the electrolyte. It forms the negative terminal.

Such a capacitor is made as described below:

A thin strip of aluminium is coated with a molecular-thin film of aluminium oxide by an electrochemical process. It is covered with a layer of gauze soaked in the electrolyte of borax. On top of this is another metal plate in contact with the electrolyte which, in reality, is the second plate of the capacitor (and not the top plate). The entire sandwhich is rolled up into a compact cylinder and placed inside a metal cylinder (Fig. 5.29) or can.

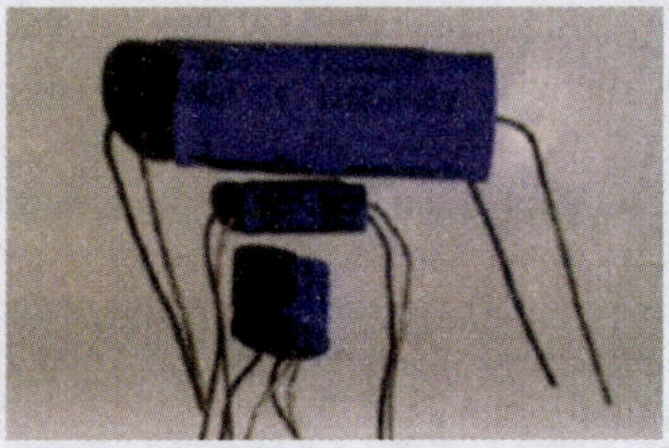

Fig. 5.29. The aluminium electrolytic capacitors.

This enclosing cylinder contacts the outside metal foil of the capacitor and serves as the negative terminal. Such capacitors are made as single capacitors or with two or three capacitors in a single container having a common negative terminal. The outside metal cylinder is usually enclosed in a paper tube in order to insulate it from components and to protect radio technicians from shock.

Because of extremely thin dielectric film, such capacitors possess very large capacitance ranging from 1 μF to 10,000 μF in very compact sizes. Hence, they are used where lot of capacitance is required in a small space. Since they use an electrolyte as the negative plate, electrolytic capacitors are classed as "*polarised*' capacitors *i.e.,* they must be connected in the circuit according to the plus (+) and minus (–) markings on the case. If this is not done, the capacitor may become short-circuited or get overheated due to excessive leakage current through its dielectric. Moreover, reversed polarity results in gas formation which may cause the capacitor to explode.

Electrolytic capacitors are frequently used in the filter circuits in order to remove the a.c. ripple from the power supply. They are also used as blocking or coupling capacitors for blocking d.c. in a circuit but allowing a.c. to pass through.

The miniature and high-voltage aluminium foil electrolytic capacitors manufactured by Bharat Capacitors Limited (BCL), Hyderabad are characterised by low leakage current, low dissipation, close tolerance in capacitance value, excellent reliability and long life. They meet all performance characteristics as laid down in JIS–C–5141 or MIL–C–62 specifications.

The two disadvantages of electrolytic capacitors are that they

1. are polarity sensitive,
2. have low leakage resistance because the oxide film is not a perfect insulator. In other words, they have high leakage current which can become troublesome.

5.41. Variable Capacitors

A variable capacitor is one whose capacitance can be varied usually by rotating a shaft. It consists of two sets of metal plates separated from each other by air. One set of plates is stationary and is called the stator [Fig. 5.30 (*a*)]. It is insulated from the frame of the capacitor upon which

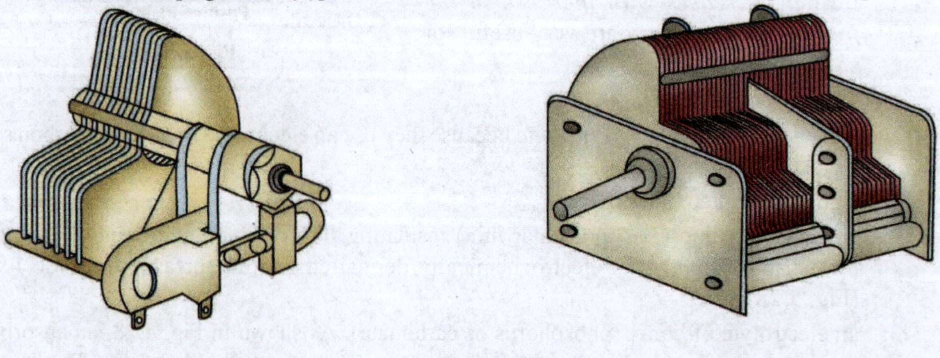

(*a*) Single section **Fig. 5.30** (*b*) Two-gang

it is mounted. The other set of plates is connected to the shaft and can be rotated. That is why it is called the rotor. By rotating the rotor with the help of a suitable knob, rotor plates can be made to move in or out of the stator plates. Capacitance is maximum when rotor plates are fully 'in' and minimum when 'out.'

If n is the total number of plates and d is the separation between any two adjacent plates, then capacitance for air dielectric is

$$= \frac{(n-1)\,\varepsilon_0\,A}{d}$$

When two or more such capacitors are operated by a single shaft, it is known as a *ganged* capacitor. Fig. 5.30 (*b*) shows a 2-gang capacitor *i.e.,* one having two separate variable capacitors in one unit rotated by a single control. Commercial receivers and those used by short-wave listeners often have a variable capacitor with three gangs.

Another type of small variable capacitor which is often used in parallel with the main variable capacitor is sometimes known as trimmer and sometimes as padder. It is used primarily for making fine adjustments on the total capacitance of the device.

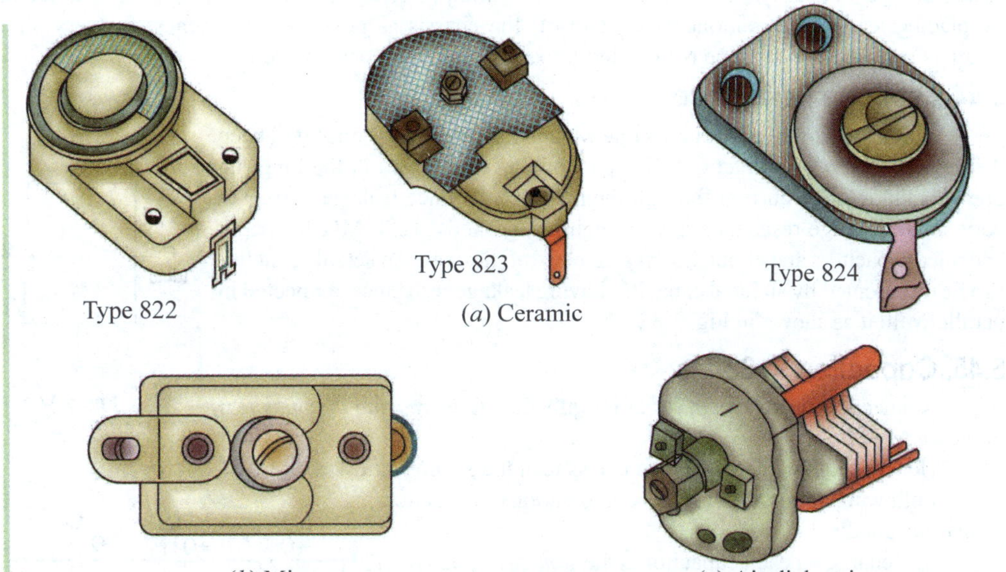

Fig. 5.31

A trimmer (Fig. 5.31) consists of two small flexible metal plates separated by air or mica or ceramic slab as the dielectric. The spacing between the plates can be changed by means of a screw adjustment. As the screw is turned inward, plates are compressed and its capacitance is increased.

Trimmers and padders are usually similar in appearance except that a padder may be somewhat larger or may have more plates. Capacitance of trimmers can be changed from 5 pF to a maximum value of 30 pF. Corresponding values for padder are 10 pF to 500 pF.

Variable capacitors are used primarily as tuning capacitors in radio receivers. When we tune two different stations, we actually vary the capacitance by moving the rotor plates in or out of the stator plates. Combined with an inductance, the variable capacitance tunes the receiver to a different resonant frequency for each transmitting station.

5.42. Voltage Ratings of Capacitors

Voltage rating of a capacitor is given by the maximum potential difference that can be applied across its plates without puncturing its dielectric. Such ratings are given for temperatures up to 60°C.

Higher temperatures result in lower voltage ratings. For general purpose paper, mica and ceramic capacitors, voltage ratings are typically 200-500 V dc. Ceramic capacitors with voltage ratings of 1 to 12 kV are also available. Electrolytic capacitors are commonly used in 25, 150 and 450 V ratings. Additionally, miniature electrolytics with 6 V and 10 V ratings are often used in transistor circuits. These ratings are for dc voltages. Those for ac voltages are less because of internal heat produced by continuous charge and discharge.

Voltage across a capacitor should not be allowed to exceed its rating. However, a capacitor with a higher voltage rating can be used in a low voltage circuit. For example, a 200 V, 0.05 μF capacitor can be replaced by a 400 V, 0.05 μF capacitor but not *vice-versa.*

5.43. Stray Circuit Capacitance

In an electronic circuit, the wiring and other components have capacitance to the metal chassis which acts as negative or grounded plate. Typical values of such stray capacitance are from 5 to 10 pF. Though at ordinary frequencies this value is quite small as compared to concentrated or lumped values of capacitance, it becomes important at high radio frequencies when one is forced to use small values of capacitance. The stray capacitance can be minimised by keeping connecting wires short and by placing leads and components as high off the chassis as possible. Sometimes, for very high frequencies, stray capacitance is included as part of the circuit design itself.

5.44. Leakage Resistance

A perfect or ideal capacitor is one which given the charge once will keep it forever. However, in practice, all capacitors get discharged in the long run due to small leakage current through their dielectric. Since leakage current is very small, leakage resistance is very high being almost 1000 MΩ for paper and mica capacitors and about 0.5 MΩ for electrolytic ones. An actual capacitor can be represented by an ideal capacitor having leakage resistance connected in parallel with it as shown in Fig. 5.32.

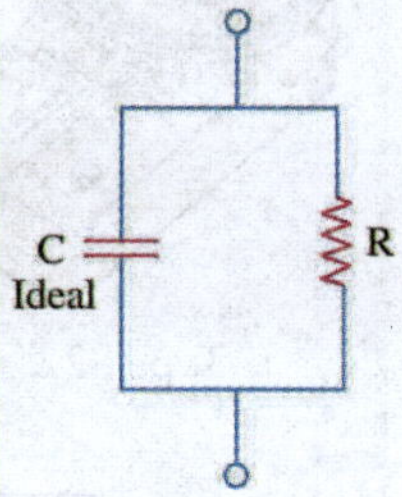

Fig. 5.32

5.45. Capacitors in Series

Connecting capacitors in series is equivalent to *increasing the thickness of the dielectric.*

Hence, combined capacitance is less than the *smallest* individual values.

Following points about series combination of capacitors should be noted :

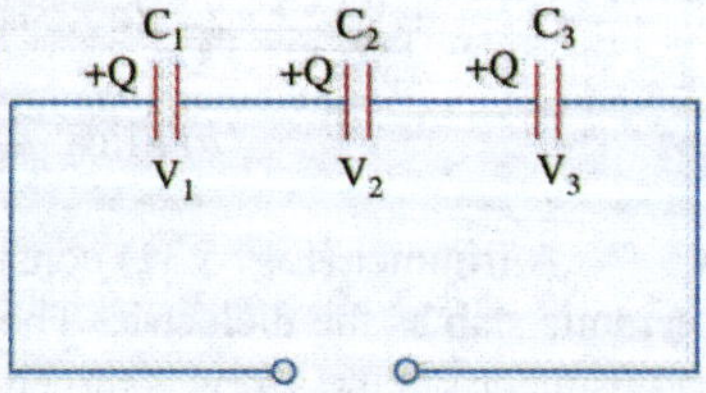

Fig. 5.33

1. charge on each capacitor is the *same* irrespective of its capacitance,
2. p.d. across each capacitor is *different* being inversely proportional to its capacitance,
3. sum of voltages across the capacitors equals the applied voltage.

   As seen from Fig. 5.33,

   $V = V_1 + V_2 + V_3$
4. combined capacitance is given by the reciprocal formula.

   In Fig. 5.33,

$$\frac{1}{C} = \frac{1}{C_1} + \frac{1}{C_2} + \frac{1}{C_3} \quad \text{or} \quad C = \frac{1}{1/C_1 + 1/C_2 + 1/C_3}$$

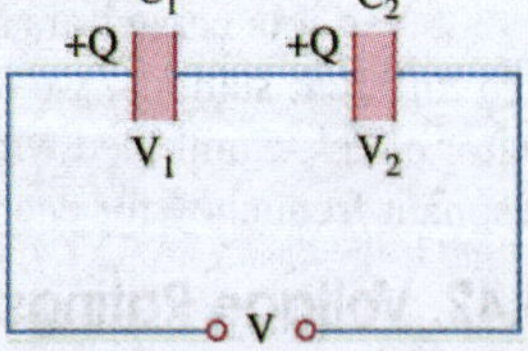

Fig. 5.34

Capacitors are used in series to provide a higher voltage breakdown rating for the combination. For example, combined voltage rating of three equal 200 V capacitors becomes 600 V when connected in series (even though overall capacitance is reduced).

5.46. Two Capacitors in Series

As seen from Fig. 5.34, charge on each capacitor is the same though voltages are different.

1. $$C = \frac{C_1 C_2}{C_1 + C_2}$$

2. $$V_1 = V \frac{C_2}{C_1 + C_2}$$

3. $$V_2 = V \frac{C_1}{C_1 + C_2}$$

5.47. Capacitors in Parallel

Connecting capacitors in parallel is equivalent to *adding their plate areas.* Hence, combined capacitance equals the sum of individual capacitances.

Following facts about parallel combination of capacitors (Fig. 5.35) should be noted :

1. charge across each capacitor is different, being directly proportional to its capacitance
   ($\because \quad Q = CV$)
2. p.d. across each capacitor is the same *i.e.*, the applied voltage V,
3. the sum of the individual charges is equal to the total charge supplied by the power source
   $$Q = Q_1 + Q_2 + Q_3$$
4. combined capacitance is equal to the sum of individual capacitances
   $$C = C_1 + C_2 + C_3$$

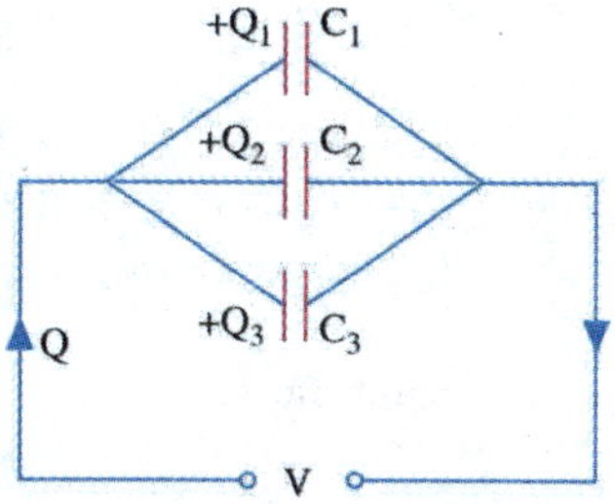

Fig. 5.35

5.48. Two Capacitors in Parallel

Consider the case when only two unequal capacitors are connected in parallel as shown in Fig. 5.36. In this case

1. since V is the same across both capacitors
   $$\because \quad V = \frac{Q_1}{C_1} = \frac{Q_2}{C_2}$$
   $$\therefore \quad \frac{Q_1}{C_1} = \frac{Q_2}{C_2} \text{ or } \frac{Q_1}{Q_2} = \frac{C_1}{C_2}$$
2. the two capacitor charges can be expressed in terms of the total charge Q taken from the power source.
   $$Q_1 = Q\frac{C_1}{C_1 + C_2}; \; Q_2 = Q\frac{C_2}{C_1 + C_2}$$

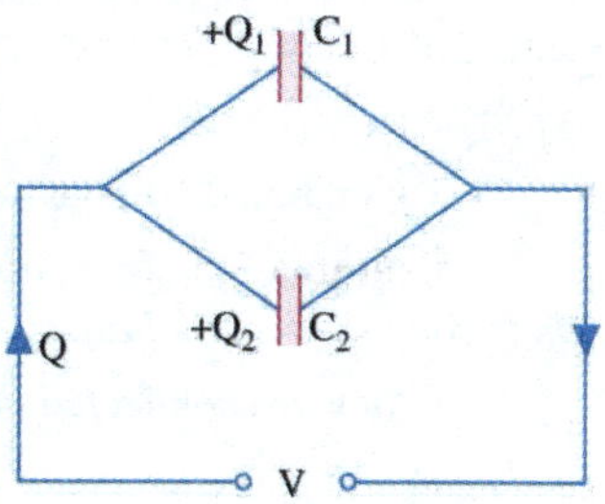

Fig. 5.36

5.49. Energy Stored in a Capacitor

Energy is required to charge a capacitor and is supplied by the charging agency *i.e.,* a battery or any other voltage source. This energy is stored in the electric field set up in the dielectric medium. On discharging the capacitor, the field collapses and the stored energy is released.

$$\text{Stored energy} = \frac{1}{2} CV^2 \text{ joules}$$

This stored energy is the reason why a charged capacitor can produce an electric shock when you touch its two leads. Stored energy of more than 1 joule can be dangerous with a capacitor charged to a voltage high enough (> 90 V) to produce electric shock.

The energy for each "flash" comes from the electrical energy stored in a capacitor.

Example 5.6. *A multiplate capacitor is made up of ten plates 4 cm × 5 cm separated by mica sheets having thickness of 1 mm and a relative permittivity of $\varepsilon_r = 6$. Find its capacitance.*

Solution. $$C = \frac{(n-1)\,\varepsilon_0\,\varepsilon_r\,A}{d}$$

Here, $n = 10,\ \varepsilon_r = 6,\ A = 4 \times 5 = 20\ \text{cm}^2 = 2 \times 10^{-3}$ m^2

$$d = 1\ \text{mm} = 10^{-3}\ \text{m},\ \varepsilon_0 = 8.854 \times 10^{-12}\ \text{F/m}$$

$$C = \frac{(10-1) \times 8.854 \times 10^{-12} \times 6 \times 2 \times 10^{-3}}{10^{-3}}$$

$$= 956 \times 10^{-12}\ \text{F} = \mathbf{956\ pF}$$

Example 5.7. *Two capacitors of 4 μF and 12 μF capacitance and each of working-voltage rating of 24 V are connected in series across a 24 V battery. Calculate*

1. *charge across each,* 2. *voltage across each,* 3. *combined voltage rating.*

Solution. 1. $$C = \frac{C_1 C_2}{C_1 + C_2} = \frac{4 \times 12}{4 + 12} = \mathbf{3\,\mu C}$$

$$Q = CV = 3 \times 24 = \mathbf{72\ \mu C}$$

Each of the two series-connected capacitors will have the same charge.

2. $$V_1 = V\frac{C_2}{C_1 + C_3} = 24 \times \frac{12}{16} = \mathbf{18\,V}$$

$$V_2 = V\frac{C_1}{C_1 + C_2} = 24 \times \frac{4}{16} = \mathbf{6\,V}$$

Alternatively, since charge across each capacitor is 72 μC and their capacitances are known, values of V_1 and V_2 can be easily found from this data.

$$V_1 = Q/C_1 = 72/4 = 18\ \text{V}\ ;\quad V_2 = Q/C_2 = 72/12 = 6\ \text{V}$$

3. Combined voltage rating is = 24 + 24 = **48 V**

Example 5.8. *Two capacitors of 0.0003 μF and 0.0006 μF are connected in series. Find their combined capacitance.*

If they are connected in parallel, what will be the new value of total capacitance?

Solution. $$C = \frac{C_1 C_2}{C_1 + C_2} = \frac{0.0003 \times 0.0006}{0.0003 + 0.0006} = \mathbf{0.0002\,\mu F}$$

Note. In the case of such fractional capacitances, an easier method is to multiply both capacitance values by a figure that will provide a whole number. Then, when the result has been obtained, it should be divided by the same number. In the present case, we will multiply both capacitances by 10,000. Hence,

$$C = \frac{3 \times 6}{3 + 6} = 2$$

Dividing this result by 10,000, we get

$$C = 2/10{,}000 = 0.0002\ \mu\text{F}$$ — as before

When connected in parallel

$$C = 3 + 6 = 9/10,000 = \mathbf{0.0009\ \mu F}$$

Example 5.9. *Find the charges on capacitors shown in Fig. 5.37 and the P.Ds. across them.*

Solution. Equivalent capacitance between points A and B is

$$C_{AB} = C_2 + C_3 = 5 + 3 = 8\ \mu F$$

This capacitance is in series with the 2 μF capacitor.

Hence, combined capacitance is

$$C = \frac{2 \times 8}{2 + 8} = 1.6\ \mu C$$

Charge taken from the battery is

$$Q = CV = 1.6 \times 100 = 160\ \mu C$$

Same would be the charge on 2 μF capacitor.

$\therefore \quad Q_1 = 160\ \mu C$

$\therefore \quad V_1 = Q_1 / C_1 = 160 / 2 = \mathbf{80\ V}$

Now, $\quad V = V_1 + V_2 \quad$ or $100 = 80 + V_2$

$\therefore \quad V_2 = 20\ V$

$\therefore \quad Q_2 = C_2V_2 = 3 \times 20 = \mathbf{60\ \mu C}$

$\quad Q_3 = C_3V_3 = 5 \times 20 = \mathbf{100\ \mu C}$

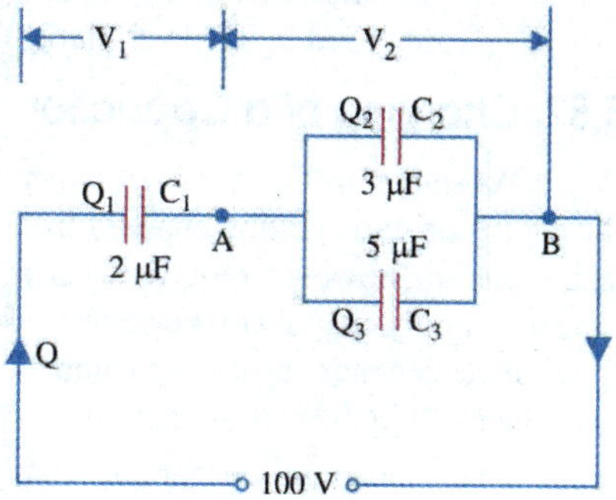

Fig. 5.37

5.50. Troubles in Capacitors

Two common problems with capacitors are (*i*) **open** and (*ii*) **short**. In either case, the capacitor is useless because it cannot store energy. A good capacitor has very high dielectric resistance in the range of 500 to 1000 MΩ for paper and ceramic capacitors and about 0.5 MΩ for electrolytic capacitors. A short-circuited capacitor has *zero* resistance *i.e.*, it shows continuity. A leaky capacitor has a resistance somewhat lower than its normal value.

5.51. Checking Capacitors with Ohmmeter

The procedure is as follows :

1. Discharge the capacitor before checking.
2. Make sure that ohmmeter voltage does not exceed the working voltage of the capacitor.
3. Disconnect one side of the capacitor from the circuit to eliminate any parallel resistance path that can lower the resistance reading.
4. Keep your fingers off the conductors because the body resistance (50 K or so) connected in parallel will lower the resistance.
5. Use highest-ohm scale.

The trouble in a capacitor is indicated by the following ohmmeter readings :

1. If meter pointer immediately goes to practically zero and stays there, the capacitor is short-circuited.
2. If the pointer first goes to low-resistance side (showing capacitor charging) and then comes up and shows a reading less than the normal, it is leaky.

3. If the pointer does not go to low-resistance side of the scale (*i.e.,* if capacitor shows no charging action) but just goes straight to infinity, the capacitor is open.
4. If the meter pointer first moves quickly towards the low-resistance side of the scale (showing capacitor charging) and then slowly goes to infinity, it shows a good capacitor which is supposed to have very large (almost infinite) resistance.

5.52. Charging of a Capacitor

As shown in Fig. 5.38 (*a*), when *S* is closed, the capacitor starts getting charged from the battery. The capacitor, however, requires a certain amount of time to get fully charged to the battery voltage *V*. The time taken depends on its capacitance (*C*) and the resistance (*R*) in the charging circuit.

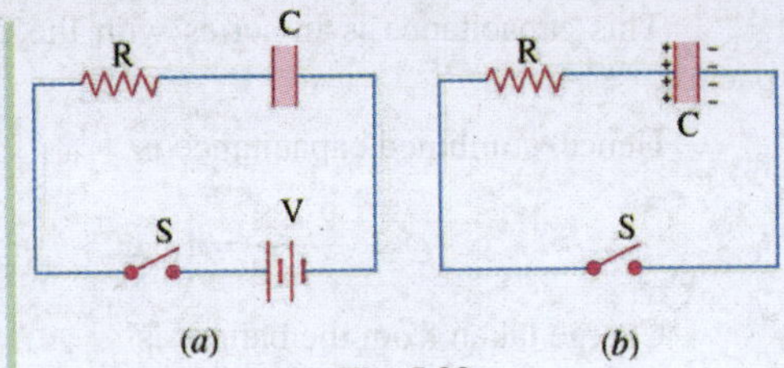

Fig. 5.38

The time required for the charge to attain 63.2% of its final value is called *time-constant* (*T*) of the circuit. Its value = *CR* and is in seconds if *C* is in farads and *R* in ohms.

A total time equal to 5 time-constants (= $5T = 5CR$) is required for the capacitor to be considered fully charged as shown in Fig. 5.39.

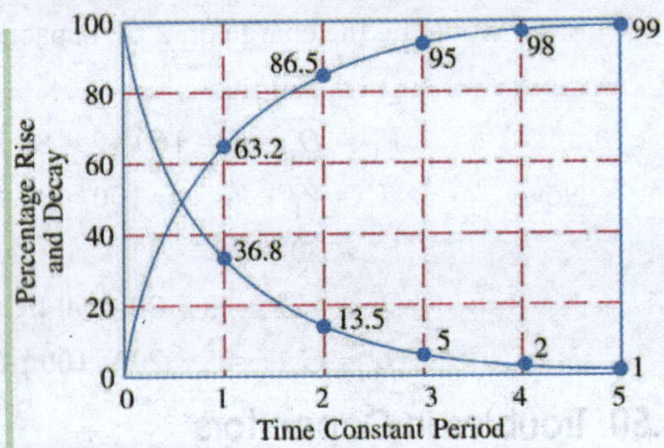

Fig. 5.39

In Fig. 5.38 (*b*), the fully-charged capacitor can be discharged through *R* by closing *S*. In one time-constant, the charge decreases to 36.8% of its fully-charged value. Again, in 5 time-constants, the capacitor is considered fully discharged.

The increase and decrease of charge across the capacitor can be found from the Universal Time Constant Chart of Fig. 5.39. It also represents rise and decay of charging and discharging current of the capacitor respectively.

Example 5.10. *A 50 μF capacitor is charged a 40 K resistor to a potential difference of 400V. Calculate*

1. *time constant of the circuit ;*
2. *value of full charge;*
3. *charge acquired in one time-constant ;*
4. *energy stored in the fully-charged capacitor.*

Solution.

1. $T = CR = 50 \times 10^{-6} \times 40 \times 10^3 = \mathbf{2\ seconds}$
2. Full charge, $Q = CV = 50 \times 20^{-6} \times 400 = \mathbf{0.02\ C}$
3. Charge acquired by the capacitor in one time-constant *i.e.,* in 2 seconds

$$= 63.2\% \text{ of full charge}$$

$$= 0.632 \times 0.02 = \mathbf{0.0126\ C}$$

4. Energy stored $= \frac{1}{2} CV^2 = \frac{1}{2} \times (50 \times 10^{-6}) \times 400^2 = \mathbf{2\ joules}$

5.53. Capacitor Connected Across an AC Source

Suppose a capacitor is connected across an ac voltage supply as shown in Fig. 5.40. It is found that the capacitor is charged first in one direction during the positive half-cycle of the applied voltage

[Fig. 5.40 (*a*)] and then in the opposite direction during the negative half-cycle [Fig. 5.40 (*b*)]. Since, it is an alternating voltage, the capacitor keeps getting charged and discharged continuously. If a lamp is included in the circuit, it will continue to glow. The reason is as follows :

During the positive half-cycle of the applied voltage, charging current flows from plate *N* to *M* through the *external* circuit containing the lamp. It cannot flow 'through' the dielectric. During the negative half-cycle, current direction is reversed but it again passes through the lamp thereby keeping it alight. It hardly makes any difference to the lamp whether current through it passes from left to right or right to left. It will be appreciated that no current whether dc or ac can pass *through* a capacitor.

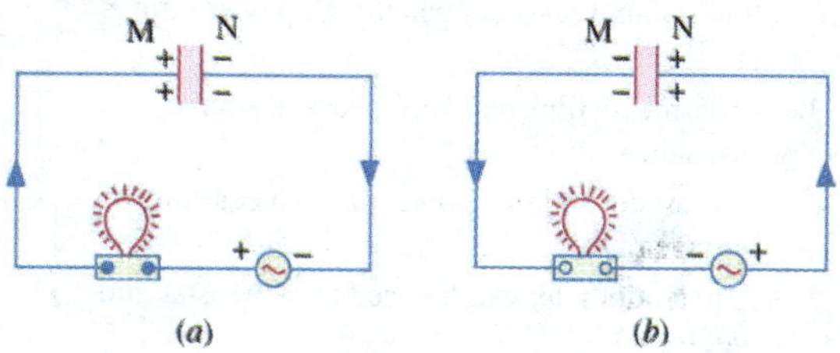

Fig. 5.40

5.54. Capacitive Reactance

A good capacitor completely blocks the passage of direct current 'through' it. However, it allows alternating current to pass 'through' it in the sense discussed above. In fact, what it actually does is that it allows current to flow through the *external* circuit first in one direction and then in the opposite direction at a frequency equal to the frequency of the applied voltage. Even then, it offers opposition to the flow of this alternating current. This opposition is called capacitive reactance (X_C) and is given by

$$X_C = \frac{1}{2\pi fC} = \frac{0.159}{f\,C}$$

Its unit is ohm.

Capacitive reactance varies *inversely* as the frequency of the applied ac voltage. Higher the frequency, lesser the reactance and *vice-versa.* For dc voltage $f = 0$, hence $X_C = 1/0 = \infty$. That is why a capacitor blocks direct current and voltage.

SELF EXAMINATION QUESTIONS

A. Fill in the blanks with most appropriate word(s) or numerical value (s).

1. Wire-wound resistors show troublesome .... effects at high frequencies.
2. The amount by which the actual resistance of a resistor may vary form its nominal value is known as the ....of the resistor.
3. A potentiometer is used for changing the value of voltage whereas a rheostat is used for changing............ .
4. The first .... colour bands of a resistor indicates the resistance value of a resistor whereas fourth one indicates its resistance ....
5. Iron cores are laminated in order to minimise .... loss.
6. Inductor cores which have the least core loss even at high frequencies are made of ....
7. Unwanted inductance in a circuit is known as inductance.
8. Opposition offered by a coil having both reactance and resistance is called ....
9. Higher the *Q*-value of a tuned circuit, greater its ....
10. As separation between the two plates of a capacitor is increased, its capacitance is ....
11. For the same size, a ceramic capacitor has .... capacitance than a mica capacitor.
12. Ceramic capacitors are preferable for use at .... frequencies.
13. Electrolytic capacitors are called .... capacitors.
14. Variable capacitors are frequently used in ....circuits of radio receivers.
15. Higher the applied frequency, ...the reactance offered by a capacitor.

B. Answer True or False

1. Wire-wound resistors are especially used where precision and stability of resistance is necessary.
2. Carbon-composition resistors are preferred to wire-wound resistors when high power ratings are required.
3. Thin-metal film resistors generally have low resistance.
4. For resistors, higher power rating means higher voltage rating.
5. A potentiometer can be used as a rheostat and *vice-versa*.
6. Air-core coils have maximum loss at high frequencies.
7. Ferrite cores are used to increase inductance without increasing core losses.
8. If number of turns over a given core are doubled, its inductance is doubled.
9. Mutual inductance is possible between two coils even if they are connected together electrically.
10. The maximum value of coefficient of coupling between two coils is unity.
11. Direct current cannot pass through a capacitor but an alternating current can.
12. Ceramic capacitors have very large capacitance even in small sizes.
13. Mylar capacitors have insulation resistance of 1000 MΩ and above.
14. Electrolytic capacitors have very large capacitance but comparatively low leakage resistance.
15. Ganged capacitors are frequently used in tuning circuits of radio receivers.
16. Voltage rating of a capacitor gives the maximum voltage it can withstand without breakdown.
17. Connecting capacitors in parallel is equivalent to increasing the voltage rating of the combination.
18. For increasing capacitance, the given capacitors should be connected in parallel.
19. When connected in series, all capacitors have the same charge but different voltages.
20. A healthy capacitor offers infinite resistance to direct current.

C. Multiple Choice Items

1. If current through a resistor is halved, wattage developed by it would be
   (*a*) halved (*b*) quadrupled
   (*c*) cut to one-fourth (*d*) doubled
2. The carbon-composition resistor shown in Fig. 5.41 has a resistance of
   (*a*) 270 K ± 27 Ω (*b*) 274 ± 5%
   (*c*) 470 K ± 47 K (*d*) 274 Ω ± 10%

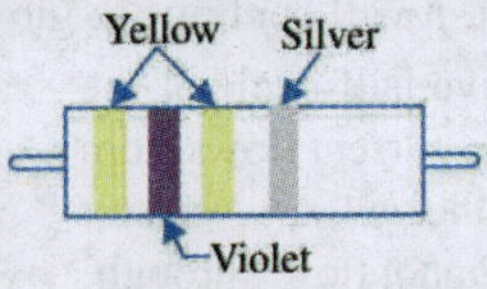

Fig. 5.41

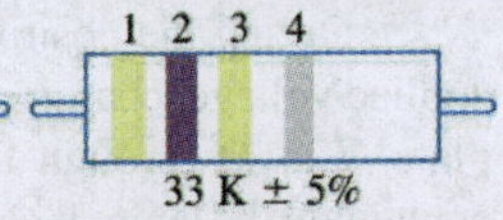

Fig. 5.42

3. The colour code for the resistor shown in Fig. 5.42 is
   (*a*) red, orange, red, gold
   (*b*) orange, orange, red, gold
   (*c*) orange, orange, orange, gold
   (*d*) red, red, violet, silver
4. The combined resistance of the two resistors connected as in Fig. 5.43, is about
   (*a*) 120 (*b*) 45
   (*c*) 165 (*d*) 35

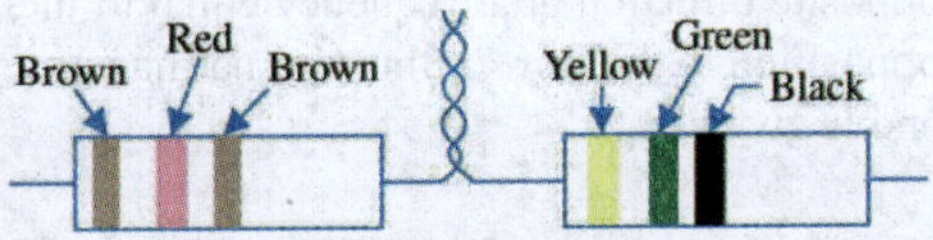

Fig. 5.43

5. When checked with an ohmmeter, an 'open' resistor reads
   (*a*) infinite (*b*) zero
   (*c*) low but not zero (*d*) high but not infinite
6. While measuring resistance with an ohmmeter, one should not touch test leads because
   (*a*) of the possibility of electric shock
   (*b*) it may lower the ohmmeter reading
   (*c*) it may ground the resistor
   (*d*) it may 'short' the resistor being checked
7. Ferrite cores commonly used at high frequencies
   (*a*) increase core losses
   (*b*) decrease core losses
   (*c*) decrease inductance
   (*d*) increase resistance
8. Point out the **incorrect** answer. Inductance of a coil depends
   (a) directly on the square of the number of its turns
   (*b*) directly on the cross-sectional area of the core
   (*c*) directly on the thickness of the wire used
   (*d*) inversely as the length of the core

9. A coil has an inductance of 1 H if current changing at the rating of 2 A/s induces
   (*a*) one volt in it
   (*b*) two volts in another coil
   (*c*) two volts in it
   (*d*) half a volt in it
10. A coil of 2 H develops a counter emf of 4V in it. The rate of change of current through it must be .... amperes/second.
   (*a*) 2 (*b*) 0.5
   (*c*) 8 (*d*) 1
11. The reactance of a coil at 100 Hz is 20Ω. Its reactance at 1 kHz would be ....ohm.
   (*a*) 2 (*b*) 10
   (*c*) 40 (*d*) 200
12. The *main* function of a capacitor is to
   (*a*) block current flow
   (*b*) help current flow
   (*c*) store energy (*d*) dissipate heat
13. A capacitor that stores 0.5 C at 10 volts has a capacitance of ....farad.
   (*a*) 5 (*b*) 20
   (*c*) 10 (*d*) 0.05

ANSWERS

A. Fill in the blanks

1. inductance **2.** tolerance **3.** current **4.** three, tolerance **5.** eddy current
6. ferrite **7.** stray **8.** impedance **9.** selectivity **10.** decreased
11. larger **12.** higher **13.** polarised **14.** tuning **15.** greater

B. True or False

1. T **2.** F **3.** F **4.** T **5.** F **6.** F **7.** T **8.** F **9.** T **10.** T **11.** F
12. T **13.** T **14.** T **15.** T **16.** T **17.** F **18.** T **19.** T **20.** T

C. Multiple Choice Items

1. *c* **2.** *c* **3.** *b* **4.** *d* **5.** *a* **6.** *b* **7.** *b* **8.** *c* **9.** *c* **10.** *a* **11.** *d*
12. *c* **13.** *d*

6 CHAPTER

Energy Sources

1. Primary and Secondary Cells
2. Cell and Battery
3. Voltage and Current of a Cell
4. Cell Life
5. Different Types of Dry Cells
6. Carbon Zinc Cell
7. Alkaline Cells
8. Manganese Alkaline Cell
9. Nickel Cadmium Cell
10. Mercury Cell
11. Silver Oxide Cell
12. Lead Acid Cell
13. Battery Rating
14. Testing Dry Cells
15. Photoelectric Devices
16. Photovoltaic Cell
17. Solar Cell

6.1. Primary and Secondary Cells

A voltaic chemical cell is a combination of materials which produce direct-current (dc) electrical energy from its *internal* chemical reactions. These cells can be sub-divided into the following two classes :

(*a*) **Primary Cell.** In this cell, chemical reactions are irreversible. During the generation of electric energy, it consumes materials which cannot be replenished by recharging. Hence, such cells cannot be recharged back to their original condition after being discharged.

(*b*) **Secondary Cell or Storage Cell or Accumulator.** In these cells, chemical reactions are reversible. After discharge, such a cell can be restored to its original condition by passing an electric current through it in a direction opposite to that of the discharge current.

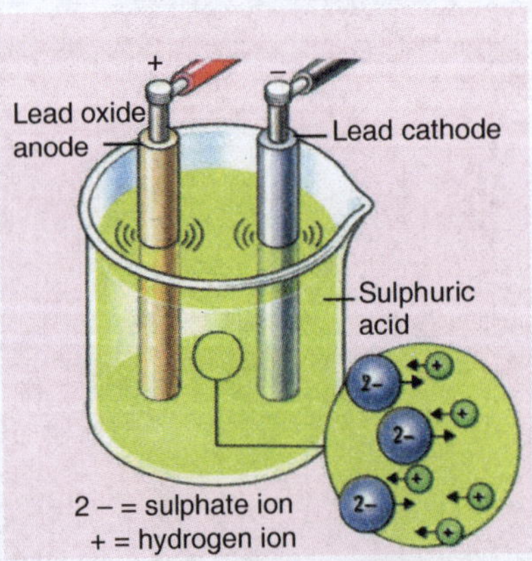

Secondary cell.

Further points of comparison between the two classes are tabulated given on next page.

	Primary Cells	Secondary Cells
1.	low cost	expensive
2.	small size	reasonably small
3.	short life	comparatively long life
4.	useless when discharged	rechargeable
5.	light weight	heavier

6.2. Cell and Battery

A battery consists of two or more cells connected either in series or parallel or both. Often, many people call a flashlight cell a flash light battery which is technically incorrect.

6.3. Voltage and Current of a Cell

The voltage rating of a cell is given by its open-circuit voltage *i.e.*, voltage it can produce when not connected to a load circuit. This voltage depends on the types of materials used and not on the physical size of the cell.

The capacity of a cell is given by the amount of current it can supply to an external load circuit. It depends on the condition and quantity of electrolyte and physical size of the electrodes. Everything else being the same, a larger cell can deliver more current for a longer period of time than a smaller one.

6.4. Cell Life

It is given by the period of time during which the cell can be stored on a shelf without losing more than approximately 10% of its original capacity. The loss of capacity occurs even when cell is not in use and is primarily due to partial drying up of paste electrolyte and to other chemical actions which change the materials within the cell. Since heat helps both these processes, it is advisable to keep a cell in a cool place in order to extend its life.

6.5. Different Types of Dry Cells

We will consider the following primary and secondary cells :

(*a*) carbon-zinc cell

(*b*) alkaline cell

1. manganese-alkaline cell
2. nickel-cadmium cell
3. mercury cell
4. silver-oxide cell
5. lead-acid cell

6.6. Carbon Zinc Cell

It is the oldest and most widely used commercial dry cell (Fig. 6.1).

(*a*) Construction

1. The zinc can functions both as a container to hold the electrolyte and as the negative electrode.
2. The positive electrode is a carbon rod down the centre but not low enough to touch the zinc bottom.
3. The electrolyte is a paste of ammonium chloride and manganese dioxide.

(*b*) Size and Voltage

Such cells are available in several sizes and have open-circuit voltage from 1.4 to 1.6 V, regardless of size. Larger cells with more zinc, electrolyte and depolarizer have a higher current rating upto 0.25 A or 250 mA. The flashlight cell has a current rating of 50 mA for nearly 60 hours of service. The cell has a height of about 57 mm and a diameter of about 32 mm.

Carbon-zinc batteries are made in many types having voltages of 3, 4, 5, 6, 9, 13.5, 22.5 and 45 V (all multiples of 1.5 V). In some batteries, cells are cylindrical in shape and in others they are flat.

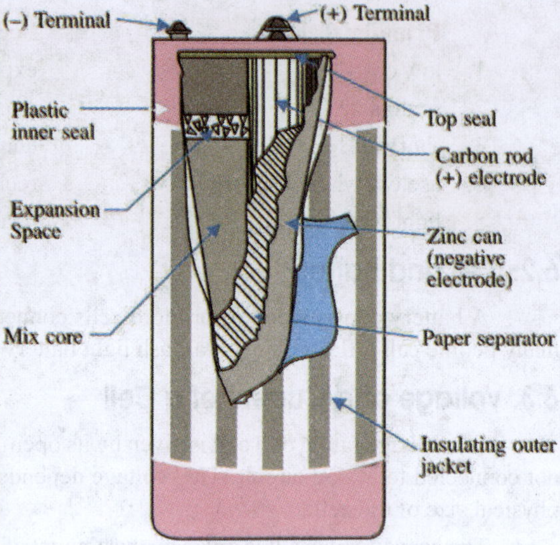

Fig. 6.1

(*c*) Operating Efficiency

Carbon-zinc cells and batteries render efficient service provided they are used *intermittently* for short periods at a time at relatively low currents. Such operation helps to keep them sufficiently depolarised.

(*d*) Recharging

The dry cell can be recharged if certain precautions are taken :

1. voltage should not be below 1 V when removed for service,
2. charging rate should be kept low and spread over 10 to 15 hours,
3. recharged cells should be used at one in view of their limited shelf life.

This walkman is powered by batteries of the primary cell type. Typically, they have to be replaced every two or three months.

6.7. Alkaline Cells

All those cells which use a caustic electrolyte are called alkaline cells. We will consider

1. manganese-alkaline cell
2. nickel-cadmium cell
3. mercury cell and
4. silver-oxide cell.

Fig. 6.2

6.8. Manganese Alkaline Cell

It is a *primary* cell having zinc as anode and manganese dioxide as cathode in a leak-proof steel can (Fig. 6.2). The electrolyte is potassium hydroxide (KOH) or sodium hydroxide (NaOH). Because of the high conductivity of the electrolyte, this cell has higher current rating than a carbon-zinc cell. It provides an output voltage of 1.5 V. Some alkaline batteries can be recharged a few times and have voltages of 4.5, 7.5, 13.5 and 15 V.

6.9. Nickel Cadmium Cell

It is a *dry* cell but is *rechargeable*. In present-day electronics, it has superseded both lead-acid cell and nickel-iron (NiFe) alkaline cell.

(*a*) Construction

The action material cadmium (a metal akin to zinc), in powder form, is pressed (*i.e.*, sintered) into perforated steel plates which then form the negative electrode of the cell (Fig. 6.3). The positive anode is a steel mesh coated with solid nickel hydroxide (NiOH). The electrolyte is potassium hydroxide (caustic potash) usually in jelly form. The container is a steel can which is sealed after the cell is placed in it. These cells are also available in flat button shape.

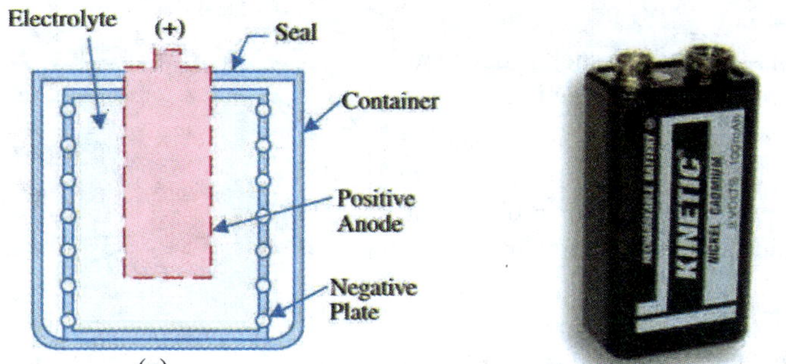

(*a*) (*b*)

Fig. 6.3

(*b*) Voltage and Current

The open-circuit voltage of such cells is from 1.25 to 1.5 V. They have much longer life when they are rapidly discharged at intervals. In fact, long periods of inactivity can cause the cells to fail. Such cells need to be charged at constant current and can be recharged upto 1000 times or more.

These cells have very low internal resistance and can deliver high currents without much loss of terminal voltage. The most common nickel-cadmium batteries have voltages of 6, 9.6 and 12 V.

Their chemical reaction is given by the following equation :

$$Cd + 2\,NiOH + 2\,KOH = Cd\,(OH)_2 + 2\,Ni + 2\,KOH$$

Fig. 6.4 shows the range of rectangular nickel-cadmium (Nicad) cells and batteries. These Nicad cells have a normal no-load voltage of 1.2 V, are capable of high current loads and 1000 cycles of discharge, can tolerate overcharge and general electrical abuse like short circuit.

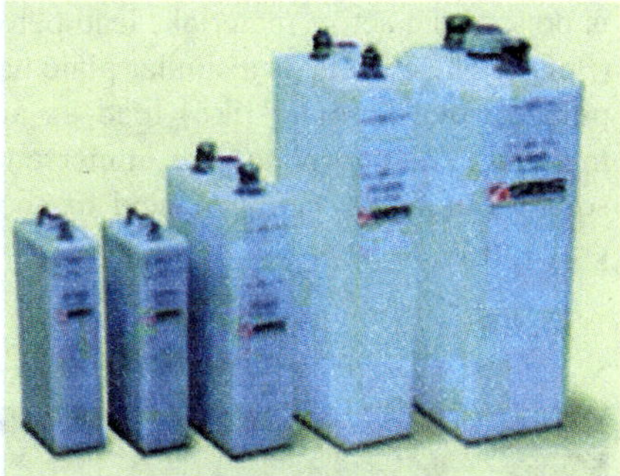

Fig. 6.4

(*c*) Uses

Since such cells can easily supply high current, they are widely used as sources of energy in radios, TV sets, tape recorders, cameras and cordless type home appliance such as electric toothbrush or electric carving knife etc. They give good service under extreme conditions of shock, vibration and temperature.

6.10. Mercury Cell

It is a *primary* cell and is widely used in miniaturised electronic gadgetry.

(*a*) Construction

The cathode is made of compressed mercuric oxide and graphite whereas anode is made of highly purified zinc powder.

The two are separated by a porous material. The electrolyte is a paste mixture of potassium hydroxide and zinc oxide. The flat pallet construction of a mercury cell is shown in Fig. 6.5.

(*b*) Size and Voltage

Such cells are available in wide range of shapes and sizes—the smallest being about 3 mm thick and 12.5 mm in diameter. Open-circuit voltage at room temperature varies from 1.35 to 1.4 V depending on the electrolyte mixture. With pure mercuric oxide for cathode, voltage output is extremely stable at 1.35 V. Mercury batteries are available in a variety of voltage ratings.

(c) Uses

Mercury cells enjoy a long shelf life without deterioration and are extremely rugged.

Since they can provide almost constant voltage under varying load conditions, mercury cells and batteries are widely used in electronic watches, hearing aids and test instruments etc.

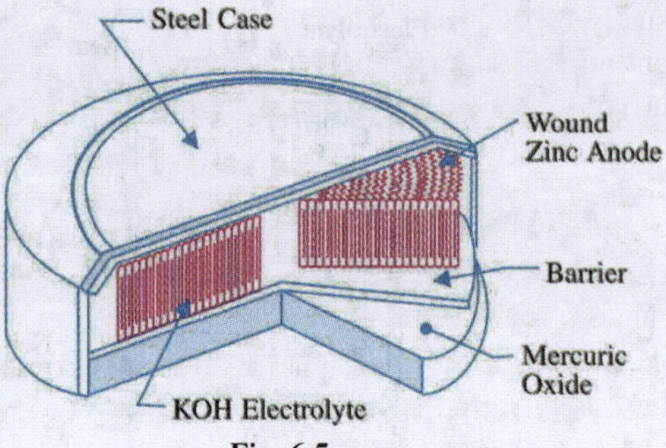

Fig. 6.5

6.11. Silver Oxide Cell

It is a *primary* cell with an open-circuit voltage of 1.5 V. It has a cathode of silver oxide and an anode of zinc in an alkaline electrolyte. As shown in Fig. 6.6, it is generally made in button size for use in cameras, hearing aids and electronic watches.

6.12. Lead Acid Cell

It is a *wet* secondary cell mostly used for making automobile batteries.

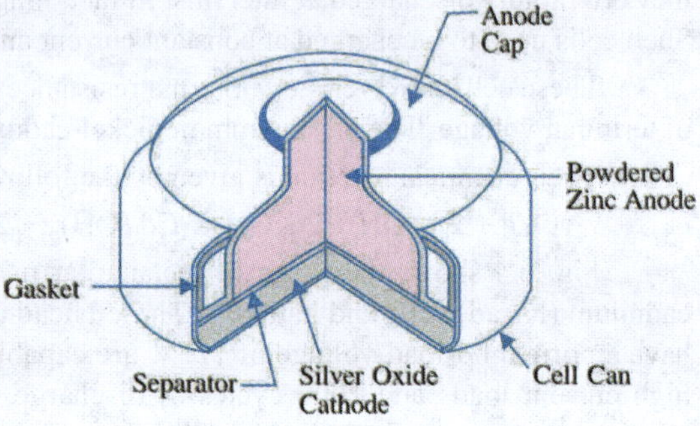

Fig. 6.6

(a) Construction

Its anode plate (brown coloured) is a grid made of a lead-antimony alloy into which is pressed the active material : lead peroxide (PbO_2). The cathode is a similar plate having pure porous (sponge like) lead as active material. The two plates are immersed in a solution of dilute sulphuric acid of specific gravity 1.21 or so.

(b) Voltage

It has an open-circuit voltage of 2 to 2.2 V when fresh. It should not be used after its voltage has fallen to 1.85 V when it should be put on charge. It has much larger current capacity as compared to dry cells, considered earlier.

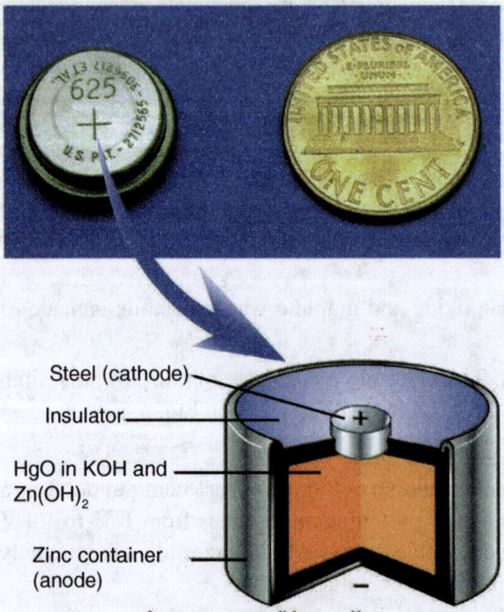

A mercury cell in section.

(c) Use

Principal use of these cells is in the manufacture of portable lead-acid batteries used in automobiles (car and motor vehicles). For car battery, 6 cells are connected in series to give a total voltage of 12 V or more. For traction batteries, the voltages vary from 12 V (6 cells) to 64 V (32 cells) for industrial trucks and from 30V (15 cells) to 72 V (56 cells) for dairy and bakery road vehicles.

6.13. Battery Rating

Lead-acid batteries are rated in terms of discharge current, they can supply continuously for a specified period of time. It is usually expressed in ampere-hours (Ah). The capacity is always given at a specified rate of discharge usually 20 hour rate. It means that a battery which is rated at 100 Ah should deliver 5 A of current continuously for 20 hours and maintain at least 1.75 V per cell. Of course, battery can supply less current for longer time or more current for a shorter time.

6.14. Testing Dry Cells

The condition of a dry cell or battery can be checked by the terminal voltage developed *when it is connected to a load.* If this voltage is less than 80% of the open-circuit voltage, it should be rejected. The basic reason is that a bad cell has high internal resistance which will become detectable when cell is tested *on load.* Due to normal current flow, there would be a large internal drop and hence terminal voltage would drop by a considerable amount. Such would not be the case when cell is tested on no-load.

6.15. Photoelectric Devices

As the name shows, these devices convert light energy directly into electric energy. We will consider the following semiconductor devices only because they are self-generating.

1. photovoltaic cell
2. solar cell

6.16. Photovoltaic Cell

In this cell, light energy is used to create a potential difference which is directly proportional to the frequency and intensity of the incident light. Fig. 6.7 shows a basic photovoltaic cell.

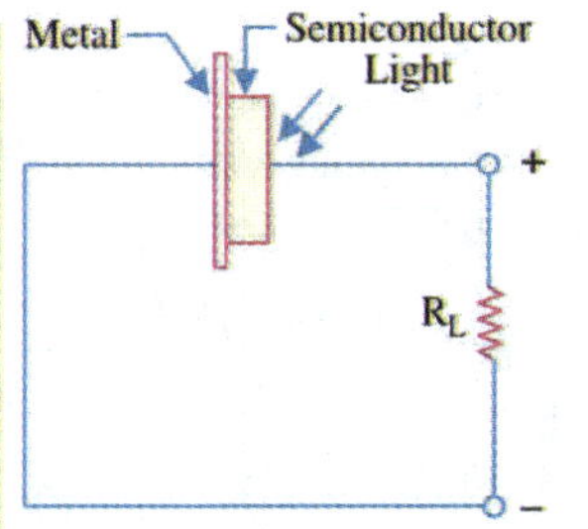

Fig. 6.7

It consists of a piece of semiconductor material such as selenium (Se), silicon (Si) or germanium (Ge) which is bonded to a metal plate.

When light falls on the semiconductor, valence electrons and holes are liberated from its crystal structure. These electrons flow out of the semiconductor into the metals whereas holes flow in the opposite direction. It creates a potential difference between the semiconductor and the metal which is sufficient to cause current flow through a load resistor as shown in Fig. 6.7.

These cells are used in devices like portable exposure meters and direct-reading illumination meters.

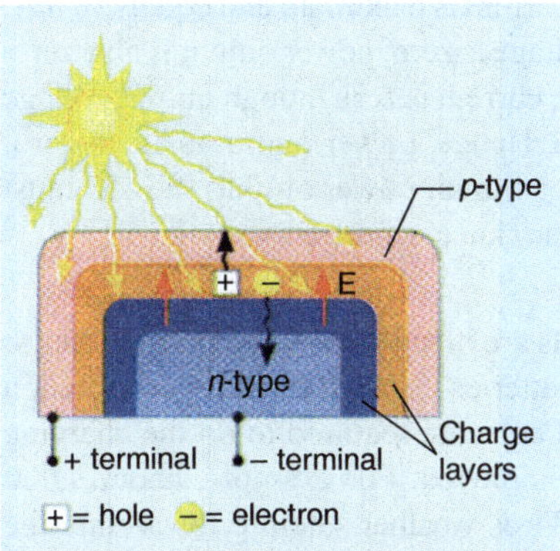

A solar cell formed from a p-n junction. When sunlight strikes the solar cell it acts like a battery, with positive and negative terminals.

6.17. Solar Cell

It is also called solar energy converter and is basically a P-N junction device which converts solar energy into electric energy.

(*a*) Construction and Working

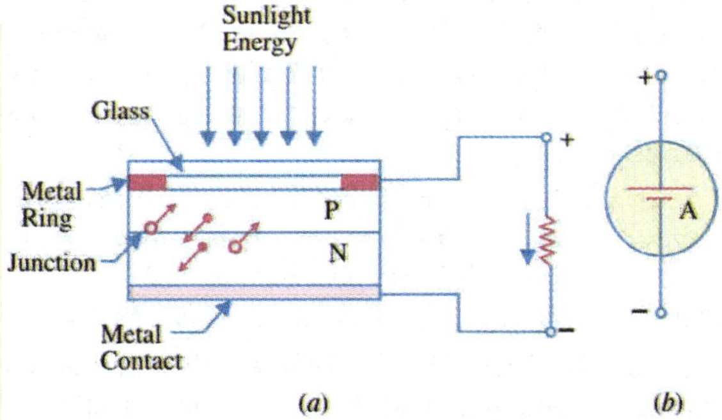

Fig. 6.8

As shown in Fig. 6.8 (*a*), it essentially consists of a silicon P-N junction diode generally packaged in TO-type can with a glass window on top. Surface layer of P-material is made extremely thin so that incident light photons may easily reach the P-N junction.

When these photons collide with valence electrons, they impart them sufficient energy as to leave their parent atoms. In this way, free electrons and holes are generated on both sides of the junction and their flow constitutes the minority current. This current is directly proportional to the illumination (lumen/metre^2 or mW/cm^2) and also depends on the size of the surface area being illuminated. The open-circuit voltage V_{oc} is a function of illumination. Consequently, power output of a solar cell depends on the level of sunlight illumination. Power cells are also available in flat strip form so as to cover sufficiently large surface area.

This calculator makes use of solar energy.

The nickel-plated ring around the P-layer acts as the positive output terminal (*i.e.*, anode) and the metal contact at the bottom serves as the negative output terminal (*i.e.*, cathode). The symbol is shown in Fig. 6.8 (*b*). Si and Ge are the most widely used semiconductor materials for solar cell although gallium arsenide (GaAs), indium arsenide (In As) and cadmium arsenide (Cd As) are also being used now-a-days.

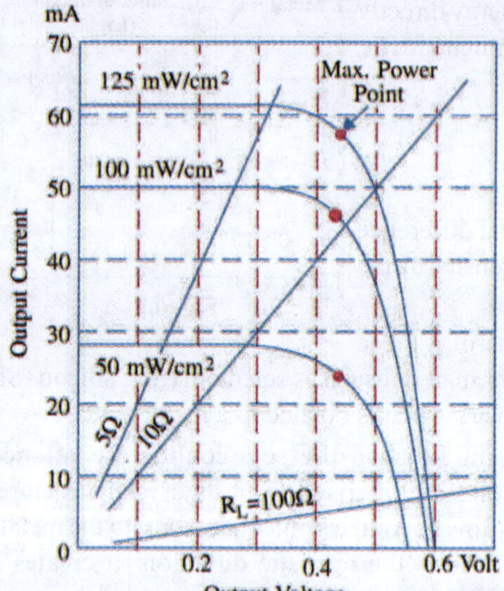

Fig. 6.9

(*b*) Characteristics

Typical V-I characteristics of a solar cell corresponding to different levels of illumination are shown in Fig. 6.9.

Consider the characteristic for incident illumination of 100 mW/cm^2 in Fig. 6.9. If we short-circuit the cell, current is maximum and equals 50 mA. Since output voltage is zero, power output is also zero. On open-circuit, current is zero though output voltage is nearly 0.57 V. Hence, power output is again zero. For obtaining maximum power output, the cell must be operated at the 'knee' of the curve.

(*c*) Uses

Solar cells are being used on board satellites to recharge their batteries. Since their sizes are small, a large number of cells are required to do the charging job for which purpose series-parallel cell combinations are employed. For example, about 9,000 solar cells were used to charge nickel-cadmium batteries of Tiros weather satellite for around the clock operation. Their present-day efficiency level is around 15% but it is hoped that higher efficiencies would be reached in the near future. Scientists are planning to orbit large banks of solar cells outside the Earth's atmosphere for converting solar energy into electricity. This energy would then be sent to Earth in the form of a powerful microwave beam which would be reconverted into electricity for terrestrial use.

Example 6.1. *An Earth satellite has 12 V nickel-cadmium batteries which supply a continuous current of 0.5 A throughout the day. Solar cells having V-I characteristics of Fig. 6.9 are employed to keep the batteries fully charged. If illumination from the Sun for 12 hours in every 24 hours is 125 mW/cm^2, determine the approximate number of solar cells required.*

Solution. The circuit for the solar cell battery charger is shown in Fig. 6.10. Let there be *n* cells connected in series to provide higher voltage and let *m* such series-groups be connected in parallel to provide the necessary current. Total number of cells required = $m \times n$.

As seen from the V-I characteristics of Fig. 6.9, each cell supplies 56 mA at 0.45 V. If we allow a drop of 1.5 V for series resistor etc., we need (12 + 1.5) = 13.5 V from each series group.

$\therefore \quad n \times 0.45 = 13.5$ Ah

or $\quad n = 13.5/0.45 = 30$

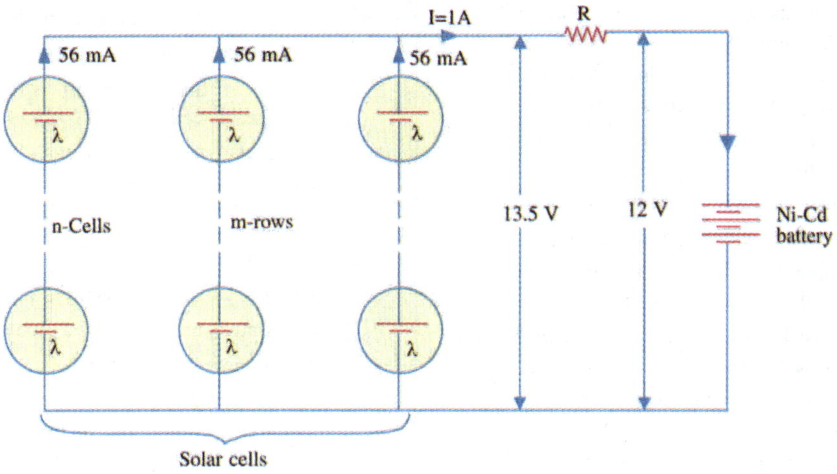

Fig. 6.10

Obviously, this much charge must be put back into the battery by the solar battery charger. Since solar battery itself is illuminated for only 12 hours per day, it must provide a charging current of 12 Ah/12 h = 1 A. Since each row can supply only 56 mA = 0.056 A.

$\therefore \quad m \times 0.056 = 1 \quad$ or $\quad m = 1 / 0.056 = 18$

Hence, total number of solar cells needed to make up the required solar battery charger is

$= m \times n = 18 \times 30 =$ **540**

CONVENTIONAL PROBLEMS

1. A 1.5 V carbon-zinc dry cell is connected across a load of 1000 Ω. Calculate the current and power supplied by it. **[1.5 mA ; 2.25 mW]**
2. A 6 V lead-acid has an internal resistance of 0.01 Ω. What would be the short-circuit current of the cell ? Will you be able to get it in practice ? **[600 A, No]**
3. A lead-acid battery discharges at the rate of 5 A for 20 hours. (*a*) How much charge must be put back into the battery to restore its original charge assuming 100% efficiency ? (*b*) How long will this charging take place if charging current is 2.5 A ? **[(*a*) 100 Ah or 36 × 10^4 (*b*) 40 h]**
4. The output voltage of a Ni-Cd battery drops from 12 V at zero load to 11.95 V with a load current of 5 A. What is the internal resistance of the battery ? **[0.01 Ω]**

SELF EXAMINATION QUESTIONS

A. Fill in the blanks with most appropriate word (s) or numerical value (s).

1. Devices which convert light energy directly into electric energy are called .... devices.
2. A............. cell is one that cannot be recharged.
3. A battery consists of a number of ....
4. For increasing the shelf life of a cell, it is advisable to keep it in a .... place.
5. Nickel-cadmium cell is a dry cell but is ....
6. Lead-acid cell is a wet ....cell because it can be recharged.
7. The working of a photovoltaic cell depends on contact between a................. and a metal.
8. A solar cell is essentially a P-N ....

B. Answer True or False

1. Chemical reactions in a secondary cell are reversible.
2. The capacity of a cell depends on the physical size of its electrodes.

3. A larger cell can deliver more current for a longer period of time than a smaller one.
4. Nickel-cadmium cell is the example of a dry but rechargeable cell.
5. Mercury cell provides almost constant voltage under varying load conditions.
6. Silver oxide cells are generally made in the button size.
7. Battery rating is generally expressed in Ah.
8. It is best to test dry cells on no-load.
9. Photoelectric devices convert light energy directly into electric energy.
10. Solar cell converts Sun's heat energy directly into electric energy.
11. The current produced by a solar cell depends only on the solar illumination.
13. Silicon and germanium are the most widely-used semiconductor materials for a solar cell.

C. Multiple Choice Items

1. The output voltage of a lead-acid cell is about .... volt.
   (*a*) 1.5 (*b*) 2.2
   (*c*) 1.25 (*d*) 6
2. The current delivering capacity of a cell depends on
   (*a*) condition of the electrolyte
   (*b*) quality of the electrolyte
   (*c*) physical size of the electrode
   (*d*) all the above
3. Which of the following is a dry but rechargeable cell ?
   (*a*) nickel-cadmium cell
   (*b*) mercury cell
   (*c*) lead-acid cell
   (*d*) solar cell
4. Rating of a battery is usually expressed in
   (*a*) watt-hour (*b*) joule
   (*c*) ampere-hour (*d*) amperes
5. Constructionally, a solar cell consists of
   (*a*) a semiconductor bonded to a metal
   (*b*) a piece of silicon
   (*c*) P-N junction
   (*d*) photoemissive material
6. For obtaining maximum power from a solar cell, it should be operated on
   (*a*) level part (*b*) knee
   (*c*) falling portion
   (*d*) any part of its V-I characteristic.

ANSWERS

A. Fill in the blanks

1. photoelectric **2.** primary **3.** cells **4.** cool
5. rechargeable **6.** secondary **7.** semiconductor **8.** junction

B. True or False

1. T **2.** T **3.** T **4.** T **5.** T **6.** T **7.** T
8. F **9.** T **10.** T **11.** F **12.** F **13.** T

C. Multiple Choice Items

1. *b* **2.** *d* **3.** *a* **4.** *c* **5.** *c* **6.** *b*.

CHAPTER 7

Magnetism and Electro-magnetism

1. Magnetic Materials
2. Ferrites
3. Types of Magnets
4. Demagnetising or Degaussing
5. Magnetic Shielding
6. Magnetic Terms and Units
7. Ohm's Law for Magnetic Circuit
8. Transformer
9. Transformer Working
10. Transformer Impedance
11. Can a Transformer Operate on DC ?
12. RF Shielding
13. Autotransformer
14. Impedance Matching

7.1. Magnetic Materials

Different materials can be classified as either magnetic materials or non-magnetic materials (air, glass, wood, paper, porcelain, plastic and rubber etc.). However, it should be noted that even though these non-magnetic materials cannot be magnetised, they allow the magnetic flux to pass through them. Magnetic materials may be further sub-divided into following three groups as regards their magnetic properties :

1. Ferromagnetic Materials

These are the most important magnetic materials used in Electricity and Electronics. They are easily and strongly magnetised in the *same* direction as the field. They have high value of relative permeability from 50 to 5000 *i.e.*, they conduct magnetic flux 50 to 5000 times more easily than air. Most commonly used ferromagnetic materials are : iron, steel, nickel, cobalt and commercial alloys such as alnico, permalloy and supermalloy.

Alnico is a trade name for an alloy of aluminium, nickel, iron and cobalt. Permanent magnets made of alnico are commonly used in motors, generators, loudspeakers, microphones and meters etc. Permalloy (nickel and iron or cobalt, nickel and iron) has a relative permeability of the order of 100,000. Similarly, supermalloy is an alloy of nickel, iron, molybdenum and manganese.

2. Paramagnetic Materials

They become only slightly- or weakly-magnetised in a

strong magnetic field in the *same* direction as the field. They conduct magnetic flux only slightly better than vacuum (or air). This group includes aluminium, chromium, platinum and manganese etc. Their relative permeability is slightly greater than one.

3. Diamagnetic Materials

These include bismuth, antimony, copper, zinc, mercury, gold and silver which have a relative permeability of less than one *i.e.*, they conduct magnetic flux less readily than air. They are also slightly magnetised but in a direction **opposite** to that of the magnetising field.

7.2. Ferrites

It is the name given to the recently-discovered ceramic materials that have the ferromagnetic properties of iron. Ferrites are made first by grinding a combination of iron oxide and an alkaline-earth material such as barium into a fine powder. This powder is then pressed into the desired shape and baked at a high temperature. It produces magnetic material which is highly magnetic having a relative permeability in the range 50 to 3000. However, unlike iron, it is an insulator so far as electric conduction is concerned. Like alloy magnets, ceramic magnets can also be shaped into any desired shape. Permanent ceramic magnets are used as gasket latches on refrigerator doors. Ferrite cores (usually adjustable) are used for RF transformers upto 20 MHz frequencies. Another application is ferrite beads. A bare wire is passed through one or more ferrite beads (Fig. 7.1). When current is passed through the wire, a magnetic field is produced. This field is concentrated by the beads into the wire which serves as a simple and economical RF choke.

Ferrite cores are used for RF transformers upto 20 MHz frequencies.

7.3. Types of Magnets

All magnets may be divided into (*i*) permanent magnets and (*ii*) electromagnets.

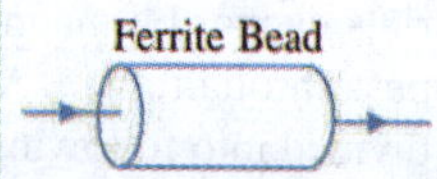

Fig. 7.1

1. Permanent Magnets (PM)

Once magnetised, they maintain their magnetic strength almost indefinitely. They are made of hard magnetic materials such as cobalt steel which is magnetised by induction in the manufacturing process. A very strong field is needed for this purpose. When magnetising field is removed, cobalt steel retains most of its induced magnetism due to its very high retentivity. Other high-retentivity materials are alnico and permalloy etc., which are used in PM loudspeakers. As the name indicates, permanent magnets will last indefinitely if not subjected to high temperature, to physical shock or to a strong demagnetising field. Moreover, they do not get exhausted with use.

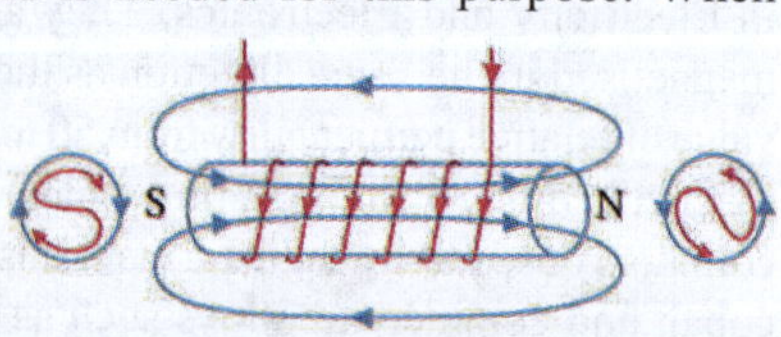

Fig. 7.2

2. Electromagnets

They consist of a coil of wire wound over a soft iron core. When current is passed through the coil, it produces a magnetic field which magnetises the core into a bar magnet with polarities as shown in Fig. 7.2. More current and more turns produce a stronger magnetic field which results in a stronger electromagnet. When current is switched off, field disappears

A buzzer uses copper wires in its electromagnets.

and the iron core is no longer a magnet. This ability of an electromagnet to provide a strong magnetic force of attraction that can be turned OFF and ON has found many applications in lifting magnets, buzzers, bells, horns, relays and magnetic circuit breakers etc. A relay is just a switch with contacts which are opened or closed by an electromagnet. Another application of electromagnets is in the magnetic tape recording.

7.4. Demagnetising or Degaussing

Though magnetism is useful, still there are times when need arises to remove magnetism from certain objects. For example, wrist watches made of magnetic material will not keep correct time if they become magnetised. Similarly, metal cutting tools such as drills and reamers become magnetised due to Earth's magnetic field and start attracting metal chips and filings. This causes them to become dull in due course of time.

Such objects can be demagnetised by using a demagnetiser which consists of a multi-turn coil carrying alternating current. When the object to be demagnetised is placed inside the coil, the alignment of its molecular magnets is destroyed by the alternating magnetic field of the demagnetiser.

A permanent magnet may be demagnetised by heating it to a high temperature or by hammering it.

7.5. Magnetic Shielding

There is no known shield against magnetism *i.e.*, there is no material which does not allow magnetic flux to pass through it. In other words, there is no magnetic insulator. However, some materials have greater permeability than others. For example, iron allows magnetic flux to pass through more easily than air. This fact is made use of in protecting a certain object against the disturbing magnetic field of a nearby component. Suppose we want to protect or shield a meter from the unwanted magnetic field of a neighbouring magnet or Earth's magnetic field. It can be done by surrounding the meter by a ring of soft iron or any other ferromagnetic material as shown in Fig. 7.3. The magnetic flux finds it easier to pass through the ring than air thereby causing no disturbance to the working of the meter. This action is called *shielding*.

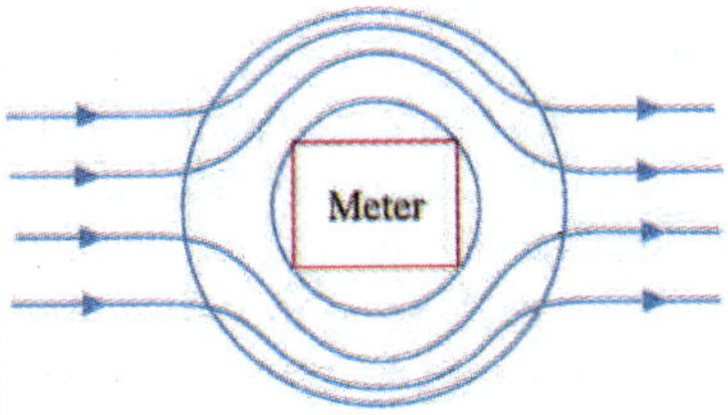

Fig. 7.3

7.6. Magnetic Terms and Units

Following terms are commonly used while discussing the subject of magnetism and electromagnetism.

1. Magnetic Flux (Φ)

The entire group of magnetic lines of force coming out of the N-pole of a magnet is called magnetic flux (Fig. 7.4).

Unit. Unit of magnetic flux is weber (Wb).

2. Flux Density (B)

It is given by the flux incident normally on a unit area. As shown in Fig. 7.5, if a magnetic flux of Φ webers falls perpendicularly on an area of A m^2, then flux density is given by

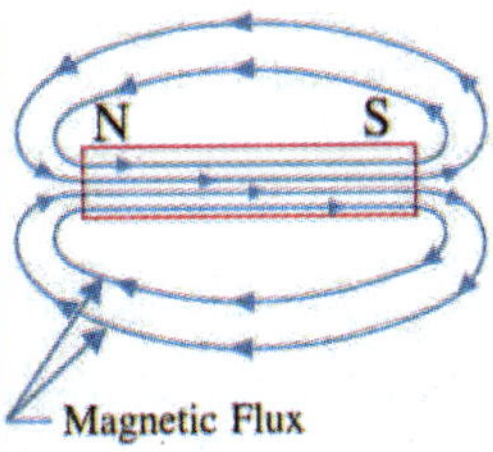

Fig. 7.4

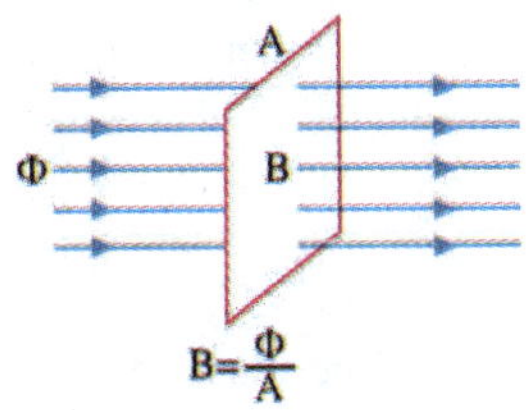

Fig. 7.5

$$B = \frac{\Phi}{A}$$

Unit. Obviously, the unit for flux density is weber/metre^2 (Wb/m^2) which is also called Tesla (T).

3. Magnetic Field Strength (H)

It is also called intensity of magnetic field or (more commonly) magnetising force. As we know, each magnet has its own magnetic field consisting of lines of force which start from its *N*-pole, pass through the surrounding medium, re-enter the *S*-pole and complete their path from *S* to *N*-pole through the body of the magnet. When a magnetic material is placed in the magnetic field, it becomes magnetised whereas non-magnetic materials remain unaffected.

The strength of a magnetic field at any point is measured by the force experienced by a *N*-pole of 1 Wb placed there. A uniform magnetic field is one whose strength remains the same everywhere (Fig. 7.7). It is represented by equally-spaced straight lines of flux.

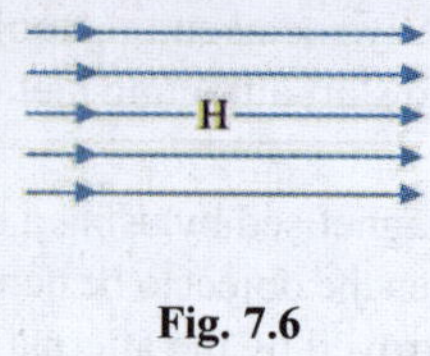

Fig. 7.6

Fig. 7.7

Unit. The unit of H is newton/weber (N/Wb). It is the same thing as an ampere/meter (A/T) which is sometimes written as ampere-turn/meter (AT/m).

4. Magnetising Force of a Solenoid

As shown in Fig. 7.7, if *L* is the length of the iron core, the value of magnetising force produced by the electromagnetic is

$$H = \frac{NI}{l} \text{ A/m or AT/m}$$

5. Permeability

It is the ability of a magnetic material to conduct magnetic flux through it. If it allows the flux to pass through more easily or readily, it is said to have greater permeability. The permeability of a substance is measured both in absolute terms and in relative terms with respect to vacuum (or approximately, air).

(a) Absolute Permeability (μ)

Suppose there is a uniform magnetic field of strength *H* established in air as shown in Fig. 7.8. Further, suppose that a bar of a magnetic material, say, iron is placed in it as shown in Fig. 7.9. The iron bar gets magnetised by *induction*. Suppose, it develops a *polarity* of *m* weber. Then, induced flux developed by it is also *m* weber. The lines of induction flux emanate from its *N*-pole, go around and re-enter its *S*-pole and then continue from *S*- to *N*-pole within the magnet as shown in Fig. 7.9.

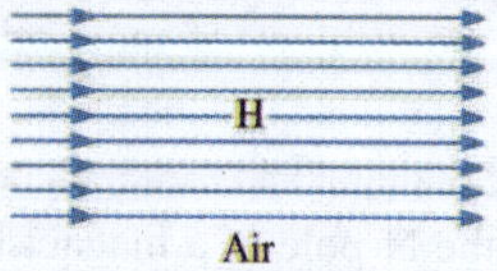

Fig. 7.8

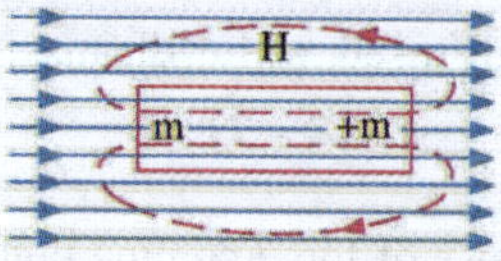

Fig. 7.9

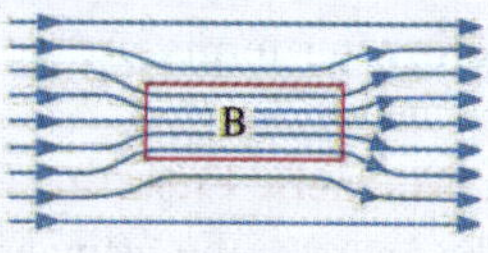

Fig. 7.10

These lines are seen to be in opposition to the lines of force of the main field *H outside* the magnet but in the same direction *within* it. The resultant field is shown in Fig. 7.10. If Φ is the *total* flux* passing through the bar and *A* is its pole area, then flux density within the bar is

$$B = \frac{\Phi}{A} \text{ tesla or Wb/m}^2$$

The absolute permeability of the bar is given by

$$\mu = \frac{B}{H} = \frac{\text{flux density}}{\text{magnetising force}}$$

* There are two fluxes : one due to *H* and the other due to induced magnetism.

Its unit is henry/metre (H/m).

Also, $B = \mu H$ tesla

(b) Relative Permeability (μ_r)

The absolute permeability of vacuum is denoted by μ_0 and has been allotted the value of $4\pi \times 10^{-7}$ H/m. Permeabilities of all other magnetic materials are expressed in terms of the absolute permeability of vacuum which has been selected (by mutual agreement) as the reference medium.

Suppose a certain medium has an absolute permeability of μ. Then, its relative permeability (μ_r) *i.e.*, permeability as compared to vacuum is given by

$$\mu_r = \frac{\mu}{\mu_0} = \frac{\text{absolute permeability of medium}}{\text{absolute permeability of vacuum}}$$

Being a mere ratio of two similar quantities, it has no unit.

Also $\mu = \mu_0 \mu_r$

As an example, suppose mild steel has a relative permeability $\mu_r = 400$. Then, its absolute permeability is given by

$$\mu = \mu_0 \mu_r = 4\pi \times 10^{-7} \times 400 \text{ H/m}$$
$$= 16\pi \times 10^{-5} \text{ H/m}$$

It is universal practice to give relative permeabilities of various media since μ can always be found by multiplying μ_r with μ_0 which is a universal constant.

6. Retentivity

It is the ability of a material to hold its magnetism after the magnetising force has been removed. Materials having high retentivity make good permanent magnets.

7. Hysteresis

Suppose the exciting coil of an electromagnet is energised by a source of alternating current (Fig. 7.11). As the current reverses its direction of flow through the coil, the flux also reverses its direction. Hence, the core also undergoes reversal of magnetisation. But it is found that magnetisation of the core does not reverse as quickly as the reversal of flux *i.e.*, the two are not in step with each other. This phenomenon is called hysteresis and is due to the retentivity of the magnetic material of the core.

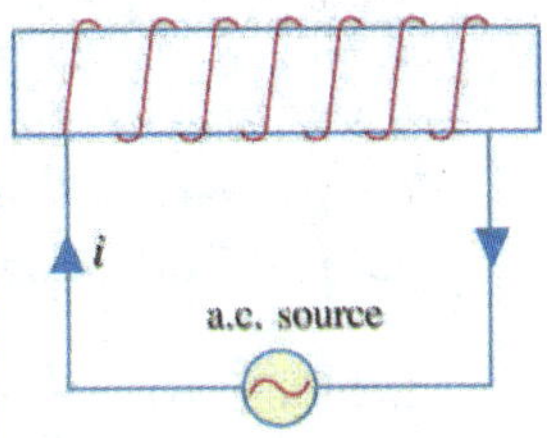

Fig. 7.11

Hysteresis leads to net loss of energy which is called hysteresis loss. This loss depends directly on

(*i*) maximum flux density B_{max} established in the core

(*ii*) frequency of reversal of magnetisation.

8. Permeance

It is the reciprocal of reluctance and resembles electrical conductance. Its unit is henry.

9. Reluctivity

It is specific reluctance and corresponds to electrical resistivity which is 'specific resistance'.

7.7. Ohm's Law for Magnetic Circuit

In Fig. 7.12 (*a*) is shown a magnetic circuit having iron path only, whereas in Fig. 7.12 (*b*) there is a small air gap in the circuit. Like electric circuit, a magnetic circuit also has three quantities interconnected by a law similar to Ohm's law.

The three quantities are :

1. Magnetomotive force (MMF)

It resembles voltage or *electromotive force* (*EMF*) in an electric circuit and is responsible for producing magnetic flux in a magnetic circuit. Its value is given by the product of current through the coil and its number of turns *i.e.*, *NI*. Its unit is ampere-turn*.

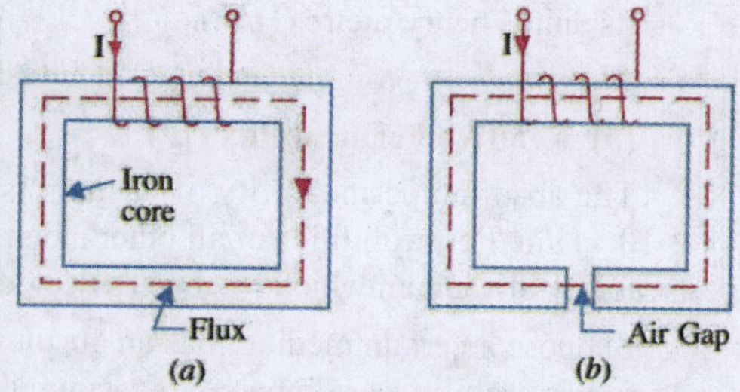

Fig. 7.12

2. Magnetic flux (Φ)

It resembles *current* in an electric circuit. It consists of magnetic lines of force and its unit is weber.

3. Reluctance (S)

It resembles *resistance* in an electric circuit. It represents the opposition which a core offers to the production of flux through it. Its value is

$$S = \frac{l}{\mu A} = \frac{l}{\mu_0 \mu_r A}$$

Its unit is 'reciprocal' henry *i.e.*, per henry.

Ohm's law for magnetic circuit is

$$\text{flux} = \frac{\text{mmf}}{\text{reluctance}} = \frac{NI}{S}\text{ weber}$$

$$= \frac{NI}{l/\mu A}\text{ weber} = \frac{\mu NAI}{l}\text{ weber} = \frac{\mu_0 \mu_r NAI}{l}\text{ weber}$$

Example 7.1. *A mild-steel ring having a cross-sectional area of 5 cm² and a mean circumference of 40 cm has a coil of 200 turns wound uniformly around it. Calculate*

(*i*) *reluctance of the ring*

(*ii*) *current required to produce a flux of 800 μ Wb in the ring.*

Take relative permeability of mild-steel as 380.

Solution. (*i*) $\frac{l}{\mu_0 \mu_r A} = \frac{0.4}{4\pi \times 10^{-7} \times 380 \times (5 \times 10^{-4})} = \mathbf{1.675 \times 10^6\ henry^{-1}}$

(*ii*) Now, $\Phi = \frac{NI}{S}$ $\quad\therefore\quad$ $\mathbf{800 \times 10^{-6}} = \frac{200 \times I}{1.675 \times 10^6}$ $\quad\therefore\quad$ $\mathbf{I = 6.7\ A}$

7.8. Transformer

It is a static (or stationary) piece of apparatus that

1. transfers electric power from one circuit to another having mutual inductance with it.

2. Does so without change of frequency.

3. Does it by electromagnetic induction.

Constructionally, transformers may be either isolation transformers (with electrically-insulated primary and secondary windings) or autotransformers (with electrically-connected primary and secondary windings). The two are shown in Figs. 7.13 and 7.14 respectively.

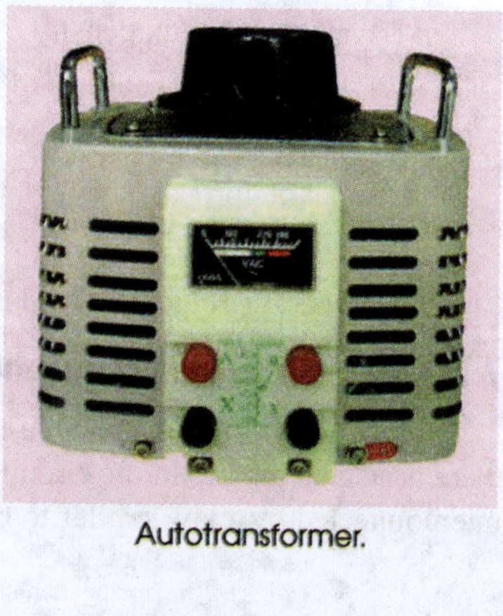
Autotransformer.

* Strictly speaking, it should be ampere only because turn has no units.

The two-winding isolation transformer may be further subdivided into

(*i*) core type transformer .... in which the windings surround a considerable part of the core (Fig. 7.15).

(*ii*) shell type transformer .... in which the core surrounds a considerable part of the windings (Fig. 7.16).

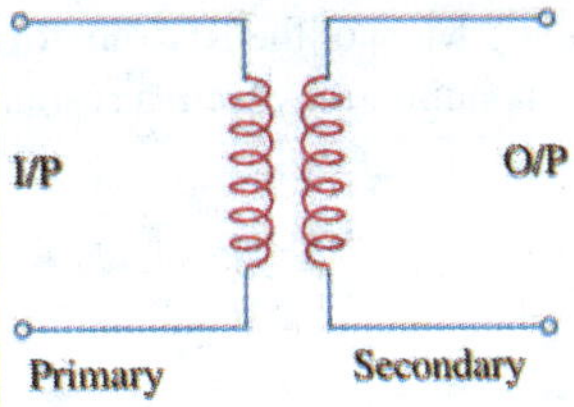

Fig. 7.13

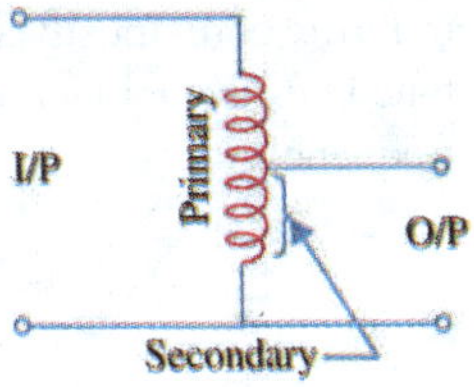

Fig. 7.14

As seen, core-type transformer is made up of a package of thin rectangular silicon steel laminations. Each lamination is coated with an insulating varnish and the total core pressed together. The primary and secondary windings are placed on each side of the common core (Fig. 7.15).

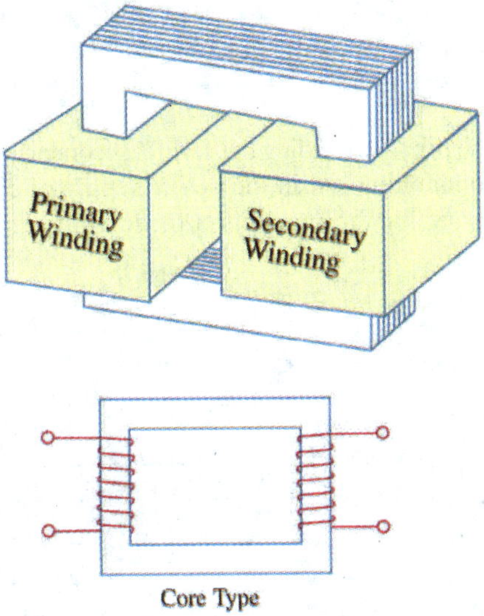

Fig. 7.15

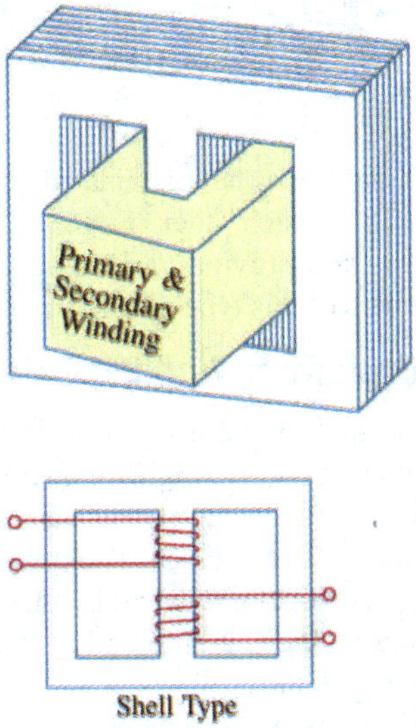

Fig. 7.16

The shell type construction also consists of similar laminations. The two windings are wound in layers and fit over the centre section of the core as shown in Fig. 7.16.

Functionally, the transformers used in electronic circuits can be classified according to the frequency range over which they operate such as :

1. Audio Frequency (AF) Transformers

They are designed to operate over the audio frequency (AF) range of 20 Hz to 20 kHz, have laminated core and are usually smaller than power transformers. They are primarily used for impedance matching and, in some cases, for voltage amplification. Two such typical transformers are shown in Fig. 7.17. Such transformers are usually designated according to their applications as input or output transformer, microphone transformer, modulation transformer and interstage transformer etc. Usually, they are rated by their primary and secondary impedances and current-carrying capability.

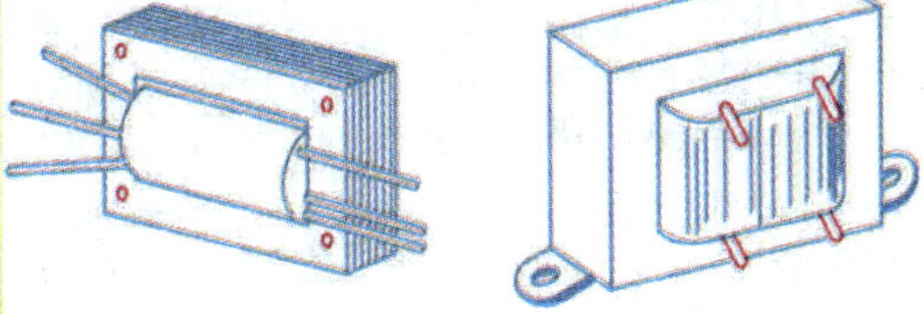
Fig. 7.17

2. Radio Frequency (RF) Transformers

They are designed to operate at high frequencies (above audio range) and are referred to either as intermediate frequency (IF) transformers or radio frequency transformers. They may have air core or ferrite core (mostly adjustable). Most of the RF transformers have either one or both of the windings tuned *i.e.*, in conjunction with capacitor, they form a resonant circuit which works best at one particular frequency.

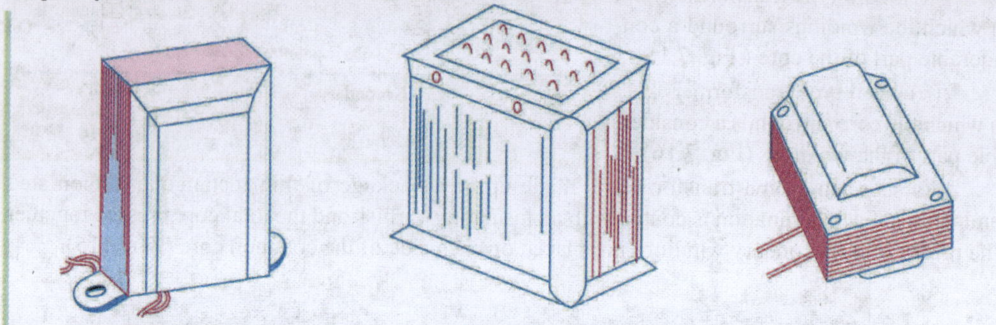

Fig. 7.18

3. Power Transformers

Usually, they have laminated core and have one primary winding but several secondary windings insulated from each other (Fig. 7.18). They are commonly used in the power supply of electronic equipment and provide various ac voltage necessary for the production of dc voltages. Typical transformers of this type are shown in Fig. 7.18.

7.9. Transformer Working

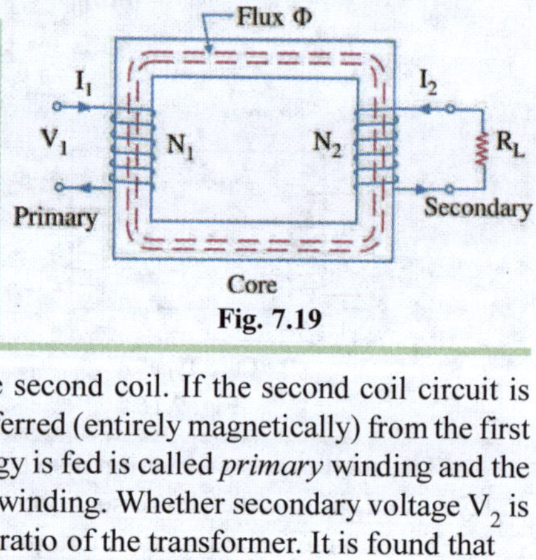

Fig. 7.19

Consider the core-type transformer shown in Fig. 7.19. It consists of two highly inductive coils which are electrically separate but magnetically linked through an iron core of low reluctance. The two coils possess high mutual inductance. If one coil is connected to source of alternating voltage, an alternating flux is set up in the laminated core most of which is linked with the other coil. Hence, mutually-induced voltage is produced in the second coil. If the second coil circuit is closed, a current flows in its and so electric energy is transferred (entirely magnetically) from the first coil to the second coil. The first coil in which electric energy is fed is called *primary* winding and the other from which energy is drawn out is called *secondary* winding. Whether secondary voltage V_2 is more or less than primary voltage V_1 depends on the turn ratio of the transformer. It is found that

$$\frac{V_2}{V_1} = \frac{N_2}{N_1}$$

If $N_2 > N_1$, then $V_2 > V_1$ and the transformer is called *step-up* transformer, since it steps up the input primary **voltage**. If $N_2 < N_1$, then $V_2 < V_1$ and the transformer is called *step-down* transformer.

Voltage transformation ratio (K) of a transformer is given by V_2/V_1.

$$\therefore \quad K = \frac{V_2}{V_1} = \frac{N_2}{N_1} \quad \text{or} \quad V_2 = KV_1$$

As seen, voltage transformation ratio equals the turn ratio.

Assuming an ideal transformer and equal power factor for both windings,

input power = output power

$$V_1 I_1 = V_2 I_2 \qquad \therefore \quad I_2 = I_1 / K$$

$$\therefore \quad \frac{I_2}{I_1} = \frac{V_1}{V_2} = \frac{1}{K}$$

It is obviously from the above that a transformer which is *step-up for voltage is step-down for current.* If voltage is increased five times, current becomes one-fifth because output power has to equal the input power (in an ideal case). It means that current ratio is reciprocal of voltage ratio.

Suppose, we have transformer with $N_1 = 100$ and $N_2 = 600$. Let $V_1 = 200$ V and $I_1 = 3$ A. Then,

$$K = \frac{N_2}{N_1} = \frac{600}{100} = 6; \qquad V_2 = KV_1 = 6 \times 200 = 1200 \text{ V}$$

$$I_2 = I_1/K = 3/6 = 0.5 \text{ A}$$

It is seen that secondary voltage is 6 times the primary voltage but, at the same time, secondary current is one-sixth of the primary current.

$$P_1 = 200 \times 3 = 600 \text{ W}; \qquad P_2 = 1200 \times 0.5 = 600 \text{ W}$$

As seen, the two powers are equal.

It is worth noting that whatever the actual value of primary and secondary volts, *the voltage/turn is the same in both windings.* In the above case

Primary volts/turn = 200/100 = 2 V; Secondary volts/turn = 1200/600 = 2 V

The two values are equal even though V_1 and V_2 are themselves unequal.

7.10. Transformer Impedance

Each transformer winding has its own resistance, inductive reactance and hence impedance. As shown in Fig. 7.20,

Primary impedance, $Z_1 = \sqrt{R_1^2 + X_1^2}$; Secondary impedance, $Z_2 = \sqrt{R_2^2 + X_2^2}$

Another very interesting thing about these impedances is that they assume different values when *viewed from the other winding.* For example, when Z_2 is viewed from primary winding, it assumes a value $Z_2' = Z_2 / K^2$. But, when Z_1 is viewed from secondary, it appears to have a value of $Z_1' = K^2 Z_1$. This fact is made use of in the working of an impedance-matching transformer (Art. 7.14).

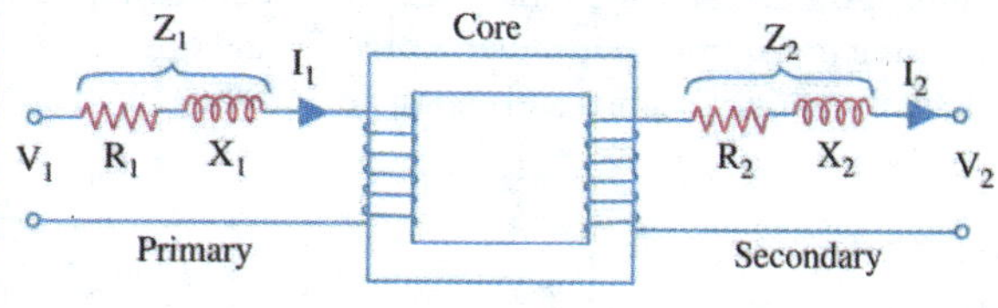

Fig. 7.20

Example 7.2. *A power transformer has 100 primary turns and 600 secondary turns. If primary voltage is 120 V and full-load primary current is 12 A, find secondary*

(i) voltage V_2 and (ii) current I_2.

Solution. Here $K = N_2/N_1 = 600/100 = 6$

(*i*) $V_2 = KV_1 = 6 \times 120 = $ **720 V**; (*ii*) $I_2 = I_1/K = 12/6 = $ **2 A**

Example 7.3. *A low-voltage soldering rod taking 40 A at 12 V is to be operated from the secondary of a 240 V transformer. Calculate*

(i) turn ratio of the transformer and (ii) primary current.

Solution. (*i*) $\frac{V_2}{V_1} = \frac{12}{240} = \frac{1}{20} \qquad \therefore \quad \frac{N_2}{N_1} = \frac{1}{20}$

Obviously, it is a step-down transformer having $K = 1/20$

(*ii*) $I_1 = KI_2 = \frac{1}{20} \times 40 = $ **2A**

7.11. Can a Transformer Operate on DC ?

A transformer cannot operate on a steady or unchanging dc voltage such as that of a battery. It requires a voltage which rises and falls. Since an ac voltage not only changes its magnitude but its direction as well (Fig. 7.21), it is used to operate the transformers.

However, a transformer will operate from dc voltage if this voltage also undergoes changes. Transformers used for audio amplifiers work on pulsating dc voltage (Fig. 7.22). Main thing which causes the transformer to work is the *change* in voltage. It is immaterial whether the voltage changes from positive to negative values as in Fig. 7.21 or from positive to zero values as in Fig. 7.22 (it could, in fact, be from minus to zero values as well).

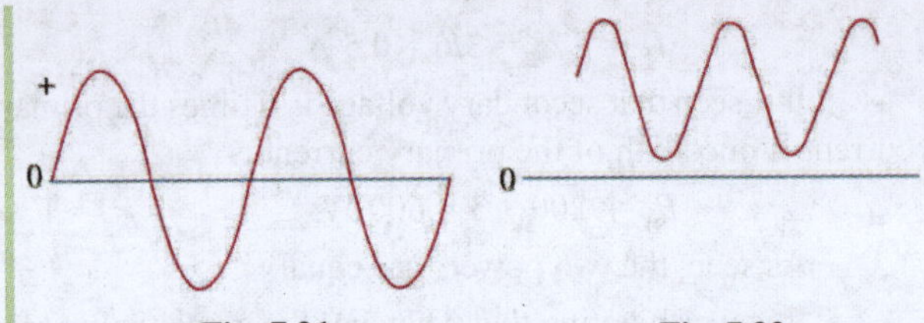

Fig. 7.21 Fig. 7.22

7.12. RF Shielding

Coils are often encased in a metal cover, usually of copper or aluminium, in order to protect them from external varying flux of RF currents. Otherwise, unwanted eddy currents would be induced in them. Purpose of RF shielding is different from magnetic shielding (Art. 7.5) which protects against steady flux only. The shield cover not only isolates the coil from external varying magnetic fields but also minimizes the effect of coil's own RF currents on other external circuits.

7.13. Autotransformer

It is a transformer with one winding only, part of it being common to both primary and secondary. Here, primary and secondary are not electrically isolated from each other as is the case in a 2-winding transformer. However, its theory and operation are similar to that of a 2-winding transformer. Because of one winding, it is compact, efficient and cheaper.

Fig. 7.23

Fig. 7.23 (*a*) shows a step-up autotransformer whereas Fig. 7.23 (*b*) shows a step-down type. As with other transformers, this step-up or step-down ratio depends on the turn ratio between the primary and secondary. Fig. 7.24 shows an audio output stage of an automobile radio that uses a step-down autotransformer.

Such a transformer is also used as an adjustable transformer for both stepping up or stepping down the input voltage (Fig. 7.25). It is often used for a light dimmer or for adjusting power to a radio transmitter.

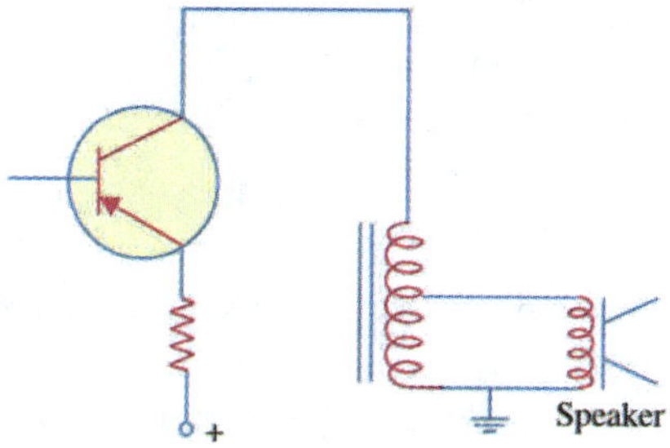

Fig. 7.24

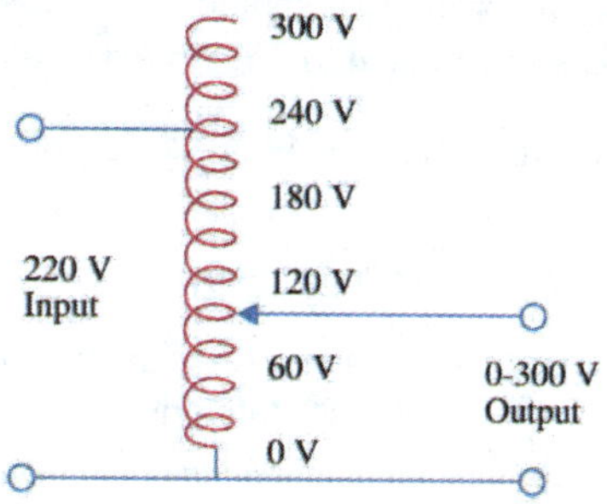

Fig. 7.25

Example 7.4. *The primary and secondary voltages of an autotransformer are 500 V and 400V respectively. Show on a diagram, the current distribution in the transformer if secondary current is 100 A.*

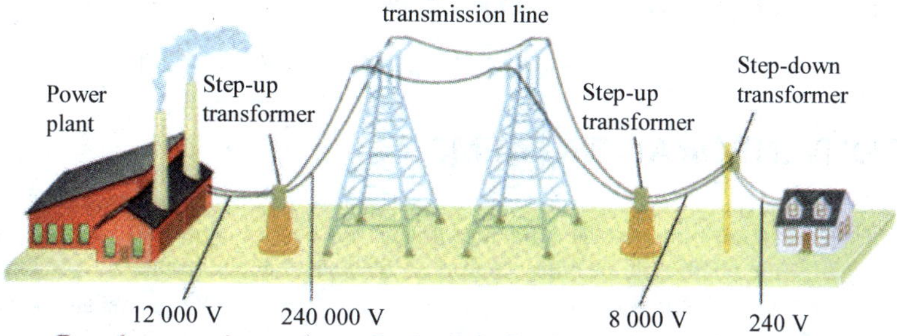

Transformers play an important role in the transmission of electric power.

Solution. $K = V_2/V_1$

$= 400/500 = 0.8$

Here, $I_2 = 100$ A —given

$I_1 = K I_2$

$= 0.8 \times 100 =$ **80 A**

As seen, current in the common portion of the winding is 20 A. Circuit diagram is shown in Fig. 7.26.

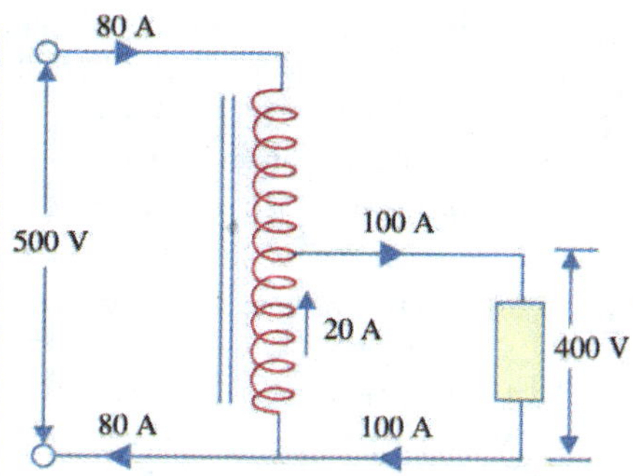

Fig. 7.26

7.14. Impedance Matching

For maximum transfer of power from one circuit to another the two should have equal impedances (Art. 4.9). If they do not have equal impedances, a transformer with suitable turn ratio can be used to achieve this impedance match. In electronic circuitry, it often becomes necessary to connect a circuit of high output impedance to one of low input impedance*. What it really means is that a certain circuit working at a high voltage but low current (hence high impedance) has sometime to be coupled to another circuit which requires lower voltages but higher current (hence low impedance). If two such circuits are coupled directly, energy transfer will not be maximum. In such cases, a transformer is used as an impedance-matching device because it can do the job of increasing or decreasing the voltages and currents very efficiently.

Suppose a circuit of *output* impedance 300 Ω is to be coupled to a circuit of *input* impedance 3 Ω. The turn-ratio (N_2/N_1) of the transformer should be such that when 3 Ω impedance in its secondary

* Usually, a higher-voltage low-current circuit is called a high impedance circuit, while a low-voltage higher-current one is referred to as low-impedance circuit.

is viewed by its primary, it should appear as 300 Ω. Now, when viewed from primary side, a 3 Ω resistance is seen as equal to $3/KC^2$ (Art. 7.9).

Hence, for equal matching $\dfrac{3}{K^2} = 300$

$$K = \sqrt{1/100} = 1/10 \qquad \therefore \quad \frac{N_2}{N_1} = \frac{1}{10}$$

It means that secondary turns should be one-tenth the primary turns. Often, autotransformers are also used for impedance matching purpose.

Example 7.5. *A transformer is required to match the 400 Ω output impedance of a transistor power amplifier to a 4 Ω speaker coil. What should be the turn ratio?*

Solution. Obviously, the transformer primary would be fed from the amplifier whereas the secondary would be connected to the speaker.

$$K = \sqrt{Z_2/Z_1} = \sqrt{4/400} = 1/10 \qquad \therefore N_2/N_1 = \mathbf{1/10}$$

CONVENTIONAL PROBLEMS

1. A coil of 500 turns and resistance 20 Ω is wound uniformly on an iron ring of mean circumference 50 cm and cross-sectional area 4 cm². It is connected to a 24 V dc supply. Under these conditions, the relative permeability of iron is 800. Calculate the values of

(*a*) mmf of the coil (*b*) magnetising force
(*c*) total flux in iron core (*d*) reluctance of the iron ring

[(*a*) 6000 AT (*b*) 1200 AT/m (*c*) 0.483 mWb (*d*) 1.24 × 10⁶ H⁻¹]

2. A magnetic circuit consists of an iron ring of mean circumference 80 cm with cross-sectional area 12 cm² throughout. A current of 2 A in the magnetising coil of 200 turns produces a total flux of 1.2 mWb in the iron. Calculate

(*i*) flux density in iron
(*ii*) absolute and relative permeability of iron
(*iii*) reluctance of the circuit **[(*a*) 1 T or Wb/m² (*b*) 0.002, 1590 (c) 3.33 × 10⁵ henry⁻¹]**

3. A step-down transformer with a voltage step-down ratio of 20 has 6 V across 0.3 Ω secondary. Calculate (*i*) secondary current and (*ii*) primary current. **[(*i*) 20 A (*ii*) 1 A]**

4. A step-down transformer with 10 : 1 turn ratio is connected to 220 V, 50 Hz ac supply mains

(*a*) What is the frequency of secondary voltage ?
(*b*) How much is secondary voltage ?
(*c*) If secondary load is 100 Ω, what is the secondary current and primary current ? Assume 100 per cent efficiency. **[(*a*) 50 Hz (*b*) 22 V (*c*) 0.22 A ; 0.022 A]**

SELF EXAMINATION QUESTIONS

A. Fill in the blanks with most appropriate word (s) or numerical value (s).

1. Even though wood is a non-magnetic material, it allows magnetic .... to pass through.

2. Ferromagnetic materials/have high value of relative ....

3. Alnico is used for making permanent magnets because it has high ....

4. Ceramic material which has ferromagnetic properties of iron is called ....

5. Unit of magnetic flux density is ....

6. Ratio of *B* and *H* is called .... permeability.

7. Unit of magneto-motive force is ....

8. .... is reciprocal of reluctance.

9. An autotransformer has only .... winding.

10. An impedance-matching transformer couples two circuits of unequal ....

B. Answer True or False

1. Cobalt is a ferromagnetic material.
2. Gold is diamagnetic whereas silver is paramagnetic.
3. Ferrites have permeability of less than one.
4. Cobalt steel is ideal for making electromagnets.
5. Demagnetising means the same thing as degaussing.
6. All transformers work on the principle of mutual inductance.
7. In autotransformers, electric energy is transferred to secondary only by induction.
8. Transformers can be used for coupling a high-output impedance circuit to a low-input impedance circuit but not *vice-versa.*

C. Multiple Choice Items

1. A permanent magnet will not attract
   (*a*) steel (*b*) nickel
   (*c*) aluminium (*d*) copper
2. Magnets made of low-retentivity but high permeability iron are called
   (*a*) electromagnets
   (*b*) permanent magnets
   (*c*) weak magnets
   (*d*) one-pole magnets
3. The unit of magnetising force is
   (*a*) ampere/metre (*b*) newton
   (*c*) tesla (*d*) ampere
4. Working of a transformer essentially depends on
   (*a*) self-inductance
   (*b*) mutual inductance
   (*c*) magnetic circuit (*d*) magnetic flux
5. Radio frequency transformers often employ air-core coils in order to
   (*a*) eliminate core loss
   (*b*) reduce coil weight
   (*c*) cut initial cost
   (*d*) eliminate winding labour
6. A transformer has 1000 primary turns and 500 secondary turns. If primary voltage is 200 V, secondary voltage would be .... volt.
   (*a*) 400 (*b*) 100
   (*c*) 800 (*d*) 50
7. If a 5 : 1 step-down audio transformer has a primary current of 20 mA, the secondary current would be .... mA.
   (*a*) 4 (*b*) 500
   (*c*) 100 (*d*) 0.8
8. Basic requirement for the operation of a transformer is that its input primary voltage must be
   (*a*) alternating (*b*) sinusoidal
   (*c*) changing (*d*) pulsating
9. If a circuit of output impedance 400 Ω is to be coupled to a circuit of input impedance 4 Ω, its primary to secondary turn ratio should be
   (*a*) 1/10 (*b*) 10
   (*c*) 100 (*d*) 1/100
10. RF shielding of a coil is primarily meant to protect it from external
    (*a*) electric fields (*b*) magnetic fields
    (*c*) varying magnetic fields
    (*d*) dust and heat
11. An autotransformer consists of 200-turn winding connected to 200 V ac supply mains. Forgetting 24 V output, the winding should be tapped at turn number
    (*a*) 24 (*b*) 12
    (*c*) 100 (*d*) 72
12. The main purpose of laminating a transformer core is to decrease its
    (*a*) electrical resistance
    (*b*) reluctance
    (*c*) eddy current loss
    (*d*) hysteresis loss

ANSWERS

A. Fill in the blanks

1. flux **2.** permeability **3.** retentivity **4.** ferrite **5.** tesla
6. absolute **7.** ampere **8.** permeance **9.** one **10.** impedances

B. True or False

1. T **2.** F **3.** F **4.** F **5.** T **6.** T **7.** F **8.** F

C. Multiple Choice Items

1. *d* **2.** *a* **3.** *a* **4.** *b* **5.** *a* **6.** *b* **7.** *c* **8.** *c* **9.** *b* **10.** *c*
11. *a* **12.** *c*

CHAPTER 8

A.C. Fundamentals

1. Introduction
2. Types of Alternating Waveforms
3. The Basic AC Generator
4. Some Definitions
5. Characteristics of a Sine Wave
6. Audio and Radio Frequencies
7. Different Values of Sinusoidal Voltage and Current
8. Phase of an AC
9. Phase Difference
10. Vector Representation of an Alternating Quantity

11, AC through Pure Resistance Only
AC through Pure Inductance Only
AC through Pure Capacitance Only

14. Non-Sinusoidal Wave-forms
15. Harmonics

8.1. Introduction

So far we have considered direct current (dc) which is produced by a voltage source whose terminals have fixed polarity *i.e.*, whose poles do not change their polarity with time. Hence, they provide a current whose direction flow does not change with time. However, this direct current may be steady (constant in value as in Fig. 8.1 (*a*) or fluctuating (pulsating as in Fig. 8.1 (*b*) or may be interrupted into short pulses as shown in Fig. 8.1 (*c*). Main point is that its direction of flow remains the same *i.e.*, from the positive terminal of the voltage source to its negative terminal. Examples of voltage sources are : electric cell, battery and d.c. generators etc.

Alternating current (a.c.) is produced by a voltage source whose terminal polarity keeps alternating (or reversing) with time. What was the positive terminal at one instant [Fig. 8.2 (*a*)] becomes negative terminal some time later and what was negative terminal at one instant [Fig. 8.2 (*b*)] becomes positive terminal at some other instant?

As a result of constantly-reversing polarity of voltage source, the direction of current flow in the circuit also keeps reversing as shown in Fig. 8.2. In addition to reversing its direction, current keeps changing in value with time–from zero to maximum in one direction and back to zero and then from zero to maximum in the opposite direction and again

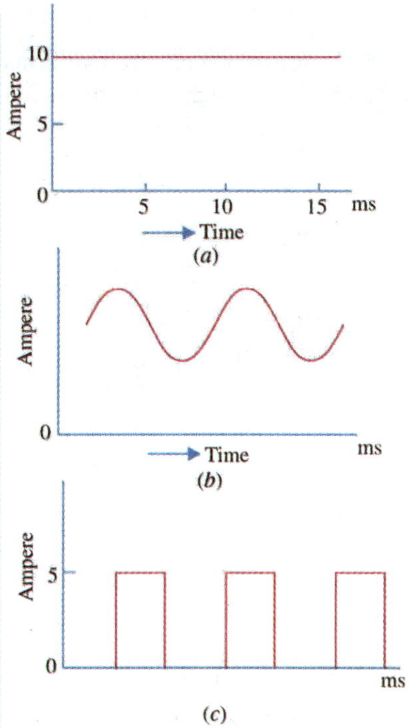

Fig. 8.1

back to zero. It is obvious that an alternating voltage source will cause an alternating current.

It may be noted in passing that 'a.c.' electricity is not 'better' than 'd.c.' electricity as some people feel. Alternating voltages and currents have their own fields of application which d.c. does not have and *vice-versa.* Any way, it is important to keep in mind that *a.c. does not replace d.c.* A.C. is commonly used in electronic circuits most of which are, however, controlled by d.c. voltages.

The most common source of alternating voltage is the alternator or a.c. generator.

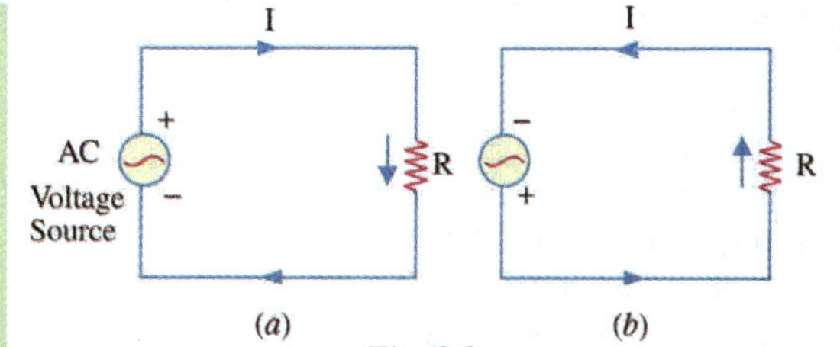

Fig. 8.2

8.2. Types of Alternating Waveforms

The alternating voltages and currents can have different waveforms or wave-shapes as shown in Fig. 8.3. These waveforms are obtained by plotting the instantaneous values of voltage or current as ordinate against time as abscissa.

Fig. 8.3 (*a*) shows a sine wave because it is obtained by plotting a sine function. Fig. 8.3 (*b*) depicts a triangular waveform whereas Fig. 8.3 (*c*) shows a square waveform and Fig. 8.3 (*d*) illustrates a complex waveform. Another point worth noting is that alternating quantities may or may not be symmetrical *i.e.*, their positive and negative, halves may or may not be identical. All the waveforms shown in Fig. 8.3 represent different types of alternating voltages which are symmetrical.

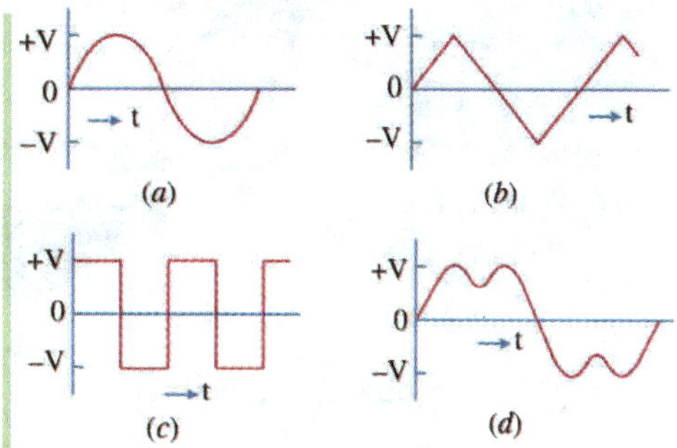

Fig. 8.3

Since sinusoidal a.c. *i.e.*, an a.c. having sine waveform is most important, we will discuss it in some detail. But it should be noted that a waveform may be sinusoidal but not symmetrical and *vice-versa.* Fig. 8.3 (*a*) shows a waveform which is both sinusoidal and symmetrical. However, the waveforms shown in Fig. 8.4 are sinusoidal but non-symmetrical. In Fig. 8.4 (*a*), negative half is different from the positive half and in Fig. 8.4 (*b*), the negative half has been completely suppressed. Such an a.c. is called half-wave rectified a.c. (Art. 16.6).

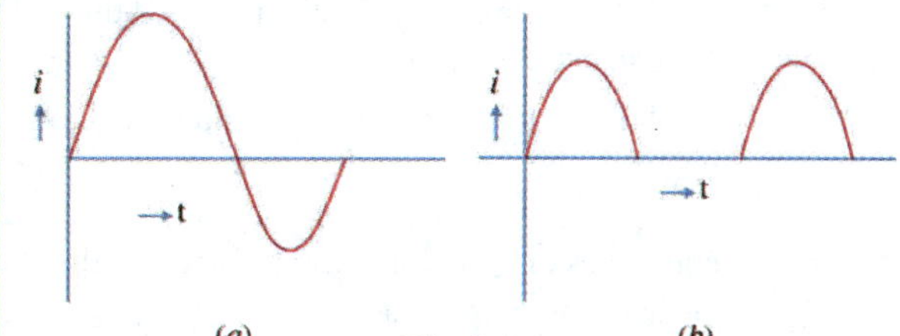

Fig. 8.4

8.3. The Basic AC Generator

Such a rotary generator is schematically shown in Fig. 8.5. It consists of a pivoted loop armature having two coil sides *A-B* and *C-D* which rotate with uniform velocity through a uniform stationary magnetic field of horse-shoe magnet. As the armature coils cut the magnetic flux, an alternating voltage is induced in them according to Faraday's Laws of Electromagnetic Induction. The magnitude of this voltage depends (among other factor) on the *rate* at which flux is cut by the coil sides. This a.c. voltage is taken out from the armature coil via sliprings and brushes (not shown in the figure) for driving current through an external load.

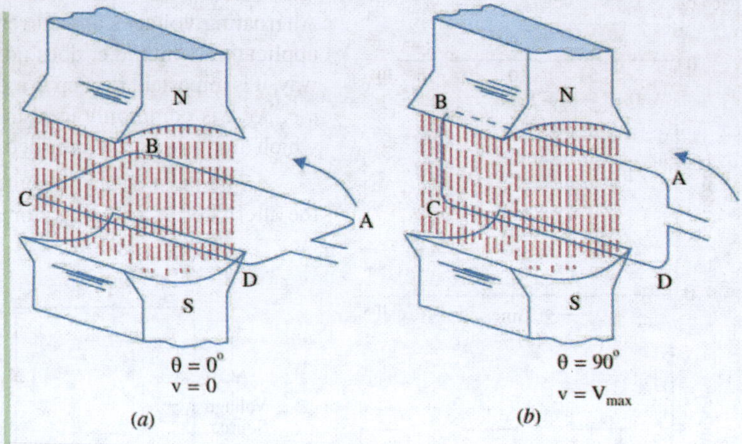

Fig. 8.5

The coil has been assumed to rotate in the anticlockwise direction in Fig. 8.5.

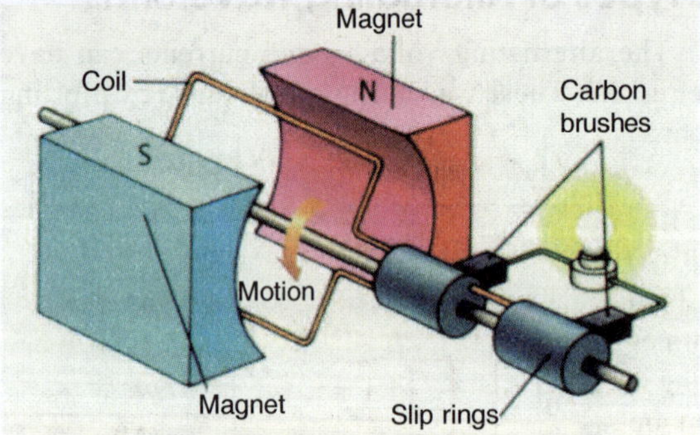

A basic AC generator.

1. When the armature coil lies in the horizontal plane, maximum flux with it but its *rate* of flux cutting (*i.e.*, $d\Phi/dt$) is zero. It is so because as the two sides *A-B* and *C-D* just start moving *upwards* and *downwards* respectively, they *slide* along or move parallel to the flux lines. They, in fact, just rub along the lines of flux without cutting them. Hence, induced voltage v at that instant is zero. We will take this horizontal position of the coil as the $\theta = 0$ position [Fig. 8.5 (*a*)].

2. As the coil rotates further up, rate of flux cutting increases till it becomes maximum when $\theta = 90°$ as shown in Fig. 8.5 (*b*). Hence, the induced voltage at $\theta = 90°$ has maximum value of V_{max}.

3. As the coil rotates further, v starts decreasing sinusoidally till it again becomes zero when $\theta = 180°$. At this stage, coil has turned through half the circle and the induced voltage has gone through half the cycle.

4. As the coil turns through the remaining half circle *i.e.*, from 180° to 360°, the induced voltages undergoes similar changes in magnitude except that its polarity is reversed. The voltage v has a maximum value of V_{max} at 270° and zero at 360°. In this way, the coil completes one full turn which corresponds to 360° of rotation and is the equivalent of one cycle of the output voltage.

The process of voltage generation is more clearly shown in Fig. 8.6. Since voltage generated in the two sides of the coil are equal (though opposite in polarity), we may consider only one coil side say, *A-B*, to explain the above process. Initially, coil side *A-B* is assumed to be positioned at point *E*.

The instantaneous values of the voltage generated at other angles of coil rotation are plotted in Fig. 8.6 (*b*). It is found that generated a.c. voltage is given by the equation

$$v = V_m \sin \theta$$

The polarity of the induced voltage can be found by applying Fleming's Right Hand rule and is shown in the figure by dots and crosses.

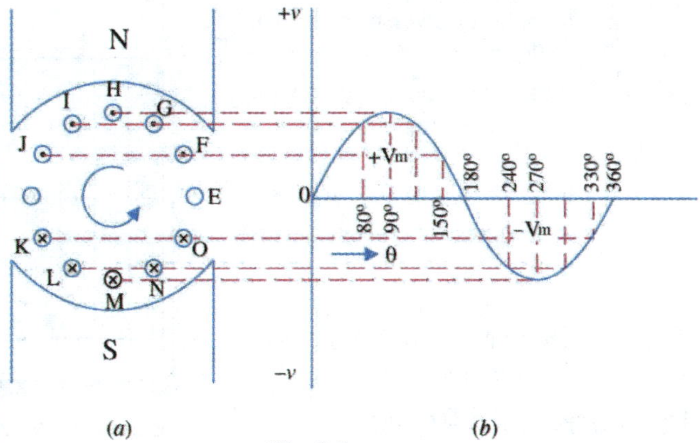

Fig. 8.6

8.4. Some Definitions

A few of the terms used in connection with alternating voltages and current are defined below. A sine wave has been taken for illustration purposes. As seen, a sinusoidal waveform is just a pattern of instantaneous changes in the value of an alternating voltage or current.

1. Cycle

One complete set of positive and negative values of an alternating quantity is known as a cycle (Fig. 8.7).

A cycle is also sometimes specified in terms of angular measure. In that case, one complete cycle is said to spread over 360° or 2π radians.

2. Time Period

It is the time taken by the alternating voltage or current to complete one cycle. For example, an alternating current making 50 cycles/second has a time period of T = 1/50 second = 0.02 second.

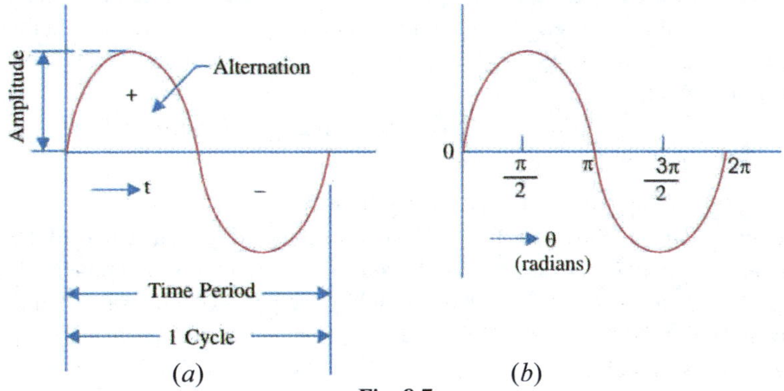

Fig. 8.7

3. Frequency

The number of cycle/second made by an alternating voltage or current is called its frequency. Its unit is cycles/second or hertz (Hz).

4. Amplitude

The maximum value, positive or negative, of an alternating quantity is known as its amplitude.

8.5. Characteristics of a Sine Wave

A sinusoidal a.c. waveform has following characteristics :

1. Its one cycle spreads over 360° or 2π radians.

2. Its polarity reverses every half cycle.
3. It has maximum positive value at 90° and maximum negative value at 270°.
4. It has zero value at 0° and 180°.
5. It changes its values fastest when it crosses the zero axis.
6. It changes its value the slowest when near its maximum value whether positive or negative.

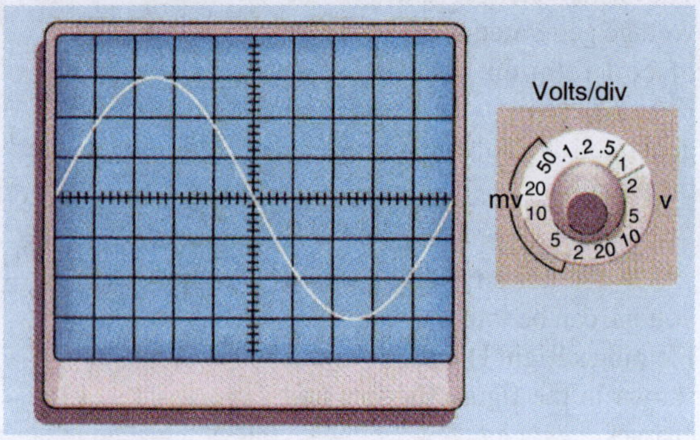

This shows one cycle of a sine wave on the scope screen with the volts/div switch set at 1V.

8.6. Audio and Radio Frequencies

The entire range of frequency from 1 Hz to several megahertz may be divided into two broad groups (*i*) audio frequencies (AF) and (*ii*) radio frequencies (RF). The audio frequencies are those which can be heard in the form of sound waves by human ear. Their range extends from 20 Hz to 20,000 Hz. Higher the frequency, higher the pitch or tone of the sound. Audio frequencies above 3000 Hz provides treble tone whereas those below 300 Hz provide bass tone.

Alternating currents above the audio range provide RF variations because electric variations of high frequency can be transmitted in the form of electromagnetic radio waves.

8.7. Different Values of Sinusoidal Voltage and Current

When referring to a sinusoidal current wave, following current values can be differentiated:

1. Instantaneous Value

It is the value of the current that exists at any instant of time measured from some reference point. It can have any value between plus maximum value $+I_m$ and negative maximum value $-I_m$ and is denoted by i. In fact, the entire sine wave is made up of instantaneous values as shown in Fig. 8.8. Mathematically, it is given by

$$i = I_m \sin\theta = I_m \sin\omega t = I_m \sin 2\pi ft = I_m \sin 2\pi\, t/T$$

2. Peak Value or Maximum Value

It is the highest value reached by the current in one cycle.

For example, current shown in Fig. 8.8 has a peak or maximum value of 1 A. Moreover, positive peak value is equal to the negative peak value. It is seen that during positive half cycle, this value occurs at $\theta = 90°$ or $t = t/4$ second where T is the time period of the wave. Similarly, negative peak value occurs at $\theta = 270°$ or $t = 3\ T/4$ second.

3. Peak-to-Peak Value

It is the sum of the positive peak and negative peak values usually written as p-p value.

The p-p value of the wave shown in Fig. 8.8 is 1 + 1 = 2 A. It is obvious that the positive and negative peak values cannot occur at the same time and may not have the same value in all waveforms.

4. Root Mean Square (RMS) Value, I

It is also called the effective value. It is the value of the current at $\theta = 45°$ which equals $0.707\ I_m$ or 70.7 per cent of the maximum value.

For example, the RMS value of the sinusoidal alternating current shown in Fig. 8.9 is = 0.707 × 10 = 7.07 A. It is called effective value because an a.c. current of peak value 10 A is as effective in producing electric power as a d.c. current of 7.07 A.

It is called root-mean-square value for the following reasons :

Suppose we take a large number of instantaneous values of the current wave shown in Fig. 8.8 and square each one of them. Further, suppose that we add these squared values and divide this total by the number of values taken, thereby getting the mean value. If we take the square root of this mean value, we get root-mean-square value. In the present case, it will come out to be 7.07 A (Fig. 8.9).

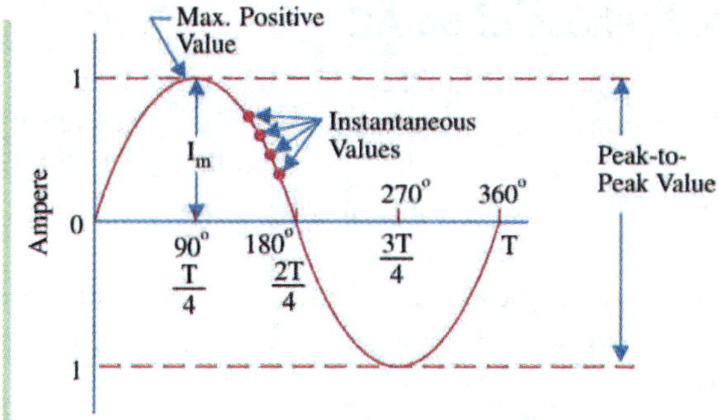

Fig. 8.8

$$\text{In general, RMS value} = \frac{\text{Maximum value}}{\sqrt{2}}$$

$$I = \frac{I_m}{\sqrt{2}} \qquad \text{— for sine waves only}$$

It should be clearly understood that whenever value of an ac voltage or current is stated, it always refers to its rms value unless stated otherwise. Usually, ac voltmeters and ammeters are calibrated to read rms values.

It is seen from above that

$$I_m = \sqrt{2}\,I = 1.414\,I$$
$$= 1.414 \times \text{rms value}$$

Also, p-p value $= 2 \times$ peak value

$$= 2 \times 1.414 \times \text{rms value}$$
$$= 2.828 \times \text{rms value}$$

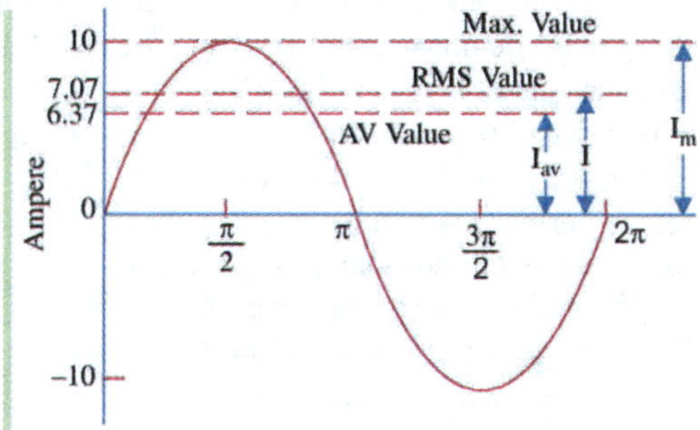

Fig. 8.9

5. Average Value (I_{av})

It is the arithmetical average of all instantaneous values in one half-cycle of the wave.

For symmetrical sine wave (like that shown in Fig. 8.9), one half cycle is taken because for full one cycle, the average would be zero. However, for non-symmetrical waves, it is *essential to take the average value over one full cycle.*

If we take the average of all instantaneous values, say, in positive half cycle of the wave shown in Fig. 8.9, the value comes out to be 6.37 per cent of the maximum value.

$$\therefore \quad I_{av} = 0.637 \times \text{max. value} = \frac{2 \times I_m}{\pi} = \frac{\text{twice max. value}}{\pi} \qquad \text{—for sine waves only}$$

$$\text{Also,} \quad I_m = \frac{I_{av}}{0.637} \quad \text{or} \quad I_m = 1.57\,I_{av}$$

6. Form Factor

It is defined as the ratio

$$K_f = \frac{\text{rms value}}{\text{average value}} = \frac{0.707\,I_m}{0.637\,I_m} = 1.11 \qquad \text{— for sine waves only}$$

7. Crest or Peak Factor

It is defined as the ratio

$$K_c = \frac{\text{maximum value}}{\text{rms value}} = \frac{I_m}{I_m/\sqrt{2}} = \sqrt{2} = 1.414 \qquad \text{— for sine wave only}$$

8.8. Phase of an AC

The fraction of a cycle or time-period that has elapsed since an ac current (or voltage) last passed a given reference point (usually the starting point) is called its phase. It may be expressed in

1. time measured in seconds or
2. fraction of a time-period or
3. angle expressed either in radians or degrees.

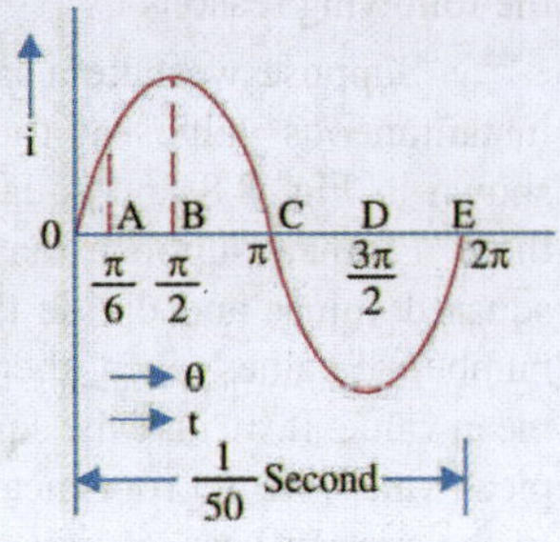

Fig. 8.10

For example, phase of the alternating current at point *A* (Fig. 8.10) is

1. $\pi/6$ radian or 30° or
2. 1/12th of a cycle or $T/12$ second or
3. 1/12 × 50 = 1/ 600 second.

Similarly, phase at point *B* is $\pi/2$ radian or 90° or $T/4$ or 1/200 second.

Phase at point *E* is 2π radians or zero because ac current is exactly in the same condition as it was at the starting point 0.

8.9. Phase Difference

In practice, the phase difference between two alternating quantities is more important than their absolute phases. In Fig. 8.11, the phase difference between two 50 Hz current waves is 60° whereas that in Fig. 8.12 is 90°.

The phase difference may be expressed in seconds by remembering that, in the present case, one cycle corresponds to 1/50 second. Its value for Fig. 8.11 is 1/50 × 6 = 1/300 second because 60° represents 60/360 = 1/6th of a cycle. Similarly, value of phase difference in Fig. 8.12 is 1/50 × 4 = 1/200 second.

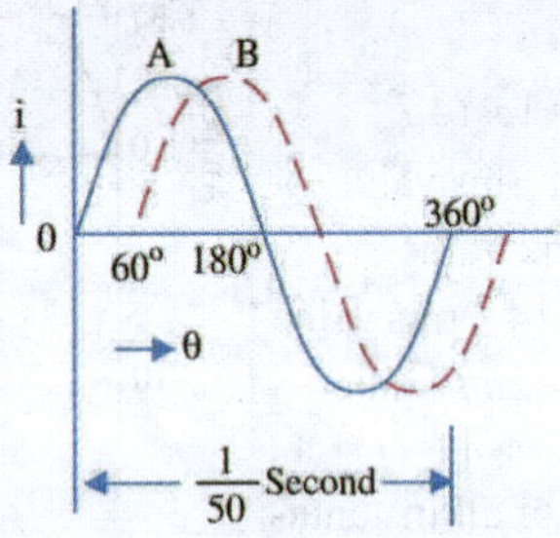

Fig. 8.11

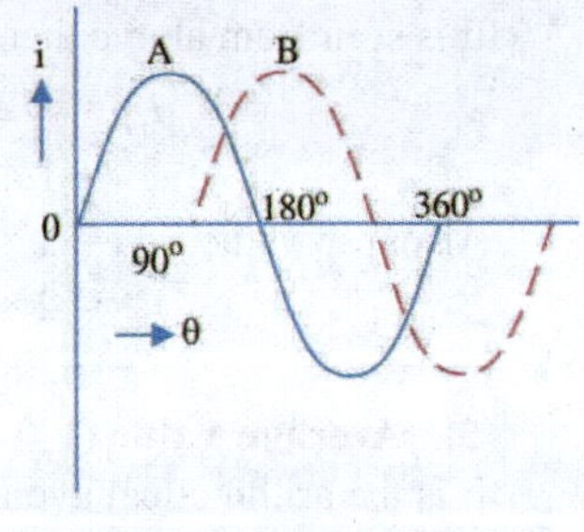

Fig. 8.12

Fig. 8.13 shows the case when phase difference is 0° *i.e.*, when two ac sine waves *A* and *B* are in phase or in step with each other. Fig. 8.14 illustrates the case when phase difference is 180°. Here, maxima of one wave fall on the minima of the other.

It has been stated above that phase difference between two sine waves *A* and *B* of Fig. 8.11 is 60°. However, this statement does not give any indication as to which current reaches its maximum value first. This deficiency is supplied by the term 'lag' or 'lead'.

A leading quantity is one which reaches it maximum (or zero) value *earlier* as compared to the other quantity. Similarly, a lagging quantity is one which reaches its maximum (or zero) value later than the other quantity.

For example, in Fig. 8.11, *B* lags behind *A* by 60° or which is the same thing, *A* leads *B* by 60°.

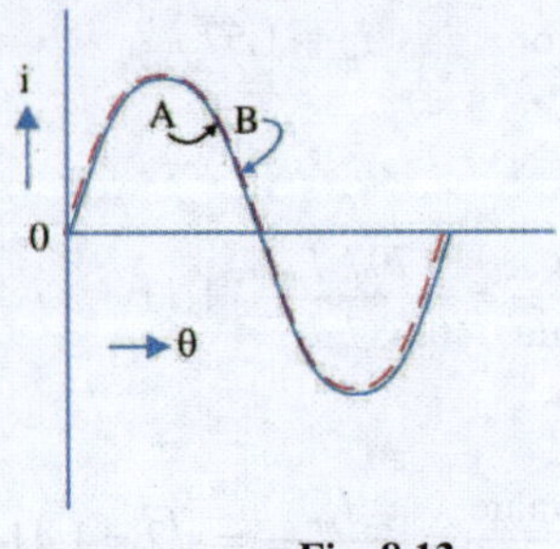

Fig. 8.13

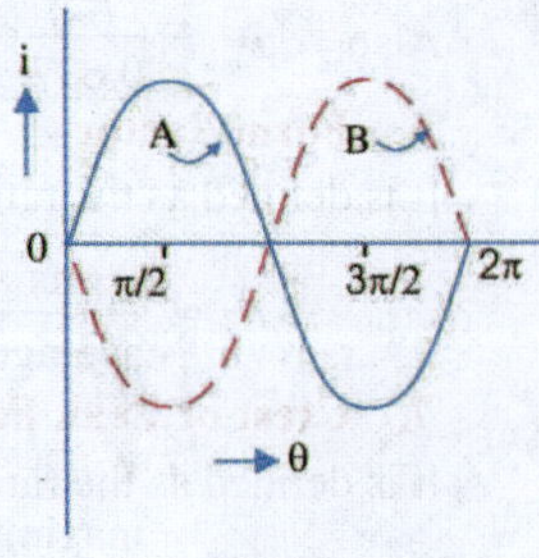

Fig. 8.14

Similarly, in Fig. 8.14, B lags A by 180° or π radians whereas they are in step or in unison or in phase with each other in Fig. 18.13.

Let us assume that the two sine waves A and B shown above have the same amplitude and same period. Taking wave A as the *reference* wave, their equations can be written as

$$i_A = I_m \sin \omega t$$

(*a*) $i_B = I_m \sin (\omega t - \pi / 6) = I_m \sin (\omega t - 30°)$ — Fig. 8.11

(*b*) $i_B = I_m \sin (\omega t - \pi / 2) = I_m \sin (\omega t - 90°)$ — Fig. 8.12

(*c*) $i_B = I_m \sin \omega t$ — Fig. 8.13

(*d*) $i_B = I_m \sin (\omega t - \pi)$ — Fig. 8.14

However, in Fig. 8.15, sine wave B leads A by 90°. Hence, again taking A as the reference wave, their equations can be written as

$$i_A = I_m \sin \omega t$$
$$i_B = I_m \sin (\omega t + \pi/2)$$

It is seen that a plus (+) sign when used in connection with phase difference denotes 'lead' whereas a minus (–) sign denotes 'lag'.

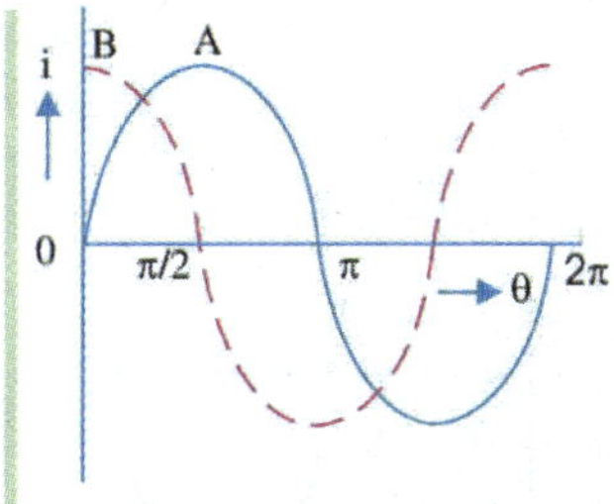

Fig. 8.15

Example 8.1. *A sinusoidal alternating current having a frequency of 50 Hz has a peak value of 10 A. What is the value of current after 1/300 second from zero ?*

Solution. The current equation is

$$i = I_m \sin \omega t = I_m \sin 2\pi f t \quad \text{— angle in radians}$$
$$= I_m \sin 360 ft \quad \text{— angle in degrees}$$

Now, $f = 50$ Hz, $t = 1/300$ second, $I_m = 10$ A

$\therefore \quad i = 10 \sin 360 \times 50 \times 1/300 = 10 \sin 60° = 10 \times 0.866 =$ **8.66 A**

Alternative Solution

Now, frequency is 50 Hz so that time for one cycle is 1/50 second. Obviously, 1/300 second represents 1/6th of a cycle spread over 2π radians or 360°. Hence, 1/300 second corresponds to $360 \times 1/6 = 60°$

$\therefore \quad i = 10 \sin 60° =$ **8.66 A**

Example 8.2. *A sinusoidal ac has a frequency of 50 Hz and a peak value of 10 A. Starting from zero, how long will it take the current to reach 5 A for the first time?*

Solution. The current equation is

$$i = 10 \sin 2\pi ft$$

Now, $i = 5$ A, $I_m = 10$ A, $f = 50$ Hz, $t = ?$

$\therefore \quad 5 = 10 \sin 360 \times 50 \times t$

or $\quad \dfrac{1}{2} = \sin 18{,}000\, t$

$\therefore \quad 18{,}000\, t = \sin^{-1}(1/2) = 30°$

$\therefore \quad t = 30/18{,}000 =$ **1/600 second**

Example 8.3. *An alternating current is represented by i = 141.4 sin 628 t. Calculate its (i) frequency (ii) rms value and (iii) average value.*

Solution. Comparing the given current equation with the standard a.c. equation ; $i = I_m \sin 2\pi ft$, we get

(*i*) $2\pi f = 628 \quad \therefore \quad f = 628 / 2\pi =$ **100 Hz** (*ii*) $I_m = 141.4$ A $\quad\therefore\quad I = \frac{I_m}{\sqrt{2}} = \frac{141.4}{\sqrt{2}} = \mathbf{100\ A}$

(*iii*) $I_{av} = 0.637 \quad I_m = 0.637 \times 141.4 = \mathbf{90\ A}$

8.10. Vector Representation of an Alternating Quantity

A sinusoidally alternating voltage or current can be graphically represented by counter-clockwise rotating vector provided it satisfies the following conditions :

1. its length is equal to the peak value (even rms value will do) of the alternating quantity,
2. it is in the horizontal position at the same instant when the alternating quantity is zero and is increasing positively,
3. its angular velocity is such that it completes one revolution in the same time as taken by the alternating quantity to complete one cycle.

In Fig. 8.16 (*a*) is shown a sinusoidal current wave lagging behind the alternating voltage wave by 90°. The same fact has been shown vectorially in Fig. 8.16 (*b*). Here, vector *OV* represents rms value of the voltage which is taken as the reference quantity. Similarly, vector *OI* represents rms. value of the alternating current which lags behind the voltage by 90°. Both vectors are supposed to be rotating in the counter-clockwise direction at the same angular velocity ω as that of the two alternating quantities.

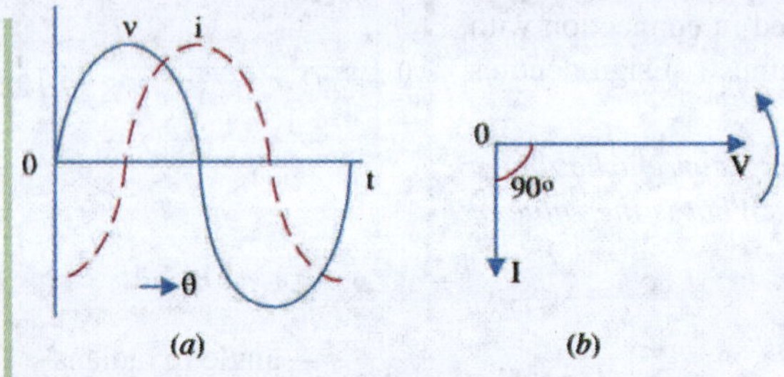

Fig. 8.16

8.11. AC through Pure Resistance Only

It is found that when a sinusoidal voltage is applied across a pure ohmic resistance, it produces a sinusoidal current

(*i*) which is in phase with the voltage as shown in Fig. 8.17, both graphically and vectorially;

(*ii*) whose rms value is given by $I = V/R$.

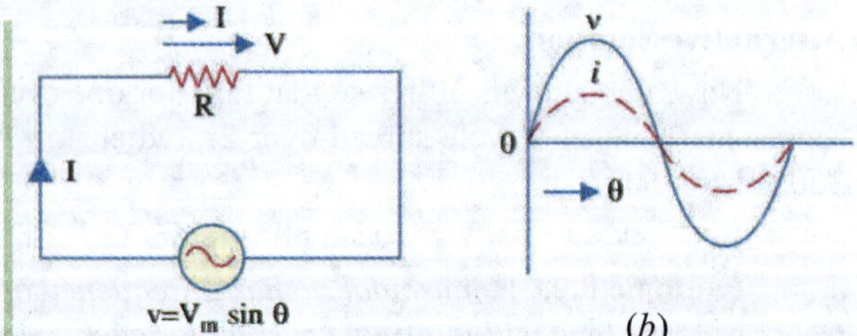

Fig. 8.17

If the equation of the applied voltage is $v = V_m \sin\theta$, then equation of the current is $i = I_m \sin\theta$.

The power wasted in the circuit in the form of heat is $P = I^2R$ where *I* is the r.m.s. value of the current.

In general, it should be noted that potential drop $v(t)$ across a resistor *R* is directly proportional to the current $i(t)$ in it. Hence, as shown in Fig. 8.18,

$$v(t) = R\,.\,i(t) \qquad \therefore \qquad i(t) = \frac{v(t)}{R}$$

and power $\quad p(t) = v(t).i(t)$

Here, $v(t)$ and $i(t)$ indicate the instantaneous values of time-varying voltage and current respectively which may have any wave form.

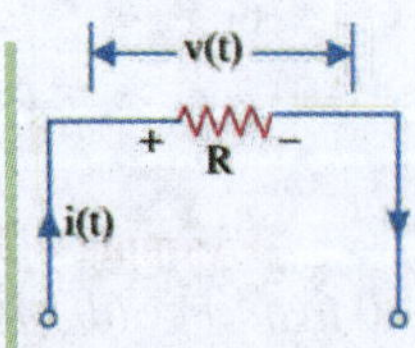

Fig. 8.18

8.12. AC through Pure Inductance Only

When a sinusoidally alternating voltage is applied across a pure inductive coil of inductance *L*, it produces a sinusoidal current (*i*) whose rms value is

$$I = \frac{V}{X_L} = \frac{V}{2\pi f L}$$

(*ii*) which lags behind the applied voltage by 90° as shown in Fig. 8.19.

If $\quad v = V_m \sin\theta$

then $\quad i = I_m \sin(\theta - \pi/2)$

It is found that power absorbed by this coil is zero. It may be noted that when current in a coil changes, the magnetic flux linked with it also changes thereby producing an induced voltage v (t) in it (Fig. 8.20). This voltage is proportional to the *rate of change of current* if permeability of the coil core is constant. The constant of proportionality is called the inductance of the coil.

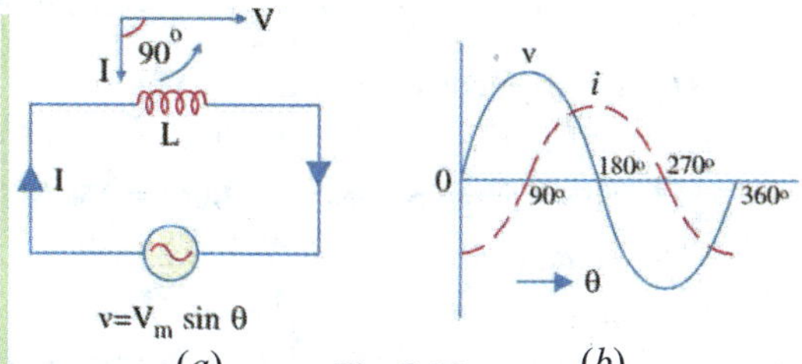

(*a*) Fig. 8.19 (*b*)

8.13. AC through Pure Capacitance Only

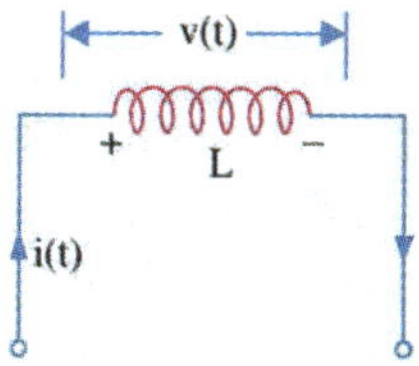

Fig. 8.20

When a sinusoidal voltage is applied across a pure capacitor,it produces a sinusoidal current

(*i*) whose magnitude is given by

$$I = \frac{V}{X_C} = \frac{V}{1/\omega C} = \omega VC$$

(*ii*) which leads the applied voltage by 90° as shown in Fig. 8.21.

Hence, if $v = V_m \sin\theta$, then current is given by

$$i = I_m \sin(\theta + \pi/2)$$

Power absorbed by such a capacitor is zero.

It may be noted that potential difference v between the terminals of a capacitor is directly proportional to the charge q on it. The constant of proportionality is the capacitance of the capacitor. As seen from Fig. 8.22,

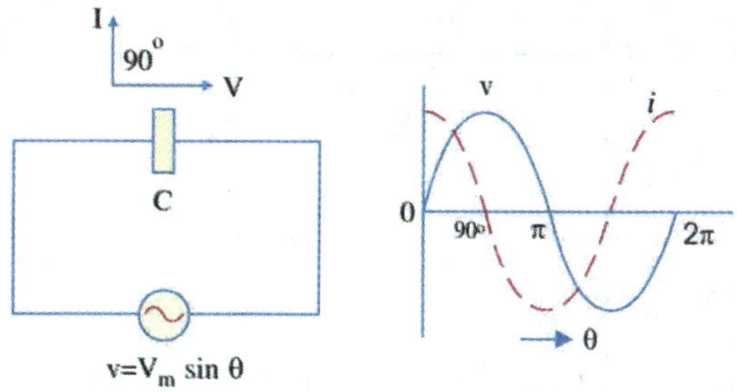

(*a*) Fig. 8.21 (*b*)

$$q(t) = C.v(t)\,; \qquad i(t) = \frac{dq}{dt} = C\frac{dv}{dt}$$

$$\therefore \qquad v(t) = \frac{1}{C}\int i\,.dt; \qquad p(t) = vi = Cv\frac{dv}{dt}$$

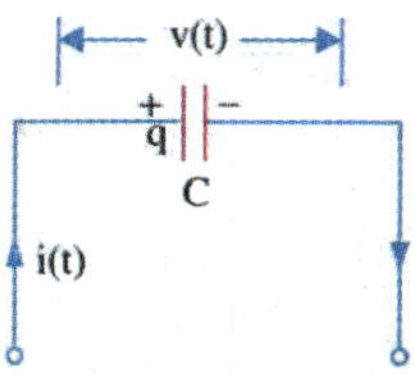

Fig. 8.22

Example 8.4. *Two capacitors of 60 μF and 30 μF are connected in series. Find the circuit current when this combination is connected across 50 Hz, 200 V supply mains.*

Solution. The equivalent capacitance of the combination is

$$C = \frac{C_1 \times C_2}{C_1 + C_2} = \frac{30 \times 60}{30 + 60} = 20\ \mu\text{F}$$

$$X_C = \frac{1}{2\pi f C} = \frac{1}{2\pi \times 50 \times 20 \times 10^{-6}} = 159\ \Omega$$

$\therefore \quad I = 200 / 159 = \mathbf{1.26\ A}$

Example 8.5. *A voltage given by v(t) = 100 sin ωt is applied across a resistor of 20 Ω. Find (a) current i(t) (b) power p(t) and (c) average power. Draw the sketches.*

Solution.

(*a*) $$i(t) = \frac{v(t)}{R} = \frac{100 \sin \omega t}{20} = \mathbf{5 \sin \omega t}$$

(*b*) Instantaneous power is given by

$$p(t) = v(t).i(t) = (100 \sin \omega t)(5 \sin \omega t) = \mathbf{500 \sin^2 \omega t}$$

(*c*) The average power is = 500/2 = **250 W.**

The sketches are shown in Fig. 8.23.

Example 8.6. *A sinusoidal voltage v (t) = 200 sin 1000 t is applied across a pure inductive coil of inductance L = 0.02 H. Determine current i(t) (b) instant power p(t) and (c) average power consumed by the coil. Also draw their waveforms.*

Solution.

(*a*) $$i(t) = \frac{1}{L}\int v\, dt = \frac{1}{0.02}\int 200 \sin 1000\, t\, dt$$

$$= \frac{200}{0.02}\left(-\frac{\cos 1000\, t}{1000}\right) = -\mathbf{10 \cos 1000}\ t$$

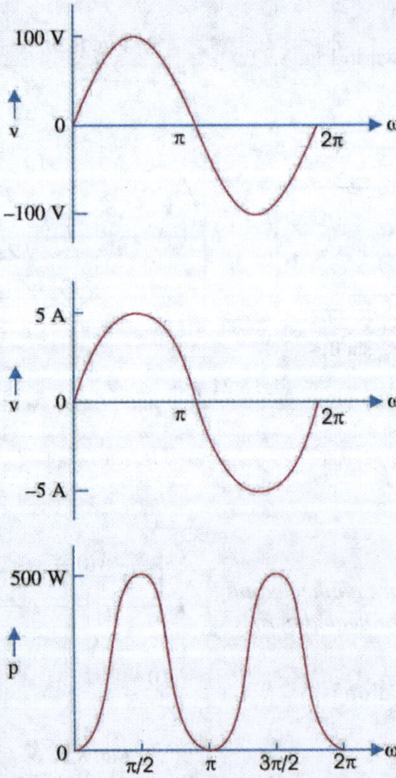

Fig. 8.23

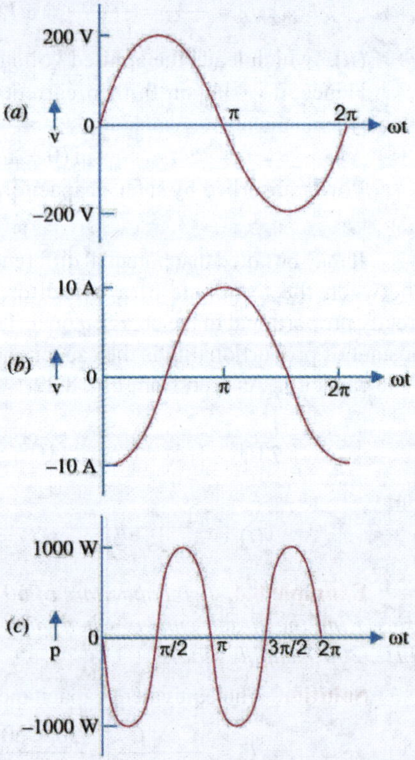

Fig. 8.24

Its wave form is shown in Fig. 8.24 (*b*).

(*b*) $p(t) = v(t).\, i(t)$

$= (200 \sin 1000\, t) \times (-10 \cos 1000\, t)$

$= -200 \sin 1000\, t.\cos 1000\, t$

$= -2000 \times \frac{1}{2} \sin 2000\, t = \mathbf{-1000 \sin 2000\, t}$

As seen from Fig. 8.24 (*c*), average power is zero.

Note. $\sin x \cos x = 1/2 \sin 2x$.

Example 8.7. *A sinusoidal voltage v(t) = 200 sin 1000 t is applied across a pure capacitor of 100 μF. Find (a) current i (t) (b) charge q (t) and (c) power p(t). Draw their sketches.*

Solution.

(*a*) $i(t) = C\frac{dv}{dt} = C\frac{d}{dt}(200 \sin 1000\, t)$

$= 100 \times 10^{-6}(200 \times 1000 \cos 1000\, t) = \mathbf{20 \cos 1000\, t}$ **amperes**

The sketch is shown in Fig. 8.25 (*b*).

(*b*) $q(t) = C.v(t) = 100 \times 10^{-6} \times 200 \sin 1000\, t = \mathbf{0.02 \sin 1000\, t}$ **coulomb.**

It is sketched in Fig. 8.25 (*c*).

(*c*) $p(t) = vi = 200 \sin 1000\, t \times 20 \cos 1000\, t$

$= 200 \times 20 \times \frac{1}{2} \sin 2000\, t = \mathbf{2000 \sin 2000} t$ **watt.**

It is shown in Fig. 8.25 (*d*). It is obvious that net power consumed by the capacitor over one cycle is zero.

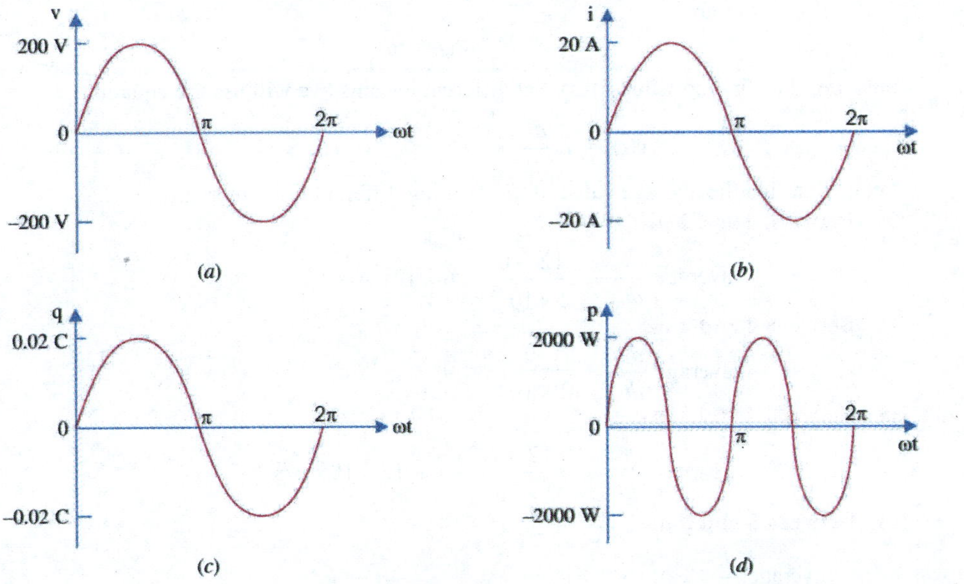

Fig. 8.25

8.14. Non-Sinusoidal Waveforms

The sine wave is considered the basic waveform because it is the one produced by alternators and electronic oscillators. Moreover, it is the only waveform which will produce a single frequency current from a single-frequency voltage when applied to resistive, inductive or capacitive load. The waves deviating from the standard sine or cosine wave are treated as *distorted* waves, they are called non-sinusoidal waveforms like the triangular and square waveforms shown in Fig. 8.3 (*b*) and (*c*) respectively and find many applications in electronic circuits. In the case of non- sinusoidal waveforms,

1. it is common practice to designate peak-to-peak value instead of peak value only because they generally have unsymmetrical peaks;
2. usually no angles are shown along the horizontal axis as is usually done in the case of sine waves. Instead, only time period is indicated;
3. the two half cycles may be non-symmetrical either in amplitude or time or both.

Non-sinusoidal waveforms are also called *complex* waveforms. A non-sinusoidal voltage may produce a current waveform similar to its own when load is resistive but it will not do so if load is either capacitive or inductive.

Example 8.8. *A non-sinusoidal current having the waveform shown in Fig. 8.26 is passed through a pure inductance of 5 mH. Sketch the voltage waveform v(t).*

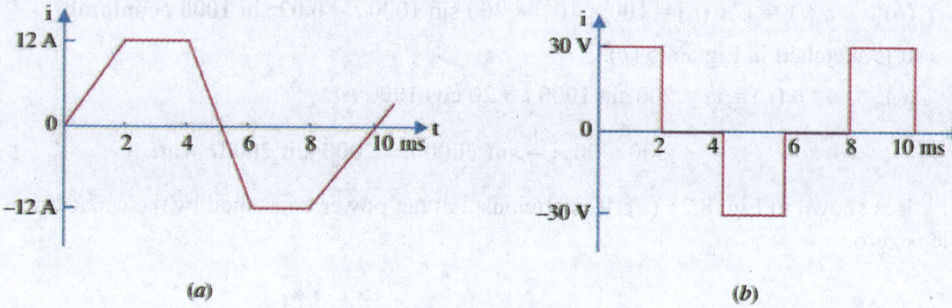

Fig. 8.26

Solution. For finding values of $v(t)$ at different instants, we will use the equation

$$v(t) = L\frac{di}{dt}$$

Let us first find the average value of *di/dt* during different time intervals.

(*a*) Between 0 and 2 ms

$$\text{average } \frac{di}{dt} = \frac{12-0}{2\times 10^{-3}} = 6\times 10^{3} \text{ A/s}$$

(*b*) Between 2 and 4 ms

$$\text{average } \frac{di}{dt} = \frac{0}{2\times 10^{-3}} = 0$$

(*c*) Between 4 and 6 ms

$$\text{average } \frac{di}{dt} = \frac{-12-(12)}{2\times 10^{-3}} = -12\times 10^{3} \text{ A/s}$$

(*d*) Between 6 and 8 ms

$$\text{average } \frac{di}{dt} = 0$$

(*e*) Between 8 and 10 ms

$$\text{average } \frac{di}{dt} = \frac{0-(-12)}{2\times 10^{-3}} = 6\times 10^{3} \text{ A/s}$$

Hence, the values of $v(t)$ during same time intervals are

1. $v(t) = L \cdot \frac{di}{dt} = 5 \times 10^{-3} \times 6 \times 10^3 = 30$ V
2. $v(t) = 5 \times 10^{-3} \times 0 = 0$
3. $v(t) = 5 \times 10^{-3} \times (-12 \times 10^3) = -60$ V
4. $v(t) = 0$
5. $V(t) = 5 \times 10^{-3} \times 6 \times 10^3 = 30$ V

The voltage wave is sketched in Fig. 8.26 (*b*).

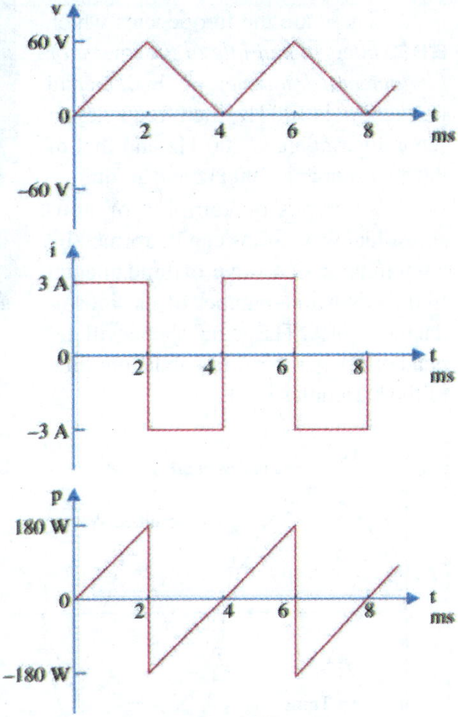

Fig. 8.27

Example 8.9. *The voltage waveform shown in Fig. 8.27 (a) is applied across the terminals of a pure capacitor of 100 μF. Sketch i (t) and p (t). Also, find maximum value of current and power.*

Solution. We will use the following two relations

$$i(t) = C\frac{dv}{dt}; \quad p(t) = Cv\frac{dv}{dt}$$

Let us find the value of dv/dt for different time intervals.

1. Between 0 and 2 ms

$$\frac{dv}{dt} = \frac{60-0}{2 \times 10^{-3}} = 30 \times 10^3 \text{ V/s}$$

2. Between 2 and 4 ms

$$\frac{dv}{dt} = \frac{0-60}{2 \times 10^{-3}} = -30 \times 10^3 \text{ V/s}$$

3. Between 4 and 6 ms

$$\frac{dv}{dt} = 30 \times 10^3 \text{ V/s} \qquad \text{— as before}$$

4. Between 6 and 8 ms

$$\frac{dv}{dt} = -30 \times 10^3 \text{ V/s} \qquad \text{— as before}$$

The value of $i(t)$ during the same time intervals are as under :

1. $i(t) = C\frac{dv}{dt} = 100 \times 10^{-6} \times 30 \times 10^3 = 3$ A

As seen from Fig. 8.27 (*b*), the magnitude of the current remains constant throughout this intervals.

2. $i(t) = 100 \times 10^{-6}(-30 \times 10^3) = -3$ A

Again, current remains constant at 3A though it flows in the opposite direction *i.e.*, it is a discharging current.

3. As before, $i(t) = 3$ A — charging current
4. As before, $i(t) = -3$A — discharging current

As seen, current is of *square* waveform with a maximum current value of ± 3 A.

Power. It is given by $p(t) = Cv\frac{dv}{dt}$

As seen, with dv/dt being constant in different time intervals, $p(t)$ varies directly as v. Hence, power waveform is exactly similar to voltage waveform except that it has a maximum value,

$$= Cv\frac{dv}{dt} = 100 \times 10^{-6} \times 60 \times 30 \times 10^3 = \mathbf{180 \text{ W}}$$

8.15. Harmonics

These are the frequencies which are an *integral multiple of the lowest or fundamental frequency.* If fundamental frequency* is 100 Hz, then frequency of second harmonic is 200 Hz and that of third harmonic is 300 Hz and so on.

All types of complex or non-sinusoidal waveforms can be formed by combining a sine wave of fundamental frequency with a number of its sinusoidal harmonics. The actual shape will depend on the number, kind and amplitude of the harmonics used.

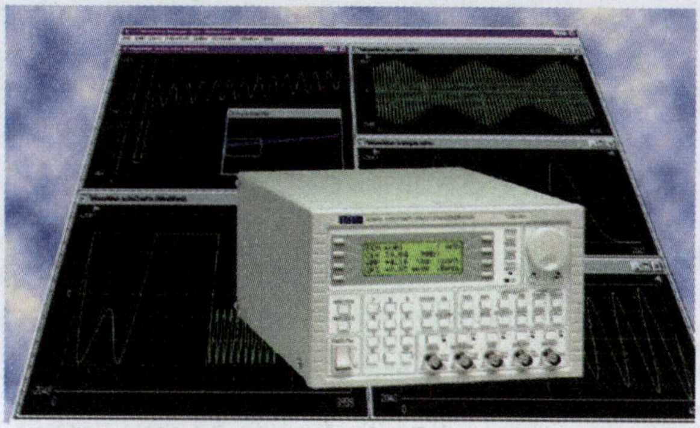

Harmonics are the multiples of a sine wave (the fundamental frequency).

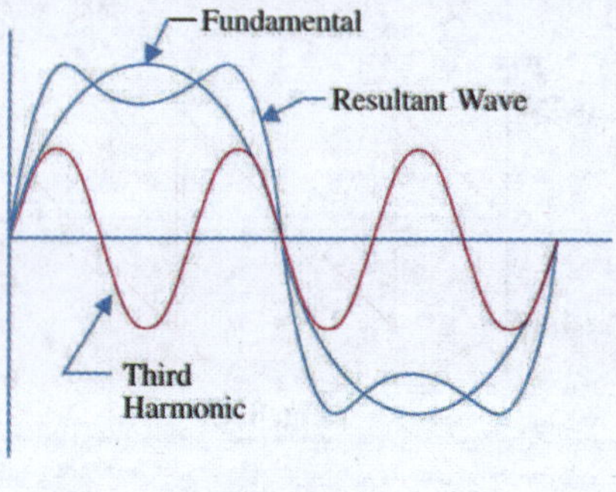

Fig. 8.28

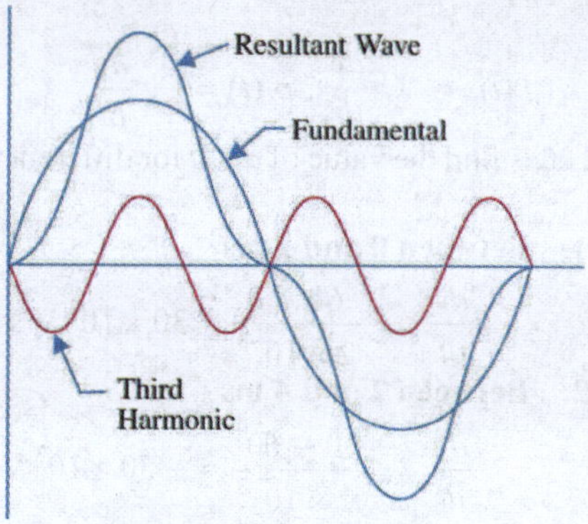

Fig. 8.29

Fig. 8.28 shows a complex wave which is obtained by adding a sine wave of fundamental frequency and its third harmonic both of which are initially *in phase with each other.*

If we shift the phase of the third harmonic by 180°, the resultant complex wave becomes as shown in Fig. 8.29. It is completely different in appearance from the one shown in Fig. 8.28. It is found that a symmetrical complex wave contains no even harmonics. But a non-symmetrical one contains both odd and even harmonics. For example, all audio waveforms of voice and music etc., contain both odd and even harmonics.

CONVENTIONAL PROBLEMS

1. Determine the time periods of the waveforms having a frequency of
   (*a*) 25 Hz (*b*) 10 kHz (*c*) 20 MHz **[(*a*) 0.04 s (*b*) 10^{-4} s (*c*) 0.05 μs]**
2. What will be the frequencies of the a.c. signals having time periods of
   (*a*) 1 μ s (*b*) 50 ns (*c*) 50 ms ?
   [(*a*) 1 MHz (*b*) 20 MHz (*c*) 20 Hz]
3. The angular frequency of a waveform is 1000 π radian/second. Find its (*a*) frequency and (*b*) time period. **[(*a*) 500 Hz (*b*) 2 ms]**
4. A sinusoidal waveform is given by
   $i = 10 \sin(6284\,t + 10°)$ ampere
   Determine its (*a*) peak value (*b*) rms value (*c*) frequency and (*d*) time period.
   [(a) 10 A (b) 7.07 A (c) 1 kHz (*d*) 1 ms]

* Fundamental frequency is also called the first harmonic.

5. Two alternating currents are given by

$$i_1 = 50 \sin\left(314t + \frac{\pi}{6}\right)$$

$$i_2 = 20 \sin\left(314\,t - \frac{\pi}{4}\right)$$

Find their phase difference. Which of the two currents leads the other ? [(*a*) 75° (*b*) I_1]

SELF EXAMINATION QUESTIONS

A. Fill in the blanks by most appropriate word (s) or numerical value (s).

1. A train of sine waves which contains 50 positive peaks and 50 negative peaks per second has a frequency of .... Hz.
2. The polarity of an ac waveform reverses every .... cycle.
3. The time period of a sine wave of 1 kHz is .... millisecond.
4. For a sinusoidal ac voltage of peak value 100 V, the rms value is ....V.
5. For a sinusoidal ac voltage of rms value 70.7 V, the p-p value is ....V.
6. The ac current drawn by a capacitor .... the applied voltage by 90°.
7. Higher the frequency of the ac current .... the reactance offered by a capacitor.
8. Complex waveforms can be formed by adding.........to the fundamental frequency.
9. Frequencies which are .... multiple of the fundamental frequency are called harmonics.
10. Net power consumed by a pure inductive coil or pure capacitor is ....

B. Answer True or False

1. Sine wave has been adopted universally as a standard waveform because it is easy to produce.
2. A waveform can be sinusoidal but non-symmetrical and *vice-versa.*
3. One complete cycle of a waveform contains two alternations which may not be identical.
4. A frequency of 15 kHz lies in the audio-frequency range.
5. Average value of a sinusoidal ac is slightly greater than its rms value.
6. Greater the capacitance of a capacitor, higher the reactance it offers to an alternating voltage.
7. A non-sinusoidal waveform consists of a fundamental frequency plus its harmonics.
8. A square waveform can be built up from sinusoidal waveforms of different frequencies.
9. Only sinusoidal voltage produces a sinusoidal current in any type of load.
10. When harmonics of a fundamental sine wave are added to it, we get a complex wave.

C. Multiple Choice Items

1. An ac current given by $i = 14.14 \sin(\omega\,t + \pi/6)$ has an rms value of .... amperes
   (*a*) 10 (*b*) 14.14
   (*c*) 1.96 (*d*) 7.07
   and a phase of .... degrees.
   (*e*) 180 (*f*) 30
   (*g*) –30 (*h*) 210
2. The rms value of a sinusoidal ac current is equal to its value at an angle of .... degrees.
   (*a*) 60 (*b*) 45 (*c*) 30 (*d*) 90
3. Two sinusoidal currents are given by the equations : $i_1 = 10 \sin(\omega t + \pi/3)$ and $i_2 = 15 \sin(\omega t - \pi/4)$. The phase difference between them is .... degrees.
   (*a*) 105 (*b*) 75 (*c*) 15 (*d*) 60
4. An ac current is given by $i = 100 \sin 100\,\pi\,t$. It will achieve a value of 50 A after .... second.
   (*a*) 1/600 (*b*) 1/300
   (*c*) 1/1800 (*d*) 1/900
5. The average value of an ac current wave given by $i = 100 \sin 100\,\pi\,t$ is .... amperes.
   (*a*) 70.7 (*b*) 141.4 (*c*) 157 (*d*) 63.7

ANSWERS

A. Fill in the blanks

1. 50 **2.** half **3.** one **4.** 70.7 **5.** 200 **6.** leads
7. greater **8.** harmonics **9.** integral **10.** zero

B. True or False

1. F **2.** T **3.** T **4.** T **5.** F **6.** F **7.** T **8.** T **9.** T **10.** T

C. Multiple Choice Items

1. *a*, *f* **2.** *b* **3.** *a* **4.** *a* **5.** *d*.

CHAPTER 9

Series A.C. Circuits

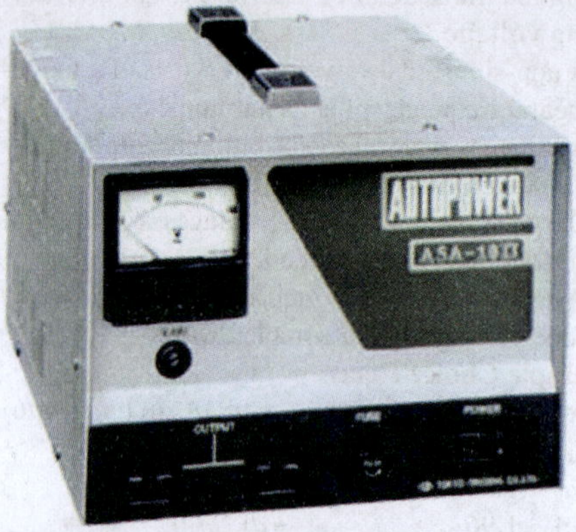

1. R-L Circuit
2. Q Factor of a Coil
3. Skin Effect
4. R.C. Circuit
5. Coupling Capacitor
6. R-L-C Circuit
7. Resonance in an R-L-C Circuit
8. Resonance Curve
9. Main Characteristics of Series Resonance
10. Bandwidth of a Tuned Circuit
11. Sharpness of Resonance
12. Tuning
13. Tuning Ratio
14. Radio Tuning Dial
15. Parallel Resonance

9.1. R-L Circuit

Suppose a *pure* coil of inductance L is connected in series with a *pure* resistance R* and is energised by a sinusoidal voltage of r.m.s. value V** as shown in Fig. 9.1 (*a*). If we take the current as the reference quantity, then it will give rise to a sinusoidal voltage drop v_R across R which will be in phase with it as shown in Fig. 9.1 (*c*). The voltage drop v_L across the coil will *lead* the current by 90° as shown in Fig. 9.1 (*d*). In other words, current through a coil lags behind the voltage across it. The resultant voltage is the *vector* (or phasor) sum of v_R and v_L and as shown in Fig. 9.1 (*e*), leads current by some angle ϕ. In other words, circuit current *lags behind* the applied voltage by an angle ϕ.

These facts have been shown in the phasor diagram of Fig. 9.2 (*a*) which is called voltage triangle. Here

V = r.m.s. value of applied voltage

I = r.m.s. value of circuit current

$V = IR$

= r.m.s. voltage drop across R (in phase with I)

$V = IX_L$

= r.m.s. voltage drop across L (at right angles to I)

* This R could also be the internal resistance of an actual coil itself.

** The equation of the alternating voltage would be $v = \sqrt{2}\ V \sin \omega t$.

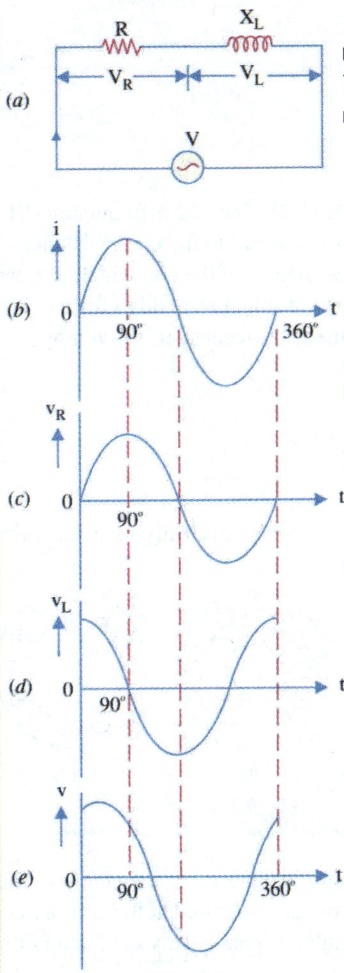

Fig. 9.1

Since in a series combination, same current flows through the two components, it has been taken as the reference quantity. That is why current vector has been taken along the horizontal or reference axis. As seen from Fig. 9.2 (*a*).

$$V = \sqrt{V_R^2 + V_L^2} = \sqrt{(IR)^2 + (IX_L)^2}$$

$$= I\sqrt{R^2 + X_L^2} = IZ \text{ or } I = \frac{V}{Z}$$

The quantity $(R^2 + X_L^2)^{1/2}$ is called *impedance* (*Z*) of the circuit and is shown separately in Fig. 9.2 (*b*). Obviously, *Z* is the vector sum of *R* and X_L as shown in the impedance triangle of Fig. 9.2 (*b*).

It is also seen that current *I* lags behind the applied voltage *V* by angle ϕ such that

$$\tan\phi = X_L/R$$

Since pure inductive coil consumes no power, the power drawn by the circuit is the same as dissipated by *R*.

$$\therefore\ P = I^2 R$$

$$= I.I.R = I.\frac{V}{R}.R$$

$$= VI\frac{R}{Z} = VI\cos\phi \quad \text{— Fig. 10.2}$$

The terms 'cos ϕ' is called the power factor of the *R-L* circuit.

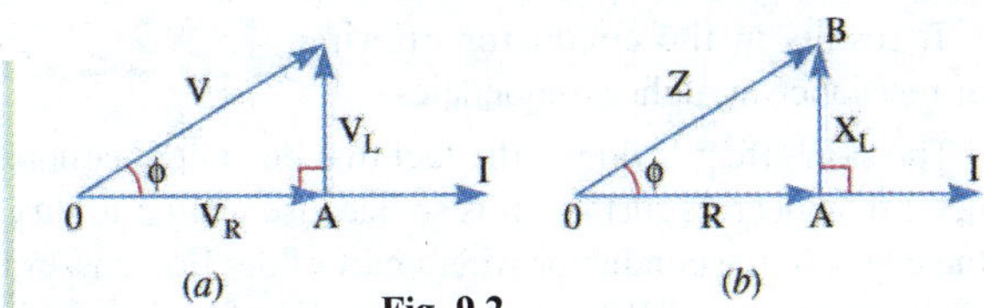

Fig. 9.2

Example 9.1. *A coil having L = 0.14 H and R = 9.43 Ω is connected across a 50 Hz, 135V supply. Compute*

(*i*) X_L (*ii*) *Z* (*iii*) *I* (*iv*) R_V (*v*) *V* (*vi*) ϕ (*vii*) power factor and (*viii*) power absorbed.

Solution. (*i*) $X_L = 2\pi fL = 2\pi \times 50 \times 0.14 = \mathbf{44\ \Omega}$

(*ii*) $Z = \sqrt{R^2 + X_L^2} = \sqrt{9.43^2 + 44^2} = \mathbf{45\ \Omega}$

(*iii*) $I = \frac{V}{Z} = \frac{135}{45} = \mathbf{3A}$

(*iv*) $V_R = IR = 3 \times 9.43 = \mathbf{28.3\ V}$

(*v*) $V_L = IX_L = 3 \times 44 = \mathbf{132\ V}$

(*vi*) $\cos\phi = R/Z = 9.43/45 = \mathbf{0.21}$

(*vii*) $\phi = \cos^{-1}(0.21) = \mathbf{77.6° (lag)}$

It means that current lags behind the applied voltage by 77.6°.

(*viii*) Power absorbed = $VI\cos\phi = 135 \times 3 \times 0.21 =$ **85 watt**

or $P = I^2R = 3^2 \times 9.43 =$ **85 W**

9.2. Q Factor of a Coil

The quality factor Q of a coil is given by

$$Q = \frac{X_L}{R} = \frac{\omega L}{R} = \frac{2\pi f L}{R}$$

where L is the inductance of the coil and R its resistance.

The Q-value of a coil may vary from less than 10 to about 1000. The radio-frequency (RF) coils usually have a Q of about 30 to 500. At low frequencies, R is just equal to the d.c. resistance of the coil but at radio-frequencies, it represents the a.c. effective resistance R_e of the coil which is much greater than R. The factors which make R_e greater than R are (*i*) skin effect (*ii*) eddy currents and (*iii*) hysteresis loss. Hence, Q of the coil is decreased at high frequencies because it is given by

$$Q = \frac{X_L}{R_e}$$

The Q of a coil can be measured with the help of a Q-meter.

9.3. Skin Effect

For d.c. and low frequency a.c., current through a conductor flows uniformly over its entire cross-section as shown in Fig. 9.3 (*a*). As the frequency is increased, current tends to flow towards the outer parts of the conductor [Fig. 9.3 (*b*)]. At radio-frequencies, the current is practically confined to the outer surface of the conductor as shown in Fig. 9.3 (*c*). The gradual shift of current flows from centre to the surface of the conductor is known as *skin effect*. It results in the conductor offering greater resistance at higher frequencies.

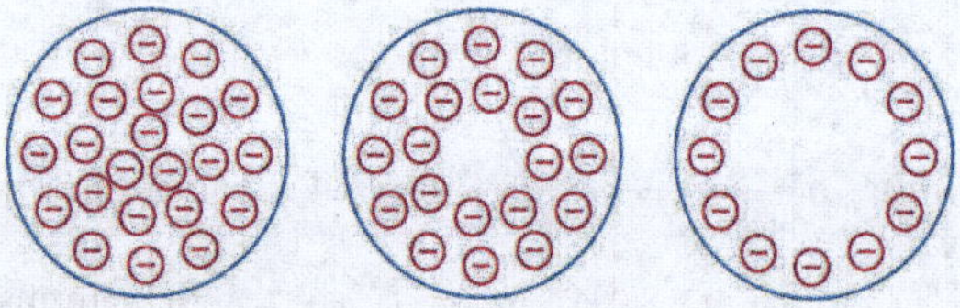
(*a*) (*b*) (*c*)

Fig. 9.3

The skin effect is due to the fact that current encounters more inductance at the centre of the conductor than near its surface. It is so because magnetic flux is more concentrated at the centre than near the edges of the conductor where part of the flux passes through air. That is why conductors for very high frequency (VHF) currents are made of hollow tubing.

Skin effect can be minimised by forming the conductor from a large number of interwoven wires connected in parallel at their ends but insulated from each other throughout the rest of their length. In this way, each conductor is linked with the same amount of magnetic flux and carries equal current thereby greatly increasing the useful cross-section of the conductor. Such a stranded conductor is called a Litz conductor.

Example 9.2. *A 100 μH coil has a Q-value of 500 at 5 MHz. What is its effective resistance?*

Solution. $R_e = \frac{2\pi f L}{Q} = \frac{2\pi \times 5 \times 10^6 \times 100 \times 10^{-6}}{500} = \mathbf{6.2\ \Omega}$

9.4. R.C. Circuit

Such a series combination connected across an a.c. voltage of rms value V is shown in Fig. 9.4. Here

$V_R = IR$ = drop across R (in phase with I)

$V_C = IX_C$ = drop across C (lagging I by 90°)

Since capacitive reactance X_C is taken negative, V_C is drawn along negative Y-axis as shown in Fig. 9.4 (*b*).

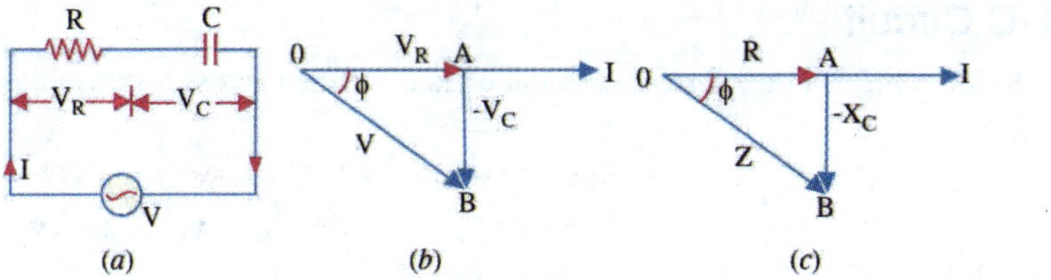

Fig. 9.4

$$\therefore \quad V = \sqrt{V_R^2 + V_C^2} = \sqrt{(IR)^2 + (-IX_C)^2} = I\sqrt{R^2 + X_C^2} = IZ$$

$$\therefore \quad I = \frac{V}{Z}$$

It is seen that current *leads* the applied voltage by an angle ϕ such that

$$\tan\phi = \frac{-X_C}{R} \qquad \text{— Fig. 10.4 (b)}$$

Since a pure capacitor consumes no power, the entire circuit power consumption is due to resistor only.

$$\therefore \quad P = I^2 R = VI\cos\phi \qquad \text{— as in an } RL \text{ circuit}$$

Example 9.3. *A capacitor having a capacitance of 10 μF is connected in series with a pure resistance of 120 Ω across a 100 V, 50 Hz supply. Calculate*

(a) current *(b) phase angle, φ* *(c) power consumed.*

Solution. $X_C = \frac{1}{2\pi f C} = \frac{1}{2\pi \times 50 \times 10 \times 10^{-6}} = \frac{10^6}{100\,\pi} = 318\ \Omega$

$$Z = \sqrt{120^2 + 318^2} = 340\ \Omega$$

(*a*) $I = \frac{V}{Z} = \frac{100}{340} = \mathbf{0.294\ A}$

(*b*) $\tan\phi = 318/100 = 3.18 \quad \therefore \phi = \tan^{-1}(3.18) = \mathbf{69.3°}$

(*c*) $P = 0.294^2 \times 120 = \mathbf{10.4\ W}$

9.5. Coupling Capacitor

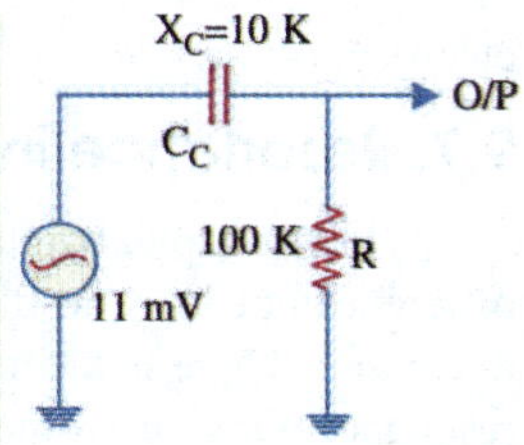

Fig. 9.5

An *R-C* circuit is often used in electronic circuits for passing on a *high-frequency signal* from one circuit to another. The capacitor used in the circuit is called coupling capacitor because it *couples* or joins the two circuits so far as a.c. signal is concerned but, by blocking d.c. keeps them isolated from each other. The low reactance C_C of the coupling capacitor (Fig. 9.5) allows practically all the a.c. signal to be dropped across R which passes it on to the next circuit. Very little signal is developed across C_C itself.

Usually, reactance of C_C is about one-tenth of R. Suppose, an a.c. signal of 11 mV is to be coupled from the signal source in Fig. 9.5 to a circuit on the right, say, the next stage of an audio amplifier.

Drop across $C_C = 11 \times \frac{10}{(10 + 100)} = 1\ \text{mV}$

Drop across $R = 11 \times \frac{100}{110} = 10\ \text{mV}$

It is seen that practically 91 per cent of the input signal appears across R and can be passed on to the next amplifier stage.

9.6. R-L-C Circuit

As shown in Fig. 9.6, the series combination is connected across a sinusoidal voltage of rms value V.

Let $V_R = IR$ — in phase with I

$V_L = IX_L$ — leading I by 90°

$V_C = IX_C$ — lagging I by 90°

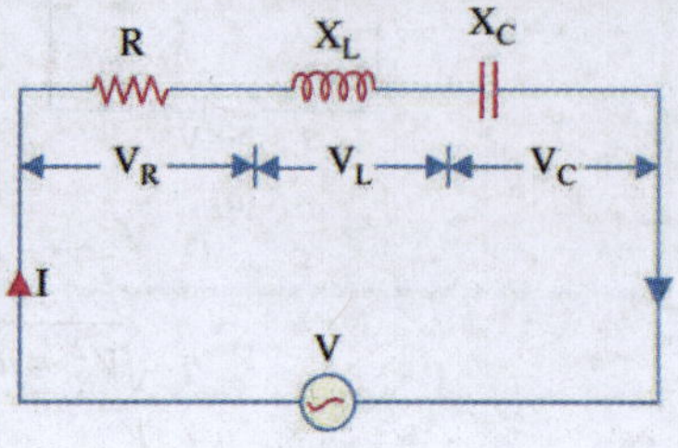

Fig. 9.6

Since V_L is ahead of I by 90° and V_C is behind it by 90°, the phase difference between the two is 180° *i.e.*, they are in direct opposition to each other as shown in Fig. 9.7 (*a*).

In the voltage triangle of Fig. 9.7 (*a*), V_L has been assumed greater than V_C which makes I lag behind V. If $Vc > V_L$, then I leads V. Subtracting AC from $AB = BD$, we get the net reactive drop

$AD = V_L - V_C = I(X_L - X_C) = IX$

From the voltage triangle OAD of Fig. 9.7 (*a*), we have

$$OD = \sqrt{OA^2 + AD^2}$$

$$\text{or} \quad V = \sqrt{(IR)^2 + (IX_L - IX_C)^2}$$

$$= I\sqrt{R^2 + (X_L - X_C)^2}$$

$$= I\sqrt{R^2 + X^2} = IZ$$

$$\therefore \quad I = \frac{V}{Z} = \frac{\text{applied voltage}}{\text{impedance}}$$

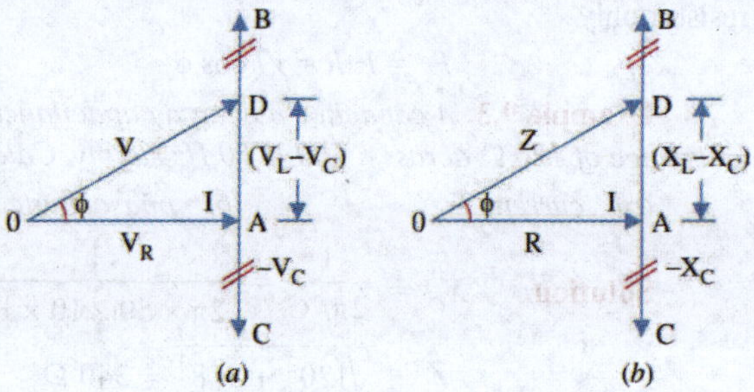

Fig. 9.7

The phase angle ϕ is given by $\tan \phi = X/R$

where X is the net reactance and is inductive in the present case. If V_C would have been greater than V_L, then X would have been capacitive. If $X_L = X_C$, then $X = 0$ and the circuit is said to be in electrical *resonance.*

9.7. Resonance in an R-L-C Circuit

If a sinusoidal voltage of variable frequency is applied across an R-L-C circuit, it encounters different impedance at different frequencies. As frequency is increased, X_L is increased but X_C is decreased. There is a certain frequency of the applied a.c. voltage for which $X_L = X_C$. It is called resonance. Obviously, the only impedance offered by the circuit is R and is the *lowest* it can offer. Hence, under resonant condition, circuit current is maximum and is given by

$$I_m = V/R$$

Moreover, this current is in phase with the applied voltage. Hence, the circuit behaves like a purely resistive circuit with a power factor of unity. The resonant frequency can be found from the condition

$$X_L = X_C \quad \text{or} \quad 2\pi f_0 L = \frac{1}{2\pi f_0 C}$$

$$\therefore \quad f_0 = \frac{1}{2\pi\sqrt{LC}} = \frac{0.159}{\sqrt{LC}} \cong \frac{0.16}{\sqrt{LC}}$$

If L is in henrys and C in farads, then f_0 is in hertz. Obviously, f_0 can be changed by changing either L or C. It should be noted that resistance R plays *no part in determining the resonant frequency* although it does limit the current at resonance and also affects the off-resonance behaviour of the circuit.

9.8. Resonance Curve

It is the curve which shows variation in circuit current with change in the frequency of the applied voltage. The shape of such a curve for various values of R is shown in Fig. 9.8. For smaller values of R, the resonance curve is not only sharply-peaked but also has very steep sides. For large values of R, the curve is flat and broad-sided.

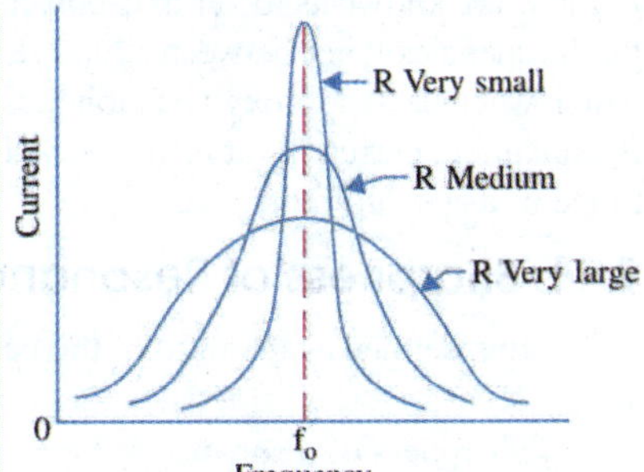

Fig. 9.8

It should be particularly noted that the value of R not only affects the value of circuit current I, it also affects the *shape* of the resonance curve itself. For low values of R, sides of the curve are very steep which means that current falls off very rapidly as the frequency changes from resonance to off-resonance value. For large values of R, curve is broad-sided which means limited change in current for resonance and off-resonance conditions.

As seen from above, smaller the R, steeper the sides of the resonance curve and consequently sharper the tuning of the circuit.

9.9. Main Characteristics of Series Resonance

Following main points about series resonance are worth noting :

1. The circuit current is maximum and is given by $I_m = V/R$.
2. The circuit offers minimum impedance $Z_{min} = R$.
3. The circuit behaves like a pure resistive circuit and has a power factor of unity.
4. Voltage drops V_L and V_C are maximum and equal in magnitude but cancel out since they are 180° out of phase with each other.
5. Resonant frequency is given by

$$f_0 = \frac{0.159}{\sqrt{LC}} \cong \frac{0.16}{\sqrt{LC}}$$

9.10. Bandwidth of a Tuned Circuit

Bandwidth of a tuned circuit is given by the band of frequencies which lie between two points on either side of its resonance curve where current is $1/\sqrt{2}$ or 0.707 of its maximum value at resonance. As shown in Fig. 9.9, bandwidth or passband is given by

$$BW = \Delta f = f_2 - f_1$$

It can be shown that

$$BW = f_2 - f_1 = \frac{R}{2\pi L} \quad \text{or} \quad BW = \frac{f_0}{Q} \text{ — how ?}$$

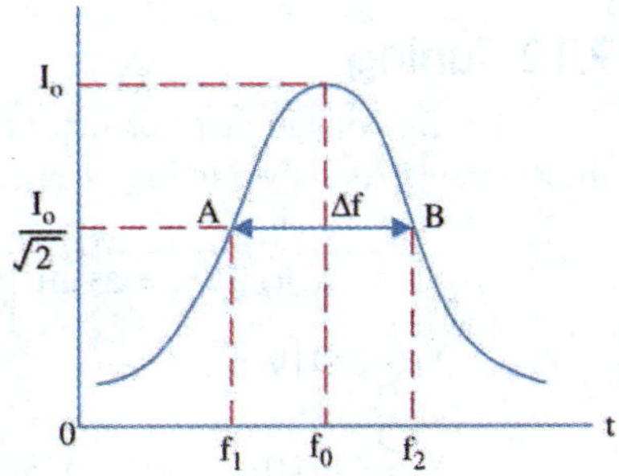

Fig. 9.9

It can also be proved that *edge* frequencies are given by

$$f_1 = f_0 - \frac{\Delta f}{2} = f_0 - \frac{BW}{2} = f_0 - \frac{R}{4\pi L}$$

Similarly, $$f_2 = f_0 + \frac{\Delta f}{2} = f_0 + \frac{BW}{2} = f_0 + \frac{R}{4\pi L}$$

It may be noted that points *A* and *B* in Fig. 9.9 are also called *half-power* points because power developed at each of these points is half the power developed at resonance. Hence, edge frequencies f_1 and f_2 are known as lower half-power frequency and upper half-power frequency respectively. All the frequencies lying between f_1 and f_2 are called *useful* frequencies because they produce currents which when passed through headphones produce a sound that is not much weaker than that produced by maximum current I_m at resonance. Hence, bandwidth of a resonant circuit, in fact, represents the range of its useful frequencies.

9.11. Sharpness of Resonance

It is defined as the ratio of the bandwidth of the circuit to its resonance frequency.

$$\text{Sharpness of resonance} = \frac{f_2 - f_1}{f_0} = \frac{\Delta f}{f_0} = \frac{\text{BW}}{f_0} = \frac{1}{Q_0}$$

It shows that as value of Q_0 increases, bandwidth decreases. As bandwidth decreases, selectivity of the circuit increases.

Fig. 9.10 (*a*) shows the resonance curve of a circuit having comparatively large value of *R*. The passband of the circuit is from 980 kHz to 1020 kHz. Its bandwidth is 40 kHz so that its sharpness of resonance is 40/1000 = 0.04. Similar value for the circuit of Fig. 9.10 (*b*) is 20/1000 = 0.02. This circuit has a selectivity of 20 kHz as compared to 40 kHz of the first circuit. The circuit shown in Fig. 9.10 (*a*) has poorer selectivity because *it does not reject frequencies close to its resonance frequency.*

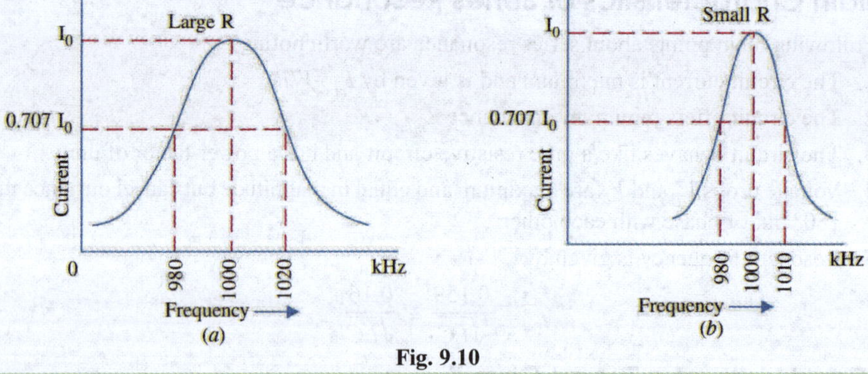

Fig. 9.10

9.12. Tuning

It means obtaining resonance at different frequencies by changing either *L* or (more often) *C* of an *LC* circuit. As shown in Fig. 9.11 (*a*), the variable capacitor *C* can be adjusted to tune the *R-L-C*

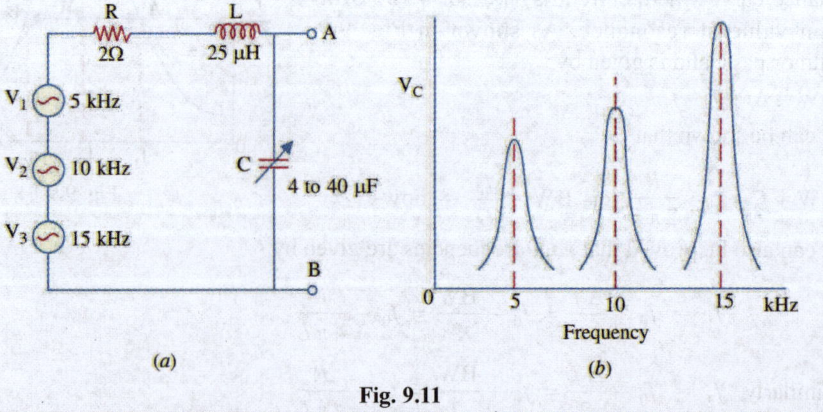

Fig. 9.11

circuit to resonance at any one of the three different frequencies produced by the three signal generators V_1, V_2 and V_3. Which of these frequencies will be selected for maximum output depends upon the resonance frequency of the LC circuit ? When C is adjusted to 40 μF, its resonance frequency becomes

$$f_0 = \frac{1}{2\pi\sqrt{25 \times 10^{-6} \times 40 \times 10^{-6}}} = 5000 \text{ Hz} = 5 \text{ kHz}$$

Hence, out of the three input voltage, only V_1 having a frequency of 5 kHz is selected to produce a resonant rise of current which results in maximum output voltage across C. Of course, some current from V_2 and V_3 will also be produced but it would insignificant.

Suppose, we now adjust C = 10 μF, then circuit resonant frequency becomes

$$f_0 = \frac{0.159}{\sqrt{25 \times 10^{-6} \times 10 \times 10^{-6}}} = \mathbf{10 \text{ kHz}}$$

Now, the circuit will offer minimum impedance to V_2 so that input signal of 10 kHz will produce maximum current and hence maximum output voltage V_C. Other signals being too off-resonance will be practically suppressed. In a similar way, by adjusting value of C to 4.44 μF, V_3 can be tuned in. Though we have considered three frequencies only, the LC circuit can be tuned to select any desired frequency.

Tuning in radio, radar and TV receivers are all examples of resonance phenomenon. Instruments like wavemeter etc., utilize the tuning properties of an LC circuit. In fact, resonant circuits are widely used in both radio transmitting and radio receiving circuits.

An inductor plays the same role in an L-C circuit that a spring like this does when connected to a mass. It provides a restoring force that always tries to return the system to equilibrium.

9.13. Tuning Ratio

For a given value of L, the resonant frequency is inversely proportional to the square root of C as seen from the relation

$$f_0 = \frac{1}{2\pi\sqrt{LC}} \quad \text{or} \quad f_0 \propto \frac{1}{\sqrt{C}}$$

$$\therefore \quad \frac{f_0}{f_0'} = \sqrt{\frac{C'}{C}} \quad \text{or} \quad \frac{f_0'}{f_0} = \sqrt{\frac{C}{C'}}$$

It means that if C is reduced to one-fourth of its previous value, frequency becomes double. In the example considered in Art. 9.12, when C changes from 40 μF to 10 μF, resonant frequency *increases* from 5 kHz to 10 kHz *i.e.*, it is doubled. Suppose, we want to tune through the whole frequency range of 5 to 15 kHz which represents a tuning ratio of 3 : 1 from the highest to the lowest frequency. Then capacitance must be varied from 40 μF to 4.44 μF—a 9 : 1 capacitance ratio.

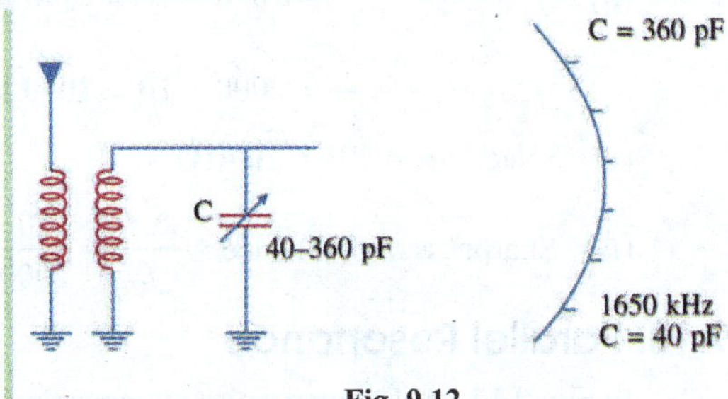

Fig. 9.12

9.14. Radio Tuning Dial

In Fig. 9.12 is shown how resonant circuits can be used in tuning a radio receiver to the carrier frequency of a broadcasting station in the medium band (550 to 1650 kHz). The tuning is done with the help of a variable capacitor *C* whose capacitance can be varied from 360 μF when plates are completely in mesh to 40 μF when they are out of mesh.

It will be seen that lowest frequency is tuned with highest capacitance and *vice-versa*. Since frequency tuning ratio is 3 : 1, the capacitance ratio is 9 : 1.

Example 9.4. *An RLC circuit consists of a capacitor of reactance 120 Ω and a coil having a resistance of 60 Ω and inductive reactance of 180 Ω. The combination is connected across a 200 V, 50 Hz source. Compute (i) current (ii) p.f. and (iii) power taken by the circuit.*

Solution. $X_L = 180\ \Omega, \quad X_C = -120\ \Omega$

net reactance $X = 180 - 120 = 60\ \Omega$ (inductive)

$$Z = \sqrt{R^2 + X^2} = \sqrt{60^2 + 60^2} = 84.8\ \Omega$$

(*i*) $I = V/Z = 200/84.8 = \mathbf{2.36\ A}$

(*ii*) $\cos\phi = R/Z = 60/84.8 = \mathbf{0.707\ (lag)} \quad \because X_L > X_C$

(*iii*) $P = VI\cos\phi = 200 \times 2.36 \times 0.707 = \mathbf{333\ kW}$

or $P = I^2R = 2.36^2 \times 60 = \mathbf{333\ kW}$

Example 9.5. *A circuit consists of a capacitor of 100 pF connected in series with a coil of resistance 5 Ω and inductance 100 μH. Calculate (i) resonance frequency (ii) Q-factor (iii) bandwidth.*

Solution. (*i*)
$$f_0 = \frac{1}{2\pi\sqrt{LC}} = \frac{0.159}{\sqrt{LC}} = \frac{0.159}{\sqrt{100 \times 10^{-6} \times 100 \times 10^{-12}}}$$
$$= \frac{0.159}{10^{-7}} = 1.59 \times 10^6 = \mathbf{1.59\ MHz}$$

(*ii*)
$$Q_0 = \frac{2\pi f_0 L}{R} = \frac{2\pi \times 1.59 \times 10^6 \times 100 \times 10^{-6}}{5} = \mathbf{200}$$

(*iii*) bandwidth,
$$\Delta f = \frac{f_0}{Q_0} = \frac{1.59 \times 10^6}{200} = \mathbf{7.95\ kHz}$$

Incidentally, $f_1 = 1590 - 7.95 = 1582.05$ kHz

$f_2 = 1590 + 7.95 = 1597.95$ kHz

Example 9.6. *An R L C resonant circuit has a resonant frequency of 2 MHz and a Q-factor of 100. Calculate (i) the bandwidth of the circuit (ii) lower and upper half-power frequencies (iii) selectivity and (iv) sharpness of resonance.*

Solution. (*i*)
$$BW = \Delta f = \frac{f_0}{Q_0} = \frac{2000\ \text{kHz}}{100} = \mathbf{20\ kHz}$$

(*ii*)
$$f_1 = f_0 - \frac{\Delta f}{2} = 2000 - 10 = \mathbf{1990\ kHz}$$
$$f_2 = f_0 + \frac{\Delta f}{2} = 2000 + 10 = \mathbf{2010\ kHz}$$

(*iii*) Selectivity = BW = **20 kHz**

(*iv*) Sharpness of resonance
$$= \frac{\Delta f}{f_0} = \frac{20}{2000} = \mathbf{\frac{1}{100}}$$

9.15. Parallel Resonance

In Fig. 9.13 (*a*) is shown a circuit consisting of a capacitor in parallel with a coil of negligible small resistance. When fed from an a.c. voltage source, the capacitor draws a leading current whereas coil draws a lagging current.* This circuit resonates to a frequency which makes $X_L = X_C$ so that the two branch currents are equal but opposite. Hence, they cancel out with the result that current drawn from the supply is zero. In practice, however, line current drawn is not zero but has minimum value due to small resistance R of the coil.

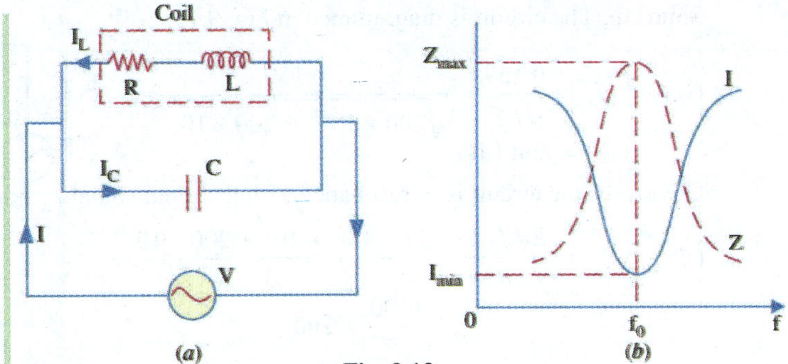

Fig. 9.13

Since current drawn by the circuit is minimum, it means it offers maximum impedance to the applied voltage under resonance condition.

If R is neglected, then

$$f_0 = \frac{1}{2\pi\sqrt{LC}} \qquad \text{— as for series circuit}$$

$$I_{min} = \frac{V}{L/CR} \qquad \text{and} \qquad Z_{max} = L/CR$$

As seen from Fig. 9.13 (*b*), current is minimum at f_0 but increases for off-resonance frequencies due to decrease in the circuit impedance.

Main application of a parallel resonant circuit is to act as a load in the output circuit of an RF amplifier. Since impedance is maximum at f_0, amplifier gain is also maximum at f_0. The advantages of using such a circuit as load are that

(*i*) it offers maximum impedance only to the a.c. signal which is required to be amplified,

(*ii*) since coil resistance is almost negligible, there is practically no d.c. voltage drop across it.

The Q-factor of the parallel circuit is essentially the same as that of the coil and is given by

$$Q = \frac{2\pi f_0 L}{R}$$

when R is coil resistance.

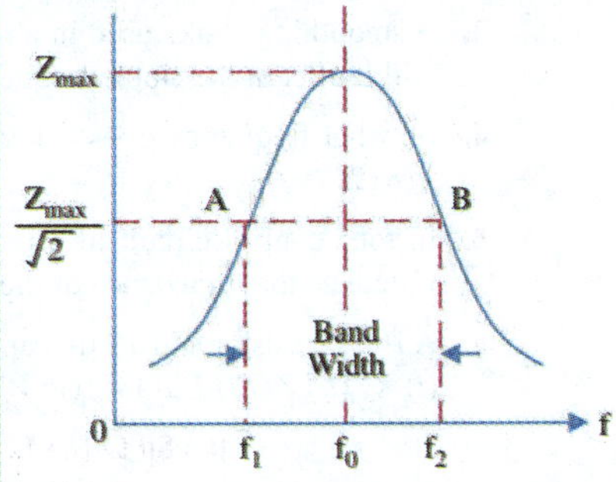

Fig. 9.14

In the case of a resonant parallel circuit, the bandwidth is defined by the two points on either side of the resonant frequency where value of impedance drops to 0.707 or $1/\sqrt{2}$ of its maximum value at resonance (Fig. 9.14).

$$BW = \Delta f = f_2 - f_1 = \frac{f_0}{Q_0}$$

Example 9.7. *A parallel circuit consisting of a 200 pF capacitor and a coil of inductance 200 μH and resistance 5 Ω is connected across a 0.2 V, 800 kHz signal source. Determine for the circuit (i) resonant frequency (ii) Q-value (iii) impedance offered (iv) current drawn and (v) bandwidth of the circuit.*

* The two branch currents are shown flowing in opposite directions in Fig. 9.13 (*a*). It is so because when coil draws in the current, the capacitor discharges out at the same time and *vice-versa*.

Solution. The circuit is diagrammed in Fig. 9.15.

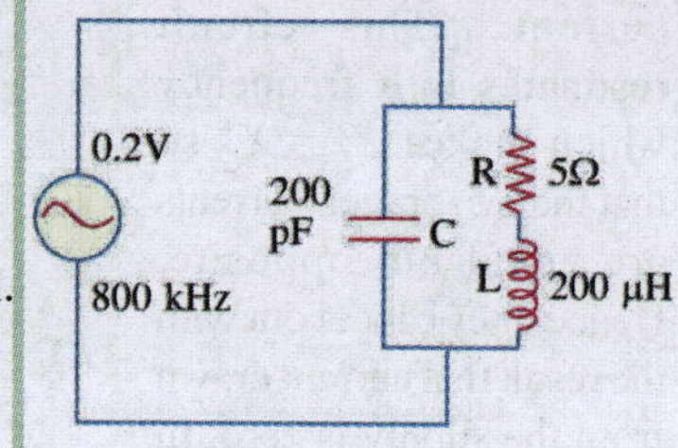

Fig. 9.15

(*i*) $$f_0 = \frac{0.159}{\sqrt{LC}} = \frac{0.159}{\sqrt{200 \times 10^{-6} \times 200 \times 10^{-12}}}$$
$$= \mathbf{800\ kHz}$$

Obviously, the circuit is in resonance with the input signal.

(*ii*) $$Q_0 = \frac{2\pi f_0 L}{R} = \frac{2\pi \times 800 \times 10^3 \times 200 \times 10^{-6}}{5}$$
$$= \frac{1000}{5} = \mathbf{200}$$

(*iii*) $$Z_{max} = \frac{L}{CR} = \frac{200 \times 10^{-6}}{200 \times 10^{-12} \times 5} = \mathbf{200{,}000\ \Omega}$$

(*iv*) $$I_{min} = \frac{V}{Z_{max}} = \frac{0.2}{200{,}000} = \mathbf{1\ \mu A}$$

(*v*) $$BW = \frac{f_0}{Q_0} = \frac{800}{200} = \mathbf{4\ kHz}$$

CONVENTIONAL PROBLEMS

1. A 30 Ω resistor and a pure 40 Ω inductor are connected in series across a 200 V supply. Determine (*i*) Z (*ii*) I (*iii*) V_R (*iv*) V_L (*v*) phase angle ϕ (*vi*) power absorbed.
[(*i*) 50 Ω (*ii*) 4 A (*iii*) 120 V (*iv*) 160 V (*v*) 53° (*vi*) 480 W]
2. An *RF* coil has an inductive reactance of 600 Ω at a certain frequency and an a.c. resistance of 10 Ω. Calculate its *Q*-factor. **[60]**
3. A smoothing choke used in a radio receiver circuit has $L = 20$ H and $R = 300\ \Omega$. What impedance will it offer at the ripple frequency of 100 Hz ? **[12,560 Ω]**
4. At what frequency does a 1 kΩ resistor in series with a 2 H choke coil offer an impedance of 1181Ω ? **[50 Hz]**
5. In tone control circuit of a receiver, a 1 kΩ resistor is joined in series with a 0.1 μF capacitor. Calculate the impedance of the combination at 2 kHz and 10 kHz. **[1261 Ω, 1012 Ω]**
6. A 100 Ω resistor and a 2 μF capacitor are connected in series across a 16 V, 1 kHz source. Determine (*i*) X_C (*ii*) Z (*iii*) I (*iv*) V_R (*v*) V_C (*vi*) p.f. and (*vii*) power absorbed.
[(*i*) 80 Ω (*ii*) 128 Ω (*iii*) 0.125 A (*iv*) 12.5 V (*v*) 10 V (*vi*) 0.78 (lead) (*vii*) 1.56 W]
7. A series *R-L-C* circuit has $R = 5\ \Omega$, $L = 200\ \mu H$ and $C = 0.4\ \mu F$. If it is connected across an a.c. voltage source of 10 V at $\omega = 10^5$ radian/second, determine (*i*) current (*ii*) p.f. and (*iii*) power absorbed.
[(*i*) 1.414 A (*ii*) 0.707 (lead) (*iii*) 10 W]
8. What is the resonance frequency of a series *L-C* circuit if $L = 200\ \mu H$ and $C = 200$ pF ?
[800 kHz]
9. A series *R-L-C* circuit consists of a capacitor of capacitance 200 μF and a coil of resistance 10 Ω and inductance 200 μF. Calculate
(*i*) resonance frequency (*ii*) *Q*-factor
(*iii*) bandwidth (*iv*) lower and upper half-power frequencies.
[(*i*) 800 kHz (*ii*) 100 (*iii*) 8 kHz (*iv*) 792 kHz, 808 kHz]

SELF EXAMINATION QUESTIONS

A. Fill in the blanks by the most appropriate word(s) or numerical value (s).

1. In a series *R-L* circuit, voltage .... the current.
2. Impedance of an *R-L* circuit is given by the .... sum of resistance and reactance.
3. Skin effect increases the resistance of a conductor at .... frequencies.
4. Skin effect may be minimised by using .... conductor.
5. If in an *R-L* circuit, $V_R = 30$ V and $V_L = 40$ V, then applied voltage must be .... V.
6. A pure inductor or capacitor dissipates ....power.
7. Power factor is given by the ratio of circuit resistance and ....
8. The power factor of an *R-L* circuit lies between .... and ....
9. A series circuit becomes resonant when algebraic sum of X_L and X_C equals ....
10. Resonance curve shows variation of circuit current with ....
11. Higher the *Q*-factor of a circuit, ... its bandwidth.
12. Lower the resistance of a resonant circuit, ... its selectivity.
13. At half-power frequencies, the current in a series *R-L-C* resonant circuit is .... times the maximum value of current.

B. Answer True or False

1. The impedance of a series *R-L* circuit is given by the algebraic sum of *R* and X_L.
2. The power factor of an a.c. circuit can lie between – 1 and + 1.
3. Skin effect at high frequencies can be neutralized by using conductors made of hollow tubes.
4. Out of the input signals having different frequencies, a series resonant circuit allows that one to pass through whose frequency equals its own lies close to it.
5. A parallel resonant circuit rejects the signal having same frequency as its own resonance frequency.
6. For sharper tuning, series resistance should be decreased.
7. Bandwidth of a circuit can be increased by decreasing inductance but increasing resistance.
8. Greater the bandwidth, higher the selectivity.

C. Multiple Choice Items

1. The power in an a.c. circuit is given by
   (*a*) $VI \cos \phi$ (b) $VI \sin \phi$
   (*c*) I^2Z (*d*) I^2X_L
2. In a series *R-L-C* circuit, $R = 100\ \Omega$, $X_L = 300\ \Omega$, and $X_C = 200\ \Omega$. The phase angle ϕ of the circuit is .... degrees.
   (*a*) 0 (*b*) 90
   (*c*) 45 (*d*) 60
3. In a series circuit with $R = 10\ \Omega$, $X_L = 25\ \Omega$ and $X_C = 35\ \Omega$ and carrying effective current of 5 A, the power dissipated is .... watt.
   (*a*) $250\sqrt{2}$ (*b*) 50
   (*c*) 100 (*d*) 250
4. A resonance curve for a series circuit is a plot of frequency versus ....
   (*a*) voltage (*b*) impedance
   (*c*) current (*d*) reactance
5. The power factor of a resonant series circuit is
   (*a*) 1 (*b*) 0
   (*c*) – 1 (*d*) 0.5
6. At half-power points of a resonance curve, the current is .... times the maximum current.
   (*a*) 2 (*b*) $\sqrt{2}$
   (*c*) $1/\sqrt{2}$ (*d*) $1/2$
7. Higher the *Q* of a series circuit,
   (*a*) greater its bandwidth
   (*b*) sharper its resonance curve
   (*c*) broader its resonance curve
   (*d*) narrower its passband
8. The selectivity of a series circuit can be increased by
   (*a*) reducing its resonance frequency
   (*b*) increasing its *Q* -value
   (*c*) increasing its resistance
   (*d*) increasing its bandwidth
9. The resonance frequency of a series resonant circuit is given by
   (*a*) $f_0 = \sqrt{CR}$
   (*b*) $f_0 = 2\pi\sqrt{LC}$
   (*c*) $f_0 = 12\pi\sqrt{LC}$
   (*d*) $f_0 = 2\pi/\sqrt{LC}$

10. Out of the many input signals of different frequencies, a series resonant circuit will accept one which has

(*a*) the highest frequency

(*b*) lowest frequency

(*c*) frequency farthest from its resonance frequency

(*d*) frequency closest to its resonance frequency.

ANSWERS

A. Fill in the blanks

1. leads **2.** vector **3.** high **4.** Litz **5.** 50
6. no **7.** impedance **8.** 0; 1 **9.** zero **10.** frequency
11. narrower **12.** better **13.** 0.707

B. True or False

1. F **2.** F **3.** F **4.** T **5.** T **6.** T **7.** T **8.** F

C. Multiple Choice Items

1. *a* **2.** *c* **3.** *d* **4.** *c* **5.** *a* **6.** *c* **7.** *d* **8.** *b* **9.** *c* **10.** *d*

10 CHAPTER

Time Constant

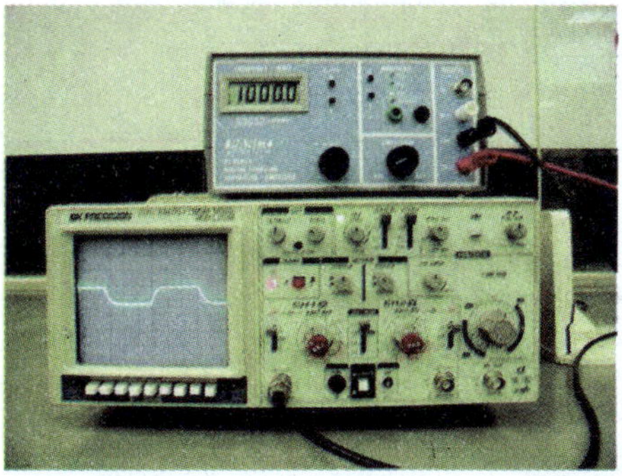

1. Rise and Fall of Current in Pure Resistance
2. Time Constant of an R-L Circuit
3. Circuit Conditions
4. Inductive Kick
5. Time Constant of an RC Circuit
6. Charging and Discharging of a Capacitor
7. Decreasing Time Constant
8. Flasher
9. Pulse Response of an RC Circuit
10. Effect of Long and Short Time Constants
11. Square Voltage Wave Applied to Short λ RC Circuit
12. Square Voltage Wave Applied to Long λ RC Circuit

10.1. Rise and Fall of Current in Pure Resistance

A pure resistor is defined as one which possesses ohmic resistance only and has neither inductance nor capacitance*. Such a resistor of $R = 6$ ohm is shown connected across a 12 V battery via a switch S in Fig. 10.1 (a). When S is closed, a current of 2 A (as given by Ohm's law) is set up in the circuit *instantly i.e.*, current changes from 0 to 2 A in no time as shown in Fig. 10.1 (b). Similarly, it is found that if S is opened, current drops to zero *instantly.* It is obvious from the above that a pure resistor offers only opposition to the flow of current but *no reaction to its change.* The reason for this is that (unlike an inductor) R has no concentrated

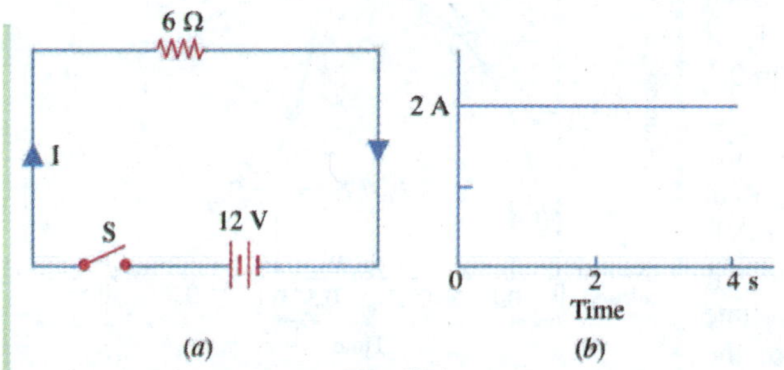

Fig. 10.1

* Actually, it is impossible to get such an ideal resistor.

magnetic field to oppose any changes in current and also (unlike a capacitor), it does not store any *charge* which opposes any change in *voltage.*

10.2. Time Constant of an R-L Circuit

Consider the circuit of Fig. 10.2 where L is in series with R (which may be either the coil resistance, external resistance or both). When switch S is closed, current starts increasing exponentially from 0 to its final steady value of 1 A as given by Ohm's Law [Fig. 10.2 (*b*)]. However, this Ohm's Law value of the current is not achieved *instantly* but after lapse of some time (theoretically, infinite time) determined by the values of R and L. The reason is that during the exponential build-up of the current from 0 to 1 A, every change of current is opposed by the induced emf in the coil. This transient response of the circuit lasts till steady-state current of 1A is reached after some time.

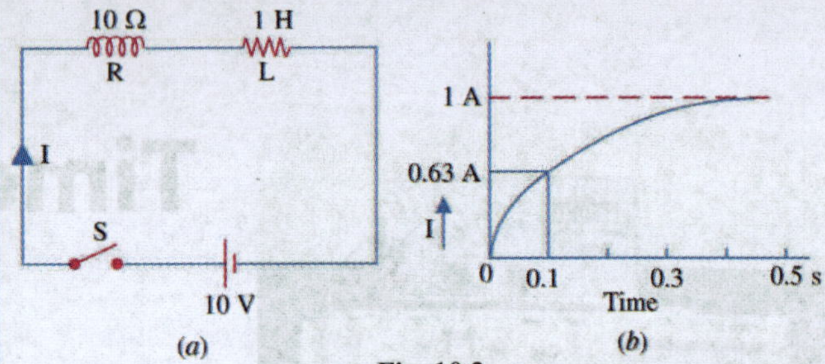

Fig. 10.2

Similarly, when S is opened, current does not drop to zero instantly but takes certain definite amount of time. This, again, is due to the fact that every change in current is opposed by the induced or counter-emf, thereby delaying its decay to ultimate value of zero.

This delay in both rise and decay of current in an inductive circuit is dependent on its time constant (λ) which is given by the ratio L/R.

Time constant, $$\lambda = \frac{L}{R} \text{ second}$$

Here, L is in henrys and R in ohms. With reference to Fig. 10.2 (*b*), it is seen that

$$\lambda = 1/10 = 0.1 \text{ second}$$

Time period (λ) may be defined as the time taken by the current to rise to 63.2% (or approximately, 63%) of its final (or maximum or steady) value.

The maximum (or Ohm's Law) value of the circuit current is 1 A. Hence, when S is closed, current will rise to a value of 0.63 A in 0.1 second.

Time constant of an R-L circuit may also be defined as the time during which current falls to 37% or 0.37% or 0.37 of its maximum value while decaying.

In the present case, it means that when S in Fig. 10.2 (*a*) is opened, current will fall from its initial value of 1 A to 0.37 A in 0.1 second. These facts have been shown in Fig. 10.3. However, it is generally accepted that the circuit current, while rising, achieves its final maximum value after a period equal to FIVE time constants *i.e.*, = 5λ seconds. Similarly, current practically decays to zero after a time interval equal to five time-constants of the circuit.

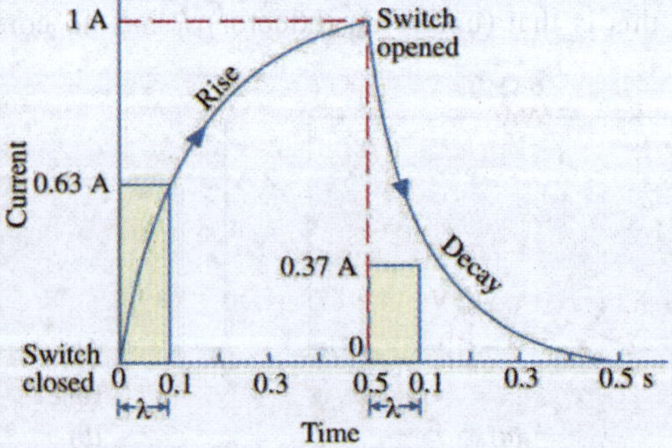

Fig. 10.3

It may be noted that λ becomes longer with larger values of L but shorter with larger values of R.

10.3. Circuit Conditions

In Fig. 10.4, the *R-L* combination becomes connected to battery of *V* volts when switch *S* is in position 1 and is short-circuited when *S* is in position 2.

When *S* is in position 1, following circuit conditions exist :

1. At the initial instant of switching on, circuit current is zero *i.e.*, $i = 0$. Hence, $v_R = 0$.

2. Initially, the rate of increase of current is maximum *i.e.*, di/dt is maximum ($= V/L$ amperes second) when $i = 0$. Hence, v_L has maximum value almost equal to V.

3. Thereafter, current rises in the circuit at progressively diminishing rate of increase. Hence, v_R increases but v_L correspondingly decreases because at all times

$$V = v_R + v_L$$

4. When after sometime ($\cong 5\ \lambda$), current reaches its maximum steady value, $v_L = 0$ and hence v_R becomes equal to V.

5. Under the steady or equilibrium conditions, the circuit appears as only a resistor. The effect of inductance has disappeared.

6. However, maximum magnetic flux exists in L and so maximum energy ($=1/2\ LI^2$) is stored in the inductor.

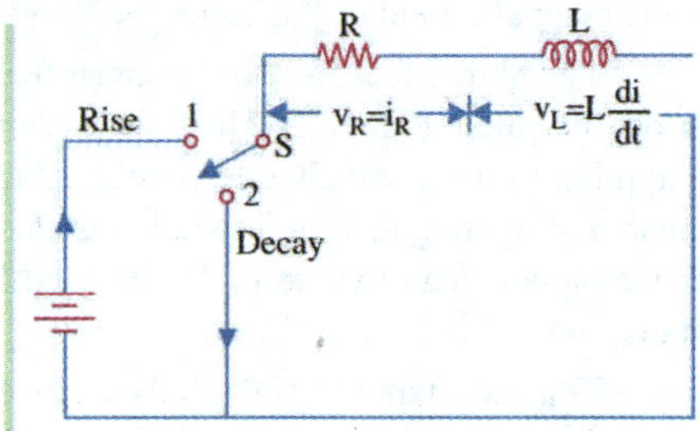

Fig. 10.4

Now, when *S* is shifted to position 2, the circuit is short-circuited and following conditions exist :

1. Initial rate of current decay is most rapid ($= V/L$ ampere/second). Hence, maximum voltage of opposite polarity is induced in *L* which would try to prolong the current decay.

2. As current decays at a progressively diminishing rate, v_R also decays to zero.

3. Voltage v_L will also decay towards zero though the rate of current decay would become progressively less.

The mathematical expression for current rise through such an inductive circuit is

$$i = I_m\,(1 - e^{-t/\lambda})$$

Similar expression for current decay is

$$i = I_m\, e^{-t/\lambda}$$

Example 10.1. *A coil having R = 120 Ω and L = 24 H is connected across a 12 V battery. Find*

(*i*) *time constant of the circuit* (*ii*) *current after 0.2 second*

(*iii*) *current after 1 second* (*iv*) *current after 0.4 second*

Solution. (*i*) $\lambda = \dfrac{L}{R} = \dfrac{24}{120} = $ **0.2 second**

(*ii*) Now, the given time of 0.2 second represents the time constant of the circuit. Hence, circuit current would be $0.63\ I_{max}$.

Now, $I_{max} = 12/120 = 0.1\text{ A} = 100\text{ mA}$

∴ Current after 0.2 s = 0.63 × 100 = **63 mA**

(*iii*) Now, the given time of 1 s = 5 λ. Hence, during this time, current would rise to its maximum value of **100 mA.**

(*iv*) The given time of 0.4 s is twice the time constant of the circuit. In the first time constant *i.e.*, first 0.2 s, current would become 63 mA. The balance of current left would be = (100 – 63) =

37 mA. In the next 0.2 s, the current will increase by 63% of 37 mA *i.e.*, by 0.63 × 37 = 23.3 mA. Hence, current after 0.4 second or after 2 time constants would be

$$= 63 + 23.3 = \mathbf{86.3\ mA}$$

10.4. Inductive Kick

When a current-carrying *R-L* circuit is opened, the time constant *L*/*R* for current decay becomes very *short because of the extremely high resistance of the open circuit.* Consequently rate of fall of current is much faster than its rate of rise when switch is closed. As a result, very high voltage (much greater than the applied voltage) is induced in the coil on opening the circuit. However, duration of this high voltage peak is proportionately smaller because there is no gain in the energy stored in the magnetic field of the coil.

The above fact can be demonstrated with the help of a neon glow bulb *N* shown connected across the coil in Fig. 10.5. The bulb presents a very high resistance until a firing voltage of about 90 is applied to its terminals after which it starts glowing. Obviously, the 10 V source by itself is not capable of lighting it. When switch *S* is closed, current rises to its maximum steady value of 10/100 = 0.1 A in 5 time constants.

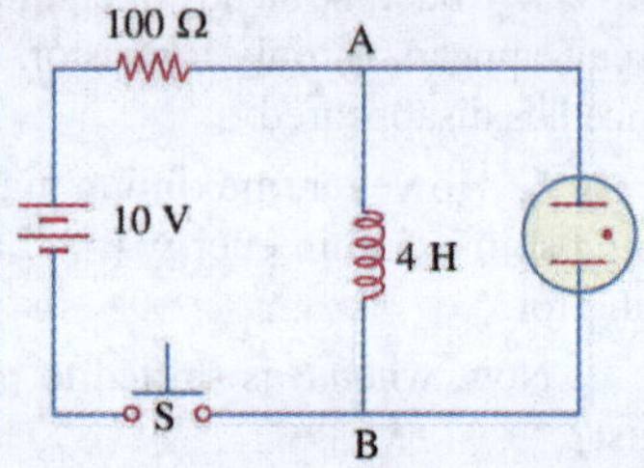

Fig. 10.5

Time constant = 4/100 = 0.04 s. Hence, current achieves Ohm's Law value in 5 × 0.04 = 0.2 second. At that time, v_R = 10 V, v_L = 0 and I_m = 0.1 A.

Now, when *S* is opened, battery is cut off but the self-induced emf in the coil tends to maintain the circuit current of 0.1 A. However, due to high resistance of the neon bulb (about 50 K), time constant becomes very small and, consequently, rate of fall of current becomes exceedingly rapid.

Now, $$\lambda = \frac{L}{R} = \frac{4}{50 \times 10^3} = 8\times10^{-5} \text{ second}$$

During this time, current will fall by 63% of its initial maximum value of 0.1 A *i.e.*, it will decrease by 0.1 × 0.63 = 0.063 A. Hence, *di* = 0.063 A and *dt* = 8 × 10^{-5} second.

$$\therefore \quad e_L = L\frac{di}{dt} = 4 \times \frac{0.063}{8 \times 10^{-5}} = 3150 \text{ V}$$

This voltage, is high enough to flash the neon bulb.

In general, whenever an inductive circuit is interrupted or opened abruptly, the high induced voltage causes

1. arcing and burning of switch contacts,
2. large heat dissipation which is likely to break down coil insulation,
3. possible danger to the persons handling the equipment.

Hence, sufficient care must be taken while opening an inductive circuit because of the 'inductive kick' *i.e.*, high induced voltage. However, this inductive kick has been usefully employed in the ignition system of an automobile. In this system, a battery circuit in series with a highly-inductive spark coil is opened by the breaker points of the distributor to produce high voltage (15 to 20 kV) to fire the spark plugs. Another application of this 'inductive kick' is the production of high voltage of 10 to 25 kV for the anode of the picture tube in a TV receiver.

10.5. Time Constant of an RC Circuit

In this case, the time constant indicates the rate of charge or discharge of the capacitor. It is given by

$$\lambda = RC \text{ seconds}$$

where R is in ohms and C in farads. If R is in MΩ and C in μF, then, again, RC is in seconds.

It is seen from above that time constant depends both on R and C. More C means that the capacitor can store more charge and hence takes longer time to get this charge. Similarly, more resistance means reduced rate of charging thus taking longer to charge the capacitor.

It may also be noted that RC merely specifies the *rate* of charge. The actual voltage across the capacitor depends both on the applied voltage and the RC time.

In this circuit also, time constant may be defined in the following two ways :

1. It is the time during which the charging current falls to 37% of initial maximum value [Fig. 10.7 (*a*)].

or

2. It is the time during which capacitor voltage rises to 63% of its final steady value [Fig. 10.7 (*b*)].

10.6. Charging and Discharging of a Capacitor

In Fig. 10.6, the *R-C* combination becomes connected to the battery of V volts when switch S is in position 1 but is short-circuited when S is shifted to position 2.

When S is in position 1, following circuit conditions exist :

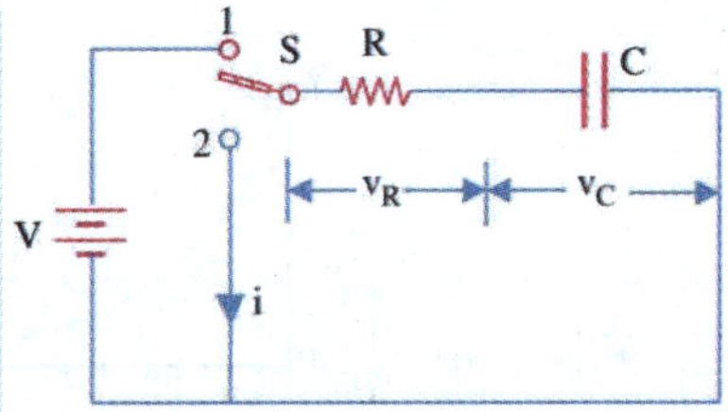

Fig. 10.6

1. At the instant of switching on $V_C = 0$ but charging current is maximum given by $I_m = V/R$ as shown in Fig. 10.7 though its rate of decrease is also maximum.

2. Hence, at the start, v_R has maximum value equal to $I_m R = 1$ V. At any other time after switching on

$$V = v_R + v_C = i_R + v_C = i_C . r + v_C$$

$\because i = i_C$

3. As charge starts collecting on C, v_C start decreasing at a progressively diminishing rate as shown in Fig. 10.7.

4. After about five time constants *i.e.*, after 5 RC seconds, charging current becomes zero and hence $v_R = iR$ becomes zero. But v_C achieves maximum value equal to V.

5. Under these steady conditions, *the circuit appears as only a capacitor.*

6. Since capacitor is fully charged to a potential difference of V volts, it has maximum energy stored in it

$$= \tfrac{1}{2} CV^2$$

7. The exponential build-up of voltage across C is given by the equation

$$v_C = V(1 - e^{-t/\lambda})$$

Similarly, the gradual tapering of charging current is given by the relation

$$i = I_m e^{-t/\lambda}$$

Now, when S is shifted to position 2, battery is cut out of the circuit but capacitor is short-circuited through R and, hence, starts discharging.

1. Initially, the discharging current is maximum $= -v/R$ because full capacitor voltage $v_C = V$ is applied across R. Moreover, its rate of discharge is maximum initially but keeps decreasing thereafter. The negative sign of current indicates that it is flowing in a direction opposite to that of the charging current.

2. Hence, $v_R = -I_m R$ but gradually tends towards zero as does v_C.

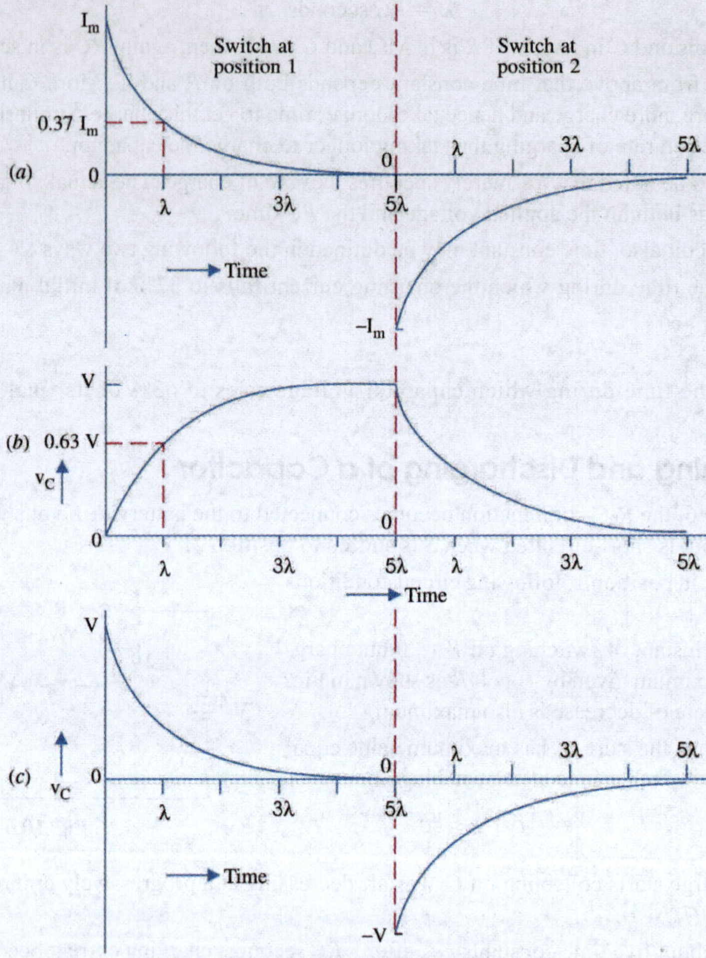

Fig. 10.7

3. After lapse of time equal to about five time-constants, v_C becomes zero and so does the discharge current. The entire circuit becomes passive by that time.

The progressive decrease of capacitor voltage is given by the equation

$$v_C = V - e^{-t/\lambda}$$

Similarly, the exponential decrease of discharge current is given by the relation

$$i = -I_m e^{-t/\lambda}$$

Example 10.2. *A series combination having $R = 2$ MΩ and $C = 0.02$ μF is connected across a dc voltage source of 100 V. Determine*

(*i*) *time constant of the circuit*

(*ii*) *capacitor voltage after 0.02 s, 0.04 s, 0.1 s and 2 hours.*

(*iii*) *charging current after 0.02 s, 0.04 s and 0.1 s.*

Solution. (*i*) $\lambda = RC = 2 \times 0.01 =$ **0.02 s**

(*ii*) (*a*) ***t* = 0.02 second**

Since this time happens to be equal to the time constant of the circuit,

$\therefore \quad v_C = 0.63\ V = 0.63 \times 100 =$ **63 V**

(*b*) $t = 0.04$ **second**

This time equals two time constants. In the first time constant, $v_C = 63$ V. In the second time constant, v_C increases by 0.63 × 37 = 23.3 V. Hence, after 0.04 s

$$v_C = 63 + 23.3 = \mathbf{86.3\ V}$$

(*c*) $t = 0.1$ **second**

This time equals 5 time constants by which time v_C becomes practically equal to the applied voltage.

$$\therefore \qquad v_C = \mathbf{100\ V}$$

(*d*) Provided circuit remains connected to the voltage source, v_C would remain constant at 100V for any length of time after 5λ.

(*iii*) (*a*) t **= 0.02 s;** $i_C = 0.37$ of I_m

Now, $I_m = 100/2\ \text{M}\Omega = 50\ \mu\text{A}$

$\therefore \qquad i_C = 0.37 \times 50 = \mathbf{18.5\ \mu A}$

(*b*) $t = 0.04$ **s**

In the first time constant of 0.02 s, the charging current becomes (as found above) 18.5 μA. In the second time constant, current further decreases by 63% of the balance of the current *i.e.*, by 0.63 × 18.5 = 11.6 μA

$\therefore\ i_C = 18.5 - 11.6 = \mathbf{6.9\ \mu A}$

(*c*) $t = 0.1$ **s**

After this time, i_C is almost zero since 0.1s represents five time constants.

10.7. Decreasing Time Constant

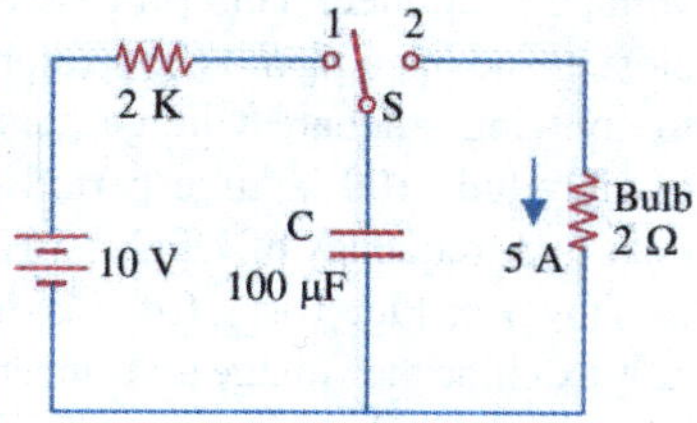

Fig. 10.8

The time constant of a given capacitor circuit can be increased by increasing *R* and can be decreased by decreasing *R*. Suppose we first charge a capacitor slowly with a small charging current through a high resistance and then discharge it through a low resistance. Since resistance during discharge would be small, we will get a momentary surge or pulse of discharge current.

The above method is employed in the operation of battery capacitor unit used for firing flash bulbs for photo cameras. Let us suppose that a flash bulb needs 5 A current to ignite. Such a heavy current would be too much of a load for a small 10 V battery (Fig. 10.8) whose normal current rating may be as low as 25 mA. By shifting switch *S* to position 1, the capacitor *C* is first charged from the battery through 2 K resistor. The charging time constant is = $RC = 2 \times 10^3 \times 100 \times 10^{-6} =$ 0.2 second. The peak charging current = 10/2 K = 5 mA which can be easily supplied by the battery. Hence, after a time lapse of 5 × 0.2 = 1 s, the capacitor is charged to 10 V. Next, *S* is shifted to position 2 for discharging the capacitor through the flash bulb. The peak discharge current = 10/2 = 5 A which is enough to fire the flash bulb.

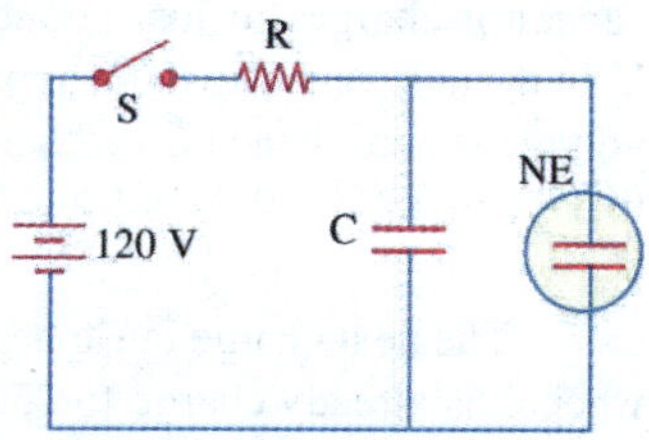

Fig. 10.9

10.8. Flasher

A practical demonstration of time constant of an *R-C* circuit can be given with the help of the circuit shown in Fig. 10.9. The neon glow lamp has a high resistance when not glowing. If a voltage of 90 V is applied, it will 'fire' *i.e.*, glow and also offer very low resistance while burning.

When switch S is closed, C starts to charge. When voltage across C reaches 90 V, neon lamp fires and starts to glow. Because of its very low resistance while glowing, C starts getting rapidly discharged through it. But soon after, the bulb goes out due to decrease in v_C and C starts to charge again. When v_C again reaches 90 V, bulb starts glowing once again but is soon extinguished when C discharges through it. In this way, neon bulb will glow each time the voltage across it reached 90 V and go out soon after due to decrease in voltage across it.

Obviously, it acts as a flasher or blinker although technically, it is called a 'relaxation oscillator'. By changing the circuit time constant, flashing rate may be controlled.

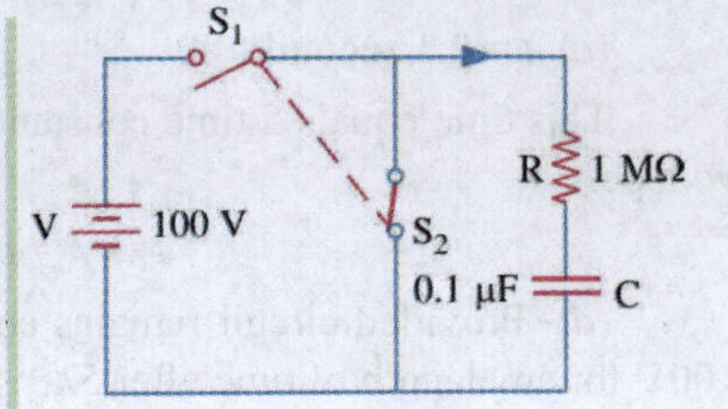

Fig. 10.10

10.9. Pulse Response of an RC Circuit

It is very instructive to investigate the response of an *RC* circuit when a square voltage pulse (of time period = 2 × circuit time constant) is applied across it. Fig. 10.10 shows a circuit in which S_1 is closed and opened at regular intervals of 0.1 second which equals the time constant of the *RC* circuit. When S_1 is closed for 0.1 s, 100 V is applied across the circuit. When S_1 is opened, then no voltage is applied. This process of closing and opening S_1 is equivalent to applying a square voltage pulse of amplitude 100 V, time period = 0.2 s and frequency of 1/0.2 = 5 Hz as shown in Fig. 10.11 (*a*). Let us now examine the voltage and current waveshapes in the *RC* circuit of Fig. 10.11. Moreover, it should be noted that when S_1 closes, S_2 gets opened up and *vice-versa*. With S_2 closed, C gets discharged through R.

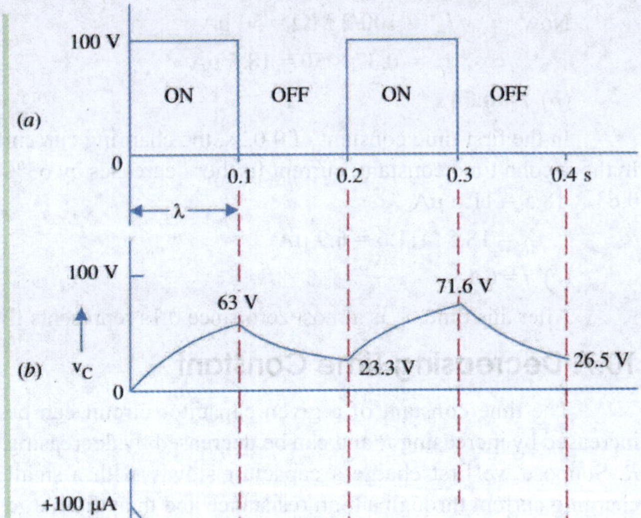

Fig. 10.11

(a) Capacitor Voltage (v_C)

As seen from Fig. 10.11 (*b*) during one time constant of 0.1 s, the capacitor charges to just about 63 V. In the next time constant, applied voltage is zero, hence C discharges to 37% of 63 V = 0.37 × 63 = 23.3 V.

The next charge cycle begins when C is already charge to 23.3 V. Consequently, the net charging voltage = 100 − 23.3 = 76.7 V. Obviously, during the next pulse, the capacitor, voltage increases by 63% of 76.7 V = 0.63 × 76.7 = 48.3 V. Hence, during second pulse, v_C becomes = 23.3 + 48.3 = 71.6 V. During the next 0.1 s, v_C drops to 37% of 71.6 V = 26.5 V and so on.

(b) Charge and Discharge Current (i_C)

As shown in Fig. 10.11 (c), capacitor charging current has its peak value at the start of charging when the full pulse voltage of 100 V is applied. Since just at the start $v_C = 0$ but $v_R = V$, maximum value of $i_C = V/R$. Similarly, at the start of discharge, i_C is again maximum because capacitor is fully charged to V volts. But this time, i_C flows in the opposite direction and its initial peak value is given by $-V/R$ because $v_C = V$. As seen, i_C has an ac waveform around the zero axis.

(c) Voltage Across R (v_R)

Since at every instant, $v_R = iR = i_C.R$, its waveshape is similar to that of the circuit current *i.e.*, i_C. The drop v_R has an ac waveshape as shown in Fig. 10.11 (d). Since waveshapes of i_C and v_R are similar, it is a general practice to connect an oscilloscope across R for studying i_C waveshape.

10.10. Effect of Long and Short Time Constants

It is easier to obtain different time constants in an RC circuit (by adjusting either R or C) than in an R-L circuit where it is difficult to change coil resistance and also impossible to eliminate capacitive effects between its different turns. Hence, RC circuits are commonly used for obtaining useful voltage and current waveshapes with required time constants.

(a) Long Time Constant

Whether an RC time constant is long or short depends on its relationship with the pulse width of the applied voltage. Any time constant which is at least five times longer than the pulse width is considered a long time constant. Suppose the pulse width of a given square voltage wave is 0.5 μs and the time constant of the RC circuit is 0.1 ms. Obviously, this time constant would be considered long because it is = 0.1 ms/0.5 μs = 200 times the pulse width. This time constant is too long for capacitor voltage v_C to rise appreciably before the applied voltage pulse drops to zero, thereby forcing C to discharge. Consequently, C takes on very little charge. Similarly, on discharges also, C discharges very little before the next pulse of the applied voltage comes on to make C charge again. Long time constants are usually employed for integrating circuits.

(b) Short Time Constant

A short RC time constant may be defined as any time constant which is no more than one-fifth the pulse width of the applied voltage. For example, a $\lambda = 0.1$ ms is considered a short time constant as compared to a pulse width of 0.02 s. As seen, λ is $0.1 \times 10^{-3}/0.02 = 1/200$ of the pulse duration. It means that voltage would remain applied across V for 200 time constants thereby allowing it to become fully charged. Similarly, when applied voltage drops to zero, C discharges completely and remains in this condition till the next voltage pulse comes on after a comparatively long time. On the next

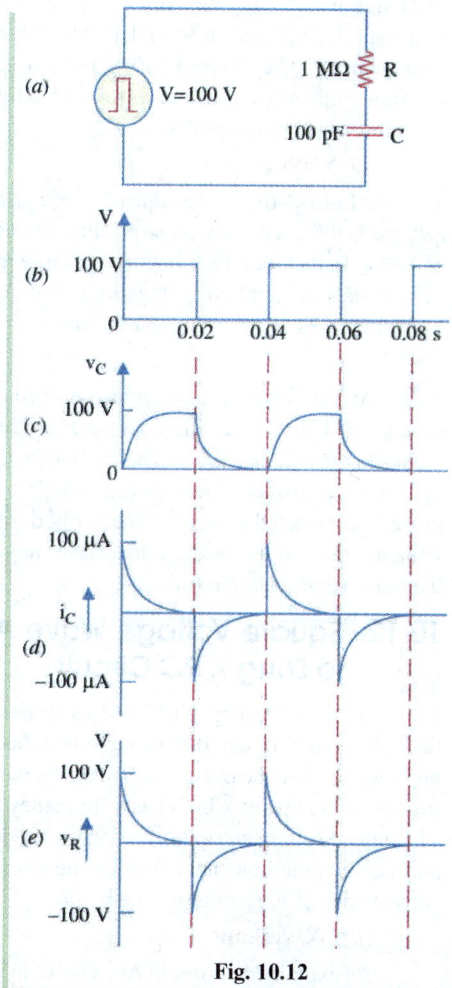

Fig. 10.12

cycle of the applied voltage pulse, C charges and discharges completely once again and so on.

Such short time constants are generally used for differentiating circuits for providing sharp pulse of v_R.

10.11. Square Voltage Wave Applied to Short λ RC Circuit

As shown in Fig. 10.12 (*b*), a square voltage wave of amplitude 100 V, time-period 0.04 s and frequency 25 Hz is applied across an *RC* circuit of $\lambda = 0.1$ ms. Obviously, time constant λ is 1/200 of the pulse width (of 0.02 s = 20 ms). Here, time axis is calibrated in seconds and not in time constants. Let us, now, examine the waveshapes of v_C, i_C and v_R one by one.

(*a*) Waveshape of v_C

Since λ is very short as compared to pulse width, C gets charged fully and, for the same reason, gets discharged also completely during each half-cycle of the applied voltage. Hence, waveshape of v_C is essentially the same as that of the applied voltage as shown in Fig. 10.12 (*c*). In fact, it very closely resembles V except for slightly rounded corners of its curve.

(*b*) Waveshape of i_C

This waveshape shows peaks both for the charge and discharge which coincide with the leading and trailing edges of the applied voltage pulse as depicted in Fig. 10.12 (*d*). Each peak value = $100/R = 100/1\ \text{M}\Omega = 100\ \mu\text{A}$. In fact, the current pulses are much sharper than shown in the figure because they are not to scale horizontally.

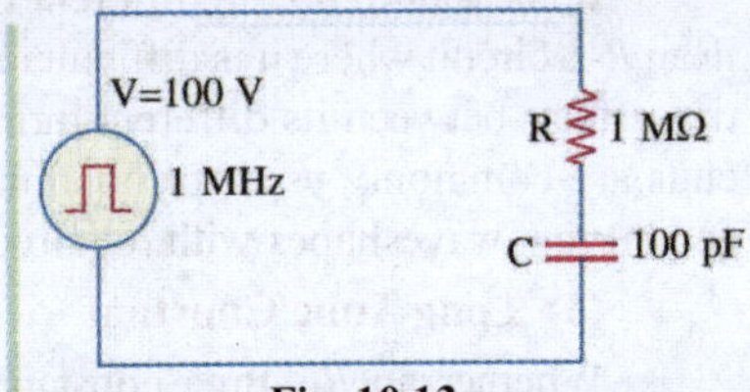

Fig. 10.13

(*c*) Waveshape of v_R

It follows the waveshape of i_C because $v_R = iR = i_CR$. Each 100 μA current pulse produces a voltage pulse of 100 V. In fact, before C begins to charge, peak value of v_R equals the applied voltage of 100 V. Afterwards, as v_C increases, v_R decreases at all times

$$V = v_R + v_C$$

At the instant of discharge, full capacitor voltage of 100 V is applied across R so that v_R peak on discharge equals –100 V. Then in five time constants, v_R drops to zero as shown in Fig. 10.12 (*e*). Such v_R pulses that match the edges of the applied square-wave voltage are used as timing pulses. We may use either the positive or negative pulses.

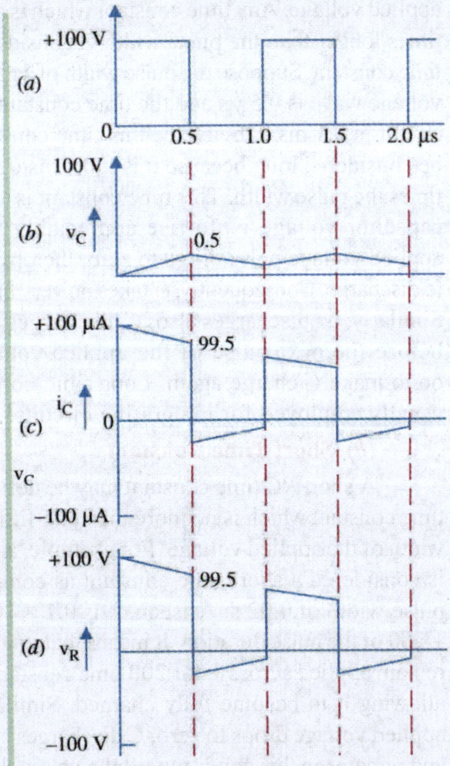

Fig. 10.14

10.12. Square Voltage Wave Applied to Long λ RC Circuit

As shown in Fig. 10.13, the *RC* time constant is the same as before *i.e.*, 0.1 ms but now it becomes 200 times longer than the pulse width of 0.5 μs of the applied square voltage wave which has a frequency of 1 MHz. The time axis has been calibrated in microseconds (μs) and not in time constants. Let us, now, examine the waveshapes of v_C, i_C and v_R one by one.

(*a*) Waveshape of v_C

As explained earlier in Art. 10.10, in view of the

extremely short duration of the applied voltage pulses, C has hardly any time to get charged. Since pulse width is 1/200 of λ, v_C would hardly achieve a value of 100/200 = 0.5 V before it is compelled to discharge as shown in Fig. 10.14 (*b*). Similarly, C is hardly given any time to discharge with the result that v_C drops by a very little amount.

(*b*) Waveshape of i_C

The initial charging current is maximum = V/R = 100/1 M = 100 μA. Hence, waveshape for i_C stays close to 100 μA throughout the duration of the pulse. It is so because C does not have much charge (*i.e.*, v_C is negligible), thereby allowing applied voltage V to keep the charging current near about 100 μA. The decrease is 100/200 = 0.5 μA. Hence, during the pulse period, the charging current falls from 100 μA to 99.5 μA as shown in Fig. 10.14 (*c*).

(*c*) Waveshape of v_R

As shown in Fig. 10.14 (*d*), it follows the waveshape for i_C. In fact, waveshapes of both i_C and v_R are essentially the same as that of the applied square-wave pulse voltage. Eventually, v_C will climb to an average value of 50 V, i_C will vary ± 50 μA above and below zero whereas v_R will vary (like V) ± 50 V above and below zero.

CONVENTIONAL PROBLEMS

1. An inductive coil has a resistance of 100 Ω and an inductance of 2 **H**. It is connected across a dc voltage of 10 V. Determine
   (*i*) time constant of the coil (*ii*) current after one time constant
   (*iii*) current after two time constants. **[(*i*) 0.2 s (*ii*) 63 mA (*iii*) 86.3 mA]**
2. A dc voltage source of 100 V is in series with a series combination of 2 MΩ and 2 μF capacitor. Determine
   (*a*) time required by v_C to become 63 V (*b*) value of v_C after 20 seconds
   [(*a*) 4 s (*b*) 100 V]
3. A 100 V d.c. source is applied to a 1 MΩ resistor connected in series with a 4 μF capacitor which is already charged to 63 V. What will be the value of v_C after 4 s ? **[86.3 V]**
4. A 0.05 μF capacitor charges through a 2 MΩ resistor but is discharged through a 20 kΩ resistor. Determine
   (*a*) charging time constant (*b*) discharging time constant
   (*c*) rate of voltage rise (*d*) rate of voltage fall.
   [(*a*) 0.1 s (*b*) 0.001 s (*c*) 1000 V/s (*d*) 100 kV/s]

SELF EXAMINATION QUESTIONS

A. Fill in the blanks by the most appropriate word(s) or numerical value (s).

1. A pure resistor does not ....any change in the current flowing through it.
2. In an *R-L* circuit, time constant is given by the ratio of .... and ....
3. In one time constant, current through a series *R-L* circuit rises to nearly .... per cent of its final steady value.
4. In one time constant, current through an *R-L* circuit decreases by about ....per cent.
5. In a series *R-L* circuit
   (*a*) initial rate of rise of current is ....
   (*b*) the initial current is ....
   (*c*) rate of rise of current keeps ....
   (*d*) sum of resistive and inductive voltage drops equals the .... voltage.
   (*e*) current reaches its maximum value in about .... time constants.
   (*f*) current decays at a progressively .... rate.
6. In a long time constant *R-C* circuit, capacitor hardly gets enough time either to .... or .... itself.

B. Answer True or False

1. A pure resistor offers only resistance to the flow

of current through it but no reaction to its change.

2. The delay in the establishment of full current in an *R-L* circuit is primarily due to its inductance.
3. Resistance of a coil plays no part in determining the rate of rise or decay of current through it.
4. The time constant of a series *R-L* circuit can be doubled by doubling the value of *R*.
5. In every time constant, current through a coil increases by 63% of the remaining current.
6. The rate of rise of current through a coil keeps decreasing though amount of current flowing through it keeps increasing.
7. In about five time constants, current through a coil is considered to have reached its final steady value.
8. In one time constant, the current through a series *R-L* circuit decays by about 37% of its initial ....value.
9. Under conditions of equilibrium, an *R-C* circuit appears as only a capacitor.
10. In a series *R-C* circuit, charging current decreases but voltage across the capacitor increases.
11. *R-C* circuits are often used to get useful voltage and current waveshapes of any desired time constant.
12. When a square wave voltage is applied to an *R-C* circuit having long time constant, it produces highly-peaked output wave shapes.

C. Multiple Choice Items

1. During one time constant, current through an *R-L* circuit
   (*a*) rises by 63% of its initial value
   (*b*) rises by 37% of its final steady value
   (*c*) decays to 63% of its initial value
   (*d*) rises to 63% of its final steady value
2. In an *R-L* circuit with $R = 100\ \Omega$ and $L = 0.2$ H and $V = 5$ volts, current will reach a steady value after about .... milliseconds.
   (*a*) 2 (*b*) 10
   (*c*) 20 (*d*) 5
3. The arcing across a switch which opens an *R-L* circuit is due to
   (*a*) very low resistance of the switch
   (*b*) high resistance of the circuit
   (*c*) high self-induced emf in the coil
   (*d*) long time constant of the circuit
4. The time constant of an *R-C* circuit is defined as the time during which capacitor charging current becomes .... per cent of its .... value.
   (*a*) 37, final (*b*) 63, final
   (*c*) 63, initial (*d*) 37, initial
5. During five time constants of an *R-C* circuit, the capacitor is usually considered to be .... per cent charged.
   (*a*) 37 (*b*) 100
   (*c*) 63 (*d*) 50
6. A square voltage wave is applied to an *R-C* circuit having a short time constant. The waveshape of the charging current
   (*a*) resembles the waveshape of the applied voltage
   (*b*) has rounded corners
   (*c*) has sharp peaks coinciding with leading edge of the applied voltage pulse
   (*d*) is triangular.

ANSWERS

A. Fill in the blanks

1. oppose **2.** inductance, resistance **3.** 63 **4.** 63 **5.** (*a*) maximum (*b*) zero (*c*) decreasing (*d*) applied (*e*) five (*f*) diminishing (or decreasing) **6.** charge, discharge.

B. True or False

1. T **2.** T **3.** F **4.** F **5.** T **6.** T **7.** T **8.** F **9.** T **10.** T
11. T **12.** F

C. Multiple Choice Items

1. *d* **2.** *b* **3.** *c* **4.** *d* **5.** *b* **6.** *c*

CHAPTER 11

Tuning Circuits And Filters

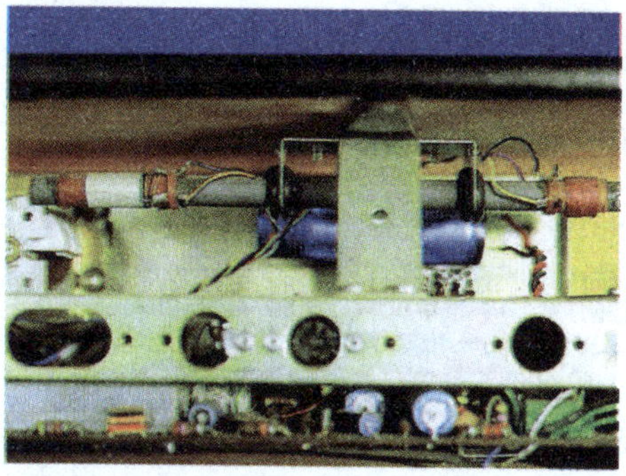

1. What is a Tuning Circuit ?
2. Operating Characteristics of a Tuning Circuit
3. Resonance
4. Actual Series Resonance
5. Is it Series or Parallel Resonance ?
6. Tuned Transformers
7. Double Tuned Transformers
8. Parallel Circuit
9. Coupled Circuits
10. Simple Coupled Circuits
11. Coefficient of Coupling
12. Filters
13. Filter Definitions
14. Types of Filter Circuits
15. Low-pass Filter
16. Highpass Filter
17. Bandpass Filter
18. Bandstop Filter
19. Multisection Filter Circuits

11.1. What is a Tuning Circuit ?

It is a circuit whose parameters can be varied in order to tune it to any desired frequency *i.e.*, to make it resonant at any particular frequency.

The process of selecting the desired frequency is called *tuning*.

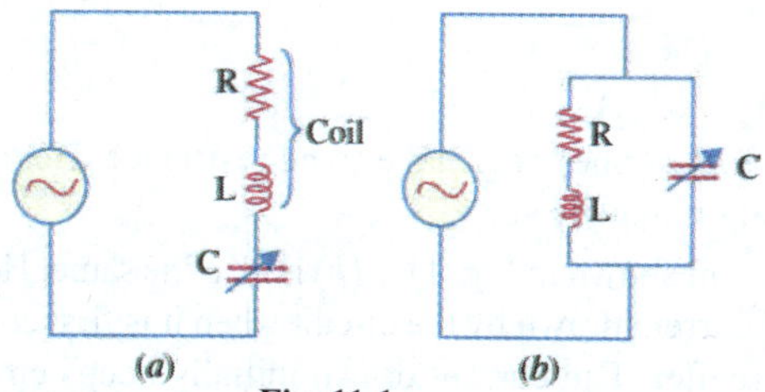

Fig. 11.1

It could be either a series or a parallel *RLC* circuit as shown in Fig. 11.1. Such circuits are the basis of all transmitter, receiver and antenna operation. No radio communication is possible without them. For example, the desired programme in a radio or TV is obtained by tuning the receiver to a particular broadcasting station. In fact, we select the desired station by adjusting the tuning circuit of the receiver so as to bring it in resonance with the carrier frequency transmitted by that station.

11.2. Tuned Circuit

It is a circuit whose parameters have values that will make it resonant to a particular frequency.

11.3. Operating Characteristics of a Tuning Circuit

A tuning circuit is required to perform the following three functions :

1. to select the desired signal out of the many available.
2. to reject all undesired signal.
3. to increase the voltage of the desired signal before passing it on to the next circuit.

The ability to perform the first function is called *sensitivity*, the second *selectivity* and the third *fidelity*.

11.4. Resonance

A circuit is said to be in electrical resonance when its inductive reactance equals its capacitive reactance. Consequently, net reactance offered by the circuit is zero. Consider the series and parallel LC circuits shown in Fig. 11.2. We will first consider the case of an ideal coil and an ideal capacitor. The coil is supposed to have only inductance but no resistance and hence no I^2R loss. Similarly, the capacitor has no leakage current *i.e.*, its dielectric is a perfect insulator.

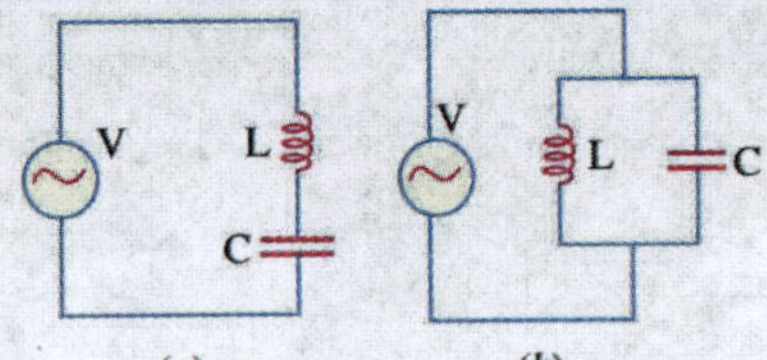

(*a*) (*b*)

Fig. 11.2

The series circuit of Fig. 11.2 (*a*) is in resonance or it resonates at that frequency for which

$$X_L = X_C$$

or $$2\pi f L = \frac{1}{2\pi f C} \quad \text{or} \quad f_0 = \frac{1}{2\pi\sqrt{LC}}$$

The subscript 0 has been added to indicate that this frequency refers to the resonant condition. It is seen that f_0 depends on the values of both L and C.

When in resonance, the circuit offers zero impedance to the applied voltage. Hence, current under resonant condition is

$$I_0 = \frac{V}{X} = \frac{V}{(X_L - X_C)} = \frac{V}{0} = \infty \; !$$

Now, every *actual* coil (as opposed to an *ideal* one) does have some resistance. Hence, actually the current is never zero though theoretically it may be so.

The resonant frequency of the parallel circuit shown in Fig. 11.2 (*b*) is also the same. However, it is interesting to note that except for the initial current drawn by the circuit when it is first connected to the supply, the circuit draws no current thereafter. The current drawn initially keeps circulating between the two branches because current requirements of L and C are 180° out of phase with each other. Hence, under parallel resonant conditions, $I_0 = 0$.

$$\therefore \qquad Z = \frac{V}{I_0} = \frac{V}{0} = \infty \; !$$

It means that impedance offered by such a circuit when in resonance is infinity ! In other words, it acts like an open switch to an input signal of a particular frequency.

However, as all actual coils do have some resistance, current is not exactly zero so that impedance is not infinite but extremely high.

11.5. Actual Series Resonance

An actual coil always possesses some resistance as shown in Fig. 11.1 (*a*). At any frequency of the applied voltage, impedance offered by the circuit is

$$Z = R + j(X_L \sim X_C) = R + JX$$

or
$$Z = \sqrt{R^2 + X^2}$$

At resonance,

$$X = 0 \text{ or } X_L \sim X_C = 0 \text{ or } X_L = X_C$$

$$\therefore \quad f_0 = \frac{1}{2\pi\sqrt{LC}} = \frac{0.159}{\sqrt{LC}} \text{ Hz}$$

It is seen that resonant frequency *does not depend on the coil resistance.*

Though main points regarding series resonance have already been discussed in Chapter 9, we will briefly touch upon the following item :

1. Q-Value

It is also called quality factor or figure of merit. It is defined as under :

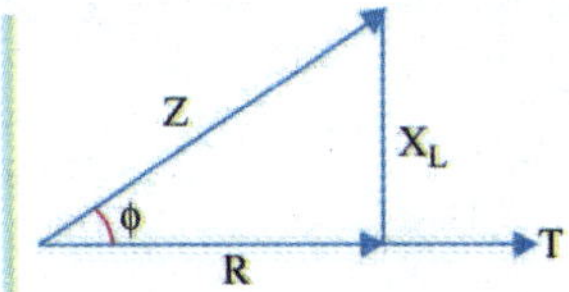

Fig. 11.3

(*a*)
$$Q = \frac{X_L}{R} = \frac{\omega_0 L}{R}$$

The Q-factor of a coil can be increased by making its R as low as possible and L as high as practicable. An ideal coil whose $R = 0$ has a Q of infinity !

Moreover, Q of a given coil will become less at high frequencies because of increase in R due to skin effect.

(*b*) Also
$$Q_0 = \frac{1}{\omega_0 CR} = \frac{1}{R}\sqrt{\frac{L}{C}} = \frac{\omega_0}{\Delta\omega}$$
$$= \frac{f_0}{f_2 - f_1} = \frac{f_0}{\Delta f}$$

When applied to a tuning circuit, it measures the selectivity or sharpness of the tuning circuit. Fig. 11.4 shows how sharpness of tuning varies with the Q-value. Values of Q_0 of the order of 200 are common in radio circuits.

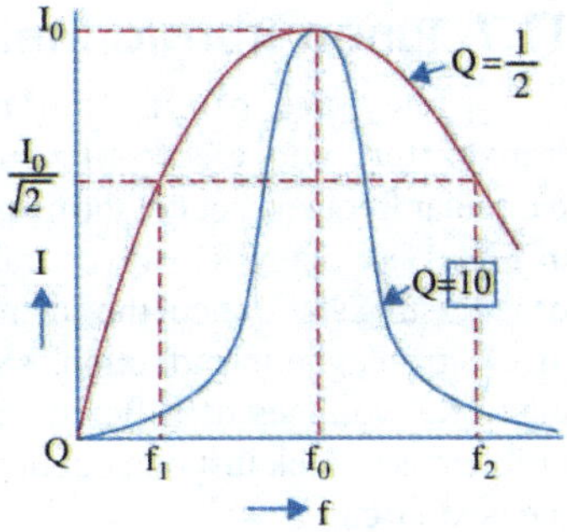

Fig. 11.4

2. Bandwidth

It has already been discussed in Art. 9.10. The bandwidth (BW) of a series resonant circuit is given by

$$BW = f_2 - f_1 = \Delta f = \frac{f_0}{Q_0} = \frac{R}{2\pi L}$$

Obviously, higher the value of Q_0, narrower the bandwidth and higher the *selectivity* of the circuit. As seen, for getting high selectivity, coil resistance should be low but inductance high. Also, bandwidth depends on the R/L ratio and not on the *individual values of R and L as such.*

It should also be noted that bandwidth *does not depend on C at all.* Value of C only affects f_0.

3. *LC* Product

It determines the value of f_0. Values of L and C may be changed but so long as their product remains constant, f_0 remains constant. If the LC product for a desired frequency is known, the capacitance or inductance required can be found.

11.6. Is it Series or Parallel Resonance ?

A tuned circuit commonly used in radio receivers is shown in Fig. 11.5. It, in fact, represents a transformer with tuned secondary which at first glance looks like a parallel resonant circuit. But, actually, it is a case of *series* resonance. It is so because transformer primary induces e.m.f. into each turn of the secondary. No e.m.f. is being induced *across* the ends of the secondary coil. Theoretically, the secondary may be considered to have an ac source inserted in series with it [Fig. 11.5 (*b*)]. Since the induced e.m.f. is in series with the coil, it behaves like a series resonant circuit.

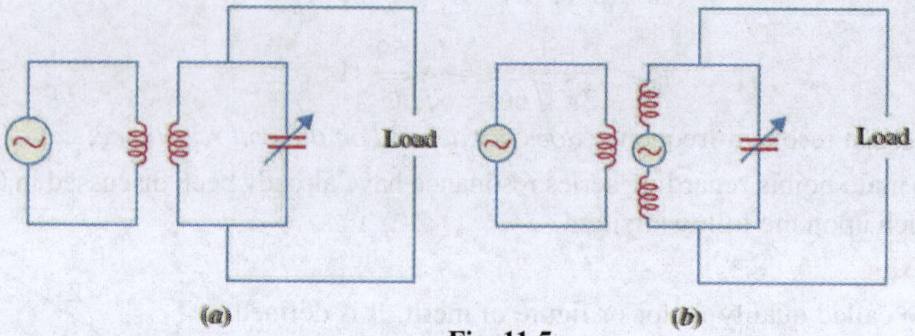

Fig. 11.5

Incidentally, if a suitable capacitor is connected across the primary, it will become a parallel resonant circuit because the ac voltage will be fed across its ends and will not be induced in series with it. Such series-parallel circuits are often used in radio receivers.

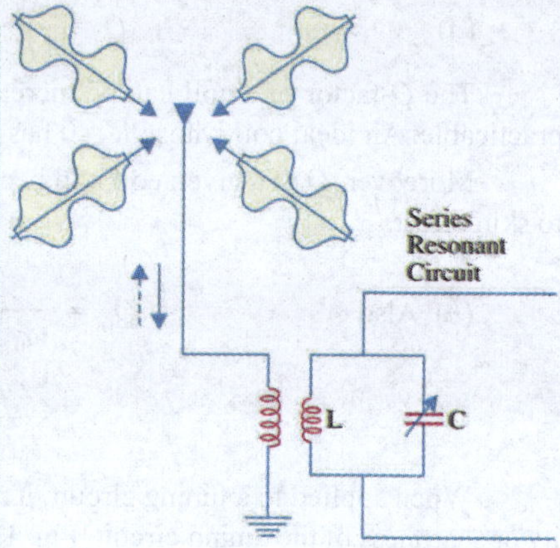

Fig. 11.6

11.7. Tuned Transformers

They are often used in radio receivers, transmitters and electronic circuits to select one particular frequency out of the many available. As shown in Fig. 11.6, many RF waves transmitted by different broadcasting stations cut the antenna and as per the laws of electromagnetic induction, simultaneously induce many RF voltages of different frequencies in it. The problem is to pick just one out of them and pass it on to the next circuit.

By varying the capacitance of C connected across the secondary coil, its resonance frequency can be varied. Since it is a series-resonant circuit, it offers least impedance to that aerial signal to which it resonates. Hence, that particular signal voltage induces large current in secondary and is then passed on to the next circuit.

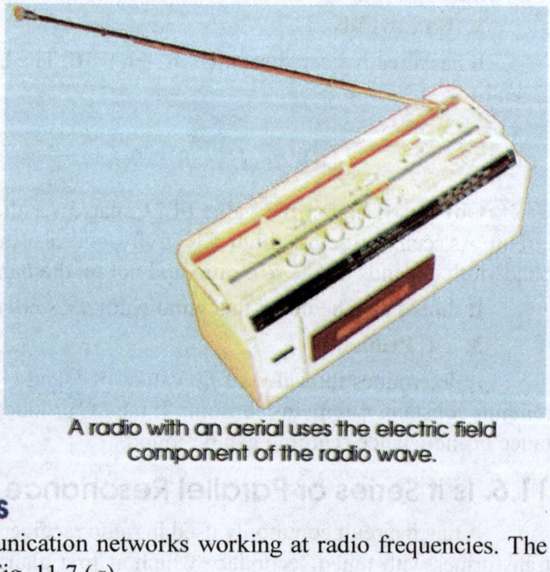

A radio with an aerial uses the electric field component of the radio wave.

To other signals having frequencies other than the resonance frequency, the circuit will offer much higher impedance. Hence currents produced by them would be almost negligible and, consequently, they would not be able to get through to the next circuit.

11.8. Double Tuned Transformers

Such circuits are often used in communication networks working at radio frequencies. The circuit and its frequency curve are shown in Fig. 11.7 (*a*).

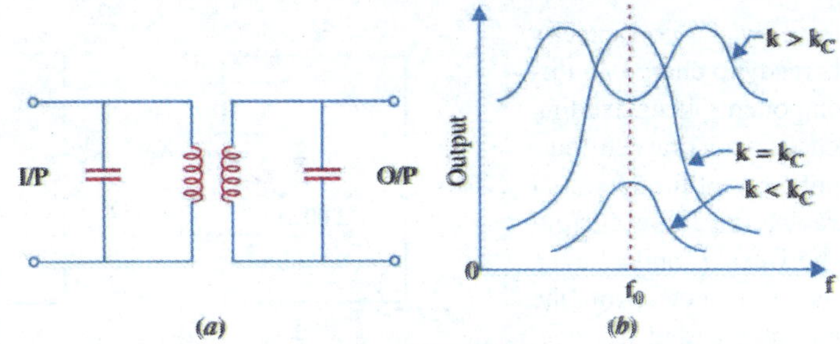

Fig. 11.7

Fig. 11.7 (*b*) shows the conditions when the tuned primary and secondary circuits are over-coupled, under-coupled and critically-coupled. For critical-coupling

$$k_c = \frac{1}{\sqrt{Q_p Q_s}}$$

where Q_p and Q_s are the *Q*-factors for the primary and secondary winding respectively.

Such transformers (often used in RF service) have coefficient of coupling of the order of 0.01 to 0.05.

The bandwidth of a double-tuned coupling transformer is given by

$$BW = k f_0$$

Example 11.1. *A series tuned circuit has a resistance of 5 ohm, an inductance of 220 μH and a capacitor of 400 pF all connected in series. Determine*

(i) resonant frequency *(ii) bandwidth* *(iii) Q-factor*

How will the bandwidth and Q-factor change if the resistance value is doubled ?

Solution. (*i*)
$$f_0 = \frac{0.159}{\sqrt{LC}} = \frac{0.159}{\sqrt{220 \times 10^{-6} \times 400 \times 10^{-12}}} = \mathbf{536\ kHz}$$

(*ii*)
$$BW = \frac{R}{2\pi L} = 0.159\frac{R}{L} = 0.159 \times \frac{5}{220 \times 10^{-6}} = \mathbf{3614\ Hz}$$

(*iii*)
$$Q_0 = \frac{f_0}{BW} = \frac{536 \times 10^3}{3614} = \mathbf{148}$$

When *R* is doubled, the bandwidth is **doubled** but *Q*-value is **halved.**

$$BW = 2 \times 3614 = \mathbf{7{,}228\ Hz}$$

and
$$Q_0 = 148/2 = \mathbf{74}$$

11.9. Parallel Circuit

A parallel circuit using ideal components is shown in Fig. 11.8 (*a*). It consists of an ideal resistanceless inductor connected in parallel with a perfect capacitor. When frequency of the applied voltage is such that X_L becomes equal to X_C, it becomes a parallel resonant circuit. It is also called *antiresonant* circuit in contrast to a series resonant circuit from which it differs in many important respects.

When the capacitor is removed leaving behind the inductor alone, let the current drawn from the supply be 1 A. When the capacitor is connected back, it also has a current of 1 A flowing through it even though *no additional current is taken from the supply* ! It is found that the current of 1 A drawn by the coil initially keeps alternating or circulating between it and the capacitor. It is so because there is a phase difference of 180° between the currents drawn by the two. It means that capacitor starts discharging exactly at the instant when coil needs current to build up its magnetic field. Similarly, current is given out by the coil exactly when capacitor is ready to charge. In this way, the two components keep feeding each other and hence do not draw any current from the supply except for the initial intake. Once drawn, the same current keeps shuttling between L and C even when the circuit is disconnected from the supply. It is sometimes called *flywheel effect* because it is similar to the action of a mechanical flywheel which, once started, tends to keep going indefinitely until stopped by friction or other losses. However, it should not be forgotten that all this happens when the *circuit is in resonance with the supply voltage*.

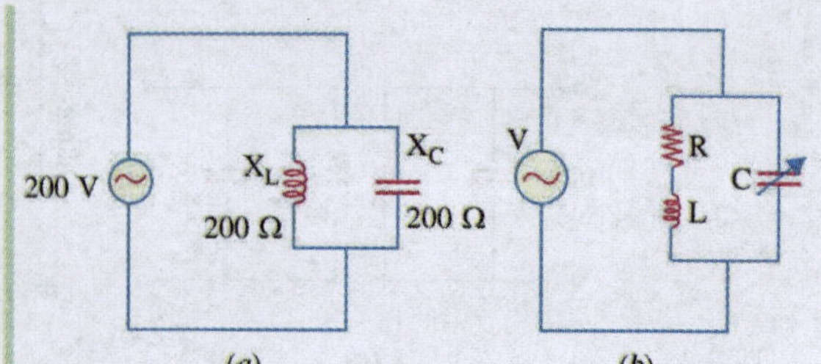

Fig. 11.8

Since current drawn from the supply is zero, such an ideal parallel resonant circuit offers infinite resistance to the applied voltage. However, because of some inherent resistance (however small) possessed by the coil, current is not exactly zero but has a certain minimum value. Consequently, impedance offered by such a circuit is maximum and not infinite.

For the practical case shown in Fig. 11.8 (*b*), following points are worth noting :

1. resonance occurs when $Z^2 = X_L \cdot X_C = L/C$ where Z is coil impedance.

2. $$f_0 = \frac{1}{2\pi}\sqrt{\frac{1}{LC} - \frac{R^2}{L^2}} = \frac{1}{2\pi\sqrt{LC}} \quad \text{— if } R \text{ is negligible}$$

3. $$I_{min} = \frac{V}{L/CR}$$

4. $$Z_{max} \text{ or } Z_0 = L/CR$$

The approximate relations for Z_0 are

$$Z_0 \cong \frac{X_L}{R} \quad \text{— if } X_L > 10\,R$$

$$\cong Q \cdot X_L \cong \frac{Q}{X_C} \cong Q^2\,R$$

5. power factor $= 1$

6. $$BW = \frac{X_L}{R} = \frac{1}{R}\sqrt{\frac{L}{C}}$$

Most significant thing about such a circuit is that it offers maximum impedance to that applied voltage whose frequency equals its own resonance frequency. Hence, current driven by that voltage is the least. Those voltages whose frequency is farther away from f_0 are offered less impedance. In fact, farther their frequency from f_0, lesser the impedance encountered by them and hence greater the current produced by them. Since it rejects (or attenuates) those applied voltages whose frequency equals its own resonance frequency, such a circuit is called a *rejector* circuit (in contrast to a series resonance circuit which is called an *acceptor* circuit).

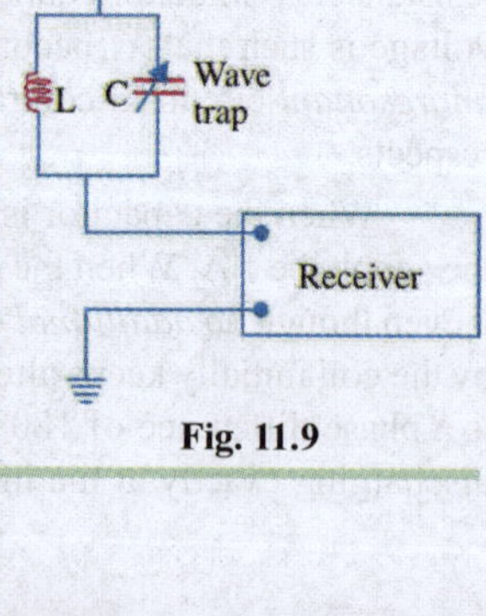

Fig. 11.9

The use of a parallel-resonant circuit to reject unwanted frequency is shown in Fig. 11.9. Here, it is used as a wave trap. When a receiver set is operated very close to a powerful transmitter, its reception of other stations on adjacent channels becomes impossible due to the strong interfering signal from this transmitter. By using a parallel-resonant circuit in series with the aerial and resonating it to the frequency of the interfering station, this particular signal can be rejected out of the receiver.

Example 11.2. *For the parallel resonant circuit shown in Fig. 11.10. Find*

(i) *resonant frequency* *(ii)* *Q of the circuit*

(iii) *circuit impedance at resonance*

(iv) *line current when a 10 V source at resonant frequency is applied*

(v) *current circulating between the two branches* *(vi)* *bandwidth.*

Solution. (*i*) $f_0 = \dfrac{1}{2\pi\sqrt{LC}} = \dfrac{1}{2\pi\sqrt{200\times10^{-6}\times500\times10^{-12}}} = \mathbf{503\ kHz}$

(*ii*) $Q_0 = \dfrac{X_L}{R}$

$X_L = 2\pi f_0 L = 2\pi\times503\times10^3\times200\times10^{-6} = \mathbf{632\ \Omega}$

$Q_0 = 632/10 = \mathbf{63.2}$

(*iii*) $Z_0 = Q_0 X_L = 63.2\times632 = \mathbf{39.94\ k\Omega}$

(*iv*) $I_0 = V/Z_0 = 10/39.94\ \text{K} = \mathbf{0.25\ mA}$

(*v*) Current circulating between branches

$= Q_0 I_0 = 63.2\times0.25 = \mathbf{15.8\ mA}$

(*vi*) $BW = \dfrac{f_0}{Q_0} = \dfrac{503\ \text{kHz}}{63.2} = \mathbf{7.96\ kHz}$

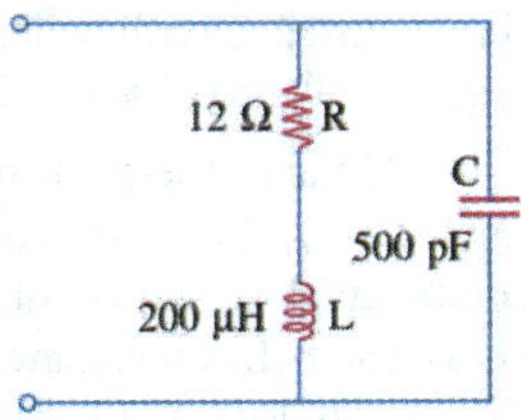

Fig. 11.10

Example 11.3. *Find the coefficient of critical coupling for a double-tuned transformer in which Q of the primary is 60 and that of the secondary is 80.*

Solution. $k = \dfrac{1}{\sqrt{Q_p Q_s}} = \dfrac{1}{\sqrt{60\times80}} = \mathbf{0.0144.}$

11.10. Coupled Circuits

Two circuits are said to be coupled to each other if they have a common *impedance* between them. This common impedance is called *coupling element* and permits the transfer of electrical energy from one circuit to the other.

Coupling element can be any one of the following :

1. a resistor,
2. a capacitor,
3. an inductor,
4. a transformer etc.

Often coupling elements also perform some filter action (Art. 11.12) in addition to transferring energy between the two circuits.

11.11. Simple Coupled Circuits

A simple coupled circuit is one in which the common impedance consists of only one element such as a resistor, capacitor or an inductor.

Fig. 11.11 shows three simple coupled circuits which are also called *direct-coupled* circuits. In these circuits, coupling is achieved by passing input circuit current through the common element where a voltage drop is

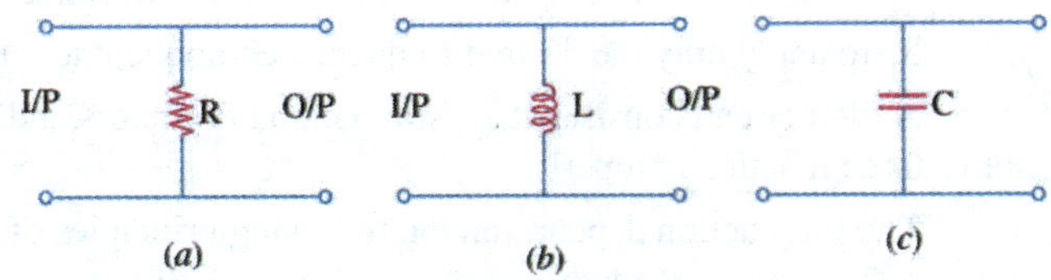

Fig. 11.11

produced. This voltage drop is applied across the output circuit thereby transferring some electric energy to it.

Indirect coupling where mutual and self- inductions are used, is illustrated in Fig. 11.12.

In Fig. 11.12 (*a*), transfer of energy is purely by magnetic induction. The input and output circuits are electrically separate. In Fig. 11.12 (*b*), energy transfer is partly by magnetic induction and partly by electric conduction.

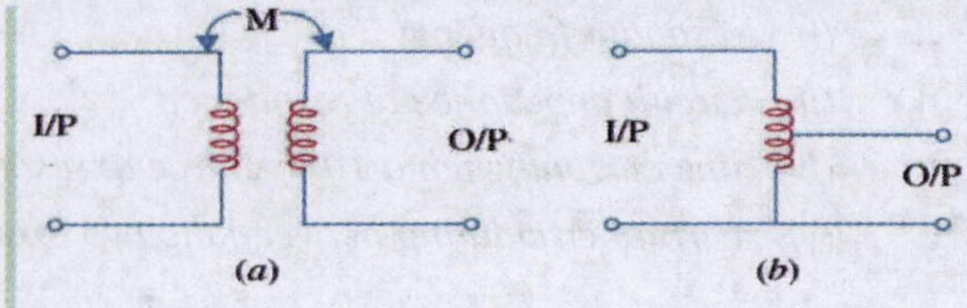

Fig. 11.12

In Fig. 11.11 (*b*), the proportion of energy transferred increases as frequency increases, whereas in Fig. 11.11 (*c*), it decreases as frequency increases. By using a combination of *L*, *R* and *C*, any desired proportion of energy can be transferred between the input and output circuits.

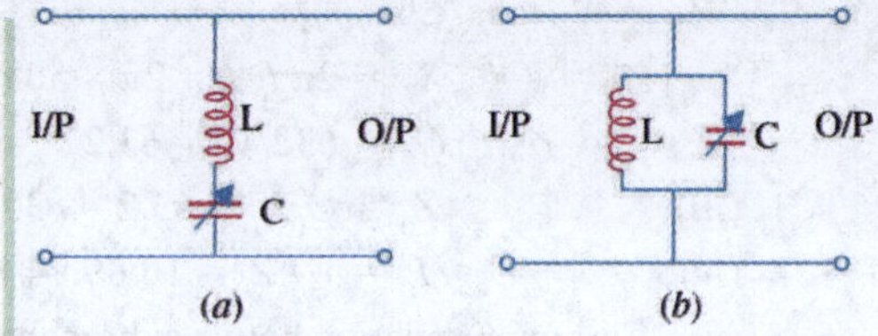

Fig. 11.13

In Fig. 11.13 (*a*), the common element is a series tuned circuit which will offer *minimum* impedance at resonant frequency.

Hence, energy transferred would be the *lowest* at f_0 because voltage drop across *LC* combination would be minimum. At frequencies above and below resonance, energy transfer would be more. In Fig. 11.13 (*b*) opposite of the above happens. Here, energy transfer would be maximum at resonance and would decrease off-resonance.

11.12. Coefficient of Coupling

The amount of energy transferred between two mutually-coupled circuits depends on their coefficient of coupling.

The response curves for tight, critical and loose coupling are shown in Fig. 11.14. With tight coupling, resonance occurs at two new frequencies different from the respective resonance frequencies of the two circuits. As coupling is decreased, the two frequencies come closer and merge into one frequency at critical coupling (also called optimum coupling). When coupling is decreased below its critical value, a single peak of reduced height is obtained as shown in Fig. 11.14 (*c*).

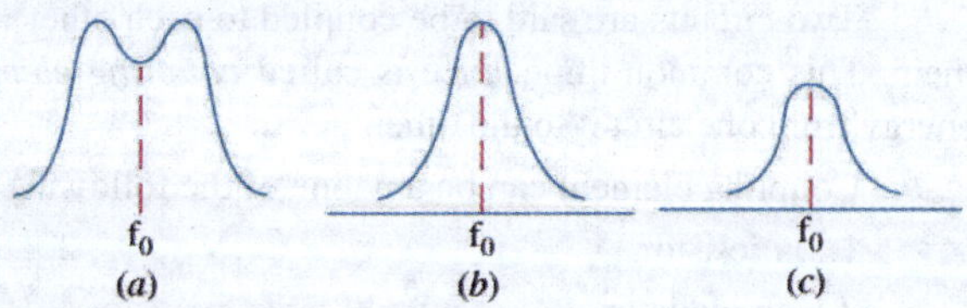

Fig. 11.14

11.13. Filters

These are frequently used in radio and other electronic circuits. Their purpose is

1. to select the desired frequency (or band of frequencies) from a complex input wave,
2. to reject the undesired frequency (or band of frequencies), and
3. to apply only the desired frequency component to the circuit where it is required.

A filter circuit consists of a combination of resistors, inductors and capacitors that perform the above three functions properly.

The filter action depends on the following principles of ac circuits :

1. The opposition offered by an inductor to the flow of ac current through it increases with increase in current frequency ($X_L = 2\pi f L$). An inductor offers very little opposition to dc current as

well as to low-frequency ac currents. But it offers increasingly higher impedance to the flow of radio frequencies through it. In other words, it attenuates *RF* currents as shown in Fig. 11.15.

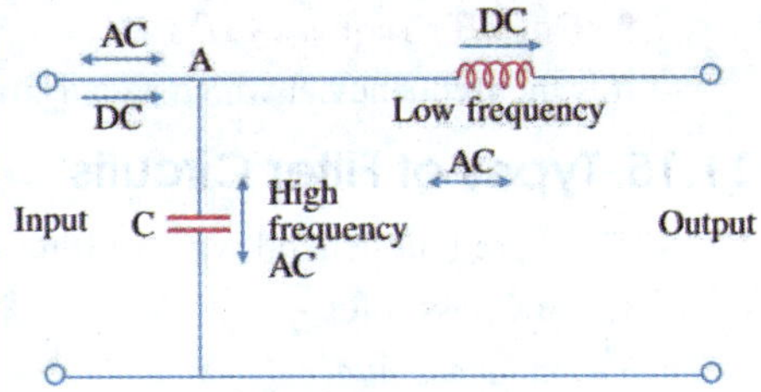

Fig. 11.15

2. The opposition offered to the flow of ac current through a capacitor decreases with increase in current frequency ($X_C = 1/2\pi fC$). A capacitor offers little opposition to *RF* currents but offers large reactance to low-frequency currents. Of course, it completely blocks the passage of dc current through it (since for dc current, $f = 0$, $X_C = 1/2\pi \times 0 \times C = \infty$!). This fact is shown in Fig. 11.15.

3. A series circuit offers minimum impedance to that current whose frequency equals its own resonance frequency. But it will offer very high opposition to all other frequencies and will hardly let them pass through as shown in Fig. 11.16.

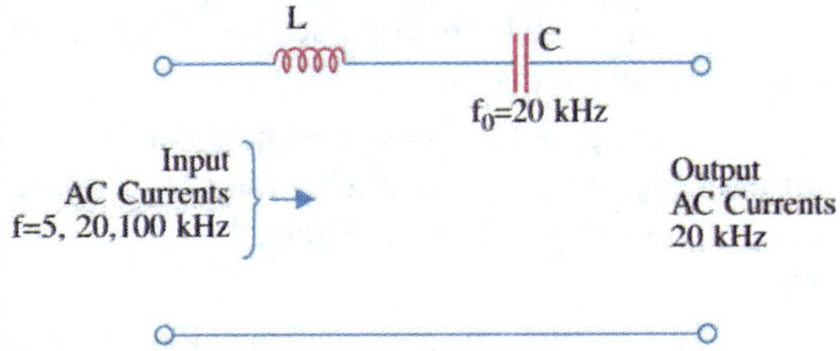

Fig. 11.16

4. A parallel circuit offers very high impedance to that ac current whose frequency equals its own resonant frequency. However, it will offer comparatively much less opposition to other off-resonance frequencies. Consequently, it will allow all other frequencies to pass through it except the one which equals its resonance frequency. This fact has been illustrated in Fig. 11.17.

5. Resistors, by themselves, do not provide any filtering action because they offer equal resistance to all currents regardless of their frequency. However, when combined with an inductor or capacitor, they increase the impedance of the circuit as a whole thereby decreasing the sharpness of filter circuit. Their presence in a resonant circuit determines the slope of its resonance curve.

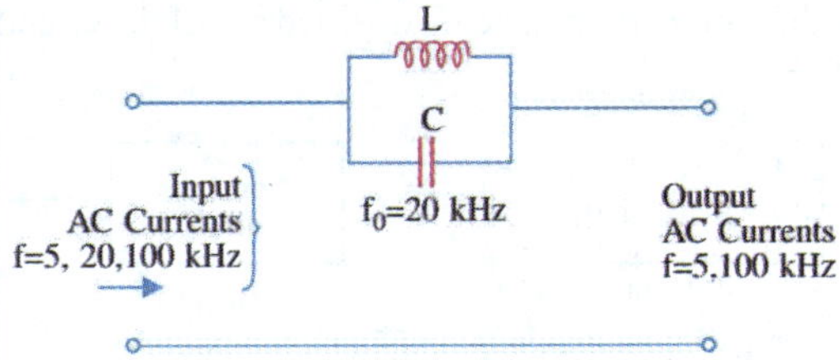

Fig. 11.17

11.14. Filter Definitions

1. Passband

It is that band (or group) of frequencies which is *passed* through the filter freely.

2. Stopband

It is that group of frequencies which is not allowed to pass through the filter *i.e.*, it is *stopped* from passing through.

3. Transmission

It is the act of carrying a frequency or a range of frequencies from the input to the output of a filter without weakening it.

In practice, –3 dB point is generally taken as the point where significant weakening of the signal begins.

4. Attenuation

It is opposite of transmission. When signal strength drops to –3 dB as compared to its original value, attenuation is said to begin.

5. Cut-off Frequency (f_C)

It is the frequency at which signal strength drops by 3 dB *i.e.*, signal power becomes half.

11.15. Types of Filter Circuits

There are four general types of filter circuits :

1. low-pass filter
2. high-pass filter
3. bandpass filter
4. bandstop filter

They are made up of inductors and capacitors which, under ideal conditions, are assumed to have no losses. In practice, effort is made to keep their losses as small as possible.

11.16. Low-pass Filter

It is a circuit which passes low frequencies but stops high ones as shown in Fig. 11.18.

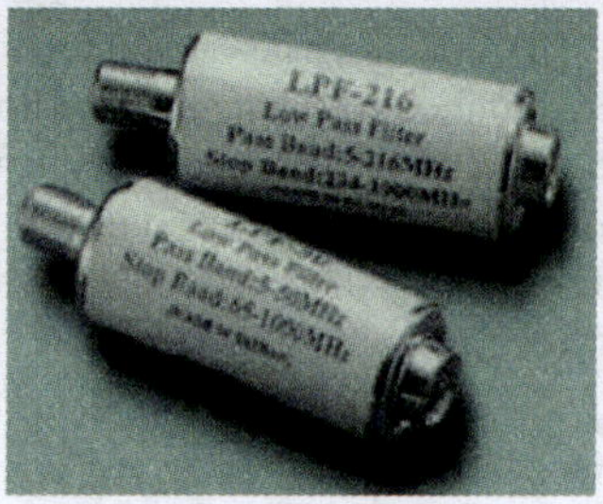

Low-pass filters.

A low-pass filter is so designed that it transmits currents of all frequencies from zero to a certain cut-off frequency f_C without attenuation. All frequencies higher than f_C are attenuated (or cut-off).

An *L*-shaped *LC* low-pass filter along with its response curve is shown in Fig. 11.19. As seen, series-connected inductor *L* offers very little opposition to the low-frequency currents but high opposition to high-frequency currents. Moreover, for diverting these high-frequency currents back to their source, a capacitor is connected in parallel to provide a low-reactance path for that purpose. In this way, a small band of frequencies ranging from 0 to f_C (where output falls to –3 dB) is passed through *L*, the rest being diverted out of the circuit through *C*. Higher the *Q*-value of *L* and *C*, sharper the cut-off.

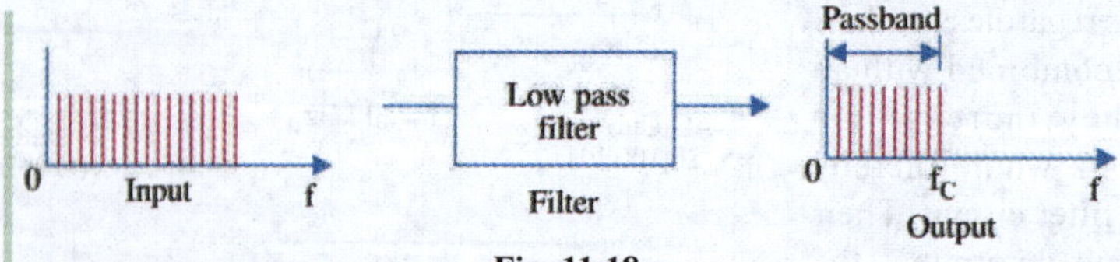

Fig. 11.18

The LP filter discussed above is one of the simplest of the group known as *constant-k* filters. This name comes from the fact that the product '$X_L \cdot X_C$' for the circuit always remains constant which is given the name 'constant-k'. For example, if current frequency is doubled, X_L is doubled but X_C is halved. Hence, their product remains the same.

The *L* and *C* values can be calculated from the formulas

$$L = \frac{R}{\pi f_C} \quad \text{and} \quad C = \frac{1}{\pi f_C R}$$

where *R* is the impedance of *both source and load*.

Uses

Low-pass filters are often used in electronic power supplies since they pass direct current but not variations of current or voltage. They are also sometimes used in voice frequency circuits which pass frequencies upto 3 kHz or so. They are also used between the transmitter proper and its antenna to prevent high frequencies (*i.e.* harmonics) from appearing in the antenna.

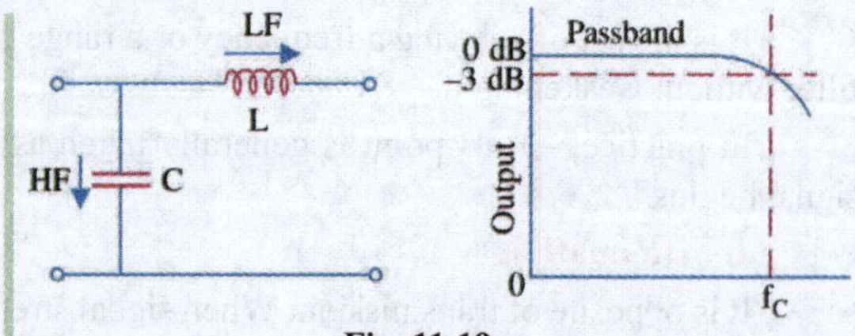

Fig. 11.19

11.17. Highpass Filter

It is a circuit which passes high frequencies but stops low ones as shown in Fig. 11.20.

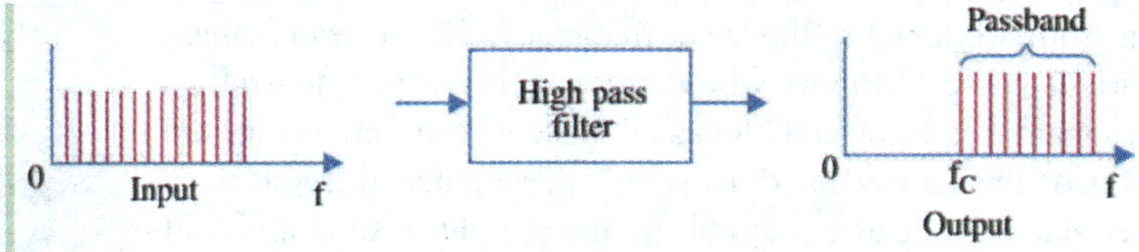

Fig. 11.20

A highpass circuit is so designed that it transmits currents of all frequencies lying between infinity and a certain cut-off frequency f_C. All other frequencies below f_C are attenuated.

As seen from Fig. 11.21 (*a*), an HP filter can be obtained from an LP filter by merely exchanging the position of its two circuit components. The response curve of such a filter is shown in Fig. 11.21 (*b*).

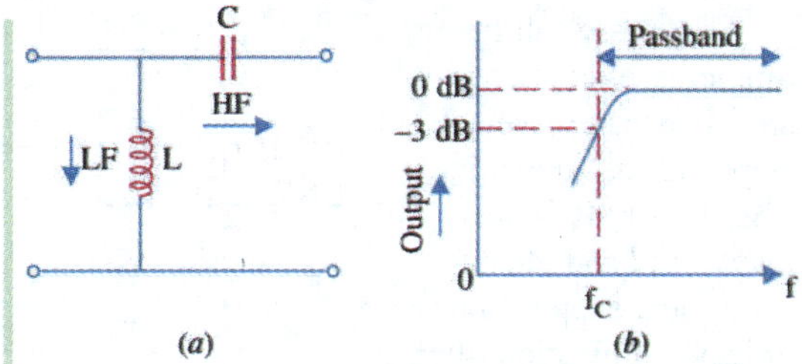

Fig. 11.21

As seen, high frequencies prefer to pass through *C* whereas low frequencies find an easier path through *L*.

The values of *L* and *C* for a highpass constant-k filter can be found from the following formulas :

$$L = \frac{R}{4\pi f_C} \quad \text{and} \quad C = \frac{1}{4\pi f_C R}$$

High pass filter.

Uses

Such HP circuits are frequently used f_{C_2}

1. in audio circuits
2. between a TV and its antenna to prevent a nearby powerful station from interfering with its reception.

11.18. Bandpass Filter

It transmits a specified band of frequencies and stops others as shown in Fig. 11.22.

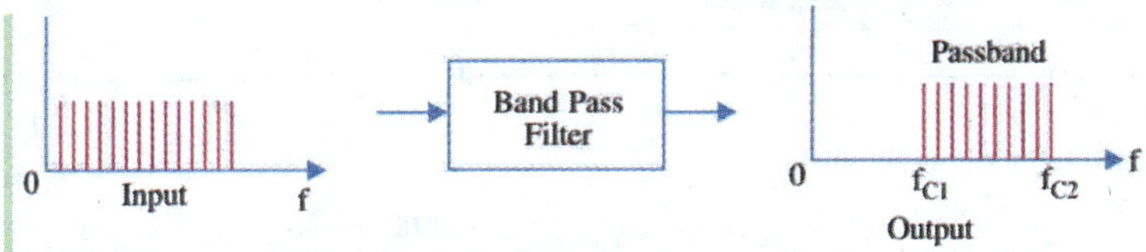

Fig. 11.22

In other words, it transmits currents of all frequencies whose values lie between the cut-off frequencies f_{C_1} and f_{C_2} and stops all frequencies lying either below f_{C_1} or above f_{C_2}.

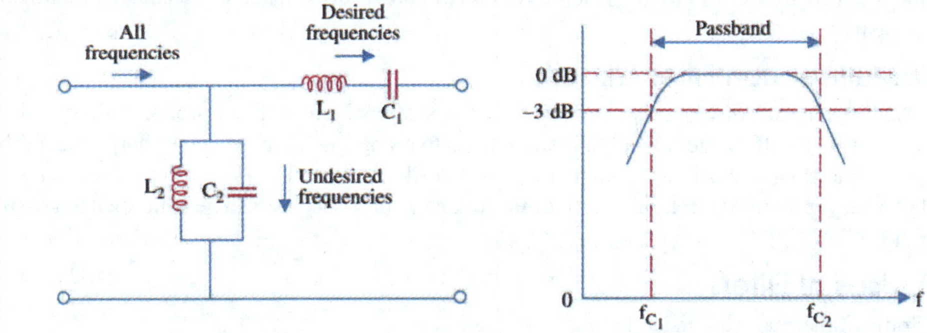

Fig. 11.23

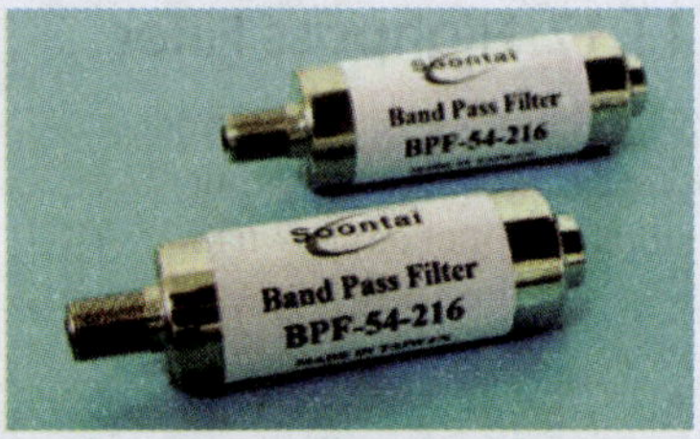

Bandpass filter.

As shown in Fig. 11.23, the filter consists of two resonant circuits, one connected in series with the line and the other across it. Both are tuned to the same frequency. The series resonant circuit passes currents whose frequencies are at or near its resonant frequency and blocks all those whose frequencies are outside this narrow band. However, these blocked frequencies get shunted out of the circuit by the parallel resonant circuit which offers them very little impedance.

11.19. Bandstop Filter

It passes all frequencies except in a specified band (called stopband) as shown in Fig. 11.24. It is also known as band-attenuation, band-suppression or band-elimination filter.

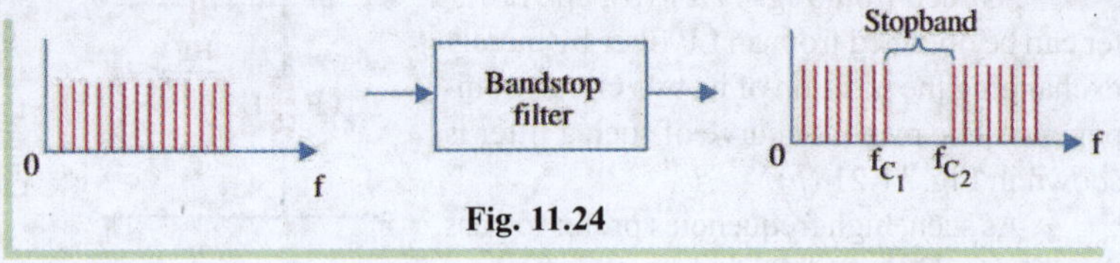

Fig. 11.24

Its function is just opposite of a bandpass filter. As shown in Fig. 11.25 (*a*), it can be obtained by interchanging the positions of circuit elements of a bandpass filter.

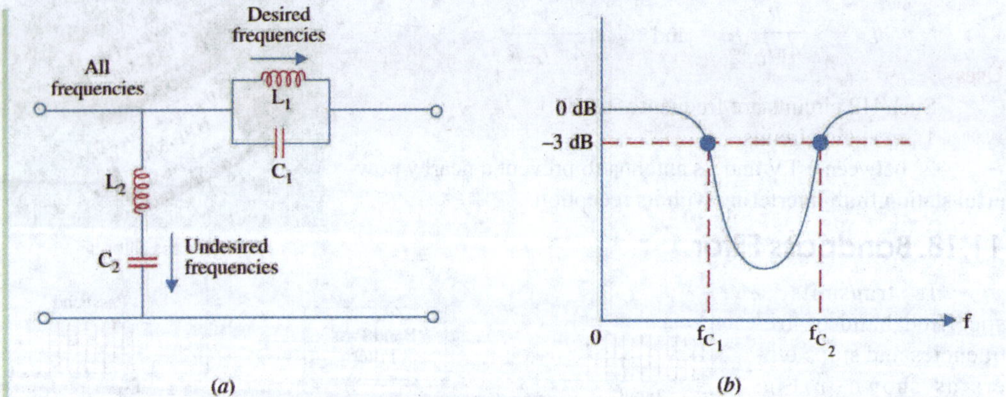

Fig. 11.25

The parallel resonant circuit stops the band of frequencies at or near its own resonant frequency but passes those lying outside this band. The frequencies blocked by parallel-resonant circuit are shunted out by the series circuit which offers them hardly any impedance because it is already tuned to them.

11.20. Multisection Filter Circuits

The filter circuits discussed so far have only one section. None of these produces sharp reduction of current at cut-off. By adding additional sections to a single-section filter, the sharpening of the edges of the bands can be achieved. Such multisection filters are called *T*- or π-type depending on which letter they resemble structurally. Two or more such *T*- or π-sections can be combined to get still sharper cut-off.

11.21. Uses of Filters

Some of their uses are listed below :

1. in decoupling amplifier stages,

2. in electronic power supply ripple smoothening,
3. as tuned circuits in RF and IF stages,
4. for selection of a particular sideband as in SSB transmitters,
5. for harmonic frequency suppression in transmitters,
6. as wave traps for eliminating undesirable signals from radio and TV receivers,
7. for improving high and low-frequency response in video amplifiers,
8. for restricting audio frequency bandpass in SSB transmitters,
9. for eliminating undesired frequencies from motors, generators and other electrical and electronic equipment.

CONVENTIONAL PROBLEMS

1. An *RLC* series circuit consists of $R = 80\ \Omega$, $L = 40\ \mu H$ and $C = 0.0783$ pF. Find its
   (*i*) resonant frequency (*ii*) *Q*-factor (*iii*) bandwidth
   [(*i*) 90 MHz (*ii*) 283 (*iii*) 318.5 kHz]
2. Calculate the resistance of series *RLC* circuit which is resonant at 50 MHz and has $Q_0 =$ 100. The coil inductance is 100 μH. **[314 Ω]**
3. A parallel circuit consists of a coil connected in parallel with a capacitor of 2.5 nF. The coil has a resistance of 4 Ω and an inductance of 40 μH. A 200 mV signal at resonant frequency is applied across the circuit. Determine
   (*i*) resonant frequency (*ii*) *Q*-factor
   (*iii*) impedance at resonance (*iv*) line current drawn
   (*v*) circulating current between branches (*vi*) bandwidth
   [(*i*) 503.2 kHz (*ii*) 31.6 (*iii*) 3996 Ω (*iv*) 50 μA (*v*) 1.58 μA (*vi*) 15.92 kHz]
4. A double-tuned transformer is required to have a resonant frequency of 1 MHz and a bandwidth of 15 kHz. Both primary and secondary have an inductance of 200 μH. The primary has a *Q* of 60 and secondary of 80. Determine
   (*i*) capacitance required across each coil
   (*ii*) coefficient of coupling required
   (*iii*) whether the circuit is over, under, or critically-coupled.
   [(*i*) 126.6 pF (*ii*) 0.015 (*iii*) overcoupled]

SELF EXAMINATION QUESTIONS

A. Fill in the blanks with most appropriate word (s) or numerical value (s).

1. A circuit whose parameters can be varied to tune in any desired frequency is called a .... circuit.
2. The process of selecting a desired frequency out of a complex wave is called ....
3. The basic principle of tuning is the phenomenon of ....
4. An *LC* circuit is said to be in resonance when $X_L =$ ....
5. The *Q*-factor of a resistanceless coil is .... high.
6. Bandwidth of an *RLC* circuit does not depend on the value of its ....
7. Higher the *Q*-factor of a resonant circuit, the .... its bandwidth.
8. The widest application of the tuned circuit is as a .... selective filter.
9. When resistance of a tuned circuit is doubled, its bandwidth is .... but *Q*-factor is ....
10. A low pass filter is that circuit which stops .... frequencies.

B. Answer True or False

1. A tuned circuit is a resonant circuit.
2. If *LC* product of a resonant circuit is doubled, its resonance frequency becomes one-half of the original.
3. Selectivity means only the ability to select a desired signal out of the many.
4. Narrow width of the resonance curve of a circuit indicates its higher selectivity.
5. The *Q* of a given coil is totally independent of the frequency of the applied voltage.
6. Coupling circuits and filter circuits often look alike.
7. Main purpose of a filter circuit is to separate currents of different frequencies at some point in the circuit.
8. Maximum energy transfer between two coupled circuits takes place when they are overcoupled.
9. Every type of filter produces signal attenuation though at a different frequency.
10. A highpass filter is no good at stopping signals of high frequency.
11. A bandpass filter can be converted into a bandstop filter simply by interchanging its two resonant circuits.
12. Frequencies within the passband of an ideal bandpass filter have zero attenuation.

C. Multiple Choice Items

1. The process of selecting the desired frequency from a composite wave is called
   (*a*) tuning (*b*) filtering
   (*c*) rectification (*d*) demodulation
2. Tuning circuits use the principle of
   (*a*) maximum energy transfer
   (*b*) mutual inductance
   (*c*) resonance (*d*) modulation
3. Without tuning circuits, no .... would be possible.
   (*a*) amplification (*b*) oscillation
   (*c*) filtering (*d*) communication
4. The main idea behind the use of a coupling circuit is to
   (*a*) connect two independent circuits
   (*b*) transfer energy from one circuit to another
   (*c*) reject unwanted frequencies
   (*d*) utilize mutual inductance
5. The basic function of an electronic filter circuit in radio communication is to
   (*a*) replace tuning circuits
   (*b*) eliminate ripples in dc power supplies
   (*c*) separate different frequencies in a composite signal
   (*d*) divert RF currents out of a given circuit
6. Signal attenuation means its
   (*a*) amplification (*b*) transmission
   (*c*) stoppage (*d*) weakening

ANSWERS

A. Fill in the blanks

1. tuning	2. tuning	3. resonance	4. X_C	5. infinitely
6. capacitance	7. narrower	8. frequency	9. doubled, halved	10. high

B. True or False

1. T 2. F 3. T 4. T 5. F 6. T 7. T 8. F 9. T 10. T 11. T 12. T

C. Multiple Choice Items

1. *a* 2. *c* 3. *d* 4. *b* 5. *c* 6. *d*.

CHAPTER 12

Solid State Physics

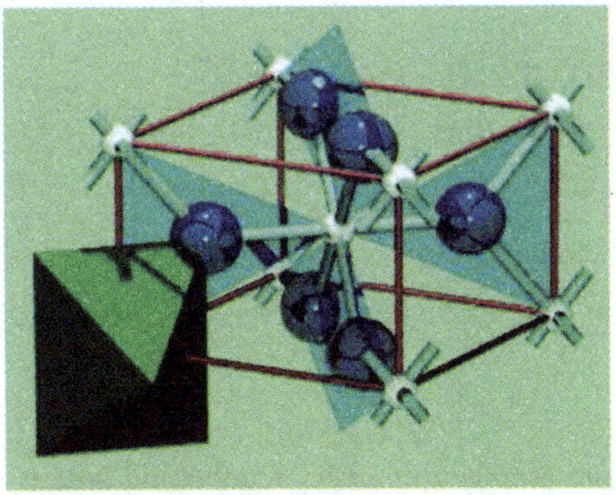

1. Crystalline Solids
2. Unit Cell
3. Forms of Matter
4. Atom and Molecule
5. Atomic Structure
6. Electron Orbits or Shells
7. Electron Distribution of Different Atoms
8. Electrons Suborbits or Subshells
9. Energy Bands in Solids
10. Bonds in Solids
11. Valence and Conduction Bands
12. Conduction in Solids
13. Conductors, Semiconductors and Insulators
14. Majority and Minority Charge Carriers
15. Drift Current in Good Conductors
16. Drift Current in Intrinsic Semiconductors
17. Intrinsic Conduction

12.1. Definition of Matter

Matter is anything which has mass and occupies space. It can exist in any of the following three states :

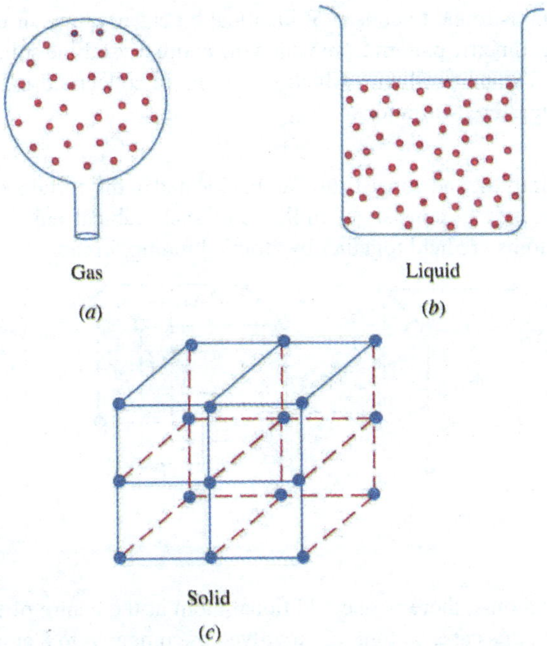

Fig. 12.1

1. solid 2. liquid 3. gas

In the gaseous state, the atoms or molecules are separated from each other by comparatively *large* distances and are *haphazardly arranged* as shown in Fig.12.1 (*a*). In the liquid state, they are somewhat closer but are still arranged haphazardly as shown in Fig. 12.1 (*b*). In the solid state, atoms are the *closest* and take on an orderly three-dimensional geometric pattern called *crystal lattice* as shown in Fig. 12.1 (*c*).

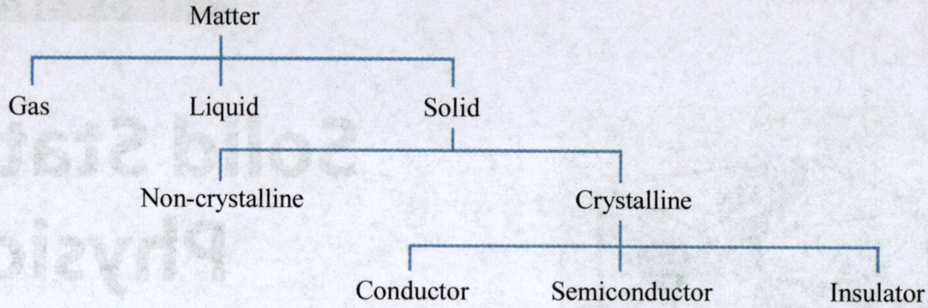

In fact, considered in the materials context, matter may be subdivided into various categories, shown above.

12.2. Crystalline Solids

These are the solids in which atoms or molecules are arranged in a very regular and orderly fashion in a three-dimensional pattern. Each atom or molecule is fixed at a definite point in space at a definite distance and in a definite direction from all others surrounding it. In fact, there is an internal spatial symmetry of atomic orientation within a crystalline solid. This spatial pattern of atoms is called *space lattice* or *lattice array*.

12.3. Unit Cell

The entire lattice structure of a crystal is found to consist of identical blocks of atoms or unit cells. This unit cell is the smallest block or geometric pattern from which the entire crystalline solid is built up by repetition in three dimensions. The unit cell with which we are more concerned in this chapter has cubic structure with the following three variations :

(*a*) Simple Cube

This unit cell or space lattice is the simplest and consists of 8 atoms located at the corners of a cube as shown in Fig. 12.2 (*a*). The entire crystal consists of millions of such cubical unit cells stacked one upon the other. The different atoms are held together by atomic binding forces.

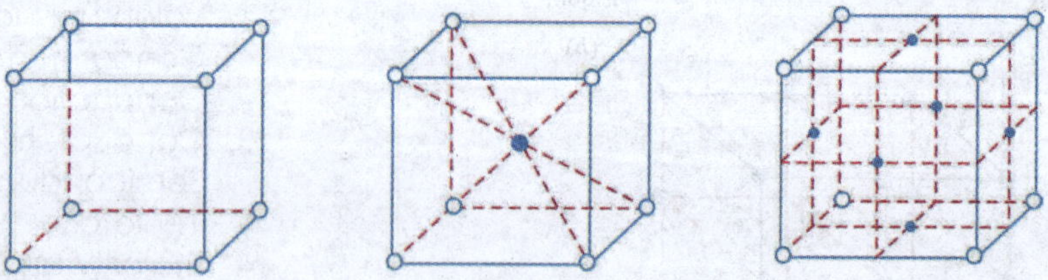

(*a*) (*b*) (*c*)

Fig. 12.2

(*b*) Body Centred Cube

In this unit cell, apart from 8 corner atoms, there is one additional atom at the centre of the cube as shown in Fig. 12.2 (*b*). Obviously, in this case, 9 atoms are involved as compared to 8 atoms in a simple cubic lattice. The two most important semiconductor materials silicon and germanium form this type of cubic crystal lattice.

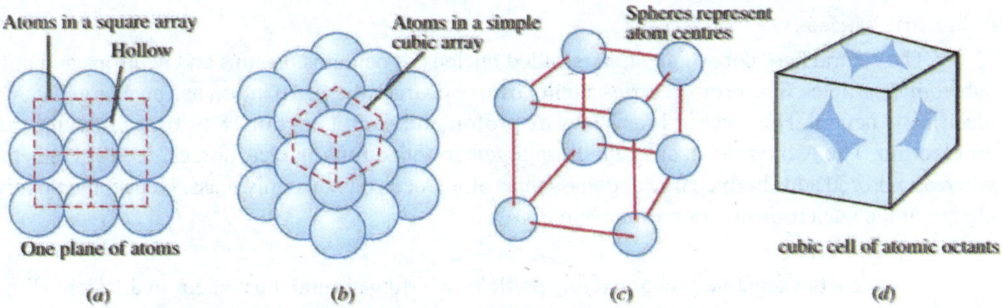

Different ways of representing a simple cubic structure.

(c) Face Centred Cube

In this arrangement, apart from the 8 corner atoms, there is one atom each at the centre of the six faces of the cube as shown in Fig. 12.2 (*c*). In all, there are (8 + 6) = 14 atoms which are bonded together by atomic forces. Copper forms this type of lattice.

12.4. Forms of Matter

In nature, matter is found in the form of either elements or compounds.

(*a*) Element

It is a substance which cannot be broken down any further by chemical methods into a simpler substance.

For example, copper can neither be decomposed into simpler substances nor can it be built up from any simpler substances. Elements are identified by such properties as colour, density and melting temperature etc. Common examples of elements are : oxygen, bromine, aluminium, copper, silicon, germanium and uranium etc. They all are made up of identical atoms which themselves consist of electrons, protons and neutrons etc. Till today, 106 elements have been discovered.

(*b*) Compound

It is a substance which consists of more than one element in chemical combination. For example, water is formed from the chemical union of two elements *i.e.*, hydrogen and oxygen. Sodium carbonate is an example of a compound formed from the union of three elements *i.e.*, sodium, carbon and oxygen. The smallest part of a compound which retains its chemical characteristics is called a *molecule*.

12.5. Atom and Molecule

The atoms of most elements cannot exist by themselves, hence they generally combine to form molecules. A molecule is the smallest particle of a substance that retains all the characteristic properties of the original substance. Molecules are composed of one or more atoms and are called monoatomic, diatomic and triatomic molecules etc.

12.6. Atomic Structure

Our present planetary atomic model was proposed by Niels Bohr in 1913. According to this model, an atom is composed of negatively-charged electrons moving in fixed circular or elliptical orbits around a heavy positively-charged nucleus made up of protons and neutrons as shown by the three-dimensional diagram of Fig. 12.3.

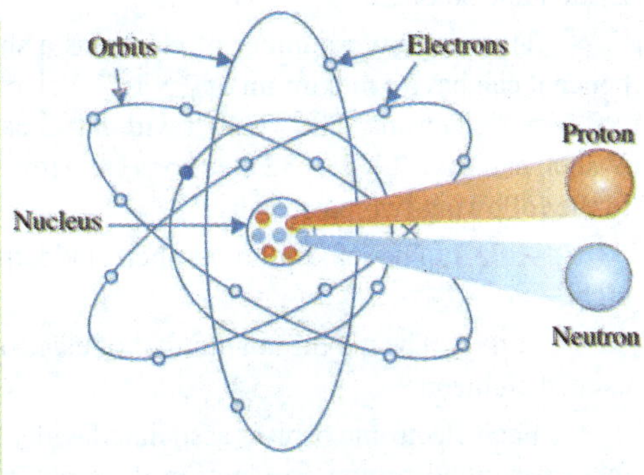

Fig. 12.3

(a) Nucleus

The central hard core of an atom is called nucleus. It contains protons and neutrons and other subatomic particles. The proton carries a unit positive charge whereas neutron has no charge *i.e.*, it is electrically neutral. However, it is as heavy as proton and each is about 1840 times heavier than an electron. The two particles are held close together with strong nuclear forces. It is obvious that whereas a proton adds both charge and mass to an atom, neutron adds only mass. Hence, the positive charge of the nucleus is due to protons only.

(b) Electrons

These are the negatively-charged tiny particles whirling around the nucleus in different elliptical orbits at fantastic speeds. Their number in each orbit is fixed. An electron carries a unit negative charge ($= -1.6 \times 10^{-19}$ C) but has negligible mass. There are as many orbital electrons as there are protons in a given atom which, therefore, makes the entire atom electrically neutral.

The centripetal force ($= mv^2/r$) necessary to keep electrons rotating round the nucleus is supplied by the force of attraction between their opposite charges as given by Coulomb's Law. It is obvious that nearer an electron is to the nucleus, greater is the force with which it is bound to it.

12.7. Atomic Number (Z)

It is equal to the number of protons (or electrons) contained in an atom.

For example, hydrogen (H) atom has one proton in the nucleus (and of course, one orbital electron). Hence, for hydrogen, $Z = 1$. For carbon (C), $Z = 6$, for copper (Cu), $Z = 29$, for silicon (Si) $Z = 14$ and for germanium (Ge), $Z = 32$.

12.8. Atomic Mass Number (A)

It gives the total number of protons and neutrons contained in the nucleus of an atom. For example, Si has 14 protons and 14 neutrons in its nucleus. Hence, its $A = 28$. Similarly, Ge has 32 protons and 41 neutrons thereby having $A = 73$.

The atomic weight is the *actual* weight of an atom and numerically differs slightly from the atomic mass number.

12.9. Electron Orbits or Shells

According to Planck's Quantum Theory, an electron cannot revolve round the nucleus in any arbitrary orbit but in only certain definite discrete and fixed orbits. These orbits are designated by alphabetical letters K, L, M, N, O, P etc., starting from the nucleus outwards. The orbit (or shell) closest to nucleus is called K-shell, the next farther one is called L-shell and so on. These shells are also known by their principal quantum number n which can have values of $n = 1$ for K-shell, $n = 2$ for L-shell and so on.

The maximum number of electrons a shell can have $= 2n^2$. For example, for K-shell, $n = 1$, hence it can have a maximum of $2 \times 1^2 = 2$ electrons. Similarly, L-shell with $n = 2$, can accommodate $2 \times 2^2 = 8$ electrons. The M-shell with $n = 3$ can have a maximum of $2 \times 3^2 = 18$ electrons whereas N-shell can have $2 \times 4^2 = 32$ electrons etc. However, this electronic distribution in all atoms is subject to the following two basic rules :

Rule 1. The maximum number of electrons in the outermost shell of an atom cannot exceed eight.

Rule 2. The maximum number of electrons in the shell just prior to the outermost shell cannot exceed eighteen.

Each electronic orbit is associated with a certain definite amount of energy. While revolving in these permitted orbits, an electron does not radiate out any energy. But it does radiate out some definite energy when *jumping from one orbit to another*. If E_2 and E_1 are the energies corresponding to the two orbits before and after the jump, the frequency of the emitted radiation is given by the relation

$$E_2 - E_1 = hf \quad \text{or} \quad \Delta E = hf$$

where h = Planck's constant

$= 6.625 \times 10^{-34}$ J-s.

12.10. Electron Distribution of Different Atoms

In Fig. 12.4 is shown the two-dimensional schematic distribution of electrons in different orbits for hydrogen, boron, silicon and germanium atoms. Following facts are clear from the diagram.

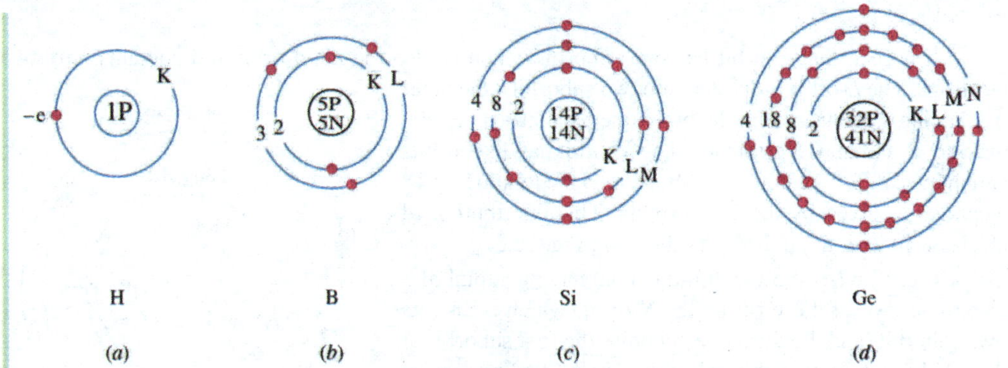

Fig. 12.4

The hydrogen atom [Fig. 12.4 (*a*)] has one positive proton in its nucleus symbolised as "IP". To balance this charge, it has one electron revolving in the extra-nuclear *K*-shell.

As shown in Fig. 12.4 (*b*), a boron atom has 5 protons and 5 neutrons inside its nucleus and 5 electrons outside it. Two electrons occupy *K*-shell which is then said to be completely- or fully-occupied because it cannot hold more electrons as per the $2n^2$ formula. The other three electrons occupy the next *L*-shell which, obviously, is not full or completely-occupied because it can hold a total of 8 electrons.

The Si atom of 12.4 (*c*) has *K*- and *L*-shells completely filled whereas *M*-shell is partially-occupied because it has only four electrons*.

Similarly, Ge atom of Fig. 12.4 (*d*) has first three orbits fully occupied whereas the fourth (and the outermost) orbit is partially filled. It could also have at least four additional electrons (Rule 1 of Art. 12.9).

12.11. Electrons Suborbits or Subshells

It has been found that each electron orbit actually contains a number of *suborbits* with the exception of the $n = 1$ or *K*-orbit which is its own suborbit. The number of suborbits in an orbit equals its principal quantum number, n. For example, *K*-orbit ($n = 1$) has one suborbit, *L*-orbit ($n = 2$) has two suborbits and *M*-orbit ($n = 3$) consists of three suborbits etc. These orbits and suborbits are also respectively known as shells and subshell. The total number of electrons (which equals $2n^2$) gets distributed amongst these subshells as shown on next page in Table 12.1.

* Being the outermost orbit, it could have at least 4 more electrons (Art. 12.9).

Table 12.1

Orbit	Total No. of electrons	No. of suborbits	Electron distribution in suborbits
K ($n = 1$)	2	1	2
L ($n = 2$)	8	2	2,6
M ($n = 3$)	18	3	2, 6, 10
N ($n = 4$)	32	4	2, 6, 10, 14

It is seen that *K*-orbit has one suborbit which is also the main orbit and contains only two electrons. The *L*-orbit is divided into two suborbits, the inner one having 2 electrons and the outer one 6 electrons thereby making up a total of 8 electrons. The *M*-orbit has 3 suborbits which have 2, 6, 10 electrons respectively, the total being 18 electrons as given by the $2n^2$ formula. This distribution of electrons in various suborbits is shown in Fig. 12.5.

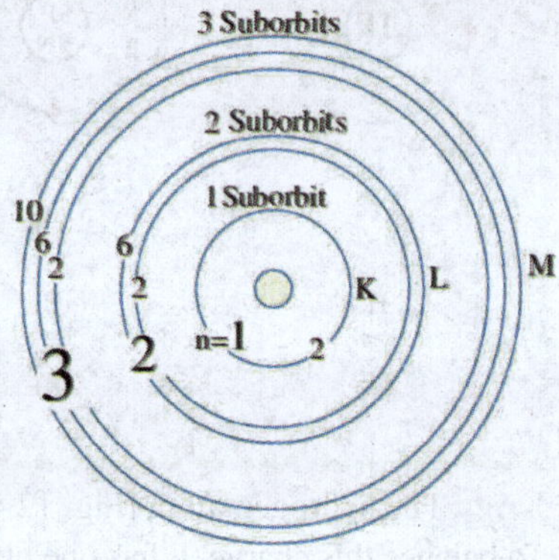

Fig. 12.5

Fig. 12.6 (*a*) shows a lithium atom having a total of 3 electrons. Two of these fill up the *K*-orbit (which is also its own suborbit) and the third one occupies the first suborbit of the *L*-orbit. Obviously, this suborbit is half-filled because it needs one more electron to fill it completely.

Fig. 12.6 (*b*) shows electron distribution in a boron atom. As seen, this atom has a total of 5 electrons. As usual, *K*-orbit has 2 electrons. The first suborbit of *L*-orbit has 2 electrons and the next suborbit has just one electron (though it could have five more). Hence, second suborbit of *L*-orbit is partially-filled.

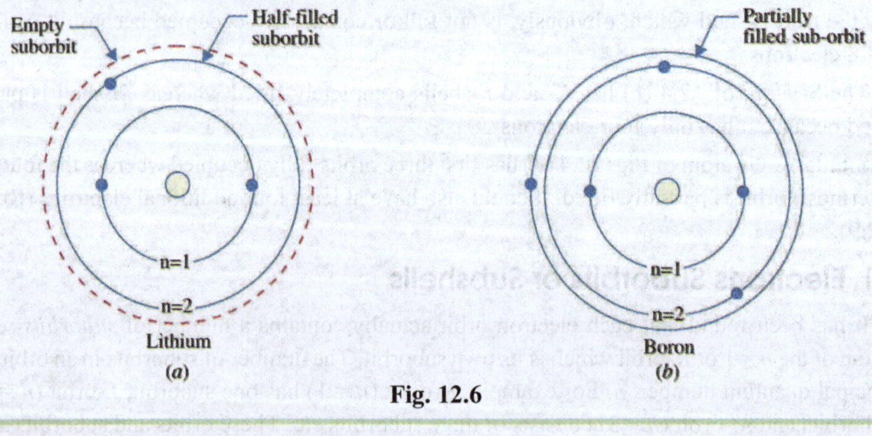

Fig. 12.6

Example 12.1. *The atomic number of copper is 29. Give its electronic distribution.*

Solution. Since atomic number represents protons, it means that the number of protons in the nucleus of Cu is 29. This is also the number of orbiting electrons which are distributed orbitwise as under :

K-orbit =	2 electrons	— full
L-orbit =	8 electrons	— full
M-orbit =	18 electrons	— full
N-orbit =	1 electron	— incomplete
Total =	29 electrons	

Example 12.2. *Silicon atom has Z = 14. Give its electronic distribution indicating which orbits are full and which are not.*

Solution. There are 14 electrons per atom. They are distributed as under :

K-orbit =	2 electrons	— full
L-orbit =	8 electrons	— full
M-orbit =	4 electrons	– incomplete

Example 12.3. *Xenon atom has Z = 54. Determine its electronic distribution. Also find out if it is chemically stable. What is the number of its valence electron?*

Solution. The 54 electrons are distributed as under :

Shell	:	*K*	*L*	*M*	*N*	*O*
No. of electrons	:	2	8	18	18	8

It is seen that it has 8 valence electrons. Hence, its valence or outermost orbit has maximum possible number of electrons as permitted by Rule 1 of Art. 12.9. Consequently, Xenon atom is chemically highly stable and is also an insulator.

12.12. Valence Electrons

Most of the atoms of different elements do not have their outermost shells completely-filled *i.e.*, they do not have eight electrons in their outermost orbit as permitted by Rule 1 of Art. 12.9. The electrons occupying the outermost orbit or shell of an atom are called *valence* electrons. They determine the chemical and electrical properties of the element. Elements deficient in valence electrons are highly active in the sense that they are always ready to chemically combine with other elements. Those elements which have one or two valence electrons are good conductors of electricity. These valence electrons are responsible for forming atomic bonds (Art. 12.18). The number of valence electrons in an atom also determines the valency of the element whose atom it is. For example, boron is trivalent whereas both Si and Ge are tetravalent *i.e.*, they have four valence electrons each.

12.13. Orbital Energy

The total energy (both kinetic and potential) possessed by an electron when it revolves in the *n*th orbit of an atom with atomic number *Z* is given by

$$E_n = -21.76 \times 10^{-19} \frac{Z}{n^2} \text{ joule}$$

Now, 1 electron-volt (eV) = 1.6×10^{-19} J

$$\therefore \quad E_n = -\left(\frac{21.76 \times 10^{-19}}{1.6 \times 10^{-19}}\right)\frac{Z^2}{n^2} \text{ eV} = -13.6 \frac{Z^2}{n^2} \text{ eV}$$

From the above equation, it is seen that

1. Total energy of the electron is *negative i.e.*, it is a *binding* energy.
2. Total energy is inversely proportional to the square of the principal quantum number.
   It means that the *binding* energy of electrons is less when they revolve in higher orbits. Consequently, it is easier to remove electrons from *M*-shell than from *L*-shell and also from *N*-shell than from the *M*-shell and so on.
3. Total energy of the electron varies as Z^2.
   It means that electrons in the atoms of heavier elements have much greater binding energy than those in the lighter elements when revolving in the same orbit.

12.14. Normal, Excited and Ionised Atom

Consider the case of the simplest atom *i.e.*, hydrogen atom which has only one orbital electron. When this electron occupies the innermost orbit *i.e.*, *K*-orbit ($n = 1$), the atom is said to be in its *normal* or *ground* or *unexcited* state as shown in Fig. 12.7 (*a*). Generally, it is this condition in which most of the free hydrogen atoms in a gas are found to exist at normal temperature and pressure. However, if a spark is passed through hydrogen gas contained in a vessel, the high-speed electrons produced by this spark collide with the hydrogen electron and may either raise it from its normal $n = 1$ orbit to higher permitted orbits with $n = 2, 3, 4$ etc. [Fig. 12.7 (*b*)] or may altogether remove it from the atom itself and make it a free electron which is no longer bond to the atom as shown in Fig. 12.7 (*c*).

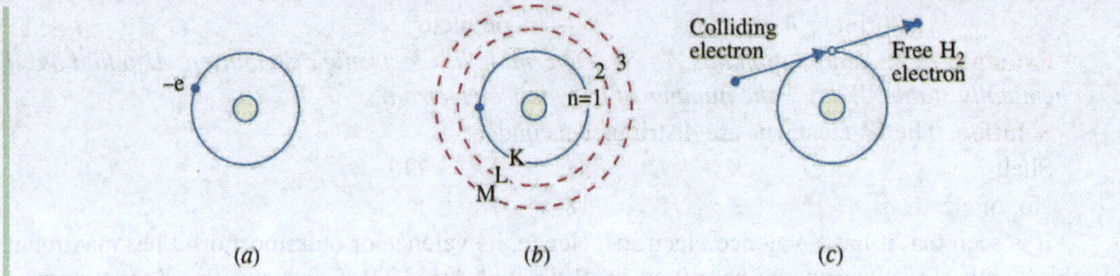

Fig. 12.7

As shown in Fig. 12.8, the hydrogen electron can be raised to higher permitted orbits provided it is given the right amount of energy. In that case, the hydrogen atom is said to be *excited* or be in an *excited* state. However, the atom does not remain in the excited state longer than 10^{-8} second because the raised electron tends to come back to its normal or ground state. In so doing, it gives out the energy (it had gained earlier during collision) in the form of radiations which may or may not be visible. The raised electron may return by several jumps instead of one thereby emitting many radiations of different frequencies (and, hence wavelengths).

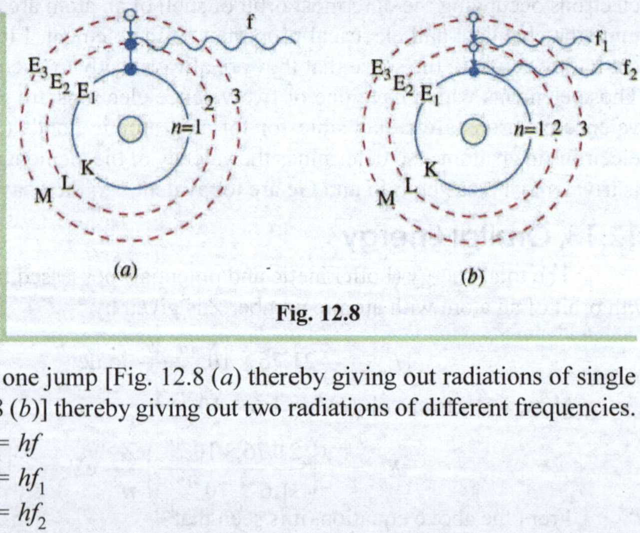

Fig. 12.8

As shown in Fig. 12.8, suppose that the electron had been raised to $n = 3$ orbit where its energy is E_3. It can, now, come back to its normal $n = 1$ orbit either directly by one jump [Fig. 12.8 (*a*) thereby giving out radiations of single frequency or by two jumps [Fig. 12.8 (*b*)] thereby giving out two radiations of different frequencies.

In Fig. 12.8 (*a*), $E_3 - E_1 = hf$

In Fig. 12.8 (*b*), $E_3 - E_2 = hf_1$

$E_2 - E_1 = hf_2$

If, as shown in Fig. 12.7 (*c*), the electron is completely removed from the atom, the atom is said to be *ionised i.e.*, it becomes a positively-charged ion. The energy required to remove the electron away from its parent atom is called *ionisation energy*.

12.15. Orbital Energies in Hydrogen Atom

The general expression for total energy of an electron in an atom is

$$E_n = -13.6 \frac{Z^2}{n^2} \text{ eV}$$

For hydrogen $Z = 1$

$$\therefore \quad E_n = -\frac{13.6}{n^2} \text{ eV}$$

(*i*) For *K*-shell, $n = 1$

$$\therefore \quad E_1 = -\frac{13.6}{1^2} = -13.6 \text{ eV}$$

(*ii*) For *L*-shell, $n = 2$

$$\therefore \quad E_2 = -\frac{13.6}{2^2} = -3.4 \text{ eV}$$

(*iii*) For *M*-shell, $n = 3$

$$\therefore \quad E_3 = -\frac{13.6}{3^2} = -1.51 \text{ eV etc.}$$

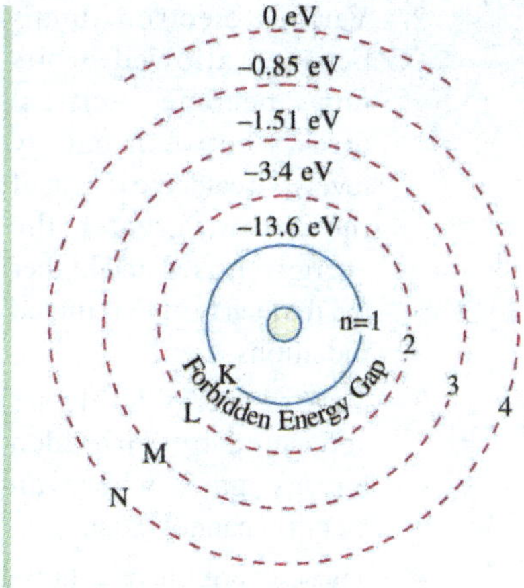

Fig. 12.9

The different permitted orbits with their associated electronic energies are shown in Fig. 12.9. It is seen that the electrons can have certain specific amounts of energy only. In the $n = 1$ orbit, it can have an energy of – 13.6 eV and in the $n = 2$ orbit, it can have – 3.4 eV. It cannot have any energy *lying in between these two values*. For example, we will have to give it an extra amount of energy = (13.6) – (3.4) = 10.2 eV to raise it from the lower $n = 1$ orbit to the higher $n = 2$ orbit. If we try to give it, say, 8 eV, it will simply not accept it because it is forbidden (by Quantum Theory) to lie anywhere in between the permitted orbits.

Example 12.4. *Determine the frequency of radiation emitted by an electron in an excited hydrogen atom when it jumps from n = 3 orbit to n = 2 orbit. Take h = 6.625 × 10⁻³⁴ J-s.*

Solution. As seen from Art. 12.5, $E_3 = -1.51$ eV and $E_2 = -3.4$ eV

$$\therefore \quad E_3 - E_2 = -1.51 - (-3.4) = 1.89 \text{ eV}$$

$$= 1.89 \times 1.6 \times 10^{-19} \text{ J}$$

Now, $\quad E_3 - E_2 = hf$

or $\quad 1.89 \times 1.6 \times 10^{-19} = 6.625 \times 10^{-34} \times f$

$\therefore \quad f = \mathbf{4.57 \times 10^{14}\ Hz}$

If required, wavelength of the radiations emitted can also be found.

Now, $\quad c = f\lambda$

where c is the velocity of light in vacuum.

$\therefore \quad 3 \times 10^8 = 4.57 \times 10^{14} \times \lambda$

$\therefore \quad \lambda = \mathbf{6563 \times 10^{-10}\ m}$

12.16. Energy Levels in an Isolated Atom

We will consider a *single isolated* atom of hydrogen *i.e.*, an atom which is so far removed from other atoms as not to be affected at all by their electric fields. In that case, energies possessed by various electrons in different orbits would remain totally unaffected by any external influence. The orbital energies have already been given in Art. 13.15 above. However, instead of drawing various orbits to the scale of their radii as in Fig. 12.9, it is customary (and also more convenient) to draw horizontal lines to an energy scale as shown in Fig. 12.10. Such a diagram is called an *energy-level diagram* (ELD) of the atom.

Each energy level is represented by a horizontal line of the same length although the length and thickness of the lines have no significance. It is seen that in this diagram

1. less negative energies are at the *top* whereas more negative ones are at the *bottom*. It means that it is easier to remove electrons from higher orbits than from lower ones.

2. Various electron jumps between allowed orbits now become vertical arrows between energy levels. Greater the length of the arrow, greater the energy released and higher the frequency of the emitted radiations.
3. Various energy levels are separated by forbidden energy gaps where an electron cannot exist.
4. These energy levels, however, carry *no indication of their occupancy by electrons* which level an electron will occupy at a given time depends on the total energy possessed by it. However, the maximum number of electrons which an energy level can have is given by $2n^2$ and has been shown in the ELD of Fig. 12.10.

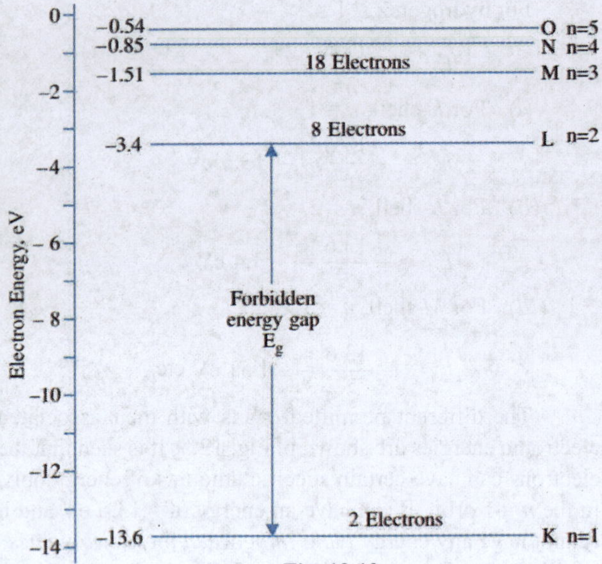

Fig. 12.10

12.17. Energy Bands in Solids

So far we have considered electron energy *levels* in a *single isolated* atom. Atoms of gases can be considered isolated because of their greater spacing under normal pressure and temperature. Hence, energy levels of such atoms are not affected by other distant atoms. But there are significant changes in the energy levels when atoms exist close together as in solids. If there are 100 atoms in a piece of solid material, then there will be 100 levels of slightly different energies for each *n*-value *i.e.*, 100 *energy-level-diagrams would be superimposed on each other*. Consequently, a single energy level of an isolated atom will become a *band* of energy in a solid as shown in Fig. 12.11.

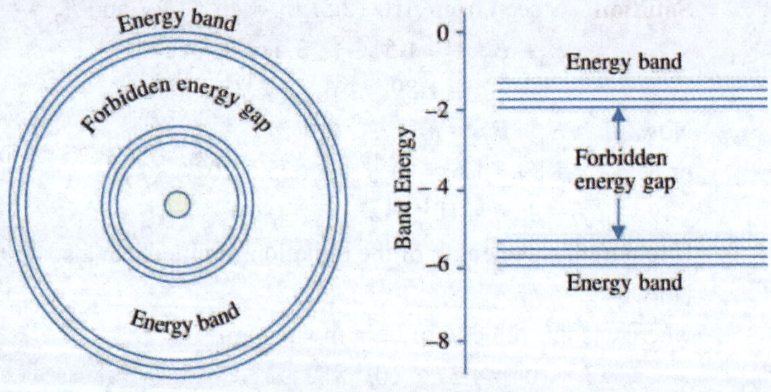

Fig. 12.11

Only four levels have been shown in each band though their number should be equal to the number of atoms which are close enough to influence each other (in the present case, 100).

Let us, for example, consider the lithium element which exists as a solid metal at ordinary room temperature. Each lithium atom has three electrons two of which occupy the *K*-shell (or *K*-level), whereas the third one occupies the first sub-shell s_1 of the *L*-shell. As explained earlier in Art. 12.11, this subshell is half-filled because it can accommodate a total of two electrons. Of course, suborbit s_2 is unoccupied. The ELD of single lithium atom along with its state of electron occupancy is shown in Fig. 12.12 (*a*).

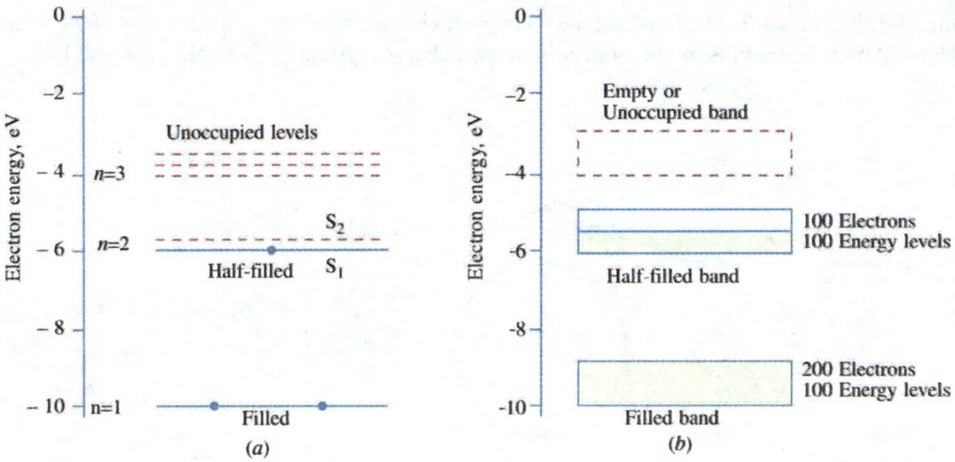

Fig. 12.12

Now, consider a small piece of lithium metal containing 100 atoms. It is obvious that there would be 100 *K*-levels differing in energy by extremely small amounts. They are, in fact, so close as to merge into one energy band as shown in Fig. 12.12 (*b*). Similarly, the 100 *L*-levels also blend into one energy band.

Now, let us consider the electron occupancy of these bands. Since each *K*-level has two electrons, there would be 2 × 100 = 200 electrons occupying the first or *K*-energy band. Since, the first sublevel s_1 of the main *L*-level of each atom has one electron, it is half-filled. Hence, the energy band corresponding to this sublevel s_1 of Fig. 12.12 (*a*) is half-filled *i.e.*, it contains only 100 electrons though it could accommodate 200 electrons. Of course, all the higher energy bands are unoccupied because Li atom has no electrons to fill them.

12.18. Bonds in Solids

There are three major types of atomic bonds created by the valence electrons. These are known as

1. ionic bonds, 2. covalent bonds, and 3. metallic bonds.

These bonds determine the electrical conductivity of the solid.

(*i*) Ionic Bonds

These occur between two *different* atoms and are due to *permanent transfer* of valence electrons from one atom to another. It is a very strong bond. Such bonds are formed between sodium and chlorine atoms during the formation of sodium chloride (NaCl) crystals.

(*ii*) Covalent Bonds

These can occur between two similar or dissimilar atoms. In this case, there is only *sharing* of one or more valence electrons between the two atoms each of which tries to fill up its outermost orbit.

Fig. 12.13 depicts two ways of showing this covalent bonding which occurs mostly in semi-conductor materials like Si and Ge. Only valence electrons have been shown in the figure. As seen, each tetravalent Si atom shares one electron each with four surrounding atom thereby completing its quota of 8 electrons in its outermost orbit (Rule 1 of Art. 12.9). Such bonds can be broken by supplying sufficient energy to the

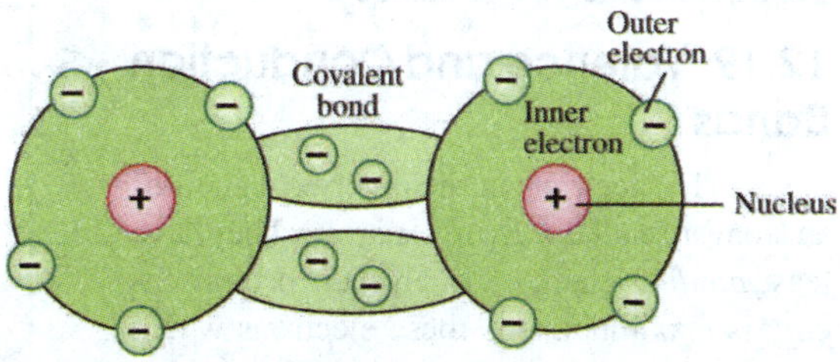

Covalent bonds of the oxygen molecule showing the sharing of the outer electrons.

solid. The electron set free by breaking the bond leaves behind a vacancy called *positive hole*. Hence, each bond breakage results in the production of two charge carriers *i.e.*, one **electron** and **hole**.

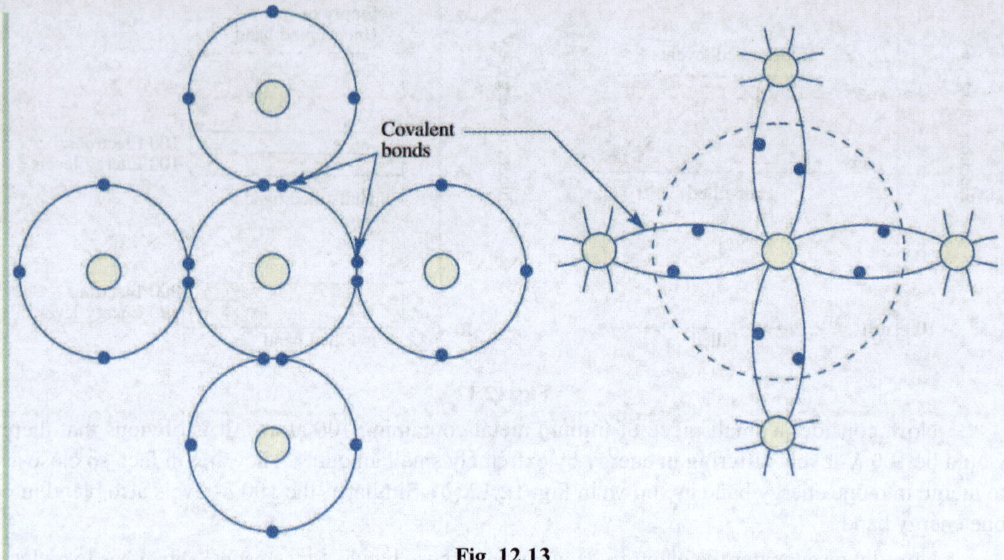

Fig. 12.13

(*iii*) Metallic Bonds

As the name indicates, such bonds, exist in metals and their alloys. They arise due to *sharing of variable number of electrons by a variable number of atoms*. It has been found that in metals (like Cu), each atom loses its valence electron at room temperature and becomes positively-charged ion. These free and mobile electrons form a kind of electron *cloud* or *gas* which permeates all atoms. In fact, one may look upon a metal as consisting of an array of closely-packed positive ions immersed in a sea of electrons. These valence electrons are not bound to any particular atom but freely move about from one atom to another. In fact, high electrical conductivity of metals is primarily due to the easy availability of such free electrons in extremely large numbers.

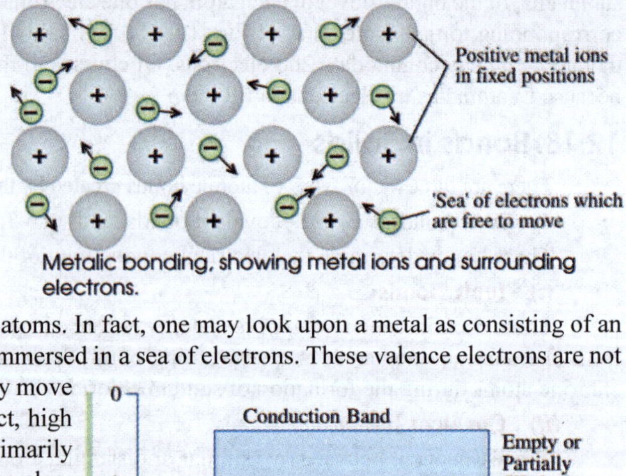

Metalic bonding, showing metal ions and surrounding electrons.

12.19. Valence and Conduction Bands

The electrons in the outermost shell of an atom are called *valence* electrons. They have least *binding* energy though their orbital energy is maximum. It is these electrons which are most affected when a number of atoms are brought *very close* together as during the formation of a solid. The states of lower-energy

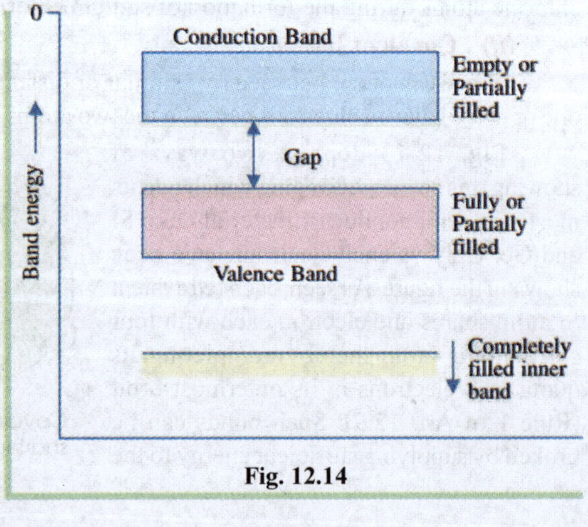

Fig. 12.14

electrons orbiting in shells nearer to the nucleus are little, if at all, affected by this atomic proximity.

The energy band occupied by the valence electrons is called the **valence band** and is, obviously, the highest occupied band. As regards occupancy, it may be either completely-filled or partially-filled with electrons but *can never be empty*.

The electrons which have left the valence band are called *conduction* electrons. They practically leave the atom or are only weakly held to the nucleus. The band occupied by these electrons is called the *conduction band*. This band lies next to the valence band. It may either be empty (when no free electrons are available) or partially filled with electrons. In fact, it may be defined as *the lowest unfilled energy band in an atom*.

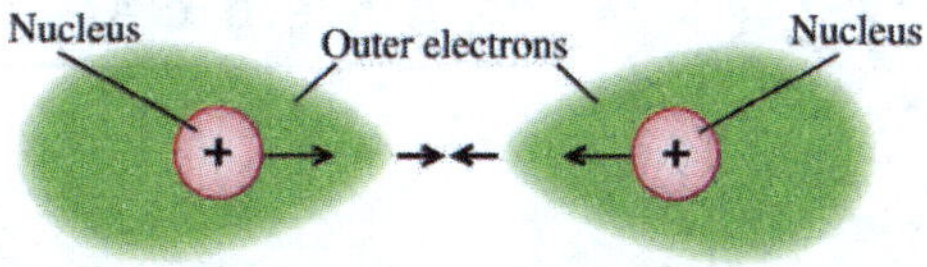

Van der Waals bonding between two neutral atoms.

In conduction band, electrons move freely and conduct electric current through the solid. That is why they are called conduction electrons.

As shown in Fig. 12.14, the valence band and conduction band are separated by a gap known as *forbidden energy gap*. Energy is required to lift an electron from the valence band, take it across the gap and put it in the conduction band.

12.20. Conduction in Solids

Current flow occurs in a given material when a voltage of suitable magnitude is applied to it which causes the charge carriers within the material to move in a desired direction. This may be due to one or both of the processes *i.e.*, electron motion and hole transfer. In electron motion, free electrons in the conduction band are moved under the influence of the electric field set up by the applied voltage. They travel from the negative terminal of the applied voltage to its positive terminal.

Hole transfer involves electrons in the *valence band* which (unlike conduction electrons) are still attached to the atom.

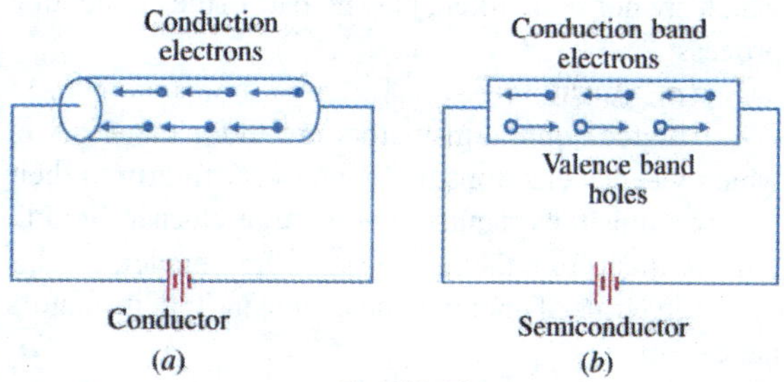

Fig. 12.15

In good conductors, current flow is due to free or conduction electrons only [Fig. 12.15 (*a*)]. There is no hole current in them (Art. 12.28). However, current flow in semiconductors consists *partly of electron current and partly of hole current* [Fig. 12.15 (*b*)].

Another point worth noting is that free electrons require much less energy than holes to move them because they are already disconnected from their atoms. It is for this reason that conduction electrons have greater mobility than holes.

12.21. Hole Formation and its Movement

As explained earlier, the covalent forces of the crystal lattice have their source in the valence band. If an electron in the valence band gets sufficient energy, it can jump across the forbidden energy gap and enter the conduction band. As shown in Fig. 12.16 (*a*), suppose covalent bond at *A* breaks due to the departure of an electron. A vacancy is left behind in the valence band which is called a positive hole. The chain of events that follows is depicted in Fig. 12.16 (*a*). An electron at *B* from an adjacent atom jumps into the hole at *A*. This fills the original hole but creates a new hole at *B*. Next, an electron at *C* jumps into the hole at *B* and so on. In this way, by a succession of electron movement, a hole appear at *G* and a negative charge would have moved from *G* to *A*. But it is more convenient to consider as if a *positive hole has moved from A to G*.

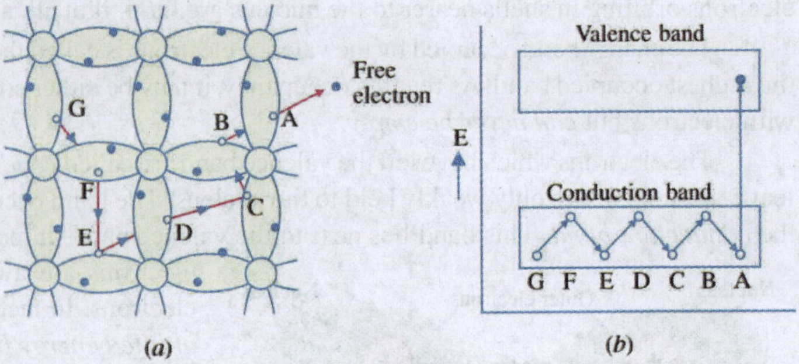

Fig. 12.16

It should be carefully noted that holes are filled by electrons which move from adjacent atoms *without* passing through the forbidden energy gap. It means that *hole movement takes place in valence band only*. No new hole is created if an existing hole is filled by a free electron and not by one from the adjacent atom.

To summarise the above, it may be repeated that

1. conduction electrons are found in and freely flow in the conduction band,
2. holes exist in and flow in the valence band,
3. electronic current is in conduction band but hole current is in valence band,
4. conduction electrons move almost twice as fast as the holes.

12.22. Conductors, Semiconductors and Insulators

The electrical conduction properties of different elements and compounds can be explained in terms of the electrons having energies in the valence and conduction bands. The electrons lying in the lower energy bands, which are normally filled, play no part in the conduction process.

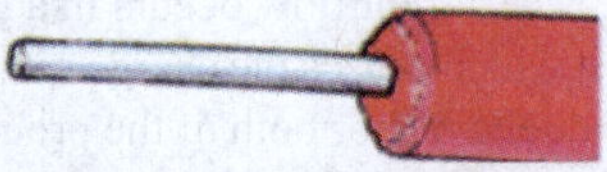

Single-core wire

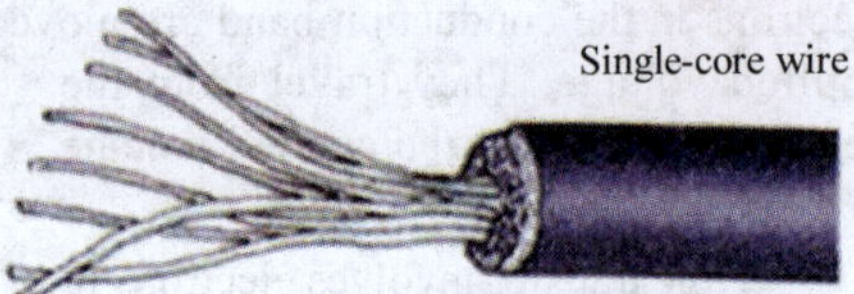

Insulated multi-strand wire

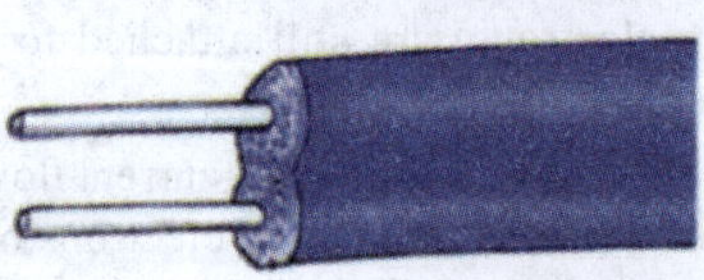

Pairs of wires with insulation

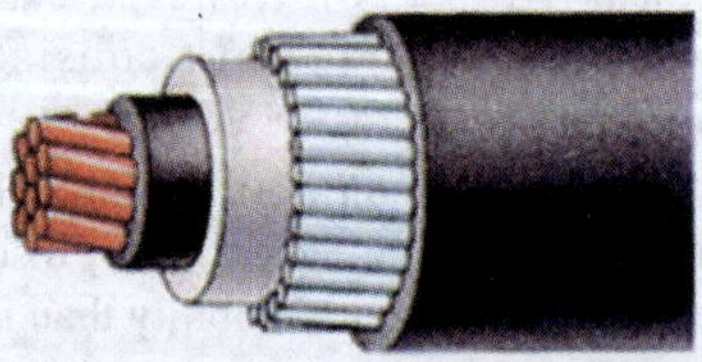

Electrical cable with three layers of insulation

By placing a rubber or plastic insulator as a sheath around a good conductor, such as a copper wire, electricity can flow along the wire without causing harm.

(*a*) Insulators

Stated simply, insulators are those materials in which valence electrons are bound very tightly to their parent atoms thus requiring very large electric field to remove them from the attraction of their nuclei.

In terms of energy bands, it means that insulators have

1. full valence band,
2. an empty conduction band, and
3. a large energy gap (of several eV) between them.

This is shown in Fig. 12.17 (*a*). For conduction to take place, electrons must be given sufficient energy to jump from the valence band to the conduction band. Increase in temperature enables some electrons to go to the conduction band which fact accounts for the negative temperature coefficient of resistance of insulators.

(*b*) Conductors

Conducting materials are those in which plenty of free electrons are available for electric conduction.

In terms of energy bands, it means that electrical conductors are those which have *overlapping valence and conduction bands* as shown in Fig. 12.17 (*b*).

In fact, there is no physical distinction between the two bands. Hence the availability of a large number of conduction electrons.

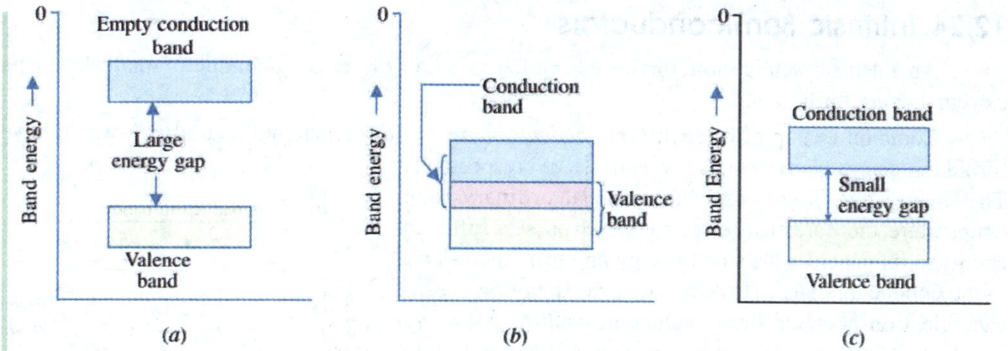

Fig. 12.17

Another point worth noting is that in the absence of forbidden energy gap in good conductors, there is no structure to establish holes. The total current in such conductors is simply a flow of electrons. It is exactly for this reason that the existence of holes was not discovered until semiconductors were studied thoroughly.

(*c*) Semiconductors

A semiconductor material is one whose electrical properties lie in between those of insulators and good conductors. Examples are : germanium and silicon.

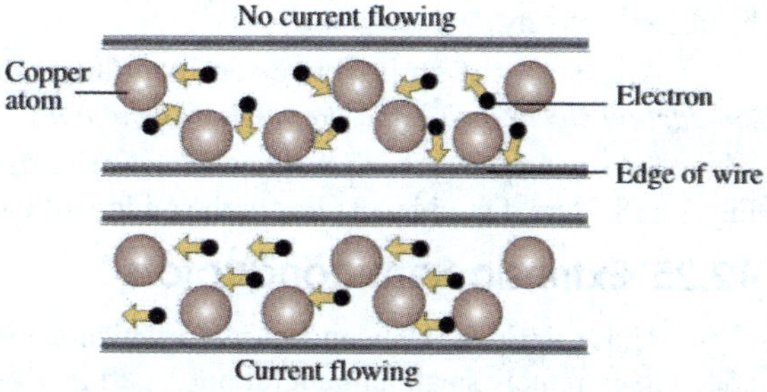

Current moves through materials that conduct electricity.

In terms of energy bands, semiconductors can be defined as those materials which *at room temperature* have

1. partially-filled conduction band,
2. partially-filled valence band, and
3. a very narrow energy gap (of the order of 1 eV) between them.

At 0°K, there are no electrons in the conduction band of semiconductors and their valence band is completely filled. It means that at absolute zero temperature, a piece of Ge or Si acts like a perfect insulator. However, with increase in temperature, width of the forbidden energy band is decreased so that some of the electrons are liberated into the conduction band. In other words, conductivity of semiconductors increases with temperature. It means that they have *negative* temperature coefficient of resistance.

Typical resistivities are of the following order :

Conductors	10^{-8} Ω-m
Semiconductors	10^{-3} Ω-m
Insulators	10^{-12} Ω-m

12.23. Types of Semiconductors

Semiconductors may be classified as under :

Semiconductors

- Intrinsic or pure semiconductors
- Extrinsic or impure semiconductors
  - *N*-type
  - *P*-type

12.24. Intrinsic Semiconductors

An intrinsic semiconductor is one which is made of the semiconductor material in its extremely pure form.

Common examples of such semiconductors are : pure germanium and silicon which have forbidden energy gaps of 0.72 eV and 1.1 eV respectively. The energy gap is so small that even at ordinary room temperature, there are many electrons which possess sufficient energy to jump across the small energy gap from the valence to the conduction band. However, it is worth noting that for each electron liberated into conduction band, *a positively-charged hole is created in the valence band*. When an electric field is applied to an intrinsic semiconductor at a temperature greater than 0°K, conduction electrons move to the anode and the holes in the valence band move to the cathode. Hence, semiconductor current consists of movement of electrons and holes in *opposite* directions in the *conduction* and *valence* bands respectively.

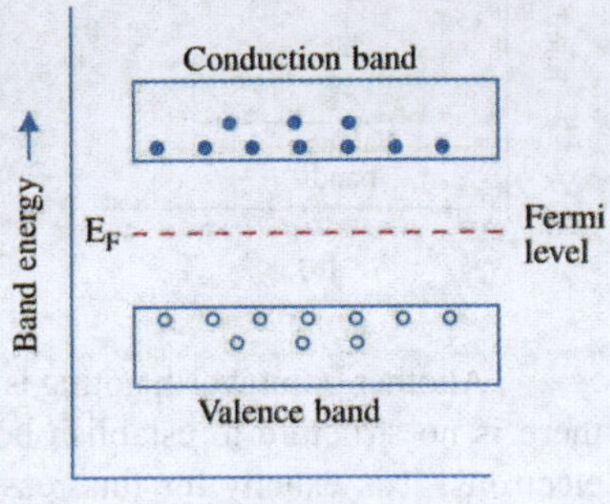

Fig. 12.18

Alternatively, an intrinsic semiconductor may be defined as one in which the *number of conduction electrons is equal to the number of holes*.

Schematic energy band diagram of an intrinsic semiconductor at room temperature is shown in Fig. 12.18. Here, Fermi level* lies in the middle of the forbidden energy gap.

12.25. Extrinsic Semiconductors

Those intrinsic semiconductors to which some suitable impurity or doping agent has been added in extremely small amount (about 1 part in 10^8) are called *extrinsic* or *impurity* semiconductors.

Usually, the doping agents are pentavalent atoms having five valence electrons (bismuth, antimony, arsenic, phosphorus) or trivalent atoms having three valence electrons (gallium, indium, aluminium, boron). Pentavalent doping atom is known as *donor atom* because it *donates* or contributes one electron to the conduction band of pure germanium. The trivalent atom, on the other hand, is called *acceptor atom* because it *accepts* one electron from the germanium atom. The reason why doping materials are called *impurities* is that they *alter the structure of pure semiconductor crystals.*

Depending on the type of doping material used, extrinsic semiconductors can be further subdivided into two classes :

(*i*) *N*-type semiconductors and (*ii*) *P*-type semiconductors.

(*i*) N-type Extrinsic Semiconductor

This type of semiconductor is obtained when a pentavalent material like antimony (Sb) is added to pure germanium crystal. As shown in Fig. 12.19 (*a*), each antimony atom forms covalent bonds with the surrounding four germanium atoms with the help of four of its five electrons. The fifth electron is superfluous and is loosely bound to the antimony atom. Hence, it can be easily excited from the valence band to the conduction band by the application of electric field or increase in its thermal energy. Thus, practically every antimony atom introduced into the germanium lattice contributes one conduction electron *without creating a positive hole*. Antimony is called donor impurity and makes the pure germanium an *N*-type (*N* for the negative) extrinsic semiconductor.

* For the present discussion, Fermi level may be defined as the energy which corresponds to the centre of gravity of conduction electrons and holes weighted according to their energies.

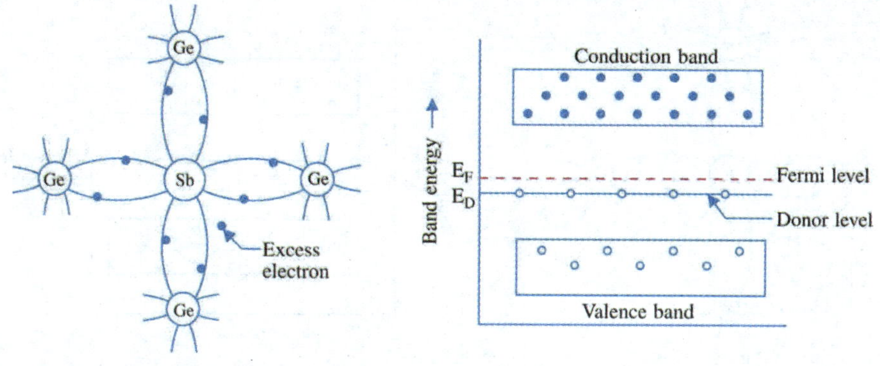

(a) (b)

Fig. 12.19

As an aid to memory, the student should associate the letter **N** in 'do**N**or' with the **N** in '**N**-type extrinsic semiconductor' and with the **N** in '**N**egative charge carrier'.

It may be noted that by giving away its one valence electron, the donor atom becomes a positively-charged ion. But it cannot take part in conduction because it is firmly fixed or tied into the crystal lattice.

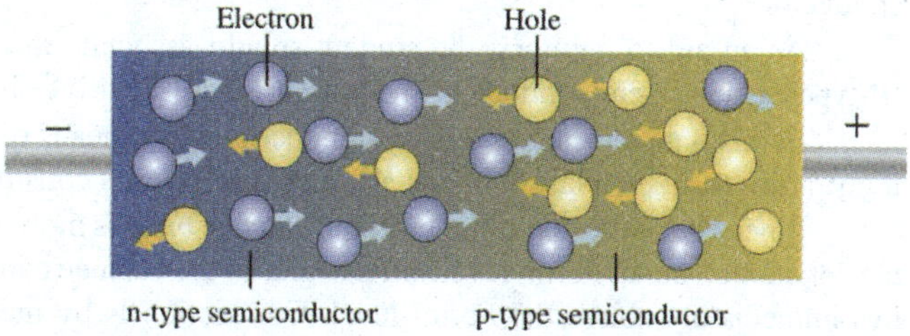

In this diagram, current flows in the preferred direction.

As seen, in addition to the electrons and holes intrinsically available in germanium, the addition of antimony greatly increases the number of conduction electrons. Hence, concentration of electrons in the conduction band is increased and exceeds the concentration of holes in the valence band. Because of this, Fermi level shifts *upwards* towards the bottom of the conduction band as shown in Fig. 12.19 (*b*).

In terms of energy levels, the fifth antimony electron has an energy level (called donor level) just below the conduction band. Usually, donor level is 0.01 eV below conduction band for germanium and 0.054 eV for silicon.

It is seen from the above description that in an *N*-type semiconductor, electrons are the majority carriers while holes constitute the minority carriers.

(*ii*) P-type Extrinsic Semiconductor

This type of semiconductor is obtained when traces of a trivalent impurity like boron (B) are added to a pure germanium crystal.

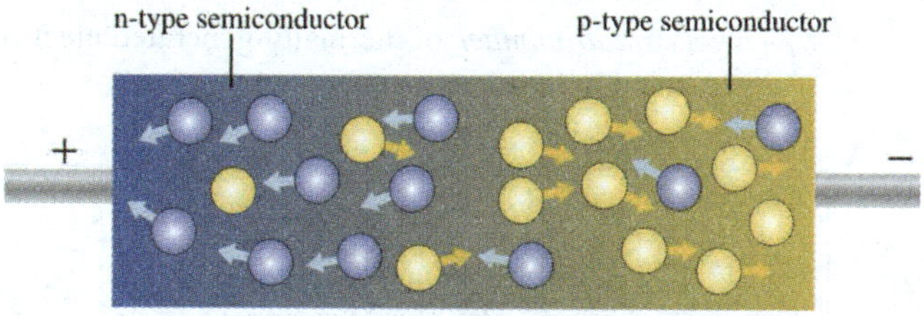

In this diagram, voltage is reversed and very little current flows.

In this case, the three valence electrons of boron atom form covalent bonds with four surrounding germanium atoms, but one bond is left incomplete and gives rise to a hole as shown in Fig. 12.20 (*a*).

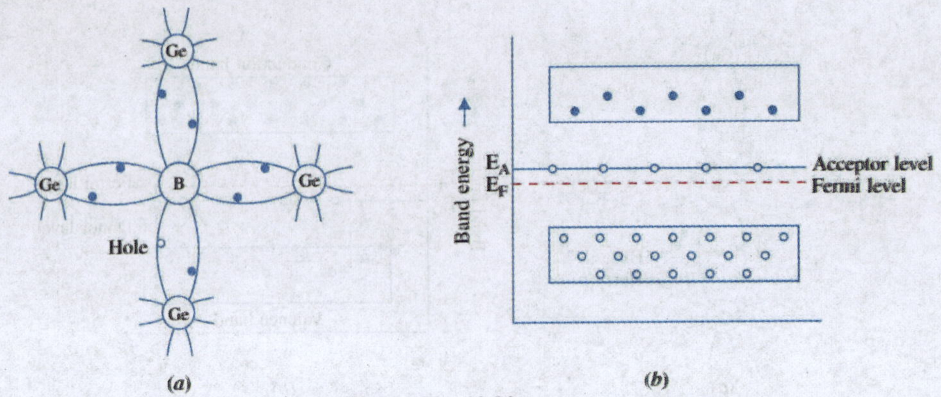

Fig. 12.20

Thus, boron which is called an acceptor impurity, causes as many positive holes in a germanium crystal as there are boron atoms thereby producing a *P*-type (*P* for the Positive) extrinsic semiconductor.

As an aid to memory, the student should associate the letter **P** in 'acce**P**tor' with the **P** in '**P**-type extrinsic semiconductor' and with the **P** in '**P**ositive charge carrier'.

In this type of semiconductor, conduction is by means of holes in the *valence* band.* Accordingly, holes form the majority carriers whereas electrons constitute minority carriers.

Since concentration of holes in the valence band is more than the concentration of electrons in the conduction band, Fermi level shifts nearer to the valence band [Fig. 12.20 (*b*)]. The acceptor level lies immediately above the Fermi level. Conduction is by means of hole movement at the top of valence band, the acceptor level readily accepting electrons from the valence band.

12.26. Majority and Minority Charge Carriers

In a piece of pure germanium or silicon, no free charge carriers are available at 0°K (Art. 12.23). However, as its temperature is raised to room temperature, some of the covalent bonds are broken by heat energy and, as a result, electron-hole pairs are produced. These are called *thermally-generated charge carriers*. They are also known as intrinsically-available charge carriers. Ordinarily, their number is quite small.

An intrinsic (or pure) germanium can be converted into a *P*-type semiconductor by the addition of an acceptor impurity which adds a large number of holes to it. Hence, a *P*-type material contains following charge carriers :

(*a*) *large number* of positive holes—most of them being the added impurity holes with only a very small number of thermally generated ones;

(*b*) *a very small number* of thermally-generated electrons (the companions of the thermally-generated holes mentioned above).

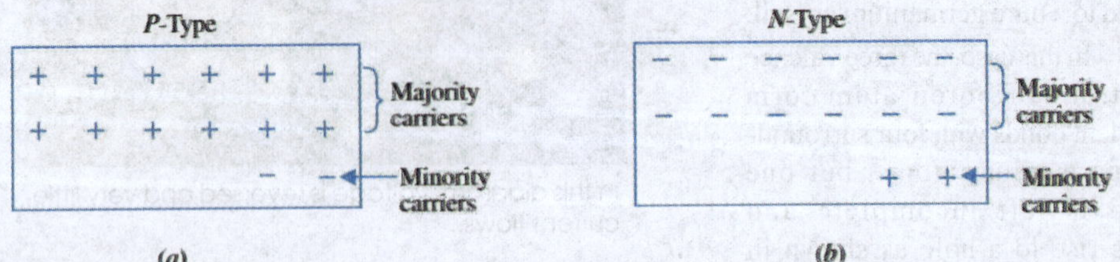

Fig. 12.21

* Hole current flows more slowly than electron current in *N*-type semiconductors.

Obviously, in a *P*-type material, the number of holes (both added and thermally-generated) is much more than that of electrons. Hence, in such a material, holes constitute majority carriers and electrons form minority carriers as shown in Fig. 12.21 (*a*) where plus (+) sign represents hole and negative (–) sign represents electron.

Similarly, in an *N*-type material, the number of electrons (both added and thermally-generated) is much larger than the number of thermally-generated holes. Hence, in such a material, electrons are majority carriers whereas holes are minority carriers as shown in the Fig. 12.21 (*b*).

12.27. Mobile Charge Carriers and Immobile Ions

As discussed in Art. 12.25, *P*-type material is formed by the addition of acceptor impurity atoms like boron to the pure Ge or Si crystals. The number of holes added is equal to the number of added boron atoms because each such atom contributes one hole. Now, when a hole moves away from its parent atom, the remaining atom becomes a negative ion. Unlike the mobile and free-moving hole, this ion cannot take part in conduction because it is fixed in the crystal lattice. In Fig. 12.22 (*a*), these immobile ions are shown by *circled* minus signs whereas free and mobile holes are shown by *uncircled* plus signs. Thermally-generated electrons (which form minority carriers) are shown by uncircled minus signs.

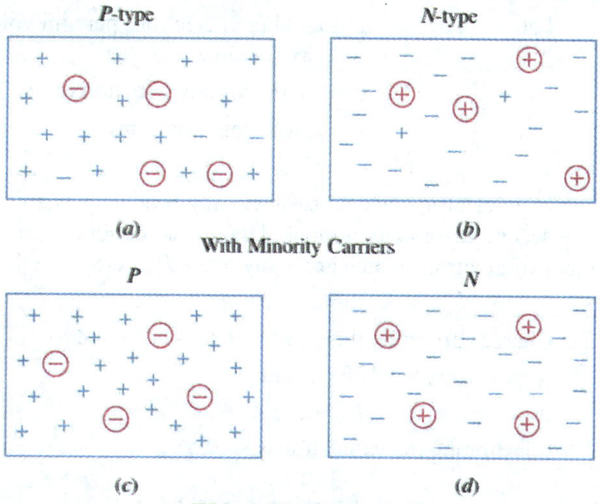

Fig. 12.22

Similarly, addition of pentavalent atoms like antimony to pure Ge or Si crystal produces *N*-type material. The number of free and mobile electrons which are added equals the number of donor Sb atoms. Again, when an electron moves away from its parent atom, it leaves behind a positive ion. This ion being fixed in crystal structure, cannot take part in conduction. As shown in Fig. 12.22 (*b*), these immobile ions are represented by *circled* plus signs whereas free and mobile electrons are represented by *uncircled* minus signs. The thermally-generated holes (which form minority carriers in this case) are shown by uncircled plus signs. In Fig. 12.22 (*c*) and (*d*), minority carriers of both types have been neglected. Hence, the figure does not show the small number of free electrons in the *P*-type material or the small number of holes in the *N*-type material.

12.28. Drift Current in Good Conductors

If an electric field is applied to an electron existing in a free space, it will accelerate the electron in a *straight line* from the negative terminal to the positive terminal of the applied voltage. But same thing does not happen in the case of electrons available in good conductors. As explained earlier, good conductors have plenty of free electrons moving randomly in between the fixed positive ion cores. If no voltage is applied, the net flow of charge in any direction is zero.

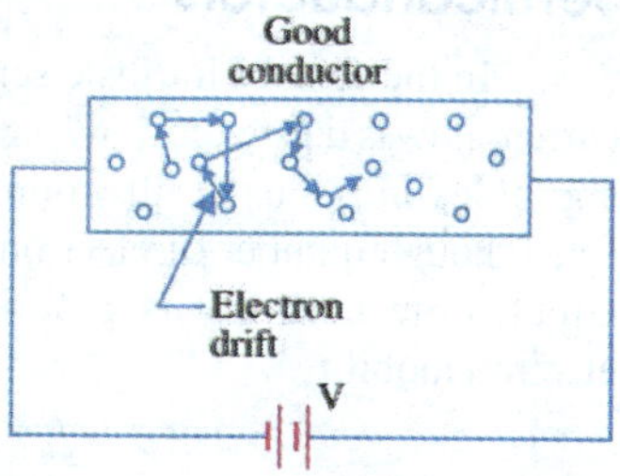

Fig. 12.23

However, as shown in Fig. 12.23, when an electric field is applied to a conductor at room temperature, electrons move towards the positive terminal of the applied voltage but they continuously collide with atoms along the way. Each time the electron collides with an atom, it rebounds in a random fashion. At each collision, the electron loses some kinetic energy, then accelerates again, gains certain component of velocity in the direction of $-E$ and loses its energy at the next collision. Obviously, the presence of the electric field *does not stop collisions and random motion* but it does cause the electrons to *drift* towards the positive terminal of the applied voltage V. Consequently, the electrons gain an average directed drift velocity v which is directly proportional to E.

$$\therefore \quad v = \mu_e E$$

where μ_e is called electron mobility*.

The resulting flow of electrons carrying negative charge at drift velocity V constitutes electric current called drift current.

Let n = number of electrons per unit volume of the conductor *i.e.*, electron density (per m^3)

A = conductor cross-section (m^2)

l = conductor length (m)

$E = V/l$ — applied electric field (V/m)

Now, electric current flowing in any conductor is given by the amount of charge which flows in one second across any normal plane of the conductor. It is seen from Fig. 12.24 that the total number of electrons which cross any plane P of cross-section A in one second

$$= n \times (v \times A)$$

Charge carried by them per second is $= e\,n\,v\,A$

It represents the drift current.

$$\therefore \quad I = v\,e\,n\,A$$

Substituting the value of v, we get

$$I = ne\,A\,\mu_e\,E = ne\,A\,\mu_e\,\frac{V}{l}$$

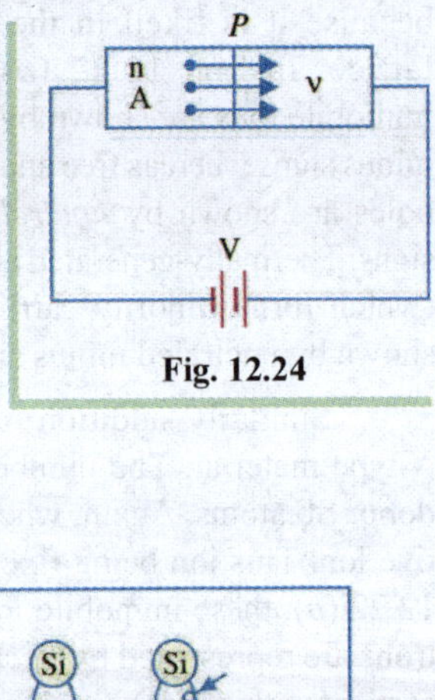

Fig. 12.24

Now, $$R = \frac{V}{I} = \frac{l}{A}\cdot\left(\frac{1}{ne\mu_e}\right)$$

Now, $$\rho = \frac{l}{A}$$

$$\therefore \text{ Resistivity, } \rho = \frac{1}{ne\mu_e}\ \Omega\text{-m}$$

and conductivity, $\sigma = ne\mu_e$ siemens/m

12.29. Drift Current in Intrinsic Semiconductors

In the case of intrinsic semiconductors, current flow is due to the movement of electrons and holes in opposite direction (Fig. 12.25). Even though number of electrons and holes is equal, hole mobility is practically half of electron mobility.

$$I = I_c + I_h = env_eA + epv_hA$$

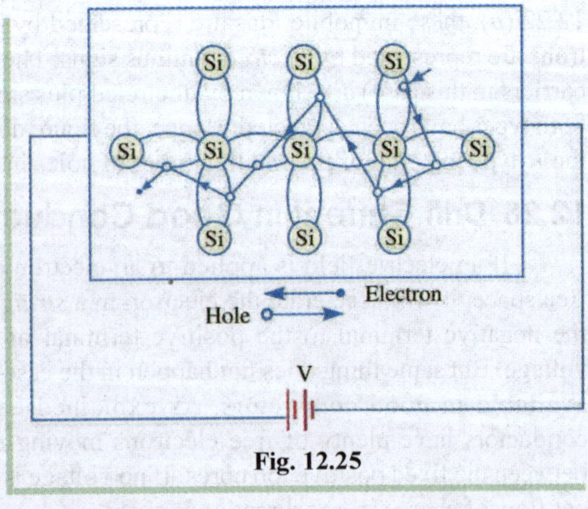

Fig. 12.25

* $\mu_e = \dfrac{V \text{ metre/second}}{E \text{ volt/metre}} = \dfrac{v}{E}\ m^2/\text{V-s}$. This also gives the unit of mobility.

where n = electron density (per m^3)
p = hole density (per m^3)
v_h = hole drift velocity (m/s)
v_e = electron drift velocity (m/s)

Since in an intrinsic semiconductor, $n = p$

$$\therefore \quad I = en(v_e + v_h) A$$

Now $v_e = \mu_e E$ and $v_h = \mu_h E$

$$\therefore \quad I = en(\mu_e + \mu_h) AE = en(\mu_e + \mu_h) A \frac{V}{l}$$

$$\text{Now,} \quad R = \frac{V}{I} = \frac{l}{A} \cdot \frac{l}{en(\mu_e + \mu_h)} = \rho \frac{l}{A}$$

$$\therefore \text{ Resistivity, } \rho = \frac{l}{\mathbf{en}(mu_e + \mu_h)} \ \mathbf{\Omega\text{-m}}$$

Conductivity, $\sigma = en(\mu_e + \mu_h)$ S/m

Example 12.5. *At room temperature, copper has free electron density of* 8.4×10^{28} *per* m^3. *Find electron drift velocity in a copper conductor having a cross-section of* 10^{-6} m^2 *and carrying a current of 5.4 A.*

Solution. $I = venA$

Now, $I = 5.4\ A$, $e = 1.6 \times 10^{-19}$ C

$n = 8.4 \times 10^{28}/m^3$, $A = 10^{-6}\ m^2$

$$\therefore \quad 5.4 = v \times 1.6 \times 10^{-19} \times 8.4 \times 10^{28} \times 10^{-6}$$

$$\therefore \quad v = 4 \times 10^{-4} \text{ m/s} = \mathbf{0.4\ mm/s}$$

Example 12.6. *Mobilities of electrons and holes in a sample of intrinsic germanium at room temperature are 0.36* m^2*/V-s and 0.17* m^2*/V-s respectively. If the electron and hole densities are each equal to* 2.5×10^{-19} m^3, *calculate germanium conductivity.*

Solution. For intrinsic germanium

$$\sigma = en(\mu_e + \mu_h) \text{ siemens/m}$$
$$= 1.6 \times 10^{-19} \times 2.5 \times 10^{19} (0.36 + 0.17) = \mathbf{2.12\ S/m}$$

Incidentally, $\rho = 1/2.12 = \mathbf{0.47\ \Omega\text{-m}}$

12.30. Intrinsic Conduction

The electric conduction that takes place in an intrinsic (or pure) semiconductor is called intrinsic conduction and has been briefly discussed in Art. 12.19. A more detailed view is given below.

As explained earlier, even at room temperature, some covalent bonds are broken in an intrinsic semiconductor because of heat energy. Each time a bond breaks, electron-hole pair is generated. Since both electron and hole are charge carriers, they move in opposite directions when an electric potential is applied. Hence, current is produced.

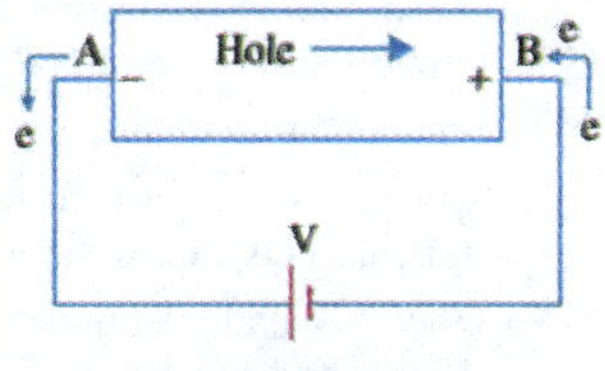

Fig. 12.26

Let us now consider the detailed mechanism of electric conduction in an intrinsic semiconductor bar *AB* shown in Fig. 12.26. Although there are many electron-hole pairs in the above sample, we will consider only one pair. Following sequence of events takes place :

1. Conduction electrons are attracted towards end *A* and travel through the connecting wire into the positive terminal of the battery. At the same time, additional electrons leave nega-

tive terminal of the battery and enter the semiconductor at *B*. Thus only electrons flow through the connecting wire in the counter-clockwise direction.

2. Positive holes in the bar are attracted towards the end *B*. On reaching there, they are neutralized by the electrons entering the bar from the wire end near *B*. It is called a *recombination process*. At the same time, some valence electrons break their covalent bonds and enter the connecting wire for going to the positive terminal of the battery. This is a charge generation process in which both electrons and holes are produced. Hence, we find that holes 'appear' at *A*, move from *A* to *B* and then 'disappear' at *B*. The net result is that there is a continuous flow of holes in a clockwise direction from *A* to *B* *inside* the bar.
3. As seen, there is continuous generation and recombination of electrons and holes. However, there are no *holes in the external circuit*. They are produced inside the bar at point *A* and are neutralised at the other end *B*. So far as the external circuit is concerned, the total current is due solely to the counter-clockwise flow of electrons as if the semiconductor bar were a simple resistor. It is only inside the semiconductor that the current has two components *i.e.*, *electron current* I_e and hole current I_h.

CONVENTIONAL PROBLEMS

1. The atomic number of silver atom is 47. Determine its orbital distribution of electrons. **[2, 8, 18, 18, 1]**
2. What is the valency of gallium whose atomic number is 31 ? **[3]**
3. Bismuth has atomic number of 83. How many valence electrons does it have ? **[5]**
4. Is tin which has $Z = 50$ tetravalent or pentavalent ? **[Tetravalent]**

SELF EXAMINATION QUESTIONS

A. Fill in the blanks with most appropriate word (s) or numerical value (s).

1. All normal atoms are electrically ....
2. The maximum number of electrons which the outermost orbit of an atom can have is ....
3. The electrons in the outermost orbit of an atom are called ....electrons.
4. Electromagnetic radiations are always emitted when an electron in an atom jumps from a .... level to any lower level.
5. There is a .... energy gap between any two permitted energy levels within an atom.
6. Whereas single isolated atoms have energy levels, solids have energy ....
7. .... electrons are responsible for forming different types of bonds in solids.
8. Semiconductors have .... bonds.
9. Covalent bonds are formed by .... of electrons between two atoms.
10. When a covalent bond breaks, usually one electron and one .... are produced.
11. Free electrons move in .... band whereas holes move in ....band.
12. *N*-type semiconductor has excess of ....
13. Elements added to semiconductor materials to produce more current are called ....
14. Antimony in a silicon matrix provides a free ....
15. The majority carriers in a *P*-type semiconductor are ....
16. All impurities that have five valence electrons are called .... impurities.
17. The two most popular semiconductor materials are germanium and ....
18. A pure semiconductor is also called .... semiconductor.
19. The most common semiconductor material is ....
20. In their pure form, germanium and silicon are not electrical ....

B. Answer True or False

1. Three-dimensional pattern in which atoms align themselves in a solid is called crystal lattice.
2. All solids have crystalline structure at low temperatures.
3. A unit cell is the smallest part of a crystal having all the structural properties of the crystal lattice.
4. Germanium has body-centred cubic lattice.

5. The maximum number of electrons in the valence shell of an atom cannot exceed 18.
6. Electrons in the higher orbits have high total energy but low binding energy.
7. When the orbiting electron of an atom jumps from a higher to a lower orbit, it emits radiations.
8. Valence electrons are those which occupy the outermost orbit of an atom.
9. Whereas isolated single atoms have energy levels, solids have energy bands.
10. Ionic bonds provide binding force in semiconductors.
11. Covalent bonds are the strongest of all the bonds formed in solids.
12. Electrons in the conduction band have higher energy than those in the valence band.
13. Holes are produced when covalent bonds in a semiconductor material get broken with loss of one electron each.
14. Holes remain confined to valence band.
15. Holes are as mobile as electrons.
16. Intrinsic semiconductors have equal number of electrons and holes.
17. *N*-type semiconductor has excess of negative charge.
18. Extrinsic semiconductors usually have 1 atom of impurity in 10,000 atoms of pure semiconductor material.
19. Adding impurity to a pure semiconductor materials is called doping.
20. The missing valence electron is called a hole.

C. Multiple Choice Items

1. In semiconductors like silicon, the unit cell is
   (*a*) a simple cube
   (*b*) body-centred cube
   (*c*) face-centred cube
   (*d*) hexagonal
2. The maximum number of electrons which the *M*-shell of an atom can contain is
   (*a*) 32 (*b*) 8
   (*c*) 18 (*d*) 50
3. The maximum number of electrons in the shell preceding the valence shell cannot exceed
   (*a*) 18 (*b*) 8
   (*c*) $2n^2$ (*d*) 32
4. Electronic distribution of an Si atom is
   (*a*) 2, 10, 2 (*b*) 2, 8, 4
   (*c*) 2, 7, 5 (*d*) 2, 4, 8
5. The electrons in the atom of an element which determine its electrical and chemical properties are called .... electrons.
   (*a*) valence (*b*) excess
   (*c*) revolving (*d*) active
6. Silicon has $Z = 14$. Its outermost orbit is
   (*a*) partially filled
   (*b*) half filled
   (*c*) completely occupied
   (*d*) empty
7. Semiconductor materials have .... bonds.
   (*a*) ionic (*b*) covalent
   (*c*) mutual (*d*) metallic
8. A single Si atom is bonded to .... other atoms in its crystal lattice.
   (*a*) 4 (*b*) 2
   (*c*) 8 (*d*) 1
9. Major part of the current in an intrinsic semiconductor is due to
   (*a*) conduction-band electrons
   (*b*) valence-band electrons
   (*c*) holes in the valence band
   (*d*) thermally-generated electrons
10. Conduction electrons have more mobility than holes because they
   (*a*) are lighter
   (*b*) experience collisions less frequently
   (*c*) have negative charge
   (*d*) need less energy to move them
11. There is no hole current in good conductors because they
   (*a*) are full of electron gas
   (*b*) have large forbidden energy gap
   (*c*) have no valence band
   (*d*) have overlapping valence and conduction bands
12. In an *N*-type semiconductor, there are
   (*a*) no minority carriers
   (*b*) immobile negative ions
   (*c*) immobile positive ions
   (*d*) holes as majority carriers
13. Doping materials are called impurities because they
   (*a*) decrease the number of charge carriers
   (*b*) change the chemical properties of semiconductors

(*c*) make semiconductors less than 100 percent pure

(*d*) alter the crystal structure of the pure semiconductors

14. When germanium crystal is doped with phosphorus atoms, it becomes

(*a*) *N*-type semiconductor

(*b*) *P*-type semiconductor

(*c*) an insulator

(*d*) photo-transistor

15. In an *N*-type semiconductor material, the majority carrier are

(*a*) holes (*b*) electrons

(*c*) ions (*d*) protons

16. Electrons existing in the outermost shell of an atom are called

(*a*) stabilizing electrons

(*b*) acceptors (*c*) donors

(*d*) valence electrons

17. When electrons leave the *N*-material to fill holes in the *P*-material, the process is called

(*a*) doping (*b*) mixing

(*c*) depletion (*d*) diffusion

ANSWERS

A. Fill in the blanks

1. neutral	**2.** eight	**3.** valence	**4.** higher	**5.** forbidden	**6.** bands
7. valence	**8.** covalent	**9.** sharing	**10.** hole	**11.** conduction; valence	
12. electrons	**13.** impurities	**14.** electron	**15.** holes	**16.** donor	**17.** silicon
18. intrinsic	**19.** silicon	**20.** conductors			

B. True or False

1. T	**2.** F	**3.** T	**4.** T	**5.** F	**6.** T
7. T	**8.** T	**9.** T	**10.** F	**11.** F	**12.** T
13. T	**14.** T	**15.** F	**16.** T	**17.** F	**18.** F
19. T	**20.** T				

C. Multiple Choice Items

1. *b*	**2.** *c*	**3.** *a*	**4.** *b*	**5.** *a*	**6.** *b*
7. *b*	**8.** *a*	**9.** *a*	**10.** *d*	**11.** *d*	**12.** *c*
13. *d*	**14.** *a*	**15.** *b*	**16.** *d*	**17.** *d.*	

CHAPTER 13

The P-N Junction

1. The *P-N* Junction
2. Formation of Depletion Layer
3. Junction or Barrier Voltage (V_B)
4. Effect of Temperature on Barrier Voltage
5. Forward Biased *P-N* Junction
6. Forward V/I Characteristics
7. Reverse Biased P-N Junction
8. Reverse Saturation Current (I_s or I_0)
9. Reverse V/I Characteristic
10. Combined Forward and Reverse V/I Characteristics
11. Junction Breakdown
12. Junction Capacitance
13. Equivalent Circuit of a P-N Junction

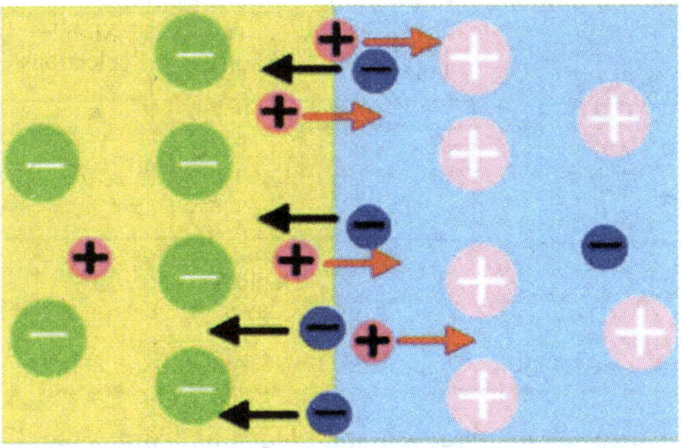

13.1. The *P-N* Junction

It is possible to manufacture a single piece of a semi conductor material (either Ge or Si) one half of which is doped by *P*-type impurity and the other half by *N*-type impurity as shown in Fig. 13.1. The plane dividing the two halves or zones is called a ***P-N* junction.** It should be noted that a useful *P-N* junction cannot be produced by connecting *P*-type material to *N*-type material by welding etc., because this would give rise to discontinuities across the crystal structure.

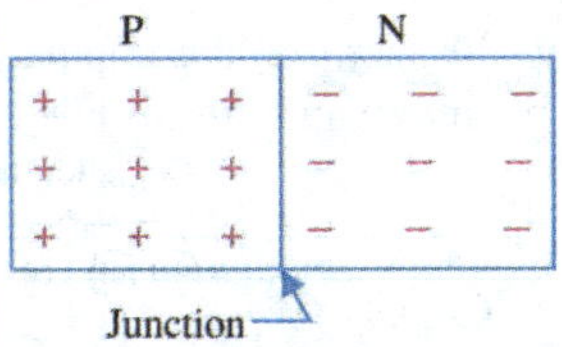

Fig. 13.1

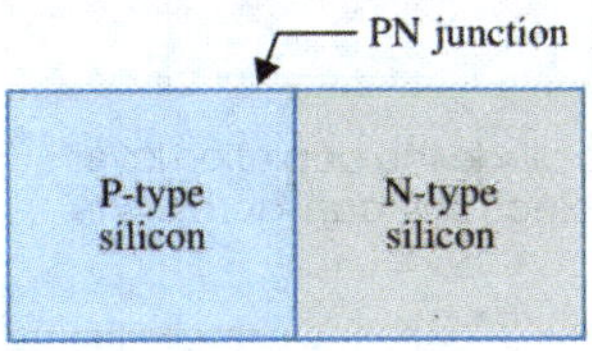

(a) A P-N junction in a diode.

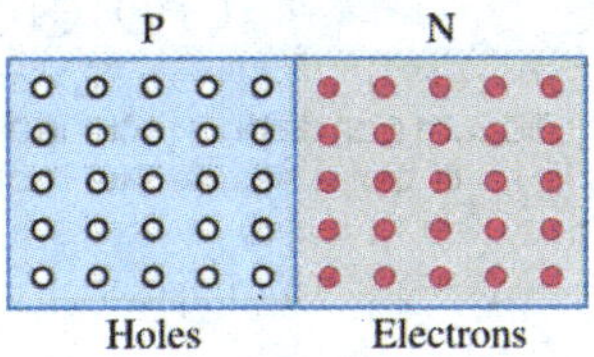

(b) The N region has in a diode many conduction electrons and the P region has many holes.

During the formation of a *P-N* junction, following two phenomena take place :

(*i*) a thin depletion layer (or region) is set up on both sides of the junction and is so-called because it is depleted or devoid of free charge carriers. Its width is about 1 μm (10^{-6} m);

(*ii*) a junction or barrier potential V_B is developed across the junction whose value is about 0.3 V for Ge and 0.7 V for Si.

When a *P-N* junction is packed as a semiconductor device, it is called a *P-N* junction diode.

13.2. Formation of Depletion Layer

Suppose that a *P-N* junction has just been formed. At that instant, holes are still in the *P*-region and electrons in the *N*-region. However, there is greater concentration of holes in *P*-region (where they form majority carriers) than in *N*-region (where they form minority carriers). Similarly, concentration of electrons is greater in *N*-region than in *P*-region (where they exist as minority carriers). This difference in concentration establishes a density gradient across the junction resulting in majority carrier diffusion. Holes diffuse from *P*- to *N*-region and electrons from *N*- to *P*-region and terminate their existence by recombination as shown in Fig. 13.2 (*a*).

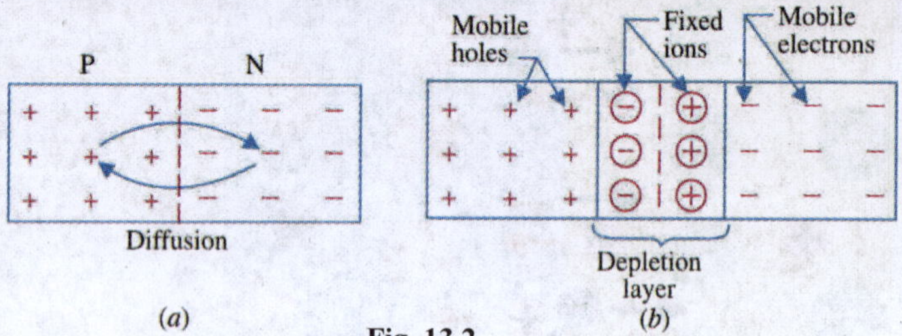

Fig. 13.2

This recombination of free and mobile holes and electrons produces the narrow region at the junction called *depletion layer*. It is so-called because this region is *devoid or depleted of free and mobile charge carriers* though it does contain **fixed** or **immobile** positive and negative ions.

This production of ions is due to the fact that the impurity atoms which provide migratory electrons and holes are themselves left behind in an ionised state bearing a charge which is opposite to that of the departed carrier. For example, when an electron migrates across the junction from *N*-region to *P*-region, it leaves behind an atom that is one electron short of its normal quota. This atom is now ionised and has a positive charge. It is said to be *uncovered.* These impurity ions so produced are fixed in their positions in the crystal lattice in the *P*- and *N*-regions. Hence, as shown in Fig. 13.2 (*b*), they form parallel rows or plates of opposite charges facing each other across the depletion layer. As seen, depletion layer contains no free and mobile charge carriers but only fixed and immobile ions. Hence, this layer (or region) behaves like an insulator and due to the presence of rows of fixed charges, it possesses capacitance as explained in Art. 13.11.

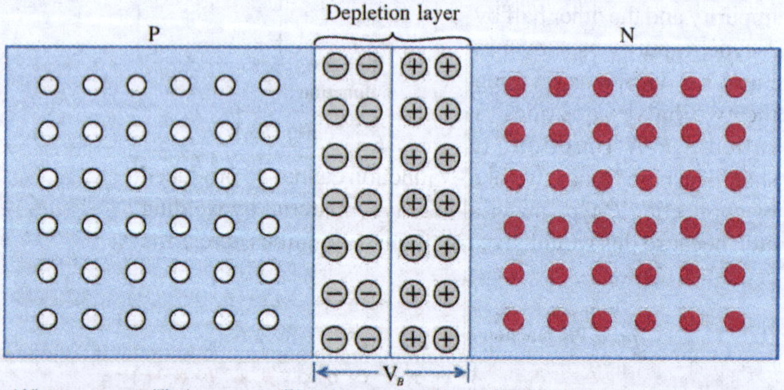

When an equilibrium condition is reached in a diode the depletion layer has widened to a point where no more electrons can across the P-N junction.

13.3. Junction or Barrier Voltage (V_B)

As explained in Art. 13.2, depletion layer of a *P-N* junction diode has no free charge carriers but only fixed rows of oppositely-charged ions on its two sides. Because of this charge separation, an electric potential V_B is established across the junction *even when the junction is not connected to any external source of e.m.f.* (Fig. 13.3). It is known as *junction or barrier* potential. It stops further flow of carriers across the junction unless supplied by energy from an external source. At room temperature of 300° K, V_B is about 0.3 V for Ge and 0.7 V for Si. Its value depends on doping density, electronic charge and temperature.

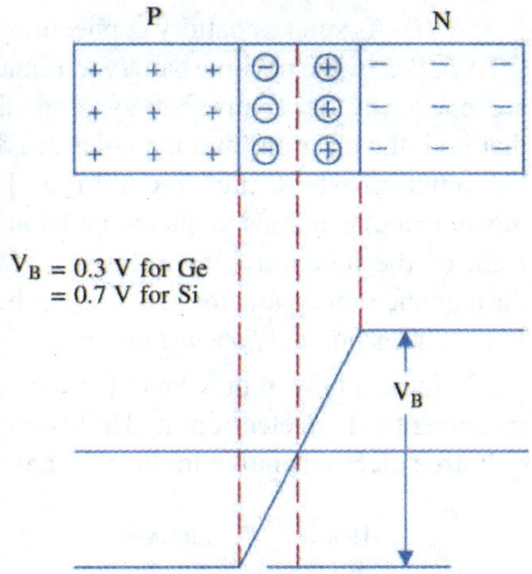

Fig. 13.3

The processes involved in the formation of a *P-N* junction are summarized below :

1. Holes from the *P*-side diffuse into the *N*-side where they combine with free electrons.
2. Free electrons from the *N*-side diffuse into the *P*-side where they combine with holes.
3. The diffusion current (also known as recombination current) decays exponentially both with time and distance from the junction.
4. Due to the departure of free and mobile carriers from both sides of the junction, a depletion layer (centred around the junction) is formed. This layer contains only immobile or fixed (also called uncovered) ions of opposite polarity.
5. These uncovered by fixed ions set up a potential barrier across the junction.
6. This potential difference opposes the diffusion of free majority charge carriers from one side of the junction to the other* till the process is completely stopped.
7. The width of depletion layer depends on the doping level. For heavy doping, depletion layer is physically thin because a diffusing charge carrier (either free electron or hole) has not to travel far across the junction for recombination (short lifetime). Opposite is the case if light doping is used.

13.4. Effect of Temperature on Barrier Voltage

Barrier voltage depends on doping density, electronic charge and temperature. For a given junction, the first two factors are constant, thus making V_B dependent on temperature. With increase in temperature, more minority charge carriers are produced leading to their increased drift across the junction. As a result, equilibrium occurs at a slightly lower barrier potential. It is found that both for Ge and Si, V_B decreases by about 2 mV/°C.

or $\Delta V_B = -0.002.\Delta t$

where Δt = change in temp. in °C.

13.5. Forward Biased *P-N* Junction

Suppose positive battery terminal is connected to *P*-region of a semiconductor and the negative battery terminal to the *N*-region as shown in Fig. 13.4. In that case, the junction is said to be biased in the **forward** direction because it permits easy flow of current across the junction. The current flow may be explained in the following two ways :

* Incidentally, this potential barrier *aids* in the transfer of thermally-generated minority charge carriers from one side of the junction to the other.

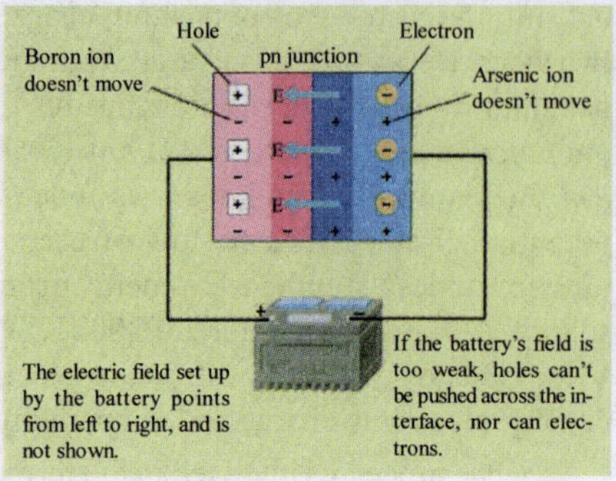

(*i*) As soon as battery connection is made, holes are *repelled* by the positive battery terminal and electrons are repelled by the negative battery terminal with the result that both the electrons and the holes are driven *towards* the junction where they recombine. This *en masse* movement of electrons to the left and that of holes to the right of the junction constitutes a large current flow through the semiconductor. Obviously, the crystal offers *low resistance in the forward direction.*

Incidentally, it may be noted that though there is movement of both electrons and holes *inside* the crystal, only free electrons move in the external circuit *i.e.*, in the battery-connected wire (Art. 12.20).

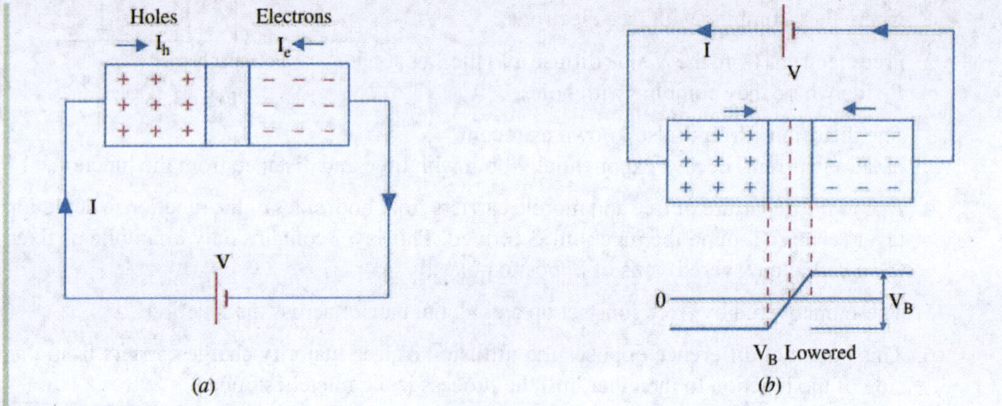

Fig. 13.4

(*ii*) Another way to explain current flow in the forward direction is to say that due to the applied external voltage, the barrier potential is reduced which now allows more current to flow across the junction [Fig. 13.4 (*b*)]. Incidentally, it may be noted that forward bias reduces the thickness of the depletion layer as shown in Fig. 13.5.

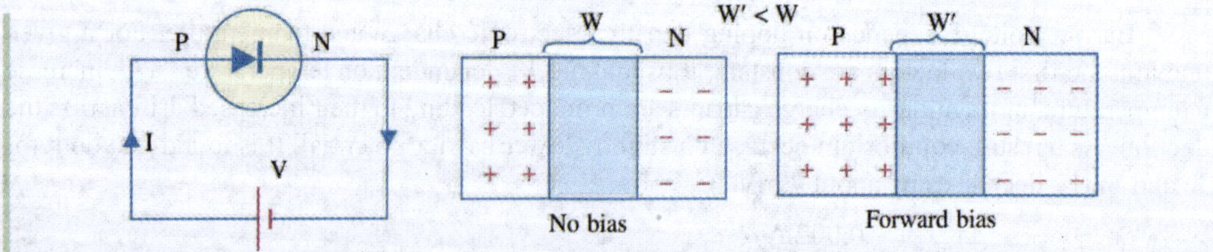

Fig. 13.5

13.6. Forward V/I Characteristics

A typical *V/I* characteristic for a forward-biased *P-N* junction is shown in Fig. 13.6. It is seen that forward current rises exponentially with the applied forward voltage. However, at ordinary room temperature, a p.d. of about 0.3 V is required before a reasonable amount of forward current starts flowing in a germanium junction. This voltage is known as *threshold voltage* (V_{th}) *or cut-in voltage or knee voltage* V_K. It is practically the same as barrier voltage V_B. Its value for silicon junction is about 0.7 volt. For $V < V_{th}$, current flow is negligible. But as applied voltage increases beyond the threshold value, the forward current increases sharply. If forward voltage is increased beyond a certain safe value, it will produce an extremely large current which may destroy the junction due to overheating. **Ge devices can stand junction temperature around 100° C whereas Si units can function upto 175°C.**

Obviously, the *forward-biased junction has a low resistance.* For point *B* in Fig. 13.6, the forward resistance for Si is

$$R_F = \frac{0.8 \text{ V}}{20 \text{ mA}} = 40\ \Omega$$

Similarly, for point *A* on the Ge curve

$$R_F = \frac{0.36 \text{ V}}{20 \text{ mA}} = 18\ \Omega$$

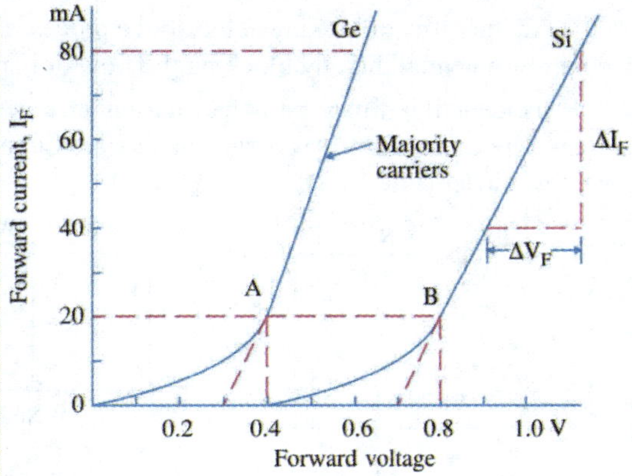

Fig. 13.6

In practice, this static forward resistance is not used. Instead, the **dynamic** resistance or **incremental** resistance or **ac resistance** of the junction is used. It is given by the reciprocal of the slope of the forward characteristic.

$$r_{ac} = \frac{1}{\Delta I_F / \Delta V_F} = \frac{\Delta V_F}{\Delta I_F}$$

Here, $\Delta V_F = 0.19\ V$ and $\Delta I_F = 37.6$ mA

$$\therefore \quad r_{ac} = \frac{0.19}{3.76 \times 10^{-3}} \cong 5\ \Omega$$

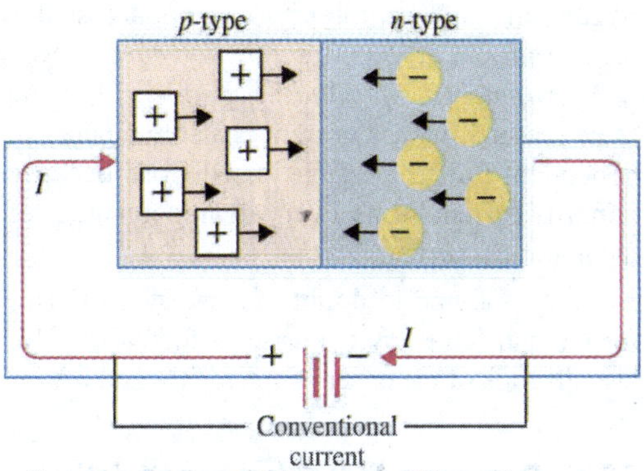

There is an appreciable current through the diode when the diode is forward biased.

13.7. Reverse Biased P-N Junction

When battery connections to the semiconductor are made as shown in Fig. 13.7 (*a*), the junction is said to be *reverse-biased.* In this case, holes are attracted by the negative battery terminal and electrons by the positive terminal so that both holes and electrons move **away** from the junction and **away** from each other. Since there is no electron-hole combination, no current flows and the junction offers high resistance.

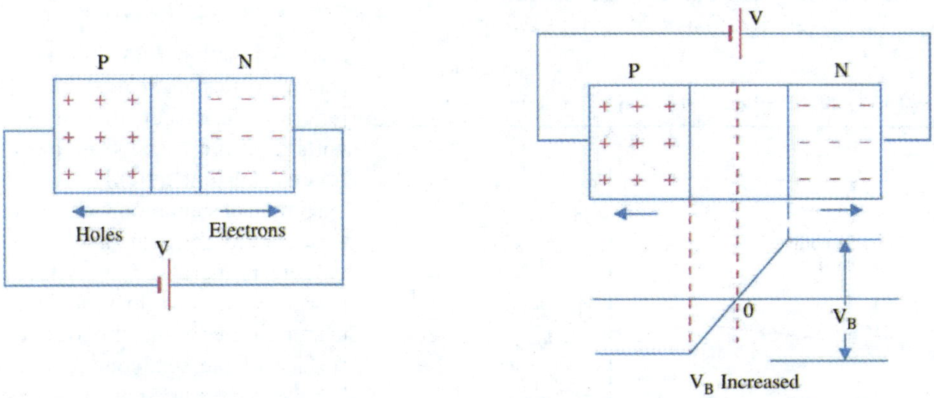

(*a*) (*b*)

Fig. 13.7

Another way of looking at the above process is that, in this case, the applied voltage increases the barrier potential thereby blocking the flow of majority carriers as shown in Fig. 13.7 (*b*).

Incidentally, it may be noted that under reverse bias conditions, width of depletion layer is increased because majority charge carriers are pulled away from the junction. It also increases the potential barrier (Fig. 13.8).

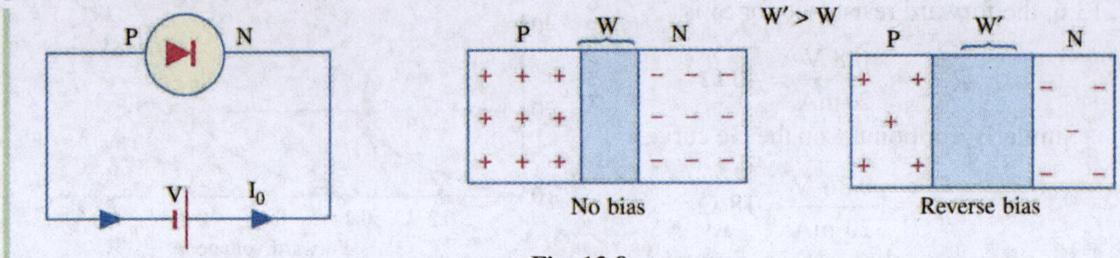

Fig. 13.8

13.8. Reverse Saturation Current (I_s or I_0)

Although under reverse-bias conditions, there is practically no current due to majority carriers, yet there is a small amount of current (a few μA only) due to the flow of minority carriers across the junction. As explained earlier in Art. 12.26, due to thermal energy, there are always generated some holes in the *N*-type region and some electrons in the *P*-type region of the semiconductor as shown in Fig. 12.21. The applied voltage acts as a forward bias for these minority carriers. The battery drives these minority carriers across the junction thereby producing a small current called reverse current or *reverse saturation current* I_s or I_0. Since minority carriers are thermally-generated, I_0 is *extremely temperature dependent.* For the same reason, forward current is also temperature dependent but to a much less degree because minority current forms a very small percentage of the majority current.

I_0 is found to double for every 10°C rise for Ge and for every 6°C rise in the case of Si. Usually, it is of the order of μA for Ge and nA for Si.

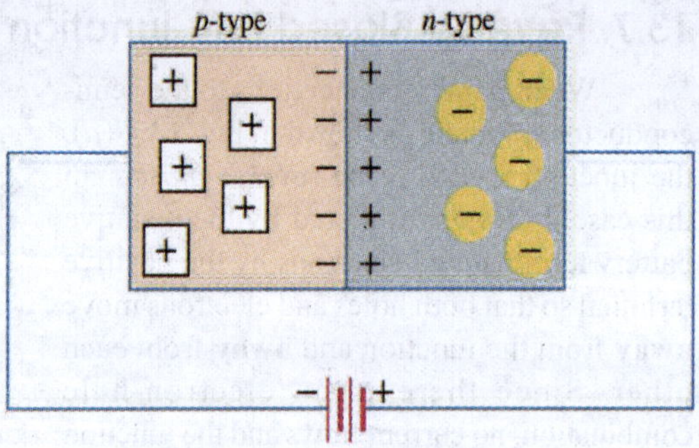

Under a reverse bias condition, there is almost no current through the diode.

13.9. Reverse V/I Characteristic

It may be noted that reverse saturation current is also referred to as *leakage current* **of** the *P-N* junction. Fig. 13.9 shows *V/I* characteristics of a reverse-biased *P-N* junction. It is seen that as reverse voltage is increased from zero, the reverse current quickly rises to its maximum or saturation value. Keeping temperature constant, as the reverse voltage is increased, I_0 is found to increase only slightly. This slight increase is due to the impurities on the surface of the semiconductor which behaves as a resistor and hence obeys Ohm's law. This gives rise to a very small current called *surface leakage*

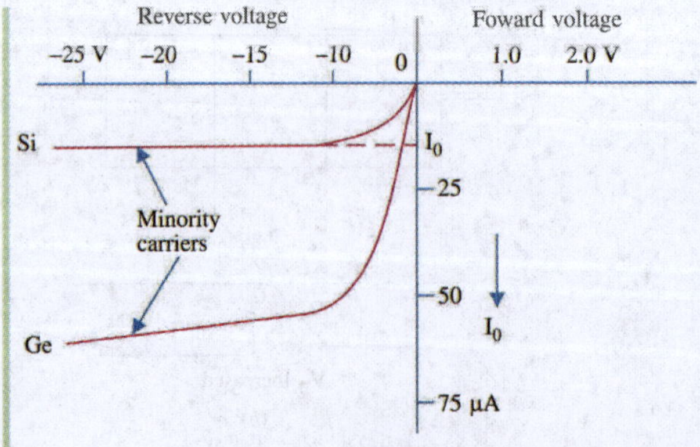

Fig. 13.9

current. Unlike the main leakage (or saturation) current, this surface leakage current is independent of temperature but depends on the magnitude of the reverse voltage.

A reverse-biased junction can be represented by a very large resistance. As seen from Fig. 13.9, in the case of Si, for a reverse voltage of about 15 V, $I_0 = 10\ \mu A$. Hence reverse resistance is

$$R_R = \frac{15\ V}{10\ \mu A} = 1.5\ M$$

13.10. Combined Forward and Reverse V/I Characteristics

Combined forward and reverse voltage-current characteristics for both Ge and Si are shown in Fig. 13.10. It is seen that leakage current of Ge junction is much more than that of Si junction. It is worthwhile to make note of the change in scale of I_0 and reverse voltage.

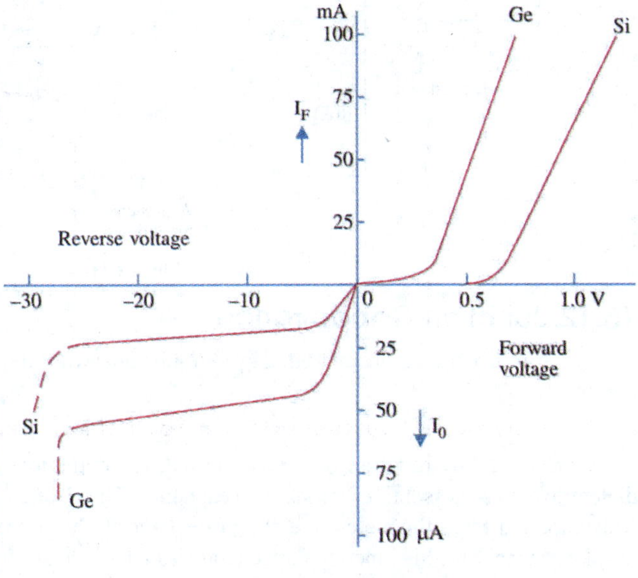

Fig. 13.10

13.11. Junction Breakdown

If the reverse bias applied to a *P-N* junction is increased, a point is reached when the junction breaksdown and reverse current rises sharply to a value limited only by the external resistance connected in series with the junction (Fig 13.10). This critical value of the voltage is known as **breakdown voltage** (V_{BR}). It is found that once breakdown has occurred, very little further increase in voltage is required to increase the current to relatively high values. The junction itself offers almost zero resistance at this point.

The breakdown voltage depends on the width of the depletion region which, in turn, depends on the doping level.

The following two mechanisms are responsible for breakdown under increasing reverse voltage :

1. Zener Breakdown

This form of breakdown occurs in junctions which, being heavily doped, have narrow depletion layers. The breakdown voltage sets up a very strong electric field (about 10^8 V/m) across this narrow layer. This field is strong enough to *break or rupture the covalent bonds* thereby generating electron-hole pairs. Even a small further increase in reverse voltage is capable of producing large number of current carriers. That is why the junction has very low resistance in the breakdown region.

2. Avalanche Breakdown

This form of breakdown occurs in junctions which, being lightly-doped, have wide depletion layers where the electric field is not strong enough to produce Zener breakdown. Instead, the minority carriers (accelerated by this field) collide with the semiconductor atoms in the depletion region. Upon collision with valence electrons, covalent bonds are broken and electron-hole pairs are generated. These newly-generated charge carriers are also accelerated by the electric field resulting in more collisions and hence further production of charge carriers. This leads to an avalanche (or flood) of charge carriers and, consequently, to a very low reverse resistance. The two breakdown phenomena are shown in Fig. 13.11.

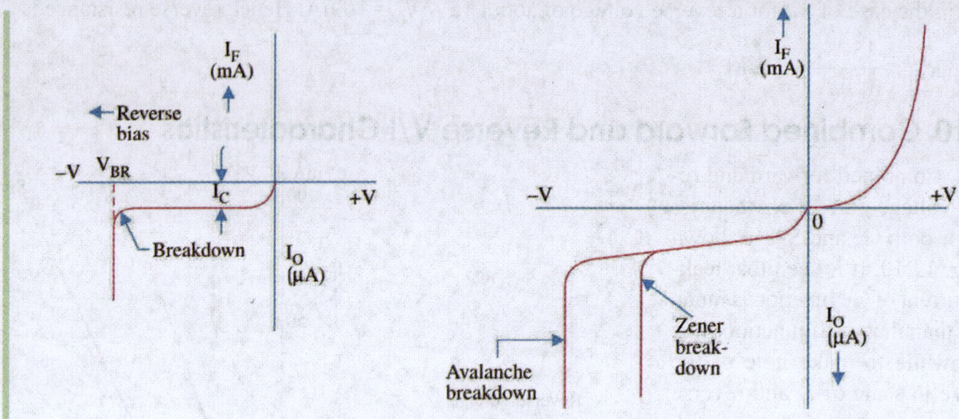

Fig. 13.11

13.12. Junction Capacitance

Capacitive effects are exhibited by *P-N* junctions when they are either forward-biased or reverse-biased.

(*a*) Transition Capacitance (C_T) or Space-charge Capacitance

When a *P-N* junction is reverse-biased, the depletion region acts like an insulator or as a dielectric material essential for making a capacitor. The *P*- and *N*-type regions on either side have low resistance and act as the plates. We, therefore, have all the components necessary for making a parallel-plate capacitor. This junction capacitance is called *transition* or *space charge* capacitance (C_{pn} or C_T). It may be calculated by the usual formula $C = \varepsilon A/d$. Its typical value is 40 pF. Since thickness of depletion (or transition) layer depends on the amount of reverse bias, capacitance C_T can be controlled with the help of applied bias. This property of variable capacitance possessed by a reverse-biased *P-N* junction is used in the construction of a device known as *varicap* or *varactor* (Art. 15.8).

This capacitance is voltage-dependent as given by the relation

$$C_T = \frac{K}{(V_K + V_R)^n}$$

where V_K = knee voltage (Art. 13.6)

V_R = applied reverse voltage

K = constant depending on semiconductor material.

$$n = \frac{1}{2} \qquad \text{— for alloy junction}$$

$$= \frac{1}{3} \qquad \text{— for diffused junction}$$

(*b*) Diffusion or Storage Capacitance (C_D)

This capacitive effect is present when the junction is *forward-biased.* It is called diffusion capacitance to account for the time delay in moving charges across the junction by diffusion process. Due to this fact, this capacitance cannot be identified in terms of a dielectric and plates. It varies directly with the magnitude of forward current as explained below in more details.

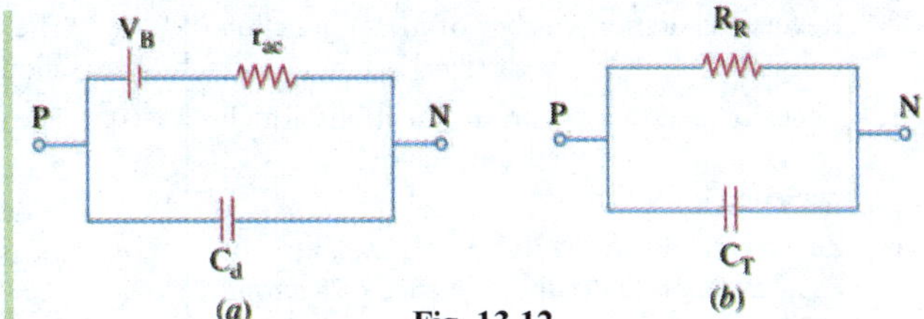

Fig. 13.12

Consider a forward-biased junction which is carrying a forward current I_F. Suppose the applied voltage is suddenly reversed, then I_F ceases suddenly but leaves lot of majority charge carriers in the depletion region. These charge carriers must get out of the region which to their bad luck, becomes wider under the reverse bias. Hence, it is seen that when a forward-biased *P-N* junction is suddenly reverse biased, a reverse current flows which is large initially but gradually decreases to the level of saturation current I_0. This effect can be likened to the discharging current of a capacitor and is, therefore, rightly represented by a capacitance called *diffusion capacitance* C_D. Since the number of charge carriers left in depletion layer is proportional to forward current, C_D is directly proportional to I_F. Its typical value is 0.02 μF which is 5000 time C_T.

This capacitance assumes great significance in the operation of devices which are required to switch rapidly from forward to reverse bias. If C_D is large, this switchover cannot be rapid. It will delay both the switch-on and the switch-off. This effect of C_D is variously known as *recovery time* or carrier storage.

13.13. Equivalent Circuit of a P-N Junction

We have seen from above that a forward-biased junction offers ac resistance r_{ac} and possesses diffusion capacitance C_D (which comes into the picture only when frequency of the applied voltage is very high). Hence, it can be represented by the equivalent circuit of Fig. 13.12 (*a*). An opposing battery has been connected in series with r_{ac} to account for the junction barrier potential (Art. 13.3).

As seen from Fig. 13.12 (*b*), a reverse-biased junction can simply represented by a reverse resistance R_R connected in *parallel* with a capacitance C_T or C_{pn}.

Example 13.1. *Calculate the barrier potential for Si junction at (a) 100°C and (b) 0°C if its value at 25°C is 0.7 V.*

Solution. As seen from Art. 13.4,

$$\Delta V = -0.002.\Delta t = -0.002\,(t_2 - t_1)$$

(*a*) $\Delta t = (100 - 25) = 75°C$

$\therefore$ $\Delta V = -0.002 \times 75 = -0.15$ V

$\therefore$ V_B at 100°C $= 0.7 + (-0.15) = \mathbf{0.55V}$

(*b*) $\Delta t = (0 - 25) = -25°C$

$\therefore$ $\Delta V = -0.002 \times (-25) = 0.05$

$\therefore$ V_B at 0° C $= 0.7 + 0.05 = \mathbf{0.75\ V.}$

SELF EXAMINATION QUESTIONS

A. Fill in the blanks with most appropriate word (s) or numerical value (s).

1. Depletion region exists on .... sides of a *P-N* junction.
2. A .... potential is developed across a *P-N* junction whose value for silicon is .... volt.
3. As soon as a *P-N* junction is formed, ... charge carriers from both sides cross-over the junction.
4. The width of the depletion layer in an unbiased *P-N* junction depends on .... level.
5. Barrier potential .... with increase in junction temperature.
6. When a junction is forward-biased for majority carriers, it is .... -biased for .... carriers.
7. The forward-biased *P-N* junction has .... forward resistance but .... reverse resistance.

8. Reverse saturation current of a *P-N* junction consists of .... carriers.
9. Zener breakdown occurs in junctions which are .... doped and have .... depletion layers.
10. Zener breakdown involves .... of covalent bond under the influence of a strong electric field.
11. Avalanche breakdown occurs in junction materials which are lightly doped and have .... depletion layers.
12. Diffusion capacitance of a *P-N* junction depends directly on forward ....
13. The area around the *P-N* junction where the ions are formed is called the ....region.
14. The barrier potential for a silicon *P-N* junction is ....volt.
15. The opposite of forward biasing a *P-N* junction is ....biasing.
16. The depletion of a P-N junction ....when reverse biased.
17. A reverse-biased diode is considered to be ....
18. Zener diodes are widely used as .... regulators.
19. A Zener diode is normally connected in the .... biased direction.
20. .... current is caused by minority carriers of a semiconductor diode.

B. Answer True or False

1. *P-N* junction can be formed by welding a *P*-type crystal to an *N*-type crystal.
2. The depletion layer of a *P-N* junction is so-called because it has no charged particles in it.
3. Temperature lowers the junction barrier potential.
4. A *P-N* junction acts like a low resistance when forward-biased but like a high capacitance when reverse-biased.
5. Reverse bias increases the barrier potential of a *P-N* junction.
6. The leakage current of a *P-N* junction varies directly as the applied reverse voltage.
7. Zener breakdown occurs in heavily-doped junction whereas avalanche breakdown occurs in lightly-doped ones.
8. Both the transition capacitance and diffusion capacitance of a *P-N* junction arise due to reverse bias.
9. The diode becomes forward-biased when its anode is more positive than the cathode.
10. The resistance of a reverse-biased diode is low.
11. Leakage current occurs when a diode is reverse-biased.
12. The average voltage drop across a forward-biased silicon diode is 0.3 V.

C. Multiple Choice Items

1. The depletion region of a *P-N* junction is formed
   (*a*) during the manufacturing process
   (*b*) when forward bias is applied to it
   (*c*) under reverse bias
   (*d*) when its temperature is reduced
2. The width of the depletion layer of a junction
   (*a*) decreases with light doping
   (*b*) increases with heavy doping
   (*c*) is independent of applied voltage
   (*d*) is increased under reverse bias
3. Point out the WRONG statement. Barrier potential of a *P-N* junction is a function of
   (*a*) diode design (*b*) temperature
   (*c*) forward bias (*d*) doping density
4. The width of the depletion region of an unbiased *P-N* junction is about a few
   (*a*) μm (*b*) mm
   (*c*) cm (*d*) nm
5. Zener breakdown occurs
   (*a*) mostly in germanium junctions
   (*b*) due to rupture of covalent bonds
   (*c*) in lightly-doped junctions
   (*d*) due to thermally-generated minority carriers
6. Avalanche breakdown is primarily dependent on the phenomenon of
   (*a*) collision (*b*) doping
   (*c*) ionization (*d*) recombination
7. Reverse current in a silicon junction nearly doubles for every ....°C rise in temperature.
   (*a*) 10 (*b*) 2
   (*c*) 6 (*d*) 5
8. Diffusion capacitance of a *P-N* junction is
   (*a*) less than its transition capacitance
   (*b*) essential for the design of varactors
   (*c*) problem for devices which are required to switch rapidly from forward to reverse bias
   (*d*) essential for the working of a tunnel diode

9. The depletion region of a semiconductor diode is due to
(*a*) absence of current carriers
(*b*) reverse-biasing
(*c*) forward-biasing
(*d*) crystal doping

10. When a *P-N* junction of a diode is forward-biased, diffusion current causes
(*a*) covalent bonding
(*b*) forward-biasing
(*c*) establishment of barrier potential
(*d*) reverse-biasing

11. The leakage current of semiconductor diode is caused by
(*a*) chemical energy
(*b*) barrier voltage
(*c*) heat energy
(*d*) doping impurity

12. An SC diode has –1.5 V on its anode and –2.5 on its cathode. The diode is
(*a*) conducting
(*b*) hot conducting
(*c*) reverse biased
(*d*) in breakdown

ANSWERS

A. Fill in the blanks

1. both **2.** barrier, 0.7 **3.** majority **4.** doping **5.** decreases **6.** reverse, minority **7.** low, high **8.** minority **9.** heavily, narrow **10.** rupture or breakdown **11.** wide **12.** bias **13.** depletion **14.** 0.7 **15.** reverse **16.** widens **17.** non-conducting **18.** voltage **19.** reverse **20.** leakage

B. True or False

1. F **2.** F **3.** T **4.** F **5.** T **6.** F **7.** T **8.** F **9.** T **10.** F **11.** T **12.** F

C. Multiple Choice Items

1. *a* **2.** *d* **3.** *c* **4.** *c* **5.** *b* **6.** *a* **7.** *c* **8.** *c* **9.** *a* **10.** *c* **11.** *c* **12.** *a*

CHAPTER 14

P-N Junction Diode

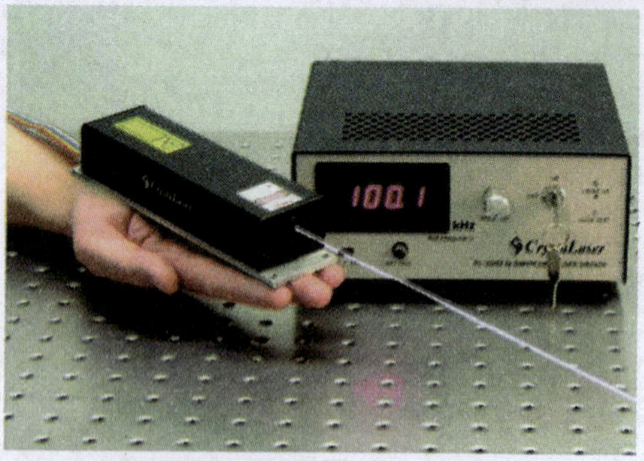

1. P-N Junction Diode
2. Diode Ratings or Specifications
3. Diode Testing
4. The Ideal Diode
5. The Real Diode
6. Diode Circuits with DC and AC Voltage Sources
7. Diode Fabrication
8. Grown Junction
9. Alloy Junction
10. Diffused Junction
11. Epitaxial Junction
12. Point Contact Junction
13. Clippers and Clampers
14. Clippers
15. Some Clipping Circuits
16. Clampers
17. Summary of Clamping Circuits

14.1. P-N Junction Diode

(a) Construction

It is a two-terminal device consisting of a *P-N* junction formed either in Ge or Si crystal. Its circuit symbol is shown in Fig. 14.1 (*a*). The *P*-and *N*-type regions are referred to as anode and cathode respectively. In Fig. 14.1 (*b*), arrowhead indicates the conventional direction of current flow when forward-biased. It is the same direction in which hole flow takes place.

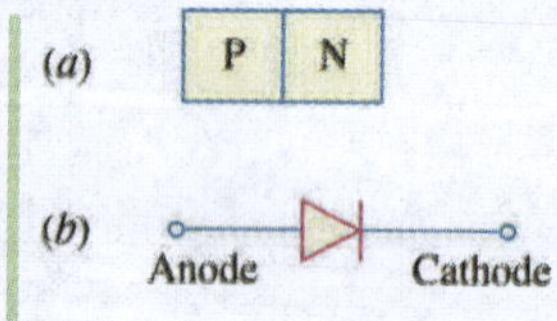

Fig. 14.1

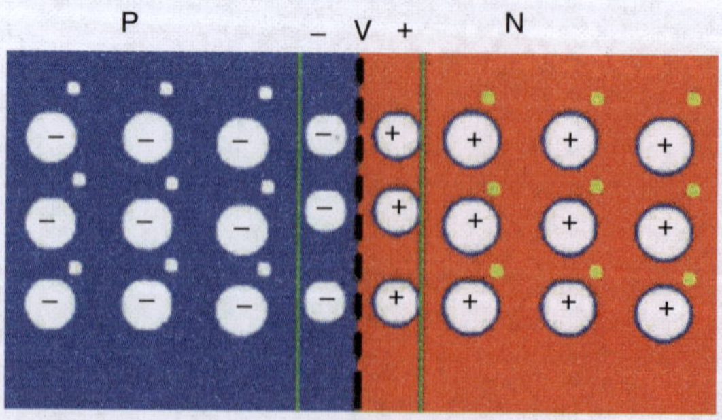

PN-junction diode.

Commercially available diodes usually have some means to indicate which lead is P and which lead is N. Standard notation consists of type numbers preceded by '*IN*' such as IN 240 and IN 1250. Here, 240 and 1250 correspond to colour bands.

The appearances of one low, two medium and one high current diodes are shown in Fig. 14.2. All the four diodes carry their own identification in the form of a small schematic painted on their bodies. The small glass-encased germanium diode has bands at one end. The end which has the black band is the cathode and the other, obviously, is the anode. Fig. 14.2 (*a*) shows low-current diode whose body is about 3 mm long. It can carry a forward current of about 100 mA, has saturation current of 5 μA at room temperature (25°C) and can stand a reverse voltage of 75 V without breaking down.

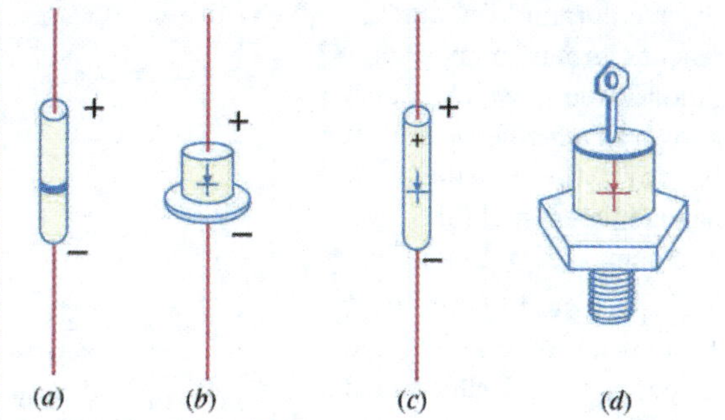

Fig. 14.2

The two medium-current diodes shown in Fig. 14.2 (*b*) and (*c*) can pass a forward current of about 500 mA and can withstand a reverse voltage of 250 V. The high-current diode or power diode shown in Fig. 14.2 (*d*) can pass a forward current of many amperes and can survive many hundred volts of reverse voltage.

(*b*) Diode Mounting

Low and medium-current diodes are usually mounted by soldering their leads to the connecting terminals. The heat generated by these diodes (when operating) is small enough to be carried away by air convection and conduction along the connecting leads. However, 'high-current stud-mounted diodes' generate large amounts of heat for which air convection is totally inadequate. For cooling, they need heat sinks made of metals such as copper or aluminium which are good conductors of heat. The sink absorbs heat from the device and then transfers it to the surrounding air by convection and radiation since it has large surface area. Such a heat sink is shown in Fig. 14.3. It has been finned to increase its surface area and hence its cooling efficiency.

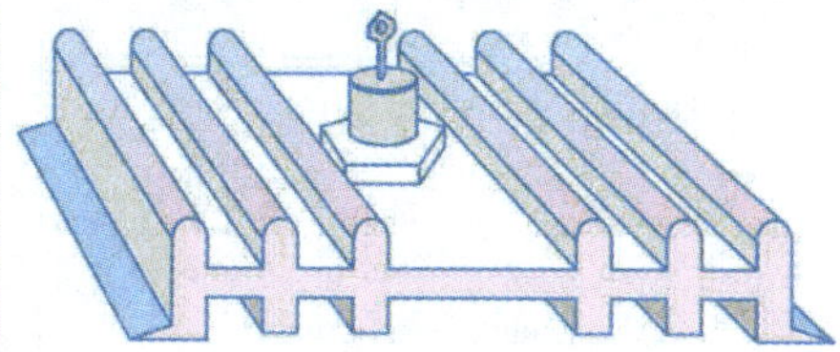
Fig. 14.3

(*c*) Working

A P-N junction diode is a one-way device offering low resistance when forward-biased and behaving almost as an insulator when reverse-biased. Hence, such diodes are mostly used as rectifiers *i.e.*, for converting alternating current into direct current.

(*d*) V/I Characteristic

Fig. 14.4 shows the static voltage-current characteristics for a low-power P-N junction diode.

(*i*) *Forward Characteristic*

When the diode is forward-biased and the applied voltage is increased from zero, hardly any current flows through the device in the beginning. It is so because the external voltage is being opposed by the internal barrier voltage V_B whose value is 0.7 V for Si and 0.3 V for Ge. As soon as V_B is neutralised, current through the diode increases rapidly with increasing applied battery voltage. It is found that as little a voltage as 1.0 V produces a forward current of about 50 mA. A burnout is likely to occur if forward voltage is increased beyond a certain safe limit.

(*ii*) *Reverse Characteristic*

When the diode is reverse-biased, majority carriers are blocked and only a small current (due to minority carriers) flows through the diode. As the reverse voltage is increased from zero, the reverse current very quickly reaches its maximum or saturation value I_0 which is also known as *leakage current*. It is of the order of nanoamperes (nA) for Si and microamperes (μA) for Ge.

As seen from Fig. 14.4, when reverse voltage exceeds a certain value called *breakdown voltage* V_{BR}, the leakage current suddenly and sharply increases, the curve indicating zero resistance at this point. When *P-N* junction diodes are employed primarily because of this porperty as voltage regulators, they are called *Zener diodes* (Art. 15.1).

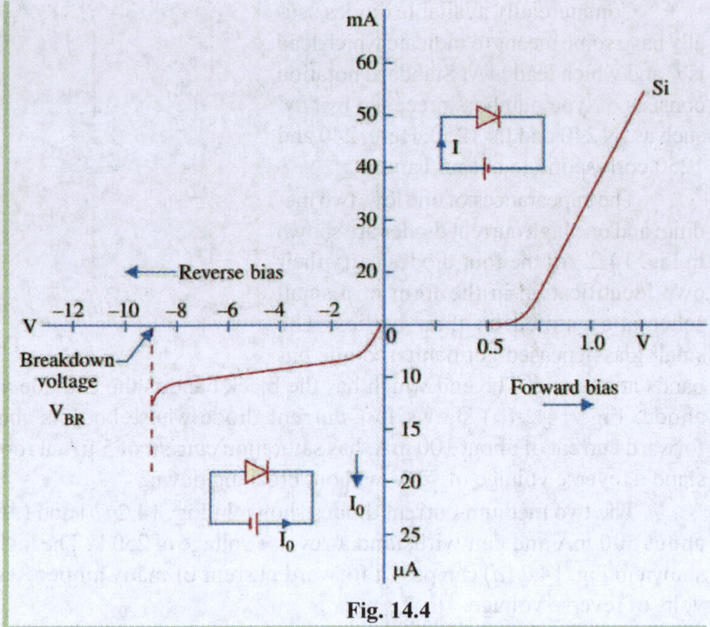

Fig. 14.4

(e) Diode Parameters

The diode parameters of greatest interest are as under :

1. Bulk resistance (r_B)

It is the sum of the resistance values of the *P*-and *N*-type semiconductor materials of which the diode is made of

$$\therefore \quad r_B = r_P + r_N \qquad \text{— Fig. 14.5 (b)}$$

Usually, it is very small. It is given by

$$r_B = \frac{V - V_K}{I_F}$$

It is the resistance offered by the diode well above the knee voltage *i.e.*, when current is large. Obviously, this resistance is offered in the *forward* direction.

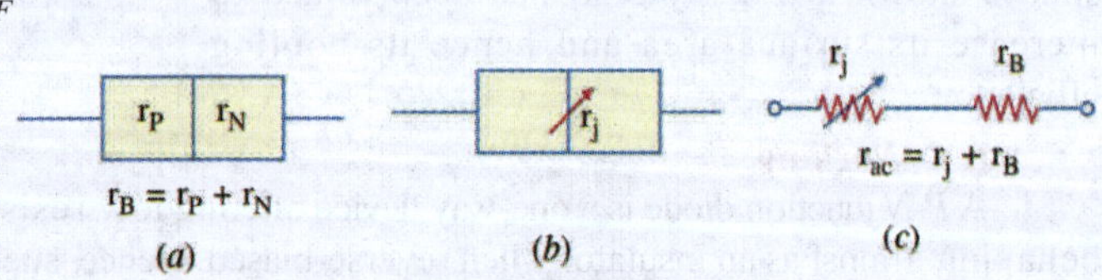

Fig. 14.5

2. Junction resistance (r_j)

Its value for forward-biased junction depends on the magnitude of *forward dc* current.

$$r_j = \frac{25 \text{ mV}}{I_F \text{ mA}} \qquad \text{— for Ge}$$

$$= \frac{50 \text{ mV}}{I_F \text{ mA}} \qquad \text{— for Si}$$

Obviously, it is a *variable* resistance.

3. Dynamic or ac resistance

$$r_{ac} \text{ or } r_d = r_B + r_j \qquad \text{— Fig. 14.5 (c)}$$

For large values of forward current, r_j is negligible. Hence, $r_{ac} = r_B$. For small values of I_F, r_B is negligible as compared to r_j.

$$\therefore \quad r_{ac} = r_j$$

4. Forward voltage drop

It is given by the relation

$$\text{forward voltage drop} = \frac{\text{power dissipated}}{\text{forward dc current}}$$

5. Reverse saturation current (I_0)

It has already been discussed in Art. 13.7.

6. Reverse breakdown voltage (V_{BR})

It has already been discussed in Art. 13.10.

7. Reverse dc resistance R_R

$$R_R = \frac{\text{reverse voltage}}{\text{reverse current}}$$

8. Equation of diode current

The analytical equation which describes both the forward and reverse characteristics is called the Boltzmann's diode equation given by

$$I = I_0\,(e^{40\text{ V}} - 1) \qquad \text{— for Ge}$$
$$= I_0\, e^{40\text{ V}} \qquad \text{if } V > 1 \text{ volt}$$

and

$$I = I_0\,(e^{20\text{ V}} - 1) \qquad \text{— for Si}$$
$$= I_0 e^{20\text{ V}} \qquad \text{if } V > 1 \text{ volt}$$

where I_0 = reverse saturation current ; V = voltage across the diode.

(f) Applications

The main applications of semiconductor diodes in modern electronic circuitry are as under :

1. As power or rectifier diodes. They convert ac current into dc current for dc power supplies of electronic circuits.
2. As signal diodes in communication circuits for modulation and demodulation of small signals.
3. As Zener diodes in voltage stabilizing circuits.
4. As varactor diodes — for use in voltage-controlled tuning circuits as may be found in radio and TV receivers. For this purpose, the diode is deliberately made to have a certain range of junction capacitance. The capacitance of the reverse-biased diode is given by where V_R is the reverse voltage.
5. In logic circuits used in computers.

Example 14.1. *A silicon diode dissipates 3 W for a forward dc current of 2 A. Calculate the forward voltage drop across the diode and its bulk resistance.*

Solution. Forward drop is $V_F = 3 / 2 =$ **1.5 V**

Now, $V_F = V_j = r_B I_F$; $1.5 = 0.7 + r_B \times 2$ $\quad\therefore\quad$ $r_B = \mathbf{0.4\ \Omega}$

Example 14.2. *A silicon diode has a forward voltage drop of 1.2 V for a forward dc current of 100 mA. It has a reverse current of 1 μA for a reverse voltage of 10 V. Calculate*

(a) bulk and reverse resistances of the diode

(b) ac resistance at forward dc current of (i) 2.5 mA and (ii) 25 mA.

Solution. (a)

$$r_B = \frac{V_F - V_j}{I_F} = \frac{1.2\text{ V} - 0.7\text{ V}}{100\text{ mA}} = \mathbf{5\ \Omega}$$

$$R_R = \frac{V_R}{I_R} = \frac{10\text{ V}}{1\ \mu\text{A}} = \mathbf{10\ M}$$

(*b*) (*i*)

$$r_j = \frac{50\text{ mV}}{2.5\text{ mA}} = \mathbf{20\ \Omega}$$

$$r_{ac} = r_B + r_j = 5 + 10 = \mathbf{15\ \Omega}$$

(*ii*)

$$r_j = \frac{50\text{ mV}}{25\text{ mA}} = \mathbf{2\ \Omega}$$

$$r_{ac} = 5 + 1 = \mathbf{6\ \Omega}$$

14.2. Diode Ratings or Specifications

In order to select a diode for a particular application, one must know the specifications as given in the data sheet provided by the device manufacturer. These specifications usually indicate the absolute ratings of the diode, typical operating conditions, mechanical data and set of curves showing typical operation etc.

The absolute maximum rating means the maximum value of voltages and current etc., that can be applied without destroying the device. It is very important that these ratings should never be exceeded. In fact, they should never be approached if reliable operation is desired. These specifications must be derated downward for temperatures exceeding 25°C.

Some of the important diode specifications are given below :

1. Peak Inverse Voltage (PIV)

It is the maximum reverse voltage that can be applied to the diode without destruction.

Other names and abbreviations used for it are

(*a*) peak reverse voltage (PRV),

(*b*) reverse breakdown voltage (V_R or V_{RB}),

(*c*) maximum reverse voltage (V_{RM} or $V_{R.max}$).

2. Average Forward Current

It is the continuous forward current which the diode can pass at the normal temperature (usually 25°C). It is written as $I_{F(av)}$.

It is also called maximum steady state forward current (I_{FM} or $I_{F.max}$).

Obviously, the current should be derated for operation at higher temperatures.

It is also known as the repetitive forward current *i.e.*, current which the diode can safely handle continuously. It is written as I_{rep}.

3. Forward Surge Current

Surge current is defined as that large current which a diode can safely take for a very short time (make upto one second). It is written as I_{FS}.

4. Maximum Forward Voltage

It is the maximum forward voltage that the diode can have without burnout. It is written as V_{FM}.

5. Forward Voltage

It is the forward voltage of the diode at a given temperature and for a specific value of forward current. It is written as V_F.

6. Reverse Current

It is the maximum reverse saturation current at the maximum reverse voltage at a given temperature.

Normal reverse current is written as I_R and its maximum value as I_{RM}.

7. Power Dissipation

It is given as the maximum power that the diode can safely dissipate on a continuous basis in free air at 25°C.

8. Reverse Recovery Time

It is the maximum time taken by the device to switch from ON to OFF state. It is written as t_{rr}. It is usually in nanoseconds (ns).

14.3. Diode Testing

For testing a diode, we may either use a diode checker or an ohmmeter (or ohmmeter scale of a multimeter). While using this resistance test, at least one lead of the diode should be disconnected from the circuit. Since an ohmmeter has its own dc source, it can bias the diode either in the forward or reverse direction and thus measure its resistance in either direction. Secondly, it also enables the technician to know which terminal is anode and which is cathode.

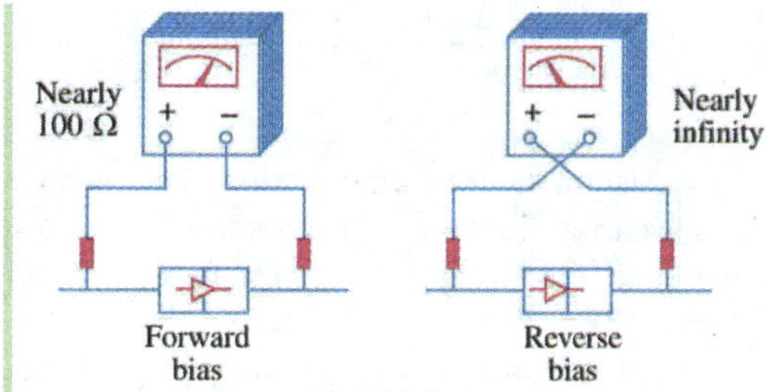

Fig. 14.6

Before checking the diode, selector switch of the ohmmeter should be placed on '$R \times 100$' and the leads connected as shown in Fig. 14.6. In the forward direction, the ohmmeter reading should be less than 100 Ω. In the reverse bias direction (when leads are reversed) it should read more than 5000 Ω. It should be kept in mind that the actual resistance values of the diode are not important as long as we get low/high readings when the ohmmeter polarity is reversed. Other points worth noting are as under :

1. a low resistance reading both ways is common when *diode is tested in-circuit*,
2. if both resistance readings are high, the diode is open,
3. if both resistance readings are low, the diode is shorted,
4. the diode is good only if there is a low resistance reading in one direction and a high reading in the other.

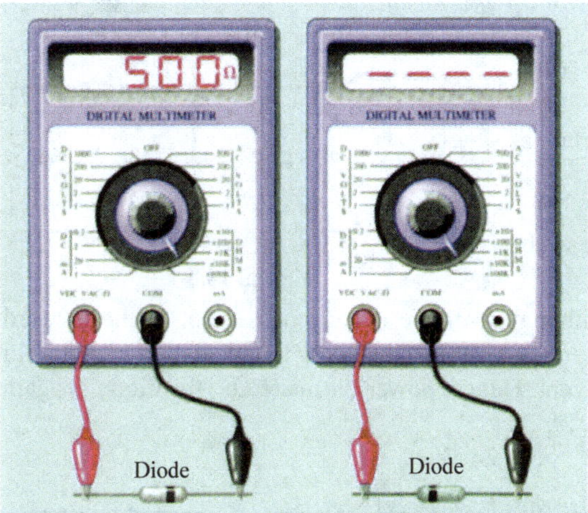

(*a*) Typical forward check (*b*) Typical reverse check

Checking a diode with a DMM set on the "Ohms" function.

The anode and cathode terminals of a good diode can also be determined if we observe the ohmmeter polarity while taking the resistance readings. The low resistance value is obtained when positive lead of the meter is connected to the anode [Fig. 14.6 (*a*)]. Obviously, the other terminal will be the cathode.

14.4. The Ideal Diode

There is no such thing as an *ideal* diode or perfect diode. The existence of such a diode is visualized simply as an aid in analysing the diode circuits.

An ideal diode may be defined as a two-terminal device which

(*a*) conducts with zero resistance when forward-biased, and

(*b*) appears as an infinite resistance when reverse-biased.

In other words, such a device acts as a short-circuit in the forward direction and as an open-circuit in the reverse direction as shown in Fig. 14.7.

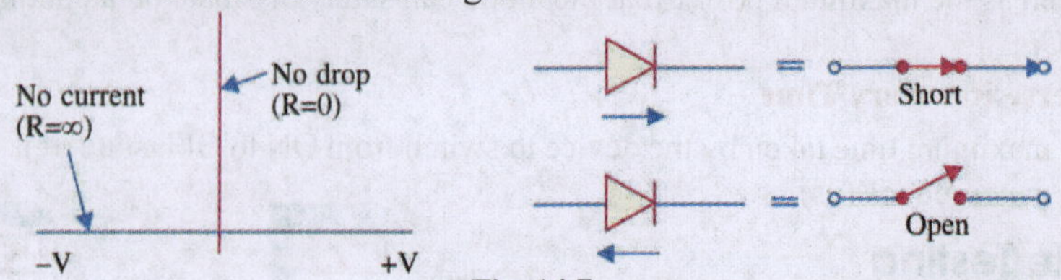

Fig. 14.7

Also in the forward direction, there is no voltage drop (even though current is there) since a short has zero resistance. On the other hand, there is no reverse current because reverse resistance is infinite.

It is helpful to think of an ideal diode as a bistable switch which is closed in the forward direction and open in the reverse direction. Hence, it has two stable states: ON or OFF.

The ideal diode.

Example 14.3. *Calculate the circuit current and power dissipated in the*

(a) ideal diode and (b) 6 Ω resistor of the circuit shown in Fig. 14.8 (a).

Solution. The diode is an ideal one and is forward-biased. Hence, it can be replaced by a short (closed switch) as shown in Fig. 14.8 (*b*). The circuit current, as given by Ohm's law, is

$$I = 12/6 = \mathbf{2\ A}$$

(*a*) Since there is no voltage drop across the diode, power consumed by it is **zero**.

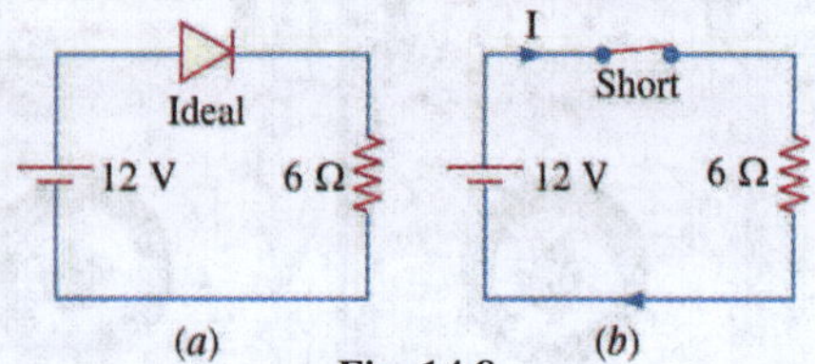

Fig. 14.8

As we know, there is no power when either the voltage or current is zero. In the forward direction, there is current but no voltage drop, hence power dissipated by the ideal diode is zero. In the reverse direction, there is voltage but no current. Hence, power dissipated by the diode is again zero. In fact, an ideal diode never *dissipates any power*.

(*b*) Power consumed by 6 Ω resistor = $2^2 \times 6$ = **24 W.**

Example 14.4. *Calculate the current in the circuit of Fig. 14.9 (a).*

Solution. As seen from the polarity of the battery terminals, the diode is reverse-biased.

Hence, it acts like an open switch as shown in Fig. 14.9 (*b*). Obviously, current in such a circuit is **zero**.

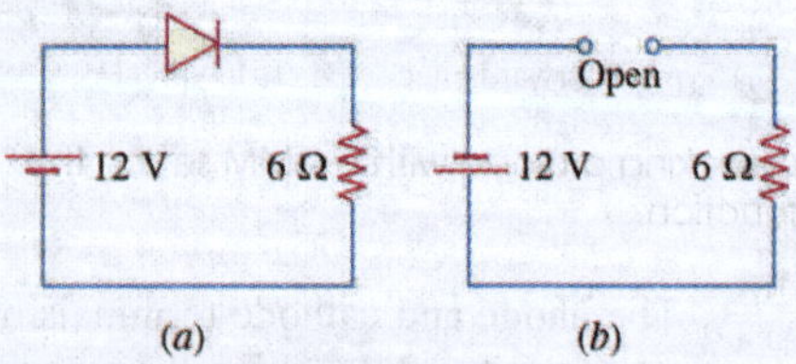

Fig. 14.9

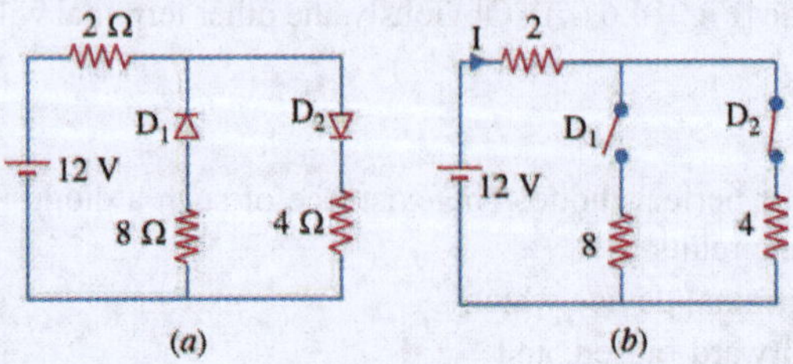

Fig. 14.10

Example 14.5. *Find the current, if any, flowing in the circuit of Fig. 14.10 (a) which uses two oppositely-connected ideal diodes in parallel.*

Solution. The diode D_1 is reverse-biased and acts as an open switch as shown in Fig. 14.10 (*b*). So, there is no current through D_1 and the 8 Ω resistor.

However, D_2 is forward-biased and acts like a short-circuit or closed switch. The current drawn is

$$I = \frac{12}{2 + 4} = \mathbf{2\ A}$$

Example 14.6. *Sketch the wave of the voltage v_0 across the 10 K resistor and current in the circuit of Fig. 14.11 if the generator voltage v_g is a sine wave of peak value 10 V.*

Solution. It should be noted that the diode will conduct only during the positive half-cycle of the input signal voltage when it is forward-biased. It will not conduct during the negative half-cycle since it would be reverse-biased.

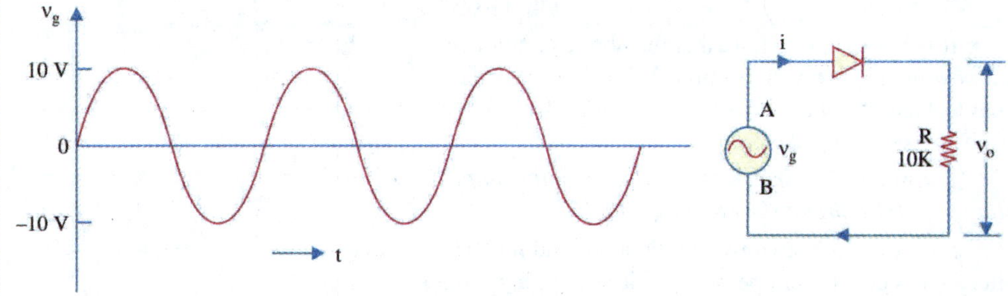

Fig. 14.11

Since during forward bias, the diode acts as a short, the entire generator voltage drops across *R*. The negative half-cycle of v_g would be suppressed because the diode would act as an open switch. Hence, the wave form of v_0 is as shown in Fig. 14.12 (*a*). It is called half-wave rectified voltage.

The wave shape of the output current is identical to that of v_0 except that peak or maximum value of *i*

$$= 10 / 10\text{ K} = 1\text{ mA}.$$

It is called half-wave rectified current and is shown in Fig. 14.12 (*b*).

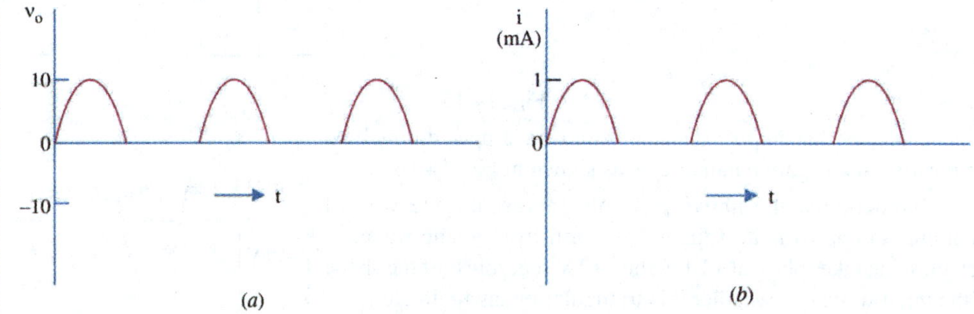

(*a*) (*b*)

Fig. 14.12

Example. 14.7. *A sine wave voltage of peak value 50 V is applied to a diode circuit of Fig. 14.13. Sketch the waveform of voltage v_0 treating the diode as an ideal one.*

Solution. First, consider the positive half-cycle of the input voltage. The diode acts as a short in the forward direction and the moment v_g exceeds battery voltage of 10 V, current will start flowing through the circuit. The value of v_0 will remain steady at 10 V, the balance of 40 V dropping across the 10 K resistor. It is seen that value of v_0 is set by the battery voltage.

Next, consider the negative input half-cycle. Now, the diode acts like an open switch and no current can flow through the circuit and hence through the 10 K resistor. In that case, v_0 follows the input voltage as shown in Fig. 14.14.

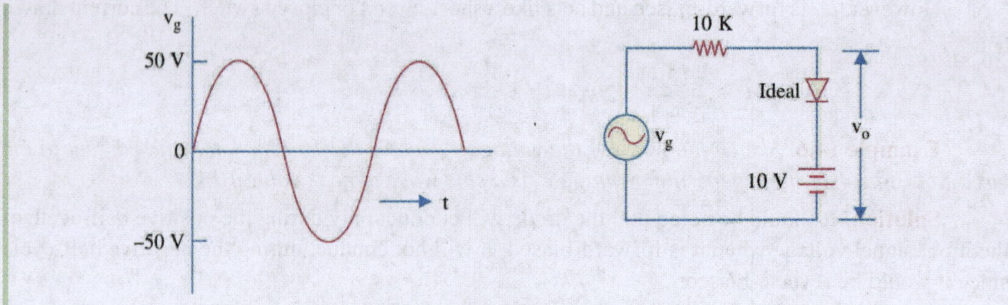

Fig. 14.13

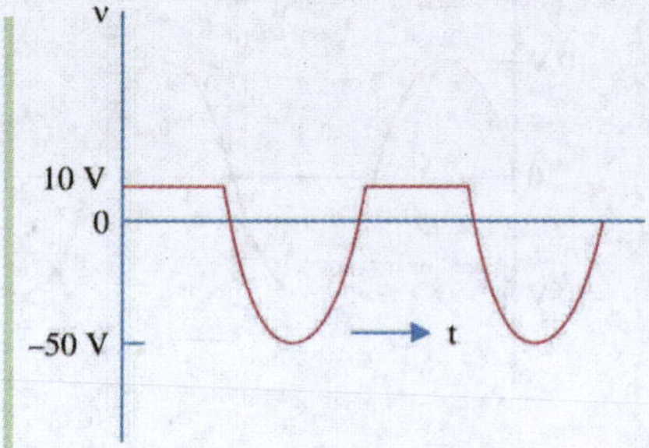

Fig. 14.14

It is interesting to note that the above circuit acts as a *positive clipper* with a clipping level at 10 V *i.e.*, equal to battery voltage. All positive parts of v_g above 10V have been clipped off.

Example 14.8. *Sketch the waveform of the output voltage v_0 of the circuit shown in Fig. 14.15.*

Solution. It is obvious that diode D_1 and its 10 V battery act as positive clipper with positive clipping level at +10 V. Similarly, D_2 and its 10 V battery act as negative clipper with a clipping level at –10 V.

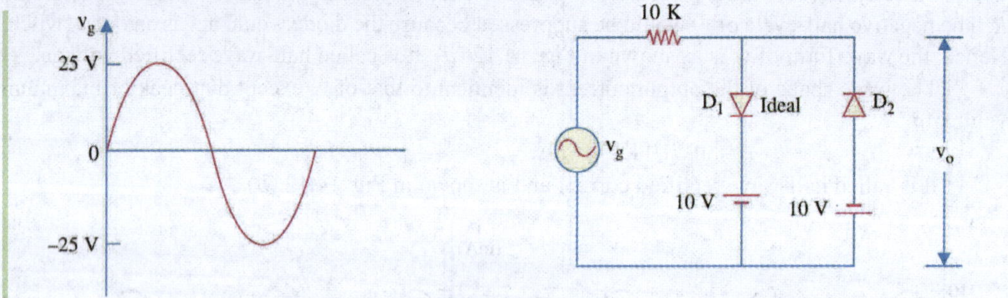

Fig. 14.15

It is clear that output voltage v_0 would be clipped during both the positive and negative half-cycles as shown in Fig. 14.16.

Incidentally, the above example represents one way of obtaining an approximate square waveform from a sine wave. In fact, clipping takes place at +10 V and –10 V regardless of the shape of the input wave *i.e.*, whether it is triangular or sawtooth etc.

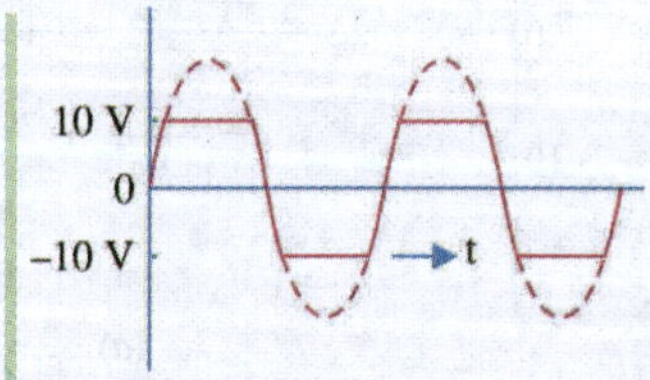

Fig. 14.16

14.5. The Real Diode

A real diode does not conduct in the forward direction with zero resistance neither it offers infinite resistance in the reverse direction.

(*a*) Forward Direction

In this case, we have to take two factors into account. One is that forward current does not start flowing until the voltage applied to the diode exceeds its threshold or knee voltage (V_K). Hence, a real diode is shown as equivalent to

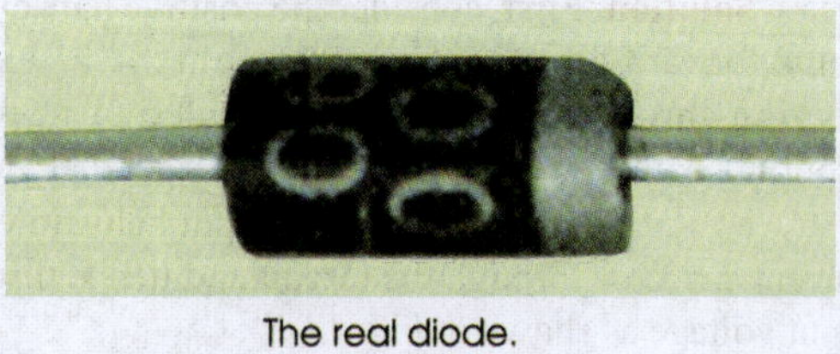

The real diode.

an ideal diode in series with a small oppositely-connected battery of e.m.f. V_K as shown in Fig. 14.17.

The second factor to be considered is the forward dynamic or ac resistance (r_{ac}) offered by the circuit. So far, we considered this resistance to be zero implying that forward characteristic is a straight vertical line (Fig. 14.17). If we take r_{ac} into account, the forward characteristic becomes as shown in Fig. 14.18. Here, the reciprocal of the slope of this characteristic represents r_{ac}.

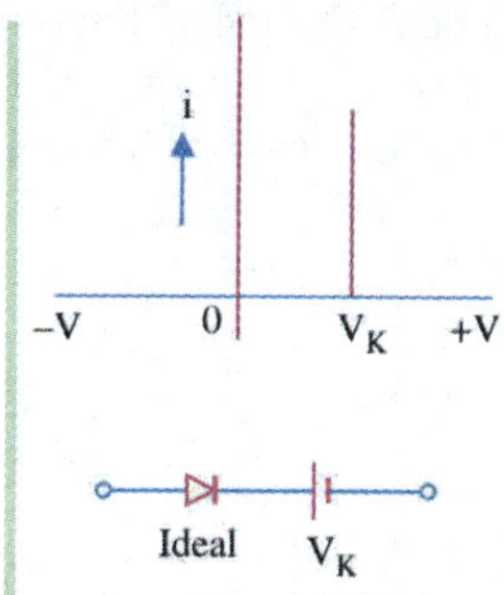

Fig. 14.17

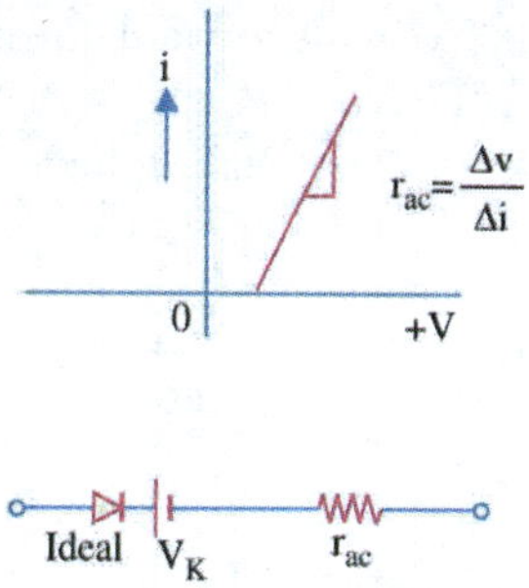

Fig. 14.18

(i) Large Signal Operation

Large signal sources are those whose voltage is *much greater* than the diode knee voltage V_K (nearly equal to barrier potential). Under such conditions, forward current would be large, so that r_j (Art. 14.1) would be negligible.

$$\therefore \qquad r_{ac} = r_j + r_B \cong r_B$$

(ii) Small Signal Operation

In this case, the signal voltage is *much smaller* than V_K (0.3 V for Ge and 0.7 for Si). Since I_F would be small, r_j would be very large as compared to r_B.

$$\therefore \qquad r_{ac} = r_j + r_B \cong r_j$$

(b) Reverse Direction

An actual or real diode does not have infinite resistance in the reverse direction because it will always have some reverse saturation current below the breakdown. For example, if with a $V_R = 50$ V, I_R is 10 μA, then $R_R = 5 \times 10^6\ \Omega = 5$ M. Silicon diodes have reverse resistance of many thousands of megohms. Hence, an actual diode in the reverse direction can be thought of as equivalent to a high resistor. This would be true only in the case of signals of low frequencies. For high-frequency signals, we will have to take into account the capacitive effects (Art. 13.11).

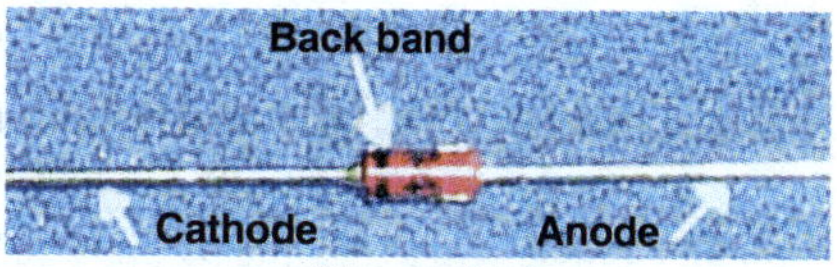

Silicon diode.

Example 14.9. *A silicon diode passes a current of 100 mA at 1 V. Find its bulk resistance. What would be its ac resistance for direct currents of (i) 0.1 mA and (ii) 25 mA ?*

Solution. As seen from Art. 14.1,

$$r_B = \frac{V - V_K}{I_F} = \frac{1 - 0.7}{100 \times 10^{-3}} = 3\ \Omega$$

(*i*) Now, $r_{ac} = r_j + r_B$; $r_j = \dfrac{50 \text{ mA}}{I_F \text{ mA}} = \dfrac{50}{0.1} = 500\ \Omega$

The bulk resistance, as calculated above, is 3 Ω

$$\therefore \quad r_{ac} = 500 + 3 = \mathbf{503\ \Omega} \quad \therefore\ r_a = 1 + 3 = \mathbf{4\ \Omega}$$

Example 14.10. *Sketch the waveform for the output voltage V_0 in Fig. 14.19. Take $V_K = 0.3$ V for germanium diode which has a forward current of 28 mA at 1 V.*

Solution. $r_B = \dfrac{1 - 0.3}{25 \text{ mA}} = 25\ \Omega$; $r_j = 25/28 = 0.9\ \Omega$

$$\therefore \qquad r_{ac} = r_j + r_B \cong r_B = 25\ \Omega$$

Hence, the equivalent circuit becomes as shown in Fig. 14.20 (*a*). Whenever v_g exceeds

0.3 V in the forward direction, diode is shorted and the circuit offers a total resistance = (25 + 75) = 100 Ω.

The peak voltage = 20 – 0.3 = 19.7 V. Hence, peak value of forward current is = 19.7 / 100 = 197 mA.

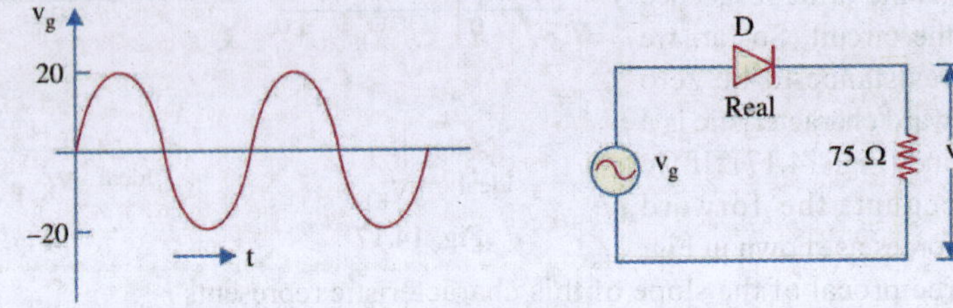

Fig. 14.19

∴ peak value of output voltage is = 197 × 10^{-3} × 75 = 14.77 V.

Hence, the peak value of the half-wave rectified voltage is 14.77 V as shown in Fig. 14.20 (*b*). During the negative half-cycle of the applied voltage, diode acts as an open switch so that $V_0 = 0$. The waveform sketch of v_0 is shown in Fig. 14.20 (*b*).

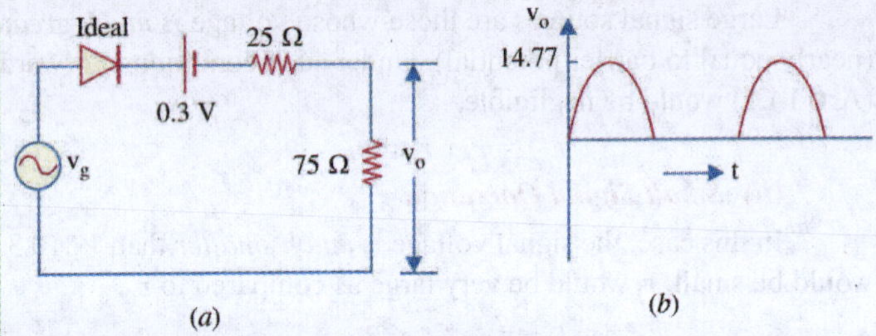

Fig. 14.20

14.6. Diode Circuits with DC and AC Voltage Sources

We will often come across diode and transistor circuits which will contain both dc and ac voltage sources. Such circuits can be easily analysed by using Superposition Theorem (Art. 4.2). We will first draw the dc equivalent circuit while neglecting ac sources and find the required current and voltage values. Next, we will draw the ac equivalent circuit neglecting the dc sources and again find the voltage and current values. Finally, we will superimpose the two sets of values to get the final result.

While drawing the equivalent circuits, following points must be remembered :

1. Direct current cannot flow 'through' a capacitor. Hence, all capacitors look *like an open to a dc source.*
2. Usually, capacitors used in most circuits are large enough to *look like a short to an ac source* particularly one of very high frequency. Similarly, dc batteries would also act as short circuits unless they have very high internal resistances.

Example 14.11. *The silicon diode shown in Fig. 14.21 has a bulk resistance of 1 Ω. The frequency of 10 mV (peak) signal is so high that the reactance of the coupling capacitor may be taken as zero. Sketch the approximate waveform of the total voltage 'v' across the diode.*

Solution. We will apply Superposition theorem to find *v*. First, voltage drop due to dc source would be found and then it would be added to the drop due to the ac source.

(i) DC Equivalent Circuit

It is seen from Fig. 14.21 (*b*) that the circuit to the left of point *A* is 'open' to the dc source of 20 V because of capacitor *C*. Hence, the equivalent dc circuit is as shown in Fig. 14.22 (*a*). As seen, diode is forward-biased by the battery. Hence only 0.7 V dc appears across the diode.

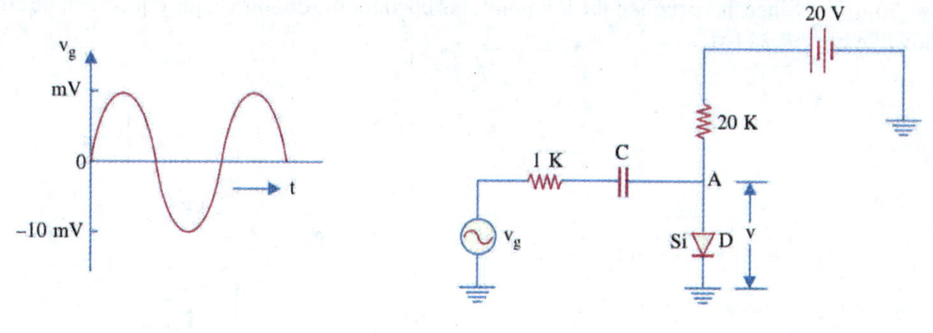

(a) (b)

Fig. 14.21

The dc current may also be calculated.

$$I = \frac{20 - 0.7}{20\text{ K}} \cong 1\text{ mA}$$

(ii) AC Equivalent Circuit

Here, the capacitor C and the 20 V battery would be treated as orts thereby giving us the ac equivalent circuit of Fig. 14.22 (*b*).

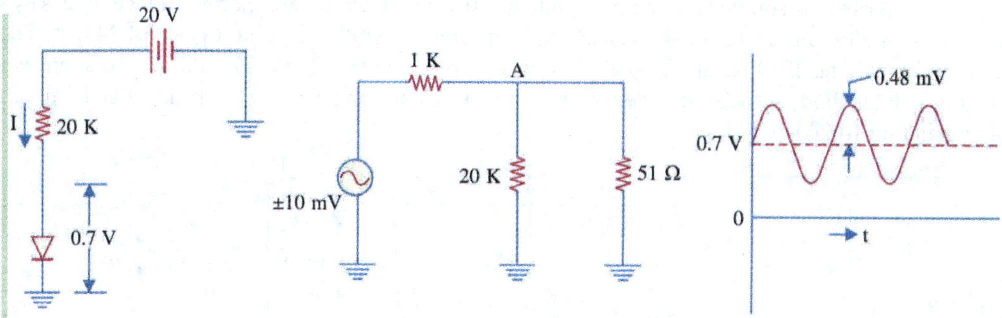

(a) (b) (c)

Fig. 14.22

Since, it is a silicon diode

$$r_j = \frac{50\text{ mV}}{1\text{ mA}} = 50\ \Omega \qquad \therefore \quad r_{ac} = r_j + r_B = 50 + 1 = 51\ \Omega$$

As shown in Fig. 14.22 (*b*), so far as the signal source is concerned, 20 K resistor and ac resistance of the diode are connected in parallel at point *A*. Now, 20 K || 51 Ω ≅ 51 Ω. Hence, 1 K and 51 Ω are put in series across the signal source of peak value 10 mV. The peak value of the ac voltage drop over 51 Ω resistance is

$$= 10 \times \frac{51}{51 + 1000} = 0.48\text{ mV}$$

The total drop across the diode is the sum of the ac and dc drops. The combined voltage waveform is shown in Fig. 14.22 (*c*). It consists of a dc voltage of 0.7 V over which rides an ac voltage of peak value ± 0.48 mV.

Example 14.12. *Neglecting the bulk resistance and knee voltage of the Ge diode shown in Fig. 14.23 (b), calculate the total current flowing through it. Also neglect the reactance of the coupling capacitor at the given signal frequency.*

Solution. Since the circuit to the left point *A* is open for the dc source, the equivalent dc circuit is shown in Fig. 14.24 (*a*).

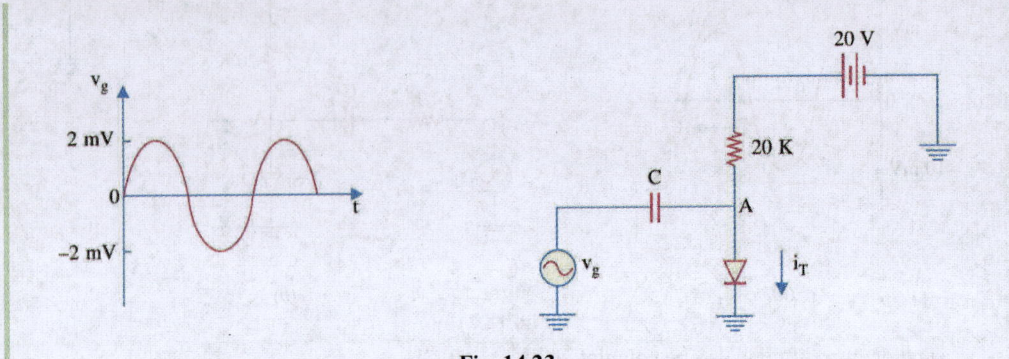

Fig. 14.23

It is seen that direct current through the diode is

$$I = \frac{20\text{ V}}{20\text{ K}} = 1\text{ mA}$$

$$\therefore \qquad r_{ac} = r_j = \frac{25\text{ mA}}{1\text{ mA}} = 25\ \Omega$$

So far as the ac source is concerned, both the 20 V battery and *C* are shorted. Hence, it sees two resistance of 20 K and 25 Ω (diode resistance) connected in parallel at point *A* [Fig. 14.24 (*b*)]. The current through the 25 Ω resistance (*i.e.*, diode) has a peak value = 2 mV/25 = 80 μA. Total current (i_T) through the diode is the sum of the dc current of 1 mA and an ac current of peak value 80 μA as shown in Fig. 14.24 (*c*).

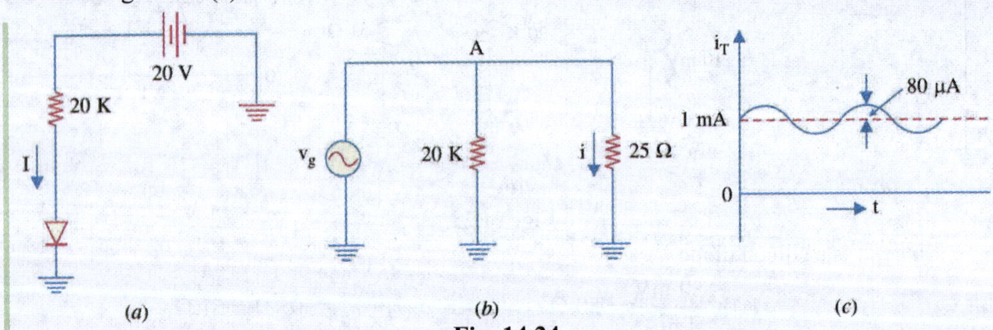

Fig. 14.24

14.7. Diode Fabrication

The electrical characteristics of a semiconductor diode depend on two factors (*i*) the material employed and (*ii*) the type of *P-N* junction used.

The two most commonly-used materials are germanium (Ge) and silicon (Si). Since Ge has higher electrical conduction than Si, it is often used in low-and medium-power diodes. On the other hand, Si has been found more suitable for high-power applications because it can be operated at higher temperatures than Ge. A new material called gallium-arsenide (GaAs) is found to combine desirable features of both Ge and Si and is finding ever-increasing use in many new applications.

The *P-N* junction may be produced by any one of the following methods :

1. grown junction,
2. alloy junction,
3. diffused junction,
4. epitaxial growth,
5. point contact junction.

The first step in the manufacture of any semiconductor device is to obtain the semiconductor material in an extremely pure form. The accepted impurity level is less than one part of impurity in one billion (10^9) parts of the semiconductor material. To begin with, the raw material is subjected to a series of chemical reactions and then to a zone refining process which employs induction heating to reduce the impurity level of the polycrystalline structure. Next, the Czochralski or floating zone technique is used to form single crystals of Ge or Si for fabrication of diodes. These crystals are then cut into wafers as thin as 0.025 mm (nearly one-fourth the thickness of this paper).

Now, we will briefly discuss the four basic processes commonly used in the manufacture of semiconductor diodes.

14.8. Grown Junction

Such junctions are produced by employing either the Czochralski or floating zone technique. The apparatus used for Czochralski technique is shown in Fig. 14.25. A single crystal **seed** of the desired impurity level is immersed in the molten semiconductor material contained in a crucible. Then, it is gradually withdrawn while the shaft holding the seed is slowly turning. When crystal is being pulled out, impurities of *P*-and *N*-type are alternately added to produce a *P-N* junction. This large area crystal is then cut into a large number of smaller-area diodes.

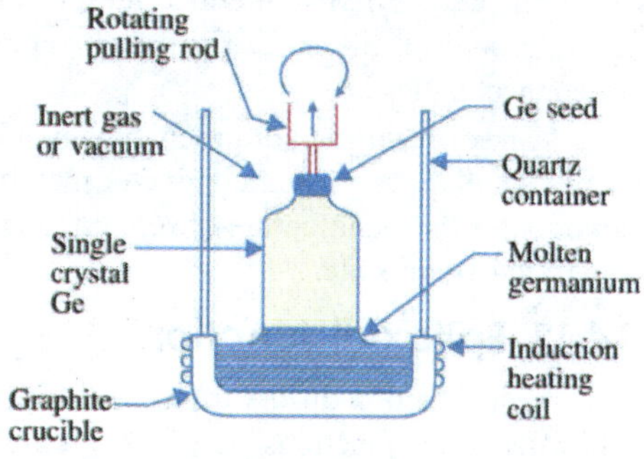

Fig. 14.25

14.9. Alloy Junction

The alloy process produces junction diodes that have high PIV and current ratings but which have large junction capacitance due to their large junction area.

In this process, a tiny dot (or pellet) of indium (or any other *P*-type impurity) is placed on the surface of an *N*-type silicon wafer and the two are heated well above the melting temperature of indium (about 150°C) as shown in Fig. 14.26 (*a*). Consequently, indium melts and dissolves some of the silicon. The temperature is then lowered and silicon refreezes to form a single crystal having a *P-N* junction as shown in Fig. 14.26 (*b*).

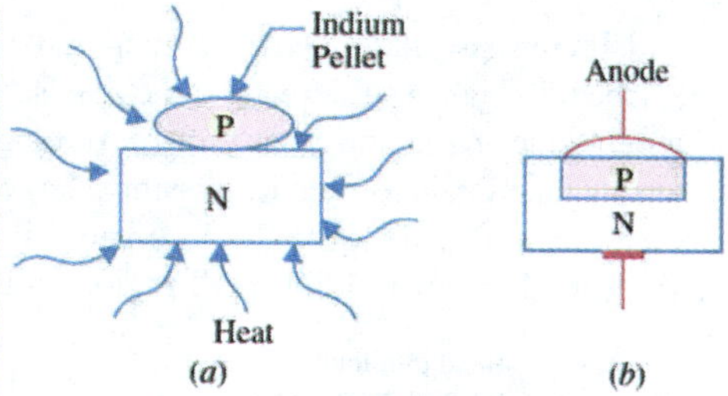

Fig. 14.26

14.10. Diffused Junction

The diffusion process employs either solid or gaseous diffusion. This process takes more time than alloy process but is relatively cheaper and more accurately controllable. In this process, particles from an area of high concentration drift to surrounding region of lesser concentration.

(*i*) Solid Diffusion

The solid diffusion process starts with the 'painting' of a *P*-type impurity (say, indium) on an *N*-type substrate and heating the two until the impurity diffuses a short distance into the substrate to form *P*-type layer (Fig. 14.27).

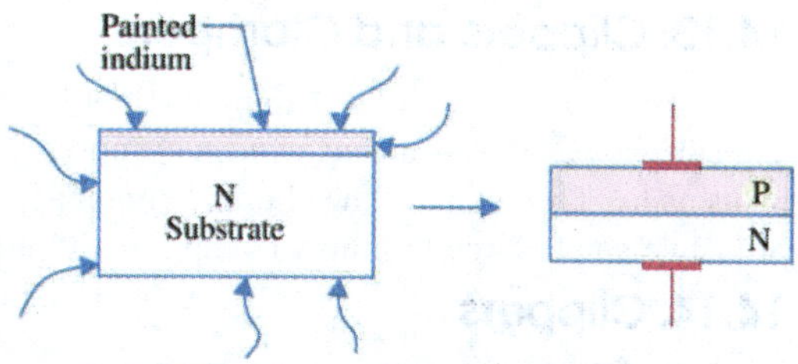

Fig. 14.27

(*ii*) Gaseous Diffusion

In the gaseous diffusion process, an *N*-type material is heated in a chamber containing a high concentration of an acceptor impurity in vapour from (Fig. 14.28). Some of the acceptor atoms are diffused (or absorbed) into the *N*-type substrate to form the *P*-type layer thus creating a *P-N* junction. By exposing only part of the *N*-type material during the diffusion process (the remainder being covered by a thin coating of SiO_2), the size of the *P*-region can be controlled. Finally, metal contacts are electroplated on the surface of each region for connecting leads.

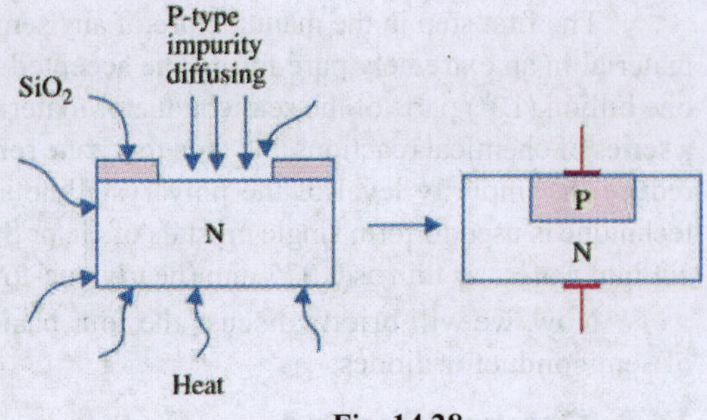

Fig. 14.28

The diffusion technique enables simultaneous fabrication of many hundreds of diodes on one small disc of a semiconductor material. That is why it is the most frequently-used technique not only for the manufacture of semiconductor diodes but also for the production of transistors and integrated circuits etc.

14.11. Epitaxial Junction

This junction differs from the diffusion junction only in the manner in which the junction is fabricated. Such junctions are grown on top of an *N*-type wafer in a high temperature chamber. The growth proceeds atom by atom and hence is exactly similar to the crystal lattice of the wafer on which it is grown. Such junctions have the advantage of low resistance.

14.12. Point Contact Junction

It consists of an *N*-type germanium or silicon wafer about 1.25 mm square by 0.5 mm thick, one face of which is soldered to a metal base by radio-frequency heating as shown in Fig. 14.29 (*b*). The other face has a phosphor bronze (or tungsten) spring (called a cat's whisker) pressed against it. The *P-N* junction is formed by passing a large current for a second or two through the wire while the crystal face with wire point is kept positive. The heat so produced drives away some of the electrons from the atoms in the small region around the point of contact thereby leaving holes behind. This small region of the *N*-type material is, consequently, converted into *P*-type material as shown in Fig. 14.29 (*a*). The small area of the *P-N* junction results in very low junction capacitance as mentioned earlier.

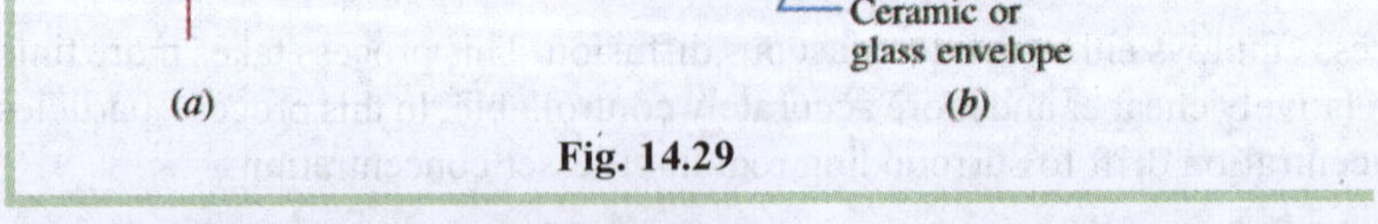

Fig. 14.29

14.13. Clippers and Clampers

These are diode waveshaping circuits *i.e.*, circuits meant to control the *shape* of the voltage and current waveforms to suit various purposes. Each performs the waveshaping function indicated by its name. The output of the clipping circuit appears as if a portion of the input signal were clipped off. But clamper circuits simply clamp (*i.e.*, lift up or down) the input signal to a different dc level.

14.14. Clippers

A clipping circuit requires a minimum of two components *i.e.*, a diode and a resistor. Often dc battery is also used to fix the clipping level. The input waveform can be clipped at different levels by simply changing the battery voltage and by interchanging the position of various elements. We will only use an ideal diode which acts like a closed switch when forward-biased and an open switch when reverse-biased.

Example 14.13. *For the simple parallel clipper of Fig. 14.30 (b), find the shape of the output voltage v_0 across the diode if the input sine wave signal is as shown in Fig. 14.30 (a). What will happen when diode and resistor are interchanged ?*

Solution. When positive half-cycle of the signal voltage is applied to the clipper *i.e.*, when *A* is positive with respect to *B*, the diode *D* is reverse-biased. Hence, it acts as an open switch. Consequently, the entire input voltage appears across it*.

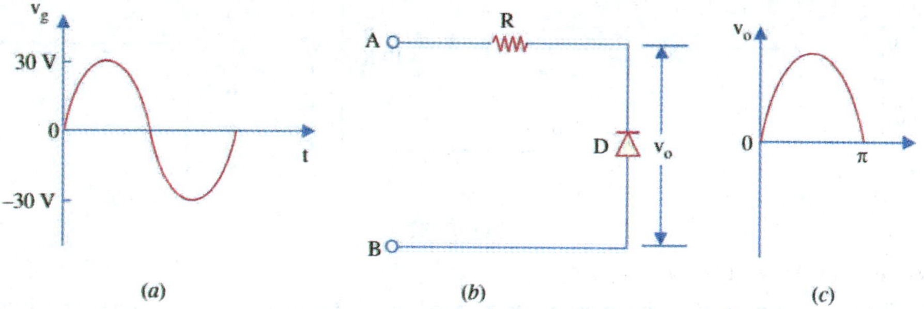

Fig. 14.30

In the negative half-cycle of the signal voltage when circuit terminal *B* becomes positive with respect to *A*, the diode is forward biased. Hence, it acts like a closed switch (or short) across which no voltage is dropped. Hence, the waveshape of v_0 is as shown in Fig. 14.30 (*c*). It is seen that the negative portion of the signal voltage has been removed. Hence, such a circuit is called a **negative** clipper.

Diode and Resistor Interchanged

In this case, the circuit becomes as shown in Fig. 14.31 (*b*). Now the output voltage v_0 is that which is dropped across *R*. During the positive half-cycle of the signal voltage, *D* acts as an open switch. Hence, all the applied voltage drops across *D* and none across *R*. So, there is no output signal voltage.

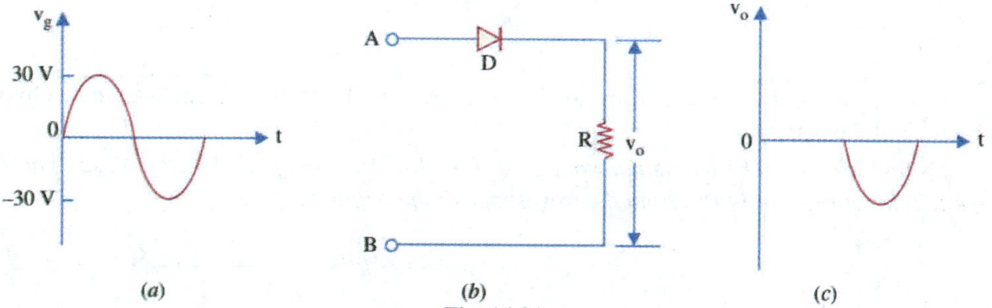

Fig. 14.31

During the negative input half-cycle, terminal *B* is positive and so it forward biases *D* which acts as a short. Hence, there is no voltage drops across *D*. Consequently, all the applied signal voltage drops across *R* and none across *D*. As a result, the negative half-cycle of the input signal is allowed to pass through the clipper circuit. Obviously, now the circuit acts as a **positive** clipper.

Example 14.14. *With the sine wave signal input of Fig. 14.32 (a), find the shape of the output signal v_0 in the biased series clipper of Fig. 14.32 (b). What would happen if battery connections were reversed?*

* Since there is no circut current, drop across *R* is zero which may, therefore, be considered a 'short' because no voltage develops across a short.

Solution. Let us consider the positive half-cycle of the signal *i.e.*, when terminal *A* of the circuit becomes positive with respect to *B*. The diode appears as a short since it is forward-biased. But no current flows till v_g exceeds the opposing battery voltage of 10 V. Hence, only upper part of the positive signal voltage passes through the clipper circuit and appears as v_0 across *R*. Its shape is shown in Fig. 14.32 (*c*). The negative half-cycle of the signal voltage is clipped off.

In fact, in this circuit, the entire input is clipped off except positive peak protions.

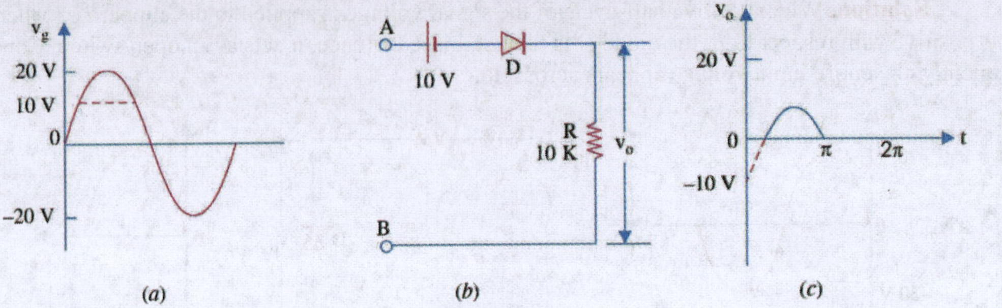

Fig. 14.32

Reversed Battery Connections

The battery connections have been reversed in Fig. 15.33. In this case, during the positive half-cycle of the signal, the voltage across *R* would be the sum of the signal voltage and the battery voltage *i.e.*, signal voltage would be lifted up by 10 V as shown in Fig. 14.33 (*c*).

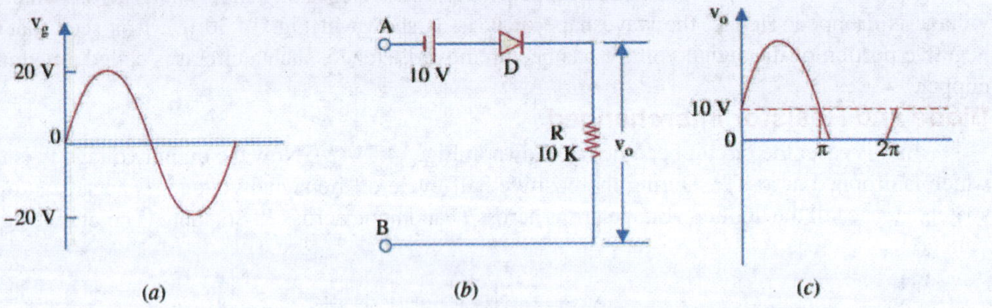

Fig. 14.33

During the negative input half-cycle, the lower peak portions of the signal would be clipped off because of the battery.

Example 14.15. *The triangular voltage of Fig. 14.34 (a) is applied to the biased parallel clipper circuit of Fig. 14.34 (b). Find the waveshape of the output voltage* v_0.

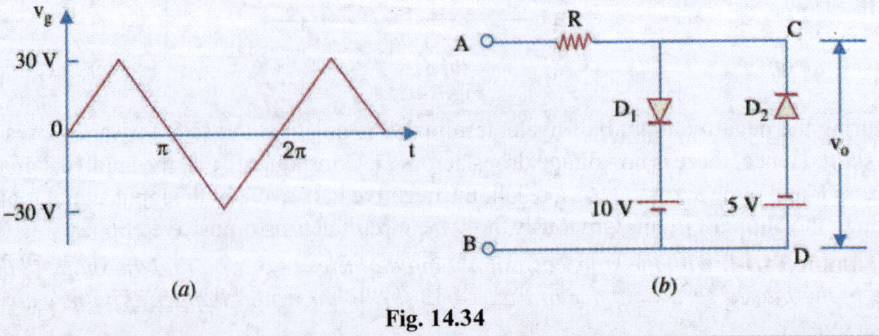

Fig. 14.34

Solution. During the positive half-cycle, D_1 would conduct but D_2 will act as an open-circuit. But value of v_0 cannot exceed 10 V because points *C* and *D* are electrically connected across the

10-V battery since D_1 is shorted. Hence, signal voltage above +10 V level would be clipped off as shown in Fig. 14.35.

During the negative half-cycle, D_1 is open but D_2 conducts. Again, v_0 cannot exceed 5 V since it is the voltage across points C and D whose value is fixed by the battery connected in that branch. Hence, signal voltage beyond 5 V is clipped off. The waveshape of v_0 is shown separately in Fig. 14.35.

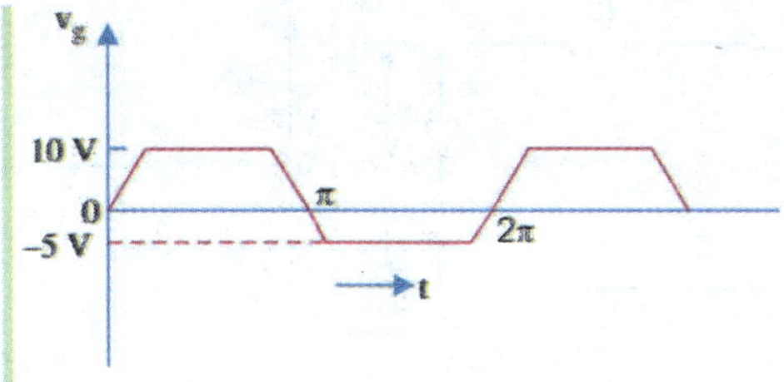

Fig. 14.35

14.15. Some Clipping Circuits

For the following circuit, a sinusoidal input signal as shown in Fig. 14.36 would be assumed. We will now discuss the biased series and parallel clipping circuits.

***(a)* Biased Series Clippers**

The output voltage has the waveform as shown in Fig. 14.37.

***(b)* Biased Parallel Clippers**

The waveforms of the output voltage are as shown in Fig. 14.38.

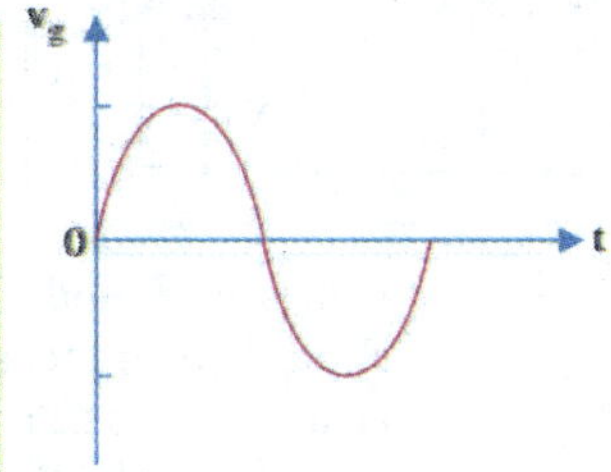

Fig. 14.36

Clipping has been changed by changing the battery and diode connections.

14.16. Clampers

To put it simply, clamping is the process *of introducing a dc level into an ac signal*.

By way of illustration, consider the signal shown in Fig. 14.39 (*a*). It is a sine wave with equal positive and negative swings of ± 5 V about 0 V. Hence, its average value over one cycle is zero (it has no dc level *i.e.*, it has zero dc level).

In Fig. 14.39 (*b*), the signal waveform has been lifted up so as to just touch the horizontal axis. It is, now said to have acquired a dc level of +5 V. This output waveform is said to be *positively clamped at* 0 V.

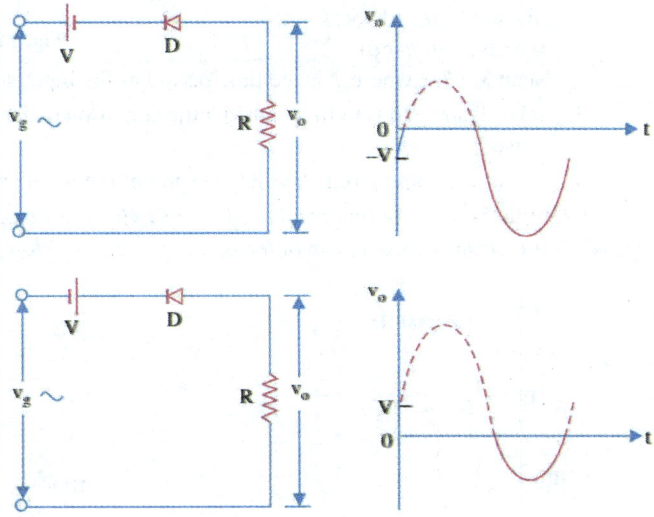

Fig. 14.37

A circuit capable of accepting the input signal shown in Fig. 14.39 (*a*) and delivering the output shown in Fig. 14.39 (*b*) is called a **clamper.** Such a circuit has a minimum requirement of three elements :

1. a diode   2. a capacitor and   3. a resistor

It will generally need a dc battery also. Following additional points regarding clamper circuit are worth keeping in mind :

1. both R and C affect the waveform;

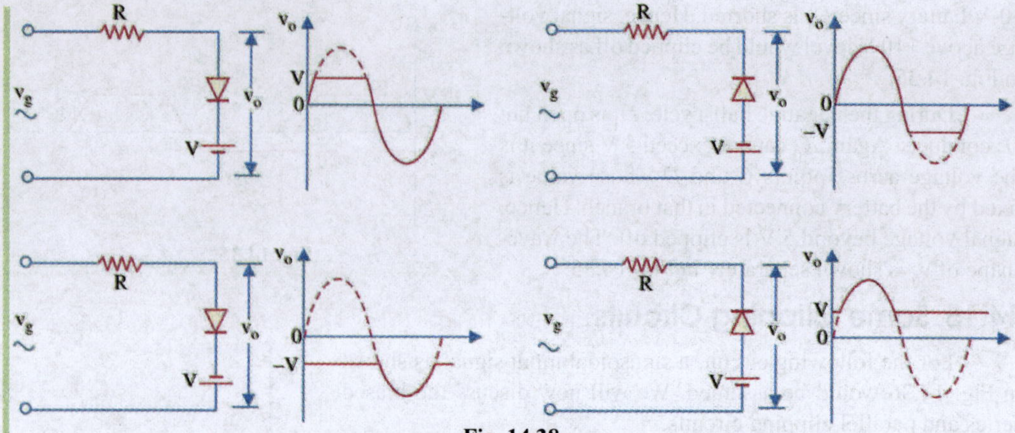

Fig. 14.38

2. values of R and C should produce a time constant ($\lambda = CR$) which is large enough to ensure that capacitor remains almost fully charged during the time period of the signal. In other words, time-constant $\lambda \gg T/2$ where T is the time period of the input signal. (Please See Example 14.16);

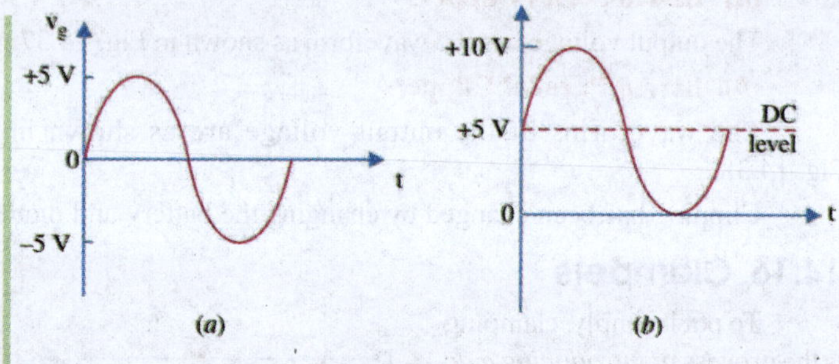

Fig. 14.39

3. it is advantageous to first consider the condition under which the diode becomes forward-biased;
4. for all clamping circuits, voltage swing of the input and output waveforms is the same.

Example 14.16. *The input signal of Fig. 14.40 (a) is applied to the clamper circuit shown in Fig. 14.40 (b). Draw the waveform of the output voltage v_0. How will it change if R is made 100 Ω ?*

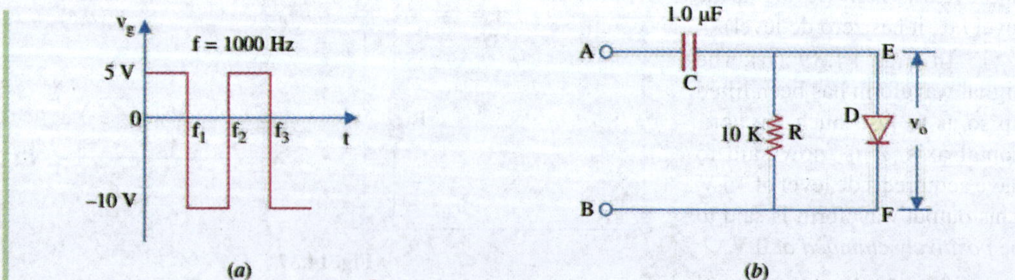

Fig. 14.40

Solution. As seen, time period of input signal is

$$T = \frac{1}{1000} \text{ second} = 1 \text{ ms}$$

$$\therefore \quad 0 \rightarrow t_1 = t_1 \rightarrow t_2 = t_2 \rightarrow t_3 = T/2 \text{ .... } = 0.5 \text{ ms}$$

$$\therefore \quad \lambda = CR = 1 \times 10^{-6} \times 10 \times 10^3 = 10 \text{ ms}$$

As seen, $\lambda \gg T/2$. Hence, once charged, the capacitor will have hardly any time to discharge by the time signal polarity reverses.

(*a*) Positive Input Half-cycle

When positive half-cycle of the input signal voltage is applied to the clamper circuit, its terminal *A* becomes positive with respect to terminal *B*. Hence, *D* acts like a short as shown in Fig. 14.41 (*a*). A steady positive voltage of 5 V remains applied to *A* for 0.5 ms. At the same time, *R* is also shorted out [Fig. 14.41 (*a*)] because it is in parallel with *D*. Hence, *C* will rapidly charge to 5 V. Being across a short, $v_0 = 0$ during positive half-cycle as shown in Fig. 14.41 (*b*).

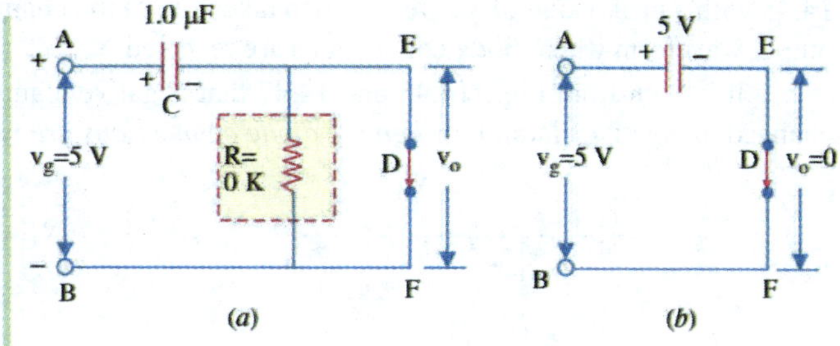

Fig. 14.41

(*b*) Negative Input Half-cycle

In this case, terminal *B* becomes positive and so reverse-biases *D* by 10 V. Hence, *D* acts like an open switch as shown in Fig. 14.42. Now, *R* and *C* get connected in series so that their $\lambda = RC = 10$ ms. As stated earlier in Art. 10.6, capacitor will take a time of $5\lambda = 50$ ms to get fully discharged. But the input signal will allow it just 0.5 ms during which to discharge. Obviously, *C* would hardly get discharged in this extremely short time interval of 0.5 ms. Hence, it can be assumed to be still fully charged with the ***original*** polarity during this negative half-cycle.

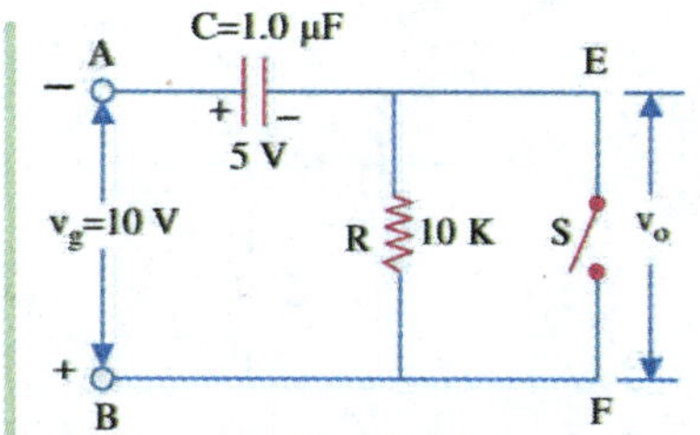

Fig. 14.42

The output voltage v_0 across the 'open' will be

= voltage from $E \rightarrow A \rightarrow B \rightarrow F$ — Fig. 14.42

= 5 + 10 = 15 V — with *E* negative

The waveform of the output voltage is shown in Fig. 14.43. It has the same frequency as that of the input signal. However, it has been clamped down in the negative region. It is seen that voltage swing of both input and output circuits is the same *i.e.*, 15 V. It is never the case in clipping circuits.

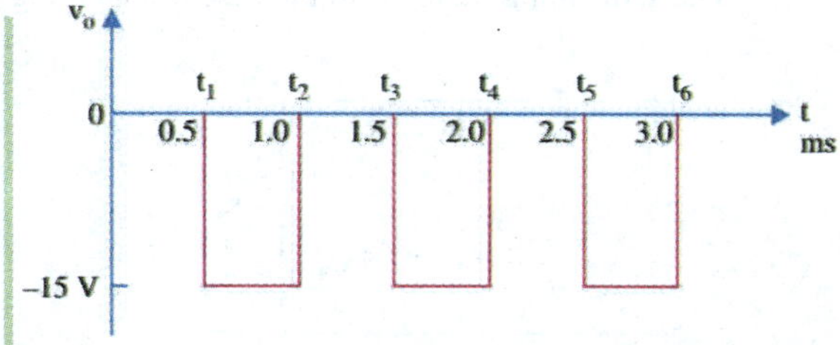

Fig. 14.43

When $R = 100\ \Omega$

Now, $\lambda = 100 \times 1 \times 10^{-6}$ ms = 0.1 ms. Hence, the capacitor which is almost *instantaneously* charged to +5 V during the positive input half-cycle, will be almost completely discharged during the negative half-cycle because, now, 5 λ (full discharge time) equals the half time-period (0.5 s) of the signal.

Hence, in this case, v_0 would be momentarily equal to –15 V at the beginning of the negative half-cycle but will fall off to almost 0 V before the signal reverses its polarity (Fig. 14.44). As seen, v_0 consists of voltage *spikes* of amplitude –15 V.

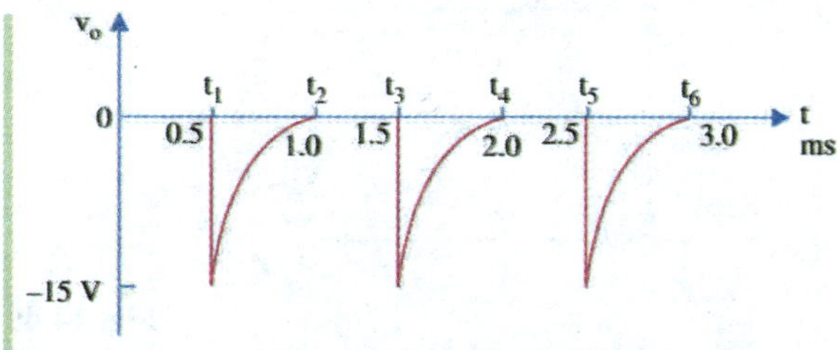

Fig. 14.44

14.17. Summary of Clamping Circuits

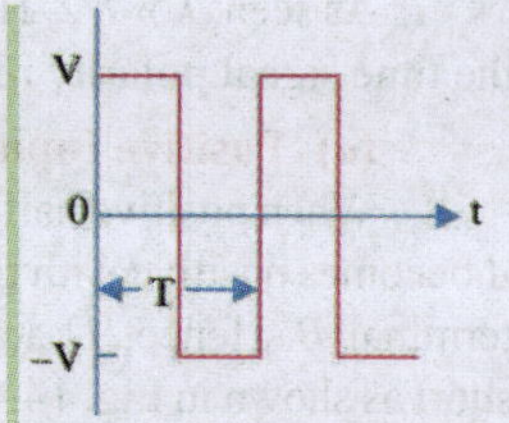

Fig. 14.45

In the following clamping circuits, it would be assumed that the amount of the time 5 λ = 5*RC* » *T*/2 where *T* is the time period of the input signal. For all circuits, we will take the same input signal shown in Fig. 14.45 with a peak value of *V*. We will also take note of the change in the output waveform when diode connections are reversed.

It is seen from Figs. 14.46 and 14.47 that negative clamping has changed to positive clamping *when the diode connections are reversed.*

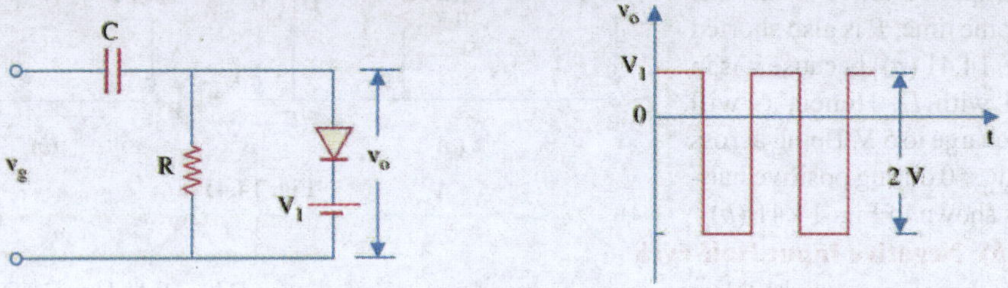

Fig. 14.46

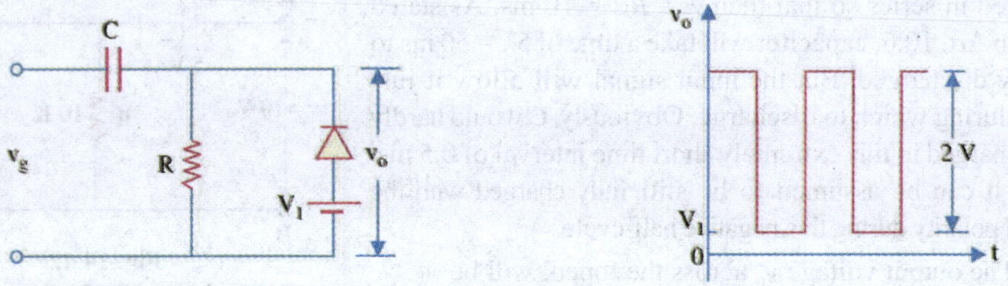

Fig. 14.47

The same thing happens in the case of clamping circuits shown in Figs. 14.48 and 14.49.

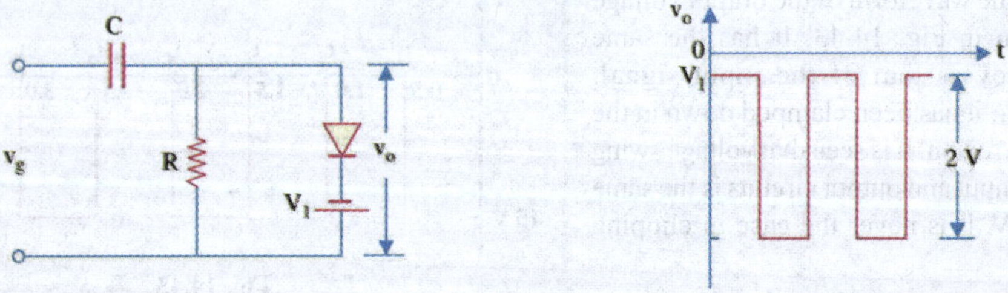

Fig. 14.48

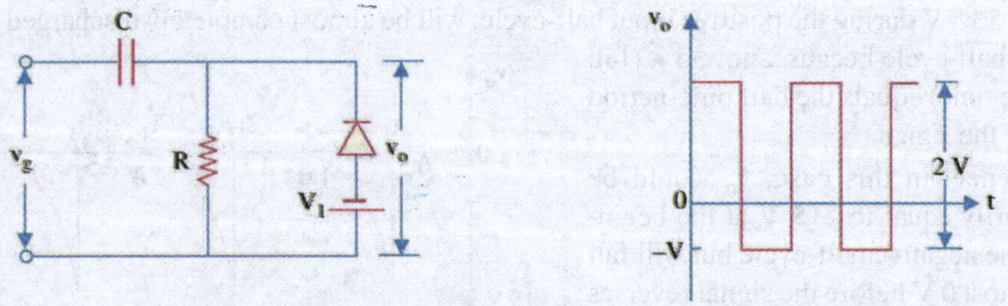

Fig. 14.49

CONVENTIONAL PROBLEMS

1. Find the bulk resistance of the silicon diode IN 3881 which has an I_F of 100 mA at a $V_F = 1$ V **[3 Ω]**

2. Find the value of V_F for a silicon diode having $r_B = 20\ \Omega$ and a forward current of 1 mA. **[0.72 V]**

3. A certain diode has a reverse saturation current of 5 μA at 25°C for a reverse voltage of 20 V. Calculate its reverse resistance. **[4 MΩ]**

4. In Fig. 14.50, which diode is forward-biased during positive input half cycle? And which during negative half-cycle ? **[D_2, D_1]**

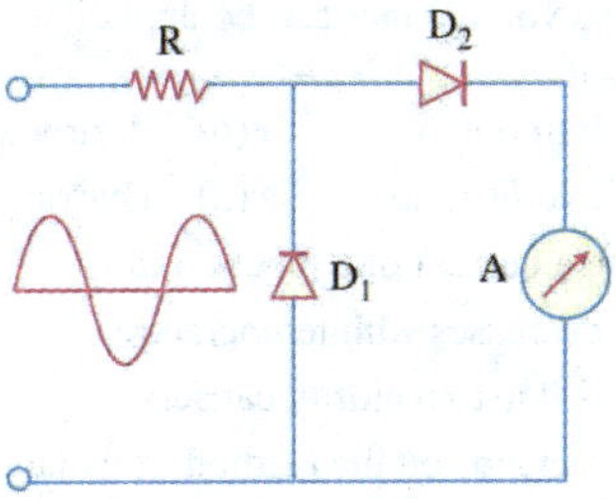

Fig. 14.50

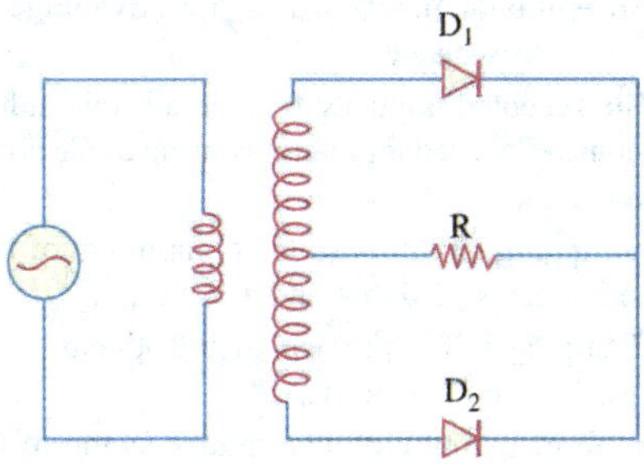

Fig. 14.51

5. In Fig. 14.51, when upper end of the secondary is positive with respect to its lower end which diode is forward biased ? Which will conduct when lower end of the secondary becomes positive ? **[D_1 D_2]**

6. In Fig. 14.52, find
   (*i*) peak forward current
   (*ii*) peak voltage across the diode when reverse-biased.
   [(*i*) 2 mV (*ii*) 10 V]

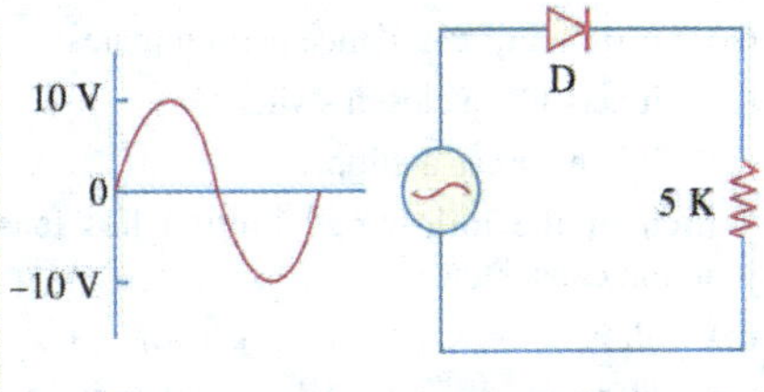

Fig. 14.52

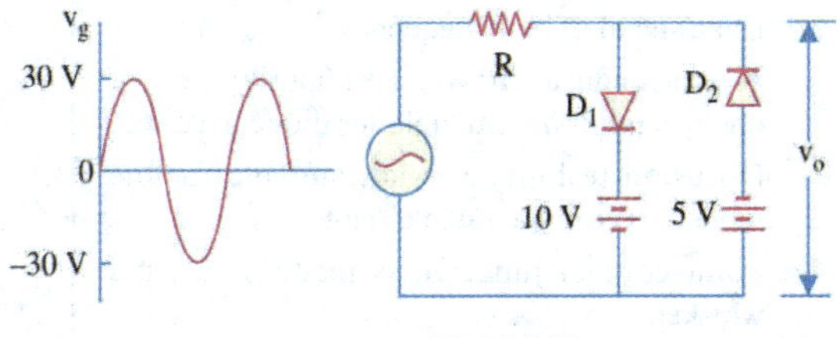

Fig. 14.53

7. In Fig.. 14.53, what would be the peak value of v_0 during
   (*i*) positive half-cycle and
   (*ii*) negative half- cycle of the input signal.
   [(*i*) 10 V (*ii*) 5 V]

SELF EXAMINATION QUESTIONS

A. Fill in the blanks with most appropriate word (s) or numerical value (s).

1. The maximum voltage that can be applied to a diode without destroying it is called peak ....voltage.
2. While testing a diode by an ohmmeter, low reading in one direction and high in the other shows a .... diode.
3. An ideal diode conducts in forward direction with .... resistance.
4. A reverse-biased ideal diode looks like an .... resistance.
5. A semiconductor diode designed to operate in the breakdown region is called a .... diode.

6. The maximum time taken by a diode to switch from ON to OFF state is called reverse .... time.
7. When a diode is tested in-circuit, it is common to get .... resistance readings in both directions.
8. An ideal diode is like a / an ....switch when forward-biased.
9. For forming a *P-N* junction, diffusion process takes .... time than the alloy process.
10. An epitaxial junction has the advantage of .... resistance.
11. The accepted impurity level in a semiconductor material used for manufacturing diode is one in a ....
12. A clipping circuit requires a minimum of two components : a diode and a ....
13. Clamping is the process of introducing a dc .... into an ac signal.
14. In all clamping circuits, voltage swing of the input and output waveforms is the ....

B. Answer True or False

1. A good diode gives high-resistance readings in both directions when tested by an ohmmeter.
2. An ideal *P-N* junction diode acts like a bistable switch.
3. During an ohmmeter test, if both resistance readings are high, the diode is 'open'.
4. In a circuit, a capacitor looks like an 'open' to an ac signal of high frequency.
5. A semiconductor having an impurity level of one in a million is suitable for diode making.
6. Diffusion technique is not suitable for the fabrication of integrated circuits.
7. Point-contact junction is made of a cat's whisker.
8. The primary function of clippers and clampers is to control the shape of the voltage and current waveforms.
9. The input signal waveforms can be clipped at any level by simply changing the battery voltage.
10. Clipping and clamping are practically similar processes under different names.
11. Clipping subtracts whereas clamping adds voltage to an ac signal.
12. In clamping circuits, time constant of the circuit must be much less than the time-period of the input signal.

C. Multiple Choice Items

1. A general purpose diode is more likely to suffer avalanche breakdown rather than Zener breakdown because
   (*a*) its leakage current is small
   (*b*) it has weak covalent bonds
   (*c*) it is lightly-doped
   (*d*) it has low reverse resistance
2. The turn-on voltage of a Ge junction diode is nearly–volt.
   (*a*) 0.7 (*b*) 0.3
   (*c*) 1.0 (*d*) 0.1
3. The peak inverse voltage (PIV) is the maximum reverse voltage that can be applied to a diode without
   (*a*) burnout (*b*) destruction
   (*c*) overheating (*d*) charring
4. Leakage current of a junction diode
   (*a*) decreases with temperature
   (*b*) is due to majority carriers
   (*c*) depends on the method of its fabrication
   (*d*) is in the range of mA or μA
5. In the forward region of its characteristic, a diode appears as
   (*a*) an OFF switch
   (*b*) a high resistance
   (*c*) a capacitor
   (*d*) an ON switch
6. Under small signal operation of a diode
   (*a*) its bulk resistance increases
   (*b*) its junction resistance predominates
   (*c*) it acts like a closed switch
   (*d*) it behaves as a clipper
7. Which of the following junction has least junction capacitance?
   (*a*) alloy (*b*) grown
   (*c*) diffused (*d*) point contact
8. A clipper circuit always
   (*a*) needs a dc source
   (*b*) clips both half-cycles of input signal
   (*c*) clips upper portions of the signal
   (*d*) clips some part of the input signal
9. Without a dc source, a clipper acts like a
   (*a*) rectifier (*b*) clamper
   (*c*) demodulator (*d*) chopper
10. The clipping level is primarily determined by
   (*a*) shape of the input waveform
   (*b*) battery voltage
   (*c*) value of the resistor
   (*d*) knee voltage of the diode

11. The primary function of a clamper circuit is to
   (*a*) suppress variations in signal voltage
   (*b*) raise positive half-cycle of the signal
   (*c*) lower negative half-cycle of the signal
   (*d*) introduce a dc level into an ac signal
12. For proper working of a clamper, time constant of the circuit should be
   (*a*) large
   (*b*) small
   (*c*) equal to signal time-period
   (*d*) greater than 5 times the signal time-period
13. Assuming a silicon diode in Fig. 14.54, the voltage of point *A* is .... volt.

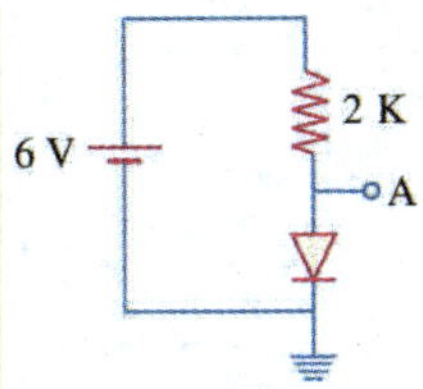

Fig. 14.54

   (*a*) 0 (*b*) 0.3
   (*c*) 0.7 (*d*) 6
14. In Fig. 14.55, voltage of point *B* is .... volt.

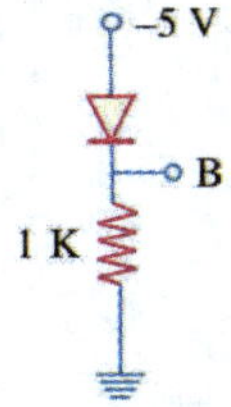

Fig. 14.55

   (*a*) 0 (*b*) –5
   (*c*) +5 (*d*) 0.7

ANSWERS

A. Fill in the blanks

1. inverse **2.** good **3.** zero **4.** infinite **5.** Zener **6.** recovery **7.** low
8. closed **9.** more **10.** low **11.** billion **12.** resistor **13.** level **14.** same

B. True of False

1. *F* **2.** *T* **3.** *T* **4.** *F* **5.** *F* **6.** *F*
7. *F* **8.** *T* **9.** *T* **10.** *F* **11.** *T* **12.** *F*

C. Multiple Choice Items

1. *c* **2.** *b* **3.** *b* **4.** *d* **5.** *d* **6.** *b* **7.** *d* **8.** *d* **9.** *a*
10. *b* **11.** *d* **12.** *a* **13.** *c* **14.** *a*.

15 CHAPTER

Special Diodes

1. Zener Diode
2. Voltage Regulation
3. Zener Diode as Peak Clipper
4. Meter Protection
5. Tunneling Effect
6. Tunnel Diode
7. Tunnel Diode Oscillator
8. Varactor
9. PIN Diode
10. Schottky Diode
11. Step Recovery Diode
12. Thermistors

15.1. Zener Diode

It is a *reverse-biased heavily-doped* silicon (or germanium) *P-N* junction diode which is operated in the breakdown region where current is limited by both external resistance and power dissipation of the diode. Silicon is preferred to Ge because of its higher temperature and current capability. As seen from Art. 13.11, when a diode breaksdown, both Zener and avalanche effects are present although usually one or the other predominates depending on the value of reverse voltage. At reverse voltages less than 6 V, Zener effect predominates whereas above 6 V, avalanche effect is predominant. Strictly speaking, the first one should be called Zener diode and the second one as avalanche diode. But the general practice is to call both types as Zener diodes.

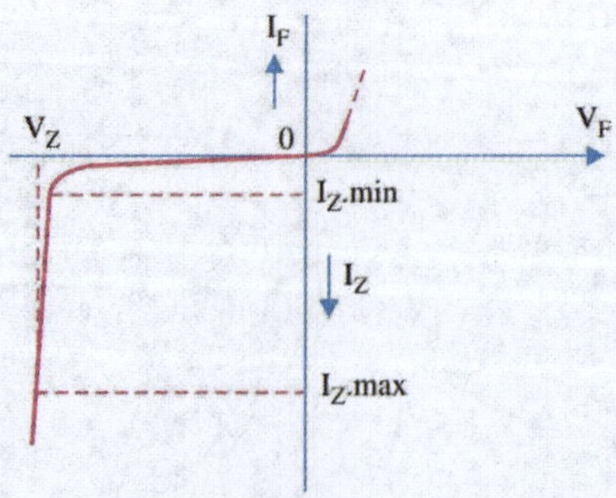

Fig. 15.1

Zener breakdown occurs due to *breaking of covalent bonds by the strong electric field set up in the depletion region by the reverse voltage*. It produces an extremely large number of electrons and holes which constitute the reverse saturation current (now called Zener current, I_z) whose value is limited only by the external resistance in the circuit. It is independent of the applied voltage. Avalanche breakdown occurs at higher reverse voltages when thermally-generated electrons acquire sufficient energy to produce more carriers by *collision*.

(a) V / I Characteristic

A typical characteristic is shown by Fig. 15.1 in the negative quadrant. The forward characteristic is simply that of an ordinary forward-biased junction diode. The important points on the reverse characteristic are :

V_Z = Zener breakdown voltage

$I_{z.min}$ = minimum current to sustain breakdown

$I_{z.max}$ = maximum Zener current limited by, maximum power dissipation.

Since its reverse characteristic is not exactly vertical, the diode possesses some resistance called *Zener dynamic impedance**. However, we will neglect it assuming that the characteristic is truly vertical. In other words, we will assume an ideal Zener diode for which voltage does not change once it goes into breakdown. It means that V_z remains constant even when I_z increases considerably.

The schematic symbol of a Zener diode and its equivalent circuit are shown in Fig. 15.2. The complete equivalent circuit is shown in Fig. 15.2 (*b*) and the approximate one in Fig. 15.2 (*c*) where it looks like a battery of V_z volts.

The schematic symbol of Fig. 15.2 (*a*) is similar to that of a normal diode except that the line representing the cathode is bent at both ends. With a little mental effort, the cathode symbol can be imagined to look like the letter *Z* for Zener.

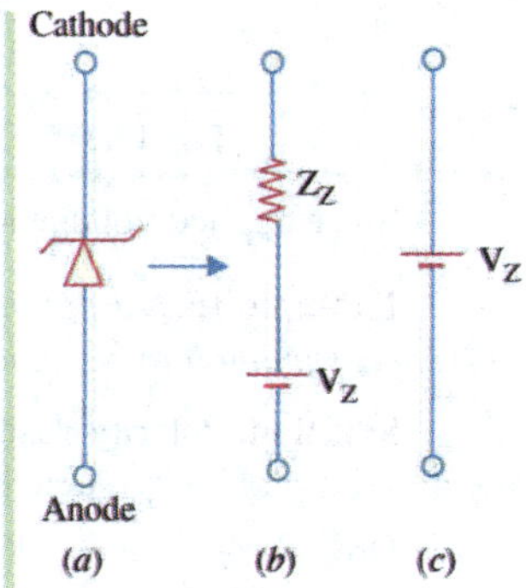

Fig. 15.2

(b) Zener Voltage

Zener diodes are available having Zener voltages of 2.4 V to 200 V. This voltage is temperature-dependent. Their power dissipation is given by the product $V_z I_z$. Maximum ratings vary from 150 mW to 50 W.

(c) Zener Biasing

For proper working of a Zener diode in any circuit, it is essential that it must

1. be reverse-biased,
2. have voltage across it greater than V_z,
3. be in a circuit where current is less than $I_{z.max}$.

(d) Diode Identification

Physically, Zener diode looks like any other diode and is recognized by its IN number such as IN 750 (10 W power) or IN 4000 (high power).

(e) Uses

Zener diodes find numerous applications in transistor circuitry. Some of their common uses

* Its value is given by $Z_z = \Delta V_z / \Delta I_z$. It is negligible as compared to large external resistance connected in the circuit.

are :

1. as voltage regulators,
2. as a fixed reference voltage in a network for biasing and comparison purposes and for calibrating voltmeters,
3. as peak clippers,
4. for meter protection against damage from accidental application of excessive voltage.

Example 15.1. *Determine whether the ideal Zener diode of Fig. 15.3 is properly biased. Explain why ?*

Solution. Since positive battery terminal is connected to its cathode, the diode is reverse-biased.

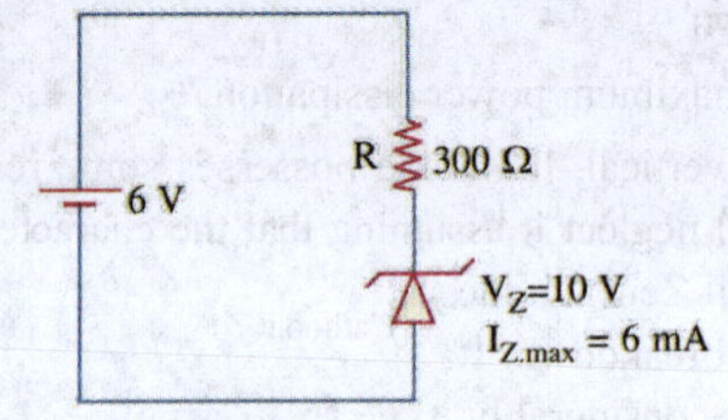

Fig. 15.3

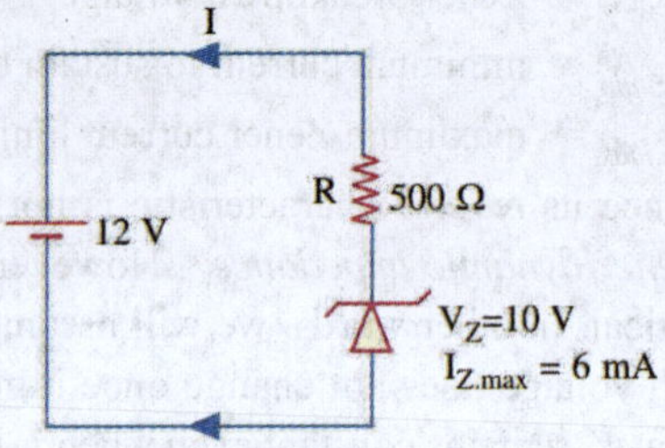

Fig. 15.4

Since applied voltage is less than V_z, the circuit is *not properly voltage-biased.*

Example 15.2. *Find out if the Zener diode of Fig. 15.4 is properly-biased. If so, find diode current assuming it to be an ideal one.*

Solution. Polarity-wise, the diode is properly-biased. Since applied voltage is greater than V_z, the diode is properly voltage biased.

Drop across $R = 12 - 10 = 2$ V $\quad\therefore\quad I = 2 / 500 = 4$ mA.

Since this current is less than the maximum diode current, the diode is properly-biased according to the criteria laid down in Art. 15.1 (*c*).

Example 15.3. *Determine if the Zener diode of Fig. 15.5 is biased properly. If so, find I_z and the power dissipated by the diode.*

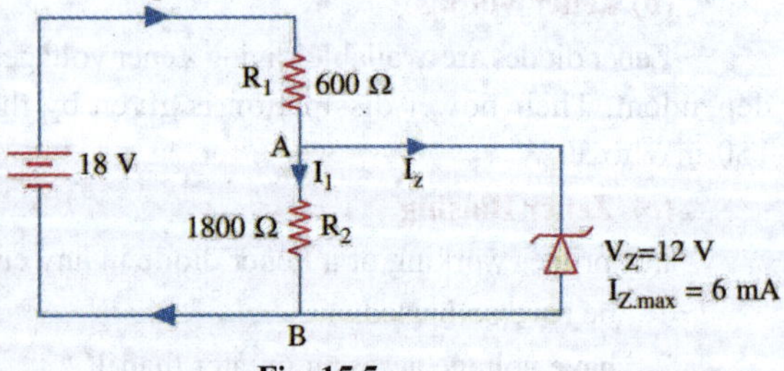

Fig. 15.5

Solution. Since its anode is connected to the negative battery terminal, it is reverse-biased.

Now, $V_{AB} = V_Z = 12$ V. Hence, drop across $R_1 = 18 - 12 = 6$ V

$$\therefore \quad I = 6 / 600 = 0.01 \text{ A} = 10 \text{ mA}$$

$$I_1 = \frac{12}{1800} = 0.0067 \text{ A} = 6.7 \text{ mA}$$

$$\therefore \quad I_Z = I - I_1 = 10 - 6.7 = 2.3 \text{ mA}$$

Since I_Z is less than $I_{Z.max}$, the diode is properly-biased in every respect as per Art. 15.1 (*c*).

15.2. Voltage Regulation

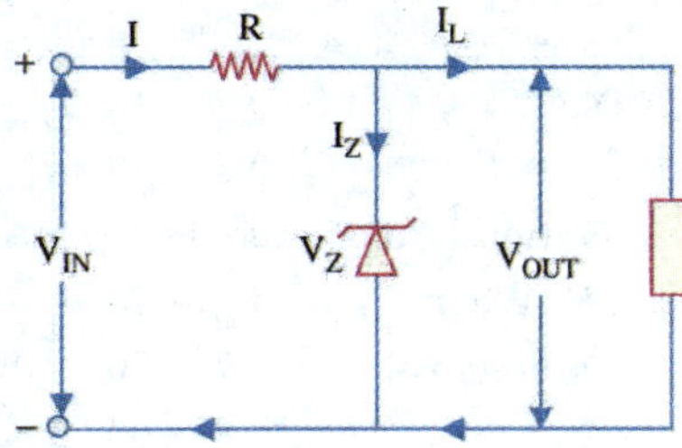

Fig. 15.6

It is a measure of a circuit's ability to maintain a constant output voltage even when either input voltage or load current varies. A Zener diode, when working in the breakdown region, can serve as a voltage regulator. In Fig. 15.6, V_{in} is the input dc voltage whose variations are to be regulated. The Zener diode is *reverse connected* across V_{in}. When p.d. across the diode is greater than V_Z, it conducts and draws relatively large current through the series resistance R. The load resistance R_L across which a constant voltage V_{out} is required, is connected in parallel with the diode. The total current I passing through R equals the sum of diode current and load current *i.e.*, $I = I_Z + I_L$.

It will be seen that under all conditions, $V_{out} = V_Z$.

Hence, $V_{in} = IR + R_{out} = IR + V_Z$.

Case 1. Suppose R is kept fixed but supply voltage V_{in} is increased slightly. It will increase I. This increase in I will be absorbed by the Zener diode without affecting I_L. The increase in V_{in} will be dropped across R thereby keeping V_{out} constant. Conversely, if supply voltage V_{in} falls, the diode takes a smaller current and voltage drop across R is reduced, thus again keeping V_{out} constant. Hence, when V_{in} changes, I and IR drop change in such a way as to keep V_{out} (= V_Z) constant.

Case 2. In this case, V_{in} is fixed but I_L is changed. When I_L increases, diode current I_Z decreases thereby keeping I and hence IR drop constant. In this way, V_{out} remains unaffected.

Should I_L decrease, I_Z would increase in order to keep I and hence IR drop constant. Again V_{out} would remain unchanged because

$$V_{out} = V_{in} - IR = V_{in} - (I_Z + I_L)\,R$$

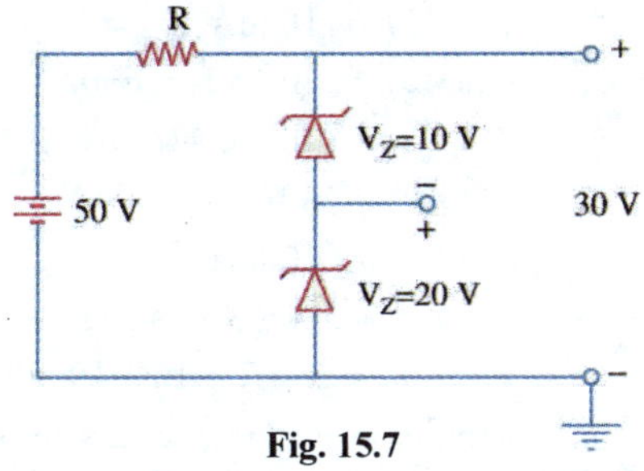

Fig. 15.7

Incidentally, it may be noted that

$$R = \frac{V_{in} - V_{out}}{I_Z + I_L}$$

It may also be noted that when diode current reaches its maximum value, I_L becomes zero. In that case

$$R = \frac{V_{in} - V_{out}}{I_{Z(max)}}$$

In Fig. 15.6, only one reference voltage level is available. Fig. 15.7 shows the circuit for establishing two reference levels. Here, two diodes having different Zener voltages have been connected in series.

Example 15.4. *Calculate the battery current I, I_Z and I_L in the circuit of Fig. 15.8. How will these values be affected if source voltage increases to 70 V? Neglect Zener resistance.*

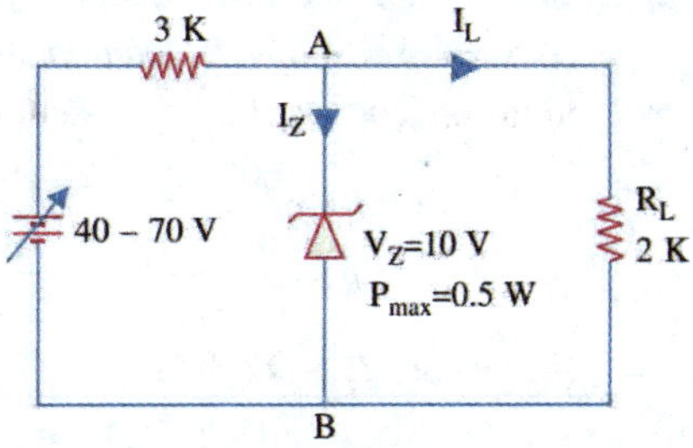

Fig. 15.8

Solution. When V_{in} = **40 V**

Now, $V_{AB} = V_Z = 10$ V

∴ drop across 3 K series (or line) resistor is

$= 40 - 10 = 30$ V

∴ $I = 30 / 3\text{ K} = \mathbf{10\ mA}$

$$I_L = \frac{V_Z}{R_L} = \frac{V_{AB}}{R_L} = \frac{10}{2\text{K}} = \mathbf{5\ mA}$$

$\therefore \quad I_Z = I - I_L = 10 - 5 = \mathbf{5\ mA}$

Now, $\quad P_{max} = V_Z \cdot I_{Z(max)}$

$\therefore \quad 0.5 = 10 \times I_{Z.(max)}$ or $I_{Z(max)} = 0.5 / 10 = 0.05\ A = 50\ mA$

Obviously, diode current of 5 mA is very much within the current range of the diode.

(*b*) When $\quad \mathbf{V_{in} = 70\ V}$

Drop across $\quad R = 70 - 10 = 60\ V$

$\therefore \quad I = 60 / 3\ K = \mathbf{20\ mA}$

$I_L = \mathbf{5\ mA}$ — as before

$I_Z = I - I_L = 20 - 5 = \mathbf{15\ mA}$

Example 15.5. *Using ideal Zener diode approximations, find current through the diode of Fig. 15.9 when load resistance R is*

(*i*) 30 K (*ii*) 5 K (*iii*) 3 K

Solution. (*i*) $\quad \mathbf{R_L = 30\ K}$

$V_{AB} = V_Z = 30\ V$

$\therefore \quad$ drop across $R = 60 - 30 = 30\ V$

$\therefore \quad I = 30/3\ K = \mathbf{10\ mA}$

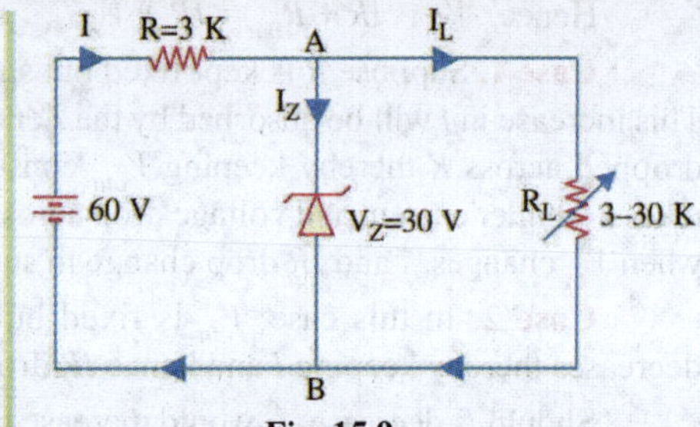

Fig. 15.9

Now, $\quad I_L = \dfrac{V_{AB}}{R_L} = \dfrac{30}{30\ K} = 1\ mA$

$\therefore \quad I_Z = I - I_L = 10 - 1 = \mathbf{9\ mA}$

(*ii*) $\quad \mathbf{R_L = 5\ K}$

$I = 10\ mA$ — as before

$I_L = 30 / 5\ K = \mathbf{6\ mA}$

$I_Z = 10 - 6 = \mathbf{4\ mA}$

(*iii*) $\quad \mathbf{R_L = 3\ K}$

$I = 10\ mA$ — as before

$I_L = 30 / 3\ K = 10\ mA$

$I_Z = I - I_L = 10 - 10 = \mathbf{0}$.

In this case, it is obvious that the diode is just on the verge of *coming out of breakdown region*. If R_L is reduced further, the diode will come out of breakdown region and would no longer act as a voltage regulator.

Example 15.6. *A 24 V, 600 mW Zener diode is to be used for providing a 24 V stabilized supply to a variable load (Fig. 15.10). If input voltage is 32 V, calculate the*

(*i*) *series resistance R required,* (*ii*) *diode current when* $R_L = 1200\ \Omega$.

Solution. (*i*) $\quad \mathbf{V_Z I_{Z(max)} = 600\ mW}$

$\therefore \quad I_{Z(max)} = 600 / 24 = 25\ mA$

$\therefore \quad R = \dfrac{V_{in} - V_{out}}{I_{Z.(max)}} = \dfrac{32 - 24}{25 \times 10^{-3}} = \mathbf{320\ \Omega}$

(*ii*) When $\mathbf{R_L = 1200\ \Omega}$

$I_L = \dfrac{V_Z}{R_L} = \dfrac{24}{1200} = 20\ mA$

$\therefore \quad I_Z = 25 - 20 = \mathbf{5\ mA}$

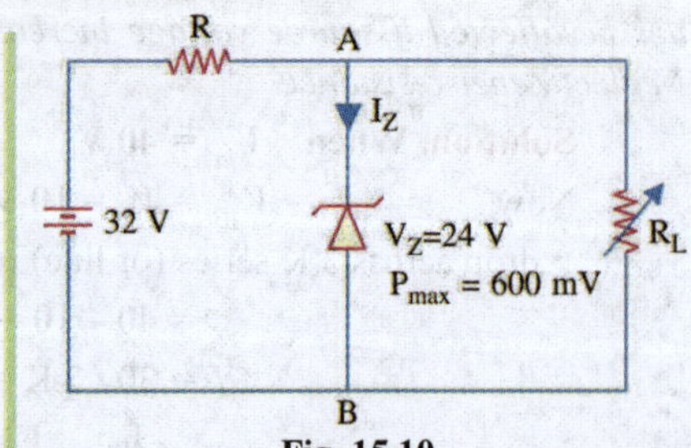

Fig. 15.10

15.3. Zener Diode as Peak Clipper

Use of Zener diodes in waveshaping circuits is illustrated in Fig. 15.11. The two similar diodes D_1 and D_2 have been joined back-to-back across the input sine wave voltage of peak value ± 25 V. Both have V_Z = 20 V. As seen, the output is a semi-square wave with a peak value of ± 20 V.

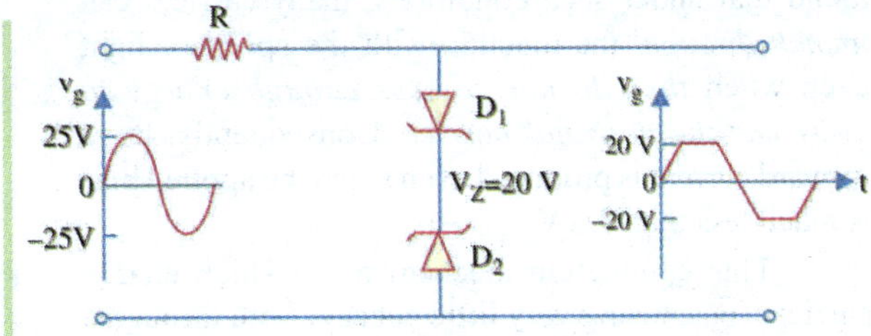

Fig. 15.11

It is well-known that a Zener diode acts like a '*short*' (or very low resistance) in the forward direction and an '*open*' in the reverse direction till it goes into breakdown at V_Z. During positive input half-cycle, D_1 is shorted (being forward-biased) but D_2 acts like an open upto 20 V. Thereafter, it goes into breakdown and holds the output voltage constant till input voltage falls below 20 V in the later part of the half-cycle. At that point, D_2 comes out of the breakdown and again acts like an open across which *the entire voltage is dropped*.

During the negative input half-cycle, roles of D_1 and D_2 are reversed. As a result, the output wave is clipped on both peaks as shown in Fig. 15.11.

If we increase the peak value of the input signal voltage and use Zener diodes of less V_Z value, we can get an almost square output voltage wave from a sinusoidal input wave as shown in Fig. 15.12.

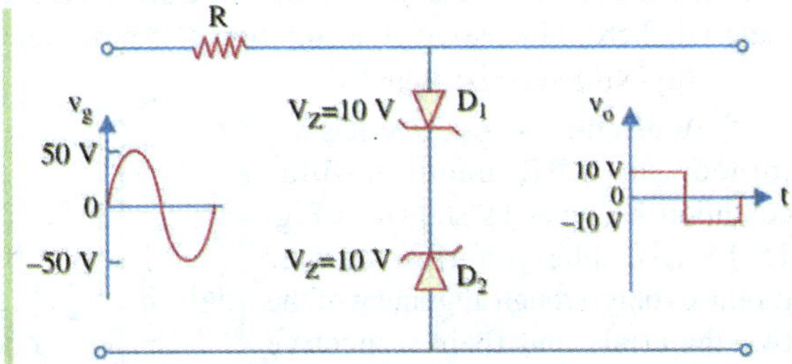

Fig. 15.12

15.4. Meter Protection

Zener diodes are frequently used in volt-ohm-miliammeters (VOM) for protecting meter movement against burn-out from accidental overloads. If VOM is set to its 2.5 V range and the test leads are accidentally connected to a 25 V circuit, an unprotected meter will be burned out or at least get severely damaged.

This hazard can be avoided by connecting a Zener diode in parallel with the meter as shown in Fig. 15.13 (*a*). In the event of an accidental overload, most of the current will pass through the diode. Two Zener diodes connected as shown in Fig. 15.13 (*b*) can provide overload protection regardless of the applied polarity.

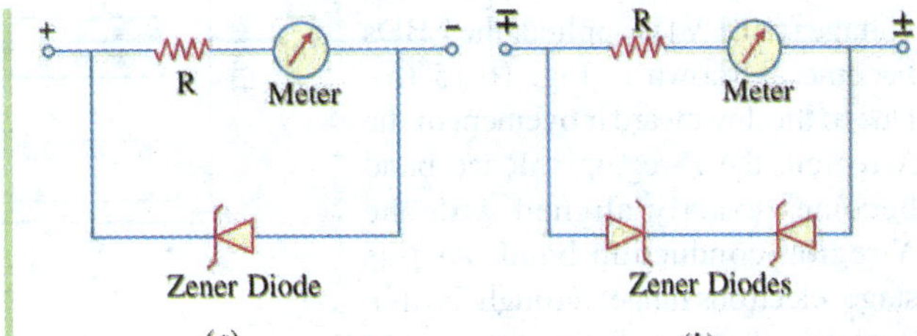

Fig. 15.13

15.5. Tunneling Effect

In a normally-doped *P-N* junction, the depletion layer is relatively wide and a potential barrier exists across the junction. The charge carriers on either side of the junction cannot cross over unless they possess sufficient energy to overcome this barrier (0.3 V for Ge and 0.7 V for Si). As is well-known, width of the depletion region depends directly on the doping density of the semiconductor. If a *P-N* junction is doped *very heavily**

* Much more than even for a Zener diode.

(about 1000 times or more), its depletion layer becomes *extremely thin* (about 0.00001 mm). It is found that under such conditions, many carriers can '*punch through*' the junction with the speed of light even when *they do not possess enough energy to overcome the potential barrier*. Consequently, large forward current is produced even when the applied bias is much less than 0.3 V.

This conduction mechanism in which charge carriers (possessing very little energy) bore through a barrier directly instead of climbing over it is called tunneling.

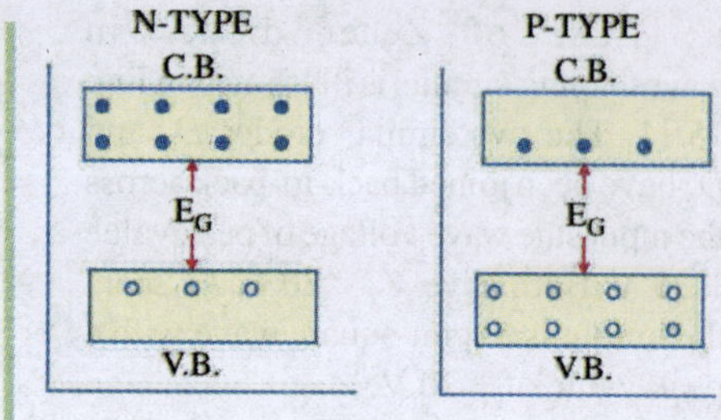

Fig. 15.14

Explanation

Energy band diagrams (EBD) of *N*-type and *P*-type semiconductor materials can be used to explain this tunneling phenomenon.

Fig 15.14 shows they energy band diagram of the two types of silicons separately. As explained earlier (Art. 12.24), in the *N*-type semiconductor, there is increased concentration of electrons in the conduction band. It would be further increased under heavy doping. Similarly, in a *P*-type material, there is increased concentration of holes in the valence band for similar reasons.

(*a*) No Forward Bias

When the *N*-type meterials are joined, the EBD under no-bias condition becomes as shown in Fig. 15.15 (*a*). The junction barrier produces only a rough alignment of the two materials and their respective valence and conduction bands. As seen, the depletion region between the two is extremely narrow due to very heavy doping on both sides of the junction. The potential hill is also increased as shown.

(*b*) Small Forward Bias

When a very small forward voltage (≅ 0.1 V) is applied, the EBDs become as shown in Fig. 16.15 (*b*). Due to the downward movement of the *N*-region, the *P*-region valence band becomes exactly aligned with the *N*-region conduction band. At this stage, electrons tunnel through the thin depletion layer with the velocity of light thereby giving rise to a large current called peak current I_P.

(*c*) Large Forward Bias

When the forward bias is increased further, the two bands get out of alignment as shown in Fig. 15.15 (*c*). Hence, tunneling of electrons stops thereby decreasing the current. Since

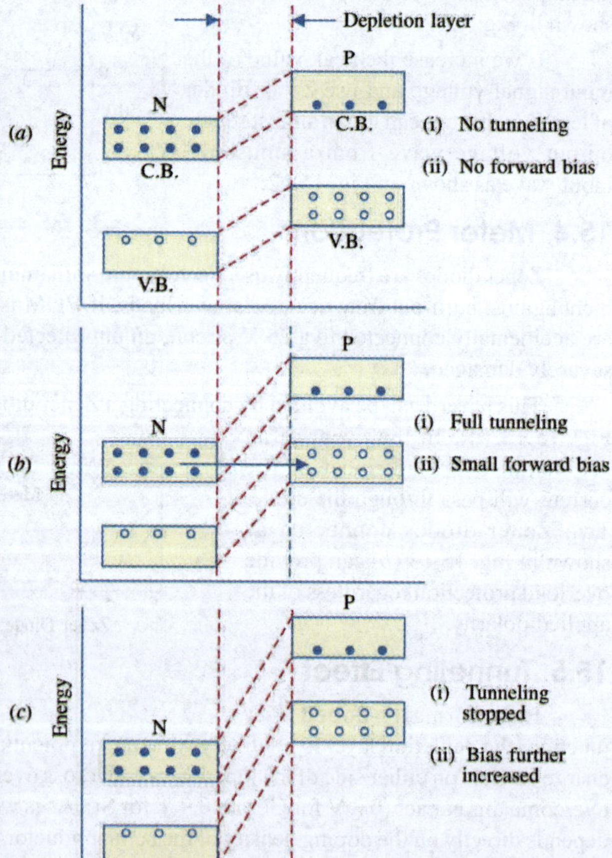

Fig. 15.15

current decreases with increase in applied voltage (*i.e.*, *dV/dI* is negative), the junction is said to possess **negative resistance** at this stage. This resistance increases throughout the negative region.

However, it is found that when applied forward voltage is increased still further, the current starts increasing once again as in a normal junction diode.

15.6. Tunnel Diode

This diode was first introducted by Dr. Leo Easki in 1958.

(*a*) Construction

It is a *high-conductivity* two-terminal *P-N* junction diode having doping density about 1000 times higher as compared to an ordinary junction diode. This heavy doping produces following three unusual effects :

1. Firstly, it reduces the width of the depletion layer to an extremely small value (about 0.00001 mm).
2. Secondly, it reduces the reverse breakdown voltage to a very small value (approaching zero) with the result that the diode appears to be broken down for any reverse voltage.
3. Thirdly, it produces a negative resistance section on the *V/I* characteristic of the diode.

It is called a tunnel diode because due to its extremely thin depletion layer, electrons are able to tunnel through the potential barrier at relatively low forward bias voltage (less then 0.05V). Such diodes are usually fabricated from germanium, gallium-arsenide (GaAs) and gallium antimonide (GaSb).

The commonly-used schematic symbols for the diode are shown in Fig. 15.16. It should be handled with caution because being a low-power device, it can be easily damaged by heat and static electricity.

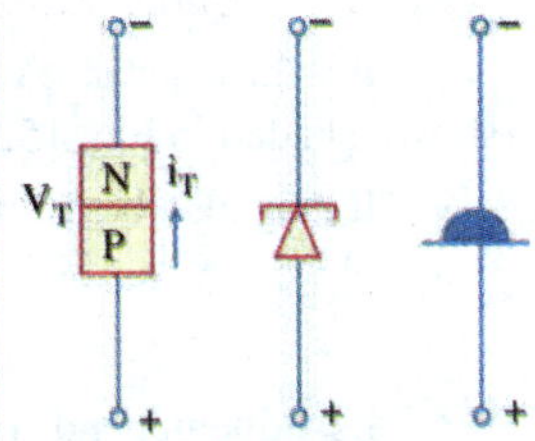

Fig. 15.16

(*b*) V /I Characteristic

It is shown in Fig. 15.17. As seen, forward bias produces immediate conduction *i.e.*, as soon as forward bias is applied, significant current is produced. The current quickly rises to its *peak value* I_P when the applied forward voltage reaches a value V_P (point *A*). When forward voltage is increased further, diode current starts decreasing till it achieves its minimum value called *valley current* I_V corresponding to valley voltage V_V (point *B*). For voltages greater than V_V, current starts increasing again as in any ordinary junction diode.

As seen from Fig. 15.17, between the peak point *A* and valley point *B*, *current decreases with increase in the applied voltage*. In other words, tunnel diode possesses negative resistance ($-R_N$) in this region. In fact, this constitutes the most useful property of the diode. Instead of *absorbing* power, a negative resistance *produces* power. By offsetting losses in *L* and *C* components of a tank circuit, such a negative resistance permits oscillations. Hence, a tunnel diode is used as a very high frequency oscillator.

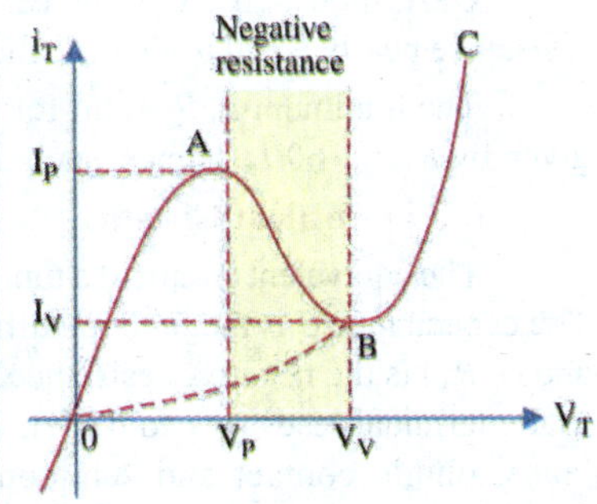

Fig. 15.17

Another point worth noting is that this resistance increases as we go from point *A* to *B* because as voltage is increased, current keeps decreasing which means that diode negative resistance keeps increasing.

(*c*) Tunneling Theory

At zero forward bias, the energy levels of conduction electrons in *N*-region of the junction are slightly out of alignment with the energy levels of holes in the *P*-region. As the forward voltage is *slightly* increased, electron levels start getting aligned with the hole levels on the other side of the junction thus permitting some electrons to cross over. This kind of junction crossing is called *tunneling* .

As voltage is increased to peak voltage (V_P), all conduction band electrons in the *N*-region are able to cross over to the valence band in the *P*-region because the two bands are exactly aligned. Hence, maximum current (called peak current I_P) flows in the circuit.

After V_P, as the applied voltage is increased, current starts decreasing because the two bands start gradually getting out of alignment. It reaches minimum value (called valley current) when the two are totally out of alignment at a forward bias of V_V (valley voltage).

For voltages greater than V_V , current starts increasing again exactly as it does in the case of an ordinary *P-N* junction diode.

Tunneling is much faster than normal crossing which enables a tunnel diode to switch ON and OFF much faster than an ordinary diode. That is why tunnel diode is extensively used in special applications requiring very fast switching speeds like high-speed computer memories and high-frequency oscillators etc.

(d) Diode Parameters

1. *Negative Resistance* ($-R_N$)

It is the resistance offered by the diode within the negative-resistance section of its characteristic (shown shaded in Fig. 15.17). It equals the reciprocal of the slope of the characteristic in this region.

It may also be found from the following relation :

$$R_N = -\frac{dV}{dI}$$

Its value depends on the semiconductor material used (varying from –10 Ω to – 200 Ω).

2. I_P/I_V *Ratio*

It is almost as important a factor (particularly for computer applications) as the negative resistance of the diode.

Silicon diodes have a low I_P/I_V ratio of 3 : 1 and their negative resistance can be approximated from $R_N = -220/I_P$. Such diodes are used mainly for switches operating in high ambient temperatures.

Germanium diodes have an I_P/I_V ratio of 6 : 1 and negative resistance formula $R_N = -120/I_P$.

GaAs diodes (used exclusively in oscillators) have an I_P/I_V ratio of about 10 : 1 and a negative resistance nearly equal to that of silicon diodes.

The minimum I_P/I_V ratio for GaSb diode is about 12 : 1 and has the lowest resistance of all given by $R_N = -60/I_P$. Hence, such diodes have the *lowest noise*.

(e) Equivalent Circuit

The equivalent circuit of a tunnel diode is shown in Fig. 15.18. The capacitance *C* is the junction diffusion capacitance (1 to 10 pF) and ($-R_N$) is the negative resistance. The inductor L_s is due mainly to the terminal leads (0.1 to 4 nH). The resistance R_s is due to the leads, ohmic contact and semiconductor materials themselves (1 – 5Ω). These factors limit the frequency at which the diode may be used. They are also important in determining the switching-speed limit.

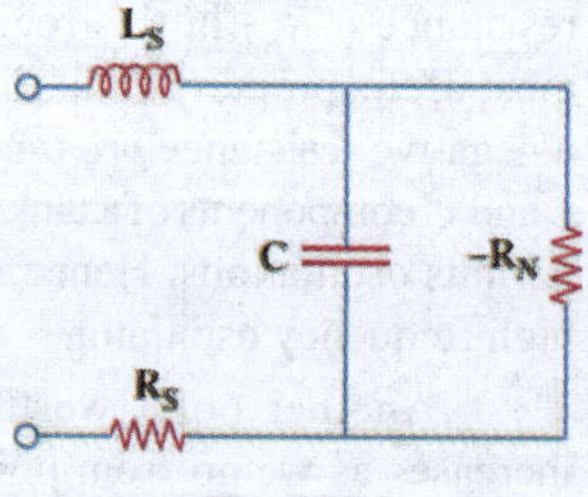

Fig. 15.18

(f) Biasing the Diode

The tunnel diode has to be biased from some dc source for fixing its *Q*-point on its characteristic when used as an amplifier or as an oscillator and modulator.

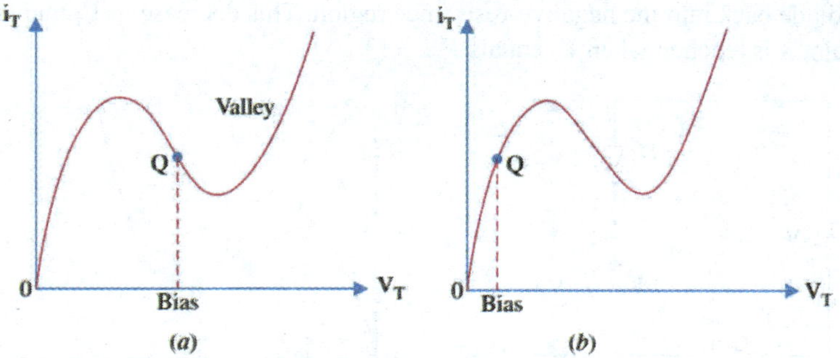

Fig. 15.19

The diode is usually biased in the negative region [Fig. 15.19 (*a*)]. In mixer and relaxation oscillator applications, it is biased in the positive resistance region nearest zero [Fig. 15.19 (*b*)].

(*g*) Applications

Tunnel diode is commonly used for the following purposes :

1. As an ultrahigh-speed switch due to tunneling mechanism which essentially takes place at the speed of light. It has a switching time of the order of nanoseconds or even picoseconds.
2. As logic memory storage device—due to triple-valued feature of its curve for current.
3. As microwave oscillator at a frequency of about 10 GHz—due to its extremely small capacitance and inductance and negative resistance.
4. In relaxation oscillator circuits—due to its negative resistance.
   In this respect, it is very similar to the unijunction transistor (Art. 27.2).
5. As an amplifier—due to its negative resistance.

(*h*) Advantages and Disadvantages

The advantages of a tunnel diode are :

1. low noise,
2. ease of operation,
3. high speed,
4. low power,
5. insensitivity to nuclear radiations.

The disadvantages are :

1. the voltage range over which it can be operated properly is 1 V or less,
2. being a two-terminal device, it provides no isolation between the input and output circuits.

15.7. Tunnel Diode Oscillator

The basic job of an oscillator is to convert dc power into ac power. Ordinarily, we do not expect an ac signal from a circuit which has no input ac source. But the circuit shown in Fig. 15.20 (*a*) does exactly that as explained below.

The value of *R* is so selected as to bias the diode *D* in the negative-resistance region *A-B*. The working or quiescent point *Q* is almost at the centre of a curve *A-B*. When S is closed, the current immediately rises to a value determined by *R* and diode resistance which are in series. The applied voltage *V* divides across *D* and *R* according to the ratio of their resistances.

However, as V_T exceeds V_P, diode is driven into the negative area and its *resistance starts to increase* (Art. 15.4). Hence, V_T increases further till it becomes equal to valley voltage V_V (point *B*). At this point, further increase in V_T drives the diode into the positive-resistance region *BC* [Fig. 15.20 (*b*)]. The resulting increase in current now increases V_R but correspondingly decreases V_T, thereby bringing the diode back into the negative-resistance region. This decrease in V_T increases the circuit current till point A is reached when V_T equals V_P.

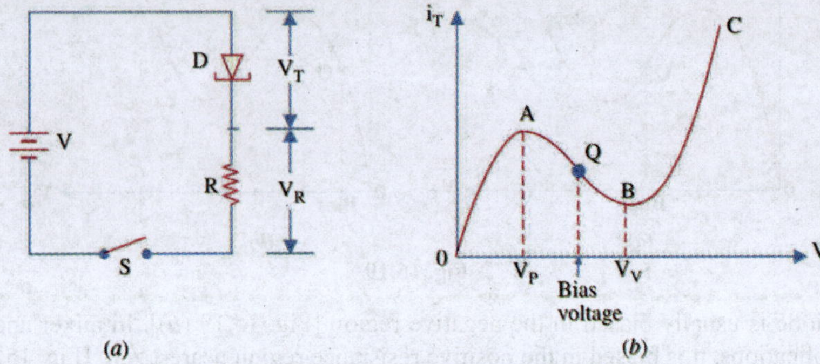

Fig. 15.20

It describes one cycle of operation. In this way, the circuit will continue to oscillate back and forth through the negative-resistance region *i.e.*, between points A and B on its characteristic. Its output across R is like a sine wave.

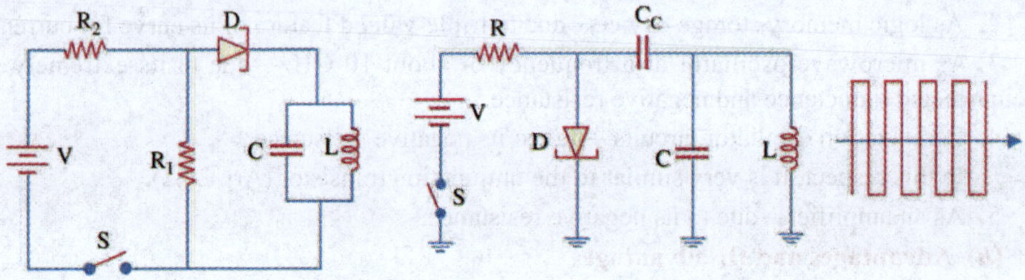

Fig. 15.21

Fig. 15.21 shows a practical circuit drawn in two slightly different ways. Here, R_2 sets the proper bias level for the diode whereas R_1 (in parallel with the LC tank circuit) sets proper current level for it. The capacitor C_C is the coupling capacitor. As the switch S is closed, the diode is set into oscillations whose frequency equals the resonant frequency of the tank.

15.8. Varactor

The varactor diode is a semiconductor, voltage-dependent variable capacitor alternatively known as *varicap or voltacap* or voltage-variable capacitor (VVC) diode. Basically, it is just a reverse-biased junction diode whose mode of operation depends on its transition capacitance (C_T). As explained earlier in Art. 13.12, reverse-biased junctions behave like capacitors whose capacitance $\propto 1/V_R^{\,n}$ where n varies from 1/3 to 1/2. As reverse voltage V_R is increased, depletion layer widens thereby decreasing the junction capacitance. Hence, we can change diode capacitance by simply changing V_R. Silicon diodes which are optimised for this variable capacitance effect are called *varactors*.

Fig. 15.22

The schematic symbol and a simple equivalent circuit for a varactor are shown in Fig. 15.22.

Varactors may be of two types as shown in Fig. 15.23.

The doping profile of the abrupt-junction diode is shown in Fig. 16.23 (a) and that of the hyperabrupt junction diode in Fig. 16.23 (b). The abrupt-junction diode has uniform doping and a capacitive tuning ratio (TR) of 4 : 1. For example, if its maximum transition capacitance is 100 pF and minimum 25 pF, then its *TR* is 4 : 1 which is not enough to tune a broadcast receiver over its entire frequency range of 550 to 1650 kHz.

The hyperabrupt-junction diode has highest impurity concentration near the junction. It results in narrower depletion layer and larger capacitance. Also, changes in V_R produce larger capacitance changes. Such a diode has a tuning range of 10 : 1, enough to tune a broadcast receiver through its frequency range of nearly 3 : 1.

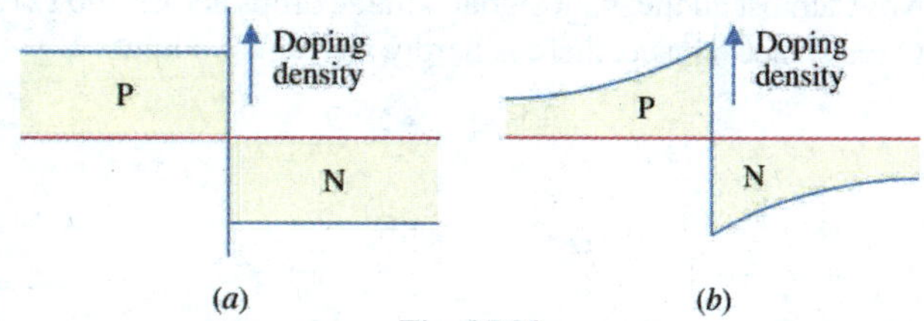

Fig. 15.23

Application

Since the junction capacitance of a varactor is in the pF range, it is suitable for use in high-frequency circuits. Its main applications are as

1. automatic frequency control device,
2. FM modulator,
3. adjustable band-pass filter,
4. parametric amplifier.

Fig. 15.24

15.9. PIN Diode

(*a*) Construction

It is composed of three sections. There are the usual *P*- and *N*-regions but sandwiched between them is an intrinsic layer or *I*-layer of pure silicon (Fig. 15.24). Being intrinsic (or undoped) layer, it offers relatively high resistance. This high-resistance region gives it two advantages as compared to an ordinary *P-N* diode. The advantages are :

1. Decrease in capacitance C_{pn} because capacitance is inversely proportional to the separation of *P*- and *N*-regions.

It allows the diode a faster response time. Hence, *PIN* diodes are used at high frequencies (more than 300 MHz).

2. Possibility of greater electric field between the *P*- and *N*- junctions.

It enhances the electron-hole pair generation thereby enabling *PIN* diode to process even very weak input signals.

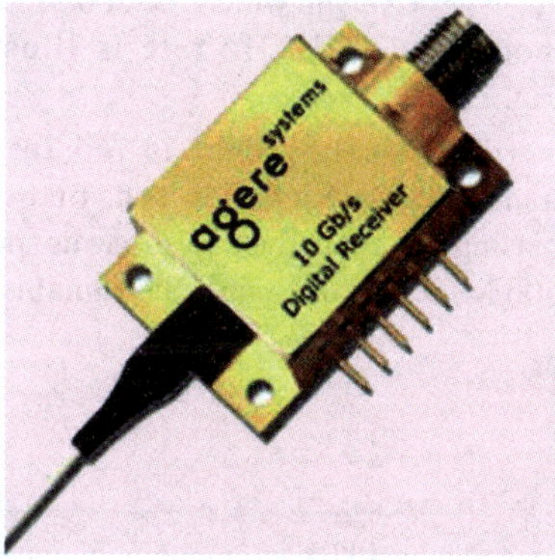

Pin diode.

(*b*) Diode Resistance

1. When forward-biased, it offers a variable resistance $r_{ac} \cong 50 / I$ where *I* is the dc current in mA (Art. 15.1). For large dc currents, it would look like a *short*.

2. When reverse-biased, it looks like an '*open*' *i.e.*, it offers infinite resistance in the reverse direction.

(*c*) Operation

(i) High Frequency Switching

Its use in electronic high frequency switching is illustrated in Fig. 15.25.

When the diode is reverse-biased, it looks like an 'open' as shown in Fig. 15.25 (*b*). The 500 MHz input signal voltage divides across the series-connected 100 Ω resistor and the diode in proportion to their resistances. Since the diode has infinite resistance (being open), the entire input signal appears across it. Hence, the whole input signal passes out via coupling capacitance C_2 without any attenuation (or loss). When the diode is forward-biased by the +25 V dc source, $I = 25/1K = 25$ mA. Hence, diode resistance $r_{ac} = 50/25 = 2\ \Omega$ as shown by its equivalent circuit in Fig. 15.26.

Now, almost all the input signal voltage drops across 100 Ω resistor and practically none across the 2 Ω resistance. Hence, there is hardly any signal output.

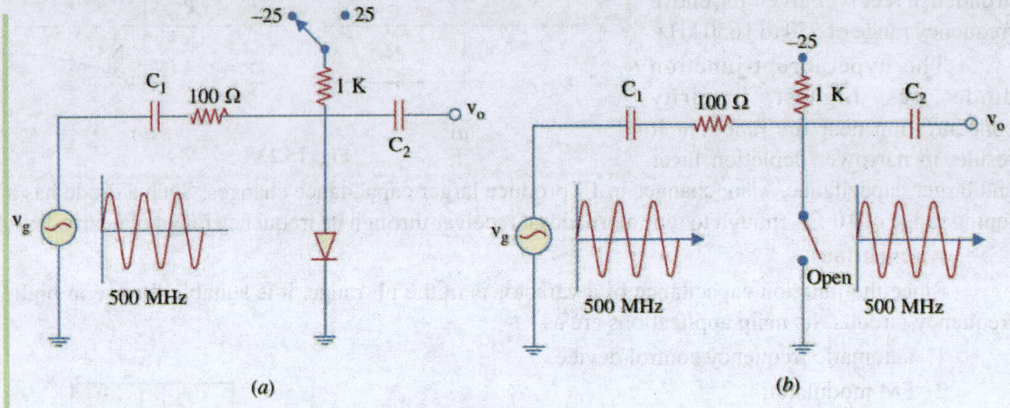

Fig. 15.25

In practice, instead of mechanically switching the diode-biasing supply from –25 V to +25V, a transistor is used to do this switching operation. In this way, we can turn a very high frequency signal (MHz range) OFF and ON with the speed of a transistor switching circuit.

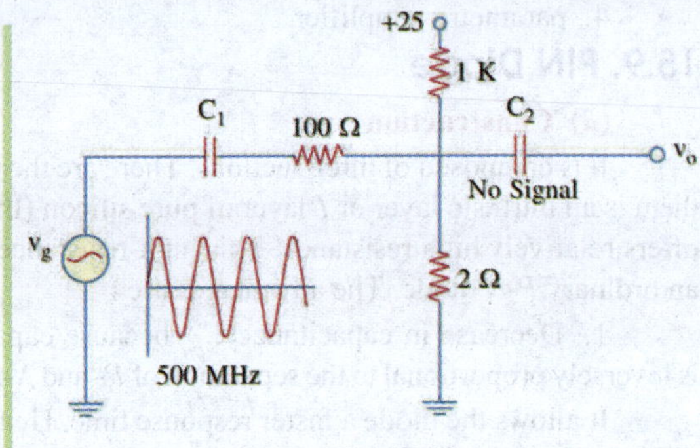

Fig. 15.26

(ii) Use as AM Modulator

The way in which the 500 MHz signal is modulated at 1 kHz rate is illustrated in Fig. 15.27.

A 1-kHz signal is fed into a *PNP* transistor where it varies its dc output current at the same rate. This varying dc current is applied as biasing current to the *PIN* diode as shown in Fig. 15.27. It varies the diode ac resistance as seen by the 500 MHz signal. Hence, the signal is modulated at 1kHz rate as shown.

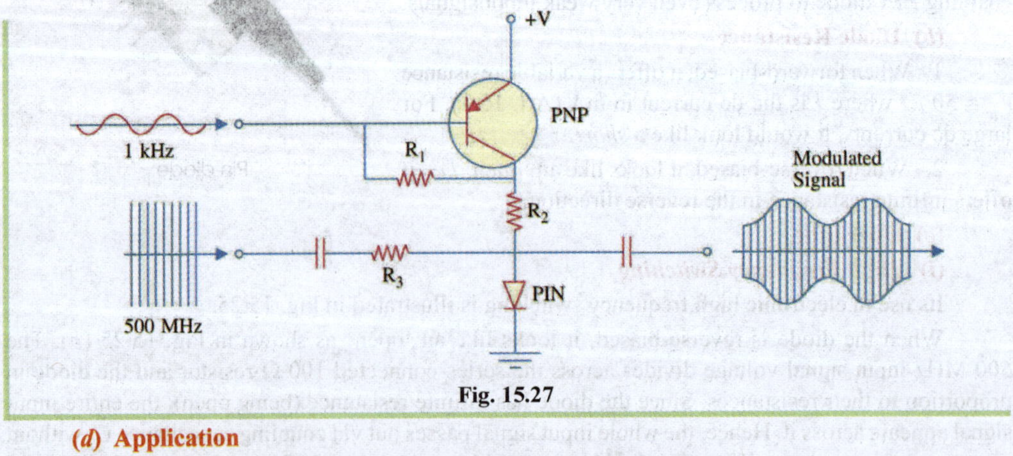

Fig. 15.27

(d) Application

1. as a switching diode for signal frequencies up to GHz range,
2. as an AM modulator of very high frequency signals.

15.10. Schottky Diode

It is also called Schottky *barrier* diode and *hot-carrier* diode. It is mainly used as a rectifier at signal frequencies exceeding 300 MHz. It has more uniform junction region and is more rugged than *PIN* code—its main rival.

(*a*) Construction

It is a metal-semiconductor junction diode with *no depletion layer*. It uses a metal (like gold, silver, platinum, tungsten etc.) on one side of the junction and usually an *N*-type doped silicon semiconductor on the other side. The diode and its schematic symbol are shown in Fig. 15.28.

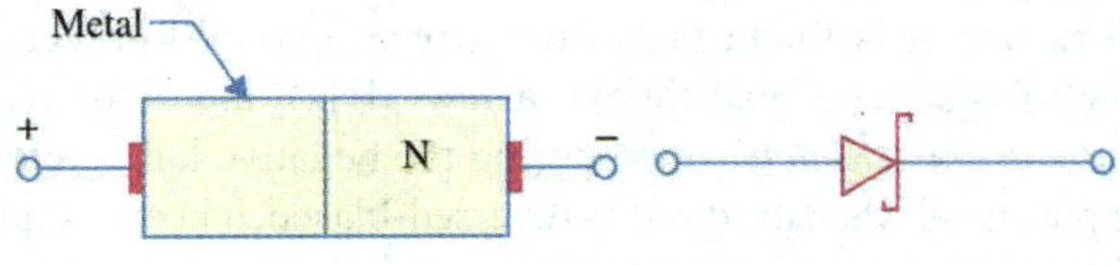

Fig. 15.28

(*b*) Operation

When the diode is unbiased, electrons on the *N*-side have lower energy levels than electrons in the metal. Hence, they cannot surmount the junction barrier (called Schottky barrier) for going over to the metals.

When the diode is forward-biased, conduction electrons on the *N*-side gain enough energy to cross the junction and enter the metal. Since these electrons plunge into the metal with very large energy, they are commonly called '*hot carriers*'. That is why this diode is often referred to as *hot carrier* diode.

(*c*) Application

This diode possesses two unique features as compared to an ordinary *P-N* junction diode :

1. It is a *unipolar* device because it has electrons as majority carriers on both sides of the junction. An ordinary *P-N* junction diode is a bipolar device because it has both electrons and holes as majority carriers.

2. Since no holes are available in metal, there is *no depletion layer* or *stored charges* to worry about. Hence, Schottky diode can switch OFF faster than a bipolar diode.

Because of these qualities, Schottky diode can easily rectify signals of frequencies exceeding 300 MHz. As shown in Fig. 15.29, it can produce an almost perfect half-wave rectified output.

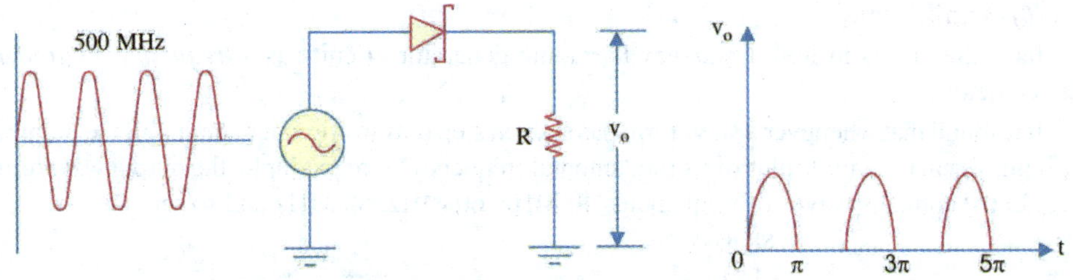

Fig. 15.29

The present maximum current rating of the device is about 100 A. It is commonly-used in switching power supplies that operate at frequencies of 20 GHz. Another big advantage of this diode is its low noise figure which is extremely important in communication receivers and radar units etc.

It is also used in clipping and clamping circuits, computer gating, mixing and detecting networks used in communication systems.

15.11. Step Recovery Diode

It is another type of VVC diode having a *graded doping profile* where doping density decreases near the junction as shown in Fig. 15.30. This results in the production of strong electric fields on *both* sides of the junction.

(a) Theory

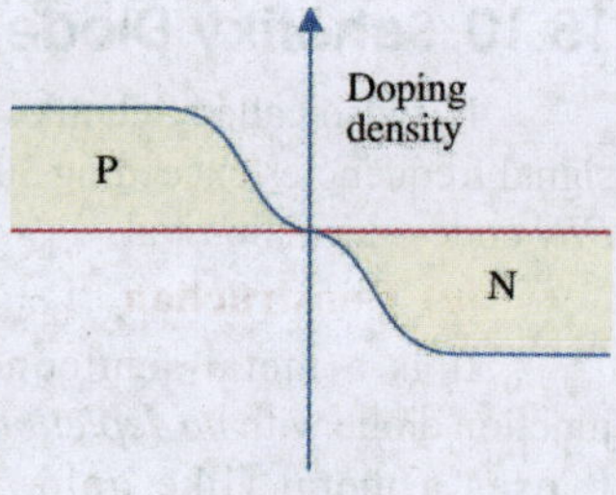

Fig. 15.30

At low frequencies, an ordinary diode acts as a rectifier. It conducts in the forward direction but not in the reverse direction *i.e.*, it recovers immediately from ON state to the OFF state. However, it is found that when driven forward-to-reverse by a high-frequency signal (above a few MHz), the diode *does not recover immediately*. Even during the negative half-cycle of the input signal when the diode is reversed-biased, it keeps conducting for a while after which the reverse current ceases abruptly in one step. This reverse conduction is due to the fact that charges stored in the depletion layer during the period of forward bias take time to drain away from the junction.

Fig. 15.31 (*a*) shows a step-recovery diode being driven by a 20 MHz signal source. As seen from Fig. 15.31 (*b*), it conducts in the forward direction like any diode. During the reverse half-cycle, we get reverse current due to the draining of the stored charge after which current suddenly drops to zero. It looks as though the diode has suddenly *snapped open* during the early part of the reverse cycle. That is why it is sometimes called a *snap diode*.

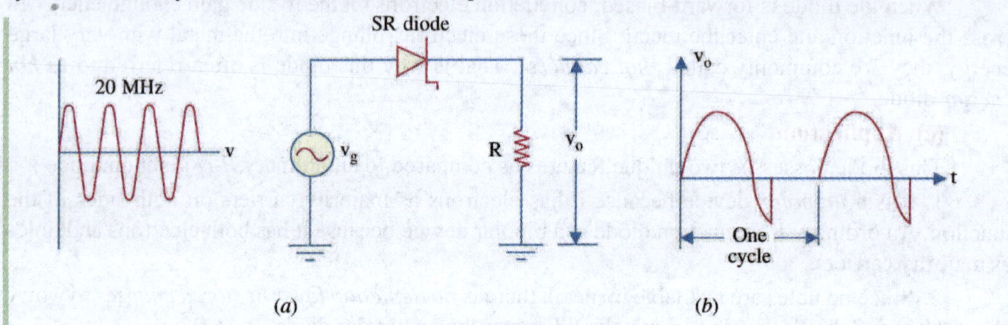

Fig. 15.31

The step or sudden recovery from reverse current ON to reverse current OFF gives the diode its name.

(b) Applications

Its mains use is in high-frequency harmonic generator circuits as a *frequency multiplier* as explained below.

It is found that whenever a waveform has sudden step or transition, it contains all the harmonics of the input signal (*i.e.*, multiples of its fundamental frequency). For example, the output waveform of Fig. 15.32 (*b*) contains waves of frequencies 40 MHz, 60 MHz, 80 MHz and so on.

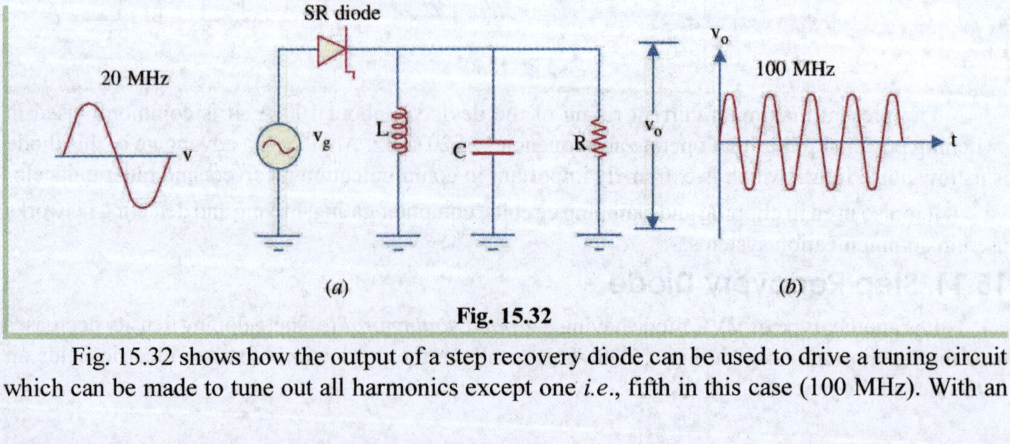

Fig. 15.32

Fig. 15.32 shows how the output of a step recovery diode can be used to drive a tuning circuit which can be made to tune out all harmonics except one *i.e.*, fifth in this case (100 MHz). With an input signal of 20 MHz, the step recovery diode generates harmonics of different multiple frequencies listed above. However, the resonant *L-C* circuit is tuned to 5th harmonic of $f = 100$ MHz. Hence, all except this harmonic are filtered out of the circuit. The signal appearing across R is almost a pure sine wave with $f = 100$ MHz as shown separately in Fig. 15.32 (*b*).

Step-recovery diodes are also used in pulse and digital circuits for generating very fast pulses with rise time of less than 1 nanosecond.

15.12. Thermistors

It is a thermal resistor *i.e.*, a temperature-sensitive resistor. Most thermistors have negative temperature coefficient of resistance though devices having positive temperature coefficient are also available.

(*a*) Construction

Thermistors are made by heating under pressure (sintering) semiconductor ceramic materials made from mixture of metallic oxides. Various mixtures used for this purpose are those of manganese, nickel, cobalt, copper and iron etc. These are pressed into desired shapes and then baked at high temperature for making different types of thermistors. Electrical connections are made either by including wires during the shaping process or by silvering the surfaces after sintering.

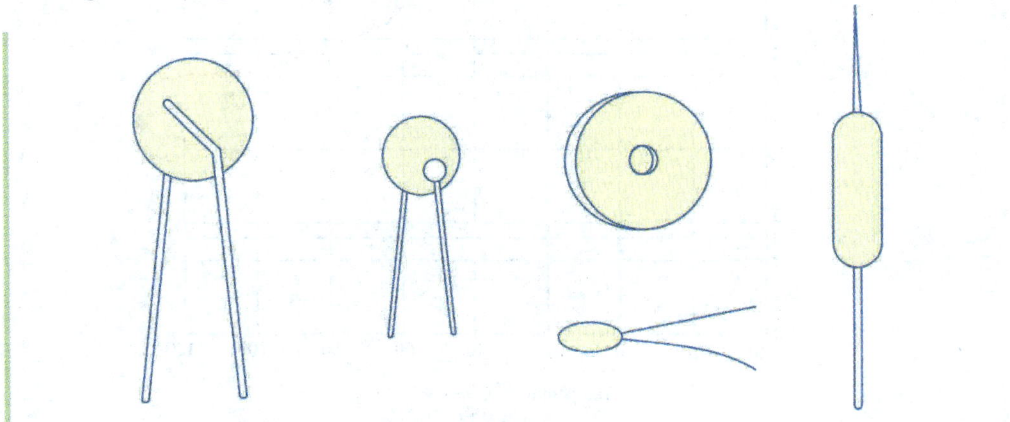

Fig 15.33

As shown in Fig. 15.33, thermistors are made in the shape of beads (with diameters ranging from 0.15 mm to 2.5 mm), probes (beads sealed in glass rods, discs or washers) upto 25 mm in diameter and rods from 6 mm to 50 mm long. Beads may be glass-coated or enclosed in evacuated envelopes for protection against corrosion.

Washer-shaped thermistors can be bolted together for series or parallel connection. The usual symbol for the device is also shown in Fig. 15.33.

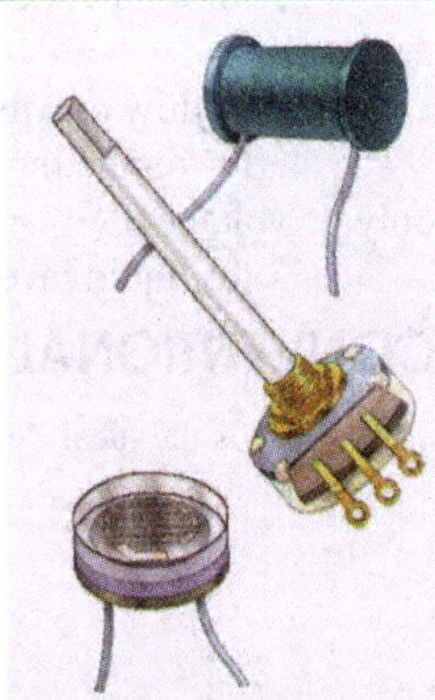

The resistance of a thermistor (top) changes with temperature. The variable resistors are adjusted by hand (middle). Others change when light strikes them (bottom).

(*b*) Temperature/Resistance Curve

A typical curve is shown in Fig. 15.34. It is seen that the thermistor has a resistance of about 3000 Ω at – 20°C and 40Ω at 80°C. Hence a change of 100°C has resulted in a 75 : 1 change in resistance. It is this very large change in resistance which makes a thermistor such a valuable temperature measurement and control device.

There are two ways of changing the resistance of the device: internally and externally.

In other words, device resistance depends on ambient temperature and self-heating.

(c) Applications

Thermistors find many applications some of which are given below :

1. for measurement and control of temperature as in ovens,

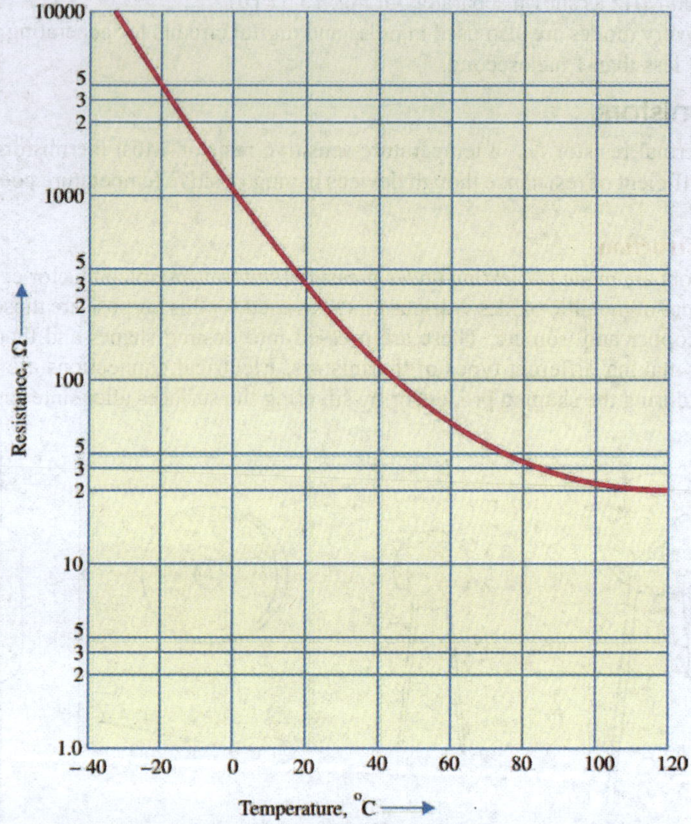

Fig. 15.34

2. for temperature compensation *i.e.*, cancelling the effect of temperature on other electronic devices,
3. as flow rate measuring devices,
4. for regulating ac voltage (unlike Zener diode which, being polarity sensitive, can regulate only dc voltages),
5. for liquid level detection etc.

CONVENTIONAL PROBLEMS

1. Is the ideal Zener diode shown in Fig. 15.35 properly biased ? If not explain why ? **[No]**

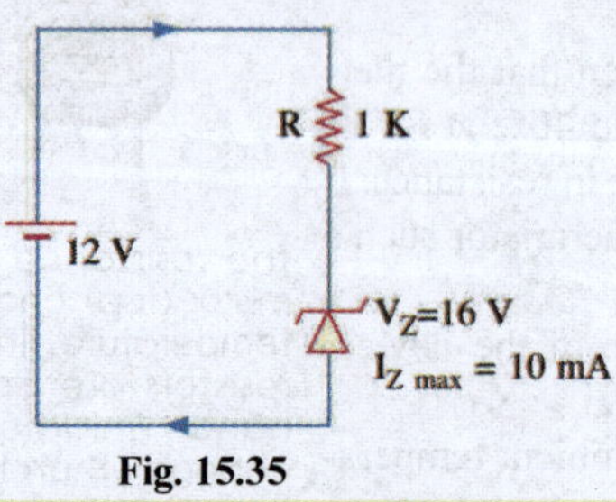

Fig. 15.35

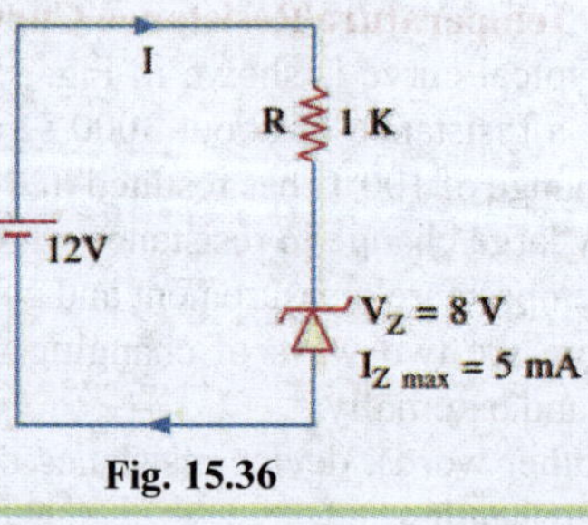

Fig. 15.36

2. Check up if the diode in Fig. 15.36 is biased properly for normal operation. What is the current taken by the diode ? **[Yes, 4 mA]**

3. Check up if Zener diode of Fig. 15.37 is reverse-biased as well as properly voltage-biased. Calculate diode current and power dissipation. **[6 mA, 96 mW]**

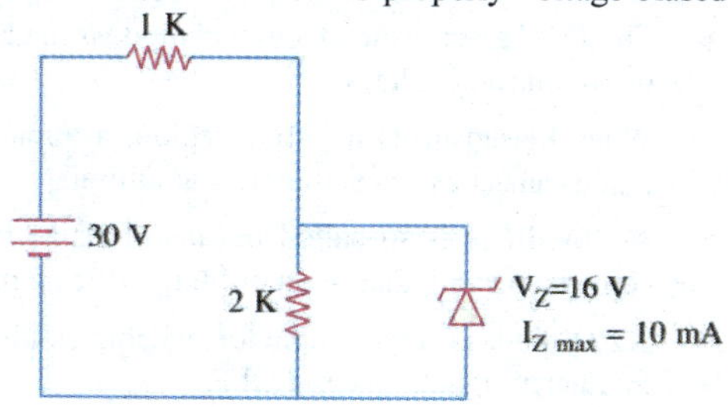

Fig. 15.37

4. Using ideal Zener diode approximations, find the minimum and maximum currents through the diode in Fig. 15.38. **[4 mA, 8 mA]**

5. A 9 V stabilized voltage supply is required to run a car stereo system from car's 12 V battery. A Zener diode with V_Z = 9 V and P_{max} = 0.25 W is used as a voltage regulator as shown in Fig. 15.39. Find the value of the series resistor R. **[120 Ω]**

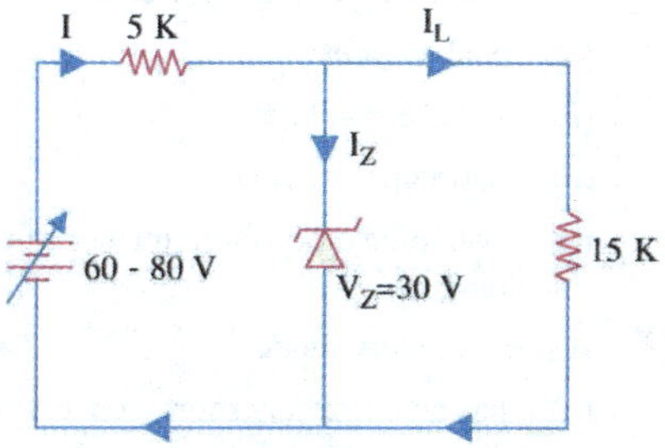

Fig. 15.38

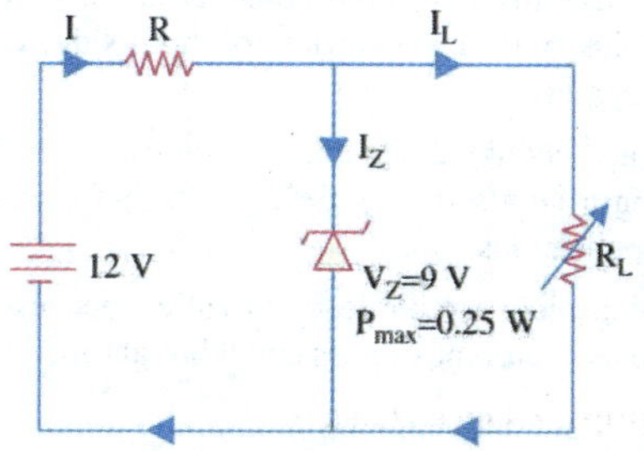

Fig. 15.39

SELF EXAMINATION QUESTIONS

A. Fill in the blanks with most appropriate word (s) or numerical value (s).

1. Zener diode is operated in the .... region.
2. While in breakdown, a Zener diode keeps the voltage across it ....
3. Zener diodes are commonly used as .... regulators.
4. An almost square output wave can be obtained by connecting two Zener diodes joined .... to .... across an input sine wave.
5. Clipping level of a Zener diode can be adjusted by changing......... of the dc source.
6. In a tunnel diode, electrons tunnel through the junction with almost the speed of ....
7. When used as an amplifier, a tunnel diode is biased in the .... region.
8. A varactor is a .... dependent variable capacitor.
9. A varactor diode is always .... biased.
10. Varactor diodes find extensive use in .... amplifiers.
11. A *PIN* diode is composed of............. sections.
12. When reverse-biased, a *PIN* diode looks like an ....
13. The main rival of Schottky diode is the .... diode.
14. .... constitute the majority carriers on both side of the junction in a Schottky diode.
15. Step-recovery diode is also a .... dependent capacitance diode.
16. The Step-recovery diode finds its main use in .... generator circuits.

B. Answer True or False

1. Zener diode is a reverse-biased lightly-doped *P-N* junction diode.
2. Once in breakdown, a Zener diode keeps its voltage practically constant.
3. Zener diodes are widely-used as voltage regulators and peak clippers.
4. A sine wave can be converted into a square wave with the help of Zener diodes.

5. A very heavily-doped *P-N* junction can be tunneled through by electrons having even low energy.
6. The *V/I* characteristic of a tunnel diode is made up of hills and valleys.
7. When biased in its negative region, a tunnel diode can act as an oscillator or amplifier.
8. A *PIN* diode is so called because its size is comparable with that of an ordinary office pin.
9. A *PIN* diode is very suitable for switching high-frequency signals ON and OFF.
10. A *PIN* diode consists of three equally-doped semiconductor sections.
11. A Schottky diode utilizes metal-metal junction.
12. A Schottky diode is unipolar because it has only holes as majority carriers on both sides of its junction.
13. The Schottky diode and *PIN* diode are the two main rivals in the field of high-frequency applications.
14. Step-recovery diode is so called because its reverse current stops is small but abrupt steps.

C. Multiple Choice Items

1. The value of Zener current is
   (*a*) limited by external circuit resistance
   (*b*) determined by Zener voltage
   (*c*) totally independent of temperature
   (*d*) always in μA range
2. When used in a circuit, the Zener diode is always
   (*a*) forward-biased
   (*b*) connected in series
   (*c*) troubled by overheating
   (*d*) reverse-biased
3. Once a Zener diode goes into breakdown, its............ does not change much.
   (*a*) voltage (*b*) current
   (*c*) dynamic impedance
   (*d*) capacitance
4. The main reason why electrons can tunnel through a *P-N* junction is that
   (*a*) they have high energy
   (*b*) barrier potential is very low
   (*c*) deplection layer is extremely thin
   (*d*) impurity level is low
5. A tunnel diode is always biased
   (*a*) by dc source
   (*b*) in the middle of its negative resistance region
   (*c*) in the positive-resistance region nearest zero
   (*d*) in the reverse direction
6. The capacitance of a varactor diode can be changed by
   (*a*) increasing its doping level
   (*b*) changing its forward bias
   (*c*) changing the barrier potential
   (*d*) changing the reverse voltage
7. A *PIN* diode is frequently used as a
   (*a*) peak clipper
   (*b*) voltage regulator
   (*c*) harmonic generator
   (*d*) switching diode for frequencies upto GHz range
8. A step-recovery diode
   (*a*) has an extremely short recovery time
   (*b*) conducts equally well in both directions
   (*c*) is mainly used as a harmonic generator
   (*d*) is an ideal rectifier of high-frequency signals
9. A semiconductor device that resembles a voltage variable capacitor is known as .... diode.
   (*a*) tunnel (*b*) varactor
   (*c*) Schottky (*d*) *PIN*
10. A semiconductor device that has only high energy electrons as carriers is called .... diode.
    (*a*) Schottky (*b*) IMPATT
    (*c*) PIN (*d*) varactor
11. A semiconductor device that has an intrinsic material between *P* and *N* materials is called .... diode.
    (*a*) Schottky (*b*) IMPATT
    (*c*) tunnel (*d*) PIN
12. A semiconductor that operates in avalanche region is called the .... diode.
    (*a*) gunn (*b*) tunnel
    (*c*) IMPATT (*d*) varactor

13. When the reverse bias voltage of a varactor increases, its
(*a*) depletion region decreases
(*b*) capacitance decreases
(*c*) leakage current increases
(*d*) negative resistance increases

14. A semiconductor device which operates with a forward biased metal-semiconductor junction is called .... diode.
(*a*) Schottky (*b*) tunnel
(*c*) Gunn (*d*) PIN

ANSWERS

A. Fill in the blanks

1. breakdown **2.** constant **3.** voltage **4.** back, back **5.** voltage
6. light **7.** negative **8.** voltage **9.** reverse **10.** parametric
11. three **12.** open **13.** *PIN* **14.** electrons **15.** voltage
16. harmonic

B. True or False

1. F **2.** T **3.** T **4.** T **5.** T **6.** T **7.** T **8.** F **9.** T **10.** F
11. F **12.** F **13.** T **14.** F

C. Multiple Choice Items

1. *a* **2.** *d* **3.** *a* **4.** *c* **5.** *a* **6.** *c* **7.** *d* **8.** *c* **9.** *b* **10.** *a*
11. *d* **12.** *c* **13.** *b* **14.** *a*

16 CHAPTER

Optoelectronic Devices

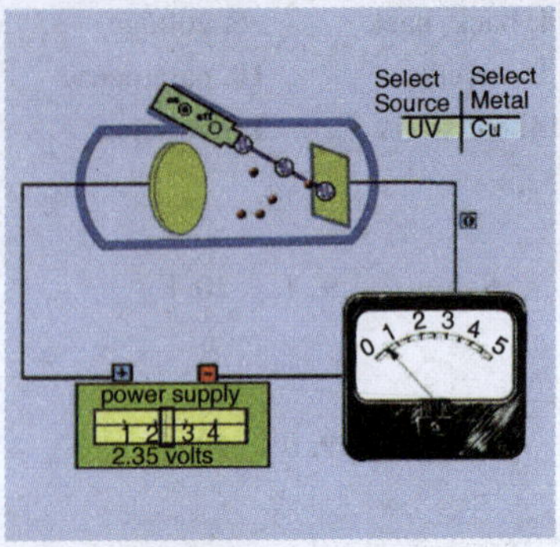

1. Introduction
2. Spectral Response of Human Eye
3. Light Emitting Diode (LED)
4. Photoemissive Devices
5. Photomultiplier Tube
6. Photovoltaic Devices
7. Bulk Type Photoconductive Cells
8. Photodiodes
9. P-N Junction Photodiode
10. PIN Photodiode
11. Avalanche Photodiode

16.1. Introduction

Optoelectronic devices are the products of a technology that combines optics with electronics.

The present-day family tree of these devices is shown in Fig. 16.1.

Such devices fall into two broad categories :

1. Devices that convert electricity into light (visible or invisible). They are known as *emitters.*
2. Devices that convert light into electricity. It includes photoemissive (non-solid-state) devices and photodetectors (solid- state devices).

Since the characteristics of these devices depend on light, some fundamental facts about light would be discussed below.

What we call light is only a small portion of the vast electromagnetic spectrum shown in Fig. 16.2. This gives a graphical representation of the different radiations that exist at different frequencies and wavelengths.

Wavelength and Frequency

The wavelength (λ) and frequency (f) of light are related to the velocity of light by the relation

$$c = f\lambda$$

where $c = 3 \times 10^8$ m/s, f is in hertz (Hz) and λ in metres (m).

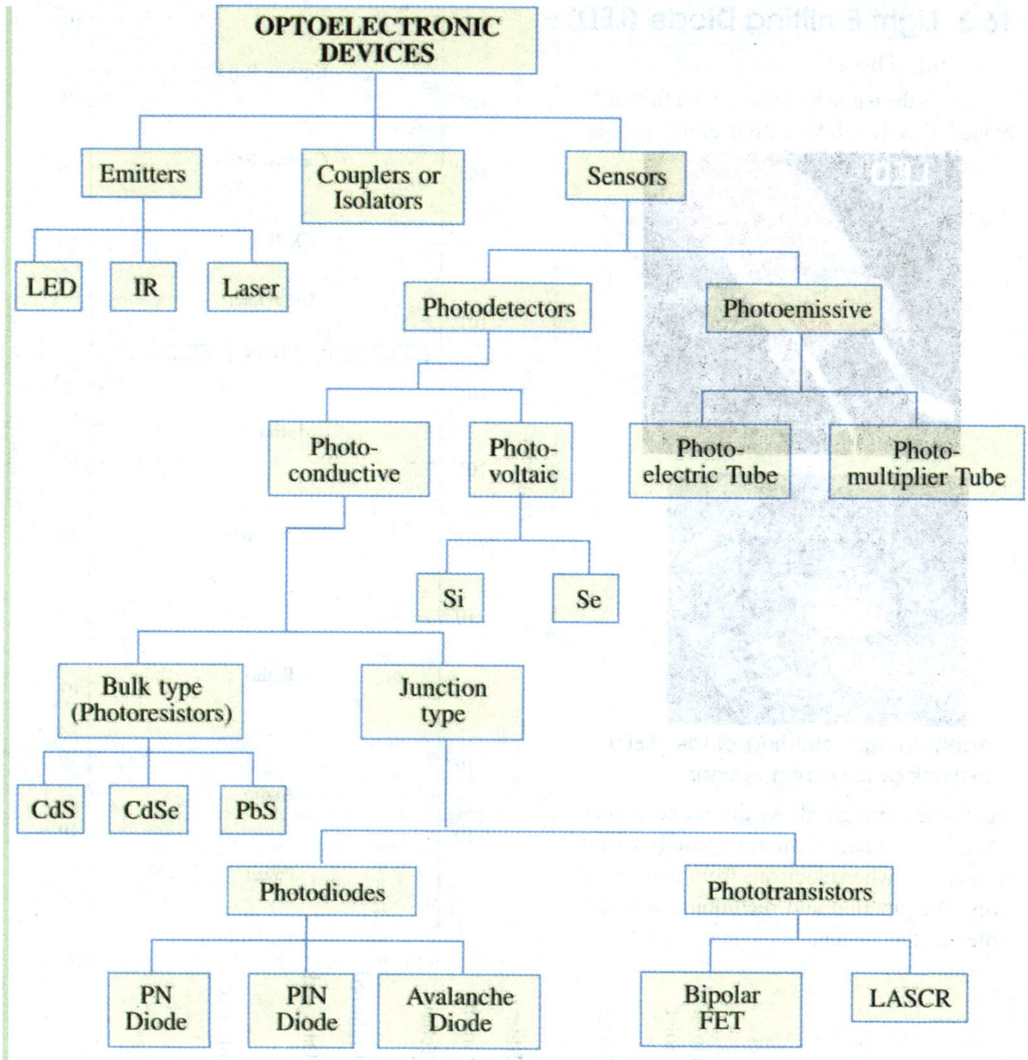

Fig. 16.1

The different units employed for measuring wavelength are

1 micrometre (μm) = 10^{-6} m

It is also known as micron (μ).

1 nanometre (nm) = 10^{-9} m

16.2. Spectral Response of Human Eye

The human eye is somewhat like a filter with a response similar to that of a tuned circuit. Fig. 16.3 shows the response of an average human eye to the visible part of the spectrum extending from 400 nm to 760 nm. It is seen that peak sensitivity occurs at 550 nm and tapers off to zero at 400 nm and 760 nm. It drops to 50% of the peak sensitivity at 610 nm. Different colours corresponding to different wavelengths have also been shown in the figure. Speaking in terms of colours, the human eye is most sensitive to green-yellow and less sensitive to violet (shorter λ) and red (longer λ).

16.3. Light Emitting Diode (LED)

(a) Theory

As the name indicates, it is a forward-biased *P-N* junction which emits visible

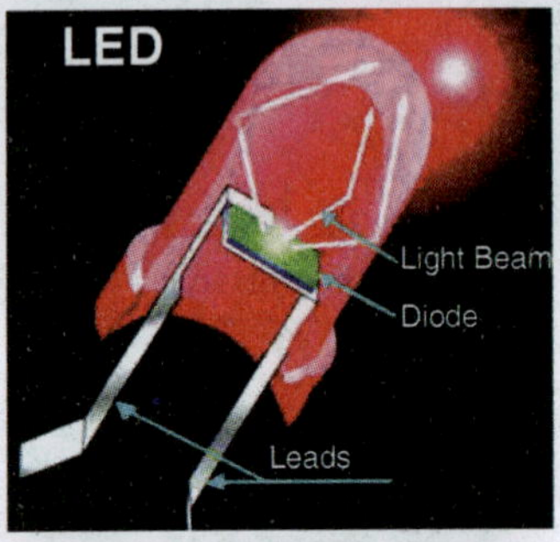

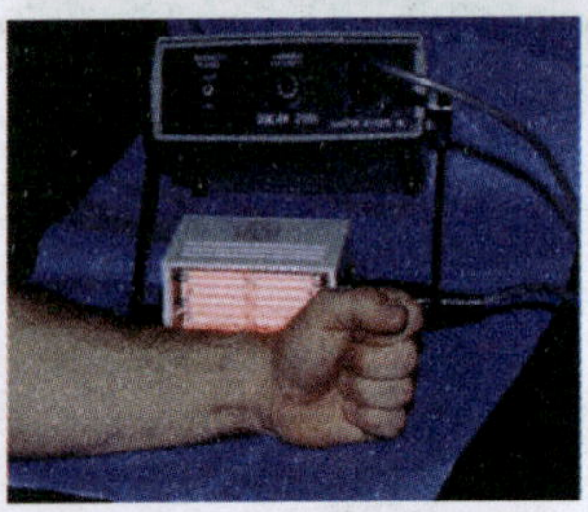

Applying light-emitting diode (LED) technology to healing wounds.

light when energised. As discussed earlier (Art. 13.2), charge carrier recombination takes place when electrons from the *N*-side cross the junction and recombine with the holes on the *P*-side.

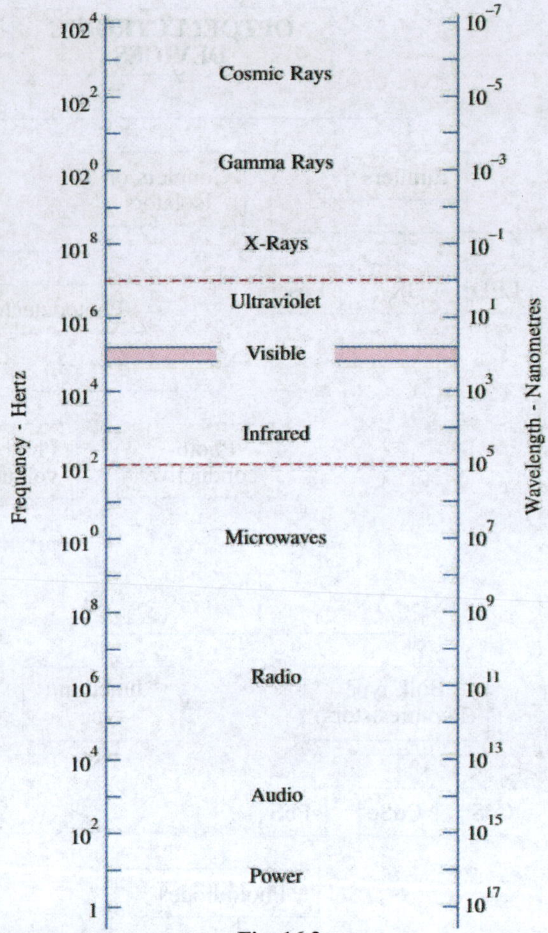

Fig. 16.2

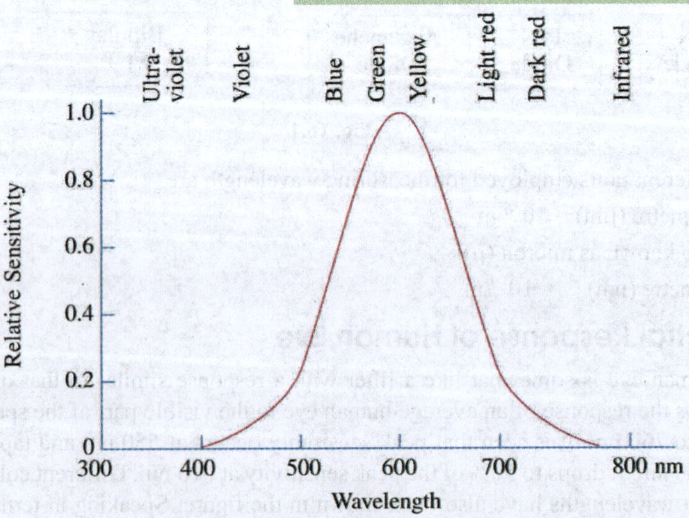

Fig. 16.3

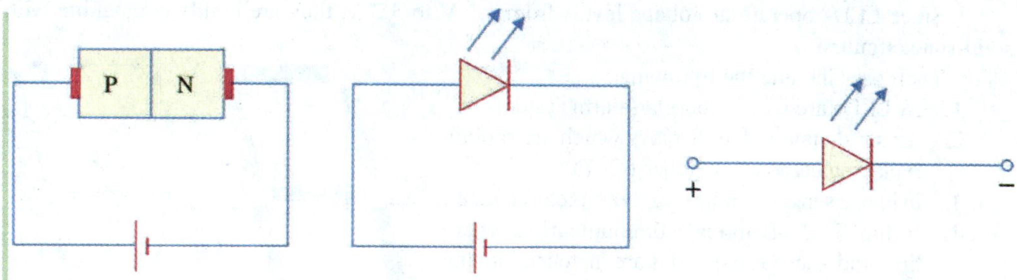

Fig. 16.4

Now, electrons are in the higher conduction band on the *N*-side whereas holes are in the lower valence band on the *P*-side. During recombination, some of this energy difference is given up in the form of heat and light (*i.e.*, photons). For Si and Ge junctions, greater percentage of this energy is given up in the form of heat so that the amount of emitted light is insignificant. But in the case of other semiconductor materials like gallium arsenide (GaAs), gallium phosphide (GaP) and gallium-arsenide-phosphide (GaAsP), a greater percentage of energy released during recombination is given out in the form of light. If the semiconductor material is transluscent, light is emitted and the junction becomes a light source *i.e.*, a light emitting diode (LED) as shown schematically in Fig. 16.5. The colour of the emitted light depends on the type of material used as given below :

1. GaAs — infrared radiation (invisible)
2. GaP — red or green light
3. GaAsP — red or yellow (amber) light

LEDs emit no light when reverse-biased. In fact, operating LEDs in reverse direction will quickly destroy them.

(*b*) Construction

An *N*-type layer is grown on a substrate and a *P*-type is deposited on it by diffusion. Since carrier recombination takes place in the *P*-layer *it is kept uppermost*. The metal anode connections are made at the outer edges of the *P*-layer so as to allow more central surface area for the light to escape. LEDs are manufactured with domed lenses in order to lessen the reabsorption problem.

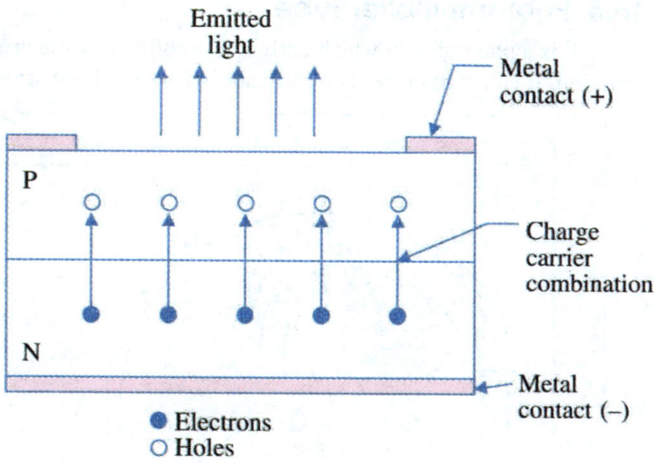

Fig. 16.5

A metal (gold) film is applied to the bottom of the substrate for reflecting as much light as possible to the surface of the device and also to provide cathode connection LEDs are always encased to protect their delicate wires.

Being made of semiconductor material, it is rugged and has a life of more than 10,000 hours.

(*c*) Applications

To choose emitting diodes for a particular application, one or more of the following points have to be considered: wavelength of light emitted, input power required, output power, efficiency, turn-on and turn-off time, mounting arrangement, light intensity and brightness etc.

Since LEDs operate at voltage levels from 1.5 V to 3.3 V, they are highly compatible with solid-state circuitry.

Their uses include the following :

1. *IR* LEDs are used in burglar-alarm systems,
2. for solid-state video displays which are rapidly replacing cathode-ray tubes (*CRT*),
3. in image sensing circuits used for 'picturephone',
4. in the field of optical communication where high-radiance GaAs diodes are matched into the silica-fibre optical cable,
5. in arrays of different types for displaying alphanumerics (letters and numbers) or supplying input power to lasers or for entering information into optical computer memories,
6. for numeric displays in hand-held or pocket calculators.

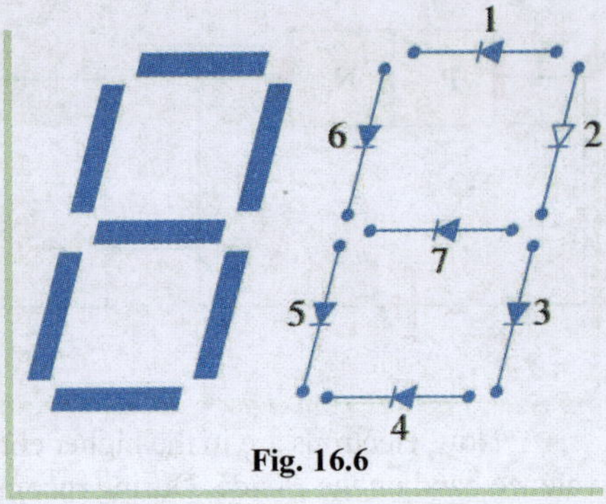

Fig. 16.6

A seven-segment array of LEDs is shown in Fig. 16.6 which can form the digits 0 to 9. Each segment contains a LED which can be turned ON or OFF to form the desired digit. For example, when all segments are ON, the digit formed is 8. If only the centre is OFF, we get zero digit.

16.4. Photoemissive Devices

There are two devices in this category :

1. photoemissive cell or phototube, 2. photomultiplier tube.

The phototube will not be discussed becasue it has been almost displaced by its solid-state counterpart discussed later. Only photomultiplier tube will be considered because it is still being used.

16.5. Photomultiplier Tube

It is a phototube in which current is amplified by electron multiplication through secondary emission. It can give current amplification. Cross-section of one form of such a tube is shown in

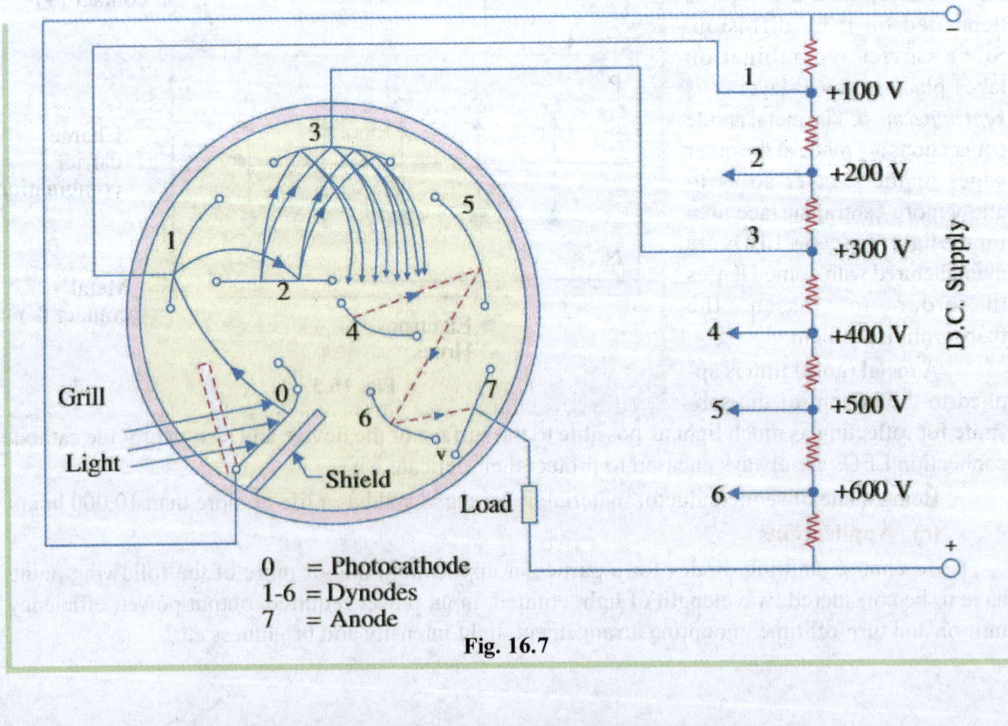

Fig. 16.7

Fig. 16.7. It consists of electrodes (called dynodes) which are maintained at increasing potentials in sequence from photocathode 0 to the anode marked 7 in the diagram. When light falls on the cathode, photoelectrons are emitted which are accelerated towards dynode 1, since it is at a higher potential. On colliding with this dynode, they liberate secondary electrons which are attracted by dynode No. 2, since it is at a relatively still higher potential. These electrons release further secondary electrons from it which are attracted by dynode 3. This process of secondary emission from different dynodes continues till the electrons are finally collected by the anode. It is obvious that at each dynode, the number of secondary keeps on multiplying. Photomultiplier tubes operate at voltages from 1 to 5 kV. Because of their ultrahigh sensitivity, low noise, extremely small dark current and very fast response time (less than 1 nanosecond), they are extensively employed in scintillators, radiation detectors and for space exploration.

16.6. Photovoltaic Devices

These are made of either silicon or selenium. We will discuss only the solid-state device that consists of a *P-N* junction which is used *without any bias*. It is worth making a distinction here regarding these devices. When a *P-N* junction is used without any bias, it is called a photovoltaic device (or solar cell already discussed in Art. 6.17) and when it is used with a reverse bias, it is called a photoconductive device *i.e.*, a photodiode (see chart of Fig. 16.1).

Photoconductive devices may be either of bulk type or junction type. The bulk type are made from a single layer of photosensitive material and have no *P-N* junction. The devices utilizing reverse biased junctions are called *photodiodes*.

16.7. Bulk Type Photoconductive Cells

It is also called a *photoresistive* device.

(*a*) Principle

It is based on the principle that the resistance of certain semiconductor materials *decreases when they are exposed to radiations* (both visible and invisible). In other words, such materials have high dark resistance and low irradiated resistance.

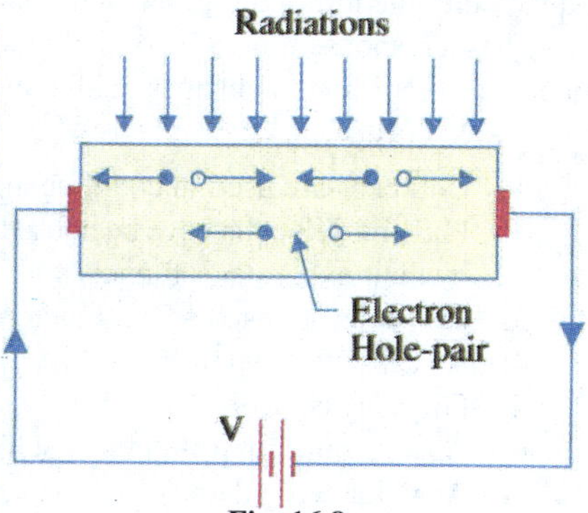

Fig. 16.8

It is found that when radiations of sufficient energy fall on such photosensitive materials, they cause the electrons to break away from their covalent bonds thereby generating electron-hole pairs (Fig. 16.8). Since these charge carriers are created within the material, they reduce its resistance.

(*b*) Construction

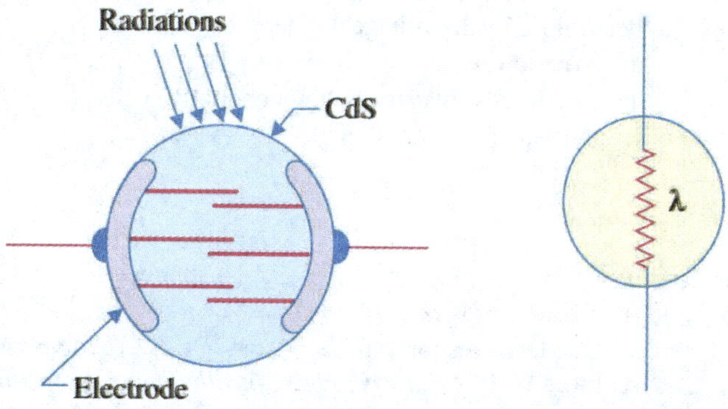

Fig. 16.9

The four materials normally employed in photoconductive cells are : cadmium sulphide (CdS), cadmium selenide (CdSe), lead sulphide (PbS) and thalmium sulphide (TlS). The simplified two-dimensional model of the commonly-used CdS cell is shown in Fig. 16.9 (*a*) and its circuit symbol in Fig. 16.9 (*b*). As shown, the two electrodes are extended in an interdigital pattern in order to increase the contact area with the sensitive material. In this way, it is possible to obtain a large ratio of 'dark-to-light' resistance.

An external power supply is necessary to generate a direction and provide a path for the current to flow. The value of applied voltage varies from a few volts to several hundred volts depending on photocell applications.

(c) Working

The illumination characteristic of the cell is shown in Fig. 16.10 where logarithmic scales have been used. It is seen that when not illuminated, the cell has a resistance of more than 100 kΩ. It is known as its dark resistance. When illuminated, the resistance falls to only a few hundred ohms.

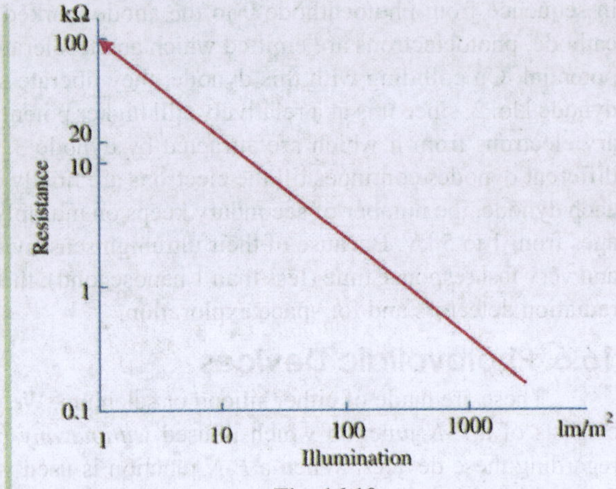

Fig. 16.10

The spectral response of CdS cell is similar to that of the human eye *i.e.*, it is sensitive to visible light. The response of CdSe cell is at the longer wavelength of the visible spectrum and extends into the infrared region. Both materials respond rather slowly to *changes* in light intensity. CdSe has a response time of about 10 ms whereas CdS takes as long as 100 ms. Moreover, CdSe is much more sensitive to changes in ambient temperature than CdS.

(d) Applications

1. CdS cells are used in counting applications.
2. PbS and TlS cells have been used for detecting ships and aircraft by the radiations given out by their exhausts or funnels.
3. CdS cells are used as light meters.
4. As ON-OFF switch.
5. For relay control.
6. As a volume control that is itself controlled by light level.
7. As voltage regulator.

As shown in Fig. 16.11, if due to any reason, V_i decreases, the brightness of lamp will also decrease. This decrease in illumination would increase photocell resistance. Hence, V_0 would be maintained at its rated level as determined by the voltage divider rule (Art. 2.8).

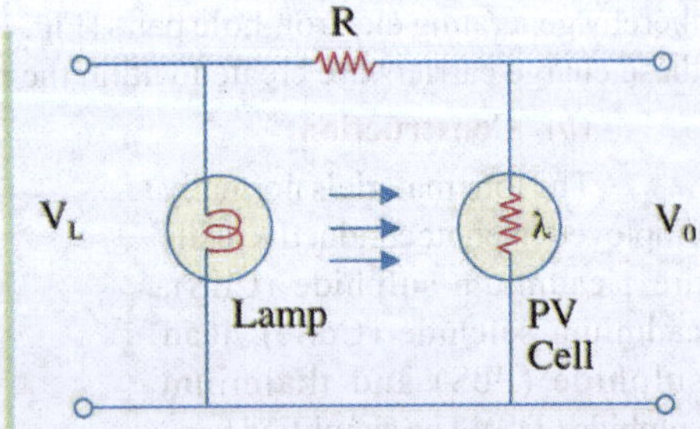

Fig. 16.11

(e) Advantages

Some of the advantages of photoconductive cells over other photodetectors are their

1. high sensitivity,
2. low cost,
3. high dark-to-light resistance ratio exceeding 100:1.

(f) Disadvantages

1. large response time and 2. relatively narrow spectral response.

Example 16.1. *A relay is to be controlled by a photoconductive cell having a dark resistance of 100 kΩ of 1 kΩ when illuminated mith 400 lm/m². The relay is to be supplied with 8 mA from a 24-V battery (Fig. 16.12) when the cell is illuminated with 400 lm/m² and is required to be de-energised when the cell is dark. Calculate the required series resistance and the dark current.*

Solution. The relay circuit is shown in Fig. 16.12 where R is included to limit circuit current. As seen

$$I = \frac{24}{R + \text{cell resistance}}$$

$$\therefore \quad 8 \times 10^{-3} = \frac{24}{R + 1000}$$

$$\therefore \quad R = 2000\ \Omega = \mathbf{2\ K}$$

Dark current is

$$= \frac{24}{2\text{K} + 100\text{K}} \cong \mathbf{0.24\ mA}$$

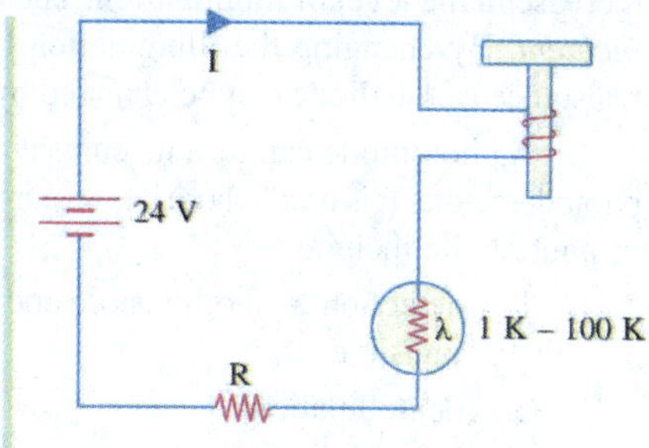

Fig. 16.12

16.8. Photodiodes

They are junction-type photoconductive devices and consists of

1. *PN* diode   2. *PIN* diode and   3. avalanche diode

We will take them up one by one in the pages to follow.

16.9. P-N Junction Photodiode

It is a two-terminal junction device which is operated by first reverse-biasing the junction and then illuminating it. A reverse-biased *P-N* junction has a small amount of reverse saturation current I_s (or I_0) due to thermally-generated electron-hole pairs. In silicon, I_s is in the range of nanoamperes. The number of these minority carriers depends on the intensity of light incident on the junction. When the diode is in a glass package, light can reach the junction and thus change the reverse current.

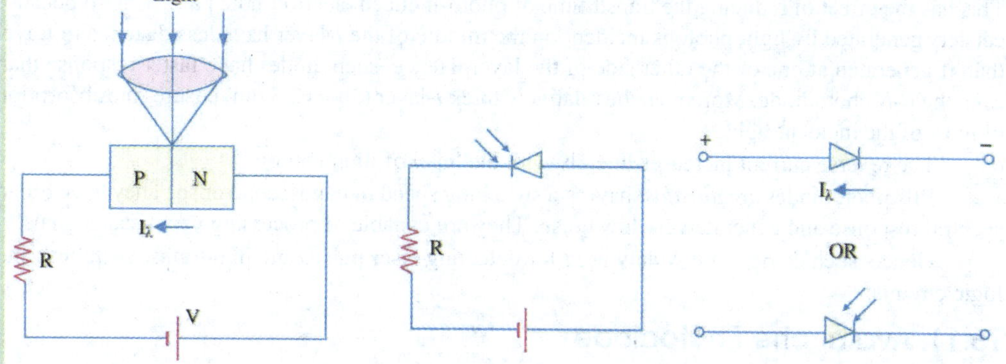

Fig. 16.13

The basic biasing arrangement, construction and symbol of a photodiode are shown in Fig. 16.13. As seen, a lens has been used in the cap of the unit to focus maximum light on the reverse-biased junction. The active diameter of these devices is about 2.5 mm but they are mounted in standard TO-5 packages with a window to allow maximum incident light.

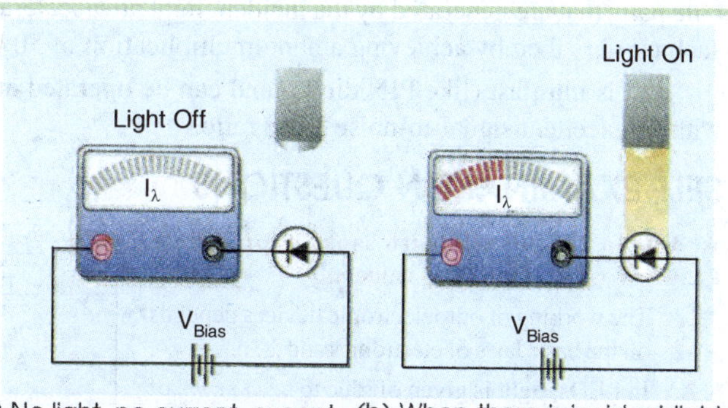

(a) No light, no current, except a negligible dark current.

(b) When there is incident light, resistence decreases and reverse (light) current flows.

The characteristics of Fig. 16.14 show that for given reverse voltage, I_λ (or I_s) increases with increase in the level of illumination. The dark current refers to the current that flows *when no light is incident*. By changing the illumination level, reverse current can be changed. In this way, reverse resistance of the diode can be changed by a factor nearly 20.

A photodiode can turn its current ON and OFF in nanoseconds. Hence, it is one of the fastest photodetectors. It is used where it is required to switch light ON and OFF at maximum rate. Applications of photodiode include

1. detection, — both visible and invisible
2. demodulation,
3. switching,
4. logic circuits that require stability and high speed,
5. character recognition,
6. optical communication equipment,
7. encoders etc.

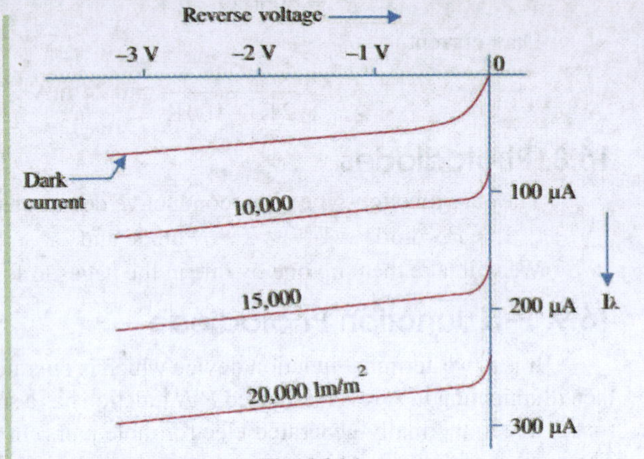

Fig. 16.14

16.10. PIN Photodiode

It is a three-region reverse-biased junction diode as shown in Fig. 16.9. A layer of intrinsic silicon is sandwiched between two heavily-doped *P*- and *N*-type silicon materials. This has the effect of reducing the transit time of photo-induced electron-hole pairs. It is so because carriers generated by light photons incident on the middle of the *I*-layer have less distance to travel than if generated at one or the other side of the layer. Hence, such diodes have faster response than even the *P-N* photodiode. Moreover, the relatively thick *I*-layer (about 2.5 μm) ensures the absorption of most of the incident light.

The reverse current increases linearly with the level of illumination.

PIN photodiodes are *ultrafast* having a switching speed of nanosecond or so. They have broad spectral response and generate very low noise. They are capable of processing very weak signals.

Hence, such devices are widely used for detecting laser pulses and in ultrafast switching and logic circuits.

16.11. Avalanche Photodiode

It is a *P-N* junction diode which operates in the avalanche breakdown region. The electron-hole pairs that are generated by the incident light are accelerated by the electric field to produce more such carriers thereby achieving a photomultiplication of 50 or so.

It is ultrafast (like PIN diode) and can be operated at a modulation frequency of a few GHz with an excellent signal-to-noise (*S/N*) ratio.

SELF EXAMINATION QUESTIONS

A. Fill in the blanks with most appropriate word (s) or numerical value(s).

1. The working of optoelectronic devices depends on the basic laws of electronics and ....
2. In LEDs, light is given off due to .... of electrons and holes.
3. LEDs emit light only when .... biased.
4. LED alphanumeric displays can indicate both .... and numbers.
5. A seven-segment LED array can display digits ....to 9.

6. Photovoltaic cells do not need an external ....supply.
7. A photodiode is essentially a .... biased *P-N* junction which is illuminated with radiations.
8. Incident radiations in a photodiode increase its .... current.

B. Answer True or False

1. A LED emits light when forward-biased.
2. Light emission by LED is due to recombination of electrons and holes.
3. A LED using GaAs material gives out infrared radiations.
4. LED emits no light when reverse-biased.
5. Photovoltaics are self-generating devices.
6. A photodiode is a junction device needing forward bias.
7. Bulk type photoconductive cells are also called photoresistors.
8. Higher its dark-to-light resistance ratio, better the photoconductive cell.
9. *P-N* photodiode has faster response than a PIN photodiode.
10. Higher the illumination level, greater the reverse current of a photodiode.

C. Multiple Choice Items

1. The colour of light emitted by a LED depends on
   (*a*) its forward bias
   (*b*) its reverse bias
   (*c*) the amount of forward current
   (*d*) the type of semiconductor material used
2. Bulk type photoconductive cells have
   (*a*) wide spectral response
   (*b*) small response time
   (*c*) high cost
   (*d*) high dark-to-light resistance ratio
3. A *P-N* junction photodiode is
   (*a*) operated in forward direction
   (*b*) encased in an opaque package
   (*c*) a very fast photodetector
   (*d*) dependent on thermally-generated minority carriers
4. Silicon is invariably used in the manufacture of junction photodiodes because
   (*a*) more electron-hole pairs are generated in it
   (*b*) its thermally-generated minority current is extremely small
   (*c*) it is more rugged than Ge
   (*d*) it needs less reverse bias
5. A PIN photodiode has ultrafast response primarily due to
   (*a*) the presence of middle *I*-layer
   (*b*) heavy doping of *P* and *N*-regions
   (*c*) higher electrical conductivity of Si
   (*d*) its wide spectral response

ANSWERS

A. Fill in the blanks

1. optics **2.** recombination **3.** forward **4.** letters **5.** 0
6. voltage **7.** reverse **8.** reverse

B. True or False

1. *T* **2.** *T* **3.** *T* **4.** *T* **5.** *T* **6.** *F* **7.** *T* **8.** *T* **9.** *F* **10.** *T*

C. Multiple Choice Items

1. *d* **2.** *d* **3.** *c* **4.** *b* **5.** *a*

CHAPTER 17

DC Power Supplies

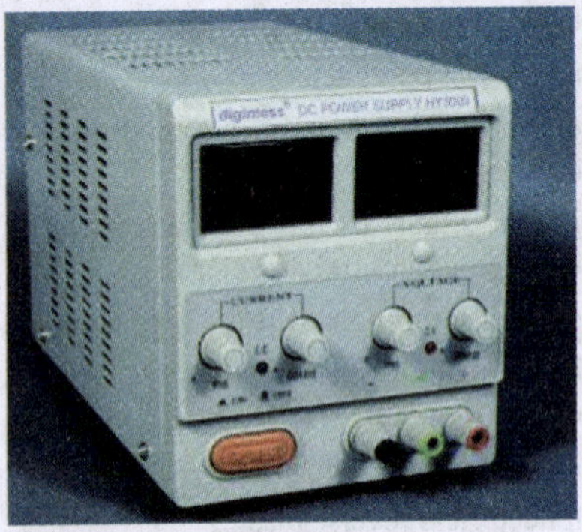

1. Unregulated Power Supply
2. Regulated Power Supply
3. Steady and Pulsating DC Voltages
4. Rectifiers
5. Full-wave Bridge Rectifier
6. Filters
7. Bleeder Resistor
8. Voltage Regulation
9. Zener Diode Shunt Regulator
10. Voltage Dividers
11. Complete Power Supply
12. Voltage Multipliers
13. Half-wave Voltage Doubler
14. Full-wave Voltage Doubler
15. Troubleshooting Power Supplies
16. Controlled Rectification
18. Silicon Controlled Rectifier (SCR)
19. UJT Controlled Circuit

17.1. Introduction

Most of the electronic devices and circuits require a dc source for their operation. Dry cells and batteries are one form of dc source. They have the advantage of being portable and ripple-free. However, their voltages are low, they need frequent replacement and are expensive as compared to conventional dc power supplies. Since the most convenient and economical source of power is the domestics ac supply, it is advantageous to convert this alternating voltage (usually, 220 V rms) to dc voltage (usually smaller in value). This process of converting ac voltage into dc voltage is called *rectification* and is accomplished with the help of

(*i*) rectifier,

(*ii*) filter; and

(*iii*) voltage regulator circuits.

These elements put together constitute dc power supply.

17.2. Unregulated Power Supply

An unregulated power supply is one whose dc terminal voltage is affected significantly by the amount of load. As the load draws more current, the dc terminal voltage becomes less.

17.3. Regulated Power Supply

It is that dc power supply whose terminal voltage remains almost constant regardless of the amount of current drawn from it. An unregulated supply can be converted into a regulated power supply by adding a voltage regulating circuit to it (Art. 17.17).

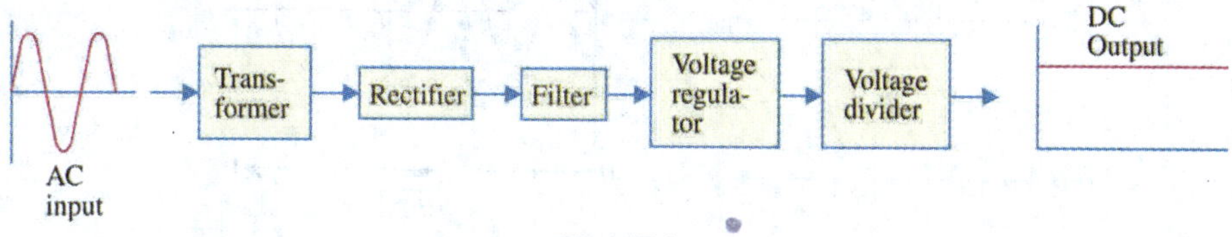

Fig. 17.1

A typical dc power supply consists of five stages as shown in Fig. 17.1.

1. Transformer

Its job is either to step up or (mostly) step down the ac supply voltage to suit the requirement of the solid-state electronic devices and circuits fed by the dc power supply. It also provides isolation from the supply line–an important safety consideration.

2. Rectifier

It is a circuit which employs one or more diodes to convert ac voltage into pulsating dc voltage.

3. Filter

The function of this circuit element is to remove the fluctuations or pulsations (called ripples) present in the output voltage supplied by the rectifier. Of course, no filter can, in practice, give an output voltage as ripple-free as that of a dc battery, but it approaches it so closely that the power supply performs as well.

4. Voltage Regulator

Its main function is to keep the terminal voltage of the dc supply constant even when

(*i*) ac input to the transformer varies (deviations from 220 V are common) or

(*ii*) the load varies.

Usually, Zener diodes and transistors are used for voltage regulation purposes. Again, it is impossible to get 100% constant voltage but minor variations are acceptable for most of the jobs.

5. Voltage Divider

Its function is to provide different dc voltages needed by different electronic circuits. It consists of a number of resistors connected in series across the output terminals of the voltage regulator. Obviously, it eliminates the necessity of providing separate dc power supplies to different electronic circuits working on different dc levels.

Comments

Strictly speaking, all that is really required for conversion from ac to dc is a transformer and rectifier (in fact, even the transformer could be eliminated if no voltage transformation is required). The filter, voltage regulator and voltage divider are mere refinements of a dc power supply though they are essential for most applications except for battery charging and running small dc motors etc.

17.4. Steady and Pulsating DC Voltages

As mentioned earlier, the filtered output of a dc power supply is not a steady dc voltage like the one shown in Fig. 17.2 (*a*). It is a fluctuating or pulsating one as shown in Fig 17.2 (*b*). It has high frequency *pulsations* or *ripples* in it. It can be looked upon as a dc voltage having some amount of ac voltage riding on it. The amount of ripple or ac voltage variations as compared to the dc voltage is a measure of the *purity* of the power supply output and is defined by a ripple factor (γ).

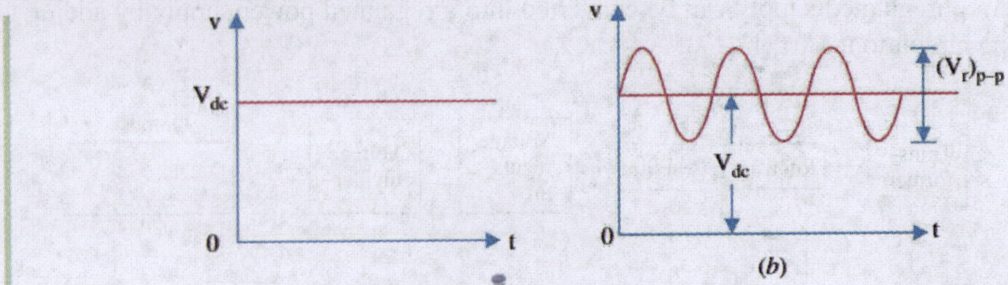

Fig. 17.2

$$\gamma = \frac{\text{ripple voltage (rms)}}{\text{circuit dc voltage}} = \frac{\text{rms value of ac component}}{\text{dc value of the wave}} = \frac{V_r\ (rms)}{V_{dc}}$$

Example 17.1. *A dc and an ac voltmeter were used to measure the output voltage from a filter circuit. If the readings of the two voltmeters are 25 V and 2.5 V respectively, calculate the ripple factor of the filter circuit.*

Solution. It should be kept in mind that an ac voltmeter measures rms value of an alternating voltage.

$$\therefore \quad V_{dc} = 25 \text{ V} \quad \text{and} \quad V_r\text{(rms)} = 2.5 \text{ V}$$

$$\therefore \quad \text{ripple factor, } \gamma = \frac{2.5}{25} = 0.1 = 0.1 \times 100 = \mathbf{10\%}$$

$$= 0.1 \times 100 = \mathbf{10\%}$$

17.5. Rectifiers

We will consider the following circuits :

1. half-wave rectifier,
2. full-wave rectifier,
3. full-wave bridge circuit,
4. voltage multiplier circuits.

Many semiconductor devices or systems (like car stereo systems) require a negative dc source or both a negative and a positive dc source. For the sake of simplicity, we will analyse only the positive dc power supplies. However, a positive dc supply can be converted into a negative one by simply reversing the two output leads in the same way as we reverse the polarity of a dry cell.

Quite a number of integrated circuits (ICs) require both a positive and negative source with common ground. In that case, the polarised components in the negative portion of the supply will have to be reversed. For example, its rectifier, filter, capacitor and voltage/current regulation devices will have to be reversed as compared to the positive supply.

Half wave rectifier.

17.6. Half-wave Rectifier

The basic half-wave diode rectifier circuit is shown in Fig. 17.3 along with its input and output waveforms. An alternating voltage is applied to a single diode connected in series with a load resistor R_L.

(a) Working

During the positive half-cycle of the input ac voltage, the diode *D* is forward-biased (ON) and conducts. While conducting, the diode acts as a short-circuit so that circuit current flows and hence, positive half-cycle of the input ac voltage is dropped across R_L. It constitutes the output voltage v_0 as shown in 17.3 (*b*). Waveform of the diode current (which equals load current) is shown in Fig. 17.3 (*c*).

During the negative input half-cycle, the diode is reverse-biased (OFF) and, so, does not conduct *i.e.*, there is no current flow. Hence, there is no voltage drop across R_L. In other words $i_D = 0$. and $v_0 = 0$. Obviously, the negative input half-cycle is suppressed

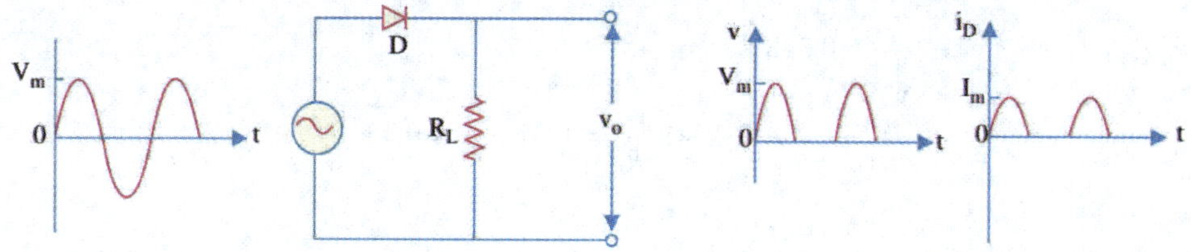

(a) (b) (c)

Fig. 17.3

i.e., it is not utilized for delivering power to the load. As seen, the output is not a steady dc but *only a pulsating dc wave* having a ripple frequency equal to that of the input voltage frequency. This wave can be observed by an oscilloscope connected across R_L. When measured by a dc meter, it will show some *average* positive value both for voltage and current. Since only one half-cycle of the input wave is used, it is called a half-wave rectifier. It should be noted that forward voltage drop across the diode has been neglected in the above discussion. We have, in fact, assumed an ideal diode. If it is required to step-up or step-down the input voltage, we will have to use a power transformer as shown in Fig. 17.4.

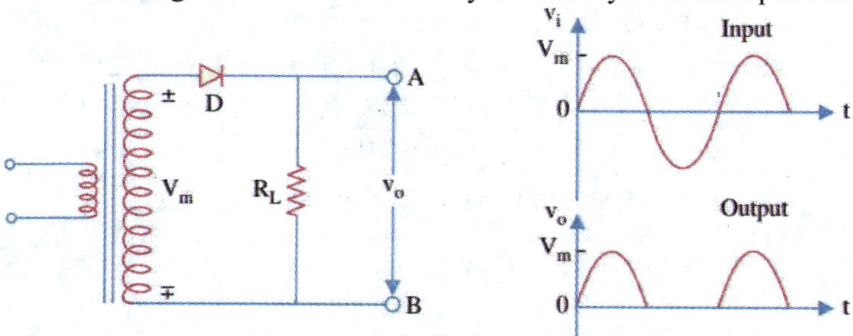

Fig. 17.4

(b) Average Values

We will now state the average values of the output voltage and current. These are shown in Fig. 17.5.

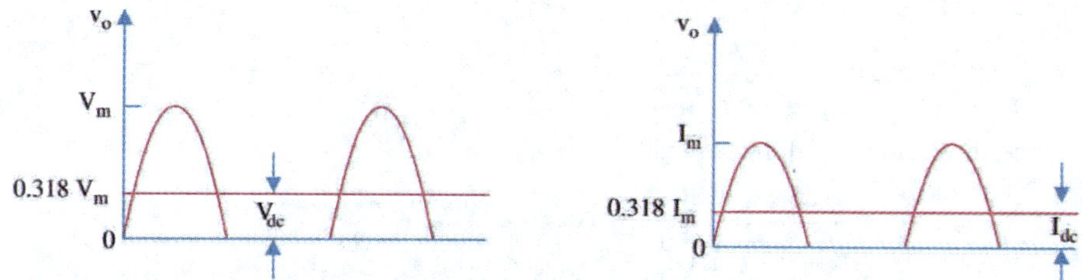

Fig. 17.5

Let the equation of the input supply voltage of Fig. 17.3 be

$$v = V_m \sin \theta$$

Same would be the equation of the secondary voltage in Fig. 17.4, except that here V_m will represent the maximum voltage across the transformer secondary.

Let V_m = maximum value of supply voltage in Fig. 17.3 = $\sqrt{2}$ V

V = rms value of supply voltage

I_m = maximum value of diode or load current

V_{dc} = average dc voltage across load

I_{dc} = average dc current through load

I_{rms} = rms current through load

Then, it can be proved that

(*i*) $$V_{dc} = \frac{V_m}{\pi} = 0.318\, V_m\, *$$

(*ii*) $$I_{dc} = \frac{V_{dc}}{R_L} = \frac{V_m}{\pi\, R_L} = 0.318\, \frac{V_m}{R_L} = 0.318\, I_m = 0.45\, \frac{V}{R_L}$$

$$I_m / 2 = 0.5\, I_m$$

(*c*) Form Factor

$$F = \frac{\text{rms value}}{\text{average value}} = \frac{0.5\, I_m}{0.318\, I_m} = 1.57 \qquad \text{— for sinusoidal ac wave}$$

(*d*) PIV

It is the maximum voltage across the diode in the reverse direction. Obviously, its value in the present case is V_m.

(*e*) Ripple Factor (γ)

The ripple factor** of a single-phase half-wave rectifier is

$$\gamma = V_r\, \frac{(rms)}{V_{dc}} = \frac{0.385\, V_m}{0.318\, V_m} = 1.21$$

The relation between V_m and V_{dc} is given by

$$\frac{V_m}{V_{dc}} = 1 + \sqrt{3}.\, \gamma$$

(*f*) Efficiency (η)

It is given by the ratio of the output dc power to the total amount of input power supplied to the circuit

$$\therefore \quad \eta = \frac{P_{out}}{P_{in}}$$

Now, $$P_{out} = I_{dc}^2\, R_L\,;\; P_{in} = I_{rms}^2\, (r_d + R_L)$$

$$\therefore \quad \eta = \frac{I_{dc}^2\, R_L}{I_{rms}^2\, (r_d + R_L)} \times 100$$

Substituting and simplifying, we get

$$\eta = \frac{0.406}{1 + \frac{r_d}{R_L}} = \frac{40.6}{1 + \frac{r_d}{R_L}} \text{ per cent}$$

If diode resistance r_d is neglected, then η = 40.6%.

Obviously, it is the maximum possible efficiency.

Example 17.2. *A 1 phase half-wave rectifier supplies power to a 1 kΩ load. The input supply voltage is 200 V rms. Neglecting forward resistance of the diode, calculate*

(*a*) V_{dc} (*b*) I_{dc} (*c*) *ripple voltage* (*rms value*).

$$* \; V_{dc} = V_{av} = \frac{1}{2\pi}\int_0^{\pi} v.d\theta = \frac{1}{2\pi}\int_0^{\pi}(V_m \sin\theta)\, d\theta = \frac{V_m}{2\pi}\left|-\cos\theta\right|_0^{\pi} = \frac{V_m}{2\pi}[+1 - (-1)]$$

$$= \frac{V_m}{\pi} = 0.318\, V_m.$$

** Apart from the definition given in Art. 17.4, ripple factor can also be defined in terms of the form factor (*F*) of the rectified output. For a H.W. rectified voltage, *F* = 1.57.

Solution. $V = 200$ V

(*a*) $V_{dc} = 0.45\text{ V} = 0.45 \times 200 = \mathbf{90\ V}$

(*b*) $I_{dc} = \dfrac{V_{dc}}{R_L} = \dfrac{90}{1\,K} = \mathbf{90\ mA}$

(*c*) $\gamma = \dfrac{V_r\text{ (rms)}}{V_{dc}}$

$\therefore \quad V_r\text{(rms)} = \gamma.V_{dc} = \dfrac{1.21}{100} \times 90 = \mathbf{1.09\ V}$

Example. 17.3. *In the unfiltered half-wave rectifier circuit of Fig. 17.6, calculate the value of the dc output voltage.*

Solution. Since the transistor ratio $K = 1/20$, the rms value of the secondary voltage is given by

$$V_s = 240 \times (1/20) = 12\text{ V}$$

Hence, maximum or peak value of the secondary voltage is

$$V_m = 12 \times \sqrt{2} = 17\text{ V}$$

Talking diode drop of 0.7 V, the voltage available across the load resistance R_L is

$$V_{RL} = 17 - 0.7 = 16.3\text{ V}$$

$$\therefore \quad V_{dc} = 0.381 \times 16.3 = \mathbf{6.2\ V}$$

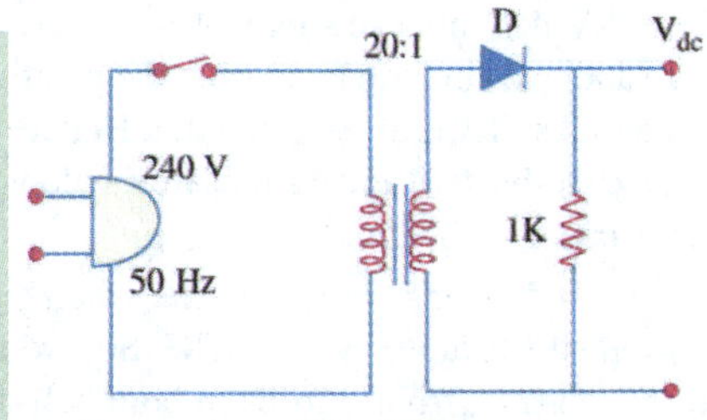

Fig. 17.6

17.7. Full-wave Rectifier

In this case, *both* half-cycles of the input are utilized with the help of two diodes working alternately. For full-wave rectification, use of a transformer is essential (though it is optional for half-wave rectification).

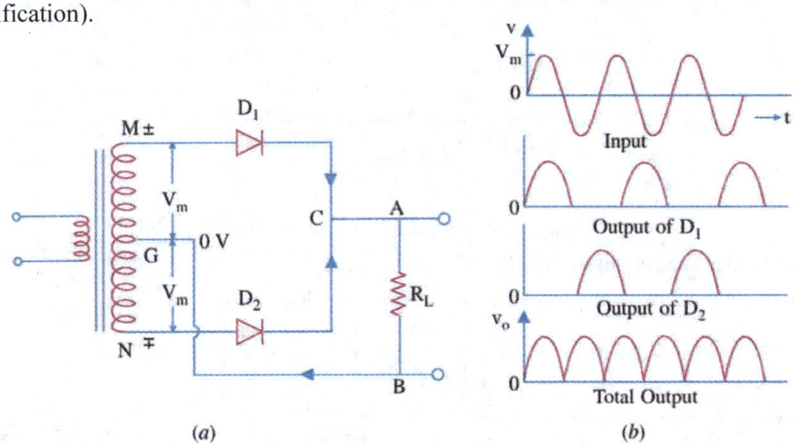

Fig. 17.7

Full-wave rectifier.

The full-wave rectifier circuit using two diodes and a centre-tapped transformer is shown in Fig. 17.7 (*a*). The centre-tap is usually taken as the ground or zero voltage reference point.

Fig. 17.8 shows two different ways of drawing the circuit. In Fig. 17.8 (*a*), R_L becomes connected to point *G* via the earth whereas in Fig. 17.8 (*b*), it is connected directly to *G*.

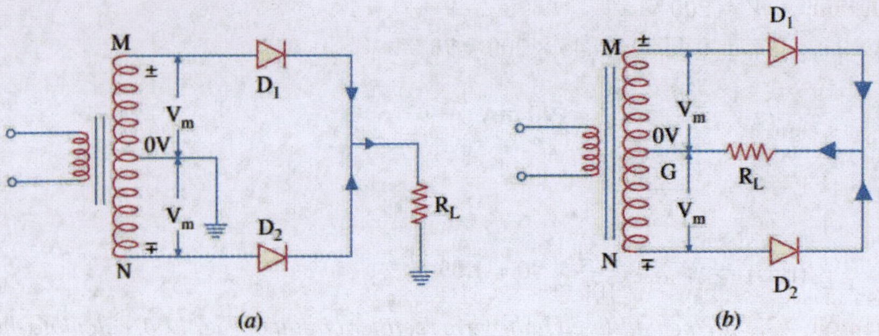

Fig. 17.8

(a) Working

When input ac supply is switched on, the ends *M* and *N* of the transformer secondary become +ve and –ve alternately. During the positive half-cycle of the ac input, terminal *M* is +ve, *G* is at zero potential and *N* is at –ve potential. Hence, being forward-biased, diode D_1 conducts (but not D_2 which is reverse-biased) and current flows along MD_1CABG. As a result, positive half-cycle of the voltage appears across R_L.

In the negative half-cycle, when terminal *N* becomes +ve, then D_2 conducts (but not D_1) and current flows along $ND_2\,CABG$. So, we find that current keeps on flowing through R_L in the same direction (*i.e.*, from *A* to *B*) in both half-cycles of the ac input. It means that both half-cycles of the input ac supply are utilized as shown in Fig. 17.7 (*b*). Also, the frequency of the rectified output voltage is twice the supply frequency.

(b) Average Values

(*i*) $$V_{dc} = \frac{2V_m}{\pi} = 0.636\ V_m \quad \text{— twice the HW rectifier value}$$

$$= \frac{2.\sqrt{2}V}{\pi} = 0.9\ V$$

where V_m = maximum voltage across each half of the secondary winding (Figs. 17.7 and 17.8)

$V = V_m/\sqrt{2}$

= rms voltage across *each* half of the secondary winding

These values are shown in Fig. 17.9.

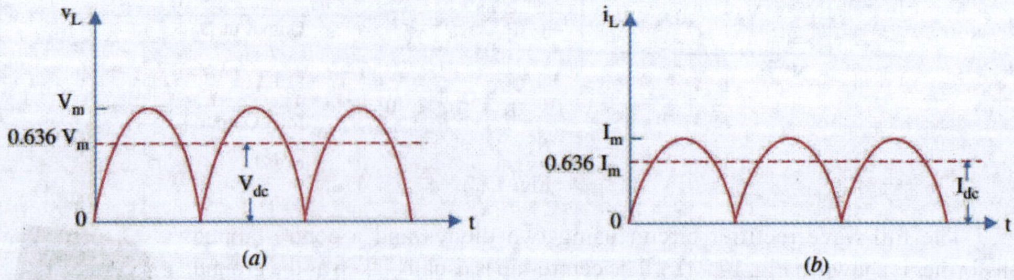

Fig. 17.9

(*ii*) $$I_{dc} = \frac{2\,I_m}{\pi} = \frac{2}{\pi}\cdot\frac{V_m}{R_L} = 0.636\ I_m$$

$$= \frac{V_{dc}}{R} = \frac{2.\sqrt{2}}{\pi}\cdot\frac{V}{R_L} = 0.9\,\frac{V}{R_L}$$

(*iii*) $I_{rms} = 0.707\, I_m$

(*iv*) form factor, $F = \frac{I_{rms}}{I_{av}} = \frac{I_{rms}}{I_{dc}} = \frac{0.707\, I_m}{0.636\, I_m} = 1.11$

(*c*) PIV

In this case, *PIV* rating of each diode is $2V_m$. It can be understood thus :

As stated earlier, it is the maximum possible voltage across a diode when it is *reverse-biased*. Consider the positive half-cycle of the input ac supply when D_1 acts as a short and D_2 as an open since it is reverse-biased. A voltage V_m develops across R_L. As seen from Fig. 17.10, voltage across D_2 is equal to the sum of voltages across the lower half *GN* of the transformer secondary and the load resistor. Hence, *PIV* of $D_2 = 2\, V_m$. Same is the value for D_1.

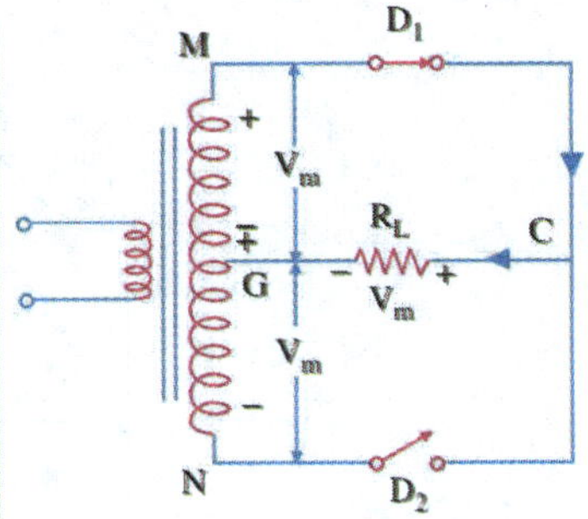

Fig. 17.10

(*d*) Ripple Factor (γ)

For full-wave rectifier circuit*

$$\gamma = \frac{V_r(rms)}{V_{dc}} = \frac{0.305\, V_m}{0.636\, V_m} = 0.482$$

It is much less than that for a HW rectifier. The relation between V_m and V_{dc} is the same as given in Art. 17.6.

(*e*) Efficiency

$$\eta = \frac{P_{out}}{P_{in}} = \frac{I_{dc}^2\, R_L}{I_{rms}^2 (r_d + R_L)}$$

Substituting and simplifying, we get

$$\eta = \frac{81.2\%}{1 + \frac{r_d}{R_L}} = 81.2\% \quad \text{— if } r_d = 0$$

It is twice the value for the half-wave rectifier for the simple reason that a full-wave rectifier utilizes both half-cycles of the input ac supply.

Example 17.4. *A full-wave rectifier supplies a load of 1 kΩ. The ac voltage applied to the diodes is 200-0-200 V rms. If diode resistance is neglected, calculate*

(*a*) *average dc voltage* (*b*) *average dc current* (*c*) *ripple voltage* (*rms*)

Solution. (*a*) $V_{dc} = 0.9\, V = 0.9 \times 200 =$ **180 V.**

(It is twice the value found in Ex. 17.2)

(*b*) $I_{dc} = \frac{V_{dc}}{R_L} = \frac{180}{1\,K} =$ **180 mA**

(*c*) $V_r(\text{rms}) = \frac{\gamma . V_{dc}}{100} = \frac{0.482 \times 180}{100} =$ **0.868 V**

17.8. Full-wave Bridge Rectifier

It is the most frequently-used circuit for electronic dc power supplies. It requires four diodes but the transformer used is not centre-tapped and has a maximum voltage of V_m. The full-wave bridge rectifier is available in three distinct physical forms.

* Since for full-wave rectifier ac, form factor is 1.11

∴ $\gamma = \sqrt{1.12^2 - 1} = 0.482.$

1. four discrete diodes,
2. one device inside a four-terminal case,
3. as part of an array of diodes in an IC.

The circuit using four discrete diodes is shown in Fig. 17.11 (*a*).

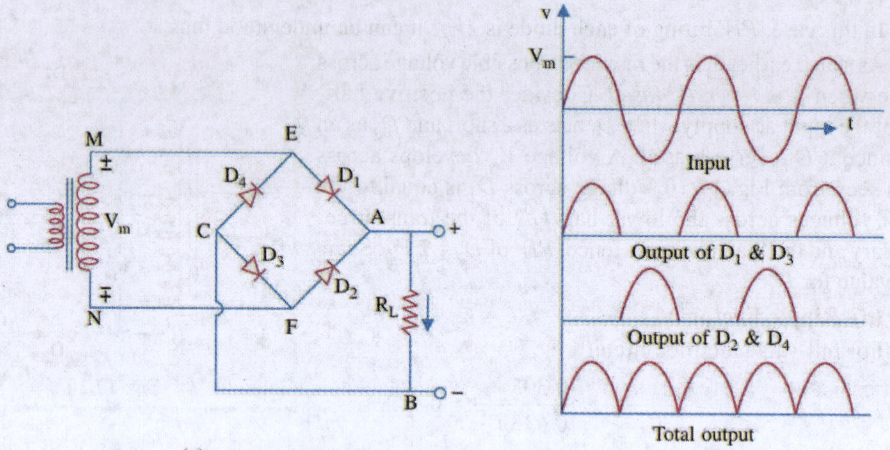

(*a*) (*b*)

Fig. 17.11

(*a*) Working

During the positive input half-cycle, terminal *M* of the secondary is positive and *N* is negative as shown separately in Fig. 17.12 (*a*). Diodes D_1 and D_3 becomes forward-biased (ON) whereas D_2 and D_4 are reverse-biased (OFF). Hence, current flows along MEABCFN producing a drop across R_L.

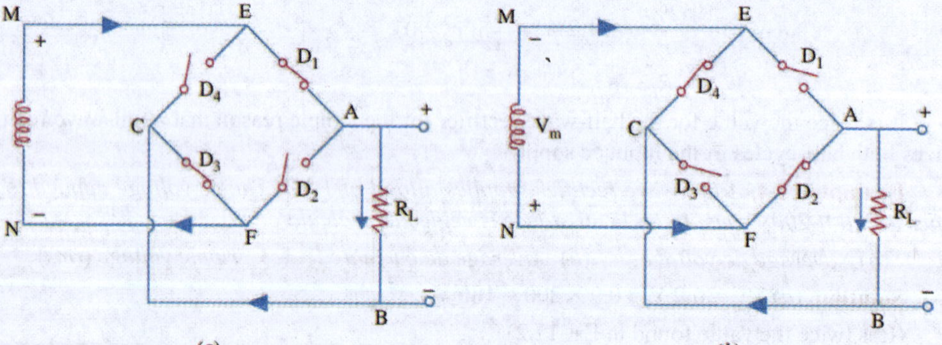

(*a*) (*b*)

Fig. 17.12

During the negative input half-cycle, secondary terminal *N* becomes positive and *M* negative. Now, D_2 and D_4 are forward-biased. Circuit current flows along NFABCEM as shown in Fig. 17.12 (*b*). Hence, we find that current keeps flowing through load resistance R_L in the *same direction AB* during both half-cycles of the ac input supply. Consequently, point *A* of the bridge rectifier always acts as an anode and point *C* as cathode. The output voltage across R_L is as shown in Fig. 17.11 (*b*). Its frequency is *twice* that of the supply frequency.

(*b*) Average Values

These are the same as for the centre-tapped full-wave rectifier discussed in Art. 17.7

(*i*) $V_{dc} = 0.636\ V_m = 0.9\ V$

(*ii*) $I_{dc} = 0.636\ \dfrac{V_m}{R_L} = 0.636\ I_m = 0.9\ \dfrac{V}{R_L} = 0.9 I_{rms}$

(iii) $I_{rms} = 0.707\ I_m$

(iv) form factor, $F = 1.11$

(c) PIV

In a full-wave bridge rectifier, the *PIV* rating of each of the four diodes is V_m (and not 2 V_m as is the case for ordinary full-wave rectifier of Art. 17.7).

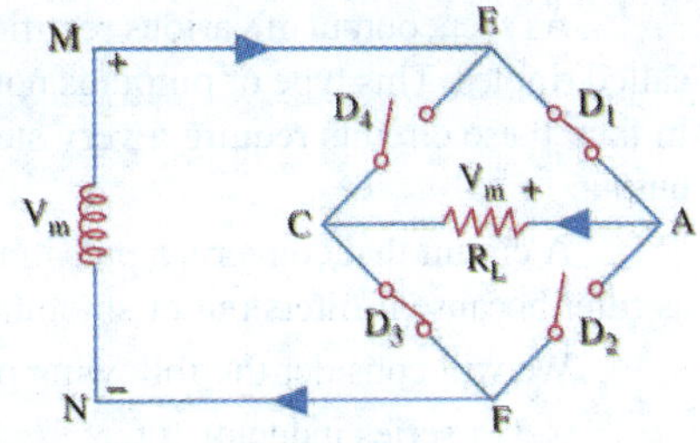

Fig. 17.13

It can be understood with reference to Fig. 17.13 which shows the condition when positive half-cycle of the supply is applied to the bridge. The circuit has been drawn in a slightly different way for ease of understanding. As seen, D_2 and D_4 are open due to reverse bias. If we go along the path MECAFN, the total voltage across the path is equal to the *sum* of secondary voltage and voltage drop across R_L due to conduction by D_1 and D_3. This sum is equal to $(V_m + V_m) = 2\ V_m$.

This voltage appears across two open diodes D_2 and D_4. Hence voltage across each diode is $2\ V_m / 2 = V_m$ which represents its *PIV* value.

Alternatively, consider the open diode D_4 alone (Fig. 17.13). The voltage across its end points *E* and *C* is the same as between *M* and *N* (remember D_3 is ON) *i.e.*, secondary voltage or peak input voltage V_m.

(d) Ripple Factor (γ)

It is the same as for ordinary FW rectifier $\gamma = 0.482$.

(e) Efficiency

It is almost the same as for ordinary FW rectifier

$$\eta = 81.2\% \qquad \text{— if diode resistance is neglected.}$$

(f) Advantages

After the advent of low-cost, highly-reliable and small-sized silicon diodes, bridge circuit has become much more popular than the centre-tapped transformer FW rectifier. The main reason for this is that for a bridge rectifier, a much smaller transformer is required for the same output because it utilizes the transformer secondary continuously unlike the 2-diode FW rectifier which uses the two halves of the secondary alternately.

So, the advantages of the bridge rectifier are

1. no centre-tap is required on transformer,
2. much smaller transformers are required,
3. it is suitable for high-voltage applications,
4. it has less *PIV* rating per diode.

The obvious disadvantage is the need for twice as many diodes as for the centre-tapped transformer version. But ready availability of low-cost silicon diodes has made it more economical despite its requirement of four diodes.

17.9. Filters

The main function of a filter circuit (Fig. 17.14) is to minimize the ripple content in the rectifier output.

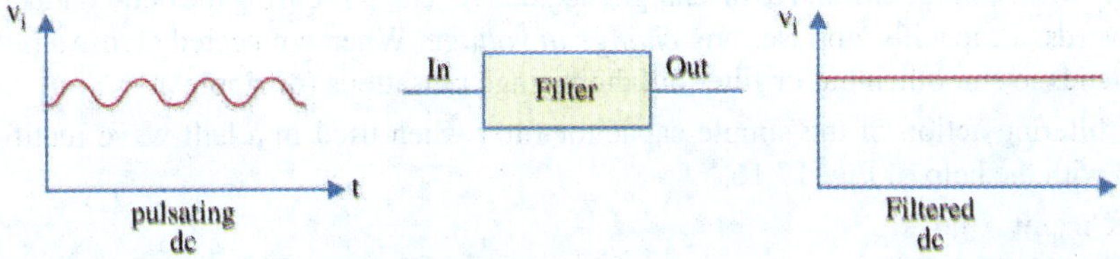

Fig. 17.14.

As seen, output of various rectifier circuits is pulsating. It has a dc value and some ac variations called ripples. This type of output is not useful for driving sophisticated electronic circuits / devices. In fact, these circuits require a very steady dc output that approaches the smoothness of a battery's output.

A circuit that converts a pulsating output from a rectifier into a very steady dc level is known as filter because it filters out or smoothens out the pulsations in the output.

We will consider the following popular filter circuits :

1. series inductor filter,
2. shunt capacitor filter,
3. *LC* filter (or *L*-type filter),
4. π-filter.

17.10. Series Inductor Filter

An inductor (also called choke) is inserted between the rectifier and the load (Fig. 17.15). This inductor presents high impedance to the ac variations (*i.e.*, ripple current) and none to dc current. This action is due to the fundamental property of an inductor *i.e.*, *to oppose any changes in the current passing through it.* Hence, the sudden changes are evened out or toned down.

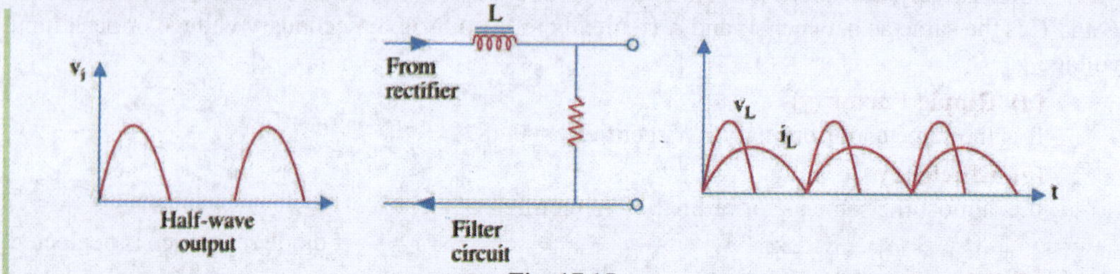

Fig. 17.15

As shown in Fig. 17.15 when output current tends to rise above the average value, magnetic energy is stored in the inductor which has the effect of toning down the sudden rise in current. However, when circuit current tends to fall below the average value, this stored energy is returned to the circuit in an effort to prevent the current from falling down too much. In this way, current variations are reduced to the minimum.

The ripple factor for such a filter is given by

$$\gamma = \frac{R_L}{3\sqrt{2}\ \omega L} = \frac{R_L}{1330L} \qquad \text{— for } f = 50 \text{ Hz}$$

Here, L is in henrys and RL in ohms.

As seen γ varies directly with load resistance R_L and inversely with inductance L.

17.11. Shunt Capacitor Filter

In this circuit, a suitable single capacitor C is connected across the rectifier and in parallel with the load R_L to achieve filtering action. This type of filter is known as *capacitor input filter.*

This filter circuit depends for its operation on the property of a capacitor to charge up (*i.e.*, store energy) when conducting and to discharge (*i.e.*, deliver energy) during the non-conduction cycle. In simple words, a capacitor opposes any *change in voltage.* When connected across a pulsating dc voltage, it tends to smoothen out or filter out the voltage pulsations (or ripples).

The filtering action of the simple capacitor filter when used in a half-wave rectifier can be understood with the help of Fig. 17.16.

(*a*) Circuit Analysis

When positive half-cycle of the ac input is applied, the diode is forward-biased and hence is turned ON. This allows C to quickly charge up to V_m [point *b* in Fig. 17.16 (*c*)] because charging time constant is almost zero. It is so because there is no resistance in the charging path except diode forward resistance which is negligible. Hence, capacitor follows the charging voltage as shown. After being fully charged, the capacitor holds the charge till input ac supply to the rectifier goes negative.

During the negative half-cycle, the capacitor attempts to discharge. However, it cannot discharge through D which, being now reverse-biased, is OFF. Hence, it discharges through R_L from point b to c in Fig. 17.16 (c) and its voltage decreases somewhat. The discharging time constant ($= CR_L$) is usually 100 times more than the charging time. Hence, C does not have sufficient time to discharge appreciably. It is seen that even during negative half-cycle of the input supply, the capacitor maintains large voltage across R_L.

During the next positive half-cycle, when rectifier voltage exceeds the capacitor voltage represented by point c in Fig. 17.16 (c), C is again charged quickly to V_m as represented by point d. Once more, input voltage goes negative, opening the diode and forcing C to discharge through R_L during the interval de. In this way, R_L *sees nearly constant dc voltage across it at all times.*

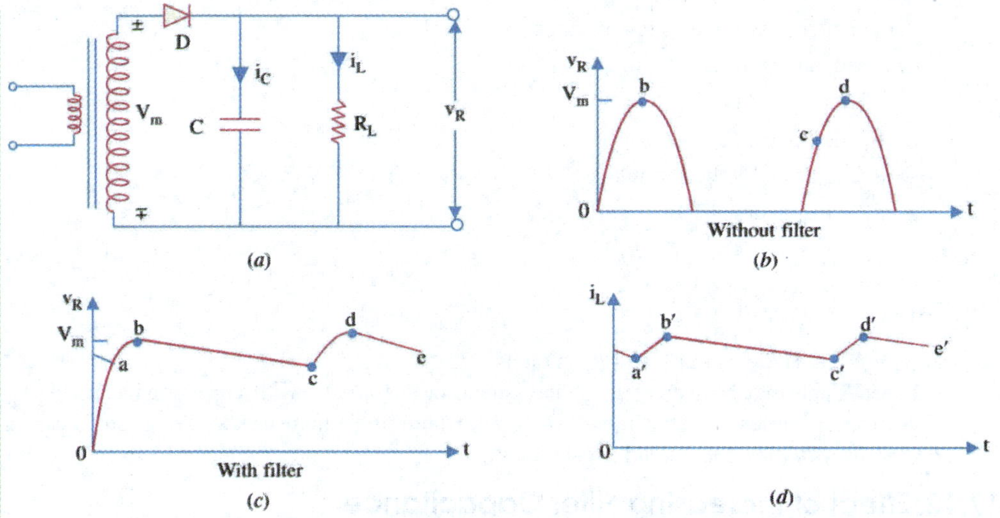

Fig. 17.16

The filtering action of this simple capacitor filter on a full-wave rectifier is shown in Fig. 17.17. It is seen that as compared to a HW rectifier

(*i*) dc load voltage increases slightly towards V_m,

(*ii*) ripple voltage has been reduced by half.

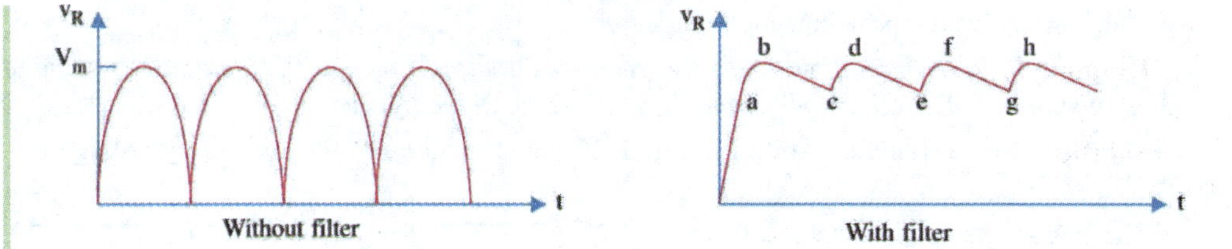

Fig. 17.17

This decreased ripple is because of shorter discharge time before the capacitor is re-energised by another pulse of current.

(*b*) Load Current

The load current has the same waveshape as v_R because load is purely resistive. It is shown in Fig. 17.16 (d). During periods $a'b'$ and $c'd'$ etc. current is supplied by the diode and during period $b'c'$ and $d'e'$ etc., by the capacitor.

(c) Diode Current

Diode current flows during short intervals of time like *ab* and *cd* etc., in Fig. 17.16 (*c*) which is reproduced in Fig. 17.18. During these intervals, diode output voltage is greater than the capacitor voltage which is also the load voltage. Hence, diode current is a surging current *i.e.*, it takes the form of short-duration pulses as shown in Fig. 17.18. A small resistor is always connected in series with the diode to limit this surge current. It is known as surge limiting resistor.

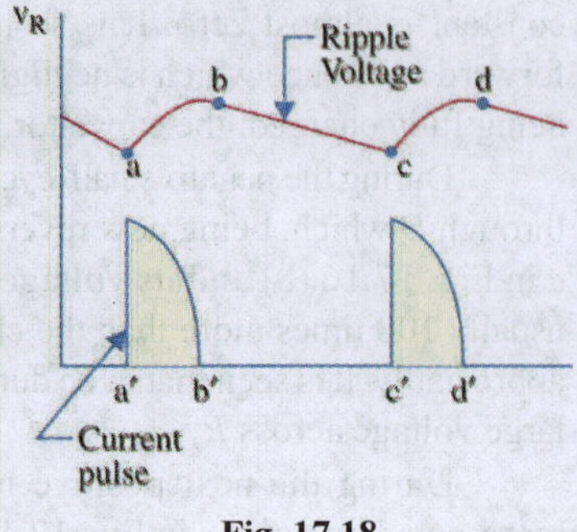

Fig. 17.18

The sole function of the diode is to recharge C and the sole function of C is to supply load current by discharge.

(d) PIV

When diode is reverse-biased, the voltage across it is the sum of

(*i*) capacitor voltage which is almost equal to V_m,

(*ii*) secondary, voltage V_m.

Hence, *PIV* is $2V_m$.

(e) Ripple Factor

Its value is given by the formula

$$\gamma = \frac{1}{4\sqrt{3}\, f\, CR_L}$$

Also, $$\gamma = \frac{2890}{CR_L}$$

where f = 50 Hz, C is in μF and R_L in ohms.

As seen, ripple may be decreased by increasing C or R_L (or both) with a resulting increase in V_{dc}. In contrast to series inductor filter (Art. 17.10) where ripple increases with R_L in shunt capacitor, it *decreases with increase* in load resistance R_L.

17.12. Effect of Increasing Filter Capacitance

A capacitor has the *basic property of opposing changes in voltage.* Hence, a bigger capacitor would tend to reduce the ripple magnitude. It has been found that increasing the capacitor size.

1. increases V_{dc} towards the limiting value V_m,
2. reduces the magnitude of ripple voltage,
3. reduces the time of flow of current pulse through the diode,
4. increases the peak current in the diode.

Example 17.5. *In the filtered half-wave rectifier circuit of Fig. 17.19, calculate the value of the dc output voltage. The component values are as shown in the figure.*

Solution. Since transformation ratio is 1/20, the rms value of the secondary voltage is = 240 × 1/20 = 12 V.

The maximum or peak value of the secondary voltage.

$V_m = 12 \times \sqrt{2} = 17$ V

Taking diode drop of 0.7 V, the peak or maximum value of the secondary voltage is

$V_m = 17 - 0.7 = 16.3$ V

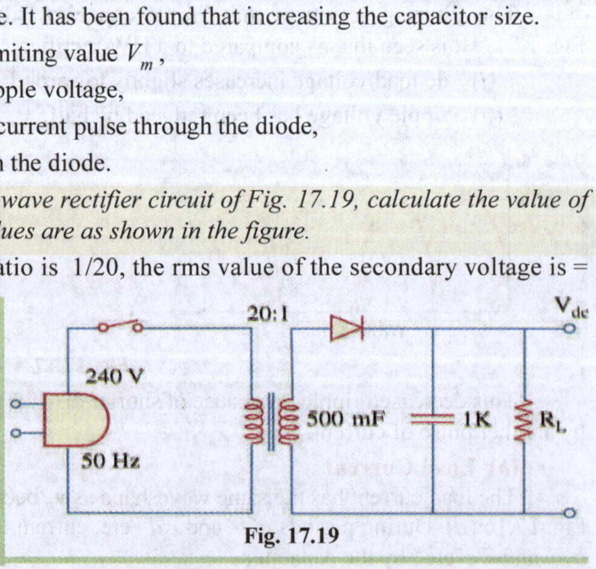

Fig. 17.19

Maximum value of the diode current is

$$I_m = 16.3 \text{ V/1 K} = 16.3 \text{ mA}$$

The peak-to-peak ripple voltage is given by

$$V_{r\,(p-p)} = \frac{I_m}{fC}$$

where f is the ripple frequency in Hz and C is the value of filter capacitor in farad. In a half-wave rectifier, ripple frequency equals the supply frequency.

$$\therefore \quad V_{r\,(p-p)} = \frac{16.3 \text{ mA}}{50 \text{ Hz} \times 500 \text{ mF}} = 652 \text{ mV}$$

Magnified view of the ripple voltage is shown in Fig. 17.20.

$$\therefore \quad V_{dc} = 16.3 - (0.652 / 2) = \mathbf{16\ V}$$

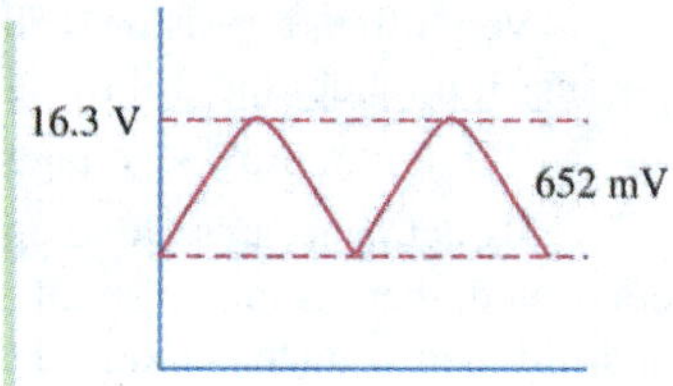

Fig. 17.20

17.13. LC Filter

It combines the advantages of series inductor and shunt capacitor filters. It is also called *L*-type choke input filter. This name due to the fact that the inductor and is capacitor are connected as an inverted *L* (Fig. 17.21).

The output of this filter has very little ripple and good regulation.

The action of this filter may be described by considering it to be a low-pass filter. The capacitor offers very low reactance to the ac ripple current and hence shunts most of it to ground. But it appears *as an open to the dc current* and force it to pass through inductor and hence R_L and finally to ground. In this way, most of the ripple voltage is eliminated from the load voltage.

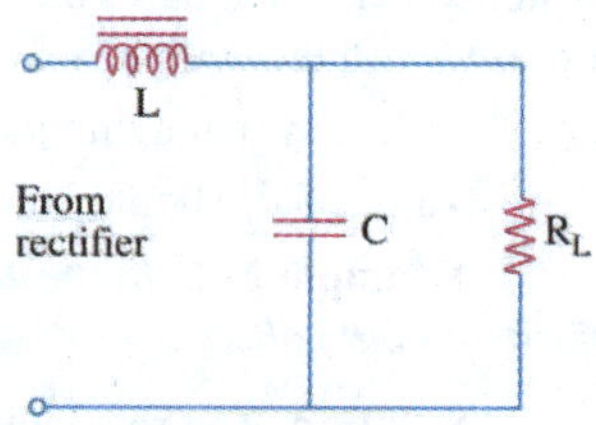

Fig. 17.21

These days small and inexpensive transistor voltage regulator circuits are easily available. Hence, need for filters using large inductors is on the decline.

The ripple factor of this circuit is

$$\gamma = \frac{\sqrt{2}}{12\,\omega^2\,LC} = \frac{1.19}{LC} \qquad \text{— for } f = 50 \text{ Hz}$$

Here, L is henrys and C is in μF.

17.14. The CLC or Pi Filter

As shown in Fig. 17.22, it consists of one inductor and two capacitors connected across its each end. The three components are arranged in the shape of the Greek letter π. It is also called capacitor input π-filter. The input capacitor C_1 is selected to offer very low reactance to the ripple frequency. Hence, major part of filtering is done by C_1. Most of the remaining ripple is removed by the combined action of L and C_2.

The charging and discharging action of C_1 is exactly the same as described in Art. 17.11. The output voltage waveform is also like that shown in Fig. 17.16 (*b*).

This circuit gives much better filtering than *LC* filter circuit. However, C_1 is still directly connected across the supply and would need high pulses of current if load current is large. Since these high peak current pulses are likely to damage the rectifier diode, this filter is used *with low-current equipment.*

Though this filter gives somewhat higher output voltage, its voltage regulation is inferior to that of the *LC* filter.

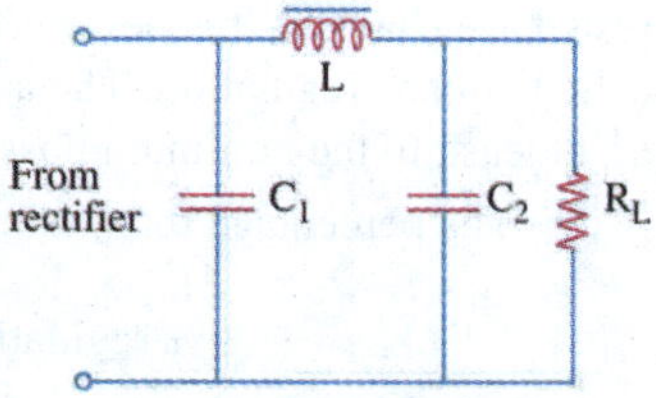

Fig. 17.22

The ripple factor of this filter is given by

$$\gamma = \frac{\sqrt{2}}{12\,\omega^3\, C_1 C_2\, LR_L} = \frac{5700}{LC_1\, C_2\, R_L} \quad \text{— when } f = 50 \text{ Hz}$$

Here, C_1, C_2 are in µF, L in henrys and R_L in ohms.

17.15. Bleeder Resistor

Very often, a resistor (called bleeder resistor) is placed across the filter output (Fig. 17.23) because it provides the following advantages :

1. It improves voltage regulation of the supply.

By acting as a pre-load on the supply, it causes an initial voltage drop. When the real load is connected, there is only a small amount of additional drop. In this way, difference between no-load and full-load voltage is reduced thereby improving the regulation.

2. It provides safety to the technicians handling the equipment.

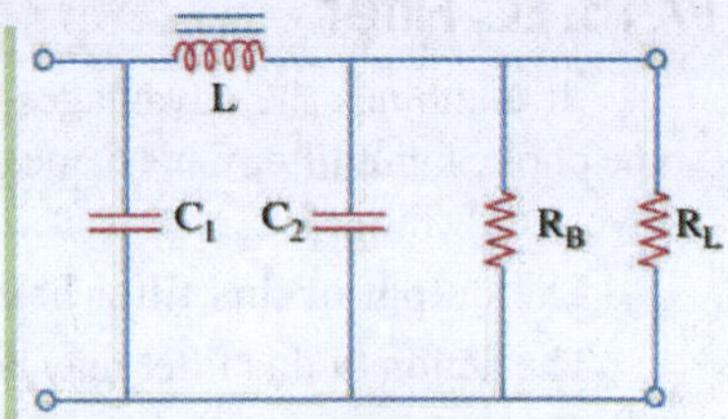

Fig. 17.23

When power supply is switched off, it provides a path for the filter capacitor to discharge through. That is why it is called *bleeder* resistor. Without it, the capacitor will retain its charge for quite sometimes even when the power supply is switched off. This high voltage can be dangerous for people working with the equipment.

3. By maintaining a minimum current through the choke, it improves its filtering action.

Value of R_B should be such as to conduct 10 per cent of the total current.

Example 17.6. *In the unfiltered full-wave rectifier circuit of Fig. 17.24, calculate the value of the dc output voltage.*

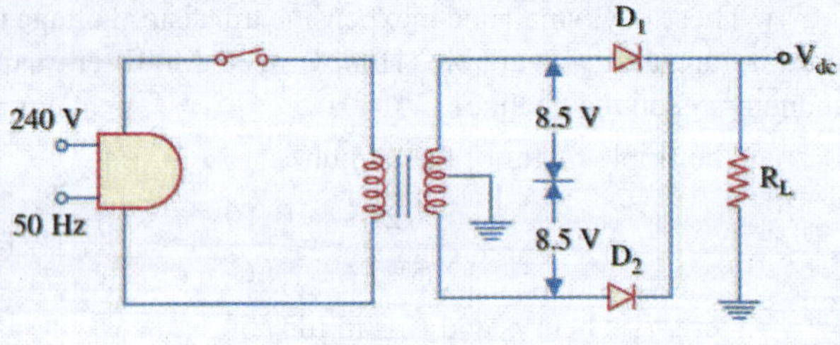

Fig. 17.24

Solution. The rms value of the secondary voltage is given by (240 × 1/20) = 12 V. Since the secondary has a centre-tap, voltage across each secondary half is 12/2 = 6 V. The peak or maximum value of the centre-tap voltage is $6 \times \sqrt{2} = 8.5$V.

Considering the diode drop of 0.7 V, the maximum value of the voltage across the load resistance is 8.5 – 0.7 = 7.8 V.

The dc output voltage is given by

$$V_{dc} = 0.636 \times 7.8 = \mathbf{4.96\ V}$$

17.16. Voltage Regulation

In an unregulated power supply, output voltage changes whenever input supply voltage or load resistance changes. It is never constant. The change in voltage from no-load to full-load condition is called voltage regulation. The aim of a voltage regulator circuit is to reduce these variations to zero or, at least, to the minimum possible value.

The percentage regulation or, simply, regulation of a power supply is given by

$$\%\ \text{regulation} = \frac{V_{max} - V_{min}}{V_{max}} \times 100$$

where V_{max} = maximum dc output voltage

V_{min} = minimum dc output voltage

When we say that 10 V regulated dc power supply has a regulation of 0.005 per cent, it means that dc output voltage will vary within an envelope 0.005 per cent of 10 V.

Now, $$0.005\% \text{ of } 10\text{ V} = \frac{0.005 \times 10}{100} = 0.0005\text{ V} = 0.5\text{ mV}$$

Hence, output voltage will vary by ± 0.25 mV. So, we see that instead of expressing voltage regulation by unwieldly expression 0.005 per cent, we can express it by a simple figure ± 0.25mV.

In general, $$\%\text{ regn.} = \frac{V_{NL} - V_{FL}}{V_{FL}} \times 100$$

where V_{NL} = no-load or open-circuit terminal voltage of the supply

V_{FL} = full-load terminal voltage of the supply

17.17. Zener Diode Shunt Regulator

A simple shunt voltage regulating system using a Zener diode is shown in Fig. 17.25. The input voltage V_{in}, in fact, is the unregulated output of a rectifier. This simple regulator restricts output voltage variations within reasonable limits around V_Z in the face of changing load current or changing input voltage. Obviously, the Zener diode will regulate *so long as it is kept in reverse conduction.*

Example 17.7. *The Zener diode of Fig. 17.25 has the following ratings :*

$V_Z = 6.8\ V$ at $I_Z = 50\ mA$

$r_z = 2\Omega$ at $I_Z = 50\ mA$

$I_{Z(min)} = 5\ mA,$ $I_{Z(max)} = 150\ mA.$

What would be the load voltage when load current I_L varies from 10 mA to 120 mA? Also calculate voltage regulation of the regulator.

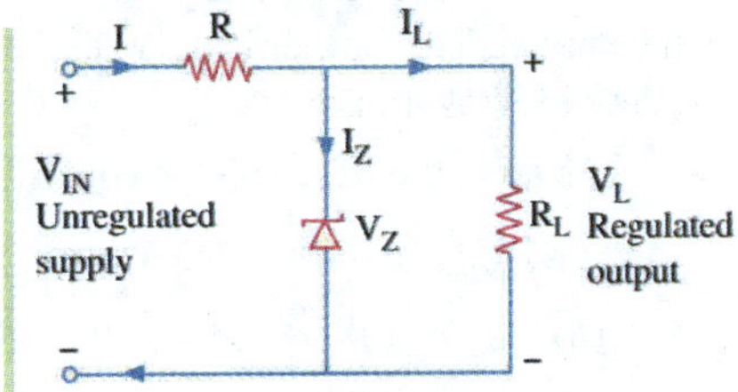

Fig. 17.25

Solution. We will call V_Z = 6.8 V and I_Z = 50 mA as *reference values* and calculate changes in voltage with respect to these values.

(*i*) $\boldsymbol{I_L}$ **= 120 mA**

Obviously, I_Z = 150 – 120 = 30 mA

Deviation from 50 mA = 30 – 50 = – 20 mA

Drop across diode $= I_Z r_Z = -20 \times 2 = -40$ mV

∴ $V_L = V_Z + I_Z r_Z = 6.8 + (-40 \times 10^{-3}) = \mathbf{6.76\ V}$

(*ii*) $\boldsymbol{I_L}$ **= 10 mA**

Now, I_Z = 150 – 10 = 140 mA

Deviation from reference value

= 140 – 50 = 90 mA

Diode drop $= I_Z + I_Z r_Z = 90 \times 10^{-3} \times 2 = \mathbf{0.18\ V}$

∴ $V_L = V_Z + I_Z r_Z = 6.8 + 0.18 = \mathbf{6.98\ V}$

$$\%\text{ regn.} = \frac{6.98 - 6.76}{6.98} \times 100 = \mathbf{3.15\%}$$

For many applications, a change in load voltage of 3.15% is acceptable but, in some, it may be intolerable. This regulation can be reduced to 1% or less with the help of circuits discussed below.

17.18. Transistor Series Voltage Regulator

The circuit is shown in Fig. 17.26. It is known as emitter follower because voltage at the emitter follows the base voltage. In this set up, the transistor behaves like a variable resistor whose resistance is determined by the base current. It is also called pass transistor because all the load current must pass through it.

Keeping in mind the polarities of different voltages, they are related by the equation

$$V_L + V_{BF} - V_Z = 0$$

— Kirchhoff's Voltage law

$$\therefore \quad V_{BE} = V_Z - V_L$$

(fixed)

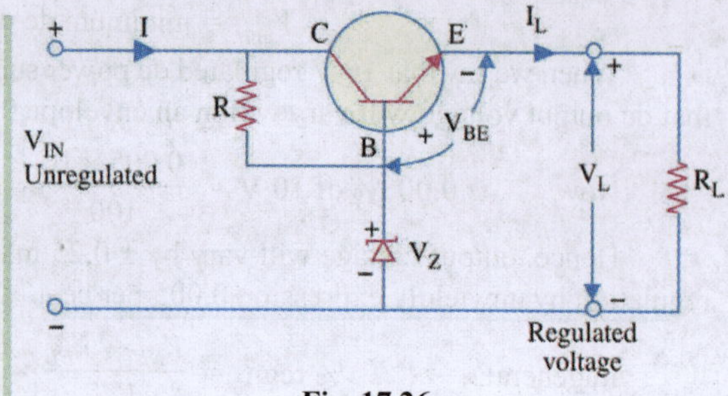

Fig. 17.26

When current demand is increased by decreasing R_L, V_L tends to decrease. As seen from the above equation, it will increase V_{BE} because V_Z is fixed. This will increase forward bias of the transistor thereby increasing its level of conduction. This will lead to decrease in the collector-emitter resistance of the transistor which will slightly increase the input current to compensate for decrease in R_L so that $V_L = (I_L R_L)$ will remain at constant value.

Example. 17.8. *In the emitter-follower regulation circuit of Fig. 17.27, calculate the values of (a) V_L (b) I_L (c) I_E (d) V_{CE} and (e) power dissipated by Q_1.*

Solution. (*a*) The load voltage V_L is equal to the voltage at the emitter. The emitter is 0.7V less than the base voltage in the case of *NPN* transistors.

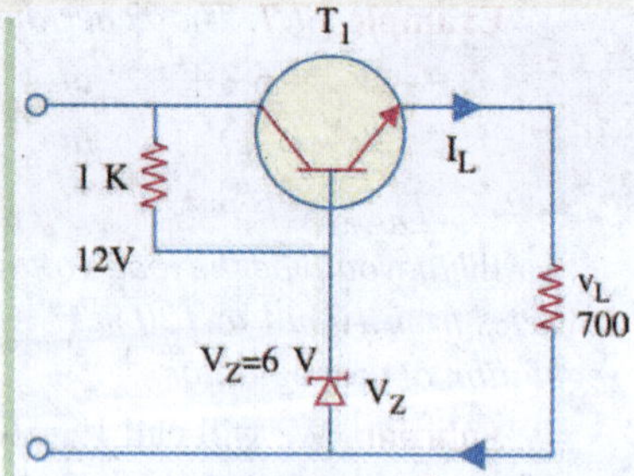

Fig. 17.27

As seen, the base voltage equals the Zener voltage.

$$\therefore \quad V_L = 6 - 0.7 = \mathbf{5.3\ V}$$

(*b*) $I_L = V_L / R_L = 5.3 / 100 = \mathbf{53\ mA}$

(*c*) $I_E = I_L = \mathbf{53\ mA}$

(*d*) $V_{CE} = V_{in} - V_L = 12 - 5.3 = \mathbf{6.7\ V}$

(*e*) Power dissipated by the passive transistor Q_1 is

$$= V_{CE} \times I_E = 6.7 \times 53 \times 10^{-3} = \mathbf{0.355\ W}$$

17.19. Controlled Transistor Series Regulator

The circuit employing a second transistor T_2 for control purposes is shown in Fig. 17.28. It has the additional feature of an output control with the help of potentiometer $R_1 - R_2$.

In the discussion to follow, it will be assumed that *I* is much greater than I_{B2}.

Now, there is a drop of V_L on $(R_1 + R_2)$ and drop of $(V_Z + V_{BE2})$ across R_2.

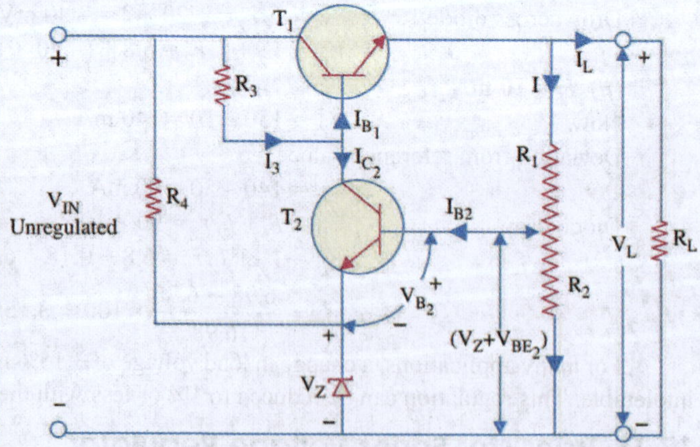

Fig. 17.28

$$\therefore \quad \frac{V_L}{V_Z = V_{BE2}} = \frac{R_1 + R_2}{R_2}$$

or $$V_L = \frac{R_1 + R_2}{R_2}(V_Z + V_{BE2})$$

Now, $(R_1 + R_2)$ and $(V_Z + V_{BE2})$ both have constant values so that $V_L \propto 1/R_2$. If the potentiometer is adjusted so that R_2 decreases, then V_L increases and *vice versa*.

Suppose R_L is decreased. Then, I_L increases but V_L decreases. Decrease in V_L decreases I_{B2} and also I_{C2}. Assuming I_3 to be relatively constant (or decreasing slightly), I_{B1} is increased thereby decreasing the terminal (collector-emitter) resistance of T_1. This leads to decrease in V_{CE1} thereby offsetting the decrease in V_L which is, therefore, returned to its original value.

In sequential, we have

$$V_L \downarrow I_{B2} \downarrow I_{C2} \downarrow I_{B1} \uparrow V_{CE1} \downarrow V_L \uparrow$$

17.20. Transistor Shunt Voltage Regulator

It employs the transistor in shunt configuration as shown in Fig. 17.29.

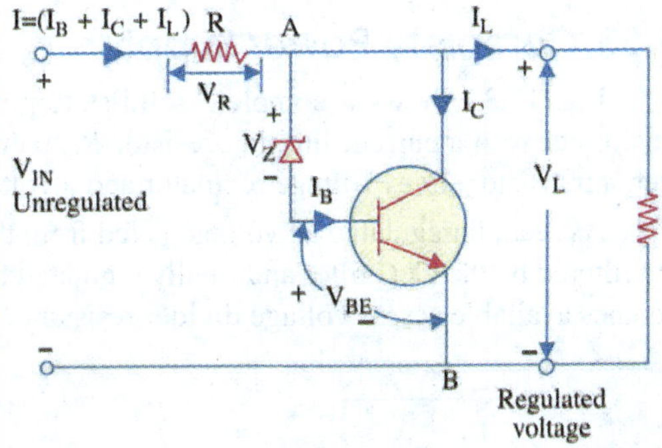

Fig. 17.29

Since path AB is in parallel across V_L, we have

$$V_L - V_Z - V_{BE} = 0$$

or $$V_{BE} = V_l - V_Z$$

— Kirchhoff 's Voltage Law

Since V_Z is fixed, any decrease or increase in V_L will have a corresponding effect on V_{BE}. Suppose, V_L decreases, then as seen from the above relation V_{BE} also decreases. As a result, I_B decreases, hence I_C (= β_{IB}) decreases, thereby decreasing I and hence V_R (= IR). Consequently, V_L increases because at all times

$$V_{in} = V_R + V_L \quad \text{or} \quad V_L = V_{in} - V_R$$

In sequential logic

$$V_L \downarrow V_{BE} \downarrow I_B \downarrow I_C \downarrow I_R \downarrow V_R \downarrow V_L \uparrow$$

Same line of logic applies in case V_L tries to increase.

17.21. Transistor Current Regulator

The main function of a current regulator is to maintain a fixed current through the load despite variations in the terminal voltage. Such a circuit employing a Zener diode and a *PNP* transistor is shown in Fig. 17.30. Suppose due to drop in V_L, current I_L (= I_C) is decreased. This will decrease I_E ($\cong I_C$). Hence, drop across R_E *i.e.*, V_{RE} will decrease. As seen

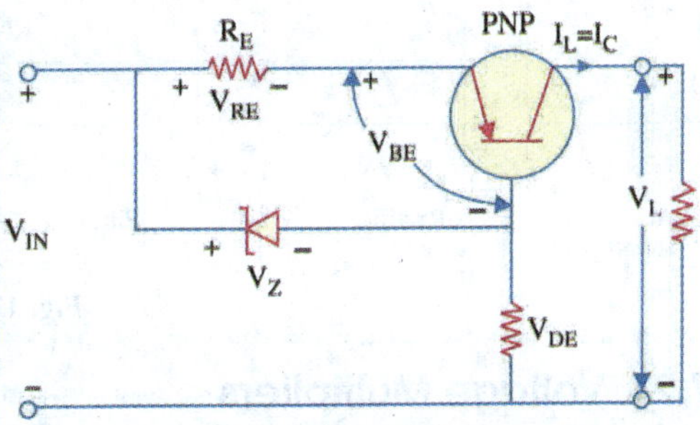

Fig. 17.30

$$-V_{RE} - V_{BE} + V_Z = 0$$

—Kirchhoff 's Voltage Law

or $$V_{BE} = V_Z - V_{RE}$$
(fixed)

Hence, a decrease in V_{RE} will increase V_{BE} and, hence, the conductivity of the transistor thereby keeping I_L at fixed level.

A similar logic applies when there is increase in V_L.

17.22. Voltage Dividers

Often more than one dc voltage is needed for the operation of electronic circuits. A single power supply can provide as many voltages as are needed by using a voltage divider. As shown in Fig. 17.31, a voltage divider is a single tapped resistor connected across the output terminals of the supply. The tapped resistor may consist of two or three resistors connected in series across the supply. In fact, bleeder resistor may also be used as a voltage divider.

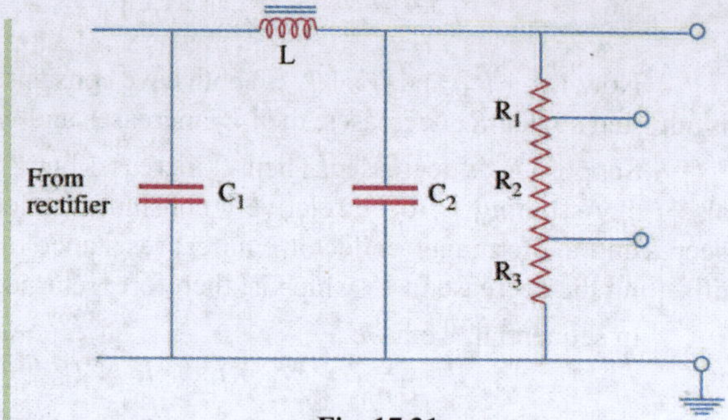

Fig. 17.31

17.23. Complete Power Supply

Fig. 17.32 shows a complete solid-state power supply. From left to right, it consists of a transformer with a current-limiting resistor R_1, rectifier diodes for full-wave rectification, a π-type filter, a transistor series voltage regulator and a voltage divider.

As seen, unregulated ac voltage is fed from the transformer through a full-wave rectifier. It is then filtered by the *CLC* filter and finally regulated by a transistor regulator. The regulated dc supply becomes available across voltage divider resistance R_B. The output is practically ripple-free.

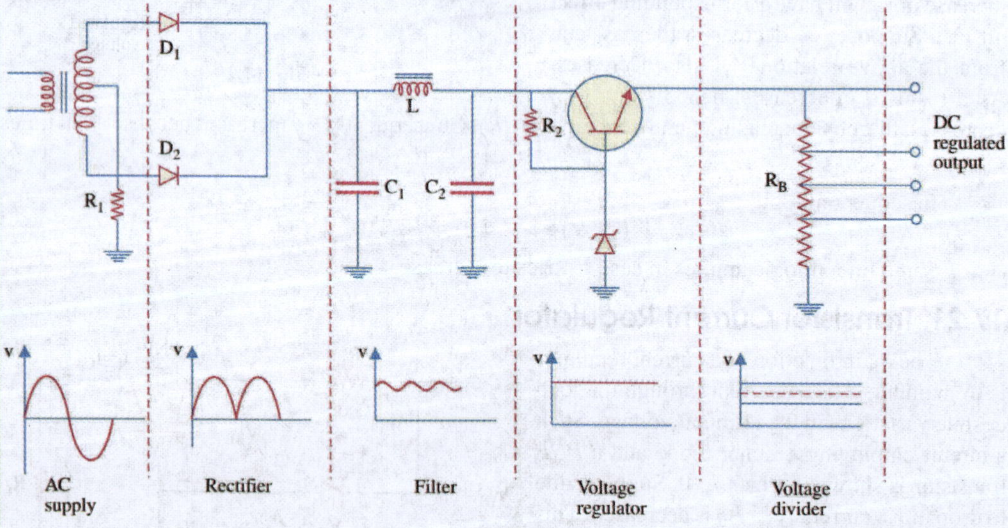

Fig. 17.32

17.24. Voltage Multipliers

A voltage multiplier is a circuit which produces a greater dc output voltage than ac input voltage to the rectifiers. Multipliers are required in many circuit applications where it is necessary to have high voltage with low current as for accelerating purposes in a cathode ray tube (CRT).

We will consider the following circuits :

1. half-wave voltage doubler,
2. full-wave voltage doubler,
3. voltage tripler,
4. voltage quadrupler.

17.25. Half-wave Voltage Doubler

It is also known as cascade voltage doubler. The circuit is shown in Fig. 17.33.

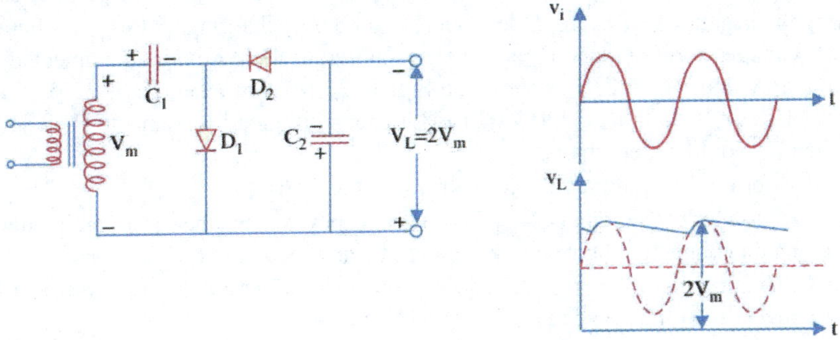

(a) (b)

Fig. 17.33

Circuit Analysis

During the positive half-cycle of the input voltage, D_1 conducts (not D_2) and charges C_1 to peak value of secondary voltage (V_m) with the polarity as shown in Fig. 17.33 (a). During the negative half cycle, D_2 conducts (not D_1) and charges C_2. The voltage across C_2 is the sum of peak supply voltage and the voltage across C_1 (Fig. 17.34). It can be proved by applying KVL to the outer loop. Starting from the bottom of the transformer secondary in Fig. 17.34 and going clockwise, we get

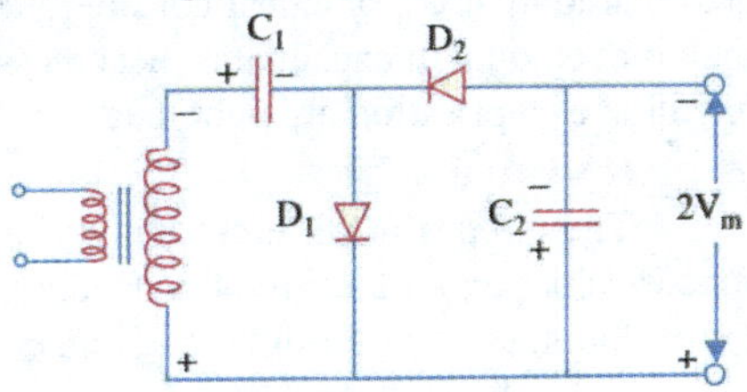

Fig. 17.34

$$-V_m - V_m + V_{C_2} = 0$$

or $$V_{C_2} = 2V_m = 2 \times \text{peak input voltage}$$

During the next positive half-cycle, D_2 is open and C_2 will discharge through the load if it is connected. If no load is connected across C_2, then both capacitors stay charged *i.e.*, C_1 to V_m and C_2 to $2V_m$. If there is a load connected across C_2, it will discharge a little bit and, as a result of it, voltage across it will drop slightly. But it will get recharged in the next half-cycle.

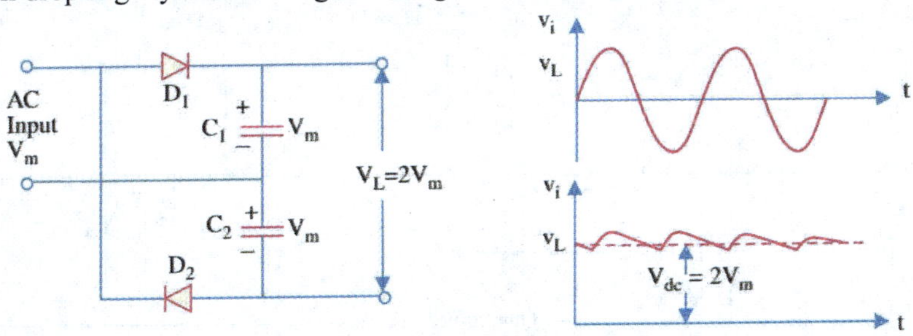

(a) (b)

Fig. 17.35

The output waveform shown in Fig. 17.33 (b) is that of a half-wave rectifier filtered by a shunt capacitor. *PIV* across each diode is 2 V_m. Ripple frequency is equal to the supply frequency.

This circuit has very poor regulation and its ripple content is also high. This circuit has a common connection between the line and load (which a full-wave doubler does not have).

17.26. Full-wave Voltage Doubler

This circuit is shown in Fig. 17.35 (*a*). During the positive half-cycle of the input voltage, D_1 conducts (but not D_2) and charges capacitor C_1 to the peak voltage V_m with the polarity as shown.

During the negative half-cycle, D_2 conducts (but not D_1) charging C_2 to V_m. As far as the load is concerned, voltages across C_1 and C_2 are in series-aiding. If there is no load connected across the output, then load voltage $V_L = 2V_m$ as shown in Fig. 17.35 (*a*). For example, if 220V, 50 Hz is the supply, then $V_{dc} = 2V_m = 2 \times \sqrt{2} \times 220 = 620$ V. Of course, if a load is connected across the output terminals, then V_L would be less than 2 V_m.

The waveform of the output voltage is shown in Fig.17.35 (*b*).

The *PIV* rating of each diode is $2V_m$. Ripple frequency is *twice* the supply frequency. As seen, there is no common connection between the supply line and the load.

It is worth noting that where the expense of a line transformer is justified, it is preferable to use the superior conventional full-wave rectifier (Art. 17.7).

17.27. Voltage Tripler And Quadrupler Circuits

(*a*) General

The half-wave voltage doubler circuit (Fig. 17.33) can be extended to obtain any multiple of the peak input voltage (V_m) *i.e.*, $3V_m$, $4V_m$, $5V_m$ etc. Theoretically speaking, there is no upper limit to the amount of voltage multiplication that can be obtained. Though voltage triplers and quadruplers are commonly used, practical considerations limit additional multiplications. The main handicap is that total amount of capacitance becomes unduly large to maintain the desired dc output voltage for anything except extremely light loads.

(*b*) Circuit

The circuit for different multipliers is shown in Fig. 17.36. It should be obvious from the repetitive pattern of the circuit connections how additional diodes and capacitors may be connected to the doubler circuit for obtaining higher multiplications of the peak output voltage V_m.

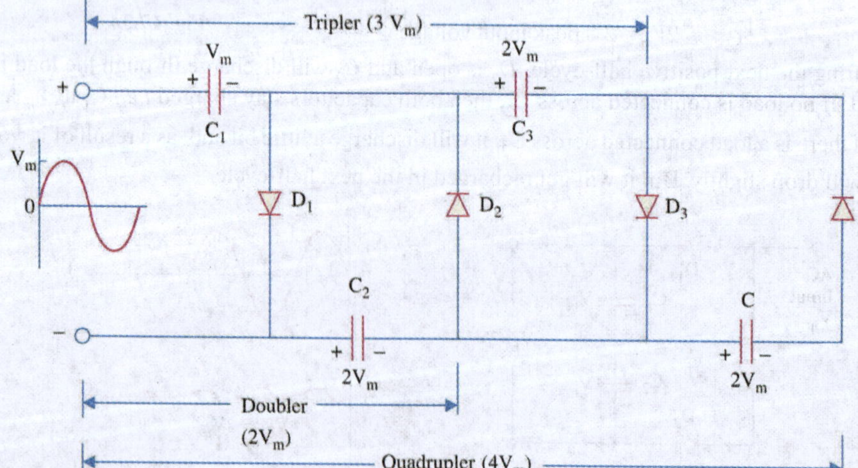

Fig. 17.36

(*c*) Analysis

During the first positive half-cycle, C_1 charges to V_m as diode D_1 conducts. During negative half-cycle, C_2 is charged through D_2 to 2 V_m *i.e.*, to the sum of voltage across C_1 and peak input voltage (Art. 17.6).

During the second positive half-cycle, D_3 conducts and voltage across C_2* charges C_3 to same voltage across V_m. During the negative half-cycle, diodes D_2 and D_4 conduct allowing C_3 to charge C_4 to the same peak voltage 2 V_m.

It is seen from Fig. 17.36 that voltage across C_2 is 2 V_m, across C_1 and C_3 is 3 V_m and across C_2 and C_4 is $4V_m$.

If additional diodes and capacitors are used, each capacitor would be charged to a peak voltage of $2V_m$.

The *PIV* rating of each diode is $2V_m$ and ripple frequency is twice the line frequency. Generally, these circuits are used where both the supply voltage and load are maintained constant.

17.28. Troubleshooting Power Supplies

There are usually two types of problems with power supplies *i.e.*, either no dc output or low dc output.

The situation of no dc output can occur due to any one of the following reasons :

1. when there is no output from the rectifiers,
2. when there is no ac input to power supply,
3. when filter choke is open,
4. when the first input capacitor shorts.

A low dc output can occur in the following situations :

1. decreased input ac voltage,
2. open input capacitor of the filter circuit,
3. partial short across the load.

17.29. Controlled Rectification

It is that rectification in which the output of a rectifier circuit can be varied by controlling the point in the ac cycle at which the circuit is turned ON. A thyristor or *SCR* can be used for the purpose because under proper firing conditions, it can control the conduction angle of the rectifier circuit.

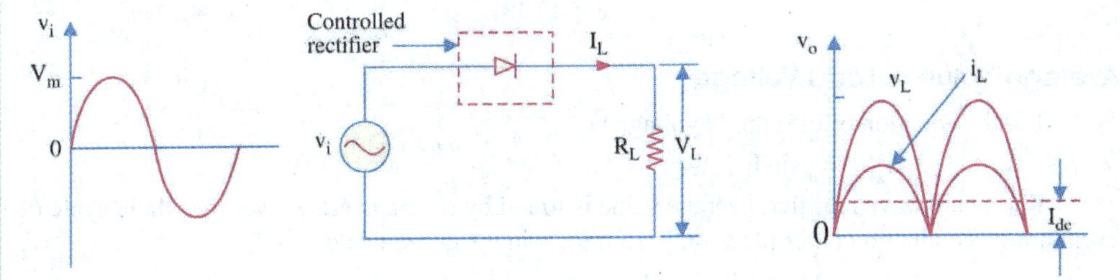

Fig. 17.37

In an ordinary diode rectifier circuit, current flows through the diode whenever *instantaneous value of the ac supply voltage is greater than the voltage across the load at that instant.*

For a resistive load shown in Fig. 17.37, load current flows at all times during the positive half-cycles of the supply.

In a controlled rectifier, on the other hand, load current flows *only when a control signal is*

* C_1 cannot charge C_3 because it is shorted by D_1.

applied to turn on the rectifier at a specific point [like A in Fig. 17.37 (*b*)] in the ac cycle. Point *A* corresponds to the angle θ_1 so that condition is delayed by this much period.

Once the HW rectifier is turned ON, it remains in conduction for the rest of the positive half-cycle *i.e.*, upto 180°. Obviously, the firing point *A* is determined by the angle of delay in applying the firing signal by the control circuit. As θ_1 increases, condition *occurs later in the cycle* thereby decreasing the load current further.

The curve for voltage drop across rectifier diode is shown in Fig. 17.38 (*c*). During the positive half-cycle when the rectifier is fired into conduction, it acts like a short and voltage across it drops to zero (neglecting forward voltage of the diode). During negative half-cycle when diode is reverse-biased (and hence open), the full supply voltage appears across it as shown.

The curve for load voltage is shown in Fig. 17.38 (*d*). Since current obeys Ohm's law, it follows the load voltage.

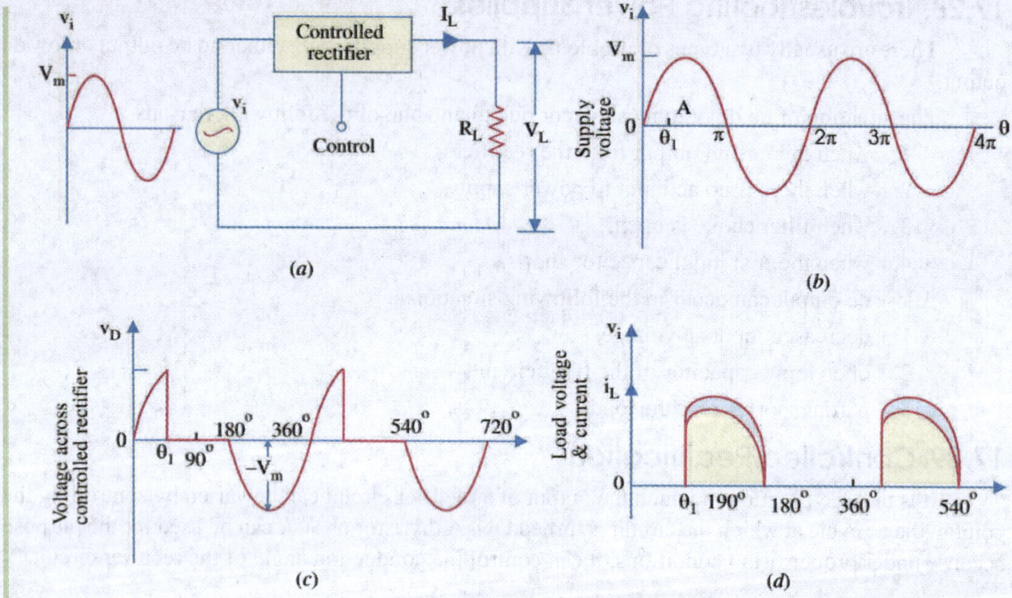

Fig. 17.38

Average Value of Load Voltage

Let the equation of the supply voltage be

$$v_i = V_m \sin \theta$$

For a half-wave rectifier, average value is found by taking average over the whole cycle (2π) even though conduction takes place only from θ_1–π in the entire cycle.

$$\therefore \quad V_L = \frac{1}{2\pi}\int_{\theta_1}^{\pi} v_i \, d\theta = \frac{1}{2\pi}\int_{\theta_1}^{\pi} V_m \sin\theta \, d\theta$$

$$= \frac{V_m}{2\pi}\int_{\theta_1}^{\pi} \sin\theta . d\theta = \frac{V_m}{2\pi}\left| -\cos\theta_1 \right.$$

$$= \frac{V_m}{2\pi}\left| (\cos\pi) - (-\cos\theta_1) \right. = \frac{V_m}{2\pi}(1 + \cos\theta_1)$$

$$\therefore \quad V_{dc} = V_L = \frac{V_m}{2\pi}(1 + \cos\theta_1)$$

For a resistive load, $I_{dc} = I_L = \dfrac{V_m}{2\pi R_L}(1 + \cos\theta_1)$

Note. If $\theta_1 = 0$, then

$$V_{dc} = \frac{V_m}{\pi} \quad \text{and} \quad I_{dc} = \frac{V_m}{\pi R_L}$$

These are the same values as for an uncontrolled or ordinary HW rectifier (Art. 17.6).

17.30. Output Waveforms For Different Firing Angles

In all waveforms given in Fig. 17.39, it has been assumed that the ac supply voltage is sinusoidal given by the equation $v_i = V_m \sin\theta$. Also a resistive load has been assumed.

Following points must always be kept in mind while drawing these diagrams. When the diode is forward-biased, it conducts and *behaves like a short.* Hence drop across it is *almost zero.* Instead, whole of the applied voltage drops across load resistance. When during the negative input half-cycle, diode is reverse-biased, it does not conduct and behaves like an 'open'. Hence, all the applied voltage appears across the rectifier diode and none across R_L. In other words, diode anode voltage and load voltage are mutually exclusive *i.e., when one is there, the other is not.*

It is seen from Fig. 17.40 that as delay angle θ_1 is increased, the output keeps decreasing till for $\theta_1 = 180°$, *the output is zero.*

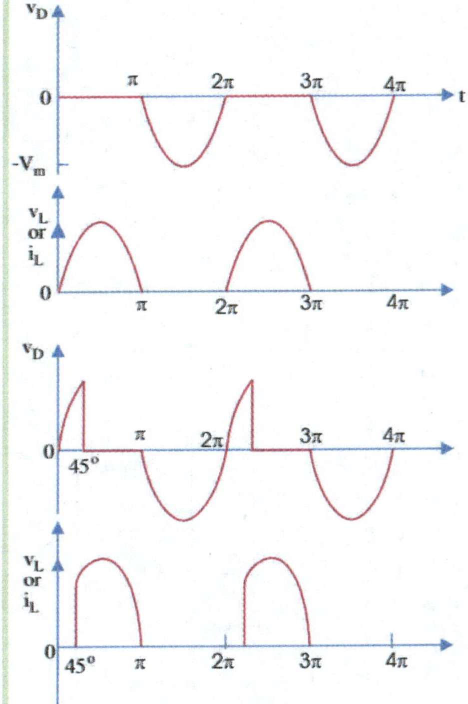

Fig. 17.39

17.31. Output Voltage and Current Values in Controlled Rectifiers

We will use the equation derived in Art. 17.29 for a half-wave rectifier

$$V_{dc} = \frac{V_m}{2\pi}(1 + \cos\theta_1)$$

$$I_{dc} = \frac{V_m}{2\pi R_L}(1 + \cos\theta_1) = \frac{V_{dc}}{R_L}$$

(a) when $\theta_1 = 0°$

$$\cos\theta_1 = \cos 0° = 1$$

$$\therefore \quad V_{dc} = \frac{V_m}{2\pi}(1 + 1) = \frac{V_m}{\pi} = 0.318\, V_m; \qquad I_{dc} = \frac{V_{dc}}{R_L} = \frac{V_m}{\pi R_L} = 0.318\,\frac{V_m}{R_L}$$

(b) when $\theta_1 = 30°$

$$\cos\theta_1 = \cos 30° = 0.866$$

$$\therefore \quad V_{dc} = \frac{V_m}{2\pi}(1 + 0.866) = 0.297\, V_m; \qquad I_{dc} = \frac{V_{dc}}{R_L} = 0.297\,\frac{V_m}{R_L}$$

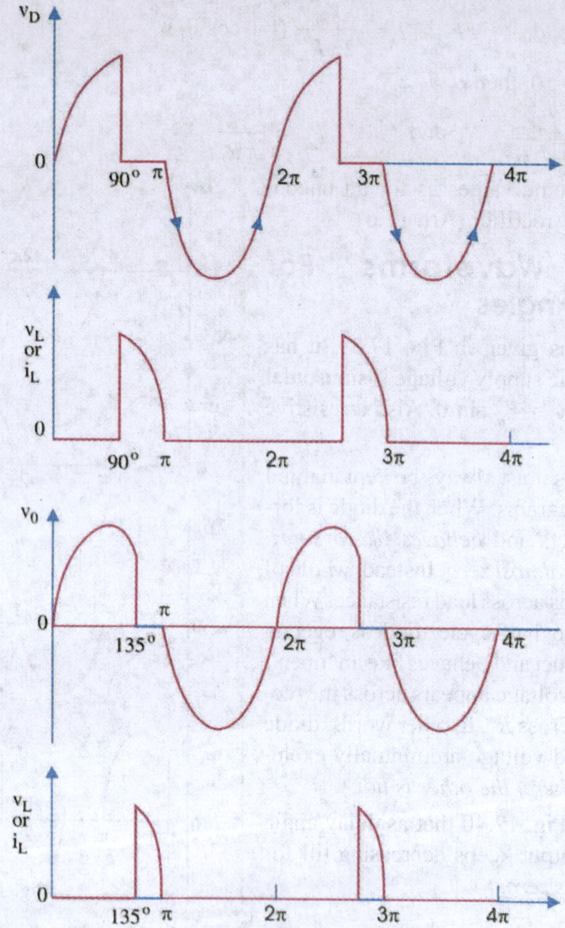

Fig. 17.40

(c) when $\theta_1 = 45°$

$$\cos \theta_1 = \cos 45° = 0.707$$

$$\therefore \quad V_{dc} = \frac{V_m}{2\pi}(1 + 0.707) = 0.27\, V_m; \qquad I_{dc} = \frac{V_{dc}}{R_L} = 0.27\,\frac{V_m}{R_L}$$

(d) when $\theta_1 = 60°$

$$\cos \theta_1 = \cos 60° = 0.5$$

$$\therefore \quad V_{dc} = \frac{V_m}{2\pi}(1.05) = 0.239\, V_m; \qquad I_{dc} = \frac{V_{dc}}{R_L} = 0.239\,\frac{V_m}{R_L}$$

(e) when $\theta_1 = 90°$

$$\cos \theta_1 = \cos 90° = 0$$

$$V_{dc} = \frac{V_m}{2\pi}(1 + 0) = \frac{V_m}{2\pi} = 0.159\, V_m; \qquad I_{dc} = \frac{V_{dc}}{R_L} = \frac{V_m}{2\pi R_L} = 0.159\,\frac{V_m}{R_L}$$

(f) when $\theta_1 = 120°$

$$\cos \theta_1 = \cos 120° = -0.5$$

$\therefore \quad V_{dc} = \frac{V_m}{2\pi}(1 - 0.5) = 0.08\ V_m; \qquad I_{dc} = \frac{V_{dc}}{R_L} = 0.08\frac{V_m}{R_L}$

(*g*) when $\theta_1 = 135°$

$\cos\theta_1 = \cos 135° = -0.707$

$V_{dc} = \frac{V_m}{2\pi}(1 - 0.707) = 0.0466\ V_m; \qquad I_{dc} = 0.0466\frac{V_m}{R_L}$

Example 17.8. *A 100 Ω load is connected to peak supply of 300 V through a controlled half-wave diode rectifier. The load power is to be varied from 25 W to 80 W. What is the angular firing control required ? Neglect forward drop of the diode.*

Solution. $P = V_{dc}.I_{dc}$

Now, $V_{dc} = \frac{V_m}{2\pi}(1 + \cos\theta)$

$$I_{dc} = \frac{V_m}{2\pi R_L}(1 + \cos\theta)$$

$$\therefore \quad P = \left(\frac{V_m}{2\pi}\right)^2 . \frac{1}{R_L}(1 + \cos\theta)^2$$

$$= \left(\frac{300}{2\pi}\right)^2 . \frac{1}{100}(1 + \cos\theta)^2$$

$$= 22.5\,(1 + \cos\theta)^2$$

(*i*) when $P = 80$ W

$\therefore \quad 80 = 22.5\,(1 + \cos\theta)^2$

or $(1 + \cos\theta) = 1.8856$

$\therefore \quad \cos\theta = 0.8856$

or $\theta =$ **27.7°**

(*ii*) when $P = 25$ W

$25 = 22.5\,(1 + \cos\theta)^2$

$\therefore \quad \cos\theta = 0.1111$ or $\theta =$ **83.6°**

Example 17.9. *In a controlled half-wave rectifier, peak supply voltage is 200 V and the value of load resistor is 1 kΩ. Calculate the power delivered to the load circuit for firing angles of (i) 0° (ii) 45° (iii) 90° and (iv) 135°.*

Solution. (*i*) $V_{dc} = 0.318 \times 200 = 63.6$ V; $\qquad I_{dc} = 63.6 / 1000 = 63.6$ mV

$P = V_{dc}.\,I_{dc} = 63.6 \times 63.6 = 4045$ mW = **4.045 W**

(*ii*) $V_{dc} = 0.27 \times 200 = 54$ V

$I_{dc} = 54 / 1000 = 54$ mA ; $\qquad P = 54 \times 54 = 2916$ mW = **2.916 W**

(*iii*) $V_{dc} = 0.15 \times 200 = 31.8$ V

$I_{dc} = 31.8$ mA; $\qquad P = 31.8 \times 31.8 = 1011$ mW = **1.011 W**

(*iv*) $V_{dc} = 0.0466 \times 200 = 9.3$ V

$I_{dc} = 9.3$ mA; $\qquad P = 9.3 \times 9.3 = 86.5$ mW = **0.0865 W**

17.32. Average Values for FW Controlled Rectifier

In this case, the average values would be doubled because there are two half sinusoids to be averaged as shown in Fig. 17.41 for $\theta_1 = 60°$. The general expression is

$$V_{dc} = 2 \times \frac{V_m}{2\pi}(1 + \cos\theta_1)$$

$$= \frac{V_m}{\pi}(1 + \cos\theta_1)$$

$$\left[I_{dc} = \frac{V_{dc}}{R_L} = \frac{V_m}{\pi R_L}(1 + \cos\theta_1) \right]$$

In the present case, $\theta_1 = 60°$, $\cos\theta_1 = 0.5$

$$\therefore \quad V_{dc} = \frac{V_m}{\pi}(1 + 0.5) = 0.477\, V_m$$

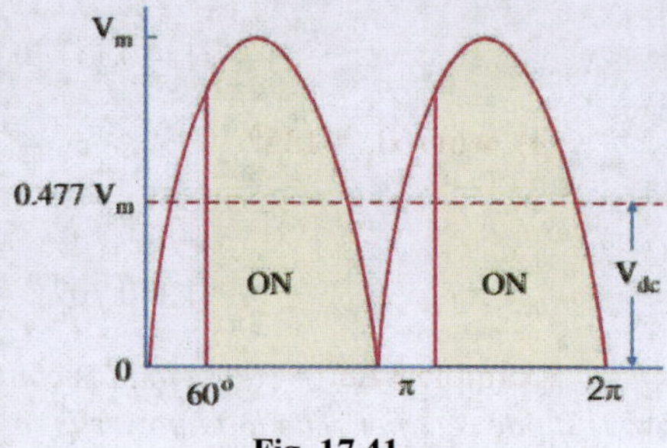

Fig. 17.41

17.33. Silicon Controlled Rectifier (SCR)

It is a trijunction *PNPN* device having three external connections : anode (*A*), cathode (*C*) and gate (*G*) as shown in Fig. 17.42.

When *A* is made positive with respect to *C*, the junctions marked 1 and 3 are forward-biased but the centremost junction marked 2 is reverse-biased. Hence, the *SCR* does not conduct. However, if a sufficiently positive voltage pulse is applied to this junction via the gate *G*, the *SCR* starts conducting like an ordinary diode. Gate has no control over *SCR* once it has been triggered into conduction. Conduction can only be terminated by removing positive voltage from its anode.

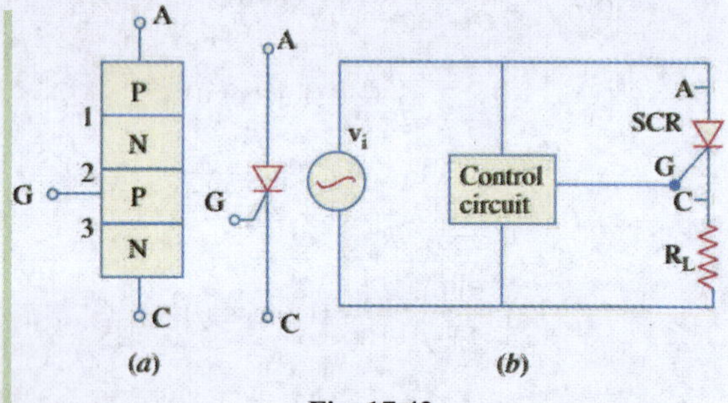

Fig. 17.42

As shown in Fig. 17.42 (*b*), *SCR* needs a control circuit which triggers it into conduction by a gate pulse. Two popular ways of turning an *SCR* ON are

1. amplitude firing — in which gate current flows for the whole period of conduction.
2. *UJT* oscillator firing — in this method, gate current is supplied only *momentarily.* This method allows control over the power delivered to *SCR*'s load from zero watt to fully ON.

Some of the simple ways of controlling the conduction of an *SCR* are discussed below.

17.34. Pulse Control of SCR

The simplest *SCR* control circuit of this type is shown in Fig. 17.43. If *SCR* were an ordinary rectifier, it would produce half-wave rectified ac voltage across R_L. However, *SCR* will not conduct even during positive half-cycle of input ac voltage unless *G* is given positive voltage to forward-bias

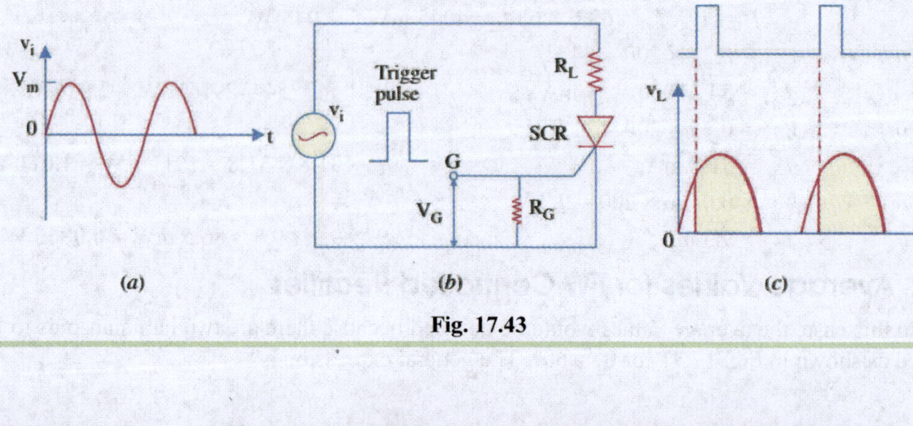

Fig. 17.43

the centremost junction. By applying a trigger pulse at any time during the positive input half-cycle, it can be fired into conduction.

The resultant load waveform across R_L will consist of a portion of the positive half-cycle commencing at any instant at which the *SCR* is triggered [Fig. 17.43].

17.35. 90° Phase Control of SCR

Fig. 17.44 shows a circuit which can trigger (or switch on) the *SCR* anywhere from the commencement of the ac cycle to the peak of its positive half-cycle *i.e.*, between 0° and 90°. As shown, gate current is derived from the ac supply (often it is from rectifier output as in Fig. 17.41) via *R*. If *R* is set at low value, the *SCR* will trigger almost at the commencement of the positive half-cycle of the ac input. On the other hand, if *R* is set at high resistance, the *SCR* may not switch ON until the peak positive half-cycle when sufficiently large gate current would become available.

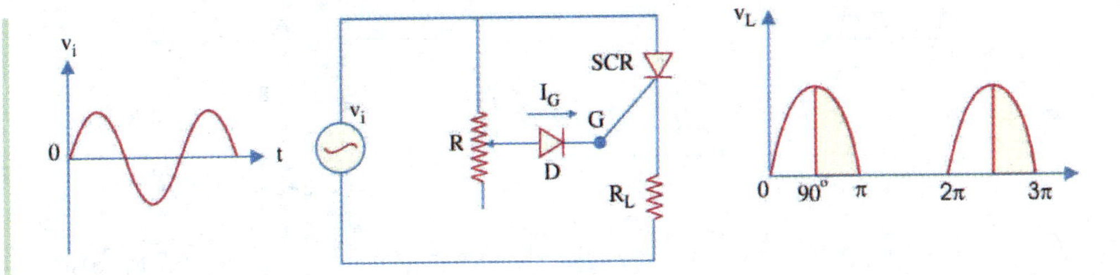

Fig. 17.44

If I_G is not large enough even at peak positive half-cycle, then *SCR* will *not trigger at all* because I_G has the greatest value at the peak input and then falls off as the voltage falls. The purpose of diode *D* is to protect the *SCR* gate from negative voltage during the negative input half-cycle. This method is also known as the *amplitude firing* of an *SCR*.

17.36. 180° Phase Control of SCR

The circuit shown in Fig. 17.45 can trigger the *SCR* from 0° to 180° of the input waveform.

The circuit is identical to that of Fig. 17.44 except for the addition of the diode D_2 and capacitor *C*.

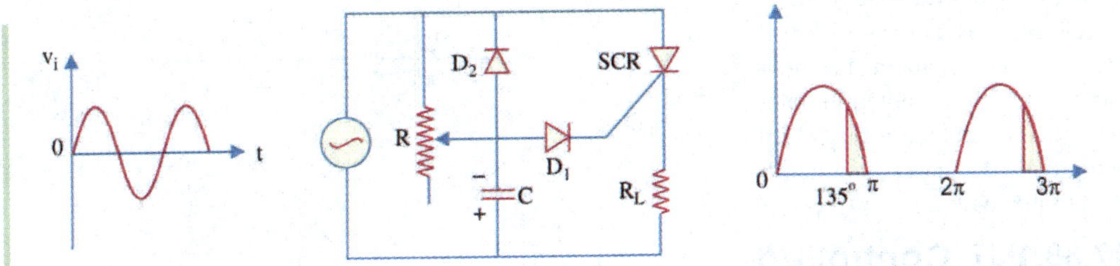

Fig. 17.45

We will start the analysis with the negative half-cycle. During the negative half-cycle of the input, *C* is charged immediately (with the polarity as shown) to the peak of the input voltage because D_2 is forward-biased. When peak of the negative half-cycle passes over, D_2 becomes reverse-biased because its anode (connected to *C*) becomes more negative than its cathode (connected to the supply). Hence, C starts to discharge through *R*.

Depending on the time constant (=*CR*), *C* may almost be completely discharged at the commencement of the oncoming positive half-cycle or may retain partial charge until almost 180° of positive half-cycle has passed. So long as *C* remains negatively-charged, D_1 is reverse-biased and the gate cannot become positive to trigger the *SCR* into conduction. Hence, *R* and/or *C* can be adjusted to trigger the *SCR* anywhere from 0° to 180° of the input ac cycle.

17.37. SCR Controlled Circuit

We will now consider a circuit where an amplitude-fired *SCR* is used to control the power in a full-wave rectifier circuit.

In Fig. 17.46 (*a*), R_L is the load resistance which may be a lamp load, heater or a small dc motor whose power we wish to control. The full-wave bridge rectifier furnishes the rectified output shown in Fig. 17.46 (*b*). Obviously, control over both half-cycles of the ac input is possible.

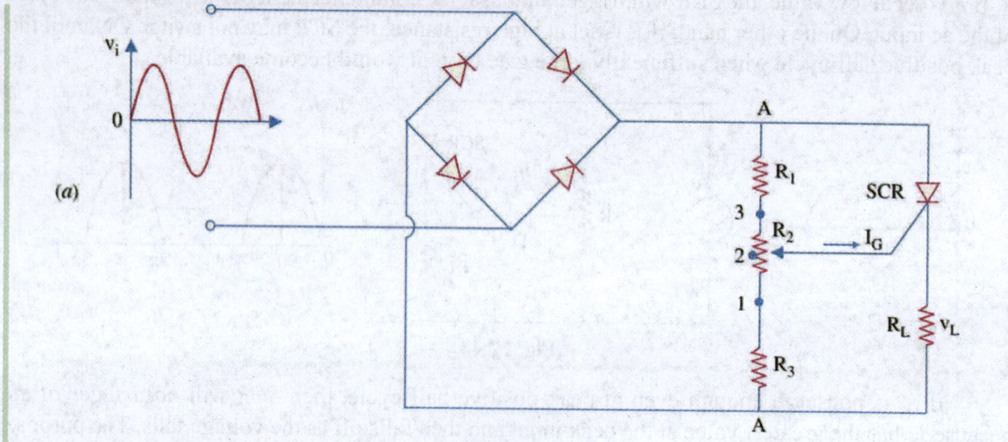

Fig. 17.46

The rectifier output voltage is dropped across the potentiometer *AA*. When potentiometer is set at point 1, drop across R_3 is not enough to provide sufficient gate current to switch ON the *SCR*. Hence, there is no load current and, consequently, no drop across the load. When potentiometer is set at point 2 and then 3, the *SCR* gets fired yielding load voltage waveforms shown in Fig. 17.46 (*c*) and (*d*). Hence, changing the moving contact on the potentiometer changes the conduction angle and hence the amount of power delivered to the load.

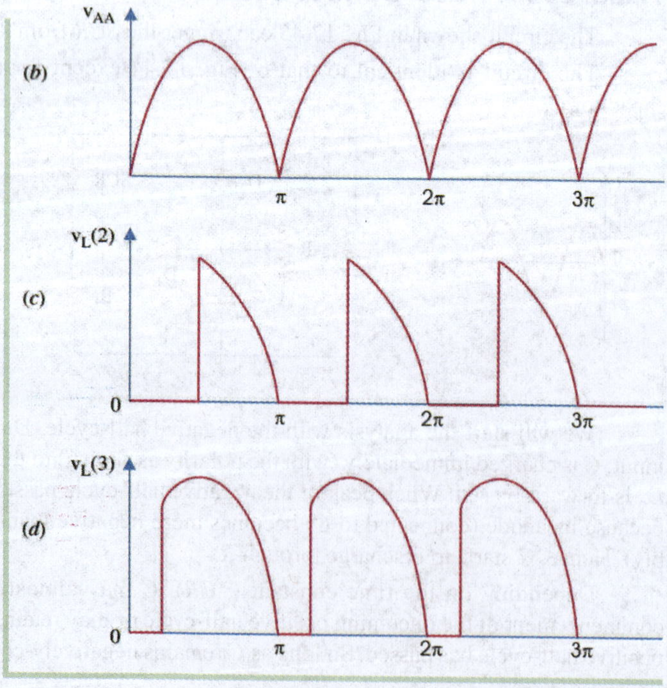

17.38. UJT Controlled Circuit

This circuit employs a *UJT* oscillator to control the firing of the *SCR* as shown in Fig. 17.47. The full-wave rectified output voltage V_{AF} is available across parallel paths *AF* and *BF*. When *SCR* is not conducting (*i.e.*, it acts as an open), then whole output voltage drops across it and none across R_L. When it conducts (*i.e.*, acts as a short), whole of V_{AF} drops across R_L and none across it.

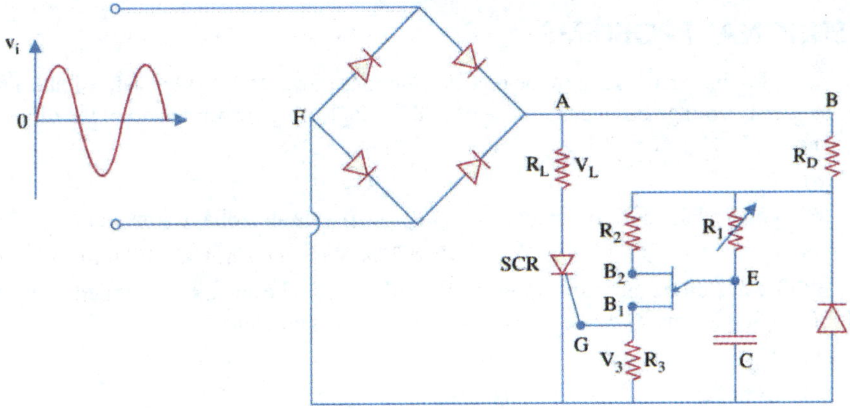

Fig. 17.47

Now, consider the other parallel circuit *BF*. R_D is the voltage dropping resistor so that a suitable voltage is applied to the *UJT* oscillator circuit for the *UJT* to work without damage. The full wave rectified voltage is clipped at a convenient level by Zener diode *D* for proper operation of *UJT*. Varying R_1 varies the time it takes for *C* to charge to the *UJT*'s firing voltage. Charging of *C* always starts when V_C is at 0 V dc or at 0° of V_{AF}.

If R_1 is very small, *C* charges very quickly ($\because$ time constant CR_1 is very short) and reaches the *UJT* firing voltage soon after V_{AF} starts to go positive. Hence, *UJT* fires early on every cycle of V_{AF}. When *UJT* fires, it becomes almost a short between its emitter (*E*) and B_1. Thus, it discharges *C* very quickly through R_3 which is in parallel with the *SCR*'s gate. This triggers the *SCR* early in the cycle of V_{AF} and the *SCR latches itself* ON until V_{AF} falls to zero. When V_{AF} goes to zero, the *SCR* is shut OFF. The cycle repeats itself with the *SCR* being turned ON early in the cycle thereby delivering nearly full-power to the load.

When R_1 is increased, v_C rises slowly and reaches the *UJT* firing voltage later during each cycle of V_{AF}. Therefore, *SCR* is turned ON much later thereby considerably cutting down in the power

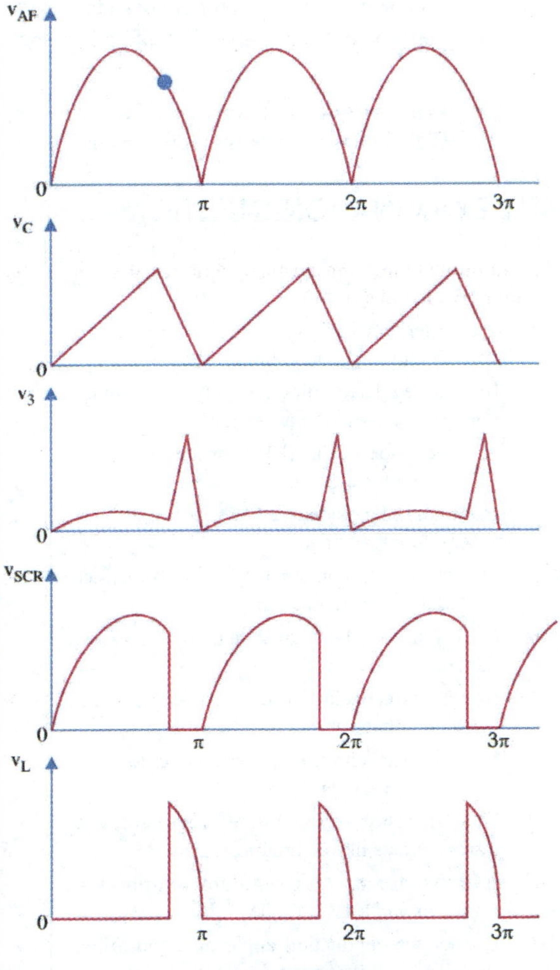

Fig. 17.48

delivered to the load as shown in Fig. 17.48. Obviously, the *SCR* is being fired during the last few degrees of the cycle.

CONVENTIONAL PROBLEMS

1. A single-phase half-wave diode rectifier supplies power to a 2 kΩ resistive load. The input ac supply voltage has a peak value of 300 V. Neglecting forward drop of the diode, calculate
   (*a*) V_{dc} (*b*) I_{dc}
   (*c*) power delivered to the load (*d*) ripple voltage (rms value)
   [(*a*) 95.4 V (*b*) 47.7 mA (*c*) 4550 mV (*d*) 115.4 V]
2. A full-wave diode rectifier supplies a load of 10 kΩ. The ac voltage applied to the diode is 300-0-300 V rms. If diode resistance is neglected, calculate
   (*a*) V_{dc} (*b*) I_{dc} (*c*) I_{rms}
   (*d*) form factor (*e*) ripple voltage.
   [(*a*) 270 V (*b*) 27 mA (*c*) 30 mA (*d*) 1.11 (*e*) 130.1 V]
3. A dc and an ac voltmeter are used to measure the output of a filter circuit. The readings of the two voltmeters are 50 V and 5 V respectively. Calculate the ripple factor of the filter.
   [10%]
4. In a controlled full-wave rectifier, peak supply voltage is 200 V and load resistance 1 kΩ. Calculate the power delivered to the load for firing angles of (*a*) 60° and (*b*) 120°.
   [(*a*) 9.1 W (*b*) 1.01 W]

SELF EXAMINATION QUESTIONS

A. Fill in the blanks most appropriate word(s) or numerical value(s).

1. A rectifier's main function is to convert an ac voltage into .... dc voltage.
2. In an unregulated voltage supply, dc terminal voltage .... with the amount of load.
3. Average value of the HW rectified voltage is .... time the peak supply voltage.
4. *PIV* rating of a diode in a FW rectifier is .... × peak supply voltage.
5. Frequency of the output voltage of a *FW* rectifier is .... the supply frequency.
6. Greater the form factor of a filter, higher its .... factor.
7. In a shunt capacitor filter, load sees a .... constant voltage across it at all times.
8. In a shunt capacitor filter, ripple decreases with .... in load resistance.
9. In a transistor series voltage regulator, the transistor acts like a variable ....
10. A transistor current regulator employs a transistor and a .... diode.
11. Greater the conduction angle of a controlled rectifier, ... its output.
12. For a firing angle of 180°, the output of a half-wave controlled rectifier is ....
13. An *SCR* needs a control circuit to .... it into conduction at any time after input cycle goes positive.
14. In amplitude firing of an *SCR*, ... current flows continuously.

B. Answer True or False

1. Ripple voltage is directly proportional to the dc value of the output wave.
2. As compared to a full-wave rectifier, a half-wave rectifier has more ripple but less efficiency.
3. In a FW diode rectifier, *PIV* rating of each diode is twice that in a HW rectifier.
4. Ripple factor of a FW rectifier is about half the value for a HW rectifier.
5. Every other factor being the same, an ordinary two-diode full-wave rectifier is preferrable to a bridge rectifier.
6. The voltage regulation of a π filter is inferior to that of an LC filter.
7. In a transistor series rgulator, the transistor behaves like a variable resistor.
8. A controlled transistor series regulator has the ability to control its own output.

9. When input filter capacitor shorts, output of a dc supply falls to zero.
10. Controlled rectification provides full control over output power.
11. Smaller the firing angle of a controlled rectifier, greater its output.
12. When firing angle of a controlled rectifier is increased from 30° to 60°, its power output is doubled.

C. Multiple Choice Items

1. The ripple factor of a power supply is a measure of
   (*a*) its filter efficiency
   (*b*) its voltage regulation
   (*c*) diode rating
   (*d*) purity of power output
2. The output of a half-wave rectifier is suitable only for
   (*a*) running car radios
   (*b*) charging batteries
   (*c*) running ac motors
   (*d*) running tape-recorders
3. A bridge rectifier is preferable to an ordinary two diode full-wave rectifier because
   (*a*) it uses four diodes
   (*b*) its transformer has no centre tap
   (*c*) it needs much smaller transformer for the same output.
   (*d*) it has higher safety factor
4. The ripple factor of a bridge rectifier is
   (*a*) 0.406 (*b*) 0.812
   (*c*) 1.21 (*d*) 1.11
5. The *PIV* of a half-wave rectifier circuit with a shunt capacitor filter is
   (*a*) $2V_m$ (*b*) V_m
   (*c*) $v_m/2$ (*d*) $3V_m$
6. The primary function of a filter is to
   (*a*) minimise ac input variations
   (*b*) suppress odd harmonics in the rectifier output
   (*c*) stabilise dc level of the output voltage
   (*d*) remove ripples from the rectified output
7. Larger the value of filter capacitor
   (*a*) larger the p-p value of ripple voltage
   (*b*) larger the peak current in the rectifying diode
   (*c*) longer the time that current pulse flows through the diode
   (*d*) smaller the dc voltage across the load
8. Which stage of a dc power supply uses a Zener as the main component ?
   (*a*) rectifier (*b*) voltage divider
   (*c*) regulator (*d*) filter
9. Which rectifier requires four diodes ?
   (*a*) half-wave voltage doubler
   (*b*) full-wave bridge circuit
   (*c*) full-wave voltage doubler
   (d) voltage quadrupler
10. For a half-wave controlled rectifier, the average value of output dc voltage is given by
   (*a*) $V_{dc} = \dfrac{V_m}{2\pi}(1 - \cos\theta_1)$
   (*b*) $V_{dc} = \dfrac{2V_m}{\pi}(1 + \cos\theta_1)$
   (*c*) $V_{dc} = \dfrac{V_m}{\pi}(\cos\theta_1 - 1)$
   (*d*) $V_{dc} = \dfrac{V_m}{2\pi}(\cos\theta_1 + 1)$

ANSWERS

A. Fill in the blanks

1. pulsating **2.** varies **3.** 0.318 **4.** 2 **5.** twice
6. ripple **7.** nearly **8.** increase **9.** resistance **10.** Zener
11. higher **12.** zero **13.** trigger **14.** gate

B. True or false

1. *F* **2.** *T* **3.** *T* **4.** *F* **5.** *F* **6.** *T* **7.** *T* **8.** *T* **9.** *T* **10.** *T*
11. *T* **12.** *F*

C. Multiple Choice Items

1. *d* **2.** *b* **3.** *c* **4.** *c* **5.** *a* **6.** *d* **7.** *b* **8.** *c* **9.** *b* **10.** *d*

18 CHAPTER

The Basic Transistor

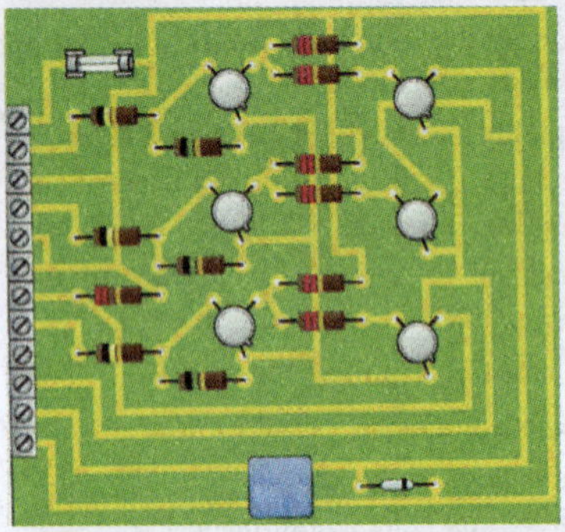

1. The Bipolar Junction Transistor
2. Transistor Biasing
3. Important Biasing Rule
4. Transistor Currents
5. Summing Up
6. Transistor Circuit Configurations
7. CB Configuration
8. CE Configuration
9. Relations between α and β
10. CC Configuration
11. Relations between Transistor Currents
12. Leakage Currents in a Transistor
13. Thermal Runaway

18.1. The Bipolar Junction Transistor*

Basically, it consists of two back-to-back *P-N* junctions manufactured in a single piece of a semiconductor crystal. These two junctions give rise to three regions called **emitter**, **base** and **collector**. As shown in Fig. 18.1, a junction transistor is simply a sandwich of *one type* of semiconductor material between two layers of the *other type*. Fig. 18.1, (*a*) shows a layer of *N*-type material sandwiched between two layers of *P*-type material. It is described as *PNP* transistor. Fig. 18.1 (*b*) shows an *NPN* transistor consisting of a layer of P-type material sandwiched between two layers of *N*-type material.

NPN-bipolar junction transistor.

* It is a contraction of two words : **transfer** and **resistor.** It is so because a transistor is basically a resistor that amplifies electrical impulses as they are transferred through it from its input to output terminals.

The emitter, base and collector are provided with terminals which are labelled as *E*, *B* and *C*. The two junctions are : emitter-base (*E*/*B*) junction and collector-base (*C*/*B*) junction.

The symbols employed for *PNP* and *NPN* transistors are also shown in Fig. 18.1. The arrowhead is *always at the emitter* (not at the collector) and in each case, its direction indicates the *conventional* direction of current flow. For a *PNP* transistor, arrowhead points from emitter to base meaning that *emitter is positive with respect to base* (and also with respect to collector)*. For *NPN* transistor, it points from base to emitter meaning that base (and collector as well) is positive with respect to emitter.

The *PNP* transistor.

1. Emitter

It forms the left-hand section of region of the transistor as shown in Fig. 18.1. It is more *heavily doped* than any of the other regions because its main function is to supply majority charge carriers (either electrons or holes) to the base.

2. Base

It forms the middle section of the transistor. It is very thin (10^{-6} m) as compared to either the emitter or collector and is very *lightly-doped*.

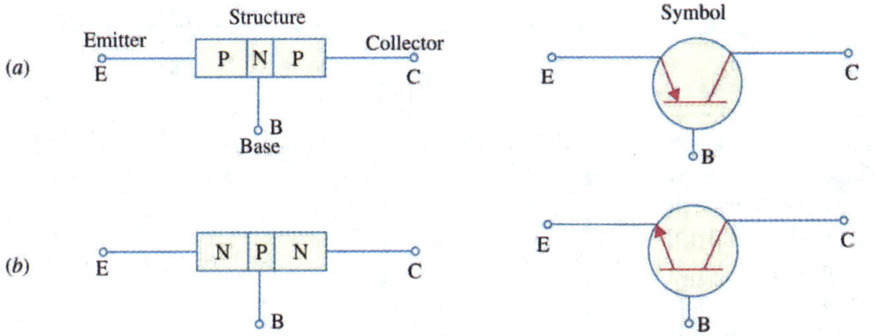

Fig.18.1

3. Collector

It forms the right-hand side section (or region) of the transistor shown in Fig. 18.1 and its main function (as indicated by its name) is to collect majority charge carriers through the base.

In most transistors, collector region is made *physically larger* than the emitter region because it has to dissipate much greater power. Because of this difference, there is no possibility of inverting the transistor *i.e.*, making its collector the emitter and its emitter the collector.

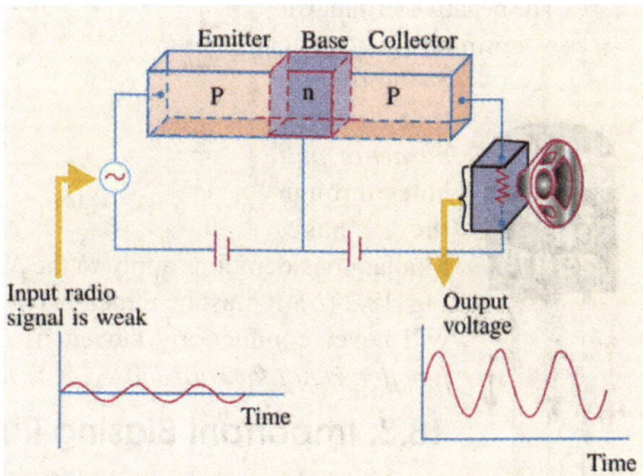

It may be noted, in passing, that transistors are made by growing, alloying or diffusing processes.

18.2. Transistor Biasing

For proper working of a transistor, it is essential to apply voltages of correct polarity across its two junctions. It is worthwhile to remember that for **normal** operation

* In a transistor, for normal operation, collector and base have the **same** polarity *with respect to emitter* (Art. 18.3).

1. emitter-base junction is always forward-biased, and
2. collector-base junction is always reverse-biased.

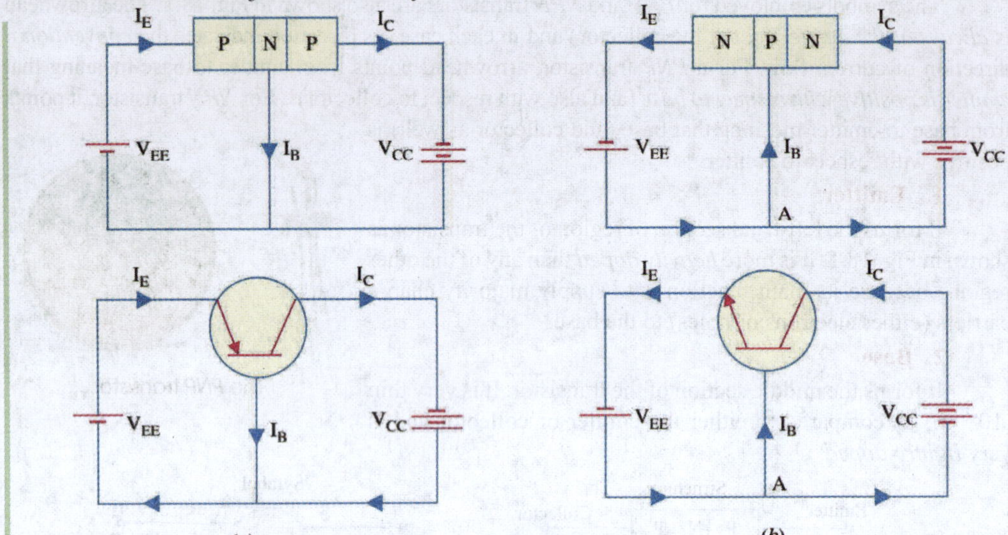

(a) (b)

Fig. 18.2

In Fig. 18.2, two batteries respectively provide the dc emitter supply voltage V_{EE} and collector supply voltage V_{CC} for properly biasing the two junctions of the transistor. In Fig. 18.2 (a), **Positive** terminal of V_{EE} is connected to **P**-type emitter in order to *repel* or **Push** holes into the base.

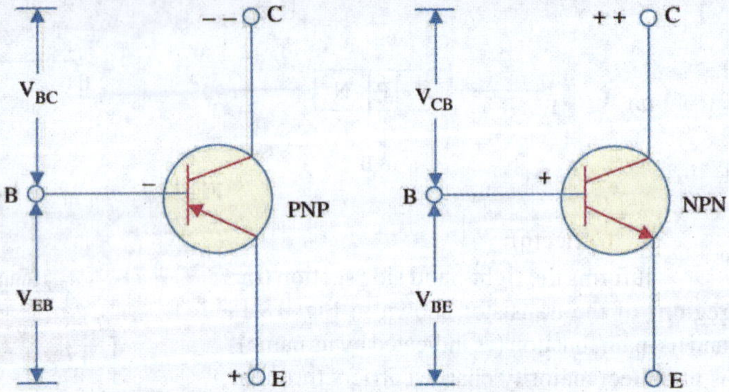

(a) **Fig. 18.3** (b)

The negative terminal of V_{CC} is connected to the collector so that it may *attract or pull* holes through the base. Similar considerations apply to the *NPN* transistor of Fig. 18.2 (*b*). It must be remembered that a transistor will never conduct any current if *its emitter-base junction is not forward-biased*.*

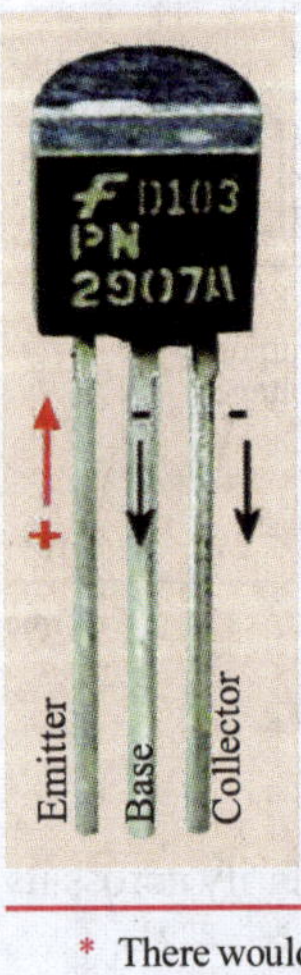

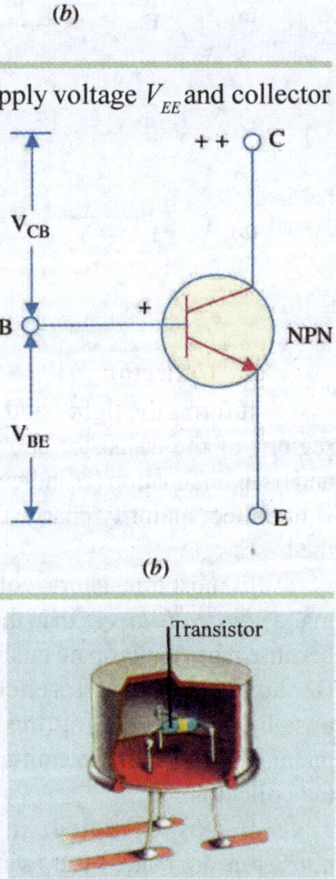

A small current flowing into the base of a *n-p-n* transistor allows a larger current to flow between its emitter and collector.

18.3. Important Biasing Rule

For a *PNP* transistor, both collector and base are negative with respect to the emitter (the letter **N** of **N**egative being the same as the middle letter of **PNP**). Of course, collector is *more negative* than base

* There would be no current due to majority charge carriers. However, there would be an extremely small current due to minority charge carriers which is called leakage current of the transistor (Art. 18.12).

[Fig. 18.3 (*a*)]. Similarly, for **NPN** transistor, both collector and base are *positive* with respect to the emitter (the letter **P** of **P**ositive being the same as the middle letter of **NPN**). Again, collector is *more positive* than the base as shown in Fig. 18.3 (*b*).

It may be noted that different potentials have been designated by double subscripts. The first subscript always represents the point or terminal which is more positive (or less negative) than the point or terminal represented by the second subscript. For example, in Fig. 18.3 (*a*) the potential difference between emitter and base is written as V_{EB} (and not V_{BE}) because emitter is positive with respect to base. Now, between the base and collector themselves, collector is more negative than base. Hence, their potential difference is written as V_{BC} and not as V_{CB}. Same is the case with voltage marked in Fig. 18.4.

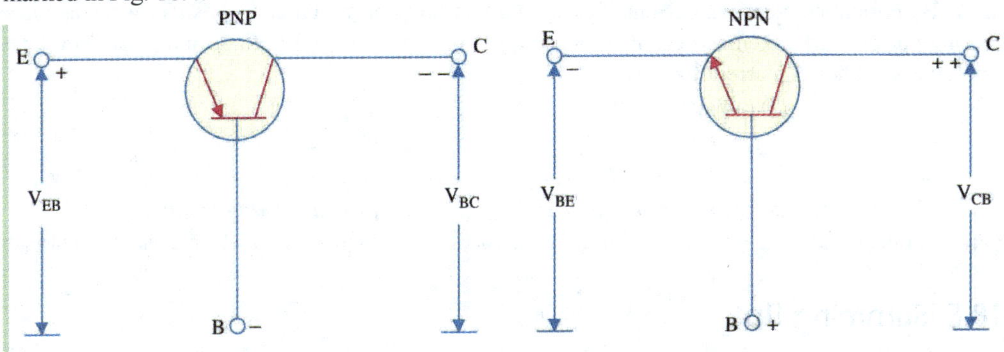

Fig. 18.4.

18.4. Transistor Currents

The three primary currents which flow in a properly-biased transistor are I_E, I_B and I_C. In

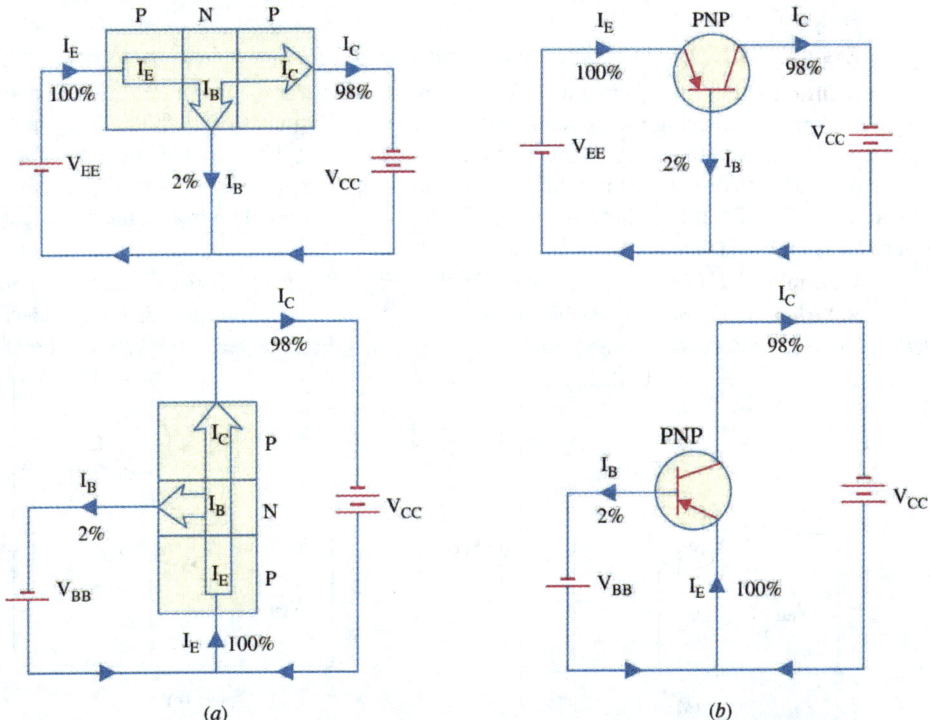

Fig. 18.5

Fig. 18.5 (*a*) are shown the directions of flow as well as relative magnitudes of these currents for a *PNP* transistor connected in the common-base mode. It is seen that

$$I_E = I_B + I_C$$

It means that a small part (about 1-2%) of emitter current goes to supply base current and the remaining major part (98 – 99%) goes to supply—collector current.

Moreover, I_E flows *into* the transistor whereas both I_B and I_C flow out of it.

Fig. 18.5 (*b*) shows the flow of currents in the same transistor when connected in the common-emitter mode. It is seen that again.

$$I_E = I_B + I_C$$

By normal convention, currents flowing *into* a transistor are taken as positive whereas those flowing out of it are taken as negative. Hence, I_E is positive whereas both I_B and I_C are negative. Applying Kirchhoff's Current Law, we have

$$I_E + (-I_B)(-I_C) = 0$$

or $$I_E - I_B - I_C = 0$$

or $$I_E = I_B + I_C$$

This statement is true *regardless of transistor type or transistor configuration.*

Note. For the time being, we have not taken into account the leakage currents which exist in a transistor (Art. 18.12).

18.5. Summing Up

The four basic guideposts about all transistor circuits are :

1. conventional current flows along the arrow whereas electrons flow against it.
2. *E/B* junction is always forward-biased.
3. *C/B* junction is always reverse-biased.
4. $I_E = I_B + I_C$.

Example 18.1. *Determine whether the lamp L in Fig. 19.6 (a) is OFF or ON and explain why?*

Solution. From the schematic symbol we find that the transistor is a *PNP* type. Now, according to the transistor biasing rule given in Art. 18.3, in a *PNP* transistor, both the base and collector have to be Negative with respect to the emitter. However, in Fig. 18.6 (*a*) both the emitter and collector are positive with respect to the emitter. Hence, the transistor is not correctly voltage-biased. Thus, the lamp will be OFF. It is so because when *EBJ* is reverse-biased, then from collector to emitter the transistor is OFF and *vice versa*.

Example 18.2. *Find out whether the lamp in Fig. 18.6 (b) is ON or OFF. Explain why?*

Solution. Since the arrowhead on the emitter points *away* from the base, the transistor is of *NPN* type. According to the biasing rule given in Art. 18.3, both the base and the collector should be

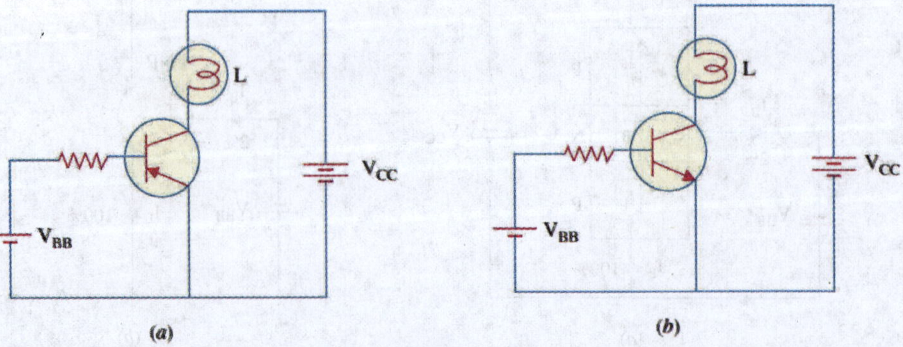

Fig. 18.6

Positive with respect to the emitter. Since the transistor is correctly-biased, the lamp will be ON. It is so because a forward-biased *EBJ* turns a transistor from collector to emitter ON.

18.6. Transistor Circuit Configurations

Basically, there are three types of circuit connections (called configurations) for operating a transistor.

1. common-base (*CB*),
2. common-emitter (*CE*),
3. common-collector (*CC*).

The term '*common*' is used to denote the electrode that is *common to the input and output circuits*. Because the common electrode is generally grounded, these modes of operation are frequently referred to as grounded-base, grounded-emitter and grounded-collector configurations as shown in Fig. 18.7 for a *PNP* transistor.

Since a transistor is a 3-terminal (and not a 4-terminal) device, one of its terminals has to be common to the input and output circuits.

18.7. CB Configuration

In this configuration, emitter current I_E is the input current and collector current I_C is the output current. The input signal is applied between the emitter and base whereas output is taken out from the collector and base as shown in Fig. 18.7 (*a*).

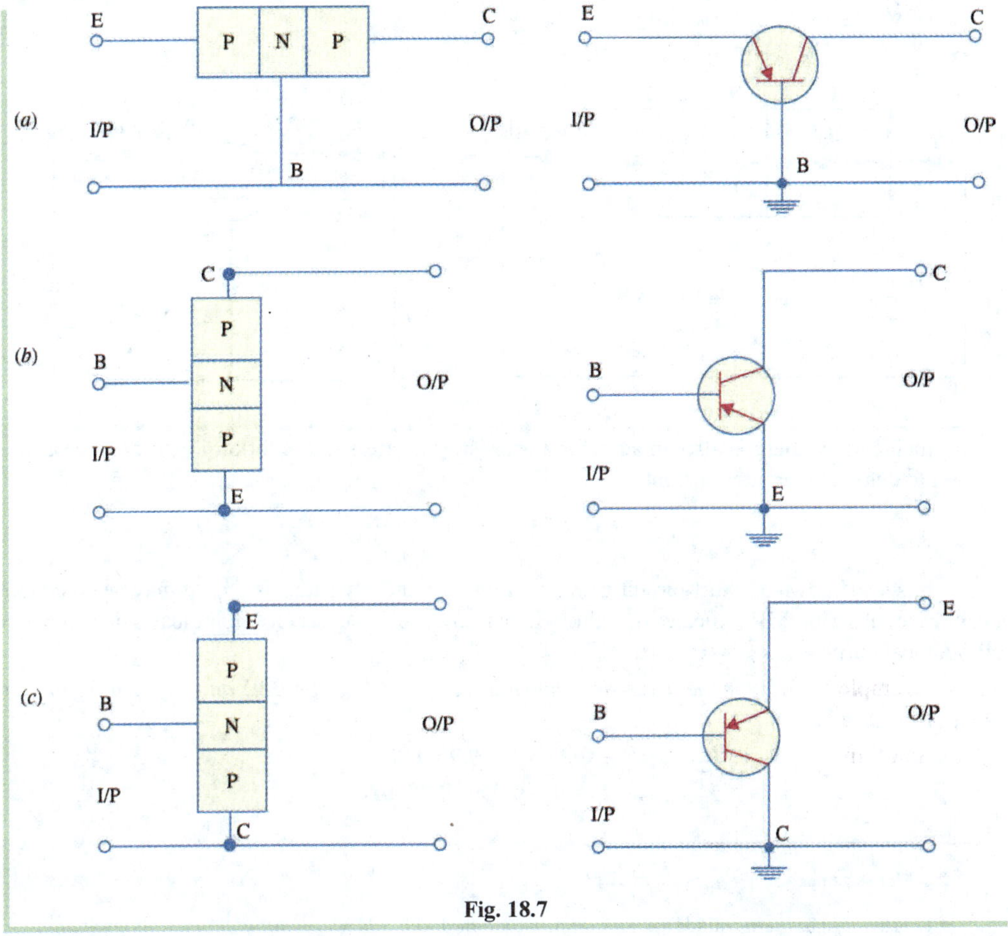

Fig. 18.7

The ratio of the collector current to the emitter current is called dc alpha (α_{dc}) of a transistor.

$$\therefore \qquad \alpha_{dc}{}^{*} = \frac{-I_C}{I_E}$$

The negative sign is due to the fact that current I_E flows into the transistor whereas I_C flows *out* of it. Hence, I_E is taken as positive and I_C as negative.

$$\text{or} \qquad I_C = -\alpha_{dc}.I_E{}^{**}$$

If we write α_{dc} simply as α, then

$$\alpha = -\frac{I_C}{I_E}$$

It is also called forward current transfer ratio ($-h_{FB}$). In h_{FB}, subscript *F* stands for forward and *B* for common-base. The negative sign merely indicates that emitter and collector currents flow in and out of the transistor. The subscript *dc* on α signifies that this ratio is defined from dc values of I_C and I_E.

The α of a transistor is a measure of the quality of a transistor : higher the value of α, better the transistor in the sense that collector current more closely equals the emitter current. Its value ranges from 0.95 to 0.999. Obviously, it applies only to *CB* configuration of a transistor. As seen from above and Fig. 18.8.

$$I_C = \alpha I_E$$

$$\text{Now,} \qquad I_B = I_E - I_C = I_E - \alpha I_E$$

$$\therefore \qquad I_B = (1-\alpha)I_E$$

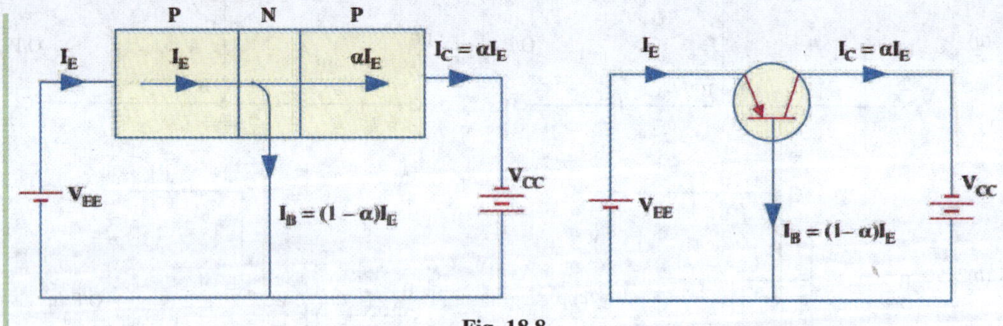

Fig. 18.8

Incidentally, there is also an ac α for a transistor. It refers to the ratio of *change* in collector current to *change* in emitter current.

$$\therefore \qquad \alpha_{ac} = -\frac{\Delta I_C}{\Delta I_E}$$

It is also known as short-circuit gain of a transistor and is written as $-h_{fb}$. It may be noted that upper case subscript 'FB' indicates dc value whereas lower case subscript '*fb*' indicates dc value. For all practical purposes, $\alpha_{dc} = \alpha_{ac} = \alpha$.

Example 18.3. *If for the transistor shown in Fig. 19.7 (a), $\alpha = 0.95$ and $I_E = 1$ mA, find the values of I_C and I_B.*

Solution. $I_C = \alpha I_E = 0.95 \times 1 = \mathbf{0.95\ mA}$

$I_B = I_E - I_C = 1 - 0.95 = \mathbf{0.05\ mA}$

* More accurately, $\alpha_{dc} = \dfrac{I_C - I_{CBO}}{I_E}$ — Art. 18.12

*Negative sign has been omitted since we are here concerned with only magnitudes of the currents involved.

Example 18.4. *Following current readings are obtained in transistor circuit of Fig. 19.7 (a): $I_E = 2$ mA and $I_B = 20$ μA. Compute the values of α and I_C*

Solution.
$$I_C = I_E - I_B$$
$$= 2 \times 10^{-3} - 20 \times 10^{6} = \mathbf{1.98\ mA}$$
$$\alpha = I_C / I_E = 1.98 / 2 = \mathbf{0.99}$$

18.8. CE Configuration

Here, input signal is applied between the base and emitter and output signal is taken out from collector and emitter circuit. As seen from Fig. 18.7 (*b*), I_B is the input current and I_C is the output current.

The ratio of dc collector current to dc base current is called dc beta (β_{dc}) or just β of the transistor.

$$\beta = \frac{I_C}{I_B} \qquad \text{or} \qquad I_C = \beta I_B$$

It is also called common-emitter dc forward transfer ratio and is written as h_{FE}. It is possible for β to have as high a value as 500.

While analysing ac operation of a transistor, we use ac β which is given by

$$\beta_{ac} = \frac{\Delta I_C}{\Delta I_B}$$

It is also written as h_{fe}.

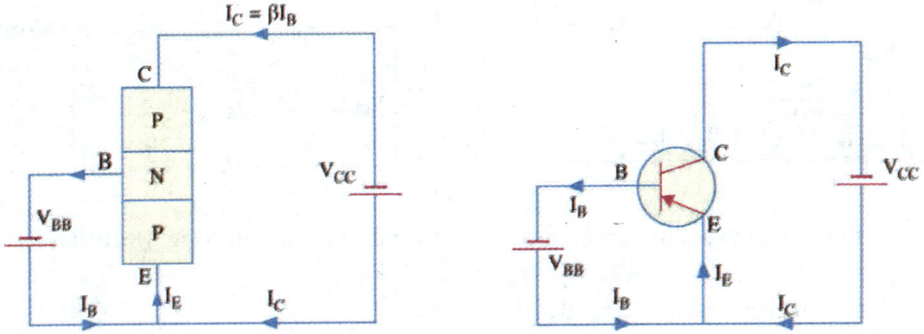

Fig. 18.9

The flow of various currents in a *CE* configuration is shown in Fig. 18.9. As seen

$$I_E = I_B + I_C = I_B + \beta I_B = (1 + \beta) I_B$$

18.9. Relations between α and β

$$\beta = \frac{I_C}{I_B} \qquad \text{and} \qquad \alpha = \frac{I_C}{I_E}$$

$$\therefore \quad \frac{\beta}{\alpha} = \frac{I_E}{I_B}$$

Now, $\quad I_B = I_E - I_C$

$$\therefore \quad \beta = \frac{I_C}{I_E - I_C} = \frac{I_C / I_E}{I_E / I_E - I_C / I_E} \qquad \text{or} \qquad \beta = \frac{\alpha}{1 - \alpha}$$

Cross-multiplying the above equation and simplifying it, we get

$$\beta(1 - \alpha) = \alpha \qquad \text{or} \qquad \beta = \alpha(1 + \beta)$$

$$\alpha = \frac{\beta}{1+\beta}$$

It is seen from the above two equations that

$$1 - \alpha = \frac{1}{1+\beta}$$

18.10. CC Configuration

In this case, input signal is applied between base and collector and output signal is taken out from emitter-collector circuit. [Fig. 18.7 (*c*)]. Conventionally speaking, here I_B is the input current and I_E is the output current as shown in Fig. 18.10. The current gain of the circuit is

$$\frac{I_E}{I_B} = \frac{I_E}{I_C} \cdot \frac{I_C}{I_B} = \frac{1}{\alpha} \cdot \beta$$

$$= \frac{\beta}{\alpha} = \frac{\beta}{\beta/(1+\beta)} = (1+\beta)$$

It means the output current is $(1 + \beta)$ times the input current.

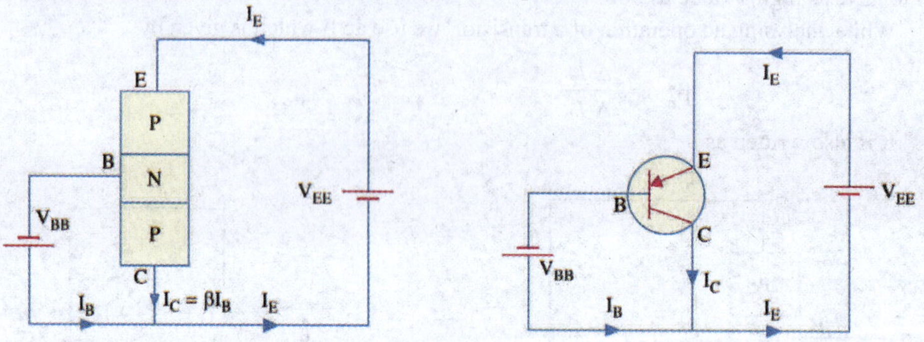

Fig. 18.10

The flow paths of various currents in a *CC* configuration are shown in Fig. 18.10. It is seen that

$$I_E = I_B + I_C = I_B + \beta\, I_B = (1+\beta) I_B$$

∴ output current = $(1 + \beta) \times$ input current.

18.11. Relations between Transistor Currents

While deriving various equations, following definitions should be kept in mind.

$$\alpha = \frac{I_C}{I_E} \quad \text{and} \quad \beta = \frac{I_C}{I_B}$$

Also, $$\alpha = \frac{\beta}{(1+\beta)} \quad \text{and} \quad \beta = \frac{\alpha}{(1-\alpha)}$$

(*i*) $$I_C = \beta I_B = \alpha I_E = \frac{\beta}{1+\beta} \cdot I_E$$

(*ii*) $$I_B = \frac{I_C}{\beta} = \frac{I_E}{1+\beta} = (1-\alpha) I_E$$

(*iii*) $$I_E = \frac{I_C}{\alpha} = \frac{1+\beta}{\beta} \cdot I_C = (1+\beta) I_B = \frac{I_B}{(1-\alpha)}$$

(*iv*) The three transistor dc currents always bear the following ratio

$$I_E : I_B : I_C :: 1 : (1-\alpha) : \alpha$$

Incidentally, it may be noted that for ac currents, small letters i_e, i_b and i_c are used.

18.12. Leakage Currents in a Transistor

(a) CB Circuit

Consider the *CB* transistor circuit shown in Fig. 18.12. The emitter current (due to majority carriers) initiated by the forward-biased emitter base junction is split into two parts :

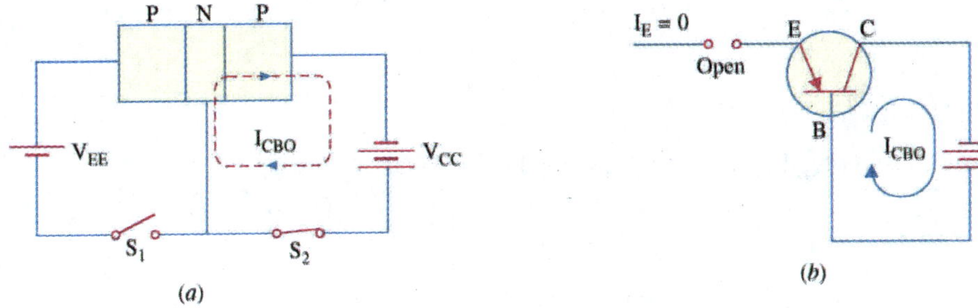

Fig. 18.11

(*i*) $(1 - \alpha) I_E$ which becomes base current I_B in the external circuit, and

(*ii*) αI_E which becomes collector current I_C in the external circuit.

As mentioned earlier (Art. 18.2) though *C/B* junction is reverse-biased for *majority* charge carriers (*i.e.*, holes in this case), it is forward-biased so far as thermally-generated *minority* charge carriers (*i.e.*, electrons in this case) are concerned. This current flows *even when emitter is disconnected from its dc supply* as shown in Fig. 18.11 (*a*) where switch S_1 is open. It flows in the *same* direction* as the collector current of majority carriers. It is called leakage current I_{CBO}. The subscripts *CBO* stand for 'current from **C**ollector to **B**ase with emitter **O**pen'. Very often, it is simply written as I_{CO}.

It should be noted that

(*i*) I_{CBO} is exactly like the reverse saturation current I_S or I_O of a reverse-biased diode discussed in Art. 14.1.

(*ii*) I_{CBO} is extremely temperature-dependent because it is made up of thermally-generated minority carriers. As mentioned earlier, I_{CBO} doubles for every 10°C rise in temperature for Ge and 6°C for Si.

If we take into account the leakage current, the current distribution in a *CB* transistor circuit becomes as shown in Fig. 18.12 both for *PNP* and *NPN* type transistors.

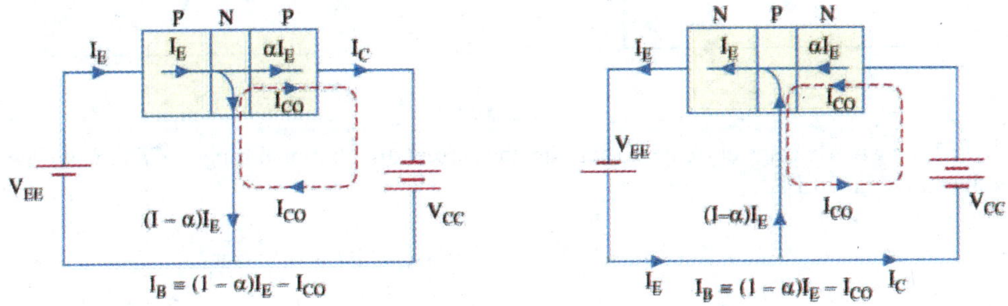

Fig. 18.12

It is seen that total collector current is actually the sum of two components:

(*i*) current produced by normal transistor action *i.e.*, component controlled by emitter current. Its value is αI_E and is due to *majority carriers*;

* Actually, electrons (which form minority charge carriers in collector) flow from negative terminal of collector battery, to collector, then to base through *C/B* junction and finally, to positive terminal of V_{CC}. However, conventional current flows in opposite direction as shown by dotted line in Fig. 18.11 (*a*).

(*ii*) temperature-dependent leakage I_{CO} due to minority carriers

$$\therefore \quad I_C = \alpha I_E + I_{CO}$$

$$= \underset{\text{majority}}{\alpha I_E} + \underset{\text{minority}}{I_{CO}} \quad ...(i)$$

$$\therefore \quad \alpha = \frac{I_C - I_{CO}}{I_E}$$

Since $I_{CO} \ll I_C$ hence $\alpha \cong \dfrac{I_C}{I_E}$

(*iii*) Substituting the value of $I_E = (I_C + I_B)$ in Eq. (*i*) above, we get

$$I_C = \alpha\,(I_C + I_B) + I_{CO} \quad \text{or} \quad I_C\,(1-\alpha) = \alpha I_B + I_{CO}$$

$$\therefore \quad I_C = \frac{\alpha\, I_B}{1-\alpha} + \frac{I_{CO}}{1-\alpha}$$

(*iv*) Eliminating I_C from Eq. (*i*) above, we get

$$(I_E - I_B) = \alpha I_E + I_{CO}$$

Rearranging the above equation, we get

$$I_B = (1-\alpha)\, I_E - I_{CO}$$

(*b*) CE Circuit

In Fig. 18.13 (*a*) is shown a grounded-emitter circuit of an *NPN* transistor whose base lead is open. It is found that despite $I_B = 0$, there is a leakage current from collector to emitter. It is called I_{CEO}, the subscripts *CEO* standing for '**C**ollector to **E**mitter with base **O**pen'.

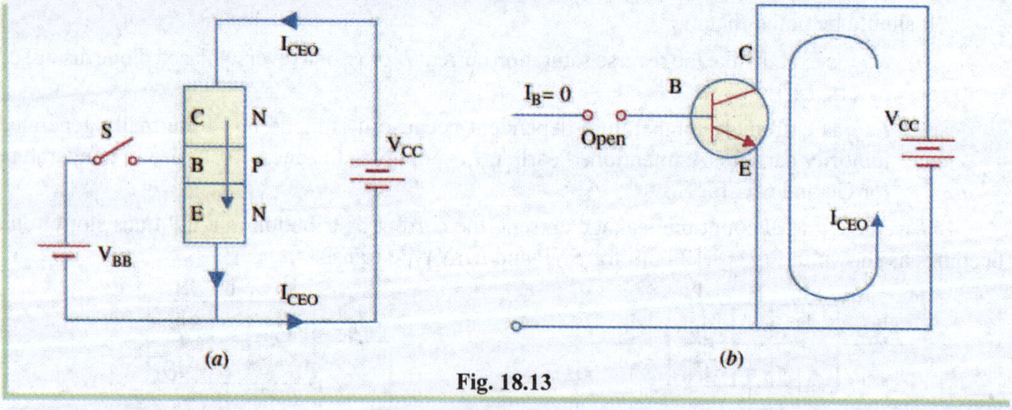

Fig. 18.13

Taking this leakage current into account, the current distribution through a *CE* circuit becomes as shown in Fig. 18.14.

$$= I_C\, \beta I_B + I_{CEO} = \beta I_B + (1+\beta)\, I_{CO} = \beta\, I_B + \frac{I_{CO}}{(1-\alpha)}$$

(*i*) $\therefore \quad I_C = \dfrac{\alpha\, I_B}{1-\alpha} + \dfrac{I_{CO}}{1-\alpha}$

Now, $\beta\, I_B = \alpha I_E$. Substituting this value above, we get

$$I_C = \alpha I_E + I_{CEO}.$$

Also, $\quad I_B = I_E - I_C$

Substituting the value of I_C from above, we have

(*ii*) $\quad I_B = I_E - \alpha I_E - I_{CEO} = (1-\alpha)\, I_E - I_{CEO}$

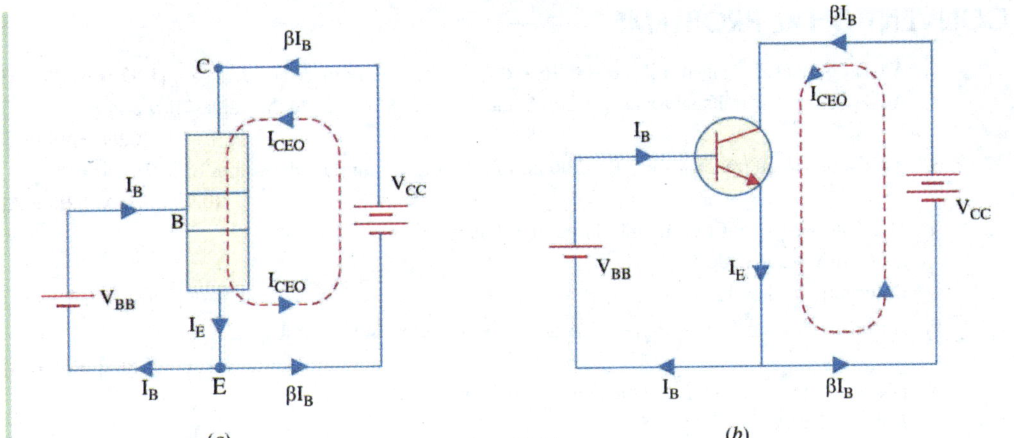

Fig. 18.14

18.13. Thermal Runaway

As seen from Art. 18.12, for a *CE* circuit

$$I_C = \beta I_B + (1+\beta) I_{CO}$$

The leakage current is extremely temperature-dependent. It almost doubles for every 6°C rise in temperature in Ge and for every 10°C rise in Si. Any increase in I_{CO} is magnified $(1 + \beta)$ times *i.e.*, 200 to 300 times. Even a slight increase in I_{CO} will affect I_C considerably. As I_C increases, collector power dissipation increases which raises the operating temperature that leads to further increase in I_C. If this succession of increases is allowed to continue, soon I_C will increase beyond safe operating value thereby damaging the transistor itself —— a condition known as *thermal runaway*. Hence, some form of stabilization is necessary to prevent this thermal runaway.

Example 18.5. *A transistor has $\alpha = 0.98$, $I_B = 100\ \mu A$ and $I_{CO} = 6\ \mu A$. Calculate I_C and I_E.*

Solution.
$$I_C = \frac{\alpha I_B}{1-\alpha} + \frac{I_{CO}}{1-\alpha} = \frac{0.98 \times 100}{1-0.98} + \frac{6}{1-0.98}$$
$$= 4900 + 300 = 52\ \mu\text{A} = \mathbf{5.2\ mA}$$
$$I_E = I_C + I_B$$
$$= 5200 + 100 = 5300\ \mu\text{A} = \mathbf{5.3\ mA}$$

Example 18.6. *The following measurements were made in a particular transistor connected in a circuit.*

$$I_C = 10.525\ mA,\ I_B = 100\ \mu A,\ I_{CBO} = 5\ \mu A$$

Determine

(i) α_{dc}, β_{dc} *and* I_E ; *(ii) the new level of* I_B *required to make* $I_C = 15 mA$.

Solution. (*i*) As seen from Art. 18.12

$$I_C = \beta I_B + (1+\beta) I_{CO}$$

or
$$\beta = \frac{I_C - I_{CBO}}{I_B + I_{CO}} = \frac{10.525 - 5}{100 + 5} = \mathbf{100}$$

$$\alpha = \frac{\beta}{1+\beta} = \frac{100}{1+100} = \mathbf{0.99}$$

CONVENTIONAL PROBLEMS

1. A CB-connected transistor has $\alpha = 0.96$ and $I_E = 2$ mA. Find its I_C and I_B. **[1.92 mA, 80 μA]**
2. A CB-connected transistor has $I_B = 20\mu A$ and $I_E = 2$ mA. Compute the values of α and I_C. **[0.99, 1.98 mA]**
3. A CE-connected transistor has $\beta = 100$ and $I_B = 50$ μA. Compute the values of α, I_C and I_E. **[0.99 ; 5 mA, 5.05 mA]**
4. The following quantities are measured in a transistor :
   $I_C = 5$ mA ; $I_B = 100\mu A$
   Determine α, β and I_E **[0.98; 50 ; 5.1 mA]**
5. A transistor has $\alpha = 0.98$, $I_{CO} = 5$ μA and $I_B = 100$ μA. Find the values of I_C and I_E. **[5.15 mA; 5.25 mA]**
6. Following measurements are made in a transistor :
   $I_C = 5.202$ mA, $I_B = 50$ μA, $I_{CO} = 2$ μA
   Compute the values of α, β and I_E. **[0.99; 100; 5.252 mA]**

SELF EXAMINATION QUESTIONS

A. Fill in the blanks by most appropriate word (s) or numerical value (s).

1. Structurally, a bipolar junction transistor consists of emitter, base and ....
2. Out of the three regions of a transistor, ... is more heavily doped than the other two.
3. For proper functioning of a transistor its *E/B* junction is always .... biased and *C/B* junction .... biased.
4. In a normally-working *PNP* transistor, both base and collector are at .... potential with respect to the emitter.
5. For all transistor configurations, emitter current is the .... of base and collector currents.
6. In an *NPN* transistor, the emitter-to-collector carrier is the ....
7. The leakage currents in a transistor are due to .... carriers.
8. I_{CO} is extremely .... dependent.
9. The ratio of I_C and I_E gives the .... of a transistor.
10. The dc beta of a transistor is given by the ratio of .... and I_B.
11. The transistor β is much .... than the α.
12. A transistor does not conduct unless its ...... base junction is forward-biased.

B. Answer True or False

1. In most transistors, collector region is physically the largest.
2. Emitter of a transistor is the most heavily-doped region.
3. In an *NPN* transistor, collector is positive with respect to base.
4. Irrespective of transistor configuration, emitter current equals the sum of base current and collector current.
5. The α of a transistor is given by the ratio of I_E and I_B.
6. The α of a transistor is always numerically less than unity.
7. The β of a transistor is always greater than its α.
8. The leakage current of transistor is extremely temperature dependent.
9. Base current is able to turn the collector OFF or ON.
10. When *EBJ* is reverse-biased, collector-to-emitter junction is OFF and *vice versa.*

C. Multiple Choice Items

1. The emitter of a transistor is generally doped the heaviest because it
   (*a*) has to dissipate maximum power
   (*b*) has to supply the charge carriers
   (*c*) is the first region of the transistor
   (*d*) must possess low resistance
2. For proper working of a transistor in normal circuits
   (*a*) *EBJ* is reverse-biased, *CBJ* forward-biased
   (*b*) *EBJ* is forward-biased, *CBJ* is forward-biased
   (*c*) *CBJ* is reverse-biased, *EBJ* is forward-biased

(*d*) *CBJ* is reverse-biased, *EBJ* is reverse-biased.

3. In a properly-biased *NPN* transistor, most of the electorns from the emitter
   (*a*) recombine with holes in the base
   (*b*) recombine in the emitter itself
   (*c*) pass through the base to the collector
   (*d*) are stopped by the junction barrier

4. The current amplification factor alpha dc (α_{dc}) is given by
   (*a*) I_C/I_E (*b*) I_C / I_B
   (*c*) I_B / I_E (*d*) I_B / I_C

5. The common-emitter forward amplification factor βdc is given by
   (*a*) I_C / I_E (*b*) I_C / I_B
   (*c*) I_E / I_B (*d*) I_B / I_E

6. The following relationships between α and β are correct EXCEPT
   (*a*) $\beta = \frac{\alpha}{1-\alpha}$ (*b*) $\alpha = \frac{\beta}{1-\beta}$
   (*c*) $\alpha = \frac{\beta}{1+\beta}$ (*d*) $1-\alpha = \frac{1}{1+\beta}$

7. The value of total collector current in a *CB* circuit is
   (*a*) $I_C = \alpha I_E$ (*b*) $I_C = \alpha I_E + I_{CO}$
   (*c*) $I_C = \alpha I_E - I_{CO}$ (*d*) $I_C = \beta I_E$

8. In the case of a bipolar transistor, α is
   (*a*) positive and > 1 (*b*) positive and < 1
   (*c*) negative and >1 (*d*) negative and <1

9. When the *E*/*B* junction of a transistor is reverse-biased, collector current
   (*a*) is reversed
   (*b*) increases
   (*c*) decreases
   (*d*) stops

10. The *EBJ* of a given transistor is forward-biased and its *CBJ* reverse-biased. If the base current is increased, then its
   (*a*) I_C will decrease
   (*b*) V_{CE} will increase
   (*c*) I_C will increase
   (*d*) V_{CC} will increase

ANSWERS

A. Fill in the blanks

1. collector 2. emitter 3. forward, reverse 4. positive 5. sum 6. electron
7. minority 8. temperature 9. alpha 10. I_C 11. larger 12. emitter

B. True or False

1. T 2. T 3. T 4. T 5. F 6. T 7. T 8. T 9. T 10. T

C. Multiple Choice Items

1. *b* 2. *c* 3. *c* 4. *a* 5. *b* 6. *b* 7. *b* 8. *d* 9. *d* 10. *c*

CHAPTER 19

Transistor Characteristics And Approximations

1. Common Emitter Test Circuit
2. Common Emitter Static Characteristics
3. Common Collector Static Characteristics
4. Different Ways of Drawing Transistor Circuits
5. Common Collector Formulas
6. The Beta Rule
7. Cut-off and Saturation Points
8. Normal DC Voltage Transistor Indications
9. Transistor Fault Location
10. Notation for Voltages and Currents
11. Applying AC to a DC Biased Transistor
12. Transistor AC/DC Analysis

19.1. Transistor Static Characteristics

These are the curves which represent relationship between different dc currents and voltages of a transistor. These are helpful in studying the operation of a transistor when connected in a circuit. The three important characteristics of a transistor are :

1. input characteristic
2. output characteristic
3. constant-current transfer characteristic.

19.2. Common Base Test Circuit

The static characteristics of an *NPN* transistor connected in common-base configuration can be determined by the use of test circuit shown in Fig. 19.1. Milliammeters are included in series with the emitter and collector circuits to measure I_E and I_C. Similarly, voltmeters are connected *across* E and B to measure voltage V_{BE}* and across C and B to measure V_{CB}.* The two potentiometer resistors R_1 and R_2 supply variable voltage from the collector and emitter dc supplies.

* In these subscripts, first letter indicates that electrode which is at higher potential as compared to the other electrode.

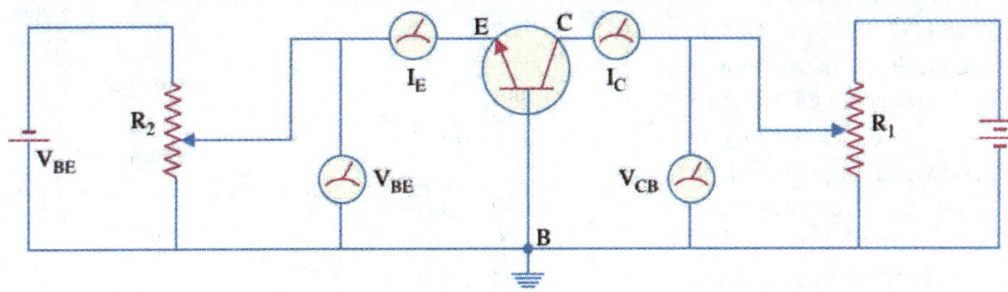

Fig. 19.1

19.3. Common Base Static Characteristics

(a) Input Characteristic

It shows how I_E varies with V_{BE} when voltage V_{CB} is held constant. The method of determining this characteristic is as follows :

First, voltage V_{CB} is adjusted to a suitable value with the help of R_1 (Fig. 19.1). Next, voltage V_{BE} is increased in a number of discrete steps and corresponding values of I_E are noted from the milliammeter connected for the purpose. When plotted, we get the input characteristics shown in Fig. 19.2, one for Ge and the other for Si. Both curves are exactly similar to the forward characteristic of a *P-N* diode which, in essence, is what the emitter-base junction is.

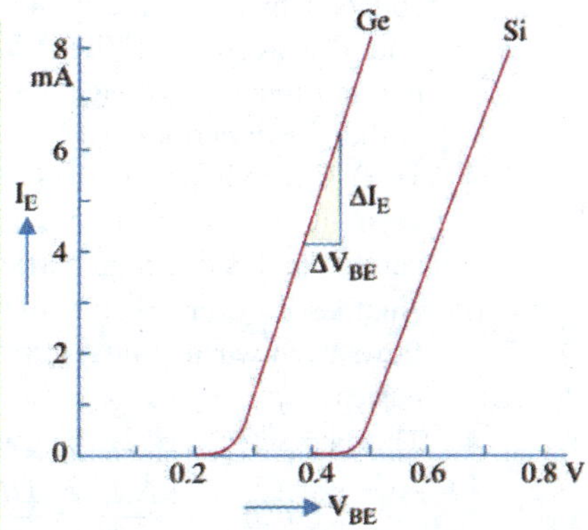

Fig. 19.2

This characteristic may be used to find the input resistance of the transistor. Its value is given by the reciprocal of its slope.

$$R_{in} = \frac{\Delta V_{BE}}{\Delta I_E} \qquad \text{— } V_{CB} \text{ constant}$$

Since the characteristic is initially non-linear, R_{in} will vary with the point of measurement. Its value over linear part of the characteristic is about 50 Ω but for low values of V_{BE}, it is considerably greater. This change in R_{in} with change in V_{BE} *gives rise to distortion of signals handled by the transistor*.

This characteristic is hardly affected by changes either in V_{CB} or temperature.

(b) Output Characteristic

It shows the way I_C varies with V_{CB} when I_E is held constant.

The method of obtaining this characteristic is as follows :

First, movable contact on R_2 (Fig. 19.1) is changed to get a suitable value of V_{BE} and hence I_E. While keeping I_E constant at this value, V_{CB} is increased from zero in a number of steps and the corresponding collector current I_C that flows is noted.

Next, V_{CB} is reduced back to zero, I_E is increased to a value a little higher than before and the whole procedure is repeated. In this way, whole family of curves is obtained, a typical family being shown in Fig. 19.3.

1. The reciprocal of the slope of the near horizontal part of the characteristic gives the output resistance R_{out} of the transistor which it would offer to an input signal. Since the characteristic is linear over most of its length (meaning that I_C is virtually independent of V_{CB}), R_{out} is very high, a typical value being 500 kΩ.

2. It is seen that I_C flows even when $V_{CB} = 0$. For examples, it has a value = 1.8 mA corresponding to $V_{CB} = 0$ for $I_E = 2$ mA as shown in Fig. 19.3. It is due to the fact that electrons are being injected into the base under the action of forward-biased E/B junction and are being collected by the collector due the action of the internal junction voltage at the C/B junction (Art. 18.2). For reducing I_C to zero, it is essential to neutralize this potential barrier by applying a small forward bias across C/B junction.

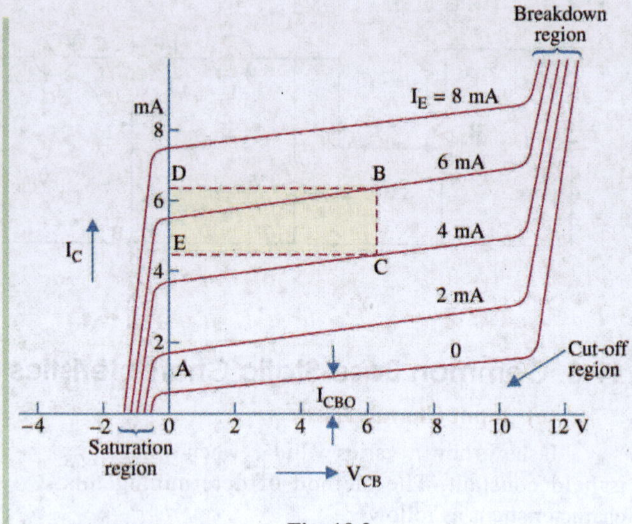

Fig. 19.3

3. Another important feature of the characteristic is that a small amount of collector current flows even when emitter current $I_E = 0$. As we know (Art. 18.12), it is collector leakage current I_{CBO}.
4. This characteristic may be used to find α_{ac} of the transistor as shown in Fig. 19.3.

$$\alpha_{ac} = \frac{\Delta I_C}{\Delta I_E} = \frac{DE}{BC} = \frac{6.2 - 4.3}{2} = 0.95$$

5. Another point worth noting is that although I_C is practically independent of V_{CB} over the working range of the transistor, yet if V_{CB} is permitted to increase beyond a certain value, I_C eventually increases rapidly due to avalanche breakdown as shown.

(c) Current Transfer Characteristic

It shows how I_C varies with changes in I_E when V_{CB} is held constant. For drawing this characteristic, first V_{CB} is set to a convenient value and then I_E is increased in steps and corresponding values of I_C noted. A typical transfer characteristic is shown in Fig. 19.4 (*a*). Fig. 19.4 (*b*) shows a more detailed view of the portion near the origin.

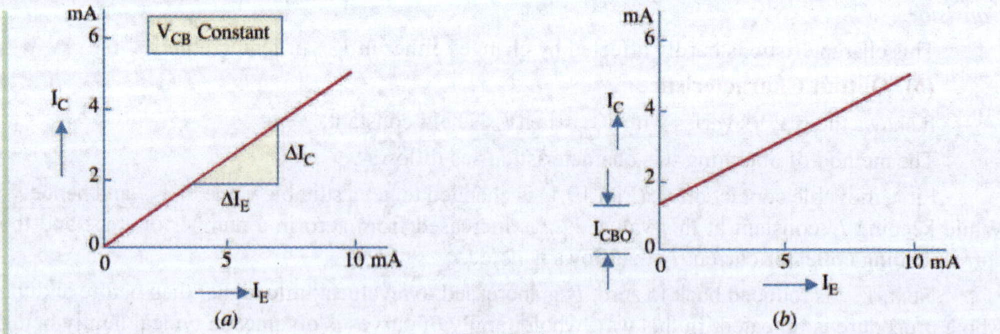

Fig. 19.4

As seen, α_{ac} may be found from the equation

$$\alpha_{ac} = \frac{\Delta I_C}{\Delta I_E}$$

Usually, α_{ac} is found from output characteristic than from this characteristic.

It may be noted in the end that *CB* connection is rarely employed for audio-frequency circuits because (*i*) its current gain is less than unity and (*ii*) its input and output resistances are so different.

19.4. Common Emitter Test Circuit

The static characteristics of a transistor connected in *CE* configuration may be determined by the use of circuit diagram shown in Fig. 19.5. A milliammeter (or a microammeter in the case of a low-power transistor) is connected in series with the base to measure I_B. Similarly, an ammeter is included in the collector circuit to measure I_C. A voltmeter with a typical range of 0-1 V is connected across base and emitter terminals for measuring V_{BE}.

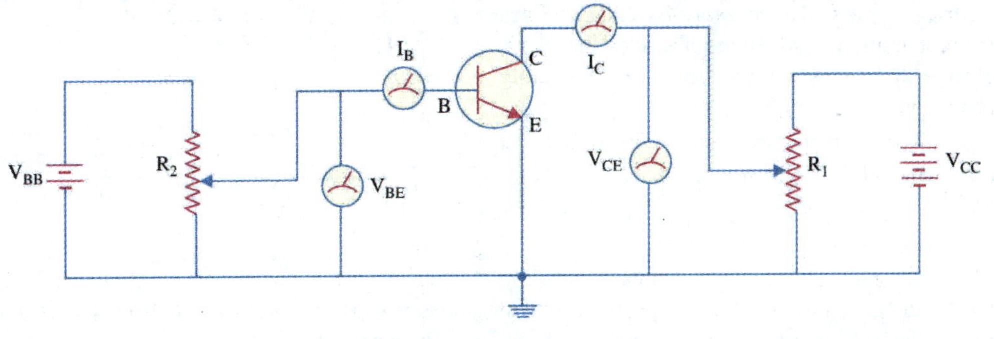

Fig. 19.5

Potentiometer R_2 connected across dc supply V_{BB} is used to vary I_B and V_{BE}. A second voltmeter with a typical range of 0-20 V is connected across collector-emitter terminals to measure the output collector-emitter voltage V_{CE}.

19.5. Common Emitter Static Characteristics

(*a*) Input Characteristic

It shows how I_B varies with changes in V_{BE} when V_{CE} is held constant at a particular value.

To begin with, voltage V_{CE} is maintained constant at a convenient value and then V_{BE} is increased in steps. Corresponding values of I_B are noted at each step. The procedure is then repeated for a different but constant value of V_{CE}. A typical input characteristic is shown in Fig. 19.6. Like *CB* connection, the overall shape resembles the forward characteristic of a *P-N* diode. The reciprocal of the slope gives the input resistance R_{in} of the transistor.

$$R_{in} = \frac{1}{\Delta I_B / \Delta V_{BE}} = \frac{\Delta V_{Be}}{\Delta I_B}$$

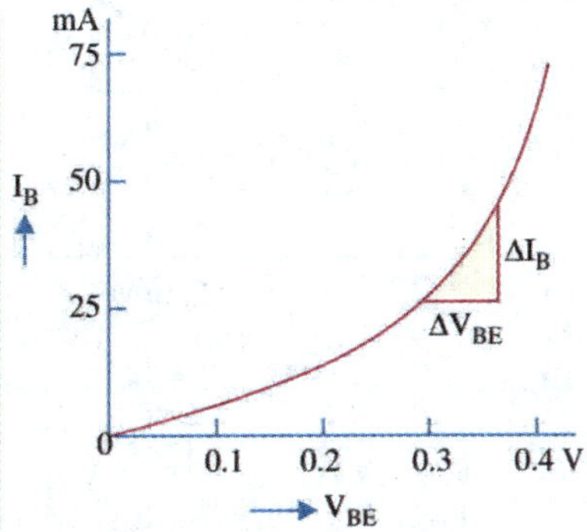

Fig. 19.6

Due to initial non-linearity of the curve, R_{in} varies considerably from a value of 4 kΩ near the origin to a value of 600 Ω over the more linear part of the curve.

(*b*) Output or Collector Characteristic

It indicates the way in which I_C varies with changes in V_{CE} when I_B is held constant.

For obtaining this characteristic, first I_B is set to a convenient value and maintained constant and then V_{CE} is increased from zero in steps, I_C being noted at each step. Next, V_{CE} is reduced to zero and I_B increased to another convenient value and the whole procedure repeated. In this way, a family of curves (Fig. 19.7) is obtained.

It is seen that as V_{CE} increases from zero, I_C rapidly increases to a near saturation level for fixed value of I_B. As shown, a small amount of collector current flows even when $I_B = 0$. It is called I_{CEO} (Art. 18.2). Since main collector current is zero, the transistor is said to be *cut-off*.

It may be noted that if V_{CE} is allowed to increase too far, *C/B* junction completely breaks down and due to this avalanche breakdown, I_C increases rapidly and may cause damage to the transistor.

When V_{CE} has very low value (ideally zero), the transistor is said to be *saturated* and it operates in the saturation region of the characteristic. Here, change in I_B does not produce a corresponding change in I_C.

This characteristic can be used to find β at a specific value of I_B and V_{CE}.

$$\beta = \frac{\Delta I_C}{\Delta I_B}$$

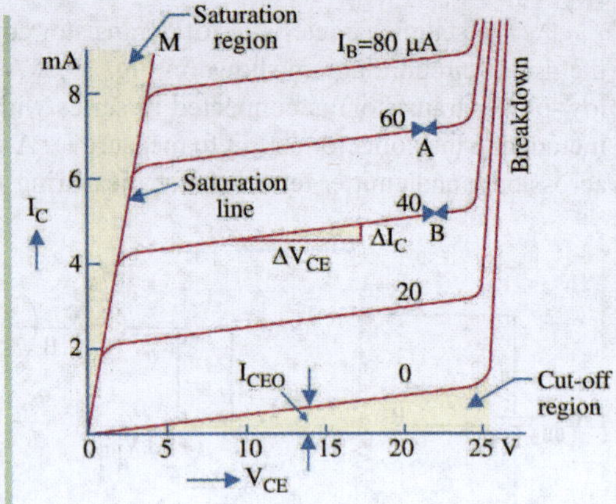

Fig. 19.7

We may select any two points *A* and *B* on the I_B = 60 μA and 40 μA lines respectively and measure corresponding values of I_C from the diagram for finding ΔI_C. Since $\Delta I_B = (60 - 40) = 20$ μA, β can be easily found.

The value of output resistance R_{out} (= $\Delta V_{CE}/\Delta I_C$) over the near horizontal part of the characteristic varies from 10 kΩ to 50 kΩ.

(c) Current Transfer Characteristic

It indicates how I_C varies with changes in I_B when V_{CE} is held constant at a given value. Such a typical characteristic is shown in Fig. 19.8 (*a*). Its slope gives

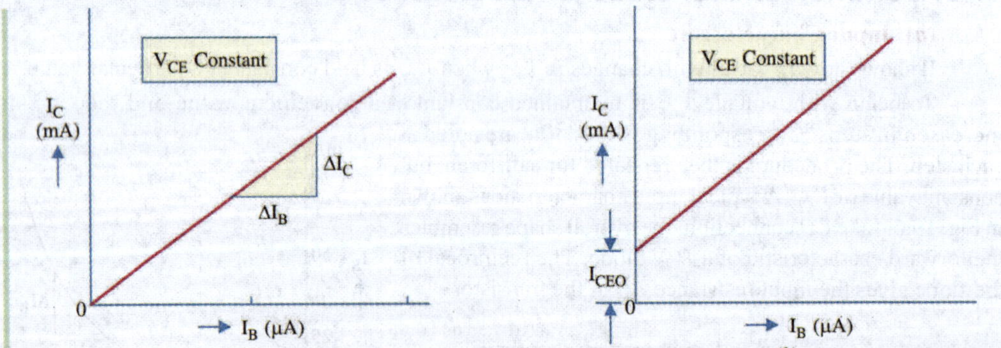

Fig. 19.8

$$\beta = \frac{\Delta I_C}{\Delta I_B}$$

From Fig. 19.8 (*b*), it is seen that a small collector current flows even when $I_B = 0$. It is the common-emitter leakage current $I_{CEO} = (1 + \beta)I_{CO}$. Like I_{CO}, it is also due to the flow of minority carriers across the reverse-biased *C/B* junction.

19.6. Common Collector Static Characteristics

As shown in Fig. 19.9, in this case, collector terminal is common to both the input (*CB*) and output (*CE*) circuits.

The output characteristic is I_E versus V_{CE} for several fixed values of I_B. Since $I_C \cong I_E$, this characteristic is practically identical to that of the *CE* circuit and is shown in Fig. 19.10 (*a*).

Similarly, its current gain characteristic I_C versus I_B for different values of V_{CE} is similar to that of a *CE* circuit because $I_C \cong I_E$.

The *CC* input characteristic is a plot of V_{CB} versus I_B for different values of V_{CE} and is shown in Fig. 19.10 (*b*). It is quite different from those for *CB* or *CE* circuits. This difference is due to the fact that input voltage V_{CB} is largely determined by the value of *CE* voltage. Consider the input characteristic for I_B = 100 μA and V_{CE} = 2 V.

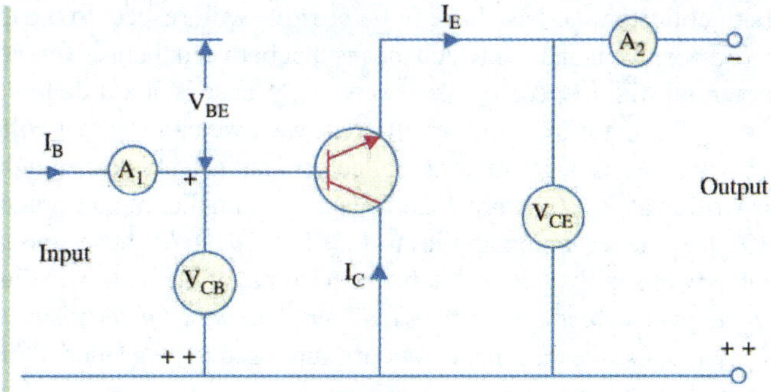

Fig. 19.9

$$V_{CB} = V_{CC} - V_{BE} = 2 - 0.7 = 1.3 \text{ V for Si material}$$

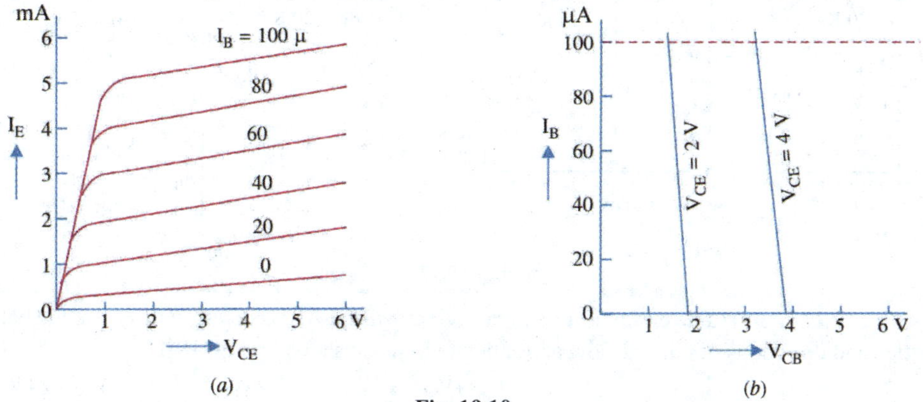

Fig. 19.10

Moreover, as V_{CB} is increased, V_{BE} is reduced thereby reducing I_B.

Now, consider the values V_{CE} = 4 V and I_B = 100 μA

$$V_{CB} = 4 - 0.7 = 3.3 \text{ V}$$

Again, as V_{CB} increases, I_B is decreased.

19.7. Different Ways of Drawing Transistor Circuits

In Fig. 19.11 (*a*) is shown a *CB* transistor circuit which derives its voltage and current requirements from two independent power sources *i.e.*, two different batteries. Correct battery connections can be done by remembering the transistor polarity rule (Art. 18.2) that in an *NPN* transistor,

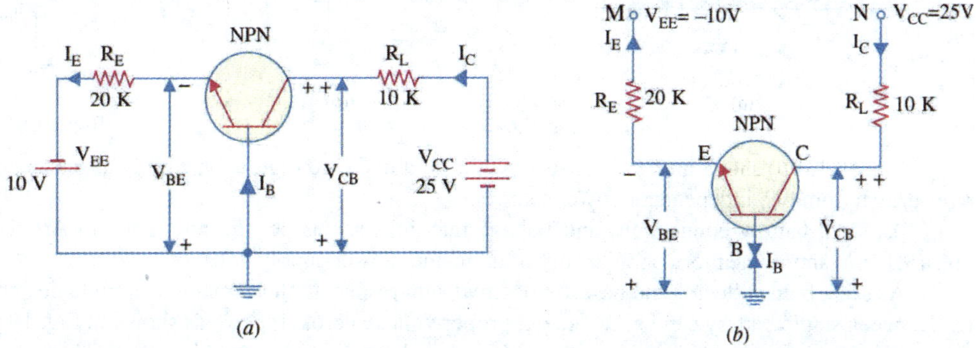

Fig. 19.11

both collector and base have to be *positive* with respect to the emitter. Of course, collector is a *little more* positive than base which means that between themselves, collector is at a *slightly higher positive* potential with respect to base. Conversely, base is at a little lower potential with respect to collector.

Putting it in a slightly different way, we can say that collector is positive w.r.t. base and conversely, base is negative w.r.t. collector. That is why, potential difference between collector and base is written as V_{CB} (and not V_{BC}) because terminal at higher potential is mentioned first [footnote Art. 19.2]. Same reasoning applies to V_{BE}. Fig. 19.11(*b*) shows another and more popular way of indicating power supply voltage. Only one terminal of the battery is shown, the other terminal *is understood to be grounded so as to provide a complete path for the current*. For example, negative terminal of V_{CC} and positive terminal of V_{EE} are supposed to be grounded (as is the base) even though not shown as such in the diagram.

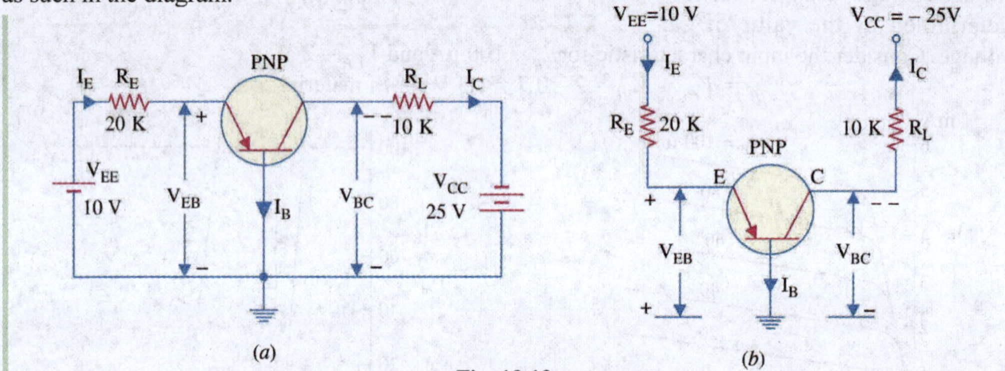

Fig. 19.12

Fig. 19.12 shows same circuit drawn for a *PNP* transistor. As expected, in this case, the battery polarities and directions of current flow are opposite to those shown in Fig. 19.11.

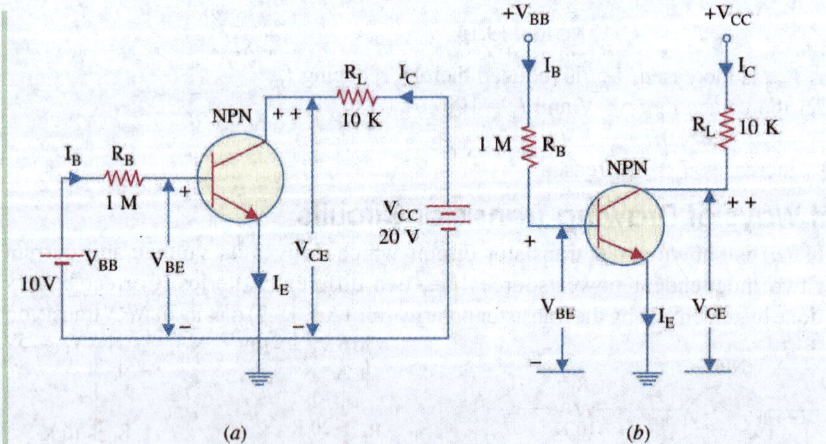

Fig. 19.13

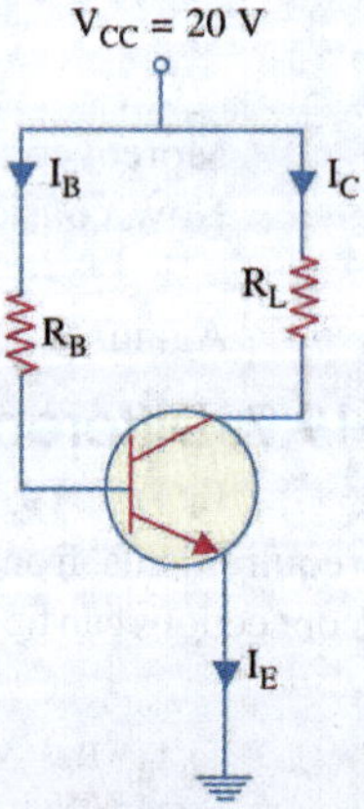

Fig. 19.14

Fig. 19.13 (*a*) shows an *NPN* transistor connected in *CE* configuration with voltages and currents drawn from two independent power sources.

As seen, battery connections and voltage markings are as per the rule given in Art. 18.2. Fig. 19.13 (*b*) shows the more popular way of indicating power-supply voltages.

As seen, both collector and base are positive with respect to the common electrode *i.e.*, emitter. Hence, a single battery can be used to get proper voltage across the two as shown in Fig. 19.14.

Fig. 19.15 (*a*) shows the *CC* configuration of an *NPN* transistor and Fig. 19.15 (*b*) shows the same circuit drawn differently.

19.8. Common Base Formulas

Let us find the values of different voltages and currents for the circuit shown in Fig. 19.11 (*b*). Consider the circuit *MEBM*. Applying Kirchhoff 's voltage law and starting from point *B* (or ground) upwards, we get

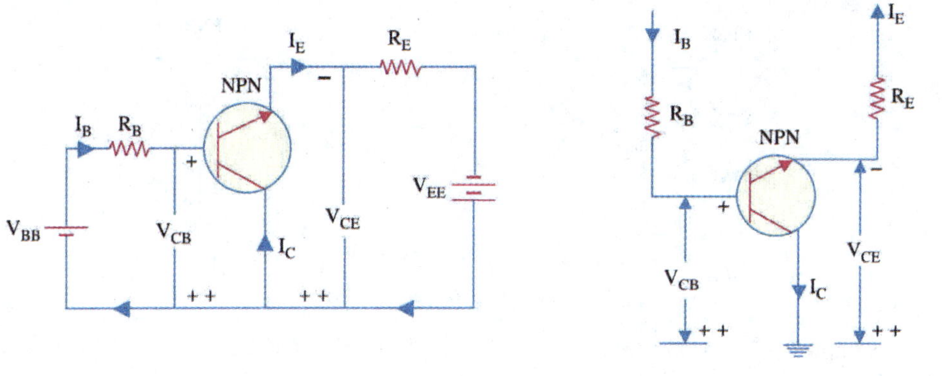

Fig. 19.15

(*a*) $-V_{BE} - I_E R_E + V_{EE}{}^* = 0$ or $I_E = \dfrac{V_{EE} - V_{BE}}{R_E}$

where $V_{BE} = 0.3 \text{ V}$ — for Ge

$= 0.7 \text{ V}$ — for Si

Since, generally, $V_{EE} \gg V_{BE}$, we can simplify the above to

$$I_{EE} \cong V_{EE} / R_E$$

$$= 10 \text{ V} / 20 \text{ K} = 0.5 \text{ mA} \qquad \text{— Fig. 19.11 } (b)$$

Taking V_{BE} into account and assuming silicon transistor

$$I_E = \frac{(10 - 0.7) \text{ V}}{20 \text{ K}} = 0.465 \text{ mA}$$

(*b*) $I_C = \alpha I_E \cong I_E$ — neglecting leakage current I_{CO}

$= 0.5 \text{ mA}$

(*c*) From circuit *NCBEN*, we get

$$V_{CB} = V_{CC} - I_C R_L$$

$$\cong V_{CC} - I_E R_L \qquad \therefore I_C \cong I_E$$

$\therefore$ $V_{CB} = 25 - 0.5 \times 10 = 20 \text{ V}$

Example 19.1. *In the circuit of Fig. 19.16 (a), what value of* R_L *causes* $V_{CB} = 5 \text{ V}$*?*

Solution. $I_E \cong \dfrac{V_{EE}}{R_E} = \dfrac{10 \text{ V}}{10 \text{ K}} = 1 \text{ mA}$

$I_C = \alpha I_E \cong I_E = 1 \text{ mA}$

Now, $V_{CC} = I_C R_L + V_{CB}$

$\therefore$ $R_L = \dfrac{V_{CC} - V_{CB}}{I_C} = \dfrac{20 - 5}{1 \text{ mA}} = \mathbf{15 \text{ K}}$

* It is taken positive because we are going from the negative to the positive terminal of the emitter battery (Art. 3.4).

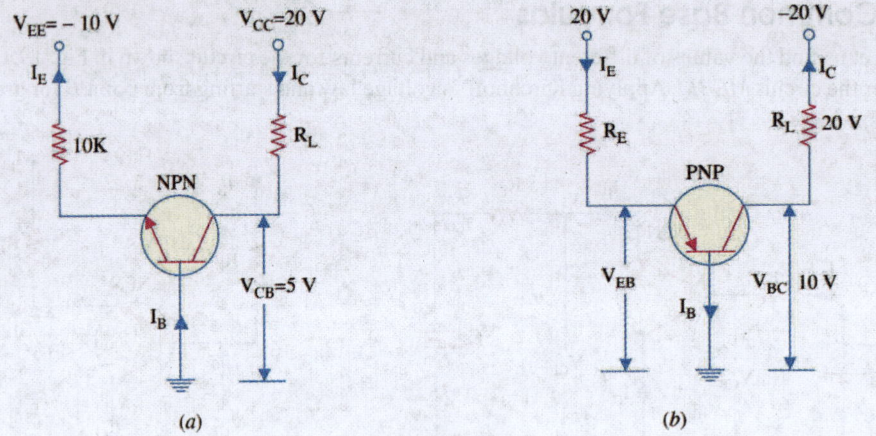

Fig. 19.16

Example 19.2. *For the circuit shown in Fig. 19.16 (b), find the value of R_E which causes $V_{BC} = 10\ V$.*

Solution. $$I_C = \frac{V_{CC} - V_{BC}}{R_L} = \frac{20 - 10}{20\ K} = 0.5\ mA$$

Now, $$I_E = I_C / \alpha \cong I_C = 0.5\ mA$$

If we neglect V_{EB}, then entire V_{EE} = 20 V has to be dropped across R_E.

$$\therefore \quad 0.5\ R_E = 20 \text{ or } R_E = 20 / 0.5\ mA = \mathbf{40\ K}$$

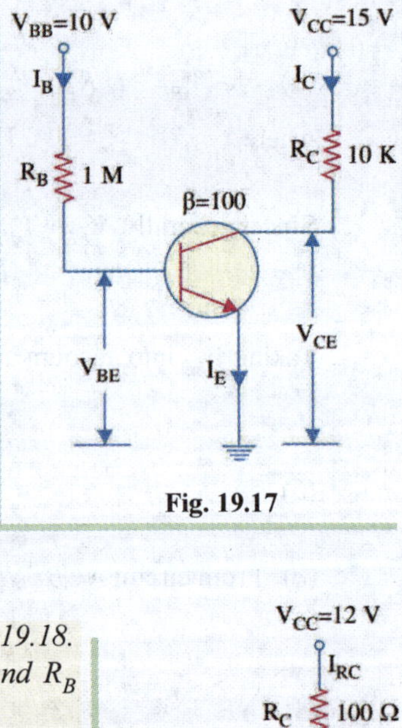

Fig. 19.17

19.9. Common Emitter Formulas

Consider the *CE* circuit of Fig. 19.17.

Taking the emitter-base circuit, we have

$$I_B = \frac{V_{BB} - V_{BE}}{R_B} \cong \frac{V_{BB}}{R_B}$$

$$I_C = \beta\ I_B \qquad \text{— neglecting}$$

Leakage current I_{CEO} (Art. 18.12)

$$V_{CE} = V_{CC} - I_C R_C$$

Example 19.3. *In the collection-feedback bias circuit of Fig. 19.18. Compute (a) the three transistor currents (b) drop across R_C and R_B (c) value of V_{CE}.*

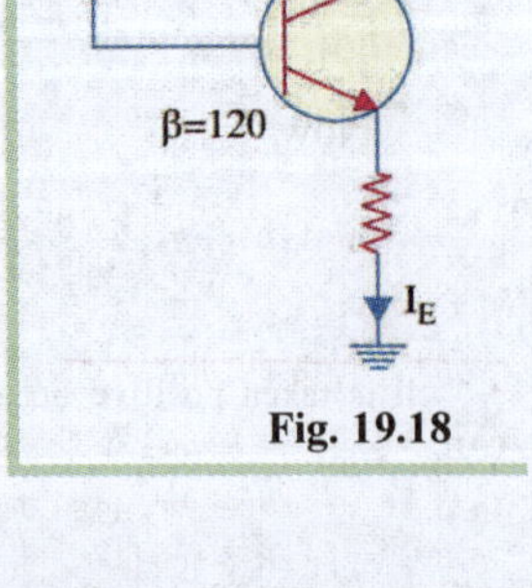

Fig. 19.18

Solution. (*i*) $$I_C = \frac{12 - 0.7}{100 + (12\ K / 120)} = \mathbf{56.5\ mA}$$

$$I_B = I_C / \beta = 56.5\ mA / 120 = \mathbf{471\ \mu A}$$

$$\therefore \quad I_E = I_C + I_B = \mathbf{57\ mA}$$

$$I_{RC} = I_C + I_B = I_E = \mathbf{57\ mA}$$

(*b*) $$V_{RC} = I_{RC} \times R_C = 57\ mA \times 100\Omega = \mathbf{5.7\ V}$$

$$V_{RB} = I_B \times R_B = 471\ \mu A \times 12\ K = \mathbf{5.65\ V}$$

(*c*) $$V_{CE} = V_{CC} - V_{RC} = 12 - 5.7 = \mathbf{6.3\ V}$$

Example 19.4. *For the circuit of Fig. 19.17, find (i) I_B (ii) I_C (iii) I_E and V_{CE}. Neglect V_{BE}.*

Solution. (*i*) $I_B = \frac{V_{BB}}{R_B} = \frac{10}{1\text{ M}} = \mathbf{10\ \mu A}$

(*ii*) $I_C = \beta I_B = 100 \times 10\ \mu\text{A} = \mathbf{1\ mA}$

(*iii*) $I_E = I_B + I_C = 1\text{ mA} + 10\ \mu\text{A} = \mathbf{1.01\ mA}$

(*iv*) $V_{CE} = V_{CC} - I_C R_C = 15 - 1 \times 10 = \mathbf{5\ V}$

Example 19.5. *Find the exact value of emitter current I_E in the two-supply emitter-bias circuit of Fig. 19.19.*

Solution. Let us apply Kirchhoff's voltage law to the loop containing R_B, R_E and V_{EE}. Starting from emitter and going clockwise, we get

$$-I_E R_E + V_{EE} - I_B R_B - V_{BE} = 0$$

$$I_E R_E + I_B R_B = V_{EE} - V_{BE} \quad ...(i)$$

Now, $\beta = \frac{I_C}{I_B} \cong \frac{I_E}{I_B} \quad \therefore \quad I_B \cong \frac{I_E}{\beta}$

Substituting this value in Eq. (*i*) above, we get

$$I_E R_E + \frac{I_E R_B}{\beta} = V_{EE} - V_{BE}$$

$$\therefore \quad I_E = \frac{V_{EE} - V_{BE}}{R_E + R_B/\beta}$$

Since, in most cases, $(R_B/\beta) \ll R_E$

$$\therefore \quad I_E = \frac{V_{EE} - V_{BE}}{R_E} \cong \frac{V_{EE}}{R_E}$$

Also, $I_B = I_E/(1+\beta) \cong I_E/\beta$.

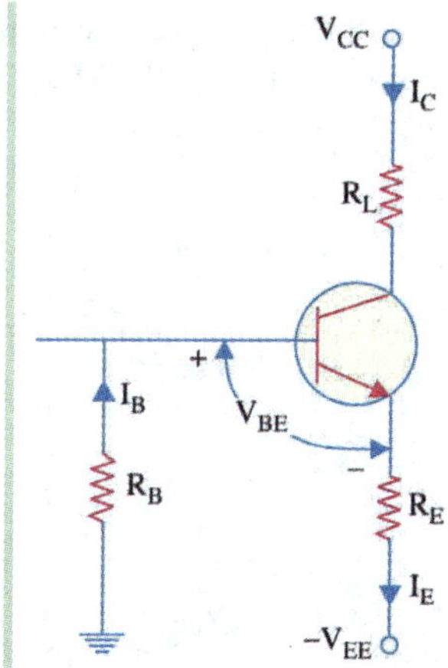

Fig. 19.19

Example 19.6. *In the circuit of Fig. 19.20, find (i) I_E (ii) I_B (iii) I_C and (iv) V_{CE}. Neglect V_{BE} and take $\alpha = 100$.*

Solution. (*i*) $I_E = \frac{V_{EE}}{R_E + R_B/\beta} = \frac{30}{30 + 20/100} \cong \mathbf{1\ mA}$

(*ii*) $I_B \cong I_E/\beta = 1/100 = \mathbf{0.01\ mA}$

(*iii*) $I_C = I_E - I_B = 1.0 - 0.01 = \mathbf{0.99\ A}$

(*iv*) $V_{CE} = V_{CC} - I_C R_L = 30 - 10 \times 0.99 = \mathbf{20.1 V}$

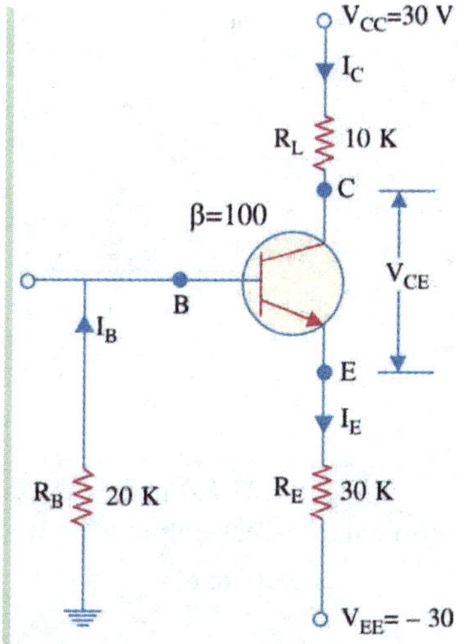

Fig. 19.20

Example 19.7. *In the emitter bias circuit of Fig. 19.21 calculate V_{RC}, V_{CE} and V_{RE}. What is the total sum of these voltage drops? Comment on the result.*

Solution. This circuit uses two supplies, one positive and the other negative. This arrangement provides a greater range of operating voltages.

In loop *A*, it will be assumed that V_{RB} is negligible. Hence, the 12 V supply voltage is used to meet two drops *i.e.*, $V_{BE} = 0.7$ V and V_{RE}. Since $V_{RE} = 12 - 0.7 = \mathbf{11.3\ V}$, hence, $I_E = V_{RE}/R_E = 11.3/9 = 1.255$ mA.

Since $I_C \cong I_E$, hence voltage drop across R_C is

$$V_{RC} = I_C R_C = 1.255 \times 4 = \mathbf{5.02\ V}$$

In the loop B, the two places where voltage can drop are : base-collector junction and R_C because V_{RB} is negligible

$$\therefore \quad V_{BC} = V_{CC} - V_{RC} = 12 - 5.02 = 6.98 \text{ V}$$

Since V_{BC} nearly equals V_C , hence, collector voltage with respect to ground is 6.98 V. Also, voltage at the emitter is – 0.7 V.

$$\therefore \quad V_{CE} = 6.98 - (-0.7) = \mathbf{7.68\ V.}$$

It will be seen that the sum total of V_{RC} , V_{CE} and V_{RE} along with the two junction drops equals 20 V. This is as it should be because the circuit has positive and negative supplies whose total range spans 20V.

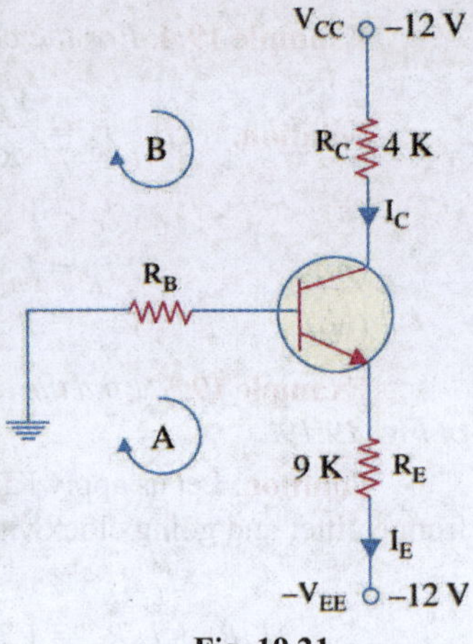

Fig. 19.21

19.10. Common Collector Formulas

The *CC* circuit with its proper dc biasing voltage sources is shown in Fig. 19.22 (*a*). The two circuits given in Fig. 19.22 represent the same thing.

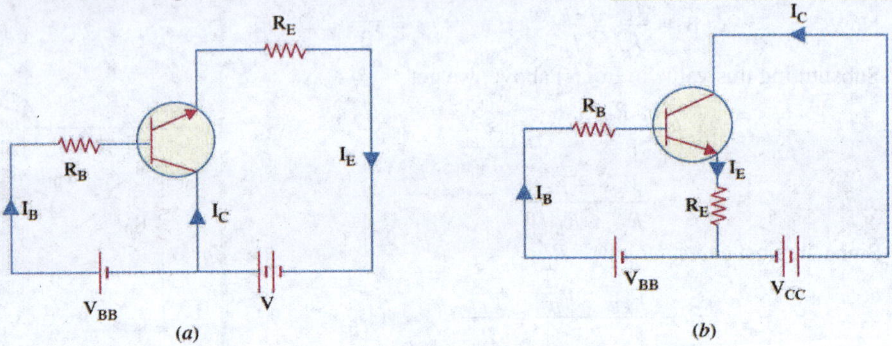

Fig. 19.22

Another way of drawing the same circuit is shown in Fig. 19.23 (*a*) where only one battery has been used. It should be noted that load resistor is not in the collector lead but in the emitter lead as shown.

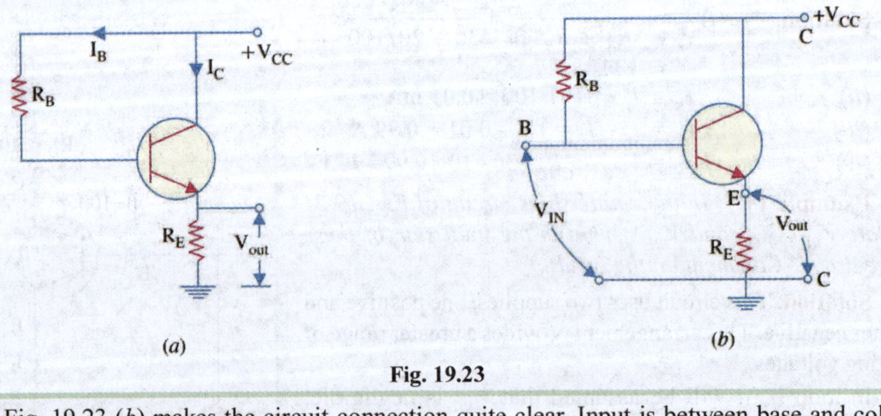

Fig. 19.23

Fig. 19.23 (*b*) makes the circuit connection quite clear. Input is between base and collector terminals whereas output is between emitter and collector terminals.

It is seen that

$$I_E = \frac{V_{CC} - V_{BE}}{R_E + R_B/\beta}; \quad V_{CC} = V_{CE} + I_E R_E$$

$$I_B = \frac{V_{CC} - V_{BE}}{R_B + \beta R_E}; \quad I_C = \beta I_B$$

Example 19.8. *In the CC circuit of Fig. 19.24, find (a) I_B (b) I_E (c) V_{CE} (d) V_E and (e) V_B Take $\beta = 49$ and $V_{BE} = 0.7$ V.*

Solution. (*a*) $I_B = \dfrac{V_{CC} - V_{BE}}{R_B + (1+\beta) R_E}$

$$= \frac{9.0 - 0.7}{100 + (1 + 49)\,2}$$

$$= \frac{8.3}{200\text{ K}} = \mathbf{41.5\ \mu A}$$

(*b*) $I_E = (1 + \beta) I_B$

$= 50 \times 41.5 = \mathbf{2.075\ mA}$

(*c*) $V_{CE} = V_{CC} - I_E R_E$

$= 9 - (2.075 \times 2) = \mathbf{5.85\ V}$

(*d*) $V_E = I_E R_E = 2.075 \times 2 = \mathbf{4.15\ V}$

(*e*) $V_B = V_{BE} + I_E R_E$

$= 0.7 + 4.15 = \mathbf{4.85\ V}$

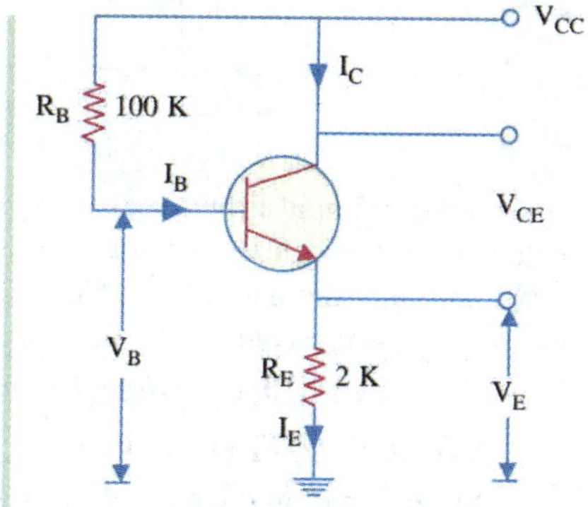

Fig. 19.24

19.11. The Beta Rule

According to this rule, resistance from one part of a transistor circuit can be referred to another of its parts (as we do with the primary and secondary winding impedances of a transformer). For example, resistance, R_L in the collector circuit can be referred to the base circuit and *vice versa*. Similarly, R_E can be referred to the base circuit and, reciprocally, R_B can be referred to emitter circuit. Since current through R_L is $I_C (= \beta I_B)$, hence β factor comes into the picture. Similarly, current through R_E is I_E which is $(1+\beta)$ times I_B, hence $(1 + \beta)$ or approximately β factor comes into picture again. Use of this 'β-rule' makes transistor circuit calculations quite quick and easy. It makes the calculation of I_B quite simple.

The 'β-rule' may be stated as under :

1. When referring R_L to base circuit, **multiply** it by β. When referring R_B to collector circuit, **divide** it by β.
2. When referring R_E to base circuit, **multiply** it by (1 + β) or just β (as a close approximation).
3. Similarly, when referring R_B to emitter circuit, **divide** it by (1 + β) or β.

Before you apply this rule to any circuit, you must remember one very important point otherwise you are likely to get wrong answers. The point is that *only those resistances are transferred which lie in the path of the current being calculated*. Not otherwise. The utility of this rule will be demonstrated by solving the following problems.

Example 19.9. *Calculate the value of V_{CE} in the collector stabilization circuit of Fig. 19.25.*

Solution. We will use β-rule to find I_C in the following two ways :

(*i*) First Method

Here, we will transfer R_L to base circuit

$$I_B = \frac{V_{CC}}{R_B + \beta R_L} = \frac{20}{1000 + 100(10)} = 10\,mA$$

$$I_C = \beta I_B$$
$$= 100 \times 10 = 1000 \text{ mA} = 1 \text{ A}$$
$$V_{CE} \cong V_{CC} - I_C R_L$$
$$= 20 - 1 \times 10 = \mathbf{10\ V.}$$

(*ii*) Second Method

Now, we will refer R_B to collector circuit.

$$I_C \cong \frac{V_{CC}}{R_L + R_B/\beta} = \frac{20}{10 + (1000/100)} = 1\text{A}$$
$$V_{CE} = V_{CC} - I_C R_L = 10 \text{ V} \quad \text{— as before}$$

It was a simple circuit because $R_E = 0$ and R_B was connected to V_{CC} through R_L and not directly (in which case, R_L would not lie in the path of I_B). Now, we will consider the case when R_E is present and R_L does not lies in the path of I_B.

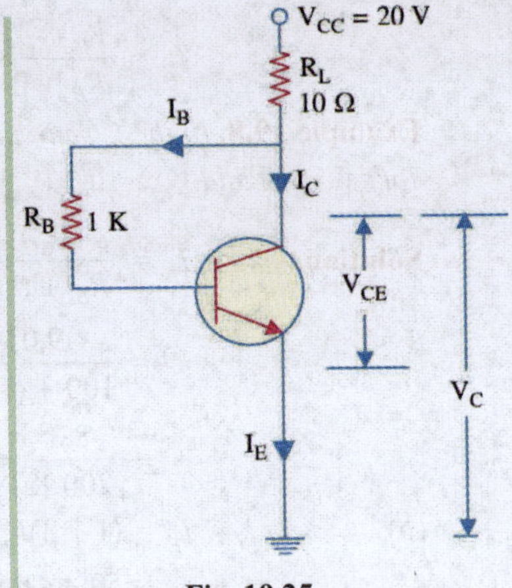

Fig. 19.25

Example 19.10. *Calculate the three transistor currents in the circuit of Fig. 18.26.*

Solution. (*i*) First Method

Since R_E lies in the path of I_B

$$\therefore \quad I_B = \frac{V_{CC}}{R_B + \beta R_E} \quad \text{— neglecting } V_{BE}$$
$$= \frac{10}{100 + 200 \times 0.5} = \mathbf{0.05\ mA}$$
$$I_C = I_E = \beta I_B$$
$$= 200 \times 0.05 = \mathbf{10\ mA}$$
$$I_E = I_B + I_C = \mathbf{10.05\ mA}$$

(*ii*) Second Method

Now, we will transfer R_B to emitter circuit and find I_E directly.

$$I_E = \frac{V_{CC}}{R_E + R_B/\beta} = \frac{10}{0.5 + 100/200} = 10 \text{ mA} \quad \text{— as before}$$
$$I_B = I_C/\beta = 10/200 = 0.05 \text{ mA}$$

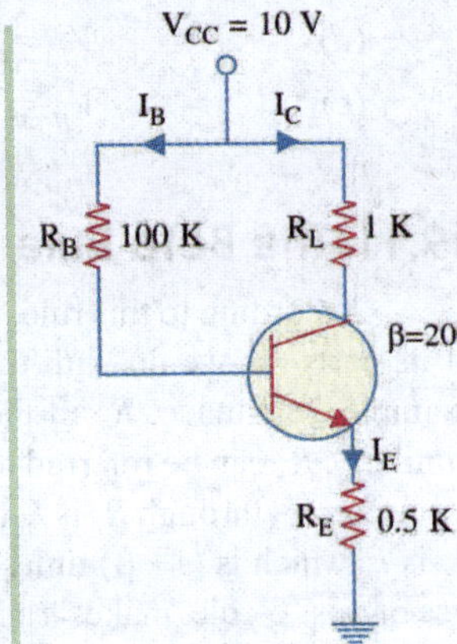

Fig. 19.26

Example 19.11. *Calculate I_E in the circuit of Fig. 19.27.*

Solution. If we neglect V_{BE}, then as seen from the circuit of Fig. 19.27,

$$I_E = \frac{V_{EE}}{R_E + R_B/\beta}$$
$$= \frac{10}{10 + 10/100} = \mathbf{0.99\ mA}$$

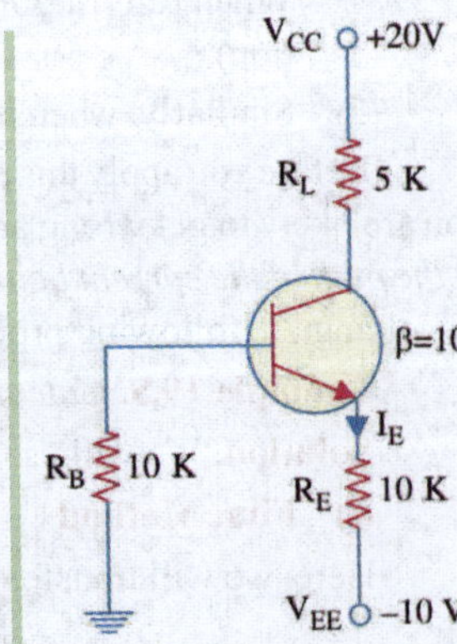

Fig. 19.27

19.12. Importance of V_{CE}

The voltage V_{CE} is very important in checking whether the transistor is

(*a*) defective,

(*b*) working in cut-off (Examples 19.19 and 19.20),

(*c*) in saturation or well into saturation (Examples 19.12 and 19.13).

When $V_{CE} = V_{CC}$, the transistor is in cut-off *i.e.*, it is turned OFF. When $V_{CE} = 0$, the transistor is in saturation *i.e.*, it is turned fully ON.

When V_{CE} is less than zero *i.e.*, negative, the transistor is said to be **well in saturation**. In practice, both these conditions are avoided. For amplifier operation, V_{CE} = 1/2 V_{CC} *i.e.*, transistor is operated at approximately 1/2 ON. In this way, variations in I_B in either direction will control I_C in both directions. In other words, when I_B increases or decreases, I_C also increases or decreases. However, if I_B is OFF, I_C is also OFF. On the other hand, if collector has been turned fully ON, maximum I_C flows. Hence, no further increase in I_B can be reflected in I_C.

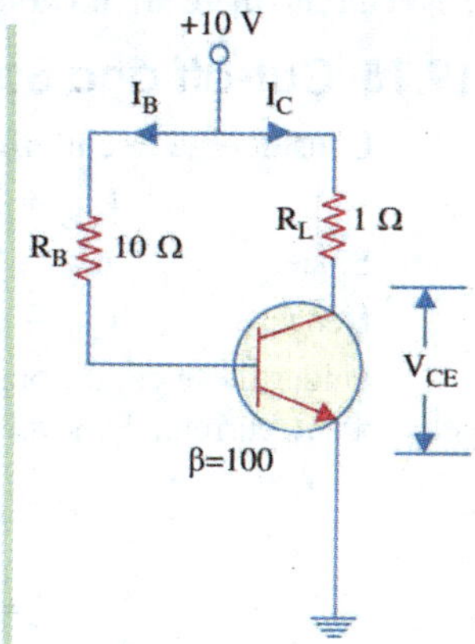

Fig. 19.28

Example 19.12. *For the CE circuit of Fig. 19.28, find the value of V_{CE}. Take β = 100. Is the transistor working in cut-off or saturation?*

Solution.

$$I_B = \frac{10}{100} = 0.1 \text{ A}$$

$$I_C = \beta I_B$$

$$= 100 \times 0.1 = 10 \text{ A}$$

$$V_{CE} = V_{CC} - I_C R_L$$

$$= 10 - 10 \times 1 = \mathbf{0}$$

Obviously, the transistor is operating just at saturation and not well into saturation.

Example 19.13. *Find out whether the transistor of Fig. 19.29 is working in saturation or well into saturation.*

Solution.

$$I_B = \frac{10}{10} = 1 \text{ A}$$

$$I_C = 100 \times 1 = 100 \text{ A}$$

Obviously, I_C cannot be that large because its maximum value is given by V_{CC} / R_L = 10/1 = 10 A. However, let us assume that I_C takes this value temporarily. Then

$$V_{CE} = V_{CC} - L_C R_L$$

$$= 10 - 100 \times 1 = \mathbf{-90 V.}$$

It means that the transistor is working *well into saturation*.

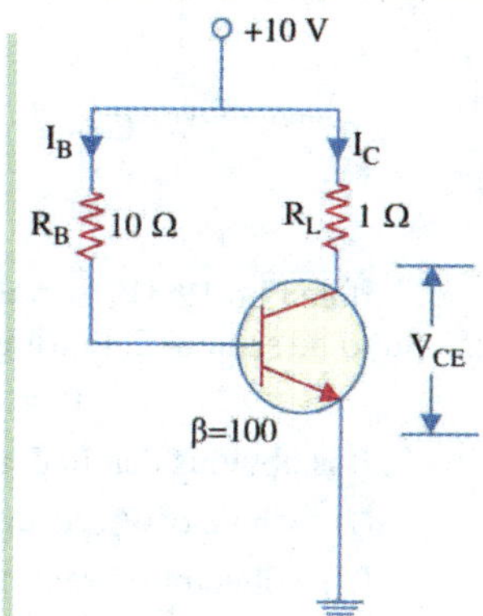

Fig. 19.29

Example 19.14. *In Fig. 19.30, determine the voltage at the conductor under (a) active (b) saturation and (c) cut-off operating conditions.*

Solution. (*a*) The collector current I_C can be found by using the relation

$$I_C = \frac{V_{CC} - V_{BE}}{R_E + (R_B/\beta)} = \frac{10 - 0.7}{100 + (160 \text{ K}/100)} = 6.65 \text{ mA}$$

Now, $V_{RC} = I_C R_C = 6.65 \times 900 = 6 \text{ V}$

∴ $V_C = 12 - 6 = \mathbf{6 V}$

(*b*) The V_{CE} of a saturated transistor is equal to zero. Hence, V_{CC} has to be divided between R_C and B_C. By finding drop across R_C , we can find the saturated voltage at the collector.

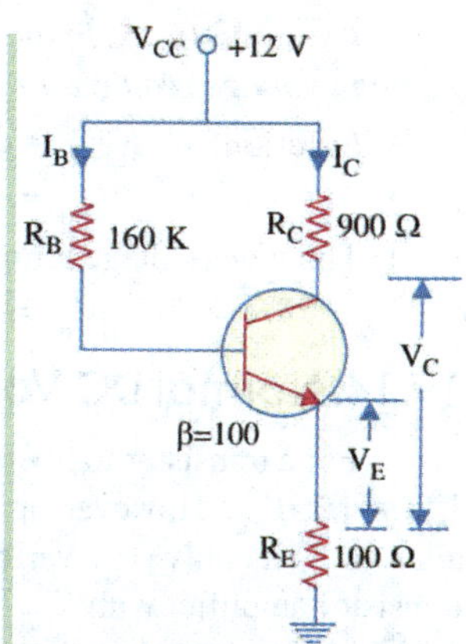

Fig. 19.30

$$I_{C(sat)} = \frac{12}{900 + 100} = 12 \text{ mA}$$

$$V_{RC(sat)} = 12 \times 900 = 10.8 \text{ V}$$

∴ $V_{C(sat)} = 12 - 10.8 = \mathbf{1.2 V}$

(*c*) A transistor that stops conducting is said to be cut-off. In that case, $V_{CE} = V_{CC}$. Since there is no current, there are no voltage drops. Hence $V_{C(cut\text{-}off)}$ = **12 V.**

19.13. Cut-off and Saturation Points

Consider the circuit of Fig. 19.31 (*a*). As seen from Art. 19.9,

$$V_{CE} = V_{CC} - I_C R_L$$

Since, $\quad I_B = 0, I_C = 0$

Hence, $\quad V_{CE} = V_{CC}$

Under these conditions, transistor is said to be cut-off for the simple reason that it does not conduct any current. This value of V_{CE} is written as $V_{CE(cut\text{-}off)}$.

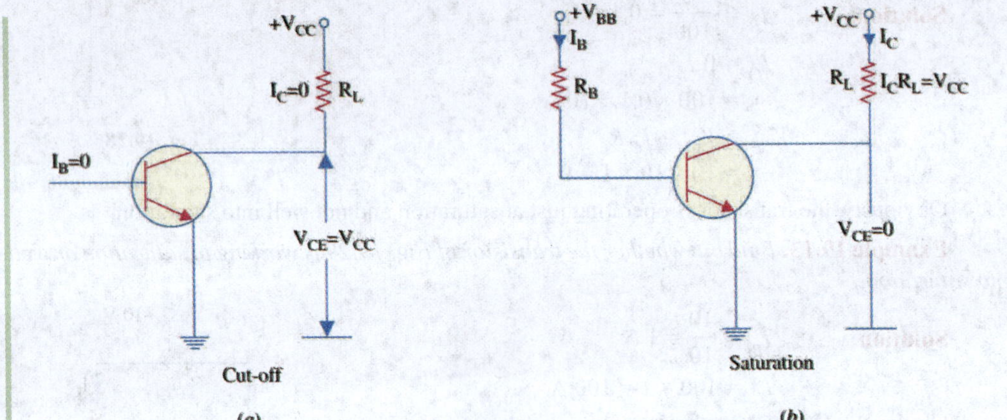

Fig. 19.31

If, in Fig. 19.31 (*b*), values of R_B and R_L are such that V_{CE} comes out to be zero, then transistor is said to be saturated. Putting $V_{CE} = 0$ in the above equation, we get.

$$0 = V_{CC} - I_C R_L \text{ or } I_C = V_{CC} / R_L$$

It is obvious that under saturation condition,

(*i*) whole of V_{CC} drops across R_L,

(*ii*) collector current has maximum possible value called $I_{C(sat)}$.

Normal operation of a transistor lies between the above two extreme conditions of cut-off and saturation.

Example 19.15. *In the silicon transistor shown in Fig. 19.32. Calculate the base current and the input voltage required to saturate the transistor.*

Solution. $\quad I_{C(sat)} = V_{CC} / R_C = 10 \text{ V} / 1 \text{ K} = 10 \text{ mA}$

$$I_B = I_C / \beta = 10 \text{ mA} / 100 = \mathbf{0.1 \text{ mA}}$$

The input voltage required at the base terminal is

$$V_{in} = V_{RB} + V_{BE} = 0.1 \text{ mA} \times 10 \text{ K} + 0.7 = \mathbf{1.7 \text{ V}}$$

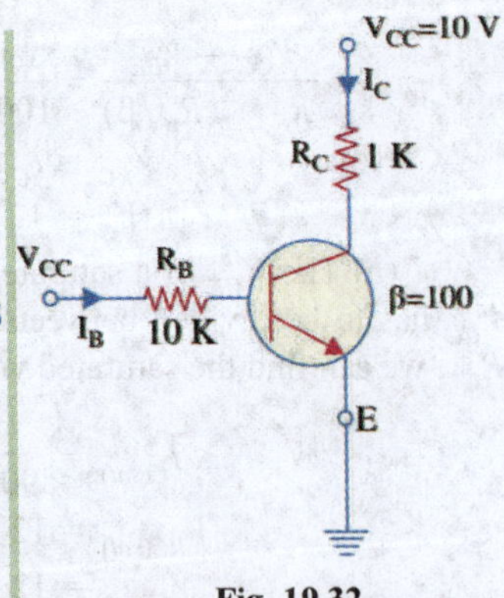

Fig. 19.32

19.14. Normal DC Voltage Transistor Indications

For a transistor to operate as an amplifier, it is desirable that $V_{CE} = 1/2\, V_{CC}$. However, in actual practice, wide tolerances are allowed. Generally, V_{CE} varies between 25% to 75% of V_{CC}. Any transistor amplifier with $V_{CE} = V_{CC}$ is either open or is operating in out-off. When operating with V_{CE} near cut-off, the amplifier causes lot of distortion. Same is the case when V_{CE} is near zero. Hence, any transistor amplifier with V_{CE} more than 75% V_{CC} or less than 25% V_{CC} should be suspected of having a problem and further investigated.

In the circuit shown in Fig. 19.33 (*a*), V_{CE} should be in the range 25-75% of V_{CC}. In the circuit of Fig. 19.33 (*b*), V_{CE} may be normal but either R_L or R_E could be shorted. Hence, V_C and V_E should be measured separately. Moreover, V_{CE} could be found by subtracting V_E from V_C.

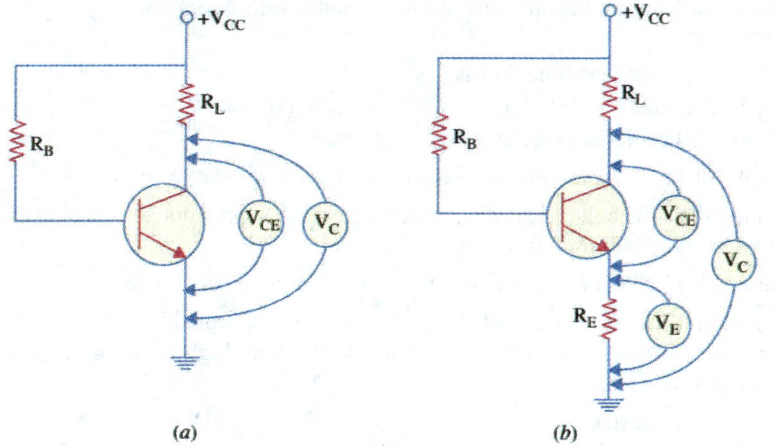

Fig. 19.33

For the circuit of Fig. 19.33 (*b*), the normal voltmeter readings are

$V_{CE} = 1/2\ V_{CC}$; $V_E = 1/4\ V_{CC}$; $V_C = 3/4\ V_{CC}$

If instead of R_L, there is a low-resistance *coil* in the circuit,

then

$V_{CE} = 1/2\ V_{CC}$; $V_E = 1/2\ V_{CC}$; $V_C = V_{CC}$.

Example 19.16. *The LED of Fig. 19.34. has a drop of 1.6 V and requires 10 mA for normal brightness. Taking V_{BE} = 0.7 V and $V_{CE(sat)}$ = 0.1 V for the transistor, calculate the value of the collector resistor R_C and R_B.*

Solution. We have to find the value of R_C that will make I_C = 10 mA and the value of R_B that will guarantee saturation of the transistor with a collector current of 10 mA.

As seen from the Fig. 19.34 drop across R_C is given by

$$V_{RC} = V_{CC} - V_{LED} - V_{CE(sat)}$$
$$= 5 - 1.6 - 0.1 = 3.3 \text{ V}$$

$\therefore$ $R_C = 3.3 \text{ V} / 10 \text{ mA} = \mathbf{330\ \Omega}$

The base current at minimum beta is given by

$$I_B = \frac{I_C}{\beta_{min}} = \frac{10 \text{ mA}}{50} = 0.2 \text{ mA}$$

Drop over $R_B = 2.5 - 0.7 = 1.8$ V

$\therefore$ $R_B = 1.8 \text{ V}/0.2 \text{ mA} = \mathbf{9\ k\Omega}$

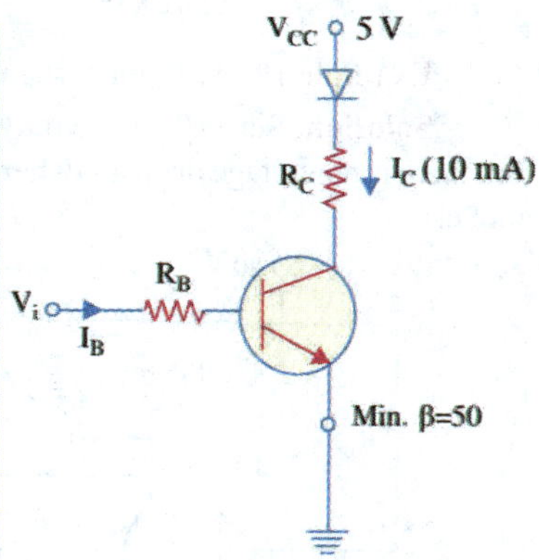

Fig. 19.34

19.15. Transistor Fault Location

Voltage measurements are employed in the vast majority of trouble situations because current measurements are comparatively difficult to make. Magnitude of V_{CE} is of great diagnostic value in finding and locating faults in a transistor circuit. Following possibilities are considered:

1. $V_{CE} = 0$

Possibilities are that the transistor is

(*a*) shorted out (*b*) operating in saturation

(*c*) disconnected from V_{CC}.

2. $V_{CE} = 1/2\ V_{CC}$

It shows that the circuit is operating normally and is well designed.

3. $V_{CE} = V_{CC}$

Possibilities are that the transistor is

(*a*) open-circuited (*b*) operating in cut-off

(*c*) having all resistors in series with V_{CE} shorted.

Example 19.17. *Compute the value of V_{CE} for the circuit shown in Fig. 19.35.*

Solution. Since the collector is disconnected from the supply due to 'open' in the circuit, V_{CE} = 0. It represents fault condition No.1 in Art. 19.15.

Example 19.18. *What is the value of V_{CE} in the circuit of Fig. 19.36 ?*

Solution. Since R_B is shorted out, I_B would increase and probably burn out the *E/B* junction. But this burn out is not indicated in the question. Hence, with high base current, the transistor is operating in saturation so that V_{CE} = 0.

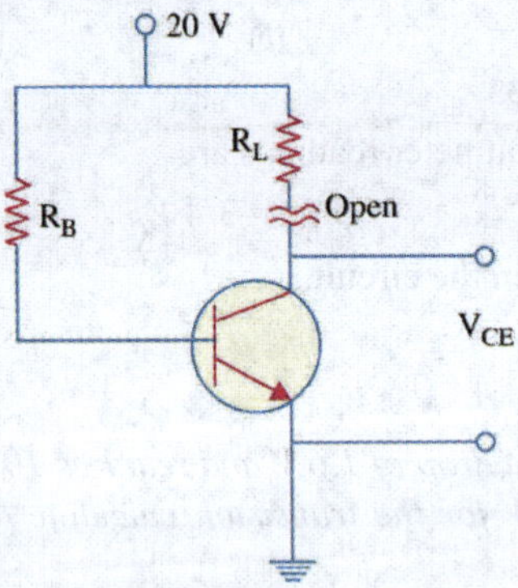

Fig. 19.35

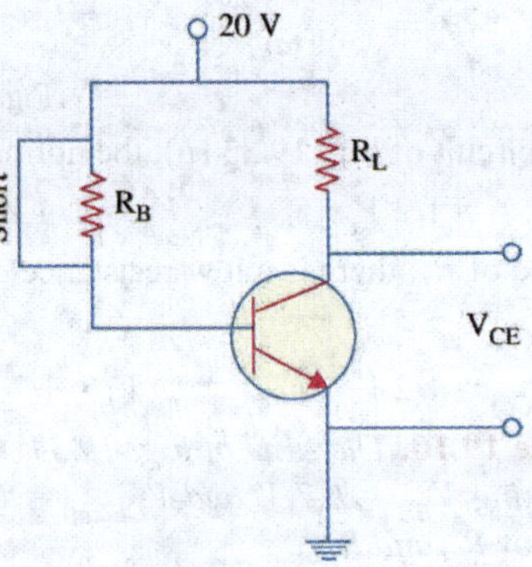

Fig. 19.36

Example 19.19. *What is the value of V_{CE} in the circuit of Fig. 19.37?*

Solution. Since R_L is the only resistor in series with the transistor and is shorted out, it means that there is no voltage drop anywhere. Hence, $V_{CE} = V_{CC}$. It represents fault No. 3 stated in Art. 19.15 above.

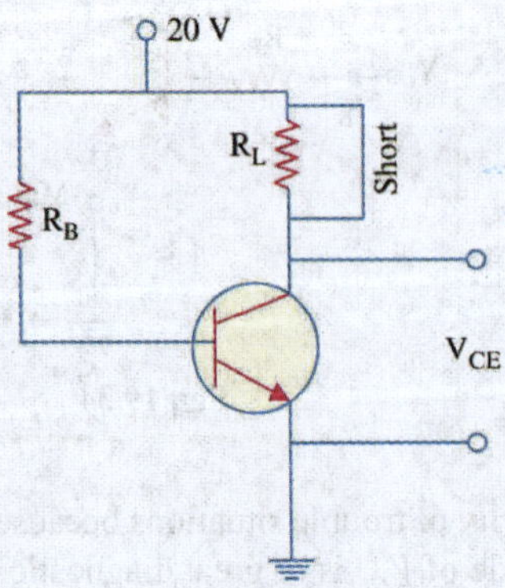

Fig. 19.37

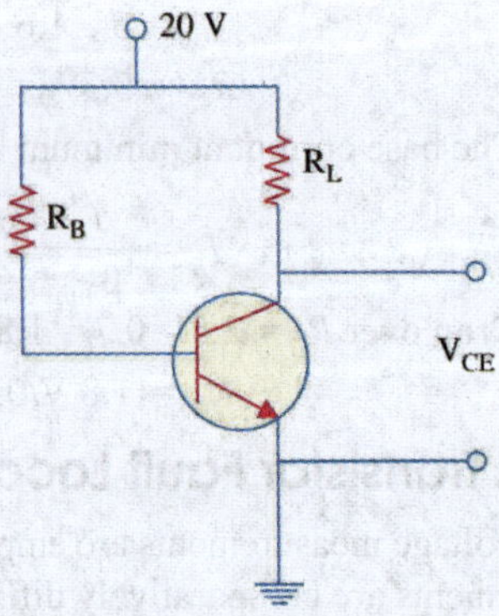

Fig. 19.38

Example 19.20. *Find the possible values of V_{CE}, V_C and V_E for the circuit shown in Fig. 19.38.*

Solution. In the circuit of Fig. 19.38, there is neither a short nor an open and the voltage polarities are correct for an *NPN* transistor. It looks like a well-designed circuit operating normally. Hence, according to Art. 19.15.

$$V_{CE} = 1/2\ V_{CC} = 1/2 \times 20 = \mathbf{10\ V}$$
$$V_E = 1/4\ V_{CC} = 1/4 \times 20 = \mathbf{5\ V}$$
$$V_C = 3/4\ V_{CC} = 3/4 \times 20 = \mathbf{15\ V}$$

Example 19.21. *Find the values of V_C, V_E and V_{CE} in the circuit of Fig. 19.39.*

Solution. Since $I_B = 0$, transistor is cut off.

Hence, $V_E = \mathbf{0}$. Also $V_C = \mathbf{-20\ V}$

and $V_{CE} = \mathbf{-20\ V.}$

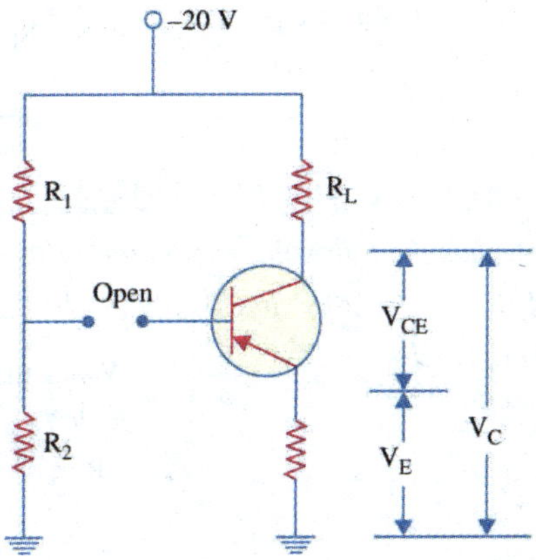

Fig. 19.39

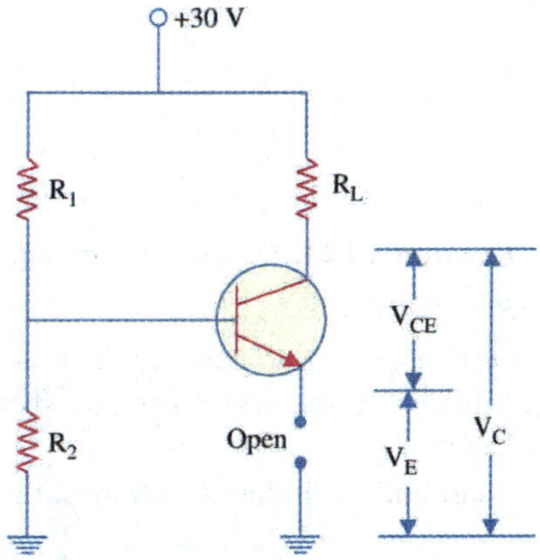

Fig. 19.40

Example 19.22. *What would be the values of V_C, V_E and V_{CE} for the circuit shown in Fig. 19.40?*

Solution. Since emitter is open, no current flows in *any part of the circuit*. The transistor is essentially cut off. Without I_R drops all points above the emitter are at 30 V.

$$\therefore \quad V_E = \mathbf{30\ V},\ V_C = \mathbf{30\ V} \text{ and } V_{CE} = \mathbf{0}$$

19.16. Solving Universal Stabilization Circuit

Such a circuit is shown in Fig. 19.41 in which R_E *appears* to be in parallel with R_2. But according to the β-rule (Art. 19.9), R_2 is actually *in parallel* with βR_E. In a well-designed circuit, the resistance βR_E is much larger than R_2. Hence, their combined resistance = $R_2 \parallel \beta R_E \cong R_2$. On this assumption as well as another that I_B is practically zero, we can find voltage drop across R_2 by the Proportional Voltage Formula (Art. 2.7). Since V_{CC} is applied across $R_1 - R_2$ potential divider circuit, drop across R_2

$$= \frac{R_2}{R_1 + R_2}$$

If we neglect V_{BE}, then this drop equals V_E.

$$\therefore \quad V_E = V_{CC} \cdot \frac{R_2}{R_1 + R_2}$$

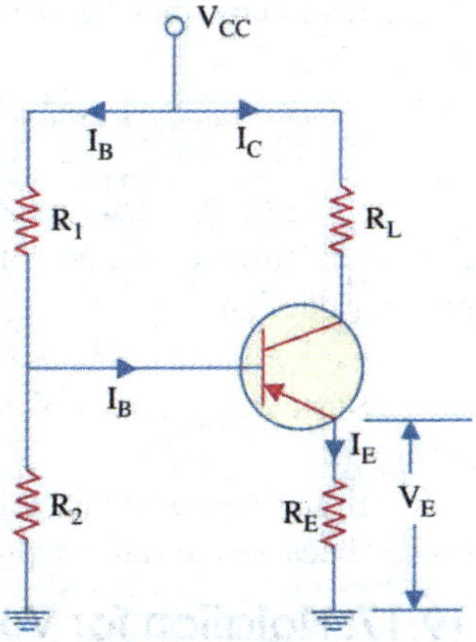

Fig. 19.41

$$I_E = \frac{V_E}{R_E}$$

Having found I_E, other currents and voltage drops can be easily found.

$$V_{CE} = V_{CC} - I_C R_L - I_E R_E$$

Since $\quad I_E \cong I_C$

$$\therefore \quad V_{CE} = V_{CC} - I_E R_L - I_E R_E = V_{CC} - I_E (R_L + R_E)$$

Example 19.23. *Find V_{CE} and V_E for the circuit shown in Fig. 19.42. Neglect V_{BE}.*

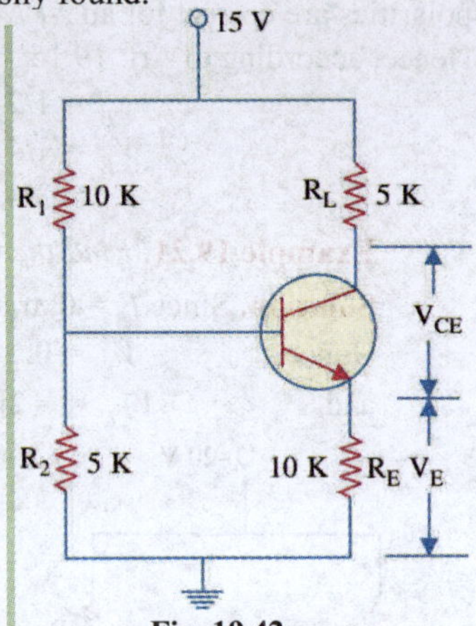

Fig. 19.42

Solution. As explained above

$$V_E = V_{CC} \frac{R_2}{R_1 + R_2} = 15 \times \frac{5}{10+5} = \mathbf{5\ V}$$

$$I_E = \frac{V_E}{R_E} = \frac{5}{10\text{ K}} = 0.5\text{ mA}$$

$$I_C \cong I_E = 0.5\text{ mA}$$

$$V_{CE} = V_{CC} - I_E(R_E + R_L) = 15 - 0.5(5+10) = \mathbf{7.5 V.}$$

Example 19.24. *The silicon transistor of Fig. 19.43 has the following maximum ratings: $P_{D(max)} = 0.6$ W, $V_{CE(max)} = 15$ V, $I_{C(max)} = 100$ mA, $V_{CB(max)} = 20$ V and $V_{EB(max)} = 10$ V.*

Determine the maximum value to which V_{CE} can be adjusted without exceeding the above ratings. Which rating would be exceeded first ?

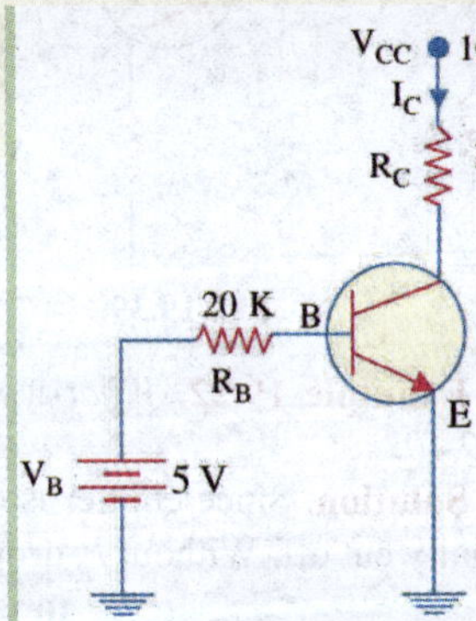

Fig. 19.43

Solution. For finding I_C, we must first calculate I_B,

$$I_B = \frac{V_{BB} - V_{BE}}{R_B} = \frac{5 - 0.7}{20 \times 10^3} = 215\text{ mA}$$

$$I_C = \beta I_B = 100 \times 215\ \mu\text{A} = 21.5\text{ mA}$$

As seen, I_C is much less than $I_{C(max)}$ and will not change with V_{CC} since it depends only on I_B and β.

Voltage drop across R_C is

$$V_{RC} = I_C R_C = 21.5\text{ mA} \times 1\text{ K} = 21.5\text{ V}$$

Let us now find the value of V_{CC} when $V_{CE} = V_{CE(max)} = 15$ V.

$$V_{RC} = V_{CC} - V_{CE} \quad \therefore \quad V_{CC(max)} = V_{CE(max)} + V_{RC}$$

$$\therefore \quad V_{CC(max)} = 15 + 21.5 = \mathbf{36.5\ V}$$

As seen, V_{CC} can be increased upto 36.5 V under the existing conditions before $V_{CE(max)}$ is exceeded. However, we do not know whether or not $P_{D(max)}$ has been exceeded at this point. Let us find this out first.

$$P_D = V_{CE(max)} I_C = 15\text{ V} \times 21.5\text{ mA} = 0.322\text{ W}$$

Since $P_{D(max)}$ is 0.6 W, it is not exceeded when $V_{CE} = 36.5$V. So, $V_{CE(max)}$ is the limiting rating in this case.

It may be noted that if I_B is removed causing the transistor to turn OFF, $V_{CE(max)}$ *will be exceeded* because the entire supply voltage $V_{CC} = 100$ V will be dropped across the transistor.

19.17. Notation for Voltages and Currents

In order to avoid confusion while dealing with dc and ac voltages and currents, following notation will be employed :

1. For dc or non-time-varying quantities

We will use capital letters with capital subscripts such as

I_E, I_B, I_C — for dc currents

V_E, V_B, V_C — for dc voltage to ground

V_{BE}, V_{CB}, V_{CE} — for dc potential differences

V_{EE}, V_{CC}, V_{BB} — for dc source or supply voltages

2. For ac quantities

We will use the following symbols

i_e, i_b, i_c — for instantaneous values of ac currents

I_e, I_b, I_c — for rms values of ac currents

v_e, v_b, v_c — for instantaneous values of ac voltage to ground

v_{be}, v_{cb}, v_{cc} — for ac voltage differences

3. Total ac and dc Voltages and Currents

In this case, we will use a hybrid notation. For example, i_E will be used to represent the *total* emitter current *i.e.*, sum of dc and ac currents in the emitter.

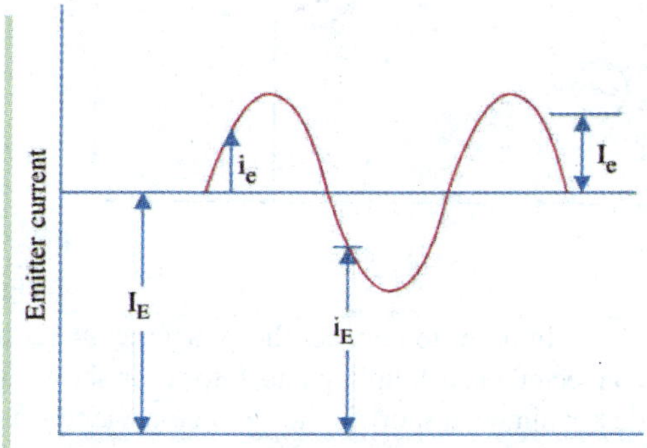

Fig. 19.44

Fig. 19.44 illustrates the notation discussed above.

19.18. Increase / Decrease Notation

This notation is very helpful in analyzing transistor operation when ac signal is applied to it. It is simply this:

↑ means increase and ↓ means decrease

As an illustration, consider the transistor circuit of Fig. 19.45.

If V_{BB} were increased (↑), I_B would increase (↑) this would increase I_C (↑) because it equals βI_B. The drop $I_C R_L$ would increase (↑) and, hence, V_{CE} will decrease (↓) because

$$V_{CE} = V_{CC} - I_C R_L$$

Using increase/decrease notation, the above sequence of changes can be written as

$$V_{BB}\uparrow, I_B\uparrow, I_C\uparrow, I_C R_L\uparrow V_{CE}\downarrow$$

At one look, we can straight away say that as input voltage is increased, output voltage is decreased.

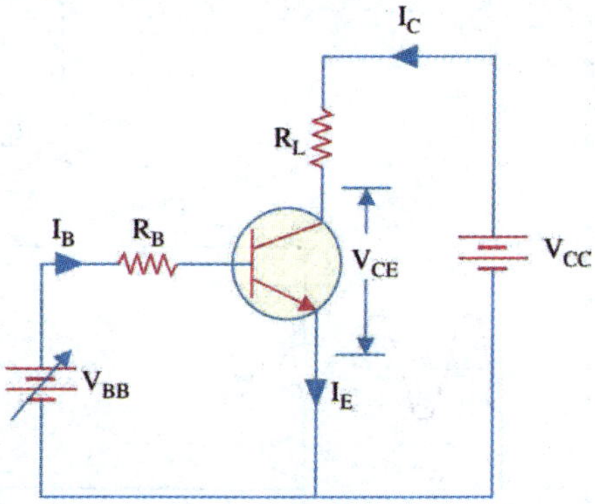

Fig. 19.45

19.19. Applying AC to a DC Biased Transistor

Suppose we want to apply an ac signal to the input emitter-base circuit of a properly-biased transistor shown in Fig. 19.46. If we apply the ac source directly across the *EBJ* as shown in Fig. 19.46 (*a*), it will upset the dc bias. It should be kept in mind that most ac signal sources *are nearly a short to dc*. Hence, nearly whole of I_B would pass through ac source rather than the base thereby spoiling the transistor bias.

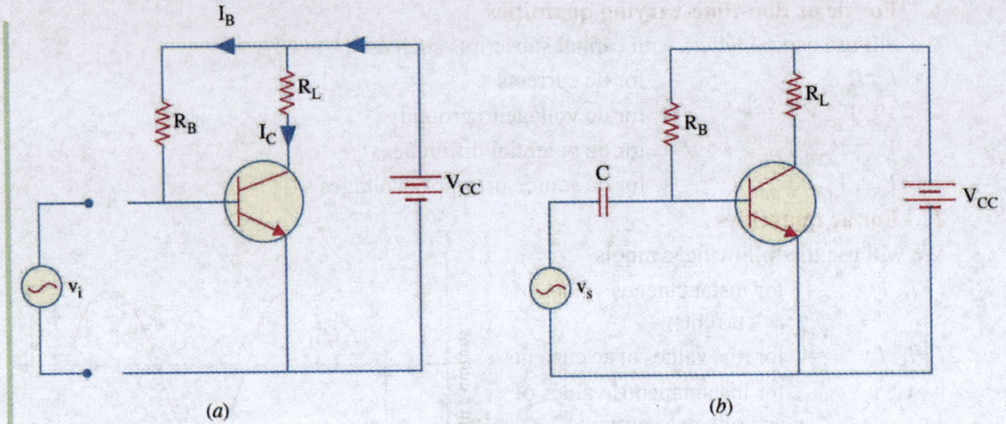

Fig. 19.46

In order to connect the ac source and at the same time not upset the dc bias, the ac source is connected via a coupling capacitor C as shown in Fig. 19.46 (*b*). This capacitor acts as an 'open' for dc but almost a short for ac source provided it is of sufficiently large capacitance.

19.20. Transistor AC/DC Analysis

In Fig. 19.47 is shown a *CE* amplifier circuit having an ac signal voltage v_{be}* applied across its E/B junction. This voltage will be added to the dc voltage V_{BE} as if the two were connected in series. The resultant voltage is shown in Fig. 19.48 (*a*), which shows ac voltage riding the dc level. The variations in the resultant output voltage v_{CE} can be expressed in terms of the increase/decrease notation. It will be assumed that V_{BE} is such as to bias V_{CE} at 1/2 V_{CC} when no ac signal is applied.

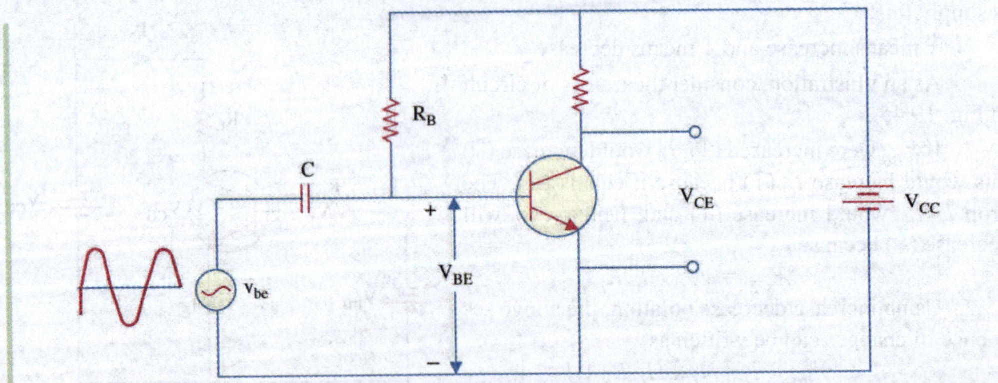

Fig. 19.47

(*i*) First Quarter Cycle

In the first quarter-cycle of the input signal, both v_{be} and v_{BE} increase thereby giving rise to the following sequence of changes

$$v_{BE}\uparrow, i_B\uparrow, i_C\uparrow, i_C R_L\uparrow, v_{CE}\downarrow$$

Hence, output voltage decreases as shown in Fig. 19.48 (*b*).

(*ii*) Second Quarter Cycle

Here, v_{bc} as well as v_{BE} decrease. Hence,

$$v_{BE}\downarrow, i_B\downarrow, i_C\downarrow, i_C R_L\downarrow, v_{CE}\uparrow$$

* Normally, we will use the notation v_i or v_{in} v_s while discussing amplifiers.

Again, v_{CE} does the opposite of v_{BE}.

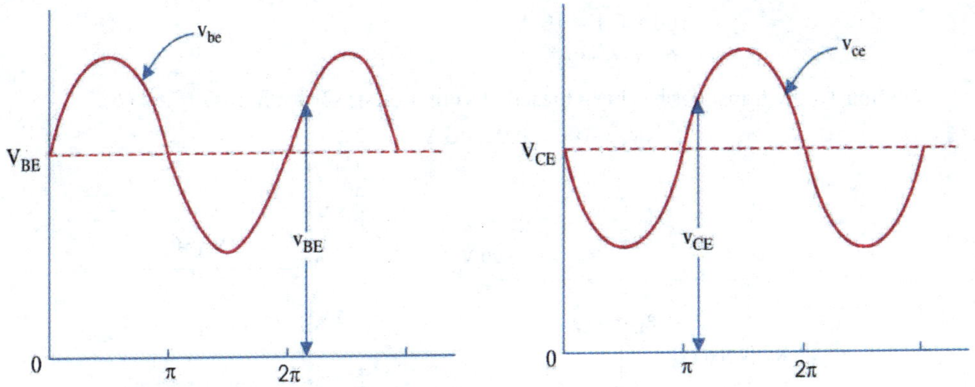

Fig. 19.48

Same changes will happen in third quarter cycle as happened in the first quarter-cycle and so on. It is seen from Fig. 19.48 that output ac voltage is 180° out of phase with the input voltage.

Example 19.25. *Calculate the value of v_{CE} in the circuit of Fig. 19.49 (a) if ac signal voltage is sinusoidal with a peak value of 0.01 V. Take voltage gain A_v of the circuit as 100 and $\beta = 100$. Depict the waveform of the output voltage separately.*

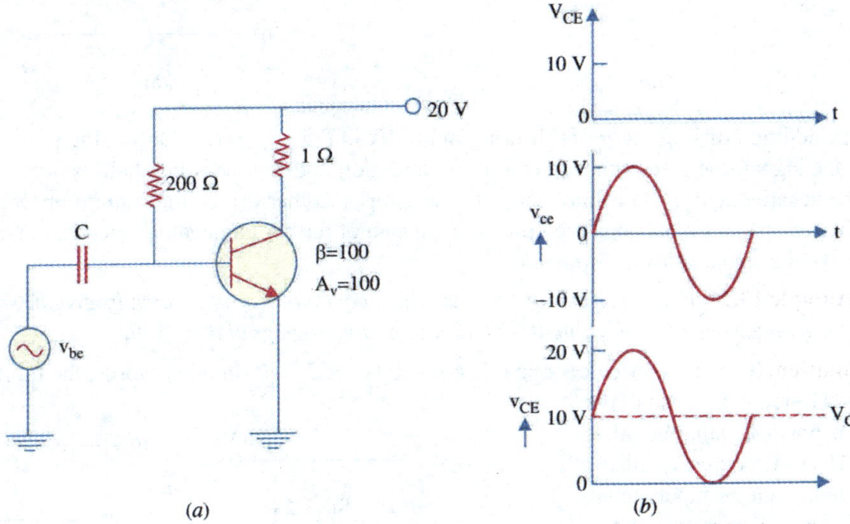

Fig. 19.49

Solution.

$$I_B = \frac{V_{CC}}{R_B} = \frac{20}{200} = 0.1 \text{ A}$$

$$I_C = \beta I_B = 100 \times 0.1 = 10 \text{ A}$$

$$V_{CE} = V_{CC} - I_C R_L = 20 - 10 \times 1 = 10 \text{ V}$$

$$v_{ce} = A_v \times v_{be} = 100 \times 0.01 = 10 \text{ V}$$

The combined output voltage v_{CE} is the sum of V_{CE} and v_{ce} and is shown graphically in Fig. 19.49 (*b*). It is seen that 100 times amplified ac signal rides the dc voltage.

Example 19.26. *find v_{CE} in the circuit of Fig. 19.50 (a) and sketch its waveform. Take $A_v = 100$ and $\beta = 100$ and peak input signal voltage as 0.05 V.*

Solution. $I_B = 20 / 200 = 0.1$ A

$I_C = 100 \times 0.1 = 10$ A

$V_{CE} = 20 - (10 \times 2) = 0$

Obviously, the transistor has been biased at saturation as shown in Fig. 19.50 (*b*).

$$v_{ce} = A_v \times v_{be} = 100 \times 0.05 = 5 \text{ V}$$

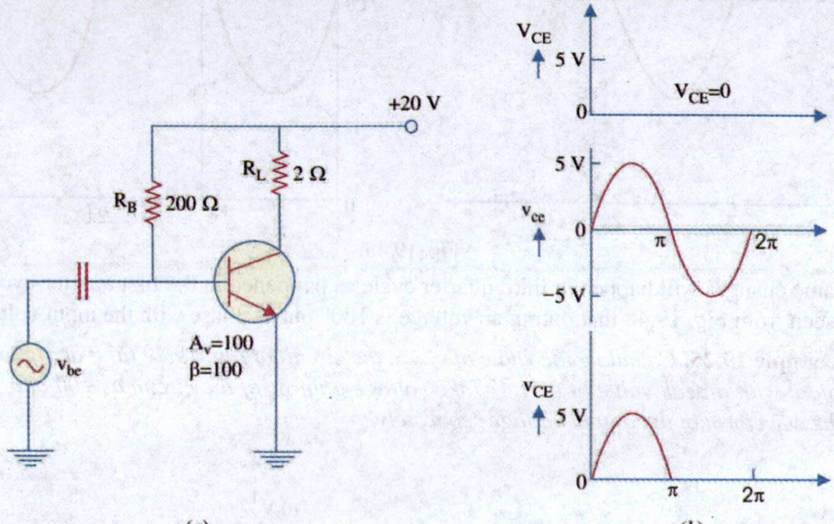

Fig. 19.50

The addition of V_{CE} and v_{ce} is shown graphically in Fig. 19.50 (*b*). During the positive half-cycle of the signal, the transistor comes out of saturation and lets pass this half-cycle. However, during the negative half-cycle of input signal, transistor is further driven into saturation. Since it is already biased at V_{CC}'s most negative limit (OV), it cannot further go negative. Hence, the negative half-cycle of the signal *is lost in saturation*.

Example 19.27. *For the circuit of Fig. 19.51 (a), compute v_{CE} and sketch its graph. The input ac signal is a sine wave of peak value 0.05 V and circuit voltage gain A_v is 100.*

Solution. Since base circuit is open, $I_B = 0$ and hence $I_C = 0$. In other words, the transistor is cut-off. Hence, V_{CE} has the maximum positive value equal to V_{CC} of 10 V. Obviously, during positive half-cycle of input signal, V_{CE} cannot be driven more positive since it already has maximum positive value. Hence, positive half-cycle of the signal is lost. During negative half-cycle of the ac signal, the transistor is brought out of cut-off. Hence, this half-cycle gets through as shown in Fig. 19.51.

Consequently, only negative half-cycles of the input signal can appear on the output side.

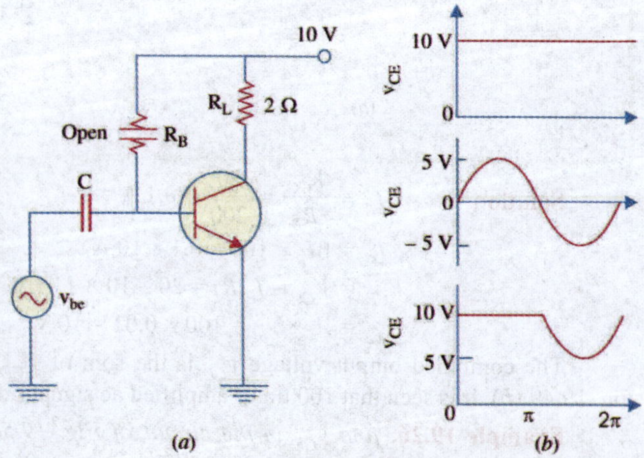

Fig. 19.51

CONVENTIONAL PROBLEMS

1. For the *CB* circuit of Fig. 19.52, find the value of V_{CB}. Neglect junction voltage V_{EB}. **[5 V]**
2. In the *CB* circuit of Fig. 19.53, what value of R_E causes V_{BC} = 10 V? Neglect V_{BE}. **[5 K]**

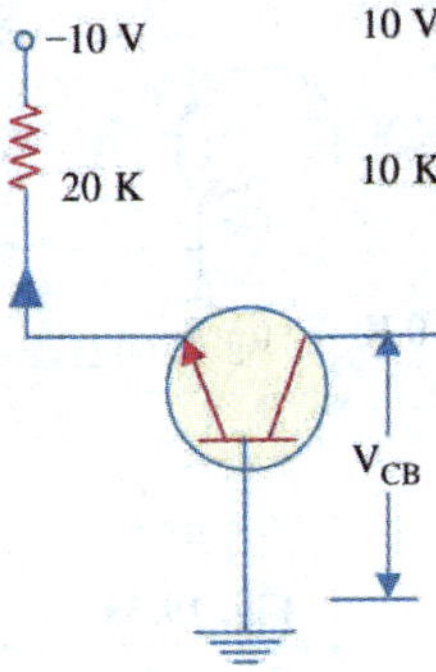

Fig. 19.52

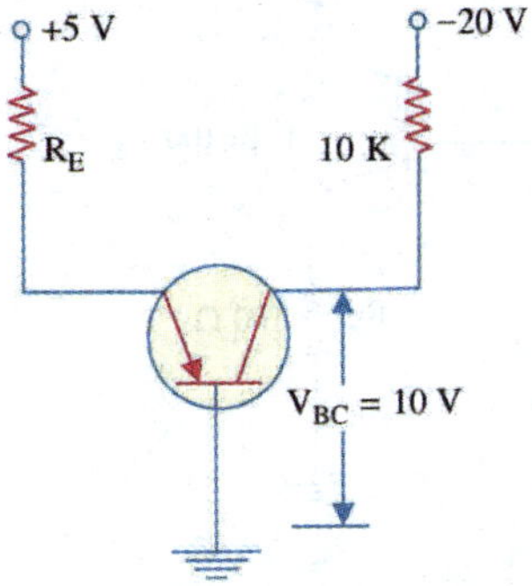

Fig. 19.53

3. For the *CE* circuit of Fig. 19.54, calculate the values of I_B, I_C, I_E and V_{CE}. Take β = 50 and neglect V_{BE}.
[100 μA, 5 mA, 5.1 mA, 7.5 V]
4. In the circuit of Fig. 19.55, calculate I_B, I_C, I_E and V_{CE}. Take transistor β = 50 and neglect V_{BE}.
[100 μA, 5 mA, 5.1 mA, 5 V]

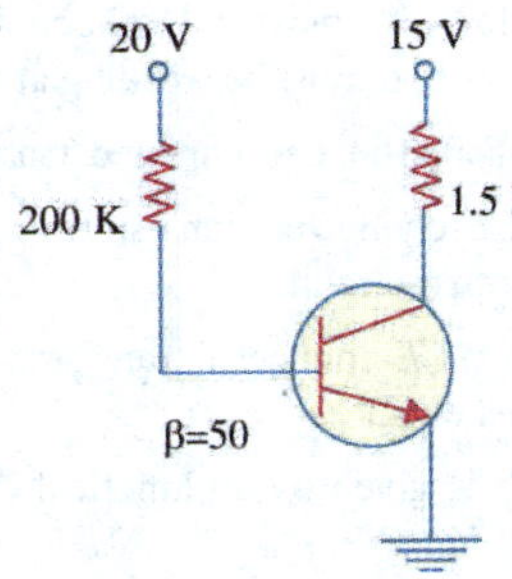

Fig. 19.54

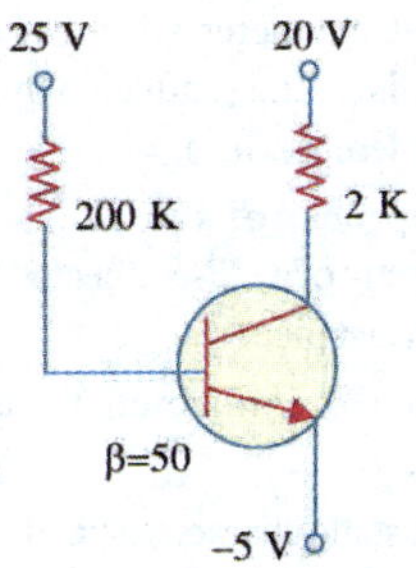

Fig. 19.55

5. In the *CC* circuit of Fig. 19.56, compute the values of I_E, I_B, I_C and V_{CE}. Neglect V_{BE}.
[0.25 mA, 2.48 μA, 0.248 mA, 5 V]
6. In the *CC* circuit of Fig. 19.57, find I_E, I_B, I_C and V_{CE}. Neglect V_{BE}.
[0.25 mA, 2.48 μA, 0.248 mA, 5 V]

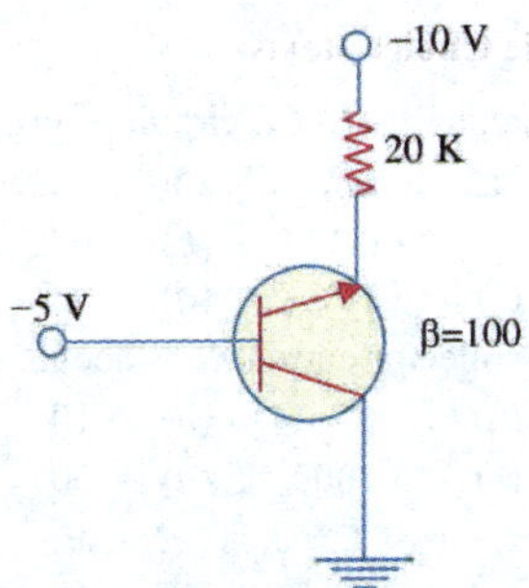

Fig. 19.56

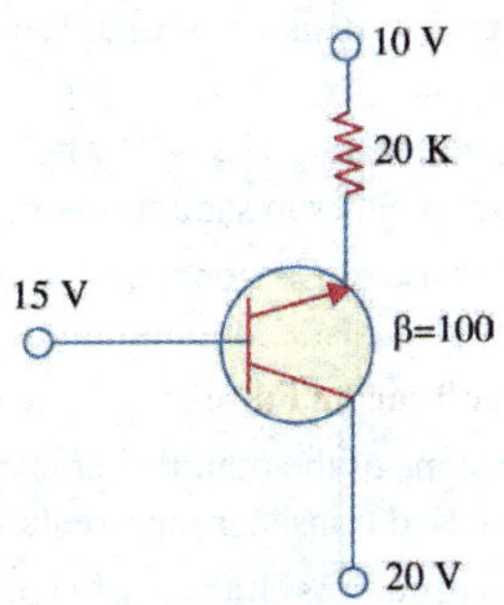

Fig. 19.57

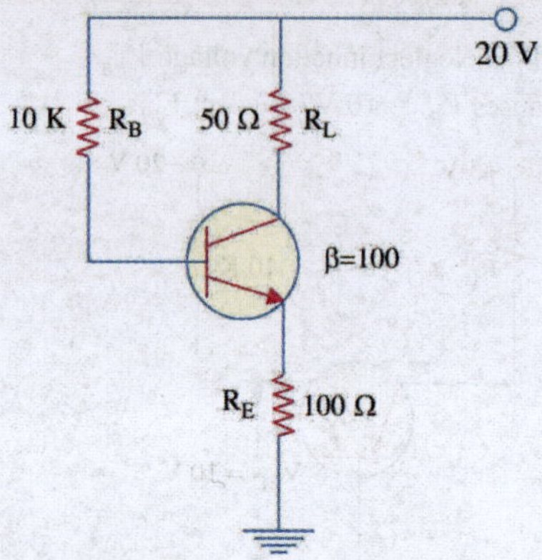

Fig. 19.58

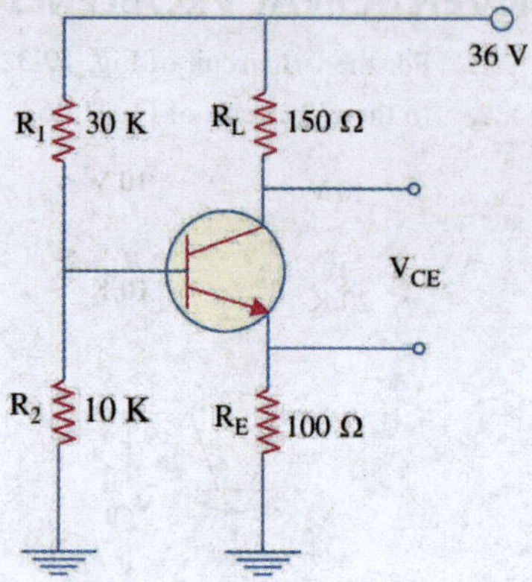

Fig. 19.59

7. In the circuit of Fig. 19.58, find the drop across emitter circuit. The transistor β = 100. **[5V]**
8. Find the value of V_{CE} for the universal stabilization circuit of Fig. 19.59. **[13.5 V]**

SELF EXAMINATION QUESTIONS

A. Fill in the blanks by most appropriate word (s) or numerical value (s).

1. The input characteristic of a *CB*-connected transistor is .... affected by changes either in V_{CB} or temperature.
2. The reciprocal of the slope of the output characteristic of a *CB*-connected transistor gives its .... resistance.
3. R_{out} of a *CB*-connected transistor is very ....
4. The input static characteristic of a *CE*-connected transistor gives its .... resistance.
5. A transistor is said to be cut-off when V_{CE} = ...
6. A transistor is said to operate at saturation when V_{CE} = ... volt.
7. Under saturation conditions, I_C has .... value.
8. When a transistor is open-circuited, V_{CE} = ...
9. In a transistor, V_{CE} = 0 when the transistor operates either in saturation or is...... .
10. If the base of *CE*-connected transistor is open, its V_{CE} equals........... .

B. Answer True or False

1. The slope of the output characteristic of a *CB*-connected transistor represents R_{out}.
2. The input resistance of a *CE* connected transistor is given by the slope of its input characteristic.
3. In a properly-biased transistor in *CE* configuration, I_C is not zero even when I_B = 0.
4. If for a transistor connected in *CE* mode, V_{CE} = 0, it must be working at or in saturation.
5. A shorted out *CE*-connected transistor has V_{CE} = 0.
6. A *CE*-connected transistor, $V_{CE} = V_{CC}$, it may be open-circuited.
7. If in a *CE*-connected transistor, I_B = 0, then V_{CE} must equal V_{CC}.
8. V_{CE} is given by arithmetical difference of V_C and V_E.
9. For a well-designed transistor circuit, V_{CE} = 1/2 V_{CC}.
10. No positive going ac signal can pass through a transistor circuit working in cut-off.
11. When viewed from base, emitter resistance appears like β R_E.

C. Multiple Choice Items

1. With reference to *CE* circuit of Fig. 19.60.

(*i*) base current I_B is about .... mA.

(*a*) 0.1 (*b*) 0.05
(*c*) 10 (*d*) 0.5

(*ii*) collector current I_C is about ......mA

(*a*) 15 (*b*) 10
(*c*) 1000 (*d*) 50

(*iii*) value of V_{CE} is ....volts

(*a*) 0 (*b*) 10
(*c*) –5 (*d*) 5

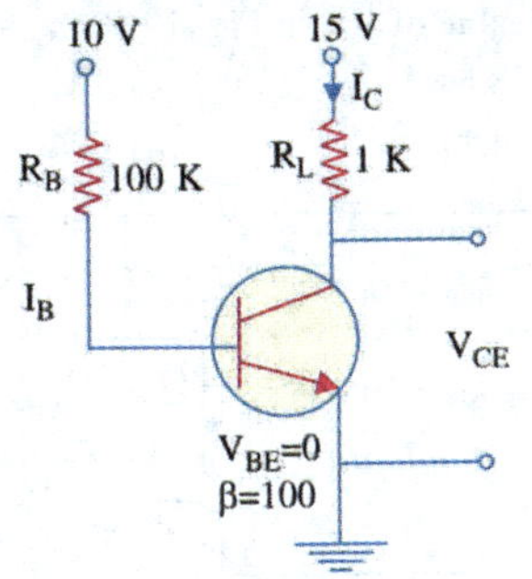

Fig. 19.60

2. If we neglect V_{EB} in Fig. 19.61, value of R_E required to make V_{BC} = 10 V is .... kilohm

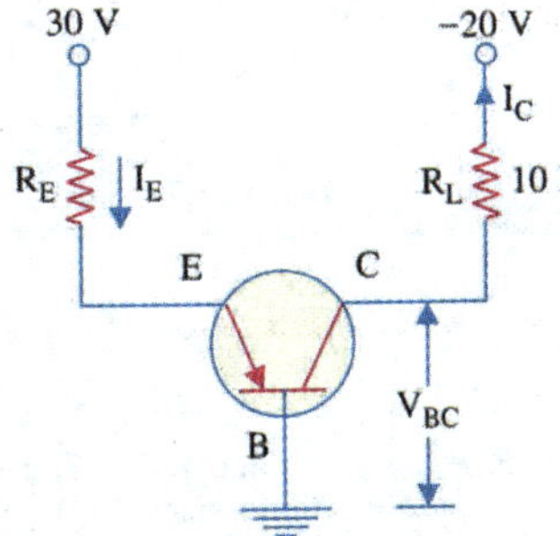

Fig. 19.61

(*a*) 30 (*b*) 20
(*c*) 15 (*d*) 5

3. In the ideal transistor shown in Fig. 19.62, the value of R_B required for making V_{CE} = 10 V is
(*a*) 1 M (*b*) 2 M
(*c*) 20 K (*d*) 10 K

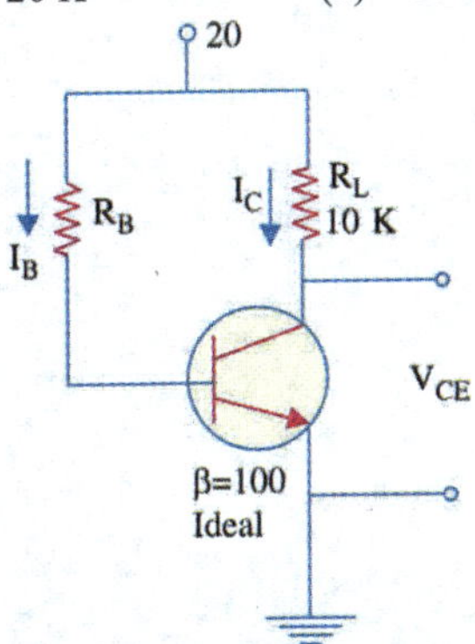

Fig. 19.62

4. The value of dc collector-ground voltage V_C in Fig. 19.63 is .... volt.
(*a*) 0 (*b*) 10
(*c*) 20 (*d*) −55

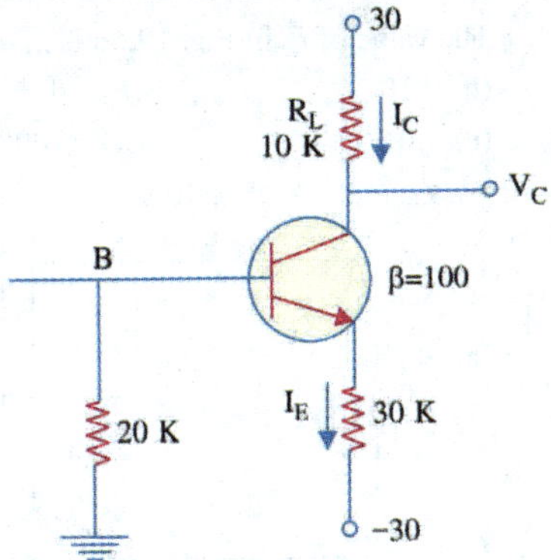

Fig. 19.63

5. In Fig. 19.64, the value of R_E which will saturate the transistor is very nearly ......kΩ.
(*a*) 5 (*b*) 10
(*c*) 20 (*d*) 25

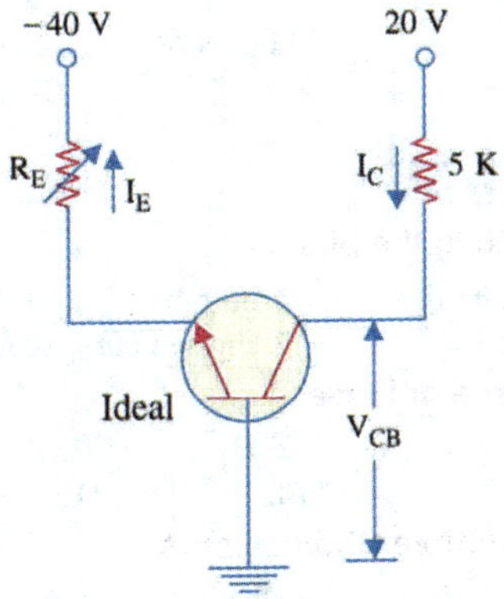

Fig. 19.64

6. The transistor shown in Fig. 19.65 is biased
(*a*) at cut-off
(*b*) at saturation
(*c*) well into saturation
(*d*) well into cut-off

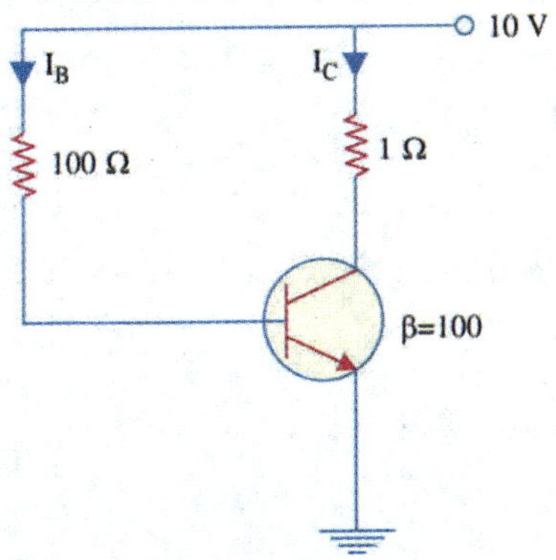

Fig. 19.65

7. The value of I_C in Fig. 19.66 is .....amperes.

(*a*) 10 (*b*) 0

(*c*) 0.1 (*d*) 1000

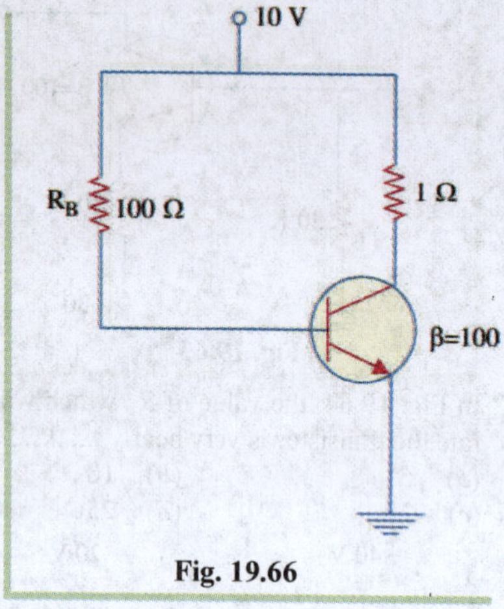

Fig. 19.66

8. The value of V_{CE} in Fig. 19.67 is ....

(*a*) 8.6 (*b*) 3

(*c*) 4.17 (*d*) –3

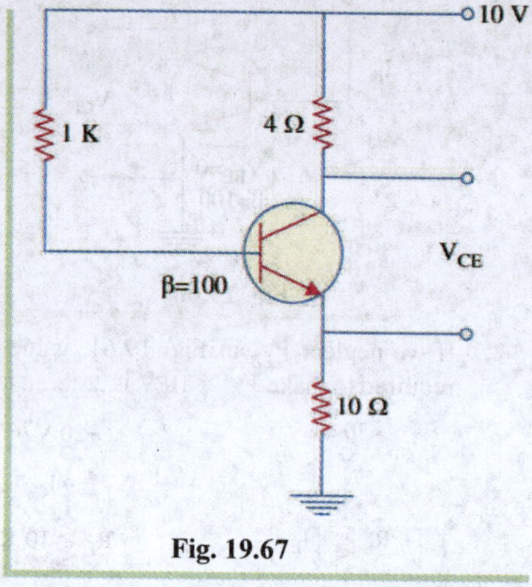

Fig. 19.67

ANSWERS

A. Fill in the blanks

1. hardly **2.** output **3.** high **4.** input **5.** V_{CC} **6.** 0 **7.** maximum **8.** V_{CC} **9.** shorted out **10.** V_{CC}

B. True or False

1. *F* **2.** *F* **3.** *T* **4.** *T* **5.** *T* **6.** *F* **7.** *T* **8.** *T* **9.** *T* **10.** *T* **11.** *T*

C. Multiple Choice Items

1. *a*, *b*, *d* **2.** *a* **3.** *b* **4.** *c* **5.** *b* **6.** *b* **7.** *b* **8.** *b*.

20
CHAPTER

Load Lines And DC Bias Circuits

1. DC Load Line
2. Q-point and Maximum Undistorted Output
3. Need for Biasing a Transistor
4. Factors Affecting Bias Variations
5. Stability Factor
6. Beta Sensitivity
7. Stability Factor for CB and CE Circuits
8. Different Methods for Transistor Biasing
9. Base Bias
10. Base Bias with Emitter Feedback
11. Base Bias with Collector Feedback
12. Base Bias with Collector and Emitter Feedbacks
13. Voltage Divider Bias
14. Load Line and Output Characteristics
15. AC Load Line

20.1. DC Load Line

For drawing the dc load line of a transistor, one need to know only its *cut-off* and *saturation points*. It is a *straight line* joining these two points. For the *CE* circuit of Fig. 19.17, the load line is drawn in Fig. 20.1. *A* is the cut-off point and *B* is the saturation point.

The voltage equation of the collector-emitter is

$$V_{CC} = I_C R_L + V_{CE}$$

$$\therefore I_C = \frac{V_{CC}}{R_L} - \frac{V_{CE}}{R_L}$$

Consider the following two particular cases :

(*i*) when $I_C = 0,\ V_{CE} = V_{CC}$ — cut-off point *A*

(*ii*) when $V_{CE} = 0,\ I_C = V_{CC}/R_L$ — saturation point *B*

Obviously, load line can be drawn if only V_{CC} and R_L are known.

Incidentally slope of the load line $AB = -1/R_L$.

Note. The above given equation can be written as

$$I_C = -\frac{V_{CE}}{R_L} + \frac{V_{CC}}{R_L}$$

It is a linear equation similar to $y = -mx + c$.

The graph of this equation is a straight line whose slope is

$$m = -1/R_L$$

Active Region

All operating points (like *C*, *D*, *E* etc., in Fig. 20.1) lying between cut-off and saturation points form the *active region* of the transistor. In this region, *E*/*B* junction is forward-biased and *C*/*B* junction is reverse-biased—conditions necessary for the proper operation of a transistor.

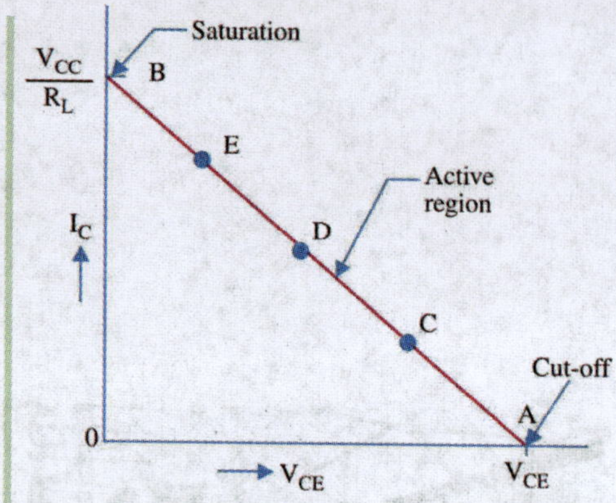

Fig. 20.1

Quiescent Point

It is a point on the dc load line, which represents the values of I_C and V_{CE} that exist in a transistor circuit when *no input signal is applied*.

It is also known as the *dc operating* point or *working point*. The best position for this point is midway between cut-off and saturation points where $V_{CE} = \frac{1}{2} V_{CC}$ (like point *D* in Fig. 20.1).

Example 20.1. *For the circuit shown in Fig. 20.2 (a), draw the dc load line and locate its quiescent or dc working point.*

Solution. The cut-off point is easily found because it lies along *X*-axis where $V_{CB} = V_{CC} =$ 20 V *i.e.*, point *A* in Fig. 20.2 (*b*). At saturation point *B*, saturation value of collector current is

$$I_{C(sat)} = V_{CC}/R_L = 20/5\text{ K} = 4\text{ mA}$$

The line *AB* represents the load line for the given circuit.

We will now find the actual operating point.

$$I_E = V_{EE}/R_E = 30\ /15\text{ K} = 2\text{ mA} \qquad \text{— neglecting } V_{BE}$$

$$I_C = \alpha I_E \cong I_E = 2\text{ mA}; \qquad \therefore \quad V_{CB} = V_{CC} - I_C R_L = 20 - 2 \times 5 = 10\text{ V}$$

Hence, *Q*-point is located at **(10 V; 2 mA)** as shown in Fig. 20.2 (*b*).

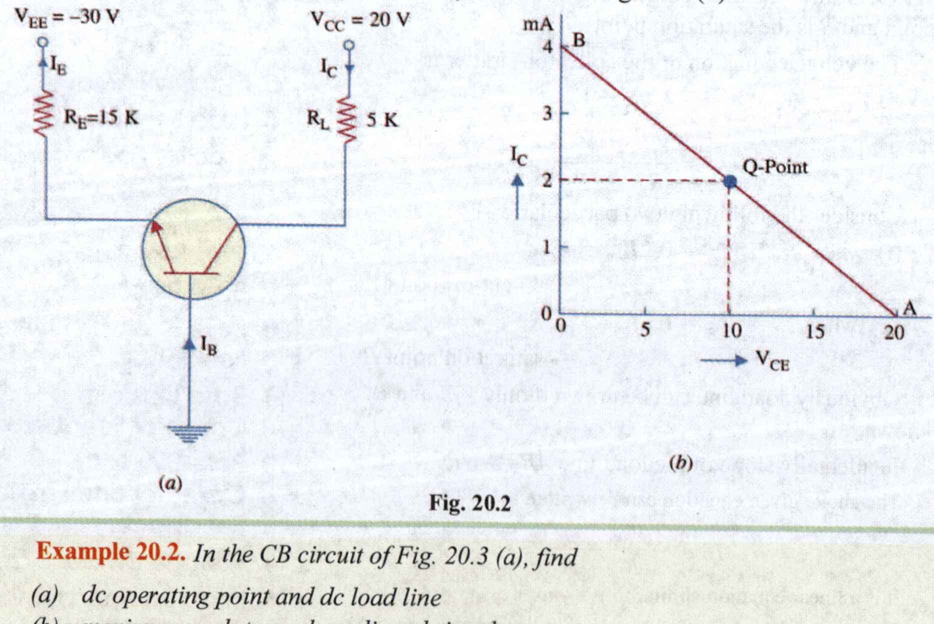

Fig. 20.2

Example 20.2. *In the CB circuit of Fig. 20.3 (a), find*

(*a*) *dc operating point and dc load line*

(*b*) *maximum peak-to-peak unclipped signal*

(*c*) *the approximate value of ac source voltage that will cause clipping.*

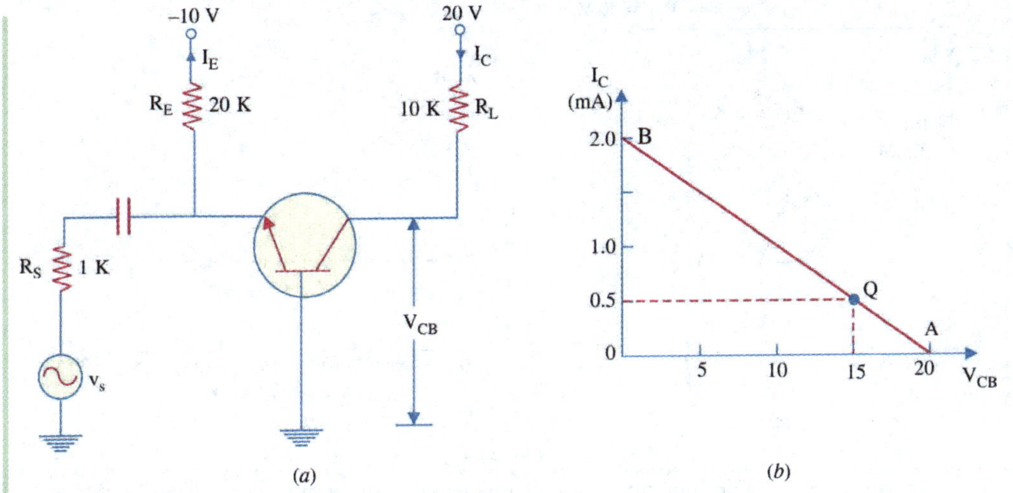

Fig. 20.3

Solution. (a) $I_{C\,(sat)} = \dfrac{V_{CC}}{R_L} = \dfrac{20}{10} = 2$ mA — point *B*

V_{CB} at cut-off $= V_{CC} = 20$ V — point *A*

Hence, *AB* is the dc load line and is shown in Fig. 20.3 (*b*).

Now, $I_E = 10/20 = 0.5$ mA ; $I_C \cong I_E = 0.5$ mA

$\therefore$ $V_{CB} = V_{CC} - I_C R_L = 20 - 0.5 \times 10 = 15$ V

The *Q*-point is located at **(15 V, 0.5 mA)**

(*b*) It is obvious from Fig. 20.3 (*b*) that maximum positive swing can be from 15 V to 20 V *i.e.*, 5 V only. Of course, on the negative swing, the output swing can go from 15 V down to zero volt. The limiting factor being *cut-off on positive half-cycle*, hence maximum unclipped peak-to-peak voltage that we can get from this circuit is 2 × 5 = **10 V.**

(*c*) The approximate voltage gain of the above circuit is

$$= \frac{v_{out}}{v_s} = \frac{R_{ac}}{R_s} = \frac{10\text{ K}}{1\text{ K}} = 10$$

It means that signal voltage will be amplified 10 times. Hence, maximum value of source voltage for obtaining unclipped or undistorted output is

$$v_s = \frac{v_{out}}{10} = \frac{10\text{ V}_{p-p}}{10} = \mathbf{1\ V_{p-p}}$$

Example 20.3. *For the CE circuit shown in Fig. 20.4 (a), draw the dc load line and mark the dc working point on it. Assume β = 100 and neglect V_{BE}.*

Solution. Cut-off point *A* is located where,

$I_C = 0$ and $V_{CE} = V_{CC} = 30$ V. Saturation point *B* is given where

$V_{CE} = 0$

and $I_{C(sat)} = 30/5\text{ K} = 6$ mA.

Line *AB* represents the load line in Fig. 20.4 (*b*).

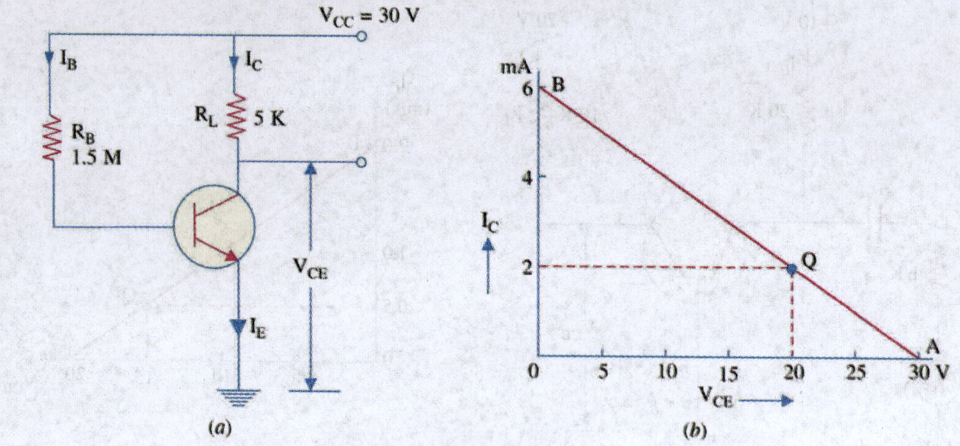

Fig. 20.4

Let us find the dc working point from the given values of resistances and supply voltage.

$$I_B = 30/1.5\text{ M} = 20\ \mu\text{A};\quad I_C = \beta I_B = 100 \times 20 = 2000\ \mu\text{A} = 2\text{ mA};$$

$$V_{CE} = V_{CC} - I_C R_L = 30 - 2 \times 5 = 20\text{ V}$$

Hence, Q-point is **(20 V; 2 mA)** as shown in Fig. 20.4. (*b*).

20.2. Q-point and Maximum Undistorted Output

Position of the Q-point on the dc load line determines the maximum signal that we can get from the circuit before clipping occurs. Consider the cases shown in Fig. 20.5.

In Fig. 20.5 (*a*), when Q, is located near cut-off point, signal first starts to clip at A. It is called *cut-off clipping* because the positive swing of the signal drives the transistor to cut-off. In fact, as seen from Fig. 20.5 (*a*), maximum positive swing is $= I_{CQ} R_{ac}$.

If the Q-point Q_2 is located near saturation point, then clipping first starts at point B as shown in Fig. 20.5 (*b*). It is caused by saturation. The maximum negative swing $= V_{CEQ}$.

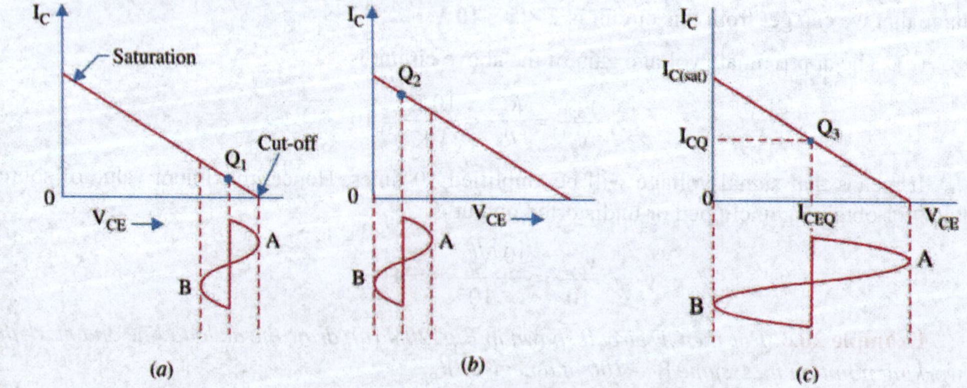

Fig. 20.5

In Fig. 20.5 (*c*), the Q-point Q_3 is located at the *centre* of the load line. In this condition, we get the *maximum possible output signal*. The point Q_3 gives the *optimum* Q-point. The maximum undistorted signal $= 2V_{CEQ}$.

In general, consider the case shown in Fig. 20.6. Since $A < B$, maximum possible peak-to-peak output signal $= 2A$.

If the operating point were so located that $A > B$, then maximum possible peak-to-peak output signal = $2B$.

When operating point is located at the centre of the load line, then maximum undistorted peak-to-peak signal is = $2A = 2B = V_{CC} = 2V_{CEQ}$.

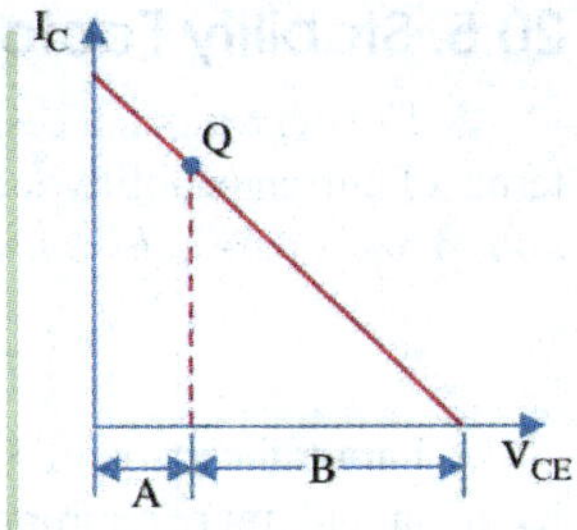

Fig. 20.6

Under optimum working conditions corresponding to Fig. 20.5 (c), I_{CQ} is half the saturation value given by V_{CC}/R_L (Art. 20.1).

$$\therefore \quad I_{CQ} = \frac{1}{2} \cdot \frac{V_{CC}}{R_L} = \frac{V_{CC}}{2R_L}$$

Example 20.4. *Determine the value of R_B required to adjust the circuit of Fig. 20.7 to optimum operating point. Take $\beta = 50$ and $V_{BE} = 0.7$ V.*

Solution. As seen from above

$$I_{CQ} = \frac{V_{CC}}{2R_L} = \frac{20}{2 \times 10} = 1 \text{ mA}$$

The corresponding base current is

$$I_{BQ} = \frac{I_{CQ}}{\beta} = \frac{1}{50} = 20\ \mu\text{A}$$

Now, $V_{CC} = I_B R_B + V_{BE}$

$$\therefore \quad R_B = \frac{V_{CC} - V_{BE}}{I_B} = \frac{20 - 0.7}{20 \times 10^{-6}}\ \mathbf{965\ K}$$

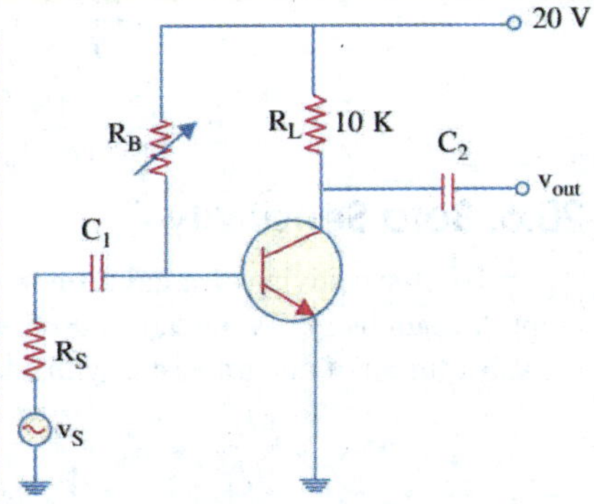

Fig. 20.7

20.3. Need for Biasing a Transistor

For normal operation of a transistor amplifier circuit, it is essential that there should be a

(*a*) forward bias on the emitter-base junction, and

(*b*) reverse bias on the collector-base junction.

In addition, *amount* of bias required is important for establishing the *Q*-point which is dictated by the mode of operation desired.

If the transistor is not biased correctly, it would

1. work inefficiently and
2. produce distortion in the output signal.

It is desirable that once selected, the *Q*-point should remain stable *i.e.*, should not shift its position due to temperature rise etc. Unfortunately, this does not happen in practice unless special efforts are made for the purpose.

20.4. Factors Affecting Bias Variations

In practice, it is found that even after careful selection, *Q*-point tends to shift its position.

This bias instability is the direct result of thermal instability which itself is produced by cumulative increase in I_C that may, if unchecked, lead to thermal runaway (Art. 19.13).

The collector current for *CE* circuit is given by

$$I_C = \beta I_B + I_{CEO} = \beta I_B + (1 + \beta) I_{CO}$$

This equation has three variables : β, I_B and I_{CO} all of which are found to increase with temperature. In particular, increase in I_{CO} produces significant increase in collector current I_C. This leads to increased power dissipation with further increase in temperature and hence I_C. Being a cumulative process, it can lead to thermal runaway which will destroy the transistor itself !

However, if by some circuit modification, I_C is made to decrease with temperature automatically, then decrease in the term βI_B can be made to neutralize the increase in the term $(1 + \beta) I_{CO}$, thereby keeping I_C constant. This will achieve thermal stability resulting in bias stability.

20.5. Stability Factor

The degree of success achieved in stabilizing I_C in the face of variations in I_{CO} is expressed in terms of current stability factor S. It is defined as the *rate of change of I_C with respect to I_{CO} when both β and I_B are held constant.*

$$\therefore \qquad S = \frac{d\,I_C}{d\,I_{CO}} \qquad \text{— } I_B \text{ constant}$$

Larger the value of S, greater the thermal instability and *vice versa* (in view of the above, this factor should, more appropriately, be called instability factor).

The stability factor may be alternatively expressed by using the well-known equation $I_C = \beta I_B + (I + \beta)\, I_{CO}$ which, on differentiation with respect to I_C, yields.

$$I = \beta \frac{dI_B}{dI_C} + (1+\beta)\frac{dI_{CO}}{dI_C} = \beta \frac{dI_B}{dI_C} + (1+\beta)\frac{1}{S}$$

$$\therefore \qquad S = \frac{(1+\beta)}{1 - \beta\,(dI_B / dI_C)}$$

20.6. Beta Sensitivity

By β-sensitivity of a circuit is meant the influence that the β-value has on its dc operating point. Variations in β-value are caused either by variations in the circuit operating conditions or by the substitution of one transistor with another. Beta sensitivity K_β is given by

$$\frac{dI_C}{I_C} = K_\beta \frac{dI_B}{\beta} \qquad \text{or} \qquad K_\beta = \frac{\beta}{I_C} \cdot \frac{dI_C}{d_B}$$

Obviously, K_β is dimensionless ratio and can have values ranging from zero to unity.

20.7. Stability Factor for CB and CE Circuits

(*i*) CB Circuit

Here, collector current is given by

$$I_C = \alpha\, I_E + I_{CO} \qquad \text{— Art. 18.12}$$

$$\therefore \qquad \frac{dI_C}{dI_{CO}} = 0 + 1 \qquad \therefore \quad S = 1$$

(*ii*) CE Circuit

$$I_C = \beta\, I_B + (I+\beta)\, I_{CO} \qquad \text{— Art. 18.13}$$

$$\therefore \qquad \frac{dI_C}{dI_{CO}} = (1+\beta) \qquad \text{— treating } I_B \text{ as a constant}$$

$$\therefore \qquad S = (1+\beta).$$

If $\beta = 100$, then $S = 101$

which means that I_C changes 101 times as much as I_{CO}.

20.8. Different Methods for Transistor Biasing

Some of the methods used for providing bias for a transistor are :

1. Base bias or fixed current bias (Fig. 19.14)

It is not a very satisfactory method because bias voltages and currents do not remain constant during transistor operation.

2. Base bias with emitter feedback (Fig. 20.7)

This circuit achieves good stability of dc operating point against changes in β with the help of emitter resistor which causes degeneration to take place.

3. Base bias with collector feedback (Fig. 20.9)

It is also known as collector-to-base bias or collector feedback bias. It provides better bias stability.

4. Voltage divider bias (Fig. 20.10)

It is most widely used in linear discrete circuits because it provides good bias stability. It is also called universal bias circuit or base bias with one supply.

Each of the above circuits will now be discussed separately.

20.9. Base Bias

It has already been discussed in Art. 19.17 and is shown in Fig. 19.44. For such a circuit, $S = 1$ and $K\beta = 1$.

20.10. Base Bias with Emitter Feedback

This circuit is obtained by simply adding an emitter resistor to the base bias circuit as shown in Fig. 20.8.

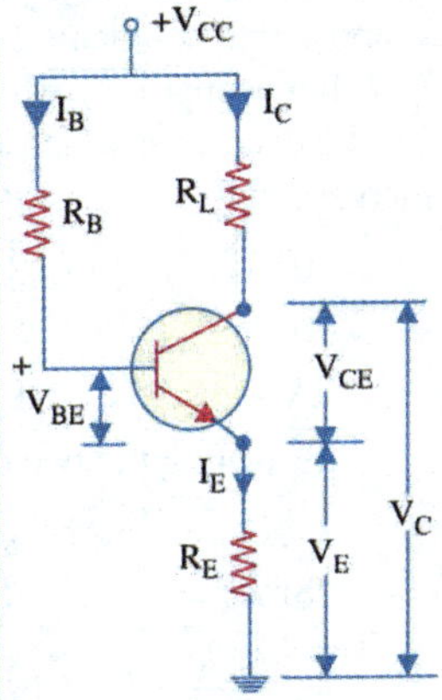

Fig. 20.8

1. At saturation, V_{CE} is essentially zero, hence V_{CC} is distributed over R_L and R_E.

$$\therefore \quad I_{C(sat)} = \frac{V_{CC}}{R_E + R_L}$$

2. I_C can be found as follows :

Consider the supply, base, emitter and ground route. Applying Kirchhoff's Voltage Law, we have

$$-I_B R_B - V_{BE} - I_E R_E + V_{CC} = 0$$

or $\quad V_{CC} = I_B R_B + V_{BE} + I_E R_E \qquad ...(i)$

Now $\quad I_B = I_C/\beta \quad$ and $\quad I_E \cong I_C$

Substituting these values in the above equation, we have

$$V_{CC} \cong \frac{I_C R_B}{\beta} + V_{BE} + I_C R_E$$

$$\therefore \quad I_C \cong \frac{V_{CC} - V_{BE}}{R_E + R_B/\beta} \cong \frac{V_{CC}}{R_E + R_B/\beta} \qquad \text{— neglecting } V_{BE}.$$

(we could have applied the β-rule given in Art. 19.11)

3. collector-to-ground voltage $\quad V_C = V_{CC} - I_C R_L$

4. emitter-to-ground voltage $\quad V_E = I_E R_E \cong I_C R_E$

The β-sensitivity of this circuit is $K_\beta = \dfrac{1}{1 + \beta R_E / R_B}$

Example 20.5. *For the circuit shown in Fig. 20.9, find (i) $I_{C(sat)}$, (ii) I_C, (iii) V_C, (iv) V_E, (v) V_{CE} and (vi) K_β.*

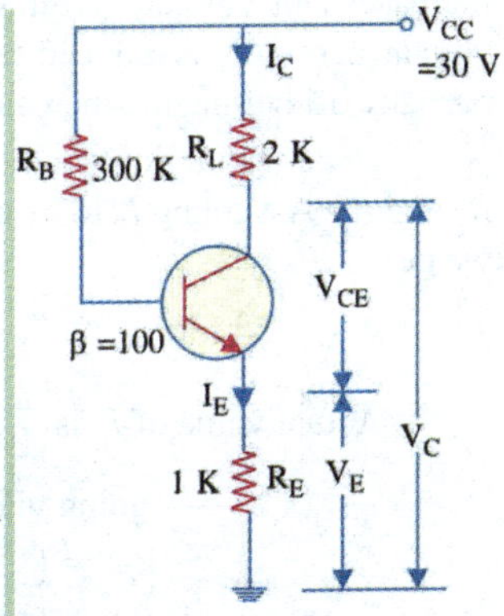

Fig. 20.9

Solution. (i) $I_{C(sat)} = \dfrac{V_{CC}}{R_E + R_L} = \dfrac{30}{1+2} = \mathbf{10\ mA}$

(ii) $\quad$ actual $I_C \cong \dfrac{V_{CC}}{R_E + R_B/\beta} = \dfrac{30}{1 + 300/100} = \mathbf{7.5\ mA}$

(iii) $\quad V_C = V_{CC} - I_C R_L = 30 - 7.5 \times 2 = \mathbf{15\ V}$

(iv) $\quad V_E \cong I_E R_E \cong I_C R_E = 7.5 \times 1 = \mathbf{7.5\ V}$

(v) $V_{CE} = V_C - V_E = 15 - 7.5 = \mathbf{7.5\ V}$

(vi) $K_\beta = \dfrac{1}{1 + 100 \times 1/300} = \mathbf{0.75}$

20.11. Base Bias with Collector Feedback

This circuit (Fig. 20.10) is like the base bias circuit except that base resistor is returned to *collector* rather than to the V_{CC} supply. It derives its name from the fact that since voltage for R_B is derived from collector, there exists a negative feedback effect which tends to stabilise I_C against changes in β. To understand this action, suppose that somehow β increases. It will increase I_C as well as $I_C R_L$ but decrease V_C which is applied across R_B. Consequently, I_B will be decreased which will partially compensate for the original increase in β.

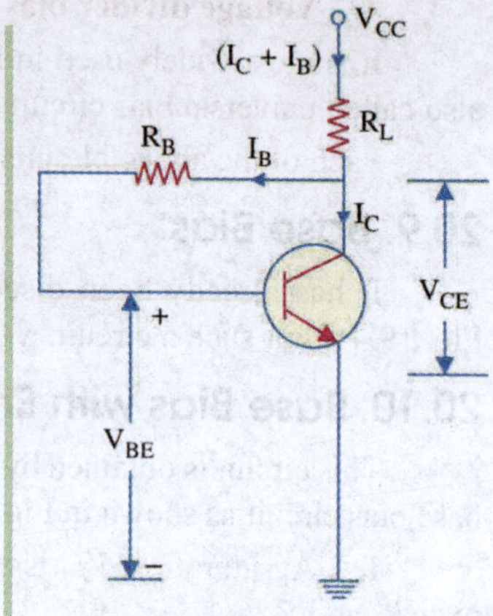

Fig. 20.10

(i) $I_{C(sat)} = V_{CC} / R_L$ — since $V_{CE} = 0$

(ii) $V_C = V_{CC} - (I_B + I_C) R_L \cong V_{CC} - I_C R_L$

Also, $V_C = I_B R_B + V_{BE}$

Equating the two expressions for V_C, we have

$$I_B R_B + V_{BE} \cong V_{CC} - I_C R_L$$

Since $I_B = I_C / \beta$, we get

$$\frac{I_C}{\beta} \cdot R_B + V_{BE} \cong V_{CC} - I_C R_L$$

$$\therefore \quad I_C = \frac{V_{CC} - V_{BE}}{R_L + R_B/\beta} \cong \frac{V_{CC}}{R_L + R_B/\beta}$$

This is also the approximate value of I_E (again, we could take the help of β-rule).

$$K_\beta = \frac{1}{1 + \beta R_L / R_B} = 1 - \frac{I_C}{I_{C(sat)}}$$

20.12. Base Bias with Collector and Emitter Feedbacks

In the circuit of Fig. 20.11, both collector and emitter feedbacks have been used in an attempt to reduce circuit sensitivity to changes in β. If β increases, emitter voltage increases but collector voltage decreases. It means that voltage across R_B is reduced causing I_B to decrease thereby partially off-setting the increase in β.

Under saturation conditions, V_{CC} is distributed over R_L and R_E. Assuming I_B to be negligible as compared to I_C, we get

$$I_{C(sat)} = V_{CC}/(R_E + R_L).$$

Actual value of I_C is $I_C \cong \dfrac{V_{CC} - V_{BE}}{R_E + R_L + R_B/\beta}$

— going via R_B because V_{CE} is unknown

$$V_C = V_{CC} - (I_C + I_B) R_L; \cong V_{CC} - I_C R_L$$

$$V_E = I_E R_E \cong I_C R_E; \quad V_{CE} = V_C - V_E$$

$$V_{CE} \cong V_{CC} - I_C (R_L + R_E)$$

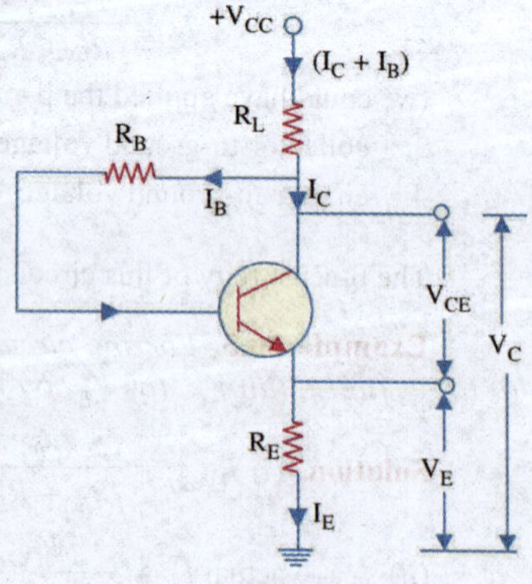

Fig. 20.11

It can be proved that

$$K_\beta = \frac{1}{1+\beta\,(R_E + R_L)/R_B} = 1 - \frac{I_C}{I_{C(sat)}}$$

Obviously, K_β will be degraded with increase in R_B.

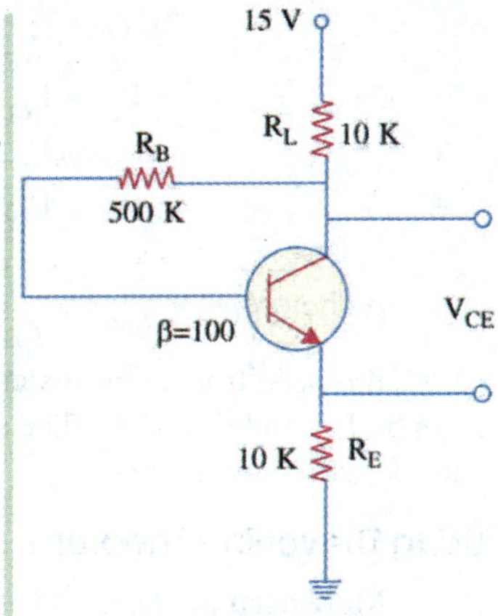

Fig. 20.12

Example 20.6. *For the circuit shown in Fig. 20.12, find (a) $I_{C\,(sat)}$ (b) V_{CE} and (c) K_β. Neglect V_{BE} and take β = 100.*

Solution. (*a*) $I_{C(sat)}$ = 15/(10 + 10) = **0.75 mA**

(*b*) $$I_C = \frac{V_{CC}}{R_E + R_L + R_B/\beta} = \frac{15}{10+10+500/100}$$

= **0.6 mA**

V_{CE} = 15 – 0.6 (10 + 10) = 3 V

(*c*) $$K_\beta = \frac{1}{1+100\,(10+10)/50} = \mathbf{0.2}$$

or $K_\beta = 1 - I_C/I_{C(sat)} = 1 - 0.6/0.75 = \mathbf{0.2}$

Obviously, K_β will be degraded with increase in R_B.

Example 20.7. *For the transistor amplifier circuit shown in Fig. 20.13, find the value of the collector bias of voltage V_C. Also, find the new value of R_B so that the maximum transistor β of 300 brings V_C down to 5 volt.*

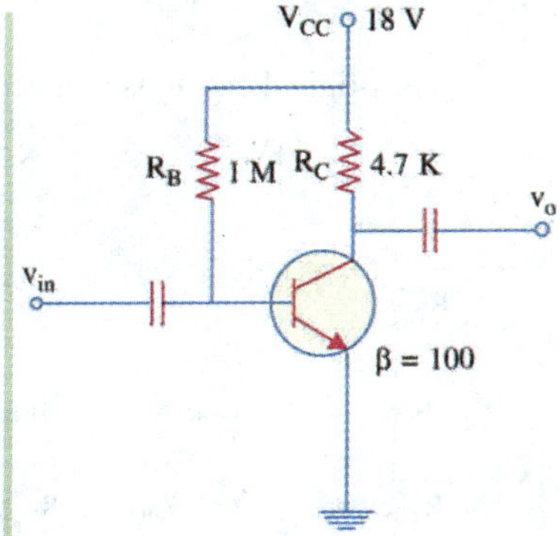

Fig. 20.13

Solution. I_B = (18 – 0.7) V/1 M = 17.3 μA

$I_C = \beta I_B$ = 100 × 17.3 = 1.73 mA

$V_{RC} = I_C R_C$ = 1.73 mA × 4.7 K = 8.1 V

∴ $V_C = V_{CC} - V_{RC}$ = 18 – 8.1 = **9.9 V**

Since, now V_C = 5 V, hence drop across the col lector load resistor is given by

$V_{RC} = V_{CC} - V_C$ = 18 – 5 = 13 V

∴ I_C = 13 V/4.7 K = 2.8 mA

$I_B = I_C/\beta$ = 2.8 mA/300 = 9.2 μA

$V_{RB} = V_{CC} - V_{BE}$ = 18 – 0.7 = 17.3 V

∴ new value of R_B = 17.3 V/9.2 μA = **1.9 M**

20.13. Voltage Divider Bias

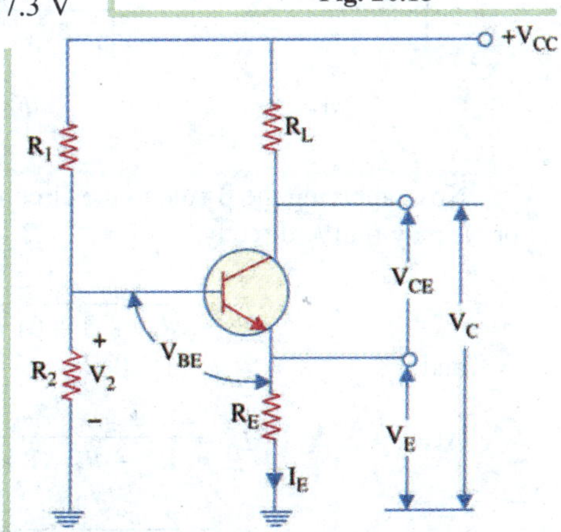

Fig. 20.14

This arrangement is commonly used for transistors incorporated in integrated circuits (ICs).

The name 'voltage divider' is derived from the fact that resistors R_1 and R_2 form a potential divider across V_{CC} (Fig. 20.14)*. The voltage drop V_2 across R_2 forward-biases the emitter whereas V_{CC} supply reverse biases the collector.

As per voltage divider theorem,

$$V_2 = V_{CC} \cdot R_2/(R_1 + R_2)$$

* It is also known as Universal Bias Stabilization Circuit.

As seen, $V_E = V_2 - V_{BE}$ — Art. 19.14

$$\therefore \quad I_E = \frac{V_E}{R_E} = \frac{V_2 - V_{BE}}{R_E} \cong \frac{V_2}{R_E}$$

Also,
$$V_C = V_{CC} - I_C R_L$$
$$V_{CE} = V_C - V_E = V_{CC} - I_C R_L - I_E R_E$$
$$\cong V_{CC} - I_C (R_L + R_E) \qquad \therefore \quad I_C \cong I_E$$

As before, $I_{C(sat)} \cong \dfrac{V_{CC}}{R_L + R_E}$; $\quad K_\beta = \dfrac{1}{1 + \beta R_E/(R_1 \parallel R_2)}$

It is seen from above calculations that value of β was never used anywhere. The base voltage is set by V_{CC} and R_1 and R_2. The dc bias circuit is independent of transistor β. That is why it is such a very popular bias circuit.

Using Thevenin's Theorem

More accurate results can be obtained by Thevenizing the voltage divider circuit as shown in Fig. 20.15. The first step is to open the base lead at point *A* and remove the transistor along with R_L and R_E, thereby leaving the voltage divider circuit behind as in Fig. 20.15 (*a*) and (*b*).

$$V_{th} = V_2 = V_{CC} \cdot \frac{R_2}{R_1 + R_2} \quad \text{and} \quad R_{th} = R_1 \parallel R_2 = \frac{R_1 R_2}{(R_1 + R_2)}$$

The original circuit is reduced to that shown in Fig. 20.15 (*c*) where $V_{th} = V_{BB'}$ and $R_{th} = R_{B'}$.

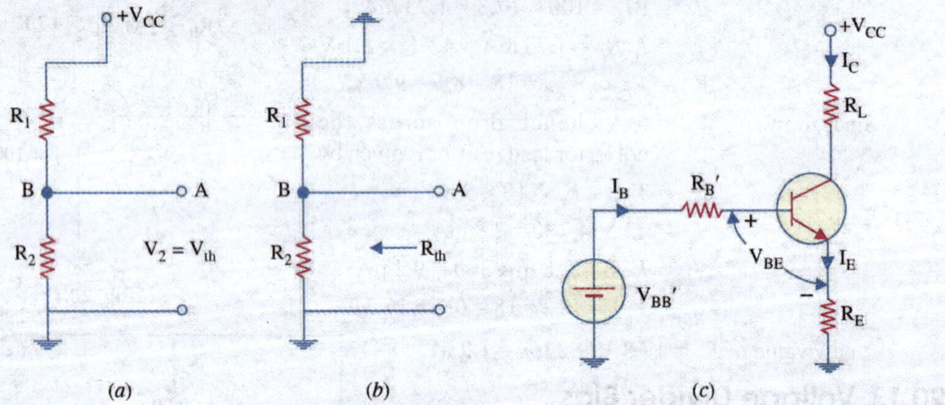

Fig. 20.15

Now, applying the β-rule to the circuit of Fig. 20.15 (*b*), we may find I_B and then $I_E = (1 + \beta) I_B$ or we may find I_E directly.

$$I_B = \frac{V_{BB'} - V_{BE}}{R_{B'} + (1+\beta) R_E} \cong \frac{V_{BB'} - V_{BE}}{R_{B'} + \beta R_E} = \frac{V_{BB'}}{R_{B'} + \beta R_E}$$

and
$$I_E = (1 + \beta) I_B$$

Also
$$I_E = \frac{V_{BB'} - V_{BE}}{R_E + R_{B'}/(1+\beta)}$$
$$\cong \frac{V_{BB'} - V_{BE}}{R_E + R_{B'}/\beta} \cong \frac{V_{BB'}}{R_E + R_{B'}/\beta}$$

$$\therefore \quad V_{CE} = V_{CC} - I_C R_L - I_E R_E$$
$$= V_{CC} - I_C (R_L + R_E)$$

Example. 20.8. *For the circuit of Fig. 20.16, find (a) $I_{C(sat)}$, (b) I_C, (c) V_{CE}, (d) K_β. Neglect V_{BE} and take $\beta = 50$.*

Solution. (*a*) $I_{C(sat)} = \dfrac{V_{CC}}{R_L + R_E} = \dfrac{20}{2+6} = \mathbf{2.5\ mA}$

(*b*) $I_C \cong I_E \cong V_2/R_E = 6/6 = \mathbf{1\ mA}$

(*c*) $V_{CE} = V_{CC} - I_C(R_L + R_E)$

$= 20 - 1\,(2+6) = \mathbf{12\ V}$

(*d*) $R_1 \parallel R_2 = 84/20 = 4.2$ K

$\therefore \quad K_\beta = \dfrac{1}{1 + 50 \times 6/4.2} = \mathbf{0.0138}$

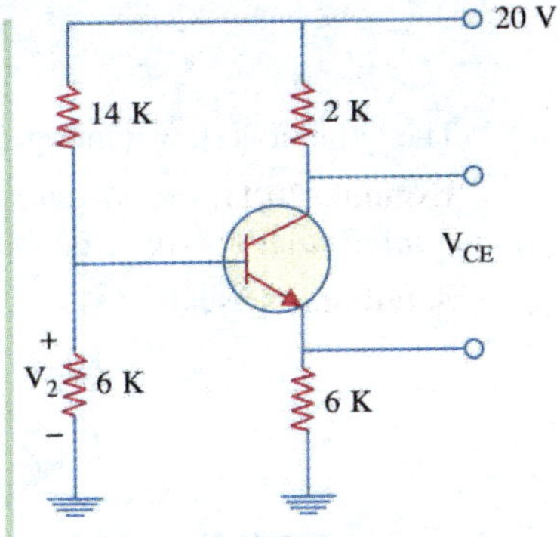

Fig. 20.16

Example 20.9. *Calculate the emitter bias current for the transistor amplifier circuit shown in Fig. 20.17.*

Solution. Let us first find the value of base voltage.

$$V_B = V_{CC}\frac{R_2}{R_1 + R_2} = 6 \cdot \frac{3}{3+6} = 2\ \text{V}$$

$$V_E = V_B - 0.7 = 2 - 0.7 = 1.3\ \text{V}$$

$$\therefore \quad I_E = V_E/R_E = 1.3/650 = \mathbf{2\ mA}$$

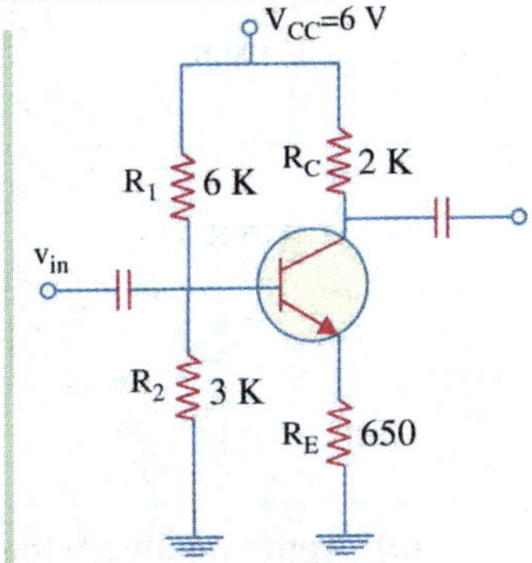

Fig. 20.17

Example 20.10. *For the circuit shown in Fig. 20.18 (a), draw the dc load line and mark in the Q-point of the circuit. Assume silicon transistor.*

Solution. Under cut-off condition ; $V_{CE} = V_{CC}$

$\therefore \quad V_{CE(cut\text{-}off)} = V_{CC} = 20$ V — point *A* in Fig. 20.18 (*b*)

$I_{C(sat)} = \dfrac{V_{CC}}{R_L + R_E} = \dfrac{20}{2+3} = 4$ mA — point *B* in Fig. 20.18 (*b*)

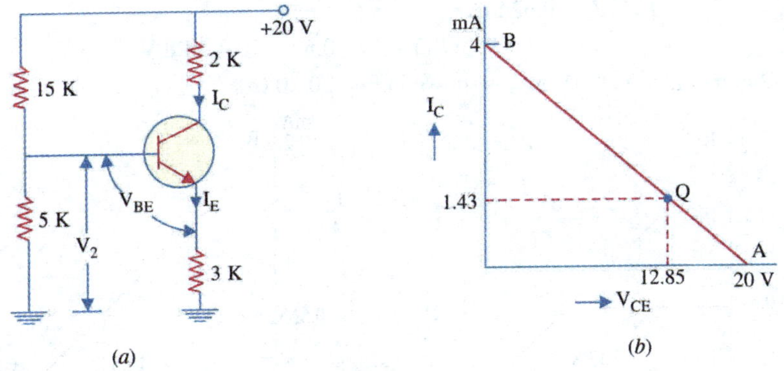

Fig. 20.18

$$V_2 = V_{CC}\frac{R_2}{R_1 + R_2} = 20 \times \frac{5}{15+5} = 5\ \text{V}$$

$$I_E = \frac{V_2 - V_{BE}}{R_E} = \frac{5 - 0.7}{3} = 1.43\ \text{mA}$$

Now, $\quad I_C \cong I_E = 1.43$ mA

$$V_{CE} = V_{CC} - I_C (R_L + R_E)$$
$$= 20 - 1.43\ (2 + 3) = \mathbf{12.85\ V}$$

The Q-point is shown in Fig. 20.18 (*b*).

Example 20.11. *For the circuit shown in Fig. 20.19 (a), draw the dc load line and mark the Q-point of the circuit. Assume germanium material with $V_{BE} = 0.3$ V and $\beta = 50$.*

Solution. As usual.

$$V_{CE(cut\text{-}off)} = V_{CC} = 20\ \text{V} \qquad \text{— point } A \text{ in Fig. 20.20 } (a)$$

$$I_{C(sat)} = \frac{V_{CC}}{R_L + R_E} = \frac{20}{4+6} = 2\ \text{mA} \qquad \text{— point } B \text{ in Fig. 20.20 } (a)$$

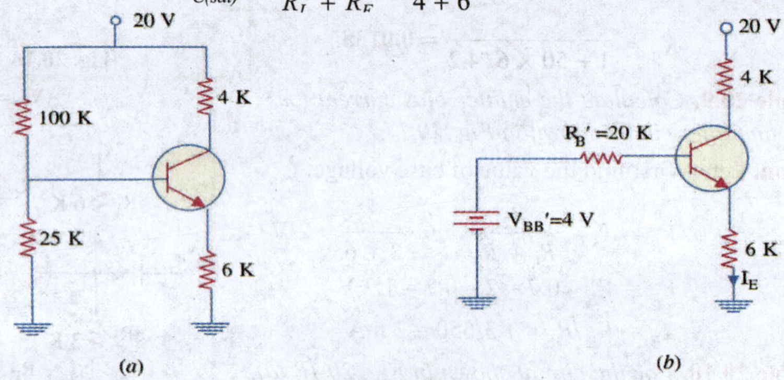

Fig. 20.19

(*a*) Approximate Method

$$V_2 = V_{CC}\frac{R_2}{R_1 + R_2} = 20 \times \frac{25}{125} = 4\ \text{V}$$

$$I_E = \frac{V_2 - V_{BE}}{R_E} = \frac{4 - 0.3}{6} = 0.62\ \text{mA}$$

Also, $\quad I_C \cong I_E = 0.62$ mA

$\therefore \quad V_{CE} = V_{CC} - I_C (R_L + R_E) = 20 - 0.62 \times 10 = \mathbf{13.8\ V}$

This Q-point **(13.8 V, 0.62 mA)** is shown in Fig. 20.20 (*a*).

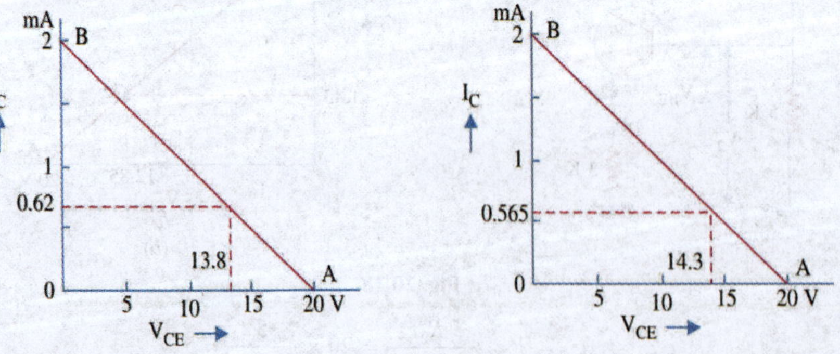

Fig. 20.20

(*b*) Accurate Method

As seen from Fig. 20.19 (*b*)

$$V_{BB'} = 20 \times 25\ /\ 125 = 4\ \text{V}$$

$$I_B = \frac{V_{BB'} - V_{BE}}{R_{B'} + (1 + \beta) R_E} = \frac{4 - 0.3}{20 + 51 \times 6} = 11.3 \ \mu A$$

$$I_C = \beta I_B = 50 \times 11.3 = 565 \ \mu A \qquad = 0.565 \text{ mA}$$

$$I_E = (1 + \beta) I_B = 576 \ \mu A = 0.576 \text{ mA}$$

$$\therefore \quad V_{CE} = V_{CC} - I_C R_L - I_E R_E = 20 - 0.565 \times 4 - 0.576 \times 6 = \mathbf{14.3 \ V}$$

The new and more accurate *Q*-point **(14.3 V, 0.565 mA)** is shown in Fig. 20.20 (*b*).

20.14. Load Line and Output Characteristics

In order to study the effect of bias conditions on the performance of a *CE* circuit, it is necessary to superimpose the dc load line on the transistor output (V_{CE}/I_C) characteristics. Consider a silicon *NPN* transistor which is connected in *CE* configuration (Fig. 20.21) and whose output characteristics are given in Fig. 20.22. Let its $\beta = 100$.

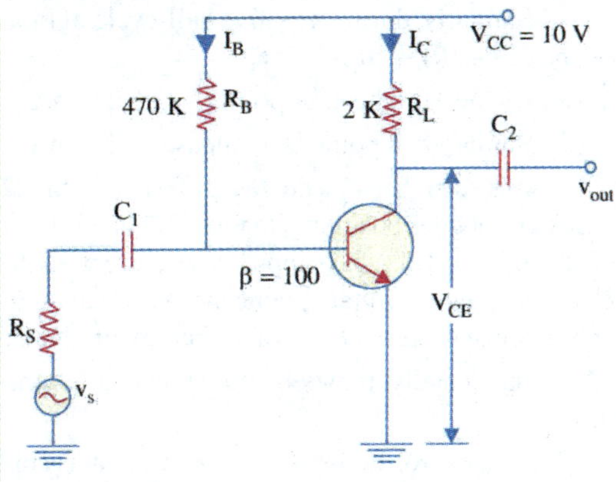

Fig. 20.21

First, let us find the cut-off and saturation points for drawing the dc load line and then mark in the *Q*-point.

$$I_{C(sat)} = 10/2 = 5 \text{ mA}$$

— point *B* in Fig. 20.22

$$V_{CE(cut\text{-}off)} = V_{CC} = 10 \text{ V}$$

— point *A* in Fig. 20.22

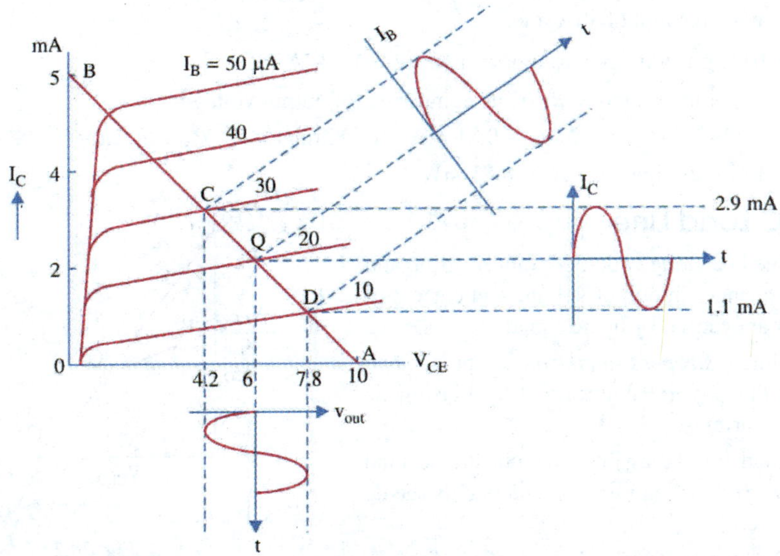

Fig. 20.22

The load line is drawn in Fig. 20.22.

Actual $$I_B = \frac{V_{CC} - V_{BE}}{R_B} = \frac{10 - 0.7}{470} = 20 \ \mu A$$

$$I_C = \beta I_B = 100 \times 20 = 2000\ \mu A$$
$$= 2\ mA$$
$$V_{CE} = V_{CC} - I_C R_L = 10 - 2 \times 2 = 6\ V$$

This locates the Q-point in Fig. 20.22.

Suppose an ac input signal voltage injects a sinusoidal base current of peak value 10 μA into the circuit of Fig. 20.21. Obviously, it will swing the operating or Q-point up and down along the load line.

When positive half-cycle of I_B is applied, the Q-point shifts to point C which lies on the (20 + 10) = 30 μA line.

Similarly, during negative half-cycle of input base current, Q-point shifts to point D which lies on the (20 – 10) = 10 μA line.

By measurement, at point C, $I_C = 2.9$ mA. Hence, $V_{CE} = 10 - 2 \times 2.9 = 4.2$ V

Similarly, at point D, I_C measures 1.1 mA. Hence, $V_{CE} = 10 - 2 \times 1.1 = 7.8$ V

It is seen that V_{CE} decreases from 6 V to 4.2 V *i.e.*, by a peak value of (6 – 4.2) = 1.8 V when base current goes positive. On the other hand, V_{CE} increases from 6 V to 7.8 V *i.e.*, by a peak value of (7.8 – 6) = 1.8 V when input base current signal goes negative. Since changes in V_{CE} represent changes in output voltage, it means that when input signal is applied, I_B varies according to the signal amplitude and causes I_C to vary, thereby producing voltage variations.

Incidentally, it may be noted that variations in voltage drop across R_L are exactly the same as in V_{CE}.

Steady drop across R_L when no signal is applied = 2 × 2 = 4 V. When base signal goes positive, drop across $R_L = 2 \times 2.9 = 5.8$ V. When base signal goes negative, $I_C = 1.1$ mA and drop across $R_L = 2 \times 1.1 = 2.2$ V.

Hence, voltage variation is = 5.8 – 4 = 1.8 V during positive input half-cycle and 4 – 2.2 = 1.8 V during negative input half-cycle.

Obviously, rms voltage variation = $1.8/\sqrt{2} = 1.27$ V

Now, power dissipated in R_L by ac component of output voltage is

$P_{ac} = 1.27^2/2 = 0.81$ mW. Similarly, $P_{dc} = I_C 2R = 2^2 \times 2 = 8$ mW

Total power dissipated in $R_L = 8.81$ mW.

20.15. AC Load Line

It is the line along which Q-point shifts up and down when changes in output voltage and current of an amplifier are caused by an ac signal.

This line is *steeper than the dc line* but the two intersect at the Q-point determined by biasing dc voltages and currents.

AC load line takes into account the ac load resistance whereas dc load line considers only the dc load resistance.

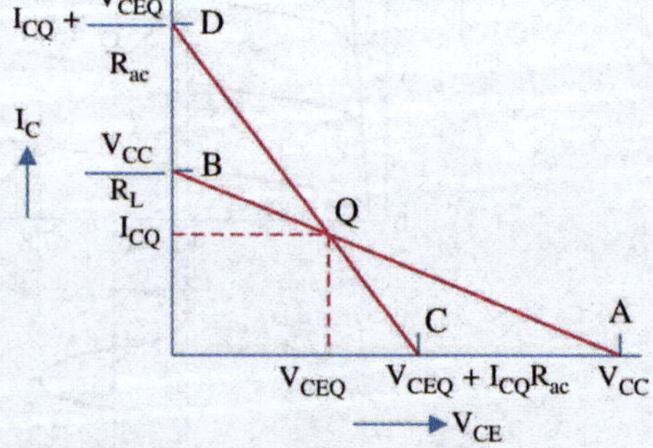

Fig. 20.23

(i) DC Load Line

The cut-off point for this line is where $V_{CE} = V_{CC}$.

It is also written as $V_{CE(cut\text{-}off)}$.

Saturation point is given by

$I_C = V_{CC}/R_L$. It is also written as $I_{C(sat)}$.

It is represented by straight line *AQB* in Fig. 20.23.

(*ii*) AC Load Line

The cut-off point is given by $V_{CE(cut\text{-}off)} = V_{CEQ} + I_{CQ} R_{ac}$ where R_{ac} is the *ac load resistance**.

Saturation point is given by

$$I_{C(sat)} = I_{CQ} + V_{CEQ}/R_{ac}. \qquad \text{— as shown in Fig. 20.23}$$

It is represented by straight line *CQD* in Fig. 20.23.

The slope of the ac load line is given by $y = -1/R_{ac}$.

It is seen from Fig. 20.23 that maximum possible positive signal swing is $= I_{CQ} R_{ac}$. Similarly, maximum possible negative signal swing is V_{CEQ}. In other words, peak-signal handling capacity is limited to $I_{CQ} R_{ac}$ or V_{CEQ} whichever is smaller.

Example 20.12. *Draw the dc and ac load lines for the CE circuit shown in Fig. 20.24 (a). What is the maximum peak-to-peak signal that can be obtained ?*

Solution. DC Load Line [Fig. 20.24 (*b*)]

$$V_{CE(cut\text{-}off)} = V_{CC} = 20 \text{ V} \qquad \text{—(point } A)$$

$$I_{C(sat)} = V_{CC}/(R_L + R_E) = 20/5 = 4 \text{ mA} \qquad \text{—(point } B)$$

Hence, *AB* represents dc load line for the given circuit.

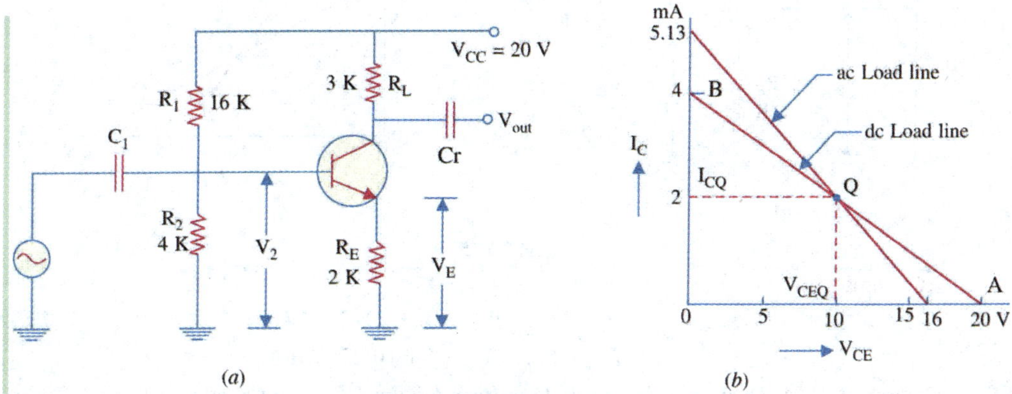

Fig. 20.24

Approximate bias conditions can be quickly found by assuming that I_B is too small to affect the base bias in Fig. 20.24 (*a*).

$$V_2 = 20 \times 4/(4 + 16) = 4 \text{ V}$$

If we neglect V_{BE}, $\quad V_2 = V_E; \quad I_E = \dfrac{V_E}{R_E} = \dfrac{V_2}{R_E} = \dfrac{4}{2} = 2 \text{ mA}$

Also, $\quad I_C \cong I_E = 2 \text{ mA}. \qquad$ Hence, $I_{CQ} = 2$ mA.

The corresponding value of V_{CEQ} can be found by drawing dotted line in Fig. 20.24 (*b*) or may be calculated as under :

$$V_{CEQ} = V_{CC} - I_{CQ}(R_L + R_E) = 20 - 2(3 + 2) = 10 \text{ V}$$

AC Load Line

Cut-off point,

$$V_{CE(cut\text{-}off)} = V_{CEQ} + I_{CQ} R_{ac}$$

* Written as r_L in Chapter 21 (Art. 21.4)

Now, for the given circuit, ac load resistance is $R_{ac} = R_C = 3$ K

Cut-off point = 10 + 2 × 3 = 16 V — [Fig. 20.24 (*b*)]

Saturation point, $$I_{C(sat)} = I_{CQ} + \frac{V_{CEQ}}{R_{ac}} = 2 + \frac{10}{3} = 5.13 \text{ mA}$$

Hence, line joining 16 V point and 5.13 mA point gives ac load line as shown in Fig. 20.24 (*b*). As expected, this line passes through the *Q*-point.

Now, $I_{CQ} \cdot R_{ac} = 2 \times 3 = 6$ V and $V_{CEQ} = 10$ V. Taking the smaller quantity, maximum peak output signal = 6 V. Hence, peak-to-peak value = 2 × 6 = **12 V.**

Example 20.13. *Find the dc and ac load lines for CE circuit shown in Fig. 20.24 (a).*

Solution. The given circuit is identical to that shown in Fig. 20.24 (*a*) except for the addition of 6 K resistor. This makes R_{ac} = 3 K ‖ 6 K = 2 K because collector feeds these two resistors in parallel. The dc load line would remain unaffected. Change would occur only in the ac load line.

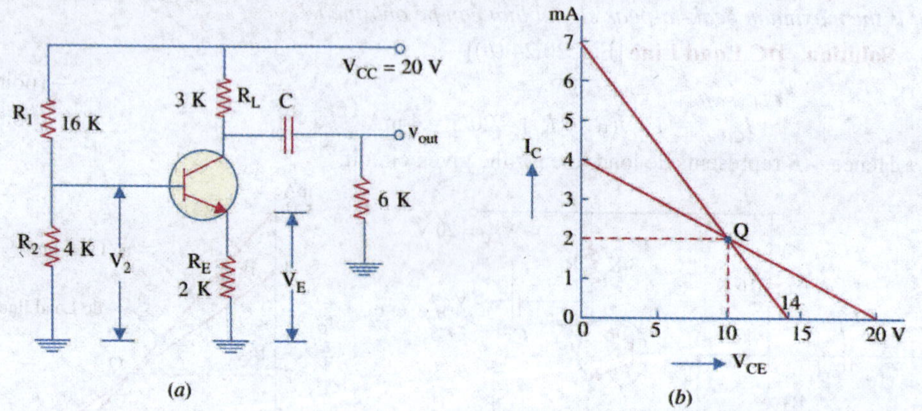

Fig. 20.25

AC Load Line

$$V_{CE(cut\text{-}off)} = V_{CEQ} + I_{CQ} R_{ac} = 10 + 2 \times 2 = \mathbf{14\ V}$$
$$I_{C(sat)} = I_{CQ} + V_{CEQ}/R_{ac} = 2 + 10/2 = \mathbf{7\ mA}$$

Example 20.14. *Draw the dc and ac load lines for the CB circuit shown in Fig. 20.26 (a). Which swing starts clipping first ?*

Solution. The dc load line passes through cut-off point of 30 V and saturation point of $V_{CC}/R_L = 1$ mA.

Now, $$I_E \cong 20/40 = 0.5 \text{ mA} ; I_C \cong I_E = 0.5 \text{ mA}; \qquad I_{CQ} = 0.5 \text{ mA}$$
$$V_{CB} = V_{CC} - I_C R_C = 30 - 0.5 \times 30 = 15 \text{ V} ; \qquad \therefore \quad V_{CBQ} = 15 \text{ V}$$

Hence, dc operating point or *Q*-point is **(15V, 0.5 mA)** as shown in Fig. 20.26 (*b*).

The cut-off point for ac load line is = $V_{CBQ} + I_{CQ} R_{ac}$.

Since collector sees an ac load of two 30 K resistors in parallel,

$$\therefore \quad R_{ac} = 30 \text{ K} \parallel 30 \text{ K} = 15 \text{ K}.$$
$$\therefore \quad V_{CBQ} + I_{CQ} R_{ac} = 15 + 0.5 \times 15 = 22.5 \text{ V}.$$

Saturation current for ac line is = $I_{CQ} + V_{CBQ}/R_{ac} = 0.5 + 15/15 = 1.5$ mA.

The line joining these two points (and also passing through *Q*) gives the ac load line as shown in Fig. 20.26 (*b*).

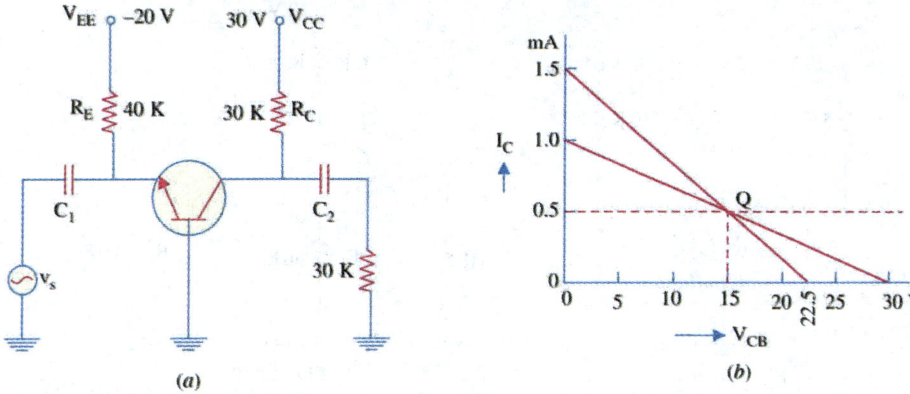

Fig. 20.26

Note. Knowing ac cut-off point and Q-point, we can draw the ac load line. Hence, we need not find the value of saturation current for this purpose.

As seen, positive swing starts clipping first because $I_{CQ}\ R_{ac}$ is less than V_{CBQ}. Obviously, maximum peak-to-peak signal that can be obtained from this circuit = 2 × 7.5 = 15 V.

Example. 20.15. *For the circuit shown in Fig. 20.26 (a), find the approximate value of source voltage v_s that will cause clipping. The voltage source has an internal resistance of 1 K.*

Solution. As found out in Ex. 20.14, the maximum swing of the unclipped output = 2 × 7.5 = 15 V.

Now,
$$A_v = \frac{v_{out}}{v_s} = \frac{R_{ac}}{R_S} = \frac{15}{1} = 15$$

$$\therefore \quad v_s = \frac{v_{out}}{A_v} = \frac{15\ V_{p-p}}{15} = 1\ \mathbf{V}_{p-p}$$

CONVENTIONAL PROBLEMS

1. For the CB circuit shown in Fig. 20.27, find the approximate location of Q-point. **[10 V, 2 mA]**

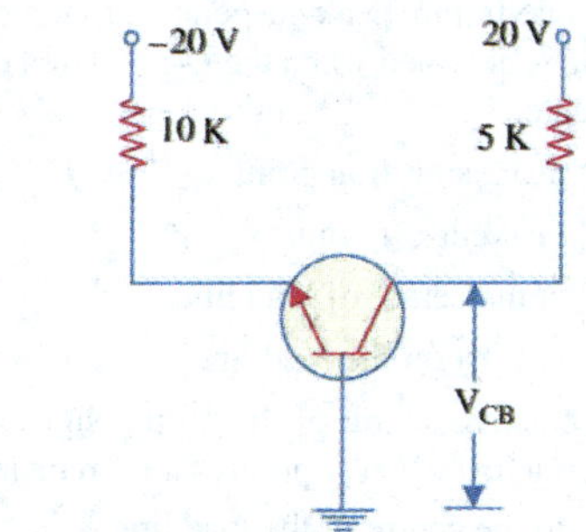

Fig. 20.27

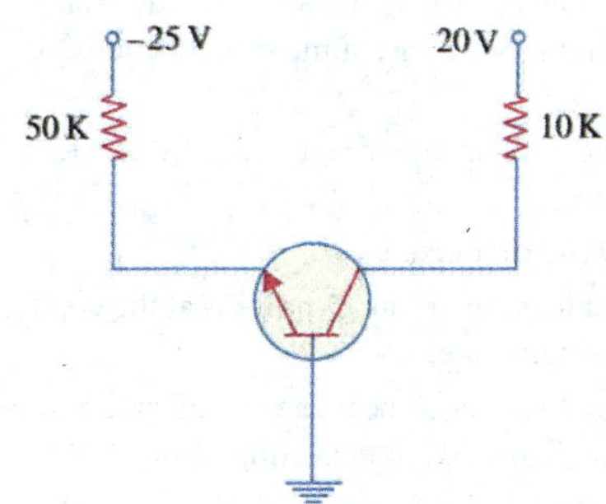

Fig. 20.28

2. For the circuit of Fig. 20.28, find
   (*a*) dc operating-point,
   (*b*) maximum peak-to-peak unclipped signal. **[15 V, 0.5 mA ; 10 V_{p-p}]**
3. What is the maximum peak-to-peak signal that can be obtained from the circuit of Fig. 20.29 ? **[20 V_{p-p}]**

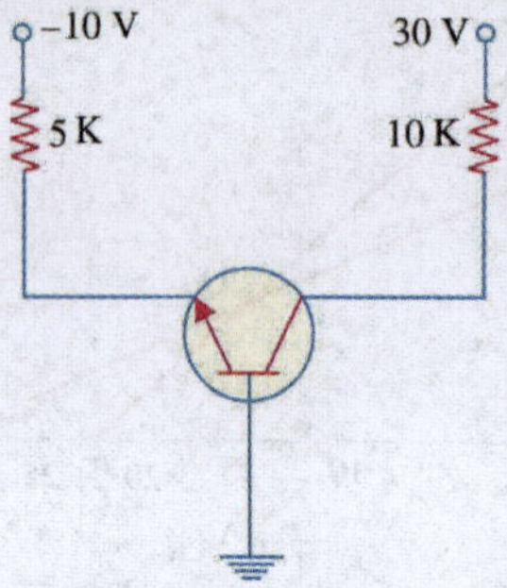

Fig. 20.29

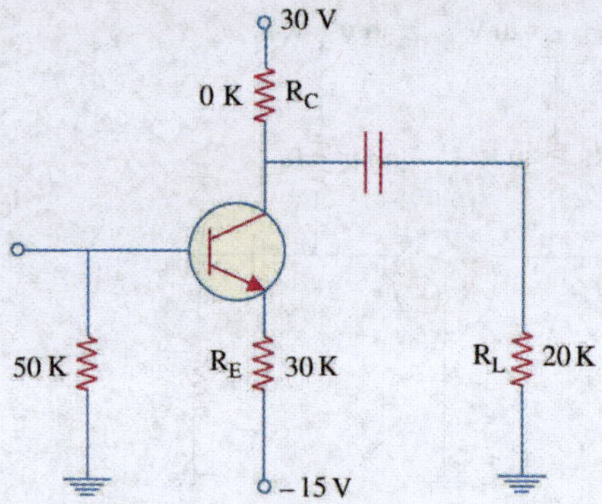

Fig. 20.30

4. Find the value of maximum peak-to-peak output of signal that can be obtained from the circuit of Fig. 20.30. **[10 V_{p-p}]**

SELF EXAMINATION QUESTIONS

A. Fill in the blanks with most appropriate word (s) or numerical value(s).

1. The dc load line of a transistor can be drawn if we know its cut-off and .... points.
2. The point on the dc load line which represents value of I_C and V_{CE} that exist in the absence of input signal is called .... point.
3. AC load line of a transistor is .... than its dc load line.
4. A load line describes graphically the relationship between possible voltage and .... values of a circuit.
5. An improperly-biased transistor produces .... in the output signal.
6. In the universal bias stabilization circuit, voltage and current values are almost independent of ....
7. AC load line is based on .... load resistance.

B. Answer True or False

1. The best location of the Q-point is at the centre of the dc load line.
2. If Q-point is located near the cut-off point, the negative signal swing gets clipped first.
3. In a CE circuit, location of Q-point is determined by V_{EE} and I_C.
4. The dc load line of a transistor is steeper than its ac load line.
5. In a transistor circuit, thermal stability leads to bias stability.
6. Larger the current stability factor of a circuit, greater its thermal instability.
7. Larger the β-value, greater the β-sensitivity factor of a transistor circuit.
8. β-sensitivity of a circuit depends inversely on the β-value of the transistor.
9. Universal bias stabilization circuit has the highest stability factor.
10. The value of K_β varies from zero to unity.

C. Multiple Choice Items

1. The dc load line of a transistor circuit
   (*a*) has a negative slope
   (*b*) is a curved line
   (*c*) gives graphic relation between I_C and I_B
   (*d*) does not contain the Q-point.
2. The maximum peak-to-peak output voltage swing is obtained when the Q-point of a circuit is located.
   (*a*) near saturation point
   (*b*) near cut-off point
   (*c*) at the centre of load line
   (*d*) at least on the load line
3. The positive swing of the output signal starts clipping first when Q-point of the circuit moves.
   (*a*) to the centre of the load line
   (*b*) two-third way up the load line
   (*c*) towards the saturation point
   (*d*) towards the cut-off point
4. Improper biasing of a transistor circuit leads to
   (*a*) excessive heat production at collector terminal
   (*b*) distortion in output signal

(*c*) faulty location of load line
(*d*) heavy loading of emitter terminal

5. The ac load line of a transistor circuit is steeper than its dc line because
(*a*) ac signal sees less load resistance
(*b*) it has steeper slope
(*c*) I_C is higher
(*d*) input signal varies in magnitude

6. The universal bias stabilization circuit is most popular because
(*a*) I_C does not depend on transistor characteristics
(*b*) its β-sensitivity is high
(*c*) voltage divider is heavily loaded by transistor base
(*d*) I_C equals I_E

ANSWERS

A. Fill in the blanks

1. saturation **2.** Q or quiescent **3.** steeper **4.** current **5.** distortion **6.** β **7.** *AC*

B. True or False

1. *T* **2.** *F* **3.** *F* **4.** *F* **5.** *T* **6.** *T* **7.** *T*
8. *F* **9.** *T* **10.** *T*

C. Multiple Choice Items

1. *a* **2.** *c* **3.** *d* **4.** *b* **5.** *a* **6.** *a*

21 CHAPTER

Transistor Equivalent Circuits and Models

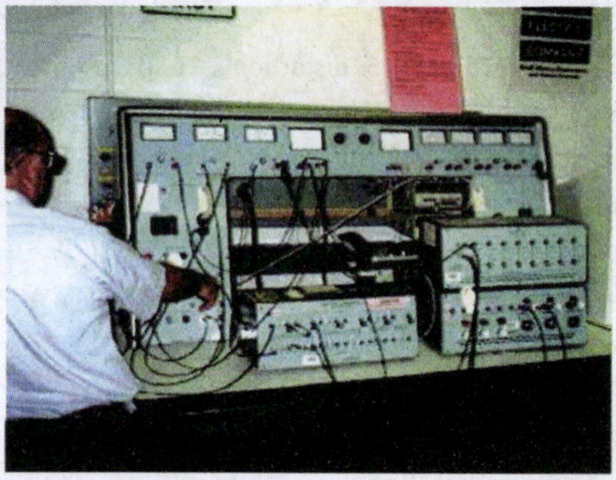

1. DC Equivalent Circuit
2. AC Equivalent Circuit
3. Equivalent Circuit of a CB Amplifier
4. Equivalent Circuit of a CE Amplifier
5. Effect of Source Resistance R_S
6. Equivalent Circuit of a CC Amplifier
7. T-Model
8. Formulas for T-Equivalent of a CB Circuit
9. Formulas for T-Equivalent of a CE Circuit
10. Formulas for T-Equivalent of a CC Circuit
11. What are *h*-parameters?

21.1. General

We will begin by idealizing a transistor with the help of simple approximations that will retain its essential features while discarding its less important qualities. These approximations will help us to *analyse transistor circuits easily and rapidly.*

We will discuss only the *small signal* equivalent circuits in this chapter. Small signal operation is that in which the ac input signal voltages and currents are in the order of ± 10 per cent of *Q*-point voltages and currents.

There are two prominent schools of thought today regarding the equivalent circuit to be substituted for the transistor. The two approaches make use of

(*a*) four *h*-parameters of the transistor and the values of circuit components,

(*b*) the beta (β) of the transistor and the values of the circuit components.

Since long, industrial and educational institutions have heavily relied on the hybrid parameters because they produce more accurate results in the analysis of amplifier circuits. In fact, hybrid-parameter equivalent circuit continues to be popular even today. But their use is beset with the following difficulties :

1. The values of *h*-parameters are not so readily or easily available.
2. Their values vary considerably with individual transistors *even of the same type number.*
3. Their values are limited to a particular set of operating conditions for reasonably accurate results.

The second method which employs transistor beta and resistance values is gaining more popularity of late. It has the following advantages :

1. The required values are easily available.
2. The procedure followed is simple and easy to understand.
3. The results obtained are quite accurate for the study of amplifier circuit characteristics.

To begin with, we will consider the second method first.

21.2. DC Equivalent Circuit

(*a*) CB Circuit

In an ideal transistor, $\alpha = 1$ which means that $I_C = I_E$.

The emitter diode acts like any *forward-biased ideal diode*. However, due to transistor action, collector diode acts as a *current source.* In other words, for the purpose of drawing dc equivalent circuit, we can view an ideal transistor as nothing more than a rectifier diode in emitter and a current source in collector. In the dc equivalent circuit of Fig. 21.1 (*b*), current arrow always points in the direction of conventional current.

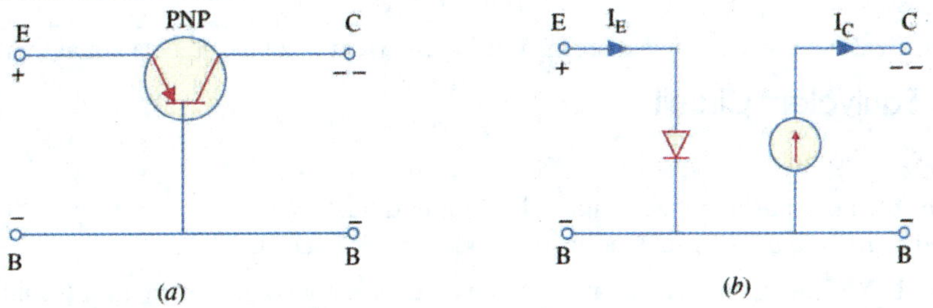

Fig. 21.1

As per the polarities of transistor terminals (Art. 18.3) shown in Fig. 21.1 (*a*), emitter current flows from *E* to *B* and collector current from *B* to *C*.

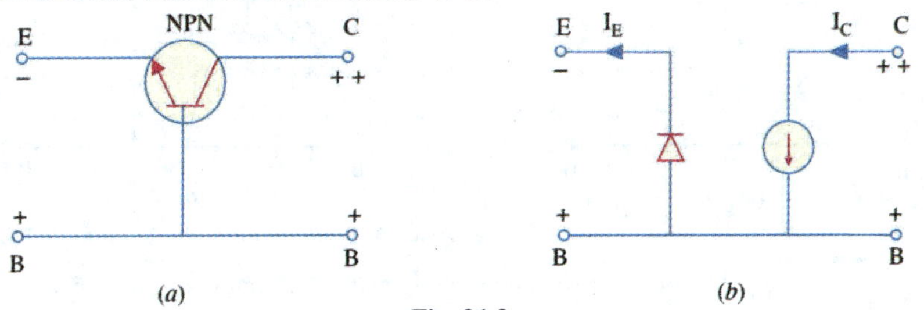

Fig. 21.2

The dc equivalent circuit shown in Fig. 21.2 for an *NPN* transistor is exactly similar except that direction of current flow is opposite.

(*b*) CE Circuit

Fig. 21.3 shows the dc equivalent circuit of an *NPN* transistor when connected in the *CE* configuration Fig. 21.4 that of a *PNP* transistor. Direction of current flow can be easily found by remembering the transistor polarity rule given in Art. 18.3.

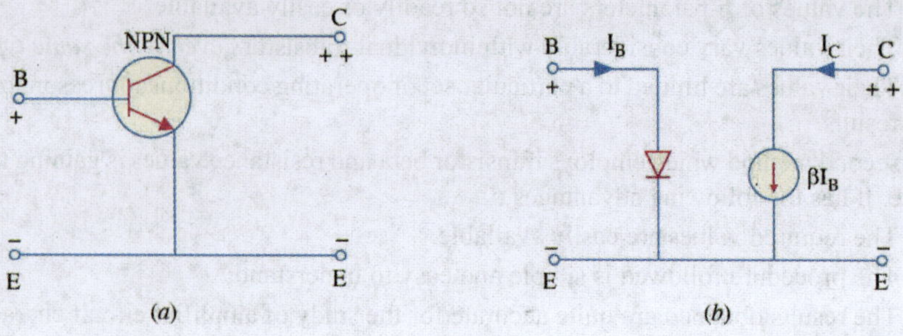

Fig. 21.3

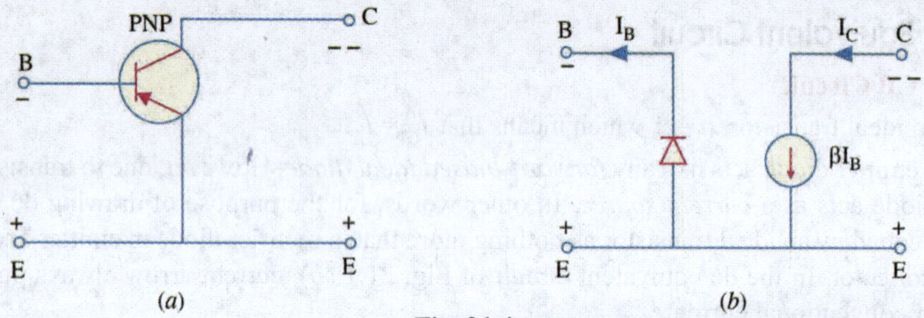

Fig. 21.4

In an ideal *CE* transistor, we disregard leakage current and take ac beta equal to dc beta.

21.3. AC Equivalent Circuit

(*a*) CB Circuit

In the case of *small* input ac signals, the emitter diode *does not rectify,* instead it offers resistance called ac *resistance*. As usual, collector diode acts as a current source.

Fig. 21.5 (*b*) shows the ac equivalent circuit of a transistor connected in the *CB* configuration.* Here, ac resistance offered by the emitter diode is

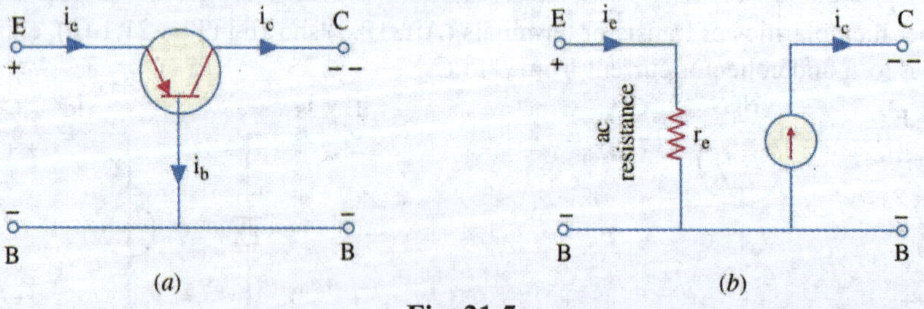

Fig. 21.5

$$r_{ac} = \text{junction resistance } (r_j) + \text{base-spreading resistance } (r_B)^{**}$$

$$= r_j \qquad \because r_B \text{ is negligible}$$

$$= \frac{25 \text{ mV}}{I_E} \qquad \text{— where } I_E \text{ is dc emitter current in mA}$$

It is written as r_e' signifying junction resistance of the emitter *i.e.*, ac resistance looking into the emitter (though shown as r_e in figures).

* This circuit is valid both for *PNP* as well as *NPN* transistors because difference in the direction of ac currents does not matter.

** Also called bulk-resistance.

$$\therefore \qquad r_e' = \frac{25 \text{ mV}}{I_E}$$

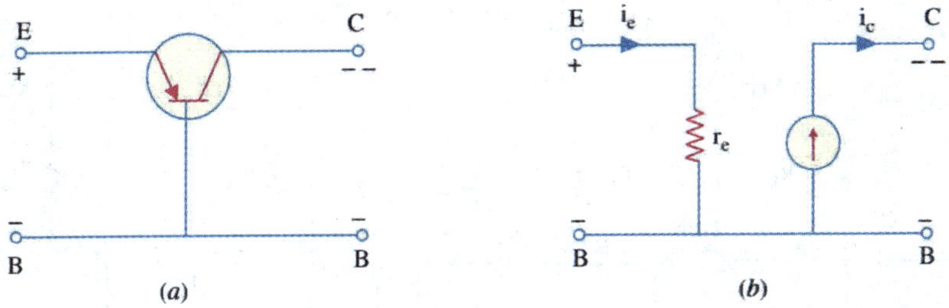

Fig. 21.6

Hence, the ac equivalent circuit of a *CB* circuit becomes as shown in Fig. 21.6. Since changes in collector current are almost equal to changes in emitter current, $\Delta i_c = \Delta i_e$.

(*b*) CE Circuit

Fig. 21.7 (*b*) shows the equivalent circuit when transistor has been connected in the *CE* configuration.

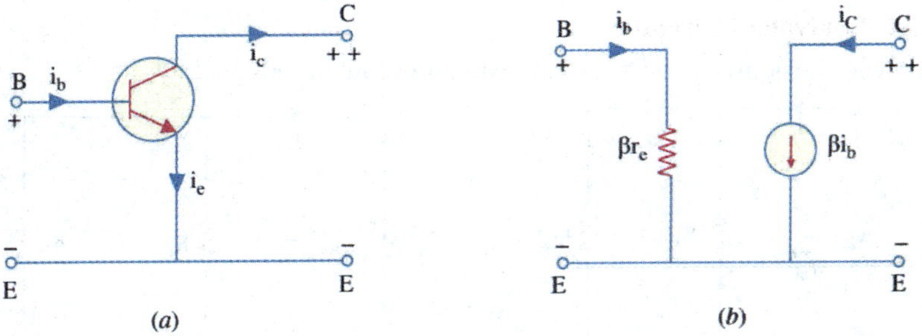

Fig. 21.7

The ac resistance *looking into the base is*

$$r_{ac} = \frac{25 \text{ mV}}{I_B} \qquad \text{— dc current is } I_B \text{ not } I_E$$

$$= \frac{25 \text{ mV}}{I_C/\beta} = \beta . \frac{25 \text{ mV}}{I_C} \cong \beta . \frac{25 \text{ mV}}{I_E} = \beta r_e'$$

Strictly speaking, $r_{ac} = (1 + \beta)\, r_e' \cong \beta\, r_e'$ — Art. 9.11

The ac collector current is β times the base current *i.e.*, $i_c = \beta\, i_b$.

21.4. Equivalent Circuit of a CB Amplifier

In Fig. 21.8 (*a*) is shown the circuit of a common-base amplifier. As seen, emitter is forward-biased by $-V_{EE}$ and collector is reverse-biased by V_{CC}. The ac signal source voltage v_s drives the emitter. It produces small fluctuations in transistor voltage and currents in the output circuit. It will be seen that ac output voltage is amplified because it is more than v_s.

(*a*) DC Equivalent Circuit

For drawing dc equivalent circuit, following procedure should be adopted :

(*i*) short all ac sources *i.e.*, reduce them to zero,

(*ii*) open all capacitors because they block dc.

If we do this, then as seen from Fig. 21.8 (*a*), neither emitter current can pass through C_1 nor collector current can pass through C_2. These currents are confined to their respective resistances R_E and R_C (earlier we had been designating it as resistance R_L).

Here, $I_C \cong I_E$ and $V_{CB} = V_{CC} - I_C R_L$

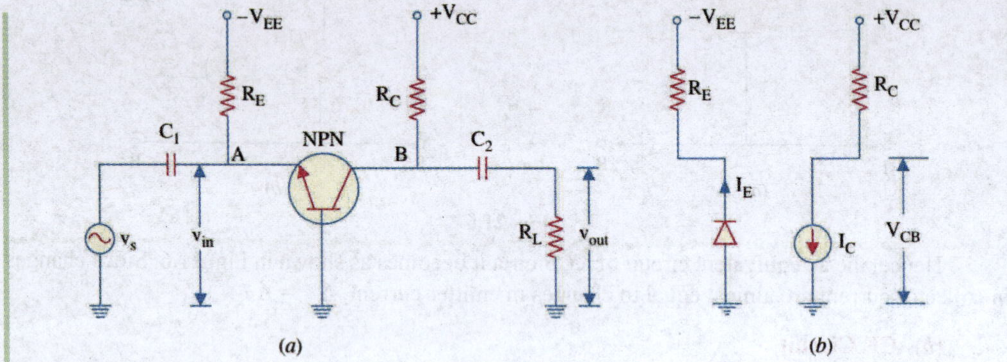

Fig. 21.8

Hence, the dc equivalent circuit becomes as shown in Fig. 21.8 (*b*).

(*b*) AC Equivalent Circuit

For drawing ac equivalent circuit, following procedure is adopted :

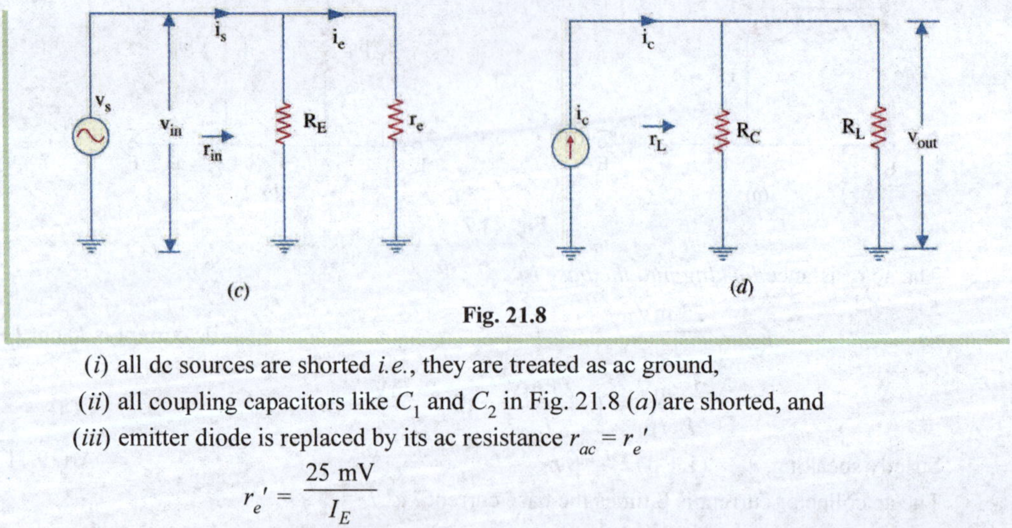

Fig. 21.8

(*i*) all dc sources are shorted *i.e.*, they are treated as ac ground,

(*ii*) all coupling capacitors like C_1 and C_2 in Fig. 21.8 (*a*) are shorted, and

(*iii*) emitter diode is replaced by its ac resistance $r_{ac} = r_e'$

$$r_e' = \frac{25 \text{ mV}}{I_E}$$

where I_E is dc emitter current in mA.

As seen by the input ac signal, it has to feed R_E and r_e' in parallel [Fig. 21.8 (*c*)]. As looked from point *A* in Fig. 21.8 (*a*), R_E is grounded through V_{EE} which has been shorted and r_e' is grounded *via* the base.

Similarly, collector has to feed R_C and R_L which are connected in parallel across it *i.e.*, at point *B* in Fig. 21.8 (*a*). The ac signal in collector sees an output load resistance of $r_L = R_C \| R_L$

Hence, ac equivalent circuit is as shown in Fig. 21.8 (*d*). Here, collector diode itself has been shown as a current source.

Following two points are worth noting :

(*i*) changes in collector signal current are very nearly equal to changes in emitter signal current. Hence, $\Delta i_c \cong \Delta i_e$.

(*ii*) directions of ac currents shown in the circuit diagram are those which correspond to postive half-cycle of the ac input voltage. That is why i_c is shown flowing upwards in Fig. 21.8 (*d*).

(c) Principal Operating Characteristics

1. Input Resistance

As seen from Fig. 21.8 (*c*), the input resistance of the circuit (or stage) is given by

$$r_{in} = R_E \parallel r_e' = \frac{r_e' R_E}{R_E + r_e'}$$

In practice, R_E is always much greater than r_e' so that parallel combination

$$R_E \parallel r_e' \cong r_e'$$

$\therefore$ $r_{in} \cong r_e'$ = input resistance of the emitter diode

2. AC Load Resistance

The collector load as seen by output ac signal consists of a parallel combination of R_L and R_C. This has already been designated as r_L (Art. 21.4).

$$\therefore \quad r_L = R_L \| R_C$$

It should be carefully noted that it is the output resistance *as seen by collector and not the ac output resistance when looking into the collector.*

Note. In case, R_L has not been connected, then $r_L = R_C$ (Ex. 21.1).

3. Current Gain

It is given by the ratio

$$A_i = \frac{i_c}{i_c} \cong \alpha$$

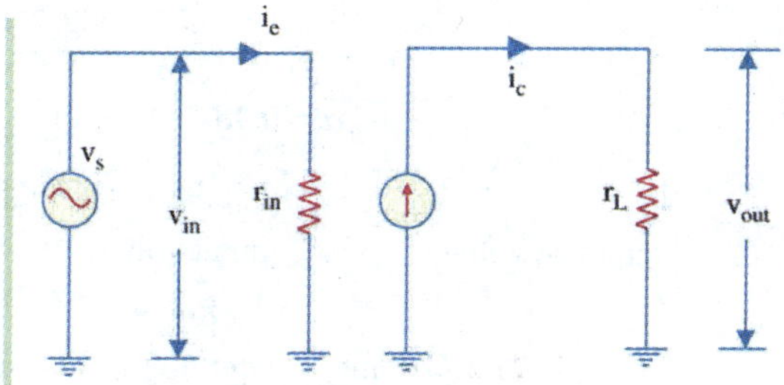

Fig. 21.9

4. Voltage Gain

It is given by the ratio

$$A_v = \frac{v_{out}}{v_{in}}$$

Now, v_{in}* = *i.e.*, r_{in}

$$= (1 + \beta)\, i_b \,.\, r_{in}$$

$$v_{out} = i_c r_L = \beta i_b r_L$$

$$\therefore \quad A_v = \frac{\beta i_b r_L}{(1 + \beta)\, i_b r_{in}}$$

Taking the ratio $\beta / (1 + \beta)$ as unity $\quad A_v \cong \dfrac{r_L}{r_{in}} = \dfrac{r_L}{r_{e'}}$

5. Power Gain

$$A_p = A_v . A_i$$

When expressed in decibels, it is written as $G_p = 10 \log_{10}{}^{Ap}$ dB.

When expressed in terms of r_{in} and r_L, the ac equivalent circuit becomes as shown in Fig. 21.9.

Example 21.1. *For the single-stage CB amplifier shown in Fig. 21.10 (a), find r_{in}, r_L, A_i, A_v and A_p.*

What would be the r.m.s. value of the signal voltage across the load if v_s has an r.m.s. value of 1.5 mV? Assume silicon material and transistor α = 0.98.

---

* Here, v_{in} equals v_s because there is no internal resistance of the source. In case it is there, the two will not be equal (Art. 21.5).

Solution. For finding r_e', let us first find I_E.

$$I_E = \frac{V_{EE} - V_{BE}}{R_E} = \frac{10 - 0.7}{5\ K} = 1.86\ \text{mA}$$

$$r_e' = \frac{25\ \text{mV}}{I_E\ \text{mA}} = \frac{25}{1.86} \cong 14\ \Omega$$

(*i*) $r_{in} = r_e' \| R_E = 14\ \Omega \| 5\ \text{K} \cong \mathbf{14\ \Omega}$ (*ii*) $r_L = R_C = \mathbf{4\ K}$

The ac equivalent circuit becomes as shown in Fig. 21.10 (*b*).

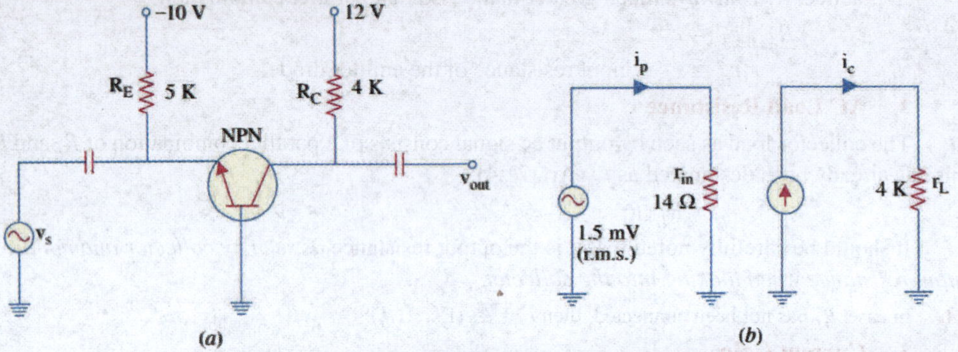

Fig. 21.10

(*iii*) $A_i = \alpha = \mathbf{0.98}$ (*iv*) $A_v = \frac{r_L}{r_{in}} = \frac{r_L}{r_e'} = \frac{4K}{14\Omega} = \mathbf{286}$

(*v*) $A_p = A_i A_v = 0.98 \times 286 = 280$ $G_p = 10 \log_{10}{}^{280} = \mathbf{24.5\ dB}$

Output ac voltage $= A_v \times$ input voltage

$\therefore$ $V_{out} = A_v \times V_{in} = 286 \times 1.5 = 429\ \text{mV} = \mathbf{0.429\ V}$

Example 21.2. *For the CB amplifier circuit shown in Fig. 21.11 (a), find*

(i) stage r_{in} (ii) r_L (iii) ac output voltage v_{out} (iv) voltage gain A_v.

Take $V_{BE} = 0.7$ V.

Solution. $I_E = (20 - 0.7)/30\ \text{K} = 0.64\ \text{mA}\ ;\ r_e' = 25/0.64 = 39\Omega$

(*i*) stage $r_{in} = r_e' \| R_E = 39\ \Omega \| 30\ \text{K} \cong \mathbf{39\ \Omega}$

(*ii*) $r_L = R_L \| R_C = 30\ \text{K} \| 15\ \text{K} = \mathbf{10\ K}$

The amplifier circuit alongwith its ac equivalent circuit is shown in Fig. 21.11.

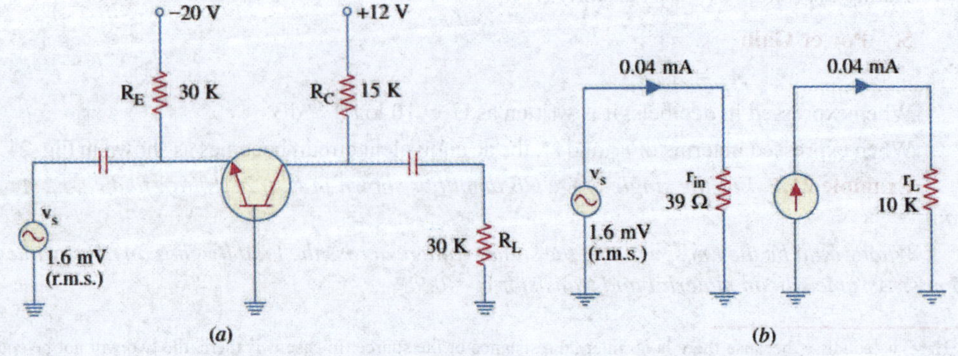

Fig. 21.11

(*iii*) We will first find the value of v_{out} as a drop across $r_L = R_C \parallel R_L$. For that purpose, we will employ r.m.s. values.

Now, $$i_e = \frac{V_{in}}{r_{in}} = \frac{V_s}{r_{in}} = \frac{1.6}{3.9} = 0.04 \text{ mA*}$$

$$I_c = I_e = 0.04 \text{ mA}; \ V_{out} = I_c r_L = 0.04 \times 10 \text{ K} = \mathbf{0.4\ V}$$

(*iv*) $$A_v = \frac{r_L}{r_{in}} = \frac{r_L}{r_e'} = \frac{10\ K}{39\ \Omega} = \mathbf{250}$$

The r.m.s. value of ac output voltage may be found with the help of A_v.

Now, $$A_v = \frac{V_{out}}{V_{in}} \quad \therefore \quad V_{out} = A_v \cdot V_{in} = 250 \times 1.6 = 400 \text{ mV} = \mathbf{0.4\ V.}$$

21.5. Effect of Source Resistance R_S on Voltage Gain

The voltage gain for the *CB* amplifier circuit has so far been calculated on the assumption that the resistance R_S of the ac signal source is negligible. The voltage gain will decrease as R_S increases because more and more of v_s will drop on R_S rather than on r_{in} and so will not appear in the output.

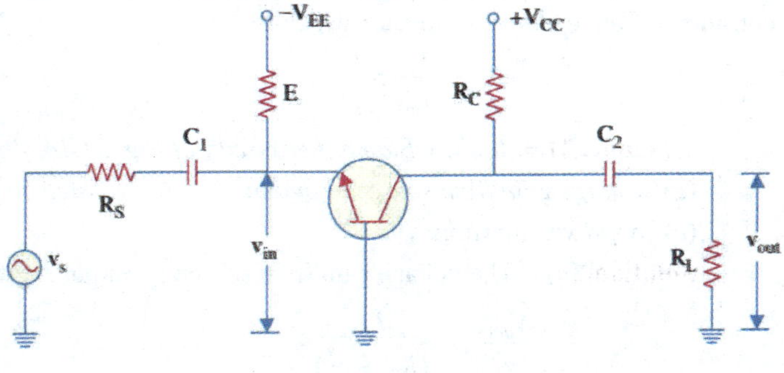

Fig. 21.12

The basic circuit is shown in Fig. 21.12 where R_S is the internal resistance of the ac signal source.

The ac equivalent circuit is shown in Fig. 21.13 (*a*). Here, R_S is in series with $(r_e' \parallel R_E)$. The input ac signal voltage v_s drops on these two series resistors. Obviously, v_{in} is the drop across $r_e' \parallel R_E$ and is less than v_s. Using Proportional Voltage Formula (Art. 2.7)

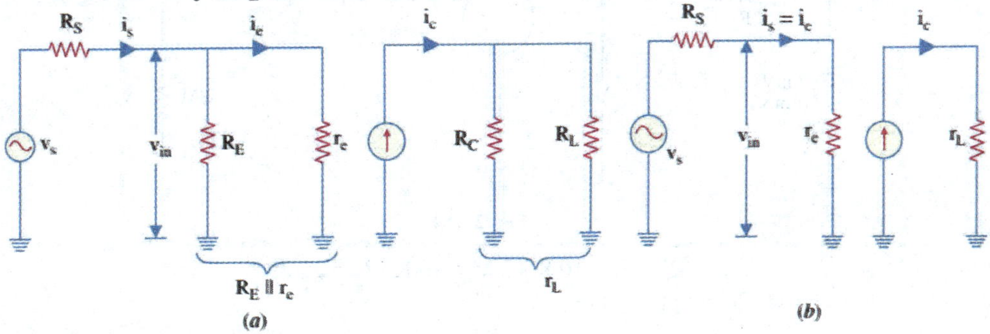

Fig. 21.13

$$v_{in} = v_s \frac{r_e' \parallel R_E}{R_S + r_e' \parallel R_E}$$

* Strictly speaking, it is not I_e but I_s. A small part of the ac source current goes through R_E = 30 K and the balance (major part) goes through parallel resistance r_e'. However, if we neglect current through 30 K, then $I_e = I_s$.

In practice, $R_E \gg r_e'$, so that $r_e' \parallel R_E \cong r_e'$

$$\therefore \quad v_{in} = v_s \frac{r_e'}{R_S + r_e'} = \frac{v_s}{I + \dfrac{R_S}{r_e'}}$$

As R_S increases in comparison to r_e', the ratio R_S/r_e' increases thereby making v_{in} increasingly less than v_s.

Moreover, as seen from the source, input resistance is

$$R_S + r_e' \parallel R_E \cong R_S + R_e'$$

$$\therefore \quad \frac{v_{out}}{v_s} = \frac{r_L}{(R_S + r_e')} \cong \frac{r_L}{R_S} \qquad \text{— if } R_S \gg r_e'$$

Making R_S much larger then r_e' is called *swamping out* the emitter diode. This makes voltage gain independent of r_e' and hence the particular transistor used. It is different from the voltage gain considered from *emitter to collector* which is

$$\frac{v_{out}}{v_{in}} = \frac{r_L}{r_{in}} \cong \frac{r_L}{r_e'}$$

Example. 21.3. *In the CB amplifier circuit of Fig. 21.14, find*

(*a*) *voltage gain from source to output* (*b*) *voltage gain from emitter to output*

(*c*) *approximate value of* v_{in}.

Solution. (*a*) The voltage gain from source to output is given by

$$\frac{v_{out}}{v_s} = \frac{r_L}{(R_S + r_e')}$$

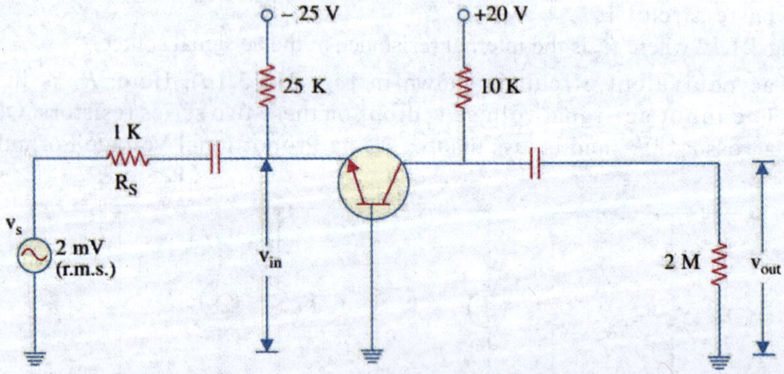

Fig. 21.14

$$r_L = R_C \parallel R_L = 10 \text{ K} \parallel 2 \text{ M} \cong 10 \text{ K} ; I_E = 25/25 \text{ K} = 1 \text{ mA}$$

$$r_e' = 25/1 = 25\ \Omega$$

$$\frac{v_{out}}{v_{in}} = \frac{10\ K}{(1\ K + 26\ \Omega)} \cong \mathbf{10} \qquad \ldots(i)$$

(*b*) The voltage gain from emitter to output is

$$\frac{v_{out}}{v_{in}} = \frac{r_L}{r_{in}}$$

Now, $\quad r_{in} = R_E \parallel r_e' = 25 \text{ K} \parallel 25\ \Omega \cong 25\ \Omega$

$$\therefore \quad \frac{v_{out}}{v_{in}} = \frac{10\ K}{25\ \Omega} = \mathbf{400} \qquad ...(ii)$$

(c) The value of v_{in} may be found by the following two ways :

(i) As seen from Eq. (ii) above, $v_{in} = \frac{v_{out}}{400}$

Now, from Eq. (i) above, we have

$$v_{out} = 10\ v_s = 10 \times 2 = 20 \text{ mV}$$

$$\therefore \quad v_{in} = \frac{20}{400} \times 10^3 = \mathbf{50\ \mu V}$$

(ii) As seen from the ac equivalent circuit of the amplifier (Fig. 21.15), v_s is dropped proportionally on the two series resistances R_S and r_e' (strictly speaking $r_e' \parallel R_E$). The drop across r_{in} is called v_{in}.

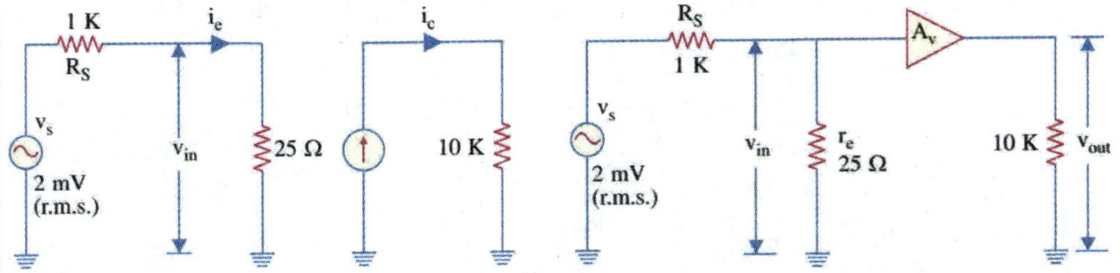

Fig. 21.15

$$\therefore \quad v_{in} = v_s \frac{r_e'}{R_S + r_e'} = 2mV \times \frac{25}{1000 + 25} = \mathbf{50\ \mu\ V}$$

Example 21.4. *For the circuit shown in Fig. 21.16, find*

(i) voltage gain A_v and (ii) output voltage v_{out}.

Solution. This circuit is exactly similar to the one shown in Fig. 21.11 (a) except for the addition of ac voltage source resistance R_S.

Since R_S and r_e' are in series so far as the input ac signal is concerned.

$\therefore$ input resistance $= R_S + r_e'$

$$\therefore \quad A_v = \frac{v_{out}}{v_s} = \frac{r_0}{R_S + r_e'} = \frac{r_L}{R_S + r_e'} = \frac{10\ K}{(100 + 39)\ \Omega} = \mathbf{72}$$

It is seen that overall voltage gain is drastically reduced *i.e.*, from 250 to 72.

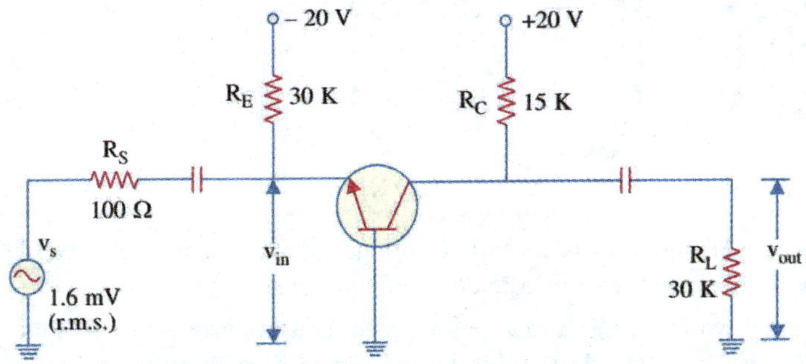

Fig. 21.16

(ii) $\qquad v_{out} = v_s \times A_v = 1.6 \times 72$

$= 115 \text{ mV} = \mathbf{0.115\ V}$

21.6. Equivalent Circuit of a CE Amplifier

Consider the simple *CE* amplifier circuit of Fig. 21.17 (*a*) in which base bias has been employed.

(a) DC Equivalent Circuit

For drawing the dc equivalent circuit, same procedure is adopted as given in Art. 21.2. It is shown in Fig. 21.17 (*b*).

As seen, $\qquad I_B = \dfrac{V_{CC} - V_{BE}}{R_B} \cong \dfrac{V_{CC}}{R_B}; \; I_C = \beta . I_B ; \; I_E \cong I_C \cong \beta \, I_B$

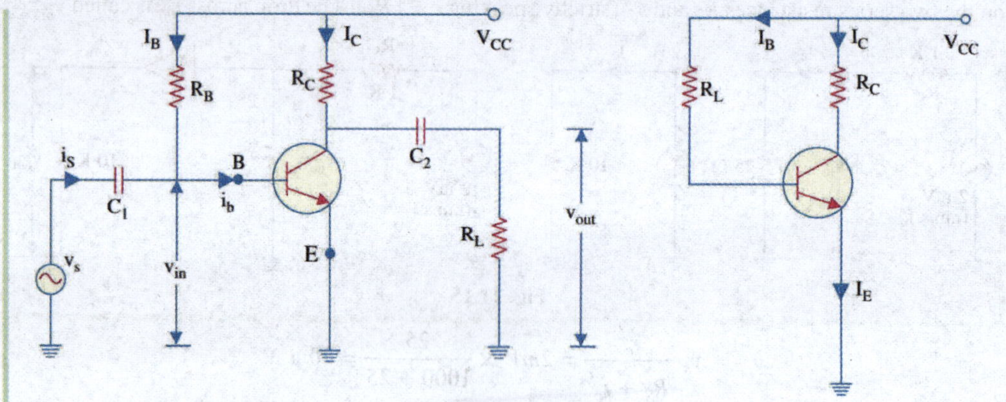

Fig. 21.17

(b) AC Equivalent Circuit

Let us now analyse the ac equivalent circuit given in Fig. 21.18.

As proved in Art. 21.3, the ac resistance as seen by the input signal when looking into the base is $= \beta \, r_e'$. It may be called $r_{in(base)}$ to distinguish it from r_{in} which is resistance of the *CE* ***stage*** as shown in Fig. 21.18 (*a*).

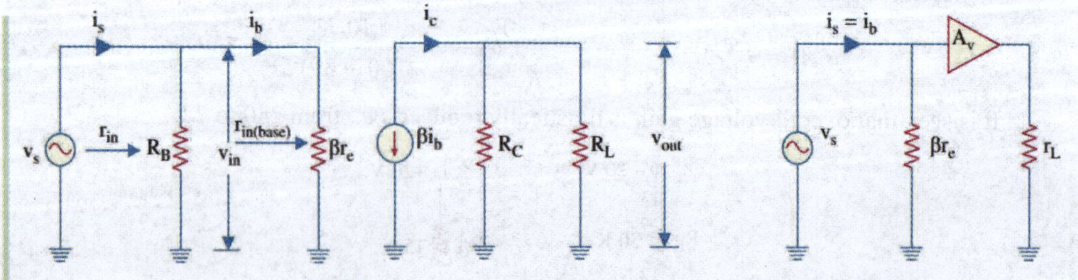

Fig. 21.18

It may be noted that in the absence of ac voltage source resistance R_S , whole of v_s acts across R_B as well as $\beta r_e'$ because the two are connected in parallel across it.

Another point worth noting is that major part of source current i_s passes through $\beta \, r_e'$ and an extremely small part ($= v_s / R_B$) passes through R_S. Since R_B is usually very large, current passing through it can be easily neglected. Hence, as shown in Fig. 21.18 (*b*), $i_b = i_s$.

Moreover, R_S being zero, whole of v_s drops over $R_B \parallel \beta\, r_e'$.

(c) Principal Operating Characteristics

1. Input Resistance

As seen from Fig. 21.18, input resistance of the *stage* is

$$r_{in} = R_B \parallel \beta\, r_e' \cong \beta\, r_e' \qquad \because R_B >> \beta\, r_e'$$

$$= \text{input resistance of the base}$$

$$r_{in(stage)} = r_{in(base)}$$

2. AC Load Resistance. $r_L = R_C \parallel R_L$.

3. Current Gain. $A_i = \dfrac{i_c}{i_b} = \beta$.

4. Voltage gain. The voltage gain of the stage or circuit is $A_v = \dfrac{v_{out}}{v_{in}}$.

Now, $\quad v_{in} = i_b\, \beta\, r_e' \quad$ and $\quad v_{out} = i_c . r_L = \beta\, i_b\, r_L$

$$\therefore \quad A_v = \frac{\beta\, i_b\, r_L}{\beta\, i_b\, r_e'} = \frac{r_L}{r_e'} \qquad \text{— if } R_B >> \beta\, r_e'$$

5. Power Gain. $A_p = A_i \,.\, A_v$ and $G_p = 10 \log_{10}^{Ap}$ dB

Example. 21.5. *If in the CE circuit of Fig. 21.17 (a), V_{CC} = 20 V, R_C = 10 K, R_B = 1 M, R_L = 1 M, v_s = 2 mV and β = 50, find*

(i) i_b and i_c (ii) r_{in} (iii) r_L (iv) A_v (v) A_p and G_p.

Solution. I_B = 20/1 M = 20 μA; $\quad I_C = \beta\, I_B$ = 50 × 20 μA = 1 mA

r_e' = 25 mV/1 mA = 25 Ω; $\quad r_{in(base)} = \beta r_e'$ = 50 × 25 = 1250 Ω

(i) As seen from Fig. 21.18 (*a*), ac base current is given by

$$i_b = \frac{v_s}{r_{in(base)}} = \frac{2\ mV}{1250\ \Omega} = \mathbf{1.6\ \mu A};\ i_c = \beta i_b = 50 \times 1.6 = \mathbf{80\ \mu A}$$

(ii) stage $\quad r_{in} = R_B \parallel \beta r_e'$ = 1 M ∥ 1250 Ω ≅ **1250 Ω**

(iii) $\quad r_L = R_L \parallel R_C$ = 1 M ∥ 10 K ≅ **10 K**

(iv) $$A_v = \frac{r_v}{r_e'} = \frac{r_L}{r_e'} = \frac{10\ K}{25\ \Omega} = \mathbf{400}$$

(v) $\quad A_p = A_i \,.\, A_v$ = 50 × 400 = 2,000 ; $G_p = 10 \log_{10}^{20000}$ = **43 dB.**

Example 21.6. *In the CE amplifier circuit of Fig. 21.19 employing emitter feedback, find :*

(i) r_{in} (ii) r_L (iii) A_v

(iv) A_p and (v) G_p

Take transistor β = 100. How will these values change if emitter bypass capacitor is removed?

Solution. It should be carefully noted that emitter bypass capacitor *C* provides *ac ground* to the signal *i.e.*, it shorts out R_E to ground so far as *ac signal is concerned.* However, it plays its normal role so far as dc quantities are concerned.

Hence, the ac equivalent circuit of the amplifier becomes as shown in Fig. 21.20.

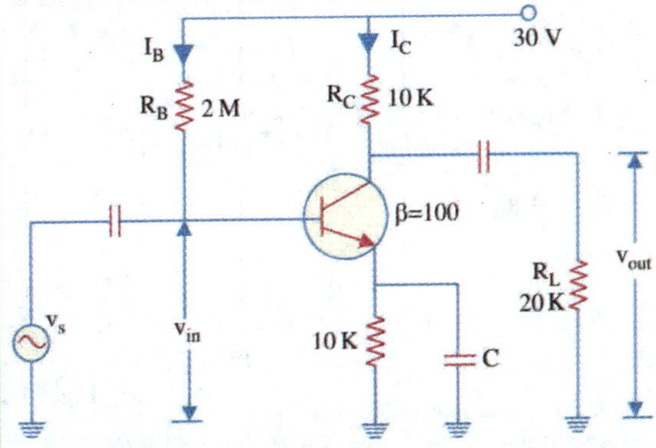

Fig. 21.19

Now, $$I_B = \frac{V_{CC}}{R_B + \beta\, R_E}$$

$$= \frac{30}{2\text{ M} + 100 \times 10\text{ K}} = 10\ \mu\text{A};$$

$$I_C = \beta\, I_B = 100 \times 10\ \mu\text{A} = 1\text{ mA}$$

$$I_E \cong I_C = 1\text{ mA},\ r_e' = 25/1 = 25\ \Omega$$

$$\beta\, r_e' = 100 \times 25 = 2500\ \Omega$$

(*i*) As seen from Fig. 21.20

$$r_{in} = R_B \parallel \beta\, r_e'$$

$$= 2\text{ M} \parallel 2500\ \Omega \cong \mathbf{2500\ \Omega}$$

(*ii*) $$r_L = R_C \parallel R_L = 10\text{ K} \parallel 20\text{ K} = \mathbf{6.67\ K}$$

(*iii*) $$A_v = \frac{v_{out}}{v_{in}} = \frac{r_L}{r_e'} = \frac{6.67\ K}{25\ \Omega} = \mathbf{267}$$

(*iv*) $$A_p = A_i \cdot A_v = 100 \times 267 = \mathbf{26{,}700};\ G_p = 10 \log_{10}{}^{26700} = \mathbf{44.3\ dB}$$

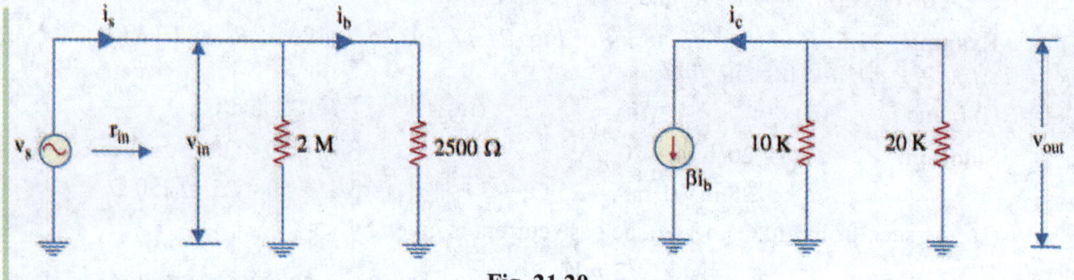

Fig. 21.20

When Emitter Bypass Capacitor is Removed

When bypass capacitor is removed, ac ground is removed. Now, the ac signal will have to pass through R_E also. According to β-rule of Art. 19.11, the total emitter resistance referred to base will become $(1 + \beta)\,(r_e' + R_E) \cong \beta\,(r_e' + R_E)$ because r_e' is in series with R_E as shown in Fig. 21.21 (*a*).

Now, the ac equivalent circuit becomes as shown in Fig. 21.21 (*b*).

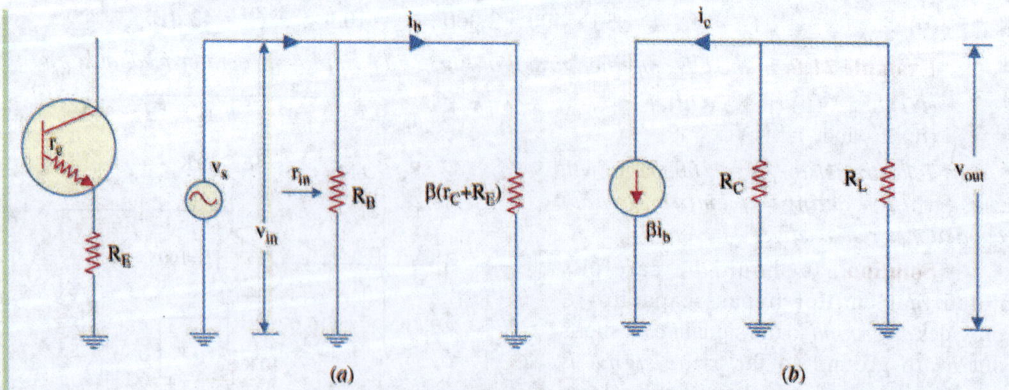

Fig. 21.21

Stage $$r_{in} = R_B \parallel \beta\,(r_e' + R_E) \cong R_B \parallel \beta\, R_E$$

It is so because R_E is much greater than r_e'.

(*i*) $r_e' = 25\ \Omega$ — as before

$\therefore \quad \beta\,(r_e' + R_E) \cong \beta\,R_E = 100 \times 10\ \text{K} = 1\ \text{M}$

$\therefore \quad r_{in} = R_B \,\|\, \beta\,R_E = 2\ \text{M} \,\|\, 1\text{M} = \frac{2}{3}\ \text{M} = \mathbf{666.7\ K}$ — much greater than before

(*ii*) $r_L = \mathbf{6.67\ K}$ — it remains unchanged

(*iii*) $v_{in} = i_b . \beta\,(r_e' + R_E);\ v_{out} = i_c . r_L = \beta\, i_b\, r_L$

$$\therefore \quad A_v = \frac{v_{out}}{v_{in}} = \frac{\beta\, i_b\, r_L}{i_b . \beta (r_e' + R_E)} = \frac{r_L}{(r_e' + R_E)} \cong \frac{r_L}{R_E}$$

$$= \frac{6.67\ K}{10\ K} = \mathbf{0.667}$$ — reduced drastically

(*iv*) $A_p = A_i A_v = 100 \times 0.667 = \mathbf{66.7}$ — reduced considerably

It is seen that by removing bypass capacitor, *excessive degeneration has occurred in the amplifier circuit.*

Example 21.7. *For the circuit shown in Fig. 21.22, compute (i) $r_{in(base)}$ (ii) v_{out} (iii) A_v (iv) r_{in}. Neglect V_{BE} and take $\beta = 200$.*

Solution. It may be noted that voltage divider bias has been used in the circuit.

$$V_2 = \frac{15}{15 + 45} \times 30 = 7.5\ \text{V}$$

$$V_E = V_2 - V_{BE} \cong V_2 = 7.5\ \text{V}$$

$$I_E = 7.5/7.5\ \text{K} = 1\ \text{mA}$$

$$r_e' = 25\ \text{mV}/1\ \text{mA} = 25\ \Omega$$

The ac equivalent circuit is shown in Fig. 21.23. Since dc source is shorted, 45 K, resistor is ac grounded. On the input side, three resistance become paralleled across v_s *i.e.*, (*i*) 15 K (*ii*) 45 K and (*iii*) $\beta\, r_e'$ or $r_{in(base)}$.

Capacitor C_2 ac grounds the emitter resistance R_E, so does C_3 to 5 K and V_{CC} to 10 **K.**

(*i*) $r_{in(base)} = \beta\, r_e' = 200 \times 25 = \mathbf{5\ K}$ (*ii*) $i_b = 5\ \text{mV}/5\ \text{K} = 1\ \mu\text{A}$

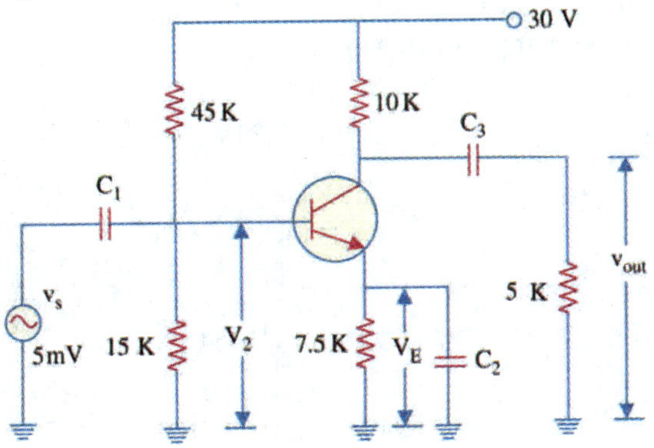

Fig. 21.22

(Obviously, i_b is not the current which *leaves* the source but that part of the source current i_s which *enters* the base).

$$i_c = \beta\, i_b = 200 \times 1 = 200\ \mu\text{A} = 0.2\ \text{mA}$$

Now, collector load resistance is

$r_L = 10\ \text{K} \,\|\, 5\ \text{K} = 10/3\ \text{K} = 3.33\ \text{K}, \qquad v_{out} = i_c\, r_L = 0.2 \times 3.33 = \mathbf{0.667\ V}$

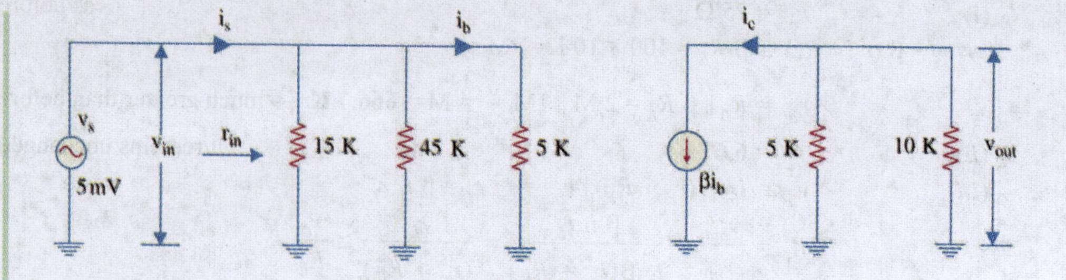

Fig. 21.23

(*iii*) $$A_v = \frac{v_{out}}{v_{in}} = \frac{0.667 \text{ V}}{5 \text{ mV}} = 133 \text{ or } A_v = \frac{r_L}{r_e'} = \frac{3.33\ K}{25\ \Omega} = \mathbf{133}$$

(*iv*) r_{in} means the input ac resistance as seen from the source *i.e.*, from point *A* in Fig. 21.21. It is different from $r_{in(base)}$. Obviously, r_{in} is equal to the equivalent resistance of three resistances connected in parallel.

$$r_{in} = 15 \text{ K} \parallel 45 \text{ K} \parallel 5 \text{ K} = \mathbf{3.54\ K}$$

Example 21.8. *For the CE amplifier circuit of Fig. 21.24, find out (i)* r_{in}, *(ii)* r_L, *(iii)* A_v, *(iv)*A_p, *and (v)* V_{out}.

Take transistor β = 50.

Solution. $$I_E \cong \frac{V_{EE}}{R_E} = \frac{20}{40} = 0.5 \text{ mA}$$

$$r_e' = 25/0.5 = 50\ \Omega$$

$$\beta r_e' = 50 \times 50 = 2500\ \Omega$$

(*i*) $$r_{in} = 10 \text{ K} \parallel 2500\ \Omega = \mathbf{2000\ \Omega}$$

(*ii*) $$r_L = 20 \text{ K} \parallel 20 \text{ K} = \mathbf{10\ K}$$

(*iii*) $$A_v = \frac{r_L}{r_e'} = \frac{10\,K}{50\,\Omega} = \mathbf{200}$$

(*iv*) $$A_p = A_v \cdot A_i$$

$$= 200 \times 50 = 10{,}000$$

$$G_p = 10 \log_{10}{}^{10000} = \mathbf{40\ dB}$$

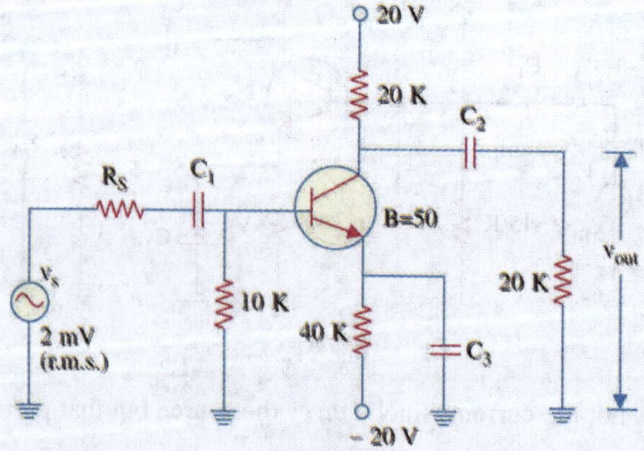

Fig. 21.24

(*v*) $$V_{out} = A_v \times V_{in} = 20 \times 2 = \mathbf{40\ mV\ (r.m.s.)}$$

Since R_S is zero, whole of v_s appears across the diode base.

Example. 21.9. *For the single-stage CE amplifier circuit of Fig. 21.25, find approximate value of (i) r_{in} (ii) r_L (iii) A_v (iv) A_p and G_p.*

Take transistor $\beta = 100$. Use $r_e' = 50\ mV/I_E$

Solution. $V_2 = 20 \times \dfrac{5}{5+45} = 2$ V;

$$I_E \cong = \frac{V_2}{R_E} = \frac{2}{1} = 2\text{mA}$$

$$r_e' = 50/2 = 25\ \Omega\ ;$$

$$\beta r_e' = 25 \times 100 = 2.5\ \text{K}$$

(*i*) $r_{in} = R_1 \parallel R_2 \parallel \beta r_e'$

$= 45 \parallel 5 \parallel 2.5 = \mathbf{1.6\ K}$

(*ii*) $r_L = 5\ \text{K} \parallel 5\ \text{K} = \mathbf{2.5\ K}$

(*iii*) $A_v = \dfrac{r_L}{r_e'} = \dfrac{2.5\ K}{25\ \Omega} = \mathbf{100}$

(*iv*) $A_v = A_v \,.\, A_i = 100 \times 100$

$= \mathbf{10{,}000}$

(*v*) $G_p = 10 \log_{10}{}^{10000} = \mathbf{40\ dB}$

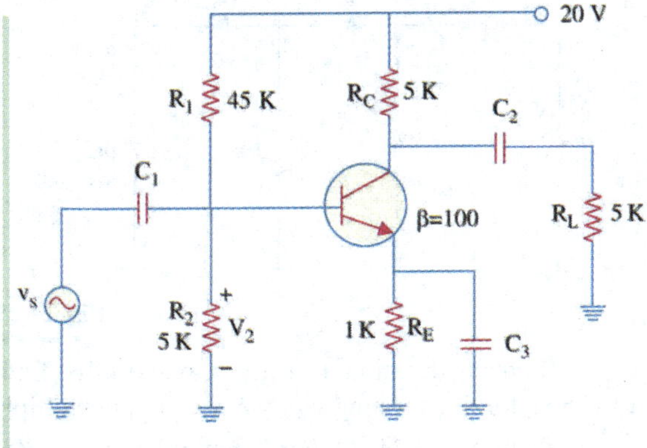

Fig. 21.25

Note. R_E did not come into picture because it was ac grounded by the bypass capacitor C_3.

Example 21.10. *Find the approximate values of the quantities of Ex. 21.9 in case bypass capacitor C_3 in Fig. 21.25 is removed.*

Solution. (*i*) In this case,

$$r_{in} = R_1 \parallel R_2 \parallel \beta\,(r_e' + R_E) = 45\ \text{K} \parallel 5\ \text{K} \parallel 100\,(25 + 1\ \text{K})$$

$$\cong 45\ \text{K} \parallel 5\ \text{K} \parallel 100\ \text{K} = \mathbf{5\ K}$$

(*ii*) $r_L = \mathbf{2.5\ K}$ — as before

(*iii*) $A_v = \dfrac{r_L}{(r_e' + R_E)} \cong \dfrac{r_L}{R_E} = \dfrac{2.5\ \text{K}}{1\ \text{K}}\ \mathbf{2.5}$ (Please note the reduction)

(*iv*) $A_p = 100 \times 2.5 = \mathbf{250}$

(*v*) $G_p = 10 \log_{10}{}^{250} = \mathbf{24\ dB}$

21.7. Effect of Source Resistance R_S

Greater the internal resistance of the ac signal source, greater the internal voltage drop and hence lesser the value of v_{in} because

$v_{in} = v_s$ — drop across R_S

Consider the *CE* circuit shown in Fig. 21.26 whose ac equivalent circuit is shown in Fig. 21.27.

As seen from Fig. 21.27, on the input side, R_S is in series with $R_B \parallel \beta\, r_e'$. Hence, v_s is divided between them in the direct ratio of their resistances.

If R_S is much less than $R_B \parallel \beta r_e'$, then $v_{in} \cong v_s$.

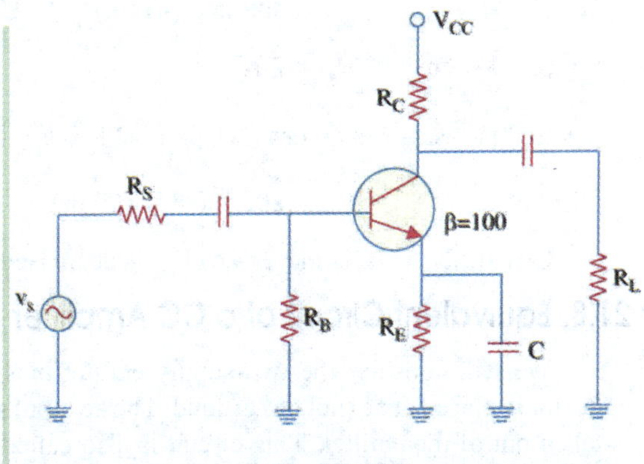

Fig. 21.26

The voltage gain from base to output is still given by .

$$\frac{v_{out}}{v_{in}} = \frac{r_L}{r_e'}$$

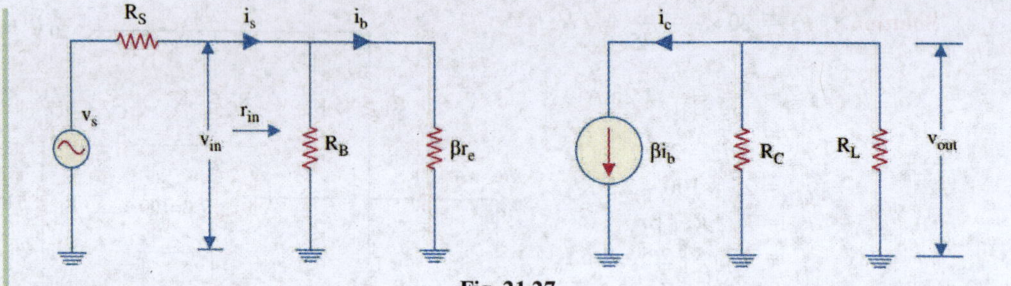

Fig. 21.27

If, in any question, v_s is given, we will first find out how much of it appears across the base as v_{in}. Then, for determining v_{out}, we will simply multiply this v_{in} by the voltage gain.

Example 21.11. *In the CE amplifier circuit of Fig. 21.28, find the r.m.s. signal output voltage when R_S is (a) 100 Ω and (b) 3 K.*

Solution. This circuit is similar to that shown if Fig. 21.24 except for the addition of R_S.

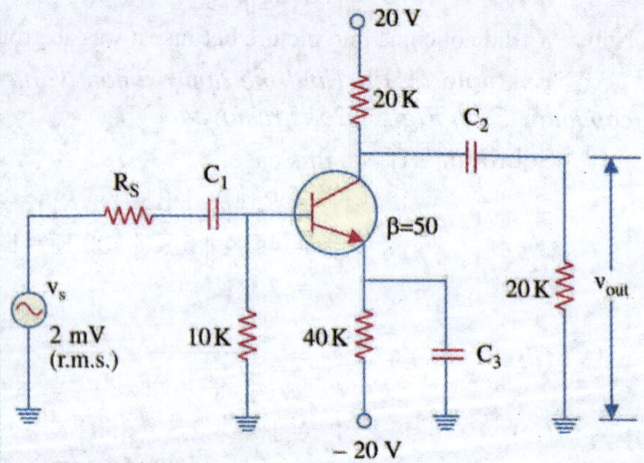

Fig. 21.28

As found in Ex. 21.8, $r_e' = 50\ \Omega$,

$\beta r_e' = 2500\ \Omega$.

$\therefore \quad R_B \parallel \beta r_e' = 10\ K \parallel 2500\ \Omega = 2000\ \Omega$

(a) When $R_S = 100\ \Omega$

In this case, 2 mV are dropped across a series combination of 100 Ω and 2000 Ω. Drop over 100 Ω is negligible as compared to that on 2000 Ω. Hence, it can be presumed that $v_s = v_{in} = 2$ mV. We have already found that $A_v = 200$.

$$\therefore \quad V_{out} = A_v \times V_{in} = 200 \times 2$$
$$= \mathbf{400\ mV\ (r.m.s.)}$$

(b) Whens $R_S = 3\ K$

In this case, drop across $R_B \parallel \beta r_e' = 2$ K is $V_{in} = V_s \dfrac{R_B \parallel \beta r_e'}{R_S + (R_B \parallel \beta r_e')} = 2 \times \dfrac{2}{3+2} = 0.8$ mV

$$\therefore \quad V_{out} = 200 \times 0.8 = \mathbf{160\ mV}$$

Obviously, as R_S is increased, V_{out} is decreased.

21.8. Equivalent Circuit of a CC Amplifier

We will consider the two-supply emitter-bias circuit shown in Fig. 21.29 (*a*) in which the collector is placed at ac (not dc) ground. The ac input signal is coupled into the base and output signal is taken out of the emitter. This circuit is also called **emitter follower circuit** because the emitter signal *follows the signal at the base both in magnitude and phase.*

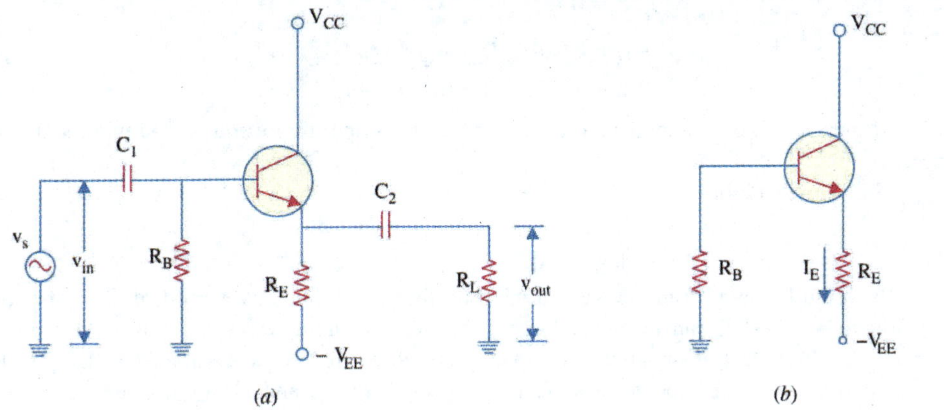

Fig. 21.29

(a) DC Equivalent Circuit

It is drawn in the usual way by opening all the capacitors and shorting all ac sources. It is shown in Fig. 21.29 (*b*).

$$I_E = \frac{V_{EE} - V_{BE}}{R_E + R_B/\beta} \cong \frac{V_{BE}}{R_E}$$

(b) AC Equivalent Circuit

It is obtained by shorting all ac sources and capacitors and is shown in Fig. 21.30.

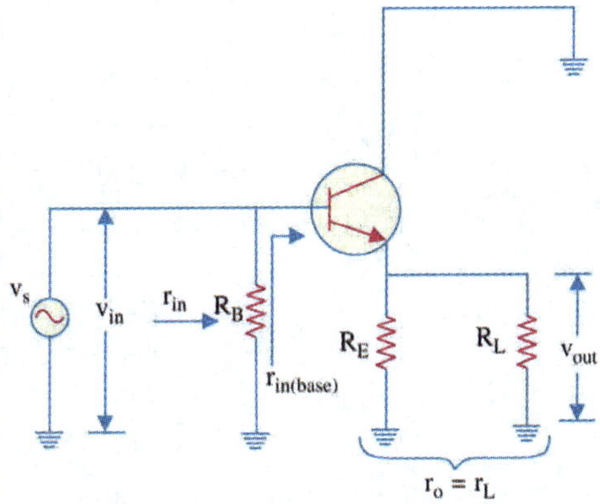

Fig. 21.30

(c) Principal Operating Characteristics

1. Input Resistance

The input resistance of the *CC* stage is given by the parallel combination of R_B and $r_{in(base)}$. Now, $r_{in(base)}$ is the input resistance looking into the base. It is found to be equal to $(1 + \beta)(r_e' + r_L) \cong \beta r_L$ *i.e.*, β times the ac load seen by the emitter.

$$\therefore \quad r_{in} \text{ or } r_{in(stage)} = R_B \parallel r_{in(base)}$$

$$= R_B \parallel \beta (r_e' + r_L)$$

$$\cong \beta (r_e' + r_L) \quad \text{— when } R_B \text{ is very large}$$

$$\cong \beta r_L \quad \text{— when } r_e' << r_L$$

$$= \beta R_E \quad \text{— if } R_L = 0$$

2. AC Load Resistance.

$$r_L = R_E \parallel R_L$$

It is the ac output resistance as *seen* by the emitter (and not the one when looking into the emitter).

3. Current Gain.

$$A_i = \frac{i_e}{i_b} = (1 + \beta) \cong \beta$$

4. Voltage Gain.

$$v_{in} = i_b \cdot r_{in} = i_b \cdot \beta (r_e' + r_L);$$

$$v_{out} = i_e \cdot r_L = \beta\, i_b \cdot r_L$$

$$A_v = \frac{v_{out}}{v_{in}} = \frac{\beta\, i_b\, r_L}{\beta\, i_b\,(r_e' + r_L)} = \frac{r_L}{r_e' + r_L} \cong \frac{r_L}{r_L} = 1 \qquad \text{— if } r_e' << r_L$$

It means that output signal from emitter has the **same magnitude** as the input signal at the base.

5. Power Gain.

$$A_p = A_v \cdot A_i = 1 \times (1 + \beta) = (1 + \beta) \cong \beta$$

$$G_p = 10 \log_{10}{}^{A_P} \text{ dB}$$

It would be noted from above that main usefulness of the emitter follower is to **step up the impedance level** *i.e.*, it transforms load impedance to a much higher value. It does not *increase the signal voltage.* Hence, primary application of emitter follower or *CC* stage is as an impedance matching device. It offers a higher impedance at the input terminals *i.e.*, r_{in} and a low output impedance (r_L)— *something opposite of typical basic transistor amplifier.*

Another point worth keeping in mind is that the above formulas are not exact because we have used ideal transistor approximations. However, these formulas do help in rapidly grasping the essential features of an emitter follower. Moreover, they are adequate for preliminary analysis and design.

Example. 21.12. *For the emitter follower shown in Fig. 21.31, find*

(i) r_{in} or $r_{in(stage)}$ *(ii)* r_L *(iii)* A_v *(iv)* A_p. *Take transistor* $\beta = 50$.

Solution.

$$I_E = 20/20 = 1 \text{ mA}$$

$$r_e' = 25/1 = 25\ \Omega$$

$$r_L = R_E \parallel R_L = 20 \parallel 5 = 4 \text{ K}$$

$$r_{in(base)} = \beta(r_e' + r_L) \cong \beta\, r_L = 50 \times 4 = 200 \text{ K}$$

(r_e' of 25 Ω has been neglected as compared to r_L of 4 K)

(i) $r_{in(stage)}$ or $r_{in} = R_B \parallel r_{in(base)} = 400 \text{ K} \parallel 200 \text{ K} = \mathbf{133.3\ K}$

(ii) $r_L = 20 \text{ K} \parallel 5 \text{ K} = \mathbf{4\ K}$

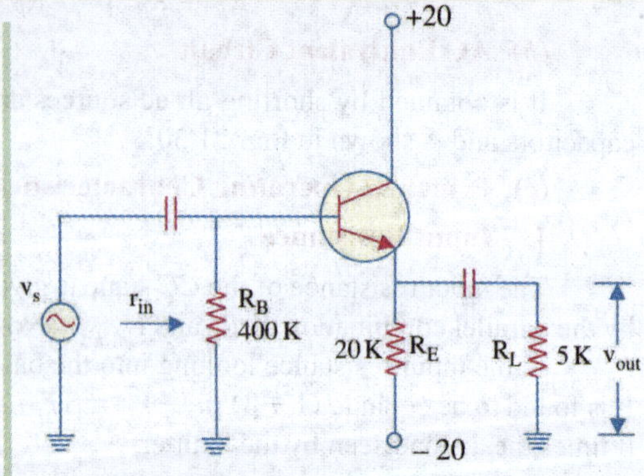

Fig. 21.31

(iii) $A_v \cong 1$ since r_e' is negligible as compared to r_L. If it is not, then A_v could be even less than 1.

(iv) $A_p = A_i \cdot A_v \cong \beta \times 1 = \beta = \mathbf{50}$;

$$G_p = 10 \log_{10}{}^{50} = \mathbf{17\ dB}$$

Example 21.13. *For the beta-stabilized emitter follower circuit of Fig. 21.31, find (i)* $r_{in(base)}$ *(ii)*r_{in} or $r_{in(stage)}$ *(iii)* r_L *and (iv)*A_p. *Take transistor* $\beta = 100$.

Solution.

$$V_2 = V_{CC}\frac{R_2}{R_1 + R_2} = 20 \times \frac{10}{40} = 5 \text{V}$$

$$I_E \cong \frac{V_2}{R_E} = \frac{5}{5} = 1 \text{ mA}$$

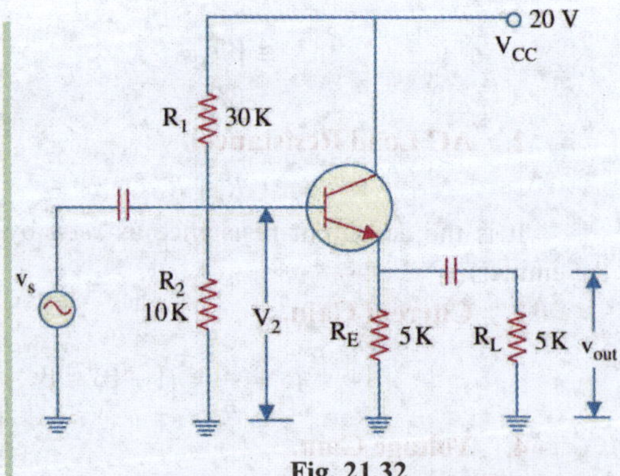

Fig. 21.32

$r_e' = 25/1 = 25\ \Omega$

$r_L = 5 \parallel 5 = 2.5$ K

(*i*) $r_{in(base)} = \beta(r_e' + r_L)$

$= 100\ (25 + 2500) \qquad = \mathbf{250\ K}$

(*ii*) $r_{in(base)} = R_1 \parallel R_2 \parallel r_{in(base)}$

$= 30\text{ K} \parallel 10\text{ K} \parallel 250\text{ K} \cong \mathbf{8K}$

(*iii*) $r_L = 5\text{ K} \parallel 5\text{ K} = \mathbf{2.5\ K}$

(*iv*) $A_p = A_v \,.\, A'_i = 1 \times \beta = 100;\ G_p = 100 \log_{10}{}^{100} = \mathbf{20\ dB}$

Example 21.14. *For the CE circuit of Fig. 21.33, find* (*i*) r_{in} or $r_{in(stage)}$ (*ii*) A_v. (*iii*) A_p. *Take transistor* $\beta = 200$.

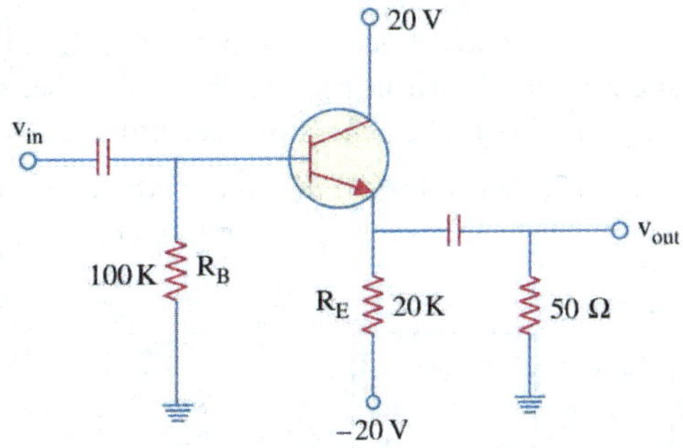

Fig. 21.33

Solution. $I_E = 20/20 = 1$ mA

$r_e' = 25/1 = 25\ \Omega$

$r_L = 20\text{ K} \parallel 50\ \Omega = 50\ \Omega$

$r_{in(base)} = \beta(r_e' + r_L)$

$= 200(25 + 50) = 15$ K

(*i*) r_{in} or $r_{in(stage)} = R_B \parallel r_{in(base)}$

$= 100\text{ K} \parallel 15\text{ K} = \mathbf{13\ K}$

(*ii*) Since r_e' is not negligible as compared to r_L, we will use the expression

$$A_v = \frac{r_L}{r_e' + r_L} = \frac{50}{25 + 50} = \mathbf{0.667}$$

(*iii*) $A_p = A_v \,.\, A_i = 0.667 \times 200 = \mathbf{133.3}$

SMALL-SIGNAL LOW-FREQUENCY MODEL OR REPRESENTATION

21.9. General

Although many transistor representations or models have been suggested and widely used, the equivalent *T*-model is the easiest to understand because, in this representation, component parts retain their identity in all configurations leading to rapid appreciation of a given network.

21.10. T-Model

Such models are not in common use today because they do not take into account any *gain between input and output.*

(*a*) CB Circuit

In Fig. 21.33 is shown the low-frequency *T*-equivalent circuit of a transistor connected in *CB* configuration. It utilizes *T*- or *r*-parameters. All these parameters are ac parameters and are measured under open-circuit conditions. Here, r_e represents the ac resistance of the forward-biased emitter-base junction. Its value is

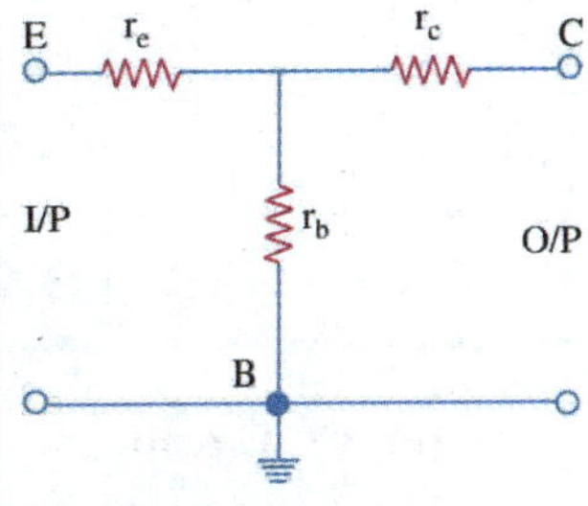

Fig. 21.34

$$r_e = \frac{25\text{ mV}}{I_E} \qquad \text{— for Ge}$$

$$= \frac{50\text{ mV}}{I_E} \qquad \text{— for Si}$$

This resistance is fairly small and depends on I_E. Also, r_c represents the ac resistance of the reverse-biased *C/B* junction. It is of the order of a few MΩ. Finally, r_b represents the resistance of the base region which is common to both junctions. Its value depends on the degree of doping. Usually, r_b is larger than r_e but much smaller than r_c.

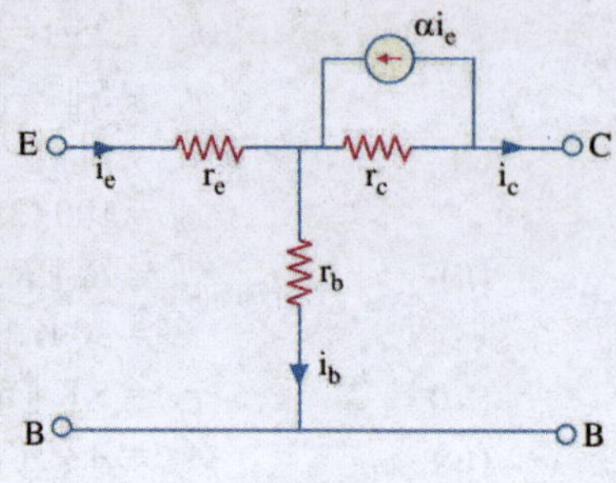

Fig. 21.35

However, circuit shown in Fig. 21.34 is not complete because it does not illustrate the forward current transfer ratio. Since current in the output of a transistor depends on the current at the input, a current-generator in parallel with r_c must be included as shown in Fig. 21.35.

The current generator may be replaced by a voltage generator with the help of Thevenin's theorem as shown in Fig. 21.36 (*a*). In that case, the *T*-equivalent circuit becomes as shown in Fig. 21.36 (*b*). The generator has a voltage of $\alpha i_e r_c = r_m i_e$ where $r_m = \alpha r_c$.

Typical values of different parameters are :

$$r_e = 25 \text{ to } 50\ \Omega; \qquad r_b = 100 \text{ to } 300\ \Omega;\ r_c = 1\ \text{M}\Omega$$

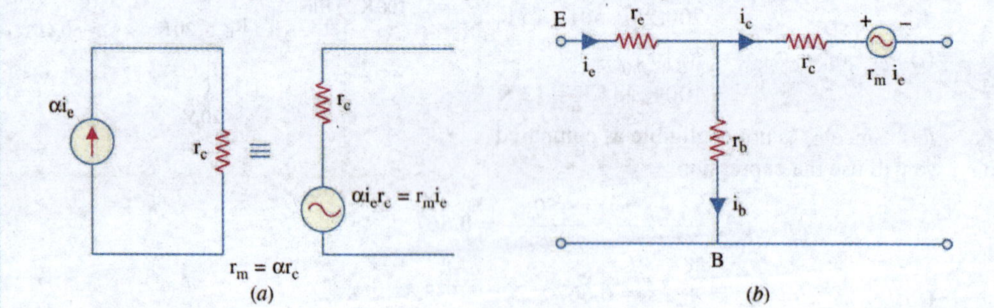

Fig. 21.36

(*b*) CE Circuit

The *T*-equivalent circuit for such a configuration is shown in Fig. 21.37. Whereas circuit shown in Fig. 21.36 contains a parallel current-generator that shown in Fig. 21.38 contains a series-voltage generator.

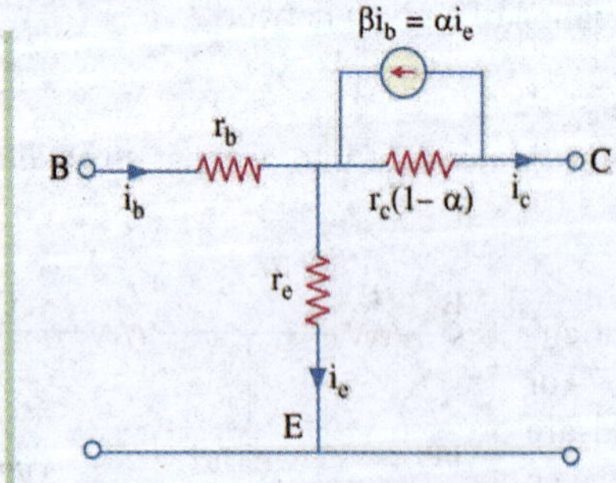

Fig. 21.37

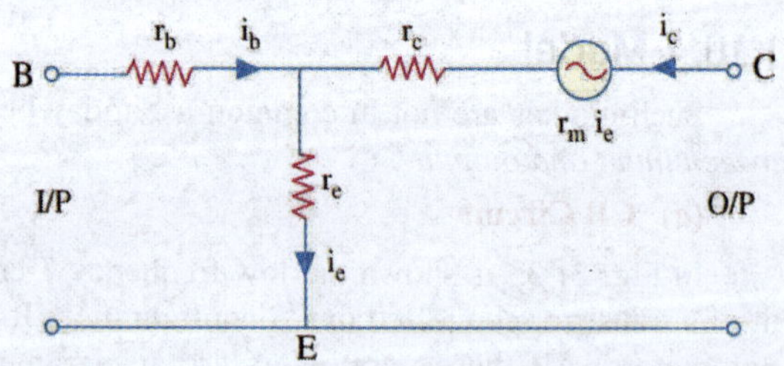

Fig. 21.38

(*c*) CC Circuit

The *T*-equivalent circuit for such a configuration is shown in Fig. 21.39.

21.11. Formulas for T-Equivalent of a CB Circuit

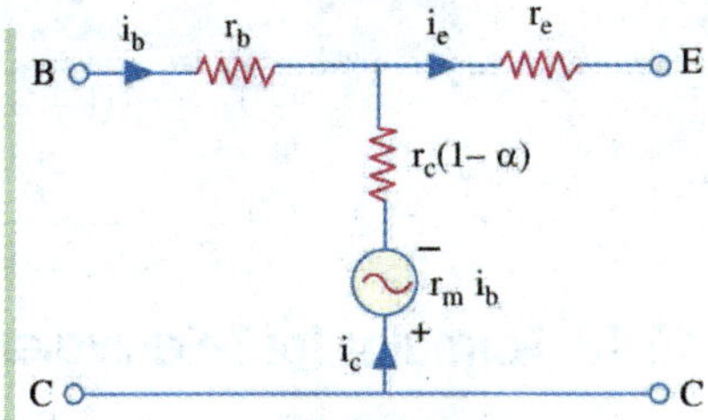

Fig. 21.39

In Fig. 21.40 is shown a small-signal low-frequency *T*-equivalent circuit for *CB* configuration. The ac input signal source has a resistance of R_S and voltage of v_s.

Of course, dc biasing circuit has been omitted and only ac equivalent shown. The approximate expressions for input and output resistance and voltage and current gains are given below without derivation :

1. $r_{in} = r_e + r_b(1-\alpha) = r_e + \dfrac{r_b}{(1+\beta)}$.

$$= r_e + r_b \frac{r_c(1-\alpha) + R_L}{r_c + R_L}.$$

2. $r_o = r_c - \dfrac{\alpha\, r_b\, r_c}{r_e + r_b + R_S}$.

3. $A_i = \alpha$

4. $A_v = \dfrac{v_{out}}{v_{in}} = \dfrac{\alpha R_L}{r_e + r_b(1-\alpha)}$.

=

$$\frac{v_{out}}{v_s} = \frac{\alpha R_L}{(r_e + R_S) + r_b(1-\alpha)} \cong \frac{\alpha R_L}{(r_e + R_S)}$$

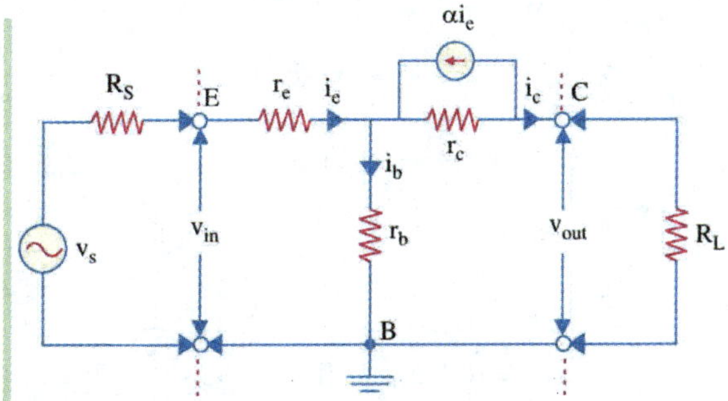

Fig. 21.40

$\log_{10}^{A_p}$

5. $A_p = A_i \cdot A_v$

$$= \frac{\alpha^2 R_L}{r_e + r_b(1-\alpha)} \quad \text{— no source resistance}$$

$$= \frac{\alpha^2 R_L}{(r_e + R_S) + r_b(1-\alpha)} \quad \text{— with source resistance}$$

$G_p = 10 \log_{10}^{A_p}$ dB

21.12. Formulas for T-Equivalent of a CE Circuit

The low-frequency small-signal *T*-equivalent circuit for such a configuration is shown in Fig. 21.41.

The approximate expressions for various resistances and gains are given below :

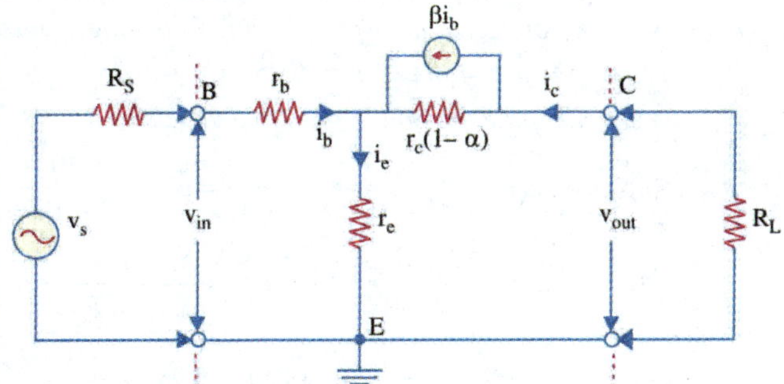

Fig. 21.41

1. $r_{in} = r_b + \dfrac{r_e}{(1-\alpha)}$

$$= r_b + (1+\beta)r_e$$

2. $r_o = r_e(1-\alpha) + \dfrac{\alpha\, r_c\, r_e}{r_b + r_e + R_S}$

3. $A_i = \beta = \dfrac{\alpha}{(1-\alpha)}$

4. $A_v = \dfrac{v_{out}}{v_{in}} = \dfrac{\alpha\, R_L}{r_e + r_b\,(1-\alpha)} = \dfrac{v_{out}}{v_s} = \dfrac{\alpha\, R_L}{r_e + (r_b + R_s)\,(1-\alpha)}$

5. $A_p = A_v . A_i = \dfrac{\alpha^2\, R_L}{(1-\alpha)[(r_e + r_b)(1-\alpha)]} = \dfrac{\alpha^2\, R_L}{(1-\alpha)\,[r_e + (r_b + R_S)\,(1-\alpha)]}$

21.13. Formulas for T-Equivalent of a CC Circuit

The low-frequency *T*-equivalent circuit for such a configuration is shown in Fig. 21.42. The approximate expressions for various resistances and gains are given below :

1. $r_{in} = \dfrac{R_L}{1-\alpha} = (1+\beta)\,R_L$

2. $r_o = r_e + (1-\alpha)\,(r_b + R_S) = r_e + \dfrac{(r_b + R_S)}{(1+\beta)}$

3. $A_i = \dfrac{1}{(1-\alpha)} = (1+\beta)$

4. $A_v = 1$

5. $A_p = \dfrac{1}{(1-\alpha)} = (1+\beta)$

$G_p = 10 \log_{10}{}^{A_P}$ dB

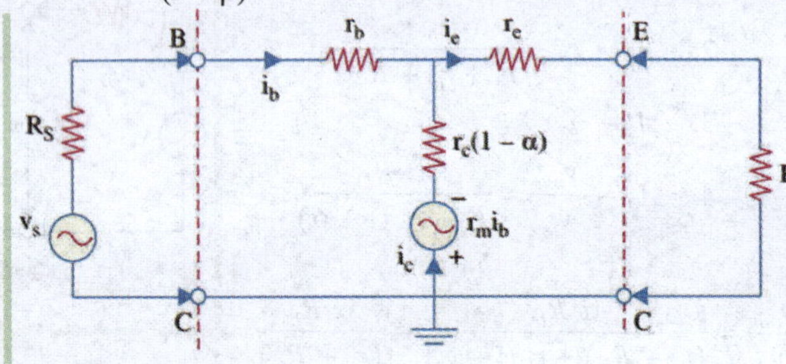

Fig. 21.42

Example 21.15. *A junction transistor has $r_e = 50\ \Omega$, $r_b = 1\ K$, $r_c = 1M$ and $\alpha = 0.98$. It is used in common-base circuit with a load resistance of 10 K. Calculate the current, voltage and power gains and the input resistance.*

Solution. (*i*) $A_i = \alpha = 0.98$

(*ii*) $A_v = \dfrac{\alpha\, R_L}{r_e + r_b\,(1-\alpha)} = \dfrac{0.98 \times 10000}{50 + 1000\,(1-0.98)} = \mathbf{140}$

(*iii*) $A_p = A_v . A_i = 140 \times 0.98 = \mathbf{137}$

(*iv*) $r_{in} = r_e + r_b\,(1-\alpha) = 50 + 1000\,(1-0.98) = \mathbf{70\ \Omega}$

Example 21.16. *Calculate the input and output resistance, overall current, voltage and power gains for a CE connected transistor having following r-parameters :*

$r_b = 30\ \Omega$, $r_e = 400\ \Omega$, $r_c = 0.75\ M$, $\alpha = 0.95$, $R_L = 10\ K$ and $R_S = 400\ \Omega$.
Also, calculate the power gain in decibels.

Solution. (*i*) $r_{in} = r_b + \dfrac{r_e}{(1-\alpha)} = 30 + \dfrac{400}{(1-0.95)} = \mathbf{8030\ \Omega}$

(*ii*) $r_o = r_e\,(1-\alpha) + \dfrac{\alpha . r_c\, r_e}{r_b + r_e + R_S}$

$= 750{,}000\,(1-0.95) + \dfrac{0.95 \times 750{,}000 \times 400}{(30 + 400 + 400)} = 380{,}870 = \mathbf{0.38\ M}$

(*iii*) $A_i = \beta = \dfrac{\alpha}{1-\alpha} = \dfrac{0.95}{(1-0.95)}\ \mathbf{19}$

(*iv*) $A_v = \dfrac{\alpha\, R_L}{r_e + (r_b + R_S)\,(1-\alpha)} = \dfrac{0.95 \times 10 \times 10^3}{400 + (30 + 400) \times 0.05} = \mathbf{22.5}$

(v) $A_p = A_v \cdot A_i = 22.5 \times 19 = \mathbf{427.5}$

Power gain in decibels, $G_p = 10 \log_{10} 427.5 = \mathbf{26.3\ dB}$

21.14. What are *h*-parameters ?

These are *four* constants which describe the behaviour of a two-port linear network. A linear network is one in which resistances, inductances and capacitances remain fixed when voltage across them is changed.

Consider an unknown linear network contained in a black box as shown in Fig. 21.43. As a matter of convention, currents flowing into the box are taken positive whereas those flowing out of it are considered negative.

Fig. 21.43

Similarly, voltages are positive from the upper to the lower terminals and negative the other way around. The electrical behaviour of such a circuit can be described with the help of four hybrid parameters or constants designated as h_{11}, h_{12}, h_{21}, h_{22}. In this type of double-number subscripts, it is implied that *the first variable is always divided by the other.* The subscript 1 refers to quantities on the *input* side and 2 to the quantities on the *output* side. The letter '*h*' has come from the word hybrid which means *mixture* of distinctly different items. These constants are hybrid because they have different units.

Out of the four *h*-parameters, two are found by short-circuiting the output terminals 2-2 and the other two by open-circuiting the input terminals 1-1 of the circuit.

(a) Finding h_{11} and h_{21} from Short-Circuit Test

As shown in Fig. 21.44, the output terminals have been shorted so that $v_2 = 0$, because no voltage can *exist on a short.* The linear circuit within the box is driven by an input voltage v_1. It produces an input current i_1 whose magnitude depends on the type of circuit within the box.

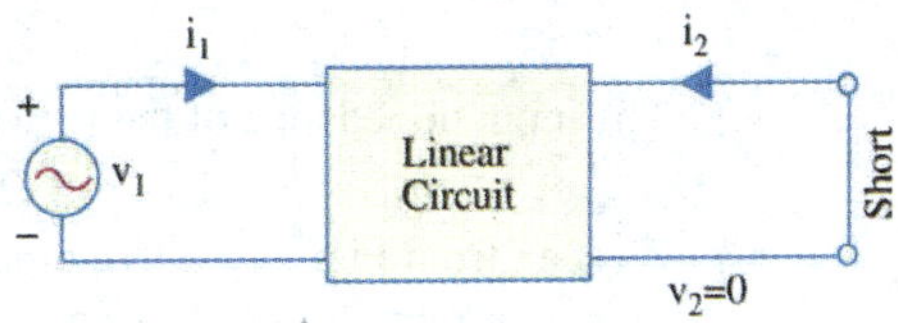

Fig. 21.44

$$h_{11} = \frac{v_1}{i_1} \quad \text{— output shorted}$$

$$h_{21} = \frac{i_2}{i_1} \quad \text{— output shorted}$$

These two constants are known as **forward** parameters.

The constant h_{11} represents input impedance with output shorted and has the unit of ohm. The constant h_{21} represents current gain of the circuit with output shorted and has no unit since it is the ratio of two *similar* quantities.

(b) Finding h_{12} and h_{22} from Open-circuit Test

As shown in Fig. 21.45, the input terminals are open so that $i_1 = 0$ but there does appear a voltage v_1 across them. The output terminals are driven by an ac voltage v_2 which sets up current i_2.

$$h_{12} = \frac{v_1}{v_2} \quad \text{— input open}$$

$$h_{22} = \frac{i_2}{v_2} \quad \text{— input open}$$

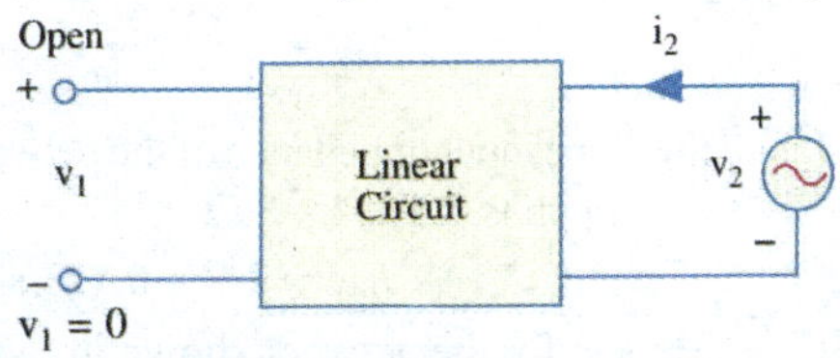

Fig. 21.45

As seen, h_{12} represents voltage gain (not forward gain which is v_2 / v_1). Hence, it has no units. The constant h_{22} represents admittance (which is reverse of resistance) and has the unit of mho or siemens, *S*. It is actually the admittance looking into the output terminals with input terminals open. Generally, these two constants are also referred to as ***reverse parameters.***

Summary of *h*-parameters

h_{11}	= input impedance	with output shorted
h_{21}	= forward current gain	
h_{12}	= reverse voltage gain	
h_{22}	= output admittance	with input open

Example. 21.17. *Find the h-parameters of the circuit shown in Fig. 21.46 (a).*

Solution. First of all, let us find the forward parameters h_{11} and h_{21}. For that purpose, a short is put across the output terminals as shown in Fig. 21.46 (*b*).

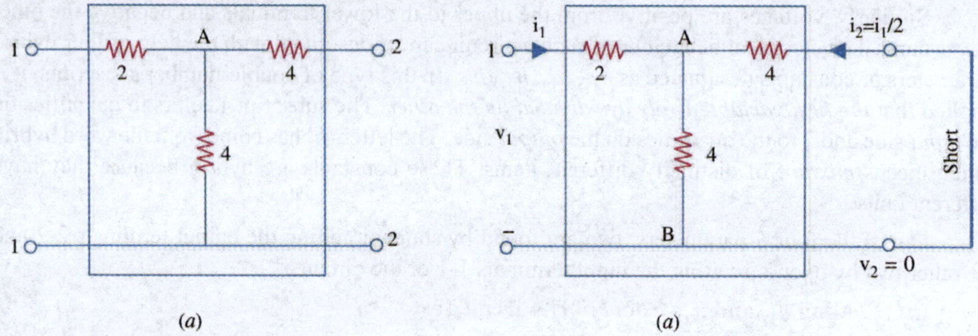

Fig. 21.46

(*i*) The input impedance of the network as viewed from input terminals is

$$h_{11} = 2 + 4 \parallel 4 = \mathbf{4\ \Omega}$$

(*ii*) As seen from Fig. 21.46 (*b*), input current i_1 divides into two equal parts at point *A*. The output current $i_2 = -i_1/2$ (negative sign has been taken because actually it is flowing *out* of the box).

$$\therefore \quad h_{21} = \frac{i_2}{i_1} = \frac{-i_1/2}{i_1} = -\frac{1}{2} = -\mathbf{0.5}$$

(*iii*) Now, for finding reverse parameters, we will keep input terminals open and apply v_2 across output terminals as shown in Fig. 21.47. It will produce a current i_2 which will produce equal drops across the two 4 Ω resistors. The voltage which appears across input terminals as v_1 is the drop across the vertical 4 Ω resistor connected at point *A*. Hence, $v_1 = v_2 / 2$.

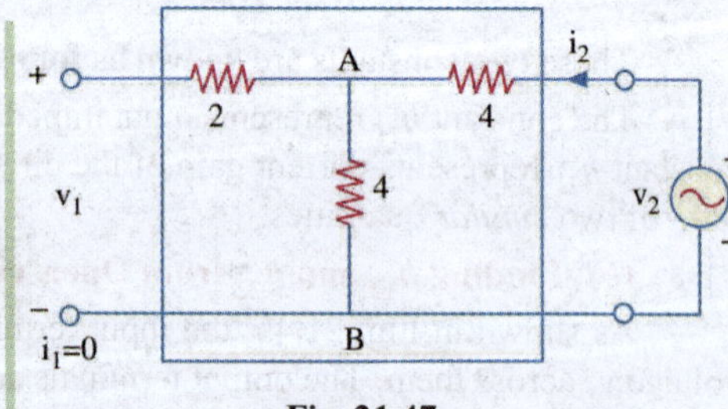

Fig. 21.47

$$\therefore \quad h_{12} = \frac{v_1}{v_2} = \frac{v_2/2}{v_2} = \mathbf{0.5.}$$

(*iv*) The input impedance of the network when viewed from output terminals with input terminals open is = 4 + 4 = 8 Ω.

$$\therefore \quad h_{22} = 1/8 = \mathbf{0.125\ siemens\ (\textit{i.e.}, mho)}$$

Hence, for the network shown in Fig. 21.46 (*a*), the *h*-parameters are as under :

$$h_{11} = 4\Omega, \quad h_{21} = -0.5, \quad h_{12} = 0.5, \quad \text{and} \quad h_{22} = 0.125\ \text{S}$$

21.15. The *h*-parameter Notation for Transistors

While using *h*-parameters for transistor circuits, their numerical subscripts are replaced by the first letters (printed black) for defining them.

h_{11}	=	h_i	=	input impedance	output shorted
h_{21}	=	h_f	=	forward current gain	
h_{12}	=	h_r	=	reverse voltage gain	input open
h_{22}	=	h_0	=	output admittance	

A second subscript is added to the above parameters to indicate the *particular configuration.* For example, for *CE* connection, the four parameters are written as :

$$h_{ie},\quad h_{fe},\quad h_{re}\quad \text{and}\quad h_{oe}$$

Similarly, for *CB* connection, these are written as h_{ib}, h_{fb}, h_{rb}, h_{op} and for *CC* connection as h_{ic}, h_{fc}, h_{rc} and h_{oe}.

21.16. The *h*-parameters of an Ideal Transistor

As stated earlier, every linear circuit has a set of parameters associated with it, which fully describe its behaviour. When small ac signals are involved, a transistor behaves like a linear device because its output ac signal *varies directly as the input signal.* Hence, for small ac signals, each transistor has its own characteristic set of *h*-parameters or constants.

The *h*-parameters depend on a number of factors such as

1. transistor type, 2. configuration, 3. operating point, 4. temperature, 5. frequency.

These *h*-parameters can be found experimentally or graphically. The parameters h_i and h_r are determined from input characteristics of the *CE* transistor whereas h_f and h_0 are found from output characteristics.

21.17. The *h*-parameters of an Ideal CB Transistor

In Fig. 21.48 (*a*), a *CB*-connected transistor has been shown contained in a black box. Fig. 21.48 (*b*) gives its equivalent circuit. It should be noted that no external biasing resistors or any signal source has been shown connected to the transistor.

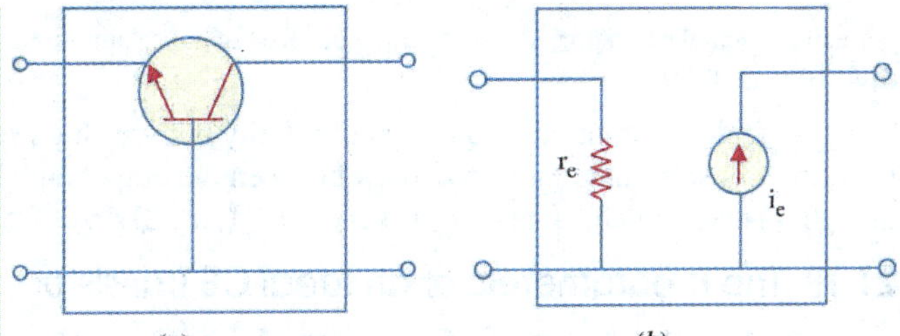

Fig. 21.48

(*i*) Forward Parameters

The two forward *h*-parameters can be found from the circuit of Fig. 21.49 (*a*) where a short has been put across the output. The input impedance is simply r_e.

$$\therefore \qquad h_{ib} = r_e$$

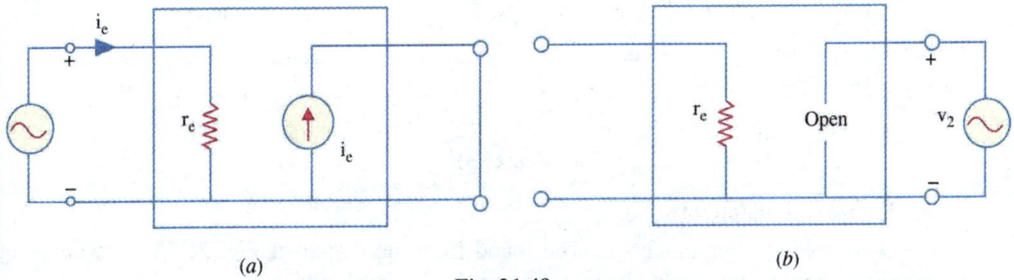

Fig. 21.49

The output current equals input current i_e. Since it flows out of the box, it is taken as negative. The forward current gain is

$$h_{fb} = \frac{-i_e}{i_e} = -1$$

(It also called the ac α of the *CB* circuit)

***(ii)* Reverse Parameters**

The two reverse parameters can be found from the circuit diagram of Fig. 21.49 (*b*). When input terminals are open, there can be no ac emitter current. It means that ac current source (inside the box) has a value of zero and so appears as an 'open'. Because of this open, no voltage can appear across input terminals, however, large v_2 may be. Hence, $v_1 = 0$.

$$\therefore \quad h_{rb} = \frac{v_1}{v_2} = \frac{0}{v_2} = 0$$

Similarly, the impedance, looking into the output terminals is infinite. Consequently, its admittance (= 1/∞) is zero.

$$\therefore \quad h_{ob} = 0$$

Summary

The four *h*-parameters of an **ideal** transistor connected in *CB* configuration are

$$h_{ib} = r_e; \quad h_{fb} = -1;$$
$$h_{rb} = 0; \quad h_{ob} = 0$$

The equivalent hybrid circuit is shown in Fig. 21.50.

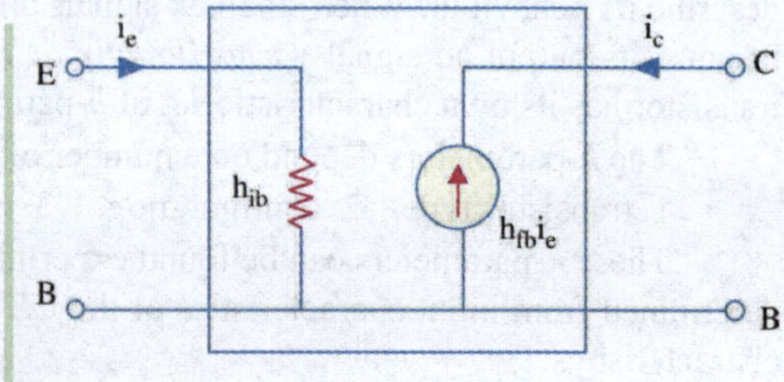

Fig. 21.50

Note. In an actual transistor, h_{rb} and h_{ob} are not zero but have some finite value though extremely small (ranging from 10^{-4} to 10^{-6}).

In reality, output impedance is not infinity but very high so that h_{ob} is extremely small. Similarly, there is some amount of feedback between the output and the input circuits (even when open) though it is very small. Hence, h_{rb} is very small (Art. 21.10).

21.18. The *h*-parameters of an Ideal CE Transistor

Fig. 21.51 (*a*) shows a *CE*-connected *ideal* transistor contained in a black box whereas Fig. 21.51 (*b*) shows its ac equivalent circuit in terms of its β and resistance values.

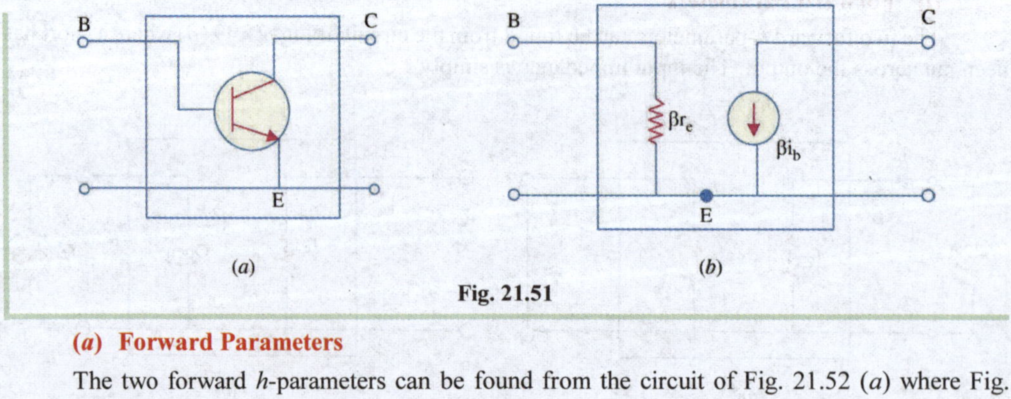

(*a*) (*b*)

Fig. 21.51

***(a)* Forward Parameters**

The two forward *h*-parameters can be found from the circuit of Fig. 21.52 (*a*) where Fig. output has been shorted. Obviously, the input impedance is simply βr_e.

$$\therefore \quad h_{ie} = \beta r_e$$

The forward current gain is given by

$$h_{fe} = \frac{i_2}{i_1} = \frac{\beta i_b}{i_b} = \beta$$

(It is also called the ac beta of the *CE circuit*)

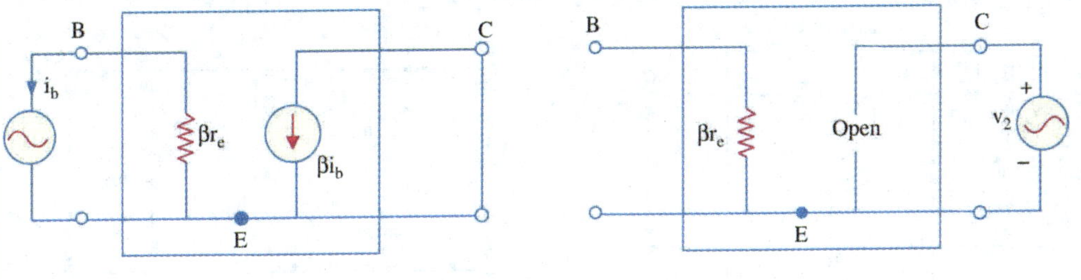

(*a*) (*b*)

Fig. 21.52

(*b*) Reverse Parameters

These can be found by reference to the circuit of Fig. 21.52 (*b*) where input terminals are open but output terminals are driven by an ac voltage source v_2. With input terminals open, there can be no base current so that $i_b = 0$. If $i_b = 0$, then collector current source has zero value and looks like an open. Hence, there can be no v_1 due to this open.

$$\therefore \qquad h_{re} = \frac{v_1}{v_2} = \frac{0}{v_2} = 0$$

Again, the impedance looking into the output terminals is infinite so that conductance is zero.

$$\therefore \qquad h_{re} = 0$$

Hence, the four *h*-parameters of an *ideal* transistor connected in *CE* configuration are :

$$h_{ie} = \beta r_e \,;\ h_{fe} = \beta \,;\ h_{re} = 0 \,;\ h_{oe} = 0.$$

The hybrid equivalent circuit of such a transistor is shown in Fig. 21.53.

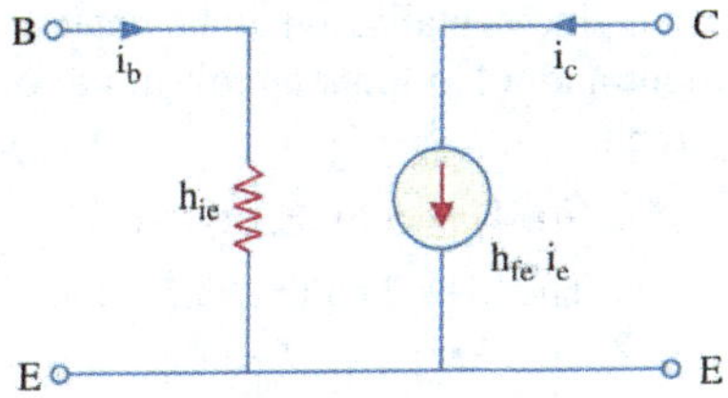

Fig. 21.53

Note. In practice, h_{re} and h_{oe} are not exactly zero but quite small for the same reasons as given in the Note to Art. 21.17.

21.19. Approximate Hybrid Equivalent Circuits

So far, we did not take into account the following two factors which exist in an actual transistor (as opposed to an ideal one):

(*i*) because of the transistor's non-unilateral behaviour, there is a 'feedback' of the output voltage into the input voltage. This feedback is represented by a voltage-controlled generator $h_r v_2$ as shown in Figs. 21.54 and 21.55.

By definition, an ideal amplifier is one which *responds only to signals applied to its input terminals.* It should not do the reverse *i.e.*, reproduce at the input any portion of the ac signal applied at the output. Such an ideal one-way device is called a unilateral device. A real transistor cannot be unilateral because of unavoidable interaction between its input and output circuits (after all, it consists of a single piece of a crystal). Therefore, not only its output responds to its input but, to a lesser degree, its input also responds to its output.

(*ii*) even when input circuit is open, there is some effective value of conductance when looking into the transistor from its output terminals. It is represented by h_0.

We will now draw *low-frequency small-signal* hybrid equivalent circuits after taking into account the 'feedback' voltage generator and output admittance.

(*a*) Hybrid CB Circuit

In Fig. 21.54 (*a*) is shown an *NPN* transistor connected in *CB* configuration. Its ac equivalent circuit employing *h*-parameters is shown in Fig. 21.54 (*b*).

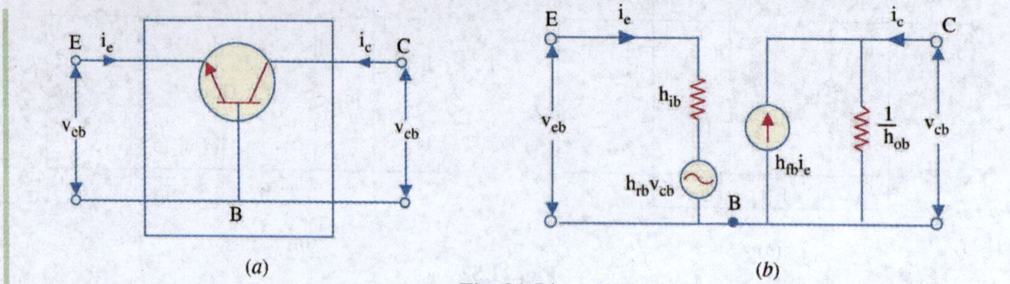

Fig. 21.54

As per current convention stated earlier (Art. 21.14), collector *i.e.*, is shown flowing *inwards* though actually this current flows *outwards* as shown by the arrow in the ac current source. Similarly, ac voltage polarities have been taken by considering upper terminal positive and lower one as negative (please remember the dc biasing rule of Art. 18.12 does not apply here).

It may be noted that no external dc biasing resistor or ac voltage sources have been connected to the equivalent circuit as yet (Art. 21.21).

Incidentally, it may be noted that the ac equivalent circuit contains a Thevenin's circuit in the input and a Norton's circuit in the output. It is all the more a reason to call it a hybrid equivalent circuit.

(*b*) Hybrid CE Circuit

The hybrid equivalent of the transistor alone when connected in *CE* configuration is shown in Fig. 21.55 (*b*).

Fig. 21.55

We may connect signal input source across its input terminals and load resistance across output terminals (Art. 21.22).

21.20. Typical Values of Transistor *h*-parameters

In the table given on next page are given typical values for each parameter for the broad range of transistors available today in each of the three configurations.

Parameter	*CB*	*CE*	*CC*
h_i	25 Ω	1 K	1 K
h_r	3×10^{-4}	2.5×10^{-4}	≅ 1
h_f	– 0.98	50	– 50
h_o	0.5×10^{-6} S	25×10^{-6} S	25×10^{-6} S

21.21. Hybrid Formulas for Transistor Amplifier

Sufficiently accurate formulas for the various gains and resistances of a transistor amplifier circuit in terms of *h*-parameters are given below. These are applicable to all transistor configurations.

(*i*) r_{in} = input resistance at *transistor* terminals

$$= h_i - \frac{h_r\, h_f}{h_o + 1/r_L}$$ — where r_L is ac load resistance

It is $r_{in(base)}$ and is different from $r_{in(stage)}$.

If either h_r or r_L is small, then $r_{in} = h_i$.

(*ii*) r_o = output resistance at transistor terminals

$$= \frac{h_i + R_S}{(h_i + R_S)\, h_o - h_f h_r} \cong \frac{h_i + R_S}{h_0 R_S + \Delta h}$$

It is worth noting that input resistance depends on ac load resistance r_L and output resistance depends on input signal source resistance R_S.

In case where R_S is very large or h_r is negligible, the above expression simplifies to $r_0 \cong 1/h_o$.

(*iii*) A_i = transistor current gain from input terminals to load

$$= \frac{h_f}{1 + h_0\, r_L}$$

where h_f is the short-circuit gain of the transistor alone *i.e.*, from input terminals to output terminals only.

Also, $A_i \cong h_f$ — if $h_0 r_L << 1$

(*iv*) A_v = voltage gain from transistor input terminals to output terminals

$$= \frac{-h_f\, r_L}{h_i\,(1 + h_0\, r_L) - h_f h_r\, r_L} = \frac{-h_f\, r_L}{h_i + \Delta h . r_L} = \frac{-h_f}{r_{in}\,(h_0 + 1/r_L)}$$

In the above formulas, $\Delta h = h_0\, h_i - h_f h_r$.

If $h_i >> \Delta h.\, r_L$, then $A_v \cong \dfrac{-h_f . r_L}{h_i}$

If overall voltage gain from **source** to output terminals is required then add R_S to h_i

$$\therefore \quad A_v' = \frac{v_{out}}{v_s} = \frac{-h_f . r_L}{(h_i + R_S)\,(1 + h_0\, r_L) - h_f h_r r_L}$$

It may be noted that ac load resistance r_L will depend on transistor connections. For *CB* and *CE* connections

$r_L = R_C$ — when load resistance R_L is not connected across output terminals.

$= r_C \| R_L$ — when R_L is connected across output terminals of the transistor.

For a *CC* connection

$r_L = R_E$ — with no R_L connected across transistor output terminals

$= R_E \| R_L$ — with R_L connected across output terminals

Note : In the above formulas, value of h_f will have to be substituted with its proper sign *i.e.*, negative sign for *CB* and *CC* configurations. Incidentally, very approximate formulas may be obtained by neglecting h_0 and h_r where they add to or subtract from an expression but not where they divide or multiply an expression.

21.22. Approximate Hybrid Formulas

The approximate hybrid formulas for the three connections are listed below. These are applicable when h_o and h_r are very small and R_S is very large. The given values refer to *transistor terminals.* The values of $r_{in(stage)}$ or r_{in} and $r_{o(stage)}$ will depend on biasing resistors and load resistance respectively.

Item	*CE*	*CB*	*CC*
r_{in}	h_{ie}	h_{ib}	$h_{ic} + h_{fc} R_L$
r_o	$\dfrac{1}{h_{oe}}$	$\dfrac{1}{h_{ob}}$	$\dfrac{h_{ic}}{h_{fc}}$
A_i	$h_{fe} = \beta$	$-h_{fb} \cong 1$	$-h_{fc} \cong \beta$
A_v	$\dfrac{h_{fe} R_C}{h_{ie}}$	$\dfrac{h_{fb}}{h_{ib}} R_C$	1

Example. 21.18. *A transistor used in CB circuit has the following set of parameters.*

$h_{ib} = 36\ \Omega,\ h_{fb} = -0.98,\ h_{rb} = 5 \times 10^{-4},\ h_{ob} = 10^{-6}$ *mho*

With $R_S = 2$ *K and* $R_C = 10$ *K, calculate (i)* r_{in} *(ii)* r_{out} *(iii)* A_i *and (iv)* A_v.

Approximate Solution

(*i*) $$r_{in} = h_{ib} = \mathbf{36\ \Omega}$$

(*ii*) $$r_o = \frac{1}{h_{ob}} = \mathbf{1\ M}$$

(*iii*) $$A_i = h_{fb} = \mathbf{-0.98\ M}$$

(*iv*) $$A_v = \frac{h_{fb}}{h_{ib}} \times R_C = \frac{0.98}{36\ \Omega} \times 10\ K = \mathbf{272}$$

More Accurate Solution

(*i*) $$r_{in(base)} = h_{ib} - \frac{h_{rb}\ h_{fb}}{h_{ob} + 1/r_L}$$ —Art. 21.22

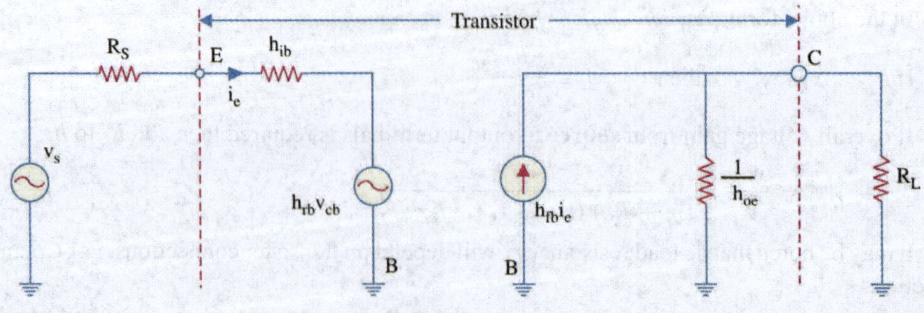

Fig. 21.56

$$= 36 - \frac{-0.98 \times 5 \times 10^{-4}}{10^{-6} + 1/10 \times 10^3}$$ $(\because r_L = R_C$ since there is no $R_L)$

$$= 36 + 4.9 = \mathbf{40.9\ \Omega}$$

It is the input resistance at transistor terminals.

(*ii*) $$r_o = \frac{h_{ib} + R_S}{h_o\,(h_{ib} + R_S) - h_{fb}.h_{rb}} = \frac{36 + 2000}{10^{-6}(36 + 2000) - (-0.98) \times 5 \times 10^{-4}} = \mathbf{0.8\ M}$$

It is the output resistance at transistor terminals.

(*iii*) $$A_i = \frac{h_{fb}}{1 + h_{ob}\,r_L} = \frac{-0.98}{1 + 10^{-6} \times 10^4} = \mathbf{-0.97}$$

(*iv*) $$A_v = \frac{-h_{fb}\,r_L}{h_{ib}\,(1 + h_{ob}\,r_L) - h_{fb}\,h_{rb}.r_L}$$

Here, ac load $r_L = R_C = 10\text{ K} = 10^4\,\Omega$

$$\therefore \quad A_v = \frac{-(-0.98) \times 10^4}{36\,(1 + 10^{-6} \times 10^4) - (-0.98) \times 5 \times 10^{-4} \times 10^4} = \mathbf{249}$$

Example 21.19. *A transistor used in CE connection (Fig. 21.57) has the following set of h-parameters : h_{ie} = 1 K, h_{fe} = 100, $h_{re} = 5 \times 10^{-4}$ and $h_{oe} = 2 \times 10^{-5}$ S. With $R_S = 2$ K and R_C = 5K, determine*

(*i*) r_{in} (*ii*) r_o (*iii*) A_i *and* (*iv*) A_{vo}

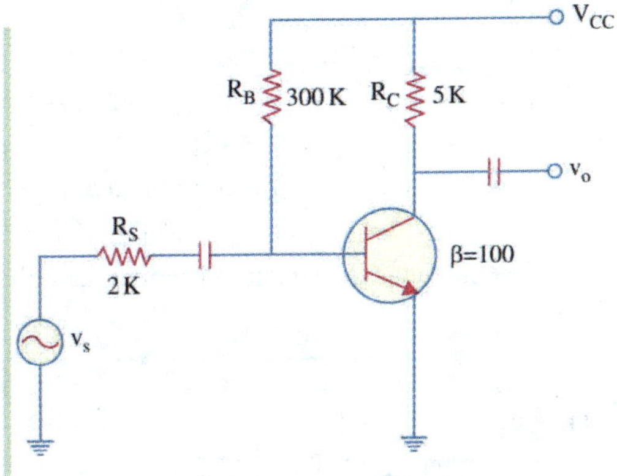

Fig. 21.57

Solution. (*i*) $$r_{in} = h_{ie} - \frac{h_{re}\,h_{fe}}{h_o + 1/r_L}$$

$$= 1000 - \frac{5 \times 10^{-4} \times 100}{2 \times 10^{-5} + 1/5 \times 10^3}$$

$$= \mathbf{723\ \Omega}$$

(*ii*) $$r_o = \frac{h_{ie} + R_s}{(h_{ie} + R_S)\,h_{oe} - h_{fe}.h_{re}}$$

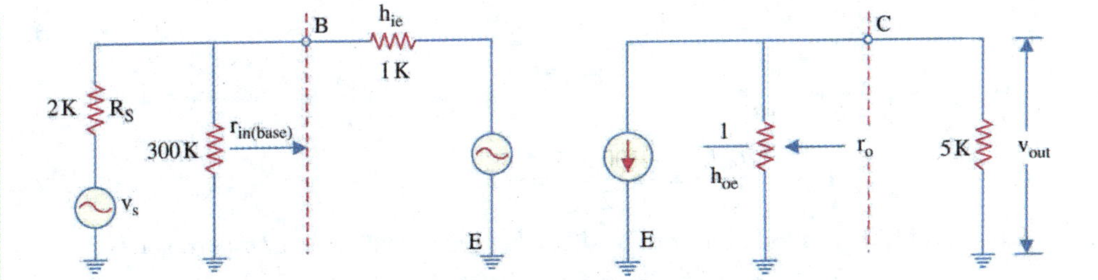

Fig. 21.58

$$= \frac{1000 + 2000}{(1000 + 2000) \times 2 \times 10^{-5} - 100 \times 5 \times 10^{-4}} = 30,000\ \Omega = \mathbf{0.03\,M}$$

(*iii*) $$A_i = \frac{h_{fe}}{1 + h_{oe}\,r_L} = \frac{100}{1 + 2 \times 10^{-5} \times 5 \times 10^3} = \mathbf{91}$$

(*iv*) $$A_v = \frac{-h_{fe}\,r_L}{(h_{ie} + R_S)\,(1 + h_{oe}.r_L) - h_{fe}\,h_{re}\,r_L}$$

$$= \frac{-100 \times 5 \times 10^3}{(1000 + 2000)\,(1 + 2 \times 10^{-5} \times 5 \times 10^3) - 100 \times 5 \times 10^{-4} \times 5 \times 10^3} = \mathbf{-164}$$

The negative sign indicates that there is 180° phase shift between the input and output ac signals. Obviously, it is the overall (or circuit) voltage gain and not the voltage gain of the transistor alone.

Example 21.20. *In the CE circuit shown in Fig. 21.59, the transistor parameters are :*

$h_{ie} = 2K,\ h_{fe} = 100,\ h_{re} = 5 \times 10^{-4},\ h_{oe} = 2 \times 10^{-5}$ S.

Calculate (i) $r_{in(base)}$, *(ii)* $r_{in(stage)}$, *(iii)* r_o, *(iv)* $r_{o(stage)}$, *(v)* A_i *and (vi)* A_v.

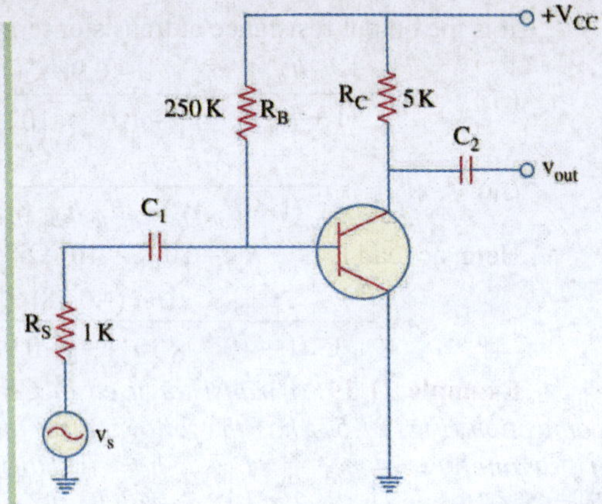

Fig. 21.59

Solution. The hybrid equivalent circuit is shown in Fig. 21.60. We will use approximate formulas given in Art. 21.22.

(i) $r_{in(base)} = h_{ie} = \mathbf{2\ K}$

(ii) $r_{in(stage)} = 2\text{ K} \parallel 250\text{ K} = \mathbf{1.98\ K}$

(iii) $r_o = 1/h_{oe} = 1/2 \times 10^{-5}$

$= \mathbf{50\ K}$

It is the *output impedance of the transistor only.*

(iv) $r_{o(stage)} = r_o' = 50\text{ K} \| 5\text{ K} = \mathbf{4.54\ K}$

The impedance takes into account the collector load.

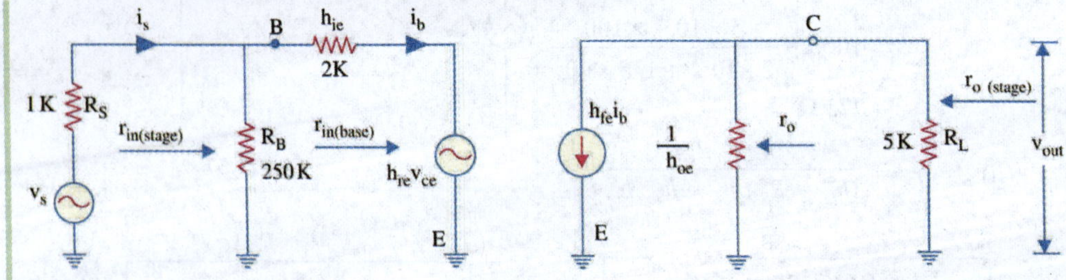

Fig. 21.60

(v) $A_i \cong h_{fe} = \mathbf{100}$

(vi) $A_v = \dfrac{-h_{fe}\, r_L}{h_{ie}} = \dfrac{-100 \times 5}{2} = \mathbf{-250}$

Example 21.21. *The transistor of Fig. 49.61 has the following set of h-parameters :*

$h_{ie} = 2K,\ h_{fe} = 100,\ h_{oe} = 5 \times 10^{-4},\ h_{oe} = 2.5 \times 10^{-5}$ S.

Find the voltage gain and the ac input impedance of the stage.

Solution. Using somewhat exact formulas given in Art. 21.21, we have

$$r_{in(base)} = h_{ie} - \frac{h_{fe}\, h_{re}}{h_{oe} + 1/r_L}$$

Now, collector load

$$r_L = 10\text{ K} \parallel 30\text{ K} = 7.5\text{ K}$$

$$\therefore\ r_{in(base)} = 2000 - \frac{100 \times 5 \times 10^{-4}}{2.5 \times 10^{-5} + 1/7.5 \times 10^3}$$

$$= 2000 - 316$$

$$= \mathbf{1684\ \Omega}$$

The ac input impedance of the stage *i.e.*, impedance when looking into point *B* is

$$r_{in(stage)} = r_{in(base)} \parallel R_1 \parallel R_2$$
$$= 1.684 \parallel 50 \parallel 25$$
$$= \mathbf{1.53\ K}$$

$$A_v = \frac{-h_{fe}}{r_{in(base)}\ (h_{oe} + 1/r_L)}$$

Now, $r_L = 10\ K \parallel 30\ K$
$= 7.5\ K$

$$\therefore \quad A_v = \frac{-100}{1684\ (2.5 \times 10^{-5} + 1/7500)} = \mathbf{-375}$$

Obviously, R_E does not come into the ac picture because it is ac grounded by the bypass capacitor.

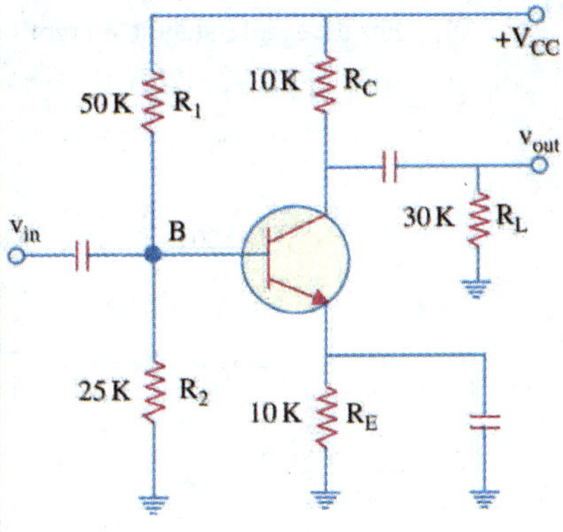

Fig. 21.61

CONVENTIONAL PROBLEMS

1. Using ideal transistor approximations for the single-stage *CB* amplifier of Fig. 21.62, find (*i*) stage r_{in} (*ii*) r_L (*iii*) A_v and (*iv*) A_p. Take transistor $\alpha = 0.99$.
[(*i*) 25 Ω (*ii*) 10 K (*iii*) 400 (*iv*) 396]

2. For the single-stage CB amplifier circuit shown in Fig. 21.63, find (*i*) r_{in} (*ii*) r_L (*iii*) A_v and (*iv*) A_p
Transistor $\alpha = -0.9$. Take Ge transistor material. **[(*i*) 25 Ω (*ii*) 5 K (*iii*) 200 (*iv*) 196]**

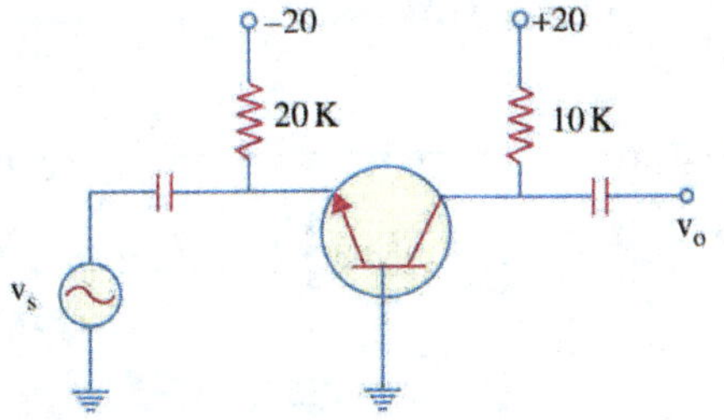

Fig. 21.62

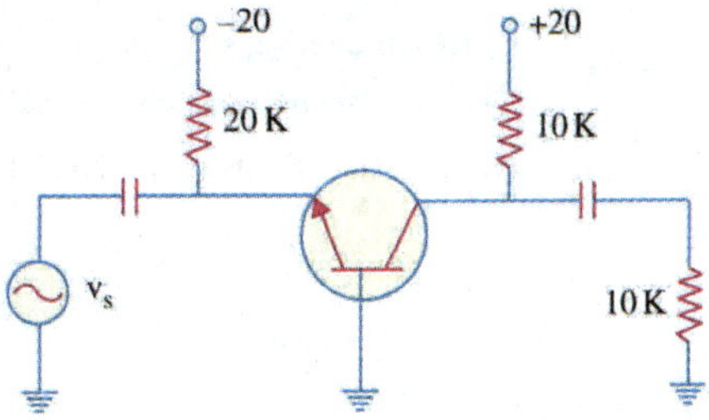

Fig. 21.63

3. For the *CB* amplifier circuit of Fig. 21.64, find approximate values of
(*i*) stage r_{in} (*ii*) r_L (*iii*) $A_v = v_{out} / v_{in}$
(*iv*) $A_{vs} = v_{out} / v_s$ (*v*) A_p (*vi*) G_p
Take transistor $\alpha = -0.98$ **[(*i*) 25 Ω (*ii*) 5K (*iii*) 200 (*iv*) 22.2 (*v*) 196 (*vi*) 22.9 dB]**

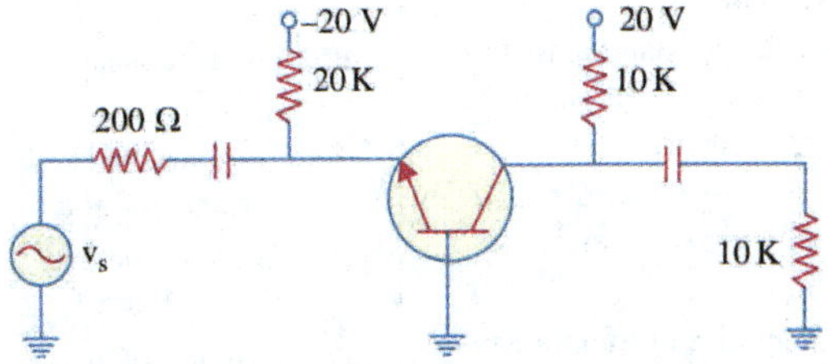

Fig. 21.64

4. Find r_{in}, r_L, A_i, A_v and G_p for the *CE* amplifier shown in Fig. 21.65.
[(*i*) 1250 Ω (*ii*) 5 K (*iii*) 50 (*iv*) 200 (*v*) 30 dB]

5. For the single-stage *CE* amplifier circuit of Fig. 21.66, find

(*i*) r_{in} (*ii*) r_L (*iii*) A_v (*iv*) A_p and (*v*) G_p. Take β = 100 and use r_e = 50 mV/I_E.

[(*i*) 160 K (*ii*) 5K (*iii*) 2.5 (*iv*) 250 (*v*) 24 dB]

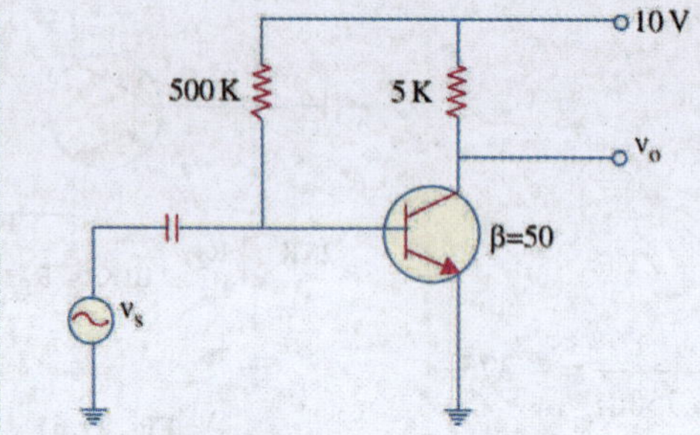

Fig. 21.65

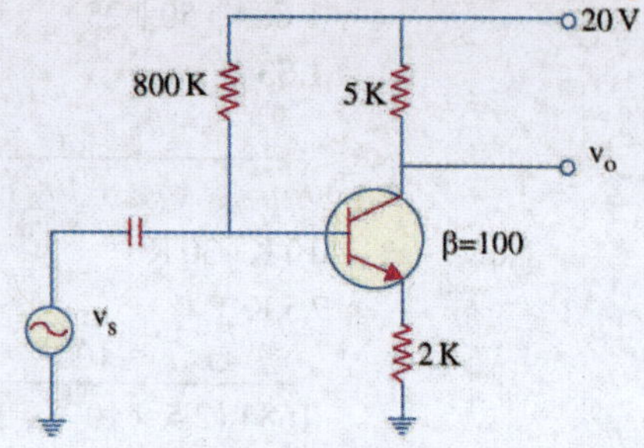

Fig. 21.66

6. For the *CE* amplifier of Fig. 21.67, find approximate values of

(*i*) r_{in} (*ii*) r_L (*iii*) A_v (*iv*) A_p and G_p.

Take transistor β = 50 and use r_e = 50 mV/I_E.

[(*i*) 2K (*ii*) 10K (*iii*) 200 (*iv*) 10,000 (*v*) 40 dB]

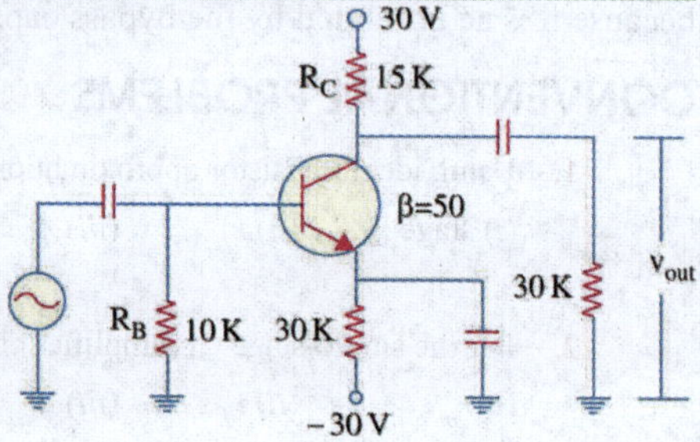

Fig. 21.67

7. For the emitter follower circuit shown in Fig. 21.68, find

(*i*) $r_{in(base)}$ (*ii*) $r_{in(stage)}$

(*iii*) stage A_v

(*iv*) A_p in decibels

Take β = 100 and use r_e = 25 mV/I_E.

[(*i*) 10K (*ii*) 9.52 K (*iii*) 0.75 (*iv*) 18.75 dB]

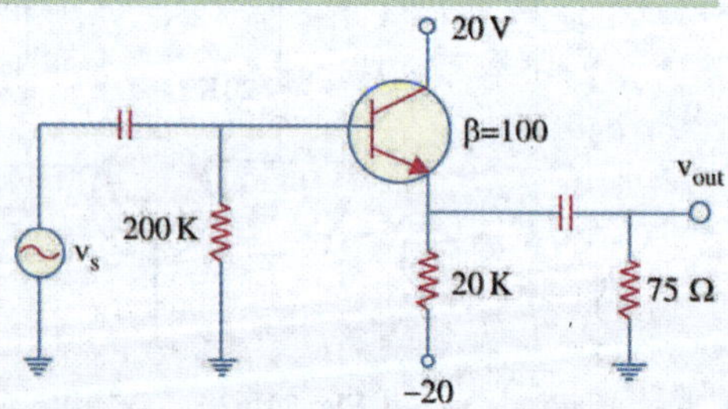

Fig. 21.68

SELF EXAMINATION QUESTIONS

A. Fill in the blanks by most appropriate word (s) or numerical value(s).

1. Small signal operation of a transistor is that in which signal excursions are = ... per cent around *Q*-point values.
2. For drawing dc equivalent circuit of a transistor, all ac sources are treated as .... and capacitors as ....
3. For drawing ac equivalent circuit of a transistor, all dc sources and capacitors are treated as ....
4. The current gain of a single-stage CE amplifier is nearly........... .
5. Voltage gain of an emitter follower is nearly ....
6. For a transistor, the number of h-parameters is ....
7. The parameters h_r and h_o are called .... parameters of a transistor.
8. The parameter h_{11} has unit of ....whereas h_{22} has that of ....
9. For an ideal transistor, both h_r and h_0 are ....
10. Typical value of............ for an actual transistor is a few μS.

B. Answer True or False

1. The current gain of a single-stage *CB* amplifier is nearly equal to transistor ac alpha.
2. Emitter bypass capacitor in a *CE* amplifier prevents signal degeneration.
3. A *CC* amplifier circuit is also called an emitter follower.
4. The voltage gain of an emitter follower is either unity or less.
5. The main use of an emitter follower is as an impedance matching device.
6. The power gain of a single-stage *CE* amplifier is β times its voltage gain.
7. Of all the *h*-parameters, h_0 has the smallest magnitude.
8. For an ideal *CB* amplifier, h_{ib} equals βr_e.
9. For an ideal *CE* amplifier, h_{ie} nearly equals r_e.
10. Amongst the four *h*-parameters, h_i has the largest value.

C. Multiple Choice Items

1. The voltage gain of a well-designed single-stage *CB* amplifier is essentially determined by ac collector load and
   (*a*) emitter resistor R_E
   (*b*) ac alpha
   (*c*) input resistance of emitter diode
   (*d*) ac beta
2. In a single-stage *CB* amplifier, a smaller load resistance R_L will produce
   (*a*) high current gain
   (*b*) low voltage gain
   (*c*) better frequency response
   (*d*) higher power gain
3. The basic reason for using an emitter bypass capacitor in a *CE* amplifier circuit is to
   (*a*) keep emitter at ac ground
   (*b*) prevent excessive degeneration from taking place
   (*c*) decrease stage voltage gain
   (*d*) bypass low-frequency signal component
4. The voltage gain of an emitter follower is
   (*a*) more than its current gain
   (*b*) greater than or equal to unity
   (*c*) less than or equal to unity
   (*d*) dependent on emitter load
5. The main use of an emitter follower is as
   (*a*) power amplifier
   (*b*) impedance matching device
   (*c*) low-input impedance circuit
   (*d*) follower of base signal
6. The *h*-parameters are called hybrid because they
   (*a*) are obtained from different characteristics
   (*b*) are mixed with other parameters
   (*c*) apply to circuits contained in a black box
   (*d*) are defined by using both open- and short-circuit terminations
7. The smallest of the four *h*-parameters of a transistor is
   (*a*) h_i (*b*) h_r
   (*c*) h_o (*d*) h_f
8. Typical value of h_{ie} is
   (*a*) 1 K (*b*) 25 K
   (*c*) 50 Ω (*d*) 100 K

ANSWERS

A. Fill in the blanks

1. 10	**2.** shorts, opens	**3.** shorts	**4.** beta	**5.** unity	**6.** four
7. reverse	**8.** ohm, siemens	**9.** zero	**10.** h_0		

B. True or False

1. T	**2.** T	**3.** T	**4.** T	**5.** T
6. T	**7.** T	**8.** F	**9.** F	**10.** T

C. Multiple Choice Items

1. *c*	**2.** *b*	**3.** *b*	**4.** *c*	**5.** *b*	**6.** *d*	**7.** *c*	**8.** *a*

22 CHAPTER

Single-Stage Transistor Amplifiers

1. Common Base (CB) Amplifier
2. Various Gains of a CB Amplifier
3. Common Emitter (CE) Amplifier
4. Various Gains of a CE Amplifier
5. Characteristics of a CE Amplifier
6. Common Collector (CC) Amplifier
7. Various Gains of a CC Amplifier
8. Characteristics of a CC Amplifier
9. Comparison of Amplifier Configurations
10. Amplifier Classification Based on Biasing Condition
10. Class-A Amplifier
11. Power Distribution in a Class-A Amplifier

22.1 Classification of Amplifiers

Linear amplifiers are classified according to their mode of operation *i.e.*, the way they operate according to a predetermined set of values. Various amplifier descriptions are based on the following factors :

1. **As based on its input**
   (*a*) small-signal amplifier
   (*b*) large-signal amplifier
2. **As based on its output**
   (*a*) voltage amplifier
   (*b*) power amplifier
3. **As based on its frequency response**
   (*a*) audio-frequency (AF) amplifier
   (*b*) intermediate-frequency (IF) amplifier
   (*c*) radio-frequency (RF) amplifier
4. **As based on its biasing conditions**
   (*a*) class-A (*b*) class-AB
   (*c*) class-B (*d*) class-C
5. **As based on transistor configuration**
   (*a*) common-base (CB) amplifier
   (*b*) common-emitter (CE) amplifier
   (*c*) common-collector (CC) amplifier

The description *small-signal, class-A, CE, voltage amplifier* means that input signal is small, biasing condition is class-A, transistor configuration is common-emitter and its output concerns voltage amplification.

We will first take up the basic working of a single-stage amplifier *i.e.*, an amplifier having one amplifying element connected in *CB*, *CE* and *CC* configuration.

22.2. Common Base (CB) Amplifier

Both Figs. 22.1 and 22.2 show the circuit of a single-stage *CB* amplifier using *NPN* transistor. As seen, input ac signal is injected into the emitter-base circuit and output is taken from the collector-base circuit. The *E/B* junction is forward-biased by V_{EE} whereas *C/B* junction is reverse-biased by V_{CC}. The *Q*-point or dc working conditions are determined by dc batteries along with resistors R_E and R_C. In other words, values of I_E, I_B and V_{CB} are decided by V_{CC}, V_{EE}, R_E and R_C. The voltage V_{CB} is given by the equation $V_{CB} = V_{CC} - I_C R_C$.

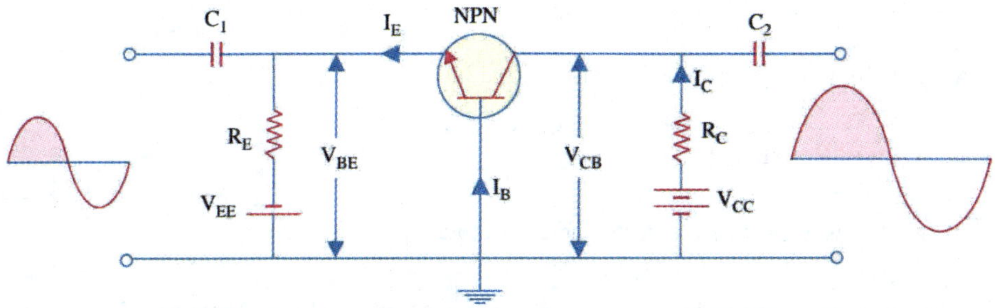

Fig. 22.1

When no signal is applied to the input circuit, the output just *sits at the Q-point so that there is no output signal*. Let us now see what happens when we apply an ac signal to the *E/B* junction via a coupling capacitor C_1 (which is assumed to offer no reactance to the signal).

Circuit Operation

When positive half-cycle of the signal is applied, then

1. forward bias is *decreased* because V_{BE} is already negative with respect to the ground as per biasing rule of Art. 18.3.
2. consequently, I_B is *decreased.*
3. I_E and hence I_C are *decreased* (because they are both nearly β times the base current).
4. the drop $I_C R_C$ is *decreased.*
5. hence, V_{CB} is *increased* as seen by the equation given above.

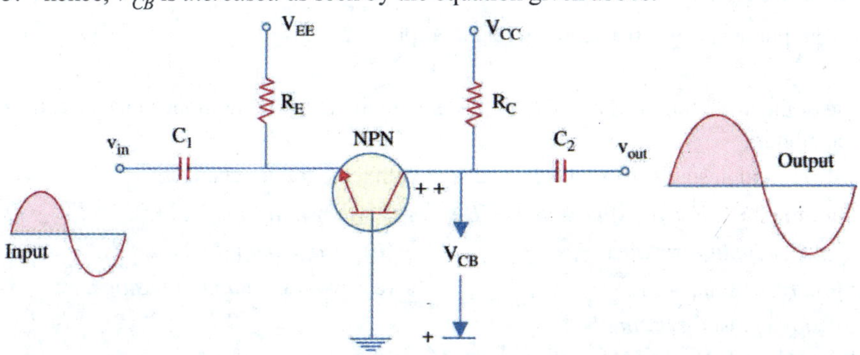

Fig. 22.2

It means that a positive output half-cycle is produced.

Since a *positive-going* input signal produces a *positive-going* output signal, there is no phase reversal between the two.

Voltage amplification in this circuit is possible by reason of relative input and output circuitry rather than current gain (α) which is always less than unity. The input circuit has low resistance whereas output circuit has very large resistance. Although changes in input and output currents are the same, the ac drop across R_L is very large. Hence, changes in V_{CB} (which is the output voltage) are much larger than changes in input ac signal. Hence, the voltage amplification.

22.3. Various Gains of a CB Amplifier

1. Input Resistance

The ac input resistance of the transistor is given by the emitter junction resistance

$$r_e = \frac{25\text{ mV}}{I_E} \quad \text{or} \quad \frac{50\text{ mV}}{I_E}$$

As seen from the ac equivalent circuit (Fig. 22.3)

$$r_{in} = r_e \,\|\, R_E$$

2. Output Resistance

$$r_o = R_C \qquad \text{— Fig. 22.2}$$

If a load resistance R_L is connected across output terminals, then

$$r_o = R_C \,\|\, R_L \qquad \text{— Fig. 22.3}$$

It is called the output resistance of the stage and is written as r_o (stage) or r_o'.

3. Current Gain. $A_i = \alpha$

4. Voltage Gain. $A_v = \frac{r_o}{r_{in}} = \frac{r_o}{r_e}$

5. Power Gain. $A_p = A_v \,.\, A_i$

The decibel gain is given by $G_p = 10 \log_{10}{}^{Ap}$ dB.

22.4. Characteristics of a CB Amplifier

Common-base amplifier has

1. very low input resistance (30 – 150 Ω),
2. very high output resistance (upto 500 K),
3. a current gain α < 1,
4. large voltage gain of about 1500,
5. power gain of upto 30 dB,
6. no phase reversal between input and output voltages.

Uses

One of the important uses of a *CB* amplifier is in matching a low-impedance circuit to a high-impedance circuit.

It also has high stability of collector current with temperature changes.

Example 22.1. *For the single-stage CB amplifier shown in Fig. 22.3 (a), find*

(*a*) *stage input resistance,* (*b*) *stage output resistance,*

(*c*) *current gain,* (*d*) *voltage gain of the stage,*

(*e*) *stage power gain in dB.*

Assume α = 1. Neglect V_{BE} and use r_e = 25 mV/I_E.

Solution. The ac equivalent circuit is shown in Fig. 22.3 (*b*).

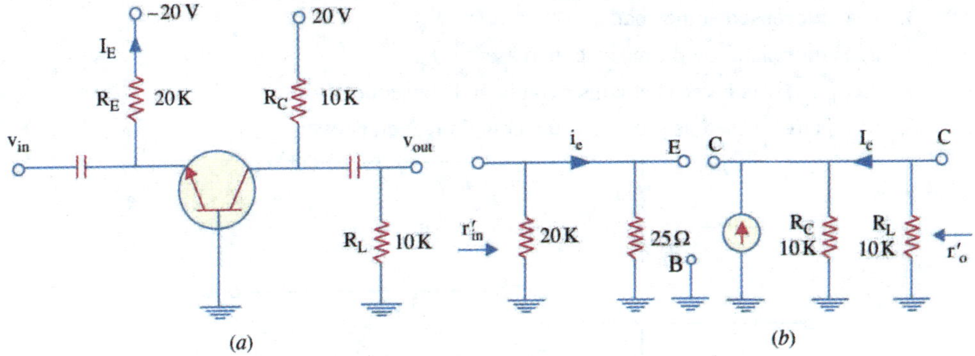

Fig. 22.3

$I_E = 20/20 = 1$ mA $\qquad r_e = 25/I_E = 25\ \Omega$

(*a*) stage $\quad r_{in} = r_e \,||\, R_E = 25\ \Omega \,||\, 20\text{K} \cong \mathbf{25\ \Omega}$

(*b*) stage $\quad r_o = R_C \,||\, R_L \quad \therefore \quad r_o' = 10\text{K} \,||\, 10\text{K} = \mathbf{5K}$

(*c*) $\quad A_i = \alpha = \mathbf{1} \qquad$ (*d*) $\quad A_v = \dfrac{r_o'}{r_e} = \dfrac{5000}{25} = \mathbf{200}$

(*e*) $\quad A_p = A_v \,.\, A_i = 200 \times 1 = \mathbf{200}$

$G_p = 10 \log_{10}{}^{Ap}$ dB $= 10 \log_{10} 200 = \mathbf{23\ dB}$

22.5. Common Emitter (CE) Amplifier

Figs. 22.4 and 22.5 show the circuit of a single-stage *CE* amplifier using an *NPN* transistor.

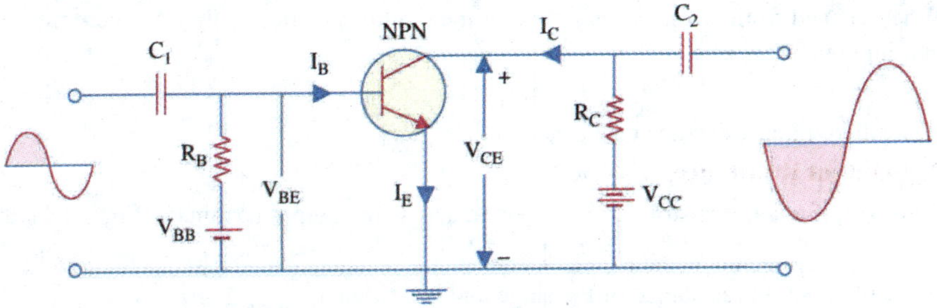

Fig. 22.4

Here, base is the driven element. The input signal is injected into the base-emitter circuit whereas output signal is taken out from the collector-emitter circuit. The *E*/*B* junction is forward-biased by V_{BB} and *C*/*B* junction is reversed-biased by V_{CC} (in fact, same battery V_{CC} provide dc power for both base and collector). The *Q*-point or working condition is determined by V_{CC} together with R_B and R_C. The dc equation is (Fig. 22.5).

$I_B \cong V_{BB}/R_B$ — neglecting V_{BE}

$I_C = \beta I_B \quad$ and $\quad V_{CE} = V_{CC} - I_C R_C$

Now, let us see what happens when an ac signal is applied at the input terminals of the circuit.

Circuit Operations

When positive half-cycle of the signal is applied (Fig. 22.4).

1. V_{BE} is **increased** because it is already positive w.r.t. the ground as per biasing rule of Art. 18.3.

2. It leads to **increase** in forward bias of base-emitter junction.
3. I_B is **increased** somewhat.
4. I_C is **increased** by β times the increased in I_B.
5. Drop $I_C R_C$ is **increased** considerably and consequently.
6. V_{CE} is **decreased** as seen from the equation given above.

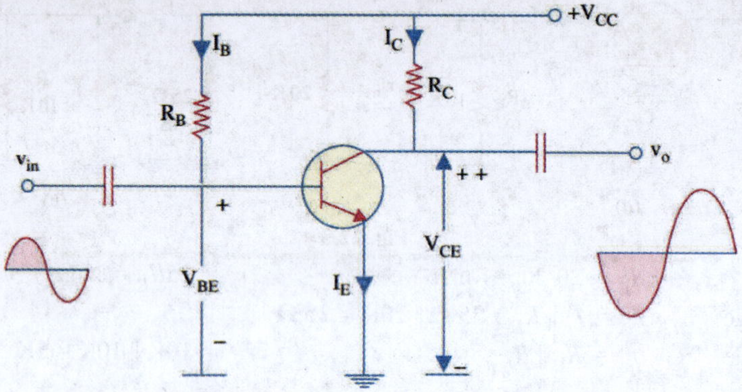

Fig. 22.5

Hence, negative half-cycle of the output is obtained. It means that a *positive-going* input signal becomes a *negative going* output signal as shown in Figs. 22.4 and 22.5.

22.6. Various Gains of a CE Amplifier

The ac equivalent of the given circuit (Fig. 22.5) is similar to the one shown in Fig. 22.6 (*b*).

1. Input Resistance

When viewed from *base*, ac resistance of the emitter junction is βr_e. As seen, from Fig. 22.6 (*b*), circuit input resistance is

$$r_{in} = R_B \parallel \beta\, r_e \cong \beta\, r_e \qquad \text{— when } R_B \gg \beta\, r_e$$

It is called input resistance of the stage *i.e.*, $r_{in(stage)}$.

2. Output Resistance. $r_o = R_C$ — Fig. 22.5

However, if a load resistor R_L is connected across the output terminals (Fig. 22.6), then

$$r_o = R_C \parallel R_L \qquad \text{— Fig. 22.6 } (b)$$

It is called output resistance of the stage and is written as $r_{o(stage)}$ or r_o'.

3. Current Gain. $A_i = \beta$.

4. Voltage Gain. $$A_v = \beta \cdot \frac{r_o}{r_{in}} = \beta \cdot \frac{r_o}{\beta r_e} \cong \frac{r_o}{r_e} \qquad \text{— if } R_B \gg \beta\, r_e$$

5. Power Gain. $$A_p = A_v \,.\, A_i = \beta \cdot \frac{r_o}{r_e} \qquad G_p = 10 \log_{10}{}^{Ap} \text{ dB}$$

22.7. Characteristics of a CE Amplifier

A *CE* transistor amplifier has the following characteristics :

1. it has moderately low input resistance (1 K to 2 K),
2. its output resistance is moderately large (50 K or so),
3. its current gain (β) is high,
4. it has very high voltage gain of the order of 1500 or so,
5. it produces very high power gain of the order of 10,000 times or 40 dB,

6. it produces *phase reversal* of input signal *i.e.*, input and output signals are 180° out of phase with each other.

Uses

Most of the transistor amplifiers are of *CE* type because of large gains in voltage, current and power. Moreover, their input and output impedance characteristics are suitable for many applications.

Example 22.2. *For the transistor amplifier shown in Fig. 22.6 (a), calculate the voltage gain without and with a load resistor of 500 Ω.*

Solution. The base current is given by

$$I_B = (V_{CC} - V_{BE})/R_B = (10 - 0.7)/186\text{ K} = 50\mu\text{A}$$

The collector current is given by

$$I_C = \beta I_B = 100 \times 50\ \mu\text{A} = 5.0\text{ mA}$$

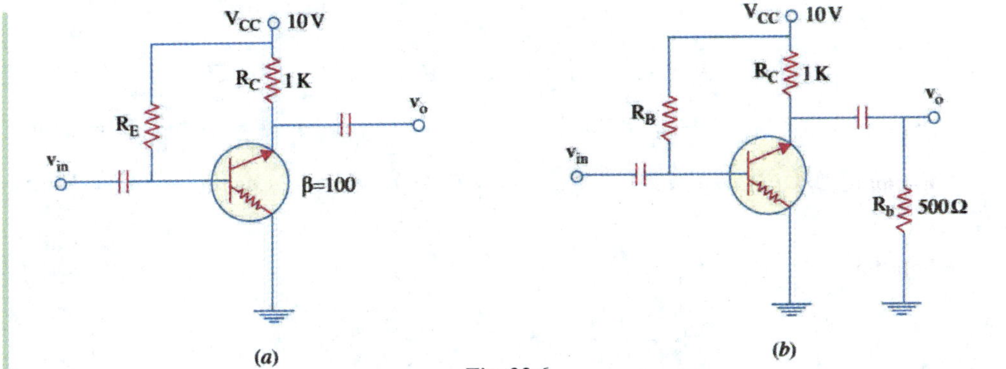

Fig. 22.6

$$\therefore \quad A_v = \frac{-1000 \parallel 500}{5} = \frac{-333}{5} = -66.6$$

The emitter current I_E is nearly the same *i.e.*, 5 mA.

The input signal appears across the internal resistance r_e of the emitter.

Now, $\quad r_e = 25\text{ mV}/5\text{ mA} = 5\Omega$

Voltage gain without load resistance of 500 Ω is given by

$$A_v = \frac{-R_C}{r_e} = \frac{-1000}{5} = -200$$

With Load Resistor

A load resistor of 500 Ω has been connected across the output as shown in Fig. 22.6 (*b*). In this case, $A_V = (-R_C \parallel R_L)/r_e$

It is seen that the addition of the 500 Ω load has caused a large drop in the amplifier gain.

Example 22.3. *For the single-stage CE amplifier circuit shown in Fig. 22.6 (a), calculate (a) r_{in} (b) r_o (c) A_i (d) A_v and (e) G_p.*

Take transistor β = 50. Neglect V_{BE} and take $r_e = 25\text{ mV}/I_E$.

Solution. $\quad I_B = \frac{20}{1\text{ M}} = 20\ \mu\text{A}\ ; \quad I_C = \beta I_B = 50 \times 20 = 1\text{ mA}$

$$r_e = 25/1 = 25\ \Omega\ ; \quad \beta r_e = 50 \times 25 = 1250\ \Omega$$

The ac equivalent circuit is shown in Fig. 22.6 (*b*).

(*a*) $\quad r_{in} = 1\text{M} \parallel 1250\ \Omega \cong \mathbf{1250\ \Omega}$

Obviously, it is the input resistance of the stage and not that of the transistor alone.

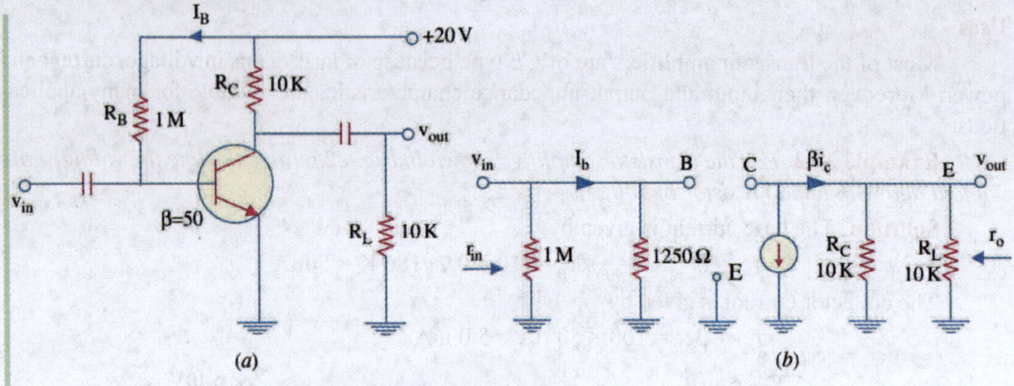

Fig. 22.7

(*b*) $r_o = R_C \| R_L = 10\text{ K} \| 10\text{ K} = \mathbf{5\ K}$ (*c*) $A_i = \mathbf{50}$

(*d*) $A_v = \dfrac{5\text{ K}}{25\ \Omega} = \mathbf{200}$ (*e*) $G_p = 10\log_{10}{}^{200} = \mathbf{23\ dB}$

Example 22.4. *Find the values of V_{CE}, A_v and Z_{in} for the transistor amplifier circuit shown in Fig. 22.8.*

Solution.

$$V_B = V_{CC}\frac{R_2}{R_1 + R_2}$$

$$= 18\cdot\frac{10}{40 + 10} = 3.6\text{ V}$$

$$V_E = 3.6 - 0.7 = 2.9\text{ V}$$

$$I_E = 2.9\text{ V}/1\text{K} = 2.9\text{ mA}$$

$$V_{RC} = I_C R_C \cong I_E R_C$$

$$= 2.9\text{ mA} \times 4\text{K}$$

$$= 11.6\text{ V}$$

$$\therefore \quad V_{CE} = V_{CC} - V_{RC} \times V_E$$

$$= 18 - 11.6 - 2.9$$

$$= \mathbf{3.5\ V}$$

$$r_e = 25\text{ mV}/2.9\text{ mA}$$

$$= 8.6\ \Omega$$

$$A_v = -R_C/r_e = -4000/8.6$$

$$= \mathbf{-4822}$$

$$r_b = \beta r_e = 200 \times 8.3 = 1.66\text{ K}$$

$$Z_{in} = R_1 \| R_2 \| r_b = 40 \| 10 \| 1.66 = \mathbf{1.37\ K}$$

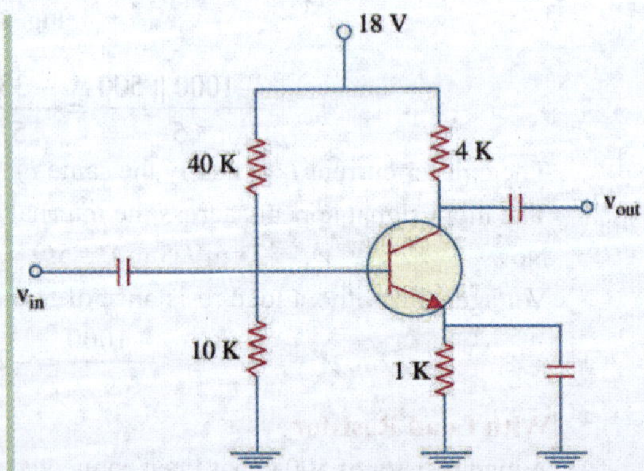

Fig. 22.8

22.8. Common Collector (CC) Amplifier

Figs. 22.8 and 22.9 show the circuit of a single-stage *CC* amplifier using an *NPN* transistor. The input signal is injected into the base-collector circuit and output signal is taken out from the emitter-collector circuit. The *E*/*B* junction is forward-biased by V_{EE} and *C*/*B* junction is reverse-biased by V_{CC}. The quiescent values of I_B and I_E are set by V_{CC} and V_{EE} together with R_B and R_E. As seen from Fig. 22.8.

$$I_E = \frac{V_{EE} - V_{BE}}{R_E + R_B/\beta} \cong \frac{V_{EE}}{R_E + R_B/\beta}$$

Let us now see what happens when an ac signal is applied across the input circuit.

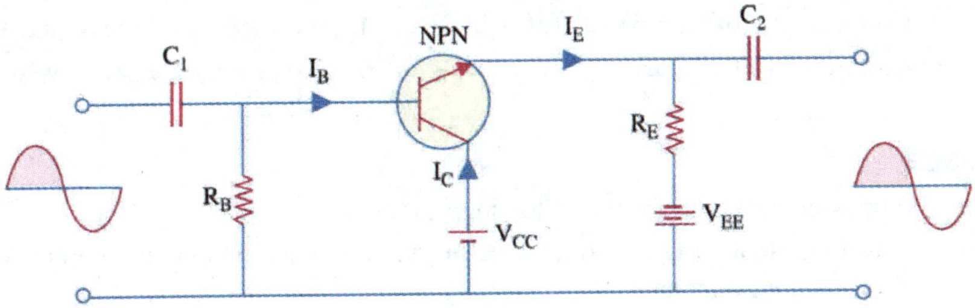

Fig. 22.9

Circuit Operation

When positive half-cycle of the signal is applied, then

1. forward bias is **increased** since V_{BE} is positive w.r.t. collector *i.e.*, ground,
2. base current is **increased**,
3. emitter current is **increased**,
4. drop across R_E is **increased**,
5. hence, output voltage (*i.e.*, drop across R_E), is **increased**.

Consequently, we get positive half-cycle of the output.

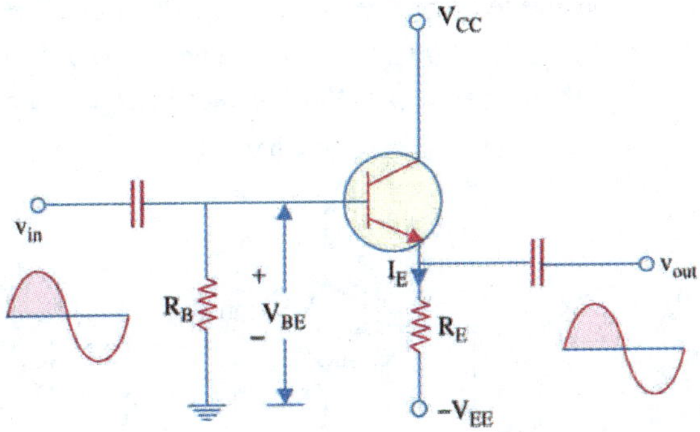

Fig. 22.10

It means that a *positive-going* input signal results in a *positive- going* output signal and, consequently, the input and output signals are in phase with each other as shown in Fig. 22.10.

22.9. Various Gains of a CC Amplifier

The ac equivalent circuit of the CC amplifier (Fig. 22.10) is given in Fig. 22.11.

1. $r_{in} = R_B \| \beta\,(r_e + r_o)$.
2. $r_o = R_E \| R_L$.   3. $A_i = \beta$.
4. $A_v = \dfrac{r_o}{r_o + r_e}$.   Since usually $r_o' >> r_e'$, $A_v \cong \dfrac{r_o'}{r_o'} = 1$.

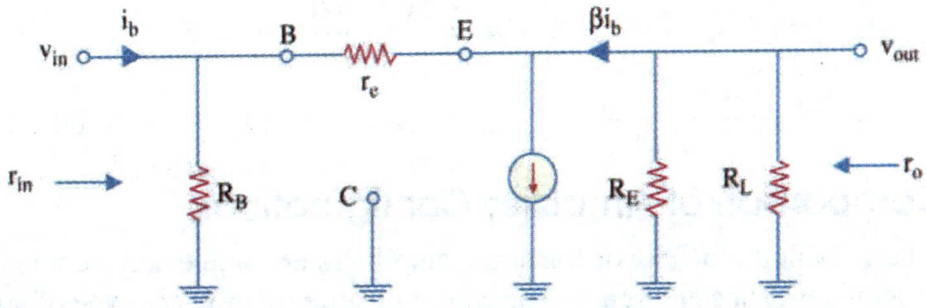

Fig. 22.11

5. $A_p = A_v \cdot A_i$, $G_p = 10 \log_{10}{}^{Ap}$ dB.

22.10. Characteristics of a CC Amplifier

A *CC* amplifier has the following characteristics :

1. high input impedance (20-500 K),
2. low output impedance (50-1000 Ω),
3. high current gain of (1 + β) *i.e.*, (50 – 300),
4. voltage gain of less than 1,
5. power gain of 10 to 20 dB,
6. no phase reversal between input and signals.

22.11. Uses

The *CC* amplifiers are used for the following purposes :

1. for impedance matching *i.e.*, for connecting a circuit having high output impedance to one having low input impedance;
2. for circuit isolation;
3. as a two-way amplifier since it can pass a signal in *either direction*;
4. for switching circuits.

Example 22.5. *For the CC amplifier circuit of Fig. 22.12 (a), compute*

(a) r_{in} (ii) r_o (iii) A_v and (iv) A_p

Take transistor β = 100. Neglect V_{BE} and use r_e = 25 mV/I_E.

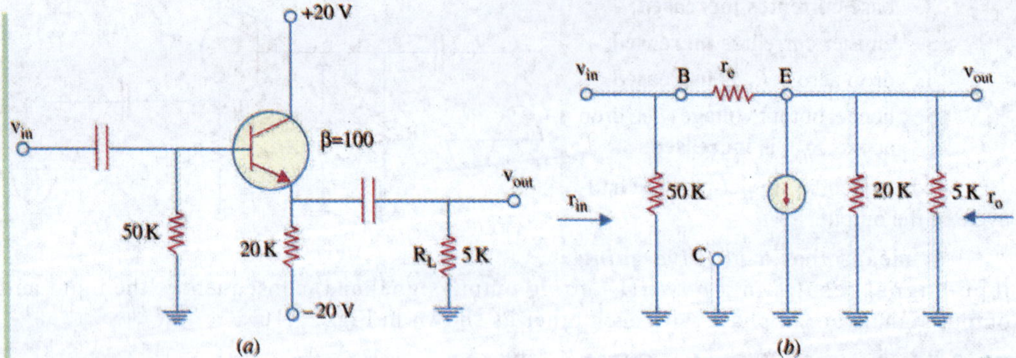

Fig. 22.12

Solution. $I_E = \frac{20}{20} = 1$ mA ; $r_e = \frac{25}{1} = 25\ \Omega$

The ac equivalent circuit is shown in Fig. 22.12 (*b*).

(*i*) $r_o = R_E \parallel R_L = 20\text{K} \parallel 5\text{K} = 4\text{K}$

(*ii*) $r_{in} = R_B \parallel \beta\,(r_e + r_o) \cong R_B \parallel \beta r_o = \frac{50 \times 400}{450} = 44.4$ K

(*iii*) $A_v = \frac{4\ \text{K}}{4\ \text{K} + 25\ \Omega} \cong 1$ (*iv*) $A_p = A_v \cdot A_i = 1 \times 100 = 100$

22.12. Comparison of Amplifier Configurations

The basic building blocks of transistor amplifiers are single-stage common-base, common-emitter and common-collecter circuits. The choice of configuration and type of transistor for a given application will depend largely upon the desired input and output impedances, voltage, current and power gains and frequency response. For convenience, these values are tabulated in Table 22.1. Here, the input and output parameters are resistances and apply only to *low-frequency conditions*.

Table 22.1

Characteristic	*Type of circuit*		
	Common base	*Common emitter*	*Common collector*
Current gain	nearly unity (α)	high (β)	highest ($1 + \beta$)
Voltage gain	high	very high	nearly unity
Power gain	moderate	highest	lowest
Input impedance	lowest	moderate	highest
Output impedance	highest	moderate	lowest
Phase reversal	No	Yes	No

In each case, the amplifying action depends on low-power input circuits controlling the high-power output circuits.

In the common-base circuit, although the input and output currents are nearly equal, the low-impedance emitter circuit absorbs for less power as compared to that which is available at the high-impedance collector.

The low base current flowing into the common-emitter circuit (where the impedance is a few kilohms) gives rise to much higher collector current flowing out of the high-impedance output circuit.

The common-collector circuit with approximately equal input and output resistances requires a low input current to control much larger output current.

22.13. Amplifier Classification Based on Biasing Condition

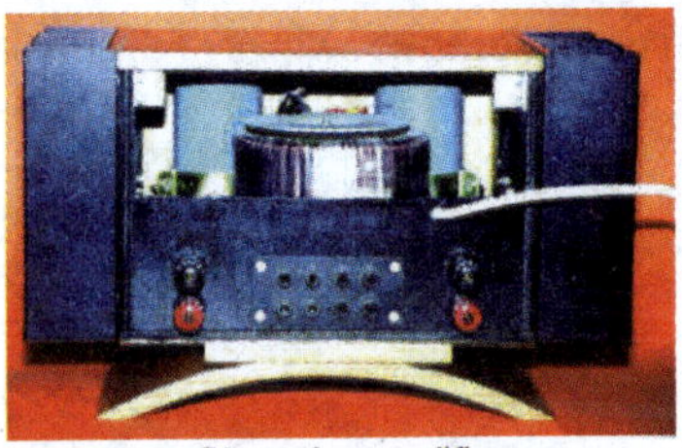

Class-A amplifier.

This classification is based on the amount of transistor bias and amplitude of the input signal. It takes into account the portion of the cycle for which the transistor conducts. The three main classifications are :

(*a*) Class-A Amplifier

In this case, the transistor is so biased that output current flows for the full-cycle of the input signal (360°) as shown in Fig. 22.13. In other words, the transistor remains FR-biased throughout the input cycle. Hence, its conduction angle is 360°.

(*b*) Class-B Amplifier

In this case, the transistor bias and the amplitude of input signal are such that output current flows for only half-cycle (180°) of the input signal. It means that transistor stays FR-biased for half the input cycle. The transistor conduction angle equals 180°.

(*c*) Class-C Amplifier

In this case, transistor bias and signal amplitude are such that output current flows for appreciably less than half-cycle of the input

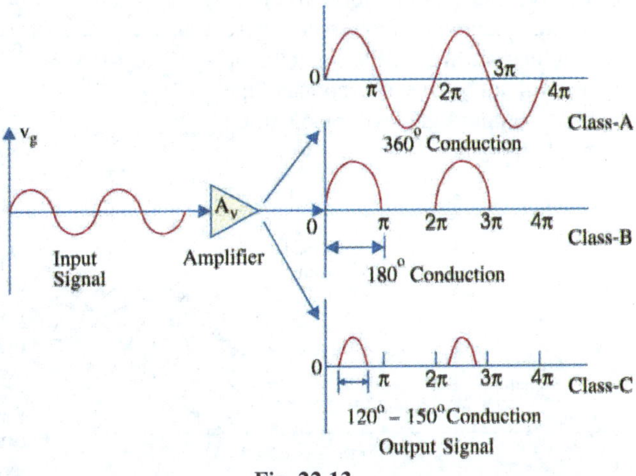

Fig. 22.13

signal *i.e.*, upto 120° or 150° angle of conduction as shown in Fig. 22.13. In other words, transistor remains FR-biased for less than half the cycle.

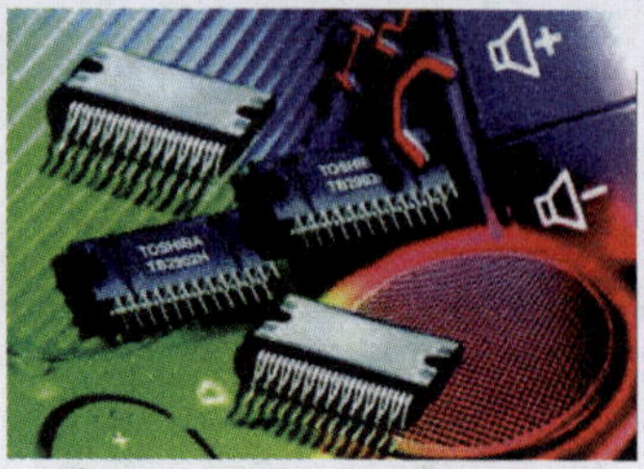

Class – AB power amplifier.

(*d*) Class-AB Amplifier

The characteristics of such an amplifier lie in-between those of class-A and class-B. Here, biasing conditions are such that output current flows for appreciably more than half but less than the entire cycle *i.e.*, current flows for more than 180° but less than 360°.

22.14. Graphic Representation

Since common-emitter is the most versatile and widely-used configuration, we will use *CE* output characteristic curves to differentiate between main classes of amplifiers.

Fig. 22.14 shows the biasing condition for class-A operation. It is seen that *Q*-point is located *at the centre of the load line* so that output (collector) current flows for the complete cycle of the input signal (conduction angle of 360°). Because of centred *Q*-point, the positive and negative swings of the input signal are confined to *linear portions of the load line.* In this linear region, equal changes in input base current produce equal changes in output (collector) current and voltage. Hence, output is an *exact replica of the input.* If the amplitude of the input signal is so large as to drive the *Q*-point closer to either cut-off or saturation region (where base current lines are not equally spaced), the output will not be an exact replica of the input *i.e.*, output signal will be distorted (Art. 22.30).

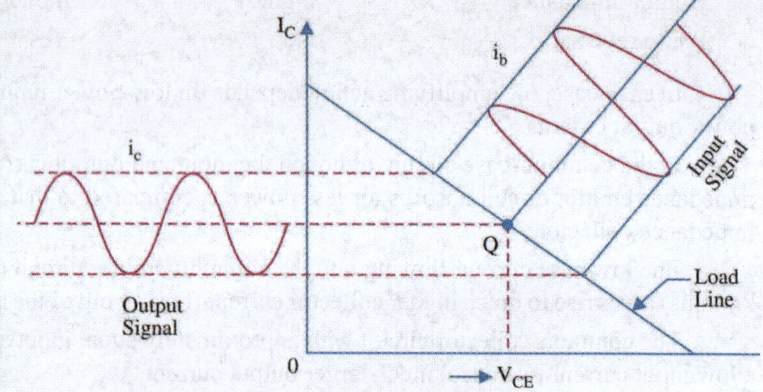

Fig. 22.14

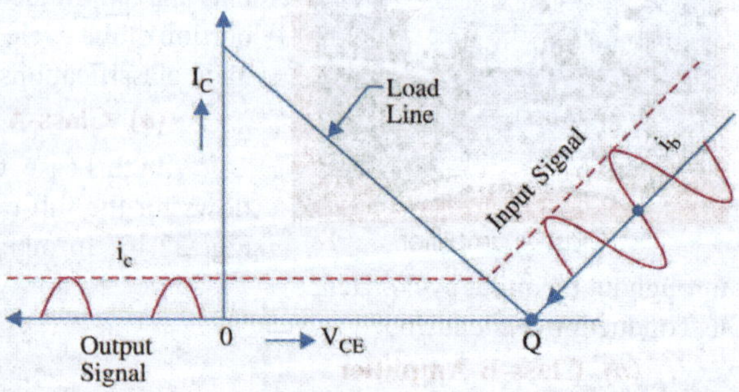

Fig. 22.15

Fig. 22.15 shows biasing conditions for class-B operation. In this case, *Q*-point is located at cut-off point or very close to it. It is seen that only for positive half-cycles of the input signal, the output current flows. For negative half-cycles, there is no output (collector) current. Obviously, output current flows only for 180° or during the positive half-cycle of the input signal.

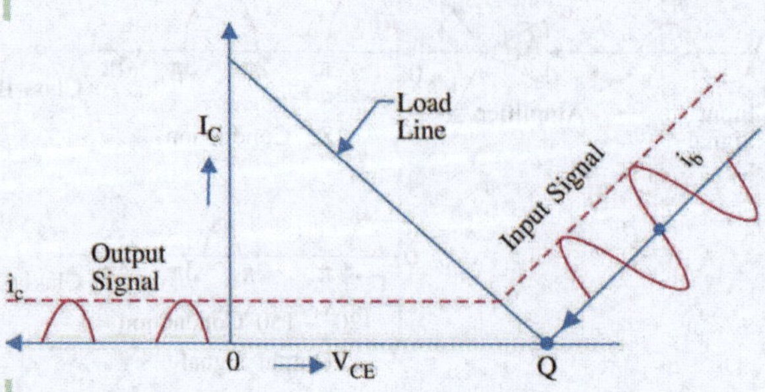

Fig. 22.16

Fig. 22.16 shows biasing condition for class-C operation. Here, the transistor is biased *well into cut-off*. In this case, (collector) current flows only for a part (upto 150°) of the positive half-cycle of the input signal when transistor comes out of the cut-off. During negative half-cycle of the input, transistor *remains deep in cut-off*.

22.15. Class-A Amplifier

It is one which has centred Q-point so that the transistor operates only over the linear region of its load line.

Another way of defining is that it is an amplifier in which the output current flows during the entire cycle of the input signal *i.e.*, it has a conduction angle of 360°. In other words, the transistor remains FR-biased throughout the input cycle.

Characteristics

1. Since the transistor operates over the *linear portion* of the load line, the output waveform is exactly similar to the input waveform. Hence, class-A amplifiers are characterised by a high fidelity of the output. They are used where linearity or *freedom from distortion is the prime requisite*.
2. Since its operation is restricted only over a small central region of the load line, this amplifier is meant only for amplifying input signals of *small amplitude*. Large signals, will shift the Q-point into non-linear regions near saturation or cut-off and produce distortion (Art. 22.30).
3. Due to the limitation of the input signal amplitude, ac power output per active device (*i.e.*, transistor) is small.
4. The overall efficiency of the amplifier circuit is
$$= \frac{\text{ac power delivered to the load}}{\text{total power delivered by dc supply}} = \frac{\text{average ac power output}}{\text{average dc power input}}$$
The maximum possible overall efficiency of a class-A amplifier with resistive load is 25%.
5. The collector efficiency of a transistor is defined as
$$= \frac{\text{average ac power output}}{\text{average dc power input to transistor}}$$
The maximum possible collector efficiency of class-A amplifier with resistive load is 50%.
6. In case an output transformer is used, the maximum possible overall efficiency and maximum possible collector efficiency for a class-A amplifier are both 50%.

22.16. Power Distribution in a Class-A Amplifier

Fig. 22.17 (*a*) shows a *CE* connected transistor which forms the active element of a single-stage class-A amplifier. Fig. 22.17 (*b*) shows its output characteristic with a centred Q-point. When ac input signal is applied, Q-point shifts up and down from its central position. The output current will also increase or decrease from its equiescent (or no-signal) value I_{CQ}.

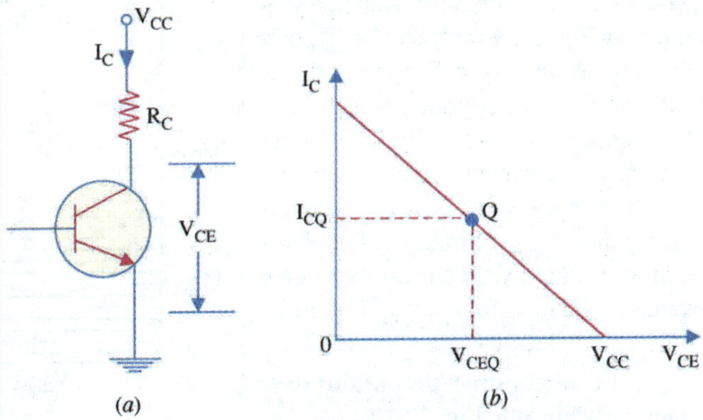

Fig. 22.17

Similarly, collector-emitter voltage V_{CE} will increase or decrease from its equiescent value V_{CEQ}. So long as signal variations are confined to linear region of the load line, average value of collector current is I_{CQ} because positive and negative input signal swings will produce equal changes in I_{CQ}.

Hence, total average dc power drawn by the *circuit* from collector battery V_{CC} is

$$P_{in\,(dc)} = V_{CC} \cdot I_{CQ}$$

Now, this power goes to supply the following :

(*i*) heat dissipated by the load resistor R_C connected to the collector

$$P_{RC(dc)} = I_{CQ}^{\,2} R_C$$

(*ii*) the balance $P_{tr(dc)}$ is given to the transistor.

It is further subdivided into :

(*a*) ac power developed across the load resistor which constitutes the ac power output

$$P_{o(ac)} = I^2 R_C = \frac{V^2}{R_C} = \frac{V_m^2}{2R_C}$$

where I is the r.m.s. value of the ac output current through the load, V is r.m.s. value of ac voltage and V_m is its maximum value.

(*b*) power dissipated (in the form of heat) by the transistor itself *i.e.*, its *collector region*. It may be called $P_{c(dc)}$.

Since, under zero-signal condition, there is no ac output power, all the power given to the transistor is wasted as heat. Hence, ***a transistor dissipates maximum power under zero-signal condition.***

The power flow diagram of a transistor is as under :

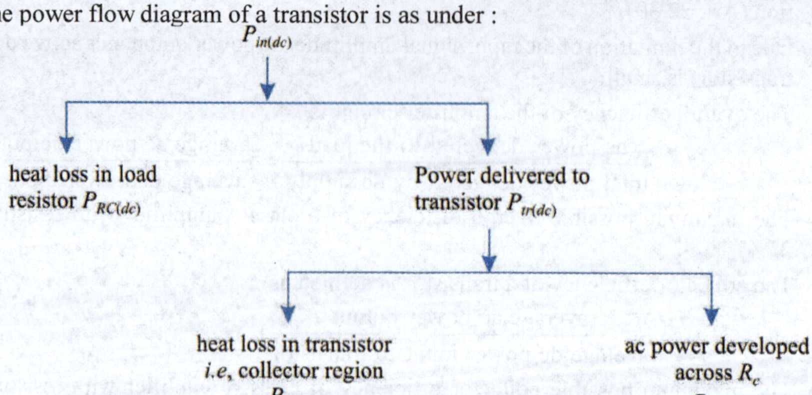

22.17. Power Rectangle

The output or collector characteristic of a *CE*-connected transistor is shown in Fig. 22.18 with centred *Q*-point for class-A operation. When input signal is applied, the *Q*-point shifts to positions Q_1 and Q_2 alternately. The output current varies around its quiescent value from maximum value of $I_{C(max)}$ to minimum value $I_{C(min)}$. Similarly, collector-emitter voltage varies from maximum value of $V_{CE(max)}$ to $V_{CE(min)}$ around its quiescent value of V_{CEQ}.

Let us construct the various rectangles as shown in Fig. 22.18.

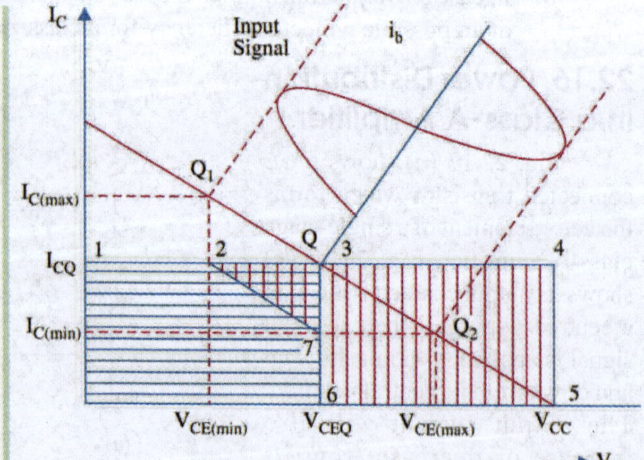

Fig. 22.18

1. rectangle 0-1-4-5 = $V_{CC} \times I_{CQ}$

$= $ total average power supplied to the circuit by V_{CC} battery

$= P_{in(dc)}$

Now, $V_{CC} - V_{CEQ}$

$= $ voltage drop across load resistor R_C.

2. rectangle 3-4-5-6 = $(V_{CC} - V_{CEQ})\, I_{CQ}$ = power lost as heat in load resistor $R_C = I_{CQ}^2 R_C$.

Now, V_{CEQ} is the voltage drop across the transistor itself *i.e.*, potential difference between its collector and emitter as shown in Fig. 22.18 (*b*).

3. rectangle 0-1-3-6 = $V_{CEQ} \cdot I_{CQ}$ = power delivered to transistor = $P_{tr(dc)}$.
4. triangle 2-3-7 = ac power across $R_C = P_{o(ac)} = I^2 R_C$.

It can also be proved that

$$P_{o(ac)} = \frac{V_{CE(max)} - V_{CE(min)}}{2\sqrt{2}} \times \frac{I_{C(max)} - I_{C(min)}}{2\sqrt{2}}$$

$$= \frac{[V_{CE(max)} - V_{CE(min)}] \times [I_{C(max)} - I_{C(min)}]}{8}$$

5. area 0-1-2-7-6 = power dissipated by the collector region of the transistor

$= P_{c(dc)}$

Obviously, when there is no input signal, there is no output signal. Hence, $P_{o(ac)} = 0$, which means that area of triangle 2-3-7 is zero. In that case, the transistor collector will have to dissipate maximum power equal to the area of the rectangle 0-1-3-6. It is called *worst-case* condition of the transistor and equals quiescent power.

22.18. Power Efficiency

Fig. 22.19 shows the power rectangles and triangles of a class-A amplifier in a very simple way.

$P_{in(dc)}$ = areas $A + B + C$

$P_{o(ac)}$ = area B

$P_{c(dc)}$ = area C

$P_{Rc(dc)}$ = area A

Amplifier efficiency or circuit efficiency or overall efficiency

$$\eta_{overall} = \frac{B}{A+B+C} = \frac{P_{o(ac)}}{P_{in(dc)}}$$

The *collector* efficiency is given by

$$\eta_{coll} = \frac{B}{B+C} = \frac{P_{o(ac)}}{P_{tr(dc)}}$$

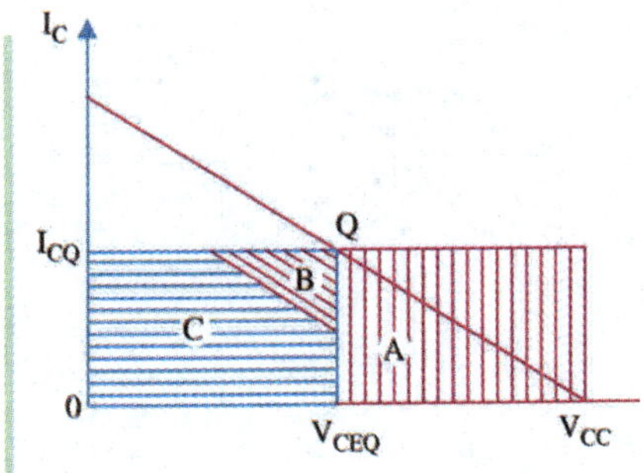

Fig. 22.19

Incidentally, it can be proved that $\eta_{overall} = 25\left[\frac{V_{CE(max)} - V_{CE(min)}}{V_{CC}}\right]$

Now, maximum value of $V_{CE} = V_{CC}$ and minimum value is = 0

$$\therefore \quad \text{maximum } \eta_{overall} = 25\left[\frac{V_{CC} - 0}{V_{CC}}\right]\% = 25\%$$

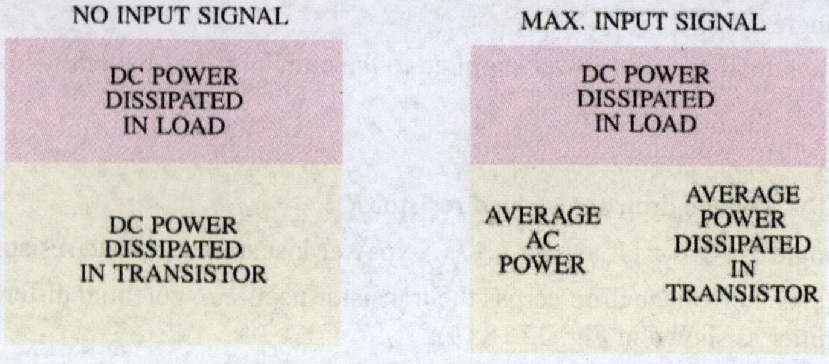

Fig. 22.20

Summary

In a direct-coupled class-A amplifier

1. maximum ac power equals half the power transistor can dissipate (or one-fourth of the total dc input power).
2. maximum ac power the transistor can dissipate is half the dc input power supplied by the battery.
3. the dc input power equals 4 times the ac output in load or twice the maximum power dissipated by the transistor.

The above facts have been illustrated with the help of power square shown in Fig. 22.20.

22.19. Maximum AC Power in Load

Fig. 22.21 (*a*) shows part of a base-driven *CE* amplifier with a collector load of R_C. Fig. 22.21 (*b*) shows its ac load line with centred *Q*-point. For the time being, we will neglect limitations of cut-off and saturation and assume (at least theoretically) that voltage and current swings down to zero are possible. Obviously, under such conditions, we will get the maximum possible ac output power.

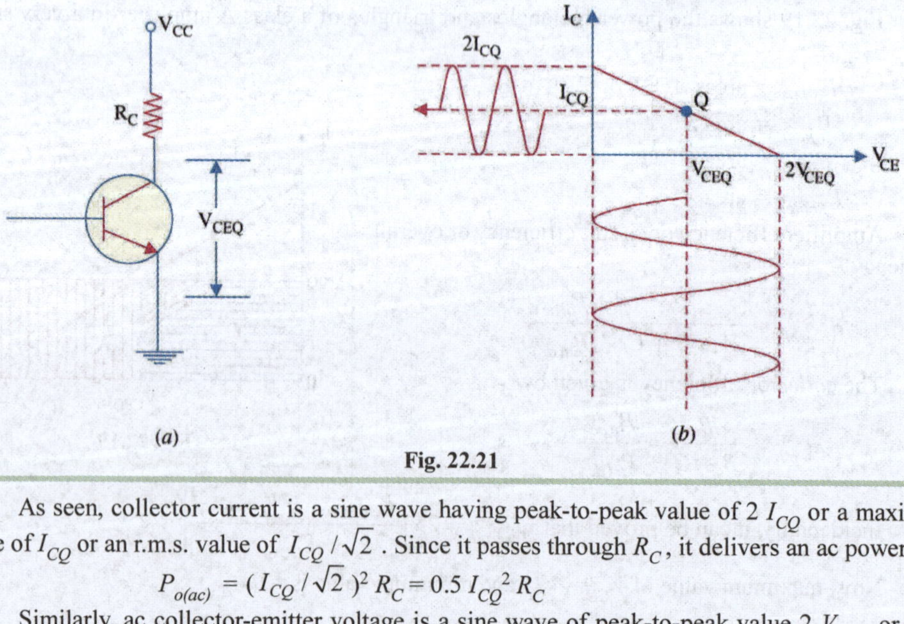

Fig. 22.21

As seen, collector current is a sine wave having peak-to-peak value of 2 I_{CQ} or a maximum value of I_{CQ} or an r.m.s. value of $I_{CQ}/\sqrt{2}$. Since it passes through R_C, it delivers an ac power of

$$P_{o(ac)} = (I_{CQ}/\sqrt{2})^2 R_C = 0.5\, I_{CQ}^2 R_C$$

Similarly, ac collector-emitter voltage is a sine wave of peak-to-peak value 2 V_{CEQ} or peak value V_{CEQ} or r.m.s. value $V_{CEQ}/\sqrt{2}$. Since this sine wave is applied across R_C, the ac power can also be expressed as

$$P_{o(ac)} = \left(\frac{V_{CEQ}}{\sqrt{2}}\right)^2 / R_C = \frac{0.5\, V_{CEQ}^{\ 2}}{R_C}$$

Note. If there had been an unbypassed resistor R_E in the emitter, then ac load R_{ac} or $r_L = R_C + R_E$. Then, we would have used this value rather than R_C alone.

As seen from power rectangle diagram of Fig. 22.22, under maximum ac power condition,

$$\text{area } B = \text{area } C$$

Also, $\quad \text{area } A = \text{area } B + \text{area } C = 2B$

$$\therefore \text{ maximum } \eta_{overall} = \frac{B}{A + B + C}$$

$$= \frac{B}{2A + 2B} = \frac{B}{4B} = 0.25$$

or $\quad$ 25%.

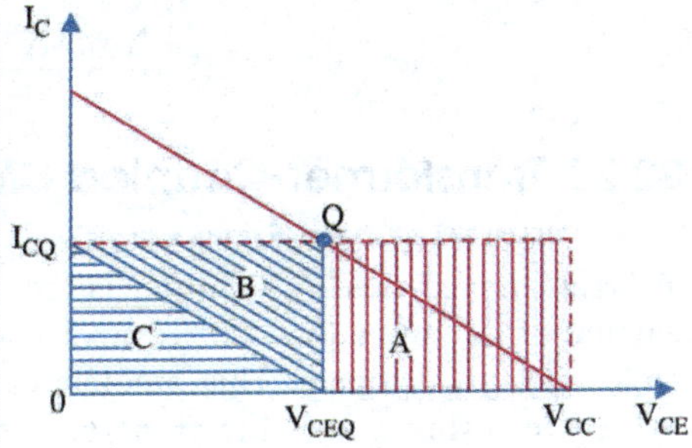

Fig. 22.22

Example 22.6. *The 2 N 1491 power transistor is used in a CE amplifier meant for class-A operation. If its zero-signal power dissipation is 10 W and ac output power is 3.5 W, find*

(a) collector efficiency and $\quad$ *(b) power rating of the transistor.*

Solution. (*i*) $\eta_{coll} = \dfrac{P_{o(ac)}}{P_{tr(dc)}} = \dfrac{3.5}{10} = \mathbf{0.35 = 35\%}.$

(*ii*) Since zero-signal condition represents worst case condition for the transistor, it means that all the 10 W power is dissipated by it. Hence, transistor power rating is **10 W**.

Example 22.7. *For the class-A, CE amplifier circuit of Fig. 22.22, V_{CEQ} = 10 V and I_{CQ} = 500 mA. If collector i.e., output current varies by ± 250 mA when an input signal is applied at the base, compute*

(*i*) *total dc power taken by the circuit,*
(*ii*) *dc power dissipated by the collector load,*
(*iii*) *ac power developed across the load,*
(*iv*) *power delivered to the transistor,*
(*v*) *dc power wasted in transistor collector,*
(*vi*) *overall efficiency,*
(*vii*) *collector efficiency.*

Solution. (*i*) $P_{in(dc)} = V_{CC} \cdot I_{CQ} = 20 \times 500$

$= 10^4 \text{ mV} = \mathbf{10\ W}$

(*ii*) $\quad P_{Rc(dc)} = I_{CQ}^{\ 2} R_C = 0.5^2 \times 20 = \mathbf{5\ W}$

(*iii*) $\quad P_{o(ac)} = I^2 R_C$

Now, maximum value of output ac current is 250 mA = 0.25A. Hence, r.m.s. value $I = 0.25/\sqrt{2}$ A.

$\therefore \quad P_{o(ac)} = (0.25/\sqrt{2})^2 \times 20$

$= \mathbf{0.625\ W}$

(*iv*) $\quad P_{tr(dc)} = 10 - 5 = \mathbf{5\ W}$

(*v*) $\quad P_{c(dc)} = 10 - 5 - 0.625 = \mathbf{4.375\ W}$

(*vi*) $\quad \eta_{overall} = \dfrac{0.625}{10} \times 100 = \mathbf{6.25\%}$

(*vii*) $\quad \eta_{coll} = \dfrac{0.625}{5} \times 100 = \mathbf{12.5\%}$

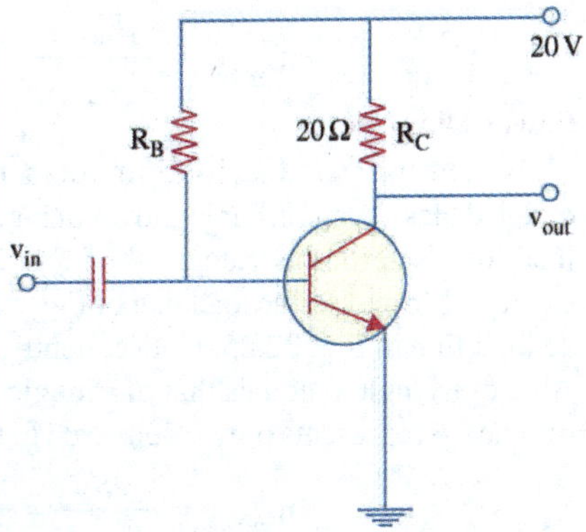

Fig. 22.23

It is seen that due to resistive load in the collector, efficiency of the circuit is very low.

Example 22.8. *For a class-A amplifier, $V_{CE(max)} = 25V$, $V_{CE(min)} = 5V$, $V_{CC} = 30V$. Find its overall efficiency.*

Solution. As seen from Art. 22.18

$$\% \eta_{overall} = 25\left[\frac{V_{CE(max)} - V_{CE(min)}}{V_{CC}}\right] = 25\left(\frac{25-5}{30}\right) = \mathbf{16.7}$$

22.20. Transformer-Coupled Class-A Amplifier

The main reason for the poor efficiency of a *direct-coupled* class-A amplifier is the large amount of dc power that the resistive load in collector must dissipate. This problem can be solved by using a suitable transformer for coupling the load (say, a speaker) to the amplifier stage as shown in Fig. 22.24. Since the load is not *directly* connected to the collector terminal, the dc collector current does not pass through it. In an ideal transformer, primary winding resistance is zero. Hence, dc power loss in the load is zero. In practice, however, there is a small dc resistance of the primary winding which does absorb some power though much less than a direct-coupled load.

In short, what the transformer does is to substitute *ac load* in place of *ohmic* or *dc load*.

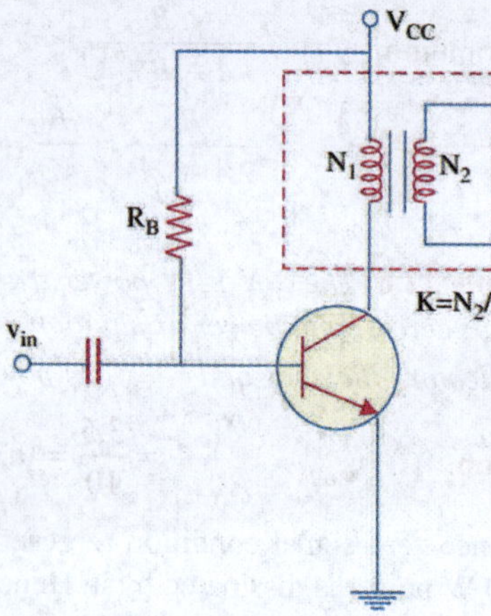

Fig. 22.24

The secondary load R_L when referred to primary becomes

$$R_L' = R_L/K^2 = a^2 R_L$$

where K = voltage transformation ratio = $N_2/N_1 = V_2/V_1$

a = turn ratio $N_1/N_2 = 1/K$

Since a is usually made much more than unity or K is much less than unity, R_L can be made to look *much bigger* than what it actually is.

In an ideal transformer, there is no primary drop, hence $V_{CC} = V_{CEQ}$. Now, all the power supplied by V_{CC} is delivered to the transistor. Hence, the overall and collector efficiencies become equal.

$$\eta_{overall} = \frac{P_{o(ac)}}{V_{CC}\, I_{CQ}} = \frac{P_{o(ac)}}{V_{CEQ} \cdot I_{CQ}}$$

Power Diagram

The power distribution for a transformer-coupled class-A amplifier is shown in Fig. 22.25. Since load loss is zero, the triangle *A* of Fig. 22.19 is non-existent. Moreover, the load line of Fig. 22.19 becomes ac load line in Fig. 22.25. Consequently, $V_{CEQ} = V_{CC}$. Area of triangle *B* equals that of triangle *C*. Also, sum of *A* and *B* represents power delivered to the transistor.

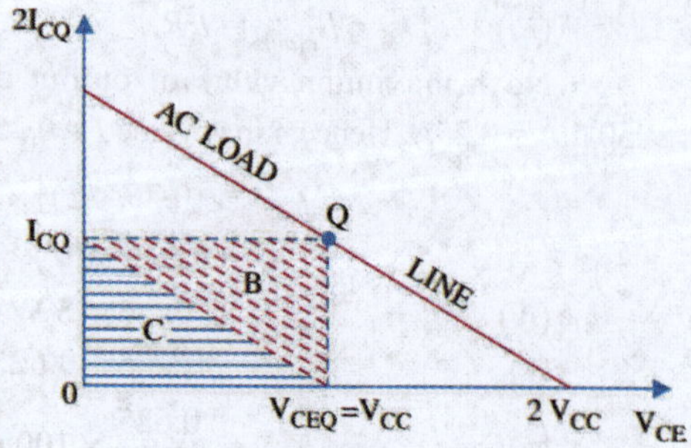

Fig. 22.25

$$\therefore \qquad \eta_{overall} = \frac{B}{B+C} = \frac{B}{2B} = 0.5$$

or $\qquad$ 50%

It can be proved that maximum possible overall efficiency and the maximum possible collector efficiency of a class-A amplifier using an output transformer are both 50%.

In general, overall efficiency of a transformer-coupled class-A amplifier is given by

$$\%\,\eta = 50\left[\frac{V_{CE(max)} - V_{CE(min)}}{2V_{CC}}\right]^2 = 50\left[\frac{V_{CE(max)} - V_{CE(min)}}{V_{CE(max)} + V_{CE(min)}}\right]^2$$

Obviously, larger the value of $V_{CE(max)}$ and smaller that of $V_{CE(min)}$, closer the efficiency to the theoretical limit of 50%.

Proof

The conditions for the development of maximum ac power are shown in Fig. 22.26. It must be kept in mind that here

$$V_{CEQ} = V_{CC}$$

Now, average power delivered by the battery is

$$P_{in(dc)} = V_{CC} \cdot I_{CQ}$$

Whole of this power is given to the transistor.

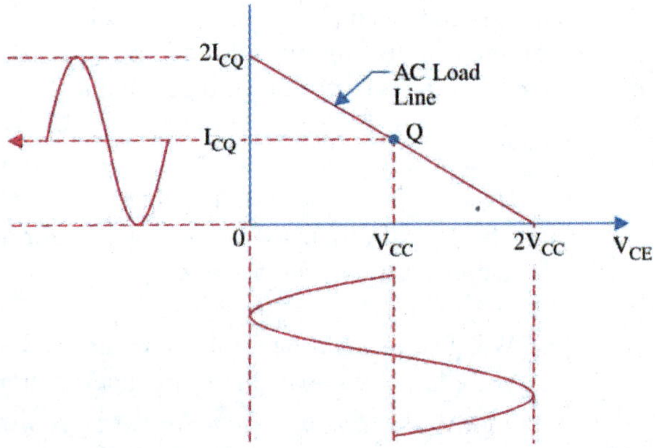

Fig. 22.26

$$\text{rms output voltage} = \frac{V_{CC}}{\sqrt{2}}$$

$$\text{rms output current} = \frac{I_{CQ}}{\sqrt{2}}$$

$$\therefore\ \text{maximum } P_{o(ac)} = \frac{V_{CC}}{\sqrt{2}} \times \frac{I_{CQ}}{\sqrt{2}}$$

$$= 1/2\ V_{CC} \cdot I_{CQ}$$

$$= \text{half the input ac power}$$

Hence, maximum value of overall efficiency

$$\eta_{overall} = \frac{P_{o(ac)}}{P_{in(dc)}} = \frac{V_{CC} \cdot I_{CC}/2}{V_{CC} \cdot I_{CQ}} = \frac{1}{2} \text{ or } 0.5 \text{ or } 50\%$$

$$\eta_{coll} = \frac{P_{o(ac)}}{P_{tr(dc)}} = \frac{1}{2} \text{ or } 50\%$$

Example 22.9. *The optimum load resistance for a certain transistor is 200Ω. What is the turn ratio (N_1/N_2) of a transformer required to couple an 8Ω loud-speaker to the transistor ?*

Solution. $R_L' = a^2 R_L \quad \therefore\ 200 = a^2 \times 8 \quad \text{or} \quad a = 5$

Hence, it should be a 5 : 1 *step-down* transformer.

Example 22.10. *In a transformer-coupled class-A amplifier $V_{CE(max)} = 27\ V$ and $V_{CE(min)} = 3V$. Compute its overall efficiency.*

Solution. $\eta_{overall} = 50\left(\frac{27-3}{27+3}\right)^2 = \mathbf{32\%}$

Example 22.11. *For the transformer-coupled optimally-biased class-A amplifier shown in Fig. 22.25, find*

(a) transformer turn ratio (b) collector current (c) transistor power rating

Solution. As shown in Fig. 22.27, if V_2 is the r.m.s. secondary voltage, then

$$P_{load} \quad \text{or} \quad P_{o(ac)} = \frac{V_2^2}{R_L} = \frac{(V_{2m}/\sqrt{2})^2}{R_L} = \frac{V_{2m}^2}{R_L}$$

where V_{2m} is the maximum value of secondary voltage.

$$\therefore \; V_{2m} = \sqrt{2RP_{load}} = \sqrt{2 \times 4 \times 4.5} = \mathbf{6\,V}$$

Now, peak, value of primary voltage is

$$V_{1m} = a \,.\, V_{2m} = 6\,a$$

This also represents the peak value of collector voltage. For optimally-biased circuit, peak value of collector-emitter voltage equals V_{CC} = 30 V.

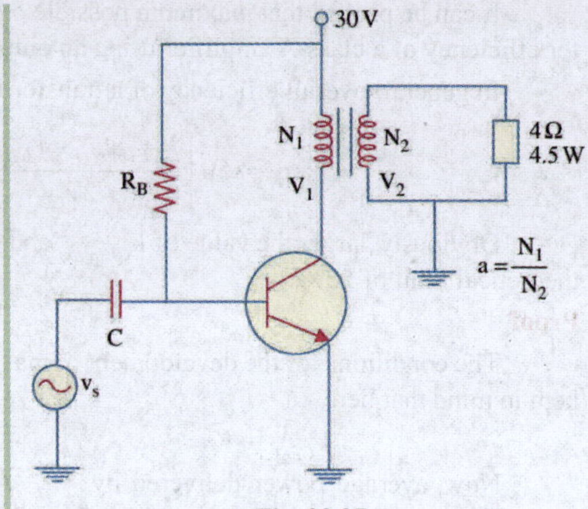

Fig. 22.27

$$\therefore \quad 6a = 30 \quad \text{or} \quad a = 5 \quad \therefore \quad N_1/N_2 = \mathbf{5}$$

(*a*) Hence, it is a step-down transformer with a turn ratio of **5 : 1**.

(*b*) The overall efficiency is 50%. Hence, for a load power of 4.5 W, total dc input power supply must be 4.5/0.5 = 9 W

$$\therefore \quad V_{CC}I_C = P_{in(dc)} = 9\text{W} \quad \text{or} \quad 30\,I_C = 9000\text{ mW} \quad \therefore \quad I_C = \mathbf{300\text{ mA}}$$

(*c*) Worst case condition for the transistor is when there is no input signal. In that case, transistor has to dissipate the entire input. ∴ transistor power rating = **9 W.**

Example 22.12. *For the optimally-biased transformer-coupled class-A amplifier, maximum collector current change is 100 mA. Find the power transferred to a 4 Ω speaker load if it is (a) directly-coupled (b) transformer-coupled to the transistor.*

Find also the turn ratio of the transformer required.

Solution. (*a*) Direct-coupled Load [Fig. 22.28 (*a*)]

It should be remembered that for an optimally-biased class-A amplifier, peak change in collector-emitter voltage equals V_{CC} = 10 V.

R.M.S. value of ac output current $I = (100/\sqrt{2})$ mA

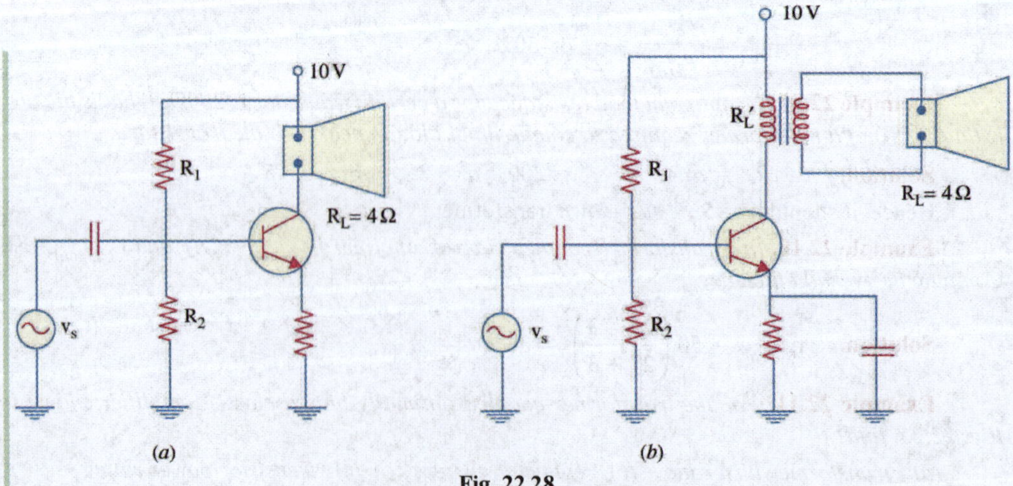

Fig. 22.28

Average power delivered to the speaker

$$= I^2 R_L = \left(\frac{100}{\sqrt{2}} \times 10^{-3}\right)^2 \times 4 = 0.02 \text{ W} = \mathbf{20 \text{ mW}}$$

(*b*) Transformer-coupled Load

Now, $\Delta V_{CE} = V_{CC} = 10$ V; $\Delta I_{CQ} = 100$ mA

If R_L' is the secondary load reflected into the primary, then

$$R_L' = \frac{\Delta V_{CE}}{\Delta I_{CQ}} = \frac{10 \text{ V}}{100 \text{ mA}} = 100 \ \Omega$$

Now, $a^2 R_L = R_L'$ $\therefore$ $a^2 \times 4 = 100$, $a = \mathbf{5}$

Hence, it is a step-down transformer with a turn ratio of 5 : 1 (Fig. 22.27).

Peak value of secondary voltage = 10 × 1/5 = 2 V

Peak value of secondary current = 2 / 4 = 0.5 A

R.M.S. value = $0.5/\sqrt{2}$ A

$$\text{Average power transferred to the speaker} = \left(\frac{0.5}{\sqrt{2}}\right)^2 \times 4 = 0.5 \text{ W} = \mathbf{500 \text{ mW}}$$

Obviously, power transferred has increased 500/20 = 25 times *i.e.*, $a^2 = (N_1/N_2)^2$ times where a is transformer primary-to-secondary turn ratio *i.e.*, $a = N_1/N_2$.

Example 22.13. *A class-A amplifier shown in Fig. 22.29 operates from V_{CC} = 20 V, draws a no-signal current of 5A and feeds a load of 40 Ω through a step-up transformer of $N_2 / N_1 = 3.16$. Find*

(*a*) *whether the amplifier is properly matched for maximum power transfer,*

(*b*) *maximum ac signal power output,* (*c*) *maximum dc power input,*

(*d*) *conversion efficiency at maximum signal input.*

Solution. (*a*) Here, $a = N_1/N_2 = 1/3.16$

$$R_L' = a^2 R_L = (1/3.16)^2 \times 40 = \mathbf{4 \ \Omega}$$

Also, $$R_L' = \frac{V_{CC}}{I_{CQ}} = \frac{20}{5} = \mathbf{4 \ \Omega}$$

Since, the two resistance values are the same, the amplifier is *properly-matched*.

(*b*)As seen from Art. 22.20,

Maximum value of

$$P_{o(ac)} = \frac{1}{2} V_{CC} I_{CQ}$$

$$= \frac{1}{2} \times 20 \times 5 = \mathbf{50 \text{ W}}$$

(*c*)The dc power input

$$= V_{CC} I_{CQ} = 20 \times 5 = \mathbf{100 \text{ W}}$$

(*d*) $$\eta = \frac{50}{100} \times 100 = \mathbf{50\%}$$

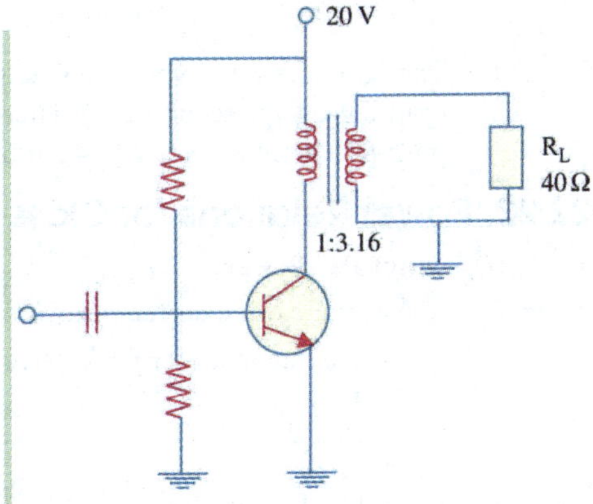

Fig. 22.29

22.21. Class-B Amplifier

Biasing condition for class-B operation has been shown both on current transfer characteristic and output characteristic for *CE* configuration in Fig. 22.30.

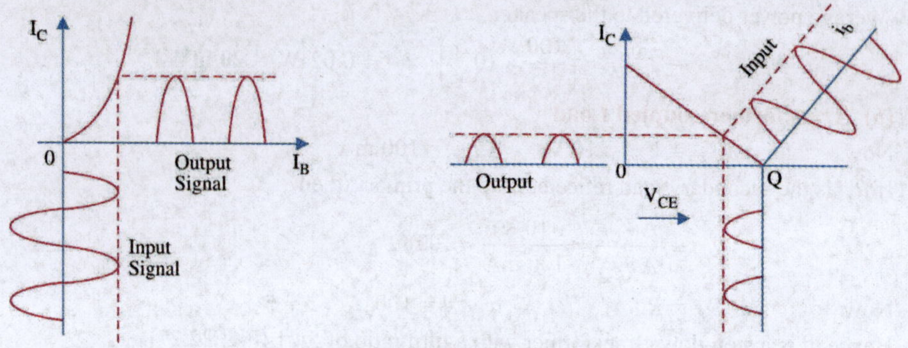

Fig. 22.30

As seen, the transistor has been biased to cut-off. It remains FR-biased for only half-cycle of the input signal. Hence, its conduction angle is only 180°. It is obvious that with zero signal, its collector current is zero.

Characteristics

1. Since negative half-cycles are totally absent from the output, the signal *distortion is high as compared to class-A amplifiers*.
2. Since input voltage is large (upto V_{CC}), voltage amplification is reduced.
3. Zero-signal input represents worst condition for class-A amplifiers but *best condition* for class-B amplifiers.
4. In class-B amplifiers, transistor dissipates more power as signal increase but opposite is the case in class-A amplifiers.
5. Average current in class-B operation is less than in class-A, hence power dissipated is less. Consequently, maximum circuit (or overall) efficiency of a class-B amplifier is 78.5% when peak signal makes $V_{CE(min)} = 0$. In general

$$\eta_{overall} = 78.5\left[1 - \frac{V_{CE(min)}}{V_{CC}}\right]\%$$

6. The half-sinewave type of collector current contains very pronounced *even harmonics* (particularly the second one). These impulses can be rendered useful either by employing push-pull circuits (Art. 22.24) or by using tuned amplifiers (Art. 22.29).

22.22. Power Relations for Class-B Operation

(*i*) Input DC Power: $P_{in(dc)} = V_{CC}\, I_{dc}$

where I_{dc} is the average or dc current drawn from the supply.

If $I_{C(max)}$ is the maximum or peak value of collector or output current, then

$$I_{dc} = \frac{I_{C(max)}}{\pi} \qquad \therefore \qquad P_{in(dc)} = V_{CC}\,\frac{I_{C(max)}}{\pi}$$

(*ii*) DC Power Loss in Load : $P_{Rc(dc)} = I_{dc}^2\, R_C$ where $I_{dc} = I_{C(max)}/\pi$

(*iii*) AC Power Output in Load : $P_{o(ac)} = I^2 R_C = \dfrac{V^2}{R_C} = \dfrac{V_m^2}{2\,R_C}$

where I = r.m.s. value of output ac current

V = r.m.s. value of output ac voltage

V_m = maximum value of output ac voltage

(*iv*) DC Power Loss in Collector Region or Transistor

$$P_{c(dc)} = P_{in(dc)} - P_{Rc(dc)} - P_{o(ac)}\ ; \quad \eta_{overall} = \frac{P_{o(ac)}}{P_{in(dc)}}$$

22.23. Maximum Values

The operation for maximum signal input is shown in Fig. 22.31.

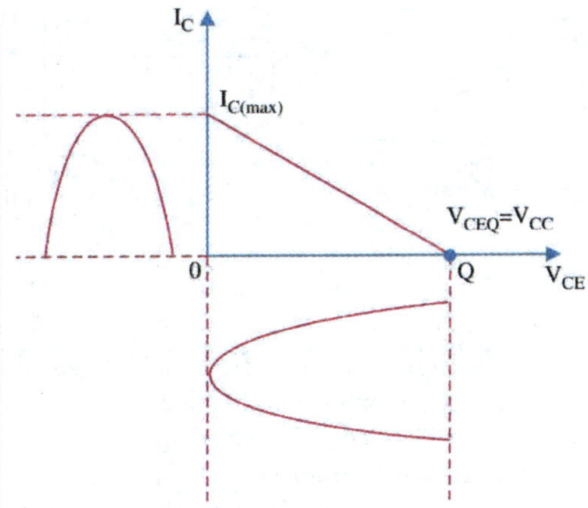

Fig. 22.31

Here, $$I_{dc} = \frac{I_{C(max)}}{\pi}$$

$$\therefore \quad P_{in(dc)} = V_{CC} \cdot I_{dc} = \frac{V_{CC} \cdot I_{C(max)}}{\pi}$$

R.M.S. value of output or collector current

$$= \frac{I_{C(max)}}{\sqrt{2}}$$

R.M.S. value of output voltage

$$= \frac{V_{CC}}{\sqrt{2}}$$

Hence,ac output power during half-cycle

$$= \frac{1}{2} \times \frac{V_{CC}}{\sqrt{2}} \cdot \frac{I_{C(max)}}{\sqrt{2}}$$

The factor $\frac{1}{2}$ comes in because power is produced during one half-cycle only.

$$P_{o(ac)} = \frac{1}{4} \cdot V_{CC} \cdot I_{C(max)}$$

$$\therefore \quad \eta_{overall} = \frac{P_{o(ac)}}{P_{in(dc)}} = \frac{\pi}{4} = 0.785 \text{ or } \mathbf{78.5\%}$$

Example 22.14. *In a class-B amplifier, $V_{CE(min)} = 2V$ and $V_{CC} = 15$ V. Find its overall efficiency.*

Solution. As seen from Art. 22.21, $\%\ \eta = 78.5\left(1 - \frac{2}{15}\right) = \mathbf{68\%}$

22.24. Class-B Push-Pull Amplifier

It employs two *identical* transistors operating as a *single-stage of amplification*. As shown in Fig. 22.32, the base of the two *CE*-connected transistors have been connected to the opposite ends of the secondary of the input transformer T_1 and collectors to the opposite ends of the primary of the output transformer T_2. For getting a balanced circuit, the two emitters have been returned to the centre tap of T_1 secondary and V_{CC} connected to the centre tap on the primary of T_2. Since zero bias is required for cut-off, the two bases have been earthed. A push-pull amplifier is sometimes referred to as *balanced amplifier.*

Principle of Operation

It is seen from Fig. 22.32 that transistors *A* and *B* are driven by two input signals which are 180° out of phase with each other. These two signals are produced by T_1.

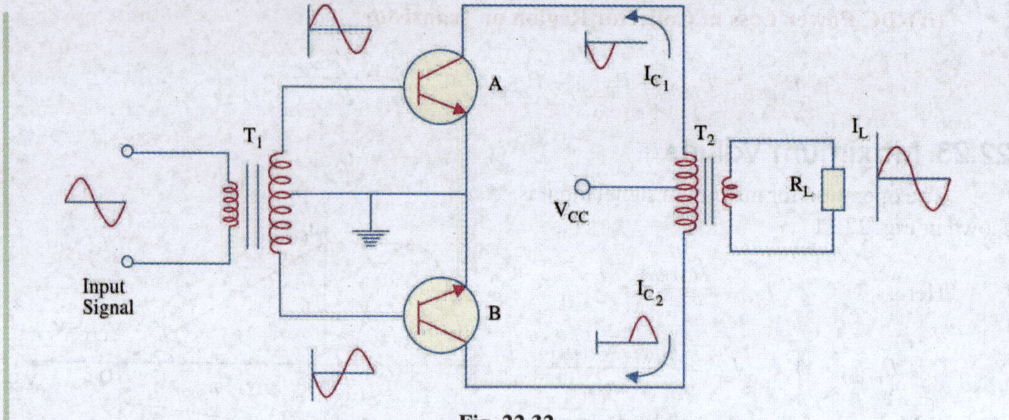

Fig. 22.32

Transistor *A* takes positive half-cycles of the signal whereas *B* handles negative half-cycles. When the two outputs are combined, an almost undistorted output waveform is produced as seen from Fig. 22.32.

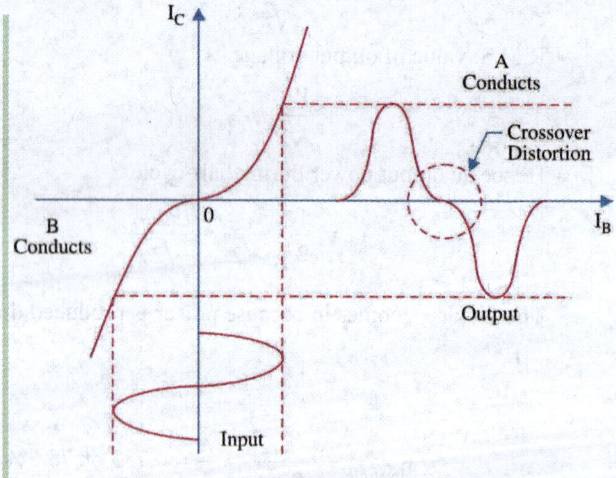

Fig. 22.33

In Fig. 22.33, the transfer characteristic of transistor *B* has been plotted upside down with respect to that of *A* in order to get the combined output.

Detailed operation is as under :

During the positive half-cycle of the signal, *A* is turned ON because its base is driven positive. It draws collector current $I_{C\text{-}1}$ in the upward direction from V_{CC}. Meanwhile, transistor *B* remains OFF because its base has negative voltage. Hence, $I_{C\text{-}2} = 0$. Obviously, one positive half-cycle of the output signal appears across secondary load R_L of T_2.

During negative half-cycle of input signal, *B* conducts whereas *A* remains OFF. Hence, $I_{C\text{-}2}$ is taken by *B* but $I_{C\text{-}1} = 0$. Now, negative output half-cycle is produced across R_L because $I_{C\text{-}2}$ is pulled down through secondary of T_2. It is obvious that in the absence of input signal, neither *A* nor *B* draws any collector current. Hence, *there is no drain on V_{CC} battery.*

Advantages

1. It has high efficiency, theoretical limit being 78.5%. It is primarily due to the fact that there is *no power drawn by the circuit under zero-signal condition.*
2. Since in push-pull arrangement, 180° phase difference exists between even-order harmonics produced by each transistor, they cancel out thereby giving an almost *distortion-free output*. This automatic cancellation of all even-order harmonics from the output current makes class-B push-pull amplifiers highly desirable for *communication sound equipment*.
3. This double-ended class-B amplifier provides practically four times the power supplied by a single-ended amplifier provided signal load resistance remains the same.
4. The dc components of the two collector currents through the two halves of the primary of T_2 flow in *opposite directions* so that net dc magnetisation of the transformer core is

almost nil. Hence, distortions due to the effect of dc magnetic saturation of the transformer core are eliminated. Moreover, smaller size core can be used due to very small net core flux.

Uses

Class-B push-pull amplifiers are extensively used for audio work in portable record players, as stereo amplifiers and in high-fidelity radio receivers.

22.25. Crossover Distortion

In class-B push-pull operation, there is severe distortion at *very low signal level* because

(*a*) bases of the transistors do not turn ON at 0 V but at 0.3 V for Ge and 0.7 V for Si.

(*b*) there is non-linearity in the low-signal area.

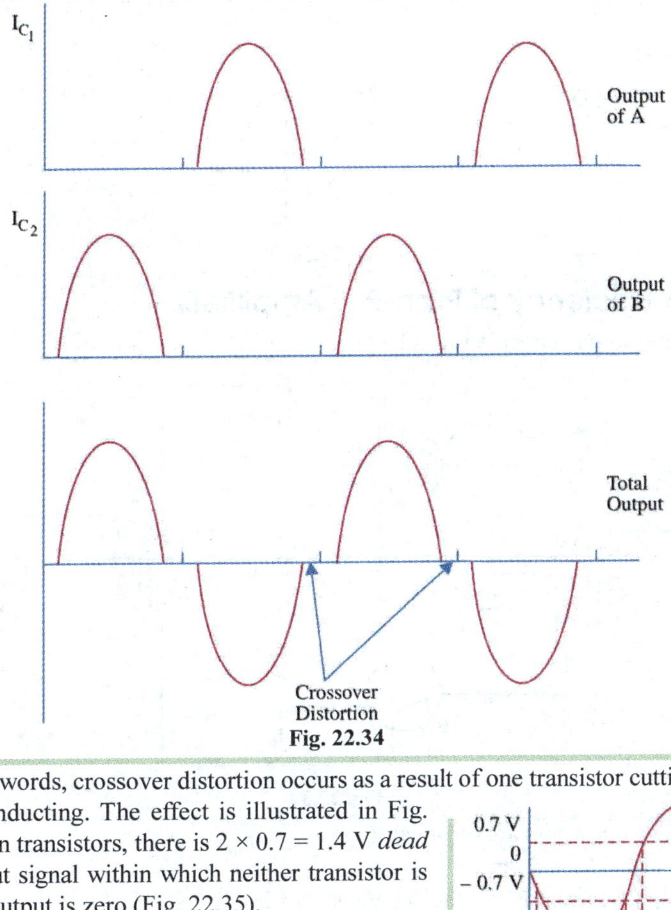

Fig. 22.34

In simple words, crossover distortion occurs as a result of one transistor cutting off before the other begins conducting. The effect is illustrated in Fig. 22.34. For silicon transistors, there is 2 × 0.7 = 1.4 V *dead zone* on the input signal within which neither transistor is turned ON and output is zero (Fig. 22.35).

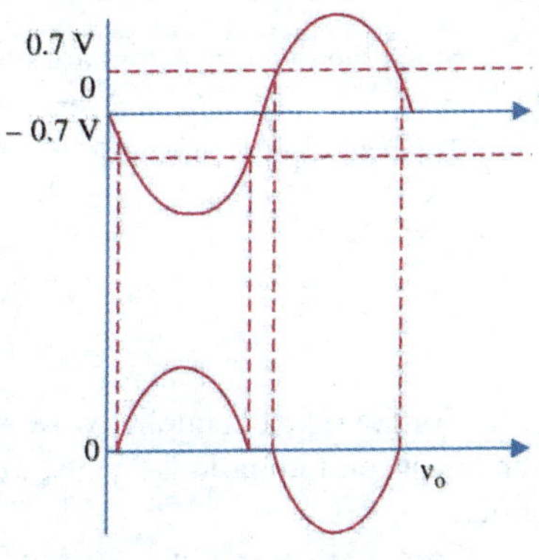

Fig. 22.35

The distortion so introduced is called *crossover distortion* because it occurs during the time operation *crosses over from one transistor to the other in the push-pull amplifier*. The same was shown earlier in Fig. 22.33 by using transfer characteristics of the two transistors.

Crossover distortion can be eliminated by applying slight forward bias to each emitter diode. It, in effect, means locating the *Q*-point of each transistor slightly above cut-off as shown in Fig. 22.36 so that each one operates for more than one half-cycle. Strictly speaking, it results in class-AB operation because each transistor may operate for about 200° (instead of 180°). However, for all practical purposes, it is still regarded as class-B push-pull operation.

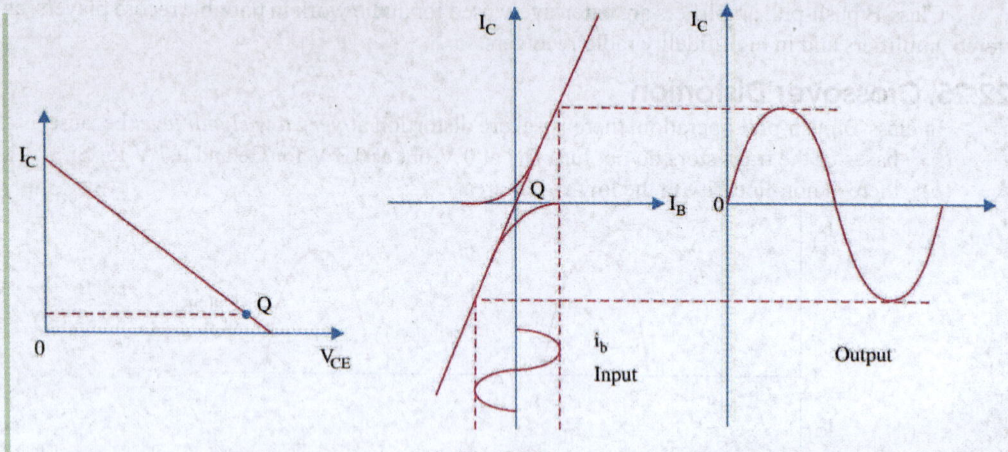

Fig. 22.36

22.26. Power Efficiency of Push-Pull Amplifiers

Consider the circuit shown in Fig. 22.37.

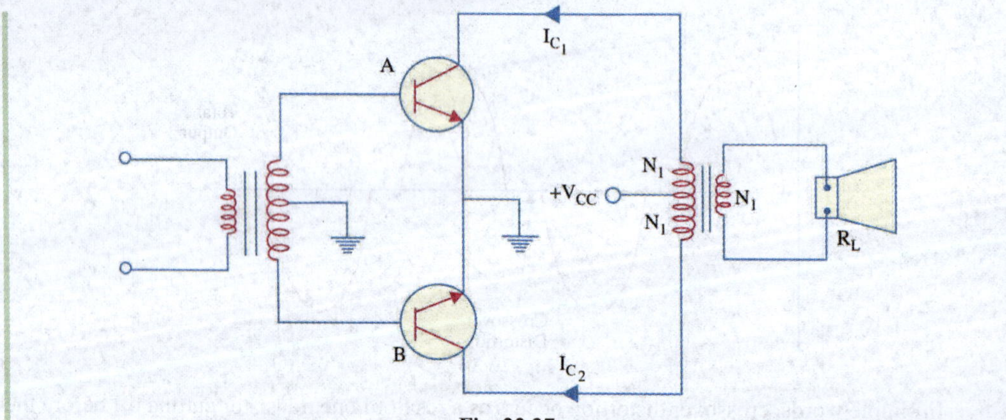

Fig. 22.37

Since the two transistors are identical, $I_{C1} = I_{C2} = I_C$

$\therefore \quad I_C = I_m / \pi$

Total dc supply current I_{dc} for the two transistors is

$$I_{dc} = 2I_C = 2\frac{I_m}{\pi}$$

$$\therefore \quad P_{in(dc)} = V_{CC} \times I_{dc} = 2\frac{I_m}{\pi} V_{CC}$$

For the sake of simplicity, we will assume transformer turn ratio as $(N_1 + N_1) : N_1$ so that for each transistor, turn ratio is $N_1 : N_1$ *i.e.*, 1 because each uses half the secondary of the output transformer.

If V_m is the peak voltage at either collector, then peak load voltage is also V_m. The peak current in the load is I_m. Then, ac power delivered to the load is

$$P_{o(dc)} = \frac{V_m}{\sqrt{2}} \cdot \frac{I_m}{\sqrt{2}} = \frac{V_m I_m}{2}$$

The total collector dissipation for the two transistors is

$$2.P_{c(dc)} = P_{in(dc)} - P_{o(dc)} = \frac{2I_m V_{CC}}{\pi} - \frac{V_m I_m}{2} = 2I_m \left(\frac{V_{CC}}{\pi} - \frac{V_m}{4} \right)$$

$$\eta_{overall} = \frac{P_{o(ac)}}{P_{in(dc)}} \times 100 = \frac{V_m I_m / 2}{2I_m / \pi} \times 100 = \frac{\pi}{4} \cdot \frac{V_m}{V_{CC}} \times 100$$

Under ideal condition of maximum power in the load, $V_m = V_{CC}$

$$\therefore \quad \text{maximum } \eta_{overall} = \frac{\pi}{4} \times 100 = 78.5\%$$

Same is the value of collector efficiency.

22.27. Complementary Symmetry Push-Pull Class-B Amplifier

The push-pull amplifier discussed in Art. 22.24 suffers from *two* disadvantages :

(*i*) it requires a bulky and expensive output transformer

(*ii*) it requires two out-of-phase input signals which necessitates an input centre-tapped transformer or phase inverter. It makes the driver circuitry quite complicated.

The complementary symmetry amplifier eliminates these two disadvantages while retaining the advantages of push-pull configuration.

As we know, a standard class-B amplifier requires two power transistors of the *same type* with closely-matched parameters. But the chief requirement of a complementary amplifier is a pair of closely-matched but *oppositely-doped* power transistors. The term 'complementary' arises from the fact that one transistor is *PNP* type and the other is of *NPN* type. They have *symmetry i.e.*, both are made with the same material and technology and have the same maximum rating.

An elementary complementary symmetry class-B push-pull amplifier is diagrammed in Fig. 22.38. The two transistors are complementary to each other and operate as emitter-follower amplifiers. The input is capacitively-coupled whereas output is direct-coupled.

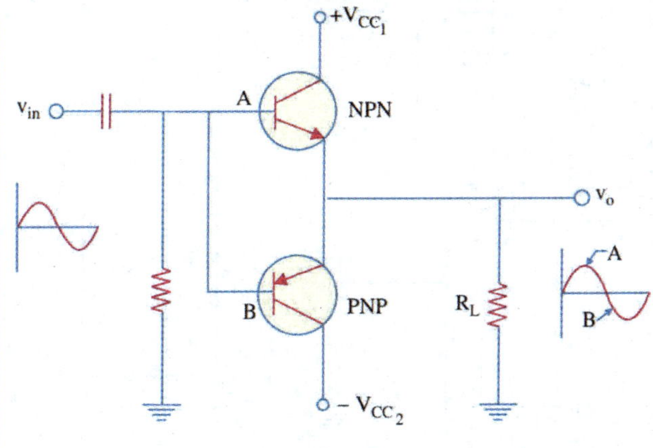

Fig. 22.38

With no input signal, neither transistor conducts and, therefore, current through R_L is zero.

When input signal is positive-going, transistor *A* is biased into conduction whereas *B* is driven into cut-off. When the signal is negative-going. *A* is turned OFF while *B* conducts. Obviously, this circuit is a push-pull amplifier because *turning one transistor ON turns the other OFF*.

The circuit possesses the essential characteristics of an emitter follower *i.e.*, unity voltage gain, no phase inversion and input impedance much higher than output impedance.

The circuit shown in Fig. 22.38 requires two dc supply batteries. Only one dc battery would be enough if '*totem-pole*' circuit configuration is used.

Due to the elimination of transformer, both the high and low-frequency responses of the circuit are extended apart from reduced cost and weight.

22.28. Class-C Amplifier

In such amplifiers, the active device *i.e.*, the transistor is biased much beyond cut-off. Hence,

1. output current flows only during *a part of the possible half-cycle* of the input signal,
2. there is no output current flow during *any part of the negative half-cycle* of the input signal,
3. output signal has hardly any resemblance with the input signal. It consists of *short pulses only*,
4. class-C amplifiers have high circuit efficiency of about 85 to 90%.

Because of high distortion, class-C amplifiers are not used for audio-frequency work. They are used for high-power output at radio frequencies (*i.e.*, RF amplifiers) where harmonic distortion can be removed by simple circuits. In reality, they are used as high-frequency power switchers in radio transmitters rather than as amplifiers.

22.29. Tuned Amplifier

The gain of a transistor amplifier depends directly on the value of its load impedance. Such a high impedance can be obtained by using a high-Q tuned or resonant LC circuit as load (Fig. 22.39). The frequency response curve of the amplifier assumes the same shape as the resonance curve of the tuned circuit. Obviously, only a narrow band of frequencies around the resonant frequency f_o would be amplified well whereas other frequencies would be discriminated against.

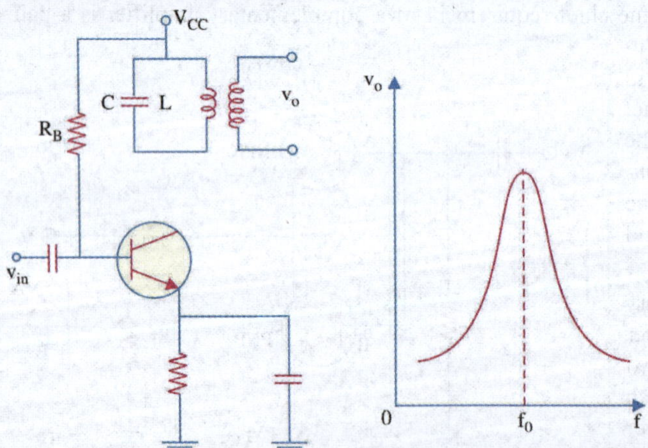

Tuned amplifier.

Fig. 22.39

Non-linear distortion is eliminated because of high selectivity of the load impedance. Hence, output is *nearly sinusoidal*. With the removal of distortion, high amplifier efficiency can be achieved by operating the transistor in its non-linear region.

22.30. Distortion in Amplifiers

Amplifiers are supposed to produce an output which does not differ from the input in any respect except amplitude *i.e.*, the output is expected *to be larger than the input*. In actual practice, it is impossible to construct such an ideal amplifier whose output is an exact duplication or replica of the input. The output is always found to differ from the input either in its waveform or frequency content. This difference between the output and input of an amplifier is called *distortion*.

The amplifier distortions may be divided into two broad categories depending on the region of the characteristic used by the transistor and the associated circuit and device reactances.

(*a*) non-linear distortion

This occurs when transistor operates in the non-linear region of its characteristic.

(*i*) when we visualize the signal in time domain, it is called *amplitude distortion or waveform distortion;*

(*ii*) when we think of the signal in the frequency domain, it is called *harmonic distortion;* This distortion occurs when input signal is of one *frequency*, say, a pure sine wave.

(*iii*) *intermodulation (IM) distortion*—when input signal has more than one frequency (like speech).

Non-linear distortion occurs in the case of *large-signal inputs* when the active device is driven into the non-linear regions of its characteristic.

(*b*) linear distortion

It occurs even when the active device is *working on linear part of its characteristic with small-signal inputs*. It is primarily due to frequency-dependent reactances associated with the circuit or active device itself and occurs when input signal is *composite i.e.*, has signals of different frequencies (say, fundamental and its harmonics). However, output contains no frequencies other than those at the input. It may be further subdivided into :

(*i*) *frequency distortion*—due to unequal amplification of different frequencies present in the input signal;

(*ii*) *phase or delay distortion*—due to unequal phase shift of various signal components.

22.31. Non-Linear Distortion

Fig. 22.40 gives time-domain view of this distortion when it is called *amplitude distortion*.

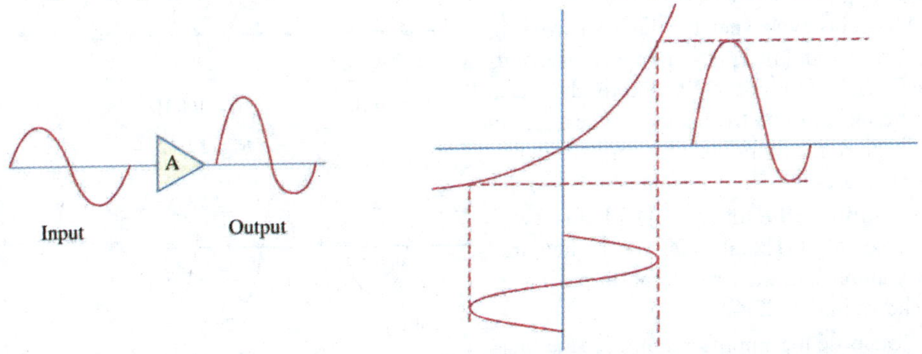

Fig. 22.40

As seen, the positive half-cycle of the input signal has been amplified more than its negative half-cycle. Consequently, waveshape of the output signal differs from that of the input signal. As shown below, it is due to the appearance of new frequencies (called harmonics) at the output which are not present in the input.

The frequency-domain view of non-linear distortion is shown in Fig. 22.41. The input is a single-frequency (f_1) signal but output signal contains dc component and different harmonics *i.e.*, frequencies which are harmonically related to each other. These harmonics are an integral multiple of the input signal frequency. Consequently, the output is distorted—the magnitude of distortion depending on the strength and number of these harmonics.

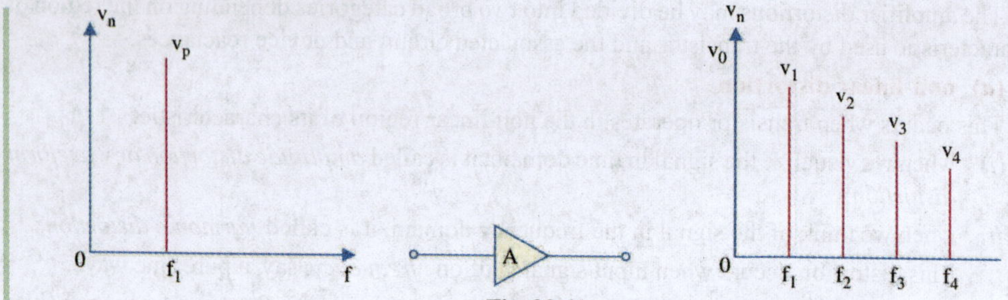

Fig. 22.41

In audio amplifiers used for amplification of speech or music, the less harmonic distortion, the better. For speech, harmonic distortion should not exceed 10% otherwise intelligibility will suffer. High-fidelity amplifiers have harmonic and *IM* distortion of less than 1 per cent.

22.32. Intermodulation Distortion

It occurs when input signal (like speech) consists of *more than one frequency*. It is also a type of non-linear distortion which generates frequency components not harmoni

cally related to the signal frequencies. Suppose, an input signal contains two frequencies f_1 and f_2. The output signal will contain their harmonics *i.e.*, f_1, $2f_1$, $3f_1$ etc., and f_2, $2f_2$, $3f_2$ etc. In addition, there would be components $(f_1 + f_2)$ and $(f_1 - f_2)$ and also the sum and difference of the harmonics. These sum and difference frequencies are called intermodulation (*IM*) frequencies and are quite undesirable in amplifiers because they subtract from original intelligence. Since *IM* frequencies *are not harmonically related to the signal*, they are easily detected by human ear as noise. Hence, great care is taken to minimize them particularly in hi-fi audio amplifiers.

22.33. Frequency Distortion

It occurs even when the device is *working with small-signal inputs over linear region of its characteristic*. It is basically due to change in the amplifier gain with frequency *i.e.*, the different input signal frequencies are amplified by different amounts.

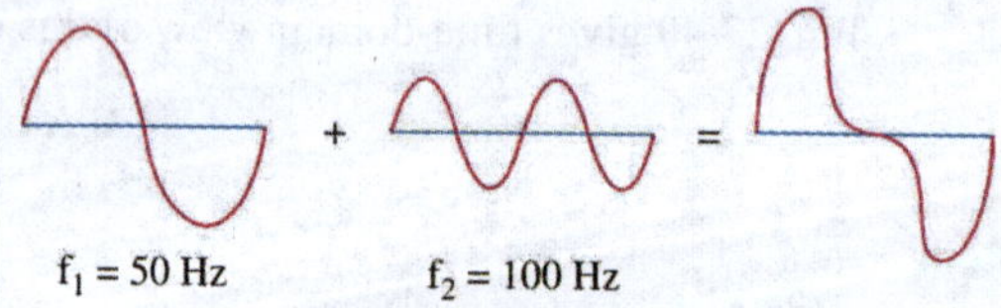

Fig. 22.42

Suppose the input signal has two frequencies of 50 Hz and 100 Hz. If they are equally amplified, their resultant output wave is as shown in Fig. 22.42.

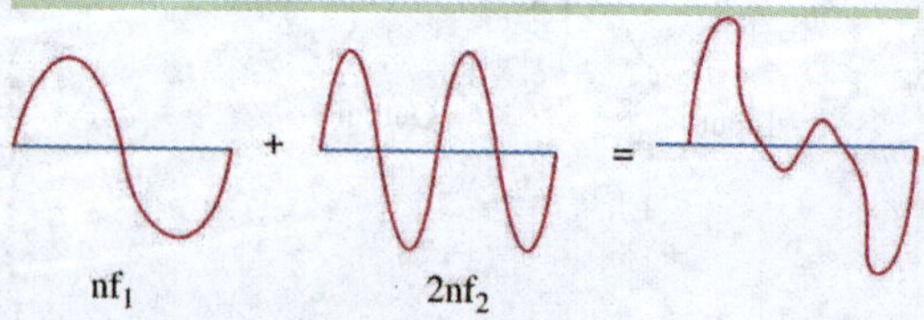

Fig. 22.43

Suppose the amplifier amplifies the first frequency n times and the second one $2n$ times. Then their frequency-distorted output would be as shown in Fig. 22.43.

This distortion is due to the various frequency-dependent reactances (both capacitive and inductive) associated either with the circuit or the active device itself. In the case of audio signals, the frequency distortion leads to a change in the quality of sound. Hence, in the design of untuned or wide-band amplifiers, special steps are taken to reduce variation of gain with signal frequency.

22.34. Phase or Delay Distortion

Phase distortion is said to take place when phase angles between the component waves of the output are not the same as the corresponding angles of the input. These changes in phase angles are also due to frequency-dependent capacitive and inductive reactances associated with the circuit and the active device of the amplifier.

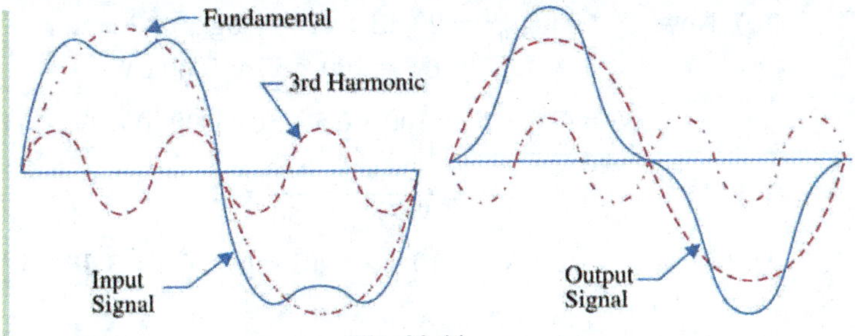

Fig. 22.44

As seen from Fig. 22.44, there has been a phase shift in the third harmonic. Hence, the resultant wave so obtained is entirely different from the input wave.

This type of distortion which is due to the non-uniform phase shift of different frequency components, is difficult to eliminate. It should be noted that if all the frequency components in a signal are shifted in phase by an integral multiple of 180°, the resultant output waveform is not changed, although the polarity of the wave may be altered but *changes in polarity do not constitute distortion.*

Fortunately, the human ear is unable to distinguish phase differences (though eye can) and is thus not sensitive to this distortion. Consequently, phase distortion is of no practical significance in audio amplifiers. But in amplifiers used in television sets (video amplifiers) and other systems where ear is not the final receiver, elimination of phase distortion is important.

22.35. Noise

In general, it may be defined as *any kind of unwanted signal not derived from or related to the input signal*. Just as distortion is the limiting factor in the amplification of large signals, noise is the limiting factor in the case of small signals.

The signal-to-noise (*S/N*) ratio should be high for good signal intelligibility. Since all amplifiers contribute some noise of their own to the signal being processed, their output *S/N* is bound to be less than their input *S/N*. High quality amplifiers are designed to have an output *S/N* as close to input *S/N* as possible. The amplifier performance is measured in terms of noise factor given by

Noise factor, $$F = \frac{\text{input } S/N}{\text{output } S/N} = \frac{S_i/N_i}{S_o/N_o};$$ Obviously, *F is greater than unity.*

When expressed in decibels, the noise factor is called noise figure (*NF*).

$$\therefore \quad NF = 10 \log_{10}{}^{F} \text{ dB}$$

If an amplifier could be built which generated no noise of its own, then

(*i*) $S_i/N_i = S_o/N_o$ (*ii*) $F = 1$ (*iii*) $NF = 0$ dB

If *G* is the power gain of the amplifier, then output and input signal powers are related by

$$G = \frac{S_o}{S_i} \quad \text{i.e.,} \quad S_o = G.S_i$$

The output noise N_o is given by

$$N_o = GN_i + N_A$$

where N_i = input noise power

N_A = noise power generated by amplifier itself.

Example 22.15. *The signal input to a small-signal amplifier consists of 50 μW of signal power and 0.5 μW of noise power. The amplifier generates an internal noise power of 50 μW and has a gain of 20 dB. For this amplifier, compute*

(a) input S/N, (b) output S/N, (c) noise factor, (d) noise figure.

Solution. (*a*) $$\frac{S_i}{N_i} = \frac{50}{0.5} = \mathbf{100}$$

(*b*) Now, $10 \log_{10}{}^{G} = 20$ dB or $\log_{10}{}^{G} = 2$ or $G = 10^2 = 100$

$$S_o = GS_i = 100 \times 50 = 5000\ \mu W$$

$$N_o = GN_i + N_A = 100 \times 0.5 + 50 = 100\ \mu W \qquad \therefore \frac{S_o}{N_o} = \frac{5000}{100} = \mathbf{50}$$

(*c*)
$$F = \frac{S_i / N_i}{S_o / N_o} = \frac{100}{50} = \mathbf{2}$$

(*d*)
$$NF = 10 \log_{10}{}^{2}\ \text{dB} = 10 \times 0.3 = \mathbf{3\ dB}$$

SELF EXAMINATION QUESTIONS

A. Fill in the blanks by most appropriate word (s) or numerical value (s).

1. In a *CB* amplifier, a positive-going input signal produces a ....going output signal.
2. There is phase .... of input signal in a *CE* amplifier.
3. .... amplifier has the highest output impedance whereas *CC* amplifier has the highest .... gain.
4. The *CC* amplifier has .... input impedance but .... output impedance.
5. The conduction angle of class-B amplifiers is .... degree.
6. The conduction angle of class-AB amplifiers lies between .... and .... degrees.
7. For class-A operation, *Q*-point is located at the .... of the load line.
8. In class-A amplifiers, worst-case condition occurs when input signal is ....
9. In a transformer-coupled class-A amplifier, the maximum overall efficiency is .... per cent.
10. Push-pull operation requires two transistors of the same type with.......... characteristics.
11. The theoretical limit for the efficiency of a class-B amplifier is .... per cent.
12. A complementary symmetry amplifier requires a pair of closely-matched but .... doped power transistors.
13. Non-linear distortion is called .... distortion when viewed in the frequency domain.
14. Frequency and phase distortions are due to the .... of the circuit elements.

B. Answer True or False

1. A *CC* amplifier has highest current gain but lowest power gain.
2. A class-B amplifier has the highest output impedance and lowest input impedance.
3. A class-B amplifier conducts only when there is input signal.
4. Lesser the no-signal conduction of an amplifier, higher its efficiency.
5. Class-A amplifiers operate with centred *Q*-point and large input signals.
6. Amongst the three amplifier configurations, class-C has the highest efficiency.
7. In amplifiers, high efficiency and high distortion go together.
8. Class-A and class-B operations have opposite worst-case conditions.
9. Small signal produces no distortion in amplifiers.
10. Class-C operation is most efficient because it hardly operates during the input cycles.
11. In push-pull arrangement, when one transistor is ON, the other is OFF and vice versa.
12. Amplifiers can be made distortion-free by just operating them over linear region of their characteristic.
13. Intermodulation distortion occurs even when input signal is of single frequency.
14. Phase distortion is objectionable in video amplifiers but not in audio amplifiers.
15. Frequency and phase distortions can occur even with single frequency inputs.
16. A noise-free amplifier has a noise factor of zero.
17. The noise figure of an ideal amplifier is 0 dB.
18. Output *S* / *N* of an actual amplifier is always less than its input *S* / *N*.

C. Multiple Choice Items

1. *CE* amplifier is characterised by
   (*a*) low voltage gain
   (*b*) moderate power gain
   (*c*) signal phase reversal
   (*d*) very high output impedance
2. A *CC* amplifier has the highest
   (*a*) voltage gain (*b*) current gain

(*c*) power gain
(*d*) output impedance

3. In a class-A amplifier, conduction extends over 360° because *Q*-point is
(*a*) located on load line
(*b*) located near saturation point
(*c*) centred on load line
(*d*) located at or near cut-off point

4. In a class-A amplifier, worst-case condition occurs with
(*a*) zero signal input
(*b*) maximum signal input
(*c*) high load resistance
(*d*) transformer coupling

5. The output of a class-B amplifier
(*a*) is distortion-free
(*b*) consists of positive half-cycles only
(*c*) is like the output of a full-wave rectifier
(*d*) comprises short-duration current pulses

6. The maximum overall efficiency of a transformer-coupled class-A amplifier is ....per cent.
(*a*) 78.5 (*b*) 25
(*c*) 50 (*d*) 85

7. A transistor audio amplifier is found to have an overall efficiency of 70 per cent. Most probably, it is a .... amplifier.
(*a*) class-B push-pull
(*b*) single-stage class-C
(*c*) transformer-coupled class-A
(*d*) direct-coupled class-A

8. A class-B push-pull amplifier has the main advantage of being free from
(*a*) any circuit imbalances
(*b*) unwanted noise
(*c*) even-order harmonic distortion
(*d*) dc magnetic saturation effects

9. Crossover distortion occurs in .... amplifiers.
(*a*) push-pull (*b*) class-A
(*c*) class-B (*d*) class AB

11. The maximum overall efficiency of a class-B push-pull amplifier cannot exceed .... per cent.
(*a*) 100 (*b*) 78.5
(*c*) 50 (*d*) 85

12. The main use of a class-C amplifier is
(*a*) as an *RF* amplifier
(*b*) as stereo amplifier
(*c*) in communication sound equipment
(*d*) as distortion generator

13. The primary cause of linear distortion in amplifiers is
(*a*) change of gain with frequency
(*b*) unequal phase shift in component frequencies
(*c*) reactances associated with the circuit and active amplifying element
(*d*) inherent limitations of the active device

14. An amplifier is said to suffer from distortion when its output is
(*a*) low
(*b*) different from its input
(*c*) noisy
(*d*) larger than its input

15. An ideal amplifier has
(*a*) noise figure of less than 1 dB
(*b*) noise factor of unity
(*c*) output *S*/*N* more than input *S*/*N*
(*d*) noise figure of more than 0 dB

ANSWERS

A. Fill in the blanks

1. positive **2.** reversal **3.** CB, current **4.** high, low **5.** 180 **6.** 180, 360 **7.** centre
8. zero **9.** 50 **10.** matching **11.** 78. 5 **12.** oppositely **13.** harmonics **14.**reactances

B. True or False

1. T **2.** T **3.** T **4.** T **5.** F **6.** T **7.** T **8.** T **9.** F
10. T **11.** T **12.** F **13.** F **14.** T **15.** F **16.** F **17.** F **18.** T.

C. Multiple Choice Items

1. *c* **2.** *b* **3.** *c* **4.** *a* **5.** *b* **6.** *c* **7.** *a* **8.** *c* **9.** *a* **10.** *a* **11.** *a* **12.** *c* **13.** *b* **14.** *b* **15.** *b*.

CHAPTER 23

Multistage Amplifiers

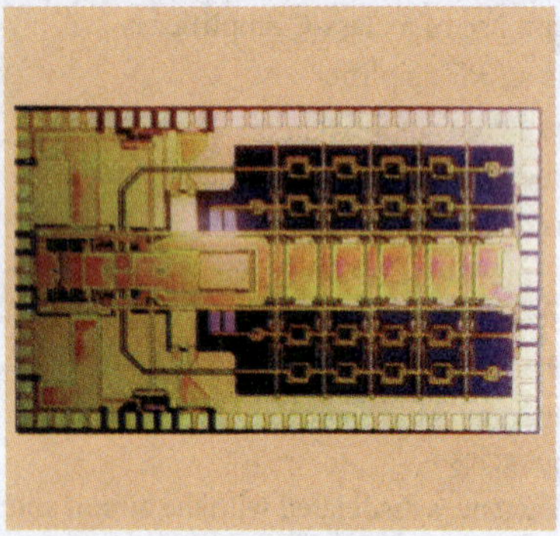

1. Amplifier Coupling
2. RC-coupled Two-stage Amplifier
3. Impedance-coupled Two-stage Amplifier
4. Advantages of Impedance Coupling
5. Transformer-coupled Two Stage Amplifier
6. Advantages of Transformer Coupling
7. Frequency Response
8. Direct-coupled Two-stage Amplifier Using Similar Transistors
9. Darlington Pair
10. Advantages of Darlington Pair
11. Comparison between Darlington Pair and Emitter Follower
12. Common Mode Input
13. Differential Amplifier

23.1. General

Often, the voltage amplification or power gain or frequency response obtained with a single stage of amplification is insufficient to meet the requirements of either a composite electronic circuit or a load device. Hence, two or more single stages of amplification are frequently used to achieve greater voltage or current amplification or both. In such cases, the output of one stage serves as input of the next stage as shown in Fig. 23.1. Such amplifiers may be divided into following two categories :

(*a*) Cascaded Amplifiers

In these amplifiers, each stage as well as the type of interstage coupling used are identical.

(*b*) Compound Amplifiers

In these amplifiers, each stage may be different from the other (one may be *CE* and the other may be *CC* stage) and also different types of interstage couplings may be employed.

As stated above, in cascaded amplifiers, the output ac voltage of the first stage becomes the input voltage of the second stage and the ac output of the second stage becomes the input of the third stage and so on. The overall voltage gain of the cascaded amplifier is equal to the *product* (not the sum) of the gain of the individual stages.

$$\therefore \quad A_v = A_{v1} \times A_{v2} \times A_{v3} \times \text{..............}$$

However, when the voltage gain is expressed in decibels (dB), then the overall decibel gain of the multistage amplifier is equal to the *sum* of the dB gains of the individual stage *i.e.*,

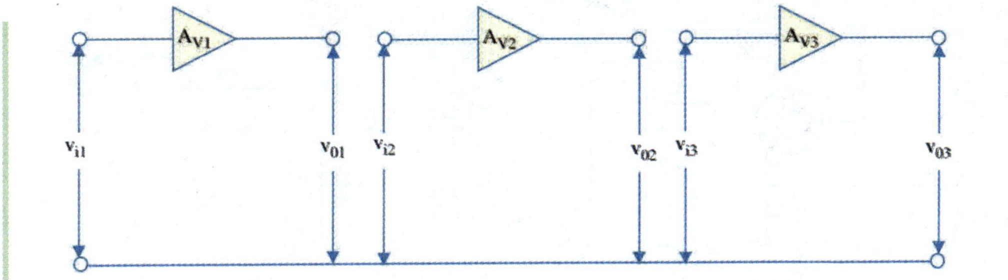

Fig. 23.1

$$G = G_1 + G_2 + G_3 + \text{..............}$$

Similarly, the overall current amplification is given by

$$A_i = A_{i1} \times A_{i2} \times A_{i3} \times \text{............}$$

The overall power gain is given by

$$A_p = A_v \,.\, A_i \qquad \text{and} \qquad G_p = 10 \log_{10}{}^{A_P} \text{ dB}$$

Suppose in a two-stage cascaded amplifier, first stage has a voltage amplification of 2000 (dB gain of the 20 $\log_{10}{}^{2000} = 66$ dB) and second stage has corresponding values of 1000 (60 dB). If the ac output of first stage is fed into the second stage, the overall amplification would theoretically become $= 1000 \times 2000 = 2 \times 10^6$ which corresponds to a dB gain of $(60 + 66) = 126$ *i.e.*, 20 $\log_{10}{}^{2 \times 106} = 20 \times 6.3 = 126$. The above result would be true only when we *neglect the loading effect of first stage by the second stage.* It would be approximately true so long as the impedance looking into the input of second stage is much greater than the output impedance of the first stage. Otherwise, the overall gain would be much less.

23.2. Amplifier Coupling

All amplifiers need some *coupling network.* Even a single-stage amplifier has to be coupled to the input and output devices. In the case of multistage systems, there is *interstage* coupling. The type of coupling used determines the characteristics of the cascaded amplifier. In fact, amplifiers are classified according to the coupling network used.

The four basic methods of coupling are :

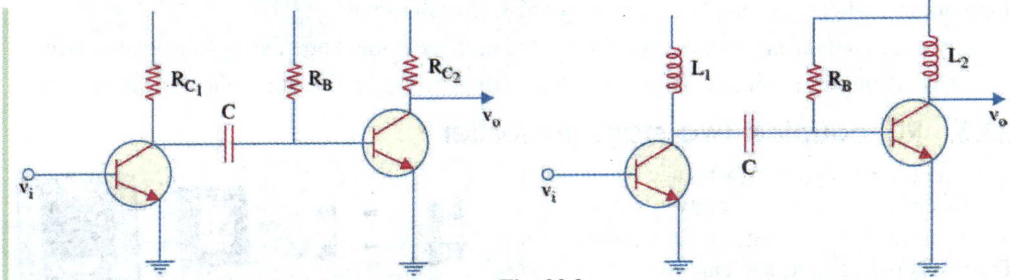

(*a*) Fig. 23.2 (*b*)

1. Resistance-Capacitance (RC) Coupling

It is also known as *capacitive* coupling and is shown in Fig. 23.2 (*a*). Amplifiers using this coupling are known as *RC*-coupled amplifiers. Here, *RC* coupling network consists of two resistors R_{C1} and R_{C2} and one capacitor *C*. The connecting link between the two stages is *C*. The function of *RC*-coupling network is two-fold :

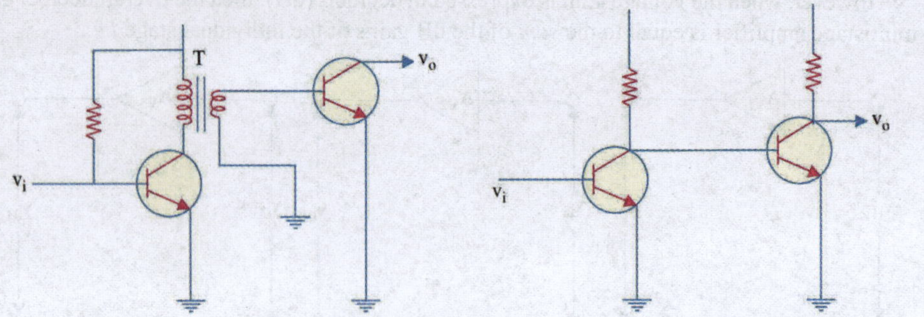

(c) (d)

Fig. 23.2

(a) to pass ac signal from one stage to the next,

(b) to block the passage of dc voltages from one stage to the next.

2. Impedance Coupling or Inductive Coupling

It is also known as choke-capacitance coupling and is shown in Fig. 23.2 (b). Amplifiers using this coupling are known as **impedance-coupled** amplifiers. Here, the coupling network consists of L_1, C and R_B. The impedance of the coupling coil depends on (i) its inductance and (ii) signal frequency.

3. Transformer Coupling

It is shown in Fig. 23.2 (c). Since secondary of the coupling transformer conveys the ac component of the signal directly to the base of the second stage, there is *no need for a coupling capacitor*. Moreover, the secondary winding also provides a base return path, hence there is *no need for a base resistance.* Amplifiers using this coupling are called *transformer-coupled amplifiers.*

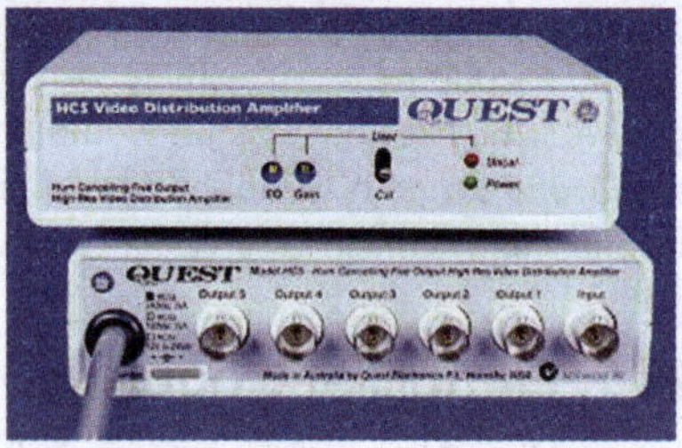

Modern coupling transformer.

4. Direct Coupling

It is shown in Fig. 23.2 (d). This coupling is used where it is desirable to connect the load directly in series with the output terminal of the active circuit element. The examples of such load devices are (i) headphones (ii) loud-speakers (iii) dc meters (iv) relays and (v) input circuit of a transistor etc. Of course, direct coupling is permissible only when

(i) dc component of the output does not disturb the normal operation of the load device,

(ii) device resistance is so low that it does not appreciably reduce the voltage at the electrodes.

23.3. RC-coupled Two-stage Amplifier

Fig. 23.3 shows a two-stage *RC*-coupled amplifier which consists of two single-stage transistor amplifiers using the *CE* configuration. The resistors R_2 and R_3 and capacitor C_2 form the coupling network. R_2 is collector load of Q_1 and R_4 is that of Q_2. Capacitor C_1 couples the input signal whereas C_3 couples out the output signal. R_1 and R_3 provide dc base bias. R_5 is the load resistor across Q_2.

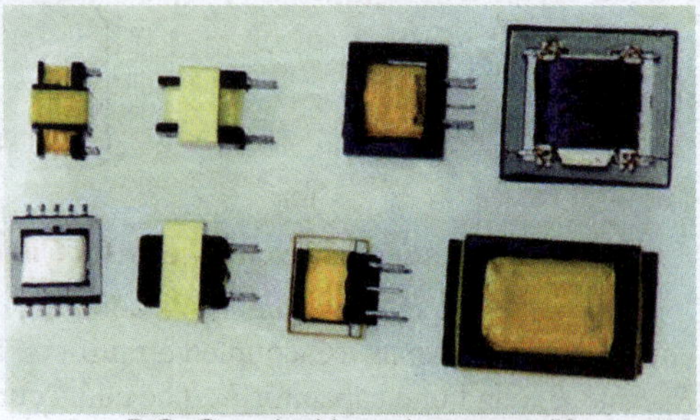

R.C. Coupled two-stage amplifier.

(i) Circuit Operation

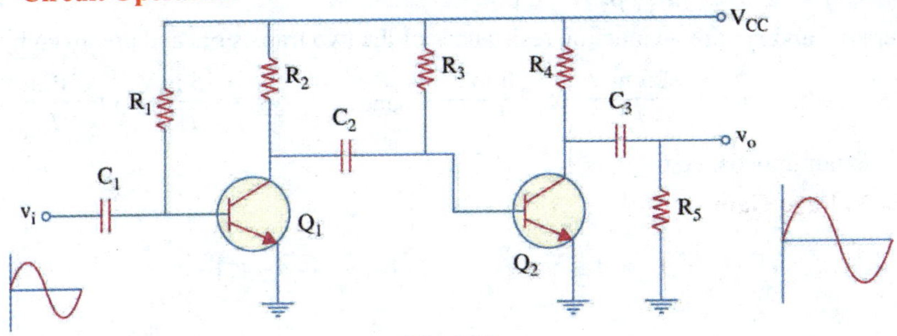

Fig. 23.3

The brief circuit operation is as under :

1. The input signal v_i is amplified by Q_1. Its phase *reversed* (usual with *CE* connection);
2. The amplified output of Q_1 appears across R_2;
3. The output of the first stage across R_2 is coupled to the input of the second stage at R_3 by coupling capacitor C_2. This capacitor is also sometimes referred to as ***blocking capacitor*** because it blocks the passage of dc voltages and currents;
4. The signal at the base of Q_2 is further amplified by Q_2 and *its phase is further reversed;*
5. The ac output of Q_2 appears across R_4;
6. The output across R_4 is coupled by C_3 to load resistor R_5;
7. The output signal v_o is the *twice-amplified replica of the input signal* v_i. It is in phase with v_i because it has been reversed twice.

(ii) AC Equivalent Circuit

The ac equivalent circuits for the two stages have been shown separately in Figs. 23.4 and 23.5 respectively.

In Fig. 23.4, $r_{i \cdot 1} = R_1 \parallel \beta_1 . r_{e \cdot 1}$

It is the input impedance of the first stage and not $r_{in(base)}$.

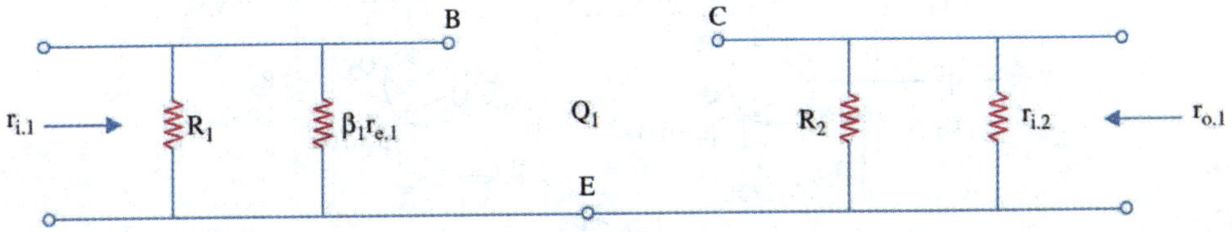

Fig. 23.4

The output impedance of the first stage is

$$r_{0 \cdot 1} = R_2 \parallel r_{i.2}$$

It is so because the input of the second stage forms a part of the output of the first stage. As seen from Fig. 23.5,

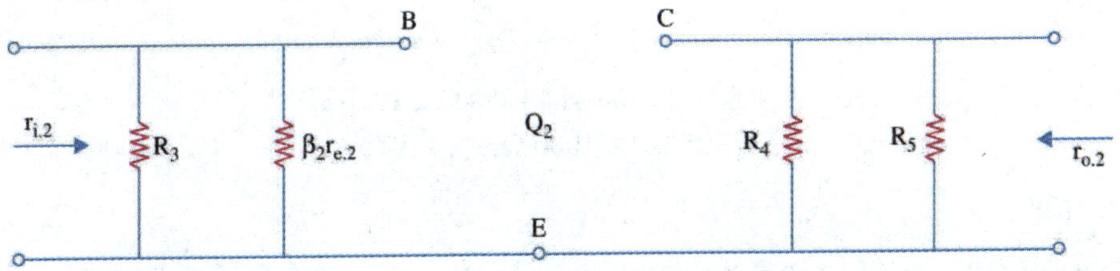

Fig. 23.5

$$r_{i.2} = R_3 \parallel \beta_{2.7}\, r_{e2} \cong \beta_2\, r_{e.2}$$

where $r_{e.1}$ and $r_{e.2}$ are ac junction resistances of the two transistors and are given by

$$r_{e.1} = \frac{25\text{ m V}}{I_{E.1}} \text{ or } \frac{50\text{ mV}}{I_{E.1}} \quad \text{and} \quad r_{e.2} = \frac{25\text{ m V}}{I_{E.2}} \text{ or } \frac{50\text{ m V}}{I_{E.2}}$$

The output impedance of Q_2 is $r_{0.2} = R_4 \parallel R_5$.

(iii) Voltage Gain

$$A_v = A_{v1} \times A_{v2} \qquad \text{Now} \quad A_{v2} = \beta_2 \cdot \frac{r_{0.2}}{r_{i.2}}$$

Also, $\quad r_{i \cdot 2} = \beta_2\, r_{e.2} \qquad \therefore \quad A_{v2} = \beta_2 \dfrac{r_{0.2}}{\beta_2 \cdot r_{e.2}} = \dfrac{r_{0.2}}{r_{e \cdot 2}}$

The voltage gain of the first stage is also given by a similar equation

$$A_{vi} = \frac{r_{0 \cdot 1}}{r_{e.1}}$$

Example 23.1. *For the two-stage RC-coupled low-level audio amplifier shown in Fig. 23.6, compute the following :*

(i) r_i *(ii)* A_{v1} *(iii)* A_{v2} and *(iv)* A_v *in dB. Neglect* V_{BE} *and take* $r_e = 25\ mV/I_E$.

Solution. The input impedance of the cascaded amplifier is

(i) $\quad r_i = R_1 \parallel \beta_1 \cdot r_{e.1}$

For finding $r_{e.1}$, we need $I_{E.1}$ which approximately equals

$$I_{C.1} = \beta \cdot I_{B.1}$$

$$I_{B1} = 12/R_1 = 12/0.6\text{ M} = 20\ \mu\text{A}$$

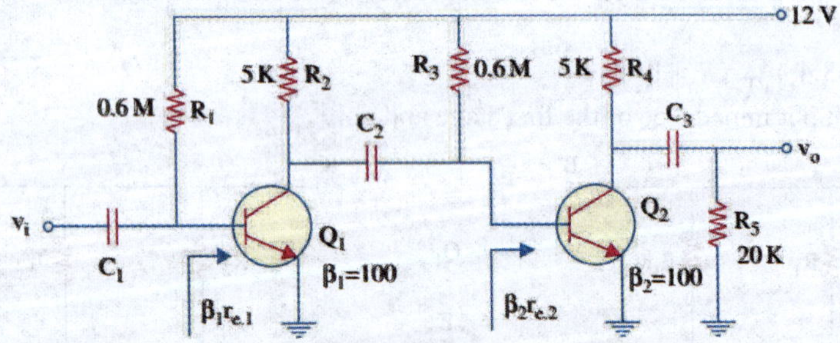

Fig. 23.6

$$I_{C1} = 100 \times 20 = 2000\ \mu\text{A} = 2\text{ mA}$$

$\therefore \quad I_{E.1} = 2\text{ mA} \qquad \therefore \quad r_{e.1} = 25/2 = 12.5\ \Omega$

$$\beta_1\, r_{e.1} = 100 \times 12.5 = 1250\ \Omega$$

$\therefore \quad r_i = R_1 \parallel \beta_1 . r_{e.1} = 0.6\text{ M} \parallel 1250\ \Omega \cong \mathbf{1250\ \Omega}$

(ii) $\quad A_{v1} = \dfrac{r_{0 \cdot 1}}{r_{e.1}}. \qquad$ Now, $r_{0 \cdot 1} = R_2 \parallel r_{i.2}$

$$r_{i.2} = R_3 \parallel \beta_2\, r_{e.2} = 0.6\text{ M} \parallel 1250\ \Omega \cong 1250\ \Omega$$

$r_{0 \cdot 1} = 5\text{ K} \parallel 1250\ \Omega = 1000\ \Omega;\ r_{e.1} = 12.5\ \Omega \qquad \therefore \quad A_{v.1} = 1000/12.5 = \mathbf{80}$

(iii) $\quad A_{v2} = \dfrac{r_{0 \cdot 2}}{r_{e.2}}$

Now, $\quad r_{0 \cdot 2} = R_4 \parallel R_5 = 5\text{ K} \parallel 20\text{ K} = \mathbf{4\ K}, \qquad I_{E.2} = 2\text{ mA} \qquad$ — same as $I_{E.1}$

$r_{e.2} = 25/2 = 12.5\ \Omega \quad \therefore \quad A_{v2} = 4000/12.5 = \mathbf{320}$

(*iv*) $A_v = A_{v.1} \times A_{v.2} = 80 \times 320 = \mathbf{25,600}$

(*v*) $G_v = 20 \log_{10} A_v \text{ dB} = 20 \log_{10} 25,600 = \mathbf{88 \text{ dB}}.$

Example 23.2. *For the two-stage RC-coupled amplifier shown in Fig. 23.7 compute the following :*

(i) r_i, *(ii)* $A_{v.1}$, *(iii)* $A_{v.2}$, *(iv)* A_v *in decibels.*

Take $\beta_1 = \beta_2 = 100$*. Neglect* V_{BE} *and use* $r_e = 25 \text{ mV}/I_E$.

Solution. (*i*) The input impedance of the stage is $r_i = R_1 \parallel \beta_1.r_{e.1}$.

It should be noted that R_6 does not come into the picture because it has been ac grounded by C_4. However, it would affect the dc emitter current.

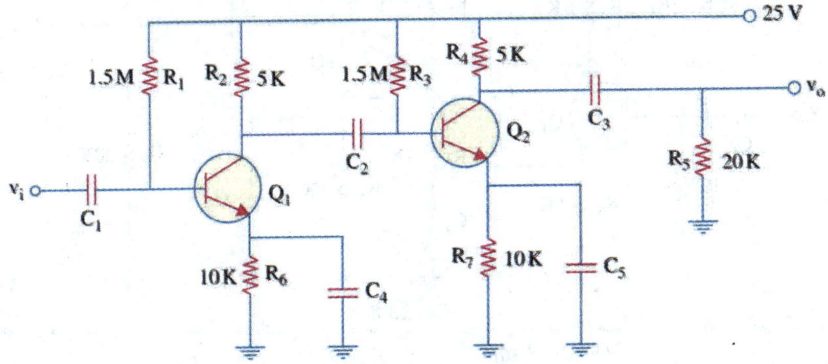

Fig. 23.7

$$I_{E1} = \frac{V_{CC}}{R_6 + R_1/\beta_1} = \frac{25}{10,000 + 1.5 \times 10^6/100} = 1 \text{ mA}$$

$$r_{e.1} = 25/1 = 25\ \Omega\ ; \beta_1.r_{e.1} = 100 \times 25 = 2500\ \Omega$$

$$r_i = 1.5 \text{ M} \parallel 2500\ \Omega \cong \mathbf{2500\ \Omega}$$

(*ii*) $$A_{v1} = \frac{r_{0.1}}{r_{e.1}}$$

Now, $$r_{0.1} = R_2 \parallel r_{i.2} \text{ and } r_{i.2} = R_3 \parallel \beta_2 r_{e.2}$$

Now, $$I_{E2} = \frac{25}{10,000 + 1.5 \times 10^6/100} = 1 \text{ mA}$$

∴ $$r_{e.2} = 25/1 = 25\ \Omega;\ \beta_2.r_{e.2} = 2500\ \Omega$$

$$r_{i.2} = 1.5 \text{ M} \parallel 2500\ \Omega \cong 2500\ \Omega$$

$$r_{o.1} = 5 \text{ K} \parallel 2.5 \text{ K} = 1,667\ \Omega$$

∴ $$A_{v1} = 1667/25 = \mathbf{66.7}$$

(*iii*) $$A_{v.2} = \frac{r_{0.2}}{r_{e.2}},$$

Now, $$r_{0.2} = R_4 \parallel R_5 = 5 \text{ K} \parallel 20 \text{ K} = 4 \text{ K}$$

∴ $$A_{v.2} = 4000/25 = \mathbf{160}$$

(*iv*) $$A_v = A_{v.1} \times A_{v.2} = 66.7 \times 160 = \mathbf{10,672}$$

$$G_v = 20 \log_{10} 10,672 = \mathbf{80.3 \text{ dB}}$$

Example 23.3. *Compute the overall voltage amplification for the two-stage RC-coupled amplifier shown in Fig. 23.8. Express the answer in decibels. Neglect* V_{BE} *and use* $r_e = 50 \text{ mV}/I_E$. *Take* $\beta_1 = \beta_2 = 100$.

Solution. For finding the overall gain, we will have to find each stage gain.

(*i*) $A_{v1} = \dfrac{r_{o.1}}{r_{e.1}}$ Now, $r_{0.1} = R_3 \parallel r_{i.2}$ and $r_{i.2} = R_5 \parallel R_6 \parallel \beta_2.r_{e.2}$

Now, $r_{e.2} = 50/I_{E.2}$

Drop across $R_6 = 20 \times \dfrac{5}{5+45} = 2\,\text{V}$

It also represents the approximate drop across R_8.

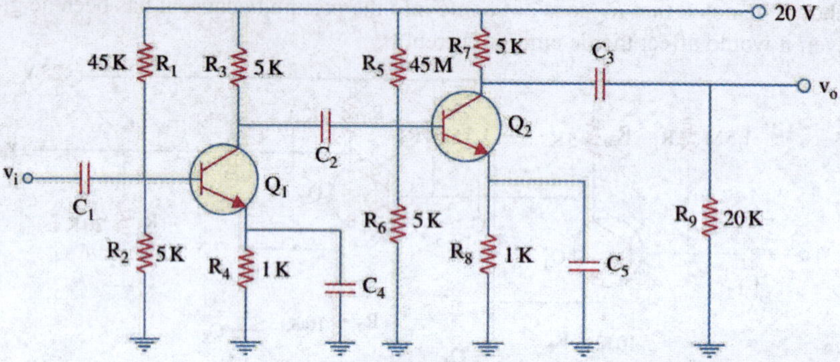

Fig. 23.8

$\therefore \quad I_{E2} = 2/1\text{ K} = 2\text{ mA}$

$\therefore \quad r_{e.2} = 50/2 = 25\ \Omega.$

$\beta_2.r_{e.2} = 100 \times 25 = 2.5\text{ K}$

$r_{i.2} = 45\text{ K} \parallel 5\text{ K} \parallel 2.5\text{ K} = 1.6\text{ K};\quad \therefore \quad r_{0.1} = 5\text{ K} \parallel 1.6\text{ K} = 1.2\text{ K}$

$\therefore \quad r_{e.1} = 50/I_{E.1} = 50/2 = 25\ \Omega$

$\therefore \quad A_{v.1} = 1200/25 = 48$

$A_{v.2} = \dfrac{r_{0.2}}{r_{e.2}};\quad r_{0.2} = R_7 \parallel R_9 = 5\text{ K} \parallel 20\text{ K} = \mathbf{4\text{ K}}$

$\therefore \quad A_{v.2} = 4000/25 = 160 \quad \therefore \; A_v = 48 \times 160 = 7{,}680;\ G_v = 20 \log_{10} 7{,}680 = \mathbf{77.6\text{ dB}}$

23.4. Advantages of RC Coupling

1. It requires no expensive or bulky components and no adjustments. Hence, it is small, light and inexpensive.
2. Its overall amplification is higher than that of the other couplings.
3. It has minimum possible non-linear distortion because it does not use any coils or transformers which might pick up undesirable signals. Hence, there are no magnetic fields to interfere with the signal.
4. As shown in Fig. 23.9, it has a very flat frequency *versus* gain curve *i.e.*, it gives uniform voltage amplification over a wide range from a few hertz to a few megahertz because resistor values are independent of frequency changes.

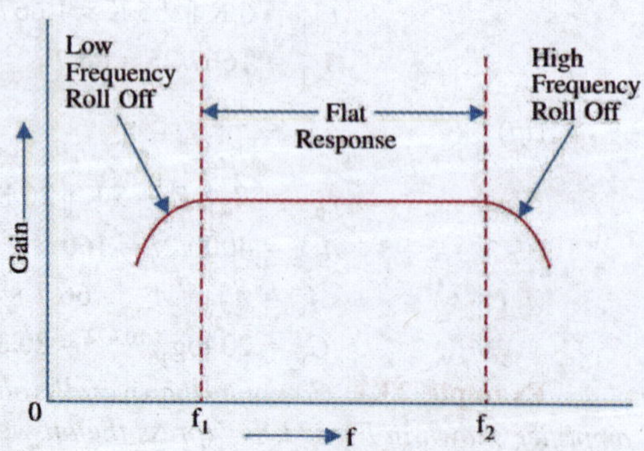

Fig. 23.9

As seen from Fig. 23.9, amplifier gain falls off at very low as well as very high frequencies. At low frequencies, the fall in gain (called *roll-off*) is due to capacitive reactance of the coupling capacitor between the two stages. The high-frequency roll-off is due to output capacitance of the first stage, input capacitance of the second stage and the stray capacitance.

The only drawback of this coupling is that due to large drop across collector load resistors, the collectors work at relatively small voltages unless higher supply voltage is used to overcome this large drop.

23.5. Impedance-coupled Two-stage Amplifier

The circuit is shown in Fig. 23.10. The coupling network consists of L, C_2 and R_3. The only basic difference between this circuit and the one shown in Fig. 23.3 is that inductor L *has replaced the resistor* R_2.

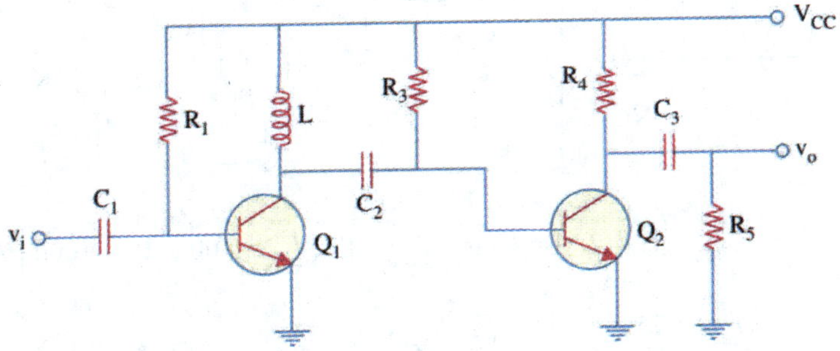

Fig. 23.10

(i) AC Equivalent Circuit

The ac equivalent circuit (at mid-frequency) of the cascaded amplifier has been shown in Fig. 23.11. Because of mid-frequency range, effects of all capacitances have been ignored.

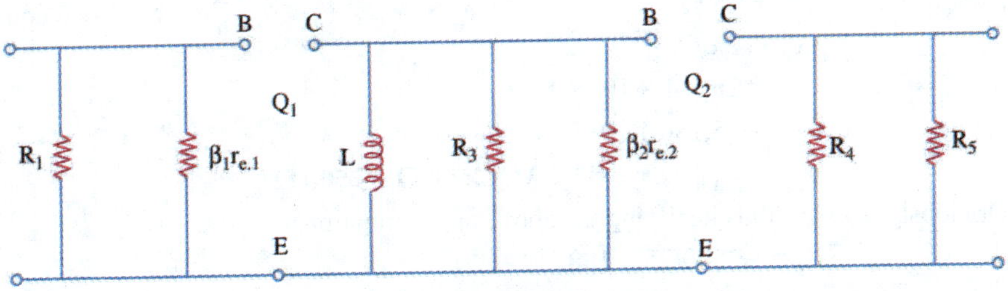

Fig. 23.11

(ii) Circuit Operation

The operation of this circuit is the same as that of the *RC*-coupled circuit described earlier.

(iii) Voltage Gain

It is given by the product of two stage gains $A_v = A_{v1} \times A_{v2}$

Now, $$A_{v.1} = \frac{Z_{o.1}}{r_{e.1}}$$ where $Z_{o.1} = X_L \parallel r_{i.2}$ and $r_{i.2} = R_3 \parallel \beta_2 . r_{e.2}$

In case, $X_L \gg r_{i.2}$, then, $Z_{o.1} \cong r_{i.2}$

$$\therefore \quad A_{v.1} = \frac{r_{i.2}}{r_{e.1}} \quad \text{and} \quad A_{v.2} = \frac{r_{o.2}}{r_{e.2}}$$

$$\therefore \quad A_v = A_{v.1} \times A_{v.2}$$

Example 23.4. *For the impedance-coupled two-stage amplifier shown in Fig. 23.12, compute the values of*

(i) $A_{v.2}$, *(ii)* $A_{v.1}$ *at 4 kHz and* *(iii)* A_v *in dB.*

Neglect V_{BE} *and use* $r_e = 25\ mV/I_E$ *Take* $\beta_1 = \beta_2 = 100$.

Solution. *(i)* $A_{v2} = \dfrac{r_{o.2}}{r_{e.2}}$

Now, $r_{o.2} = R_4 \parallel R_5 = 8\ K \parallel 24\ K = 6\ K$

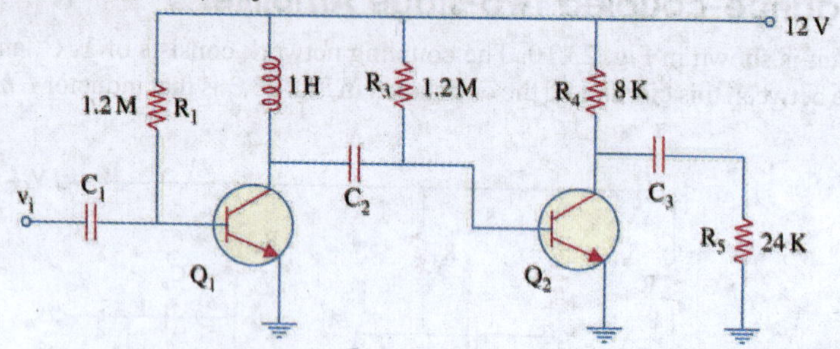

Fig. 23.12

$I_{B2} = 12/1.2\ M = 10\ \mu A,\ I_{C.2} = \beta_2 \,.\, I_{B.2} = 100 \times 10 = 1000\ \mu A = 1\ mA$

∴ $I_{E2} \cong 1\ mA$

∴ $r_{e.2} = 25/1 = 25\ \Omega$

$A_{v.2} = \dfrac{6{,}000}{25} = \mathbf{240}$

(ii) $A_{v.1} = \dfrac{Z_{o.1}}{r_{e.1}} \cong \dfrac{r_{o.2}}{r_{e.1}}$.

Now, $r_{e.1} = 25\ \Omega$ — equal to $r_{e.2}$

$X_L = 2\pi f L$

$= 2\pi \times 4 \times 10^3 \times 1$

$= 25{,}130\ \Omega$

$r_{i.2} = R_3 \parallel \beta_2 . r_{e.2} = \mathbf{1.2\ M \parallel 2500\ \Omega \cong 2500\ \Omega}$

Obviously, $X_L \gg r_{i.2}$ thus justifying the above approximation.

∴ $A_{v.1} = 2500/25 = \mathbf{100}$

(iii) $A_v = 100 \times 240 = 24{,}000,\ G_v = 20 \log_{10} 24{,}000 = \mathbf{87.6\ dB}$

23.6. Advantages of Impedance Coupling

The biggest advantage of this coupling is that there is hardly any dc drop across L so that low collector supply voltages can be used.

However, it has many disadvantages :

1. It is larger, heavier and costlier than *RC* coupling.
2. In order to prevent the magnetic field of the coupling inductor from affecting the signal, the inductor turns are wound on a closed core and are also shielded.

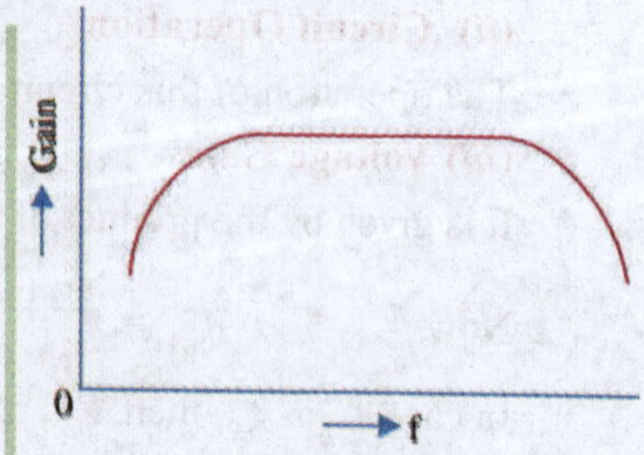

Fig. 23.13

3. Since inductor impedance depends on frequency, the frequency characteristics of this coupling are not as good as those of *BC* coupling. The flat part of the frequency *versus* gain curve is small (Fig. 23.13).

At low frequencies, the gain is low due to large capacitance offered by the coupling capacitor just as in *RC*-coupled amplifiers. The *gain increases with frequency* till it levels off at the middle frequencies of the audio range.

At relatively high frequencies, gain drops off again because of the increased reactance.

Hence, impedance coupling is rarely used beyond audio range.

23.7. Transformer-coupled Two Stage Amplifier

The circuit for such a cascaded amplifier is shown in Fig. 23.14. T_1 is the coupling transformer whereas T_2 is the output transformer. C_1 is the input coupling capacitor whereas C_2, C_3 and C_4 are the bypass capacitors. Resistors R_1 and R_2 as well as R_4 and R_5 form voltage-divider circuits whereas R_3 and R_6 are the emitter-stabilizing resistors.

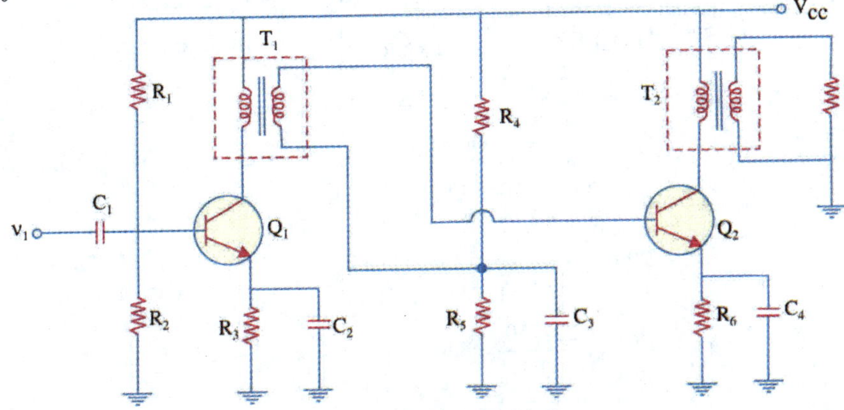

Fig. 23.14

(i) Circuit Operation

When input signal is coupled through C_1 to the base of Q_1, it appears in an amplified form in the primary of T_1. From there, it is passed on to the secondary by magnetic induction. Moreover, T_1 provides dc isolation between the input and output circuits. The secondary of T_1 applies the signal to the base of Q_2 which appears in an amplified form in the primary of T_2.

From there, it is passed on to the secondary by magnetic induction and finally appears across the matched load R_7.

(ii) Voltage Gain

$$A_{v.1} = \frac{r_{o.1}}{r_{e.1}}, \qquad r_{o.1} = a^2 r_{i.2} \qquad \text{— where } a = N_1/N_2 \text{ for } T_1$$

$$r_{i.2} = R_4 \parallel R_5 \parallel \beta_2 . r_{e.2}$$

Similarly, $$A_{e.2} = \frac{r_{o.2}}{r_{e.2}} \quad \text{where} \quad r_{e.2} = a^2 R_7$$

Example 23.5. *For the transformer-coupled two-stage amplifier shown in Fig. 23.15, calculate (i) $A_{e.1}$ (ii) $A_{v.2}$ and (iii) A_v in dB.*

Neglect V_{BE} and use $r_e = 50\ mV/I_B$. Take $\beta_1 = \beta_2 = 50$ and treat the transformers as ideal ones.

Solution. Here, for each transformer $a = 5$

Drop across $R_2 = 9 \times 4/24 = 1.5$ V

Drop across $R_3 \cong 1.5$ V and $I_{E.1} = 1.5/1 = 1.5$ mA

$r_{e.1} = 50/1.5 = 33.3\ \Omega$

$r_{e\cdot 2} = 33.3\ \Omega$ — same as above

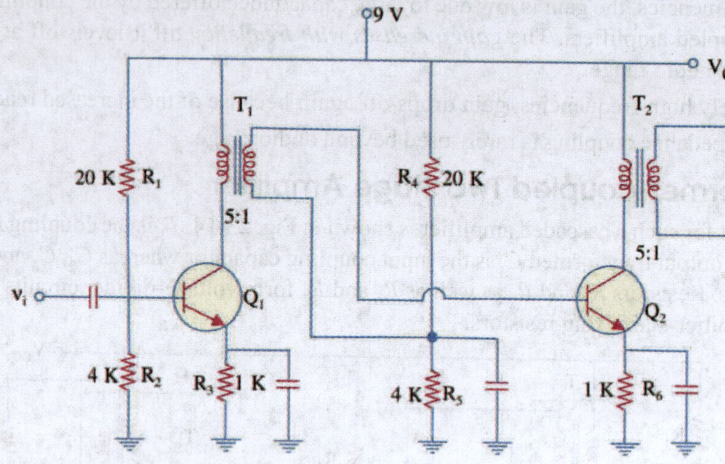

Fig. 23.15

$$\beta_2 . r_{e.2} = 50 \times 33.3 = 1665\ \Omega$$

$$r_{1.2} = R_4 \parallel R_5 \parallel \beta_2 . r_{e.2}$$

$$= 20\text{ K} \parallel 4\text{ K} \parallel 1665\ \Omega = 1110\ \Omega$$

$$r_{o.1} = a^2 . r_{i.2} = 5^2 \times 1110 = 27{,}750\ \Omega$$

(*i*) $$A_{v.1} = \frac{r_{o.1}}{r_{e.1}} = \frac{27{,}750}{33.3} = \mathbf{830}$$

(*ii*) $$A_{v.2} = \frac{r_{o.2}}{r_{e.2}} = \frac{a^2 R^7}{r_{e.2}} = \frac{25 \times 1000}{33.3} = \mathbf{750}$$

(*iii*) $$A_v = 830 \times 750 = 622{,}500\ ;\ G_v = 20 \log_{10} 622{,}500 = \mathbf{116\ dB}$$

23.8. Advantages of Transformer Coupling

1. The operation of a transformer-coupled system is basically *more efficient* because of low dc resistance of the primary connected in the collector circuit.
2. It provides a *higher* voltage gain.
3. It provides *impedance matching* between stages which is desirable for maximum power transfer. Typically, the input impedance of a transistor stage is less than its output impedance. Hence, secondary impedance of the interstage (or coupling) transformer is typically lower than the primary impedance.

This coupling is effective when the final amplifier output is fed to a low-impedance load. For example, the impedance of a typical loud-speaker varies from 4 Ω to 16 Ω whereas output impedance of a transistor stage is several hundred ohms. Use of an output audio transformer can avoid the bad effects of such a mismatch.

Disadvantages

1. The coupling transformer is costly and bulky particularly when operated at audio frequencies because of its heavy iron core.
2. At radio frequencies, the inductance and winding capacitance present lot of problems.
3. It has *poor frequency response* because the transformer is frequency sensitive. Hence, the frequency range of the transformer-coupled amplifiers is limited.
4. It tends to introduce '*hum*' in the output.

23.9. Frequency Response

The characteristics of a coupling transformer are that (*i*) it introduces inductances in both the input and output circuits (*ii*) leakage inductance exists between the primary and secondary windings and (*iii*) both windings introduce shunting (distributed) capacitance especially at high frequencies.

A typical gain *versus* frequency curve for a transformer-coupled amplifier is shown in Fig. 23.16.

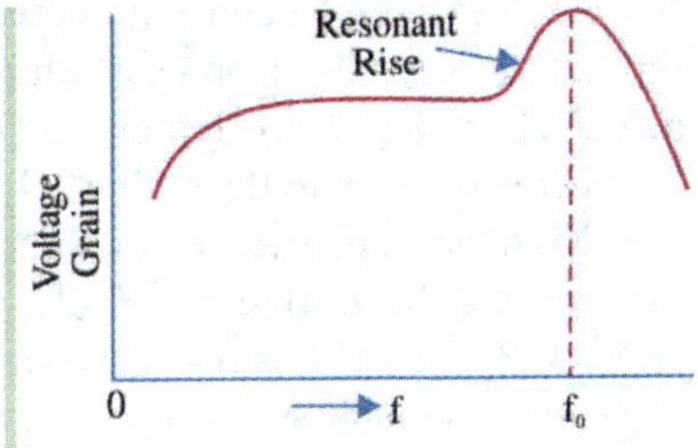

Fig. 23.16

It is seen that

1. there is decrease in gain at low frequencies and
2. also there is decrease in gain at high frequencies except for the *resonant rise* in gain at resonant frequency of the tuned circuit formed by inductance and winding capacitance in the circuit.

The output voltage is equal to ac collector current multiplied by the primary reactance of the coupling transformer. Since at low frequencies, primary reactance is small, the gain is less.

At high frequencies, the distributed capacitance existing between different turns of the winding acts as a *bypass capacitor* and so reduces the output voltage and hence the gain. The peak or exaggerated gain occurs due to resonance or tuning effect of inductance and distributed capacitance which *form a tuned circuit.*

Moreover, there is frequency distortion *i.e.*, all frequencies are not amplified equally. In fact, the flat response part of the curve is small as compared to *RC* coupling. However, transformer-coupled amplifiers can be designed to have a flat frequency response curve and excellent fidelity over the entire audio frequency range.

23.10. Applications

Transformer coupling is often employed in the last stage of a multistage amplifier where concerted effort is made to maximise power transfer by perfect impedance matching.

Example 23.6. *In a multistage transformer-coupled amplifier, the output impedance of the first stage is 5 K and the input impedance of the second stage is 1 K. Determine the primary and secondary inductances of the transformer for perfect impedance matching at f = 2000 Hz. If one turn gives an inductance of 10 μH, find the number of primary and secondary turns.*

Solution. It should be clearly understood that primary has to match with the output impedance of the first stage and secondary with the input of the second stage.

$\therefore \quad X_{Lp}$ = output of 1st stage or $2\pi f L_p = 5000$

$\therefore \quad L_p = 5000/2\pi \times 2000 = \mathbf{0.4\ H}$

Also, $\quad X_{Ls}$ = input of 2nd stage

$\therefore \quad 2\pi f_{Ls} = 1000\ ;\ L_s = 1000/2\pi \times 2000 = \mathbf{0.08\ H}$

Now, inductance of a coil varies as the square of its turns.

$\therefore \quad L \propto N^2 = kN^2$

When, $\quad N = 1, L = 10\ \mu H \quad \therefore \quad 10 \times 10^{-6} = k \times 1^2 \quad$ or $\quad k = 10^{-5}$

For primary winding $\quad 0.4 = 10^{-5} N_p^{\ 2} \quad$ or $\quad N_p = \mathbf{632}$

For secondary winding, $\quad L_s = kN_s^{\ 2} \quad$ or $\quad 0.08 = 10^{-5} N_s^{\ 2} \quad$ or $\quad N_s = \mathbf{89}$

As seen, it is a nearly 7 : 1 step-down transformer.

23.11. Direct-coupled Two-stage Amplifier Using Similar Transistors

These amplifiers operate without the use of frequency-sensitive components like capacitors, inductors and transformers etc. They are especially suited for amplifying (*a*) ac signals with frequencies as low as a fraction of a hertz,

(*b*) change in dc voltages.

Fig. 23.17 shows the circuit of such an amplifier which uses two *similar* transistors each connected in the *CE* mode. Both stages employ direct coupling (*i*) collector of Q_1 is connected directly to the base of Q_2 and (*ii*) load resistor R_2 is connected directly to the collector of Q_2. The resistor R_1 establishes the forward bias of Q_1 and also indirectly that of Q_2.

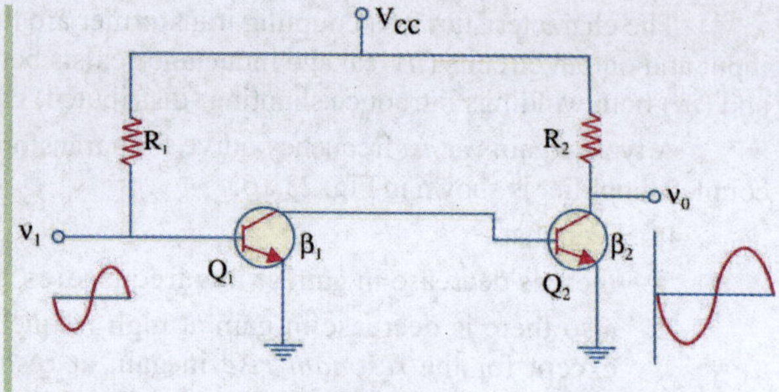

Fig. 23.17

Any signal current at the base of Q_1 is amplified β_1 times and appears at the collector of Q_1 and becomes base signal for Q_2. Hence, it is further amplified β_2 times. Obviously, signal current gain of the amplifier is

$$A_i = \beta_1 \times \beta_2 = \beta^2 \qquad \text{— if transistors are identical}$$

(*i*) AC Equivalent Circuit

The ac equivalent circuit is shown in Fig. 23.18.

(*ii*) Voltage Gain

The first stage *i.e.*, Q_1 does not produce any voltage gain *i.e.*, $A_{v.1} = 1$ as proved below :

$$A_{v.1} = \frac{r_{o.1}}{r_{e.1}} = \frac{\beta_2 \cdot r_{e.2}}{r_{e.1}} \qquad \text{...(i)}$$

As seen from Fig. 23.17, $I_{E.2} \cong \beta_2 \cdot I_{B.2}$

Also $\qquad I_{B.2} = I_{C.1} = I_{E.1}$

$\therefore \qquad I_{E.2} = \beta_2 \cdot I_{E.1}$ or $\beta_2 = \dfrac{I_{E.2}}{I_{E.1}}$

Also $\qquad r_{e.1} = \dfrac{25}{I_{E.1}}$

and $\qquad r_{e.2} = \dfrac{25}{I_{E.2}}$

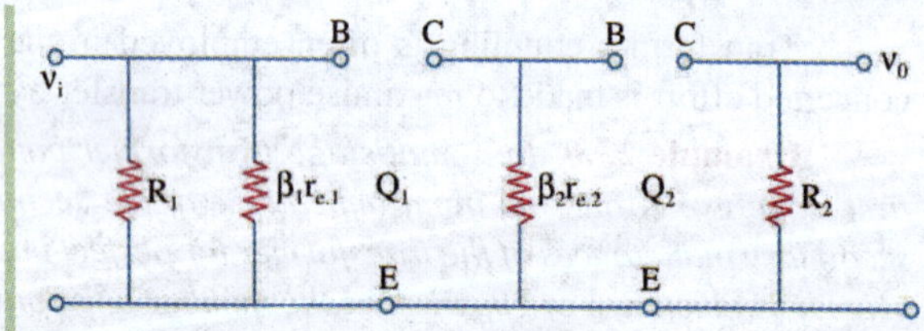

Fig. 23.18

Substituting all these values in Eq. (*i*) above, we get

$$A_{v.1} = \frac{I_{E.2}}{I_{E.1}} \times \frac{25}{I_{E.2}} \times \frac{I_{E.1}}{25} = 1$$

It proves the statement made above.

Now, $\qquad A_{v.2} = \dfrac{r_{o.2}}{r_{e.2}} \qquad$ and $\qquad A_v = A_{v.1} \times A_{e.2} = A_{v.2}$

(*iii*) Advantages

1. The circuit arrangement is very simple since it uses *minimum number of components.*
2. It is quite *inexpensive.*
3. It has the outstanding ability *to amplify direct current* (*i.e.*, as dc amplifier) and ***low-frequency signals.***
4. It has *no coupling or bypass capacitor* to cause a drop in gain at low frequencies. As seen from Fig. 23.19, the frequency-response curve is flat upto upper cut-off frequency determined by stray wiring capacitance and internal transistor capacitance.

(iv) Disadvantages

1. It cannot amplify high-frequency signals.
2. It has poor temperature stability.

It is due to the fact that any variation in base current (due to temperature changes) in one stage is amplified in the following stage (or stages) thereby shifting the Q-point. However, stability can be improved by using emitter-stability resistor (Fig. 23.21).

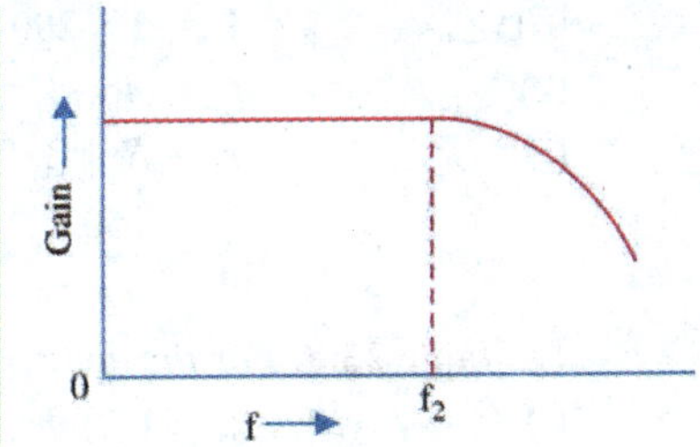

Fig. 23.19

(v) Applications

Some of the applications of direct-coupled amplifiers are in

1. regulator circuits of electronic power supplies,
2. pulse amplifiers,
3. differential amplifiers,
4. computer circuitry,
5. electronic instruments.

Example 23.7. *For the direct-coupled amplifier of Fig. 23.20, calculate*

(a) current gain, *(b) voltage gain of first stage,*
(c) voltage gain of second stage, *(d) overall voltage gain in dB,*
(e) overall power gain in dB, *(f) input resistance.*

Neglect V_{BE} and use $r_e = 50\ mV/I_E$.

Solution. (a) $A_i = \beta_1\beta_2 = 100 \times 50 = \mathbf{5000}$

(b) $A_{v.1} = 1$ — Art. 23.11

(c) $A_{v2} = \dfrac{r_{o.2}}{r_{e.2}}$. Now, $r_{e.2} = \dfrac{50}{I_{E.2}}$

Let us find the value of $I_{E.2}$ starting from the value of $I_{B.1}$. As seen from Fig. 23.20,

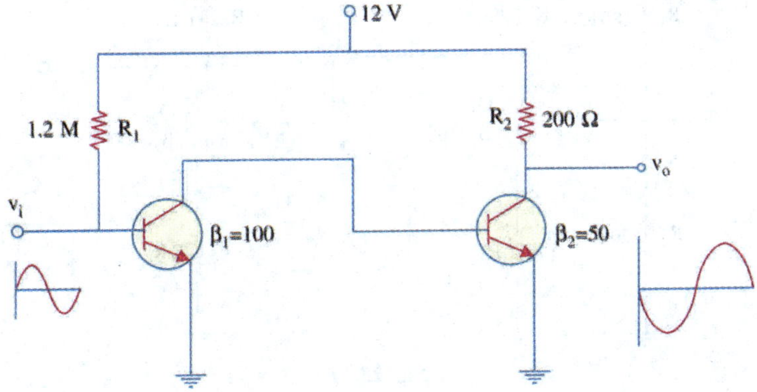

Fig. 23.20

$$I_{B\cdot 1} = 12/1.2\ M = 10\ \mu A$$
$$I_{C.1} = \beta_1 . I_{B.1} = 100 \times 10 = 1000\ \mu A$$
$$I_{E.1} = I_{C.1} = 1000\ \mu A = 1\ mA$$
$$I_{B.2} = I_{C.1} = 1\ mA$$
$$I_{C.2} = \beta_2 . I_{C.1} = 50 \times 1 = 50\ mA$$

$\therefore$ $I_{E.2} = 50\ mA$ and $r_{e.2} = 50/50 = 1\ \Omega$

Also, $r_{e2} = R_2 = 200\ \Omega$ $\therefore$ $A_{v.2} = \dfrac{200}{1} = \mathbf{200}$

(*d*) $A_v = 1 \times 200 = 200$; $G_v = 20 \log_{10} 200 = \mathbf{46\ dB}$

(*e*) $A_p = A_v . A_i = 200 \times 5{,}000 = 10^6$; $G_p = 10 \log_{10} 10^6 = \mathbf{60\ dB}$

(*f*) $r_i = R_1 \parallel \beta_1 . r_{e.1}$ — Fig. 23.18

$$r_{e.1} = \frac{50}{I_{E.1}} = \frac{50}{1} = 50\ \Omega \quad \therefore \quad r_i = 1.2\ \text{M} \parallel (50 \times 100) \cong \mathbf{5\ K}$$

Example 23.8. *For the emitter-stabilized direct-coupled amplifier of Fig. 23.21, find (i) $A_{v.1}$, (ii) $A_{v.2}$, (iii) A_v and (iv) r_i. Neglect V_{BE} and use $r_e = 50\ mV/I_E$.*

Solution. As seen, emitter resistors R_3 and R_5 have been used to improve temperature stability. The voltage divider $R_1 - R_2$ together with R_3 determines the emitter current of Q_1. The resistor R_4 has dual function (*i*) it acts as load resistor for Q_1 and (*ii*) it establishes base bias of Q_2.

(*i*) Since unbypassed resistor R_3 is present

$$A_{v.1} = \frac{r_{o.1}}{r_{e.1} + R_3}$$

As seen from the ac equivalent diagram of Fig. 23.22

$$r_{o.1} = R_4 \parallel \beta_2 . (r_{e.2} + R_5)$$

Now, $r_{e.2} = 50 / I_{E.2}$

For finding $I_{E.2}$, let us first find $I_{E.1}$

Drop across $R_2 = 20 \times \dfrac{20}{180 + 20} = 2\ \text{V}$

Same is the drop across R_3 since V_{BE} has been neglected.

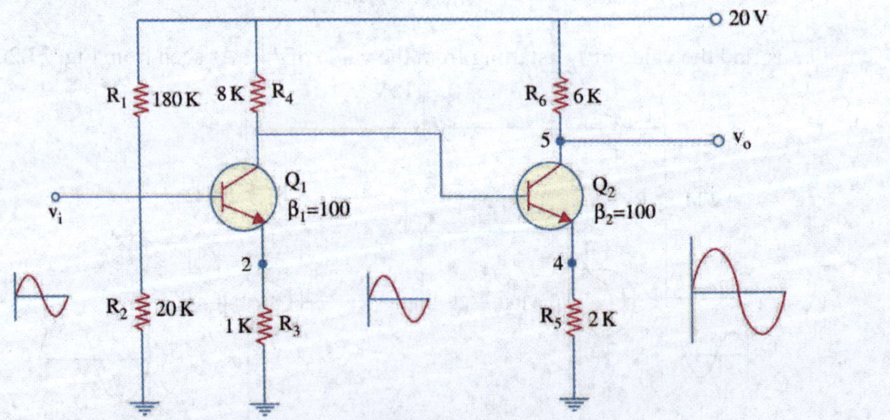

Fig. 23.21

$\therefore \quad I_{E.1} = 2/1\ \text{K} = 2\ \text{mA}$

Now, $$I_{E.2} = \frac{V_{CC} - I_{E.1} R_4}{R_5} = \frac{20 - 2 \times 8}{2} = 2\ \text{mA}$$

$\therefore \quad r_{e.2} = 50/2 = 25\ \Omega$ and $\beta_2 (r_{e.2} + R_5) \cong 200\ \text{K}$

$r_{o.1} = 8\ \text{K} \parallel 200\ \text{K} = 7.7\ \text{K}$; $r_{e.1} = 50/2 = 25\ \Omega$

$$\therefore \quad A_{v.1} = \frac{7700}{(25 + 1000)} = \mathbf{7.5}$$

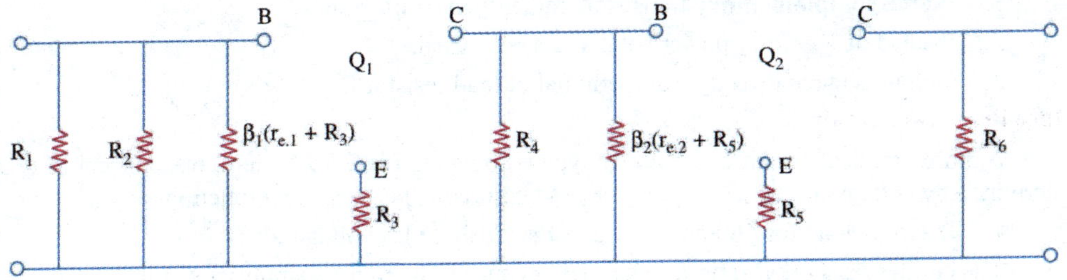

Fig. 23.22

(*ii*) $A_{v.2} = \dfrac{r_{o.2}}{(r_{e.2} + R_5)}$; $r_{o.2} = R_6 = 6000\ \Omega \therefore A_{v.2} = \dfrac{6000}{(25 + 2000)} \cong \mathbf{3}$

(*iii*) $A_v = 7.5 \times 3 = \mathbf{22.5}$

(*iv*) As seen from Fig. 23.22

$$r_i = R_1 \parallel R_2 \parallel \beta_1 (r_{e.1} + R_3) = 180\text{ K} \parallel 20\text{ K} \parallel 100\text{ K} \quad \text{— neglecting } r_{e.1}$$
$$= \mathbf{15.25\ K}$$

Example. 23.9. *In the circuit of Fig. 23.21, find dc voltage at points marked 1, 2, 3 and 4. Neglect* V_{BE}.

If a 1 V dc signal at input 1 changes by 0.01 V, what would be the voltage variation at the output?

Solution. V_1 = drop across R_2 = **2 V**

$V_2 \cong V_1 = \mathbf{2\ V}$; $V_3 = V_{CC} - I_{C.1} R_3 = 20 - 2 \times 8 = \mathbf{4\ V}$

$V_4 \cong V_3 = \mathbf{4\ V}$; $V_5 = V_{CC} - I_{C.2} R_6 = 20 - 2 \times 6 = \mathbf{8\ V}$

$\Delta v_o = A_v . \Delta v_i = 22.5 \times 0.01 = 0.225\text{ V} = \mathbf{225\ mV.}$

23.12. Direct-coupled Amplifier Using Complementary Symmetry of Two Transistors

In this case, an *NPN* transistor is directly-coupled to its complementary *i.e.*, a *PNP* transistor. Fig. 23.23 shows a two-stage cascaded amplifier using two complementary transistors connected in *CE* configuration.

The circuit differs from that shown in Fig. 23.17 in the following three ways :

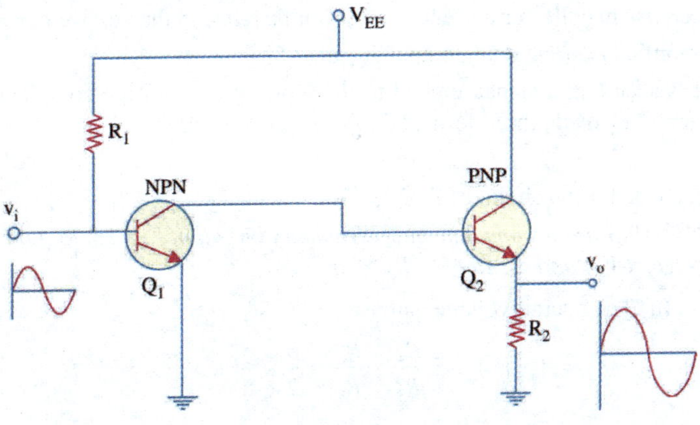

Fig. 23.23

1. It uses **complementary** transistors rather than **similar** ones.
2. Instead of V_{CC}, V_{EE} power battery has been used.
3. Output is taken directly from terminal of load resistor R_2.

More Practical Circuit

A more practical circuit of the above type is shown in Fig. 23.24. Here, bias current of Q_1 is determined by voltage divider $R_1 - R_2$ and R_3. As before, R_4 performs two functions :

1. It acts as load for Q_1 and
2. Establishes bias voltage for Q_2

The emitter resistors R_3 and R_6, as usual, are meant to improve amplifier stability.

(*i*) Circuit Operation

When a positive-going signal is applied to the base of Q_1, then

1. its base current *increases,*
2. hence, its collector current *increases* ($I_C = \beta I_B$),
3. voltage drop across R_4 *increases,*
4. consequently, voltage at the collector of Q_1 and the base of Q_2 becomes *less positive* or in other words, *more negative,*
5. hence, a negative-going signal is applied to the base of Q_2.

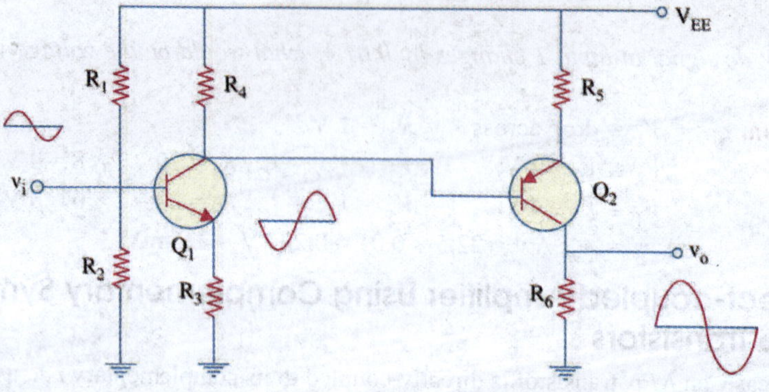

Fig. 23.24

The negative-going signal applied to the base of Q_2 causes

1. an *increase* in its forward bias (remember, it is a *PNP* transistor),
2. an *increase* in collector current,
3. an *increase* in the voltage developed across R_6,
4. an amplified *positive going* output signal at R_6.

Hence, it is seen that a signal applied to the input of a two-stage complementary amplifier appears at the output in an amplified form and *of the same polarity.*

Voltage Gain

It is the same as for the circuit of Fig. 23.17.

Example 23.10. *For the complementary symmetry circuit of Fig. 23.25, find (a) A_v (b) V_1, V_2, V_3, V_4 and V_5. Neglect V_{BE} and assume R_3, $R_5 >> r_e$.*

Solution. (*a*) The overall voltage gain is

$$A_v = A_{v.1} \times A_{v.2} \cong \frac{R_4}{R_3} \times \frac{R_6}{R_5} = \frac{6}{2} \times \frac{6}{8}$$

(*b*) $$V_1 = V_{EE} \frac{R_2}{R_1 + R_2} = 2 \text{ V} \quad \therefore \quad V_2 \cong V_1 = \mathbf{2\ V}$$

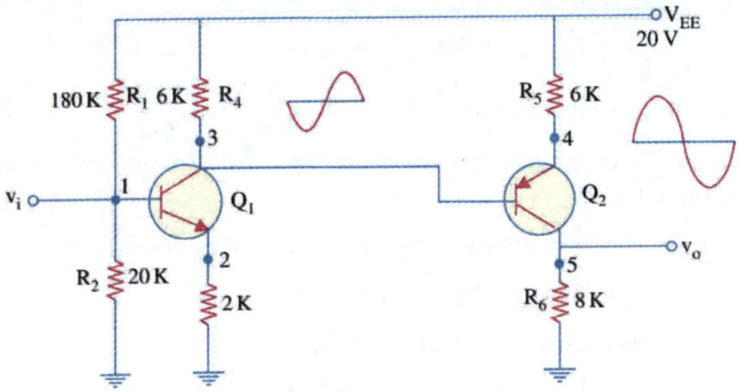

Fig. 23.25

$$\therefore \quad I_{E.1} = \frac{2\text{ V}}{R_3} = \frac{2\text{ V}}{2\ K} = 1\text{mA}, \qquad I_{C1} \cong I_{E.1} = 1\text{ mA}$$

$$V_3 = V_{EE} - I_{C.1}\, R_3 = 20 - 1 \times 6 = \mathbf{14\ V}, \quad V_4 \cong V_3 = \mathbf{14\ V}$$

$$I_{C.2} = \frac{V_{EE} - V_4}{R_5} = \frac{20 - 14}{6\text{K}} = 1\text{ mA}$$

$$\therefore \quad I_{E.2} \cong I_{C.2} = 1\text{ mA}$$

$$\therefore \quad V_5 = I_{E.2} \times R_6 = 1 \times 8 = \mathbf{8\ V}$$

23.13. Darlington Pair

It is the name given to a *pair of similar* transistors so connected that emitter of one is directly joined to the base of the other as shown in Fig. 23.26 (*a*). Obviously, the emitter current of Q_1 becomes the base current of Q_2.

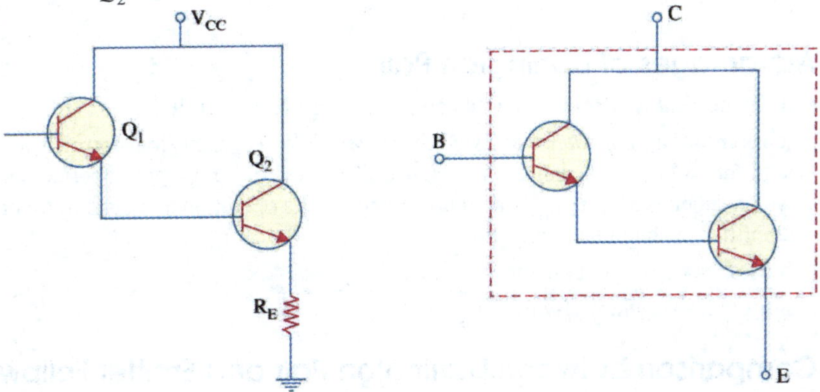

(*a*) Fig. 23.26 (*b*)

Darlington pairs are commercially mounted in a single package that has only three leads : base, collector and emitter as shown in Fig. 23.26 (*b*). It often forms a *double CC stage* in multistage amplifiers. It is so because a Darlington connection can be considered equivalent to *two cascaded emitter followers*.

Main Characteristics

(*i*) Current Gain

It can be proved that current gain of a Darlington pair is $(1 + \beta_1)(1 + \beta_2) = (1 + \beta)^2 \cong \beta^2$ if the transistors are *identical* (*i.e.*, $\beta_1 = \beta_2$).

Proof

$$I_{B.2} = I_{E.1} = (1 + \beta_1) I_{B.1} \cong \beta_1 I_{B.1}$$

$$I_{E.2} \cong \beta_2 . I_{B.2} = \beta_1 \beta_2 I_{B.1} \qquad \therefore \qquad A_i = \frac{I_{E.2}}{I_{B.1}} = \beta_1 \beta_2 = \beta^2$$

It means that a Darlington pair behaves like a *single transistor having a beta of* β^2.

(*ii*) Input Impedance

In Fig. 23.26 (*a*), the input impedance seen from the base of Q_2 is

$$r_{i.2} = \beta_2 (r_{e.2} + R_E) \cong \beta_2 . R_E$$

Input impedance as seen from the base of Q_1 is

$$r_{i.1} = \beta_1 (r_{e.1} + r_{i.2}) = \beta_1 r_{e.1} + \beta_1 r_{i.2} = \beta_1 r_{e.1} + \beta_1 \ \beta_2 R_E \cong \beta_1 \beta_2 R_E$$

or $\quad r_{in(base)}$ of $Q_1 = \beta_2 R_E$.

Note. If there is a load resistance R_L coupled to the emitter of Q_2, then

$$r_{i.1} - \beta^2 (R_E \parallel R_L) = \beta^2 . r_E$$

As seen, load impedance R_E has been transformed into $\beta^2 R_E$. Obviously, a Darlington pair is capable of high input impedance. In fact, whenever a load causes a severe loss in voltage gain (loading effect), it is usual to step up load impedance *via* a FET stage, a single *CC* stage or Darlington pair when much greater impedance transformation is required.

(*iii*) Voltage Gain

Assuming $r_{e.1} = r_{e.2} = r_e$, we have

$$A_v \cong \frac{R_E}{r_e + R_E} = \frac{1}{1 + \dfrac{r_e}{R_E}} \cong 1 \qquad \text{— as in an emitter follower}$$

23.14. Advantages of Darlington Pair

1. It can be readily formed from two adjacent transistors in an IC.
2. It has enormous *impedance transformation capability i.e.*, it can transform a low-impedance load into a high impedance load. Hence, it is used in a high-gain operational amplifier which depends on very high input impedance for its operation as an integrator or summing amplifier in analogue applications.
3. It uses very few components.
4. It provides very high β-value.

23.15. Comparison between Darlington Pair and Emitter Follower

We will refer to Figs. 21.29 and 23.26.

1. Input impedance of Darlington pair is $\beta^2 R_E$ whereas that of emitter follower is βR_E (Art. 21.8).
2. Current gain of Darlington pair is β^2 whereas that of emitter follower is β.
3. However, voltage gains of the two are identical.

Example 23.11. *For the Darlington pair shown in Fig. 23.27, calculate the value of*

1. β, 2. input impedance, 3. voltage amplification.

Assume $\beta_1 = \beta_2 = 100$ *and* $R_L \gg (r_{e \cdot 1} + r_{e.2})$.

Solution. The approximate values are as under :

1. β of Darlington pair

$= \beta_1 \times \beta_2$

$= 100 \times 100$

$= \mathbf{10{,}000}$

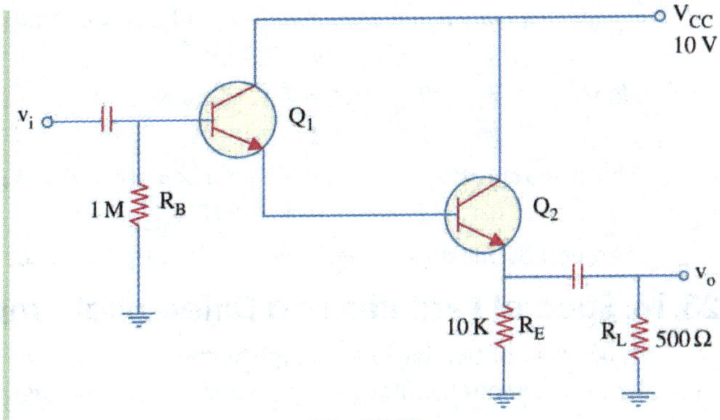

Fig. 23.27

2. $r_E = R_E \| R_L$

$= 10\text{ K} \| 500\ \Omega$

$= 475\ \Omega$

$r_{in(base)}$ of $Q_1 \cong \beta^2 r_E$

$= 10{,}000 \times 475$

$= 4.75$ M

r_i = input impedance of the pair

$= R_B \| r_{in(base)}$ of Q_1

$= 1\text{ M} \| 4.75\text{ M} = 0.826\text{ M} = \mathbf{826\ K}$

3. $A_v \cong 1$

Example 23.12. *A CE amplifier stage shown in Fig. 23.28 is to drive a 100 Ω load to 10 V_{p-p} level. An input signal of 1 V_{p-p} is available. Find out if the stage is overloaded or not. Also, find out if this overloading has been avoided by using a Darlington pair as a buffer between Q_1 and the load as shown in Fig. 23.29. Take $\beta_1 = \beta_2 = 50$.*

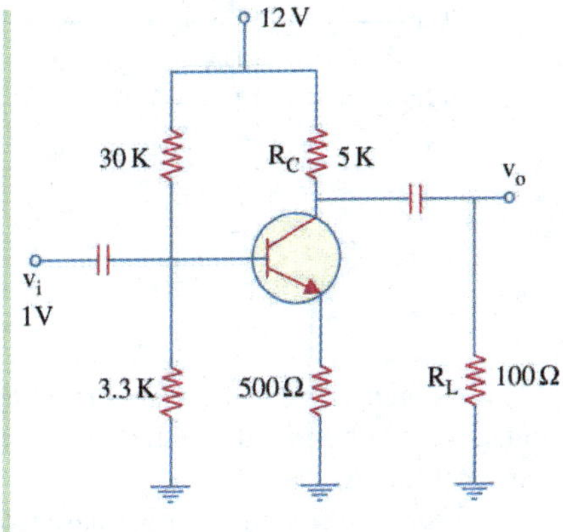

Fig. 23.28

Solution. The approximate voltage gain of Q_1 is given by

$$A_v = \frac{r_L}{r_E} = \frac{R_C \| R_L}{R_E} = \frac{5\ K \| 100\Omega}{500\ \Omega} \cong \frac{1}{5}$$

$$\therefore \quad v_o = 1 \times \frac{1}{5} = 0.2\text{ mV}$$

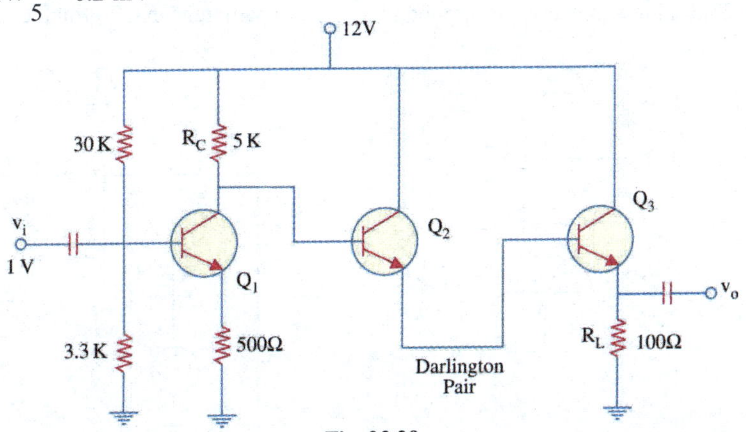

Fig. 23.29

Obviously, *due to overloading of the stage,* severe attenuation has occurred. Hence, the amplifier stage cannot, *by itself,* drive the heavy load.

By using a Darlington pair, a 100 Ω load has been transformed to

$$\beta^2 R_L = 2500 \times 100 = 250 \text{ K}$$

Now, $\quad r_L = R_C \parallel \beta^2 R_L = 5 \text{ K} \parallel 250 \text{ K} \cong 5 \text{ K}$

$\therefore \quad A_v = 5 \text{ K} / 500\ \Omega = 10$

This also represents total voltage gain because voltage gain of Darlington pair is one.

$\therefore \quad v_o = A_v \times v_i = 10 \times 1 = \mathbf{10\ V_{p\text{-}p}}$

As seen, now the stage would be able to drive the load since the design goal of 10 $V_{p\text{-}p}$ has been met.

23.16. Special Features of a Differential Amplifier

It consists of two basic *CE* amplifiers having their *emitters directly-coupled to each other.* Fig. 23.30 shows both the block diagram and the circuit diagram of such an amplifier.

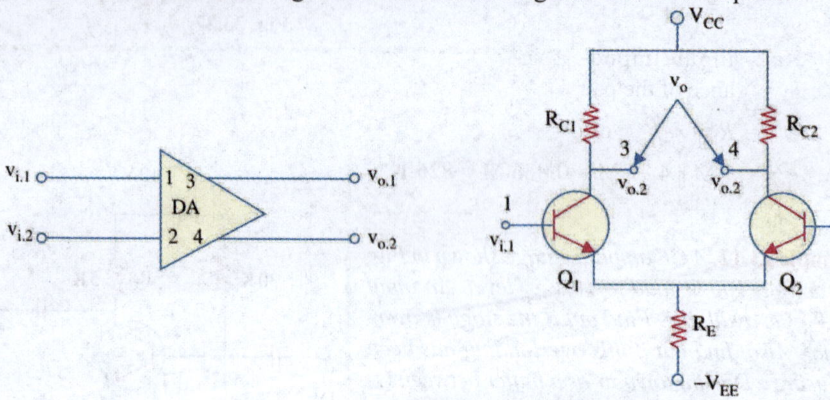

Fig. 23.30

As seen, it has two separate input terminals 1 and 2 and two separate output terminals 3 and 4. Voltages may be applied to either or both input terminals and output may be taken from either or both output terminals.

There are certain specific phase relationships between both input and both output terminals as discussed below.

(*a*) Single-ended Operation

In Fig. 23.31, input signal $v_{i.1}$ is applied to terminal 1 with terminal 2 grounded.

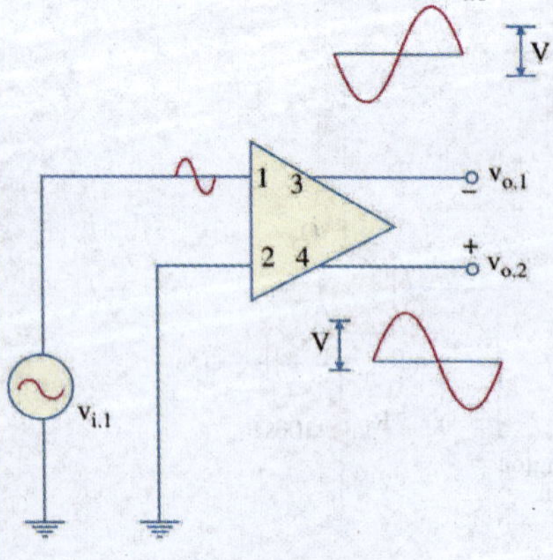

Fig. 23.31

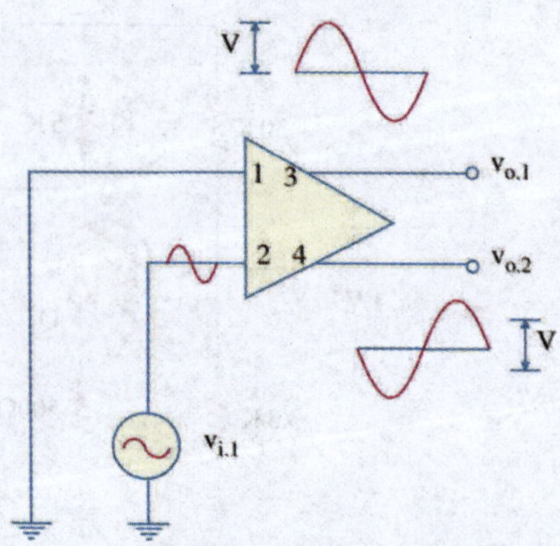

Fig. 23.32

It is seen that an *amplified* and *inverted* output signal is obtained at terminal 3 (phase inversion of a *CE* amplifier) but an *equally-amplified* and *in-phase* signal appears across output terminal 4. The differential output voltage has the polarity shown in the figure.

As shown in Fig. 23.32, when input signal v_{1-2} is applied to input terminal 2, an *amplified* and *inverted* signal appears at outputs terminal 4 whereas *equally-amplified* but *in-phase* signal appears at terminal 3.

In summary, we can say that input at any of the two terminals causes outputs at both terminals 3 and 4. The two output are opposite in phase but of equal amplitude.

(*b*) Double-ended Operation

Fig. 23.33 illustrates the double ended mode of operation when two input signals of opposite phase are applied to the two input terminals.

Input signal at each input terminal causes signals to appear at both output terminals. The resultant output signals have a peak value of 2 V–twice the value for single-ended operation.

However, if two in-phase and equal signals were applied at the two input terminals, the resultant output signal at each output terminal would be zero as shown in Fig. 23.34. It means that output *between* the collectors would be zero.

If $v_{i.1}$ and $v_{i.2}$ change by *exactly* the same amount, even then output voltage between terminals 4 and 3 remains zero because of symmetry. Only when $v_{i.1}$ and $v_{i.2}$ differ from each other, we get an output voltage. When $v_{i.1}$ is more positive than $v_{i.2}$, the output terminal 4 is more positive than terminal 3.

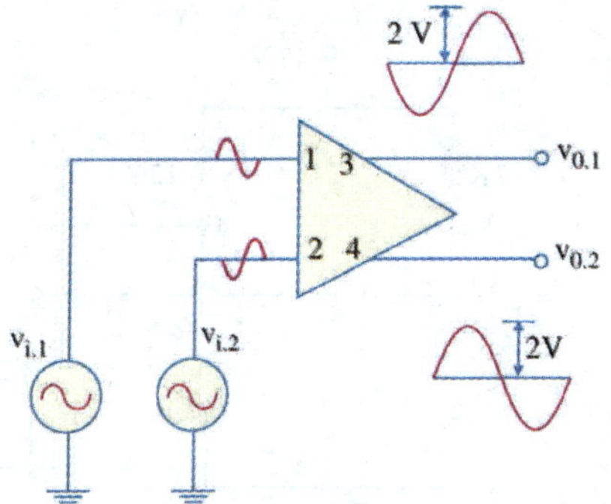

Fig. 23.33

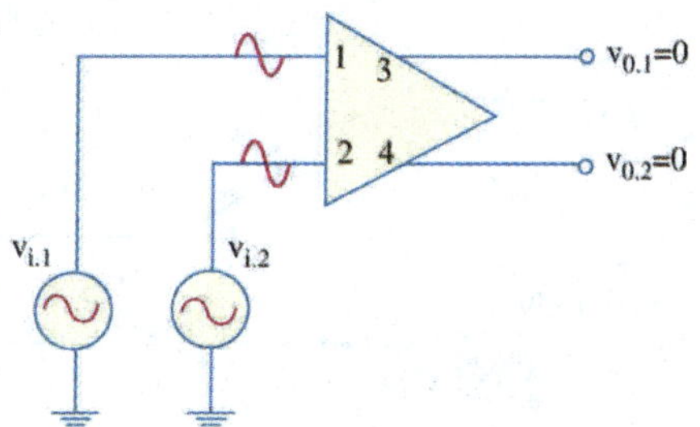

Fig. 23.34

(*c*) Inverting and Non-inverting Inputs

When positive $v_{i.1}$ acts alone, it produces a differential output voltage with terminal 4 positive with respect to terminal 3 as shown in Fig. 23.35. That is why the input terminal 1 is called non-inverting input terminal. However, when positive $v_{i.2}$ acts alone, the output voltage is inverted *i.e.*, terminal 3 becomes positive with respect to terminal 4. That is why input terminal 2 is called *inverting terminal.*

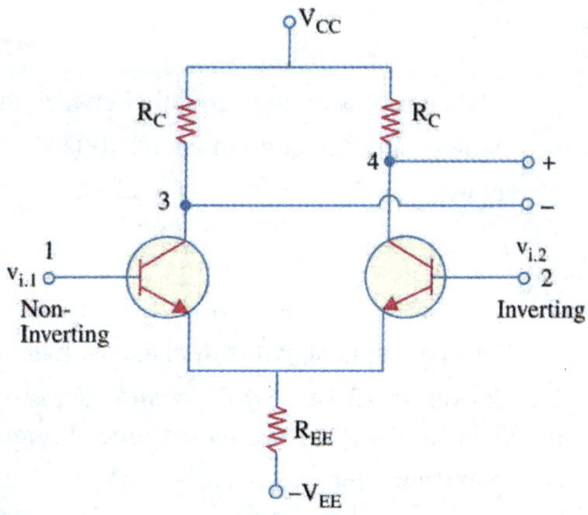

Fig. 23.35

23.17. Common Mode Input

Fig. 23.36 illustrates the common-mode input of a differential amplifier *i.e.*, when similar or same input signal is applied to both inputs. If the two halves of the diff-amp are identical, the ac output voltage will be zero. The diff-amp is then said to be perfectly balanced.

The common-mode rejection ratio (*CMMR*) is defined as

$$CMMR = \frac{A \times v_{i.cm}}{v_{o.cm}}$$

Suppose, $v_{i.cm} = 1$ V. In a perfectly-balanced and symmetrical diff-amp, the output should be zero. But, in practice, there is always a small output signal because of *non-symmetry*. Suppose, $A = 100$ and $v_{o.em} = 0.01$ V. Then

$$CMMR = \frac{100 \times 1\text{ V}}{0.01\text{ V}} = 10,000$$

$$= 20 \log_{10}{}^{10,000} = 80 \text{ dB}$$

Larger the value of *CMMR* (*i.e.*, smaller the value of $v_{o.cm}$), better the diff-amp. Ideally, a common-mode input should produce zero output voltage. Hence, an ideal diff-amp has a *CMMR* of *infinity.*

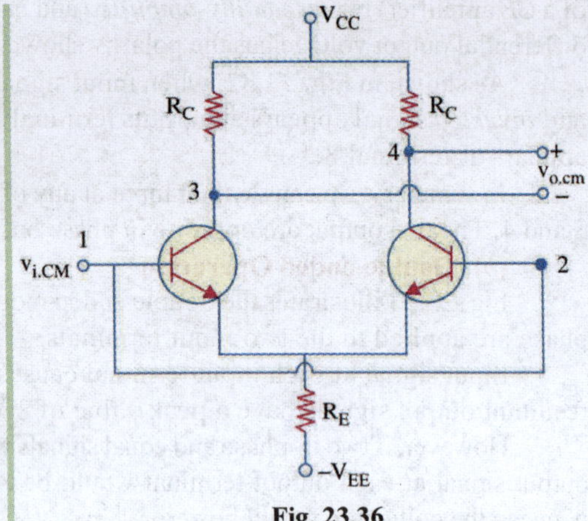

Fig. 23.36

Example 23.13. *Calculate the approximate output voltage for the diff-amp of Fig. 23.37 which uses only a single-ended non-inverting input of $v_{i.1}$ = 1 mV. Take r_e = 25 mV/I_E and neglect V_{BE}.*

Solution. $I_E \cong V_{EE}/R_E = 12/6 = 2$ mA

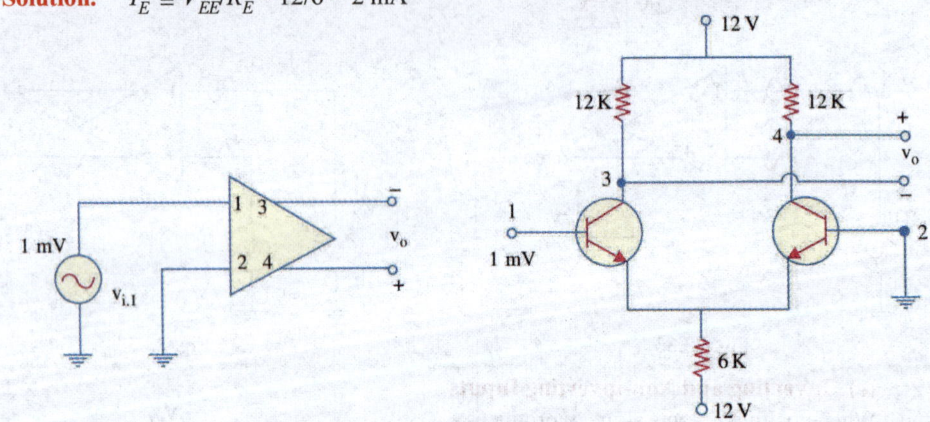

Fig. 23.37

The current is divided equally between the two transistors so that $I_{E1} = I_{E2} = 1$ mA.

Voltage amplification of each half is approximately given by $A = R_C / r_e$.

Now, $\quad r_e = 25/1 = 25\ \Omega$

$$\therefore \quad A = \frac{12,000}{25} = 480$$

$\therefore \quad v_o = A \times v_{i.1} = 480 \times 1 =$ **480 mV.**

The ac output signal is in phase with the input.

Example 23.14. *If a differential input signal of 1 mV is applied to the diff-amp shown in Fig. 23.38, calculate the output voltage. Neglect V_{BE} and take r_e = 25 mV/I_E.*

Solution. Here, $v_{i.1} - v_{i.2} = 1$ mV

As seen from Ex. 23.13, $\quad A = 480$

$\therefore \quad v_o = A\,(v_{i.1} - v_{i.2}) = 480 \times 1 = 480$ mV

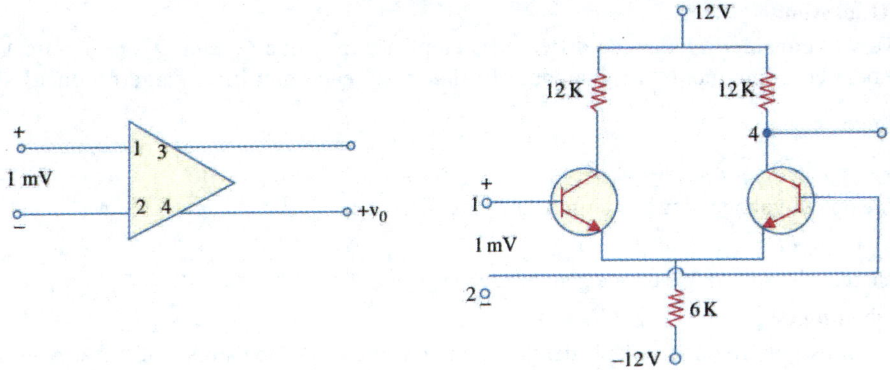

Fig. 23.38

Example 23.15. *A differential input signal of 1 mV is applied to the diff-amp of Fig. 23.39 when used in single-ended output mode. Calculate the approximate value of output voltage. Neglect V_{BE} and take $r_e = 25\ mV/I_E$*

Solution. Here, again 1 mV is the differential input signal *i.e.*, $v_{i.1} - v_{i.2} = 1$ mV

Since single-ended mode is being used, $A = 480/2 = 240$

$$\therefore \qquad v_o = A\,(v_{i.1} - v_{i.2}) = 240 \times 1 = \mathbf{240}$$

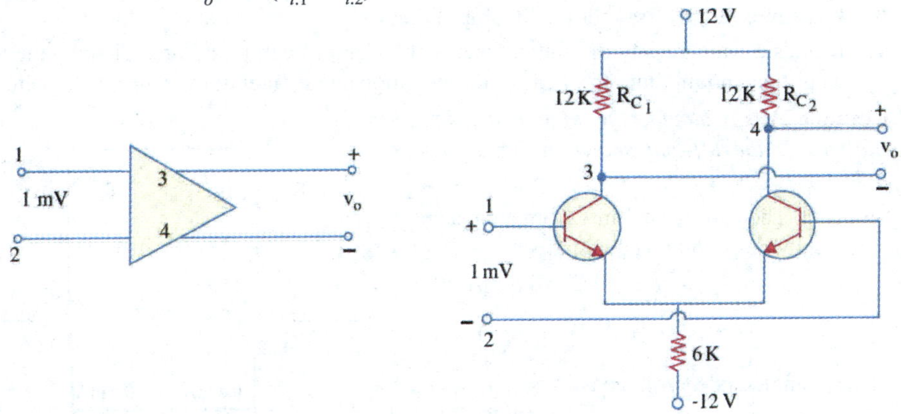

Fig. 23.39

23.18. Differential Amplifier

Fig. 23.40 shows the circuit of a differential or difference amplifier. As seen

1. it contains two *CE* amplifiers,
2. it only uses resistors and transistors.
3. it is a directly-coupled (emitter-to-emitter) amplifier,
4. it can accept two inputs by means of T_1 and ground and also T_2 and ground,
5. it can provide two separate outputs by means of T_3 and ground and T_4 and ground,
6. it can provide a single output between T_3 and T_4 *i.e.*, differential output.

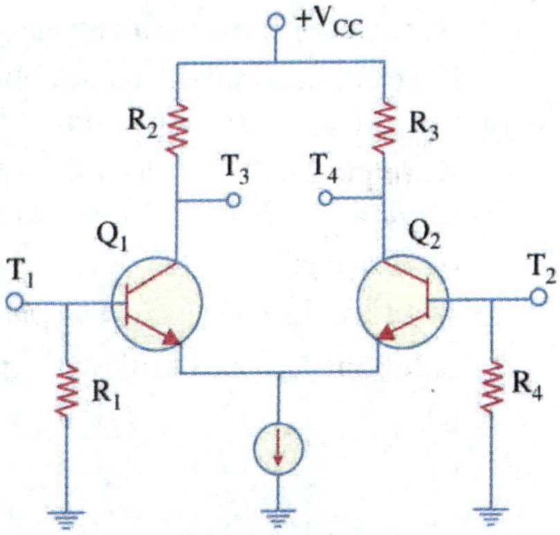

Fig. 23.40

Circuit Operation

We will consider a balanced differential amplifier in which Q_1 and Q_2 are identical and their associated components are matched. In that case, each amplifier stage produces same voltage gain. $A = \frac{r_o}{r_e}$.

In Fig. 23.40, $r_o = R_2$ or R_3

The output voltage between terminals T_3 and T_4 is

$$v_o (T_3 - T_4) = A (v_{i.1} - v_{i.2})$$

where, A = voltage gain of each stage

Advantages

1. It uses no frequency-dependent coupling or bypassing capacitors. All that it requires is resistors and transistors both of which can be easily integrated on a chip. Hence, it is extensively used in linear ICs.
2. It can compare any two signals and detect any difference. Thus, if two signals are fed into its inputs, identical in every respect except that one signal has been slightly distorted, then only the difference between the two signals *i.e.*, distortion will be amplified.
3. It gives higher gain than two cascaded stages of ordinary direct coupling.
4. It provides very uniform amplification of signal from dc upto very high frequencies.
5. It provides isolation between input and output circuits.
6. It is almost a universal choice for amplifying dc.
7. It finds a wide variety of applications such as amplification, mixing, signal generation, amplitude modulation, frequency multiplication and temperature compensation etc.

Example 23.16. *Calculate the single-ended and differential gain of the diff-amp shown in Fig. 23.41. Use $r_e = 25\ mV/I_E$.*

Solution. The current of 1 mA from a constant-current source divides into two equal parts.

$$I_{E.1} = I_{E.2} = 1/2 = 0.5 \text{ mA}$$

$$r_{e.1} = r_{e.2} = \frac{25}{0.5} = 50\ \Omega$$

Hence, single-ended voltage gain is

$$A = \frac{r_o}{r_e} = \frac{R_C}{r_e} = \frac{10\ K}{50\ \Omega} = \mathbf{200}$$

Hence, each stage has a voltage gain of 200.

If we consider differential (double-ended) gain, its value is twice *i.e.*, 2 × 200 = **400**.

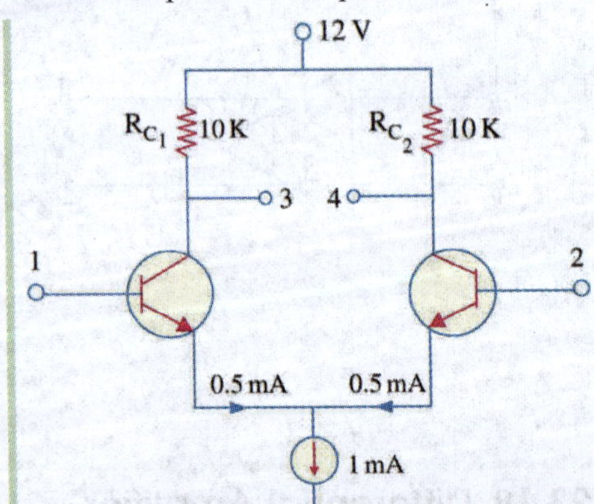

Fig. 23.41

Example 23.17. *For the diff-amp shown in Fig. 23.41, voltage gain of each stage is 200, $v_{i.1}$ = 30 mV and $v_{i.2}$ = 20 mV. Find the voltages between*

(i) T_3 and ground, (ii) T_4 and ground, (iii) T_3 and T_4.

What are the polarities of output terminals T_3 and T_4?

Solution. As stated earlier in Art. 23.18,

(*i*) $v_o (T_3) = A_1 \times v_{i.1} = 200 \times 30 \text{ mV} = 6 \text{ V}$

(*ii*) $v_o (T_4) = A_2 \times v_{i.2} = 200 \times 20 \text{ mV} = 4 \text{ V}$

(*iii*) $v_o (T_3 - T_4) = A (v_{i.1} - v_{i.2}) = 200 (30 - 20) \text{ mV} = 2 \text{ V}$

Since $v_{i.1} > v_{i.2}$; $i_{c.1} R_2 > i_{c.2} R_3$, hence, T_4 will be positive with respect to T_3.

CONVENTIONAL PROBLEMS

1. For the two-stage *RC*-coupled amplifier shown in Fig. 23.42 calculate the approximate values of
   (*i*) voltage gain for the first stage (*ii*) voltage gain for the second stage
   (*iii*) voltage gain for the amplifier (*iv*) ac input resistance of the amplifier
   Neglect V_{BE} and use $r_e = 25$ mV/I_E. **[(*i*) 76.2 (*ii*) 256 (*iii*) 19,500 or 85.8 dB (*iv*) 1250 Ω]**

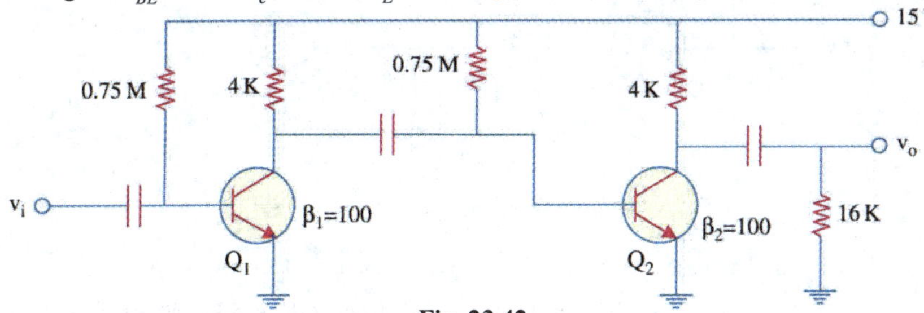

Fig. 23.42

2. For the two-stage *RC*-coupled amplifier using base and emitter bias and shown in Fig. 23.43, calculate approximate values of
   (*i*) A_{v1} (*ii*) A_{v2} (*iii*) A_v (*iv*) r_i
   Neglect V_{BE} and use $r_e = 25$ mV/I_E. **[(*i*) 61.6 (*ii*) 128 (*iii*) 7,900 or 78 dB (*iv*) 2.5 K]**

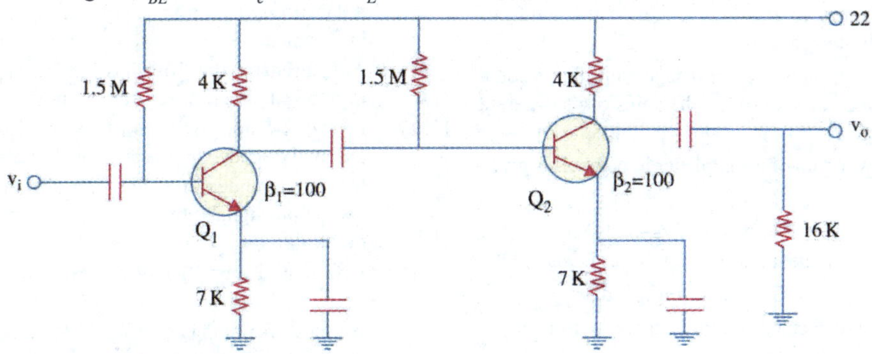

Fig. 23.43

3. Fig. 23.44 shows the circuit of a two-stage impedance-coupled amplifier. Compute the approximate values of
   (*i*) voltage gain of 1st stage
   (*ii*) voltage gain of the 2nd stage ; both at $f = 2$ kHz
   (*iii*) voltage gain of the amplifier
   Neglect V_{BE} and take $r_e = 25$ mV/I_E. **[(*i*) 80 (*ii*) 420 (*iii*) 33,600 or 90.5 dB]**

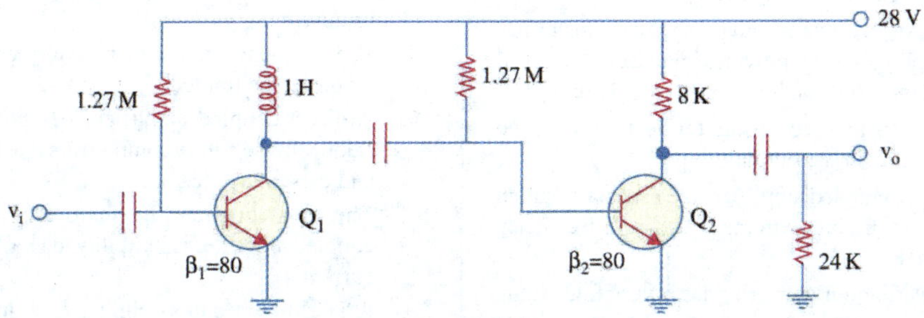

Fig. 23.44

4. For the direct-coupled amplifier shown in Fig. 23.46, calculate the approximate value of

(i) A_i (ii) A_{v1} (iii) A_{v2} (iv) A_v (v) A_p and (vi) r_4

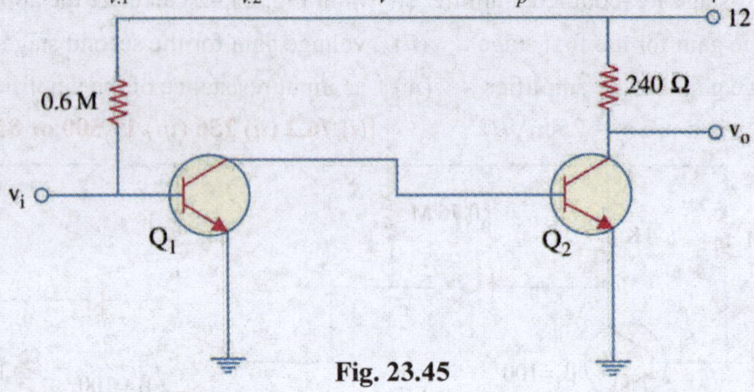

Fig. 23.45

Neglect V_{BE} and assume $r_e = 25$ mV/I_E. Also, $\beta_1 = 50$ and $\beta_2 = 25$.

[(i) 1250 (ii) 1 (iii) 240 (iv) 240 (v) 300,000 or 55 dB (vi) 1250 Ω]

SELF EXAMINATION QUESTIONS

A. Fill in the blanks with most appropriate word (s) or numerical value (s).

1. The overall voltage gain of a cascaded amplifier is equal to the .... of the gains of individual stages.
2. The dB gain of a multistage amplifier is equal to the .... of the dB gains of its different stages.
3. Power gain of an amplifier in decibels is given by

$$G_p = \text{........} \log_{10}{}^{Ap} \text{ dB}$$

4. The voltage gain of an amplifier is given by

$$G_v = \text{........} \log_{10}{}^{Ap} \text{ dB}$$

5. In transformer coupling, there is no need for coupling...............
6. *RC* coupling has minimum ....distortion.
7. *RC* coupled amplifier has a very .... frequency-response curve.
8. Impedance coupling is used, if at all, only in the ....range of frequencies.
9. Because of low dc resistance, transformer coupling is basically more.............
10. One of the biggest advantages of transformer coupling is that it provides impedance ....so essential for maximum power transfer.
11. Transformer coupling tends to introduce .... in the output.
12. Direct-coupled amplifiers are especially suited for amplifying extremely .... frequency signals.
13. A Darlington pair provides a very high value of .... not provided by any single transistor.
14. In a Darlington pair, emitter current of one transistor becomes .... current of the next one.
15. A Darlington pair has very .... input impedance.
16. A differential amplifier has .... separate inputs and .... separate outputs.
17. In a diff-amp, signal applied at one input terminal provides signals at both .... terminals.
18. In a single-ended operation of a diff-amp, the two output signals are out of ....with each other.
19. Differential amplifier is extensively used in .... ICs.
20. An ideal diff-amp has a *CMMR* of ....

B. Answer True or False

1. A voltage gain of 1000 corresponds to a dB gain of 30.
2. The chief merit of *RC* coupling is that it is inexpensive.
3. An *RC*-coupled amplifier has a very flat frequency response curve in the audio range.
4. Transformer coupling needs no input coupling capacitor.
5. The frequency range of transformer-coupled amplifiers is limited.
6. A direct-coupled amplifier does not use any frequency-sensitive components like inductors and capacitors etc.
7. The most outstanding feature of a direct-coupled amplifier is its ability to amplify direct current.
8. A Darlington pair is similar to a double *CC* stage.

9. A diff-amp basically consists of two *CE* amplifiers with their emitters directly-coupled to each other.
10. In a diff-amp, the common-mode signal is ignored and is not amplified.

C. Multiple Choice Items

1. Multistage amplifiers are used in order to achieve greater
   (*a*) voltage amplification
   (*b*) power gain
   (*c*) frequency response
   (*d*) all of the above
2. The decibel gain of a cascaded amplifier equals the
   (*a*) product of individual gains
   (*b*) sum of individual gains
   (*c*) ratio of stage gains
   (*d*) product of voltage and current gains
3. If two stages of a cascaded amplifier have decibel gains of 60 and 30, then overall gain is .... dB.
   (*a*) 90 (*b*) 1800
   (*c*) 2 (*d*) 0.5
4. *RC* coupling is popular in low-level audio amplifiers because it
   (*a*) has better low frequency response
   (*b*) is inexpensive and needs no adjustments
   (*c*) provides an output signal in phase with the input signal
   (*d*) needs low voltage battery for collector supply
5. The most desirable feature of transformer coupling is its
   (*a*) higher voltage gain
   (*b*) wide frequency range
   (*c*) ability to provide impedance matching between stages
   (*d*) ability to eliminate hum from the output
6. For matching a circuit of output impedance 200 Ω with a load of 8 Ω, the turn ratio of the two-winding transformer should be
   (*a*) 25 (*b*) 1/25
   (*c*) 1/5 (*d*) 5
7. The outstanding characteristic of a direct-coupled amplifier is its
   (*a*) utmost economy
   (*b*) temperature stability
   (*c*) avoidance of frequency-sensitive components
   (*d*) ability to amplify direct current and low-frequency signals
8. Darlington pairs are frequently used in linear ICs because they
   (*a*) do not require any capacitors or inductors
   (*b*) have enormous impedance transformation capability
   (*c*) can be readily formed from two adjacent transistors
   (*d*) resemble emitter followers
9. When same input signal is applied to both the inputs of an ideal diff-amp, the output
   (*a*) is zero
   (*b*) depends on its *CMMR*
   (*c*) depends on its voltage gain
   (*d*) is determined by its symmetry
10. The common-mode rejection ratio of an ideal diff-amp is
    (*a*) zero (*b*) infinity
    (*c*) less than unity (*d*) greater than unity

ANSWERS

A. Fill in the blanks

1. product **2.** sum **3.** 10 **4.** 20 **5.** capacitor **6.** non-linear **7.** flat
8. audio **9.** efficient **10.** matching **11.** hum **12.** low **13.** beta **14.** base
15. high **16.** two, two **17.** output **18.** phase **19.** linear **20.** infinity

B. True or False

1. F **2.** F **3.** T **4.** F **5.** T **6.** T **7.** T **8.** T **9.** T **10.** T

C. Multiple Choice Items

1. *d* **2.** *b* **3.** *a* **4.** *b* **5.** *c* **6.** *d* **7.** *d* **8.** *c* **9.** *a* **10.** *b*

24 CHAPTER

Decibels And Frequency Response

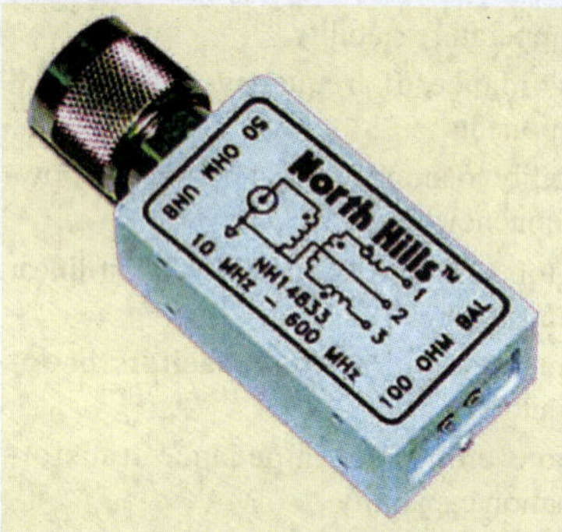

1. The Decibel System
2. Voltage and Current Levels
3. Characteristics of the Decibel System
4. Zero Decibel Reference Level
5. Variations in Amplifier Gain with Frequency
6. Changes in Voltage and Power Levels
7. Causes of Gain Variations
8. Miller Effect
9. Cut-off Frequencies of Cascaded Amplifiers
10. Transistor Cut-off Frequencies
11. Alpha Cut-off Frequency
12. Beta Cut-off Frequency
13. The f_T of a Transistor

24.1. The Decibel System

The decibel system of measurement is widely used in audio, radio, TV and instrument industries for comparing two voltage or power levels. It is based on the established fact that an individual's response to *seeing* or *hearing is non-linear*. It has been found that the changes in power and audio levels are related logarithmically. For example, when power is increased from 4W to 16 W, the audio level does not increase 16/4 = 4 times but by a factor of 2 ($\because 4^2 = 16$). Similarly, for a power change from 4 W to 64 W, the audio level changes by a factor of 3 and not by 64/4 = 16. In logarithmic form, the relationship can be written as

$$\text{increase in audio level} = \log_4^{64} = 3.$$

While comparing two powers, it is common practice to choose the log base of 10. Suppose, we want to compare any two powers P_1 and P_2. The simple method is to take their ratio P_2/P_1 and to state how many times P_2 is bigger or smaller than P_1 which is used as reference power. In the decibel system, we take the log of this ratio.

$$\therefore \quad \text{power level} = 10 \log_{10}^{P_2/P_1} \text{ bel}$$

Since bel is too large, we use decibel (dB) instead. Remembering that 1 bel = 10 decibel, the power level becomes

$$= 10 \log_{10}^{P_2/P_1} \text{ dB}$$

Similarly, if P_i is the input power of an amplifier and P_o its power output, then power gain of the amplifier in decibels is

$$= 10 \log_{10}{}^{P_o/P_i} \text{ dB}$$

Also, if the output power of an amplifier changes from P_1 to P_2, the power level change is

$$= 10 \log_{10}{}^{P_2/P_1} \text{ dB}$$

It is obvious that dB is the unit of power **change** (*i.e.*, increase or decrease) and not of **power** itself.

Another point worth remembering is that 20 dB is not twice as much power as 10 dB.

Example 24.1. *Find the decibel increase in the output power level of an amplifier when its output power increases from (i) 4W to 8W and (ii) 8W to 16W. What is the change in power in each case ?*

Solution. (*i*) Increase in power level is

$$= 10 \log_{10}{}^{8/4} = 10 \log_{10}{}^{2}$$
$$= 10 \times 0.3 = \mathbf{3\ dB}$$

absolute change in power $= 8.4 = \mathbf{4\ W}$

(*ii*) Increase in power level is

$$= 10 \log_{10}{}^{16/8} = \log_{10}{}^{2} = \mathbf{3\ dB}$$

absolute change in power $= 16 - 8 = \mathbf{8W}$

Example 24.2. *(a) Express the power gain of + 8 in dB. (b) Which power ratio is expressed by – 23 dB? (c) What voltage ratio is represented by 2 dB?*

Solution. (*a*) $G_p = 10 \log_{10} P_2/P_1$, here, $P_2/P_1 = 8$

$\therefore$ $G_p = 10 \log_{10}{}^{8} = 10 \times 0.9 = \mathbf{9}$

(*b*) $G_p = 10 \log^{P_2/P_1}$ or $-23 = 10 \log_{10}{}^{P_1/P_2}$

$\therefore$ $\dfrac{P_2}{P_1} = 10^{(-2.3/10)} = 10^{-2.3} = \dfrac{1}{10^{2.3}} = \dfrac{\mathbf{1}}{\mathbf{200}}$

(*c*) $G_v = 20_{10}{}^{A_v}$ *or* $A_v = 10^{(G_v/20)} = 10^{(2/20)}$

$\therefore$ $A_v = 10^{1/10} = 10^{0.1} = \mathbf{1.26}$

Example 24.3. *An amplifier has an input signal of 16 V peak-to-peak and an input impedance of 320 K. It gives an output voltage of 8 V peak-to-peak across a load resistor of 4 Ω. Calculate the dB power gain of the amplifier.*

Solution. $P_i = \dfrac{V_i^2}{R_i} = \dfrac{(16/2\sqrt{2})^2}{320\text{ K}} = 100\ \mu\text{W}$

where V_i = r.m.s. value of input voltage $= \dfrac{V_{i.p-p}}{2\sqrt{2}}$

If V_o is the r.m.s. value of the output voltage, then

$$P_o = \frac{V_o^2}{R_o} = \frac{(8/2\sqrt{2})^2}{4} = 2\text{ W}$$

$\therefore$ power amplification, $\dfrac{P_o}{P_i} = \dfrac{2\text{ W}}{100\ \mu\text{ W}} = 20{,}000$

decibel power gain $= 10 \log_{10}{}^{P_o/P_i} = 10 \log_{10}{}^{20,000} = 10 \times 4.3 = \mathbf{43\ dB}$

24.2. Other Expressions for Power Gain

Suppose P_o and P_i are the respective output and input powers of an amplifier, R_o and R_i are its output and input resistances and V_o and V_i the r.m.s. values of output and input voltages, then

$$P_o = \frac{V_o^2}{R_o} \quad \text{and} \quad P_i = \frac{V_i^2}{R_i}$$

Power amplification, $A_p = P_o/P_i$. But power gain is given by

$$G_p = 10 \log_{10}^{P_o/P_i} = 10 \log_{10}^{\frac{V_o^2/R_o}{V_i^2/R_i}} = 10 \log_{10}^{(V_o/V_i)^2 \cdot R_i/R_o}$$

$$= 10\left[\log_{10}^{(V_o/V_i)^2} + \log_{10}^{R_i/R_o}\right] = 10 \log_{10}^{(V_o/V_i)^2} + 10 \log_{10}^{R_i/R_o}$$

$$= 20 \log_{10}^{V_o/V_i} + 10 \log_{10}^{R_i/R_o}$$

Also, $\quad P_o = I_o^2 R_o$ and $P_i = I_i^2 R_i$

$$\therefore \text{ power gain, } G_p = 10 \log_{10}^{P_o/P_i} = 10 \log_{10}^{I_o^2 R_o/I_i^2 R_i} = 10\left[\log_{10}^{(I_o/I_i)^2 (R_o/R_i)}\right]$$

$$= 10\left[2 \log_{10}^{I_o/I_i} + \log_{10}^{R_o/R_i}\right] = 20 \log_{10}^{I_o/I_i} + 10 \log_{10}^{R_o/R_i} \text{ dB}$$

Of course, if $\quad R_o = R_i$, then $\log_{10}^{R_o/R_i} = \log_{10}^{1} = 0$

Hence, in that case, power gain, $G_p = 20 \log_{10}^{I_o/I_i}$ dB

24.3. Voltage and Current Levels

Though decibel was initially defined as the unit of power level, it can also be used with voltage and current levels.

Suppose, an amplifier has an input voltage of V_i and gives an output voltage of V_o. Then, its voltage amplification is $A_v = V_o/V_i$ but its decibel voltage gain is

$$G_v = 20 \log_{10}^{A_v} \text{ dB} = 20 \log_{10}^{V_o/V_i} \text{dB}$$

It should be carefully noted that multiplying factor of 20 has been used and not of 10 as is done for finding power level. In fact, it implies that R_o has been taken equal to R_i in the power gain equation derived in Art. 24.2.

Similarly, current amplification is $A_i = I_o/I_i$. However, current gain of the amplifier is

$$G_i = 20 \log_{10}^{I_o/I_i} \text{dB}$$

Example 24.4. *Convert each of the following ratio into their decibel values (i) $P_o/P_i = 250$ (ii) $P_o/P_i = 0.5$ (iii) $V_o/V_i = 10$ and (iv) $V_o/V_i = 0.707$.*

Solution. (*i*) $\quad A_p = 250, G_p = 10 \log^{A_p}$ dB $= 20 \log^{250}$ = **24 dB**

(*ii*) $\quad A_p = 0.5, G_p = 10 \log^{0.5}$ = **– 3dB**

(*iii*) $\quad A_v = 10, G_v = 20 \log^{10}$ = **20 dB**

(*iv*) $\quad A_v = 0.707, G_v = 20 \log^{0.707}$ = **– 3dB**

Example 24.5. *The input and output voltages of a network are 16 V and 8V respectively. If input impedances are equal, find the voltage gain.*

Solution. $\quad G_v = 20 \log_{10}^{V_o/V_i} = 20 \log_{10}^{8/16}$

$$= 20\,[\log_{16}^{8} - \log_{10}^{16}] = 20\,(0.90 - 1.2) = 20 \times (-0.3) = \mathbf{-6\ dB.}$$

Alternative Method

Whenever the voltage ratio is less than 1, *its log is negative* which is often difficult to handle. In such cases, it is best to invert the fraction and then make the result negative. The above problem could be solved thus :

$$G_v = -20 \log_{10}^{16/8} = -20 \times 0.3 = \mathbf{-6\ dB}$$

Example 24.6. *A microphone delivers 30 mV to the 300 Ω input of an amplifier. The ac power delivered to an 8 Ω speaker is 18 W. What is the power gain of the amplifier ?*

Solution. We may use any one of the equations given above.

$$G_p = 20 \log_{10}^{V_o/V_i} + 10 \log_{10}^{R_i/R_o}$$

Now, $P_o = \dfrac{V_o^2}{R_o}$ or $18 = \dfrac{V_o^2}{8}$ $\therefore$ $V_o = \sqrt{8 \times 18} = 12\ V$

$$\therefore \quad G_p = 20 \log_{10}^{12\text{ V}/30\text{ mV}} + 10 \log_{10}^{300/8} = 20 \log_{10}^{400} + 10 \log_{10}^{37.5}$$

$$= 20 \times 2.6 + 10 \times 1.57 = \mathbf{67.7\ dB.}$$

Example 24.7. *The output power of an amplifier is 100 mW when the signal frequency is 5 kHz. When the frequency is increased to 25 kHz, the output power falls to 50 mW. Calculate the dB change in power.*

Solution. The decibel change in power level

$$= 10 \log_{10}^{50/100} = 10 \log_{10}^{1/2} = -10 \log_{10}^{2} = -10 \times 0.3 = \mathbf{-3\ dB.}$$

Example 24.8. *The output voltage of an amplifier is 10 V at 5 kHz and 7.07 V at 25 kHz. What is the decibel change in output power level ?*

Solution. Since changes in voltage are across the same output resistance, the decibel change in power is

$$= 20 \log_{10}^{V_2/V_1} = 20 \log_{10}^{7.07/10}\text{ dB} = -20 \log_{10}^{10/7.07} = -20 \log_{10}^{1.414}\text{ dB}$$

$$= -20 \times 0.15 = \mathbf{-3\ dB}$$

It is seen from Ex. 24.5 and Ex. 24.16 that output decibel power falls by a 3 dB when

(*i*) absolute value of power falls to half its original value

or

(*ii*) output voltage falls to 0.707 or $1/\sqrt{2}$ of its original value.

Example 24.9. *A certain radio receiver delivers an output power of 3.6 W.*

(i) What would be the decibel gain if power output is increased to 7.2 W ?

(ii) What power output would be required to produce a power gain of 10 dB ?

Solution. (*i*) $G_p = 10 \log_{10}^{P_2/P_1} = 10 \log_{10}^{7.2/3.6} = \mathbf{3\ dB}$

(*ii*) Here, $G_p = 10$, $P_1 = 3.6$ W, $P_2 = ?$

$\therefore$ $10 = 10 \log_{10}^{P_2/3.6}$ $\therefore$ $\log_{10}^{P_2/3.6} = 1$ or $\dfrac{P_2}{3.6} = 10^1 = 10$

$\therefore$ $P_2 = 3.6 \times 10 = \mathbf{36\ W}$

24.4. Characteristics of the Decibel System

1. A decibel is a measure of **ratio** and not of **an amount.** It tells us how many times one quantity is greater or lesser with respect to another or the reference quantity. It does not measure the actual (or absolute) voltage or power but only their **changes;**

2. *Decibel is non-linear i.e.*, 20 dB is not twice as much power or voltage as 10 dB;
3. This log-based system allows a tremendous range of power ratios to be encompassed by using only two-digit numbers. For example

$$1 \text{ dB} = 1.26 : 1 \text{ power ratio and } 50 \text{ dB} = 100{,}000 : 1 \text{ power ratio;}$$

4. Total dB of a cascaded amplifier can be found by simply *adding* the stage dBs.

24.5. Value of 1 dB

It can be proved that 1 dB represents the log of two powers which *have a ratio of 1.26.*

$$1 \text{ dB} = 10 \log_{10}{}^{P_2/P_1}$$

$$\therefore \quad \log_{10}{}^{P_2/P_1} = \frac{1}{10} = 0.1 \qquad \text{or} \qquad \frac{P_2}{P_1} = 10^{0.1} = 1.26$$

Hence, +1 dB represents an increase in power of 26%.

24.6. Zero Decibel Reference Level

By now, it should be clear that decibel does not measure any physical quantity like voltage or power etc., but merely ratio of two physical quantities. For determining the power levels of various powers like P_1, P_2, P_3 etc., it is essential to fix some reference power with which their ratios can be taken. If we fix P_o as the reference power, then different power levels would be

$$10 \log_{10}{}^{P_1/P_0} \quad \text{and} \quad 10 \log_{10}{}^{P_2/P_0} \text{ etc.}$$

Obviously, P_o *cannot be 0 watt i.e.*, 0 watt cannot be taken as 0 dB because, in that case, power level of P_1 would be

$$= 10 \log_{10}{}^{P_1/0} = \infty$$

In fact, any power compared with zero power is infinity. Hence, decibels *would not be defined* if zero watt is taken as zero reference level.

Following three zero reference levels are in common use :

1. Zero dB refers to 6 mW dissipated in a 500 Ω resistive load.

   The reference voltage value corresponding to 0 dB is

$$\text{P} = V^2/R \text{ or } 6 \times 10^{-3} = V^2/500 \text{ or } V = 1.73 \text{ V}$$

2. Zero dB refers to 1 mW dissipated in 600 Ω.

   Here, $1 \times 10^{-3} = V^2/600$; V = 0.774 V

3. Zero dB refers to a 1 mW dissipation.

   It is written as dBm indicating that it uses 1 mW as a reference.

   This reference does not depend on any particular load impedance value.

   Calculations are performed by using the relation

$$G_p = 10 \log_{10}{}^{P_2/0.001} \text{ dBm}$$

24.7. Variations in Amplifier Gain with Frequency

If the input voltage of an amplifier is kept constant but its frequency is varied, it is found that the amplifier gain,

1. remains practically constant over a sizable range of mid-frequencies,

2. decreases at low as well as high frequencies.

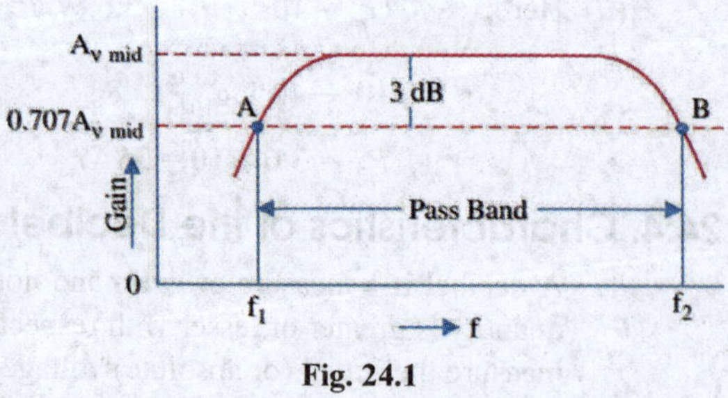

Fig. 24.1

A typical frequency-versus-gain curve is shown in Fig. 24.1. While analyzing this curve, three values of frequency are important

1. mid-frequency range,
2. lower cut-off frequency, f_1,
3. upper cut-off frequency, f_2.

The lower and upper cut-off frequencies are defined as those frequencies.

(*a*) where voltage gain of the amplifier decreases to 0.707 times the mid-frequency gain *i.e.*, to

$$\frac{1}{\sqrt{2}} A_{v.(mid)} = 0.707\ A_{v(mid)}$$

or

(*b*) in terms of power, where power amplification of the amplifier decreases to half its value at mid-frequencies

i.e., $A_{p.1}$ or $A_{p.2} = \frac{1}{2} A_{p.(mid)}$

(*c*) in terms of decibels, where power gain **falls by 3 dB.**

That is why the two cut-off frequencies are referred to as

1. –3 dB frequencies or
2. down 3 dB frequencies or
3. 3 dB loss frequencies.

The points *A* and *B* in Fig. 24.1 are called 3 dB points or (sometime) as *minus 3 dB points*.

The cut-off frequencies are also called *roll-off* frequencies because at these frequencies, the amplifier gain starts rolling down from its mid-band value or maximum value.

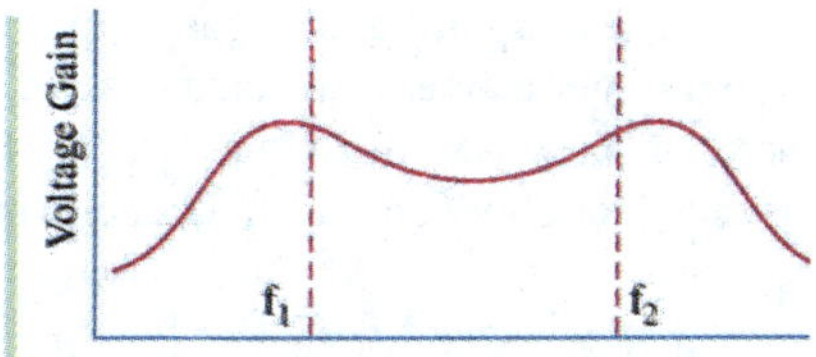

Fig. 24.2

The frequency span between these two cut-off frequencies is called the passband or bandwidth (BW) of the amplifier.

$$\therefore \qquad \Delta f = f_2 - f_1 = \text{bandwidth (BW)} = \text{passband}$$

All frequencies lying between f_1 and f_2 are amplified almost equally.

For maximum bandwidth, the stray capacitances in the amplifier must be kept to the minimum.

24.8. Changes in Voltage and Power Levels

Many amplifiers produce a maximum gain over a certain range of frequencies and a reduced gain at frequencies below and above this range as shown in Fig. 24.2.

This maximum gain is called the mid-range gain $A_{v(mid)}$ and may be assigned a zero dB value. Any other gain below or above the mid-range gain is expressed as a negative dB value. For example, if the mid-range voltage gain of a certain amplifier is 100 and the gain at a certain frequency below mid-

Table 24.1

Az	*dB change*
16	$20 \log^{16} = 24$ dB
8	$20 \log^{8} = 18$ dB
4	$20 \log^{4} = 12$ dB
2	$20 \log^{2} = 6$ dB
1	$20 \log^{1} = 0$ dB
0.5	$20 \log^{0.5} = -6$ dB
0.25	$20 \log^{0.25} = -12$ dB
0.125	$20 \log^{0.125} = -18$ dB

Table 24.2

Power Ratio (A_p)	*dB change*
8	$10 \log^{8} = 9$ dB
4	$10 \log^{4} = 6$ dB
2	$10 \log^{2} = 3$ dB
1	$10 \log^{1} = 0$ dB
0.5	$10 \log^{0.5} = -3$ dB
0.25	$10 \log^{0.25} = -6$ dB
0.125	$10 \log^{0.125} = -9$ dB

range is 50, then the reduced gain = 20 $\log^{50/100}$ = – 6 dB. It means that it is 6 dB below the 0 dB reference *i.e.*, mid-range gain. In other words, if output voltage is halved (keeping input voltage constant), the reduction in gain is 6 dB. Similarly, doubling the output voltage produces a 6 dB increase in gain. Fig. 24.2 shows gain-versus-frequency curve of an amplifier giving several dB points. Table 24.1 shows how different voltage gains translate into dB values. It will be seen that each step in the doubling or halving of the voltage ratio increase or decrease the dB value by 6 dB.

Similarly, it can be shown that every increase or decrease of the power ratio by half changes the dB level by 3 dB as shown in the Table 24.2.

24.9. Causes of Gain Variations

The primary cause of gain variation in amplifiers is the presence of capacitances, some of which are connected in series along the signal path and some in parallel.

The different types of external capacitances present in an amplifier circuit are :

1. coupling and bypass capacitors—usually of large capacitance value,
2. internal or inter-element capacitances of the transistor [Fig. 24.2 (*a*)] and stray wiring capacitance [Fig. 24.2 (*b*)].

The coupling and bypass capacitors [C_1, C_2 and C_3 in Fig. 24.3 (*a*)] are *series-connected* whereas inter element capacitances and stray capacitances are *parallel-connected* [Fig. 24.3 (*b*)] to the signal path. At mid-frequencies, C_1, C_2 and C_3 act almost as 'shorts' but C_{bc} and C_{be} act as 'open'. Hence, their effect on mid-frequencies is negligible.

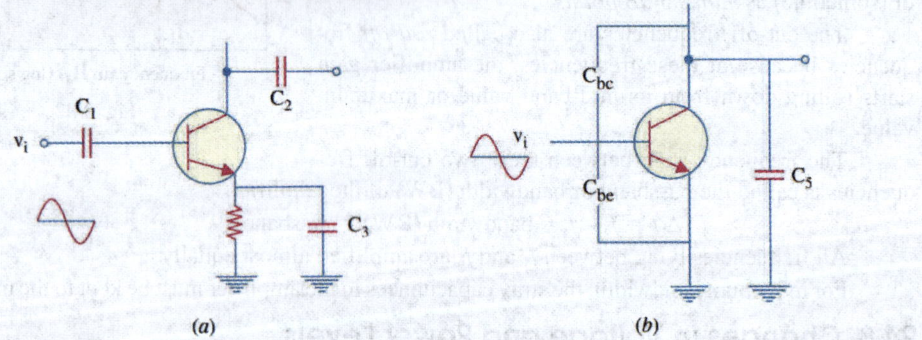

Fig. 24.3

However, at low frequencies, series-connected capacitors *i.e.*, coupling and by pass capacitors offer relatively large reactance thereby dropping off a large part of the input signal. Hence, amplifier gain starts decreasing as frequency is lowered.

The internal transistor capacitances and stray capacitances are (*i*) small and (*ii*) parallel-connected to the signal path. At low frequencies, they offer very high reactance and so act as effective '*open*'. But with increase in frequency, their reactance keeps decreasing till at very high frequencies, they almost short or shunt the ac signal to ground both at the input and output ends. This explains why amplifier gain starts decreasing at high frequencies.

In summary

(*i*) series-connected coupling and bypass capacitors cause *decrease in amplifier gain at low frequencies,*

(*ii*) parallel-connected internal transistor capacitances and stray wiring capacitances cause *decrease in amplifier gain at high frequencies.*

24.10. Miller Effect

According to this effect, when viewed from input base terminal of the *CE*-connected transistor, the capacitance C_{bc} appears as $(1 + A_v)\ C_{bc}$ *i.e.*, it is amplified by a factor of $(1 + A_v)$. In fact, Miller effect takes into account the feedback from the collector to base and *vice-versa* due to C_{bc}.

Proof

When V_i is applied to the transistor's base in Fig. 24.3 (*b*), the change in collector voltage is

$$\Delta V_c = A_v . V_i$$

The negative sign is due to phase shift inherent in a *CE* amplifier. As seen, V_C is reduced by $(A_v . V_i)$ when base voltage is *increased* by V_i. Hence, total reduction in collector base voltage is

$$\Delta V_{CB} = V_i + A_v V_i = V_i (1 + A_v)$$

This also represents the change of voltage across C_{bc} because it is connected across collector and base. Using $Q = CV$, the charge supplied to the input of the circuit is

$$Q = C_{bc} \times \Delta V_{CB} = C_{bc} \times (1 + A_v)\ V_i = (1 + A_v)\ C_{bc} \times V_i$$

Hence, C_{bc} appears as $(1 + A_v)\ C_{bc}$ when looked from the input of the circuit.

Incidentally, the total input capacitance to the transistor is

$$C_{in} = C_{be} \parallel (1 + A_v)\ C_{bc} = C_{be} + (1 + A_v)\ C_{bc}$$

At high frequencies, C_{in} reduces the input impedance of the circuit and affects the frequency response (Ex. 24.11).

Example 24.10. *A CE-connected amplifier has $C_{bc} = 4$ pF, $C_{be} = 10$ pF and $r_e = 50\ \Omega$. If circuit load resistor is 10 K, calculate the value of C_{in}.*

Solution. $A_v \cong \dfrac{R_E}{r_e} = \dfrac{100\ \text{K}}{50\ \Omega} = 200$ $\therefore$ $C_{in} = 10 + (1 + 200)\ 4 = \mathbf{814\ pF}$

Example 24.11. *Calculate the upper cut-off frequency of the CE amplifier shown in Fig. 24.4. Given the input wiring capacitance $C_{wi} = 40$ pF, $C_{bc} = 8$ pF, $C_{be} = 10$ pF and $\beta = 100$.*

Solution. $f_2 = \dfrac{1}{2\pi\ R_{eq}\ C_{in}}$

The voltage amplification of the amplifier is

$$A_v \cong \frac{R_C \parallel R_E}{R_L} = \frac{20\ \text{K} \parallel 20\ \text{K}}{400\ \Omega} = 25$$

The total capacitance from base to ground is

$$C_{in} = C_{wi} + C_{be} + (1 + A_v)\ C_{bc}$$
$$= 40 + 10 + (1 + 25) \times 8 = 258\ \text{pF}$$

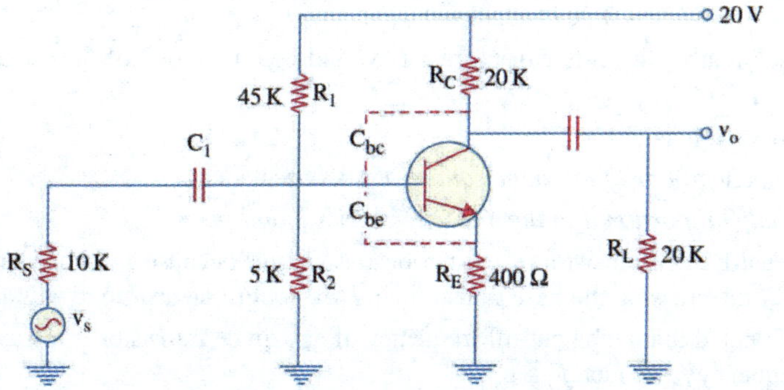

Fig. 24.4

Now, let us determine the resistance 'seen' by C_{in}. If we look to the right, a resistance of $\beta (r_e + R_E) \cong \beta R_E$ is seen while to the left $R_1 \parallel R_2 \parallel R_S$ is seen, all in parallel.

$$\therefore \quad R_{eq} = R_1 \parallel R_2 \parallel R_S \parallel \beta R_E$$
$$= 45 \text{ K} \parallel 5 \text{ K} \parallel 10 \text{ K} \parallel 40 \text{ K} = 2.88 \text{ K}$$

$$\therefore \quad f_2 = \frac{1}{2\pi \times 2.88 \times 10^3 \times 258 \times 10^{-12}} = \mathbf{214\ kHz}$$

24.11. Cut-off Frequencies of Cascaded Amplifiers

Cascading of stages gives higher amplification but narrow bandwidth because the product of the two remains almost constant. The approximate values of composite lower and upper cut-off frequencies of the cascaded amplifier having *n* identical stages are

$$f_{1.n} = 1.1\sqrt{n} \times f_1 \text{ per stage} = \frac{f_1 \text{ per stage}}{\sqrt{(2^{1/n} - 1)}}$$

$$f_{2.n} = \frac{f_2 \text{ per stage}}{1.1\sqrt{n}} = \sqrt{(2^{1/n} - 1)} \times f_2 \text{per stage}$$

The reduced bandwidth of the cascaded amplifier is $= f_{2.n} - f_{1.n}$.

24.12. Transistor Cut-off Frequencies

Even if no external stray capacitances were present in an amplifier, there would still be an upper limit on its frequency response due to

(*i*) internal or inter-element capacitances of the transistor and

(*ii*) transit time of charge carriers across the transistor junctions and through the semiconductor material.

This limitation is expressed in terms of

(*i*) alpha cut-off frequency (f_α) and (*ii*) beta cut-off frequency (f_β)

The two are defined below.

24.13. Alpha Cut-off Frequency

It is found that at high frequencies, the value of transistor α begins to fall. This decrease in α is related to the transit time effect of the charge carriers as they move from the emitter to collector.

The alpha cut-off frequency f_α is that high frequency at which the α of a *CB*-connected transistor becomes 0.707 of its low-frequency value (usually 1 kHz).

For example, if value of α at 1 kHz is 0.98, then its value at f_α would be

$$= 0.707 \times 0.98 = 0.693$$

It means that at f_α, the collector current I_C would be only 0.693 of the emitter current rather than 0.98 I_E.

It is found that f_α is

(a) inversely proportional to the square of the base width,

(b) directly proportional to the minority carrier mobility.

In this regard, *NPN* transistors are superior to *PNP* type because electrons have greater mobility than holes. For decreasing the base transit time, *base should be as thin as possible.*

It may be noted that alpha cut-off frequency of any given transistor is always greater than its beta cut-off frequency f_β. In fact, $f_\alpha \cong \beta f_\beta$.

24.14. Beta Cut-off Frequency

It is that high frequency at which the β of a *CE*-connected transistor drops to 0.707 of its low-frequency (1 kHz) value.

24.15. The f_T of a Transistor

It is another high-frequency characteristic of a transistor.

It is that high frequency of a CE-connected transistor where its β drops to unity *i.e.*, β = 1. This frequency is much larger than f_β but less than f_α. For example, for a typical transistor, their values may be $f_\beta = 6$ MHz, $f_T = 300$ MHz and $f_\alpha = 345$ MHz.

24.16. Relation Between f_α , f_β and f_T

(*i*) For simple junction transistors $f_\alpha \cong 1.2 f_T$ and (*ii*) $f_\beta = \dfrac{f_T}{\beta}$, where β is the low-frequency value of β.

24.17. Gain-Bandwidth Product

As its name indicates, it is the product of the gain and bandwidth of an amplifier. It is very useful in comparing the performance capability of various circuits.

For any amplifier, gain-bandwidth product (*GBP*) is constant and is equal to f_T.

If, for example, β = 1, at a frequency of 6 MHz, then $f_T = 6$ MHz.

∴ Gain-bandwidth product = 6 MHz.

Hence, for a given f_T in an amplifier, increased gain may be obtained only at the expense of its bandwidth.

Example 24.12. *A transistor has $f_\alpha = 8$ MHz and β = 80. When connected as an amplifier, it has stray capacitance of 100 pF at the output terminal. Calculate its upper 3 dB frequency when R_L is (a) 10 K and (b) 100 K.*

Solution. It should be remembered that stray capacitance would reduce the amplifier gain by 3 dB when

capacitive reactance = output resistance

i.e.,
$$\frac{1}{2\pi f_s C_s} = R_L \quad \text{or} \quad f_s = \frac{1}{2\pi C_s R_L}$$

First, let us find the value of f_β for comparison with f_s.

$$f_\beta = \frac{f_\alpha}{80} = \frac{8 \text{ MHz}}{80} = 100 \text{ kHz}$$

(*a*) $R_L = 10$ K

$$f_s = \frac{1}{2\pi\, C_s\, R_L} = \frac{1}{2\pi \times 100 \times 10^{-12} \times 10 \times 10^3} = 159 \text{ kHz}$$

Obviously, before this frequency is reached, cut-off would have been achieved at $f_\beta = 100$ kHz. Hence, $f_2 = f_\beta =$ **100 kHz.**

(*b*) $R_L = 100$ K

In this case,
$$f_s = \frac{10}{100} \times 159 = 15.9 \text{ kHz}$$

In this case, cut-off would be achieved much earlier at 15.9 kHz before f_β is reached.

∴
$$f_2 = \mathbf{15.9 \text{ kHz.}}$$

CONVENTIONAL PROBLEMS

1. An amplifier raises the power level of its 5 μW input signal by 30 dB. What is the output power ? **[5 mW]**
2. An attenuation network provides an output of 5 μW with an input of 5 mW. Calculate the decibel loss of the network. **[–30 dB]**
3. What is the decibel difference between a 100 kW and 500 kW radio transmitters ? **[6.98 dB]**
4. The noise level of a certain tape recording is 30 dB below the signal level. If the signal power is 5 mW, calculate the noise power. **[5 mW]**
5. An amplifier rated at 72 W output is connected to an 8 Ω speaker.
   (*a*) What input power is required for full power output if power gain is 30 dB ?
   (*b*) What is the input voltage for rated output if amplifier voltage gain is 40 dB ?
   [(a) 72 mW (b) 240 mV]
6. The characteristics of a certain audio amplifier are such that it gives a voltage amplification of 10 at 100 Hz, 30 at 3 kHz and 60 at 10 kHz. Taking the amplification at 3 kHz as the reference level, calculate the loss or gain in decibels at the other two frequencies. **[–9.54 dB, 6.02 dB]**
7. An amplifier with full power rating of 100 W drives a speaker load of 16 Ω. The hum-level rating of the amplifier is 80 dB below its full-power rating. Calculate
   (*i*) hum-level in the load,
   (*ii*) voltage produced by the hum across the load. **[(*i*) 1 μW (*ii*) 4 mW]**

SELF EXAMINATION QUESTIONS

A. Fill in the blanks by most appropriate word (s) or numerical value (s).

1. Decibel system is a ....system of comparing two physical quantities.
2. The decibel is the unit of power ....
3. One decibel represents a power ratio of ....
4. Decibel system is............. because 20 dB is not twice as much power as 10 dB.
5. Amplifier gain decreases at low as well as at .... frequencies.
6. Lower and upper cut-off frequencies are also called minus.......... dB frequencies.
7. The bandwidth of an amplifier is the band of frequencies lying between the two minus 3 .... points on its frequency response curve.
8. Lower roll-off frequency is imposed by............ and bypass capacitors.
9. Internal transistor capacitances and stray wiring capacitance determine the............ cut-off frequency of an amplifier.
10. The α of a transistor .... with increase in frequency.
11. Alpha cut-off frequency of a transistor is .... than its beta cut-off frequency.
12. Greater the minority carrier mobility, ... the alpha cut-off frequency of a transistor.

B. Answer True or False

1. Decibel is a measure of power change and not of power.
2. 20 dB represents twice as much power as 10 dB.
3. The zero dB reference level is zero watt.
4. Human response to hearing is logarithmic.
5. Negative decibels indicate loss.
6. When power output of a radio receiver is doubled, its decibel level is also doubled.
7. When power is halved, its level falls by 3 dB.
8. Decibel system is linear.
9. Amplifier gain rolls off both at lower and higher frequencies.
10. Low-frequency roll-off in amplifier gain is due to its stray wiring capacitances.
11. Internal capacitances of a transistor determine its high alpha and beta cut-off frequencies.
12. Bandwidth of an amplifier can be maximised by minimizing its stray capacitance between wiring and earth.
13. Coupling and bypass capacitors of an amplifier determine its lower cut-off frequency.
14. Cascading of amplifiers gives broader bandwidth.
15. Gain-bandwidth product of a transistor equals its f_T.

16. The f_β of a transistor can never be greater than its f_α.

C. Multiple Choice Items

1. The decibel is a measure of
   (a) power (b) voltage
   (c) current (d) power level
2. When power output of an amplifier doubles, the increase in its power level is .... decibels.
   (a) 2 (b) 20
   (c) 3 (d) 10
3. When output power level of a radio receiver increases by 3 dB, its absolute power changes by a factor of
   (a) 2 (b) 10
   (c) 1/2 (d) 3
4. Zero watt cannot be chosen as zero decibel level because
   (a) it is impossible to measure zero watt
   (b) it is too small
   (c) every power compared with it would be zero
   (d) it would be impossible to define a decibel
5. The bandwidth of an amplifier may be increased by
   (a) decreasing the capacitance of its bypass capacitors
   (b) minimizing its stray capacitances
   (c) increasing input signal frequency
   (d) cascading it
6. Lower cut-off frequency of an amplifier is primarily determined by the
   (a) internal capacitances of the active device used
   (b) stray capacitance between its wiring and ground
   (c) ac β value of its active devices
   (d) capacitances of coupling and bypass capacitors
7. The main reason for the variation of amplifier gain with frequency is
   (a) the presence of capacitances, both external and internal
   (b) due to interstage transformers
   (c) the logarithmic increase in its output power
   (d) the Miller effect
8. Series-connected external capacitors in an amplifier determine its cut-off frequency.
   (a) alpha (b) lower
   (c) upper (d) beta

ANSWERS

A. Fill in the blanks

1. logarithmic **2.** level **3.** 1.26 **4.** non-linear **5.** high **6.** 3
7. dB **8.** coupling **9.** upper **10.** decreases **11.** higher **12.** higher.

B. True or False

1. T **2.** F **3.** F **4.** T **5.** T **6.** F **7.** T **8.** F
9. T **10.** F **11.** T **12.** T **13.** T **14.** F **15.** T **16.** T.

C. Multiple Choice Items

1. *d* **2.** *c* **3.** *a* **4.** *d* **5.** *b* **6.** *d* **7.** *a* **8.** *b*.

25 CHAPTER

Feedback Amplifiers

1. Feedback Amplifiers
2. Principle of Feedback Amplifiers
3. Advantages of Negative Feedback
4. Gain Stability
5. Decreased Distortion
6. Increased Bandwidth
7. Forms of Negative Feedback
8. Shunt-derived Series-fed Voltage Feedback
9. Current-series Feedback Amplifier
10. Voltage-shunt Negative Feedback Amplifier
11. Current-shunt Negative Feedback Amplifier

25.1. Feedback Amplifiers

A feedback amplifier is that in which a *fraction of the amplifier output is fed back to the input circuit*. This partial dependence of amplifier input on its output helps to control the output. A feedback amplifier consists of two parts : an amplifier and a feedback circuit.

(*i*) Positive feedback

If the feedback voltage (or current) is so applied as to *increase the input voltage* (*i.e.*, it is in phase with it), then it is called *positive feedback*. Other names for it are : *regenerative* or *direct* feedback.

Since positive feedback produces excessive distortion, it is seldom used in amplifiers. However, because it increases the power of the original signal, it is used in oscillator circuits.

(*ii*) Negative feedback

If the feedback voltage (or current) is so applied as to reduce the amplifier input (*i.e.*, it is 180° out of phase with it), then it is called negative feedback. Other names for it are : *degenerative* or *inverse* feedback.

Negative feedback is frequently used in amplifier circuits.

25.2. Principle of Feedback Amplifiers

For an ordinary amplifier *i.e.*, one without feedback, the voltage gain is given by the ratio of the output voltage V_o

and input voltage V_i. As shown in the block diagram of Fig. 25.1, the input voltage V_i is amplified by a factor of A to the value V_o of the output voltage.

$$\therefore \quad A = V_o/V_i$$

This gain A is often called **open-loop** gain.

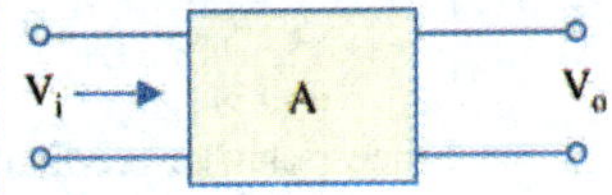

Fig. 25.1

Suppose a feedback loop is added to the amplifier (Fig. 25.2). If V_o' is the output voltage **with feedback**, then a fraction β of this voltage is applied to the input voltage which, therefore, becomes $(V_i \pm \beta V_o')$ depending on whether the feedback voltage is in phase or antiphase with it. Assuming positive feedback, the input voltage will become $(V_i + \beta V_o')$. When amplified A times, it becomes $A\,(V_i + \beta V_o')$.

$$\therefore \quad A\,(V_i + \beta V_o') = V_o'$$

$$\text{or} \quad V_o'(1 - \beta A) = AV_i$$

The amplifier gain A' **with feedback** is given by

$$A' = \frac{V_o'}{V_i} = \frac{A}{1-\beta A}$$

$$\therefore \quad A' = \frac{A}{1-\beta A} \quad \text{— positive feedback}$$

$$= \frac{A}{1-(-\beta A)} = \frac{A}{1+\beta A} \quad \text{— negative feedback}$$

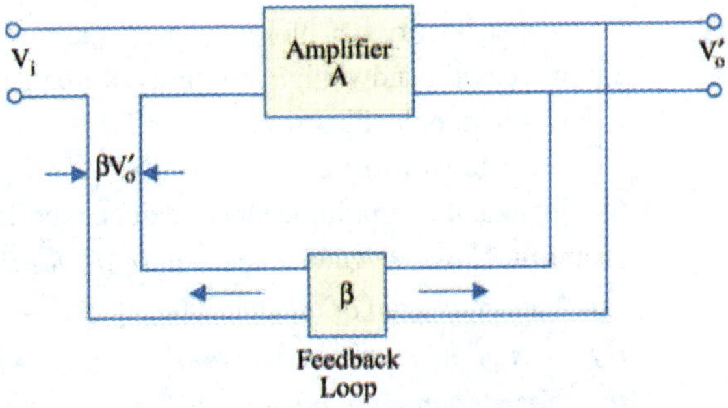

Fig. 25.2

The term 'βA' is called **feedback factor** whereas β is known as **feedback ratio.** The expression $(1 \pm \beta A)$ is called **loop gain.** The amplifier gain A' with feedback is also referred to as **closed-loop gain** because it is the gain obtained after the feedback loop in closed.

(*a*) Negative Feedback

The amplifier gain with negative feedback is given by $A' = \dfrac{A}{1+\beta A}$

Obviously, $A' < A$ because $|1 + \beta A| > 1$. Suppose, $A = 90$ and $\beta = 1/10 = 0.1$. Then, gain **without** feedback is 90 and **with** negative feedback is

$$A' = \frac{A}{1+\beta A} = \frac{90}{1+0.1\times 90} = 9$$

As seen, negative feedback reduces the amplifier gain. That is why it is called **degenerative feedback.** A lot of voltage gain is sacrificed due to negative feedback. When $|\beta A| \gg 1$, then

$$A' \cong \frac{A}{\beta A} \cong \frac{1}{\beta}$$

It means that A' depends only on β. But it is very stable because it is not affected by changes in temperature, device parameters, supply voltage and from the ageing of circuit components etc. Since resistors can be selected very precisely with almost zero temperature-coefficient of resistance, it is possible to achieve highly precise and stable gain with negative feedback.

(*b*) Positive Feedback

The amplifier gain with positive feedback is given by

$$A' = \frac{A}{1-\beta A} \qquad \text{Since } |1 - \beta A| < 1,\ A' > A.$$

Suppose gain **without feedback** is 90 and $\beta = 1/100 = 0.01$, then gain with positive feedback is

$$A' = \frac{90}{1 - (0.01 \times 90)} = 900$$

Since positive feedback *increases* the amplifier gain. It is called **regenerative** feedback. If $\beta A = 1$, then mathematically, the gain becomes infinite which simply means that **there is an output without any input.** However, electrically speaking, this cannot happen. What actually happens is that the amplifier becomes an oscillator which **supplies its own input.** In fact, the two important and necessary conditions for circuit oscillation are

1. the feedback must be positive,
2. feedback factor must be unity *i.e.*, $\beta A = +1$.

25.3. Advantages of Negative Feedback

The numerous advantages of negative feedback outweigh its only disadvantage of reduced gain. Among the advantages are :

1. higher fidelity *i.e.*, more linear operation,
2. highly stabilized gain,
3. increased bandwidth *i.e.*, improved frequency response,
4. less amplitude distortion,
5. less harmonic distortion,
6. less frequency distortion,
7. less phase distortion,
8. reduced noise,
9. input and output impedances can be modified as desired.

Example 25.1. *In the series-parallel (SP) feedback amplifier of Fig. 25.3, calculate*

(a) open-loop gain of the amplifier,

(b) gain of the feedback network,

(c) closed-loop gain of the amplifier,

(d) sacrifice factor, S.

Solution. (*a*) Since 1 mV goes into the amplifier and 10 V comes out

$$\therefore \quad A = 10 \text{ V}/1 \text{ mV} = \mathbf{10{,}000}$$

(*b*) The feedback network is being driven by the output voltage of 10 V.

$\therefore$ gain of the feedback network

$$= \frac{\text{output}}{\text{input}} = \frac{250 \text{ mV}}{10 \text{ V}} = \mathbf{0.025}$$

(*c*) So far as the feedback amplifier is concerned, input is (250 + 1) = 251 mV and final output is 10 V. Hence, gain **with feedback** is

$$A' = 10 \text{ V}/251 \text{ mA} = \mathbf{40}$$

(*d*) The sacrifice factor is given by

$$S = \frac{A}{A'} = \frac{10{,}000}{40} = \mathbf{250}$$

By sacrificing so much voltage gain, we have improved many other amplifier quantities (Art. 25.3).

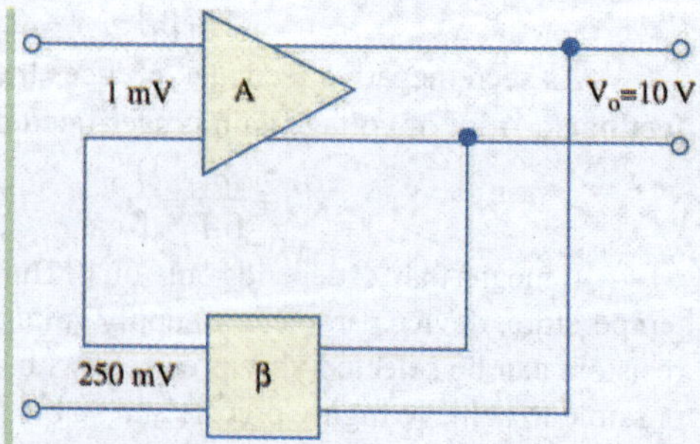

Fig. 25.3

Example 25.2. *Calculate the gain of a negative feedback amplifier whose gain without feedback is 1000 and $\beta = 1/10$.*

Solution. Since $|\beta A| >> 1$, the closed-loop gain is

$$A' = \frac{1}{\beta} = \frac{1}{1/10} = \mathbf{10}$$

Example 25.3. *In a negative-feedback amplifier, $A = 100$, $\beta = 0.04$ and $V_i = 50$ mV. Find*
(a) *gain with feedback,* (b) *output voltage,*
(c) *feedback factor,* (d) *feedback voltage.*

Solution. (a)
$$A' = \frac{A}{1+\beta A} = \frac{100}{1+0.04\times 100} = \mathbf{20}$$

(b) $$V_o' = A'V_i = 20\times 50\text{ mV} = \mathbf{1\ V}$$

(c) feedback factor $= \beta A = 0.04 \times 100 = \mathbf{4}$

(d) feedback voltage $= \beta V_o' = 0.04 \times 1 = \mathbf{0.04\ V}$

Example 25.4. *In Example 25.2, to what value should the input voltage be increased in order that the output voltage with feedback equals the output voltage without feedback?*

Solution. The new increased input voltage is given by
$$V_i' = V_i(1+\beta A) = 50\,(1+0.04\times 100) = \mathbf{250\ mV}$$

Example 25.5. *An amplifier having a gain of 500 without feedback has an overall negative feedback applied which reduces the gain to 100. Calculate the fraction of output voltage fed back. If due to ageing of components, the gain without feedback falls by 20%, calculate the percentage fall in gain without feedback.*

Solution.
$$A' = \frac{A}{1+\beta A} \qquad \therefore \qquad 1+\beta A = \frac{A}{A'}$$

$$\therefore \qquad \beta = \frac{1}{A'} - \frac{1}{A} = \frac{1}{100} - \frac{1}{500} = \mathbf{0.008}$$

Now, gain without feedback = 80% of 500 = 400

$$\therefore \text{ New} \qquad A' = \frac{400}{1+0.008\times 400} = 95.3$$

Hence, change in the gain with feedback in the two cases = 100 – 95.3 = 4.7

$\therefore$ percentage fall in gain *with* feedback is $= \frac{4.7}{100}\times 100 = \mathbf{4.7\%}$

25.4. Gain Stability

The gain of an amplifier with negative feedback is given by $A' = \frac{A}{1+\beta A}$

Taking logs of both sides, we have $\log A' = \log A - \log(1+\beta A)$

Differentiating both sides, we get
$$\frac{dA'}{A'} = \frac{dA}{A} - \frac{\beta . dA}{1+\beta A} = dA\left(\frac{1}{A} - \frac{\beta}{1+\beta A}\right) = \frac{1}{1+\beta A}\cdot\frac{dA}{A}$$

If $\beta A \gg 1$, then the above expression becomes
$$\frac{dA'}{A'} = \frac{1}{\beta A}\cdot\frac{dA}{A}$$

Example 25.6. *An amplifier has an open-loop gain of 400 and a feedback of 0.1. If open-loop gain changes by 20% due to temperature, find the percentage change in closed-loop gain.*

Solution. Here, $A = 400$, $\beta = 0.1$, $dA/A = 20\% = \mathbf{0.2}$

Now,
$$\frac{dA'}{A'} = \frac{1}{\beta A}\cdot\frac{dA}{A} = \frac{1}{0.1\times 400}\times 20\% = \mathbf{0.5\%}$$

It is seen that while the amplifier gain changes by 20%, the feedback gain changes by only 0.5% *i.e.*, an improvement of 20/0.5 = 40 times.

25.5. Decreased Distortion

Let the harmonic distortion voltage generated within the amplifier change from D to D' when negative feedback is applied to the amplifier.

Suppose $D' = x D$ ...(*i*)

The fraction of the output distortion voltage which is fed back to the input is

$$= \beta D' = \beta x D$$

After amplification, it becomes $\beta x DA$ and is antiphase with original distortion voltage D. Hence, the new distortion voltage D' which appears in the output is

$$D' = D - \beta x DA \quad \text{...(ii)}$$

From (*i*) and (*ii*), we get

$$x D = D - \beta x DA \quad \text{or} \quad x = \frac{1}{1 + \beta A}$$

Substituting this value of x in Eq. (*i*) above, we have $D' = \dfrac{D}{1 + \beta A}$

It is obvious from the above equation that $D' < D$. In fact, negative feedback reduces the amplifier distortion by the amount of loop gain *i.e.*, by a factor of $(1 + \beta A)$.

However, it should be noted that improvement in distortion is possible only when the distortion is produced by the **amplifier itself**, not when it is already present in the input signal.

Example 25.7. *An amplifier has a gain of 100 and 5 per cent distortion with an input signal of 1 V. When an input signal of 1 V is applied to the amplifier, calculate*

(i) output signal voltage, (ii) distortion voltage, (iii) output voltage.

Solution. (*i*) Signal output voltage $V_{o.s} = AV_i = 100 \times 1 = \mathbf{100\ V}$

(*ii*) distortion voltage $= DV_o = 0.05 \times 100 = \mathbf{5\ V}$

(*iii*) amplifier output voltage $V_o = V_{os} + D = 100 + 5 = \mathbf{105\ V}$

25.6. Increased Bandwidth

The bandwidth of an amplifier without feedback is equal to the separation between the 3 dB frequencies f_1 and f_2.

$\therefore \quad BW = f_2 - f_1$

where f_1 = lower 3 dB frequency

and f_2 = upper 3 dB frequency

If A is its gain, then the gain-bandwidth product is $A \times BW$.

Now, when negative feedback is applied, the amplifier gain is reduced. Since the **gain-bandwidth product has to remain the same in both cases,** it is obvious that the bandwidth must increase to compensate for the decrease in gain. It can be proved that with negative feedback, the lower and upper 3 dB frequencies of an amplifier become

$$f_1' = \frac{f_1}{(1 + \beta A)}$$

and $f_2' = f_2 (1 + \beta A)$

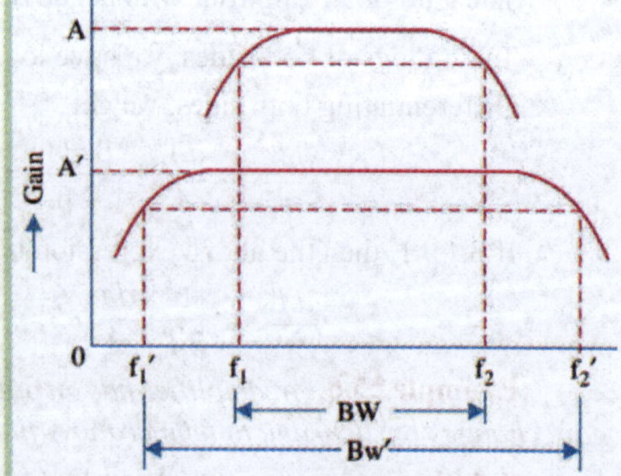

Fig. 25.4

As seen from Fig. 25.4, f_1' has decreased whereas f_2' has increased thereby giving a wider separation or bandwidth. Since gain-bandwidth product is the same in both cases.

$$\therefore \quad A \times BW = A' \times BW'$$

$$\text{or} \quad A(f_2 - f_1) = A'(f'_2 - f'_1)$$

Example 25.8. *An RC-coupled amplifier has a mid-frequency gain of 200 and a frequency response from 100 Hz to 20 kHz. A negative feedback network with β = 0.02 is incorporated into the amplifier circuit. Determine the new system performance.*

Solution.

$$A' = \frac{A}{1 + \beta A} = \frac{200}{1 + 0.02 \times 200} = \mathbf{40\ Hz}$$

$$f_1' = \frac{f_1}{1 + \beta A} = \frac{100}{1 + 0.02 \times 200} = \mathbf{20\ Hz}$$

$$f_2' = f_2 (1 + \beta A) = 20 (1 + 0.02 \times 200) = \mathbf{100\ Hz}$$

$$\therefore \quad dW' = f_2' - f'_1 \cong \mathbf{100\ kHz}$$

Incidentally, it may be proved that gain-bandwidth product remains constant in both cases.

$$dW = f_2 - f_1 \cong 20\ \text{kHz}$$

$$A \times dW = 200 \times 20 = 4000\ \text{kHz}$$

$$\therefore \quad A' \times dW' = 40 \times 100 = 4000\ \text{kHz}$$

As expected, the two are equal.

25.7. Forms of Negative Feedback

The four basic arrangements for using negative feedback are shown in the block diagram of Fig. 25.5. As seen, both voltage and current can be fed back to the input either in series or in parallel. The output voltage provides input in Fig. 25.5 (*a*) and (*b*). However, the input to the feedback network is derived from the output current in Fig. 25.5 (*c*) and (*d*).

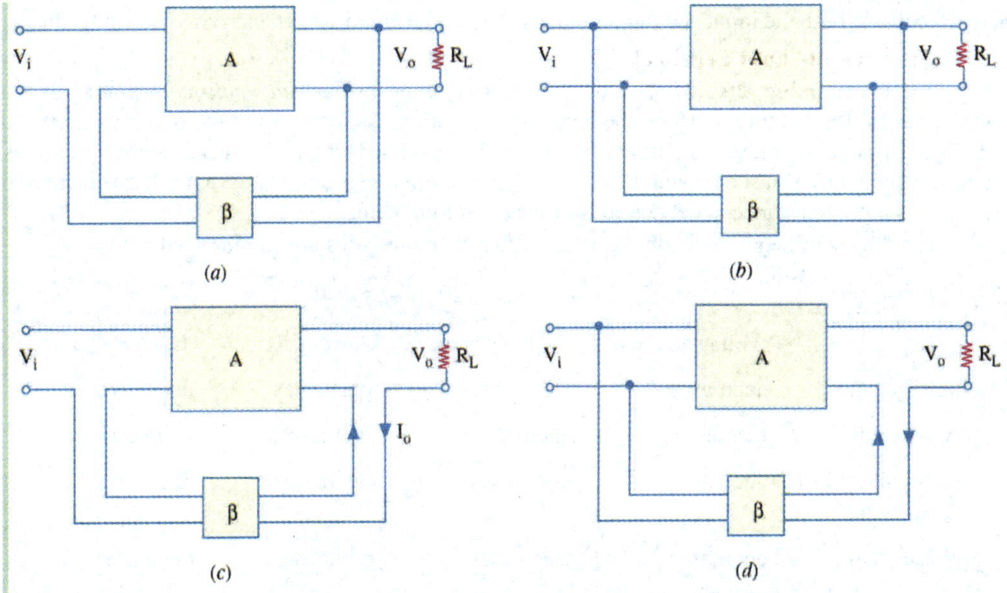

Fig. 25.5

(*a*) Voltage-series Feedback

It is shown in Fig. 25.5 (*a*). It is also called *shunt-derived series-fed feedback*. The amplifier and feedback circuit are connected series-parallel. Here, a fraction of the output voltage is applied in *series* with the input voltage *via* the feedback. As seen, the input to the *feedback network* is in *parallel* with the output of the amplifier. Therefore, so far as V_o is concerned, output resistance of the amplifier is *reduced by the shunting effect* of the input to the feedback network. It can be proved that

$$R'_o = \frac{R_o}{(1 + \beta A)}$$

Similarly, V_i sees two circuit elements in series :

(*i*) the input resistance of the amplifier and

(*ii*) output resistance of the feedback network.

Hence, input resistance of the amplifier as a *whole is increased* due to feedback. It can be proved that

$$R'_i = R_i(1 + \beta A)$$

In fact, *series feedback always increases the input impedance by a factor of* $(1 + \beta A)$.

(*b*) Voltage-shunt Feedback

It is shown in Fig. 25.5 (*b*). It is also known as *shunt-derived shunt-fed feedback i.e.*, it is parallel-parallel (*PP*) prototype. Here, a small portion of the output voltage is coupled back to the input voltage parallel (shunt).

Since the feedback network shunts both the output and input of the amplifier, it *decreases* both its output and input impedances by a factor of $1/(1 + \beta A)$.

A shunt feedback *always decreases input impedance.*

Shunt voltage.

(*c*) Current-series Feedback

It is shown in Fig. 25.5 (*c*). It is also known as *series-derived series-fed feedback*. As seen, it is a series-series (SS) circuit. Here, a part of the output current is made to feedback a proportional voltage in series with the input. Since it is a series pick-up and a series feedback, both the input and output impedances of the amplifier are *increased* due to feedback.

(*d*) Current-shunt Feedback

It is shown in Fig. 25.5 (*d*). It is also referred to as *series-derived shunt-fed feedback*. It is a parallel-series (PS) prototype. Here, the feedback network picks up a part of the output current and develops a feedback voltage in parallel (shunt) with the input voltage. As seen, feedback network shunts the input but is in series with the output. Hence, output resistance of the amplifier is increased whereas its input resistance is decreased by a factor of loop gain.

The effects of negative feedback on amplifier characteristics are summarized below :

Characteristic	***Type of Feedback***			
	Voltage series	Voltage shunt	Current series	Current shunt
Voltage gain	decreases	decreases	decreases	decreases
Bandwidth	increases	increases	increases	increases
Harmonic Distortion	decreases	decreases	decreases	decreases
Noise	decreases	decreases	decreases	decreases
Input Resistance	increases	decreases	increases	decreases
Output Resistance	decreases	decreases	increases	increases

25.8. Shunt-derived Series-fed Voltage Feedback

The basic principle of such a voltage-controlled feedback is illustrated by the block diagram of Fig. 25.6. Here, the feedback voltage is derived from the voltage divider circuit formed of R_1 and R_2.

As seen, the voltage drop across R_1 forms the feedback voltage V_f.

$$\therefore \quad V_f = V_o \frac{R_1}{R_1 + R_2} = \beta V_o$$

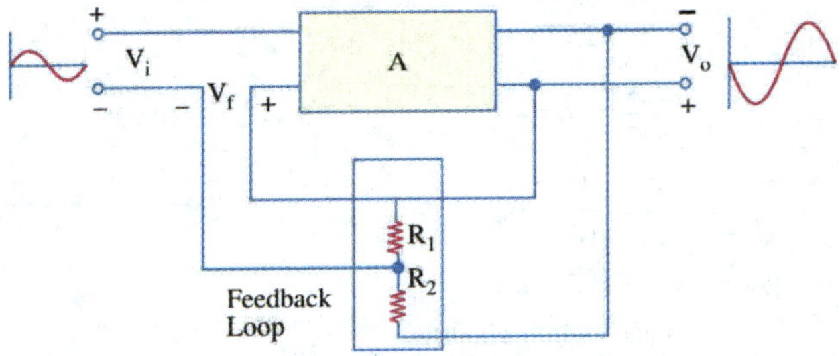

Fig. 25.6

Example 25.9. *In the voltage-controlled negative feedback amplifier of Fig. 25.7, calculate (a) voltage gain without feedback, (b) feedback factor, (c) voltage gain with feedback. Neglect* V_{BE} *and use* $r_e = 25\ mV/I_E$.

Solution. (*a*) $A = \frac{r_L}{r_e} = \frac{R_3}{r_e}$

Now, $I_B = \frac{15\text{ V}}{1.5\text{ M}} = 10\ \mu\text{A}$

$I_E = \beta I_B = 100 \times 10 = 1\text{ mA}$

$r_e = 25/1 = 25\ \Omega$

$A = \frac{10\text{ K}}{25\ \Omega} = \mathbf{400}$

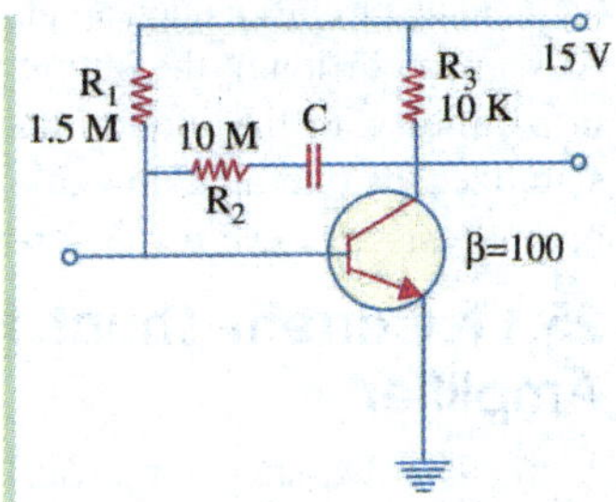

Fig. 25.7

(*b*) $\beta = \frac{R_1}{R_1 + R_2} = \frac{1.5 \times 10^6}{(1.5 + 10) \times 10^6} = 0.13$

$\therefore \quad \beta A = 0.13 \times 400 = \mathbf{52}$

(*c*) $A' = \frac{A}{1 + \beta A'} = \frac{400}{1 + 52} = \mathbf{7.55}$

25.9. Current-series Feedback Amplifier

Fig. 25.8 shows a series-derived series-fed feedback amplifier circuit. Since the emitter resistor is unbypassed, it effectively provides current-series feedback. When I_E passes through R_E, the feedback voltage drop $V_f = I_E R_E$ is developed which is applied in phase opposition to the input voltage V_i. This negative feedback reduces the output voltage V_o. This feedback can, however, be eliminated by either removing or bypassing the emitter resistor.

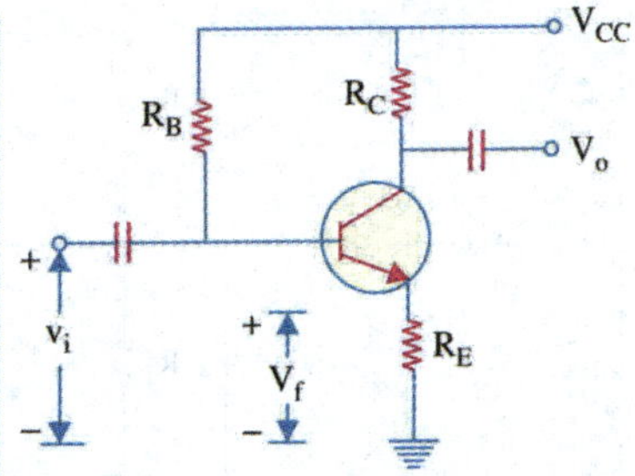

Fig. 25.8

It can be proved that

$$\beta = \frac{R_E}{R_C}; \quad A' = \frac{R_C}{r_e + R_E}; \quad A = \frac{R_C}{r_e}$$

Example 25.10. *For the current-series feedback amplifier of Fig. 25.9, calculate*

(i) voltage gain without feedback, (ii) feedback factor, (iii) voltage gain with feedback. Neglect V_{BE} and use $r_e = 25\ mV/I_E$

Solution. (*i*) $$A = \frac{R_C}{r_e}$$

Now, $$I_E = \frac{V_{CC}}{R_E + R_B/\beta} = \frac{10}{1 + 900/100} = 1\ \text{mA}$$

$\therefore$ $$r_e = 25/1 = 25\ \Omega$$

$\therefore$ $$A = \frac{10\ \text{K}}{25\ \Omega} = \mathbf{400}$$

(*ii*) $$\beta = \frac{R_E}{R_C} = \frac{1}{10} = 0.1$$

$$\beta A = 0.1 \times 400 = \mathbf{40}$$

(*iii*) $$A' = \frac{R_C}{r_e + R_E} = \frac{10,000}{20 + 1000} = \mathbf{9.756}$$

or $$A' = \frac{A}{1 + \beta A} = \frac{400}{1 + 400} = \mathbf{9.756}$$

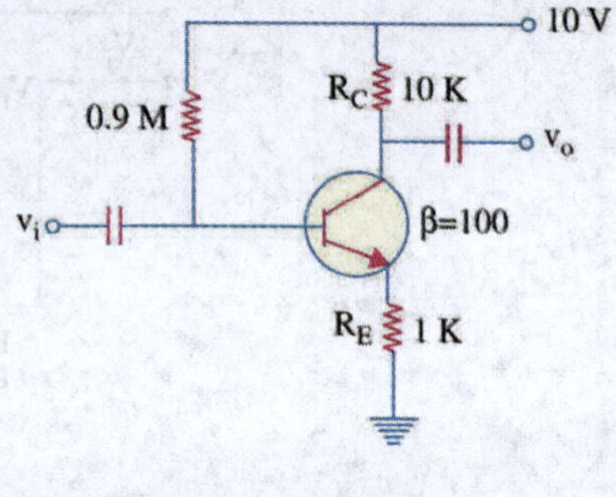

Fig. 25.9

25.10. Voltage-shunt Negative Feedback Amplifier

The circuit of such an amplifier is shown in Fig. 25.10. As seen, a portion of the output voltage is coupled through R_E in parallel with the input signal at the base. This feedback stabilizes the overall gain while decreasing both the input and output resistances. It can be proved that $\beta = R_C/R_F$.

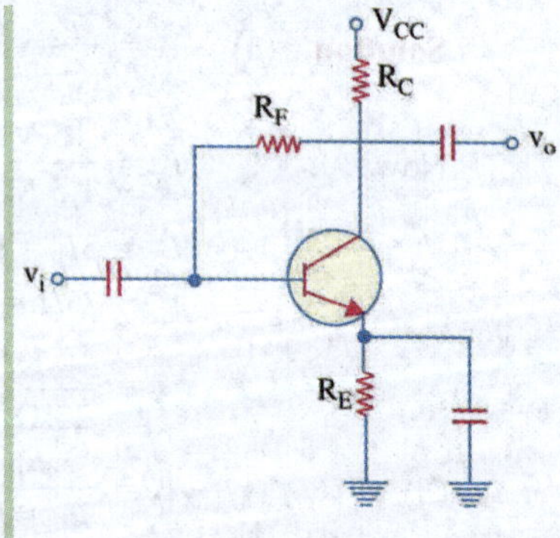

Fig. 25.10

25.11. Current-shunt Negative Feedback Amplifier

The two-stage amplifier employing such a feedback is shown in Fig. 25.11. The feedback circuit (consisting of C_F and R_F) samples the output current and develops a feedback voltage in parallel with the input voltage. The unbypassed emitter resistor of Q_2 provides current sensing.

The polarity of the feedback voltage is such that it provides the negative feedback.

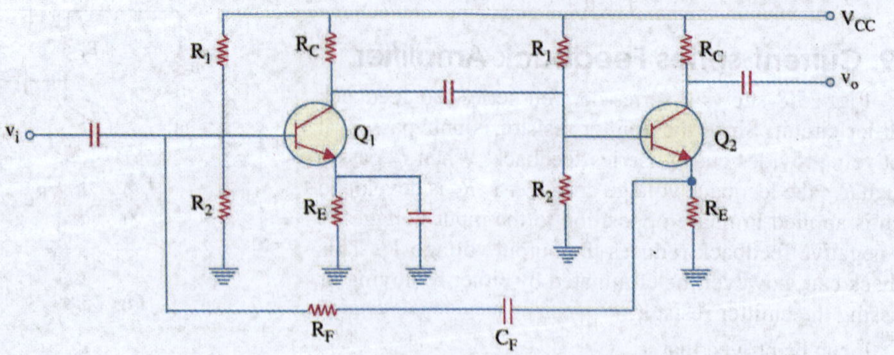

Fig. 25.11

CONVENTIONAL PROBLEMS

1. For the series-parallel feedback amplifier shown in Fig. 25.12, calculate
   (*i*) open-loop gain (*ii*) gain of feedback loop
   (*iii*) closed-loop gain (*iv*) sacrifice factor **[(*i*) 10^6 (*ii*) 0.025 (*iii*) 40 (*iv*) 25.000]**

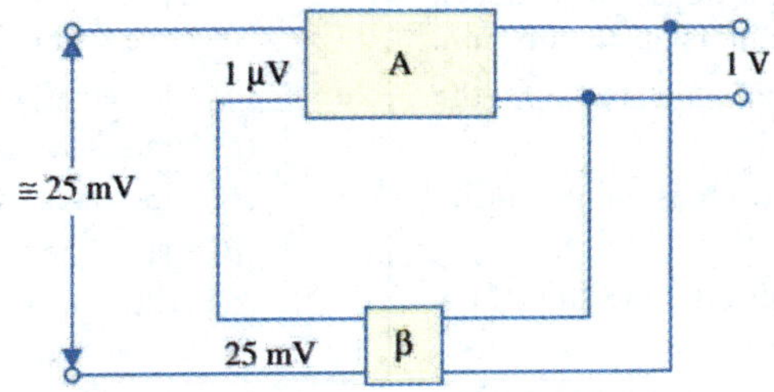

Fig. 25.12

2. A negative-feedback amplifier has the following parameters :
   A = 200, β = 0.02 and V_i = 5 mV
   Compute the following :
   (*i*) gain with feedback (*ii*) output voltage
   (*iii*) feedback factor (*iv*) feedback voltage **[(*i*) 40 (*ii*) 200 mV (*iii*) 4 (*iv*) 8 mV]**
3. An amplifier has an open-loop gain of 500 and a feedback of 0.1. If open-loop gain changes by 25% due to temperature etc., find the percentage change in closed-loop gain. **[0.5%]**
4. An *RC*-coupled amplifier has a mid-frequency gain of 400 and lower and upper 3 dB frequencies of 100 Hz and 10 kHz. A negative feedback network with β = 0.01 is incorporated into the amplifier circuit. Calculate
   (*i*) gain with feedback (*ii*) new bandwidth **[(*i*) 80 (*ii*) 75 kHz]**
5. In an amplifier with constant signal input of 1 volt, the output falls from 50 to 25 V when feedback is applied. Calculate the fraction of the output which is fed back. If, due to ageing, the amplifier gain fell to 40, find the percentage reduction in stage gain
   (*i*) without feedback (*ii*) with the feedback connection **[0.02% (*i*) 20% (*ii*) 11.12%]**
6. When voltage feedback is applied to an amplifier of gain 100, the overall stage gain falls to 50. Calculate the fraction of the output voltage fed back. If this fraction is maintained, calculate the value of the amplifier gain required if the overall stage gain is to be 75. **[β = 0.01 ; 300]**

SELF EXAMINATION QUESTIONS

A. Fill in the blanks with most appropriate word(s) or numerical value(s).

1. In feedback amplifiers, a .... of the output voltage is fed back to the input circuit.
2. Positive feedback occurs when feedback voltage and input voltage are in .... with each other.
3. Positive feedback is often used in .... circuits.
4. A feedback amplifier essentially consists of an amplifier and a .... circuit.
5. .... feedback is also called degenerative feedback.
6. Negative feedback is frequently used in .... circuits.
7. The term 'βA' is called feedback................. .
8. In a positive feedback amplifier when 'βA' becomes equal to unity, it starts ....
9. Negative feedback in amplifiers gives reduced gain but .... bandwidth.
10. One of the advantages of negative feedback is a highly .... amplifier gain.
11. Negative feedback reduces the amplifier distortion by the amount of gain.
12. Feedback does not change the gain-bandwidth................

B. Answer True or False

1. Positive feedback increases amplifier gain.
2. Negative feedback is called degenerative because it reduces the amplifier gain.
3. Closed-loop gain of a negative-feedback amplifier is greater than its open loop gain.
4. Negative feedback always reduces the amplifier gain.
5. Negative feedback provides gain stability and higher fidelity.
6. Negative feedback reduces distortion of the input signal.
7. Amongst the advantages of negative feedback is the reduced bandwidth.
8. Larger the feedback factor, higher the gain stability of an amplifier.
9. Increased bandwidth of a negative feedback amplifier is due to its decreased gain.
10. With negative-feedback, upper 3 dB frequency of an amplifier is decreased.

C. Multiple Choice Items

1. Feedback in an amplifier always helps to
   (*a*) control its output
   (*b*) increase its gain
   (*c*) decrease its input impedance
   (*d*) stabilize its gain
2. Closed-loop gain of a feedback amplifier is the gain obtained when
   (*a*) its output terminals are closed
   (*b*) negative feedback is applied
   (*c*) feedback loop is closed
   (*d*) feedback factor exceeds unity
3. A large sacrifice factor in a negative feedback amplifiers leads to
   (*a*) inferior performance
   (*b*) increased output impedance
   (*c*) characteristics impossible to achieve without feedback
   (*d*) precise control over output
4. Negative feedback in an amplifier
   (*a*) lowers its lower 3 dB frequency
   (*b*) raises its upper 3 dB frequency
   (*c*) increases its bandwidth
   (*d*) all of the above
5. Regarding negative feedback in amplifiers which statement is WRONG ?
   (*a*) It widens the separation between 3 dB frequencies.
   (*b*) It increases the gain-bandwidth product.
   (*c*) It improves gain stability.
   (*d*) It reduces distortion.
6. Negative feedback reduces distortion in an amplifier only when it
   (*a*) comes as part of input signal
   (*b*) is part of its output
   (*c*) is generated within the amplifier
   (*d*) exceeds a certain safe level
7. An amplifier with no feedback has a gain-bandwidth product of 4 MHz. Its closed-loop gain is 40. The new bandwidth is
   (*a*) 100 kHz (*b*) 160 MHz
   (*c*) 10 MHz (*d*) 20 kHz
8. The shunt-derived series-fed feedback in an amplifier
   (*a*) increases its output impedance
   (*b*) decreases its output impedance
   (*c*) increases its input impedance
   (*d*) both (*b*) and (*c*)

ANSWERS

A. Fill in the blanks

1. portion	2. phase	3. oscillator	4. feedback	5. negative	6. amplifier
7. factor	8. oscillating	9. increased	10. stabilized	11. loop	12. product

B. True or False

1. T	2. T	3. F	4. T	5. T
6. F	7. F	8. T	9. T	10. F

C. Multiple Choice Items

1. *a*	2. *c*	3. *c*	4. *d*	5. *b*	6. *c*	7. *a*	8. *d*

26
CHAPTER

Field Effect Transistors

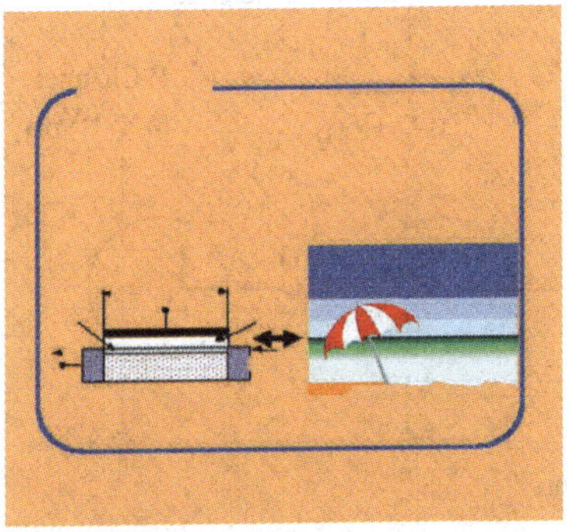

1. Junction FET (JFET)
2. JFET Drain Characteristic with $V_{GS} = 0$
3. JFET Characteristic with External Bias
4. Transfer Characteristic
5. Small Signal JFET Parameters
6. DC Biasing of a JFET
7. Common Source JFET Amplifier
8. JEFT on an IC Chip
9. MOSFET or IGFET
10. DE MOSFET
11. Schematic Symbols for a DE MOSFET
12. Enhancement-only N-Channel MOSFET
13. Transfer Characteristic
14. MOSFET Handling

26.1. What is a FET ?

The acronym 'FET' stands for **field effect transistor.** It is a three-terminal unipolar solid-state device in which current is controlled *by an electric field* as is done in vacuum tubes. Broadly speaking, there are two types of FETs:

(*a*) junction field effect transistor (JFET),

(*b*) metal-oxide semiconductor FET (MOSFET).

It is also called insulated-gate FET (IGFET).

It may be further subdivided into :

(*i*) depletion-enhancement MOSFET *i.e.*, DE MOSFET,

(*ii*) enhancement-only MOSFET *i.e.*, E-only MOSFET.

Both of these can be either P-channel or N-channel devices.

The FET family tree is shown on next page.

26.2. Junction FET (JFET)

(*a*) Basic Construction

As shown in Fig. 26.1, it can be fabricated with either an N-channel or P-channel though N-channel is generally preferred. For fabricating an N-channel JFET, first a narrow bar of N-type semiconductor material is taken and then two P-type junctions are diffused on opposite sides of its middle part [Fig. 26.1 (*a*)]. These junctions form two P-N diodes or **gates** and the area between these gates is called **channel.** The two P-regions are internally connected and a single lead is brought out which is called **gate terminal.** Ohmic contacts (direct

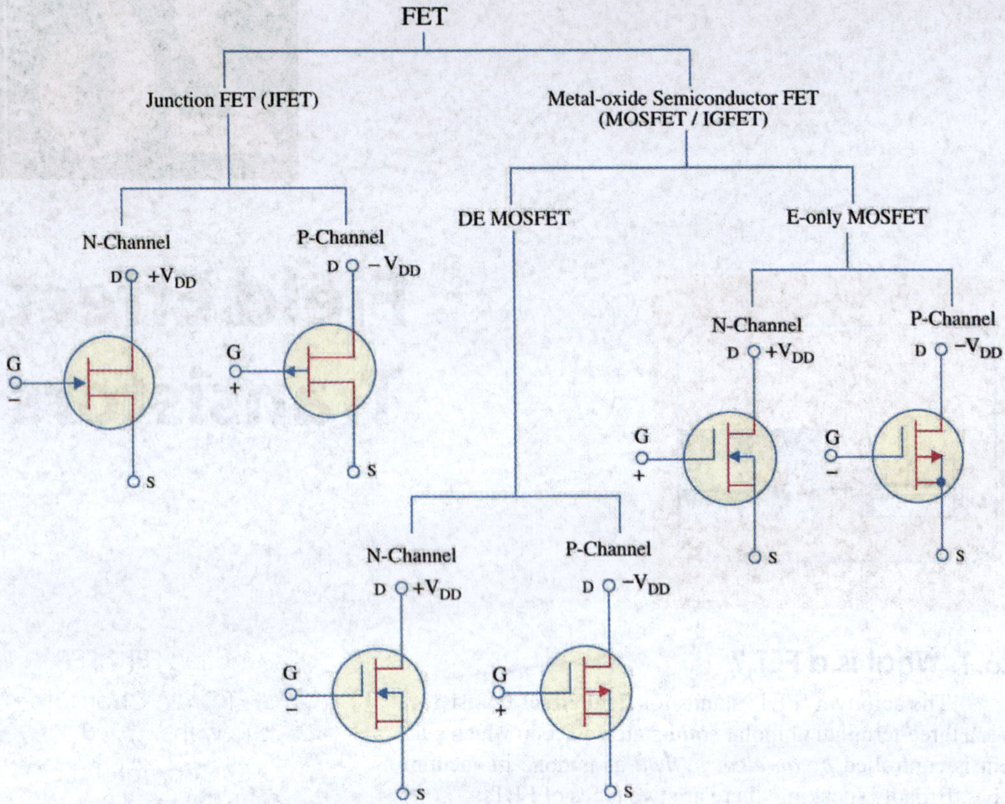

electrical connections) are made at the two ends of the bar–one lead is called **source terminal** *S* and the other **drain terminal** *D*. When potential difference is established between drain and source, current flows along the length of the 'bar' through the channel located between the two P-regions. The current consists of **only majority carriers** which, in the present case, are electrons. P-channel FET is similar in construction except that it uses P-type bar and two N-type junctions. The majority carriers are holes which flow through the channel located between the two N-regions or gates.

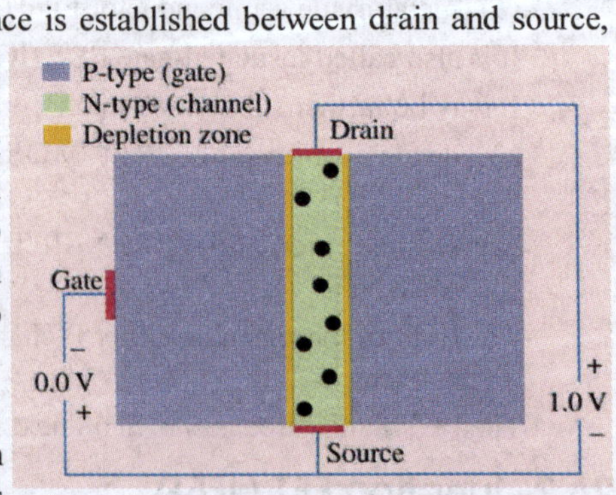

Electrons flow from source to drain when we apply a drain source voltage.

Following FET notation is worth remembering:

1. **Source.** It is the terminal through which majority carriers *enter* the bar. Since carriers come from it, it is called the source.
2. **Drain.** It is the terminal through which majority carriers *leave* the bar. The drain-to-source voltage V_{DS} drives the drain current I_D.
3. **Gate.** These are two internally-connected heavily-doped regions which form two P-N junctions. The gate-source voltage V_{GS} reverse-biases the gates.
4. **Channel.** It is the space between two gates through which majority carriers pass from

source to drain when V_{DS} is applied.

Schematic symbols for N-channel and P-channel JFET are shown in Fig. 26.1 (*c*). It must be kept in mind that **gate arrow always points to N-type material.**

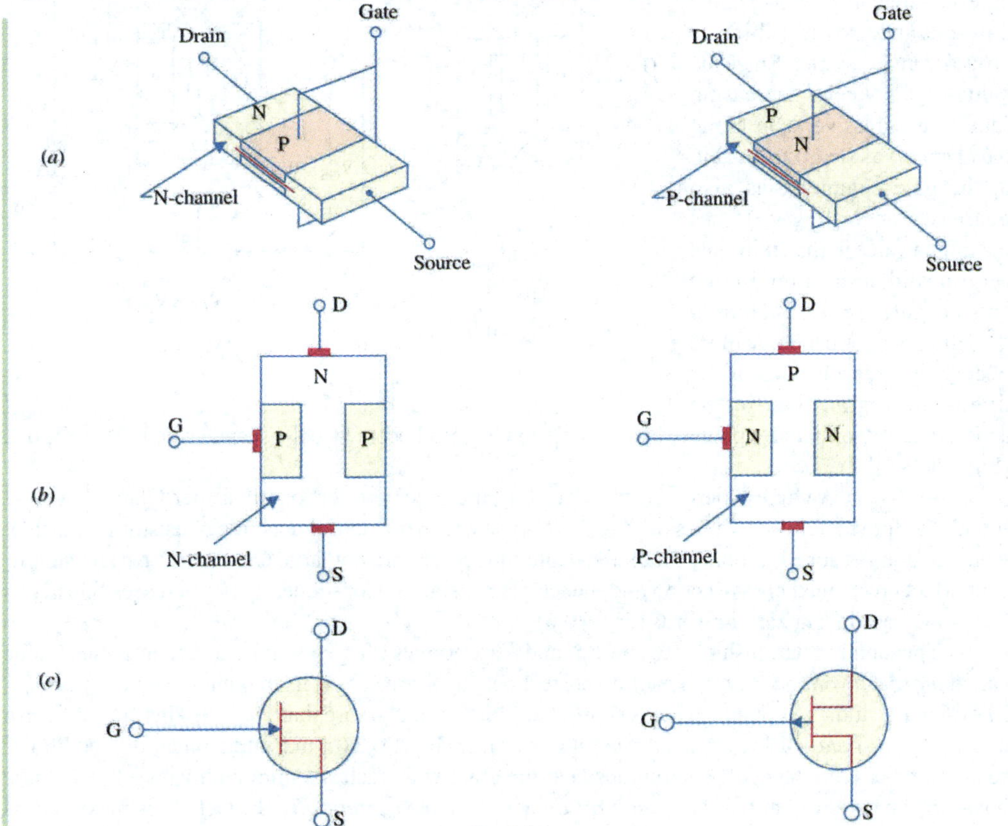

Fig. 26.1

(*b*) Theory of Operation

While discussing the theory of operation of a JFET, it should be kept in mind that:

1. Gates are always reverse-biased. Hence, gate current I_G is practically zero.
2. The source terminal is always connected to that end of the drain supply which provides the necessary charge carriers. In an N-channel JFET, source terminal *S* is connected to the negative end of the drain voltage supply (for obtaining electrons). In a P-channel JFET, *S* is connected to the positive end of the drain voltage supply for getting holes which flow through the channel.

Let us now consider an *N-channel* JFET and discuss its working when either V_{GS} or V_{DS} or both are changed.

(*i*) *When* $V_{GS} = 0$ *and* $V_{DS} = 0$

In this case, drain current $I_D = 0$, because $V_{DS} = 0$. The depletion regions around the P- junctions are of equal thickness and symmetrical as shown in Fig. 26.2 (*a*).

(*ii*) *When* $V_{GS} = 0$ *and* V_{DS} *is increased from zero*

For this purpose, the JFET is connected to the V_{DD} supply as shown in Fig. 26.2 (*b*). The electrons (which are the majority carriers) flow from *S* to *D*, whereas conventional drain current I_D flows through the channel from *D* to *S*. Now, the gate-to-channel bias at any point along the channel is = | V_{DS} | + | V_{GS} | *i.e.*, the numerical sum of the two voltages. In the present case, external bias V_{GS} = 0. Hence gate-channel reverse bias is *provided by V_{DS} alone.* Since the value of V_{DS} keeps decreasing (due to progressive drop along the channel) as we go from *D* to *S*, the gate-channel bias also decreases accordingly. It has maximum value in the drain-gate region and minimum in the source-gate region. Hence, depletion regions penetrate more deeply into the channel in the drain-gate region than in the source-gate region. This explains why the depletion regions become wedge shaped when V_{DS} is applied [Fig. 26.2 (*b*)].

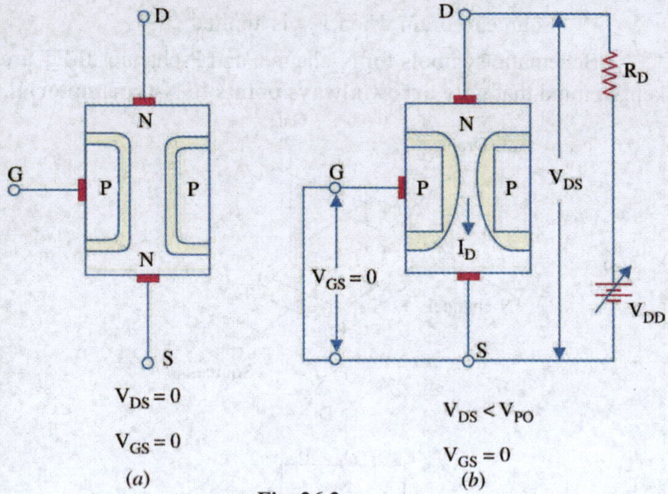

Fig. 26.2

As V_{DS} is gradually increased from zero, I_D increases proportionally as per Ohm's law. It is found that for small initial values of V_{DS}, the N-type channel material acts like a resistor of constant value. It is so because V_{DS} being small, the depletion regions are not large enough to have any significant effect on channel cross-section and, hence, its resistance. Consequently, I_D increases linearly as V_{DS} is increased from zero onwards (Fig. 26.6).

This ohmic relationship between V_{DS} and I_D continues till V_{DS} reaches a certain critical value called *pinch-off voltage* V_{PO}* when drain current becomes constant at its maximum value called I_{DSS}. The *SS* in I_{DSS} indicates that the gate is shorted to source to make sure that V_{GS} = 0. This current is also known as *zero-gate-voltage drain current.* It is seen from Fig. 26.3 (*a*) that under pinch-off conditions, separation between the depletion regions near the drain end reaches a minimum value *W*. It should, however, be carefully noted that *pinch-off does not mean 'current-off'*. In fact, I_D is maximum at pinch off.

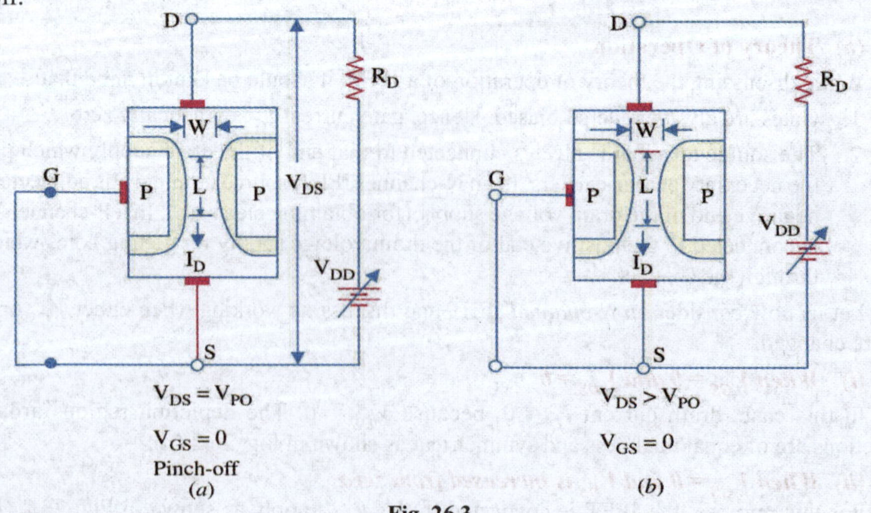

Fig. 26.3

* The subscript *O* has been added to indicate zero gate-source voltage conditions.

When V_{DS} is increased beyond V_{PO}, I_D remains constant at its maximum value I_{DSS} upto a certain point. It is due to the fact that further increase in V_{DS} (beyond V_{PO}) causes more of the channel on the source end to reach the minimum width as shown in Fig. 26.3 (*d*). It means that the channel width does not increase, instead its length L increases. As more of the channel reaches the minimum width, the resistance of the channel *increases at the same rate at which V_{DS} increases.* In other words, increase in V_{DS} is neutralized by increase in R_{DS}. Consequently, $I_D = (V_{DS} / R_{DS})$ remains unchanged even though V_{DS} is increased. Ultimately, a certain value of V_{DS} (called V_{DSO})** is reached when JFET breaks down and I_D increases to an excessive value as seen from drain characteristic of Fig. 26.6.

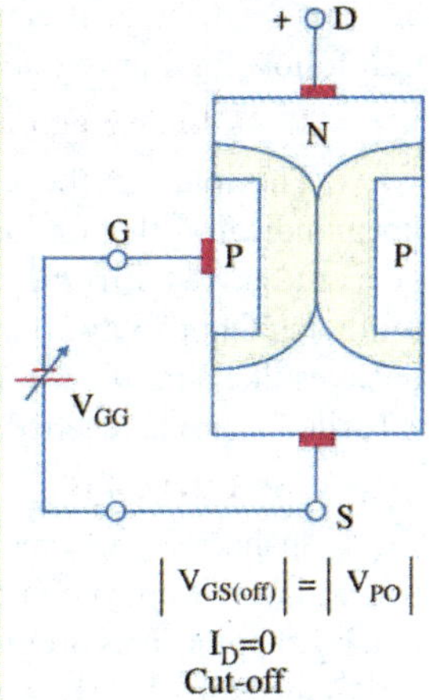

Fig. 26.4

(*iii*) When $V_{DS} = 0$ and V_{GS} is decreased from zero

In this case, as V_{GS} is made more and more negative, the gate reverse bias increases which increases the thickness of the depletion regions. As negative value of V_{GS} is increased, a stage comes when the two depletion regions touch each other as shown in Fig. 26.4. In this condition, the channel is said to be *cut-off.* This value of V_{GS} which cuts off the channel and hence the drain current is called $V_{GS}(off)$.***

It may be noted that

$$V_{GS(off)} = - V_{PO} \text{ or } |V_{PO}| = |V_{GS(off)}|.$$

As seen from Fig. 26.7 because $V_{PO} = 4$ V, $V_{GS(off)} = -4$ V. Obviously, their absolute values are equal.

(iv) When V_{GS} is negative and V_{DS} is increased

As seen from Fig. 26.7, as V_{GS} is made more and more negative, values of V_P as well as breakdown voltage V_B are decreased.

Since gate voltage controls the drain current, JFET is called a voltage-controlled device. A P-channel JFET operates exactly in the same manner as an N-channel JFET except that current carriers are holes and polarities of both V_{DD} and V_{GS} are reversed.

Since only one type of majority carrier (either electron or holes) is used in JFETs, they are called *unipolar devices* unlike bipolar junction transistors (BJTs) which use both electrons and holes as carriers (and, hence are called *bipolar devices*).

26.3. Static Characteristics of a JFET

We will consider the following two characteristics:

(*i*) Drain characteristic

It gives relation between I_D and V_{DS} for different values of V_{GS} (which is called running variable).

(*ii*) Transfer characteristic

It gives relation between I_D and V_{GS} for different values of V_{DS}.

We will analyse these characteristics for an N-channel JFET connected in the common-source mode as shown in Fig. 26.5. We will first consider the drain characteristic when $V_{GS} = 0$ and then when V_{GS} has any negative value upto $V_{GS(off)}$.

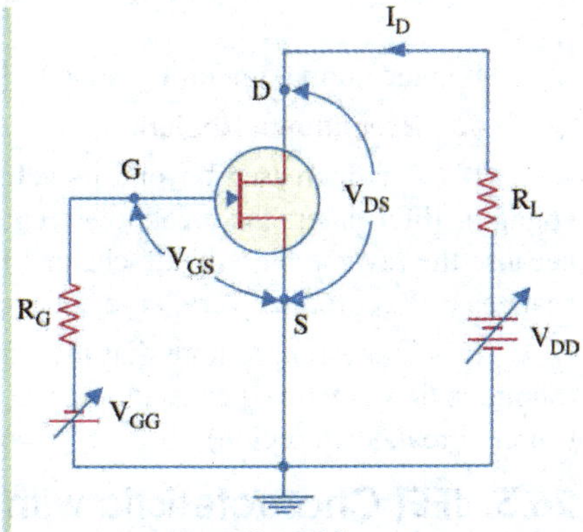

Fig. 26.5

** The subscript O has been added to indicate zero gate-source voltage conditions.

*** It has negative value for an N-channel JFET but a positive value or a P-channel JFET.

26.4. JFET Drain Characteristic with $V_{GS} = 0$

Such a characteristic is shown in Fig. 26.6 and has been already discussed briefly in Art. 26.2. It can be subdivided into following four regions :

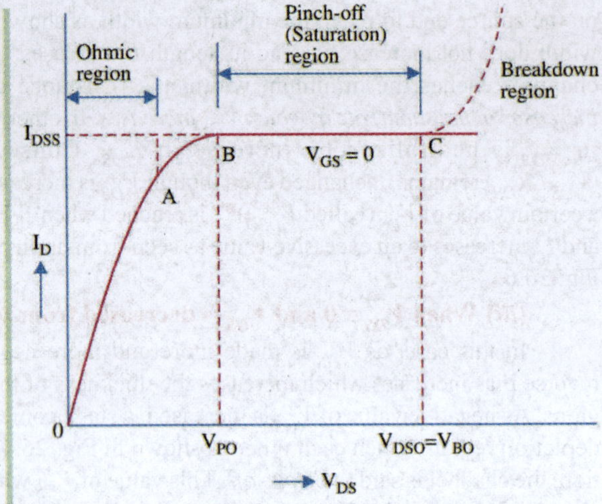

Fig. 26.6

1. Ohmic Region OA

This part of the characteristic is linear indicating that for low values of V_{DS}, current varies directly with voltage following Ohm's Law. It means that JFET behaves like an ordinary resistor till point *A* (called knee) is reached.

2. Curve AB

In this region, I_D increases at inverse square-law rate upto point *B* which is called pinch-off point. This progressive fall in the rate of increase of I_D is caused by the square law increase in the depletion region at each gate upto point *B* where the two regions are closest without touching each other. The drain-to-source voltage V_{DS} corresponding to point *B* is called *pinch-off voltage* V_{PO} (the subscript 0 indicates that $V_{GS} = 0$). But it is essential to remember that "*pinch-off*" *does not mean* "*current-off*".

3. Pinch-off Region BC

It is also known as **saturation region** or '**amplifier**' **region**. Here, JFET operates as a **constant-current** device because I_D is relatively independent of V_{DS}. It is due to the fact that as V_{DS} increases, channel resistance also increases proportionally thereby keeping I_D practically constant at I_{DSS}. It should also be noted that the reverse bias required by the gate-channel junction is supplied entirely by V_{DS} (Art. 26.2) and none by external bias because $V_{GS} = 0$.

Drain current in this region is given by Shockley's equation

$$I_D = I_{DSS}\left(1 - \frac{V_{GS}}{V_p}\right)^2 = I_{DSS}\left(1 - \frac{V_{GS}}{V_{GS(off)}}\right)^2$$

It is the normal operating region of the JFET when used as an amplifier.

4. Breakdown Region

If V_{DS} is increased beyond its value corresponding to point *C* (called avalanche breakdown voltage), JFET enters the breakdown region where I_D increases to an excessive value. This happens because the reverse-biased gate-channel P-N junction undergoes avalanche breakdown when small changes in V_{DS} produce very large changes in I_D.

It is interesting to note that increasing values of V_{DS} make a JFET behave first as a resistor (ohmic region), then as a constant-current source (pinch-off region) and finally, as a constant-voltage source (breakdown region).

26.5. JFET Characteristic with External Bias

Fig. 26.7 shows a family of I_D versus V_{DS} curves for different values of V_{GS}. It is seen that as the negative gate bias voltage is increased:

(*i*) pinch off voltage V_p is reached at a lower value of V_{DS} than when $V_{GS} = 0$.

(*ii*) value of V_{DS} for breakdown is decreased.

It is seen that with $V_{GS} = 0$, I_D saturates at I_{DSS} and the characteristic shows $V_{PO} = 4$ V. When an external bias of –1 V is applied, gate-channel junctions still require – 4 V to achieve pinch-off (remember, $|V_{GS}| = |- V_P|$). It means that a value of $V_{DS} = 3$V is now required instead of the previous 4 V. Consequently, the value of I_D is reduced. Similarly, when V_{GS} is = – 2V and – 3 V, pinch-off is achieved with V_{DS} equal to 2V and 1 V respectively. This further reduces the value of I_D. As seen, when $V_{GS} = -$ 4V (*i.e.*, numerically equal to V_{PO}), V_{DS} is not required at all. Accordingly, I_D also becomes zero.

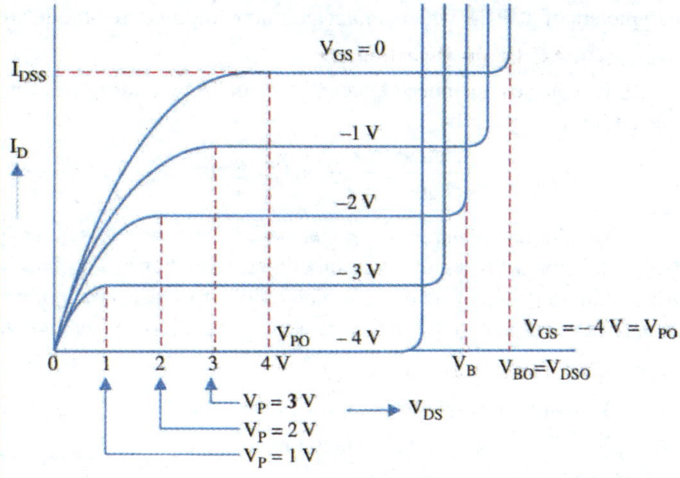

Fig. 26.7

This micrograph shows a mock field effect transistor with a layer of crystalline strontium titanate instead of silicon dioxide as the gate electrode.

The above fact could be verified from the equation

$$V_{DS(P)} = V_P = V_{PO} + V_{GS}$$

Since V_{PO} is constant for a given JFET, value of V_P decreases as V_{GS} becomes progressively more negative.

The value of breakdown voltage for any value of V_{GS} is given by

$$V_B = V_{DSO} + V_{GS} = V_{BO} + V_{GS}$$

As seen, as V_{GS} becomes more and more negative, V_B keeps decreasing. Obviously, when $V_{GS} = 0$, $V_B = V_{BO} = V_{DSO}$.

26.6. Transfer Characteristic

It is a plot of I_D versus V_{GS} for a constant value of V_{DS} and is shown in Fig. 26.8. It is similar to the transconductance characteristic of a vacuum tube or a transistor. It is seen that when $V_{GS} = 0$, $I_D = I_{DSS}$ and when $I_D = 0$, $V_{GS} = V_{PO}$. The transfer characteristic approximately follows the equation

$$I_D = I_{DSS}\left(1 - \frac{V_{GS}}{V_{PO}}\right)^2 = I_{DSS}\left(1 - \frac{V_{GS}}{V_{GS(off)}}\right)^2$$

The above equation can be written as

$$V_{GS} = V_{GS(off)}\left(1 - \sqrt{\frac{I_D}{I_{DSS}}}\right)$$

This characteristic can be obtained from the drain characteristic by reading off V_{GS} and I_{DSS} values for different values of V_{DS}.

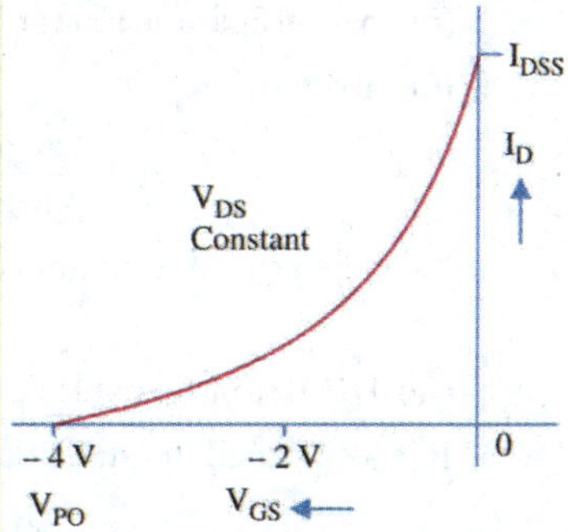

Fig. 26.8

26.7. Small Signal JFET Parameters

The various parameters of a JFET can be obtained from its two characteristics. The main parameters of a JFET when connected in common-source mode are as under:

(i) AC Drain Resistance, r_d

It is the ac resistance between drain and source terminals when JFET is operating *in the pinch-off region.* It is given by

$$r_d = \frac{\text{change in } V_{DS}}{\text{change in } I_D} - V_{GS} \text{ constant or } r_d = \left.\frac{\Delta V_{DS}}{\Delta I_D}\right|_{V_{GS}}$$

An alternative name is *dynamic drain resistance.* It is given by the slope of the drain characteristic in the pinch-off region. It is sometimes written as r_{ds} emphasizing the fact that it is the resistance from drain to source. Since r_d is usually the output resistance of a JFET, it may also be expressed as an output admittance y_{os}. Obviously, $y_{os} = 1/r_d$. It has a very high value.

(ii) Transconductance, g_m

It is simply the slope of transfer characteristic.

$$g_m = \frac{\text{change in } I_D}{\text{change in } V_{GS}} - V_{DS} \text{ constant or } g_m = \left.\frac{\Delta I_D}{\Delta V_{GS}}\right|_{V_{DS}}$$

Its unit is siemens (S) earlier called mho. It is also called *forward transconductance* (g_{fs}) *or forward transadmittance* Y_{fs}.

The transconductance measured at I_{DSS} is written as g_{mo}.

Mathematical Expression for g_m

The Shockley equation is $I = I_{DSS}\left(1 - \frac{V_{GS}}{V_P}\right)^2$

Differentiating both sides, we have

$$\frac{dI_D}{dI_{DSS}} = 2I_{DSS}\left(1 - \frac{V_{GS}}{V_P}\right)\left(-\frac{1}{V_P}\right) \quad \text{or} \quad g_m = -\frac{2I_{DSS}}{V_P}\left(1 - \frac{V_{GS}}{V_P}\right)$$

When $V_{GS} = 0, \quad g_m = g_{mo}$; $\quad \therefore \quad g_{mo} = -\frac{2I_{DSS}}{V_P}$

From the above two equations, we have

$$g_m = g_{mo}\left(1 - \frac{V_{GS}}{V_P}\right)$$

(iii) Amplification Factor, μ

It is given by

$$\mu = \frac{\text{change in } V_{DS}}{\text{change in } V_{GS}} - I_D \text{ constant} \quad \text{or} \quad \mu = \left.\frac{\Delta V_{DS}}{\Delta V_{GS}}\right|_{I_D}$$

It can be proved from above that

$$\mu = g_m \times r_d = g_{fs} \times r_d$$

(iv) DC Drain Resistance, R_{DS}

It is also called the static or ohmic resistance of the channel. It is given by

$$R_{DS} = \frac{V_{DS}}{I_D}$$

Example 26.1. *For an N-channel JFET, $I_{DSS} = 8.7\ mA$, $V_P = -3\ V$, $V_{GS} = -1V$. Find the values of (i) I_D (ii) g_{mo} (iii) g_m.*

Solution. (*i*) $I_D = I_{DSS}\left(1 - \frac{V_{GS}}{V_P}\right)^2 = 8.7\left(1 - \frac{-1\,V}{-3\,V}\right)^2$ = **3.87 mA**

(*ii*) $g_{mo} = \frac{-2I_{DSS}}{V_P} = \frac{-2 \times 8.7}{-3}$ = **5.8 mS**

(*iii*) $g_m = g_{mo}\left(1 - \frac{V_{GS}}{V_P}\right) = 5.8\left(1 - \frac{-1}{-3}\right)$ = **3.87 mS**

26.8. DC Biasing of a JFET

A JFET may be biased by using either

(*i*) a separate power source V_{GG} as shown in Fig. 26.9 (*a*), or

(*ii*) some form of self-bias as shown in Fig. 26.9 (*b*).

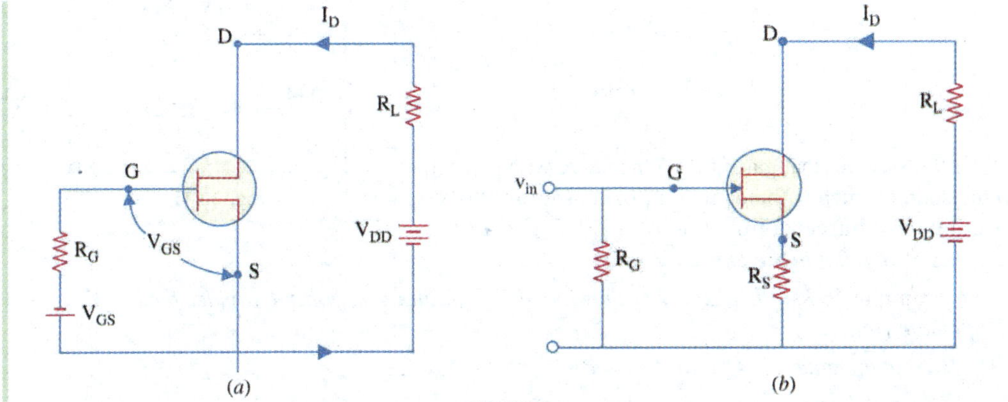

Fig. 26.9

The circuit of Fig. 26.9 (*b*) is called self-bias circuit because the V_{GS} bias is obtained from the JFET's own source current (*i.e.*, drain current I_D) through R_S.

$\therefore \qquad V_{GS} = -I_D R_S$

The gate is kept at this much negative potential with respect to the ground.

The addition of R_G in Fig. 26.9 (*b*), does not upset this dc bias for the simple reason that no gate current flows through it (the gate leakage current is almost zero). Hence, gate is essentially at *dc ground.* Without R_G, gate would be kept 'floating' which could collect charge and ultimately cut-off the JFET.

The resistance R_G has been put in the circuit for avoiding short-circuiting of the ac input voltage, v_{in}. Moreover, in case, leakage current is not totally negligible, R_G would provide it an escape route. Otherwise, the leakage current would build up static charge (voltage) at the gate which could change the bias or even destroy the JFET.

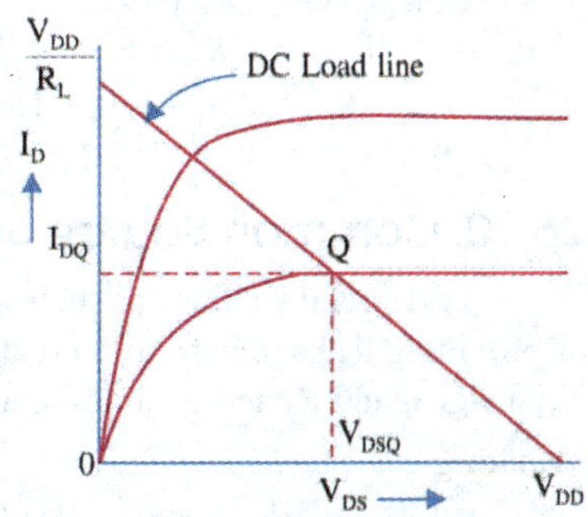

Fig. 26.10

26.9. DC Load Line

The dc load line for a JFET can be easily drawn by remembering the following two points:

(*i*) At $\qquad I_D = 0,\ V_{DS} = V_{DD}$

(*ii*) At $\qquad V_{DS} = 0,\ I_D = \frac{V_{DD}}{R_L}$

The Q-point is generally situated at the middle point of the load line (for class-A operation) so that

$$V_{DSQ} = 1/2$$

Also,
$$I_{DSQ} = \frac{1/2\ V_{DD}}{R_S + R_L}$$

Example. 26.2. *For the circuit of Fig. 26.11, find the values of V_{DSQ} and V_{GS}, assuming centrally-located Q-point and zero gate current.*

Solution. Since $I_G = 0$, dc circuit is not disturbed.

$$V_{DS} = \frac{1}{2}V_{DD} = \frac{12}{2} = \mathbf{6\,V}$$

The balance of 6 V drops across series combination of R_L and R_S.

$$\therefore \quad I_D = \frac{6}{150 + 450} = 10 \text{ mA}$$

$$\therefore \quad V_{GS} = -I_D R_S = -10 \times 150$$
$$= \mathbf{-1.5\ V}$$

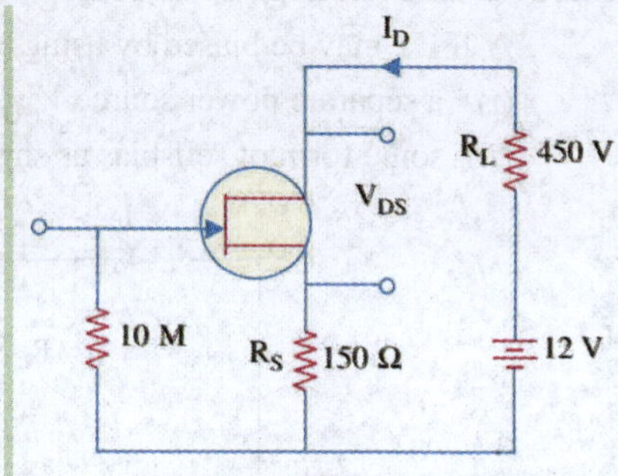

Fig. 26.11

It is obvious that gate is 1.5 V negative with respect to the source which is the common point. Incidentally, in common source connection of a JFET, *gate is the most negative point in the entire circuit.*

Example 26.3. *In Fig. 26.12 is shown a JFET circuit using voltage-divider bias. If $V_{GS} = -2$ V, calculate voltage*

(*a*) *at the gate* (*b*) *at the source*

(*c*) at the drain (*d*) *from drain to source.*

Solution. (*a*) The voltage across R_2 is the same as gate voltage. This voltage can be found by using voltage-divider formula.

$$V_G = V_{DD} \times \frac{R_2}{R_1 + R_2} = 15 \times \frac{4}{12 + 4} = \mathbf{3.75\ V}$$

(*b*) The voltage at the source is 2 V more positive than the gate voltage due to $-V_{GS}$.

$$\therefore \quad V_S = 3.75 \times (-2) = \mathbf{-5.75\ V}$$

(*c*) The drain current equals source current.

$$\therefore \quad I_D = V_S / R_S = 5.75 / 1\text{ K} = 5.75 \text{ mA}$$

Voltage across R_D is

$$V_{RD} = I_D R_D = 5.75 \times 10^{-3} \times 500 = 2.88 \text{ V}$$

$$\therefore \quad V_D = V_{DD} - V_{RD} = 15 - 2.88 = \mathbf{12.1\ V}$$

(*d*) $V_{DS} = V_D - V_S = 12.1 - 5.75 = \mathbf{6.35\ V}$

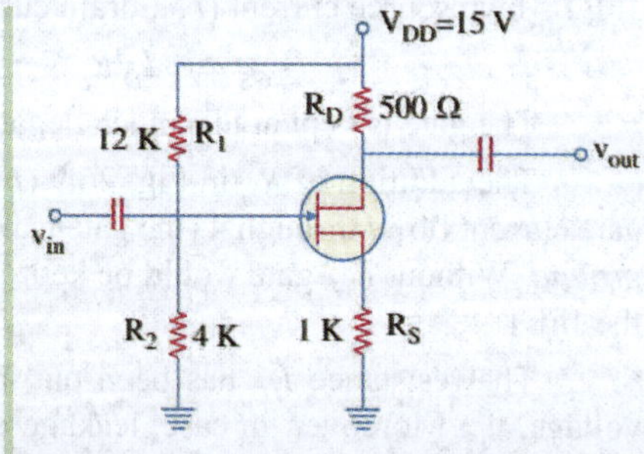

Fig. 26.12

26.10. Common Source JFET Amplifier

Two circuit versions of such an amplifier are shown in Fig. 26.13. Here, R_G serves the purpose of providing leakage path to the gate current, R_S develops gate bias, C_S provides ac ground to the input signal and R_L acts as drain load.

Working

When *negative-going* signal is applied to the input

1. gate bias is *increased,*
2. depletion regions are *widened,*
3. channel resistance is *increased,*
4. I_D is *decreased,*
5. drop across R_L is decreased,
6. consequently, a *positive-going* signal becomes available at the output through *C* point M_2 in Fig. 26.13.

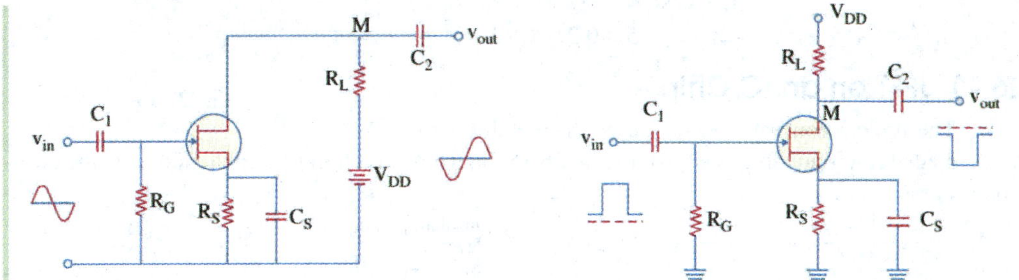

Fig. 26.13

When *positive-going* signal is applied at the input, then in a similar way, a *negative-going* signal becomes available at the output.

It is seen that there is a *phase inversion* between the input signal at the gate and output signal at the drain.

(i) Input Impedence

$$Z_{in} = R_G \| R_{GS}$$

In an ideal JFET, R_{GS} is infinite because $I_G = 0$. In an actual device, however, R_{GS} is not actually infinite but extremely high as compared to R_G.

$$\therefore \quad Z_{in} \cong R_G$$

(ii) Output Impedance

$$Z_o = r_d \| R_L \cong R_L \qquad \text{—if } r_d \gg R_L$$

(iii) Voltage Gain

Output voltage,

$$v_o = i_d \times (r_d \| R_L) \qquad \text{Now, } i_d = -g_m v_i$$

$$\therefore \quad v_o = -g_m . v_i \times (r_d \| R_L)$$

$$\therefore \quad A_v = \frac{v_o}{v_i} = \frac{-g_m v_i \times (r_d \| R_L)}{v_i}$$

$$\text{or} \quad A_v = \frac{-g_m r_d R_L}{r_d + R_L} = \frac{-\mu R_L}{r_d + R_L}$$

Since, $r_d >> R_L$, we have

$$A_v = -g_m R_L$$

Example 26.4. *For a common-source JFET voltage amplifier*

$$g_m = 2\ mS \text{ and } R_L = 10\ K$$

Find its voltage amplification assuming $r_d >> R_L$.

Solution. $A_v = g_m R_L$

$$= 2 \times 10^{-3} \times 10 \times 10^3 = \mathbf{20}$$

Example 25.5. *For the JFET amplifier circuit shown in Fig. 26.14 calculate the approximate value of (i)A_v and (ii)v_o.*

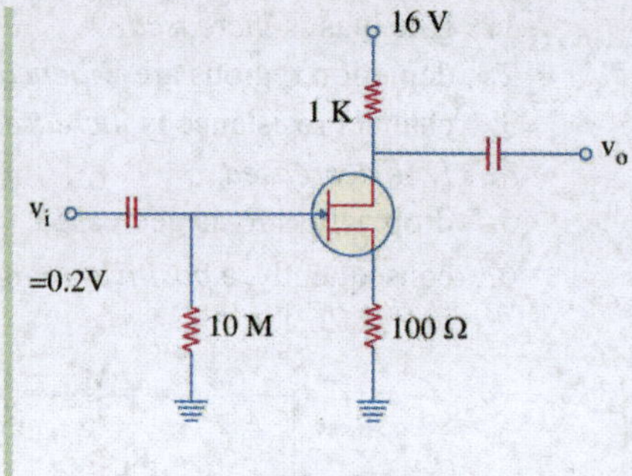

Fig. 26.14

Solution. (*i*) An approximate expression for voltage gain of an amplifier (assuming class A operation) is

$$A_r = \frac{1}{2}\left(\frac{R_L}{R_S}\right) = \frac{1}{2} \times \frac{1000}{100} = \mathbf{5}$$

(*ii*) $v_o = A_v \cdot v_i = 5 \times 0.2 = \mathbf{1.0\ V}$

26.11. JEFT on an IC Chip

The typical construction for an N-channel JFET on an integrated-circuit chip is shown in Fig. 26.15. Such construction is normally done by the diffusion process.

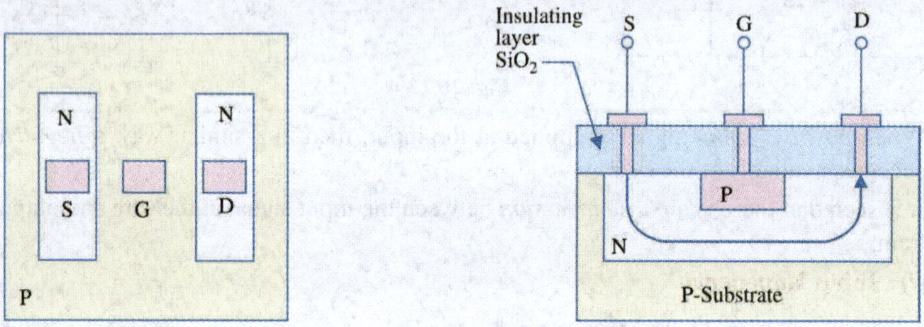

Fig. 26.15

Starting with a P-type substrate, first an N-channel is diffused. Then, P-type impurities are diffused into the N-channel to form the gate. Finally, a thin insulating layer of SiO_2 is deposited in place to make terminals.

26.12. Advantages of FETs

FETs combine the many advantages of both BJTs and vacuum tubes. Some of their main advantages are:

1. high input impedance,
2. small size,
3. ruggedness,
4. long life,
5. high frequency response,
6. low noise,
7. negative temperature coefficient, hence better thermal stability,
8. high power gain,
9. a high immunity to radiations,
10. no offset voltage when used as a switch (or chopper),
11. square law characteristics.

The only disadvantages are :

1. small gain-bandwidth product,
2. greater susceptibility to damage in handling them.

26.13. MOSFET or IGFET

It could be further subdivided as follows :

(*i*) Depletion-enhancement MOSFET or DE MOSFET

This MOSFET is so called because it can be operated in both *depletion* mode and *enhancement* mode by changing the polarity of V_{GS}. When negative gate-to-source voltage is applied, the N-channel DE MOSFET operates in the depletion mode [Fig. 26.17 (*c*)]. However, with positive gate voltage, it operates in the enhancement mode [Fig. 26.17 (*b*)]. Since a channel exists between drain and source, I_D flows even when $V_{GS} = 0$. That is why DE MOSFET is known as normally-ON MOSFET.

(ii) Enhancement-only MOSFET

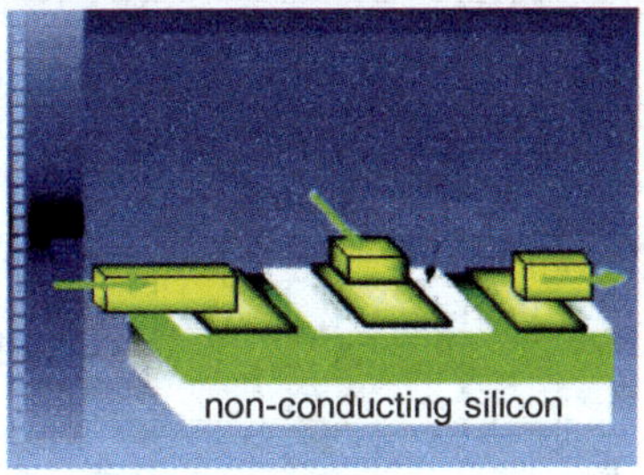

The MOSFET.

As its name indicates, this MOSFET operates *only* in the enhancement mode and has no depletion mode. It works with *large values of* V_{GS} only. It differs in construction from the DE MOSFET in that structurally there exists no channel between the drain and source (Fig. 26.20). Hence, it does not conduct when $V_{GS} = 0$. That is why it is called normally-OFF MOSFET.

In a DE MOSFET, I_D flows even when $V_{GS} = 0$. It operates in depletion mode with negative values of V_{GS}. As V_{GS} is made more negative, I_D decreases till it ceases when $V_{GS} = V_{GS(off)}$. It works in enhancement mode when V_{GS} is positive as shown in Fig. 26.17 (*b*).

In the case of E-only MOSFET, I_D flows only when V_{GS} exceeds $V_{GS(th)}$ as shown in Fig. 26.22 (*c*).

26.14. DE MOSFET

(a) Construction

Like JFET, it has source, gate and drain. However, as shown in Fig. 26.16, its gate is insulated from its conducting channel by an ultra-thin metal-oxide insulating film (usually of silicon dioxide SiO_2). Because of this insulating property, MOSFET is alternatively known as insulated-gate field-effect transistor (IGFET or IGT). Here also, gate voltage controls drain current but main difference between JFET and MOSFET is that, in the latter case, we can apply *both positive and negative voltages to the gate because it is insulated from the channel.* Moreover, the gate, SiO_2 insulator and channel from a parallel-plate capacitor. Unlike JFET, a DE MOSFET has only one P-region or N-region called substrate. Normally, it is shorted to *the source internally.* Fig. 26.16 shows both P-channel and N-channel DE MOSFETs.

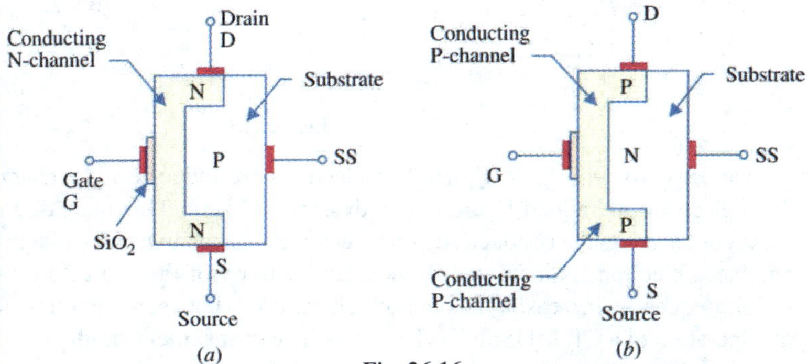

Fig. 26.16

(b) Working

(i) Depletion Mode of N-channel DE MOSFET

When $V_{GS} = 0$, electrons can flow freely from source to drain through the conducting channel which exists between them. When gate is given negative voltage, it **depletes** the N-channel of its electrons by inducing *positive charge* in it as shown in Fig. 26.17 (*a*). Greater the negative voltage on the gate, greater is the reduction in the number of electrons in the channel and, consequently, lesser its conductivity. In fact, too much negative gate voltage called $V_{GS(off)}$ can cut-off the channel. Hence, with negative gate voltage, a DE MOSFET behaves like a JFET.

For obvious reasons, negative-gate operation of a DE MOSFET is called its depletion mode operation.

(ii) Enhancement Mode of N-channel DE MOSFET

The circuit connections are shown in Fig. 26.17 (*b*). Again, drain current flows from source to drain even with zero gate bias. When positive voltage is applied to the gate, the input gate capacitor is able to create free electrons in the channel which increases I_D. As seen from the enlarged view of

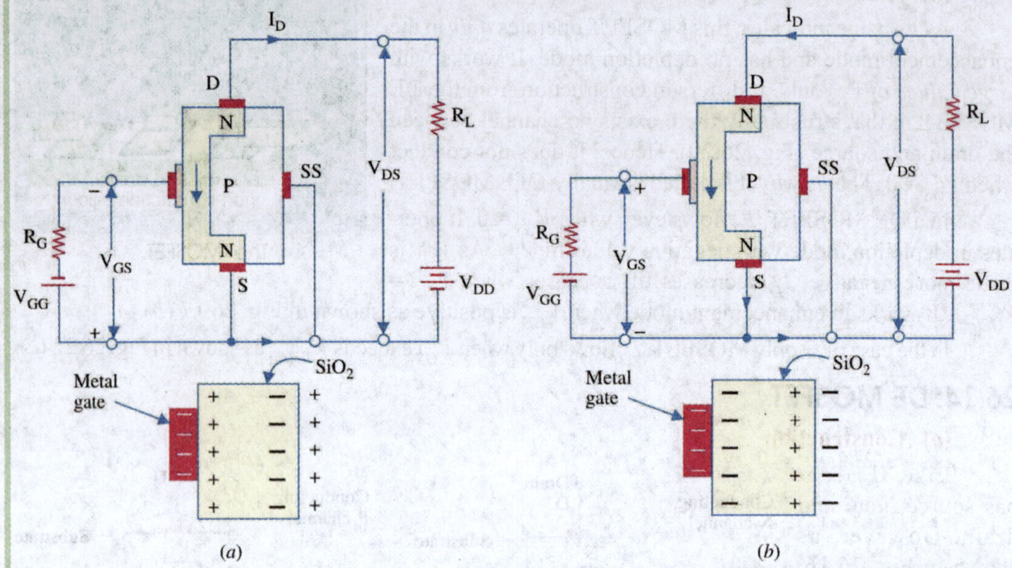

Fig. 26.17

the gate capacitor in Fig. 26.17 (*b*), free electrons are induced in the channel *by capacitor action.* These electrons are added to those already existing there. This increased number of electrons *increases* or enhances the conductivity of the channel. As positive gate voltage is increased, the number of induced electrons is increased. Hence, conductivity of the source-to-drain channel is increased and, consequently, increasing amount of current flows between the terminals. That is why, positive gate operation of a DE MOSFET is known as its **enhancement mode** operation.

26.15. Schematic Symbols for a DE MOSFET

Schematic symbol of an N-channel DE MOSFET is shown in Fig. 26.18. The gate looks like a metal plate. The arrow is on the substrate and **towards the N-channel.** When *SS* is connected to an external load, we have a 4-terminal device as shown in Fig. 26.18 (*a*) but when it is internally shorted to *S*, we get a 3-terminal device as shown in Fig. 26.18 (*b*). Fig. 26.18 (*c*) shows the symbol for a P-channel DE MOSFET.

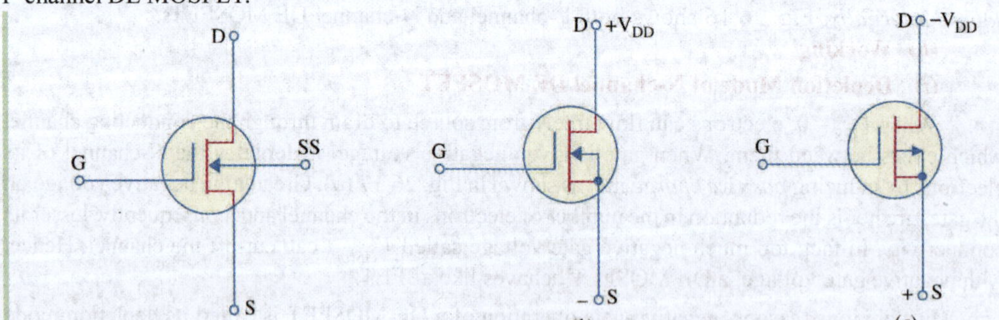

Fig. 26.18

26.16. Static Characteristics of a DE MOSFET

In Fig. 26.19 are shown the drain current and transfer characteristics of a common-source N-channel DE MOSFET for V_{GS} varying from +2 V to $V_{GS(off)}$.

It acts in the enhancement mode when gate is *positive* with respect to source and in the depletion mode when gate is *negative.* As usual, $V_{GS(off)}$ represents the gate-source voltage which cuts off the source-to-drain current. The transfer characteristic is shown in Fig. 26.19 (*b*). For a given V_{DS}, I_D flows even when $V_{GS} = 0$. However, keeping V_{DS} constant, as V_{GS} is made more negative, I_D keeps decreasing till it becomes zero at $V_{GS} = V_{GS(off)}$. When used in the enhancement mode, I_D increases as V_{DS} is increased positively.

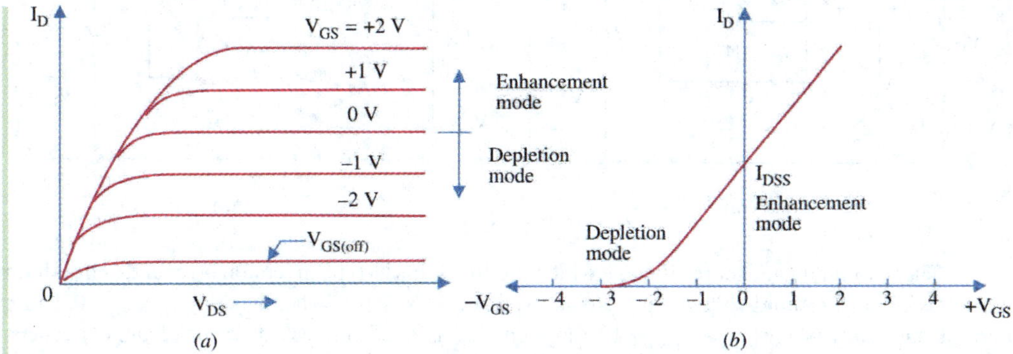

Fig. 26.19

26.17. Enhancement-only N-Channel MOSFET

This N-channel MOSFET (also called NMOS) finds wide application in digital circuitry. As shown in Fig. 26.20 (*a*) in the NMOS, the P-type substrate extends all the way to the metal-oxide layer. Structurally, *there exists no channel between the source and the drain.* Hence, an NMOS can never operate with a negative gate voltage because it will induce *positive* charge in the space between the drain and source which will not allow the passage of electrons between the two [Fig. 26.20 (b)]. Therefore, it operates *with positive gate voltage only.*

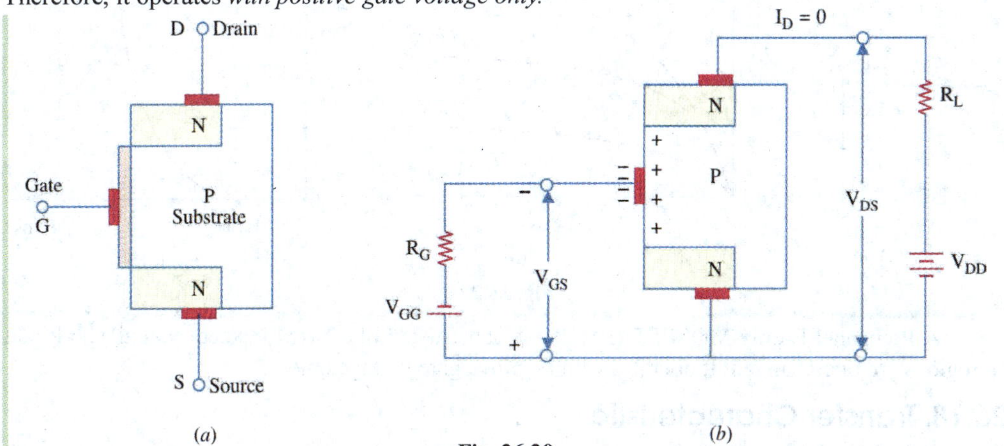

Fig. 26.20

The normal biasing polarities of this E-only MOSFET (both N-channel and P-channel) are shown in Fig. 26.21. With $V_{GS} = 0$, I_D is non-existent even when some positive V_{DD} is applied. It is found that for getting significant amount of drain current, we have to apply sufficiently high positive gate voltage. This voltage is found to produce a thin layer of free electrons very close to the metal-oxide (SiO_2) film which stretches all the way from source to drain. The thin layer of P-substrate touching the metal-oxide film which provides channel for electrons (and hence acts like N-type material) is called N-type *inversion layer or virtual* N-channel [Fig. 26.21 (*a*)].

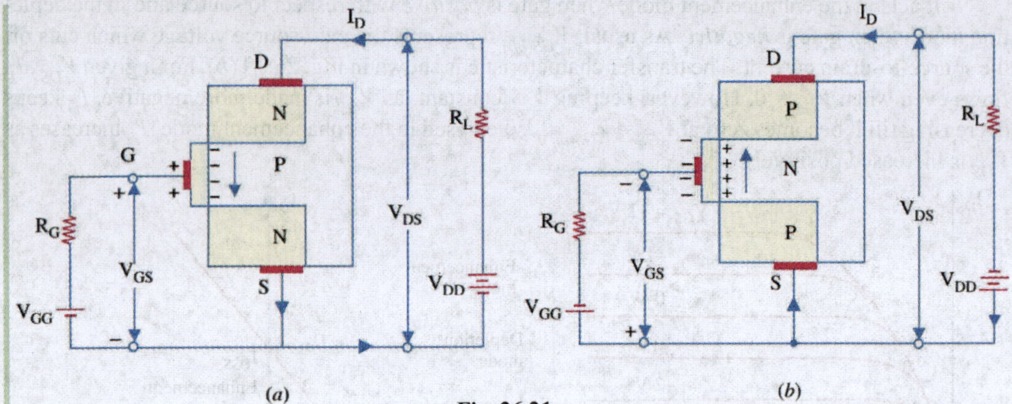

Fig. 26.21

The minimum gate-source voltage which produces this N-type inversion layer and hence drain current is called threshold voltage $V_{GS(th)}$ as shown in Fig. 26.22 (*c*). When $V_{GS} < V_{GS(th)}$, $I_D = 0$. Drain current starts only when $V_{GS} > V_{GS(th)}$. For a given V_{DS}, as V_{GS} is increased, virtual channel deepens and I_D increases. The value of I_D is given by $I_D = K(V_{GS} - V_{GS(th)})^2$ where K is a constant having a typical value of 0.3 mA/V^2. Fig. 26.22 (*a*) shows the schematic symbol for an E-only N-channel MOSFET whereas Fig. 26.22 (*b*) shows its typical drain characteristics. As usual, arrow on the substrate points to the N-type material and the vertical line (representing-channel) is broken as a reminder of the normally-OFF condition.

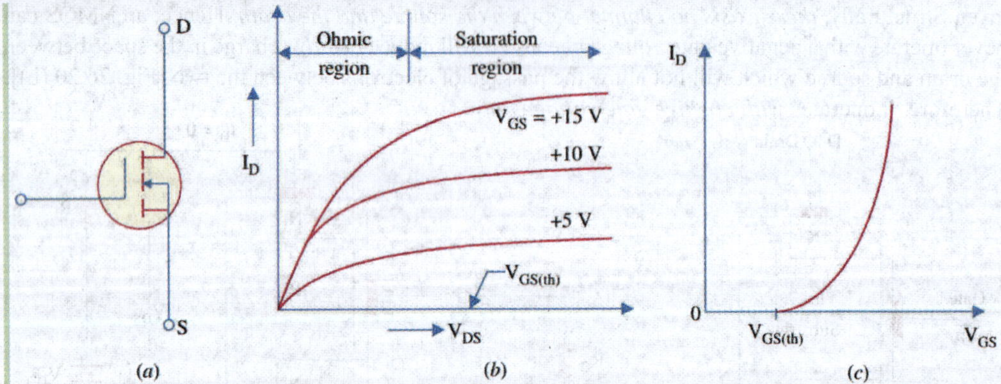

Fig. 26.22

A P-channel E-only MOSFET (PMOS) is constructed like NOMS except that all the P-and N–regions are interchanged. It operates with *negative* gate voltage only.

26.18. Transfer Characteristic

It is shown in Fig. 26.22 (*c*). I_D flows only when V_{GS} exceeds threshold voltage $V_{GS(th)}$. This MOSFET does not have an I_{DSS} parameter as do the JFET and DE MOSFET.

Example 26.6. *The data sheet for a certain E-MOSFET gives $I_D = 4.5$ mA at $V_{GS} = 12$ V and $V_{GS(th)} = 6$ V. Determine the value of I_D for $V_{GS} = 10$ V.*

Solution. $I_D = K(V_{GS} - V_{GS(th)})^2$

$\therefore$ $K = 5/(V_{GS} - V_{GS(th)})^2 = \dfrac{4.5}{(12-6)^2} = 6.125$ mA/V

Now, when $V_{GS} = 10$ V

$$I_D = K(V_{GS} - V_{GS(th)})^2 = 0.125\,(10-6)^2 = \mathbf{2\ mA}$$

Example 26.7. *The date sheet for a certain JFET indicates that I_{DSS} = 16 mA and $V_{GS(off)}$ = –5 V. Determine the drain current I_D for V_{GS} = 0 V, – 1 V and – 4 V.*

Solution. (*i*) For V_{GS} = 0 V, $I_D = I_{DSS}$ = **16 mA**

(*ii*) When V_{GS} = –1V, we can find I_D by using the following relation

$$I_D = I_{DSS}\left[1 - \frac{V_{GS}}{V_{GS(off)}}\right]^2 = 16\text{ mA}\left(1 - \frac{-1}{-5}\right)^2 = \mathbf{10.2\ mA}$$

(*iii*) When V_{GS} = – 4 V, $I_D = 16\text{ mA}\left(1 - \frac{-4}{-5}\right)^2 = \mathbf{0.6\ mA}$

26.19. FETs as Switches

Both JFETs and MOSFETs can be used as switches in electronic circuits. When the signal applied to the gate stops drain current, MOSFET acts like an open switch and when the gate signal drives the circuit into saturation, the MOSFET acts like a closed switch as shown in Figs. 26.23 and 26.24.

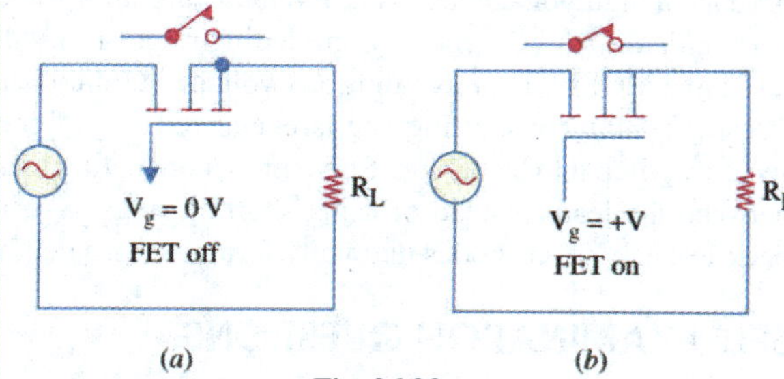

Fig. 26.23

E-MOSFETs are excellent devices for use as switches because with no gate signal, they are in the OFF state. However, with positive gate signal, the device is ON. Fig. 26.23 (*a*) shows the E-MOSFET switch used in series with the load, whereas Fig. 26.24 shows the switch in parallel with the load.

When MOSFET is in series with the load and is OFF, all the signal voltage is dropped across the high impedance R_{DS} and the MOSFET and output voltage across load resistance R_L is zero. When the MOSFET is ON, the signal voltage divides between R_L and R_{DS}. Since, $R_L >> R_{DS}$, most of the signal voltage drops across R_L.

In Fig. 26.24, MOSFET is connected in parallel with R_L. If R_S is much smaller than R_L but much larger than R_{DS}, the circuit can control the signal voltage dropped on the load. When MOSFET is OFF, most of the voltage will be delivered to the load because $R_L \gg R_S$. However, when MOSFET is ON, it shunts R_L thereby causing the input voltage to be dropped across R_S.

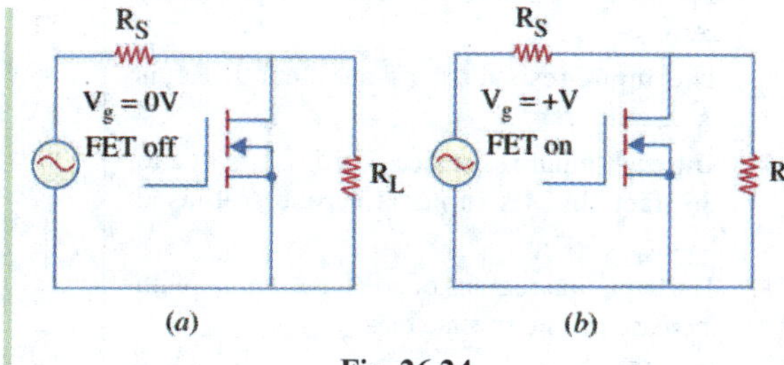

Fig. 26.24

26.20. FET Applications

FETs can be used in almost every application in which bipolar transistors can be used. However, they have certain applications which are exclusive to them :

1. As input amplifiers in oscilloscopes, electronic voltmeters and other measuring and testing equipment because their high r_{in} *reduces loading effect to the minimum.*
2. In logic circuits where it is kept OFF when there is zero input while it is turned ON with very little power input.

3. For mixer operation of FM and TV receivers.
4. As voltage-variable resistor (VVR) in operational amplifiers and tone controls etc.
5. Large-scale integration (LSI) and computer memories because of very small size.

26.21. MOSFET Handling

MOSFETs require very careful handling particularly *when out of circuit.* In circuit, a MOSFET is as rugged as any other solid-state device of similar construction and size.

It is essential not to permit any stray or static voltage on the gate otherwise the ultrathin SiO_2 layer between the channel and the gate will get ruptured. Since gate-channel junction looks like a capacitor with extremely high resistance, it requires only a few electrons to produce a high voltage across it. Even *picking up a* MOSFET *by its leads can destroy it.* Generally, grounding rings are used to short all leads of a MOSFET for avoiding any voltage build up between them. These grounding or shorting rings are removed only after MOSFET is securely wired into the circuit. Sometimes, conducting foam is applied between the leads instead of using shorting rings. As shown in Fig. 26.25, some MOSFETs have back-to-back Zener diodes internally formed to protect them against stray voltages.

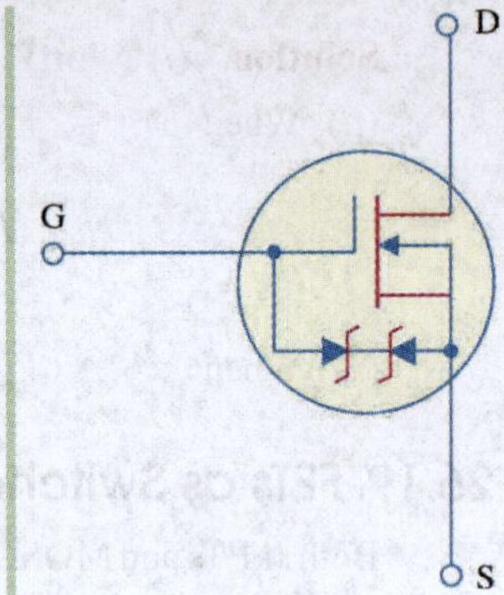

Fig. 26.25

SELF EXAMINATION QUESTIONS

A. Fill in the blanks by most appropriate word(s) or numerical value(s).

1. In a JFET, the gate-channel junctions are normally never .... biased.
2. The depletion regions in a JFET act like a .... to reduce the drain current.
3. A JFET is a .... controlled device.
4. Unlike a BJT, JFET is a .... transistor.
5. The .... current of a JFET is practically zero.
6. The input resistance of an ideal JFET is ....
7. The high input resistance of a JFET is due to the fact that its input is a reverse-biased ....
8. The depletion regions of a JFET .... with increase in gate reverse bias.
9. In a JFET, drain current is .... when $V_{GS} = 0$.
10. With no voltage applied to the gate, channel of JFET acts like a ....
11. In MOSFETs, gate and channel are .... from each other.
12. With .... gate voltage, a DE MOSFET operates in the depletion mode.
13. In depletion-mode, and N-channel DE MOSFET conducts with V_{GS} is ....
14. An N-channel E-only MOSFET works only with .... gate voltages.
15. A MOSFET can be easily destroyed by any .... voltage on its gate.
16. Any voltage build-up between different leads of a MOSFET can be avoided by using .... rings.

B. Answer True or False

1. A FET is a unipolar device.
2. FETs have high input impedance but low output impedance.
3. The gate-channel junctions of a JFET is due to its reverse-biased gate.
4. High input impedance of a JFET is due to its reverse-biased gate.
5. A JFET does not conduct with zero gate bias voltage.
6. Gate current of JFET is practically nil.
7. FETs are unipolar.
8. In a JFET, pinch-off means current-off.
9. As compared to BJTs, FETs are less noisy.
10. Input impedance of a MOSFET is extremely high due to extremely thin layer of SiO_2.
11. E-only N-channel MOSFET works with only large positive voltages.
12. Even picking up a MOSFET by hand can destroy it.

C. Multiple Choice Items

1. A FET consists of a
   (*a*) source (*b*) drain
   (*c*) gate (*d*) all of the above
2. FETs have similar properties to
   (*a*) PNP transistors
   (*b*) NPN transistors
   (*c*) thermionic valves
   (*d*) unijunction transistors
3. For small values of drain-to-source voltage, JFET behaves like a
   (*a*) resistor
   (*b*) constant-current source
   (*c*) constant-voltage source
   (*d*) negative resistance
4. In a JFET, the primary control on drain current is exerted by
   (*a*) channel resistance
   (*b*) size of depletion regions
   (*c*) voltage drop across channel
   (*d*) gate reverse bias
5. After V_{DS} reaches pinch-off value V_P in a JFET, drain current I_D becomes
   (*a*) zero (*b*) low
   (*c*) saturated (*d*) reversed
6. In a JFET, drain current is maximum when V_{GS} is
   (*a*) zero (*b*) negative
   (*c*) positive (*d*) equal to V_P
7. A JFET has the disadvantage of
   (*a*) being noisy
   (*b*) having small gain-bandwidth product
   (*c*) possessing positive temperature coefficient
   (*d*) having low input impedance
8. For the operation of enhancement only N-channel MOSFET, value of gate voltage has to be
   (*a*) high positive (*b*) high negative
   (*c*) low positive (*d*) zero
9. The extremely high input impedance of a MOSFET is primarily due to the
   (*a*) absence of its channel
   (*b*) negative gate-source voltage
   (*c*) depletion of current carriers
   (*d*) extremely small leakage current of its gate capacitor
10. The main factor which makes a MOSFET likely to breakdown during normal handling is its
    (*a*) very low gate capacitance
    (*b*) high leakage current
    (*c*) high input resistance
    (*d*) both (*a*) and (*c*)
11. In a DE MOSFET, drain current flows when gate voltage is
    (*a*) positive (*b*) negative
    (*c*) zero (*d*) all the above
12. The positive-gate operation of an N-channel DE MOSFET is known as .... mode.
    (*a*) depletion (*b*) enhancement
    (*c*) E-only (*d*) normal
13. In an E-only MOSFET, drain current starts only when $V_{GS(th)}$ is
    (*a*) positive (*b*) negative
    (*c*) zero
    (*d*) greater than $V_{GS(th)}$

ANSWERS

A. Fill in the blanks

1. forward	**2.** throttle	**3.** voltage	**4.** unipolar	**5.** gate	**6.** infinite
7. junction	**8.** increase	**9.** maximum	**10.** resistance	**11.** insulated	**12.** negative
13. zero	**14.** positive	**15.** stray	**16.** shorting/grounding		

B. True or False

1. T	**2.** F	**3.** T	**4.** T	**5.** F	**6.** T
7. T	**8.** F	**9.** T	**10.** F	**11.** T	**12.** T

C. Multiple Choice Items

1. *d*	**2.** *c*	**3.** *a*	**4.** *d*	**5.** *c*	**6.** *a*	**7.** *b*
8. *a*	**9.** *d*	**10.** *d*	**11.** *d*	**12.** *b*	**13.** *d*	

27 CHAPTER

Breakdown Devices

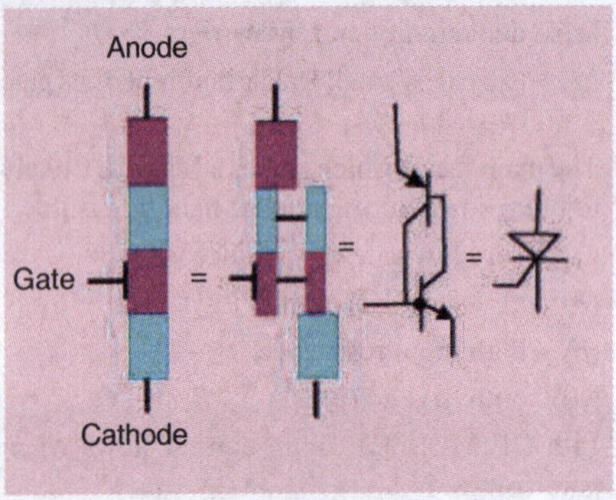

1. What are Breakdown Devices ?
2. Unijunction Transistor
3. UJT Relaxation Oscillator
4. Silicon Controlled Rectifier
5. 90° Phase Control
6. Theft Alarm
7. Triac
8. Diac
9. Silicon Controlled Switch (SCS)

27.1. What are Breakdown Devices ?

These are solid-state devices whose working depends on the *phenomenon of avalanche breakdown*. They are sometime referred to by the generic name of *thyristor* which is a semiconductor switch whose bistable action depends on P-N-P-N regenerative feedback. We will discuss the following devices :

1. Unijunction Transistor (UJT),
2. Silicon Controlled Rectifier (SCR),
3. Triac (short for 'triode ac'),
4. Diac (short for 'diode ac'),
5. Silicon Controlled Switch (SCS).

These devices have two or more junctions and can be switched ON or OFF *at an extremely fast rate*. They are also referred to as **latching** devices. A latch is a kind of switch which initially once closed, remains closed until someone opens it.

27.2. Unijunction Transistor

Basically, it is a three-terminal silicon diode. As its name indicates, it has only one P-N junction. It differs from an ordinary diode in that it has **three** leads and it differs from a FET in that it has **no ability to**

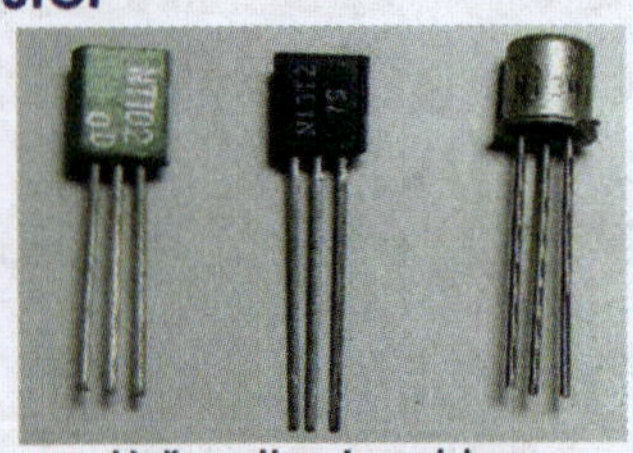

Unijunction transistors.

amplify. However, it has the ability *to control a large ac power with a small signal.* It also exhibits a negative resistance characteristic which makes it useful as an oscillator.

(*a*) Construction

It consists of a *lightly-doped* silicon bar with a heavily-doped P-type material alloyed to its one side (closer to B_2) for producing single P-N junction. As shown in Fig. 27.1 (*a*), there are three terminals: one emitter, *E* and two bases B_2 and B_1 at the top and bottom of the silicon bar. The emitter leg is drawn at an angle to the vertical and arrow points in the direction of *conventional* current when UJT is in the conducting state.

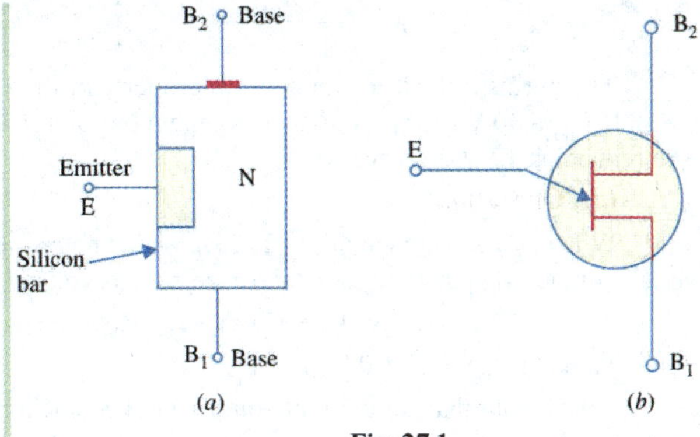

Fig. 27.1

(*b*) Interbase Resistance (R_{BB})

It is the resistance between B_2 and B_1 *i.e.*, it is the total resistance of the silicon bar from one end to the other with emitter terminal open [Fig. 27.2 (*a*)].

From the equivalent circuit of Fig. 27.2 (*b*), it is seen that $R_{BB} = R_{B2} + R_{B1}$.

It should also be noted that point *A* is such that

$R_{B1} > R_{B2}$. Usually, R_{B1} = 60% of R_{BB}.

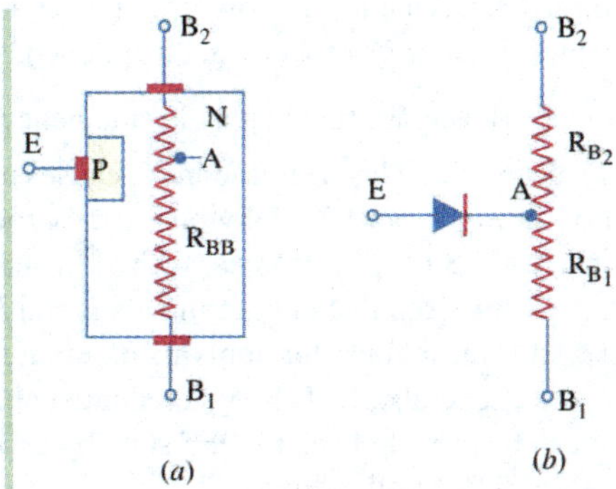

Fig. 27.2

(*c*) Intrinsic Stand-off Ratio

As seen from Fig. 27.3 (*a*), when a battery of 30 V is applied across B_2 B_1, there is a progressive fall of voltage over R_{BB} provided *E* is open. It is obvious from Fig. 27.3 (*b*) that emitter acts as a voltage-divider tap on fixed resistance R_{BB}.

With emitter open, $I_1 = I_2$. The interbase current is given by Ohm's Law.

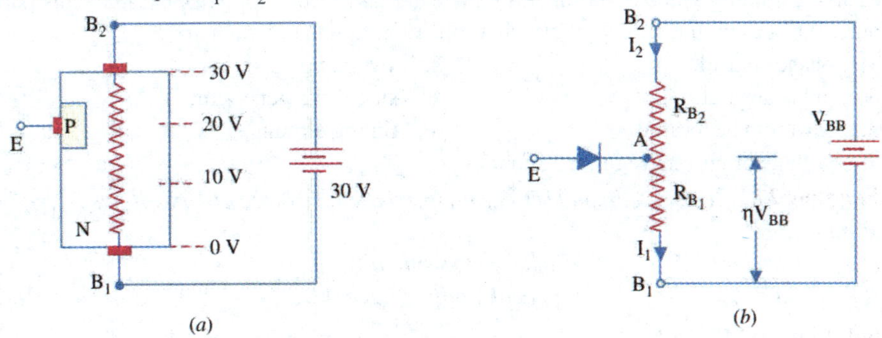

Fig. 27.3

$$I = I_2 = V_{BB}/R_{BB}$$

For example, if V_{BB} = 30 V and R_{BB} = 20 K, $I_1 = I_2$ = 2 mA.

It may be noted that part of V_{BB} is dropped over R_{B2} and part on R_{B1}. Let us call the voltage drop across R_{B1} as V_A. Using simple voltage divider relationship,

$$V_A = V_{BB} \frac{R_{B_1}}{R_{B_1} + R_{B2}}$$

The voltage division factor is given by a special symbol (η) and the name of **'intrinsic stand-off ratio'.**

$$\eta = \frac{R_{B1}}{R_{B_1} + R_{B_2}} \qquad \therefore \qquad V_A = \eta\, V_{BB}$$

The intrinsic stand-off ratio is the property of the UJT and is always less than unity (0.5 to 0.85). If $V_{BB} = 30$ V and $\eta = 0.6$, then potential of point A with respect to point $B_1 = 0.6 \times 30 = 18$ V. The remaining 12 V drop across R_{B_2}.

(*d*) Operation

When V_{BB} is switched on, V_A is developed and *reverse-biases the junction*. If V_B is the barrier voltage of the P-N junction, then total reverse bias voltage is

$$= V_A + V_B = \eta V_{BB} + V_B$$

Value of V_B for Si is 0.7 V.

It is obvious that emitter junction will not become forward-biased unless its applied voltage V_E exceeds $(\eta V_{BB} + V_B)$. This value of V_E is called *peak-point voltage* V_P. When $V_E = V_P$, emitter (peak current) I_P starts to flow through R_{B_1} to ground (*i.e.*, B_1). The UJT is then said to have been *fired* or turned ON. Due to the flow of I_E (= I_P) through R_{B1}, number of charge carriers in R_{B1} is increased which *reduces its resistance*. As η depends on R_{B_1}, *its value is also decreased.*

Hence, we find that as V_E and hence I_E increases (beyond I_P), R_{B_1} decreases, η decreases and V_A decreases. This decrease in V_A causes more emitter current to flow which causes a further reduction in R_{B1}, η and V_A. Obviously, the process is regenerative. V_A as well as V_E quickly drop as I_E increases. Since, V_E decreases when I_E increases, the UJT *possesses negative resistance*.

It is seen that only terminals E and B_1 are the active terminals whereas B_2 is the bias terminal *i.e.*, it is meant only for applying external voltage across the UJT.

Generally, UJT is triggered into conduction by applying a suitable positive pulse at its emitter. It can be brought back to OFF state by applying a negative trigger pulse.

(*e*) Applications

One unique property of UJT is that it can be triggered by (or an output can be taken from) *any one of its three terminals*. Once triggered, the emitter current I_E of the UJT increases regeneratively till it reaches a limiting value determined by the external power supply. Because of this particular behaviour, UJT is used in a variety of circuit applications some of which are :

1. phase control,
2. switching,
3. pulse generation,
4. sine wave generator,
5. sawtooth generator,
6. timing circuits,
7. voltage or current regulated supplies.

Example 27.1. *A given silicon UJT has an interbase resistance of 10 K. It has $R_{B1} = 6$ K with $I_E = 0$. Find*

(a) UJT current if $V_{BB} = 20$ V and V_E is less than V_P,

(b) η and V_A, *(c) peak point voltage, V_P.*

Solution. (*a*) Since $V_E < V_P$, $I_E = 0$ because P-N junction is reverse-biased.

$$\therefore \qquad I_1 = I_2 = \frac{V_{BB}}{R_{BB}} = \frac{20}{10\text{ K}} = \mathbf{2mA}$$

(*b*)
$$\eta = \frac{R_{B1}}{R_{BB}} = \frac{6}{10} = \mathbf{0.6}$$

$$V_A = \eta V_{BB} = 0.6 \times 20 = \mathbf{12\ V}$$

(*c*)
$$V_P = \eta V_{BB} + V_B = 12 + 0.7 = \mathbf{12.7\ V}$$

27.3. UJT Relaxation Oscillator

The relaxation oscillator shown in Fig. 27.4 consists of a UJT and a capacitor C which is charged through R as V_{BB} is switched on.

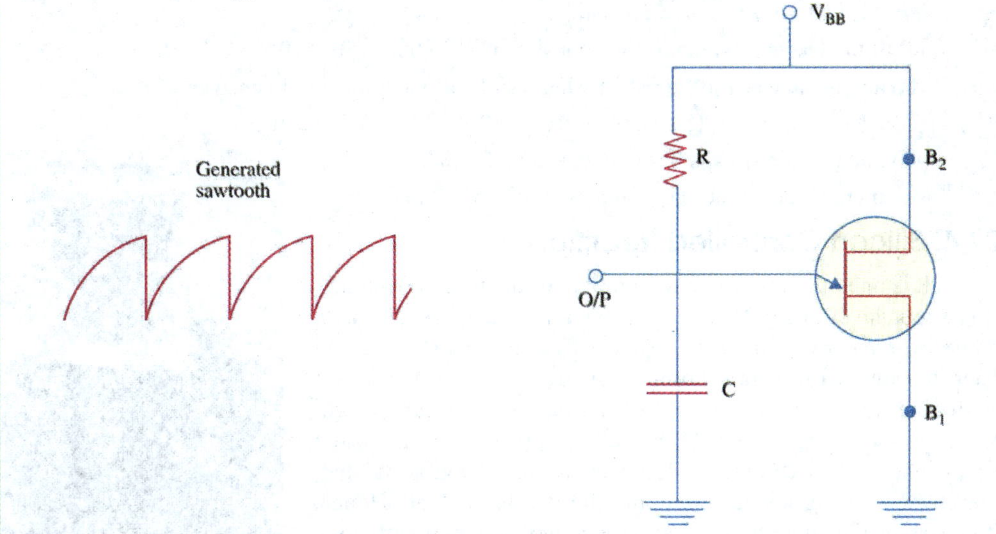

Fig. 27.4

When the capacitor voltage V_C reaches the value of V_P, the UJT fires and rapidly discharges C *via* B_1. The device then cuts off and C starts to charge again. This cycle is repeated continuously generating a sawtooth waveform across C.

The inclusion of external resistances R_2 and R_1 in series with B_2 and B_1 (Fig. 27.5) provides *spike waveforms*. When the UJT fires, the sudden surge of current through B_1 causes a drop across R_1 which produces positive going spikes. Also, at the time of firing, fall of V_{BB1} causes I_2 to increase rapidly which generates negative-going spikes across R_2.

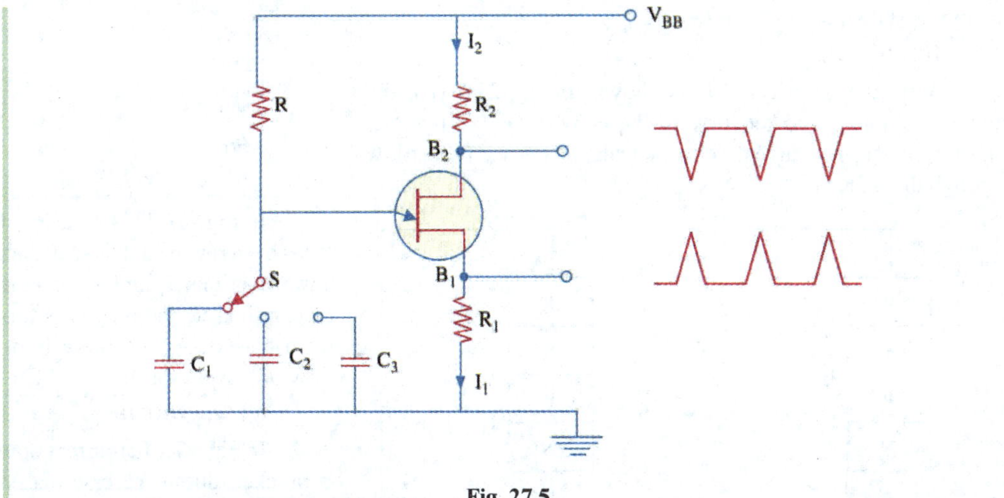

Fig. 27.5

By switching over to different capacitors, frequency of the output waveform can be changed as desired.

Example 27.2. *The windshield wiper motor of an automobile is controlled by a UJT with η = 0.6. The capacitor has a value of 50 μF and the charging resistor is a series combination of 50 K resistor and a 500 K potentiometer. Determine the minimum and maximum number of blade strokes per minute possible with this arrangement.*

Solution. The least value of time constant is = $50,000 \times 50 \times 10^{-6} = 2.5$ second.

Maximum value of time constant when whole of potentiometer resistance is used is

$$= (50 + 500) \times 10^3 \times 50 \times 10^{-6} = 27.5 \text{ second}$$

Maximum blade strokes per minute = 60/2.5 = **24.**

Minimum blade strokes per minute = 60/27.5 = **2.2.**

Silicon controlled rectifier.

27.4. Silicon Controlled Rectifier

It is one of the prominent members of the thyristor family. It is a four-layer or PNPN device. Basically, *it is a rectifier with a control element*. In fact, it consists of *three diodes* connected back-to-back with a gate connection. It is widely used as a switching device in power control applications. It can control loads by switching current OFF and ON up to many thousand times a second. It can switch ON for variable lengths of time, thereby delivering selected amount of power to the load. Hence, it possesses the advantages of a rheostat and a switch with none of their disadvantages.

(*a*) Construction

As shown in Fig. 27.6 (*a*), it is a three-junction three terminal four-layer transistor, the layers being alternately of P-type and N-type silicon. The three junctions are marked J_1, J_2 and J_3 whereas the three terminals are : anode (A), cathode (C) and gate (G) which is connected to the *inner* P-type layer. The function of the gate is to control the firing of SCR. The schematic symbol is shown in Fig. 27.6 (*b*).

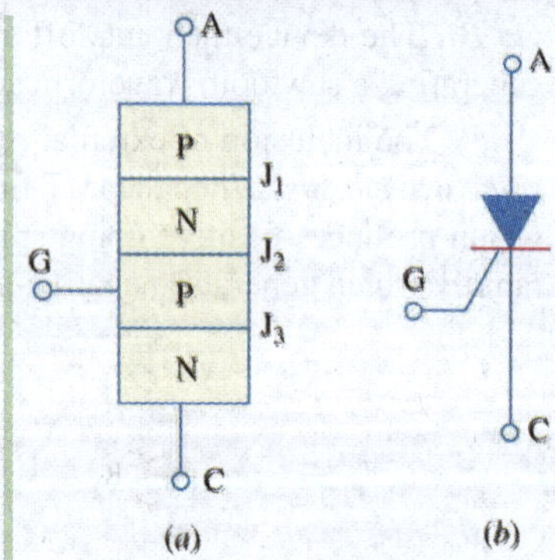

Fig. 27.6

(*b*) Biasing

With the polarity of V as shown in Fig. 27.7 (*a*), the junctions J_1 and J_3 become forward-biased whereas J_2 is reverse-biased. Hence, no current (except leakage current) can flow through the SCR.

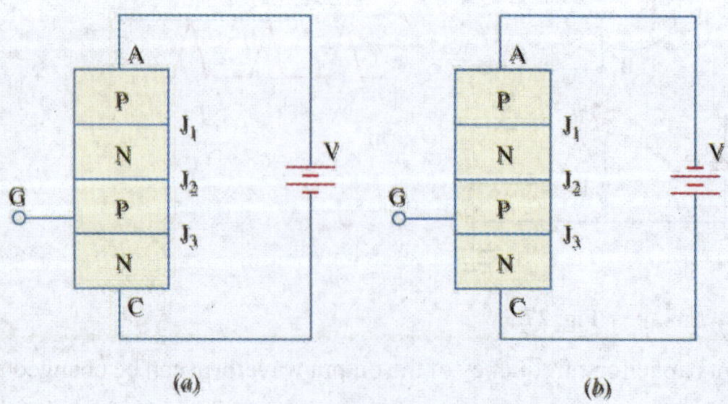

Fig. 27.7

In Fig. 27.7 (*b*) polarity of V has been reversed. It is seen that, now, junctions J_1 and J_3 become reverse-biased and only J_2 is forward-biased. Again, there is no flow of current through the SCR.

(*c*) Operation

In Fig. 27.7 (*a*) current flow is blocked due to reverse-biased junction J_2. However, when anode voltage is increased, a certain critical value called forward breakover voltage V_{BO} is reached when J_2 breaks down and SCR switches suddenly to a highly conducting state. Under this condition, SCR offers very little forward resistance (0.01 Ω – 1.0 Ω) so that voltage across it drops to a low value (about 1 V) as shown in Fig. 27.8 and current is limited only by the power supply and the load resistance. Current keeps flowing indefinitely until the circuit is opened briefly.

With supply connection as in Fig. 27.7 (*b*), the current through the SCR is blocked by the two reverse-biased junctions J_1 and J_3. When V is increased, a stage comes when Zener breakdown occurs which may destroy the SCR (Fig. 27.8). Hence, it is seen that SCR is a unidirectional device unlike triac *which is bi-directional.*

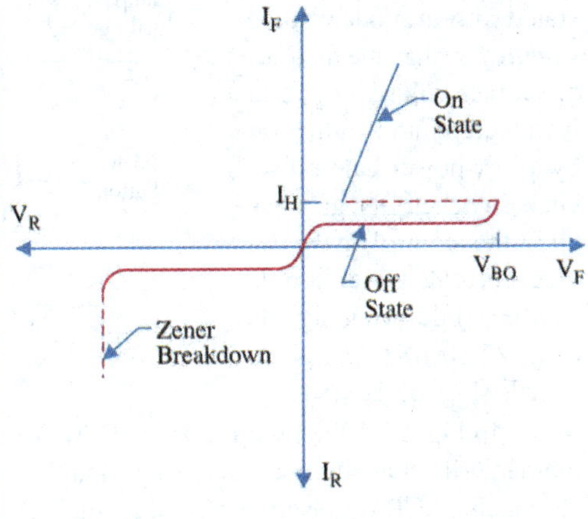

Fig. 27.8

(*d*) Two Transistor Analogy

The basic operation of an SCR can be described by using two transistor analogy. For this purpose, SCR is split into two 3-layer transistor structures as shown in Fig. 27.9 (*a*). As seen, transistor Q_1 is a PNP transistor whereas Q_2 is an NPN device interconnected together. It will also be noted from Fig. 27.9 (*b*) that

(*i*) collector current of Q_1 is also the base current of Q_2 and

(*ii*) base current of Q_1 is also the collector current of Q_2.

Suppose that the supply voltage across terminals *A* and *C* is such that reverse-biased junction J_2 starts breaking down. Then, current through the device begins to rise. It means that I_{E1} begins to increase. Then,

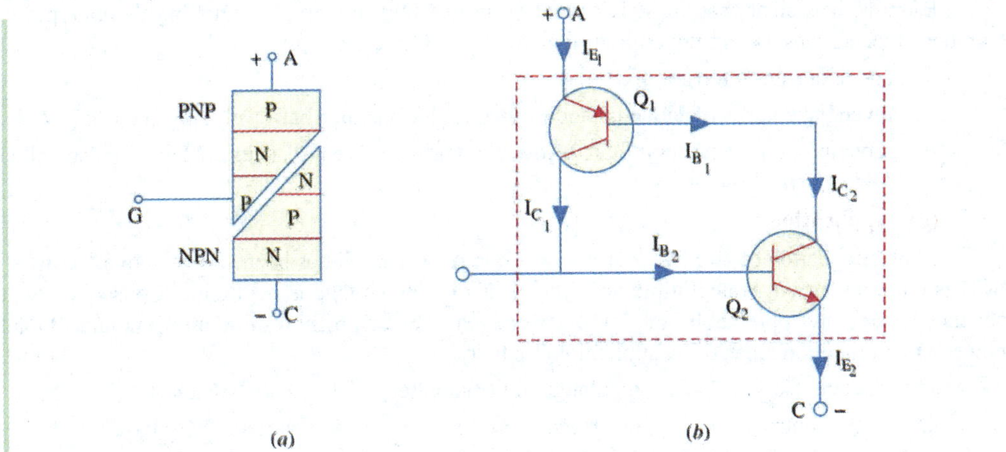

Fig. 27.9

1. I_{C1} increases (remember $I_C = \alpha\, I_E$);
2. since $I_{C1} = I_{B2}$, I_{B2} also increases;
3. hence, I_{C2} increases (remember $I_C = \beta\, I_B$);
4. now, $I_{C2} = I_{B1}$, hence I_{B1} increases;
5. consequently, both I_{C1} and I_{E1} increase.

As seen, a regenerative action takes place whereby an initial increase in current produces further increase in the same current. Soon, maximum current is reached limited by external resistances.

The two transistors are fully turned ON and voltage across the two transistors falls to a very low value. Typical turn-ON times for an SCR are 0.1 to 1.0 μs.

(*e*) Firing and Triggering

Usually, SCR is operated with an anode voltage *slightly less* than the forward breakover voltage V_{BO} and is triggered into conduction by a low-power gate pulse. Once switched ON, gate has no further control on the device current. Gate signals can be (*a*) dc firing signals [Fig. 27.10 (*a*)], (*b*) pulse signals [Fig. 27.10 (*b*)].

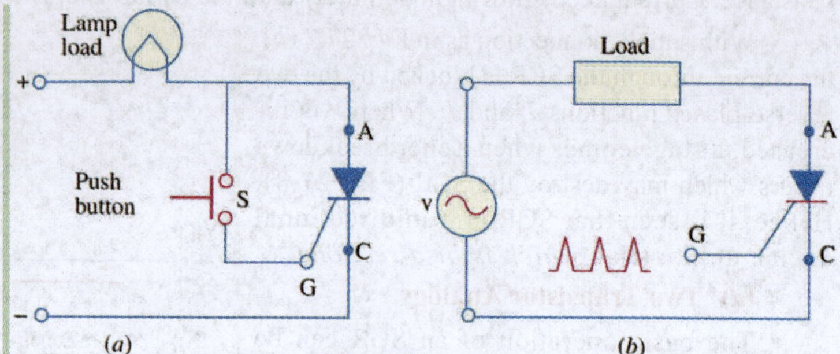

Fig. 27.10

In Fig. 27.10 (*a*) with *S* open, SCR does not conduct and the lamp is out. When *S* is closed momentarily, a positive voltage is applied to the gate which forward-biases the centre P-N junction. As a result, SCR is pulsed into conduction and the lamp lights up. SCR will remain in the conducting state until the supply voltage is removed or reversed. Fig. 27.10 (*b*) shows triggering by timed pulses obtained from a pulse source.

(*f*) Turning OFF

As stated earlier, once 'fired', SCR remains ON *even when triggering pulse is removed*. This ability of the SCR to remain ON even when gate current is removed is referred to as **latching.** In fact, SCR belongs to a class of devices known as *latching devices*.

By now, it is clear that an SCR cannot be turned OFF by simply removing the gate pulse. Number of techniques are employed to turn an SCR off. These are:

1. anode current interruption;
2. reversing polarity of anode-cathode voltage as is done each half-cycle by *v* in Fig. 27.14 (*b*);
3. reducing current through SCR below the holding current I_H (Fig. 27.8). It is also called *low-current dropout.*

(*g*) Applications

Main application of an SCR is as a *power control device*. It has been shown above that when SCR is OFF, its current is negligible and when it is ON, its voltage is negligible. Consequently, it never dissipates any appreciable amount of power even when controlling substantial amounts of load power. Other common areas of its application include

1. relay controls,
2. regulated power supplies,
3. static switches,
4. motor controls,
5. inverters,
6. battery chargers,
7. heater controls,
8. phase control.

SCRs have been designed to control powers upto 10 MW with individual ratings as high as 2000 A at 1.8 kV. Its frequency range of application has been extended to about 50 kHz.

27.5. 90° Phase Control

In the phase control circuit of Fig. 27.11, gate triggering current is derived from the supply itself. The variable resistance *R* limits the gate current during positive half-cycles of the supply. If *R* is adjusted to a low value, SCR will trigger almost immediately at the commencement of the positive half-cycle of the input.

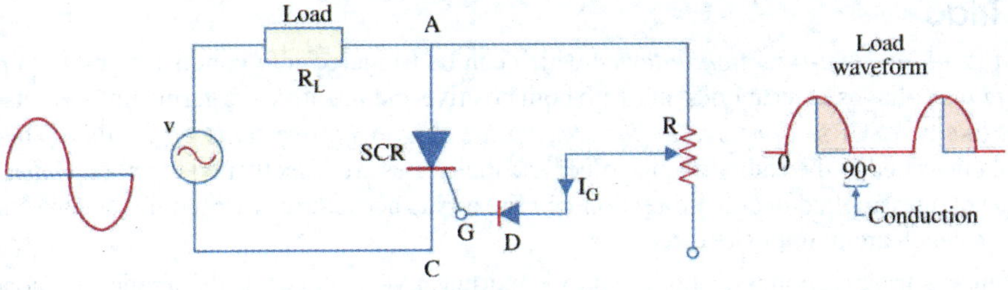

Fig. 27.11

If, on the other hand, R is set to a high resistance, SCR may not switch ON until the peak of the positive half-cycle. By adjusting R between these two extremes, SCR can be made to switch ON somewhere between the commencement and peak of the positive half-cycle *i.e.*, between 0° and 90°.

It is obvious that if I_G is not enough to trigger the SCR at 90°, then the device will not trigger at all because I_G has maximum value then. This operation is sometimes referred to as *half-wave variable-resistance phase control*. It is an effective method of controlling the load power.

The purpose of diode is to protect the gate from negative voltage which would otherwise be applied to it during the negative half-cycle of the input.

It is seen from Fig. 27.11 that at the instant of SCR switch-ON, gate current flows through R_L, R and D. Hence, at that *instant*

$$v = I_G R_L + I_G R + V_D + V_G \qquad \therefore \qquad R = \frac{v - V_D - V_G - I_G R_L}{I_G}$$

Example 27.3. *The circuit of Fig. 27.11 is connected to an ac supply $v = 50 \sin \theta$ and R_L = 50 Ω. Gate current is 100 μA and $V_G = 0.5$ V. Determine the range of adjustment of R for the SCR to be triggered between 30° and 90°. Take $V_D = 0.7$ V.*

Solution. (*i*) $\theta = 30°$

Now, $v = 50 \sin \theta = 50 \sin 30° = 25$ V

$$\therefore \qquad R = \frac{25 - 0.7 - 0.5 - (100 \times 10^{-6} \times 50)}{100 \times 10^{-6}} = \mathbf{238\ K}$$

(*ii*) $\theta = \mathbf{90°}$

$v = 50 \sin 90° = 50$ V

$$R = \frac{50 - 0.7 - 0.5 - 0.005}{100 \times 10^{-6}} = \mathbf{488\ K}$$

27.6. Theft Alarm

The circuit shown in Fig. 27.12 can be used to protect a car tape deck or a radio receiver from theft. The switch S is located at some concealed point in the car and is kept *closed*. Since gate G is grounded through the tape deck, the SCR is OFF and the horn is silent. If the tape deck is removed, G is no longer grounded. Instead, it gets connected to the car battery through R. Consequently, gate current is set up which fires the SCR. As a result, the horn starts blowing and continues to do so until S is opened.

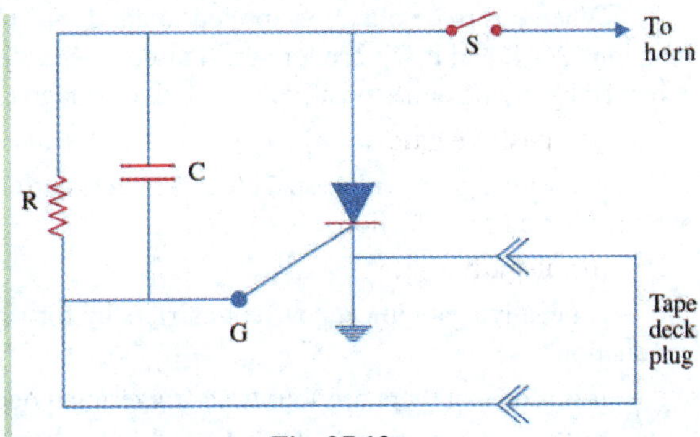

Fig. 27.12

27.7. Triac

It is a 5-*layer bi-directional device* which can be triggered into conduction by both *positive and negative* voltages at its anodes and with both positive and negative triggering pulses at its gate. It behaves like two SCRs *connected in parallel, upside down with respect to each other*. That is, the anode of one is tied to the cathode of the other and their gates are directly tied together. Hence, anode and gate voltages applied in either direction will fire a triac because they would fire at least one of the two SCRs which are in opposite directions.

Since a triac responds to both positive and negative voltages at the anode, the concept of cathode used for an SCR is dropped. Instead, the two electrodes are called anodes A_1 and A_2.

1. Construction

As shown in Fig. 27.13 (*a*), a triac has three terminals A_1, A_2 and G. As seen, gate G is closer to anode A_1. It is clear from Fig. 27.13 (*b*), that a triac is nothing but *two inverse parallel-connected SCRs with a common gate terminal*. As seen, it has six doped regions. Fig. 27.14 shows the schematic symbol which consists of two inverse-connected SCR symbols.

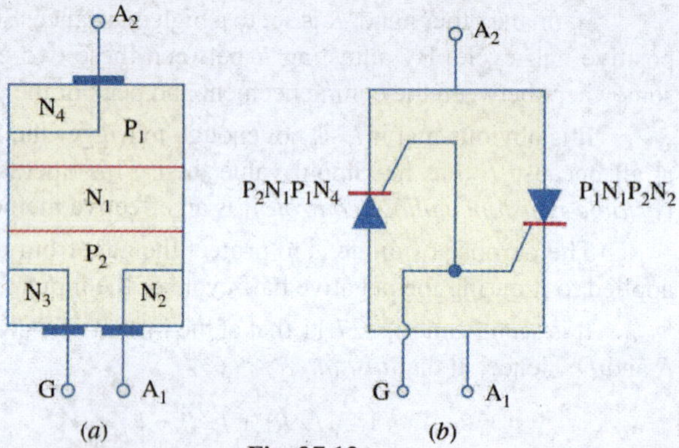

Fig. 27.13

2. Operation

(*a*) When A_2 is Positive

When positive voltage is applied to A_2, path of current flow is P_1-N_1-P_2-N_2. The two junctions P_1-N_1 and P_2-N_2 are forward-biased whereas N_1-P_2 junction is blocked. The gate can be given either positive or negative voltage to turn ON the triac as explained below.

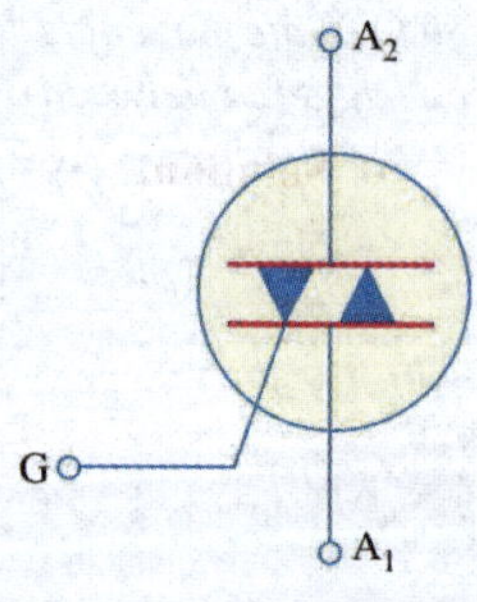

Fig. 27.14

(*i*) **positive gate**

A positive gate (with respect to A_1) forward-biases the P_2-N_2 junction and the breakdown occurs as in a normal SCR.

(*ii*) **negative gate**

A negative gate forward-biases the P_2-N_3 junction and current carriers injected into P_2 turn on the triac.

(*b*) When A_1 is Positive

When positive voltage is applied to anode A_1, path of current flow is P_2-N_1-P_1-N_4. The two junctions P_2-N_1 and P_1-N_4 are forward-biased whereas junction N_1-P_1 is blocked. Conduction can be achieved by giving either positive or negative voltage to G as explained below.

(*i*) **positive gate**

A positive gate (with respect to A_1) injects current carriers by forward-biasing P_2-N_2 junction and thus initiates conduction.

(*ii*) **negative gate**

A negative gate injects current carriers by forward-biasing P_2-N_3 junction thereby triggering conduction.

It is seen that there are four triac-triggering modes, two each for the two anodes.

Low-current dropout is the only way to open a triac.

3. V/I Characteristics

Typical characteristics of a triac are shown in Fig. 27.15.

As seen, triac exhibits same forward blocking and forward conducting characteristics as an SCR but *for either polarity of voltage applied to the main terminal.* Obviously, a triac has latch current in either direction.

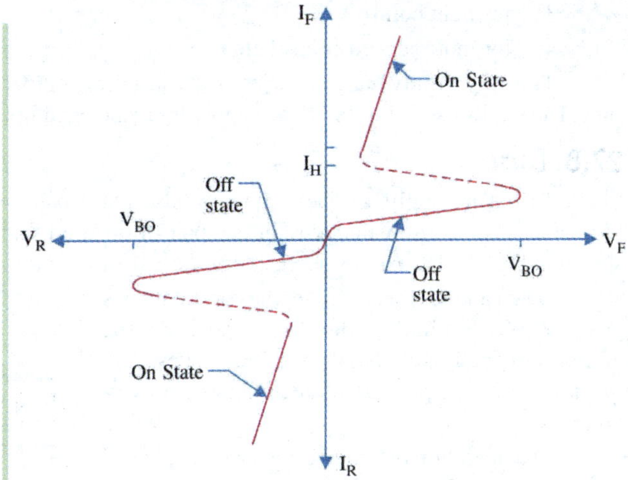

Fig. 27.15

4. Applications

One fundamental application of triac is shown in Fig. 27.16. Here, it is used to control ac power to a load by switching ON and OFF during positive and negative half-cycles of the input ac power.

During positive half-cycle of the input, diode D_1 is forward biased, D_2 is reverse-biased and gate is positive with respect to A_1. During the negative half-cycle, D_1 is reverse-biased. D_2 is forward-biased and gate is positive with respect to A_2. By adjusting R, the point at which conduction commences can be varied.

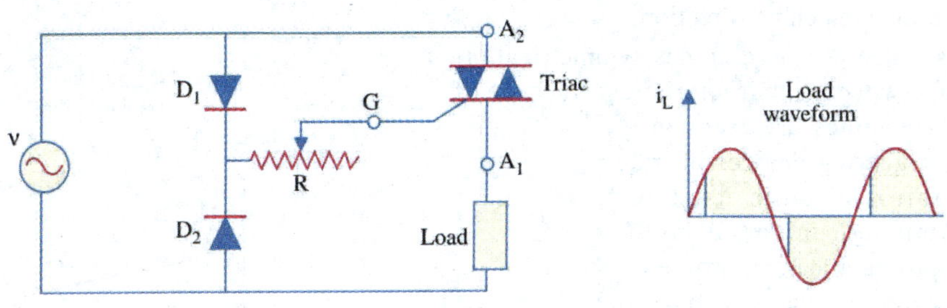

Fig. 27.16

Diac-triac combination for ac load power control is shown in Fig. 27.17. Firing control of diac is achieved by adjusting R.

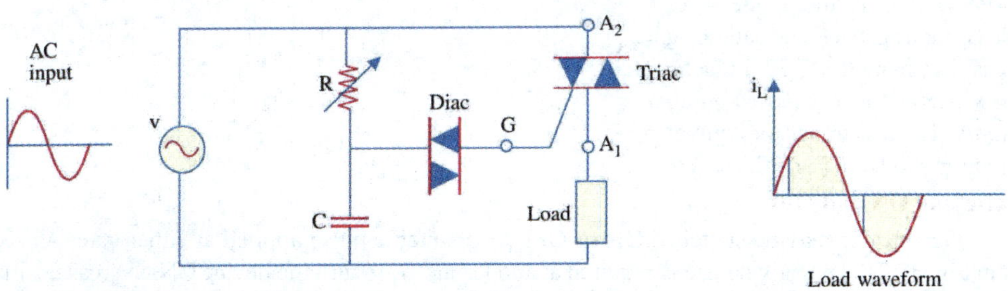

Fig. 27.17

Other applications of a triac include

1. as static switch to turn ac power OFF and ON;
2. for minimizing radio interference;

3. for light control;
4. for motor speed control etc.

The only disadvantage of triac is that it takes comparatively longer time to recover to OFF state. Hence, its use is limited to ac supply frequencies of upto 400 Hz.

27.8. Diac

To put it simply, a diac is nothing else but a *triac without its gate terminal* as shown in Fig. 27.18 (*a*). Its equivalent circuit is a pair of inverted four layer diodes. Its schematic symbol is shown in Fig. 27.18 (*b*). As seen, *it can breakdown in either direction.*

When anode A_1 is positive, the current path is P_2-N_2-P_1-N_1. Similarly, when A_2 is positive, the current flow path is P_1-N_2-P_2-N_3. Diac is designed to trigger triacs or provide protection against over-voltages.

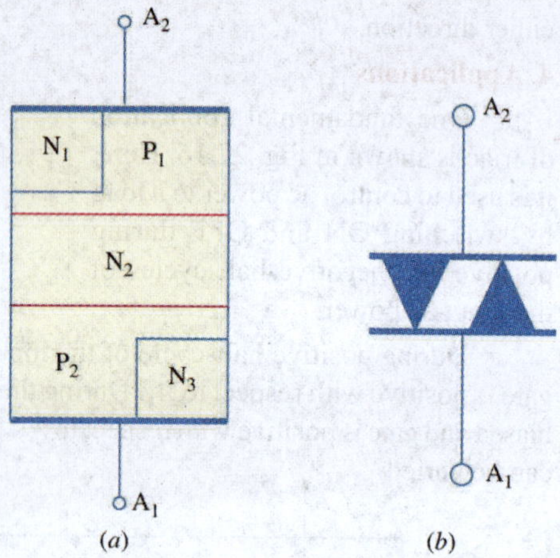

Fig. 27.18

The operation of a diac can best be explained by imaging it is **two-diodes connected in series.** Voltage applied across it in either direction turns ON one diode, reverse-biasing the other. Hence, it can be switched from OFF to ON state for either polarity of the applied voltage.

The characteristic curve of a typical diac is shown in Fig. 28.19. It resembles the letter *Z* since diac breaksdown in either direction.

As stated above, diac has symmetrical bi-directional switching characteristics. Because of this feature, diacs are frequently used as triggering devices in triac phase control circuits used for light dimming, universal motor speed control and heat control etc.

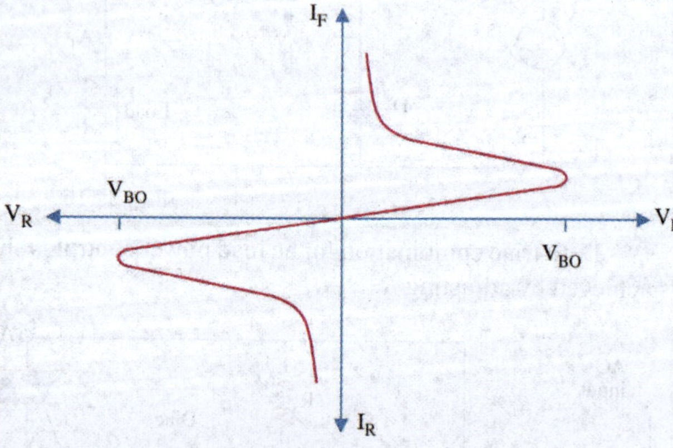

Fig. 27.19

27.9. Silicon Controlled Switch (SCS)

It is a four-layer, four-terminal PNPN device having anode *A*, cathode *C*, anode gate G_1 and cathode gate G_2 as shown in Fig. 27.20. In fact, *it is a low-current SCR with two gate terminals*. The two transistor equivalent circuit is shown in Fig. 27.21.

Switching ON and OFF

The device may be switched ON or OFF by a suitable pulse applied at either gate. As seen from Fig. 27.21, a negative pulse is required at anode gate G_1 to turn the device ON whereas positive pulse is needed to turn it OFF as explained below.

Similarly, at cathode gate G_2, a negative pulse is required to switch the device OFF and a positive pulse to turn it ON.

As seen from Fig. 27.21, when a negative pulse is applied to G_1, it forward-biases Q_1 (being PNP) which is turned ON. The resulting heavy collector current I_C, being the base current of Q_2, turns it ON. Hence, SCS is switched ON. A positive pulse at G_1 will reverse bias E/B junction of Q_1 thereby switching the SCS OFF.

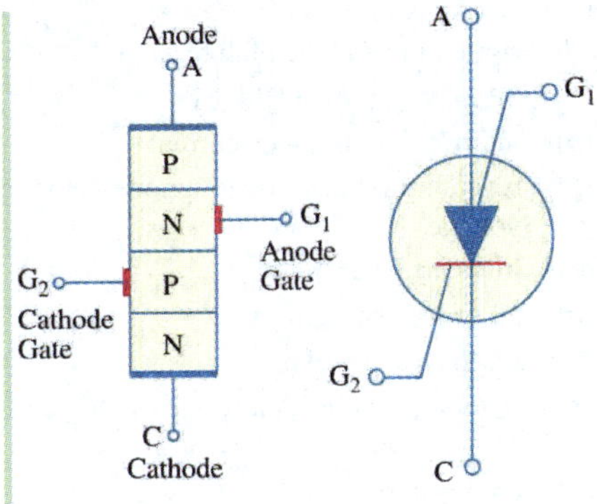

Fig. 27.20

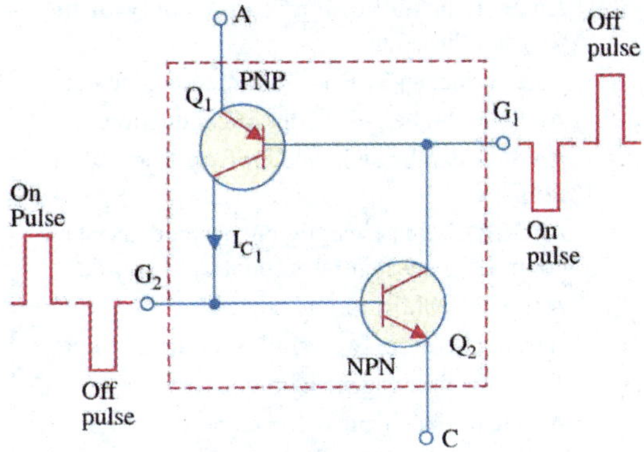

Fig. 27.21

V/I Characteristics

The *V/I* characteristics of an SCS are essentially the same as **those for the SCR** (Fig. 27.8).

As compared to an SCR, an SCS has *much reduced turn*-OFF *time*. Moreover, it has higher control and triggering sensitivity and a more predictable firing situation.

Applications

The more common areas of SCS application are as under :

1. in counters, register and timing circuits of computers,
2. pulse generators,
3. voltage sensors,
4. oscillators etc.

SELF EXAMINATION QUESTIONS

A. Fill in the blanks with most appropriate word(s) or numerical value(s).

1. A UJT has only .... P-N junction.
2. The intrinsic stand-off ratio of a UJT is the ratio of two ....
3. A UJT fires when its emitter voltage exceeds the value ....
4. Once fired, the current of a UJT increases ....
5. A UJT can be triggered into conduction by a .... pulse at its emitter.
6. A UJT is frequently used as a .... oscillator.
7. Basically, SCR is a .... with a control element.
8. Once turned ON, an SCR becomes nearly a .... circuit from anode to cathode.
9. An SCR can be imagined to consist of .... interconnected transistors of opposite polarity.
10. A triac has two .... and one gate terminal.
11. A triac can be imagined to consist of two .... parallel-connected SCRs with a common gate terminal.
12. A triac has .... doped regions.
13. Low current dropout is the only way to .... a triac.
14. A diac may be thought of as a triac without its .... terminal.
15. A diac is equivalent to .... diodes connected in series.
16. An SCS has one anode, one cathode and .... gates.
17. An SCS is just an .... with an extra gate terminal.
18. A .... pulse is required at the anode gate of an SCS to turn it ON.

B. Answer True or False

1. A latching device is like a bistable switch.
2. A UJT is a four-terminal diode.
3. A UJT is like a variable resistor.
4. The intrinsic stand-off ratio of a UJT is always less than unity.
5. A UJT can be triggered by any one of its three terminals.
6. A UJT fires when its emitter voltage just exceeds the peak point voltage.

7. An SCR is a 3-junction, 3-terminal, 4-layer device.
8. An SCR has a holding current only in the forward direction.
9. Like a triac, an SCR is a bidirectional device.
10. An SCR can be pulsed out of conduction.
11. An SCR can be switched OFF by low-current dropout.
12. An SCR dissipates negligible power even when controlling substantial amount of ac power.
13. A triac is a bidirectional device.
14. A triac consists of two parallel-connected SCRs.
15. There are four triggering modes for a triac.
16. A triac has latch current in either direction.
17. A diac consists of a pair of inverted four-layer diodes.
18. An SCS is a three-terminal PNPN device.
19. An SCS has two gates and two anodes.
20. Whereas an SCR can be turned OFF only via its anode, an SCS can turned OFF via its gates.

C. Multiple Choice Items

1. Which semiconductor device acts like a diode and two resistors ?
   (*a*) SCR (*b*) triac
   (*c*) diac (*d*) UJT
2. A UJT has $R_{BB} = 10$ K and $R_{B2} = 4$ K. Its intrinsic stand-off ratio is
   (*a*) 0.6 (*b*) 0.4
   (*c*) 2.5 (*d*) 5/3
3. An SCR conducts appreciable current when its .... with respect to cathode.
   (*a*) anode and gate are both negative
   (*b*) anode and gate are both positive
   (*c*) anode is negative and gate is positive
   (*d*) gate is negative and anode is positive
4. After firing an SCR, the gating pulse is removed. The current in the SCR will
   (*a*) remain the same
   (*b*) immediately fall to zero
   (*c*) rise up
   (*d*) rise a little and then fall to zero
5. An SCR may be turned OFF by
   (*a*) interrupting its anode current
   (*b*) reversing polarity of its anode-cathode voltage
   (*c*) low-current dropout
   (*d*) all of the above
6. A triac behaves like two
   (*a*) inverse parallel-connected SCRs with common gate
   (*b*) diodes in series
   (*c*) four-layer diodes in parallel
   (*d*) resistors and one diode
7. A triac can be triggered into conduction by
   (*a*) only positive voltage at either anode
   (*b*) positive or negative voltage at either anode
   (*c*) positive or negative voltage at gate
   (*d*) both (*b*) and (*c*)
8. A diac is equivalent to a
   (*a*) pair of SCRs
   (*b*) pair of four-layer SCRs
   (*c*) diode and two resistors
   (*d*) triac with two gates
9. An SCS has
   (*a*) four layers and three terminals
   (*b*) three layers and four terminals
   (*c*) two anodes and two gates
   (*d*) one anode, one cathode and two gates
10. An SCS may be switched ON by a
    (*a*) positive pulse at its anode
    (*b*) negative pulse at its cathode
    (*c*) positive pulse at its cathode gate G_2
    (*d*) positive pulse at its anode gate G_1

ANSWERS

A. Fill in the blanks.

1. one	**2.** resistances	**3.** $(\eta V_B + V_{BB})$	**4.** regeneratively	**5.** positive	**6.** relaxation
7. rectifier	**8.** short	**9.** two	**10.** anodes	**11.** inverse	**12.** 6
13. open	**14.** gate	**15.** two	**16.** two	**17.** SCR	**18.** negative

B. True or False

1. T	**2.** F	**3.** T	**4.** T	**5.** T	**6.** T	**7.** T
8. T	**9.** F	**10.** F	**11.** T	**12.** T	**13.** T	**14.** T
15. T	**16.** T	**17.** T	**18.** F	**19.** F	**20.** T	

C. Multiple Choice Items

1. *d*	**2.** *a*	**3.** *b*	**4.** *a*	**5.** *d*	**6.** *a*	**7.** *d*	**8.** *b*	**9.** *b*	**10.** *c*

28 CHAPTER

Sinusoidal Oscillators

1. Classification of Oscillators
2. Damped and Undamped Oscillations
3. Frequency of Oscillatory Current
4. Frequency Stability of an Oscillator
5. Essentials of a Feedback LC Oscillator
6. Tuned Base Oscillator
7. Tuned Collector Oscillator
8. Tuned Drain Oscillator (FET)
9. Hartley Oscillator
10. FET Hartley Oscillator
11. Colpitts Oscillator
12. Clapp Oscillator
13. FET Colpitts Oscillator
14. Crystals
15. Crystal Controlled Oscillator

28.1. What is an Oscillator ?

An electronic oscillator may be defined in any one of the following four ways :

1. It is a circuit which converts dc energy into ac energy at a very high frequency;
2. It is an electronic source of alternating current or voltage having sine, square or sawtooth or pulse shapes;
3. It is a circuit which generates an ac output signal without requiring any externally-applied input signal;
4. It is an unstable amplifier.

These definitions exclude electromechanical alternators producing 50 Hz ac power or other devices which convert mechanical or heat energy into electric energy.

28.2. Comparison between an Amplifier and an Oscillator

As discussed in Chapter 22, an amplifier produces an output signal whose waveform is similar to the input signal but whose power level is generally high. This additional power is supplied by the external dc source. Hence, an amplifier is essentially an energy converter *i.e.*, it takes energy from the dc power source and converts it into ac energy at signal frequency. The process of energy conversion is controlled by the input signal. If there is no input signal, there is no energy conversion and hence there is no output signal.

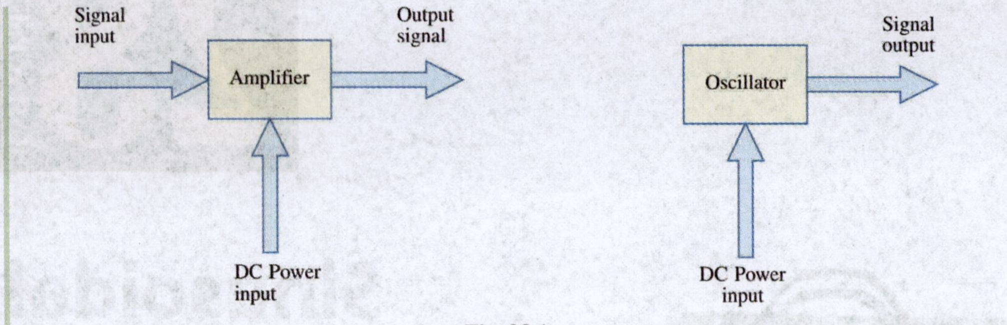

Fig. 28.1

An oscillator differs from an amplifier in one basic respect : the oscillator *does not require an exernal signal* either to start or maintain energy conversion process (Fig. 28.1). It keeps producing an output signal so long as the dc power source is connected.

Moreover, the frequency of the output signal is determined by the passive components used in the oscillator and can be varied at will.

28.3. Classification of Oscillators

Electronic oscillators may be broadly divided into following two groups :

(*i*) sinusoidal (or harmonic) oscillators—which produce an output having sine waveform;

(*ii*) non-sinusoidal (or relaxation) oscillators—they produce an output which has square, rectangular or sawtooth waveform or is of pulse shape.

Sinusoidal oscillators may be further subdivided into :

(*a*) tuned-circuits or LC feedback oscillators such as Hartley, Colpitts and Clapp etc.;

(*b*) RC phase-shift oscillators such as Wien-bridge oscillator;

(*c*) negative-resistance oscillators such as tunnel diode oscillator;

(*d*) crystal oscillators such as Pierce oscillator;

(*e*) heterodyne or beat-frequency oscillator (BFO).

The active devices (bipolars, FETs or unijunction transistors) in the above mentioned circuits may be biased class-A, B or C. Class-A operation is used in high-quality audio frequency oscillators. However, radio frequency oscillators are usually operated class-C.

28.4. Damped and Undamped Oscillations

Sinusoidal oscillations produced by oscillators may be (*i*) damped or (*ii*) undamped.

(*i*) Damped Oscillations

Oscillations whose amplitude keeps decreasing (or decaying) with time are called **damped** or **decaying** oscillations. The waveform of such oscillations is shown in Fig. 28.2 (*a*). These are produced by those oscillator circuits in which I^2R losses take place continuously during each oscillation without any arrangement for compensating the same. Ultimately, the amplitude of the oscillations decays to zero when there is not enough energy to supply circuit losses. However, the frequency or time-period remains constant because *it is determined by the circuit parameters*.

(*ii*) Undamped Oscillations

Oscillations whose amplitude remains constant *i.e*,. does not change with time are called undamped oscillations. These are produced by those oscillator circuits which have no losses or if they have, there is provision for compensating them. The constant-amplitude and constant-frequency sinusoidal waves shown in Fig. 28.2 (*b*) are called *carrier waves* and are used in communication transmitters for transmitting low-frequency audio information to far off places.

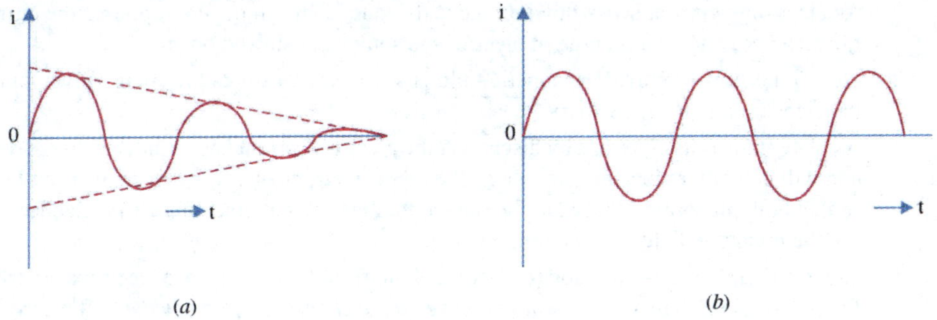

Fig. 28.2

28.5. The Oscillatory Circuit

It is also called LC circuit or tank circuit. The oscillatory circuit (Fig. 28.3) consists of two reactive elements *i.e.*, an inductor and a capacitor. Both are capable of storing energy. The capacitor stores energy in its electric field whenever there is potential difference across its plates. Similarly, a coil or an inductor stores energy in its magnetic field whenever current flows through it. Both *L* and *C* are supposed to be loss-free (*i.e.*, their *Q*-factors are infinite).

As shown in Fig. 28.3 (*a*), suppose the capacitor has been fully-charged from a dc source. Since *S* is open, it cannot discharge through *L*. Now, let us see what happens when *S* is closed.

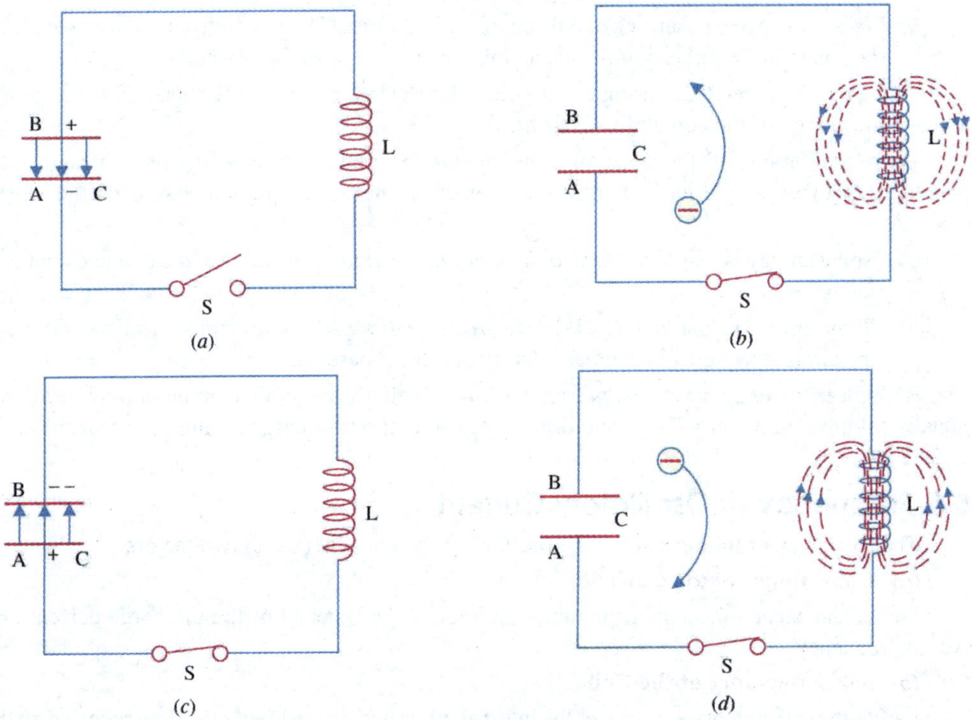

Fig. 28.3

1. When S is closed [Fig. 28.3 (b)] *electrons* move from plate A to plate B through coil L as shown by the arrow (or conventional current flows from B to A). This electron flow reduces the strength of the electric field and hence the amount of energy stored in it.
2. As electronic current starts flowing, the self-induced emf in the coil opposes the current flow. Hence, rate of discharge of electrons is somewhat slowed down.
3. Due to the flow of current, magnetic field is set up which stores the energy given out by the electric field [Fig. 28.3 (b)].
4. As plate A loses its electrons by discharge, the electron current has a tendency to die down and will actually reduce to zero when all excess electrons on A are driven over to plate B so that both plates are reduced to the same potential. At that time, there is no electric field but the magnetic field has maximum value.
5. However, due to self-induction (or electrical inertia) of the coil, more electrons are transferred to plate B than are necessary to *make up the electron deficiency there*. It means that now plate B has more electrons than A. Hence, capacitor becomes charged again though in opposite direction as shown in Fig. 28.3 (c).
6. The magnetic field L collapses and the energy given out by it is stored in the electric field of the capacitor.
7. After this, the capacitor starts discharging in the opposite direction so that, now, the electrons move from plate B to plate A [Fig. 28.3 (d)]. The electric field starts collapsing whereas magnetic field starts building up again though in the opposite direction. Fig. 28.3 (d) shows the condition when the capacitor becomes fully discharged once again.
8. However, these discharging electrons overshoot and again an excess amount of electrons flow to plate A, thereby charging the capacitor once more.
9. This sequence of charging and discharging continues. The to and fro motion of electrons between the two plates of the capacitor constitutes an oscillatory current.

It may be also noted that during this process, the electric energy of the capacitor is converted into magnetic energy of the coil and *vice-versa*.

These oscillations of the capacitor discharge are damped because energy is dissipated away gradually so that their amplitude becomes zero after sometime. There are two reasons for the loss of the energy :

(a) Some energy is lost *in the form of heat produced in the resistance of the coil* and connecting wires;

(b) Some energy is lost *in the form of electromagnetic (EM) waves* that are radiated out from the circuit through which an oscillatory current is passing.

Both these losses subtract energy from the circuit with the result that circuit current decreases gradually till it becomes zero. The waveform of the oscillatory discharge is similar to that shown in Fig. 28.2 (b).

28.6. Frequency of Oscillatory Current

The frequency or time-period of the oscillatory current depends on two factors :

(a) Capacitance of the Capacitor

Larger the capacitor, greater the time required for the reversal of the discharge current *i.e.*, lower its frequency.

(b) Self-inductance of the Coil

Larger the self-inductance, greater the internal effect and hence longer the time required by the current to stop flowing during discharge of the capacitor.

The frequency of this oscillatory discharge current is given by

$$f = \frac{1}{2\pi\sqrt{LC}} = \frac{159}{\sqrt{LC}} \text{ kHz}$$

where L = self-inductance in μH and C = capacitance in μF

It may, however, be pointed out here that damped oscillations so produced are no good for radio transmission purposes because of their *limited range and excessive distortion*. For good radio transmission, we need undamped oscillations which can be produced if some additional energy is supplied in correct phase and correct direction to the LC circuit for making up the I^2R losses continually occurring in the circuit.

28.7. Frequency Stability of an Oscillator

The ability of an oscillator to maintain a constant frequency of oscillation is called its frequency stability. Following factors affect the frequency stability :

1. Operating Point of the Active Device

The Q-point of the active device *i.e.*, transistor is so chosen as to confine the circuit operation on the linear portion of its characteristic. Operation on non-linear portion varies the parameters of the transistor which, in turn, affects the frequency stability of the oscillator.

2. Inter-element Capacitances

Any changes in the inter-element capacitances of a transistor particularly the collector-to-emitter capacitance cause changes in the oscillator output frequency, thus affecting its frequency stability. The effect of changes in inter-element capacitances, can be neutralized by adding a swamping capacitor across the offending elements—the added capacitance being made part of the tank circuit.

3. Power Supply

Changes in the dc operating voltages applied to the active device shifts the oscillator frequency. This problem can be avoided by using regulated power supply.

4. Temperature Variations

Variations in temperature cause changes in transistor parameters and also change the values of resistors, capacitors and inductors used in the circuit. Since such changes take place slowly, they cause a slow change (called drift) in the oscillator output frequency.

5. Output Load

A change in the output load may cause a change in the Q-factor of the LC tuned circuit thereby affecting the oscillator output frequency.

6. Mechanical Vibrations

Since such vibrations change the values of circuit elements, they result in changes of oscillator frequency. This instability factor can be eliminated by isolating the oscillator from the source of mechanical vibrations.

28.8. Essentials of a Feedback LC Oscillator

The essential components of a feedback LC oscillator shown in Fig. 28.4 are :

1. A resonator which consists of an LC circuit. It is also known as frequency-determining network (FDN) or tank circuit.
2. An amplifier whose function is to amplify the oscillations produced by the resonator.
3. A positive feedback network (PFN) whose function is to transfer part of the output energy to the resonant LC circuit *in proper phase*. The amount of energy fed back is sufficient to meet $I^2 R$ losses in the LC circuit.

The essential condition for maintaining oscillations and for finding the value of frequency is

$$\beta A = 1 + j0 \qquad \text{or} \qquad \beta A \angle \phi = 1 \angle 0$$

It means that

(*i*) the feedback factor or loop gain $|\beta A| = 1$,

(*ii*) the net phase shift around the loop is 0° (or an integral multiple of 360°). In other words, *feedback should be positive*.

The above conditions form Barkhausen criterion for maintaining a steady level of oscillation at a specific frequency.

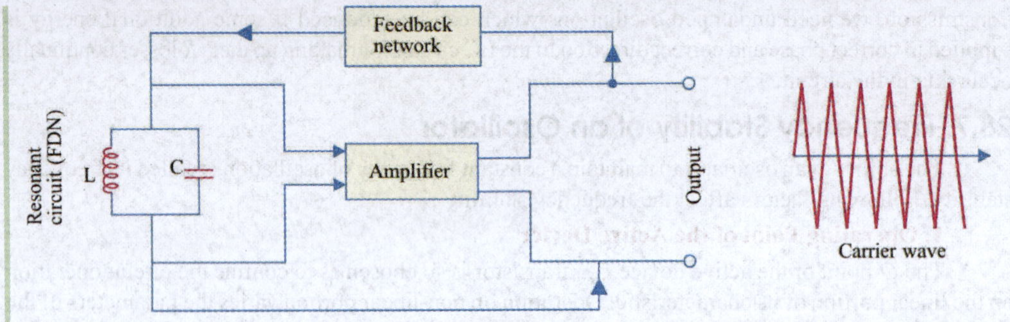

Fig. 28.4

Majority of the oscillators used in radio receivers and transmitters use tuned circuits with positive feedback. Variations in oscillator circuits are due to the different ways in which the feedback is applied. Some of the basic circuits are :

1. *Armstrong or Tickler or Tuned-base Oscillator*—It employs inductive feedback from collector to the tuned LC circuit in the base of a transistor.
2. *Tuned Collector Oscillator*—It also employs inductive coupling but the LC tuned circuit is in the collector circuit.
3. *Hartley Oscillator*—Here feedback is supplied inductively.
4. *Colpitts Oscillator*—Here feedback is supplied capacitively.
5. *Clapp Oscillator*—It is a slight modification of the Colpitts oscillator.

28.9. Tuned Base Oscillator

Such an oscillator using a transistor in CE configuration is shown in Fig. 28.5. Resistors R_1, R_2 and R_3 determine the dc bias of the circuit. The parallel R_3—C_2 network in the emitter circuit is a stabilizing circuit to prevent signal degeneration. As usual, C_1 is the dc blocking capacitor. The mutually-coupled coils L_1 and L forming primary and secondary of an *RF* transformer provide the required feedback between the collector and base circuits. The amount of feedback depends on the coefficient of coupling between the two coils. The CE connected transistor itself provides a phase shift of 180° between its input and output circuits. The transformer provides another 180° phase shift thus producing a *total phase shift of 360° which is an essential condition for producing oscillations.*

The parallel-tuned LC circuit connected between base and emitter is the frequency-determining network (FDN) *i.e.*, it generates the oscillations at its resonant frequency.

Circuit Action

The moment switch S is closed, collector current is set up which tends to rise to its quiescent value. This increase in I_C is accompanied by :

1. an expanding magnetic field through L_1 which links with L and
2. an induced e.m.f. called feedback voltage in L.

Two immediate reactions of this feedback voltage are :

(*i*) increase in emitter-base voltage (and base current) and

(*ii*) a further increase in collector current I_C.

It is followed by a succession of cycles of

1. an increase in feedback voltage,
2. an increase in emitter-base voltage and
3. an increase in I_C until saturation is reached.

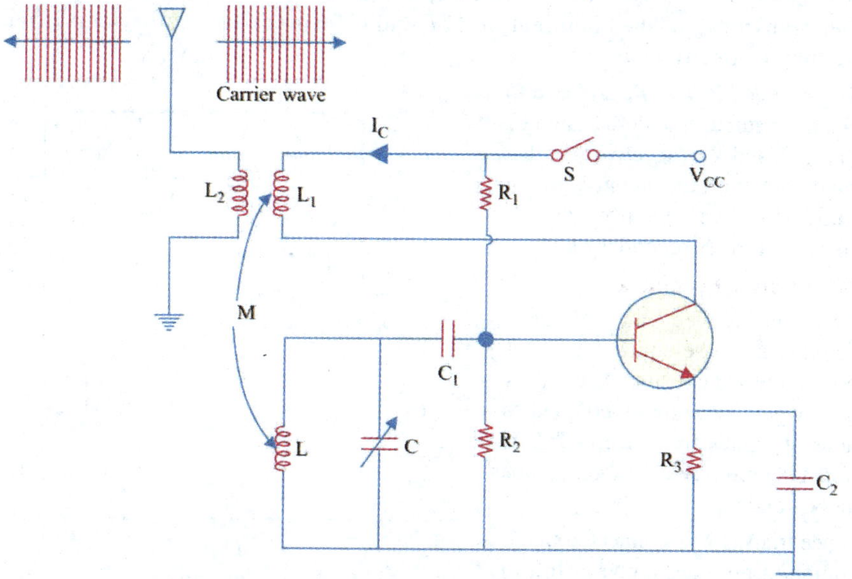

Fig. 28.5

Meanwhile, C gets charged. As soon as I_C ceases to increase, magnetic field of L_1 ceases to expand and thus no longer induces feedback voltage in L. Having been charged to maximum value, C starts to discharge through L. However, decrease in voltage across C causes the following sequence of reactions :

1. a decrease in emitter-base bias and hence in I_B, 2. a decrease in I_C,
3. a collapsing magnetic field in L_1,
4. an induced feedback voltage in L though, this time, in *opposite direction*,
5. further decrease in emitter-base bias and so on till I_C reaches its cut-off value.

During this time, the capacitor having lost its original charge, again becomes fully charged though with *opposite* polarity. Transistor being in cut-off, the capacitor will again begin to discharge through L. Since polarity of capacitor charge is opposite to that when transistor was in saturation, the sequence of reactions now will be

1. an increase in emitter-base bias,
2. an increase in I_C,
3. an expanding magnetic field in L_1,
4. an induced feedback voltage in L,
5. a further increase in emitter-base bias and
6. so on till I_C increases to its saturation value.

This cycle of operation keeps repeating so long as enough energy is supplied to meet losses in the LC circuit.

The output can be taken out by means of a third winding L_2 magnetically coupled to L_1. It has approximately the same waveform as collector current.

The frequency of oscillation is equal to the resonant frequency of the LC circuit.

28.10. Tuned Collector Oscillator

Such an oscillator using a transistor in CE configuration is shown in Fig. 28.6.

(i) Frequency Determining Network (FDN)

It is made up of a variable capacitor C and a coil L which forms primary winding of a step-down transformer. The combination of L and C forms an oscillatory tank circuit to set the frequency of oscillation.

Resistors R_1, R_2 and R_3 are used to dc bias the transistor. Capacitors C_1 and C_2 act to bypass R_3 and R_2 respectively so that they have no effect on the ac operation of the circuit. Moreover, C_2 provides ac ground for transformer secondary L_1.

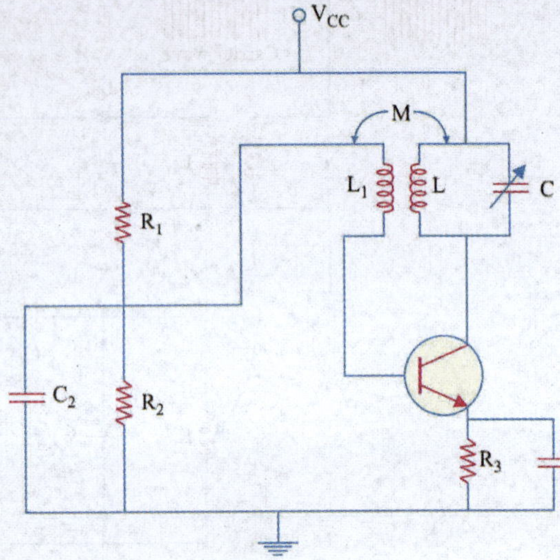

Fig. 28.6

(ii) Positive Feedback

Feedback between the collector-emitter circuit and base-emitter circuit is provided by the transformer secondary winding L_1 which is mutually-coupled to L. As far as ac signals are concerned, L_1 is connected to emitter via low-reactance capacitors C_2 and C_1.

Since transistor is connected in CE configuration, it provides a phase shift of 180° between its input and output circuits. Another phase shift of 180° is provided by the transformer thus producing a total phase shift of 360° between the output and input voltages resulting in *positive feedback* between the two.

(iii) Amplifying Action

The transistor amplifier provides sufficient gain for oscillator action to take place.

(iv) Working

When the supply is first switched on, a transient current is developed in the tuned LC circuit as the collector current rises to its quiescent value. This transient current initiates natural oscillations in the tank circuit. These natural oscillations induce a small emf into L_1 by mutual induction which causes corresponding variations in base current. These variations in I_B are amplified β times and appear in the collector circuit. Part of this amplified energy is used to meet losses taking place in the oscillatory circuit and *the balance is radiated out in the form of electromagnetic waves*.

The frequency of oscillatory current is almost equal to the resonant frequency of the tuned circuit.

$$\therefore \qquad f_o = \frac{1}{2\pi\sqrt{LC}}$$

28.11. Tuned Drain Oscillator (FET)

The basic circuit is illustrated in Fig. 28.7. It is similar to the tuned collector oscillator of Fig. 28.6. Because of its high input impedance and high voltage amplification, a FET can be used to construct very simple and efficient oscillator circuit.

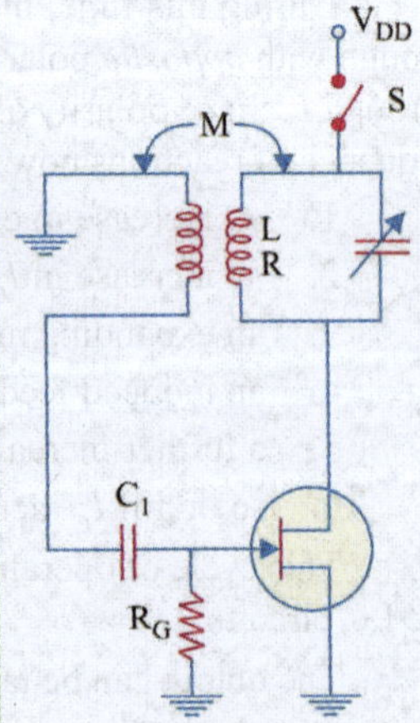

Fig. 28.7

This frequency of oscillation is given by

$$f_o = \frac{1}{2\pi\sqrt{LC}} = \sqrt{\left(1+\frac{R}{r_d}\right)}$$

where r_d = ac drain resistance.

The value of mutual inductance required for maintaining oscillations is

$$M = \frac{r_d\, RC + L}{\mu}.$$

Example 28.1. *A tuned-collector oscillator has a fixed inductance of 100 μH and has to be tunable over the frequency band of 500 kHz to 1500 kHz. Find the range of variable capacitor to be used.*

Solution. Resonant frequency is given by

$$f_o = 1/2\pi\sqrt{LC} \quad \text{or} \quad C = 1/4\pi^2 f_o^2 L$$

where L and C refer to the tank circuit.

When f_o = 500 kHz

$$C = 1/4\pi^2 \times (500 \times 10^3)^2 \times 100 \times 10^{-6} = 1015 \text{ pF}$$

When f_o = 1500 kHz

$$C = 1015/(1500/500)^2 = 113 \text{ pF}$$

Hence, capacitor range required is **113 – 1015 pF**

Example 28.2. *The resonant circuit of a tuned-collector transistor oscillator has a resonant frequency of 5 MHz. If value of capacitance is increased by 50%, calculate the new resonant frequency.*

Solution. Using the equation for resonant frequency, we have

$$5 \times 10^6 = 1/2\pi\sqrt{LC} \qquad \text{— 1st case}$$

$$f_o = 1/2\pi\sqrt{L \times 1.5C} \qquad \text{— 2nd case}$$

$$\therefore \quad \frac{f_o}{5 \times 10^6} = \frac{1}{\sqrt{1.5}} \quad \text{or} \quad f_o = \mathbf{4.08\ MHz}$$

28.12. Hartley Oscillator

In Fig. 28.8 (*a*) is shown a transistor Hartley oscillator using CE configuration. Its general principle of operation is similar to the tuned-collector oscillator discussed in Art. 28.10.

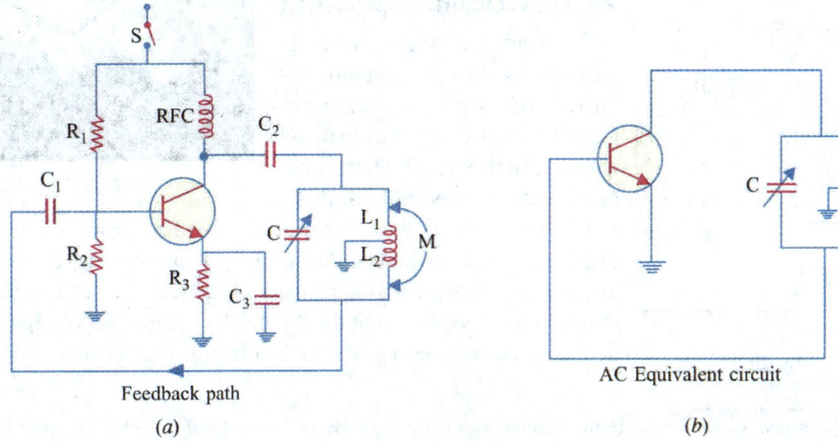

Fig. 28.8

It uses a single tapped-coil having two parts marked L_1 and L_2 instead of two separate coils. So far as ac signals are concerned, one side of L_2 is connected to base *via* C_1 and the other to emitter *via* ground and C_3. Similarly, one end of L_1 is connected to collector *via* C_2 and the other to common emitter terminal *via* C_3. In other words, L_1 is in the output circuit *i.e.*, collector-emitter circuit whereas L_2 is in the base-emitter circuit *i.e.*, input circuit. These two parts are inductively-coupled and form an auto-transformer or a split-tank inductor. Feedback between the output and input circuits is accomplished through autotransformer action which also introduces a phase reversal of 180°. This phase reversal between two voltages occurs because they are taken from *opposite ends* of an inductor ($L_1 — L_2$ combination) with respect to the tap which is tied to common transistor terminal *i.e.*, emitter which is ac grounded *via* C_3. Since transistor itself introduces a phase shift of 180°, the total phase shift becomes 360°, thereby making the *feedback positive* or *regenerative* which is essential for oscillations (Art. 28.8). As seen, positive feedback is obtained from the tank circuit and is coupled to the base *via* C_1. Fig. 28.8 (*b*) shows the ac equivalent circuit of Hartley oscillator.

Resistors R_1 and R_2 form a voltage divider for providing the base bias and R_3 is an emitter swamping resistor to add stability to the circuit. Capacitor C_3 provides ac ground thereby preventing any signal degeneration while still providing temperature stabilisation. Radio-frequency choke (RFC) provides dc load for the collector and also keeps ac currents out of the dc supply V_{CC}.

When V_{CC} is first switched on through *S*, an initial bias is established by $R_1—R_2$ and oscillations are produced because of positive feedback from the *LC* tank circuit (L_1 and L_2 constitute *L*). The frequency of oscillation is given by

$$f_o = \frac{1}{2\pi\sqrt{LC}} \qquad \text{where} \qquad L = L_1 + L_2 + 2M$$

The output from the tank may be taken out by means of another coil coupled either to L_1 or L_2.

28.13. FET Hartley Oscillator

The basic circuit is shown in Fig. 28.9. R_G is the gate biasing resistor. There is mutual induction between the two parts L_1 and L_2 of the coil.

$$f_o = \frac{1}{2\pi\sqrt{LC}}$$

where $L = L_1 + L_2 + 2M$

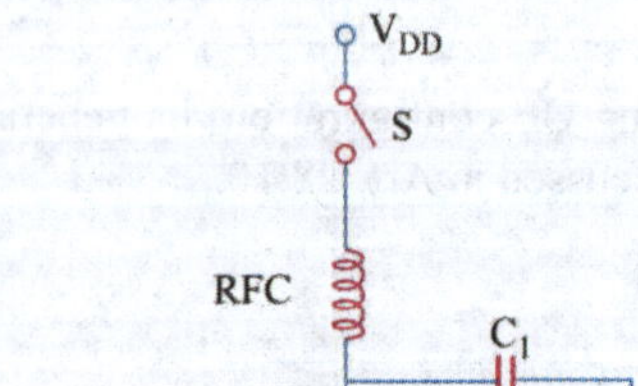

Fig. 28.9

Hartley oscillator.

28.14. Colpitts Oscillator

This oscillator is essentially the same as Hartley oscillator except for one difference. Colpitts oscillator uses **tapped capacitance** whereas Hartley oscillator uses **tapped inductance***. Fig.28.10 (*a*) shows the complete circuit with its power source and dc biasing circuit whereas Fig. 28.10 (*b*) shows its ac equivalent circuit. The two series capacitors C_1 and C_2 form the voltage divider used for providing the feedback voltage (the voltage drop across C_2 constitutes the feedback voltage). The feedback fraction

* As an aid to memory, remember that Hartley begins with letter *H* for Henry *i.e.*, coil and Colpitts begins with *C* for Capacitor.

is C_1/C_2. The minimum value of amplifier gain for maintaining oscillations is

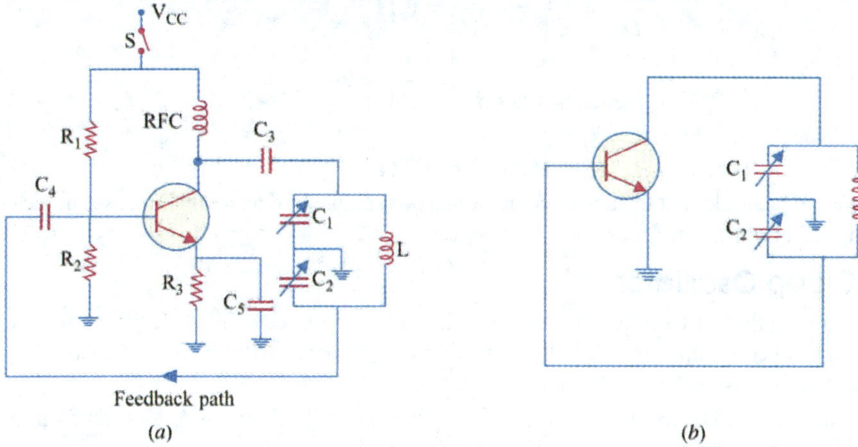

Fig. 28.10

$$A_{v\,(min)} = \frac{1}{C_1/C_2} = \frac{C_2}{C_1}$$

The tank circuit consists of two ganged capacitors C_1 and C_2 and a single fixed coil. The frequency of oscillation (which does not depend on mutual inductance) is given by

$$f_o = \frac{1}{2\pi\sqrt{LC}} \qquad \text{where} \qquad C = \frac{C_1\,C_2}{C_1 + C_2}$$

Transistor itself produces a phase shift of 180°. Another phase shift of 180° is provided by the capacitive feedback thus giving a total phase shift of 360° between the emitter-base and collector-base circuits.

Resistors R_1 and R_2 form a voltage divider across V_{CC} for providing base bias, R_3 is for emitter stabilisation and RFC provides the necessary dc load resistance R_C for amplifier action. It also prevents ac signal from entering supply dc V_{CC}. Capacitor C_5 is a bypass capacitor whereas C_4 conveys feedback from the collector-to-base circuit.

When S is closed, a sudden surge of collector current **shock-excites** the tank circuit into oscillations which are sustained by the feedback and the amplifying action of the transistor.

Colpitts oscillator is widely used in commercial signal generators upto 1 MHz. Frequency of oscillation is varied by gang-tuning the two capacitors C_1 and C_2.

Example 28.3. *For the Colpitts oscillator circuit shown in Fig. 28.11, find the values of*

(a) feedback fraction,

(b) minimum gain to sustain oscillations,

(c) emitter resistor R_E.

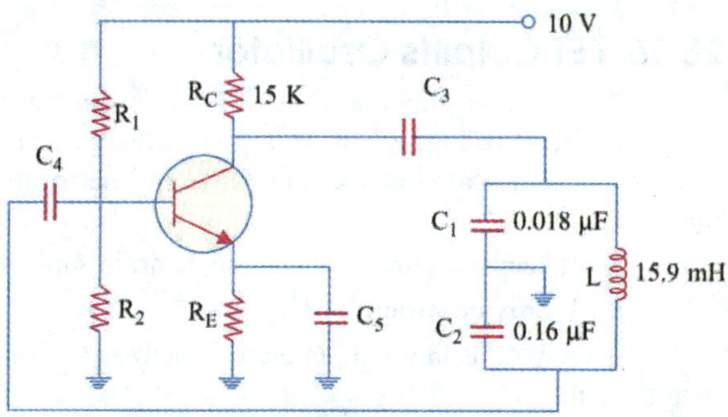

Fig. 28.11

Solution. (*a*)The voltage drop across C_2 is fed back to the input circuit.

Feedback fraction

$$= \frac{\text{drop acros } C_2}{\text{drop across } C_1}$$

$$= \frac{IX_2}{IX_1} = \frac{C_1}{C_2} = \frac{0.018}{0.16} = \mathbf{0.11}$$

(b) $$A_{v(min)} = \frac{1}{\text{feedback fraction}} = \frac{1}{0.11} = \mathbf{9}$$

(c) Now, $$A_v \cong R_C / R_E$$

$\therefore$ $$R_E = R_C / A_v = 1500 / 9 = \mathbf{167\ \Omega}$$

It should be noted that the above calculations do not take into account the losses in the coil and the loading effect of the amplifier input impedance.

28.15. Clapp Oscillator

It is a variation of Colpitts oscillator and is shown in Fig. 28.12 (*a*). It differs from Colpitts oscillator in respect of capacitor C_3 only which is joined in series with the tank inductor. Fig. 28.12 (*b*) shows the ac equivalent circuit.

Addition of C_3 (*i*) improves frequency stability and (*ii*) eliminates the effect of transistor's parameters on the operation of the circuit.

The operation of this circuit is the same as that of the Colpitts oscillator.

The frequency of oscillation is given by

$$f_o = \frac{1}{2\pi\sqrt{LC}} \quad \text{where} \quad \frac{1}{C} = \frac{1}{C_1} + \frac{1}{C_2} + \frac{1}{C_3}$$

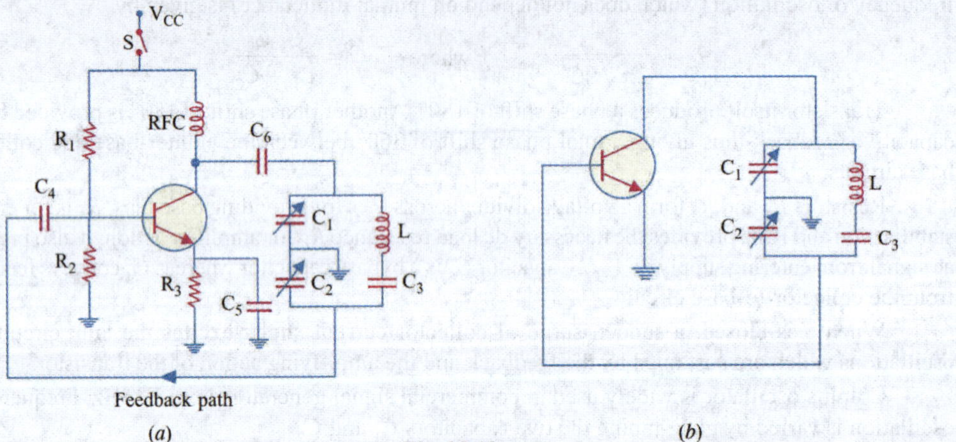

Fig. 28.12

28.16. FET Colpitts Oscillator

The circuit is shown in Fig. 28.13. It is similar to the transistor circuit of Fig. 28.10. Here, R_G is the gate biasing resistor. The radio-frequency coil choke (RFC) performs two functions :

(*i*) It keeps ac current out of the dc drain supply and

(*ii*) It provides drain load.

As seen, C_1 is in the input circuit whereas C_2 is in the output circuit.

The frequency of oscillation is given by

$$f_o = \frac{1}{2\pi\sqrt{LC}};$$

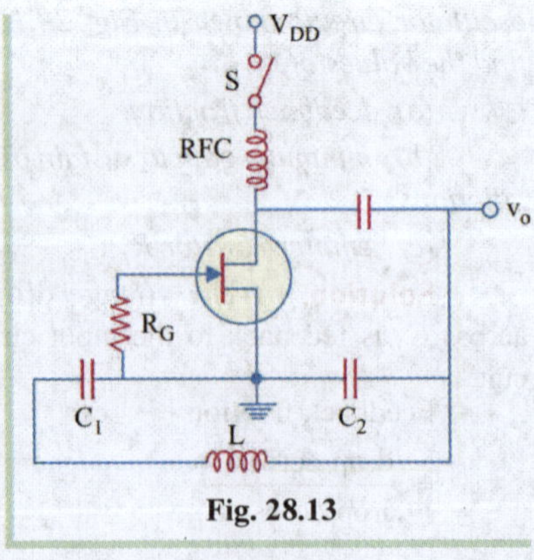

Fig. 28.13

where, $$C = \frac{C_1 C_2}{C_1 + C_2}$$

28.17. Crystals

For an exceptionally high degree of frequency stability, use of crystal oscillators is essential. The crystal generally used is a finely-ground wafer of transluscent quartz (or tourmaline) stone held between two metal plates and housed in a package about the size of a postal stamp. The crystal wafers are cut from the crude quartz in two different ways. The method of 'cutting' determines the crystal's natural resonant frequency and its temperature coefficient. When the wafer is cut so that its flat surfaces are perpendicular to its **electrical axis**, it is called an **X-cut** crystal (Fig. 28.14). But if the wafer is so cut that its flat surfaces are perpendicular to its **mechanical axis**, it is called **Y-cut** crystal.

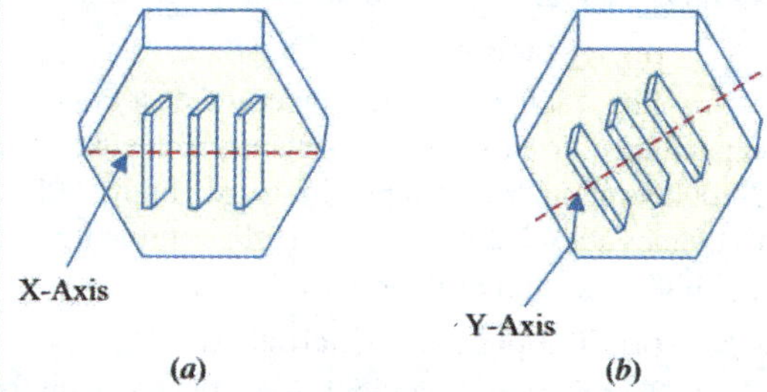

Fig. 28.14

(a) Piezoelectric Effect

The quartz crystal described above has peculiar properties. When mechanical stress is applied across its two opposite faces, **a potential difference is developed across them.** It is called **piezoelectric effect.** Conversely, when a potential difference is applied across its two opposite faces, it causes the crystal to *either expand or contract*. If an alternating voltage is applied, the crystal wafer is set into vibration. The frequency of vibration is equal to the resonant frequency of the crystal as determined by its structural characteristics. Where the frequency of the applied ac voltage equals the natural resonant frequency of the crystal, the amplitude of vibration will be maximum. As a general rule, thinner the crystal, *higher its frequency of vibration*.

(b) Equivalent Electrical Circuit

The electrical equivalent circuit of the crystal is shown in Fig. 28.15 (*b*). It consists of a series RLC_1 circuit in parallel with a capacitor C_2.

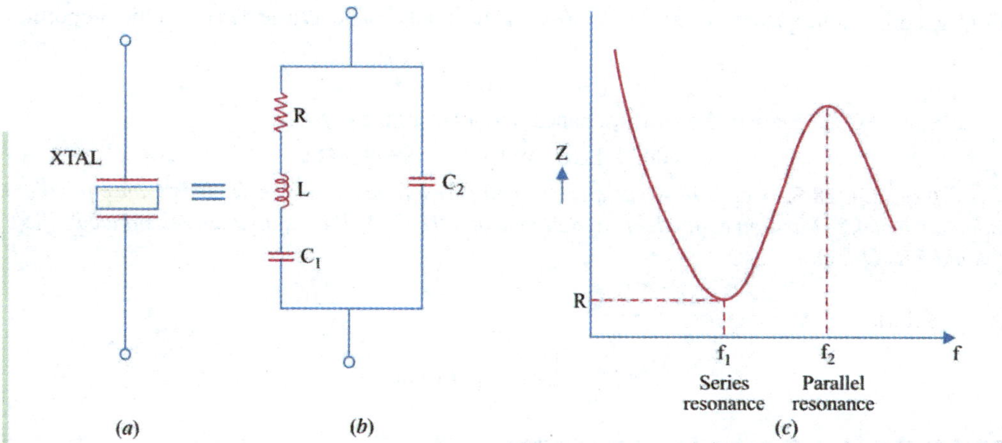

Fig. 28.15

The circuit has two resonant frequencies :

(*i*) One is the lower series resonance frequency f_1 which occurs when $X_L = X_{C1}$. In that case, $Z = R$ as shown in Fig. 28.15 (*c*).

$$f_1 = \frac{1}{2\pi\sqrt{LC_1}}$$

(*ii*) the other is the parallel resonance frequency f_2 which occurs when reactance of the series leg equals the reactance of C_2. At this frequency, the crystal offers very high impedance to the external circuit.

$$f_2 = \frac{1}{2\pi\sqrt{LC}} \quad \text{where} \quad C = \frac{C_1 C_2}{C_1 + C_2}$$

The impedance versus frequency graph of the crystal is shown in Fig. 28.15 (*c*). Crystals are available at frequencies of 15 kHz and above. However, at frequencies above 100 MHz, they become so small that handling them becomes a problem.

(*c*) *Q*-factor

The equivalent inductance of a crystal is very high as compared to either its equivalent capacitance or equivalent resistance. Because of high *L*/*R* ratio, the *Q*-factor of a crystal circuit is 20,000 as compared to a maximum of about 1000 for high-quality *LC* circuits. Consequently, greater frequency stability and frequency discrimination are obtained because of extremely high *Q* (upto 10^6) and high *L*/*R* ratio of the series RLC_1 circuit.

(*d*) Temperature Coefficient

Temperature variations affect the resonant frequency of a crystal. The number of cycles change per million cycles for a 1°C change in temperature is called the **temperature coefficient** (TC) of the crystal. It is usually expressed in parts per million (ppm) per 0°C. For example, a TC of 10 ppm per °C means that frequency variation is 0.001 per cent* per °C change in temperature. It can also be expressed as 10 Hz/°C/MHz. When kept in temperature-controlled ovens, crystal oscillators have frequency stability of about ± 1 ppm.

Usually, *X*-cut crystals have negative TC whereas *Y*-cut crystals have positive TC.

Example 28.4. *A 600 kHz X-cut crystal when calibrated at 50°C has a TC of –20 ppm/°C. What will be its resonant frequency when its temperature is raised to 60°C?*

Solution. A TC of 20 ppm/°C is the same thing as 20 Hz/MHz/°C. Since temperature rise is 10°C and original frequency is 600 kHz = 0.6 MHz, the total change as calculated by ratio proportion is

$$= 20 \times 10 \times 0.6 = 120 \text{ Hz.}$$

Since TC is negative, the new resonance frequency of the crystal is

$$= 600{,}000 - 120 = 599{,}880 = \mathbf{599.88\ kHz.}$$

Example 28.5. *A certain X-cut quartz crystal resonates at 450 kHz. It has an equivalent inductance of 4.2 H and an equivalent capacitance of 0.0297 pF. If its equivalent resistance is 60 Ω, calculate its Q-factor.*

Solution.
$$Q = \frac{\omega L}{R} = \frac{2\pi f L}{R}$$

$$= \frac{2\pi \times 450 \times 10^3 \times 4.2}{600} = \mathbf{19{,}790}$$

28.18. Crystal Controlled Oscillator

Fig. 28.16, shows the use of a crystal to stabilise the frequency of a tuned-collector oscillator which has a crystal (usually quartz) in the feedback circuit.

* Since, 10 ppm means a change of 10 in 10^6, hence percentage change is

$$= \frac{10}{10^6} \times 100 = 10^{-3} = 0.001.$$

The *LC* tank circuit has a frequency of oscillation

$$f_o = 1/2\pi\sqrt{LC}$$

The circuit is adjusted to have a frequency near- about the desired operating frequency but the exact frequency is set by the crystal and stabilized by the crystal. For example, if natural frequency of vibration of the crystal is 27 MHz, the *LC* circuit is made to resonate at this frequency.

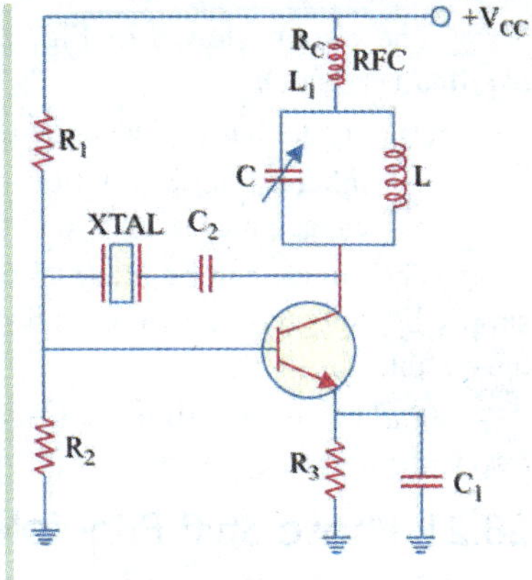

Fig. 28.16

As usual, resistors R_1, R_2 and R_3 provide a voltage-divider stabilised dc bias circuit. Capacitor C_1 by passes R_3 in order to maintain large voltage gain. *RFC* coil L_1 prevents ac signals from entering dc line whereas R_C is the required dc load of the collector. The coupling capacitor C_2 has negligible impedance at the operating frequency but prevents any dc link between collector and base. Due to extreme stability of crystal oscillations, such oscillators are widely used in communication transmitters and receivers where frequency stability is of prime importance.

28.19. Transistor Pierce Crystal Oscillator

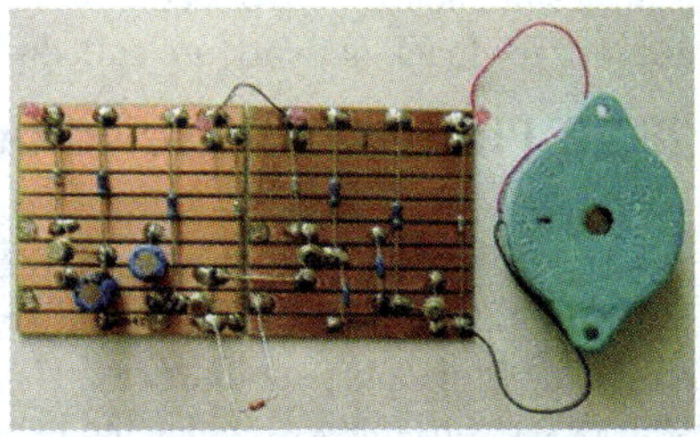
Crystal controlled oscillator.

A typical circuit originally suggested by Pierce is shown in Fig. 28.17. Here, the crystal is excited in the series-resonance mode because it is connected as a series element in the feedback path from collector to the base. Since, in series resonance, crystal impedance is the smallest, the amount of positive feedback is the largest. The crystal not only provides the feedback but also the necessary phase shift.

As usual, R_1, R_2 and R_3 provide a voltage-divider stabilised dc bias circuit. C_2 bypasses R_3 to avoid degeneration. The *RFC* coil provides dc collector load and also prevents any ac signal from entering the dc supply. The coupling capacitor C_1 has negligible reactance at circuit operating frequency but *blocks any dc between collector and base*. The oscillation frequency equals the series-resonance frequency of the crystal and is given by

$$f_o = \frac{1}{2\pi\sqrt{LC_1}} \quad \text{— Fig 28.15 } (b)$$

Advantages

1. It is a very simple circuit because no tuned circuit other than the crystal itself is required.
2. Different oscillation frequencies can be obtained by simply replacing one crystal with another. It makes it easy for a radio transmitter to work at different frequencies.
3. Since frequency of oscillation is set by the crystal, it remains unaffected by changes in supply voltage and transistor parameters etc.

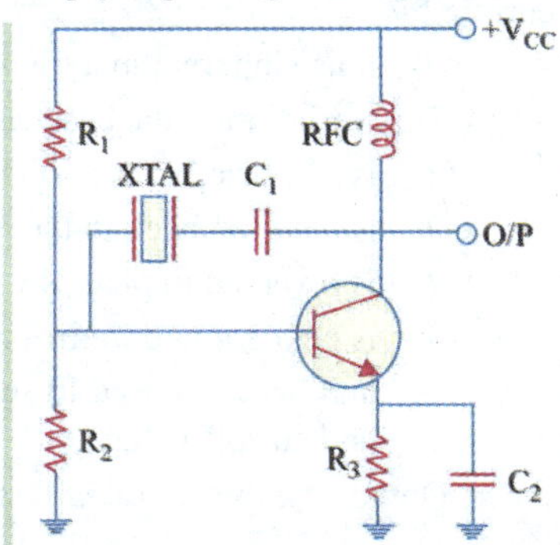

Fig. 28.17

28.20. FET Pierce Oscillator

It is shown in Fig. 28.18. Use of FET is desirable because its high input impedance results in light loading of the crystal which

1. results in good stability and
2. does not lower the *Q*-value.

The circuit shown in Fig. 28.18 is essentially a Colpitts oscillator in which

(*i*) crystal has replaced the inductor and

(*ii*) inherent FET junction capacitance provides the split capacitance as shown in the figure.

The circuit can be made to operate at different frequencies simply by plugging in different crystals between points *A* and *B* of the circuit.

The circuit oscillation frequency is given by the series-resonance frequency of the crystal (Art. 28.17).

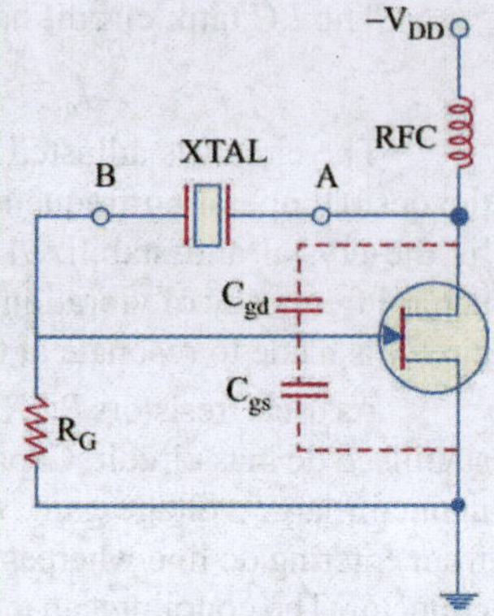

Fig. 28.18

28.21. Phase Shift Principle

Tuned circuits are *not an essential requirement for oscillation*. What is essential is that there should be a 180° phase shift around the feedback network* and loop gain should be greater than unity. The 180° phase shift in the feedback signal can be achieved by using a suitable *RC* network consisting of three or four *R-C* sections. The sine wave oscillators which use *R-C* feedback network are called **phase-shift oscillators.**

28.22. Phase Shift Oscillator

Fig. 28.19, shows a transistor phase-shift oscillator which uses a three-section *R-C* feedback network for producing a total phase shift of 180° (*i.e.*, 60° per section) in the signal is fed back to the base. Since *CE* amplifier produces a phase reversal of the input signal, total phase shift becomes 360° or 0° which is essential for **regeneration and hence for sustained oscillations.**

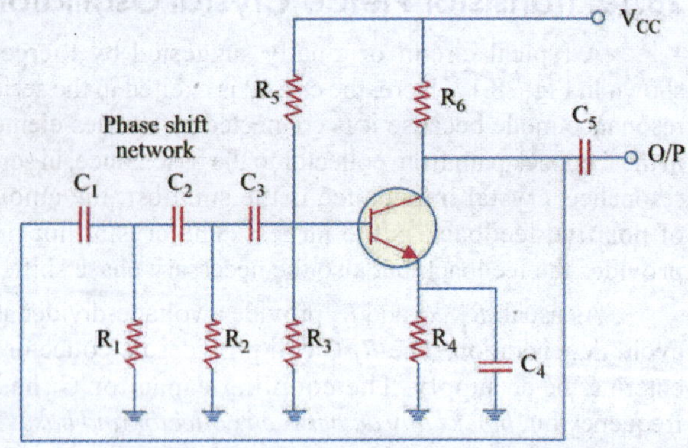

Fig. 28.19

Values of *R* and *C* are so selected that each *RC* section produces a phase advance of 60°. Addition of a fourth section improves oscillator stability. It is found that phase shift of 180° occurs **only at one frequency** which becomes the oscillator frequency.

(*a*) Circuit Action

The circuit is set into oscillations by any random or chance variation caused in the base current by

(*i*) noise inherent in a transistor or (*ii*) minor variation in the voltage of the dc source.

This variation in the base current

1. is amplified in the collector circuit,
2. is then fed back to the *RC* network R_1C_1, R_2C_2 and R_3C_3,
3. is reversed in phase by the *RC* network,
4. is next applied to the base in phase with initial change in base current,
5. and hence is used to sustain cycles of variations in collector current between saturation and cut-off values.

Obviously, the circuit will stop oscillating the moment phase shift differs from 180°.

* Total phase shift required is 360°. However, the balance of 180° is provided by the active device of the amplifier itself.

As is the case with such transistor circuits (*i*) voltage divider $R_5 - R_3$ provides dc emitter-base bias, (*ii*) R_6 controls collector voltage and (*iii*) R_4, C_4 provide temperature stability and prevent ac signal degeneration. The oscillator output voltage is capacitively coupled to the load by C_5.

(*b*) Frequency of Oscillation

The frequency of oscillation for the three-section *RC* oscillator when the three *R* and *C* components are equal is roughly given by

$$f_o = \frac{1}{2\pi\sqrt{6}.RC}\text{ Hz} = \frac{0.065}{RC}\text{ Hz}$$

Moreover, it is found that value of $\beta = 1/29$. It means that amplifier gain must be more than 29 for oscillator operation.

(*c*) Advantages and Disadvantages

1. Since they do not require any bulky and expensive high-value inductors, such oscillators are well-suited for frequencies below 10 kHz.
2. Since only one frequency can fulfil Barkhausen phase-shift requirement, positive feedback occurs only for one frequency. Hence, pure sine wave output is possible.
3. It is not suited to variable frequency usage because a large number of capacitors will have to be varied. Moreover, gain adjustment would be necessary every time frequency change is made.
4. It produces a distortion level of nearly 5% in the output signal;
5. It necessitates the use of a high β transistor to overcome losses in the *RC* network.

28.23. Wien Bridge Oscillator

It is a low-frequency (5 Hz–500 kHz), low-distortion, tunable, high-purity sine wave generator often used in laboratory work. As shown in the block diagrams of Figs. 28.20 and 28.21, this oscillator uses two *CE*-connected *RC*-coupled transistor amplifiers and one *RC*-bridge (called Wien bridge) network to provide feedback. Here, Q_1 serves as amplifier-oscillator and Q_2 provides phase reversal and additional amplification. The bridge circuit is used to control the phase of the feedback signal at Q_1.

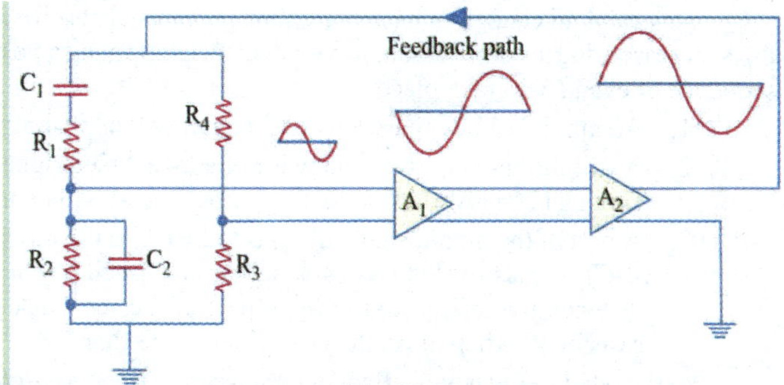

Fig. 28.20

(*a*) Phase Shift Principle

Any input signal at the base of Q_1 appears in the amplified but phase-reversed form across collector resistor R_6 (Fig. 28.21). It is further inverted by Q_2 in order to provide a total phase reversal of 360° for positive feedback. Obviously, the signal at R_{10} is an amplified replica of the input signal at Q_1 and is of the same phase since it has been inverted twice. We could feed this signal back to the base of Q_1 directly to provide regeneration needed for oscillator operation. But because Q_1 will amplify signals over a wide range of frequencies, direct coupling would result in poor frequency stability. By adding the Wien bridge, oscillator becomes sensitive to a signal of *only one particular frequency*. Hence, we get an oscillator of good frequency stability.

(*b*) Bridge Circuit Principle

It is found that the Wien bridge would become balanced at the signal frequency for which phase shift is exactly 0° (or 360°).

The balance conditions are

$$\frac{R_4}{R_3} = \frac{R_1}{R_2} + \frac{C_2}{C_1} \quad \text{and} \quad \omega_o = \frac{1}{\sqrt{R_1 C_1 R_2 C_2}} \quad \text{or} \quad f_o = \frac{1}{2\pi\sqrt{R_1 C_1 R_2 C_2}}$$

$$\text{If} \quad R_1 = R_2 = R \quad \text{and} \quad C_1 = C_2 = C, \quad \text{then} \quad f_o = \frac{1}{2\pi RC} \text{ and } \frac{R_4}{R_3} = 2$$

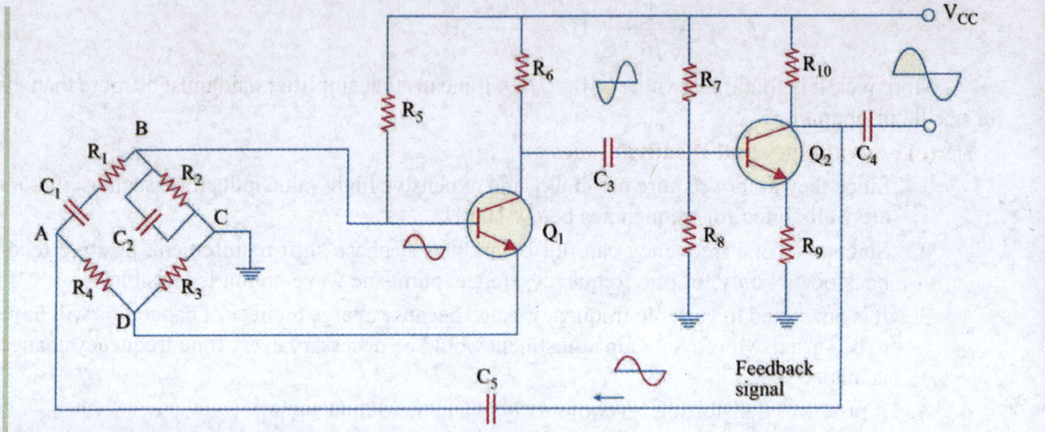

Fig. 28.21

(c) Circuit Action

Any random change in base current of Q_1 can start oscillations. Suppose, the base current of Q_1 is increased due to some reason. It is equivalent to applying a positive-going signal to Q_1. Following sequence of events will take place :

1. An amplified but phase-reversed signal will appear at the collector of Q_1.
2. A still further amplified and twice phase-reversed signal will appear at the collector of Q_2. Having been inverted twice, this output signal will be in phase with the input signal at Q_1.
3. A part of the output signal at Q_2 is fed back to the input points of the bridge circuit (point *A-C*). A part of this feedback signal is applied to emitter resistor R_3 where it produces degenerative effect. Similarly, a part of the feedback signal is applied across base-bias resistor R_2 where it produces regenerative effect.

At the rated frequency f_o, effect of *regeneration is made slightly more than that of degeneration* in order to maintain continuous oscillations.

By replacing R_3 with a thermistor, amplitude stability of the oscillator output voltage can be increased.

(d) Advantages

Such a circuit has

1. highly stabilized amplitude and voltage amplification,
2. exceedingly good sine wave output,
3. good frequency stability.

SELF EXAMINATION QUESTIONS

A. Fill in the blanks with most appropriate word (s) or numerical value (s).

1. An electronic oscillator is a circuit which converts dc energy into .... energy.
2. An electronic oscillator is an amplifier with .... feedback.
3. Temperature variations cause .... in the tuned *LC* oscillator output frequency.
4. An oscillator circuit must satisfy .... criterion.
5. Hartley oscillator uses .... feedback.
6. .... feedback is used in Colpitts oscillator.

7. Clapp oscillator is nothing but a .... form of Colpitts oscillator.
8. Conversion of mechanical stress into electric potential by a crystal is called .... effect.
9. Q-factor of a quartz wafer is very ....
10. Y-cut quartz crystals have .... temperature coefficient.
11. Phase shift oscillators do not use .... circuits.
12. Phase shift oscillators are well-suited for comparatively .... frequencies.

B. Answer True or False

1. Like an amplifier, an oscillator needs an input signal with proper phase.
2. Oscillators are used to produce high-frequency undamped waves for radio transmission.
3. An electronic oscillator needs an active device and an ac source.
4. The discharge current of a capacitor is oscillatory in nature.
5. A tuned circuit is an essential component of all types of oscillator circuits.
6. According to Barkhausen criterion, the oscillations can be sustained in a circuit if the net phase shift around the feedback loop is π or $n\pi$.
7. In Colpitts oscillator, one part of the tapped capacitor is in the output circuit and the other in the input circuit.
8. Crystal oscillators have zero temperature coefficient.
9. A crystal oscillator needs no tuned circuit.
10. Barkhausen criterion can be met by using RC phase-shift feedback network.
11. Phase-shift oscillators are not suitable for variable frequency work.
12. Wien bridge oscillator produces an exceedingly good sine wave output.

C. Multiple Choice Items

1. An electronic oscillator is
   (*a*) just like an alternator
   (*b*) nothing but an amplifier
   (*c*) an amplifier with feedback
   (*d*) a converter of ac to dc energy
2. The frequency of oscillation of an elementary LC oscillatory circuit depends on
   (*a*) coil resistance (*b*) coil inductance
   (*c*) capacitance (*d*) both (*b*) and (*c*)
3. For sustaining oscillations in an oscillator
   (*a*) feedback factor should be unity
   (*b*) phase shift should be 0° or $n\pi$
   (*c*) feedback should be negative
   (*d*) both (*a*) and (*b*)
4. If Barkhausen criterion is not fulfilled by an oscillator circuit, it will
   (*a*) stop oscillating
   (*b*) produce damped waves continuously
   (*c*) become an amplifier
   (*d*) produce high-frequency whistles
5. In a transistor Hartley oscillator
   (*a*) inductive feedback is used
   (*b*) untapped coil is used
   (*c*) entire coil is in the output circuit
   (*d*) no capacitor is used
6. A Colpitts oscillator uses
   (*a*) tapped coil
   (*b*) inductive feedback
   (*c*) tapped capacitance
   (*d*) no tuned LC circuit
7. In RC phase-shift oscillator circuits
   (*a*) there is no need for feedback
   (*b*) feedback factor is less than unity
   (*c*) pure sine wave output is possible
   (*d*) transistor parameters determine osci llation frequency
8. Wien bridge oscillator is most often used whenever
   (*a*) wide range of high purity sine waves are to be generated
   (*b*) high feedback ratio is needed
   (*c*) square output waves are required
   (*d*) extremely high resonant frequencies are required.

ANSWERS

A. Fill in the blanks

1. *ac*	**2.** positive	**3.** drift	**4.** Barkhausen	**5.** inductive	**6.** capacitive
7. modified	**8.** piezoelectric	**9.** high	**10.** positive	**11.** tuned	**12.** low

B. True or False

1. F	**2.** T	**3.** F	**4.** T	**5.** F	**6.** F
7. T	**8.** F	**9.** T	**10.** T	**11.** T	**12.** T

C. Multiple Choice Items

1. *c*	**2.** *a*	**3.** *d*	**4.** *a*	**5.** *a*	**6.** *c*	**7.** *c*	**8.** *a*

29 CHAPTER

Nonsinusoidal Oscillators

1. Nonsinusoidal Waveforms
2. Classification of Nonsinusoidal Oscillators
3. Pulse Definitions
4. Basic Requirements of a Sawtooth Generator
5. UJT Sawtooth Generator
6. Multivibrators (MV)
7. Uses of Multivibrators
8. Astable Multivibrator
9. Monostable Multivibrator (MMV)
10. Bistable Multivibrator (BMV)
11. Schmitt Trigger
12. Transistor Blocking Oscillator

29.1. Nonsinusoidal Waveforms

Any waveform whose shape is different from that of a standard sine wave is called *non-sinusoidal waveform.* Examples are : square, rectangular, sawtooth, triangular waveforms and pulses as shown in Fig. 29.1.

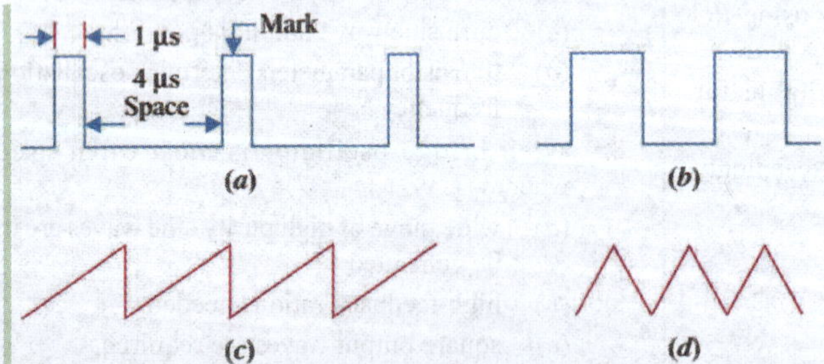

Fig. 29.1

(*a*) Pulses

Fig. 29.1 (*a*) shows a pulse train *i.e.*, a stream of pulses at regular intervals. A pulse may, in general, be defined as a voltage or current that changes rapidly from one level of amplitude to another *i.e.*, **it is an abrupt discontinuity in voltage or current.** These pulses are extensively used in digital electronics.

1. *Mark-to-Space Ratio (MSR)*

$$MSR = \frac{\text{pulse width}}{\text{time between pulses}} = \frac{1\ \mu s}{4\ \mu s} = 0.25 \qquad \text{— Fig. 29.1 } (a)$$

Hence, mark-to-space ratio of the pulse shown in Fig. 29.1 (*a*) is 1 : 4.

This name has come from early morse-code transmission systems where a pulse was used to cause a pen to mark the paper.

2. *Pulse Repetition Time (PRT)*

It may be defined as *the time between the beginning of one pulse and that of the other.*

As seen from Fig. 29.1 (*a*),

$$PRT = 5\ \mu s$$

3. *Pulse Repetition Frequency (PRF)*

It is given by *the number of pulses per second.*

$$PRF = \frac{1}{PRT} = \frac{1}{5\ \mu s} = \frac{10^6}{5} = 200{,}000 \text{ Hz} = 200 \text{ kHz}$$

Pulse circuits find applications in almost all electronic-based industries. Various types of pulse code modulations are employed in communication systems, whereas radars utilize pulses to track targets. Digital computers require circuits that can be switched very rapidly between two states by using appropriate pulses.

(*b*) Square Wave

It is shown in Fig. 29.1 (*b*) and is, in fact, a pulse waveform with a mark-to-space ratio of 1: 1.

Such square waves or pulses are used

1. for audio frequency note generation,
2. for digital electronic switching as in computers,
3. in radars,
4. as synchronizing pulses in TV,
5. for switching of high-power electronic circuits such as thyristor circuits.

(*c*) Sawtooth Wave

It is shown in Fig. 29.1 (*c*). Such waves are used

1. in the scanning circuits of cathode-ray tubes (CRT),
2. in timing circuits where the time for the wave to proceed from one level to another is measured, such as that produced in an integrating circuit.

(*d*) Triangular Wave

It is shown in Fig. 29.1 (*d*). Such waves are often used

1. in scanning circuits where a uniform left-to-right scan is required as in computer displays,
2. for audio frequency note generation, 3.in timing circuits for electronic applications.

29.2. Classification of Nonsinusoidal Oscillators

Those oscillators which generate waveforms other than sine waveform are called **nonsinusoidal oscillators** or **relaxation oscillators.** Nonsinusoidal waveforms include: square, rectangular, sawtooth and pulse-shaped waves as shown in Fig. 29.1.

A relaxation oscillator may be defined as a circuit in which voltage or current changes abruptly from one value to another and which continues to oscillate between these two values as long as dc power is supplied to it.

We will consider the following three types of such oscillators:

1. sawtooth generators 2. blocking oscillators 3. multivibrators (MV)

29.3. Pulse Definitions

Due to capacitive effects in a transistor (or to circuit elements external to it), its **output does not directly follow its input.** For example, if we apply a square input pulse to its E/B junction, some amount of time lapses before I_C starts to rise. Similarly, when the input becomes zero, there is some time lapse before I_C starts to decrease. There is always sometime delay between the application of input and change in the output. In order to measure how quickly the output changes *i.e.*, in order to define the switching (*i.e.*, OFF/ON) characteristics of a transistor, we will define the following few terms. It will be assumed that perfect square wave, as shown in Fig. 29.2 (*a*), has been applied at the input. The output wave and the various time delays are shown in Fig. 29.3 (*b*).

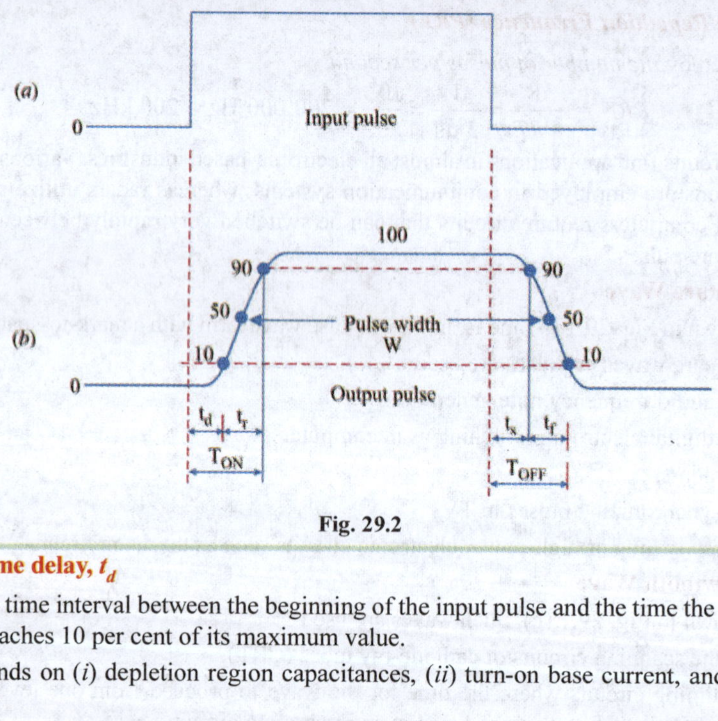

Fig. 29.2

1. Time delay, t_d

It is the time interval between the beginning of the input pulse and the time the output voltage (or current) reaches 10 per cent of its maximum value.

It depends on (*i*) depletion region capacitances, (*ii*) turn-on base current, and (*iii*) value of transistor β.

2. Rise time, t_r

It is the time taken by the output voltage (or current) to rise from 10% to 90% of its maximum value. It primarily depends on diffusion capacitance C_D of the transistor (Art. 13.12).

3. Turn-on time T_{ON}

It is equal to the sum of the delay time and rise time

$$T_{ON} = t_d + t_r$$

4. Storage time, t_s

It is the time interval between the end of the input pulse (trailing edge) and the time when output voltage (or current) falls to 90% of its initial maximum value.

It depends on the degree of saturation. Deeper the transistor is driven into saturation, more the stored charge that has to be removed and hence longer the storage time. That is why non-saturated switching is often preferred.

5. **Fall time, t_f**

It is the time interval during which the output voltage (or current) falls from 90% of its maximum value to 10%.

In simple words, it is the time interval between 90% and 10% levels of the output pulse.

6. **Turn-off time, T_{OFF}**

It is equal to the sum of storage time and fall time

$$T_{OFF} = t_s + t_f$$

For a fast switching transistor, T_{ON} and T_{OFF} must be of the order of nanoseconds.

7. **Pulse Width, *W***

It is the time duration of the output pulse measured between two 50% levels of the rising and falling waveform.

29.4. Basic Requirements of a Sawtooth Generator

The essential requirements of a sawtooth generator are:

1. a dc power source,
2. a switching device (neon tube, thyratron, thyristor, UJT etc.),
3. a capacitor,
4. a resistor.

Circuit Action

The *V/I* characteristics of the *RC* circuit play an important role in the operation of such a generator. By restricting the time interval equal to the time constant $\lambda = CR$, only the rising portion of the characteristic (*OA* in Fig. 29.3) which is almost a straight line, is utilized. For periods of time greater than λ, the rising portion of the characteristic is no longer a straight line and hence cannot be utilized.

The frequency of the wave is given by the reciprocal of time which elapses between the two waves. In Fig. 29.3, $f = 1/\lambda$.

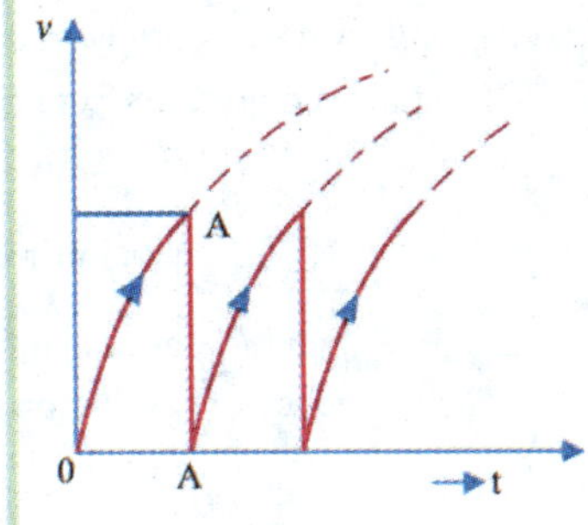

Fig. 29.3

29.5. UJT Sawtooth Generator

The circuit is shown in Fig. 29.4. It consists of a power source, a unijunction transistor and an *R-C* network.

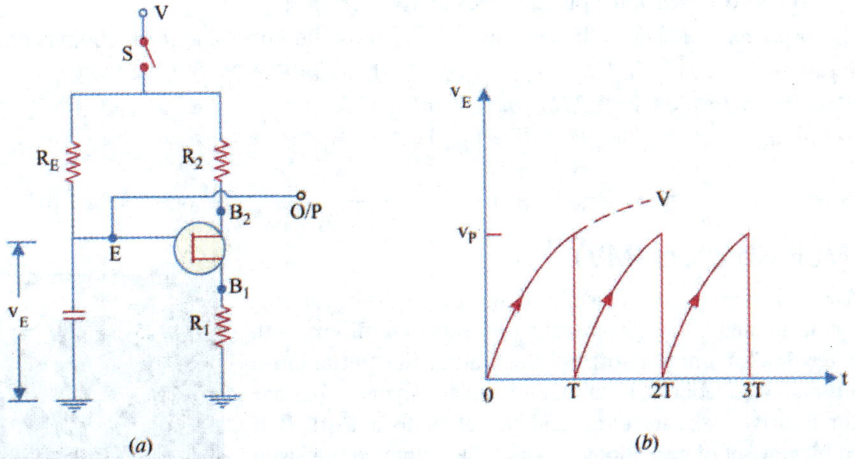

Fig. 29.4

Circuit Action

When S is initially closed, following chain of events takes place:

1. a small current is set up through R_2 and R_1 *via* B_2 and B_1 and an initial reverse bias is established across the E/B_1 junction;
2. at the same time, C begins to get charged through R_E and voltage across it increases exponentially with time towards the target voltage V;
3. when capacitor voltage equals the emitter firing (or peak point) voltage V_P, E/B_1 junction becomes forward-biased and the emitter goes into the negative region of its characteristic;
4. being forward-biased, E/B_1 junction offers very low resistance. Hence, C starts discharging through B_1 and R_1 at a rate determined mainly by E/B_1 junction resistance and R_1;
5. as capacitor voltage approaches zero, the E/B_1 junction again becomes reverse-biased and so stops conducting;
6. we revert to the initial state where C begins to charge and the whole cycle of circuit actions is repeated.

The emitter voltage waveform is shown in Fig. 29.4 (*b*). As seen, v_E rises exponentially towards the target voltage V but drops to a very low value after it reaches the value V_P due to sudden conduction through E/B_1 junction. Since R_E is large (10 K or so) charging rate is comparatively slow but discharge is much quicker since R_1 is very small (50 Ω or less). *This slow charge and fast discharge produces a sawtooth wave.*

The time required for v_E to rise to V_P is given by

$$T = kRC \qquad \text{where} \qquad k = \log_e{}^{1/(1-\eta)} \text{ and}$$

$$\eta = \text{intrinsic stand-off ratio} \quad = \frac{R_{B1}}{R_{B1} + R_{B2}} \qquad \text{...Art. 27.2}$$

Here, R_{B1} and R_{B2} are the internal inter-base resistances.

The frequency of oscillation of the UJT or of the output sawtooth wave is

$$f_o = \frac{1}{T} = \frac{1}{kRC}$$

Its amplitude is determined primarily by applied voltage V and V_P.

Applications

Sawtooth voltage waves are commonly used as

1. sweep voltages at the picture tubes of TV receivers,
2. as sweep voltages of the viewing screens of oscilloscopes and radar equipment.

Example 29.1. *For the UJT oscillator circuit shown in Fig. 29.4, R_E = 10 K, η = 0.75. If the required oscillator frequency is 1 kHz, find the value of C.*

Solution. $$k = \log_e{}^{1/(1-0.75)} = \log_e{}^4 = 2.3\log_{10}{}^4 \cong 1.4$$

Now, $$f_o = \frac{1}{kRC} \quad \therefore \quad 1 \times 10^3 = \frac{1}{1.4 \times 10 \times 10^3 \times C} \quad \text{or } C = \mathbf{0.07\ \mu F}$$

29.6. Multivibrators (MV)

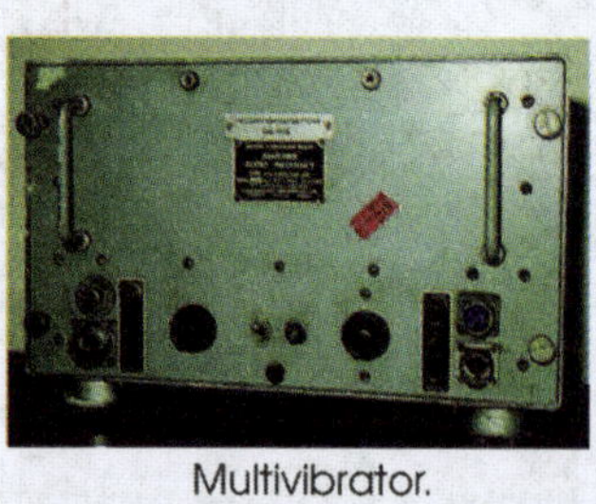

Multivibrator.

These devices are very useful as pulse generating, storing and counting circuits. They are basically **two-stage amplifiers with positive feedback from the output of one amplifier to the input of the other.** This feedback is supplied in such a manner that one transistor is driven to saturation and the other to cut-off. It is followed by new set of conditions in which the saturated transistor is driven to cut-off and the cut-off transistor is driven to saturation.

There are three basic types of MVs distinguished by the type of coupling network employed.

1. astable multivibrator (AVM), 2. monostable multivibrator (MMV),
3. bistable multivibrator (BMV).

The first one is the *non-driven* type whereas the other two are the *driven type* (also called triggered oscillators).

1. Astable Multivibrator (AMV)

It is also called *free-running relaxation oscillator.* It has no stable state but only two quasi-stable (half-stable) states between which it keeps oscillating continuously of its own accord without any external excitation.

In this circuit, neither of the two transistors reaches a stable state. When one is ON, the other is OFF and they continuously switch back and forth at a rate depending on the *RC* time constant in the circuit. Hence, it oscillates and produces pulses of certain mark-to-space ratio. Moreover, two outputs (180° out of phase with each other) are available.

It has two energy-storing elements *i.e.* **two capacitors.**

2. Monostable Multivibrator (MMV)

It is also called a **single-shot** or **single swing** or a **one-shot** multivibrator. Other names are: *delay multivibrator* and *univibrator.* It has

(*i*) one absolutely stable (stand-by) state and (*ii*) one quasi-stable state.

It can be switched to the quasi-stable state by an external trigger pulse but it returns to the stable condition after a time delay determined by the value of circuit components. It supplies a single output pulse of a desired duration for every input trigger pulse.

It has one energy-storing element *i.e.*, **one-capacitor.**

3. Bistable Multivibrator (BMV)

It is also called Eccles-Jordan or flip-flop multivibrator. It has **two absolutely stable states.** It can remain in either of these two states unless an external trigger pulse switches it from one state to the other. Obviously, *it does not oscillate.* It has **no energy storage element.**

Detailed discrete circuits for the above MVs are discussed below after listing their uses.

30.7. Uses of Multivibrators

Some of their uses are:

1. as frequency dividers, 2. as sawtooth generators,
3. as square wave and pulse generators,
4. as a standard frequency source when synchronized by an external crystal oscillator,
5. for many specialised uses in radar and TV circuits,
6. as memory elements in computers.

29.8. Astable Multivibrator

Fig. 29.5 shows the circuit of a symmetrical collector-coupled AMV using two similar transistors. It, in fact, consists of two *CE* amplifier stages, each providing a feedback to the other. The feedback ratio is unity and positive because of 180° phase shift in each stage. Hence, the circuit oscillates. Because of the very strong feedback signal, the transistors are driven either to saturation or to cut-off (they do not work on the linear region of their characteristics).

The transistor Q_1 is forward-biased by V_{CC} and R_1, whereas Q_2 is forward-biased by V_{CC} and R_2. The collector-emitter voltages of Q_1 and Q_2 are determined respectively by R_{L1} and R_{L2} together with V_{CC}. The output of Q_1 is coupled to the input of Q_2 by C_2, whereas output of Q_2 is coupled to Q_1 by C_1.

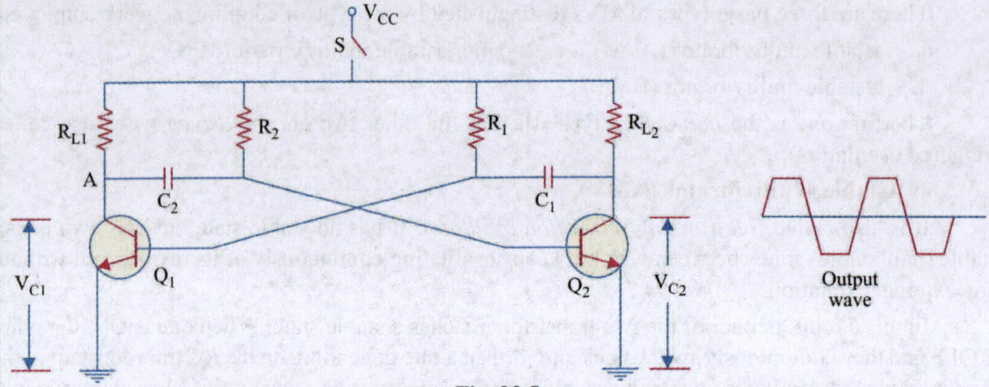

Fig. 29.5

Note that it is not essential to draw the coupling leads at 45° to the vertical as shown but it is usually done because it help to identify the circuit immediately as MV.

The output can be taken either from point *A* or *B* though these would be phase-reversed with respect to each other as shown in Fig. 29.5.

Circuit Operation

The circuit operation would be easy to understand if it is remembered that due to feedback (*i*) when Q_1 is ON, Q_2 is OFF and (*ii*) when Q_2 is ON, Q_1 is OFF.

When the power is switched on by closing *S*, one of the transistors will start conducting before the other (or slightly faster than the other). It is so because characteristics of no two seemingly similar transistors can be **exactly** alike. Suppose that Q_1 starts conducting before Q_2 does. The feedback system is such that Q_1 will be very rapidly driven to saturation and Q_2 to cut-off.

The following sequence of events will occur:

1. Since Q_1 is in saturation, whole of V_{CC} drops across R_{L1}. Hence, $V_{C1} = 0$ and point *A* is at zero or ground potential.
2. Since Q_2 is in cut-off *i.e.*, it conducts no current, there is no drop across R_{L2}. Hence, point *B* is at V_{CC}.
3. Since *A* is at 0 V, C_2 starts to charge through R_2 towards V_{CC}.
4. When voltage across C_2 rises sufficiently (*i.e.*, more than 0.7 V), it biases Q_2 in the forward direction so that it starts conducting and is soon driven to saturation.
5. V_{C2} decreases and becomes almost zero when Q_2 gets saturated. The potential of point *B* decreases from V_{CC} to almost 0 V. This potential decrease (negative swing) is applied to the base of Q_1 through C_1. Consequently, Q_1 is pulled out of saturation and is soon driven to cut-off.
6. Since, now, point *B* is at 0 V, C_1 starts charging through R_1 towards the target voltage V_{CC}.

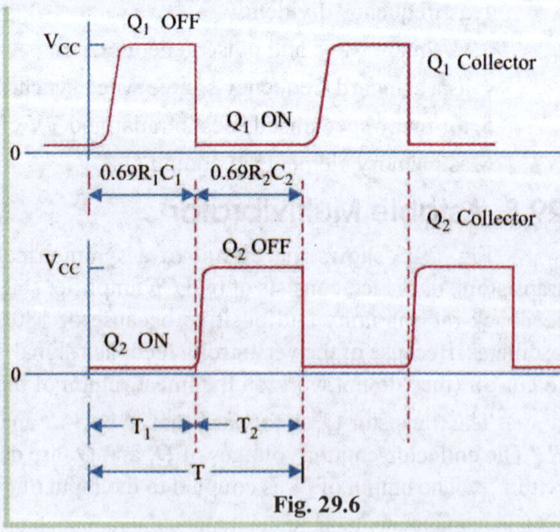

Fig. 29.6

7. When voltage of C_1 increases sufficiently, Q_1 becomes forward-biased and starts conducting. In this way, the whole cycle is repeated.

It is seen that the circuit alternates between a state in which Q_1 is ON and Q_2 is OFF and a state in which Q_1 is OFF and Q_2 is ON. The time in each state depends on RC values. Since each transistor is driven alternately into saturation and cut-off the voltage wavefrom at either collector (points A and B in Fig. 29.5) is essentially a square waveform with a peak amplitude equal to V_{CC} (Fig. 29.6).

Switching Times

It can be proved that off-time for Q_1 is $T_1 = 0.69\ R_1C_1$ and that for Q_2 is $T_2 = 0.69\ R_2C_2$. Hence, total time-period of the wave is

$$T = T_1 + T_2$$
$$= 0.69\ (R_1C_1 + R_2C_2)$$

If $R_1 = R_2 = R$ and $C_1 = C_2 = C$ *i.e.*, the two stages are symmetrical, then $T = 1.38\ RC$.

Frequency of Oscillation

It is given by the reciprocal of time period,

$$\therefore \quad f = \frac{1}{T} = \frac{1}{1.38\ RC} = \frac{0.7}{RC}$$

Minimum Values of β

To ensure oscillations, the transistors must saturate for which minimum values of β are as under :

$$\beta_1 = \frac{R_1}{R_{L1}} \quad \text{and} \quad \beta_2 = \frac{R_2}{R_{L2}}$$

If $R_1 = R_2 = R$ and $R_{L1} = R_{L2} = R_{L2}$ then

$$\beta_{min} = \frac{R}{R_L}$$

Example 29.2. *In the AMV circuit of Fig. 29.5, $R_1 = R_2 = 10\ K$, $C_1 = C_2 = 0.01\ \mu F$ and $R_{L1} = R_{L2} = 1K$. Find*

(*a*) *frequency of circuit oscillation,* (*b*) *minimum value of transistor* β.

Solution. (*a*) $T_1 = T_2 = 0.69 \times 10 \times 10^3 \times 0.01 \times 10^{-6} = 69\ \mu s$

$$\therefore \quad T = T_1 + T_2 = 2 \times 69 = 138\ \mu s \qquad \therefore \quad f_o = \frac{1}{138 \times 10^{-6}} = \mathbf{7.25\ kHz}$$

(*b*)
$$\beta_{min} = \frac{R}{R_L} = \frac{10 \times 10^3}{1 \times 10^3} = \mathbf{10}$$

29.9. Monostable Multivibrator (MMV)

A typical MMV circuit is shown in Fig. 29.7. Here, Q_1 is coupled to Q_2 base as in an AMV but the other coupling is different. In this multivibrator, a single narrow input trigger pulse produces a single rectangular pulse whose amplitude, pulse width and waveshape depend upon the values of circuit components rather than upon the trigger pulse.

Initial Condition

In the absence of a triggering pulse at C_2 and with S closed,

1. V_{CC} provides reverse bias for C/B junctions of Q_1 and Q_2 but forward-bias for E/B junction of Q_2 only. Hence, Q_2 conducts at saturation.

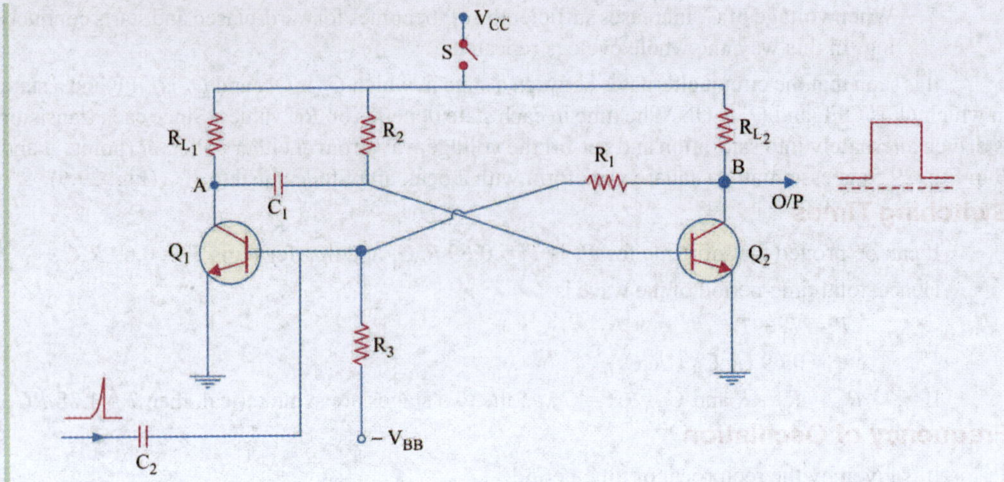

Fig. 29.7

2. V_{BB} and R_3 reverse bias Q_1 and keep it cut-off.
3. C_1 charges to nearly V_{CC} through R_{L1} to ground by the low-resistance path provided by saturated Q_2.

As seen, the initial stable state is represented by

(*i*) Q_2 conducting at saturation and (*ii*) Q_1 cut-off.

When Trigger Pulse is Applied

When a trigger pulse is applied to Q_1 through C_2, MMV will switch to its opposite unstable state where Q_2 is cut-off and Q_1 conducts at saturation. The chain of circuit actions is as under:

1. If positive trigger pulse is of sufficient amplitude, it will override the reverse bias of the E/B junction of Q_1 and give it a forward bias. Hence, Q_1 will start conducting.
2. As Q_1 conducts, its collector voltage falls due to voltage drop across R_{L1}. It means that potential of point *A* falls (negative-going signal). This negative-going voltage is fed to Q_2 *via* C_1 where it decreases its forward bias.
3. As collector current of Q_2 starts decreasing, potential of point *B* increases (positive-going signal) due to lesser drop over R_{L2}. Soon, Q_2 comes out of conduction.
4. The positive-going signal at *B* is fed *via* R_1 to the base of Q_1 where it increases its forward bias further. As Q_1 conducts more, potential of point *A* approaches 0 V.
5. This action is cumulative and ends with Q_1 conducting at saturation and Q_2 cut-off.

Return to Initial Stable State

1. As point *A* is at almost 0 V, C_1 starts to discharge through saturated Q_1 to ground.
2. As C_1 discharges, the negative potential at the base of Q_2 is decreased. As C_1 discharges further, Q_2 is pulled out of cut-off.
3. As Q_2 conducts further, a negative-going signal from point *B via* R_1 drives Q_1 into cut-off.

Hence, the circuit reverts to its original state with Q_2 conducting at saturation and Q_1 cut-off. It remains in this state till another trigger pulse comes along when the entire cycle repeats itself.

As shown in Fig. 29.7, the output is taken from the collector of Q_2 though it can also be taken from point *A* of Q_1. The width of this pulse is determined by the time constant of C_1R_2. Since this MV *produces one output pulse* for every input trigger pulse it receives, it is called **mono-or one-shot multivibrator.**

The width or duration of the pulse is given by

$$T = 0.69\, C_1 R_2$$

It is also known as the one-shot period.

Uses

1. The falling part of the output pulse from MMV is often used to trigger another pulse generator circuit thus producing a pulse delayed by a time *T* with respect to the input pulse.
2. MMV is used for regenerating or rejuvenating old and worn out pulses. Various pulses used in computers and telecommunication systems become somewhat distorted during use. An MMV can be used to generate new, clean and sharp pulses from these distorted and used ones.

29.10. Bistable Multivibrator (BMV)

The basic circuit is shown in Fig. 29.8. As stated earlier, it has **two absolutely stable** states. It can stay *in one of its two states indefinitely* (as long as power is supplied) changing to the other state only when it receives a trigger pulse from outside. When it receives another triggering pulse, only then it goes back to its original state. Since one trigger pulse causes the MV to 'flip' from one state to another and the next pulse causes it to 'flop' back to its original state, the BMV is also popularly known as **'flip-flop'** circuit.

The BMV circuit shown in Fig. 29.8 differs from the AMV circuit of Fig. 29.5 in the following respects:

1. the base resistors are not joined to V_{CC} but to a common source—V_{BB},
2. the feedback is coupled through two resistors (not capacitors).

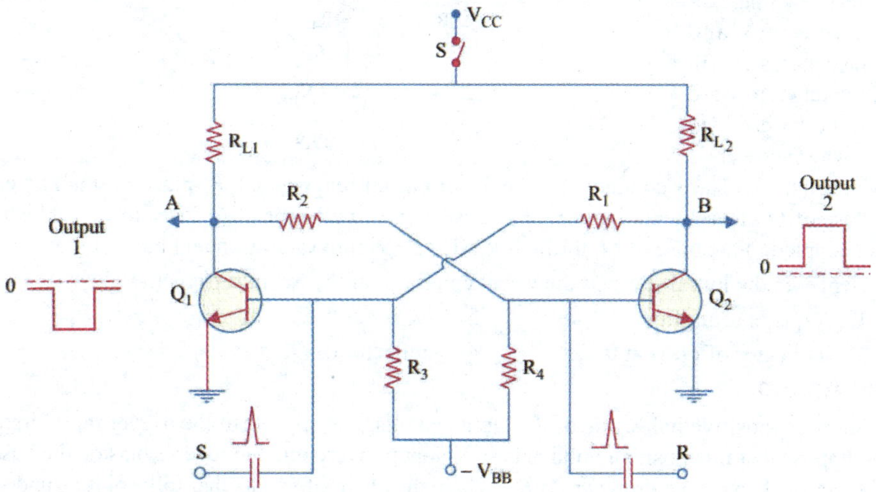

Fig. 29.8

Circuit Action

If Q_1 is conducting, then the fact that point *A* is at nearly 0 V makes the base of Q_2 negative (by the potential divider $R_2 - R_4$) and holds Q_2 off.

Similarly, with Q_2 OFF, the potential divider from V_{CC} to $-V_{BB}$ (R_{L2}, R_1, R_3) is designed to keep base of Q_1 at about 0.7 V ensuring that Q_1 conducts. It is seen that Q_1 holds Q_2 OFF and Q_3 holds Q_1 ON.

Suppose, now, a positive pulse is applied momentarily to *R*. It will cause Q_2 to conduct. As collector of Q_2 falls to zero, it cuts Q_1 OFF and, consequently, the BMV switches over to its other state.

Similarly, a positive trigger pulse applied to S will switch the BMV back to its original state.

Uses

1. in timing circuits as a frequency divider,
2. in counting circuits,
3. in computer memory circuits.

29.11. Schmitt Trigger

The Schmitt trigger (after the name of its inventor) is a **binary circuit** and **closely resembles an MV.** It has two bistable states and the magnitude of the input voltage determines which of the two is possible. It is also called emitter-coupled binary oscillator because positive feedback occurs by coupling through emitter resistor R_E.

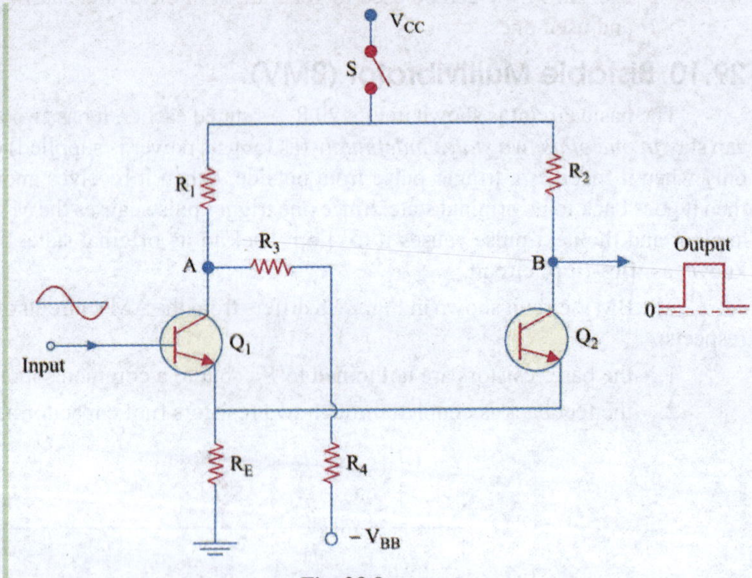

Fig. 29.9

The Quiescent Condition

As shown in Fig. 29.9, it consists of two similar transistors Q_1 and Q_2 coupled through R_E.

Resistors R_1, R_3 and R_4 form a voltage divider across V_{CC} and $-V_{BB}$ which places a small positive voltage (forward bias) on the base of Q_2. Hence, when power is first switched on, Q_2 starts conducting. The flow of its current through R_E places a small reverse bias on the base of Q_1, thereby cutting it OFF. Consequently, collector of Q_1 rises to V_{CC}. This positive voltage, coupled to the base of Q_2 through R_1, drives Q_2 into saturation and holds it there.

Hence, in the initial static or quiescent condition of the Schmitt trigger,

1. Q_2 is in saturation,
2. Q_1 is cut-off,
3. collector of Q_2 is at 0 V,
4. collector of Q_1 is at V_{CC}.

Circuit Action

Suppose, positive half-cycle of the input ac voltage is applied to the trigger input first. Let us further suppose that this positive voltage is sufficient to overcome the reverse bias on the base of Q_1 placed there by the voltage drop across R_E. Then, the chain of events that follows is as under:

1. Q_1 comes out of CUT-OFF and starts to conduct;
2. as it does so, its collector voltage drops (swings negative);
3. this negative-swinging voltage coupled to the base of Q_2 *via* R_3 reduces its forward bias and hence its emitter current;
4. with reduced emitter current, voltage drop across R_E is reduced;
5. consequently, reverse bias of Q_1 is further lowered and it conducts more heavily;
6. as a result, collector voltage of Q_1 falls further, thereby driving Q_2 still closer to cut-off.

This process is cumulative and ends up with

(*a*) Q_1 conducting at saturation with its collector voltage almost zero;

(*b*) Q_2 becoming cut-off with its collector voltage nearly V_{CC}.

Negative Half-cycle of the Input Voltage

Now, when the negative half-cycle of the input voltage is applied

1. Q_1 becomes reverse-biased. Consequently, its collector current falls and collector voltage rises (*i.e.*, potential of point *A* increases towards V_{CC});
2. this positive-swinging voltage is coupled to the base of Q_2 through R_3 and, as a result, Q_2 is driven to saturation;
3. this re-establishes the original conditions of
   (*a*) Q_1 cut-off with collector voltage at V_{CC} ,and
   (*b*) Q_2 at saturation with collector voltage at 0 V.

It completes one cycle. This cycle is repeated as the input voltage rises and falls again. Hence, each cycle of the Schmitt trigger produces a positive-going pulse at its output which is taken from the collector of Q_2 *i.e.*, from point *B* in Fig. 29.9.

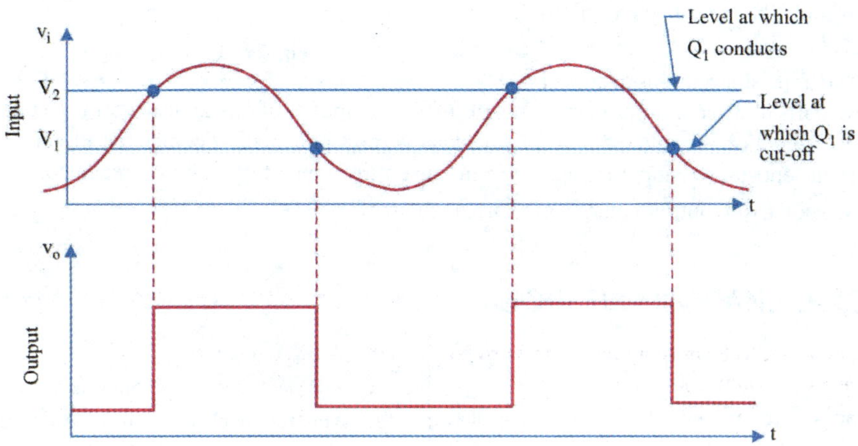

Fig. 29.10

Output Pulse Width

It depends on the time during which Q_2 is conducting. It, in turn, depends on the input voltage, within the limits imposed by emitter resistor R_E.

Uses

1. It is frequently used for waveshaping purposes.
   As shown in Fig. 29.10, it can convert inputs with any waveshape into output pulses having rectangular or square waveshapes. That is why Schmitt trigger is often called a **'squaring' circuit** or a **'squarer'** circuit.
2. It can reshape worn-out pulses by giving them sharp leading and trailing edges.
3. Since a change of state occurs whenever the input crosses a trigger point, the Schmitt trigger is often used as a **level detector** *i.e.*, as a pulse height discriminator.

29.12. Transistor Blocking Oscillator

The basic circuit is shown in Fig. 29.11. When *S* is closed, base current rises rapidly due to forward-bias placed by V_{CC} on the E/B junction of *Q*. It causes a corresponding increase in its collector current. This rising flow of I_C through L_1 produces an induced e.m.f. in L_2. Coupling between L_1 and L_2 is such that lower end of L_2 becomes positive and the grounded end negative.

The positive voltage from L_2 is applied to the base of Q through C. It further increases the forward-bias of the E/B junction which leads to further increase in I_C. Since this process is cumulative, Q is quickly driven to saturation. At that point, there is no further increase in I_C and hence, no induced e.m.f. in L_2 to be applied to Q.

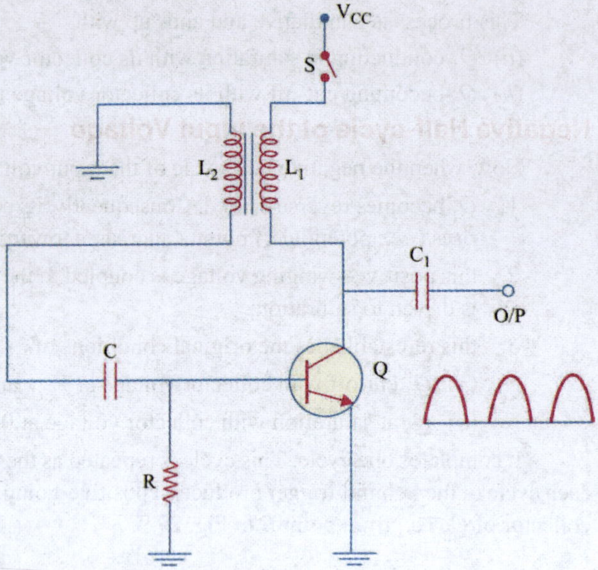

Fig. 29.11

Now, C which had been charged earlier, places a negative charge on the base of Q which reverse-biases its E/B junction and ultimately drives it to cut-off.

The transistor remains at cut-off as C now starts to discharge through R. When sufficient amount of charge leaks off C so that reverse bias of E/B junction is removed and forward bias is re-established, Q comes out of cut-off and its collector current starts rising once again. Then, the entire cycle of operation is repeated.

As shown, the output consists of sharp and narrow pulses.

SELF EXAMINATION QUESTIONS

A. Fill in the blanks by most appropriate word (s) or numerical value (s).

1. Both square and sawtooth waves can be called .... waves.
2. A pulse is an abrupt .... in voltage or current.
3. Pulse repetition frequency is given by the .... of pulse repetition rate.
4. Rise time of a wave depends mainly on the .... capacitance of the transistor.
5. Multivibrator circuits are basically two-stage amplifiers with .... feedback from the output of one to the input of the other.
6. An AMV has two .... stable states.
7. An AMV .... of its own accord.
8. In an AMV, when one stage is ON, the other is ....
9. The two output waves of an AMV are .... degrees out of phase with each other.
10. A BMV does not ....
11. A BMV has two absolutely .... states.
12. The feedback ratio in an AMV circuit is ....
13. In an MMV, a single narrow input trigger pulse produces a single .... pulse.
14. An MMV is often used for .... old and distorted pulses.
15. A BMV can remain in any one of its two states ....
16. A BMV circuit uses no energy .... element.
17. A BMV circuit is the .... of memory and counting circuits.
18. The Schmitt trigger is often called a .... circuit.
19. The Schmitt trigger is frequently used as a pulse height ....
20. A blocking oscillator produces sharp and .... pulses.

B. Answer True or False

1. There is no difference between square and rectangular waves except their mark-to- space ratio.
2. Greater the PRT of a pulse, higher its PRF.
3. Higher the MSR of a pulse, greater its width.
4. Rise time of a pulse is the time required by a wave to rise from its zero value to its final maximum value.
5. Pulse width of a pulse is defined as the time interval between its leading and trailing edges.
6. A multivibrator circuit is essentially an amplifier with 100 per cent positive feedback.
7. An AMV produces only unstable output waves.
8. In an MV circuit, when one transistor is ON, the other is OFF.
9. A symmetrical AMV uses two energy storage elements.
10. An MMV needs an external trigger pulse for its operation.
11. An MMV is called one-shot multivibrator because it oscillates only once.
12. Strictly speaking, a BMV is not an oscillator.
13. A BMV cannot change state of its own accord whereas an MMV can.
14. In an MV, one stage feeds another.
15. To ensure oscillations of an AMV, the transistor β must have a certain minimum value.
16. An MMV is well-suited for restoring the shape of waves which become distorted by use.
17. A BMV is the heart of memory and timing circuits.
18. The Schmitt trigger is called a 'squarer' circuit because it produces square output waves whatever the input.

C. Multiple Choice Items

1. Nonsinusoidal waveforms
   (*a*) are departures from sine waveform
   (*b*) have low mark-to-space ratio
   (*c*) are much easier to generate
   (*d*) are unfit for digital operation
2. A relaxation oscillator is one which
   (*a*) has two stable states
   (*b*) relaxes indefinitely
   (*c*) produces nonsinusoidal output
   (*d*) oscillates continuously
3. A square pulse has a mark-to-space ratio of
   (*a*) 1 : 1 (*b*) 1 : 2
   (*c*) 2 : 1 (*d*) 1 : 4
4. Apart from a dc power source, the essen- tial requirements of a sawtooth generator are
   (*a*) a resistor
   (*b*) a capacitor
   (*c*) a switching device
   (*d*) all of the above
5. An MMV circuit
   (*a*) has no stable state
   (*b*) gives two output pulses for one input trigger pulse
   (*c*) returns to its stand-by states auto- matically
   (*d*) has no energy-storage element
6. A BMV circuit
   (*a*) has two unstable states
   (*b*) has one energy-storage element
   (*c*) switches between its two states automatically
   (*d*) is not an oscillator
7. The frequency of oscillation of an AMV depends mainly on the
   (*a*) value of collector load resistors
   (*b*) *RC* values of the circuit
   (*c*) value of transistor β
   (*d*) width of the input pulse
8. An MMV is frequently used
   (*a*) in memory and timing circuits
   (*b*) for producing triangular waves
   (*c*) in counting circuits
   (*d*) for regeneration of distorted waves
9. A BMV is a .... oscillator.
   (*a*) triggered
   (*b*) free-running
   (*c*) sine-wave
   (*d*) sawtooth
10. In an astable multivibrator
    (*a*) $\beta = 1$ (*b*) $\beta A = 1$
    (*c*) $\beta > 1$ (*d*) $\beta < 1$

ANSWERS

A. Fill in the blanks

1. nonsinusoidal 2. discontinuity 3. reciprocal 4. diffusion 5. positive
6. quasi 7. oscillates 8. OFF 9. 180 10. oscillate
11. stable 12. unity 13. rectangular 14. regenerating 15. indefinitely
16. storage 17. heart 18. squaring 19. discriminator 20. narrow

B. True or False

1. T 2. F 3. T 4. F 5. F 6. T 7. F 8. T 9. T
10. T 11. F 12. T 13. F 14. T 15. T 16. T 17. T 18. T

C. Multiple Choice Items

1. *a* 2. *c* 3. *a* 4. *d* 5. *c* 6. *d* 7. *b* 8. *d* 9. *a* 10. *a*

30 CHAPTER

Modulation and Demodulation

1. What is a Carrier Wave?
2. Radio Frequency Spectrum
3. Sound
4. Radio Broadcasting
5. Modulation
6. Amplitude Modulation
7. Mathematical Analysis of a Modulated Carrier Wave
8. Power Relations in an AM Wave
9. Forms of Amplitude Modulation
10. Methods of Amplitude Modulation
11. Modulating Amplifier Circuit
12. Frequency Modulation
13. Modulation Index
14. Deviation Ratio
15. Per cent Modulation

30.1. Introduction

For successful transmission and reception of intelligence (code, voice, music etc.) by the use of radio waves, two processes are essential:

(*i*) modulation and (*ii*) demodulation.

Speech and music etc., are sent thousands of kilometres away by a radio transmitter. The scene in front of a television camera is also sent many kilometres away to viewers. Similarly, a moon probe or Venus probe checking its environments, sends the information it gathers millions of kilometres through space to receivers on earth. In all these cases, the carrier is the high-frequency radio wave. The intelligence *i.e.,* sight, sound or other data collected by the probe is impressed on the radio wave and is carried along with it to the destination.

Modulation is the process of **combining the low-frequency signal with a very high-frequency radio wave called**

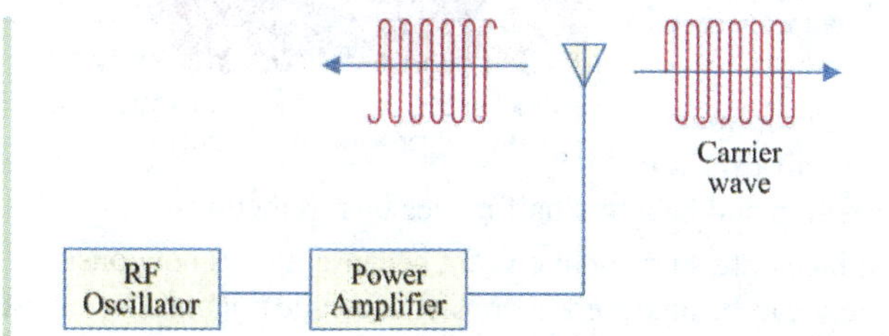

Fig. 30.1

carrier wave (CW). The resultant wave is called **modulated carrier wave.** This job is done at the transmitting station.

Demodulation is the process of separating or **recovering the signal from the modulated carrier wave.** It is just the opposite of modulation and is performed at the receiving end.

30.2. What is a Carrier Wave?

It is a high-frequency undamped radio wave produced by radio-frequency oscillators (Chapter 28). As seen from Fig. 30.1, the output of these oscillators is first amplified and then passed on to an antenna. This antenna radiates out these high-frequency (electromagnetic) waves into space. These waves have constant amplitude and travel with the velocity of light. They are inaudible *i.e.,* by themselves they cannot produce any sound in the loudspeaker of a receiver. As their name shows, their job is to **carry the signal** (audio or video) from transmitting station to the receiving station. The resultant wave is called **modulated** carrier wave.

30.3. Radio Frequency Spectrum

Radio frequencies used by different communication systems extend from very low frequencies to extra high frequencies as tabulated below along with their acronym abbreviations.

Table 30.1

Frequency	*Designation*	*Abbreviation*
3-30 kHz	very low frequency	VLF
30-300 kHz	low frequency	LF
300 kHz-3 MHz	medium frequency	MF
3-30 MHz	high frequency	HF
30-300 MHz	very high frequency	VHF
300 MHz-3 GHz	ultra high frequency	UHF
3-30 GHz	super high frequency	SHF
30-300 GHz	extra high frequency	EHF

Generally, none of the frequencies above 300 GHz is classified as radio waves.

30.4. Sound

It is a sort of disturbance which requires some physical medium for its propagation. Human voice consists of a series of compressions and rarefactions which travel through air with a velocity of about 345 m/s. The frequency range of human voice is from 20-4000 Hz which lies within the audible range of 20 to 20,000 Hz. Variations in human voice can be converted into corresponding variations in electric current with the help of a microphone as shown in Fig.30.2

Sound wave strikes the microphone and it produces AF sound.

When a sound wave strikes the microphone, it produces AF sound current. The positive half-cycles of sound current are produced by the compression and negative half-cycles by rarefactions.

As seen, human voice does not produce pure sinusoidal current because it does not consist of one frequency alone. It is quite complex and can be analysed to consist of a fundamental (or lowest frequency) and its integral multiple frequencies called **overtones** or **harmonics.**

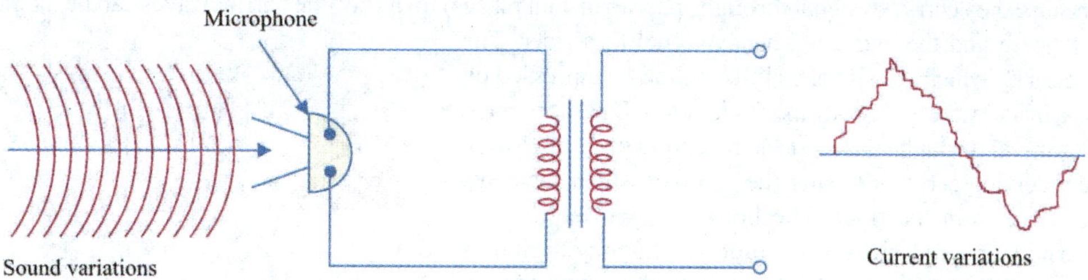

Fig. 30.2

We are interested in two main characteristics of sound:

(*i*) **intensity**—it is the energy content of the wave. It depends on its amplitude. In fact, intensity of a wave is directly proportional to the square of its amplitude *i.e.*, $I \propto a^2$. Sensation of loudness felt by a listener depends directly on the intensity of the wave falling on his ears.

(*ii*) **frequency**—it produces the sensation called pitch. Audible sounds have a frequency range from 20 Hz to 20,000 Hz.

Though every sound has complex frequency structure, we will consider only single-frequency sound whose current wave is a pure sine wave as shown in Fig. 30.3. It will be used as the modulating signal when discussing the process of modulation.

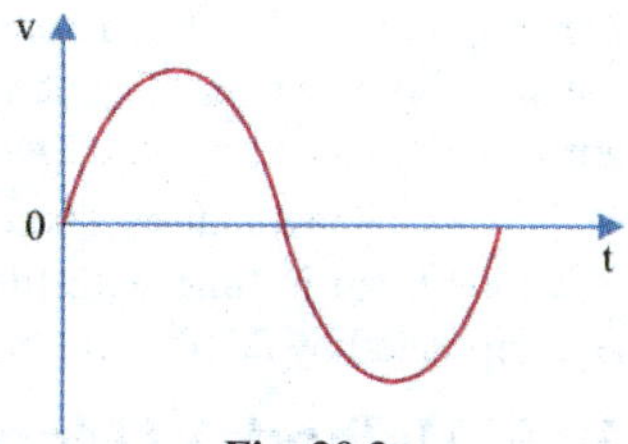

Fig. 30.3

30.5. Need for Modulation

Sometimes, beginners question the necessity of modulation *i.e.*, using a carrier wave to carry the low-frequency signal from one place to another. Why not transmit the signals directly and save lot of botheration? Unfortunately, there are three main hurdles in the process of such direct transmission of audio-frequency signals:

1. they have relatively short range,
2. if everybody started transmitting these low-frequency signals directly, mutual interference will render all of them ineffective,
3. size of antennas required for their efficient radiation would be large *i.e.*, about 75 km !

Hence, the solution lies in modulation which enables a low-frequency signal to travel very large distances through space with the help of a high-frequency carrier wave (CW). These carrier waves need reasonably-sized antennas and produce no interference with other transmitters operating in the same area.

30.6. Radio Broadcasting

Let us see how radio broadcasting stations broadcast speech or music etc., from their broadcasting studios.

First, the speech or music which consists of a series of compressions and rarefactions is translated into a tiny varying electric current with the help of a crystal microphone. The frequency of variations of this current lies in the audio-range, hence it is known as audio-frequency signal. The audio-frequency signal cannot be radiated out from the antenna directly because transmission at audio-frequencies is not practical. For this purpose, oscillations of very high frequency or radio-frequency are produced with the help of any one of the oscillators discussed in Chapter 28. The electromagnetic waves so produced are of constant amplitude but of extremely high frequency. These waves, which are neither seen nor heard, travel through space with the velocity of light *i.e.*, 3×10^8 m/s (approx). The audio-frequency signal which is to be broadcast, is then superimposed on the RF waves, which are known as carrier waves (because they carry AF signal through space to distant places). In a way, the carrier waves can be likened to a horse and the audio-frequency signal to a rider. The process by which AF signal or information is impressed on the carrier wave is known as modulation. The horse and rider travel through space. At the receiving end, they strike the receiving aerial and enter the receiver which separates the horse from the rider. The horse *i.e.,* carrier wave is returned and the rider *i.e.,* audio-frequency signal is converted back into sound. This process by which the RF waves and AF. waves are separated, is known as **detection** or **demodulation** (because it is the reverse of modulation).

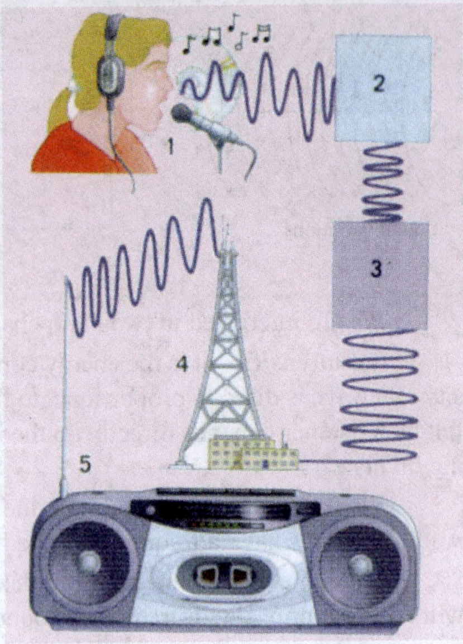

Broadcasting through waves.

30.7. Modulation

It is the process of combining *an audio-frequency (AF) signal with a radio frequency (RF) carrier wave.* The AF signal is also called a *modulating wave* and the resultant wave produced is called *modulated wave.*

During modulation, some characteristic of the carrier wave is varied in time with the modulating signal and is accomplished by combining the two.

30.8. Methods of Modulation

The mathematical expression for a sinusoidal carrier wave is

$$e = E_C \sin(\omega_c t + \phi) = E_C \sin(2\pi f_c t + \phi)$$

Obviously, the waveform can be varied by any of its following three factors or parameters:

1. E_C — the amplitude, 2. f_c — the frequency, 3. ϕ — the phase.

Accordingly, there are three types of sine-wave modulations known as:

1. **Amplitude Modulation (AM)**

   Here, the information or AF signal changes the amplitude of the carrier wave without changing its frequency or phase.

2. **Frequency Modulation (FM)**

   In this case, the information signal changes the frequency of the carrier wave without changing its amplitude or phase.

3. **Phase Modulation (PM)**

   Here, the information signal changes the phase of the carrier wave without changing its other two parameters.

30.9. Amplitude Modulation

In this case, the **amplitude** of the carrier wave is varied in proportion to the **instantaneous amplitude** of the information signal or AF signal. Obviously, the amplitude (and hence the intensity) of the carrier wave is changed *but not its frequency.* Greater the amplitude of the AF signal, greater the fluctuations in the amplitude of the carrier wave.

The process of amplitude modulation is shown graphically in Fig. 30.4. For the sake of simplicity, the AF signal has been assumed sinusoidal [Fig. 30.4 (*a*)]. The carrier wave by which it is desired to transmit the AF signal is shown in Fig. 30.4 (*b*). The resultant wave called modulated wave is shown in Fig. 30.4 (*c*).

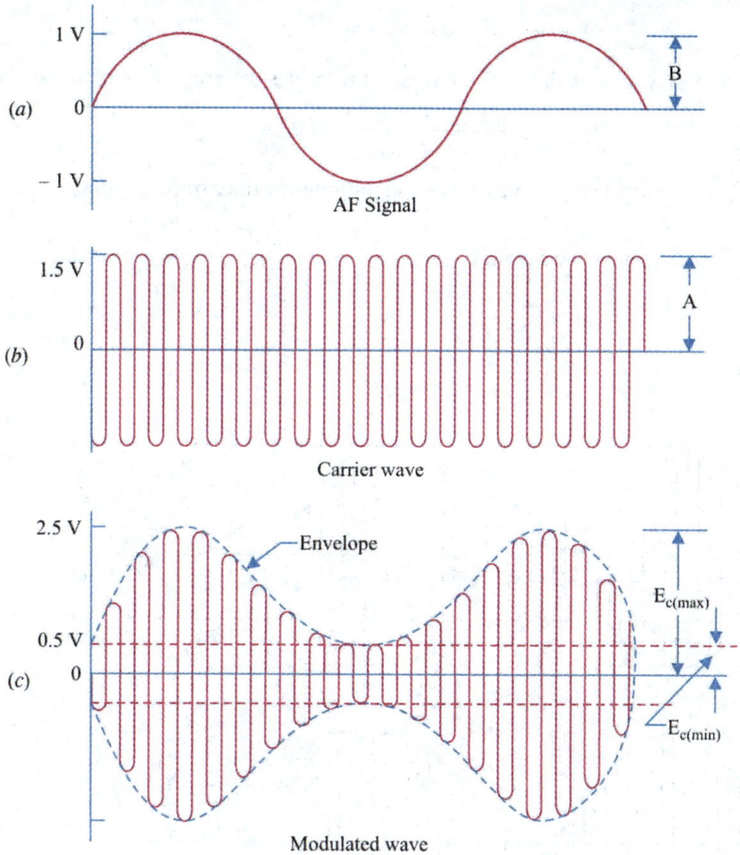

Fig. 30.4

The function of the modulator is to mix these two waves in such a way that (*a*) is transmitted along with (*b*). All stations broadcasting on the standard broadcast band (550-1550 kHz) use AM modulation.

If you observe the *envelope* of the modulated carrier wave, you will realize that it is an exact replica of the AF signal wave.

In summary

(*i*) **fluctuations** in the amplitude of the carrier wave depend on the **signal amplitude,**

(*ii*) **rate** at which these fluctuations take place depends on the **frequency** of the audio signal.

30.10. Per cent Modulation

It indicates the degree to which the AF signal modulates the carrier wave.

$$m = \frac{\text{maximum value of signal wave}}{\text{maximum value of carrier wave}} \times 100 = \frac{\text{signal amplitude}}{\text{carrier amplitude}} \times 100$$

$$= \frac{B}{A} \times 100 \qquad \text{— Fig. 30.4}$$

The ratio B/A expressed as a fraction is called **modulation index (MI)**

$$m = MI \times 100$$

From Fig. 30.4, it is seen that $B = 1$ V and $A = 1.5$ V

$$\therefore \qquad m = \frac{1}{1.5} \times 100 = 66.7\%$$

Modulation may also be defined in terms of the values referred to the modulated carrier wave.

$$m = \frac{E_{c\,(max)} - E_{c\,(min)}}{E_{c\,(max)} + E_{c\,(min)}} \times 100$$

where $E_c(max)$ and $E_c(min)$ are the maximum and minimum values of the amplitude of the modulated carrier wave.

Again, from Fig. 30.4 we see that

$$m = \frac{2.5 - 0.5}{2.5 + 0.5} \times 100$$

$$= \frac{2}{3} \times 100 = 66.7\%$$

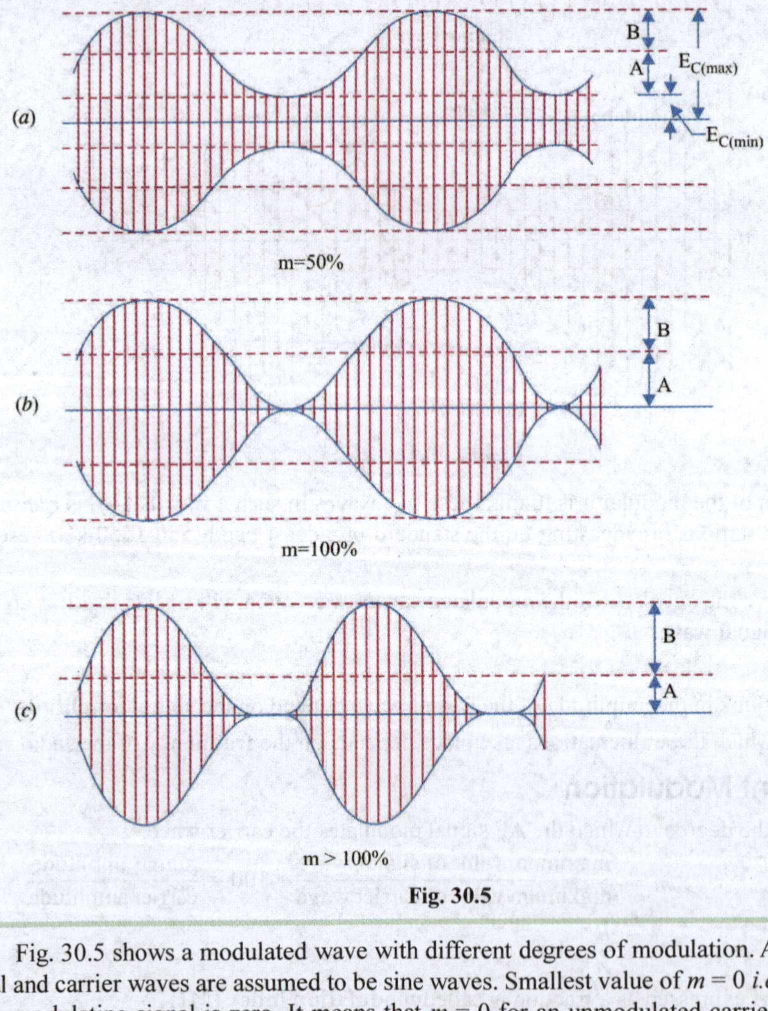

Fig. 30.5

Fig. 30.5 shows a modulated wave with different degrees of modulation. As before, both the signal and carrier waves are assumed to be sine waves. Smallest value of $m = 0$ *i.e.,* when amplitude of the modulating signal is zero. It means that $m = 0$ for an unmodulated carrier wave. Maximum value of $m = 1$ when $B = A$. Value of m can vary anywhere from 0 to 100% without introducing distortion. Maximum undistorted power of a radio transmitter is obtained when $m = 100\%$. If m is less than 100 per cent, power output is reduced though *the power content of the carrier is not.* Modulation in excess of 100 per cent produces severe distortion and interference (called splatter) in the transmitter output.

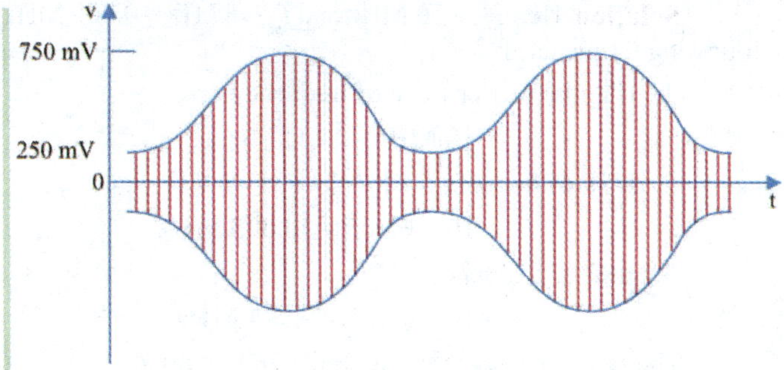

Fig. 30.6

Example 30.1. *A modulated carrier wave has maximum and minimum amplitudes of 750 mV and 250 mV. Calculate the value of percentage modulation.*

Solution. The modulated wave is shown in Fig. 30.6.

Here, $E_{c(max)} = 750$ mV and

$E_{c(min)} = 250$ mV

$$\therefore \quad m = \frac{E_{c\,(max)} - E_{c\,(min)}}{E_{c\,(max)} + E_{c\,(min)}} \times 100 = \frac{750 - 250}{750 + 250} \times 100 = \mathbf{50\%}$$

30.11. Upper and Lower Side Frequencies

An unmodulated carrier wave consists of only one single-frequency component of frequency f_c. When it is combined with a modulating signal-frequency f_m, heterodyning action takes place.

As a result, two additional frequencies called **side frequencies** are produced. The AM wave is found to consist of **three** frequency components:

1. The original carrier frequency component, f_c.
2. A higher frequency component $(f_c + f_m)$. It is called the sum component.
3. A lower frequency component $(f_c - f_m)$. It is called the difference component.

The two new frequencies are called the **upper-side frequency** (USF) and **lower-side frequency** (LSF) respectively and are symmetrically located around the carrier frequency. The modulating frequency remains unchanged but does not appear in the amplifier output because the amplifier's load presents practically zero impedance to this low frequency.

These are shown in time domain in Fig. 30.7 (*a*) and in frequency domain in Fig. 30.7 (*b*). The amplitude of the side frequencies depends on the value of *m*. The amplitude of each side frequency = $mA/2$ where *A* is the amplitude of unmodulated carrier wave.

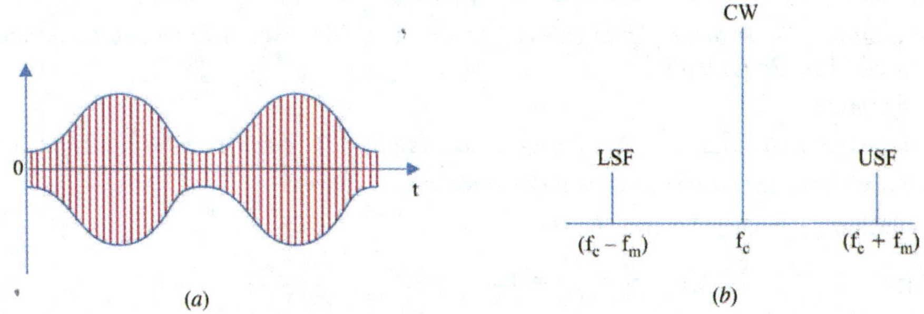

Fig. 30.7

Example 30.2. *A 10 MHz sinusoidal carrier wave of amplitude 10 mV is modulated by a 5kHz sinusoidal audio signal wave of amplitude 6 mV. Find the frequency components of the resultant modulated wave and their amplitudes.*

Solution. Here, $f_c = 10$ MHz and $f_m = 5$ kHz = 0.005 MHz. The modulated carrier contains the following frequencies:

1. original carrier wave of frequency

$$f_c = \mathbf{10\ MHz}$$

2. USF of frequency

$$= 10 + 0.005 = \mathbf{10.005\ MHz}$$

3. LSF of frequency

$$= 10 - 0.005 = \mathbf{9.995\ MHz}$$

The frequency spectrum is shown in Fig. 30.8.

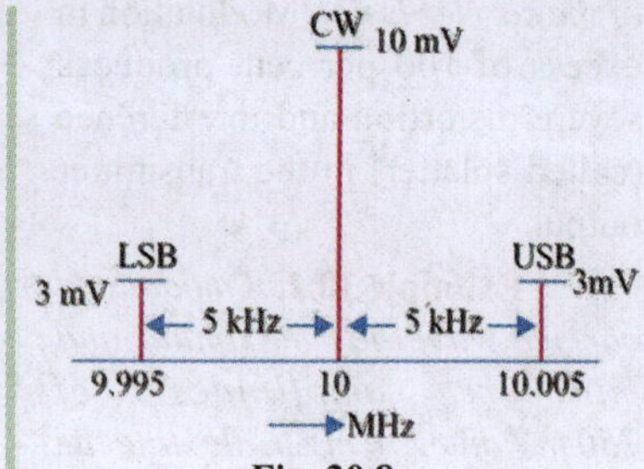

Fig. 30.8

Here, $m = \frac{B}{A} = \frac{6}{10} = 0.6$

Amplitude of LSF = USF = $mA/2 = 0.6 \times 10/2 =$ **3 mV** as shown.

30.12. Upper and Lower Sidebands

In Art. 30.11, it was assumed that the modulating signal was composed of one frequency component only. However, in a broadcasting station, the modulating signal is the human voice (or music) which contains waves with a frequency range of 20-4000 Hz. Each of these waves has its own LSF and USF. When combined together, they give rise to an **upper-side band** (USB) and a **lower-side band** (LSB) as shown in Fig. 30.9 The USB, in fact, contains all sum components of the signal and carrier frequency whereas LSB contains their difference components.

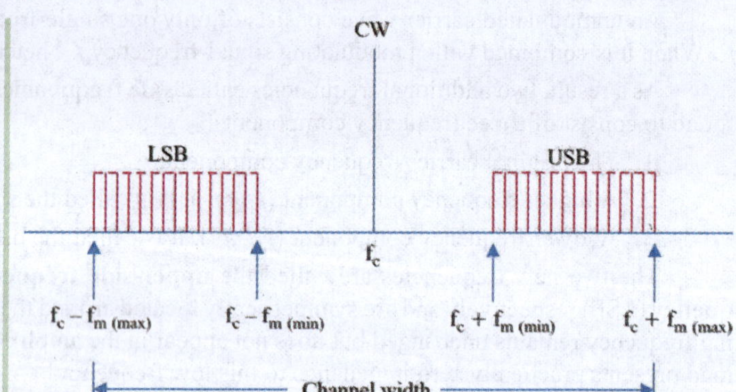

Fig. 30.9

The channel width (or bandwidth) is given by the difference between extreme frequencies *i.e.*, between maximum frequency of USB and minimum frequency of LSB. As seen

Channel width = 2 × maximum frequency of modulating signal = $2 \times f_m(max)$.

Example 30.3. *An audio signal given by 15 sin 2π (2000 t) amplitude-modulates a sinusoidal carrier wave 60 sin 2π (100,000) t.*

Determine :

(*a*) *modulation index,* (*b*) *per cent modulation,* (c) *frequencies of signal and carrier,* (*d*) *frequency spectrum of the modulated wave.*

Solution. Here, $B = 15$ and $A = 60$

(*a*) $MI = \frac{B}{A} = \frac{15}{60} = \mathbf{0.25}$

(*b*) $m = MI \times 100 = 0.25 \times 100 =$ **25%**

(*c*) f_m = **2000 Hz** —by inspection of the given equation

f_c = **100,000 Hz** —by inspection of the given equation

(*d*) the three frequencies present in the modulated CW are

(*i*) 100,000 Hz = **100 kHz**

(*ii*) 100,000 + 2000 = 102,000 Hz = **102 kHz**

(*iii*) 100,000 – 2000 = 98,000 Hz = **98 kHz**

Example 30.4. *A bandwidth of 15 MHz is available for AM transmission. If the maximum audio signal frequency used for modulating the carrier is not to exceed 15 kHz, how many stations can broadcast within this band simultaneously without interfering with each other?*

Solution. BW required by each station

$$= 2 f_{m(max)} = 2 \times 15 = 30 \text{ kHz}$$

Hence, the number of station which can broadcast within this frequency band without interfering with one another is

$$= \frac{15 \text{ MHz}}{30 \text{ kHz}} = \mathbf{500}$$

Example 30.5. *In a broadcasting studio, a 1000 kHz carrier is modulated by an audio signal of frequency range, 100 – 5000 Hz. Find (i) width or frequency range of sidebands (ii) maximum and minimum frequencies of USB (iii) maximum and minimum frequencies of LSB and (iv) width of the channel.*

Solution. (*i*) Width of sideband

= 5000 – 100 = **4900 Hz**

(*ii*) Max. frequency of USB

= 1000 + 5 = **1005 kHz**

Min. frequency of USB

= 1000 + 0.1 = **1000.1 kHz**

(*iii*) Max. frequency of LSB

= 1000 – 0.1 = **999.9 kHz**

Min. frequency of LSB

= 1000 – 5 = **995 kHz**

(*iv*) Width of channel = 1005 – 995 = **10 kHz**

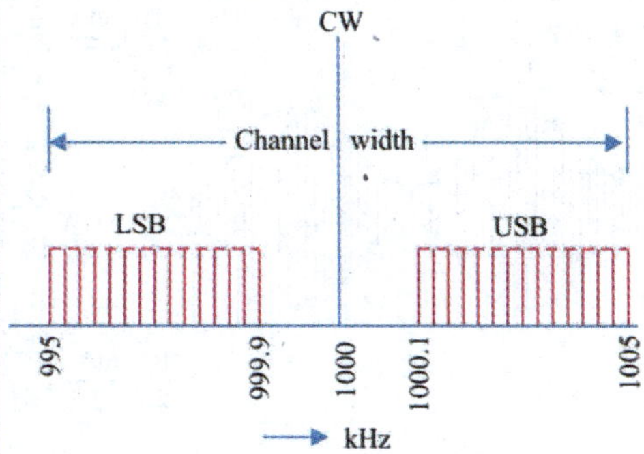

Fig. 30.10

30.13. Mathematical Analysis of a Modulated Carrier Wave

The equation of an unmodulated carrier wave is

$$e_c = E_c \sin 2\pi f_c t = A \sin 2\pi f_c t = A \sin \omega t$$

where A is the constant amplitude of the carrier wave and

$$\omega = 2\pi f_c.$$

Let the equation of the single-frequency sinusoidal modulating signal be

$$e_m = E_m \sin 2\pi f_m t = B \sin 2\pi f_m t$$

$$= B \sin pt \qquad \text{— where } p = 2\pi f_m$$

As seen from Fig. 30.11, the amplitude of the modulated carrier wave at any instant is

$$= A + e_m \qquad (\because A \text{ is constant})$$

$$= (A + B \sin pt)$$

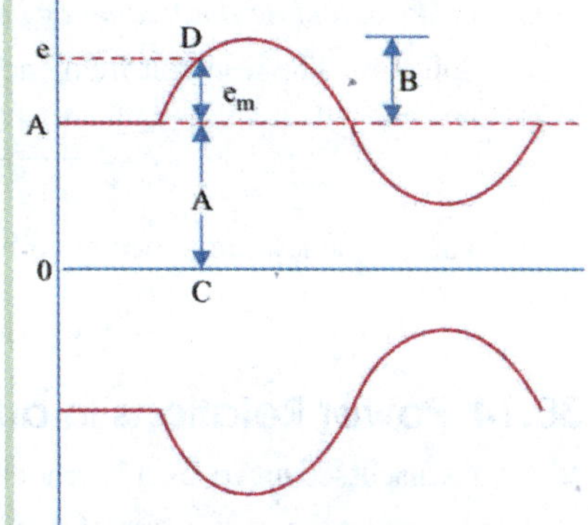

Fig. 30.11

Hence, its *instantaneous* value is given by

$$e = (A + B \sin pt) \sin \omega t = A \sin \omega t + B \sin \omega t . \sin pt$$

$$= A \sin \omega t + \frac{B}{2} . 2 \sin \omega t . \sin pt = A \sin \omega t + \frac{B}{2} [\cos (\omega - p) t - \cos (\omega + p) t]^*$$

$$= A \sin \omega t + \frac{B}{2} \cos (\omega - p) t - \frac{B}{2} \cos (\omega + p)t$$

$$= A \sin 2 \pi f_c t + \frac{B}{2} \cos 2\pi (f_c - f_m)t - \frac{B}{2} \cos 2\pi (f_c + f_m)t$$

As seen from Art. 30.10, $m = B/A$ or $B = mA$

$$\therefore \quad e = A \sin 2\pi f_c t + \left(\frac{mA}{2}\right) \cos 2\pi (f_c - f_m) t - \left(\frac{mA}{2}\right) \cos 2\pi (f_c + f_m) t$$

It is seen that the modulated wave contains three components:

(*i*) $A \sin 2\pi f_c t$ — the original carrier wave

(*ii*) $\frac{mA}{2} \cos 2\pi (f_c + f_m)t$ — upper side frequency

(*iii*) $\frac{mA}{2} \cos 2\pi (f_c - f_m)t$ — lower side frequency

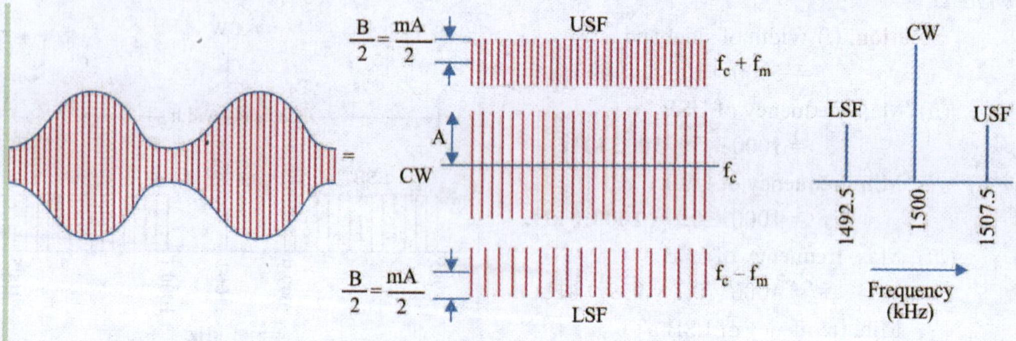

Fig. 30.12

These three frequencies are not a mathematical fiction but they actually exist. In fact, with the help of a narrow band filter, we can separate side frequencies from the carrier wave.

Example 30.6. *The tuned circuit of the oscillator in an AM transmitter uses a 40 μH coil and a 1 nF capacitor. If the carrier wave produced by the oscillator is modulated by audio- frequencies upto 10 kHz, calculate the frequency band occupied by the sidebands and channel width.*

Solution. The resonant frequency is given by

$$f_c = \frac{1}{2\pi\sqrt{LC}} = \frac{1}{2\pi\sqrt{40 \times 10^{-6} \times 1 \times 10^{-9}}} = 796 \text{ kHz}$$

The frequency range occupied by the sidebands is from **786 kHz to 806 kHz.** The channel width

$$= 2 \times 10 = \mathbf{20\ kHz}$$

30.14. Power Relations in an AM Wave

As discussed in Art. 30.13, a modulated carrier wave consists of the following three components:

1. original carrier wave of amplitude *A*,
2. USF wave of amplitude (*mA*/2),

* $2 \sin A . \sin B = \cos (A - B) - \cos (A + B)$.

3. LSF wave of amplitude ($mA/2$).

Now, power radiated out by a wave through an antenna is proportional to (amplitude)^2.

Carrier power, $P_C \propto A^2 = KA^2$

USB power, $P_{USB} \propto \left(\frac{B}{2}\right)^2 = \frac{KB^2}{4}$; LSB power, $P_{LSB} \propto \left(\frac{B}{2}\right)^2 = \frac{KB^2}{4}$

Total sideband power $P_{SB} = 2 \times \frac{KB^2}{4} = \frac{KB^2}{2}$

Total power radiated out from the antenna is $P_T = P_C + P_{SB} = KA^2 + \frac{KB^2}{2}$

Substituting $B = mA$, we get

$$P_T = KA^2 + \frac{K}{2}(mA)^2 = KA^2\left(1 + \frac{m^2}{2}\right)$$

Now $P_C = KA^2$

(*i*) $\therefore \quad P_T = P_C\left(1 + \frac{m^2}{2}\right) = P_C\left(\frac{2 + m^2}{2}\right)$

(*ii*) $P_C = P_T\left(\frac{2}{2 + m^2}\right)$

(*iii*) $P_{SB} = P_T - P_C = P_C\left(1 + \frac{m^2}{2}\right) - P_C = \frac{m^2}{2}P_C = \left(\frac{m^2}{2 + m^2}\right)P_T$

(*iv*) $P_{USB} = P_{LSB} = \frac{1}{2}P_{SB} = \frac{m^2}{4}P_C = \frac{1}{2}\left(\frac{m^2}{2 + m^2}\right)P_T$

Let us consider the case when $m = 1$ or 100%

(*i*) $P_T = 1.5\ P_C = 1.5 \times$ carrier power

(*ii*) $P_C = 2/3\ P_T = 2/3 \times$ total power radiated

(*iii*) $P_{USB} = P_{LSB} = 1/4\ P_C = 25\%$ of carrier power

(*iv*) $P_{SB} = 1/2\ P_C = 50\%$ of carrier power

(*v*) $\frac{P_{USB}}{P_T} = \frac{0.25\ P_C}{1.5\ P_C} = \frac{1}{6}$

It means that **single sideband contains 1/6th of the total power radiated out by the transmitter.** That is why single sideband (SSB) transmission is more power efficient.

Example 30.7. *The total power content of an AM wave is 1500 W. For a 100 per cent modulation, determine*

(*i*) *power transmitted by carrier,* (*ii*) *power transmitted by each sideband.*

Solution. $P_T = 1500$ W, $P_C = ?$, $P_{USB} = ?$, $P_{LSB} = ?$

(*i*) $P_C = P_T\left(\frac{2}{2 + m^2}\right) = P_T\left(\frac{2}{2 + 1^2}\right) = \frac{2}{3}P_T = \frac{2}{3} \times 1500 = \mathbf{1000\ W}$

(*ii*) $P_{USB} = P_{LSB} = \frac{m^2}{4}P_C = \frac{1^2}{4} \times 1000 = \mathbf{250\ W}$

As seen, $P_{USB} = \frac{1}{6}P_T$

Example 30.8. *The total power content of an AM wave is 2.64 kW at a modulation factor of 80%. Determine the power content of*

(*i*) *carrier, and* (*ii*) *each sideband.*

Solution. (*i*) $P_C = P_T\left(\dfrac{2}{2+m^2}\right) = 2.64 \times 10^3\left(\dfrac{2}{2+0.8^2}\right) = \mathbf{2000\ W}$

(*ii*) $P_{USB} = P_{LSB} = \dfrac{m^2}{4}P_C = \dfrac{0.8^2}{4} \times 2000 = \mathbf{320\ W}$

30.15. Forms of Amplitude Modulation

As shown in Art. 30.11, **one carrier** and **two sidebands** are produced in AM generation. It is found that it is not necessary to transmit all these signals to enable the receiver to reconstruct the original signal. Accordingly, we may attenuate or altogether remove the carrier or any one of the sidebands without affecting the communication process. The advantages would be

1. less transmitted power, and 2. less bandwidth required.

The different suppressed component systems are :

(*a*) DSB-SC

It stands for *double-sideband suppressed carrier* system [Fig. 30.13 (*a*)]. Here, carrier component is suppressed thereby saving enormous amount of power. As seen from Art. 30.14, carrier signal contains 66.7 per cent of the total transmitted power for $m = 1$. Hence, power saving amounts to 66.7% at 100% modulation.

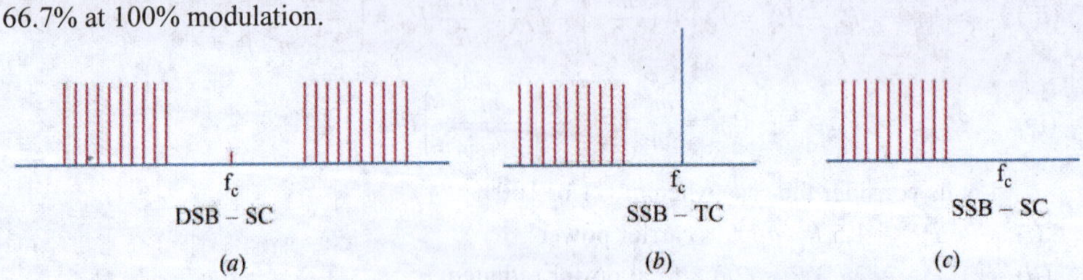

Fig. 30.13

(*b*) SSB-TC

As shown in Fig. 30.13 (*b*), in this case, one sideband is suppressed but the other sideband and carrier are transmitted. It is called *single sideband transmitted carrier* system. For $m = 1$, power saved is 1/6 of the total transmitted power.

(*c*) SSB-SC

This is the **most dramatic suppression of all** because it suppresses one sideband and the carrier and transmits *only the remaining sideband* as shown in Fig. 30.13 (*c*). In the standard or double-sideband full-carrier (DSB-FC) AM, carrier conveys **no information but contains maximum power.** Since the two sidebands are exact images of each other, they carry the same audio information. Hence, **all information is available in one sideband only.** Obviously carrier is *superfluous* and one sideband is *redundant*. Hence, one sideband and the carrier can be discarded with no loss of information. The result is SSB signal. The advantage of SSB-SC system are as follows:

1. Total saving of 83.3% in transmitted power (66.7% due to suppression of carrier wave and 16.6% due to suppression of one sideband). Hence, power is conserved in an SSB transmitter.
2. Bandwidth required is reduced by half *i.e.,* 50%. Hence, twice as many channels can be multiplexed in a given frequency range.
3. The size of power supply required is very small. This fact assumes vital importance particularly in a spacecraft.

4. Since the SSB signal has narrower bandwidth, a narrower passband is permissible within the receiver, thereby limiting the noise pick up.

However, the main reason for wide-spread use of DSB-FC (rather than SSB-SC) transmission in broadcasting is the relative simplicity of its modulating equipment.

30.16. Generation of SSB

Three main systems are employed for the generation of SSB

1. filter method, 2. phase cancellation method, 3. the 'third' method.

All these methods use some form of balanced modulator to suppress the carrier.

Example 30.9. *In an AM wave, calculate the power saving when the carrier and one sideband are suppressed corresponding to*

(*i*) $m = 1$ *and* (*ii*) $m = 0.5$

Solution. (*i*) When *m* = 1

$$P_T = P_C\left(1+\frac{m^2}{2}\right) = 1.5P_C;\ P_{LSB} = P_{USB} = \frac{m^2}{4}P_C = 0.25P_C$$

$\therefore$ saving $= P_T - P_{USB} = 1.5\,P_C - 0.25\,P_C = 1.25\,P_C$

% saving $= \dfrac{1.25P_C}{1.5P_C}\times 100 = \mathbf{83.3\%}$

(*ii*) When *m* = 0.5

$$P_T = P_C\left(1+\frac{m^2}{2}\right) = P_C\left(\frac{1+0.5^2}{2}\right) = 1.125P_C$$

$$P_{LSB} = P_{USB} = \frac{m^2}{4}P_C = \frac{0.5^2}{4}P_C = 0.0625P_C$$

$\therefore$ saving $= \dfrac{1.125\,P_C - 0.0625\,P_C}{1.125\,P_C}\times 100 = \mathbf{94.4\%}$

30.17. Methods of Amplitude Modulation

There are two methods of achieving amplitude modulation:

(*i*) amplifier modulation, (*ii*) oscillator modulation.

Block diagram of Fig. 30.14 illustrates the basic idea of amplifier modulation.

Here, carrier and AF signal are fed to an amplifier and the result is an AM output. The modulation process takes place in the active device used in the amplifier.

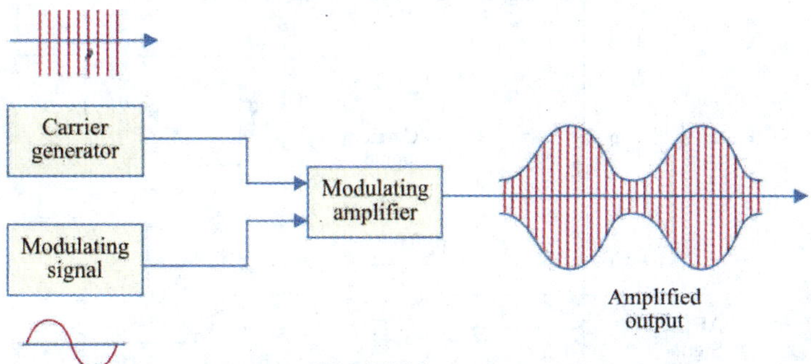

Fig. 30.14

30.18. Block Diagram of an AM Transmitter

Fig. 30.15 shows the block diagram of a typical transmitter. The carrier wave is supplied by a crystal-controlled oscillator at the carrier frequency. It is followed by a tuned buffer amplifier and an RF output amplifier.

The source of AF signal is a microphone. The audio signal is amplified by a low level audio amplifier and, finally, by a power amplifier. It is then combined with the carrier to produce a modulated carrier wave which is ultimately radiated out in the free space by the transmitter antenna as shown.

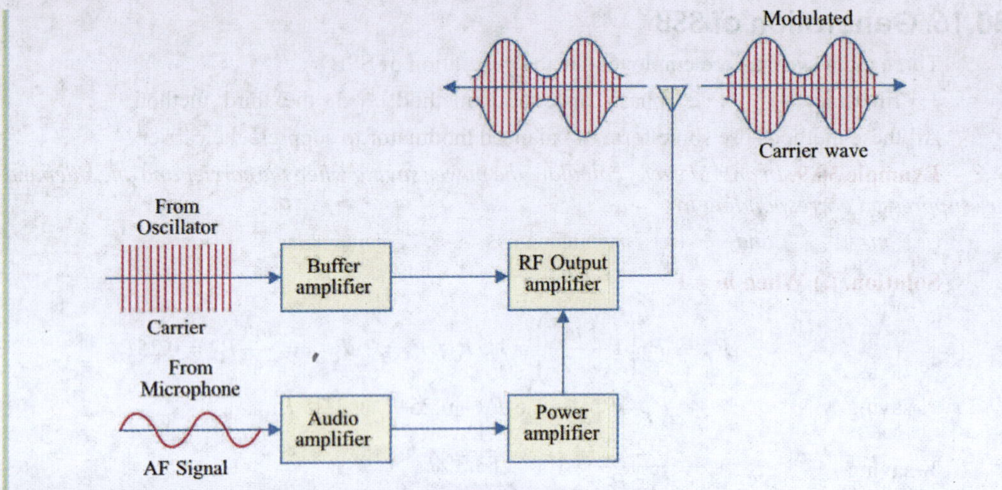

Fig. 30.15

30.19. Modulating Amplifier Circuit

It is shown in Fig. 30.16. The carrier signal from a crystal oscillator is coupled to the base of Q through transformer T_1. The low-frequency modulating signal is also coupled to the base of Q by capacitor C_1. The voltage divider R_1-R_2, as usual, provides proper forward bias for the transistor whereas C_2 and C_3 are bypass capacitors.

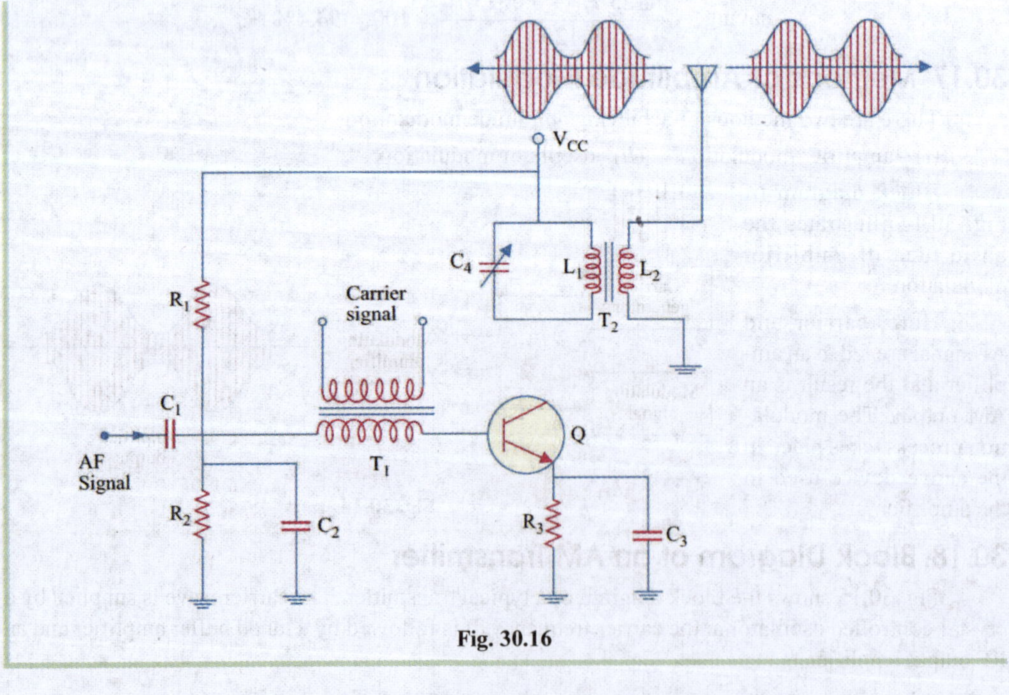

Fig. 30.16

The modulating signal applied across R_2 causes variations in the base bias in step with its own variations. The amplitude of the transistor RF output changes according to the changes in AF modulating signal. The L_1C_4 circuit is kept tuned to the carrier frequency. The amplitude-modulated carrier or RF current in L_1 induces similar current in L_2 which is connected to an antenna. Finally, the AM carrier wave is radiated out in space by the transmitting antenna.

30.20. Frequency Modulation

As the name shows, in this modulation, it is only the **frequency of the carrier which is changed and not its amplitude.**

The **amount of change** in frequency is determined by the **amplitude** of the modulating signal whereas **rate of change** is determined by the **frequency** of the modulating signal. As shown in Fig. 30.17, in an FM carrier, information (or intelligence) is carried as variations in its frequency. As seen, frequency of the modulated carrier increases as the signal amplitude increases but decreases as the signal amplitude decreases. It is at its highest frequency (point *H*) when the signal amplitude is at its maximum positive value and is at its lowest frequency (point *L*) when signal amplitude has maximum negative value. When signal amplitude is zero, the carrier frequency is at its normal frequency f_0 (also called *resting* or *centre* frequency).

The fact that amount of change in frequency depends on signal amplitude is illustrated in Fig. 30.18 where *R* stands for resting frequency. Here, signal amplitude is almost double of that in Fig. 30.17 *though its frequency is the same*. This **louder** signal causes **greater** frequency change in modulated carrier as indicated by *increased bunching and spreading* of the waves as compared with relatively weaker signal of Fig. 30.17.

The **rate** at which frequency shift takes place depends on the signal frequency as shown in Fig. 30.19. For example, if the modulating signal is 1 kHz, then the modulated carrier will swing between its maximum frequency and lowest frequency 1000 times per second. If f_m = 2 kHz, the rate of frequency swing would be twice as fast.

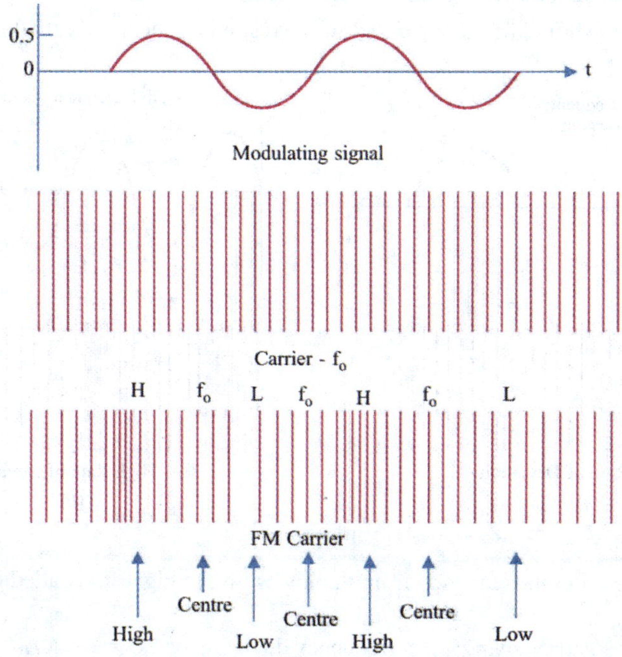

Fig. 30.17

In short, we have established two important points about the nature of frequency modulation:

(*i*) the **amount** of frequency deviation (or shift or variation) depends on the **amplitude** (loudness) of the audio signal. *Louder the sound, greater the frequency deviation and vice-versa.*

However, for the purposes of FM broadcasts, it has been internationally agreed to restrict maximum deviation to **75 kHz** on each side of the centre frequency for sounds of maximum loudness. Sounds of lesser loudness are permitted *proportionately less frequency deviation.*

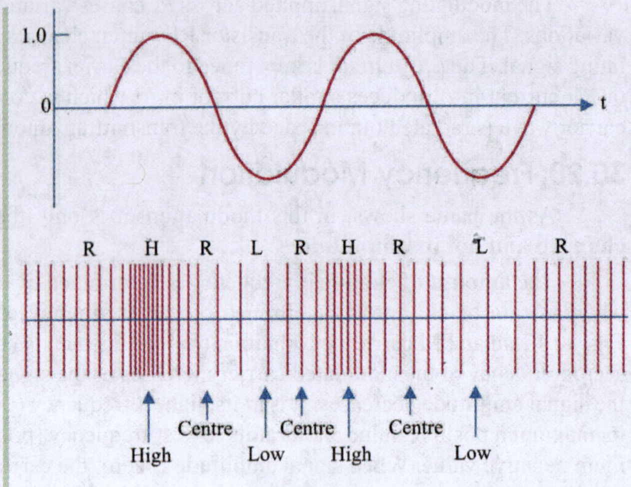

Fig. 30.18

(*ii*) the **rate** of frequency deviation depends on the **signal frequency.**

30.21. Frequency Deviation and Carrier Swing

The frequency of an FM transmitter without signal input is called the **resting frequency** or **centre frequency** (f_0) and is the allotted frequency of the transmitter. In simple words, it is the carrier frequency on which a station is allowed to broadcast. When the signal is applied, the carrier frequency deviates up and down from its resting value f_0.

This change or shift either above or below the resting frequency is called frequency deviation (Δf).

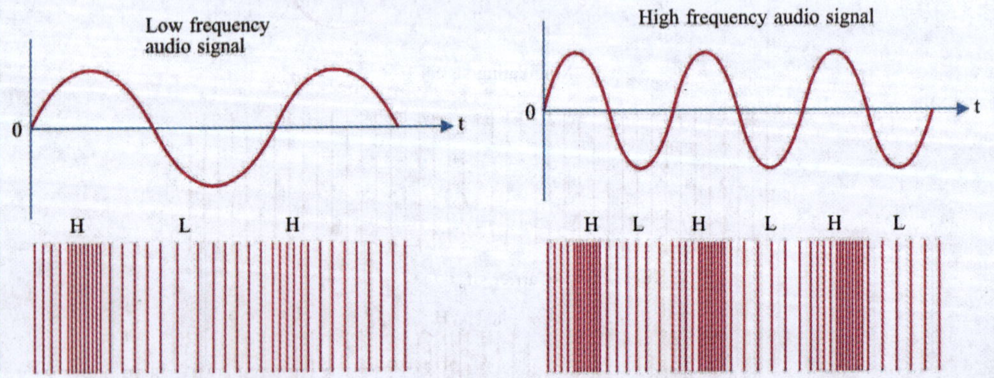

Fig. 30.19

The total variation in frequency from the lowest to the highest is called carrier swing (CS). Obviously,

$$\text{carrier swing} = 2 \times \text{frequency deviation or } CS = 2 \times \Delta f$$

A maximum frequency deviation of 75 kHz is allowed for commercial FM broadcast stations in the 88 to 168 MHz VHF band. Hence, FM channel width is 2 × 75 = 150 kHz. Allowing a 25 kHz guard band on either side, the channel width becomes = 2 (75 + 25) = 200 kHz (Fig. 30.20). This guardband is meant to prevent interference between adjacent channels. However, a maximum frequency deviation of 25 kHz is allowed in the sound portion of the TV broadcast.

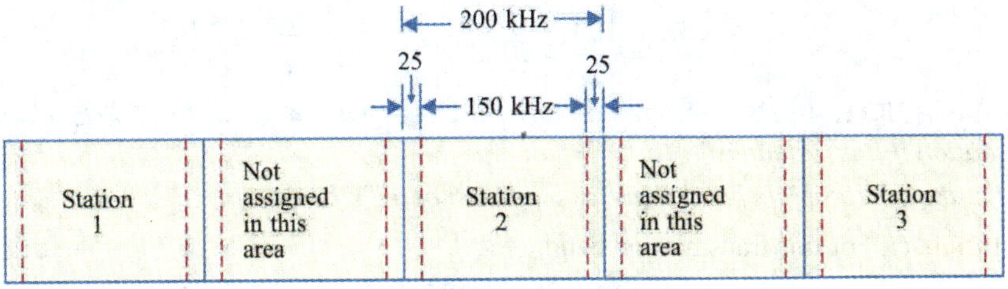

Fig. 30.20

In FM, the highest audio frequency transmitted is **15 kHz.**

Consider an FM carrier of resting frequency 100 MHz. Since $(\Delta f)_{max}$ = 75 kHz, the carrier frequency can swing from the lowest value of 999.925 MHz to the highest value of 100.075 MHz. Of course, deviations lesser than 75 kHz corresponding to relatively softer sounds are always permissible.

30.22. Modulation Index

It is given by the ratio $m_f = \dfrac{\text{frequency deviation}}{\text{modulating frequency}} = \dfrac{\Delta f}{f_m}$

Unlike amplitude modulation, this **modulation index can be greater than unity.** By knowing the value of m_f, we can calculate the number of significant sidebands and the bandwidth of the FM signal.

30.23. Deviation Ratio

It is the worst-case modulation index in which maximum permitted frequency deviation and maximum permitted audio-frequency are used.

$$\therefore \quad \text{deviation ratio} = \frac{(\Delta f)_{max}}{f_{m(max)}}$$

Now, for FM broadcast stations, $(\Delta f)_{max}$ = 75 kHz and maximum permitted frequency of modulating audio signal is 15 kHz.

$$\therefore \quad \text{deviation ratio} = \frac{75 \text{ kHz}}{15 \text{ kHz}} = 5$$

For sound portion of commercial TV deviation ratio $= \dfrac{25 \text{ kHz}}{15 \text{ kHz}} = 1.67.$

30.24. Per cent Modulation

When applied to FM, this term has slightly different meaning than when applied to AM. In FM, it is given by the ratio of actual frequency deviation to the maximum allowed frequency deviation.

$$m = \frac{(\Delta f)_{actual}}{(\Delta f)_{max}}$$

Obviously, 100% modulation corresponds to the case when actual deviation equals the maximum allowable frequency deviation. If, in some case, actual deviation is 50 kHz, then

$$m = \frac{50}{75} = \frac{2}{3} = 0.667 = \mathbf{66.7\%}$$

Value of $m = 0$ corresponds to zero deviation *i.e.,* unmodulated carrier wave. It is seen from the above equation that $m \propto (\Delta f)_{actual}$. It means that when **frequency deviation (*i.e.,* signal loudness) is doubled, modulation is doubled.**

Example 30.10. *What is the modulation index of an FM carrier having a carrier swing of 100 kHz and a modulating signal of 5 kHz ?*

Solution. $CS = 2 \times \Delta f$

$$\therefore \quad \Delta f = \frac{CS}{2} = \frac{100}{2} = 50 \text{ kHz} \quad \therefore \; m_f = \frac{\Delta f}{f_m} = \frac{50}{5} = \mathbf{10}$$

Example 30.11. *An FM transmission has a frequency deviation of 18.75 kHz. Calculate per cent modulation if it is broadcast*

(*i*) *in the 88-108 MHz band* , (*ii*) *as a portion of a TV broadcast.*

Solution. (*i*) For this transmission band,

$$(\Delta f)_{max} = 75 \text{ kHz} \quad \therefore \quad m = \frac{18.75}{75} \times 100 = \mathbf{25\%}$$

(*i*) In this case, $(\Delta f)_{max} = 25$ kHz $\quad \therefore \quad m = \frac{18.75}{25} \times 100 = \mathbf{75\%}$

Example 30.12. *An FM signal has a resting frequency of 105 MHz and highest frequency of 105.03 MHz when modulated by a signal of frequency 5 kHz. Determine*

(*i*) *frequency deviation,* (*ii*) *carrier swing,*
(*iii*) *modulation index,* (*iv*) *per cent modulation,*
(*v*) *lowest frequency reached by the FM wave.*

Solution. (*i*) $\Delta f = 105.03 - 105 = 0.03$ MHz $=$ **30 kHz**

(*ii*) $CS = 2 \times \Delta f = 2 \times 30 =$ **60 kHz**

(*iii*) $m_f = 30/5 = \mathbf{6}$

(*iv*) $m = 30/75 \times 100 = \mathbf{40\%}$

(*v*) lowest frequency $= 105 - 0.03 =$ **104.97 kHz**

30.25. FM Sidebands

In FM, when a carrier is modulated, a number of sidebands are formed.* Though theoretically their number is infinite, their strength becomes negligible after a few sidebands. They lie on both sides of the centre frequency spaced f_m apart as shown in Fig. 30.21. Sidebands at equal distances from f_0 have equal amplitudes. If f_0 is the centre frequency and f_m the frequency of the modulating signal, then FM carrier contains the following frequencies:

(*i*) f_0 (*ii*) $f_0 \pm f_m$

(*iii*) $f_0 \pm 2 f_m$ (*iv*) $f_0 \pm 3 f_m$ and so on.

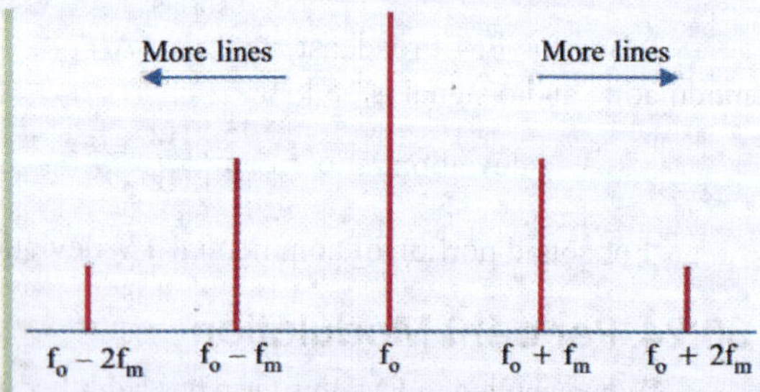

Fig. 30.21

The bandwidth occupied by the spectrum is $BW = 2\, n f_m$ where *n* is the highest order of the significant sideband.

Another approximate expression for spectrum bandwidth is

$$BW = 2\,(1 + m_f) f_m$$

Now, $m_f = \Delta f / f_m$, hence

$$BW = 2\,(\Delta f + f_m)$$

* An AM signal has only two side frequencies for each modulating frequency (Art. 30.11).

This expression is based on the assumption that sidebands having amplitudes less than 5% of the unmodulated carrier wave are negligble or when m_f is at least 6.

30.26. Modulation Index and Number of Sidebands

It is found that the number of sidebands

1. depends *directly* on the amplitude of the modulating signal,
2. depends *inversely* on the frequency of the modulating signal.

Since frequency deviation is directly related to the amplitude of the modulating signal, the above two factors can be combined in one factor called **modulation index.**

Hence, number of sidebands depends on $m_f = \Delta f / f_m$.

Obviously, the number of pairs of sidebands

(*i*) **increases** as frequency deviation (or amplitude of modulating signal) **increases,**

(*ii*) **increases** as the modulating signal frequency **decreases.**

Table 30.2

m_f	0.5	1	2	3	4	5	6	7
n	2	3	4	6	7	8	9	11
$BW = 2nf_m$	$4f_m$	$6f_m$	$8f_m$	$12f_m$	$14f_m$	$16f_m$	$18f_m$	$22f_m$
m_f	8	9	10	11	12	13	14	15
n	12	13	14	15	16	17	18	19
$BW = 2nf_m$	$24f_m$	$26f_m$	$28f_m$	$30f_m$	$32f_m$	$34f_m$	$36f_m$	$38f_m$

Table 30.2 summarizes the relation between the number (n) of sidebands on either side of the carrier and the corresponding bandwidth of the spectrum.

Example 30.13. *A 5 kHz audio signal is used to frequency-modulate a 100 MHz carrier causing a frequency deviation of 20 kHz. Determine*

(*i*) *modulation index and* (*ii*) *bandwidth of the FM signal.*

Solution. $$m_f = \frac{\Delta f}{f_m} = \frac{20}{5} = 4$$

As seen from Table 30.2 BW = $14f_m = 14 \times 5 =$ **70 kHz**

Note. We cannot use the alternate expression for BW given in Art. 30.25 above because $m_f < 6$.

Example 30.14. *In an FM circuit, the modulation index is 10 and the highest modulation frequency is 20 kHz. What is the approximate bandwidth of the resultant FM signal ?*

Solution. Since the value of m_f is more than 6, we will use the expression

$$\text{BW} = 2(\Delta f + f_m)$$

Now, $$m_f = \frac{\Delta f}{f_m} \quad \text{or} \quad 10 = \frac{\Delta f}{20} \quad \therefore \Delta f = 200 \text{ kHz}$$

$\therefore$ $$\text{BW} = 2(200 + 20) = \mathbf{440\ kHz}$$

30.27. Mathematical Expression for FM Wave

The unmodulated carrier is given by $e_c = A \sin 2\pi f_0 t$

The modulating signal frequency is given by $e_m = B \sin 2\pi f_m t$

The modulated carrier frequency f swings around the resting frequency f_0, thus

$$f = f_0 + \Delta f. \sin 2\pi f_m t$$

Hence, equation for the frequency-modulated wave becomes

$$e = A \sin 2\pi ft = A \sin [2\pi (f_0 + \Delta f. \sin 2\pi f_m t)t]$$

$$= A \sin \left(2\pi f_0 t + \frac{\Delta f}{f_m} \cos 2\pi f_m t\right) = A \sin (2\pi f_0 t + m_f \cos 2\pi f_m t)$$

30.28. Demodulation or Detection

When the RF modulated waves, radiated out from the transmitter antenna, after travelling through space, strike the receiving aerials, they induce very weak RF currents and voltages in them. If these high-frequency currents are passed through headphones or loudspeakers, they produce no effect on them because all such sound-producing devices are unable to respond to such high frequencies due to large inertia of their vibrating discs etc. Neither will such RF currents produce any effect on human ear because their frequencies are much beyond the audible frequencies (20 to 20,000 Hz approximately). Hence, it is necessary to demodulate them first in order that the sound-producing devices may be actuated by audio-frequency current similar to that used for modulating the carrier wave at the broadcasting station.

This process of *recovering AF signal from the modulated carrier wave is known as demodulation or detection.*

The demodulation of an AM wave involves two operations:

(*i*) rectification of the modulated wave, and

(*ii*) elimination of the RF component of the modulated wave.

However, demodulation of an FM wave involves three operations (*i*) conversion of frequency changes produced by modulating signal into corresponding amplitude changes, (*ii*) rectification of the modulating signal and (*iii*) elimination of RF component of the modulated wave.

30.29. Essentials of AM Detection

For recovering the AF waveform from modulated wave (a mixture of AF wave and RF carrier wave), it is essential to find some way of reducing (or better, eliminating) one half of the modulated wave. The result of this elimination (or rectification) would be that the average value of the wave would not be zero because, now, the impulse would be all in one direction as shown in Fig. 30.22. If this new wave is now passed through a headphone shunted by a suitable capacitor, then AF wave will pass through the headphone whereas the RF wave will be by-passed by the capacitor (because the high inductance of magnet coils of the headphones will offer tremendous impedance to RF currents). Hence, the two will become separated.

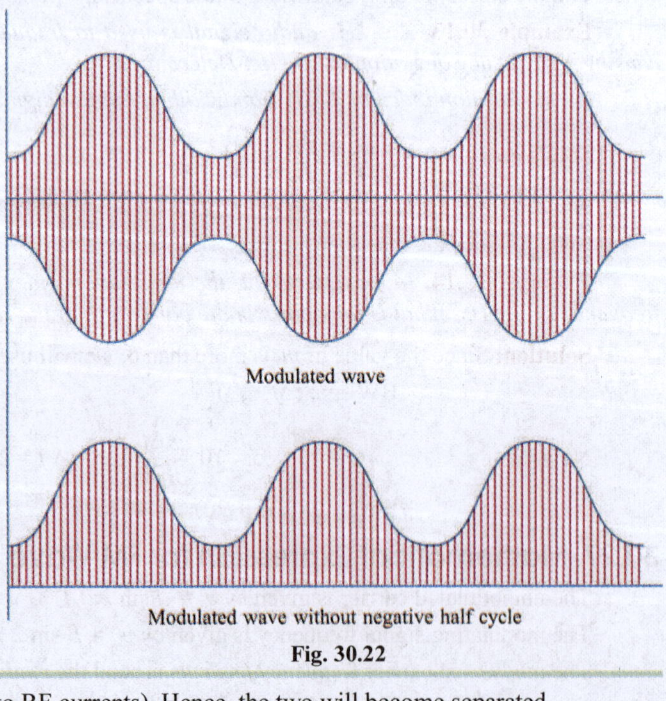

Fig. 30.22

30.30. Diode Detector for AM Signals

Diode detection is also known as **envelope-detection** or **linear detection.** In appearance, it looks like an ordinary half-wave rectifier circuit with capacitor input as shown in Fig. 30.23. It is called **envelope detection** because it *recovers the AF signal envelope* from the composite signal. Similarly, diode detector is called **linear detector** because its output is *proportional to the voltage of the input signal**.

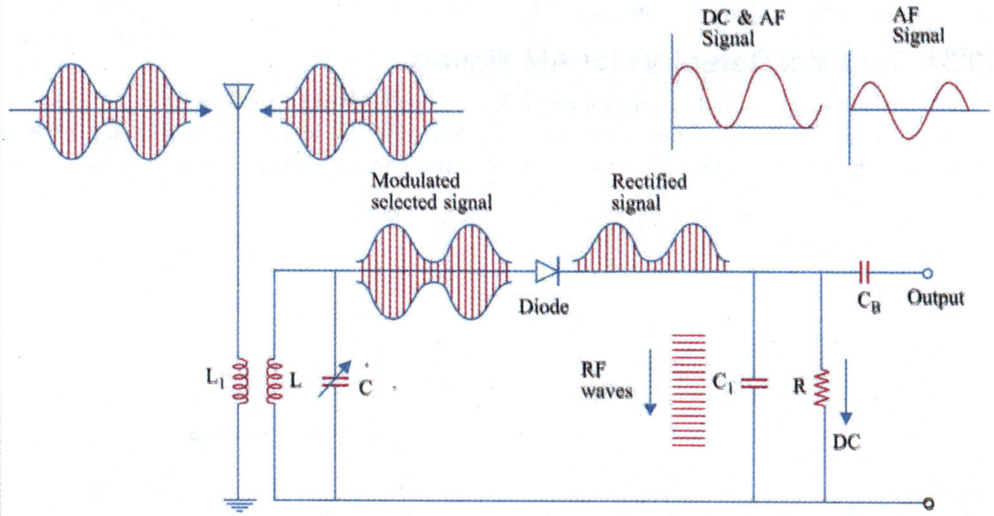

Fig. 30.23

Circuit Action

Of the various RF voltages induced in the receiver aerial, only those having the *same* frequency as the resonant frequency of *LC* circuit are tuned in due to electromagnetic induction between coils L_1 and L. By varying C, the resonant frequency of the LC circuit can be varied and hence RF signal of any desired frequency can be tuned in. This input signal is rectified by the diode and passed on to the low-pass filter RC_1.

The RF carrier wave is filtered out by low-reactance capacitor C_1 which is called RF filter capacitor or RF by-pass capacitor. The dc component of the remaining signal is shunted out through R because it cannot pass through blocking capacitor C_B. But the low-frequency AF signal can easily get through C_B and becomes available across the output. When passed through a suitable device, say, a headphone, the original sound can be heard.

Advantages

Diode detectors are extensively used in AM broadcast receivers because they have the following advantages:

1. they can handle comparatively large input signals,
2. they can be operated as linear or power detectors,
3. they rectify with negligible distortion and, hence, have good linearity,
4. they are well-adopted for use in simple automatic-gain control circuits.

* A detector circuit in which the rectified output is proportional to the square of the input signal voltage is called square-law or non-linear detector.

Disadvantages

However, the disadvantages are that

1. they do not have the ability to amplify the rectified signal by themselves as is done by a transistor detector (Art. 30.31). However, it is not a very serious drawback since signal amplification can be affected both before and after rectification;
2. while conducting, the diode consumes some power which reduces the Q of its tuned circuit as well as its gain and selectivity.

30.31. Transistor Detectors for AM Signals

Transistors can be used as detector amplifiers *i.e.,* both for rectification and amplification. As shown in Fig. 30.24, the RF signal is applied at the base-emitter junction where rectification takes place. The amplification of the recovered signal takes place in the emitter-collector circuit.

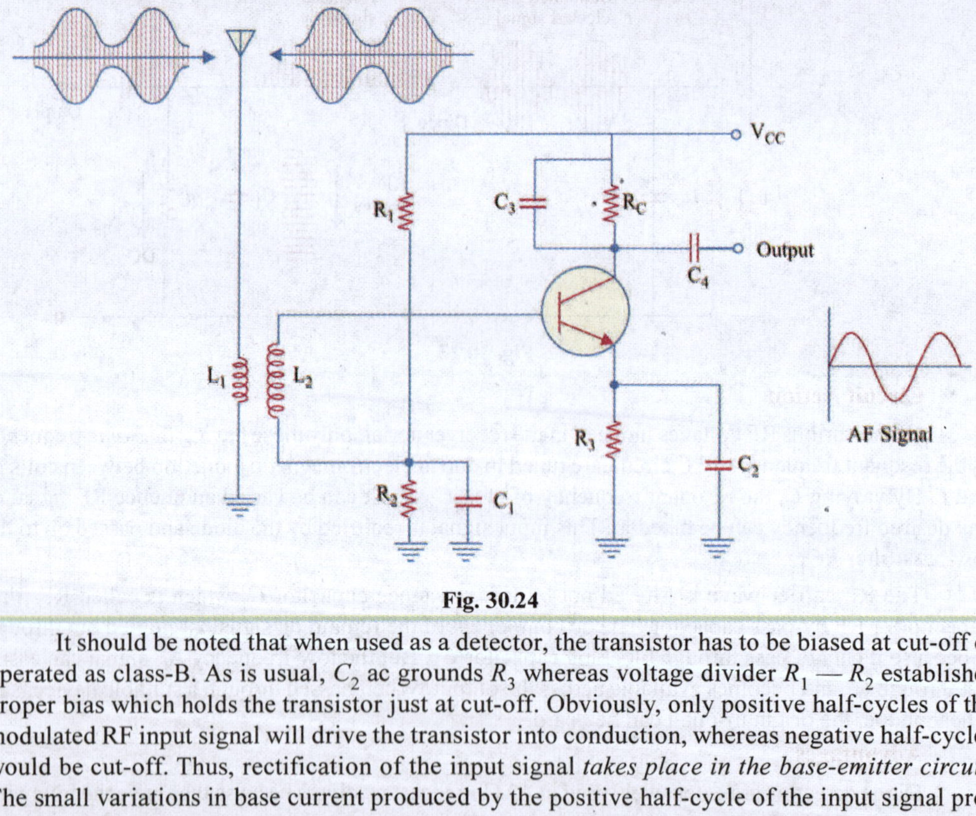

Fig. 30.24

It should be noted that when used as a detector, the transistor has to be biased at cut-off or operated as class-B. As is usual, C_2 ac grounds R_3 whereas voltage divider $R_1 — R_2$ establishes proper bias which holds the transistor just at cut-off. Obviously, only positive half-cycles of the modulated RF input signal will drive the transistor into conduction, whereas negative half-cycles would be cut-off. Thus, rectification of the input signal *takes place in the base-emitter circuit.* The small variations in base current produced by the positive half-cycle of the input signal produce large (β time) variations in the collector current. Hence, an amplified version of the rectified signal appears across $R_C — C_3$ combination. C_3 eliminates RF component so that only AF signal voltage drops across R_C. The capacitor C_4 permits AF signal to pass through but blocks its dc component.

30.32. FM Detection

As discussed earlier, an FM carrier signal contains information (or intelligence we wish to convey) in the form of frequency variations above and below the centre frequency of the carrier. For recovering the information, we must first convert the FM signal in such a way that it appears as a modulated RF voltage across the diode. A simple method of converting frequency variations into voltage variations is to make use of the principle that reactance (of coil or capacitor) varies with frequency. When an FM signal is applied to an inductor, the current flowing through it varies in amplitude according to the changes in frequency of the applied signal. Now, changes in frequency of the FM signal depend on the amplitude of the modulating AF signal. Hence, the current in the inductor varies as per the amplitude of the original modulating signal. In this way, frequency changes in FM signal are converted into amplitude changes in current. These changes in current when passed through a resistor produce corresponding changes in voltage.

Hence, we find that, ultimately, frequency variations in FM signal are converted into voltage changes. Also, there exists a linear relation between the two – something essential for distortion less demodulation.

FM demodulation may be carried out with the help of (*i*) ratio detector and (*ii*) quadrature detector.

30.33. Quadrature Detector

This detector depends on the frequency/phase relationship of a tuned circuit. It uses only one tuned circuit and is becoming increasingly popular in the integrated FM strips.

Theory

Let us first consider the general principle. A sinusoidal current is given by the equation

$$i = I_m \sin\theta = I_m \sin\omega t$$

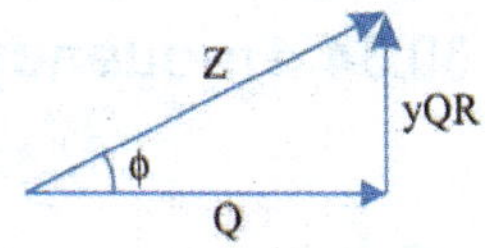

Fig. 30.25

Suppose, it flows through a circuit shown in Fig. 30.26 (*a*). The voltage v_L across the inductor (assumed pure) leads the current I by 90°

$$\therefore \quad v_L = V_L \cos\omega t$$

The voltage v_Z across the parallel tuned circuit will be in phase with I at resonance. However, at frequencies slightly different (± 1%) from the resonant frequency, the phase angle ϕ will be given by

$$\tan\phi = \frac{yQR}{R} = yQ \qquad \text{— Fig. 30.25}$$

assuming $f_0 / f \cong 1$.

Here $$y = \frac{\omega}{\omega_0} - \frac{\omega_0}{\omega} = \frac{f}{f_0} - \frac{f_0}{f} \cong \frac{2\,\Delta_f}{f_0}$$

where $\Delta f = f - f_0$

f_0 = resonant frequency

f = slightly off-resonance frequency

Hence, equation for v_Z is given by $v_Z = V_Z \sin(\omega t - \phi)$.

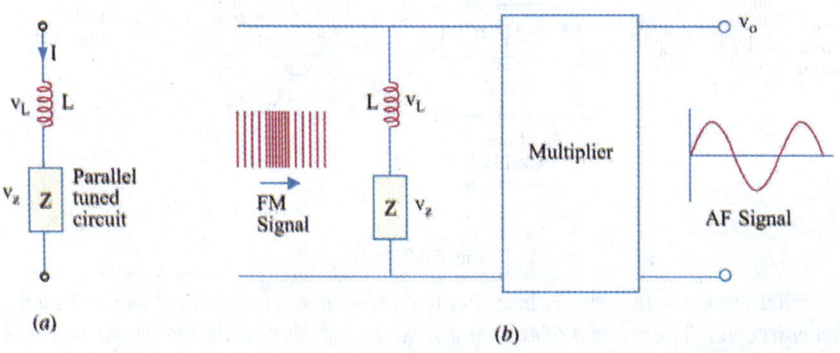

Fig. 30.26

When the two voltages v_L and v_Z are applied as inputs to a multiplier, the output voltage v_0 is found tobe proportional to their product as shown in Fig. 30.26 (*b*).

$$v_0 \propto v_L . v_Z \propto \cos \omega t . \sin (\omega t - \phi)$$
$$\propto \sin (2\,\omega t + \phi) + \sin \phi$$

A low-pass filter is used to reject the double frequency component sin (2 ωt + φ) and select only the low-frequency sin φ component.

$$\therefore \quad v_0 \propto \sin \phi$$
$$\propto \tan \phi \qquad \text{— since } \phi \text{ is very small}$$
$$\propto y\phi \propto y$$
$$\propto \frac{2\,.\,\Delta f}{f_0}$$

Now, $\quad \Delta f = f - f_0 \propto k\, v_m$

where v_m is the modulating AF voltage.

$$\therefore \quad v_0 \propto v_m$$

It shows that output voltage v_0 is proportional to the original modulating signal voltage.

30.34. Frequency Conversion

Let us first consider its necessity and basic principle of heterodyning action.

1. Need

It is very difficult to design amplifiers which give uniformly high gain over a wide range of radio frequencies used in commercial broadcast stations. However, it is possible to design amplifiers which can provide high-gain uniform amplification over a narrow band of comparatively lower frequencies called intermediate frequencies (IF). Hence, it is necessary to convert the modulated RF carrier into modulated IF carrier by using a frequency converter.

This IF signal is then amplified by narrow-band IF amplifiers and passed on to the AM detectors.

2. Basic Principle

The frequency conversion can be achieved by utilizing the heterodyne principle. For this purpose, the modulated RF signal is mixed (in a mixer) with an unmodulated RF signal produced by local oscillator as shown in Fig. 30.27.

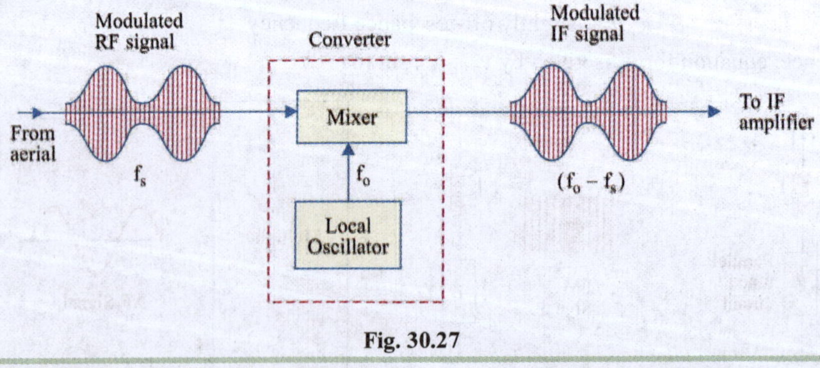

Fig. 30.27

The oscillator and the mixer may be either two separate devices or may be combined into one device called **converter.** The process of combining two ac signals of different frequencies in order to obtain a signal of new frequency is called **heterodyning action.**

3. Heterodyning Action

Suppose the carrier signal of frequency f_s is heterodyned with another signal of frequency f_0, then two additional signals are produced whose frequencies are:

(*i*) $f_0 + f_s$ —the sum component

(*ii*) $f_0 - f_s$ —the difference component

Usually, the sum frequency is removed by bandpass filtering. The difference frequency (also called *beat frequency*) is retained and forms the IF frequency in AM receivers.

Suppose, an AM carrier signal of frequency 1500 kHz is mixed with an unmodulated carrier (produced by local oscillator) of frequency 1955 kHz. Then, following two frequencies are produced

(*i*) 3455 kHz and (*ii*) 455 kHz

The **higher frequency** is generally filtered out leaving behind the difference frequency of 455 kHz which forms the IF frequency.

30.35. Superheterodyne AM Receiver

It is also referred to as superhet and is extensively used in modern AM receivers. Its block diagram with signals is shown in Fig. 30.29.

In the circuit of Fig. 30.27, the broadcasting station of a particular carrier frequency is tuned in by a suitable tuning circuit. Then, the incoming modulated signal is mixed with an unmodulated signal produced by the local oscillator (also called beat frequency oscillator, BFO). The *two heterodyning signals produce an IF signal.*

The frequency of the local oscillator is so varied that it always produces a constant frequency difference of 455 kHz which is passed on to the fix-tuned IF amplifiers. To ensure this, the variable tuning capacitor and the oscillator tuning capacitor are ganged together. In fact, both are turned by the same knob. They both track together and maintain the correct IF at 455 kHz. The oscillator frequency is **always higher than the frequency of the incoming signal.**

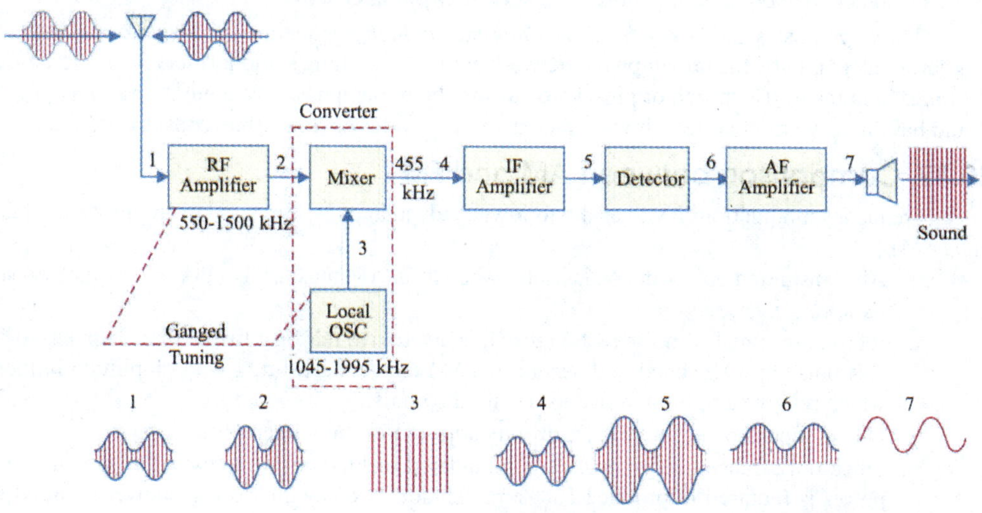

Fig. 30.28

The output of the IF amplifier is demodulated and then amplified by AF amplifiers for being applied to a loudspeaker which reproduces the original sound recorded and transmitted from the recording station.

30.36. FM Receiver

The block diagram for such a receiver is shown in Fig. 30.29. These receivers operate in the VHF band of 88-108 MHz and have an IF of 10.7 MHz with a bandpass of 2 × 75 kHz = 150 kHz.

Like AM receivers, such receivers also have a frequency converter and IF amplifiers.

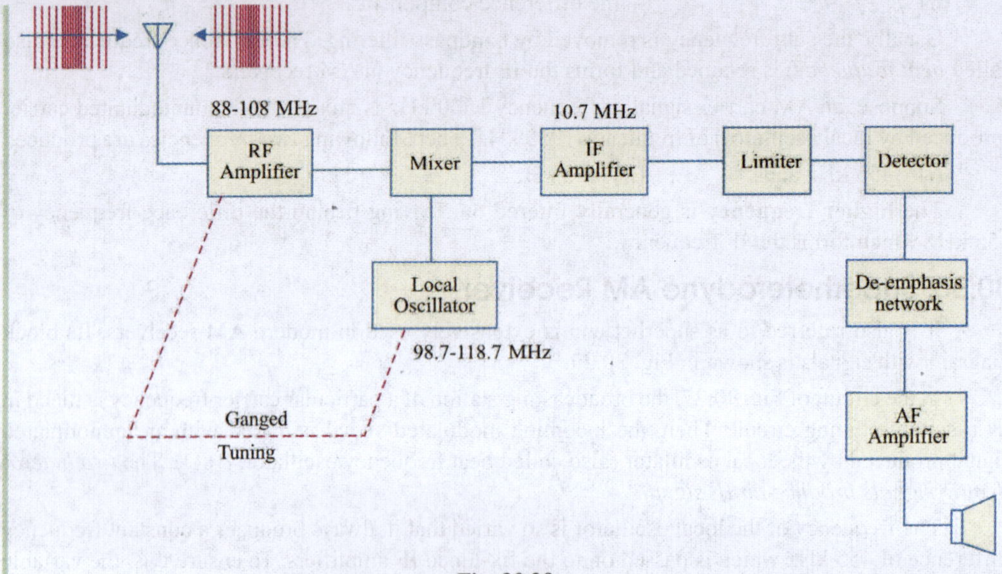

Fig. 30.29

The function of the **limiter** is to remove all *amplitude variations* (caused by noise) from IF signal which might have crept into the FM signal. This removal of amplitude variations is necessary for distortionless demodulation. **Limiter is a sort of clipping circuit.**

The de-emphasis network *reduces the amplitude of high frequencies in the audio signal* which was earlier increased by the pre-emphasis network at the transmitting station. It serves to re-establish the **tonal balance** of the speech or music etc., lost in the pre-emphasis. Without it, the sound signal would have a heavy treble effect. It is, in fact, a low-pass filter having a time constant of 75 μs.

30.37. Comparison between AM and FM

Frequency modulation (FM) has the following advantages as compared to amplitude modulation (AM):

1. All transmitted power in FM is useful whereas in AM most of it is in carrier which *serves no useful purpose.*
2. It has high signal-to-noise (S/N) ratio. It is due to two reasons: firstly, there happens to be less noise at VHF band and secondly, FM receivers are fitted with amplitude limiters which remove amplitude variations caused by noise.
3. Due to 'guard-band' there is hardly any adjacent-channel interference.
4. Since only transmitter *frequency* is modulated in FM, only fraction of a watt of audio power is required to produce 100% modulation as compared to high power required in AM.

However, FM has the following disadvantages:

1. It requires **much wider channel**—almost 7 to 15 times as large as needed by AM.
2. It requires complex and expensive transmitting and receiving equipment.
3. Since AM reception is limited to only line of sight, area of reception for FM is much smaller than for AM.

30.38. The Four Fields of FM

There are four major areas of application for FM transmission:

1. First use is in FM broadcast band 88-108 MHz with 200 kHz channels in which commercial FM stations broadcast programmes to their listeners.
2. Second use is in TV. Though video signal is amplitude-modulated, sound is transmitted by a separate transmitter which is frequency-modulated.
3. Third use is in the mobile or emergency services which transmit voice frequencies (20-4000 Hz) only.
4. Fourth use is in the amateur bands where again only voice frequencies are transmited.

CONVENTIONAL PROBLEMS

1. A carrier wave has an amplitude of 500 mV. A modulating signal causes its amplitude to vary from 200 mV top 800 mV. What is the percentage modulation? **[60%]**

2. A 100 MHz carrier having an amplitude of 50 V is amplitude-modulated by a 5 kHz audio signal having an amplitude of 20 V. Find

(*i*) modulation index, (*ii*) per cent modulation,
(*iii*) components of modulated wave, (*iv*) the amplitudes of sidebands.

[(*i*) 0.4 (*ii*) 40 (*iii*) 100 MHz, 100.005 MHz, 99.995 MHz (*iv*) 10 V]

3. An audio signal given by 30 sin ($2\pi \times 2500\, t$) is used for modulating a carrier wave given by the equation 60 sin ($2\pi \times 200{,}000\, t$). Find

(*i*) per cent modulation, (*ii*) frequencies of the signal and the carrier,
(*iii*) frequency spectrum of the modulated wave.

[(*i*) 50 (*ii*) 2500 Hz , 200,000 Hz (*iii*) 200; 202.5; 197.5 kHz)]

4. How many AM broadcast stations can be accommodated in a 5 MHz bandwidth if each station transmits a signal modulated by an audio signal having a maximum frequency of 5 kHz. **[500]**

5. A 100 kHz bandwidth is to accommodate 5 AM broadcasts simultaneously. What is the maximum modulating frequency permissible for each station? **[5 kHz]**

6. Total power content of an AM signal is 3000 W. For 100 per cent modulation, calculate

(*i*) carrier power (*ii*) power in each sideband

[(*i*) 2 kW (*ii*) 0.5 kW]

7. The total power content of an AM wave is 1320 W. What is the per cent modulation if each sideband contains 160 W ? **[0.8]**

8. An SSB-SC signal contains 2 kW. Find power contained in

(*i*) carrier, (*ii*) sideband. **[(*ii*) 0 (*iii*) 2 kW]**

9. As compared to DSB-FC. Calculate the power saving in SSB-SC when per cent modulation is

(*i*) 1 (*ii*) 0.8 (*iii*) 0.5

[(*i*) 83.3% (*ii*) 87.87% (*iii*) 94.4%]

10. What is the modulation index of an FM carrier having a carrier swing of 120 kHz and a modulating signal of 10 kHz. **[6]**

11. An FM signal has a resting frequency of 100 MHz and the highest frequency of 100.05 MHz when modulated by an audio signal of 5 kHz. Determine

(*i*) frequency deviation, (*ii*) carrier swing,
(*iii*) modulation index, (*iv*) per cent modulation.

[(*i*) 50kHz (*ii*) 100 kHz (*iii*) 10 (*iv*) 66.7%]

12. A 10 kHz audio signal is used to frequency-modulate a 100 MHz carrier causing a frequency deviation of 50 kHz. Determine

(*i*) modulation index, (*ii*) bandwidth of FM signal. **[(*i*) 5 (*ii*) 160 kHz]**

SELF EXAMINATION QUESTIONS

A. Fill in the blanks with most appropriate word (s) or numerical value (s).

1. Modulation is the process of .... low-frequency signal with high-frequency .... wave.
2. Demodulation is the process of .... the signal from the..............modulated carrier wave.
3. Intensity of a sound wave is directly proportional to the square of its ....
4. Amplitude modulation produces .... additional frequencies.
5. In amplitude modulation, carrier .... is changed but not its frequency.
6. Per cent modulation of an unmodulated carrier wave is ....
7. Modulation in excess of .... per cent produces distortion.
8. In AM transmission, each sideband contains .... the intelligence that is being transmitted.
9. In SSB-SC transmission, bandwidth required is .... of that required for normal AM transmission.
10. Bandwidth of a normal AM signal is twice the .... modulating frequency.
11. In an AM transmission with m = 100%, suppression of carrier cuts the power dissipation, by a factor of ....
12. In an AM transmission with m = 100%, suppression of carrier and one sideband cuts the power dissipation by a factor ....
13. Lower the per cent modulation, ... the power saving in SSB-SC system.
14. Maximum undistorted power from an AM transmitter is obtained with .... per cent modulation.
15. In AM transmission when modulation is less thatn 100%, power output is ....
16. In an AM transmission, all transmitted intelligence is in ....
17. In FM, louder the sound .... the deviation of carrier frequency from its centre frequency.
18. In FM, carrier .... remains constant.
19. In FM, carrier swing is twice the frequency ....
20. The modulation index of an FM transmission can never .... its deviation ratio.
21. Deviation ratio of standard FM broadcast stations is ....
22. Theoretically speaking, the number of sidebands in an FM transmission is ....
23. In FM, greater the frequency deviation, ... the number of sidebands.
24. An FM broadcast station has much .... bandwidth than an AM Station.

B. Answer True or False

1. In AM, amplitude of the carrier varies while its frequency remains constant.
2. Modulation is essential for communication at long distances.
3. An AM wave consists of only two frequency components.
4. Modulation index of an AM wave may vary between 0 and 1.
5. In an AM wave, variations in amplitude represent the intelligence to be transmitted.
6. A carrier is modulated 100% if its positive peak rises to a value twice its own amplitude.
7. Over-modulation gives rise to distorted and harsh-sounding transmitted signal.
8. Modulation index of an AM wave may vary from +1 to −1.
9. In AM, all transmitted intelligence is in sidebands and none in the carrier.
10. In AM, number of sidebands depends on the per cent modulation.
11. In AM, carrier contains most of the power but none of the intelligence.
12. As modulation increases, power in a sideband increases.
13. SSB-SC transmisison dissipates the least power.
14. In AM transmission, bandwidth is determined by carrier frequency.
15. In AM system, we can transmit all intelligence by transmitting only one sideband.
16. In an AM transmission, maximum undistorted power is transmitted when m = 1.
17. In an AM system for m = 1, each sideband contains one-sixth of the total transmitted power.
18. In an AM system, carrier suppression saves large power.
19. FM transmission is insensitive to amplitude variations.
20. In FM, louder the sound, greater the frequency deviation.

21. In FM broadcasts, 75 kHz frequency deviation corresponds to 100% modulation.
22. In FM, modulating index can be more than one.
23. FM needs much less audio power for 100 per cent modulation than AM.
24. The number of sidebands in an FM depends only on the frequency of the modulating signal.
25. FM has much higher bandwidth than AM.
26. In FM, sidebands are symmertrically located around the carrier.
27. All transmitted power in FM is useful.
28. In FM, the different high-order sidebands have progressively decreasing amplitudes.

C. Multiple Choice Items

1. The main purpose of modulation is to
   (a) combine two waves of different fre quencies
   (*b*) achieve waveshaping of the carrier wave
   (c) transmit low-frequency information over long distances efficiently
   (*d*) produce sidebands
2. Demodulation
   (*a*) is performed at the transmitting station
   (*b*) removes sidebands
   (*c*) rectifies modulated signal
   (*d*) is opposite of modulation
3. In amplitude modulation
   (*a*) carrier frequency is changed
   (*b*) carrier amplitude is changed
   (*c*) three sidebands are produced
   (*d*) fidelity is improved
4. 100% modulation is produced in AM when carrier
   (*a*) frequency equals signal frequency
   (*b*) frequency exceeds signal frequency
   (*c*) amplitude equals signal amplitude
   (*d*) amplitude exceeds signal amplitude
5. For a given carrier wave, maximum undistorted power is transmitted when value of modulation is
   (*a*) 1 (*b*) 0.8
   (*c*) 0.5 (*d*) 0
6. In an AM wave with 100 per cent modulation, each sideband carries...........of the total transmitted power.
   (*a*) one-half (*b*) one-sixth
   (*c*) one-third (*d*) two-third
7. Given a carrier frequency of 100 kHz and a modulating frequency of 5 kHz, the bandwidth of AM transmission is .... kHz.
   (*a*) 5 (*b*) 200
   (*c*) 10 (*d*) 20
8. In AM, power content of the carrier is maximum when m equals
   (*a*) 0 (*b*) 1
   (*c*) 0.8 (*d*) 0.5
9. In AM transmission with $m = 1$, suppression of carrier cuts power dissipation by a factor of
   (*a*) 6 (*b*) 2
   (*c*) 3 (*d*) 4
10. As compared to DSB-FC 100% modulated transmission, power saving in SSB-SC system is .... per cent.
   (*a*) 94.4 (*b*) 50
   (*c*) 100 (*d*) 83.3
11. In an AM transmission with 100% modulation, 66.7% of power is saved when.......is/are suppressed.
   (*a*) carrier (*b*) carrier and USB
   (*c*) carrier and LSB (*d*) USB and LSB
12. In an AM system, full information can be conveyed by transmitting only
   (*a*) the carrier
   (*b*) the upper sideband
   (*c*) the lower sideband
   (*d*) any one sideband
13. In FM, amplitude of the modulating signal determines
   (*a*) rate of frequency variations
   (*b*) amplitude of frequency shift
   (*c*) tonal balance of transmission
   (*d*) distance of broadcast
14. In FM, when frequency deviation is doubled
   (*a*) modulation is doubled
   (*b*) modulation is halved
   (*c*) carrier swing is halved
   (*d*) modulation index is decreased
15. In an FM broadcast in VHF band, channel width is ....
   (*a*) 75 (b) 25
   (*c*) 88 (*d*) 200
16. Modern FM receivers use de-emphasis circuit for
   (*a*) reducing high frequency noise

(b) reducing the amplitude of high frequencies in the audio signal
(c) increasing the amplitude of higher modulating frequencies
(d) making demodulation easy

17. One of the serious disadvantages of FM transmission is its
(a) high static noise
(b) limited line-of-sight range
(c) expensive equipment
(d) adjacent channel interference

ANSWERS

A. Fill in the blanks

1. combining, carrier	2. recovering	3. amplitude	4. two
5. amplitude	6. zero	7. 100	8. all
9. half	10. highest	11. 3	12. 6
13. greater	14. 100	15. reduced	16. sidebands
17. greater	18. amplitude	19. deviation	20. exceed
21. 5	22. infinite	23. greater	24. larger

B. True or False

1. T	2. T	3. F	4. T	5. T	6. T
7. T	8. F	9. T	10. F	11. T	12. T
13. T	14. F	15. T	16. T	17. T	18. T
19. T	20. T	21. T	22. T	23. T	24. F
25. T	26. T	27. T	28. F		

C. Multiple Choice Items

1. *c*	2. *d*	3. *d*	4. *c*	5. *a*	6. *a*
7. *c*	8. *a*	9. *c*	10. *d*	11. *a*	12. *d*
13. *b*	14. *a*	15. *d*	16. *b*	17. *b*	

31

CHAPTER

Integrated Circuits

1. What is an Integrated Circuit ?
2. Scale of Integration
3. Classification of ICs by Structure
4. Comparison between Different ICs
5. Classification of ICs by Function
6. Linear Integrated Circuits (LICs)
7. Digital Integrated Circuits
8. IC Terminology
9. How Monolithic ICs are Made ?
10. Fabrication of IC Components
11. Complete Monolithic Integrated Circuits
12. Popular Applications of ICs
13. MOS Integrated Circuits

31.1. Introduction

Electronic circuitry has undergone tremendous changes since the invention of a triode by Lee De Forest in 1907. In those days, the active components (like triode) and passive components (like resistors, inductors and capacitors etc.) of the circuits were **separate and distinct units** connected by soldered leads.

With the invention of the transistor in 1948 by W.H. Brattain and I. Bardeen, the electronic circuits became considerably reduced in size. It was due to the fact that a transistor was not only cheaper, more reliable and less power consuming but was also much smaller in size than an electron tube. To take advantage of small transistor size, the passive components too were greatly reduced in size thereby making the *entire circuit very small.* Development of printed circuit boards (PCBs) further reduced the size of electronic equipment by eliminating bulky wiring and tie points.

In the early 1960s, a new field of *microelectronics* was born primarily to meet the requirements of the Military which wanted to reduce the size of its electronic equipment to approximately one-tenth of its then existing volume. This drive for extreme reduction in the size of electronic circuits has led to the development of microelectronic circuits called **integrated circuits (ICs)** which are so small that their actual construction is done by technicians using microscopes.

31.2. What is an Integrated Circuit ?

To put it very briefly, an integrated circuit (IC) is **just a packaged electronic circuit.**

A more detailed definition is as under :

An IC is a complete electronic circuit in which both the active and passive components are fabricated on an extremely tiny single chip of silicon.

Active Components are those which have the ability to produce gain. Examples are: transistors and FETs.

Passive Components or devices are those which do not have this ability. Examples are: resistors, capacitors and inductors.

ICs are produced by the same processes as are used for manufacturing individual transistors and diodes etc. In such circuits, different components are isolated from each other by isolation diffusion within the crystal chip and are interconnected by an aluminium layer that serves as wires.

Integrated circuits.

A discrete circuit, on the other hand, is one *that is built by connecting separate components.* In this case, each component is produced separately and then all are assembled together to make the electronic circuit.

J.S. Kilby was the first to develop (in 1958) an integrated circuit—a single monolithic silicon chip in which active and passive circuit elements were fabricated by successive diffusions and depositions. He was soon followed by Robert Noyce who successfully fabricated a complete IC including the interconnections on a single silicon.

31.3. Advantages of ICs

As compared to **standard printed circuits** which use discrete components, ICs have the following advantages:

1. Extremely Small Physical Size

Often the size is thousands of times smaller than a discrete circuit. The various components and their interconnections are distinguishable only under a powerful microscope.

2. Very Small Weight

Since manufacturers can pack much more circuitry into an IC package, it drastically cuts down the weight of the circuit.

Weight and size are of great importance in military and space applications, though in consumer applications they are always the primary consideration.

3. Reduced Cost

It is a major advantage of ICs. The reduction in cost per unit is due to the fact that all circuit components are fabricated **in** or **on** the wafer at the same time and also because *hundreds of similar wafers are produced simultaneously.* Due to mass production, **an IC costs as much as an individual transistor.**

4. Extremely High Reliability

It is perhaps the **most important** advantage of an IC and is due to many factors. Most significant factor is the absence of soldered connections. Another is the need for fewer interconnections—the major cause of circuit failures. Small temperature rise due to low power consumptions of ICs also improves their reliability. In fact, an IC logic gate has been found to be 100,000 times more reliable than a vacuum tube logic gate and 100 times more reliable than a transistor logic gate.

Obviously, higher reliability means that ICs will work for longer periods without giving any trouble—something most desirable from both military and consumer application standpoints.

5. Suitability for Small-signal Operation

Since various components of an IC are located very close to each other **in** or **on** a silicon wafer, their chance of stray electrical pickup is practically nil. Hence, it makes them very suitable for small signal operation.

6. Low Power Consumption

Because of their small size, ICs are more suitable for low power operation than bulky discrete circuits.

7. Easy Replacement

ICs are hardly ever repaired because in case of failure, it is more economical to replace them than to repair them.

31.4. Drawbacks of ICs

The integrated circuits suffer from the following drawbacks:

1. coils or inductors cannot be fabricated,
2. ICs function at fairly low voltages,
3. they can handle only limited amount of power,
4. they are quite delicate and cannot withstand rough handling or excessive heat.

However, the advantages of ICs far outweigh their disadvantages or drawbacks.

31.5. Scale of Integration

Level of integration in ICs has been increasing ever since they were developed some two and a half decades back. The number of electronic circuits or components that can be fitted into a standard size IC has been dramatically increasing with each passing year. In fact, whole **electronic systems** rather than just **a circuit** are incorporated in one package.

An approximate method of classifying the amount of circuit or component density is as follows:

1. SSI—*small scale integration*

In this case, the number or **circuits** contained in one IC package is less than 30 (or number of **components** is less than 50).

2. MSI—*medium scale integration*

Here, number of circuits per package is between 30 and 100 (or number of components is between 50 and 500).

3. LSI—*large scale integration*

In this case, circuit density is between 100 and 100,000 (or component density is between 500 and 300,000).

4. VLSI—*very large scale integration*

VLSI is a major **electronic system** constructed on a single wafer of silicon containing circuits in excess of 100,000.

In summary,

$$\left.\begin{array}{ll} \text{SSI} < & 30 \\ \text{MSI} & 30-100 \\ \text{LSI} & 100-100{,}000 \\ \text{VLSI} > & 100{,}000 \end{array}\right] \quad \text{circuits per chip}$$

31.6. Classification of ICs by Structure

Structurally speaking, ICs can classified into the following three types :

1. Monolithic Integrated Circuits

The word 'monolithic' means 'single stone' or more appropriately 'a single-solid structure'. In this IC, all circuit components (both active and passive) are fabricated inseparably within a single continuous piece of silicon crystalline material called **wafer** (or **substrate**). All components are **atomically** part of the same chip. Transistors and diodes and other passive components are fabricated at appropriate spots in the substrate using epitaxial diffusion techniques.

Component interconnections are provided on the surface of the structure and external connecting wires are taken out to the terminals. It is a complete circuit requiring no 'add ons'.

Despite some of its distinct disadvantages, monolithic ICs are in wide use because for mass production, monolithic process has been found to be the most economical.

2. Thick and Thin Film ICs

The essential difference between thick-film and thin-film ICs *is not their relative thickness but* the **method** of depositing the film. Both have similar appearance, properties and general characteristics though they both differ in many respects from monolithic ICs. These ICs are not formed **within** a silicon wafer but **on** the surface of an insulating substrate such as glass or a ceramic material. Moreover, **only passive components** (resistors, capacitors) are formed through thick or thin-film techniques on the insulating surface. The active elements (transistors, diodes) are added externally as **discrete elements** to complete a functional circuit. These discrete active components are frequently produced by using the monolithic process.

As stated above, the primary difference between the thick and thin-film techniques is the process used for forming passive components and the metallic conduction pattern.

(*a*) *Thin-film ICs*

Such circuits are constructed by depositing films of conducting material through a mask on the surface of a substrate made of glass or ceramic. Resistors and conductors are formed by varying the width and thickness of the film and by using materials of different resistivity. Capacitors are produced by sandwiching an insulating oxide film between two conducting films. Small inductors can be made by depositing a spiral formation of film. The active components like transistors and diodes etc., are externally added and inter-connected by wire bonds.

Following two methods are used to produce thin-films :

(*i*) vacuum evaporation

In this method, the vaporised material is deposited through a set of masks on the glass or ceramic substrate contained in vacuum.

(*ii*) cathode sputtering

In this method, atoms from a cathode made of the desired film material are deposited on the substrate which is located between the cathode and the anode.

(*b*) *Thick-film ICs*

Such type of integrated circuits are sometimes referred to as **printed** thin-film circuits. They are so called because silk-screen printing techniques are employed to create the desired circuit pattern on the surface of the substrate. The screens are made of fine stainless steel wire mesh and the `inks' are pastes (of pulverised glass and aluminium) which have conductive, resistive or dielectric properties. After printing, the circuits are high-temperature fired in a furnace to fuse the films to the insulating substrate. As with thin-film ICs, active elements are added externally as discrete components.

Hybrid or Multichip ICs.

3. Hybrid or Multichip ICs

As the name implies, such circuits are formed either by inter-connecting a number of individual chips or by a combination of film and monolithic IC techniques. In such ICs, active components are first formed within a silicon wafer (using monolithic technique) which is subsequently covered with an insulating layer such as SiO_2. Film techniques are then employed to form passive components on the SiO_2 surface. Connections are made from the film to the monolithic structure through 'windows' cut in the SiO_2 layer.

31.7. Comparison between Different ICs

Each type of IC has its own advantages and disadvantages.

Monolithic circuits have *the advantage of lowest cost and highest reliability.* However, they have the following disadvantages :

1. isolation between components is poor,
2. range of values of passive components used in the circuits is comparatively small,
3. inductors cannot be fabricated,
4. they afford no flexibility in circuit design because for making any changes in the circuit, **a new set of masks is required.**

The film circuits have the advantage of forming passive components with broader range of values and reduced tolerances as compared to monolithic circuits. Isolation between their components is also better since they are deposited on a substrate that is an insulator. Use of external discrete active components allows greater flexibility in circuit design. These circuits also give better high frequency performance than monolithic circuits.

However, they suffer from the disadvantages of

1. not being able to fabricate active components,
2. comparatively higher cost, and
3. larger physical size.

The chief advantage of multichip ICs is their greater flexibility but they are **too expensive** for mass production and **have least reliability.** Hence, such circuits are generally used as prototypes for monolithic ICs.

The various integrated circuits are arranged in the form of an 'IC' tree of Fig. 31.1.

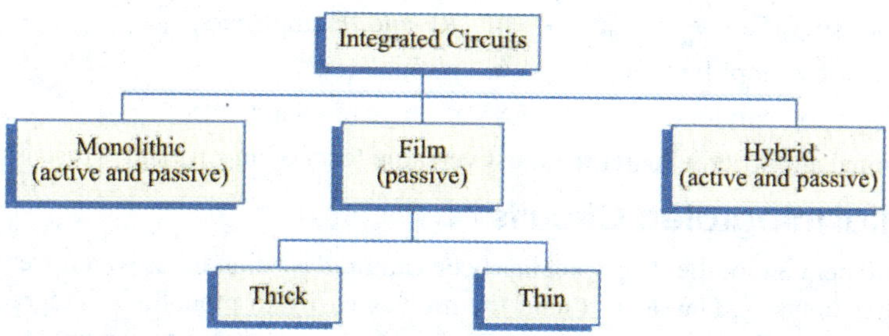

Fig. 31.1

31.8. Classification of ICs by Function

The earlier classification of ICs was based on *their method of construction.* However, the integrated circuits can also be classified according to their general function. The two most important categories are:

1. linear and
2. digital

The same fact has been shown in Fig. 31.2.

Examples of linear ICs are:

1. BEL CA-3020—used as multipurpose wide-band power amplifier.
2. BEL CA-3065—it is a monolithic IC which combines a multistage IF amplifier, limiter, an

FM detector, an electronic attenuator, a Zener diode regulated power supply and an audio amplifier. In fact, this IC provides a high performance multistage sub-system of a TV receiver. It is available in 14-pin dual-in-line package.

Both are manufactured by Bharat Electronics Ltd., Bangalore.

3. SSD 710—is a linear IC used as Differential Comparator.

It is manufactured by Solid State Devices, Syed Abdullah Road, Mumbai.

An example of a digital IC is

BEL 7400—is a TTL IC gate which provides designer with one of the gating logic necessary to design medium-speed digital control and data processing systems. It is available in 14-pin dual-in-line plastic package.

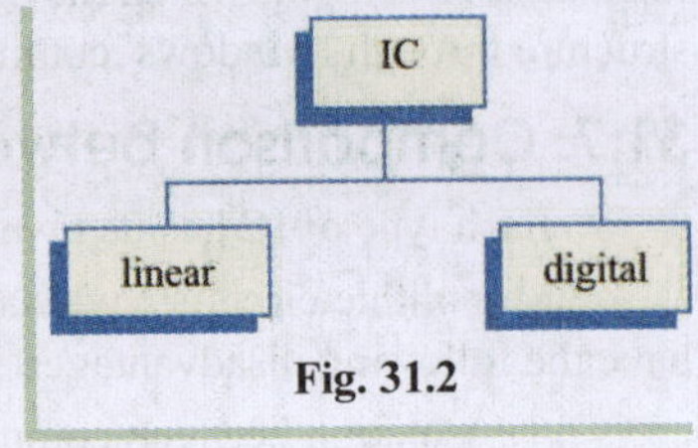

Fig. 31.2

31.9. Linear Integrated Circuits (LICs)

LICs are also referred to as analog ICs because their inputs and outputs can take on a continuous range of values and the outputs are generally proportional to the inputs. As compared to digital ICs, LICs are used much less. But LICs are quickly displacing their discrete circuit counterparts in many applications as their cost becomes competitive. They also possess much higher reliability because so many external connections (major source of circuit failure) are eliminated. LICs find wide use in military and industrial applications as well as in consumer products. They are frequently used in

Linear ICs.

1. operational amplifiers,
2. small-signal amplifiers,
3. power amplifiers,
4. RF and IF amplifiers,
5. microwave amplifiers,
6. multipliers,
7. voltage comparators,
8. voltage regulators etc.

Operational amplifier is by far the most versatile form of an LIC and is discussed separately.

31.10. Digital Integrated Circuits

About 80 per cent of the IC market has been captured by digital ICs which are mostly utilized by the computer industry. Digital ICs lend themselves easily to monolithic integration because a computer *uses a large number of identical circuits.* Moreover, such circuits employ relatively few capacitors and values of resistances, voltages and currents are low.

Digital ICs contain circuits whose input and output voltages are limited to **two** possible levels—low or high. It is so because *digital signals are usually binary.* Sometimes, digital circuits are referred to as switching circuits. Digital ICs include circuits such as

1. logic gates
2. flip-flops
3. counters
4. clock chips
5. calculator chips
6. memory chips
7. microprocessors (μP).

31.11. IC Terminology

Some of the common terms are defined below:

1. **Bonding**—attachment of wires to an IC.

2. **Chip**—an extremely small part of a silicon wafer on which IC is fabricated. One Si wafer of 2 cm diameter may contain upto 1000 IC chips (Fig. 31.3).

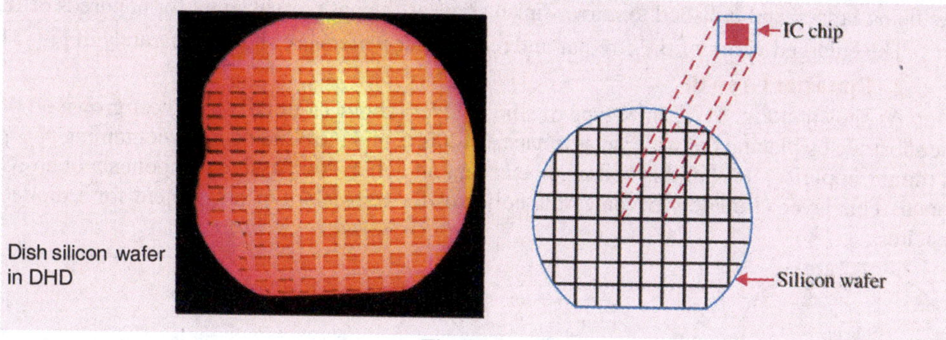

Fig. 31.3

3. **Circuit probing**—to check the proper electrical performance of each IC with the help of probes.
4. **Die**—Same as Chip
5. **Diffusion**—introduction of controlled small quantities of a material into the crystal structure for modifying its electrical characteristics.
6. **Diffusion mask**—it is a glass plate with the circuit pattern drawn on it. Impurities can diffuse through its light areas but not through its dark ones.
7. **Encapsulation**—putting a cap over the IC and sealing it in an inert atmosphere.
8. **Epitaxy**—physical placement of materials on a given surface.

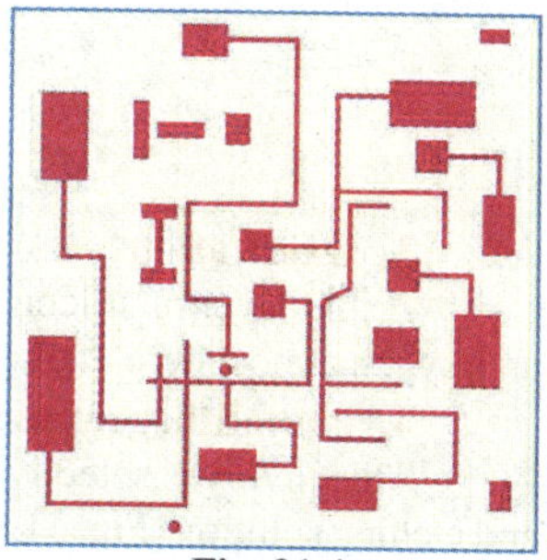

Fig. 31.4

9. **Etching**—removal of surface material from a chip by chemical means.
10. **Metallization**—providing ohmic contacts and inter-connections by evaporating aluminium over the chip (Fig. 31.4).
11. **Photoresist**—a photo-sensitive emulsion which hardens when exposed to ultraviolet light.
12. **Scribing**—incising or cutting with a sharp point.
13. **Wafer**—a thin slice of a semiconductor material either circular or rectangular in shape in which a number of ICs are fabricated simultaneously.

31.12. How Monolithic ICs are Made ?

The normal components of an IC are : transistors, diodes, resistors and capacitors. The transistor is the most complicated component and all other parts are fabricated in conjunction with the transistor processes. Following steps are involved in the fabrication of a monolithic IC by the expitaxial diffusion process.

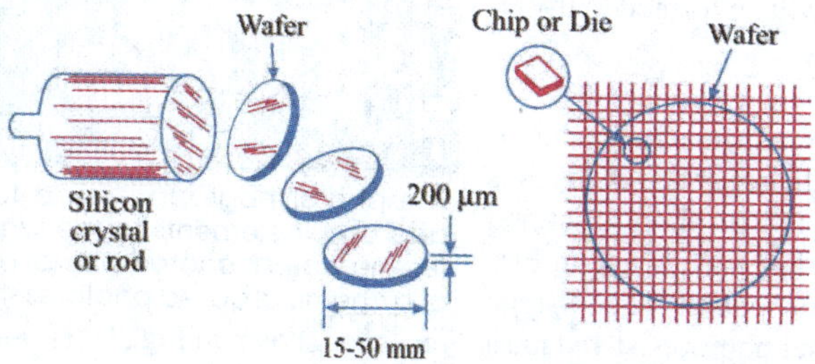

Fig. 31.5

Monolithic IC instrument.

1. Wafer Preparation

A P-type silicon bar is taken and cut into thin slices called wafers (Fig. 31.5). These wafers after being lapped and polished to mirror finish serve as the base or substrate for hundreds of ICs.

The enlarged views of the circular and rectangular wafers are shown separately in Fig. 31.6.

2. Epitaxial Growth

As shown in Fig. 31.7, an N-type of silicon layer (about 15 μm thick) is now grown on the P-type substrate by placing the wafer in a furnace at 1200°C and introducing a gas containing phosphorus (donor impurity). It is in this epitaxial layer that all active and passive components of an IC are formed. This layer ultimately becomes the collector for transistors or an element for a diode or a capacitor.

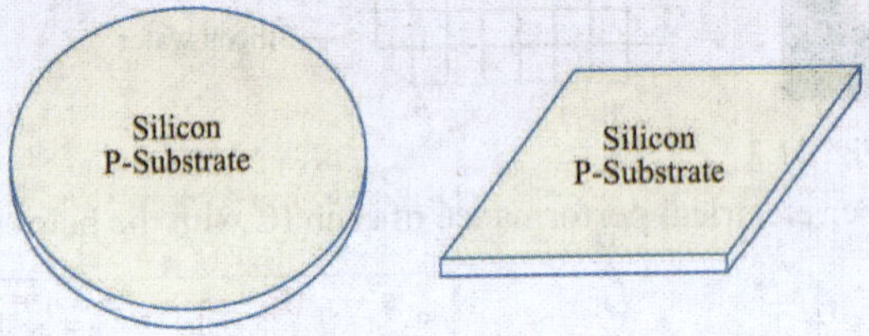

Fig. 31.6

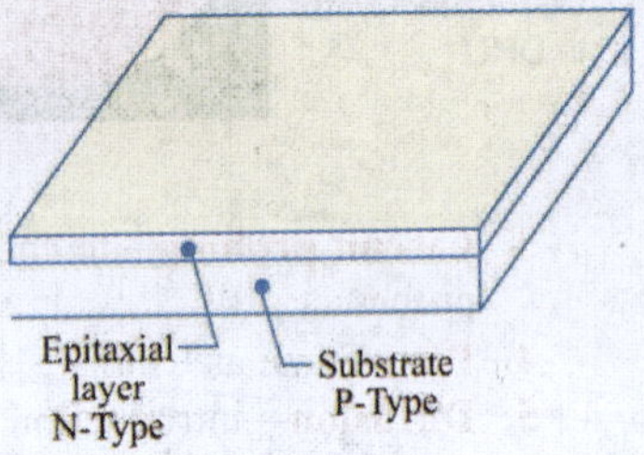

Fig. 31.7

3. Oxidisation

A thin layer of silicon dioxide (SiO_2) is grown over the N-type layer by exposing the wafer to an oxygen atmosphere at about 1000°C (Fig. 31.8).

4. Photolithographic Process

This involves selective etching of SiO_2 layer with the help of photographic mask, photoresist and etching solution [Fig. 31.9 (*a*)]. It helps to select particular areas of the N-layer which are subjected to an isolation diffusion process.

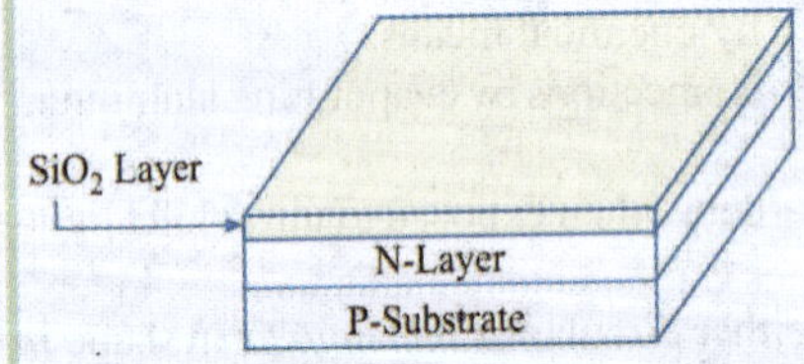

Fig. 31.8

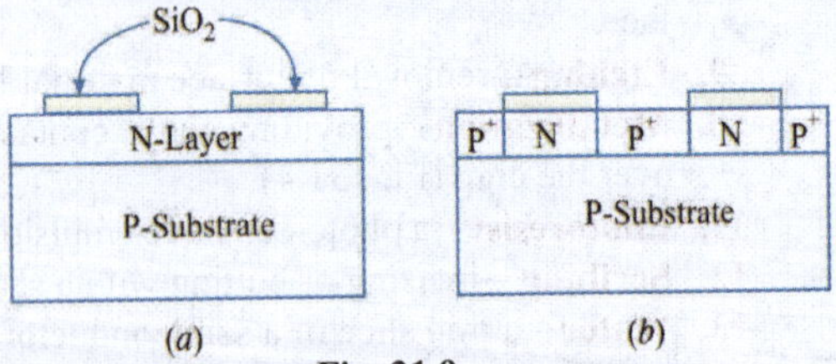

Fig. 31.9

5. Isolation Diffusion

The wafer is next subjected to a P-type diffusion process by which N-type layer is isolated into islands on which transistor or some other component is fabricated [Fig. 31.9 (*b*)]. The heavily-doped regions marked P^+ result in improved isolation between the active and passive components which will be formed in the N-type islands of the epitaxial layer.

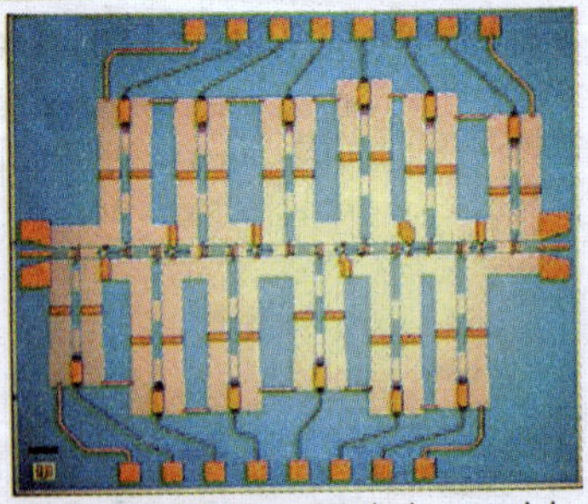

The photolithograph is used to add circuit elements to the wafer. The wafer is coated with layer of a chemical called photoresist.

6. Base and Emitter Diffusion

The P-type base of transistor is now diffused into the N-type layer which itself acts as collector. The sequence of steps is the same as in 4 and 5 above *i.e.*, by the use of photoresist and mask which creates windows in the SiO_2 layer as shown in Fig. 31.10. It is carried out in a phosphorus atmosphere. Next, N-type emitter is diffused into the base after another photoresist and mask process as shown in Fig. 31.11. Of course, no further N-type diffusion is required for the resistor which uses the resistivity of the P-type material itself for the purpose.

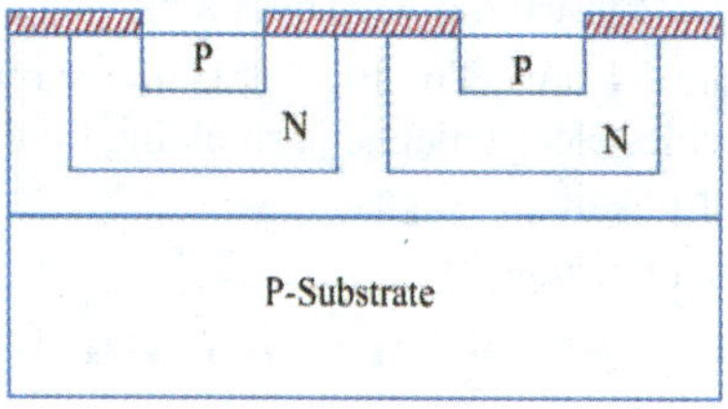

Fig. 31.10

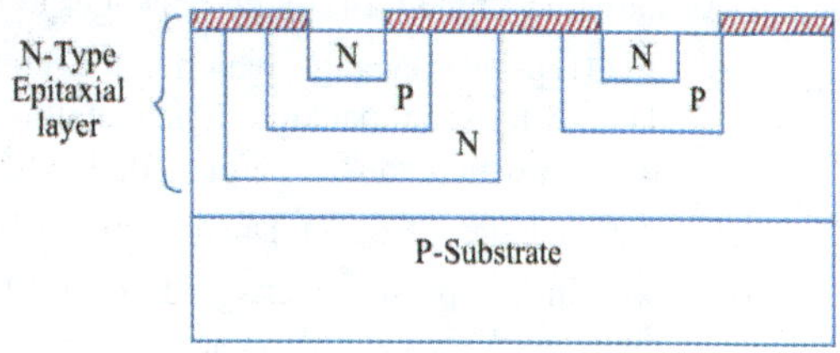

Fig. 31.11

In this way, we have fabricated one NPN transistor and one resistor simultaneously as shown in Fig. 31.11.

7. Pre-ohmic Etch

For good metal ohmic (non-rectifying) contact with diffused layers, N^+ regions are diffused into the structure as shown in Fig. 31.12. This is done once again by the SiO_2 layer, photoresist and masking process.

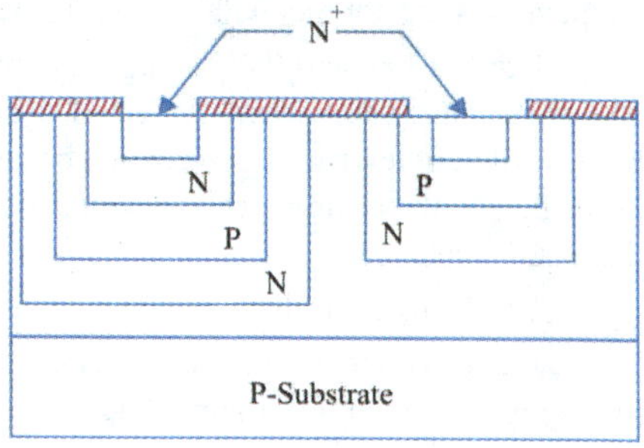

Fig. 31.12

8. Metallization

It is done for making interconnections and providing bonding pads around the circumference of the chip for later connection of wires. It is carried out by evaporating aluminium over the entire surface and then selectively etching away the aluminium to leave behind the desired inter-connecting conduction pattern and bonding pads. The final circuit alongwith its schematic is shown in Fig. 31.13.

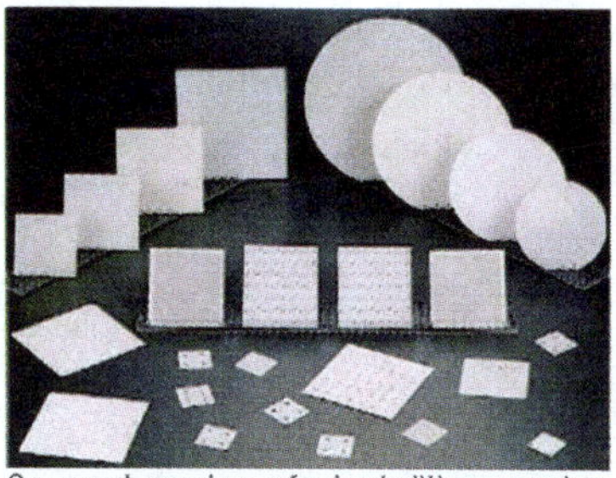

Several cycles of photolithography etching and doping are performed producing multiple layers of circuit elements on the wafer.

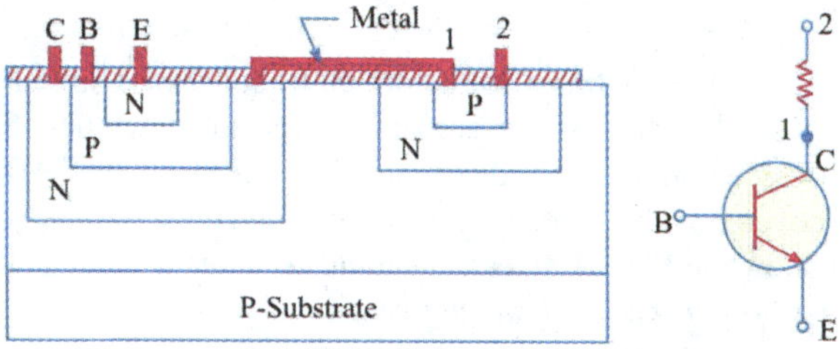

Fig. 31.13

9. Circuit Probing

Each IC on the wafer is checked electrically for proper performance by placing probes on the bonding pads. Faulty chips are marked and discarded after the wafer has been scribed and broken down into individual chips.

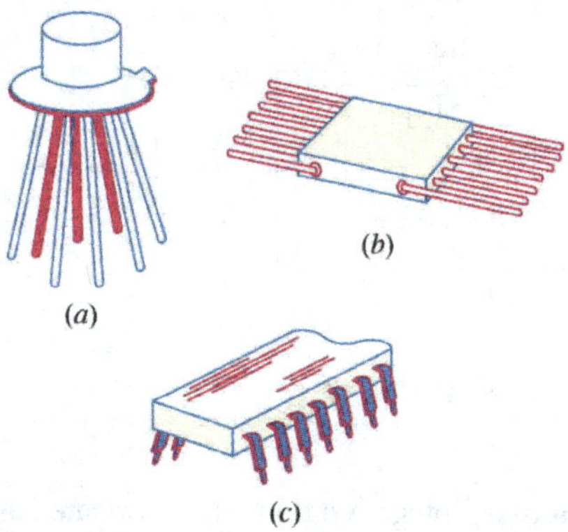

Fig. 31.14

10. Scribing and Separating into Chips

Once the metallization is complete, wafer is broken down into individual chips containing the integrated circuits. For this purpose, wafers are first scribed with a diamond tipped tool and then separated into single chips.

11. Mounting and Packing

The individual chip is very small and brittle. Hence, it is cemented or soldered to a gold-plated header through which leads have already been connected.

As shown in Fig. 31.14, many packages exist to hermetically seal the IC. These are :

(*i*) TO (top-hat) package which can accommodate upto 14 pins. The Fig. 31.14 (*a*) shows TO-5 8-terminal package. The metallic TO-can provides electromagnetic shielding for the IC chip which cannot be provided by the ceramic or plastic packages.

(*ii*) flat package—Fig. 31.14 (*b*) shows a 14-terminal flat package.

(*iii*) dual-in-line plastic package. Fig. 31.14 (*c*) shows a 14-terminal plastic DIL package (ceramic packages are also used).

These are very cheap and are widely used for industrial and consumer applications where high temperatures are not met with. Ceramic packages are used where IC is subjected to high temperatures.

Dual-in-line and flat packages are more convenient for circuit board use than TO-cans, because of their lead arrangement and because, being flatter, they allow greater circuit densities.

12. Encapsulation

A cap is now placed over the circuit and sealing is done in an inert atmosphere.

31.13. IC Symbols

These are very simple. Usually, these consist of either a triangle or a rectangle as shown in Fig. 31.15. It does not show its internal circuitry.

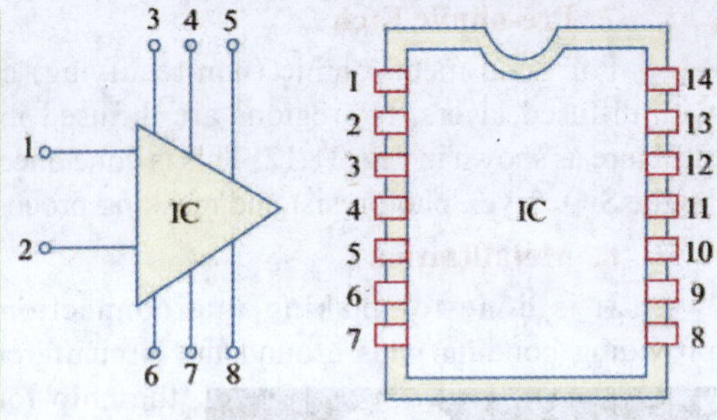

Fig. 31.15

31.14. Fabrication of IC Components

(*a*) Transistors

They are formed by the epitaxial planar diffusion process described in Art. 31.13 above.

The NPN transistor is formed by successive impurity diffusion into the N-type epitaxial layer deposited on the top surface of a P-type substrate. The SiO_2 layer is appropriately etched to allow diffusion into the desired portion of the epitaxial layer. After the diffusion is complete, the top surface is covered with a protective SiO_2 layer and metal contacts are made to the different transistor regions as shown in Fig. 31.16 (a). It should be noted that the P-substrate is only a supporting structure and *does not form part of any active device.*

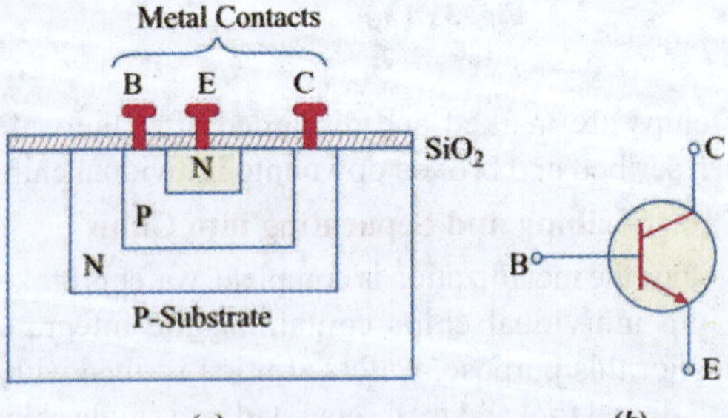

Fig. 31.16

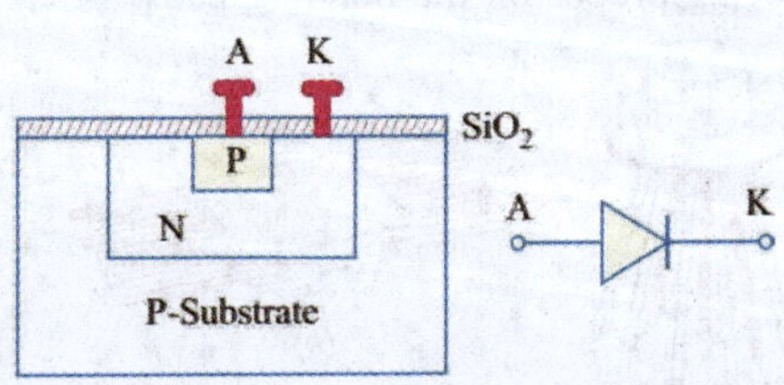

Fig. 31.17

(*b*) Diodes

They are formed by simply utilizing the P-N junctions of an integrated transistor fabricated as in (*a*) above.

In Fig. 31.17, C/B junction of the transistor is used as a diode. Anode of the diode is formed during the base diffusion of the transistor. The collector region becomes the cathode of the diode. When high speed switching is required, E/B junction is used as a diode.

(c) Diffused Resistors

IC resistors utilize the resistivity of doped silicon. By controlling the concentration of dopping impurity and depth of diffusion, the resistance value can be controlled.

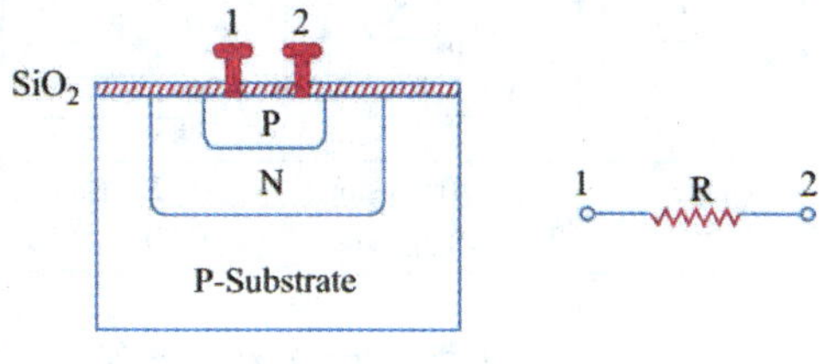

Fig. 31.18

As shown in Fig. 31.18, most resistors are formed during the base diffusion of the integrated transistor discussed in (*a*) above, because it is the highest-resistivity region. For low resistance values, emitter region is used because it has much lower resistivity.

(d) Capacitors

Monolithic IC capacitors are formed by utilizing the transition capacitance (Art. 13.12) of a reverse biased P-N junction.

As shown in Fig. 31.19 (*a*), P- and N-regions form the capacitor plates and depletion region between them is the dielectric. This capacitance is dependent on the reverse voltage across the junction and its value is limited to about 100 pF.

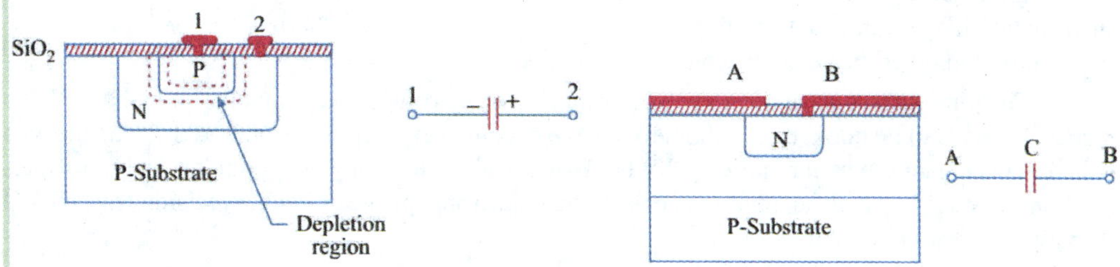

(*a*) (*b*)
Fig. 31.19 Fig. 31.20

A capacitor may also be formed by using SiO_2 layer as dielectric with metallization serving as the two plates as shown in Fig. 31.20.

(e) Inductors

Presently, there is no feasible process for fabricating inductors as part of monolithic structure. If required, it must be added externally as a discrete component.

31.15. Complete Monolithic Integrated Circuits

By way of illustration, we will now show a few transistor-diode-resistor-capacitor combinations and their schematic diagrams.

(a) TRC Combination

The IC consisting of a series combination of transistor, resistor and capacitor is shown in Fig. 31.21.

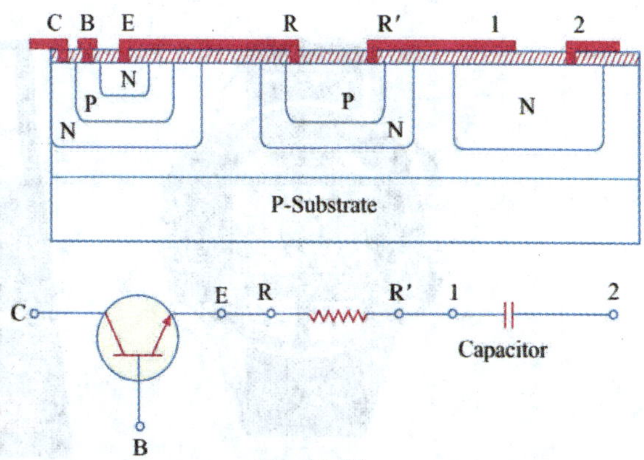

Fig. 31.21

The schematic diagram for the circuit is shown below it.

(*b*) Diode-Transistor-Resistor Combination

Such a combination is shown in Fig. 31.22 along with its interconnections.

The circuit schematic is shown alongwith.

31.16. Popular Applications of ICs

There are plenty of electronic products in the market which use integrated circuits extensively. One of the widely-accepted application is the digital watch which has an IC as its brain (Fig. 31.23). It can tell the hour, minute, second, day and month. Digital watch readouts are either LED or LCD.

Another popular application of microelectronic technology is the electronic calculator (Fig. 31.24) which can perform various functions like addition, subtraction, multiplication and division. There are also advanced scientific pocket calculators which are programmable. Modern electronic test instruments like digital meters, frequency counters and logic probes etc., use integrated circuits.

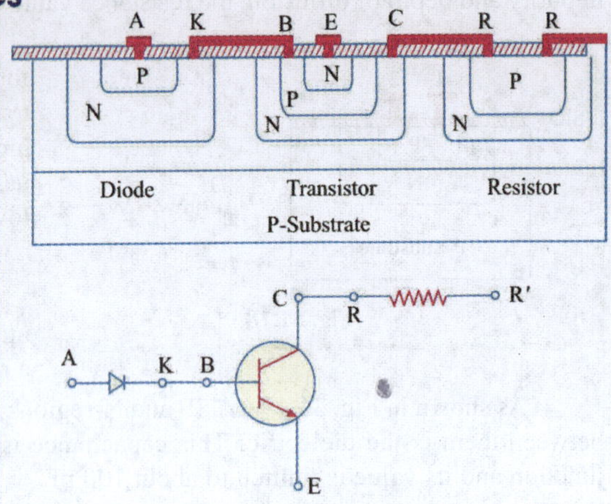

Fig. 31.22

Electronic games have been evolved from microelectronics and computer know-how. Video games that have to be hooked up to home TV have become very popular. Hand-held calculator type self-contained games which require no TV hook-up are also becoming very popular. They can operate from a battery or an ac adaptor and consist of an electronic printed circuit, precision lens, a LED display and a variety of sound effects.

31.17. MOS Integrated Circuits

The ICs discussed in Art. 31.12 are known as *monolithic bipolar* ICs because they are based on bipolar transistor fabrication and have bipolar transistor as their major circuit component. In addition to these, we also have monolithic ICs based on MOSFET structure which are widely used in many applications especially in the digital field.

Fig. 31.23

Fig. 31.24

The MOS IC wafers are processed almost in the same way as bipolar ICs. However, there are certain points of difference between the two.

1. MOS process is less expensive than biopolar process. It is so because it requires only one diffusion step to form both the source and drain regions as compared with from two to four diffusion steps needed for bipolar ICs.

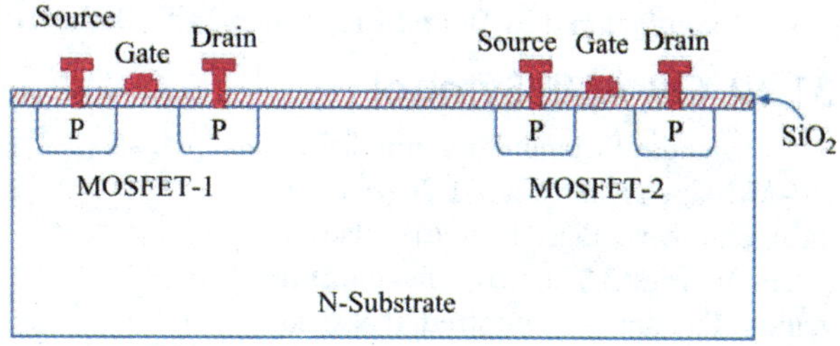

Fig. 31.25

2. There is no need for isolation technique for enhancement MOSFET devices since each source and drain region is isolated from the other by the P-N junctions formed within the N-type substrate. This fact is shown in Fig. 31.25 where two P-channel E-MOSFETs have been fabricated from the same substrate.
3. The packing density of MOS ICs is at least ten times more than that for bipolar ICs. Also, a MOS resistor occupies less than 1 per cent of the area of a conventional diffused resistor. This high packing density makes MOS ICs especially suited for LSI and VLSI.

The main disadvantage of MOS ICs is *their slower operating speed as compared to bipolar ICs*. Hence, they do not compete with bipolar ICs in ultrahigh-speed applications.

However, due to their (*i*) low cost (*ii*) low power consumption and (*iii*) high packing density, MOS ICs are widely used for LSI and VLSI chips such as calculator chips, memory chips and microprocessors (μP).

31.18. What is an OP-AMP?

It is a very high-gain, high-r_{in} directly-coupled negative-feedback amplifier which can amplify signals having frequency ranging from **0 Hz to a little beyond 1 MHz.** They are made with different internal configurations in linear ICs. An OP-AMP is so named because it was originally designed to perform mathematical operations like summation, subtraction, multiplication, differentiation and integration etc., in analog computers. Present day usage is much wider in scope but the popular name OP-AMP continues.

Typical uses of OP-AMP are : scale changing, analog computer operations, in instrumentation and control systems and a great variety of phase-shift and oscillator circuits. The OP-AMP is available in three different packages (*i*) standard dual-in-line package (DIP) (*ii*) TO-5 case and (*iii*) the flat-pack.

Although an OP-AMP is a complete amplifier, it is so designed that external components (resistors, capacitors etc.) can be connected to its terminals to change its external characteristics. Hence, it is relatively easy to tailor this amplifier to fit a particular application and it is, in fact, due to this versatility that OP-AMPs have become so popular in industry.

An OP-AMP IC may contain two dozen transistors, a dozen resistors and one or two capacitors.

Example of OP-AMPs

1. μA 709—is a high-gain operational amplifier constructed on a single silicon chip using planar epitaxial process.

   It is intended for use in dc servo systems, high-impedance analog computers and in low-level instrumentation applications.

   It is manufactured by Semiconductors Limited, Pune.
2. [LM 108 – LM 208] —Manufactured by Semiconductors Ltd., Mumbai.

3. CA 741 CT and CA 741 T—these are high-gain operational amplifiers which are intended for use as (*i*) comparator, (*ii*) integrator, (*iii*) differentiator, (*iv*) summer, (*v*) dc amplifier, (*vi*) multivibrator, and (*vii*) bandpass filter.

Manufactured by Bharat Electronics Ltd. (BEL), Bangalore.

31.19. OP-AMP Symbol

Standard triangular symbol for an OP-AMP is shown in Fig. 31.26 (*a*) though the one shown in Fig. 31.26 (*b*) is also used often. In Fig. 31.26 (*b*), **the common ground line has been omitted.** It also does not show other necessary connections such as for dc power and feedback etc.

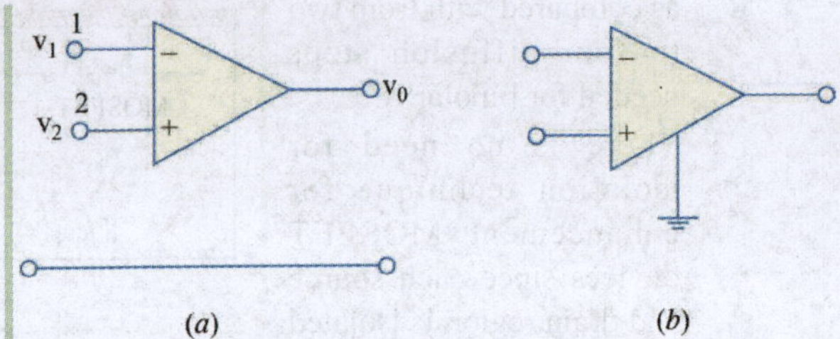

Fig. 31.26

The OP-AMP's input can be single-ended or double-ended (or differential input) depending on whether input voltage is applied to one input terminal only or to both. Similarly, amplifier's output can also be either single-ended or double-ended. The most common configuration is *two input terminals and a single output.*

All OP-AMPs have a minimum of five terminals

1. inverting input terminal,
2. non-inverting input terminal,
3. output terminal,
4. positive bias supply terminal,
5. negative bias supply terminal.

31.20. Polarity Conventions

In Fig. 31.26 (*b*), the input terminals have been marked with minus (–) and plus (+) signs. These are meant to **indicate the inverting and non-inverting terminals only** [Fig. 31.27 (*a*)]. It simply means that a signal applied at negative input terminal will appear amplified but **phase-inverted at the output terminal** as shown in Fig. 31.27 (*b*). Similarly, signal applied at the positive input terminal will appear amplified and **in-phase** at the output. Obviously, these plus and minus polarities **indicate phase reversal only.** It does not mean that voltage v_1 and v_2 in Fig. 31.27 (*a*) are negative and positive respectively. Additionally, it also does not imply that *a positive input voltage has to be connected to the plus-marked non-inverting terminal 2 and negative input voltage to the negative-marked inverting terminal 1.* In fact, the amplifier can be used 'either way up' so to speak. It may also be noted that all input and output voltages are referred to a common reference usually the ground shown in Fig. 31.26 (*a*).

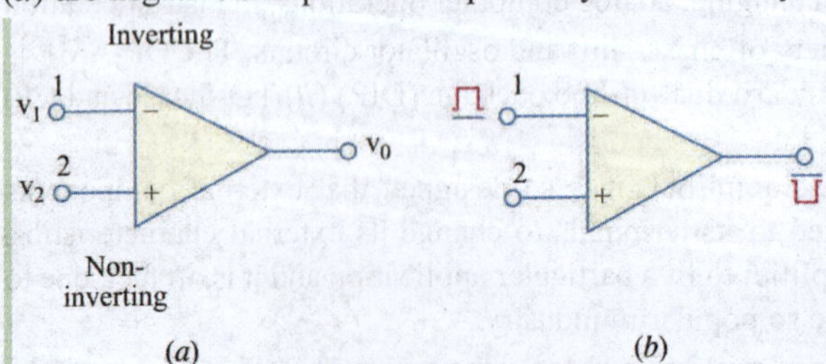

Fig. 31.27

31.21. Ideal Operational Amplifier

When an OP-AMP is operated without connecting any resistor or capacitor from its output to any one of its inputs (*i.e.*, without feedback), it is said to be in the **open-loop condition.** The word 'open loop' means that *feedback path or loop is open.* The specifications of an OP-AMP under such condition are called open-loop specifications.

An ideal OP-AMP (Fig. 31.28) has the following characteristics:

1. its open-loop gain A_v is infinite *i.e.*, $A_v = -\infty$;
2. its input resistance R_i (measured between inverting and non-inverting terminals) *is infinite i.e.*, $R_i = \infty$ ohm;
3. its output resistance R_0 (seen looking back into output terminals) *is zero i.e.*, $R_0 = 0\ \Omega$;
4. it has *infinite bandwidth i.e.*, it has flat frequency response from dc to infinity.

   Though these characteristics cannot be achieved in practice, yet an ideal OP-AMP serves as a convenient reference against which real OP-AMPs may be evaluated.

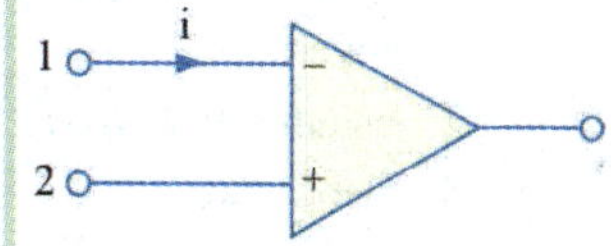

Fig. 31.28

Following additional points are worth noting:

1. Infinite input resistance means that input current $i = 0$ as indicated in Fig. 31.28.

   It means that an ideal OP-AMP is a voltage-controlled device.
2. $R_o = 0$ means that v_o is not dependent on the load resistance connected across the output.
3. Though for an ideal OP-AMP $A_v = \infty$, for an actual one, it is extremely high *i.e.*, about 10^6. However, it does not mean that 1 V signal will be amplified to 10^6V at the output. Actually, the maximum value of v_0 is limited by the bias supply voltage, typically ± 15 V. With $A_v = 10^6$ and $v_0 = 15\ V_2$ the maximum value of input voltage is limited to $15/10^6 = 15\ \mu V$. Though 1V cannot become 1 million volt in the OP-AMP, 1 μV can certainly become 1 V.

31.22. Virtual Ground and Summing Point

In Fig. 31.29 is shown an OP-AMP which *employs negative feedback* with the help of resistor R_f which feeds a portion of the output to the input.

Since input and feedback currents are algebraically added at point *A*, it is called the **summing point.**

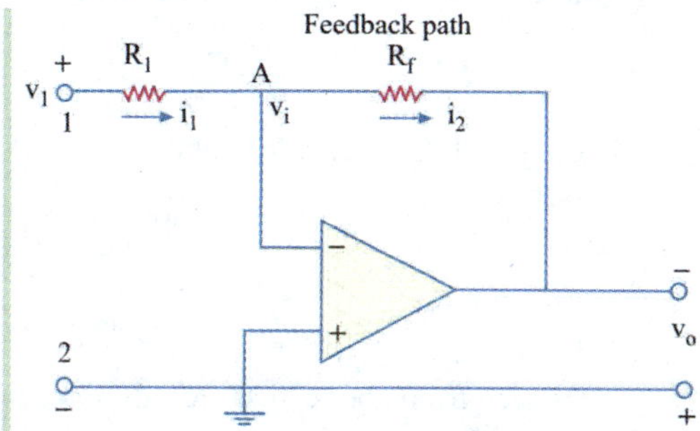

Fig. 31.29

The concept of **virtual** ground arises from the fact that input voltage v_i at the inverting terminal of the OP-AMP is forced to such a small value that, for all practical purposes, it may be assumed to be zero. Hence, point *A* is essentially at ground voltage and is referred to as *virtual ground.* Obviously, it is not the *actual ground,* which, as seen from Fig. 31.29, is situated below.

31.23. Why V_i is Reduced to almost Zero ?

When v_1 is applied, point *A* attains some positive potential and at the same time v_o is brought into existence. Due to negative feedback, some fraction of the output voltage is fed back to point *A* antiphase with the voltage already existing there (due to v_1).

The algebraic sum of the two voltages is almost zero so that $v_i = 0$. Obviously, v_i will become exactly zero when **negative feedback voltage at A is exactly equal to the positive voltage produced by v_1 at A.**

Another point worth considering is that there exists a virtual short between the two terminals of the OP-AMP because $v_i = 0$. It is virtual because no current flows (remember $i = 0$) despite the existence of this short.

31.24. OP-AMP Applications

We will consider the following applications:

1. as scaler or linear (*i.e.*, small-signal) constant-gain amplifier; both inverting and non-inverting,
2. as unity follower,
3. Adder or Summer,
4. Subtractor,
5. Integrator,
6. Differentiator,
7. Comparator.

Now, we will discuss the above circuits one by one assuming an ideal OP-AMP.

31.25. Linear Amplifier

We will consider the functioning of an OP-AMP as a constant-gain amplifier both in the inverting and non-inverting configurations.

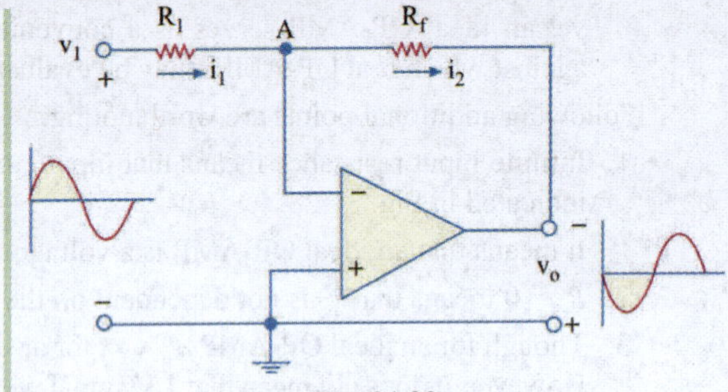

Fig. 31.30

(a) Inverting Amplifier or Negative Scale

As shown in Fig. 31.30, non-inverting terminal has been grounded whereas R_1 connects the input signal v_1 to the inverting input. A feedback resistor R_f has been connected from the output to the inverting input.

Gain

Since point A is at ground potential*, $i_1 = \frac{v_{in}}{R_1} = \frac{v_1}{R_1}$

$$i_2 = \frac{-v_o}{R_f} \quad \text{— please note – ve sign}$$

Using KCL (Art. 3.2) for point A,

$$i_1 - (-i_2) = 0 \quad \text{or} \quad \frac{v_1}{R_1} + \frac{v_o}{R_f} = 0 \quad \text{or} \quad \frac{v_0}{R_f} - \frac{v_1}{R_1} \quad \text{or} \quad \frac{v_0}{v_1} = -\frac{R_f}{R_1}$$

$$\therefore \quad A_v = -\frac{R_f}{R_1} \quad \text{or} \quad A_v = -K \quad \text{Also,} \quad v_o = -Kv_{in}$$

It is seen from above, that closed-loop gain of the inverting amplifier depends on the ratio of the two external resistors R_1 and R_f and is independent of the amplifier parameters.

It is also seen that the OP-AMP works as a negative scaler. It scales the input *i.e.*, it multiplies the input by a minus constant factor K.

(b) Non-inverting Amplifier or Positive Scaler

This circuit is used when there is need for an output which is equal to the input multiplied by a positive constant. Such a positive scaler circuit which uses negative feedback but provides an output that equals the input multiplied by a positive constant is shown in Fig. 31.31.

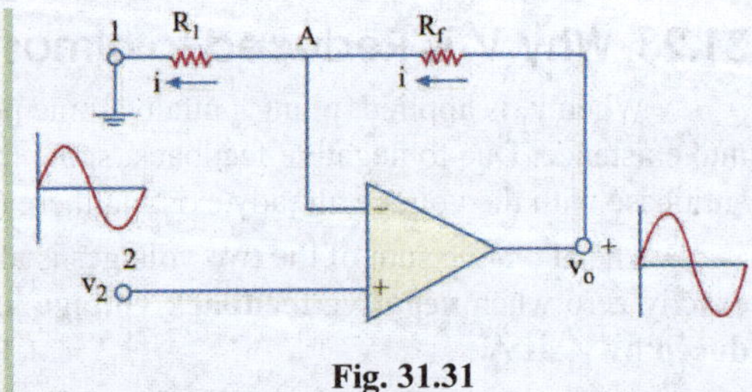

Fig. 31.31

Since input voltage v_2 is applied to the non-inverting terminal, the circuit is also called **non-inverting amplifier.**

* If not, then $i_1 = \frac{v_1 - v_i}{R_1}$ and $i_2 = \frac{v_0 - v_1}{R_f}$

Here, polarity of v_o is the same as that of v_2 *i.e.*, both are positive.

Gain

Because of virtual short between the two OP-AMP terminals, voltage across R_1 is the input voltage v_2. Also, v_0 is applied across the series combination of R_1 and R_f

$$\therefore \quad v_{in} = v_2 = iR_1 \; ; \; v_0 = i(R_1 + R_f)$$

$$\therefore \quad A_v = \frac{v_o}{v_{in}} = \frac{i(R_1 + R_f)}{iR_1} \text{ or } A_v = \frac{R_1 + R_f}{R_1} = \left(1 + \frac{R_f}{R_1}\right)$$

Alternative Derivation

As shown in Fig. 31.32, let the currents through the two resistors be i_1 and i_2.

The voltage across R_1 is v_2 and that across R_f is $(v_o - v_2)$.

$$\therefore \quad i_1 = \frac{v_2}{R_1} \text{ and } i_2 = \frac{v_o - v_2}{R_f}$$

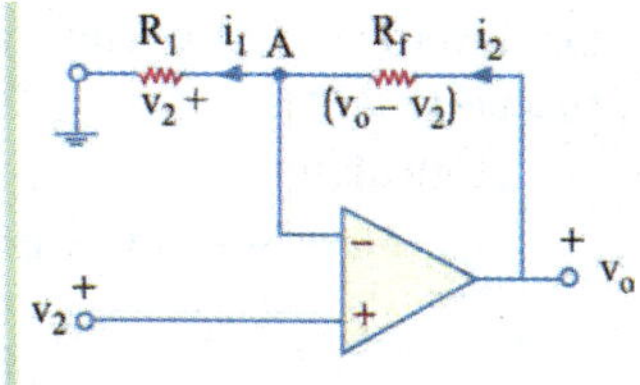

Fig. 31.32

Applying KCL to junction A, we have

$$(-i_1) + i_2 = 0 \text{ or } \frac{-v_2}{R_1} + \frac{(v_o - v_2)}{R_f} = 0$$

$$\therefore \quad \frac{v_o}{R_f} = v_2\left(\frac{1}{R_1} + \frac{1}{R_f}\right) = v_2 \frac{R_1 + R_f}{R_1 R_f}$$

$$\therefore \quad \frac{v_o}{v_2} = \frac{R_1 + R_f}{R_1} \text{ or } A_v = 1 + \frac{R_f}{R_1} \qquad \text{—as before}$$

Example 31.1. *Calculate (i) input impedance and (ii) the voltage gain of the OP-AMP amplifier circuit of Fig. 31.33.*

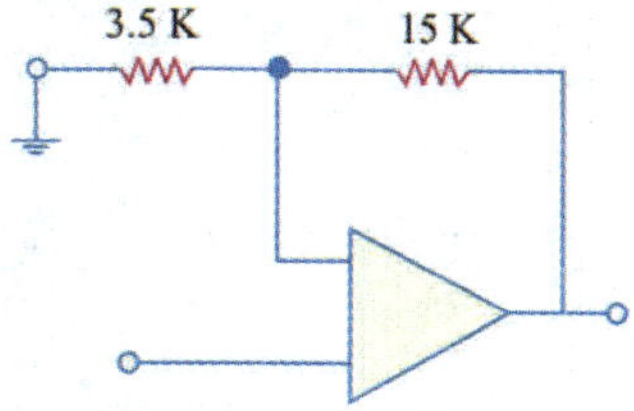

Fig. 31.33

Solution. (*i*) The input impedance of the OP-AMP amplifier is very high and when negative feedback is used, the impedance is increased even further. Hence, input impedance of a non-inverting OP-AMP amplifier can be thought of as infinite.

(*ii*) The voltage gain is given by

$$A_v = 1 + \frac{R_f}{R_1} = 1 + \frac{15}{3.5} = \mathbf{5.3}$$

Example 31.2. *For the inverting amplifier of Fig. 31.30, R_1 = 1K and R_f = 1M. Assuming an ideal OP-AMP amplifier, determine the following circuit values :*

(*a*) *voltage gain,* (*b*) *input resistance ,* (*c*) *output resistance.*

Solution. It should be noted that we will be calculating values *of the circuit* and not *for the* OP-AMP *proper.*

(*a*) $$A_v = -\frac{R_f}{R_1} = -\frac{1000 \text{ K}}{1 \text{ K}} = \mathbf{-1000.}$$

(*b*) Because of virtual ground at A, $R_{in} = R_1 =$ **1 K.**

(*c*) Output resistance of the circuit equals the output resistance of the OP-AMP *i.e.*, **zero ohm.**

31.26. Unity Follower

It provides a gain of unity without any phase reversal. It is very much similar to the emitter follower (Art. 21.8) except that its gain is very much closer to being exactly unity.

This circuit (Fig. 31.34) is useful as a buffer or isolation amplifier because it allows input voltage v_{in} to be transferred as output voltage v_o while at the same time preventing load resistance R_L from loading down the input source. It is due to the fact that its $R_i = \infty$ and $R_o = 0$.

In fact, circuit of Fig. 31.34 can be obtained from that of Fig. 31.31 by putting

$$R_1 = R_f = 0$$

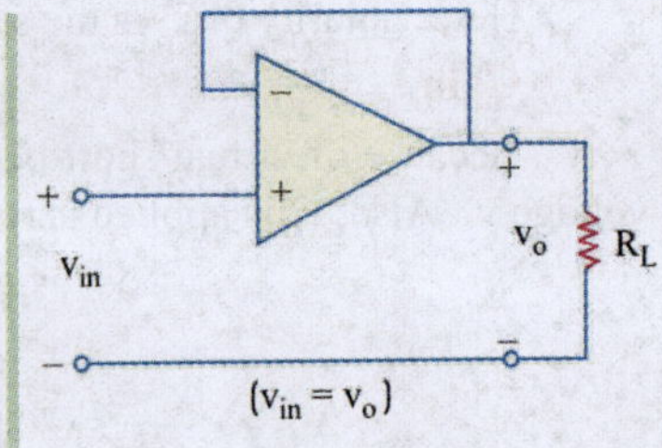

Fig. 31.34

31.27. Adder or Summer

The adder circuit provides an output voltage proportional to or equal to the algebraic sum of two or more input voltages each multiplied by a constant gain factor. It is basically similar to a scaler (Fig. 31.30) except that it has more than one input. Fig. 31.35 shows a three-input inverting adder circuit. As seen, *the output voltage is phase-inverted.*

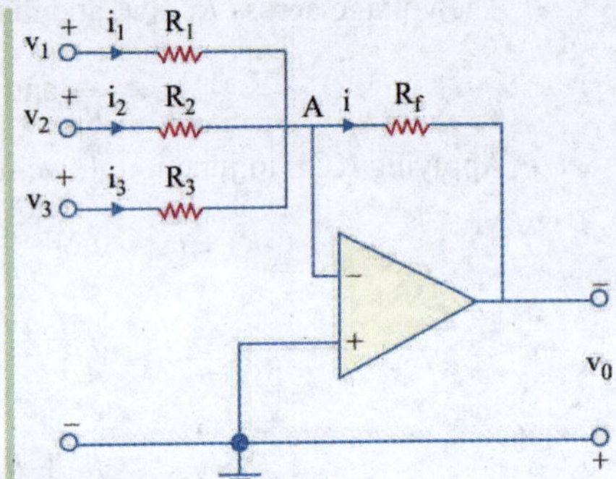

Fig. 31.35

Calculations

As before, we will treat point A as virtual ground

$$i_1 = \frac{v_1}{R_1} \quad \text{and} \quad i_2 = \frac{v_2}{R_2}$$

$$i_3 = \frac{v_3}{R_3} \quad \text{and} \quad i = -\frac{v_o}{R_f}$$

Applying KCL to point A, we have

$$i_1 + i_2 + i_3 + (-i) = 0$$

or
$$\frac{v_1}{R_1} + \frac{v_2}{R_2} + \frac{v_3}{R_3} - \left(\frac{-v_o}{R_f}\right) = 0$$

$\therefore$
$$v_o = -\left(\frac{R_f}{R_1}v_1 + \frac{R_f}{R_2}v_2 + \frac{R_f}{R_3}v_3\right)$$

or
$$v_o = -(K_1v_1 + K_2v_2 + K_3v_3)$$

The overall negative sign is unavoidable because we are using the inverting input terminal.

If $R_1 = R_2 = R_3 = R$, then

$$v_o = -\frac{R_f}{R}(v_1 + v_2 + v_3) = -K(v_1 + v_2 + v_3)$$

Hence, output voltage is proportional to (*not equal to*) the **algebraic sum** of the three input voltages.

If $R_f = R$, then output exactly equals the sum of inputs. However, if $R_f = R/3$,

then
$$v_o = -\frac{R/3}{R}(v_1 + v_2 + v_3) = -\frac{1}{3}(v_1 + v_2 + v_3)$$

Obviously, the output is equal to the *average of the three inputs.*

31.28. Subtractor

The function of a subtractor is to provide an output proportional to or equal to the difference of two input signals. As shown in Fig. 31.36, we have to apply the inputs at the inverting as well as non-inverting terminals.

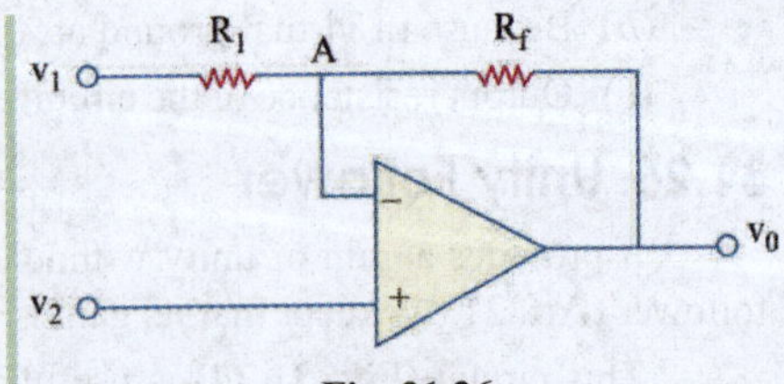

Fig. 31.36

Calculations

According to Superposition Theorem (Art. 4.2)

$$v_o = v_o' + v_o''$$

where v_o' is the output produced by v_1 and v_o'' is that produced by v_2.

Now, $$v_o' = -\frac{R_f}{R_1}.v_1 \qquad \text{— Art. 31.25 } (a)$$

$$v_o'' = \left(1 + \frac{R_f}{R_1}\right) v_2 \qquad \text{— Art. 31.26 } (b)$$

$$\therefore \quad v_o = \left(1 + \frac{R_f}{R_1}\right) v_2 - \frac{R_f}{R_1} v_1$$

Since $R_f \gg R_1$ and $R_f/R_1 \gg 1$, hence

$$v_o \cong \frac{R_f}{R_1}(v_2 - v_1) = K(v_2 - v_1)$$

Further, if $R_f = R_1$, then

$$v_o = (v_2 - v_1) = \text{difference of the two input voltages.}$$

Obviously, if $R_f \neq R_1$, then a scale factor is introduced.

Example 31.3. *Find the output voltage of an OP-AMP inverting adder for the following sets of input voltages and resistors. In all cases, $R_f = 1M$.*

$$v_1 = -3\ V,\ v_2 = +3\ V.\ v_3 = +2\ V;\ R_1 = 250\ K,\ R_2 = 500\ K,\ R_3 = 1\ M$$

Solution. $$v_o = -(K_1 v_1 + K_2 v_2 + K_3 v_3)$$

$$K_1 = \frac{R_f}{R_1} = \frac{1000\ K}{250\ K} = 4,\ K_2 = \frac{1000}{500} = 2,\ K_3 = \frac{1\ M}{1\ M} = 1$$

$$\therefore \quad v_o = -[(4 \times -3) + (2 \times 3) + (1 \times 2)] = +\mathbf{4\ V}$$

Example 31.4. *In the subtractor circuit of Fig. 31.36, $R_1 = 5\ K$, $R_f = 10\ K$, $v_1 = 4\ V$ and $v_2 = 5\ V$. Find the value of output voltage.*

Solution. $$v_o = \left(1 + \frac{R_f}{R_1}\right) v_1 - \frac{R_f}{R_1} v_2 = \left(1 + \frac{10}{5}\right) 4 - \frac{10}{5} \times 5 = +\mathbf{2\ V}$$

31.29. Integrator

The function of an integrator is to provide an output voltage which is proportional to the integral of the input voltage.

A simple example of integration is shown in Fig. 31.37 where input is dc level and its integral is *a linearly-increasing ramp output.*

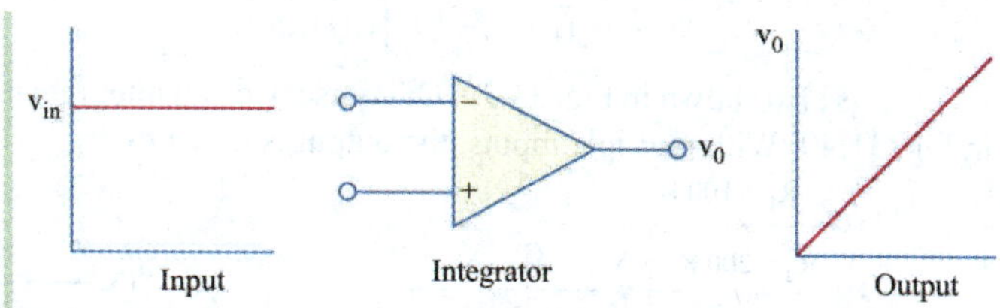

Fig. 31.37

The actual integrating circuit is shown in Fig. 31.38. This circuit is similar to the scaler circuit of Fig. 31.29 except that the **feedback component is a capacitor C instead of a resistor R_f.**

Calculations

As before, point A will be treated as virtual ground.

$$i_1 = \frac{v_1}{R};\quad i_2 = -\frac{v_o}{X_C} = -\frac{v_o}{1/j\omega C} = -\frac{v_o}{1/sC} = -sC\,v_o$$

where $s = j\,\omega$ in the Laplace notation.

Now $\qquad i_1 = i_2$ — Art. 31.20 (*a*)

$$\therefore \qquad \frac{v_1}{R} = -\,sCv_o$$

$$\therefore \qquad \frac{v_o}{v_{in}} = \frac{v_o}{v_1} = -\frac{1}{sCR} \qquad \text{—(i)}$$

$$\therefore \qquad A_v = -\frac{1}{sCR}$$

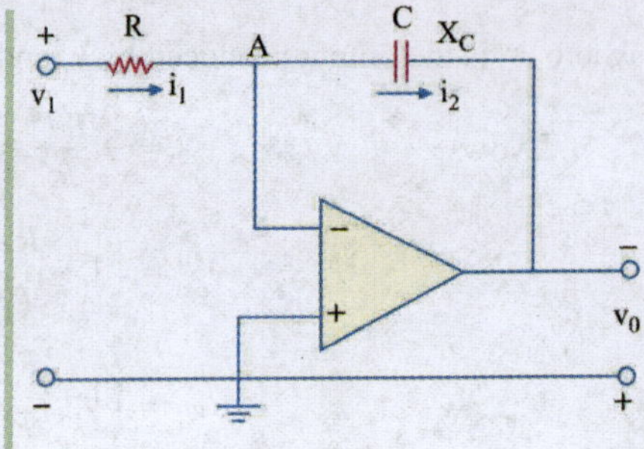

Fig. 31.38

Now, the expression of Eq. (*i*) can be written in time domain as

$$v_o(t) = -\frac{1}{CR}\int v_1(t)\,.\,dt$$

It is seen from above that output (right-hand side expressions) is an integral of the input, with an inversion and a scale factor of $1/CR$.

This ability to integrate a given signal enables an analog computer to solve differential equations and to set up a wide variety of electrical circuit analogs of physical system operations. For example, let

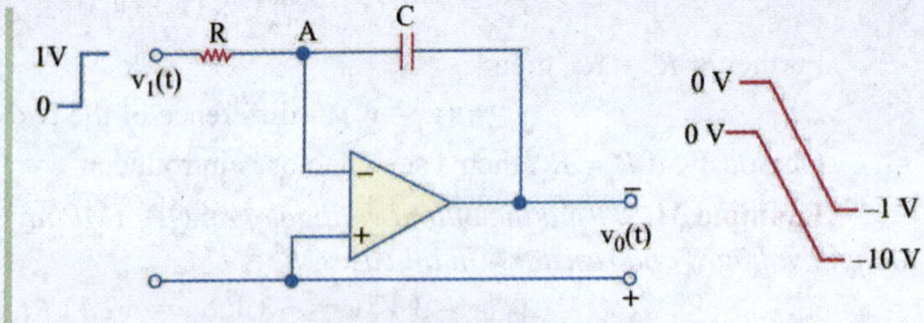

Fig. 31.39

$$R = 1\text{ M} \quad \text{and} \quad C = 1\ \mu\text{F. Then}$$

$$\text{scale factor} \qquad = -\frac{1}{CR} = -\frac{1}{10^6 \times 10^{-6}} = -1$$

As shown in Fig. 31.39, the input is a step voltage whereas output is a ramp (or linearly-changing voltage) with a scale multiplier of −1.

However, when $R = 100$ K, then

$$\text{scale factor} \qquad = -\frac{1}{10^5 \times 10^{-6}} = -10$$

$$\therefore \qquad v_o(t) = -10\int v_1(t).dt$$

It is also shown in Fig. 31.39. Of course, we can integrate more than one input as shown below in Fig. 31.40. With multiple inputs, the output is given by

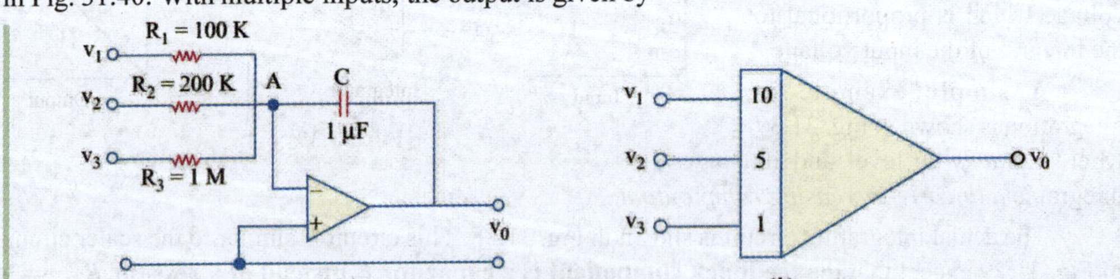

Fig. 31.40

$$v_o(t) = -[K_1 \int v_1(t)\,dt + K_2 \int v_2(t)\,dt + K_3 \int v_3(t)\,dt]$$

$$\text{where} \qquad K_1 = \frac{1}{CR_1},\ K_2 = \frac{1}{CR_2} \text{ and } K_3 = \frac{1}{CR_3}$$

Fig. 31.40 (*a*) shows a summing integrator as used in an analog computer. It shows all the three resistors and the capacitor. The analog computer representation of Fig. 31.40 (*b*) indicates only the scale factor for each input.

Example 31.5. *A 5 mV, 1 kHz sinusoidal signal is applied to the input of an OP-AMP integrator of Fig. 31.38 for which R = 100 K and C = 1 μF. Find the output voltage.*

Solution. Scale factor $= -\frac{1}{CR} = \frac{1}{10^5 \times 10^{-6}} = -10$

The equation for the sinusoidal voltage is

$$v_1 = 5 \sin 2\pi f t = 5 \sin 2000 \pi t$$

Obviously, it has been assumed that at $t = 0$, $v_1 = 0$

$$\therefore \quad v_o(t) = -10 \int_0^t 5 \sin 2000 \pi t = -50 \left| \frac{-\cos 2000 \pi t}{2000} \right|_0^t$$

$$= \frac{1}{40\pi} (\cos 2000 \pi t - 1)$$

31.30. Differentiator

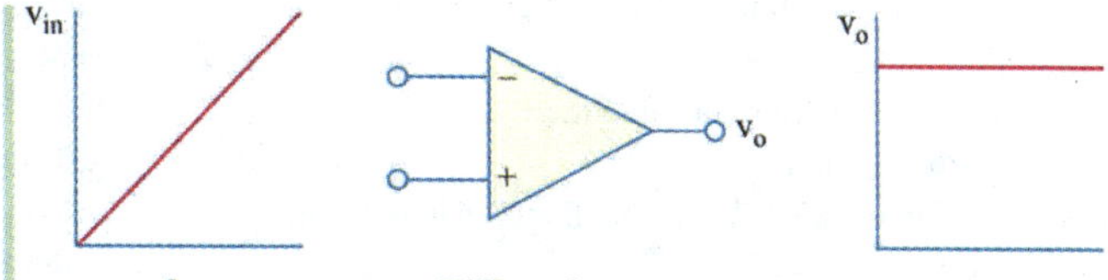

Fig. 31.41

Its function is to provide an output voltage which *is proportional to the rate of change of the input voltage.* It is an inverse mathematical operation to that of an integrator. As shown in Fig. 31.41, when we feed a differentiator, we get a constant dc output.

Circuit

Differentiator circuit can be obtained by *interchanging the resistor and capacitor of the integrator circuit of* Fig. 31.38.

Let $\quad i = \text{rate of change of charge} = \frac{dq}{dt}$

Now, $\quad q = Cv_c$

$$\therefore \quad i = \frac{d}{dt}(Cv_c) = C\frac{dv_c}{dt}$$

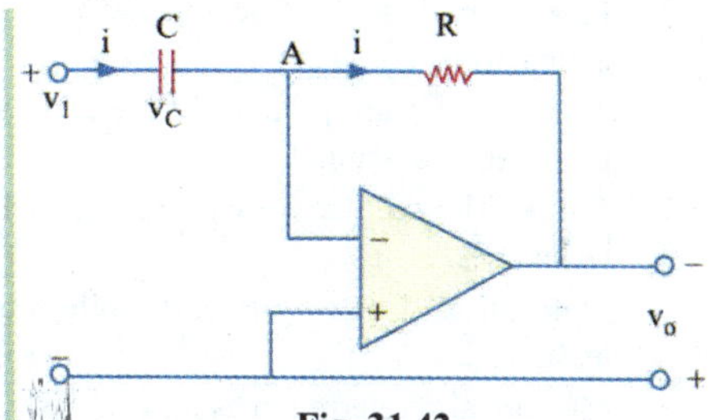

Fig. 31.42

Taking point *A* as virtual ground

$$v_o = -iR = -\left(C \cdot \frac{dv_c}{dt}\right) R$$

$$= -CR.\frac{dv_c}{dt}$$

As seen, output voltage is proportional to the derivate of the input voltage, the constant of proportionality (i.e., scale factor) being (– *CR*).

Example 31.6. *The input to the differentiator circuit of Fig. 31.42 is a sinusoidal voltage of peak value 5 mV and frequency 1 kHz. Find out the output if R = 100 K and C = 1 μF.*

Solution. The equation of the input voltage is

$$v_1 = 5 \sin 2\pi \times 1000t = 5 \sin 2000 \pi t \text{ mV}$$

scale factor $= CR = 10^{-6} \times 10^5 = 0.1$

$$v_0 = 0.1 \frac{d}{dt}(5 \sin 2000 \pi t) = (0.5 \times 2000 \pi) \cos 2000 \pi t$$

$$= 1000 \pi \cos 2000 \pi t \text{ mV}$$

As seen, output is a cosinusoidal voltage of frequency 1 kHz and peak value 1000 π mV.

31.31. Comparator

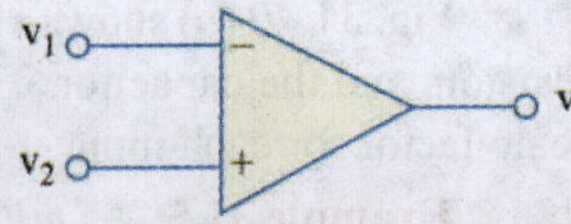

Fig. 31.43

It is a circuit which compares two signals or voltage levels. The circuit is shown in Fig. 31.43 and (like that of the follower) is the simplest because it needs no additional external components.

If v_1 and v_2 are equal, then v_o should ideally be zero. Even if v_1 differs from v_2 by a very small amount, v_o is large because of amplifier's high gain. Hence, circuit of Fig. 31.43 can detect very small level changes which is another way of saying that it compares two signals.

SELF EXAMINATION QUESTIONS

A. Fill in the blanks by most appropriate word(s) or numerical values (s).

1. In an integrated circuit, all active and passive components are .... on a single chip of silicon.
2. A major advantage of ICs is their reduced ....
3. The most important advantage of an IC is its extremely high ....
4. One drawback of ICs is that they can handle only a limited amount of ....
5. Functionally, ICs can be classified as linear and ....ICs.
6. For mass production of ICs, monolithic process is most ....
7. .... fabrication employs planar epitaxial process.
8. Whereas monolithic ICs are formed within a silicon wafer, thin film ICs are formed on its ....
9. In thin-film ICs, ... components are added externally.
10. Thin-film ICs are produced by using vacuum evaporation or cathode.
11. Thick-film ICs employ .... printing techniques.
12. Multichip ICs possess greater flexibility but have least............ .
13. LICs are also referred to as .... ICs.
14. As compared to digital ICs, linear ICs are used much............... .
15. Digital ICs are mostly utilized by the .... industry.
16. An OP-AMP is an extremely high-gain, high-r_{in} .... coupled amplifier.
17. MOS ICs are .... expensive than bipolar ICs.
18. As compared to bipolar ICs, MOS ICs have .... operating speed.
19. An OP-AMP is a .... IC.
20. The open-loop gain of an ideal OP-AMP is ....
21. An ideal OP-AMP has an input resistance of .... and an output resistance of ....
22. An adder circuit provides an output voltage which is the .... sum of two or more input voltages.
23. An integrator converts a dc level into a linearly-increasing .... ouput.
24. A differentiator converts a linear ramp into a .... dc output.
25. Differentiator can be obtained by .... the resistor and capacitor of an integrator circuit.

B. Answer True or False

1. At present, only silicon is used to fabricate ICs.
2. In monolithic ICs, all components are atomically part of the same chip.
3. Monolithic IC is a complete circuit requiring no ADD-ONs.
4. For mass production of ICs monolithic process is most economical.
5. Monolithic ICs are produced by vacuum evaporation process.
6. Thick-film ICs need active components externally.
7. Logic gates and flip-flops are examples of linear ICs.
8. VLSI is a major electronic system fabricated on a single wafer of silicon.
9. Functionally, ICs may be either linear or digital.
10. OP-AMP is a well-known example of a digital IC.
11. Electronic watches and calculators use digital ICs.
12. Metallisation of an IC means covering it with a fine layer of conducting aluminium.
13. Monolithic resistors utilize the resistivity of doped silicon.
14. Speedwise, MOS ICs are faster than bipolar ICs.
15. MOS ICs are ideal for LSI and VLSI.
16. An OP-AMP represents a linear IC.

17. The gain and input resistance of an OP-AMP are high.
18. An ideal OP-AMP has infinite bandwidth.

C. Multiple Choice Items

1. First integrated circuit chip was developed by
   (a) C.V. Raman (b) W.H. Brattain
   (c) J.S. Kilby (d) Robert Noyce
2. An integrated circuit is
   (a) a complicated circuit
   (b) an integrating device
   (c) much costlier than a single transistor
   (d) fabricated on a tiny silicon chip
3. Most important advantage of an IC is its
   (a) easy replacement in case of circuit failure
   (b) extremely high reliability
   (c) reduced cost
   (d) low power consumption
4. In monolithic ICs, all components are fabricated by .... process.
   (a) evaporation (b) sputtering
   (c) diffusion (d) oxidisation
5. Monolithic ICs are fabricated within a
   (a) soft stone (b) single stone
   (c) silicon layer (d) ceramic base
6. Monolithic technique is ideally suited for fabricating ....ICs.
   (a) thin-film (b) digital
   (c) linear (d) thick-film
7. In the context of IC fabrication, metallisation means
   (a) connecting metallic wires
   (b) forming interconnecting conduction pattern and bonding pads
   (c) depositing SiO_2 layer
   (d) covering with a metallic cap
8. The major component of a MOS IC is a/an
   (a) FET (b) MOSFET
   (c) bipolar (d) SCR
9. An OP-AMP can be classified as .... amplifier.
   (a) linear (b) low-r_{in}
   (c) positive feedback (d) *RC*-coupled
10. An ideal OP-AMP has
    (a) infinite A_v (b) infinite R_i
    (c) zero R_o (d) all the above
11. OP-AMPs have become very popular in industry mainly because
    (a) they are dirt cheap
    (b) their external characteristics can be changed to suit any application
    (c) of their extremely small size
    (d) they are available in different packages
12. Since input resistance of an ideal OP-AMP is infinite
    (a) its output resistance is zero
    (b) its output voltage becomes independent of load resistance
    (c) its input current is zero
    (d) it becomes a current-controlled device
13. An inverting amplifier has R_f = 2 M and R_1 = 2 K. Its scale factor is
    (a) 1000 (b) – 1000
    (c) 10^{-3} (d) – 10^{-3}
14. When in a negative scaler, both R_1 and R_f are reduced to zero, the circuit functions as
    (a) integrator (b) subtractor
    (c) comparator (d) unity follower

ANSWERS

A. Fill in the blanks

1. fabricated	2. cost	3. reliability	4. power	5. digital
6. economical	7. monolithic	8. surface	9. active	10. sputtering
11. screen	12. reliability	13. analog	14. less	15. computer
16. directly	17. less	18. slower	19. linear	20. infinity
21. infinity, zero	22. algebraic	23. ramp	24. constant	25. interchanging

B. True or False

1. T	2. T	3. T	4. T	5. F	6. T
7. F	8. T	9. T	10. F	11. T	12. F
13. T	14. F	15. T	16. T	17. T	18. T

C. Multiple Choice Items

1. *c*	2. *d*	3. *b*	4. *c*	5. *c*	6. *b*
7. *b*	8. *b*	9. *a*	10. *d*	11. *b*	12. *c*
13. *b*	14. *d*				

CHAPTER 32

Number Systems

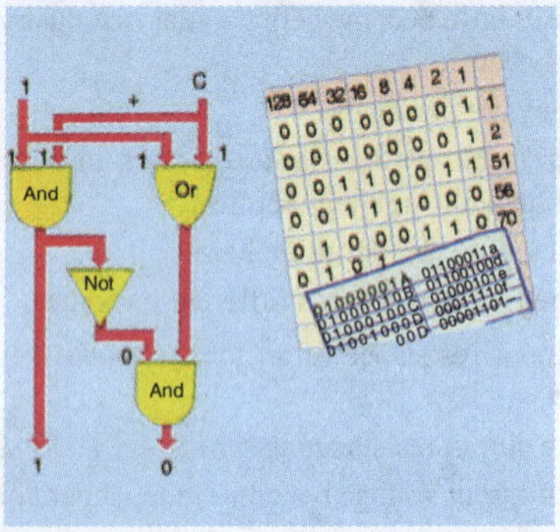

1. Number of Systems
2. Binary System
3. Binary to Decimal Conversion
4. Double-Dadd Method
5. Shifting the Place Point
6. Binary Operations
7. Binary Addition
8. Binary Subtraction
9. Complement of a Number
10. Binary Multiplication
11. Binary Division
12. Shifting a Number to Left or Right
13. Representation of Binary Numbers as Electrical Signals
14. Octal Number System
15. Octal to Decimal Conversion
16. Decimal to Octal Conversion

32.1. Number of Systems

There are four systems of arithmetic which are often used in digital circuits. These systems are:

1. **Decimal**—it has a base (or radix) of 10 *i.e.*, it uses 10 different symbols to represent numbers.
2. **Binary**—it has a base of 2 *i.e.*, it uses only two different symbols.
3. **Octal**—it has a base of 8 *i.e.*, it uses eight different symbols.
4. **Hexadecimal**—it has a base of 16 *i.e.*, it uses sixteen different symbols.

All these systems use the same type of **positional notation** except that

— decimal system uses powers of 10
— binary system uses powers of 2
— octal system uses powers of 8
— hexadecimal system uses powers of 16

Decimal numbers are used to represent quantities which are outside the digital system. Binary system is extensively used by digital systems like digital computers which operate on binary information. Octal system has certain advantages in digital work because it requires less circuitry to get information into and out of a digital system. Moreover, it is easier to read, record and print out octal numbers than binary numbers. Hexadecimal number system is particularly suited for microcomputers.

32.2. The Decimal Number System

We will briefly recount some important characteristics of this more-familiar system before taking up other systems. This system has a base of 10 and is a **position-value system** (meaning that value of a digit depends on its **position**). It has following characteristics :

(*i*) Base or Radix

It is defined as *the number of different digits which can occur in each position in the number system.*

The decimal number system has a base of 10 meaning that it contains ten unique symbols (or digits). These are : 0, 1, 2, 3, 4, 5, 6, 7, 8, 9. Any one of these may be used in each position of the number.

Incidentally, it may be noted that we call it a decimal (10's) system although *it does not have a distinct symbol of 10*. As is well-known, it expresses 10 and any number above 10 as a combination of its ten unique symbols.

(*ii*) Position Value

The absolute value of each digit is fixed but its **position value** (or place value or weight) is determined by its position in the overall number. For example, position value of 3 in 3000 is not the same as in 300. Also, position value of each 4 in the number 4444 is different as shown in Fig. 32.1.

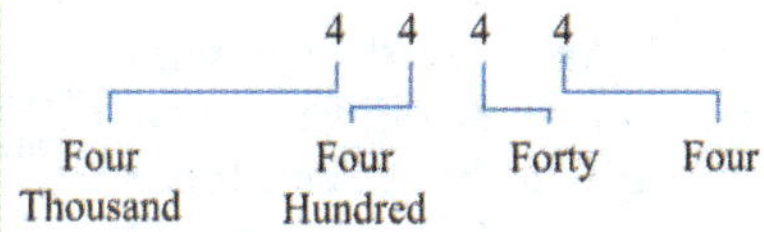

Fig. 32.1

Similarly, the number 2573 can be broken down as follows :

$$2573 = 2 \times 10^3 + 5 \times 10^2 + 7 \times 10^1 + 3 \times 10^0$$

It will be noted that in this number, 3 is the **least significant digit** (*LSD*) whereas 2 is the **most significant** *digit* (*MSD*).

Again, the number 2573.469 can be written as

$$2573.469 = 2 \times 10^3 + 5 \times 10^2 + 7 \times 10^1 + 3 \times 10^0 + 4 \times 10^{-1} + 6 \times 10^{-2} + 9 \times 10^{-3}$$

It is seen that position values are found by raising the base of the number system (*i.e.*, 10 in this case) to the power of the position. Also, powers are *numbered to the left of the decimal point starting with 0 and to the right of the decimal point starting with –1.*

Binary numbers represent all values within computers.

32.3. Binary System

Like decimal (or denary) system, it has a radix and it also uses the same type of position value system.

(*i*) Radix

Its base or radix is **two** because it uses only two digits 0 and 1 (the two binary digit are contracted to **bit**). All binary numbers consist of a string of 0s and 1s. Examples are 10, 101 and 1011 which are read as one-zero, one-zero-one and one-zero-one-one to avoid confusion with decimal numbers. Another way to avoid confusion is to add a subscript of 10 for decimal numbers and of 2 for binary numbers as illustrated below.

10_{10}, 101_{10}, 5742_{10} — decimal number and 10_2, 101_2, 110001_2 — binary numbers.

It is seen that the subscript itself is in decimal. It may be noted that binary numbers need *more places for counting because their base is small.*

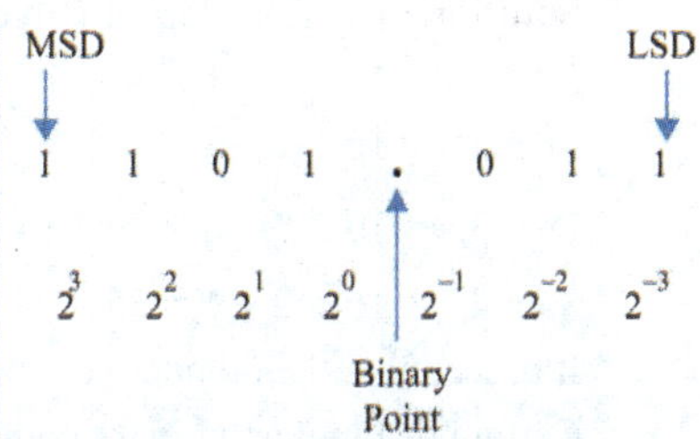

Fig. 32.2

(ii) Position Value

Like the decimal system, binary system is also positionally-weighted. However, in this case, the position value of each bit corresponds to some power of 2. In each binary number, the value increases in powers of 2 starting with 0 to the left of the binary point and decreases to the right of the binary point starting with power of –1. The position value (or weight) of each bit along with a 7-bit binary number 1101.011 is shown in Fig. 32.2.

As seen, the fourth bit to the left of binary point **carries the maximum weight** (*i.e.*, has the highest value) and is called most significant digit (MSD). Similarly, the third bit to the right of the binary point is called least significant digit (LSD). The decimal equivalent of the binary number may be found as under

$$1101.011_2 = (1 \times 2^3) + (1 \times 2^2) + (0 \times 2^1) + (1 \times 2^0) + (0 \times 2^{-1}) + (1 \times 2^{-2}) + (1 \times 2^{-3})$$

$$= 8 + 4 + 0 + 1 + 0 + \frac{1}{4} + \frac{1}{8} = 13.375_{10}$$

As stated earlier, position values of different bits are given by **ascending powers of 2** to the **left** of binary point and by **descending power of 2** to the **right** of binary point. The different digit positions of a given binary number have the following decimal weight (Fig. 32.3).

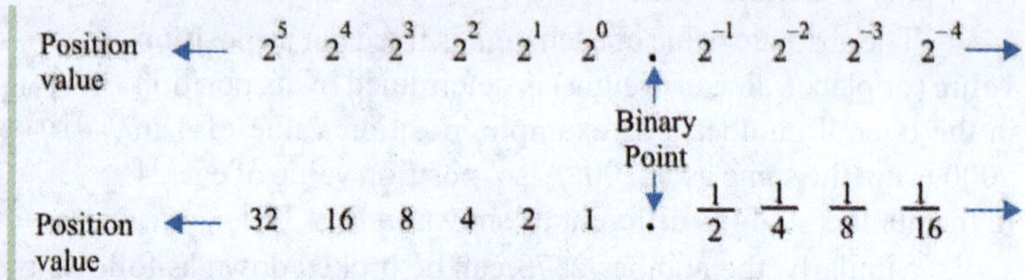

Fig. 32.3

Binary numbers are used extensively by all digital systems primarily due to the nature of electronics itself. The bit 1 may be represented by a saturated (fully-conducting) transistor, a light turned ON, a relay energised or a magnet magnetised in a particular direction. The bit 0, on the other hand, can be represented as a cut-off transistor, a light turned OFF, a relay de-energised or a magnet magnetised in the opposite direction. In such cases, there are only two values which a device can assume.

32.4. Binary to Decimal Conversion

Following procedure should be adopted for converting a given binary integer (whole number) into its equivalent decimal number :

Step 1. Write the binary number *i.e.*, all its bits in a row.

Step 2. Directly under the bits, write 1, 2, 4, 8, 16, ...starting from *right to left.*

Step 3. Cross out the decimal weights which lie under 0 bits.

Step 4. Add the remaining weights to get the decimal equivalent.

Example 32.1. *Convert 11001_2 to its equivalent decimal number.*

Solution. The four steps involved in the conversion are as under

Step 1.	1	1	0	0	1
Step 2.	16	8	4	2	1
Step 3.	16	8	~~4~~	~~2~~	1

Step 4. 16 + 8 + 1 = 25 $\quad\therefore\quad 11001_2 = 25_{10}$

It is seen that the number contains 1 sixteen, one eight, 0 four's, 0 two's and 1 one.

Certain decimal and binary equivalent numbers are tabulated on next page.

Table 32.1

Decimal	*Binary*	*Decimal*	*Binary*
1	1	6	110
2	10	7	111
3	11	8	1000
4	100	9	1001
5	101	10	1010

32.5. Binary Fractions

Here, procedure is the same as for binary integers except that the following weights are used for different bit positions.

$$\bullet \quad 2^{-1} \quad 2^{-2} \quad 2^{-3} \quad 2^{-4} \quad \rightarrow$$

$$\uparrow \quad \frac{1}{2} \quad \frac{1}{4} \quad \frac{1}{8} \quad \frac{1}{16} \quad \rightarrow$$

Binary Point

Example 32.2. *Convert the binary fraction 0.101 into its decimal equivalent.*

Solution. The following four steps will be used for this purpose.

Step 1. $\quad 0 \quad 1 \quad 0 \quad 1$

Step 2. $\quad \frac{1}{2} \quad \frac{1}{4} \quad \frac{1}{4}$

Step 3. $\quad \frac{1}{2} \quad \not{\frac{1}{4}} \quad \frac{1}{8}$

Step 4. $\quad \frac{1}{2} + \frac{1}{8} = \mathbf{0.625}$

$\therefore \quad 0.101_2 = 0.625_{10}$

Example 32.3. *Find the decimal equivalent of the 6-bit binary number 101.101.*

Solution. $1 \quad 0 \quad 1 \quad 1 \quad 0 \quad 1$

$$4 \quad \not{2} \quad 1 \quad \frac{1}{2} \quad \not{\frac{1}{4}} \quad \frac{1}{8} = 5 + \frac{1}{2} + \frac{1}{8} = 5.625$$

$\therefore \quad 101.101_2 = 5.625_{10}$

32.6. Double-Dadd Method

This method of converting binary integers into decimal equivalents is much simpler and quicker than the method given in Art. 32.4 especially in the case of large numbers. Following three steps are involved :

1. Double the first bit to the extreme left and add this doubled value to the next bit on the right.
2. Double the sum obtained and add the doubled value to the next bit.
3. Continue step 2 until the last bit has been added to the previously-doubled sum.

The conversion of 11001_2 is shown in Fig. 32.4. It is seen that $11001_2 = 25_{10}$

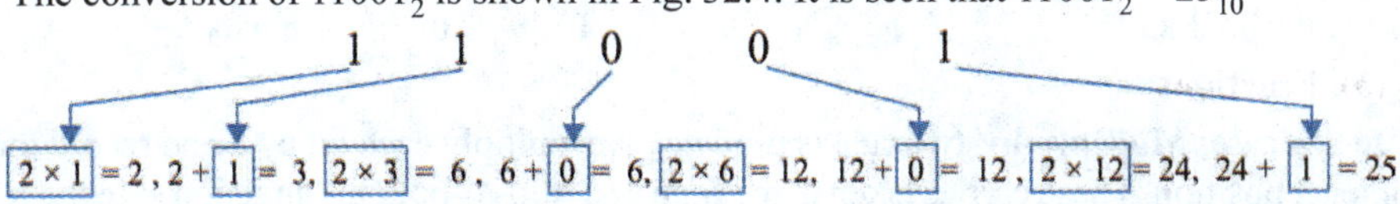

Fig. 32.4

Using double-dadd method, let us convert 111010_2 into its decimal equivalent.

1. $2 \times 1 = 2$, add next bit 1 so that $2 + 1 = 3$
2. $2 \times 3 = 6$, add next bit 1 so that $6 + 1 = 7$
3. $2 \times 7 = 14$, add next bit 0 so that $14 + 0 = 14$
4. $2 \times 14 = 28$, add next bit 1 so that $28 + 1 = 29$
5. $2 \times 29 = 58$, add next bit 0 so that $58 + 0 = 58$

$\therefore \quad 111010_2 = 58_{10}$

32.7. Decimal to Binary Conversion

(*a*) Integers

Such conversion can be achieved by using the so-called **double-dabble method.** It is also known as **divide-by-two** methods. In this method, we progressively divide the given decimal number by 2 and write down the remainders after each division. These remainders taken in the *reverse* order (*i.e.*, from bottom-to-top) form the required binary number. As an example, let us convert 25_{10} into its binary equivalent.

$25 \div 2 = 12$ + remainder of 1 (TOP)
$12 \div 2 = 6$ + remainder of 0
$6 \div 2 = 3$ + remainder of 0
$3 \div 2 = 1$ + remainder of 1
$1 \div 2 = 0$ + remainder of 1 (BOTTOM)

$\therefore \quad 25_{10} = 11001_2$

The above process may be simplified as under :

Successive Divisions	*Remainders*
2) 25	
2) 12	1
2) 6	0
2) 3	0
2) 1	1
2) 0	1

Reading the remainders from bottom to top, we get $25_{10} = 11001_2$

It may also be put in the following form :

$25 \div 2 = 12 + 1$
$12 \div 2 = 6 + 0$
$6 \div 2 = 3 + 0$
$3 \div 2 = 1 + 1$
$1 \div 2 = 0 + 1$

$\therefore$ decimal 25 = 1 1 0 0 1 binary

(*b*) Fractions

In this case, **Multiply-by-two** rule is used *i.e.*, we multiply each bit by 2 and record the carry in the integer position. These carries taken in the **forward** (top-to-bottom) direction give the required binary fraction.

Let us convert 0.8125_{10} into its binary equivalent.

$$
\begin{aligned}
0.8125 \times 2 &= 1.625 &&= 0.625 \text{ with a carry of } 1\\
0.625 \times 2 &= 1.25 &&= 0.25 \text{ with a carry of } 1\\
0.25 \times 2 &= 0.5 &&= 0.5 \text{ with a carry* of } 0\\
0.5 \times 2 &= 1.0 &&= 0.0 \text{ with a carry of } 1\\
\therefore 0.8125_{10} &= 0.1101_2
\end{aligned}
$$

Please note that *we have to add the binary point from our side*. Let us now convert 0.77_{10} into its binary equivalent.

$$
\begin{aligned}
0.77 \times 2 &= 1.54 &&= 0.54 \text{ with a carry of } 1\\
0.54 \times 2 &= 1.08 &&= 0.08 \text{ with a carry of } 1\\
0.08 \times 2 &= 0.16 &&= 0.16 \text{ with a carry of } 0\\
0.16 \times 2 &= 0.32 &&= 0.32 \text{ with a carry of } 0\\
0.32 \times 2 &= 0.64 &&= 0.64 \text{ with a carry of } 0\\
0.64 \times 2 &= 1.28 &&= 0.28 \text{ with a carry of } 1
\end{aligned}
$$

We may stop here but the answer would be approximate.

$$\therefore \quad 0.77_{10} \cong .110001_2$$

Example 32.4. *Convert 25.625_{10} into its binary equivalent.*

Solution. We will do the conversion in two steps (i) first for the integer and (ii) then for the fraction.

(*a*) Integer

$$
\begin{aligned}
25 \div 2 &= 12 + 1\\
12 \div 2 &= 6 + 0\\
6 \div 2 &= 3 + 0\\
3 \div 2 &= 1 + 1\\
1 \div 2 &= 0 + 1\\
\therefore \quad 25_{10} &= 1100_{12}
\end{aligned}
$$

(*b*) fraction

$$
\begin{aligned}
0.625 \times 2 &= 1.25 &&= 0.25 + 1\\
0.25 \times 2 &= 0.5 &&= 0.5 + 0\\
0.5 \times 2 &= 1.0 &&= 0.0 + 1\\
\therefore \quad 0.625_{10} &= 0.101_2
\end{aligned}
$$

Considering the complete number, we have $25.625_{10} = 11001.101_2$

Obviously, binary system needs more bits to express the same number than decimal system.

32.8. Shifting the Place Point

In a decimal number if the decimal point is moved one place to the right, **the number is multiplied by 10.** For example, when decimal point in 7.86 is shifted one place to the right, it becomes 78.6 *i.e.*, it increases the value of the number 10 times. Moving the decimal point one place to the left reduces its value to one-tenth.

In binary numbers, shifting the binary point by one place multiplies or divides the number by 2. For example, 111.0_2 is equal to 7_{10} but 1110.0_2 is 14_{10}. As seen, 7 is doubled to 14 by moving the binary point one place to the right.

Similarly, 11.1_2 is $(2 + 1 + \frac{1}{2}) = 3.5_{10}$. Hence, 111.0_2 is halved to 3.5_{10} by moving its binary point one place to the left.

32.9. Binary Operations

We will now consider the following four binary operations :

1. addition 2. subtraction 3. multiplication 4. division

* It is so because there is 0 at the integer position *i.e.*, to the left of the decimal point.

Addition is the most important of these four operations. In fact, by using **'complements'**, subtraction can be reduced to addition. Most digital computers subtract by complements. It leads to reduction in hardware because only adding type of circuits are required. Similarly, multiplication is nothing but repeated addition and, finally, division is nothing but repeated subtraction.

32.10. Binary Addition

Addition is simply the manipulation of numbers for combining physical quantities. For example, in the decimal number system, 2 + 3 = 5 means the combination of • • with • • • to give a total of • • • • •. Addition of binary numbers is similar to the decimal addition.

Following points will help in understanding the rules of binary addition.

1. When 'nothing' is combined with 'nothing', we get nothing.
   Binary representation of the above statement is : 0 + 0 = 0
2. When nothing is combined with •, we get •.
   In binary language 0 + 1 = 1
3. Combining • with nothing, gives •.
   The binary equivalent is 1 + 0 = 1
4. When we combine • with •, we get • •.

The binary representation of the above is 1 + 1 = 10

It should be noted that the above sum is not 'ten' but 'one-zero' *i.e.*, it represents • • and not • • • • • • • • • •. In other words, it is 10_2 which represents decimal 2. *It is not decimal ten.*

The last rule is often written as 1 + 1 = 0 with a carry of 1.

The above rules for binary addition can be summarized as under :

0 + 0 = 0
0 + 1 = 1
1 + 0 = 1
1 + 1 = 0 with a carry of 1
or = 10_2

It is worth noting that 'carry-overs' are performed in the same manner as in decimal arithmetic. The rules of binary addition could also be expressed in the form of a table as shown below

	0	1
0	0	1
1	1	10_2

or

	0	1
0	0	1
1	1	0 with a carry 1

As an illustration, let us add 101 and 110.

$$\begin{array}{r} 1\,0\,1 \\ +\,1\,1\,0 \\ \hline 1\,0\,1\,1 \end{array}$$

— first column 1 + 0 = 1
— second column 0 + 1 = 1
— third column 1 + 1 = 10 (*i.e.*, 0 with carry 1)

Similarly,

$$\begin{array}{rr} & 1 \text{ carry} \\ & 1\,1\,1 \\ + & 1\,1\,0 \\ \hline & 1\,1\,0\,1 \end{array}$$

— 1st column 1 + 0 = 1
— 2nd column 1 + 1 = 0 with carry 1
— 3rd column 1 + 1 + carry of 1 = 10 + 1 = 11_2

Let us consider one more example :

$$\begin{array}{cccc}
 & \leftarrow & \leftarrow & \leftarrow \\
\text{1 carry} & \text{1 1 carry} & \text{1 1 carry} & \text{1 1 carry} \\
1\,0\,1\,1 & 1\,0\,1\,1 & 1\,0\,1\,1 & 1\,0\,1\,1 \\
\underline{+1\,0\,0\,1} & \underline{+1\,0\,0\,1} & \underline{+1\,0\,0\,1} & \underline{+1\,0\,0\,1} \\
0 & 0\,0 & 1\,0\,0 & 1\,0\,1\,0\,0
\end{array}$$

Hence, we find from the above examples that the only two possible combinations with a carry are:

(*a*) 1 + 1 = sum of 0 with a carry of 1.

It is binary 10 *i.e.*, 10_2 which equals *decimal 2.*

(*b*) 1 + 1 + carry of 1 = a sum of 1 with a carry of 1.

It equals binary 1.1 *i.e.*, 11_2 or decimal 3.

Example 32.5. *Add 110011_2 to 101101_2.*

Solution.

$$\begin{array}{r}
1\,1\,0\,0\,1\,1 \\
\underline{1\,0\,1\,1\,0\,1} \\
1\,1\,0\,0\,0\,0\,0
\end{array}$$

1. **first column :** 1 + 1 = 0 with a carry of 1. Hence, we put down zero there and carry 1 to the second column.
2. **second column :** 1 + 0 = 1. But combined with carry 1 from first column, it gives 1 + 1 = 0 and a carry of 1. Hence, we put down 0 there and carry 1 further to the third column.
3. **third column :** 0 + 1 = 1. Again, when it is combined with carry of 1 from the second column, we get 1 + 1 = 0 with a carry of 1. Hence, again, we put down 0 and carry 1 to the fourth column.
4. **fourth column :** Here, 0 + 1 = 1. When combined with carry 1 from third column, we get 1 + 1 = 0 with a carry of 1. Hence, we put down 0 there and carry 1 to the fifth column.
5. **fifth column :** Here, 1 + 0 = 1. When combined with the carry 1 from fourth column, we get 1 + 1 = 0 with a carry of 1. Hence, we put down 0 there and carry 1 to the sixth column.
6. **sixth column :** Here, it is a case of 1 + 1 + carry of 1 = 11_2 as stated earlier in (*b*) above.

32.11. Binary Subtraction

It is also performed in a manner similar to that used in decimal subtraction. Because binary system has only two digits, binary subtraction requires more borrowing operations than decimal subtraction. The four rules for binary subtraction are as under :

1. 0 – 0 = 0, 2. 1 – 0 = 1, 3. 1 – 1 = 0,

4. 0 – 1 = 1 with a borrow of 1 from the next column of the minuend

or 10 – 1 = 1

The last result represents • • – • = • which makes sense.

While using Rule 4, it should be borne in mind that borrow reduces the remaining minuend by 1. It means that a borrow will cause a 1 in the next column to the left in the minuend to become 0. If the next column also happens to contain 0, it is changed to a 1 and the succeeding 0s in the minuend are changed to 1s until a 1 is found which is then changed to a 0.

Example 1. Let us subtract 0101_2 from 1110_2. The various steps are explained below :

$$\begin{array}{cccc}
\text{0 1 borrow} & 0\,1 & 0\,1 & 0\,1 \\
1\,1\,\not{1}\,0 & 1\,1\,\not{1}\,0 & 1\,1\,\not{1}\,0 & 1\,1\,\not{1}\,0 \\
\underline{-0\,1\,0\,1} & -\,\underline{0\,1\,0\,1} & \underline{-\,0\,1\,0\,1} & -\,\underline{0\,1\,0\,1} \\
1 & 0\,1 & 0\,0\,1 & 1\,0\,0\,1
\end{array}$$

Explanation

1. In the first column, since we cannot subtract 1 from 0, we borrow 1 from the next column to the **left.** Hence, we put down 1 in the answer and change the 1 of the next left column to a 0.
2. We apply Rule 1 to next column *i.e.*, 0 – 0 = 0.
3. We apply Rule 3 to the 3rd column *i.e.*, 1 – 1 = 0.
4. Finally, we apply Rule 2 to the last *i.e.*, fourth column *i.e.*, 1 – 0 = 1.

As a check, it may be noted that talking in terms of decimal numbers, we have subtracted 5 from 14. Obviously, the answer has to be 9 (1001_2).

Example 2

Let us now try subtracting 0001_2 from 1000_2.

Step 1	*Step 2*	*Step 3*	*Step 4*
1	1 1	0 1 1	0 1 1
1 0 $\not{0}$ 0	1 $\not{0}$ $\not{0}$ 1	$\not{1}$ $\not{0}$ $\not{0}$ 1	$\not{1}$ $\not{0}$ $\not{0}$ 1
– 0 0 0 1	– 0 0 0 1	– 0 0 0 1	– 0 0 0 1
1	1	1	0 1 1 1

Since there happened to be a 0 in the second column, it was changed to 1. Again, there was a 0 in the third column, so it was also changed to a 1. Finally, we met a 1 in the forth column which was changed to a 0 and the final answer was written down as shown above.

Example 32.6. *Subtract* 0111_2 *from* 1001_2

Solution.

1 0 0 1	1st column	: 1 – 1 = 0
– 0 1 1 1	2nd column	: 0 – 1 = 1 with a borrow of 1
0 0 1 0	3rd column	: 1 (after borrow) – 1 = 0
	4th column	: 0 (after borrow) – 0 = 0

Example 32.7. *Subtract* 01011_2 *from* 10110_2.

Solution. Step-wise the solution is as under :

Step 1	*Step 2*	*Step 3*	*Step 4*	*Step 5*
0	0 0	0 0	0 0 0	0 0 0
1 0 1 $\not{1}$ 0	1 0 $\not{1}$ $\not{1}$ 0	1 0 $\not{1}$ $\not{1}$ 0	$\not{1}$ 0 $\not{1}$ $\not{1}$ 0	$\not{1}$ 0 $\not{1}$ $\not{1}$ 0
– 0 1 0 1 1	– 0 1 0 1 1	– 0 1 0 1 1	– 0 1 0 1 1	– 0 1 0 1 1
1	1 1	0 1 1	1 0 1 1	0 1 0 1 1

32.12. Complement of a Number

In digital work, two types of complements of a binary number are used for *complemental subtraction*:

(*a*) 1's complement

The 1's complement of a binary number is obtained by changing its each 0 into a 1 and each 1 into a 0. It is also called *radix-minus-one complement*. For example, 1's complement of 100_2 is 011_2 and of 1110_2 is 0001_2.

(*b*) 2's complement

The 2's complement of a binary number is obtained by adding 1 to its 1's complement.

2's complement = 1's complement + 1

It is also known **as true** complement. Suppose we are asked to find 2's complement of 1011_2. Its 1's complement is 0100_2. Next, add 1 to get 0101_2. Hence, 2's complement of 1011_2 is 0101_2.

The complement method of subtraction reduces **subtraction to an addition process.**

This method is popular in digital computers because

1. only adder circuits are needed thus simplifying the circuitry,
2. it is easy with digital circuits to get the complements.

32.13. 1's Complemental Subtraction

In this method, instead of subtracting a number, we add its 1's complement to the minuend. The last carry (whether 0 or 1) is then added to get the final answer. The rules for subtraction by 1's complement are as under :

1. compute the 1's complement of the subtrahend by changing all its 1s to 0s and all its 0s to 1s,
2. add this complement to the minuend,
3. perform the end-around carry of the last 1 or 0,
4. if there is no end-around carry (*i.e.*, 0 carry), then the answer must be recomplemented and a negative sign attached to it,
5. if the end-around carry is 1, no recomplementing is necessary.

Suppose we want to subtract 101_2 from 111_2. The procedure is as under :

```
     1 1 1
   + 0 1 0   ←  1's complement of subtrahend 101
  ---------
  [1] 0 0 1
   └→     1  ←  end-round carry
  ---------
     0 1 0
```

As seen, we have removed from the addition sum the 1 carry in the last position and added it onto the remainder. It is called *end-around carry*.

Let us now subtract 1101_2 from 1010_2.

```
    1 0 1 0
  + 0 0 1 0   ←  1's complement of 1101
  ---------
  ↑ 1 1 0 0
  └NO CARRY
```

As seen, there is no end-around carry in this case. Hence, as per Rule 4 given above, answer must be recomplemented to get 0011 and a negative sign attached to it. Therefore, the final answer becomes – 0011.

Finally, consider the complemental subtraction of 1110_2 from 0110_2.

```
    0 1 1 0
  + 0 0 0 1   ←  1's complement of 1110₂
  ---------
  ↑ 0 1 1 1
  └NO CARRY
```

As seen, there is no carry. However, we may add an extra 0 from our side to make it a 0 carry as shown below.

```
      0 1 1 0
    + 0 0 0 1   ←  1's complement
  -----------
  [0] 0 1 1 1
   └→       0   ←  end-around carry
  -----------
      0 1 1 1
```

After recomplementing, it becomes 1000. When negative sign is attached, the final answer becomes -1000_2.

Example 32.8. *Using 1's complemental method, subtract 01101_2 from 11011_2.*

Solution.

```
      1 1 0 1 1
    + 1 0 0 1 0   ←  1's complement of subtrahend
    -----------
  [1] 0 1 1 0 1
   └──→       1   ←  end-around carry
    -----------
      0 1 1 1 0
```

Since end-around carry is 1, we take the final answer as it is (Rule 5).

Example 32.9. *Use 1's complement to subtract 11011_2 from 01101_2.*

Solution.

```
      0 1 1 0 1
    + 0 0 1 0 0   ←  1's complement of 11011₂
    -----------
     ↑1 0 0 0 1   →  – 01110
     └─NO CARRY
```

Since there is no final carry, we recomplement the answer and attach a minus sign to get the final answer—01110_2.

32.14. 2's Complemental Subtraction

In this case, the procedure is as under:

1. find the 2's complement of the subtrahend,
2. add this complement to the minuend,
3. drop the final carry,
4. if the carry is 1, the answer is positive and needs no recomplementing,
5. if there is no carry, recomplement the answer and attach minus sign.

Example 32.10. *Using 2's complement, subtract 1010_2 from 1101_2*

Solution. The 1's complement of 1010 is 0101. The 2's complement is 0101 + 1 = 0110. We will add it to 1101.

```
        1 1 0 1
      + 0 1 1 0   →  2's complement of 1010₂
      ---------
    [1] 0 0 1 1
     └──→ DROP
```

The final answer is 0011_2.

Example 32.11. *Use 2's complement to subtract 1101_2 from 1010_2.*

Solution. The 1's complement of 1101 is 0010. The 2's complement is 0011.

```
        1 0 1 0
      + 0 0 1 1   →  2's complement of 1101₂
      ---------
      ↑ 1 1 0 1
      └─ NO CARRY
```

In this case, there is no carry. Hence, we have to recomplement the answer. For this purpose, we first subtract 1 from it to get 1100. Next, we recomplement it to get 0011. After attaching the minus sign, the final answer becomes -0011_2.

Talking in terms of decimal numbers, we have subtracted 13 from 10. Obviously the answer is -3.

32.15. Binary Multiplication

The procedure for this multiplication is the same as for decimal multiplication though it is comparatively much easier. The four simple rules are as under :

1. $0 \times 0 = 0$, 2. $0 \times 1 = 0$, 3. $1 \times 0 = 0$, 4. $1 \times 1 = 1$.

The rules of binary multiplication could be summarized in the form of a table as shown.

	0	1
0	0	0
1	0	1

As in the decimal system, the procedure is

1. copy the multiplicand when multiplier digit is 1 but not when it is 0
2. shift as in decimal multiplication
3. add the resulting binary numbers according to the rules of binary addition.

Example 32.12. *Multiply 111_2 by 101_2, using binary multiplication method.*

Solution.

```
     1 1 1
   × 1 0 1
   -------
     1 1 1
   0 0 0        — shift left, no add
 1 1 1          — shift left and add
 ---------
 1 0 0 0 1 1
```

Example 32.13. *Multiply 1101_2 by 1100_2.*

Solution.

```
         1 1 0 1
       × 1 1 0 0
       ---------
         0 0 0 0
       0 0 0 0
     1 1 0 1
   1 1 0 1
   -------------
   1 0 0 1 1 0 0
```

Example 32.14. *Multiply 1111_2 by 0111_2.*

Solution. This example has been included for the specific purpose of explaining how to handle the addition if multiplication results *in columns with more than two 1s*.

1. Result of the first column is 1.
2. In the second column, addition of $1 + 1 = 10_2$. Hence, we put down 0 there and carry 1 to the third column.
3. In the third column, $1 + 1 + 1 + 1 = 100_2$ (decimal 4). We keep one 0 there, put the second 0 in fourth column and pass on 1 to the fifth column.
4. In the fourth column, $1 + 1 + 1 + 0 = 11_2$, (decimal 3). Hence, one 1 is kept there and the other 1 is passed on to the fifth column.

```
             1  1  1  1
          ×  0  1  1  1
          -------------
             1  1  1  1
          1  1  1  1
       1  1  1  1
    -------------------
                      1
                1  0      ← 1 + 1 = 10_2
          1  0  0         ← 1 + 1 + 1 + 1 = 100_2
          1  1            ← 1 + 1 + 1 = 11_2
    1  0  0               ← 1 + 1 + 1 + 1 = 100_2
       1
    -------------------
    1  1  0  1  0  0  1
```

5. In the fifth column, $1 + 1 + 1 + 1 = 100_2$ (decimal 4). Again, one 0 is retained there, second 0 is passed on to the sixth column and 1 to the seventh column.
6. In the sixth column, $1 + 0 = 1$.
7. The seventh column already has 1 given by the addition of the fifth column.

32.16. Binary Division

It is similar to the division in the decimal system. As in that system, here also division by 0 is meaningless. Rules are :

1. $0 \div 1 = 0$ or $\frac{0}{1} = 0$, 2. $1 \div 1 = 1$ or $\frac{1}{1} = 1$.

Example 32.15. *Carry out the binary division 11001 ÷ 101.*

Solution.

```
         1 0 1
      ----------
1 0 1) 1 1 0 0 1
       1 0 1
      ----------
           1 0 1
           1 0 1
           -----
```

After we bring down the next 0 bit, the number 10 so formed is not divisible by 101. Hence, we put a 0 in the quotient. Therefore, the answer is 101_2 (decimal 5).

Incidentally, it may be noted that the dividend 25_{10} and divisor is 5_{10} so that the result of division, as expected, is 5_{10}.

32.17. Shifting a Number to Left or Right

Shifting binary numbers one step to the left or right corresponds respectively to multiplication or division by decimal 2.

When binary number 101100_2 (44_{10}) is shifted one step **to the left,** it becomes 1011000_2 which is 88_{10} *i.e.*, it is doubled. If the given number is shifted one step **to the right,** it becomes 10110_2 which is 22_{10}. Obviously, the number is halved.

32.18. Representation of Binary Numbers as Electrical Signals

As seen from above, any binary number can be represented as a string of 0 s and 1 s. However, it is fine for paper and pencil calculations only. Practical poblem is how to apply the desired binary information to logic circuits in digital computers. For that purpose, two types of electrical signals are selected to represent 1 and 0. Since speed and accuracy are of primary importance in digital circuits, the two electrical signals chosen to represent 1 and 0 must meet very rigid requirements.

1. they must be suitable for use in high-speed circuitry,
2. the signals should be very easy to tell apart,
3. they must be hard to confuse with each other.

The second and third statements may look alike but they, in fact, are not so. It is found that all transistor circuits distort, to some extent, the electrical signals that pass through them. Sometime, these distorted signals can look confusingly alike. Hence, this effect of distortion or degradation has to be kept in mind while selecting the two signals.

In Fig. 32.5 are shown several signal pairs that meet the above requirements. It will be noted that it is impossible to distort a positive pulse (representing 1) to look like the no pulse or negative pulse (representing 0).

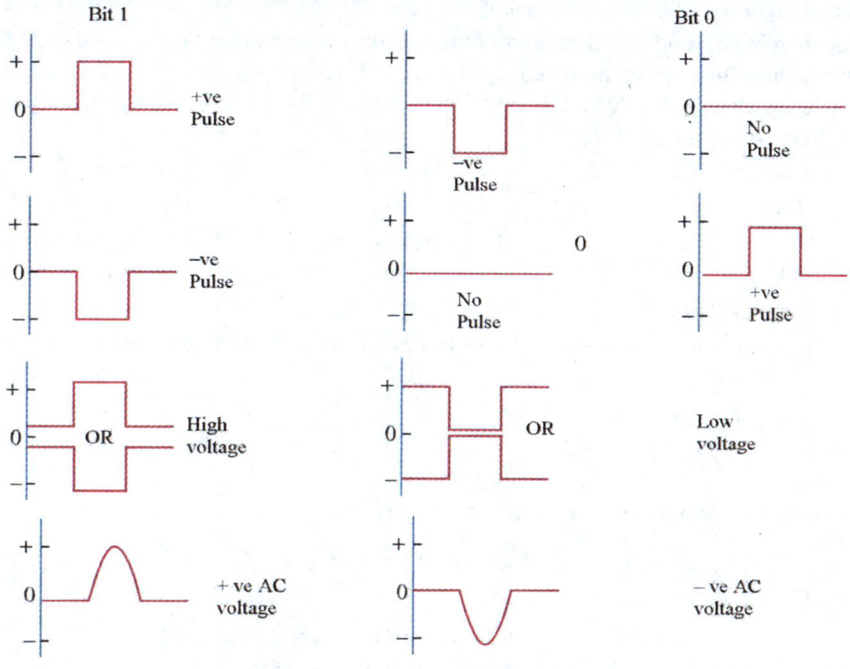

Fig. 32.5
Fig. 32.5

Fig. 32.6 shows how signal pairs can be used to represent different binary numbers.

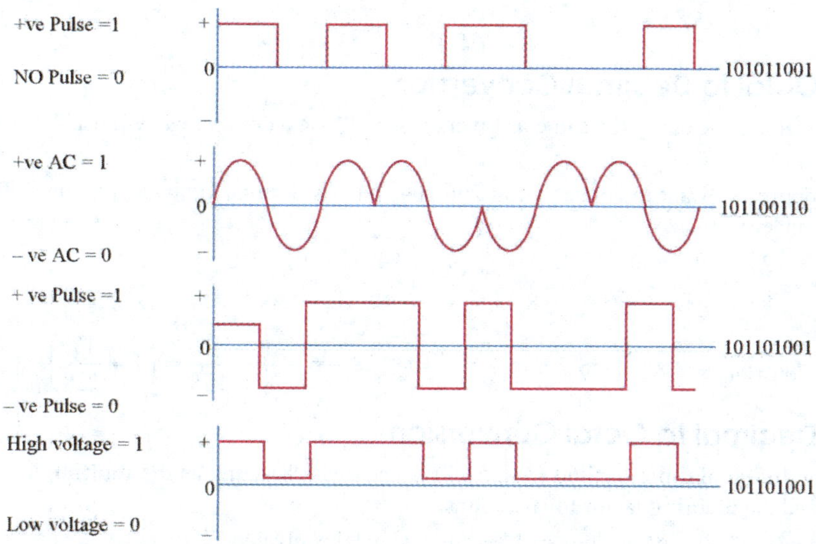

Fig. 32.6

32.19. Octal Number System

(*i*) Radix or Base

It has a base of 8 which means that it has eight distinct counting digits :

0, 1, 2, 3, 4, 5, 6, and 7

These digits 0 through 7, have exactly the same physical meaning as in decimal system.

For counting beyond 7, **2-digit combinations are formed taking the second digit followed by the first, then the second followed by the second and so on.** Hence, after 7, the next octal number is 10 (second digit followed by first), then 11 (second digit followed by second) and so on. Hence, different octal numbers are:

0,	1,	2,	3,	4,	5,	6,	7,
10,	11,	12,	13,	14,	15,	16,	17,
20,	21,	22,	23,	24,	25,	26,	27,
30,	31,	32,	...	...	...	...	...

***(ii)* Position Value**

The position value (or weight) for each digit is given by different powers of 8 as shown below :

$$\leftarrow \quad 8^3 \quad 8^2 \quad 8^1 \quad 8^0 \quad . \quad 8^{-1} \quad 8^{-2} \quad 8^{-3} \quad \rightarrow$$

↑ octal point

For example, decimal equivalent of octal 352 is

$$\begin{array}{cccc} 3 & 5 & 2 & . \; 0 \\ 8^2 & 8^1 & 8^0 & \\ 64 & 8 & 1 & = 3 \times 64 + 5 \times 8 + 2 \times 1 = 234_{10} \end{array}$$

or $352_8 = 3 \times 8^2 + 5 \times 8^1 + 2 \times 8^0 = 192 + 40 + 2 = 234_{10}$

Similarly, decimal equivalent of octal 127.24 is

$$127.24_8 = 1 \times 8^2 + 2 \times 8^1 + 7 \times 8^0 + 2 \times 8^{-1} + 4 \times 8^{-2}$$

$$= 64 + 16 + 7 + \frac{2}{8} + \frac{4}{64} = 97.3125_{10}$$

32.20. Octal to Decimal Conversion

Procedure is exactly the same as given in Art. 32.4 except that we will use digit of 8 rather than 2.

Suppose, we want to convert octal 206.104_8 into its decimal equivalent number. The procedure is as under :

$$\begin{array}{cccccc} 2 & 0 & 6 & 1 & 0 & 4 \\ 8^2 & \not{8}^1 & 8^0 & 8^{-1} & \not{8}^{-2} & 8^{-3} \end{array}$$

$$\therefore\ 206.104_8 = 2 \times 8^2 + 6 \times 8^0 + \frac{1}{8} + \frac{4}{8^3} = 128 + 6 + \frac{1}{8} + \frac{1}{128} = \left(134\frac{17}{128}\right)_{10}$$

32.21. Decimal to Octal Conversion

The **double-dabble** method (Art. 32.7) is used with 8 acting as the **multiplying** factor for **integers** and the **dividing** factor for **fractions.**

Let us see how we can convert $\mathbf{175_{10}}$ into its octal equivalent.

$175 \div 8 = 21$ with 7 remainder ↑
$21 \div 8 = 2$ with 5 remainder
$2 \div 8 = 0$ with 2 remainder

Taking the remainders in the **reverse order,** we get 257_8. $\therefore \quad 175_{10} = 257_8$

Let us now take decimal fraction 0.15. Its octal equivalent can be found as under :

$$0.15 \times 8 = 1.20 = 0.20 \quad \text{with a carry of 1}$$
$$0.20 \times 8 = 1.60 = 0.60 \quad \text{with a carry of 1}$$
$$0.60 \times 8 = 4.80 = 0.80 \quad \text{with a carry of 4 etc.} \downarrow$$

$\therefore \quad 0.15_{10} \cong 114_8$

As seen, here carries have been taken in the **forward direction** *i.e.*, from top to bottom.

Using positional notation, the first few octal numbers and their decimal equivalents are shown in Table 32.2.

Table 32.2

Octal	*Decimal*	*Octal*	*Decimal*
0	0	12	10
1	1	13	11
2	2	14	12
3	3	15	13
4	4	16	14
5	5	17	15
6	6	20	16
7	7	21	17
10	8	22	18
11	9	23	19

32.22. Binary to Octal Conversion

The simplest procedure is to use **binary-triplet** method. In this method, the given binary number is arranged into groups of 3 bits starting from the octal point and then each group is converted to its equivalent octal number. Of course, where necessary, extra 0s can be added **in front** (*i.e.*, left end) of the binary number to complete groups of three.

Suppose, we want to convert 101011_2 into its octal equivalent. Converting the bits into groups of three, we have

101 011

Now, 101_2 is 5 octal and 011_2 is 3 octal.

$$\therefore \quad \begin{matrix} 101 & 011 \\ \downarrow & \downarrow \\ 5 & 3 \end{matrix}$$

$\therefore \quad 101\ 011_2 = 53_8$

Now, take 111110111_2. We will first split it into groups of **three bits** (space is left between the groups for easy reading). Then, each group is given its octal number as shown below.

$$\begin{matrix} 111 & 110 & 111 \\ \downarrow & \downarrow & \downarrow \\ 7 & 6 & 7 \end{matrix}$$

$\therefore \quad 111\ 110\ 111_2 = 767_8$

Finally, take the example of a mixed binary number 10101.11_2. Here, we will have to add one 0 in front of the integral part as well as to the fractional part

$$\begin{array}{ccc} 010 & 101 & 110 \\ \downarrow & \downarrow & \downarrow \\ 2 & 5 & 6 \end{array}$$

$\therefore \quad 10101.11_2 = 25.6_8$

The equivalence between binary triplets and octal numbers is given in Table 32.3.

Table 32.3

Binary	*Octal*	*Binary*	*Octal*
000	0	1010	12
001	1	1011	13
010	2	1100	14
011	3	1101	15
100	4	1110	16
101	5	1111	17
110	6	10000	20
111	7	10001	21
1000	10	10010	22
1001	11	10011	23

32.23. Octal to Binary Conversion

The procedure for this conversion is just the opposite of that given in Art. 32.22. Here, each digit of the given octal number is converted into its equivalent binary triplet. For example, to change 75_8 into its binary equivalent, proceed as under :

$$\begin{array}{cc} 7 & 5 \\ \downarrow & \downarrow \\ 111 & 101 \end{array}$$

$\therefore \quad 75_8 = 111\ 101_2$

Similarly, 74.562_8 can be converted into binary equivalent as under :

$$\begin{array}{cccccc} 7 & 4 & . & 5 & 6 & 2 \\ \downarrow & \downarrow & . & \downarrow & \downarrow & \downarrow \\ 111 & 100 & & 101 & 110 & 010 \end{array}$$

$\therefore \quad 74.562_8 = 111\ 100.101\ 110\ 010_2$

Incidentally, it may be noted that number of digits in octal numbers is **one-third of that in equivalent binary numbers.** In the present case, it is five versus fifteen.

32.24. Advantages of Octal Number System

The chief advantage of this system is that its numbers *are one-third in length* as compared to binary numbers. Hence, from the user's *i.e.*, computer operator's point of view, it would be comparatively much easier to handle the input and output data of a digital computer in octal form. Moreover, the print-outs are more compact and easy to read. Additionally, conversion from binary-to-octal and octal-to-binary is quick and simple. In fact, octal numbers are used to represent binary numbers because of ease of conversion and compactness.

32.25. Hexadecimal Number System

The characteristics of this system are as under :

1. it has a base of 16. Hence, it uses sixteen distinct counting digits 0 through 9 and A through F as detailed below :

   0, 1, 2, 3, 4, 5, 6, 7, 8, 9, A, B, C, D, E, F
2. place value (or weight) for each digit is in **ascending powers** of 16 for integers and **descending powers** of 16 for fractions.

The chief use of this system is in connection with **byte-organised machines.** It is used for specifying addresses of different binary numbers stored in computer memory.

32.26. How to Count beyond F in Hex Number System ?

As usual, we resort to **2-digit combinations.** After reaching F, we take the second digit followed by the first digit, then second followed by second, then second followed by third and so on. The first few 'hex' numbers and their decimal equivalents are given in Table 32.4.

Table 32.4

Hexadecimal	*Decimal*	*Hexadecimal*	*Decimal*	*Hexadecimal*	*Decimal*
0	0	B	11	16	22
1	1	C	12	17	23
2	2	D	13	18	24
3	3	E	14	19	25
4	4	F	15	1A	26
5	5	10	16	1B	27
6	6	11	17	1C	28
7	7	12	18	1D	29
8	8	13	19	1E	30
9	9	14	20	1F	31
A	10	15	21	20	32
				21	33

32.27. Binary to Hexadecimal Conversion

The simple method is to split the given binary number into **4-bit groups** (supplying 0s from our own side if necessary) and then give each group its `hex' value as found from Table 32.5.

Table 32.5

Binary	*Hex.*	*Binary*	*Hex.*	*Binary*	*Hex.*
0000	0	0110	6	1100	C
0001	1	0111	7	1101	D
0010	2	1000	8	1110	E
0011	3	1001	9	1111	F
0100	4	1010	A	10000	10
0101	5	1011	B	10001	11

Let us see how we would convert 10001100_2 into its hexadecimal equivalent number. We will first split the given binary number into 4-bit groups and then give each group its proper value from Table 32.5.

1000   1100
↓   ↓
8   C $\therefore\ 1000\ 1100_2 = 8C_{16}$

Let us now consider 1011010111_2. Following the above procedure, we have

0010   1101   0111
↓   ↓   ↓
2   D   7

$\therefore \quad 10\ 1101\ 0111_2 = 2D7_{16}$

It is seen that two 0s have been added to complete the 4-bit groups.

32.28. Hexadecimal to Binary Conversion

Here, the procedure is just the reverse of that given in Art. 32.27. Each hexadecimal digit is converted into its equivalent 4-bit binary.

Suppose, we want to convert $23A_{16}$ into its binary equivalent. It can be done as given below :

2   3   A
↓   ↓   ↓
0010   0011   1010 $\therefore\ 23A_{16} = 0010\ 0011\ 1010_2$

Incidentally, hex numbers contain **one-fourth** the number of bits contained in the equivalent binary number. Optionally, we could drop off the two 0s in front of the binary equivalent.

$\therefore \quad 23A_6 = 10\ 0011\ 1010_2$

Example 32.16. *A computer is transmitting the following groups of bytes (each consisting of 8-bits) to some output device. Give the equivalent octal and hexadecimal listings.*

1000   1100   0011   1010
0010   1110   1001   0101
0101   1111   1011   0110
0111   1011   0101   1011

Solution.

Binary		*Octal*			*Hexadecimal*	
1000	1100	10	001	100	1000	1100
		2	1	4	8	C
0010	1110	00	101	110	0010	1110
		0	5	6	2	E
0101	1111	01	011	111	0101	1111
		1	3	7	5	F
0111	1011	01	111	011	0111	1011
		1	7	3	7	B
0011	1010	00	111	010	0011	1010
		0	7	2	3	A
1001	0101	10	010	101	1001	0101
		2	2	5	9	5
1011	0110	10	110	110	1011	0110
		2	6	6	B	6
0101	1011	01	011	011	0101	1011
		1	3	3	5	B

Hence, the groups of given memory bytes when expressed in different number systems become as under :

Binary	*Octal*	*Hexadecimal*
10001100	214	8C
00101110	056	2E
01011111	137	5F
01111011	173	7B
00111010	072	3A
10010101	225	95
10110110	266	B6
01011011	133	5B

It is clear from above that so far as the computer operator (or programmer) is concerned, it is much easier to handle this data when expressed in octal or hexadecimal system than in the binary system. For example, it is much easier and less error-prone to write the hexadecimal 8C than the binary 10001100 or 6AF than 011010101111. Of course, when the need arises, the operator can easily convert from octal or hexadecimal to binary.

CONVENTIONAL PROBLEMS

1. Find the decimal equivalents of the following binary numbers :

(*a*) 101 (*b*) 1001 (*c*) 10.011 **[(a) 5_{10} (b) 9_{10} (c) 3.0375_{10}]**

2. What are the decimal equivalents of the following binary numbers ?

(*a*) 1111 (*b*) 10100 (*c*) 1101101 (*d*) 10011001

[(a) 15_{10} (b) 20_{10} (c) 109_{10} (d) 153_{10}]

3. Express the following binary numbers into their equivalent decimal numbers :

(*a*) 11.01 (*b*) 101.11 (*c*) 110.01

[(a) 3.25_{10} (b) 5.75_{10} (c) 6.25_{10}]

4. Convert the following decimal numbers into their binary equivalents :

(*a*) 25 (*b*) 125 (*c*) 0.85 **[(a) 1100_{12} (b) 111110_{12} (c) 0.110110_2]**

5. What are the binary equivalents of the following decimal numbers ?

(*a*) 27 (*b*) 92 (*c*) 64 **[(a) 11011_2 (b) 1011100_2 (c) 1000000_2]**

6. Perform the following binary additons :

(*a*) 1011 + 1001 (*b*) 1011 + 110 (*c*) 101101 + 1101101 (*d*) 1011.01 + 1001.11

(*e*) 0.0011 + 0.1110 (*f*) 1111 + 111 + 1111

[(a) 10100_2 (b) 10001_2 (c) 1100011_2 (d) 10101_2 (e) 1.0001_2 (f) 1101001_2]

7. Perform the following binary subtractions :

(*a*) 1101 – 1011 (*b*) 111 – 101 (*c*) 1000 – 11 (*d*) 101011 – 10010

[(a) 0010_2 (b) 010_2 (c) 01001_2 (d) 011001_2]

8. Carry out the following subtractions using binary number system :

(*a*) $64_{10} - 32_{10}$ (*b*) $128_{10} - 64_{10}$ (*c*) $93.5_{10} - 42.75_{10}$ (*d*) $7_{10} - 11_{10}$

[(a) 100000_2 (b) 1000000_2 (c) 110010.11_2 (d) – 100_2]

9. Find the 1's complements of the following binary numbers :

(*a*) 01101 (*b*) 1101 (*c*) 1001 (*d*) 1010

[(a) 10010_2 (b) 0010_2 (c) 0110_2 (d) 0101_2]

10. What are 2's complements of the following binary numbers ?

(*a*) 1011 (*b*) 11011 (*c*) 11011.01 (*d*) 10011.11

[(a) 0101_2 (b) 00101_2 (c) 00100.11_2 (d) 01100.01_2]

11. Multiply the following binary numbers :

(a) 1100 × 101 (b) 10101 × 101 (c) 10111 × 101 (d) 1110 × 111

[(a) 111100_2 (b) 1101001_2 (c) 1110011_2 (d) 1100010_2]

12. Perform the following binary divisions :

(a) 11011 ÷ 100 (b) 1110011 ÷ 101 (c) 1100010 ÷ 111

[(a) 110.11_2 (b) 10111_2 (c) 1110_2]

13. Convert the following binary numbers into their octal equivalents :

(a) 1001 (b) 11011 (c) 10 101 111 (d) 1101.0110111

(e) 11 111 011 110 101 **[(a) 11_8 (b) 33_8 (c) 25_8 (d) 15.334_8 (e) 37365_8]**

14. Convert the undergiven octal numbers into their binary equivalents :

(a) 13 (b) 11 (c) 713 (d) 3674

[(a) 1011_2 (b) 1001_2 (c) $111\ 001\ 011_2$ (d) $11\ 110\ 111\ 100_2$]

15. Convert the following numbers :

(a) 357_8 to decimal (b) 6421_8 to decimal

(c) 1359_{10} to octal (d) 7777_{10} to octal **[(a) 239_{10} (b) 3345_{10} (c) 2517_8 (d) 17141_8]**

16. Convert the following binary numbers into their equivalent hexadecimal numbers :

(a) 1101 0111 (b) 1010 0110 (c) 1001 1110 (d) 1100 1111

[(a) $D7_{16}$ (b) $A6_{16}$ (c) $9E_{16}$ (d) CF_{16}]

17. What are the binary equivalents of the following hexadecimal numbers ?

(a) 9F (b) A2 (c) ED (d) 37

[(a) $1001\ 1111_2$ (b) $1010\ 0010_2$ (c) $1110\ 1101_2$ (d) $0011\ 0111_2$]

SELF EXAMINATION QUESTIONS

A. Fill in the blanks with most appropriate word(s) or numerical value(s).

1. The bits 1 and 0 are used in the .... number system.
2. The double-dabble method is used for .... to binary conversion.
3. The decimal fraction 0.75 corresponds to the binary fraction of ....
4. In the binary system, 1 + 1 = ...
5. $1_{10} + 1_{10}$ = ...
6. $1_2 + 1_2$ = ...
7. According to the rules of binary addition, 1 + 1 + 1 = ...
8. The binary addition 1 + 1 + 1 + 1 = ...
9. According to the rules of binary subtraction, 0 – 1 = 1 with .... of 1.
10. Binary 111 111 is the same as decimal ....
11. The 2's complement of the binary number 1011 is ....
12. As per binary multiplication rules, $11_2 \times 10_2$ = ...
13. In binary number system, $101_2 \times 2$ = ...
14. $11_2 + 111_2$ = ...
15. The octal 11_8 represents decimal .... and binary ....
16. $101\ 011_8$ represents decimal ....
17. Each memory byte of a minicomputer usually consists of .... bits.
18. The hexadecimal number system has a base of ....
19. The number D_{16} corresponds to decimal ....
20. The number 20_{16} corresponds to decimal ....
21. The hex number $8C_{16}$ corresponds to the binary number ....
22. The hexadecimal equivalent of binary number 1010 is ....

B. Answwer True or False

1. In digital systems, information is most often processed in binary number system.
2. In binary system, the place value of a digit increases or decreases by power of 2.
3. In 1110_2 the 0 carrier the highest weight.
4. All binary numbers consist of a string of 1s and 0s.
5. In binary numbers, shifting the binary point one place multiplies or divides it by 2.
6. $11_2 + 01_2 = 100_2$.

7. $11_2 - 01_2 = 10_2$.
8. The number 111_2 is read as one hundred and eleven.
9. $1_2 + 1_2 = 2_2$.
10. $10_2 \times 10_2 = 100_2$.
11. $1101_2 \div 10_2 = 110.1_2$.
12. The two's complement of 100_2 is 100_2.
13. $1101_2 \times 100_2 = 110100_2$.
14. The number 77 is octal but not 78.

C. Multiple Choice Items

1. The digital systems usually operate on ....system.
   (*a*) binary (*b*) decimal
   (*c*) octal (*d*) hexadecimal
2. The binary system uses powers of ....for positional values.
   (*a*) 2 (*b*) 10
   (*c*) 8 (*d*) 16
3. After counting 0, 1, 10, 11, the next binary number is
   (*a*) 12 (*b*) 100
   (*c*) 101 (*d*) 110
4. The number 1000_2 is equivalent to decimal number
   (*a*) one thousand (*b*) eight
   (*c*) four (*d*) sixteen
5. The binary addition 1 + 1 + 1 gives
   (*a*) 111 (*b*) 10
   (*c*) 110 (*d*) 11
6. The 2's complement of 1000_2 is
   (*a*) 0111 (*b*) 0101
   (*c*) 1000 (*d*) 0001
7. The chief reason why digital computers use complemental subtraction is that it
   (*a*) simplifies their circuitry
   (*b*) is a very simple process
   (*c*) can handle negative numbers easily
   (*d*) avoids direct subtraction
8. The number 12_8 is equivalent to decimal
   (*a*) 12 (*b*) 20
   (*c*) 10 (*d*) 4
9. The number 100101_2 is equivalent to octal
   (*a*) 54 (*b*) 45
   (*c*) 37 (*d*) 25
10. The number 17_8 is equivalent to binary
    (*a*) 111 (*b*) 1110
    (*c*) 10000 (*d*) 1111
11. Which of the following is NOT an octal number ?
    (*a*) 19 (*b*) 77
    (*c*) 15 (*d*) 101
12. Hexadecimal number system is used as a shorthand language for representing ....numbers.
    (*a*) decimal (*b*) binary
    (*c*) octal (*d*) large
13. The binary equivalent of A_{16} is
    (*a*) 1010 (*b*) 1011
    (*c*) 1000 (*d*) 1110

ANSWERS

A. Fill in the blanks

1. binary	**2.** decimal	**3.** 0.11	**4.** 10	**5.** 2
6. 10	**7.** 11	**8.** 100	**9.** borrow	**10.** 63
11. 0101	**12.** 110	**13.** 1010	**14.** 1010	**15.** 9, 1001
16. 53	**17.** 8	**18.** 16	**19.** 13	**20.** 32
21. 10001100	**22.** A			

B. True or False

1. T	**2.** T	**3.** F	**4.** T	**5.** T	**6.** T
7. T	**8.** F	**9.** F	**10.** T	**11.** F	**12.** T
13. T	**14.** T				

C. Multiple Choice Items

1. *a*	**2.** *a*	**3.** *b*	**4.** *b*	**5.** *d*	**6.** *c*
7. *a*	**8.** *c*	**9.** *b*	**10.** *d*	**11.** *a*	**12.** *b*
13. *a*					

33 CHAPTER

Logic Gates

1. The OR Gate
2. Transistor OR Gate
3. OR Gate Symbolizes Logic Addition
4. Three Input OR Gate
5. Exclusive OR Gate
6. The AND Gate
7. Equivalent Relay Circuit of an AND Gate
8. Diode AND Gate
9. Transistor AND Circuit
10. AND Gate Symbolizes Logic Multiplication
11. The NOT Gate
12. Equivalent Circuits for a NOT Gate
13. The NOT Operation
14. Bubbled Gates
15. The NOR Gate
16. NOR Gate is a Universal Gate
17. The NAND Gate

33.1. Definition

A logic gate is an electronic circuit **which makes logic decisions.** It has one output and one or more inputs. The output signal appears only for certain combinations of input signals. Logic gates are the basic building blocks from which most of the digital systems are built up. They implement the hardware logic function based on the logical algebra developed by George Boole, which is called Boolean algebra in his honour. A unique characteristic of the Boolean algebra is that variables used in it **can assume only one of the two values** *i.e.*, either 0 or 1. Hence, every variable is either a 0 or a 1.

These gates are available today in the form of various IC families. The most popular families are: transistor-transistor logic (TTL), emitter-coupled logic (ECL), metal-oxide semiconductor (MOS) and complementary metal-oxide semiconductor (CMOS).

A reasearcher holds up a chip containing the molecular logic gates.

In this chapter, we will consider the OR, AND, NOT, NOR, NAND and exclusive OR (XOR) gates along with their truth tables.

33.2. Positive and Negative Logic

In computing systems, the number symbols 0 and 1 represent two possible states of a circuit or device. It makes no difference if these two states are referred to as ON and OFF, CLOSED and OPEN, HIGH and LOW, PLUS and MINUS or TRUE and FALSE depending on the circumstances. Main point is that **they must be symbolized by two opposite conditions.**

In positive logic, a 1 represents

1. an ON circuit 2. a CLOSED switch 3. a HIGH voltage
4. a PLUS sign 5. a TRUE statement

Consequently, a 0 represents

1. an OFF circuit 2. an OPEN switch 3. a LOW voltage
4. a MINUS sign 5. a FALSE statement.

In negative logic, just opposite conditions prevail.

Suppose, a digital system has two voltage levels of 0 V and 5 V. If we say that symbol 1 stands for 5 V and symbol 0 for 0 V, then we have positive logic system. If, on other hand, we decide that 1 should represent 0 V and 0 should represent 5 V, then we will get negative logic system.

Main point is that in *positive* logic, **the more positive** of the two voltage levels represents the 1 while in negative logic, **the more negative** voltage represents the 1. Moreover, it is not essential that a 0 has to be represented by 0 V although in some cases the two may coincide. Suppose, in a circuit, the two voltage levels are 2 V and 10 V. Then for positive logic, the 1 represents 10 V and the 0 represents 2 V (*i.e.*, lesser of the two voltages). If the voltage levels are – 2 V and – 8 V, then, in positive logic, the 1 represents – 2 V and the 0 represents – 8 V (*i.e.*, lesser of the two voltages).

Unless stated otherwise, we will be using only *positive logic* in this chapter.

33.3. The OR Gate

The electronic symbol for a two-input OR gate is shown in Fig. 33.1 (*a*) and its equivalent switching circuit in Fig. 33.1 (*b*). The two inputs have been marked as *A* and *B* and the output as *C*. It is worth reminding the reader that as per Boolean algebra, the three variables *A*, *B* and *C* can have only one of the two values *i.e.*, either 0 or 1.

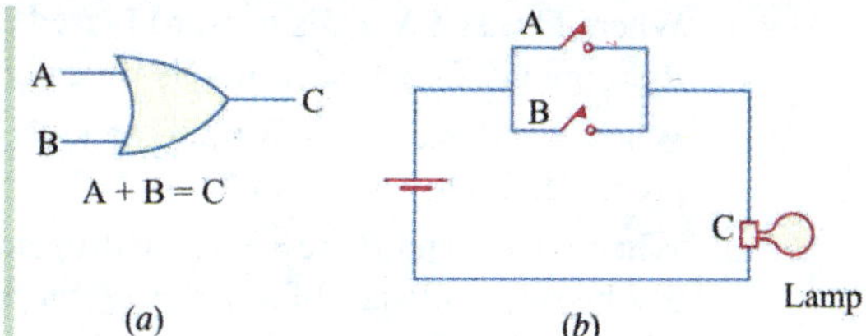

Fig. 33.1

Logic Operation

The OR gate has an **output of 1 when either *A* or *B* or both are 1.**

In other words, it is an **any-or-all** gate because an output occurs when any or all the inputs are present.

As seen from Fig. 33.1 (*b*), the lamp will light up (logic 1) when either switch *A* or *B* or both are closed.

Table 33.1

A	*B*	*C*
0	0	0
0	1	1
1	0	1
1	1	1

Obviously, the output would be 0 **if and only if both its inputs are** 0. In terms of the switching conditions, it means that lamp would be OFF (logic 0) only when both switches *A* and *B* are OFF.

The OR gate represents the Boolean equation $A + B = C$.

The meaning of this equation is that *C* is true when either *A* is true or *B* is true or both are true. Alternatively, it means that output *C* is 1 when either *A* or *B* or both are 1.

The above logic operation of the OR gate can be summarised with the help of the truth table. A truth table may be defined as a table which *gives the output state for all possible input combinations.* The OR Table 33.1 gives outputs for all possible *AB* inputs of 00, 01, 10 and 11.

We may interpret the truth table as follows:

When both inputs are 0 (switches are OPEN), output *C* is 0 (lamp is OFF). When *A* is in logic state 0 (switch *A* is OPEN) but *B* is in logic state 1 (switch *B* is CLOSED), the output *C* is logic state 1 (lamp is ON). Lamp would be also ON when *A* is CLOSED and *B* is OPEN Of course, lamp would be ON when both switches are CLOSED. It is so because an OR gate is equivalent to **a parallel** *circuit in its logic function.*

Another point worth remembering is that the above OR gate is called **inclusive** OR gate because **it includes the case when both inputs are true.**

33.4. Equivalent Relay Circuit of an OR Gate

In Fig. 33.2, the relay contacts have been wired in *parallel.* When +5 V is applied to *A*, relay K_1 is energised and pulls *M* down, thereby closing the contact. Hence, supply voltage of +5 V appears at the output *C*.

Similarly, when +5 V are applied to input *B*, K_2 is energised and pulls *N* down thereby bringing *C* in contact with *S*. Of course, when both *A* and *B* are at +5 V, *C* is at +5 V. Incidentally, when inputs at *A* and *B* are 0, *C* is also 0.

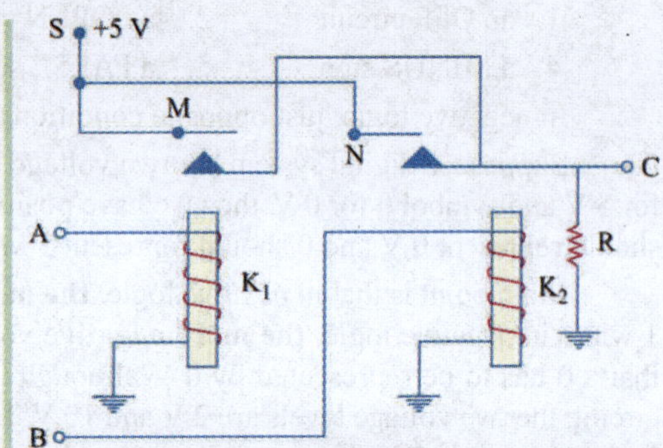

Fig. 33.2

33.5. Diode OR Gate

Fig. 33.3 shows the diode OR gate consisting of two ideal diodes D_1 and D_2 connected in parallel across the output *C*.

1. When *A* is at +5 V, D_1 is forward-biased and hence conducts. The circuit current flows *via* *R* dropping 5 V across it. In this way, point *C* achieves potential of +5 V.
2. When + 5 V is applied to *B*, D_2 conducts causing point *C* to go to +5 V.
3. When both *A* and *B* are +5 V, the drop across *R* is 5 V because voltages of *A* and *B* *are in parallel.* Again, point *C* is driven to +5 V.
4. Obviously, when there is no voltage either at *A* or *B*, output *C* remains 0.

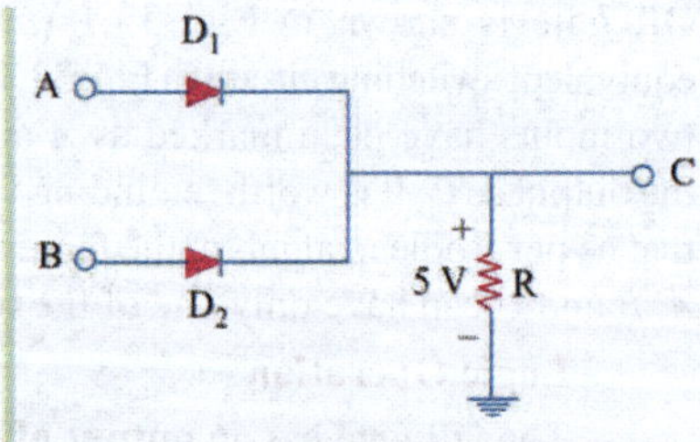

Fig. 33.3

33.6. Transistor OR Gate

Fig. 33.4 illustrates a possible transistor OR gate consisting of three interconnected transistors Q_1,Q_2, and Q_3 supplied from a common supply V_{CC} = +5 V.

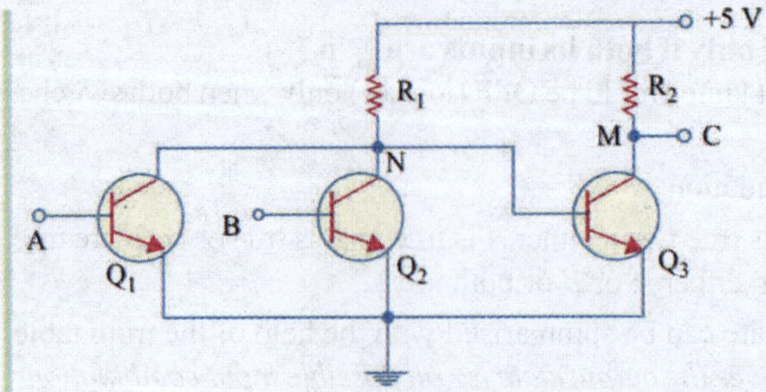

Fig. 33.4

1. When +5 V is applied to *A*, Q_1 is forward-biased and so conducts. Assuming that Q_1 is saturated, entire V_{CC} = 5 V drops across R_1 thus causing *N* to go to ground. This, in turn, cuts off Q_3 thereby causing *C* to go to V_{CC} *i.e.*, +5 V.
2. When +5 V is applied to *B*, Q_2 conducts thereby driving *N* to ground *i.e.*, 0 V. With no forward bias on its base, Q_3 is cut-off thus driving *C* again to V_{CC} *i.e.*, +5 V.

3. If both inputs A and B are grounded, Q_1 and Q_2 are cut-off driving N to + 5 V. As a result, Q_3 becomes forward-biased and conducts fully. In that case, entire V_{CC} drops across R_2 driving M and hence C to ground.

33.7. OR Gate Symbolizes Logic Addition

According to Boolean algebra, OR gate performs **logical addition.** Its truth table can be written as given below:

$$0 + 0 = 0$$
$$0 + 1 = 1$$
$$1 + 0 = 1$$
$$1 + 1 = 1$$

It must be clearly understood that '+' sign in Boolean algebra **does not stand for the addition** as understood in the ordinary or numerical algebra. In symbolic logic, the '+' sign indicates OR operation whose rules are given along side. In logic algebra, $A + B = C$ means that if A is true **OR** B is true, then C will be true. *It does* **not** *mean here that sum of A and B equals C.* The other symbols used for '+' sign are U and V. Hence, the above equation could also be written as $AUB = C$ or $AVB = C$.

The meaning of the last three logic additions is that output is 1 when either input A or B or both are 1. The first addition implies that output is 0 *only when both inputs are 0.*

The meaning of the '+' sign often becomes clear from the context as shown below:

$1 + 1 = 2$ — decimal addition

$1 + 1 = 10$ — binary addition

$1 + 1 = 1$ — OR addition

We can put the above OR laws in more general terms

$A + 1 = 1$

$A + 0 = A$

$A + A = A$ — not $2A$

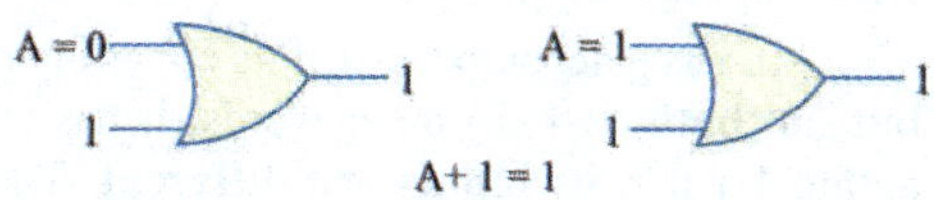

(a) (b)
Fig. 33.5

(*i*) $\boldsymbol{A + 1 = 1}$

As we know, A can have two values: 0 or 1.
When A is 0, then we have $0 + 1 = 1$ as shown in Fig. 33.5 (*a*).

When $A = 1$, then the above expression becomes : $1 + 1 = 1$ as shown in Fig. 33.5 (*b*). Hence, we find that **irrespective of the value of A.**

$A + 1 = 1$

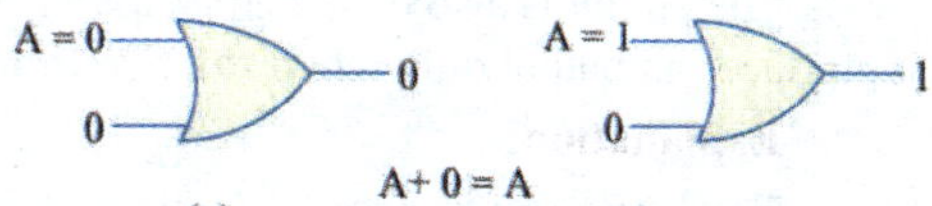

(a) (b)
Fig. 33.6

(*ii*) $\boldsymbol{A + 0 = A}$

If $A = 0$, then $0 + 0 = 0$ *i.e.*, output is 0 which is correct and is shown in Fig. 33.6 (*a*). The output is what the value of A is.

As shown in Fig. 33.6 (*b*), when $A = 1$, output is 1, because $1 + 0 = 1$. Again, output is what the value of A is.

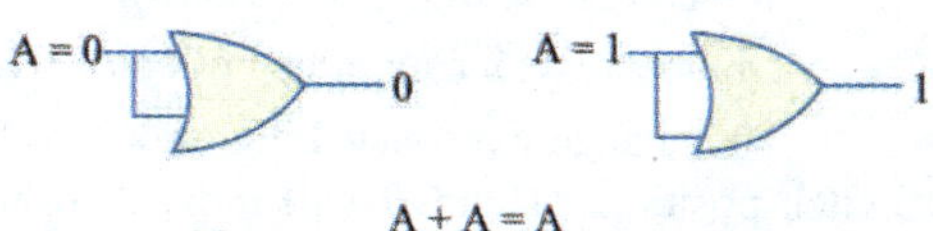

(a) (b)
Fig. 33.7

(*iii*) $\boldsymbol{A + A = A}$

With A set to 0, the output is 0 because $0 + 0 = 0$ as shown in Fig. 33.7 (*a*).

With A set to 1, the output is 1 because $1 + 1 = 1$ as shown in Fig. 33.7 (*b*). Obviously, the output in both cases is A.

33.8. Three Input OR Gate

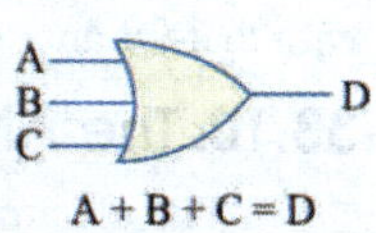

Fig. 33.8

The electronic symbol for a 3-input (fan-in of 3) *inclusive* OR gate is shown in Fig. 33.8. As is usual in logic algebra, the inputs A, B, C as well as the output D can have only one of the two values *i.e.*, 0 or 1.

Truth Table

It is shown in Table 33.2. Following points are worth noting:

1. The number of rows in the table is $2^3 = 8$ *i.e.*, there are eight ways of combining the three inputs. In general, the number of horizontal rows is 2^n where n is the number of inputs.
2. In first column A, logic values alternate between 0 and 1 every four rows twice.
3. The second input column B alternates between 0 and 1 every two rows four times.
4. The third input column C alternates between 0 and 1 every other row eight times.

Table 33.2

A	B	C	D
0	0	0	0
0	0	1	1
0	1	0	1
0	1	1	1
1	0	0	1
1	0	1	1
1	1	0	1
1	1	1	1

The truth table symbolizes the Boolean equation

$$A + B + C = D$$

which means that output D is 1 when *either* A or B or C is 1 or *all are* 1. Alternatively, D is true when either A or B or C is true or all are true.

33.9. Exclusive OR Gate

Its electronic symbol is shown in Fig. 33.9 (*a*) and its equivalent switching circuit in Fig. 33.9 (*b*).

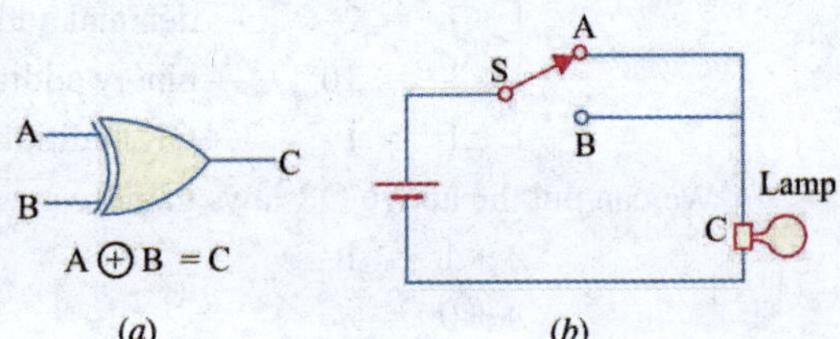

Fig. 33.9

In this gate, output is 1 if its either input **but not both,** is 1. In other words, it has an output 1 **when its inputs are different.** The output is 0 only when inputs **are the same.**

To put it a bit differently, this logic gate has output 0 **when inputs are either all 0 or all 1.**

This gate works on the Boolean equation $A \oplus B = C$.

The circle around plus (+) sign is worth noting.

The circuit is also called an *inequality comparator* or detector because it produces an output only **when the two inputs are different.**

Table 33.3

A	B	C
0	0	0
0	1	1
1	0	1
1	1	0

Explanation

The *inclusive* OR gate exemplifies the everyday usage of the word OR which stands for one or the other or both. Take the following statement :

To qualify for a competition, you might have to subscribe to a magazine OR belong to a club. Obviously, there is no bar on your doing both. But now take **exclusive** statement :

You can be rich OR you can be poor.

Obviously, you cannot be both at the same time.

The change-over switching circuit of Fig. 33.9 (*b*) simulates the exclusive OR (XOR) gate. Switch positions A and B will individually light up the lamp but a combination of A and B is not possible.

The truth table for a 2-input XOR gate is given in Table 33.3. It is instructive to compare it with that for an *inclusive* OR gate (Table 33.1).

33.10. The AND Gate

The electronic (or logic) symbol for a 2-input AND gate is shown in Fig. 33.10 (*a*) and its equivalent switching circuit in Fig. 33.10 (*b*). It is worth reminding the readers once again that the three variables *A, B, C* can have a **value of either 0 or 1.**

Logic Operation

1. The AND gate gives an output only **when all its inputs are present.**
2. The AND gate has a 1 output when both *A* **and** *B* are 1. Hence, this gate is an **all-or-nothing** gate whose output occurs only when all its inputs are present.
3. In True/False terminology, the output of an AND gate will be **true** only if **all** its **inputs are true.** Its output would be false if **any** *of its inputs is false.*

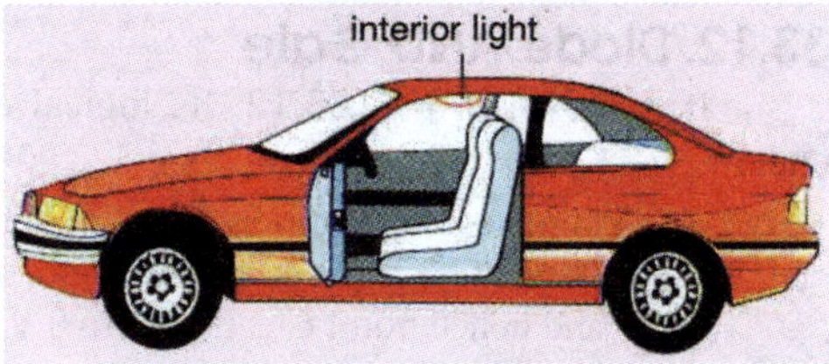

The NAND gate is used to design an interior lighting system of a car such that the light is switch off only when both doors are shut.

The AND gate works on the Boolean algebra

$$A \times B = C \quad \text{or} \quad A \,.\, B = C \quad \text{or} \quad AB = C$$

It is a **logical** multiplication and is different from the **arithmetic** multiplication. Often the sign '×' is replaced by a dot which itself is generally omitted as shown above. The logical meaning of the above equation is that

1. output *C* is 1 only when both *A* and *B* are 1.
2. output *C* is true only when both *A* and *B* are true.

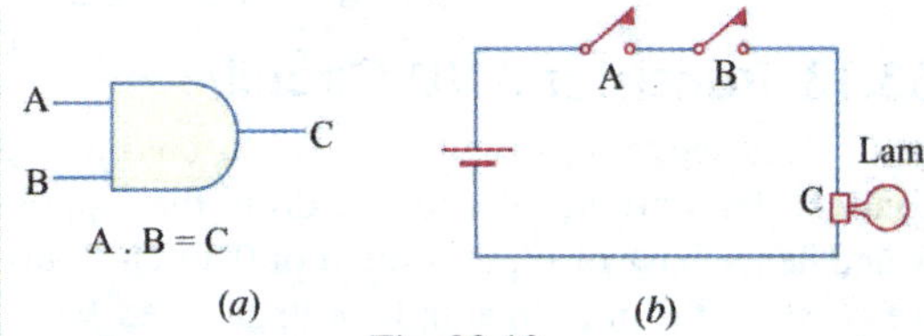

Fig. 33.10

As seen from Fig. 33.10 (*b*), the lamp would be ON when both switches *A* and *B* are closed. Even when one switch is open, the lamp would be OFF. Obviously, an AND gate is equivalent **to a series switching circuit.**

Truth Table

Table 33.4 shows truth table for a 2-input AND gate and Table 33.5 gives the same for a 3-input AND gate.

Table 33.4

A	*B*	*C*
0	0	0
0	1	0
1	0	0
1	1	1

AB = C

Table 33.5

A	*B*	*C*	*D*
0	0	0	0
0	0	1	0
0	1	0	0
0	1	1	0
1	0	0	0
1	0	1	0
1	1	0	0
1	1	1	1

As seen, *C* is at logic 1 only when *all* inputs are at logic 1, not otherwise. The procedure for writing down the first three columns is the same as explained in Art. 33.8 earlier.

33.11. Equivalent Relay Circuit of an AND Gate

The AND gate can be physically realized with the help of relay circuit shown in Fig. 33.11. Here, the two relay contacts have been **wired in series.**

When +5V is applied to both input circuits, relays K_1 and K_2 are energized thereby pulling *M* and *N* downwards which brings *C* in contact with the supply point *S*. Hence, *C* goes to +5 V.

It is obvious that energizing only *one relay* will not make *C* go to + 5 V.

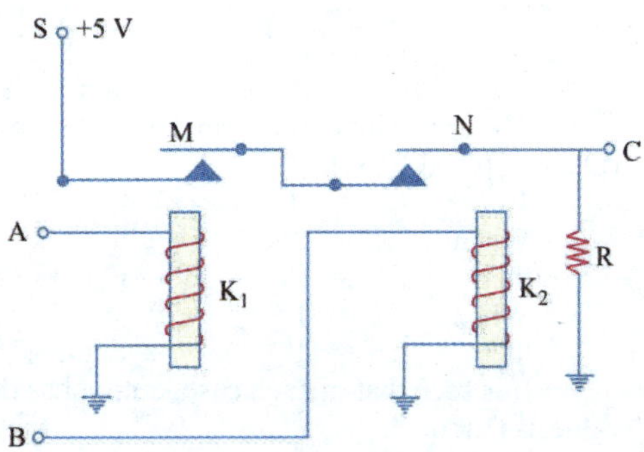

Fig. 33.11

33.12. Diode AND Gate

It is shown in Fig. 33.12. Its logical operation is as under:

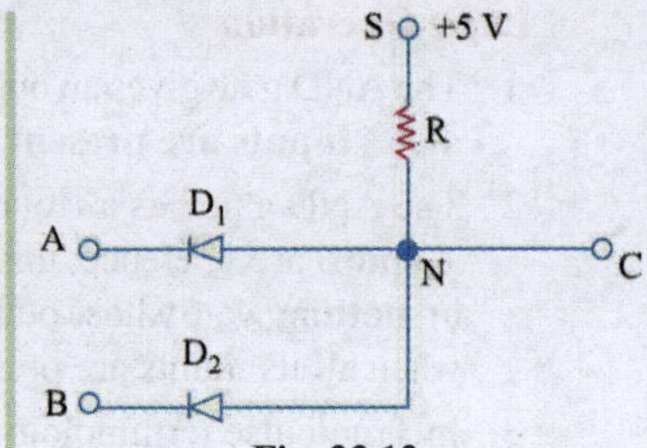

Fig. 33.12

1. When A is at 0 V, diode D_1 conducts and the supply voltage of +5 V drops across R. Consequently, point N and hence point C is driven to 0 V. Therefore, the output C is 0.
2. Similarly, when B is at 0 V, D_2 conducts thereby driving N and hence C to ground.
3. Obviously, when both A and B are at 0 V, both diodes conduct and, again, the output C is 0.
4. There is no supply current and hence no drop across R *only when both A and B are at* +5 *V*. Only in that case, the output C goes to supply voltage of +5 V.

33.13. Transistor AND Circuit

It is shown in Fig. 33.13. When both A and B are at +5 V, the two transistors Q_1 and Q_2 conduct. The current so produced drops the supply voltage of +5 V across R_1 thereby driving point N and hence base of Q_3 to ground or 0 V. This cuts off Q_3 so that C goes to supply voltage of +5 V.

Obivously, there is an output at C only when there is an input at A and B.

If either A or B is at 0 V, then Q_1 or Q_2 will be cut-off and no drop will take place across R_1. Hence, point N will go to supply voltage of +5 V. Consequently, Q_3 will conduct and whole of supply voltage will be dropped across R_2. As a result, point M and hence output C will go to 0 V.

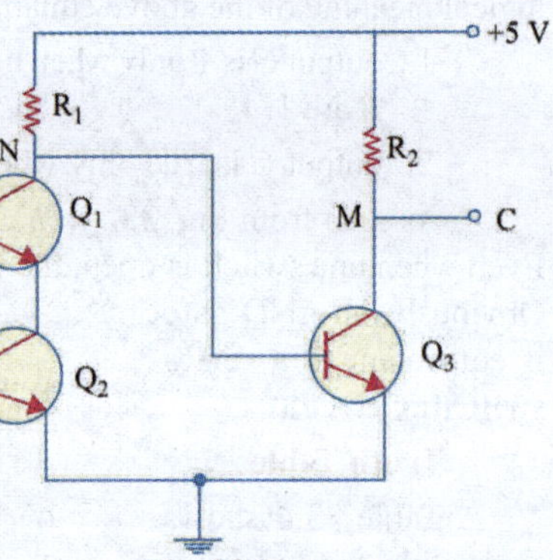

Fig. 33.13

33.14. AND Gate Symbolizes Logic Multiplication

According to Boolean algebra, the AND gate performs **logical** multiplication on its inputs as given below:

0.0 = 0
0.1 = 0
1.0 = 0
1.1 = 1

In general, we can put the laws of Boolean multiplication in the following form:

$A.1 = A$
$A.0 = 0$
$A.A = A$ — not A^2

The above indentities can be verified by giving values of 0 and 1 to A.

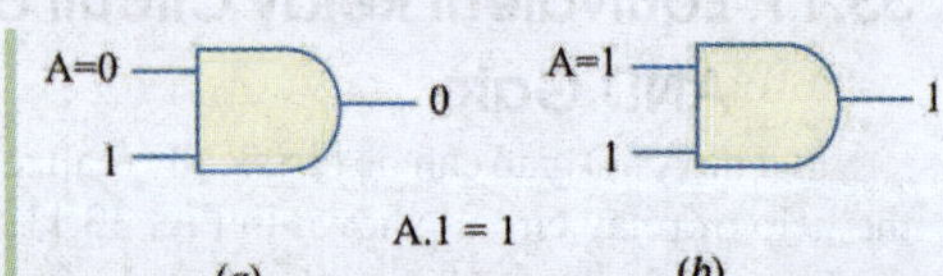

Fig. 33.14

1. $A.1 = A$

When $A = 0$
then $0.1 = 0$ —Fig. 33.14 (*a*)
When $A = 1$
then $1.1 = 1$ —Fig. 33.15 (*b*)

It is seen that in each case, output has the same value as that of A.

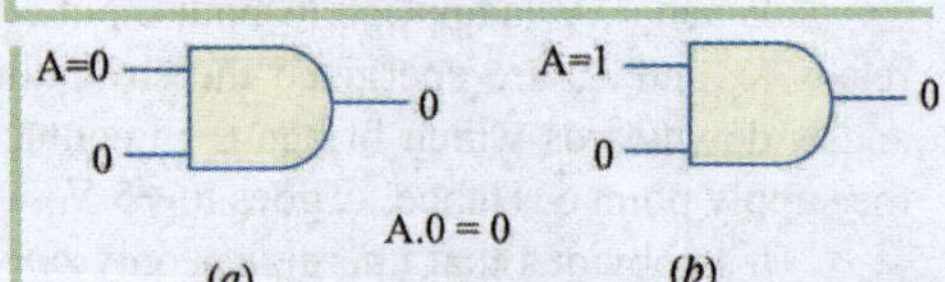

Fig. 33.15

2. A.0 = 0

When $A = 0$
then $0.0 = 0$ —Fig. 33.15 (*a*)

When $A = 1$

then $1.0 = 0$ — Fig. 33.15 (*b*)

It is seen that output is always 0 *whatever the value of A.*

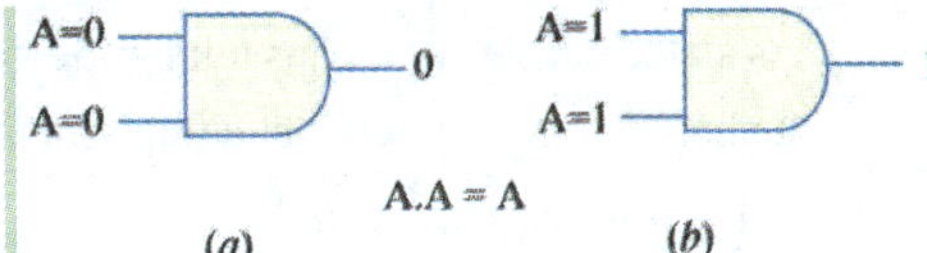

(*a*) (*b*)

Fig. 33.16

3. A.A = A

When $A = 0$, then $0.0 = 0$ —Fig. 33.16 (*a*)

When $A = 1$, then $1.1 = 1$ —Fig. 33.16 (*b*)

It is seen that output always *takes on the value of A.*

33.15. The NOT Gate

It is so called because *its output is* NOT *the same as its input.* It is also called an *inverter* because it inverts the input signal. It has **one** input and **one** output as shown in Fig. 33.17 (*a*). All it does is to *invert* (or complement) the input as seen from its truth table of Fig. 33.17 (*b*).

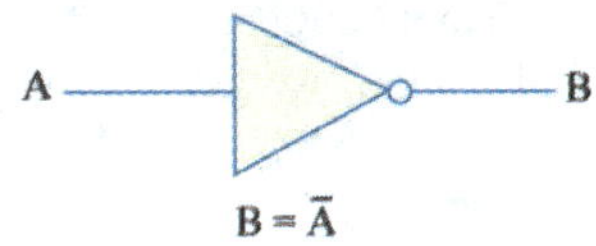

(*a*)

A	B
0	1
1	0

(*b*)

Fig. 33.17

The schematic symbol for inversion is a small circle as shown in Fig. 33.17 (*a*). The logical symbol for inversion or negation or complementation is a bar over the function to indicate the opposite state.

Sometimes, a primer is also used as A'. For example, $\overline{A}$ means not-A. Similarly, $\overline{(A + B)}$ means the complement of $(A + B)$.

33.16. Equivalent Circuits for a NOT Gate

The relay circuit of Fig. 33.18 (*a*) is a physical realization of the complementation operation of the Boolean algebra.

When +5 V is applied to input *A*, *K* is energised and opens the *normally-closed* contact thereby driving output *B* to 0 V. Of course, when *A* is at 0 V, *B* has the supply voltage of +5 V applied to it because the relay contact is normally closed.

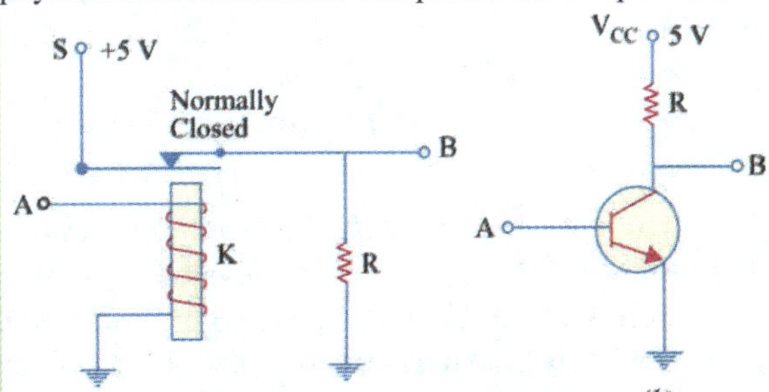

(*a*) (*b*)

Fig. 33.18

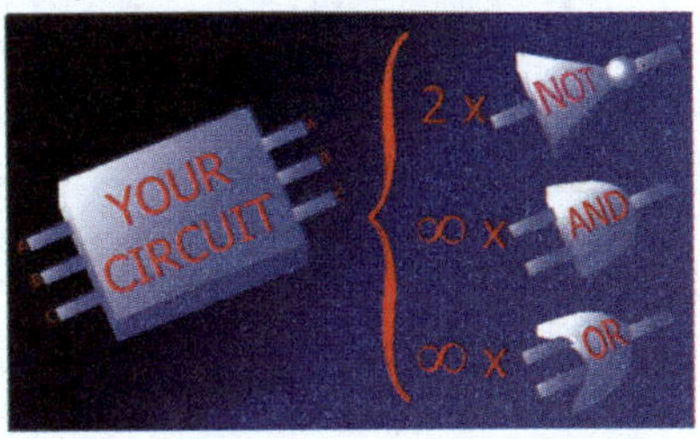

In the equivalent transistor circuit of Fig. 33.18 (*b*) when +5 V is applied to *A*, the transistor will be fully turned ON, drawing maximum *collector* current. Hence, whole of $V_{CC} = 5$ V will drop across *R* thereby sending *B* to 0 V. With 0 V applied at *A*, the transistor will be cut OFF and the output *B*, therefore, will go to V_{CC} *i.e.*, +5 V. Obviously, in each case, **output is the opposite of input.**

33.17. The NOT Operation

It is a *complementation* operation and its symbol is an **overbar.** It can be defined as under:

$\overline{0} = 1$ or $\overline{1} = 0$

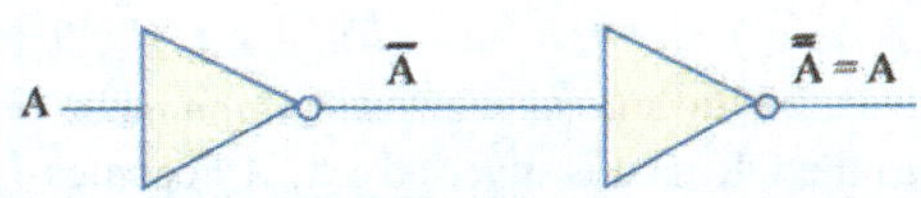

Fig. 33.19

As stated earlier, $\bar{0}$ means taking the negation or complement of 0 which is 1.

It should also be noted that complement of a value can be taken repeatedly. For example,

$\bar{\bar{1}} = \bar{0} = 1$ or $\bar{\bar{0}} = \bar{1} = 0$

As seen, double complementation gives the original value as shown in Fig. 33.19.

Example 33.1. *Find the Boolean equation for the output D of Fig. 33.20 (a). Evaluate D when*
(i) A = 0, B = 1, C = 1   (ii) A = 1, B = 1, C = 1

Solution. The output of the AND gate is *AB*. It then becomes one of the inputs for the 2 input OR gate. When *AB* is ORed with *C*, we get (*AB* + *C*).

$\therefore \quad D = AB + C$ — Fig. 33.20 (*b*)

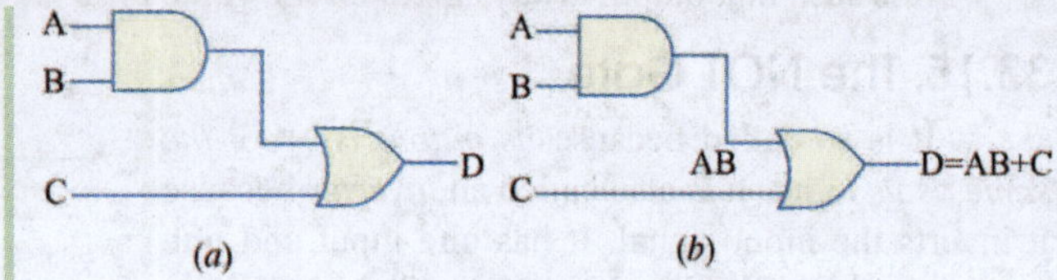

Fig. 33.20

(*i*) $D = 0.1 + 1 = 0 + 1 = \mathbf{1}$ — Art. 33.7

(*ii*) $D = 1.1 + 1 = 1 + 1 = \mathbf{1}$

Example 33.2. *Find the Boolean expression for the output of Fig. 33.21 (a) and evaluate it when*

(*i*) *A = 0, B = 1, C = 1,*
(*ii*) *A = 1, B = 1, C = 0*

Solution. The output of the OR gate is (*A* + *B*). Afterwards, it becomes the input of the AND gate. When ANDed with *C*, it becomes (*A* + *B*).*C*.

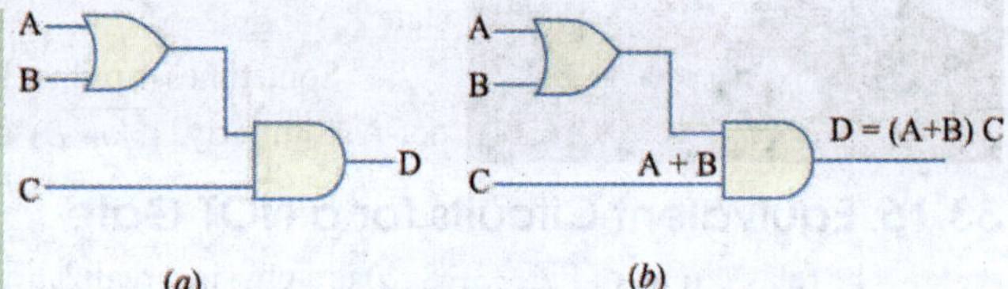

Fig. 33.21

$\therefore \quad D = (A + B).C$ — Fig. 33.21 (*b*)

(*i*) $D = (0 + 1).1 = 1.1 = \mathbf{1}$

(*ii*) $D = (1 + 1).0 = 1.0 = \mathbf{0}$

Example 33.3. *Find the Boolean expression for the output of Fig. 33.22 and compute its value when A = B = C = 1 and D = 0.*

Solution. The circuit is made up of three AND gates. Obviously, it is equivalent to a single 4-input AND gate *i.e.*, an AND gate with a fan-in of four.

Output of the first gate is *AB*, that of the second is *ABC* and that of the third is *ABCD*. Hence, final output is *E* = *ABCD*.

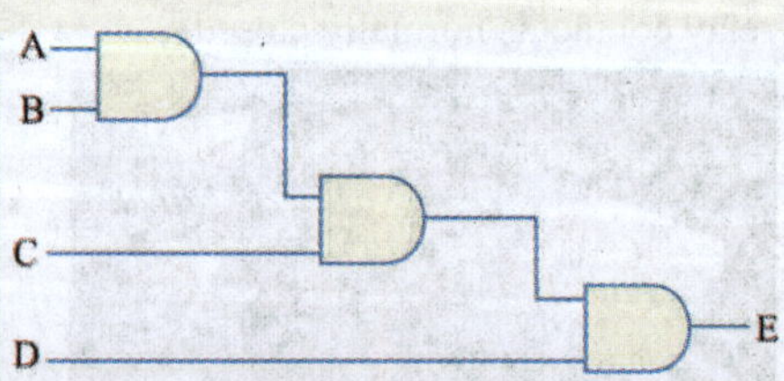

Fig. 33.22

Substituting the given values, we get

$$E = 1.1.1.0 = 1.1.0 = 1.0 = \mathbf{0}$$

Example 33.4. *Find the Boolean expression for the output C of Fig. 33.23 (a) and compute its value when*

(*i*) $A = 0, B = 1$ (*ii*) $A = 1, B = 0$

Solution. As seen, one of the inputs to the OR gate is inverted i.e., *A* becomes $\overline{A}$ as shown in Fig. 33.23 (b). Hence, output is given by $C = \overline{A} + B$

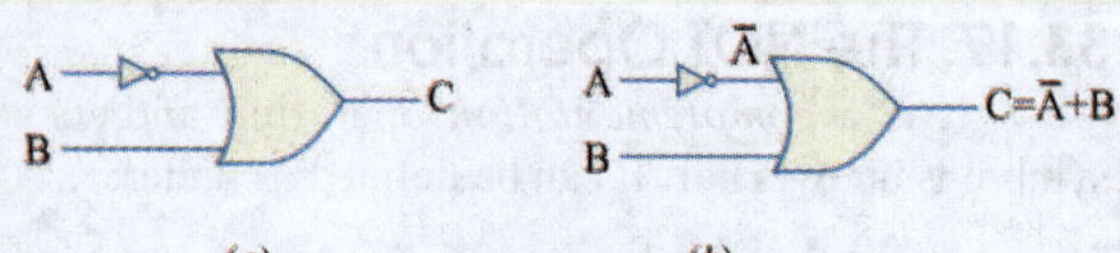

Fig. 33.23

(*i*) $C = \bar{0} + 1 = 1 + 1 = \mathbf{1}$

(*ii*) $C = \bar{1} + 0 = 0 + 0 = \mathbf{0}$

Example 33.5. *What is the Boolean expression for the output C of Fig. 33.24 (a) ? Compute the value of C when*

(*i*) $A = 0, B = 0$ (*ii*) $A = 1, B = 1$

Solution. *As seen, in this case, both inputs to the OR gate have been inverted. Hence, as shown in Fig. 33.24 (b), the inputs become $\bar{A}$ and $\bar{B}$. Therefore, Boolean expression for the output becomes $C = \bar{A} + \bar{B}$.*

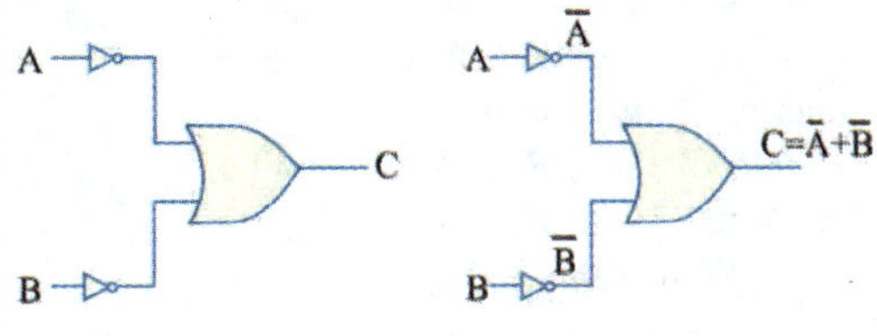

(*a*) (*b*)

Fig. 33.24

(*i*) $C = \bar{0} + \bar{0} = 1 + 1 = \mathbf{1}$

(*ii*) $C = \bar{1} + \bar{1} = 0 + 0 = \mathbf{0}$

Example 33.6. *Write down the Boolean equation for the output C of Fig. 33.25 (a). Compute its value when*

(*i*) $A = 0, B = 0$ (*ii*) $A = 0, B = 1$

(*iii*) $A = 1, B = 0$ (*iv*) $A = 1, B = 1$

Solution. As seen, inputs to the AND gate are $\bar{A}$ and B [Fig. 33.25 (*a*)]. The output of the AND gate is $\bar{A}.B$. However, this output is inverted by the second inverter connected in the output. Hence, final output equation is $C = \overline{\bar{A}\,B}$

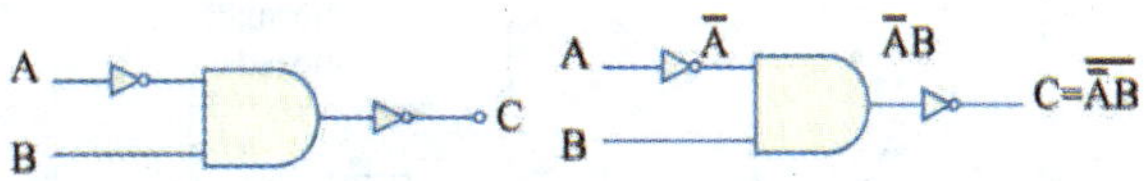

(*a*) (*b*)

Fig. 33.25

(*i*) $C = \overline{\bar{0}.0} = \overline{1.0} = \bar{0} = \mathbf{1}$ (*ii*) $C = \overline{\bar{0}.1} = \overline{1.1} = \bar{1} = \mathbf{0}$

(*iii*) $C = \overline{\bar{1}.0} = \overline{0.0} = \bar{0} = \mathbf{1}$ (*iv*) $C = \overline{\bar{1}.1} = \overline{0.1} = \bar{0} = \mathbf{1}$

While evaluating expressions of the above type, you must remember the following two points:

1. take the NOT *i.e.*, inversion of the individual term *first*.
2. When a NOT or *inversion* is applied to more than one term (like $\overline{1.0}$), you should work out the OR (or AND) operation first and then take the NOT of the result so obtained.

33.18. Bubbled Gates

A bubbled gate is one whose inputs are NOTed or inverted *i.e.*, it is a negated gate. Fig. 33.26 (*a*) shows an AND gate whose both input are inverted.

In practice, instead of this logic symbol, the one shown in Fig. 33.26 (*b*) is widely used. As seen, the inverter triangles have been eliminated and the two small circles or bubbles have been moved to the inputs of the gate. Such a gate is called a *bubbled* AND gate, the bubbles acting as a reminder of the inversion or complementation that takes place before ANDing the inputs.

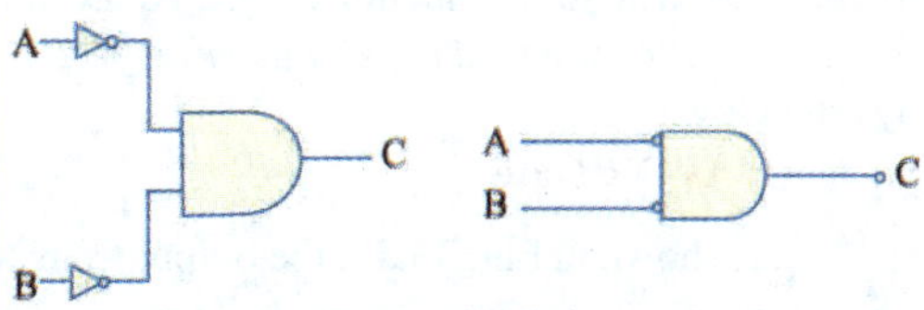

(*a*) (*b*)

Fig. 33.26

It would be shown later that a bubbled AND gate is equivalent to a NOR gate.

Similarly, a bubbled OR gate is equivalent to a NAND gate.

33.19. The NOR Gate

In fact, it is a NOT-OR gate. It can be made out of an OR gate by connecting an inverter in its output as shown in Fig. 33.27 (*a*).

The output equation is given by

$C = \overline{A + B}$

A NOR function is just the reverse of the OR function.

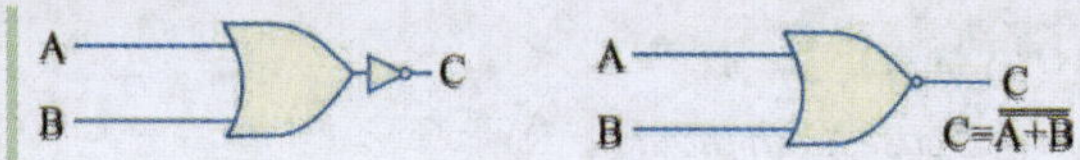

(*a*) (*b*)

Fig. 33.27

Table 33.6

A	*B*	*C*
0	0	1
0	1	0
1	0	0
1	1	0

Logic Operation

A NOR gate will have an output of 1 **only when all its inputs are 0.** Obviously, if any output is 1, the output will be 0. Alternatively, in a NOR gate, output is **true** only **when all inputs are false.**

The truth table for a 2-input NOR gate is shown in Table 33.6. It will be observed that the output *C* is just the reverse of that shown in Table 33.1.

The equivalent relay circuit for a NOR gate is shown in Fig. 33.28.

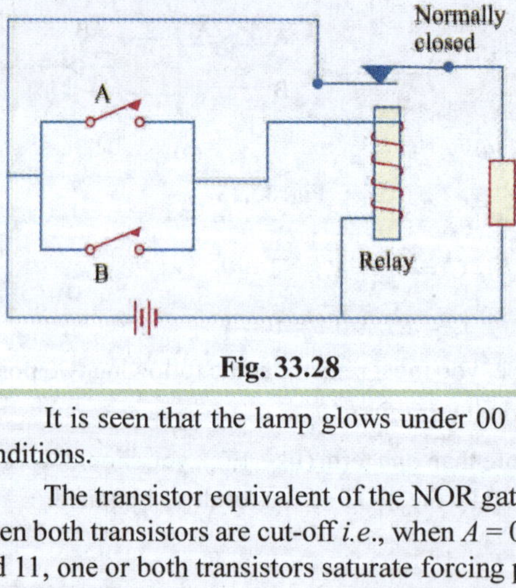

Fig. 33.28

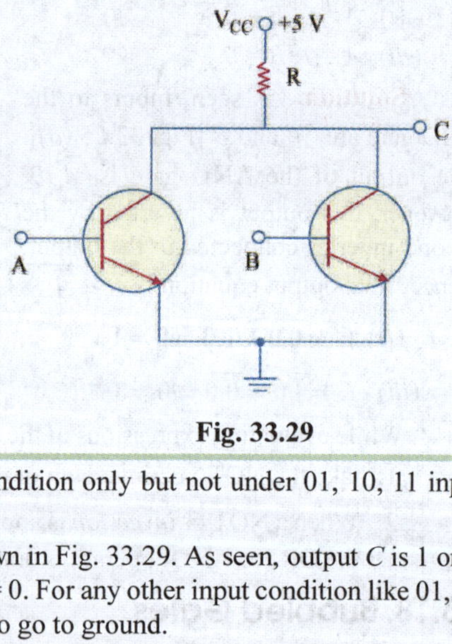

Fig. 33.29

It is seen that the lamp glows under 00 input condition only but not under 01, 10, 11 input conditions.

The transistor equivalent of the NOR gate is shown in Fig. 33.29. As seen, output *C* is 1 only when both transistors are cut-off *i.e.*, when $A = 0$ and $B = 0$. For any other input condition like 01, 10 and 11, one or both transistors saturate forcing point *C* to go to ground.

33.20. NOR Gate is a Universal Gate

It is interesting to note that a NOR gate can be used to realize the basic logic functions : OR, AND and NOT. That is why it is often referred to as a *universal* gate. Examples are given below:

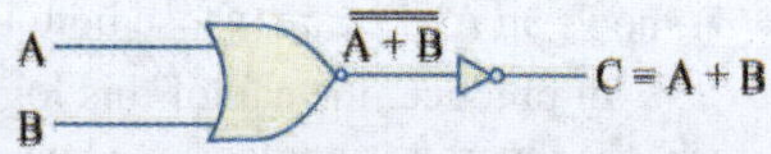

Fig. 33.30

1. As OR Gate

As shown in Fig. 33.30, the output from NOR gate is $\overline{A + B}$. By using another inverter in the output, the final output is inverted and is given by $C = A + B$ which is the logic function of a normal OR gate.

2. As AND Gate

Here, two inverters have been used, one for each input (Fig. 33.31). The inputs have, thus, been inverted **before they are applied** to the NOR gate. The output is $\overline{A} + \overline{B}$ which can be proved (with the help of De Morgan's theorem) to be equal to AB.

Incidentally, we could have used a bubbled NOR gate for the above purpose.

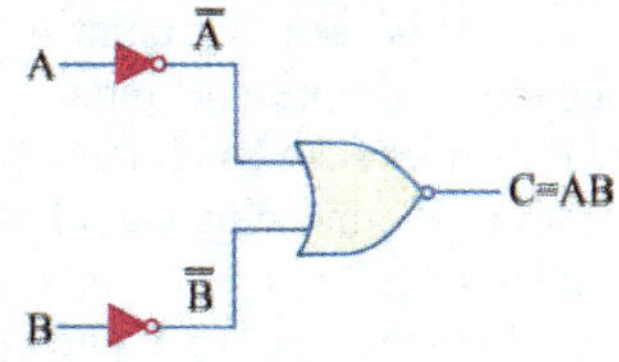

Fig. 33.31

3. As NOT Gate

The two inputs have been tied together as shown in Fig. 33.32 (*a*). The output is $\overline{A + A}$ which can be proved to be equal to $\overline{A}$ with the help of De Morgan's theorem. Instead of the first symbol, the second symbol shown in Fig. 33.32 (*b*) is widely used where only single input has been used.

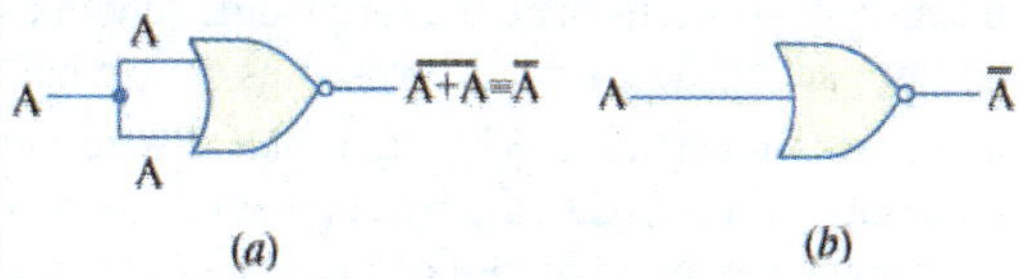

Fig. 33.32

Here is a different way of making OR and AND gates. Fig. 33.33 (*a*) shows how we can use NOR gates to produce an OR gate. Similarly, Fig. 33.33 (*b*) shows the formation of an AND gate from three NOR gates. Knowledge of De Morgan's theorem is needed to understand their logic operation.

33.21. The NAND Gate

It is, in fact, a NOT-AND gate. It can be obtained by connecting a NOT gate in the output of an AND gate as shown in Fig. 33.34. Its output is given by the Boolean equation

$$C = \overline{AB}.$$

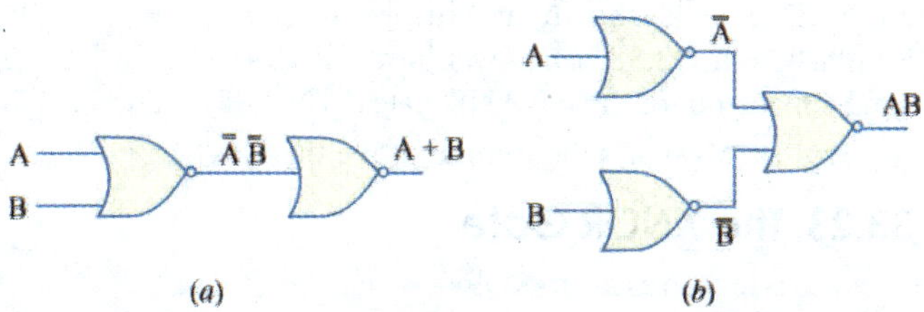

Fig. 33.33

This gate gives an output of 1, if its *both inputs are* **not** 1. In other words, it gives an output 1 if **either A or B or both are 0.**

The truth table for a 2-input NAND gate is given in Table 33.7. It is just the opposite of the truth for AND gate. It is so because NAND gate performs reverse function of an AND gate.

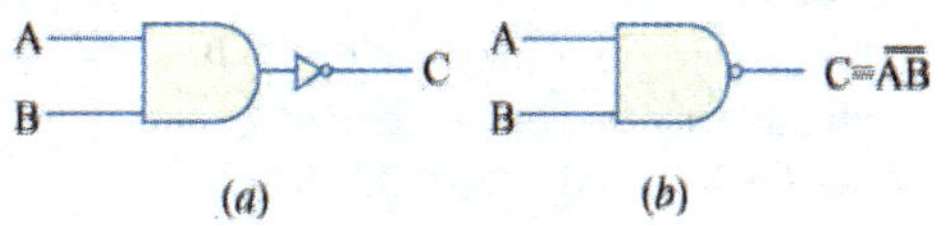

Fig. 33.34

The equivalent relay circuit of a NAND gate is shown in Fig. 33.35.

Table 33.7

A	*B*	*C*
0	0	1
0	1	1
1	0	1
1	1	0

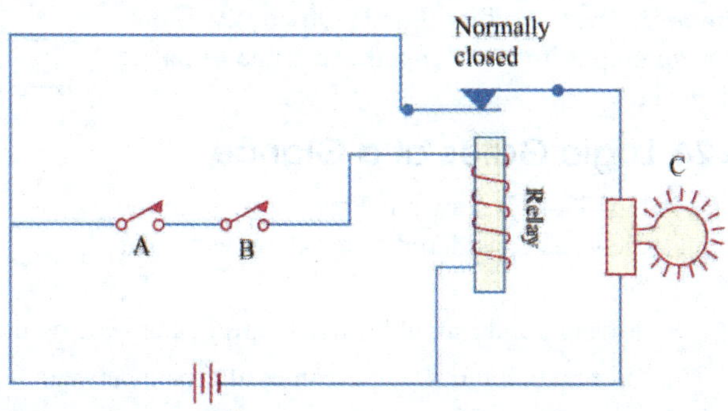

Fig. 33.35

It is seen that lamp glows under input conditions of 00, 01, 10 but not under 11 input condition when both switches *A* and *B* are ON. The diode-transistor equivalent of a NAND gate in shown in Fig. 33.36.

It is seen that point N would be driven to ground when either D_1 or D_2 or both D_1 and D_2 conduct. It represents input conditions of 10, 01 and 11*. Under such conditions, Q is cut-off and hence C goes to V_{CC} meaning logic 1 state. Only time C is 0 is when $A = 1$ and $B = 1$ (*i.e.*, input voltages at A and B are 0) so that N is +5 V and Q is saturated.

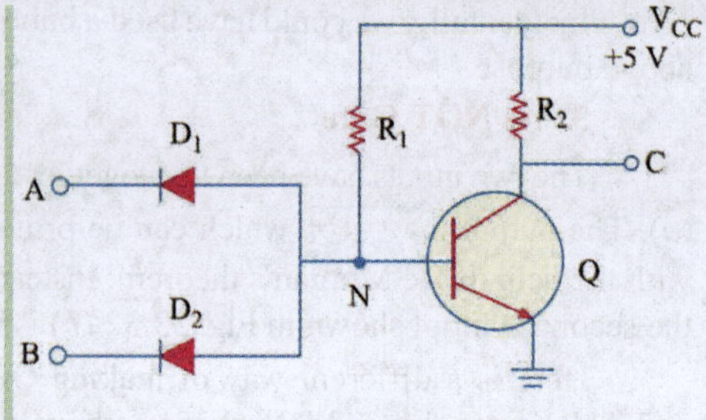

Fig. 33.36

33.22. NAND Gate is a Universal Gate

NAND gate is also called universal gate because it can perform all the three logic functions of an OR gate, AND gate and inverter as shown below.

As shown in Fig. 33.37 (*a*), a NOT gate can be made out of a NAND gate by connecting its two inputs together. When a NAND gate is used as a NOT gate, the logic symbol of Fig. 33.37 (*b*) is employed instead.

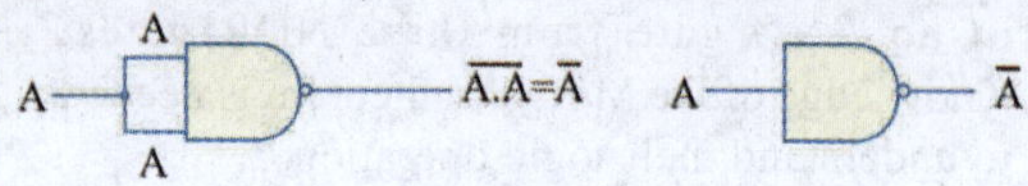

(*a*) (*b*)

Fig. 33.37

The use of two NAND gates to produce an AND gate is shown in Fig. 33.38 (*a*). Similarly, Fig. 33.38 (*b*) shows how OR gate can be made out of three NAND gates. The OR function may not be clear from the figure because we need De Morgan's theorem to prove that $\overline{A}\,\overline{B} = A + B$.

33.23. The XNOR Gate

It is known as a not-XOR gate *i.e.*, $\overline{\text{XOR}}$ gate. Its logic symbol and truth table are shown in Fig. 33.39.

Its logic function and truth table are *just the reverse of those for* XOR *gate* (Art. 33.9).

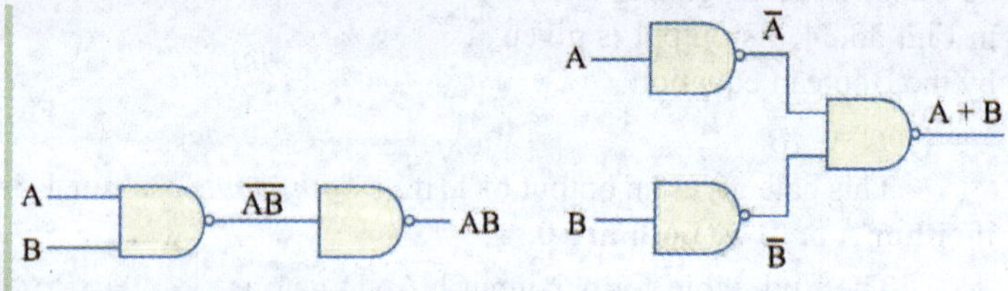

(*a*) (*b*)

Fig. 33.38

This gate has an output 1 if **its both inputs are either 0 or 1**. In other words, for getting an output, its both inputs should be *at the* **same** *logic level* of either 0 or 1. Obviously, it produces **no** output if its two inputs are at the **opposite** logic level.

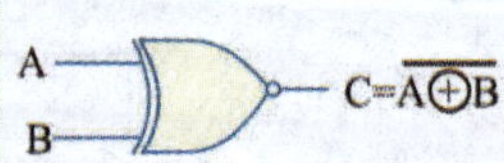

A	*B*	*C*
0	0	1
0	1	0
1	0	0
1	1	1

Fig. 33.39

33.24. Logic Gates at a Glance

In Fig. 33.40 is shown the summary of all the 3-input logic gates considered so far along with their truth tables.

Following points should prove helpful when writing these truth tables:

1. In first column A, logic values alternate between 0 and 1 every two rows.
2. In second column, logic values alternate every other row.
3. Column C is filled up as per the logic function it performs.

* In this case, $V_A = V_B = 0$ V.

A + B = C

A	B	C
0	0	0
0	1	1
1	0	1
1	1	1

AB = C

A	B	C
0	0	0
0	1	0
1	0	0
1	1	1

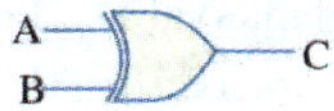

$A \oplus B = C$

A	B	C
0	0	0
0	1	1
1	0	1
1	1	0

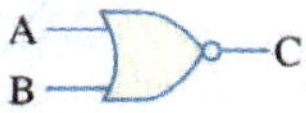

$\overline{A + B} = C$

A	B	C
0	0	1
0	1	0
1	0	0
1	1	0

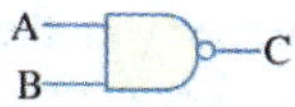

$\overline{AB} = C$

A	B	C
0	0	1
0	1	1
1	0	1
1	1	0

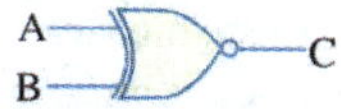

$\overline{A \oplus B} = C$

A	B	C
0	0	1
0	1	0
1	0	0
1	1	1

Fig. 33.40

4. Truth tables for NOR, NAND and XNOR (or XOR) gates are *just the opposite* of those for OR, AND and XOR gates.

Fig. 33.41

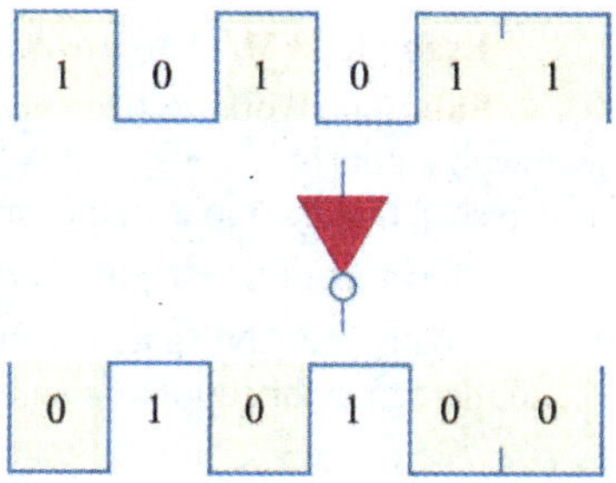

Fig. 33.42

Example 33.7. *An electrical signal is expressed as 101011. Explain its meaning. If this signal is applied to a NOT gate, what would be the output signal ?*

Solution. The signal represents binary number 101011_2. It is electrically represented as a train of pulses. Taking positive logic, 1 will represent high voltage and 0 will represent low (or zero) voltage as as shown in Fig. 33.41.

When such a signal is applied to a NOT gate, it would be inverted or complemented as shown in Fig. 33.42.

The NOT output will represent the binary number 010100_2.

Example 33.8. *Two electrical signals represented by A = 101101 and B = 110101 are applied to a 2-input AND gate. Sketch the output signal and the binary number it represents.*

Solution. The pulse trains corresponding to *A* and *B* are shown in Fig. 33.43.

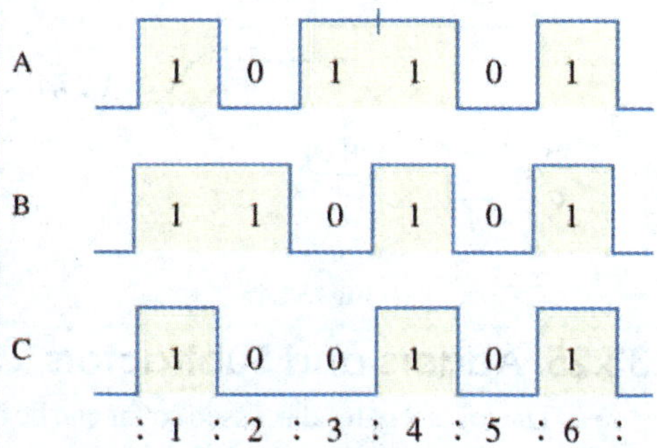

Fig. 33.43

Remember that in an AND gate, *C* is 1 only when both *A* and *B* are 1. It is an *all-or-nothing gate.* The output can be found in different time intervals as under :

1. 1st interval : 1 + 1 = 1
2. 2nd interval : 0 + 1 = 0

3. 3rd interval : $1 + 0 = 0$
4. 4th interval : $1 + 1 = 1$
5. 5th interval : $0 + 0 = 0$
6. 6th interval : $1 + 1 = 1$

Hence, ouput of the AND gate is 100101_2. It is sketched in Fig. 33.43.

Example. 33.9. *Convert the Boolean expression (AB + C) into a logic circuit using different logic gates.*

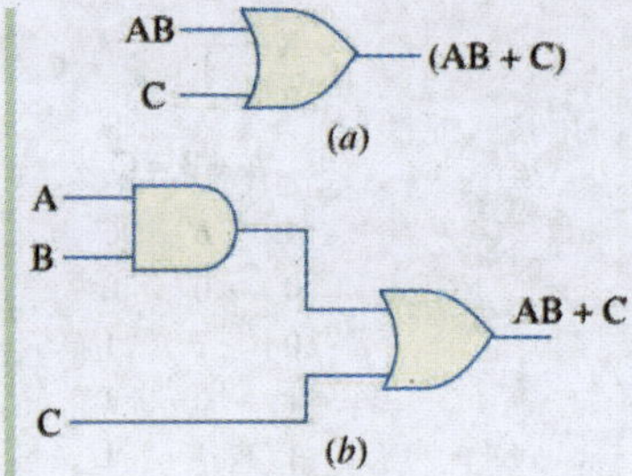

Fig. 33.44

Solution. In such cases, it is best to start with the *output and work towards the input.* As seen, *C* has been ORed with *AB*. Hence, the output gate must be a 2-input OR gate as shown in Fig. 33.44 (*a*).

Now, term *AB* is an AND function. Hence, we need an AND gate with inputs *A* and *B*. The complete logic circuit is shown in Fig. 33.44 (*b*).

Example 33.10. *Design logic hardware based on the Boolean expression* $(A + \overline{B}C)$.

Solution. We will work from *output to input.* It is seen that the last gate is a 2-input OR gate with inputs of *A* and $\overline{B}C$. It is shown in Fig. 33.45 (*a*).

Since *B* has been ANDed with *C*, it requires an AND gate as shown in Fig. 33.45 (*b*). For inversion of *B*, a NOT gate has been used as shown in Fig. 33.45 (*c*).

Example 33.11. *Design a logic circuit whose output is given by the Boolean expression* $(A+ B).\ \overline{AB}$.

Solution. Working from output to input, we find that the output gate has to be a 2-input AND gate with inputs of (A + B) and $\overline{AB}$. The first step of the circuit design is shown in Fig. 33.46 (*a*). It is also seen that the input to the entire circuit consists of *A* and *B* only.

The input of $(A + B)$ has been obtained with the help of an OR gate as shown in Fig. 33.46 (*b*).

Finally, a NAND gate is connected **in parallel** with the OR gate for getting its inputs of *A* and *B* and thereafter for supplying an output of $\overline{AB}$. The complete circuit is shown in Fig. 33.46 (*c*).

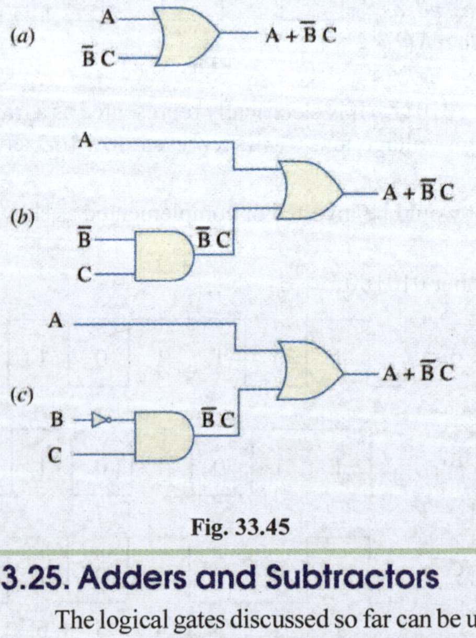

Fig. 33.45

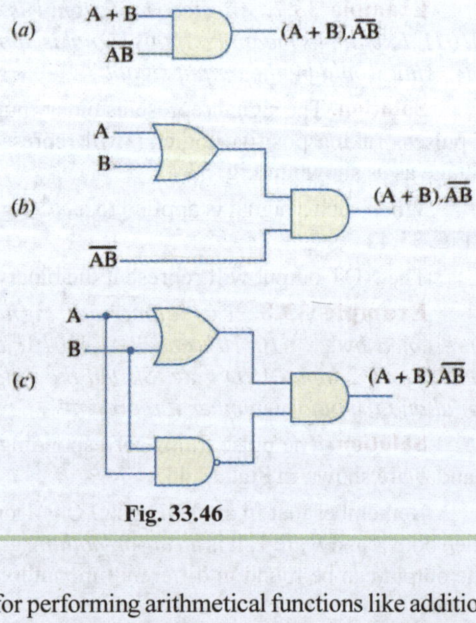

Fig. 33.46

33.25. Adders and Subtractors

The logical gates discussed so far can be used for performing arithmetical functions like addition, subtraction, multiplication and division in electronic calculators and digital instruments. In the central processing unit (CPU) of a computer, these arithmetic functions are carried out by the arithmetic and logic unit (ALU). The logic functions used generally are XOR, OR and AND. We will consider the following:

1. Half Adder—it is 1-bit adder and carries out binary addition with the help of XOR and AND gates. It has **two** inputs and **two** outputs.
2. Full Adder—it has **three** inputs and can add three bits at a time. It is made up of *two half adders and one OR gate.*

These adders can also perform subtraction by the method of 1's and 2's complements.

3. Half Subtractor—it uses one XOR and one AND gate .
4. Full Subtractor—it employs two half subtractors and one OR gate.

33.26. Half Adder

It can add 2 binary digits **at a time** and produce a 2-bit data *i.e.*, sum and carry according to the binary addition rules (Art. 33.10).

Block Diagram

It is shown in Fig. 33.47. As can be seen, it has two inputs for applying the two binary digits to be added. As is well known, binary addition of two bits always produces 2-bit output data *i.e.*, one SUM and one CARRY. For example, (1 + 1) gives a sum of 0 and a carry of 1. Also, (0 + 0) gives the sum 0 and carry 0. That is why the adder has two outputs : one for SUM and the other for CARRY.

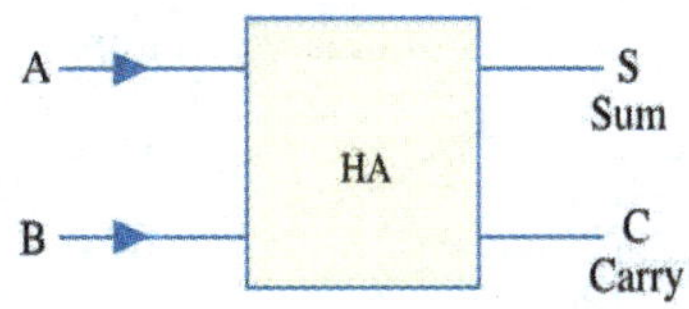

Fig. 33.47

Table 33.8

Input		Output	
A	*B*	*S*	*C*
0	0	0	0
0	1	1	0
1	0	1	0
1	1	0	1

Truth Table 33.8 lists the two columns of input, one of SUM and one of CARRY. The SUM output has the same logic pattern as when *A* is XORed with *B*. In fact, *to add it to* XOR. Also, the CARRY output has the same logic pattern as when *A* is ANDed with *B*. That is why a half-adder can be formed from a combination of one XOR gate and one AND gate as shown in Fig. 33.48.

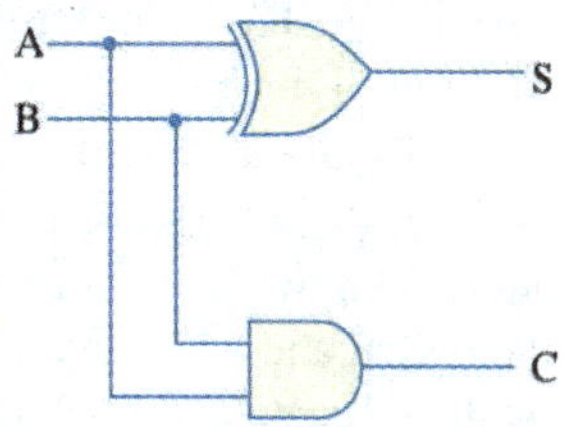

Fig. 33.48

The circuit is called *half-adder* because it cannot accept a carry-in from previous additions. For that purpose, we need a *3-input* adder called full-adder.

Incidentally, the logical equations for the SUM and CARRY are $S = A \oplus B$ and $C = A \,.\, B$.

33.27. Full Adder

As shown in the block diagram of Fig. 33.49, it has *three inputs and two outputs.* It can add 3 digits (or bits) **at a time.** The bits *A* and *B* which are to be added come from the two registers and the third input comes from the carry generated by the previous addition. It produces two outputs; SUM and CARRY-OUT.

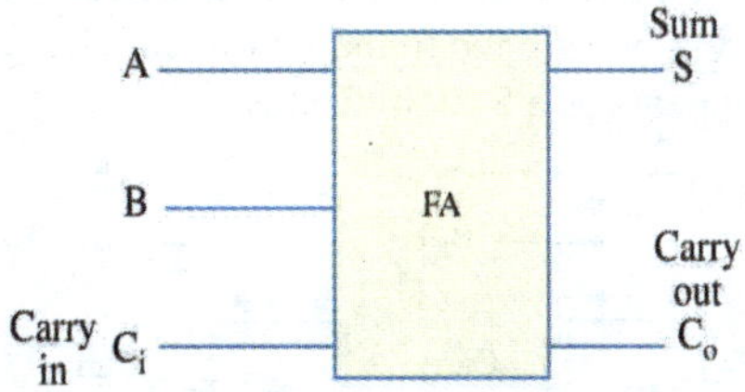

Fig. 33.49

Table 33.9

Input			Output	
A	*B*	C_i	*S*	C_o
0	0	0	0	0
0	0	1	1	0
0	1	0	1	0
0	1	1	0	1
1	0	0	1	0
1	0	1	0	1
1	1	0	0	1
1	1	1	1	1

The truth table gives all possible input/output relationships for the full adder. A and B are the inputs from the respective digits of the registers to be added and C_i is the input for any carry generated by the previous stage. The SUM output gives binary addition of A, B and C_i. The other output generates the carry C_o to be added to the next stage.

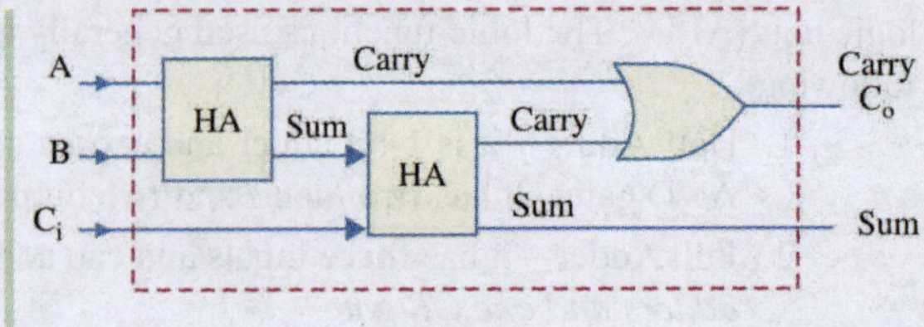

Fig. 33.50

The full-adder can be constructed from two half-adders and one OR gate (Fig. 33.50).

Working. Let us illustrate, with the help of two examples, how this full adder adds three bits.

(*i*) **A = 1, B = 1, C_i = 0**

The full adder with these three inputs is shown in Fig. 33.51. First half adder gives a sum of 0 and a carry of 0. The second HA gives a sum of 0 with a carry of 0. The final output is : SUM 0, CARRY 1. As we know from the rules of binary addition, $1 + 1 + 0 = 10_2$ (*i.e.*, decimal 2).

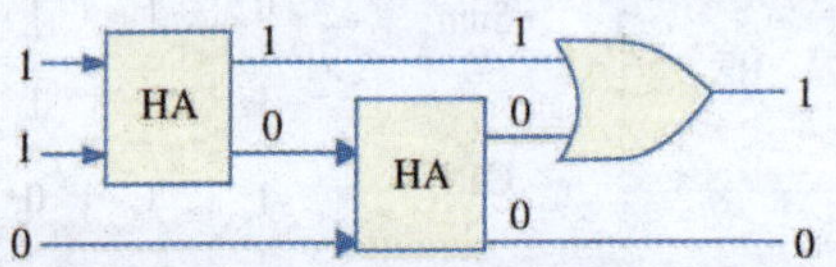

Fig. 33.51

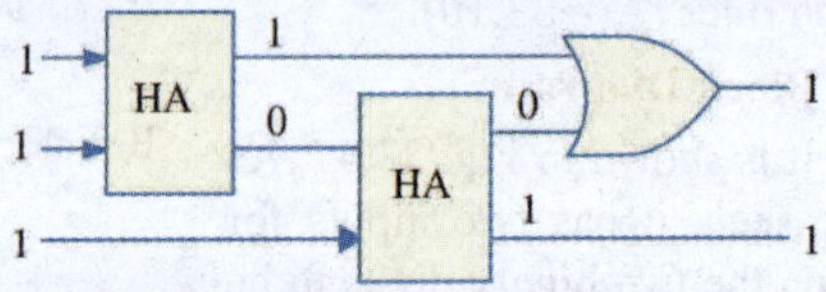

Fig. 33.52

(*ii*) **A = 1, B =1, C = 1**

As detailed in Fig. 33.52, we get a final SUM 1 with a CARRY 1. The result conforms to the binary addition : $1 + 1 + 1 = 11_2$ (*i.e.*, decimal 3).

Detailed Circuit

Fig. 33.53 shows more circuit details of a full adder. The two half adders have been replaced by their XOR and AND gates. The final carry is given by the OR gate and final sum by the XOR gate of the second adder.

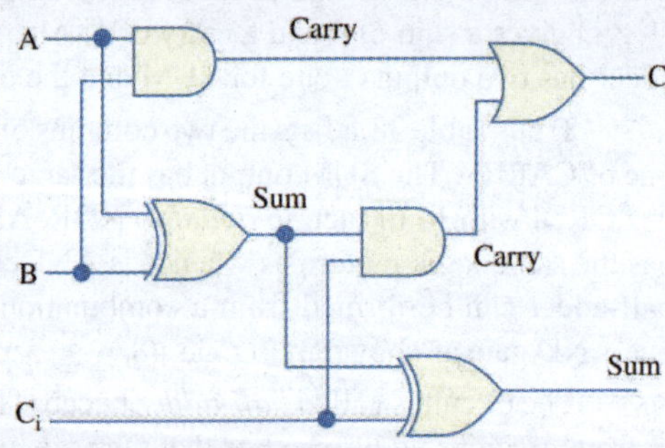

Fig. 33.53

33.28. Parallel Binary Adder

For adding two 4-bit numbers, we need 4 full adders *connected in parallel* as shown in Fig. 33.54. The two numbers being added are $A_3\ A_2\ A_1\ A_0$ and $B_3\ B_2\ B_1\ B_0$ and their sum is $S_4\ S_3\ S_2\ S_1\ S_0$.

The first adder could be a half adder though we may use a full adder but leave its CARRY-IN lead unconnected. As seen, different bits are fed to the four adders from two parallel registers which hold these bits. The final SUM appears as a 5-digit display.

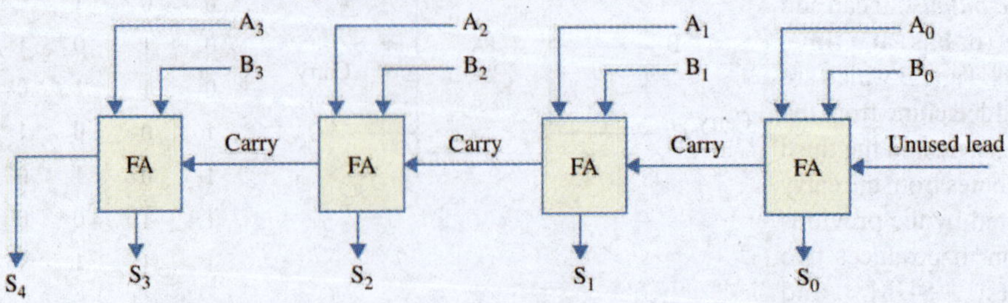

Fig. 33.54

Operation

The actual operation may be better understood with the help of the diagram of Fig. 33.55. Suppose, we want to add the following two 4-bit numbers.

$$\begin{array}{r} 1010 \\ +\,1011 \\ \hline 10101 \end{array} \qquad\qquad \begin{array}{r} 10 \\ +\,11 \\ \hline 21 \end{array}$$

The first adder performs 0 + 1 binary addition, giving a sum of 1 and a carry of 0. The two bits 0 and 1 are supplied *simultaneously* from the two registers *A* and *B*. The sum 1 appears on the display panel and carry 0 is passed on to the next full adder.

The next adder adds 1 + 1 +0 carry = sum 1 with a carry 1. The third adder performs 0 + 0 + 1 carry = 1 with carry 0. The fourth adder adds 1+ 1 + 0 carry = sum 0 with carry 1 both of which appear on the display unit. Hence, the final addition of the two numbers appears as 10101.

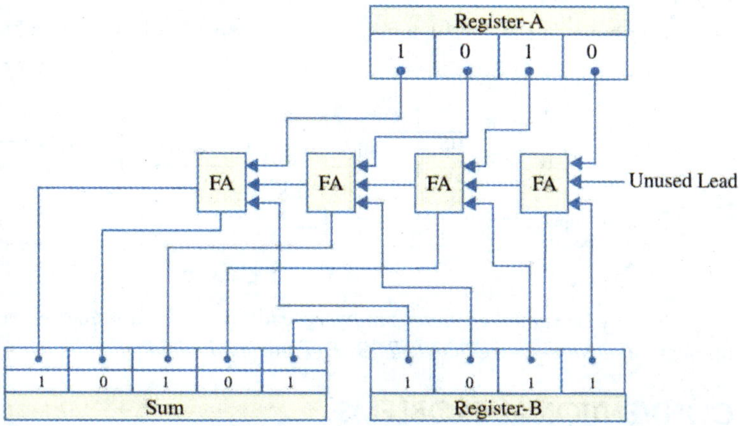

Fig. 33.55

It may be noted that the largest binary numbers that can be added by this parallel adder are 1111 and 1111 which give a sum of 11110_2 (decimal 30). To increase its capacity, more full adders may be connected at the left end of Fig. 33.55. For example, for adding 6-bit numbers, we will have to add two more adders thus making a total of **six.**

33.29. Half Subtractor

It can subtract only two binary digits *at a time* and produce an output of a difference and a borrow. As shown in the block diagram of Fig. 33.56, it has two inputs and two outputs.

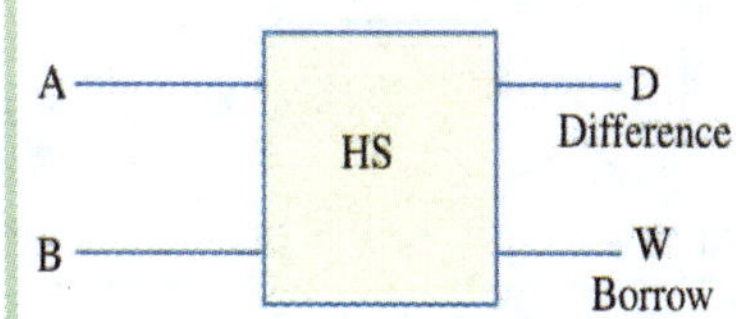

Fig. 33.56

Table 33.10

Input		*Output*	
A	*B*	*W*	*D*
0	0	0	0
0	1	1	1
1	0	0	1
1	1	0	0

The operation of a half subtractor is based on the rules of binary subtraction illustrated in the truth Table 33.10 for all possible input/output combinations. The difference output in the fourth column has the same logic pattern as when *A* is XORed with *B* (same was the case for SUM in Art. 33.26). Hence, we can use an XOR gate to get the difference of two bits. The borrow output in the third column can be obtained by ANDing $\overline{A}$ with *B*.

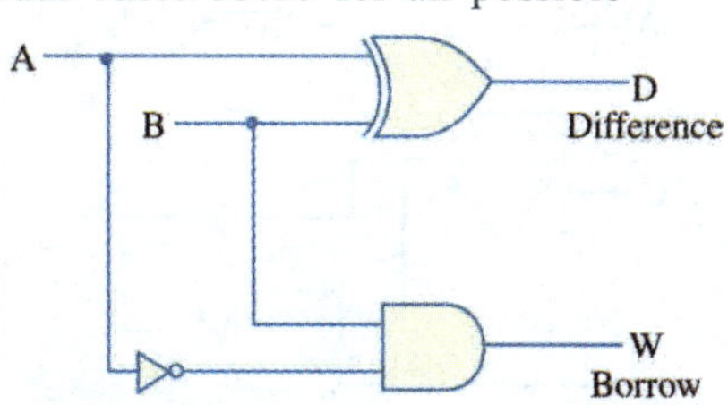

Fig. 33.57

The circuit for a half subtractor is shown in Fig. 33.57.

As mentioned earlier, the logical equations for the difference and borrow are given by

$$D = A \oplus B \text{ and } W = \overline{A}\,B$$

33.30. Full Subtractor

As shown in the block diagram of Fig. 33.58, it has three inputs and two outputs. As explained above, half subtractor can handle only 2 bits at a time and can be used for the least significant column of a subtraction problem. A full subtractor can, however, take care of higher-order columns.

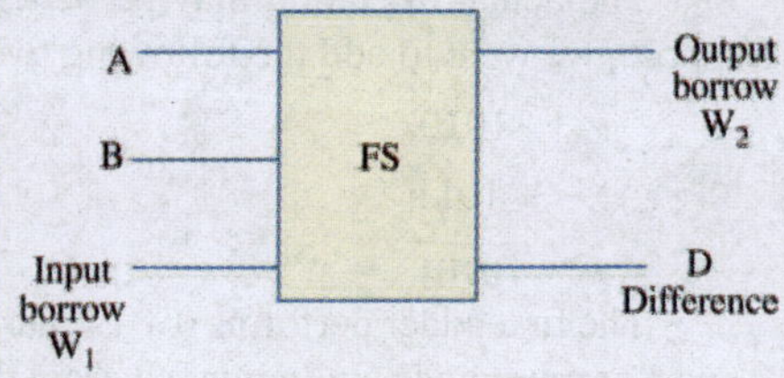

Fig. 33.58

As shown in Fig. 33.59, a full subtractor consists of two half subtractors and one OR gate.

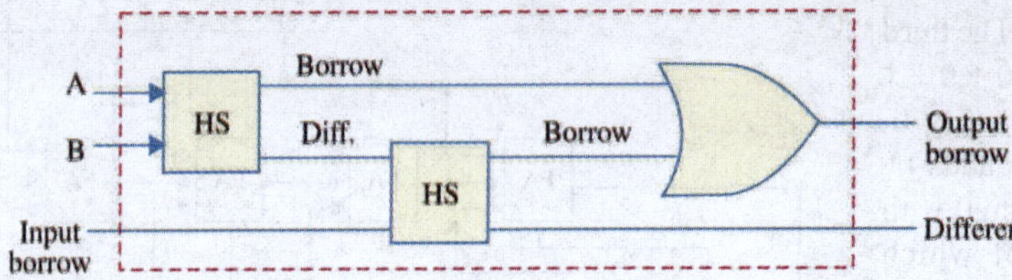

Fig. 33.59

It may be remarked here that by cascading 4 full subtractors, we can directly subtract 4-bit numbers *i.e.*, we can subtract $B_3\,B_2\,B_1\,B_0$ from $A_3\,A_2\,A_1\,A_0$.

CONVENTIONAL PROBLEMS

1. Find the Boolean equation for the output of the logic circuit shown in Fig. 33.60. What would be the output if $A = 1$, $B = 0$, $C = 1$, $D = 1$? **[Y = AB + CD; 1]**

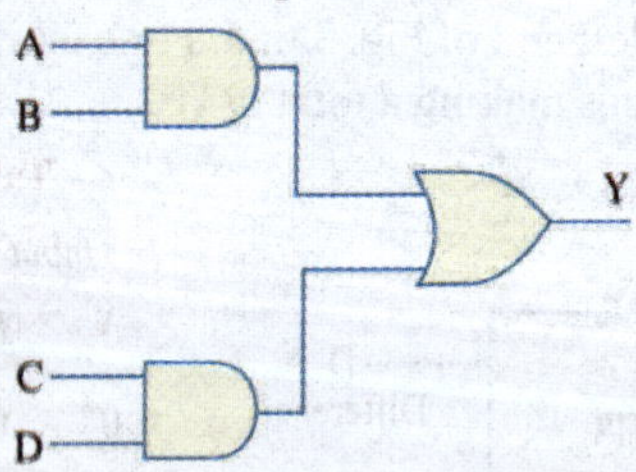

Fig. 33.60

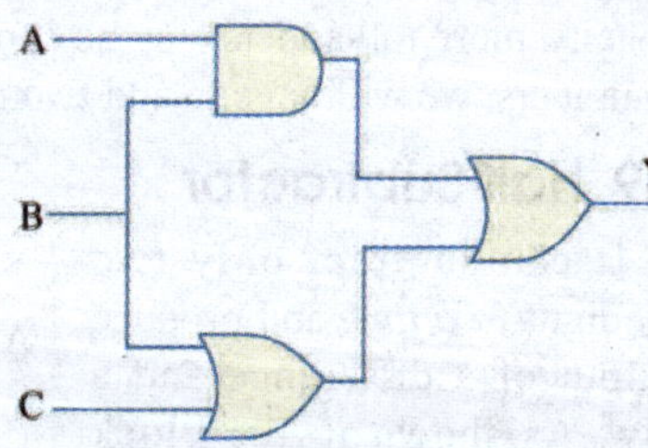

Fig. 33.61

2. What is the output equation of the logic circuit shown in Fig. 33.61 ? Evaluate the output if $A = 1$, $B = 1$, $C = 0$. **[Y = AB + (B + C); 1]**
3. After finding the Boolean equation for the circuit shown in Fig. 33.62, compute the output if $A = 1$, $B = 0$, $C = 1$, $D = 0$. **[Y = (A + B) (C + D); 1]**

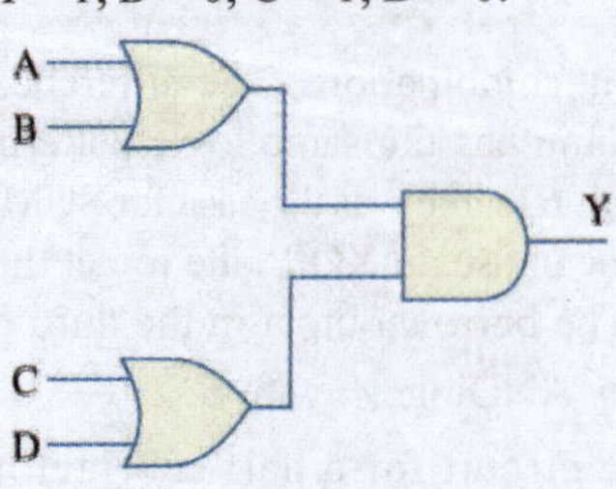

Fig. 33.62

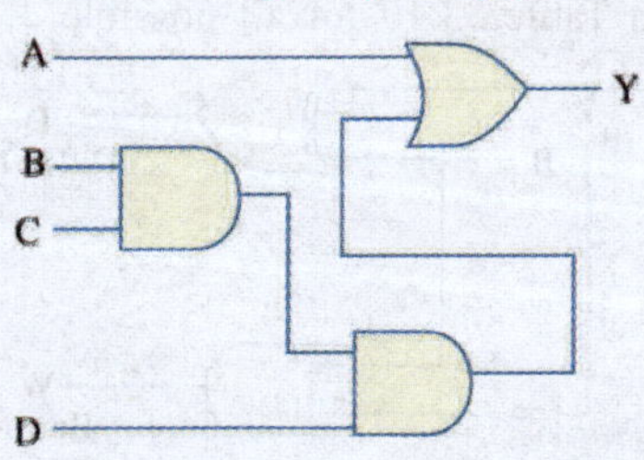

Fig. 33.63

4. Translate the hardware shown in Fig. 33.63 into a Boolean expression. Compute the value of the output if $A = 0, B = 1, C = 0, D = 1$. **[Y = A + BCD; 0]**
5. What is the Boolean expression for the logic diagram shown in Fig 33.64? Evaluate its output if $A = 1$, $B = 1$, and $C = 1$. **[Y = AB + C; 1]**

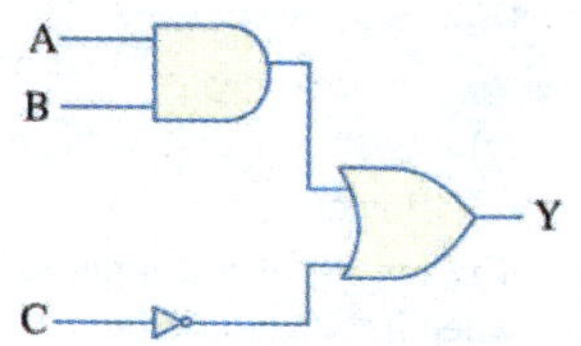

Fig. 33.64

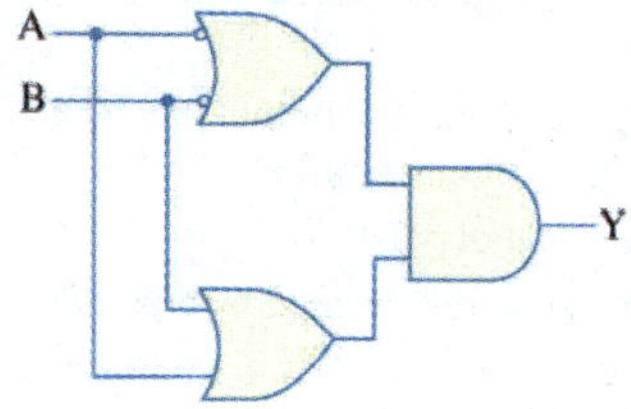

Fig. 33.65

6. Find Boolean expression for the logic circuit of Fig. 33.65. What is the output if $A = 1$, $B = 1$? **[$Y = (\overline{A} + \overline{B}).\ (A + B)$; 0]**
7. What would be the output signal if two signals $A = 101011_2$ and $B = 110101_2$ are applied to the inputs of an AND gate? **[$100\ 001_2$]**
8. What would be the output signal if two input binary signals given by $A = 100101$ and $B = 110110$ are applied to (*a*) OR gate (*b*) NAND gate and (*c*) XNOR gate? **[(*a*) 110111_2 (*b*) 011011_2 (*c*) 101100_2]**

SELF EXAMINATION QUESTIONS

A. Fill in the blanks with most appropriate word(s) or numerical value(s).

1. A logic gate is a circuit that makes.......... decisions.
2. In logic algebra, variables can assume only two values : either .... or 1.
3. The .... gate is also called any-or-all gate.
4. In a 2-input OR gate, output is 0 if and only if................ input are 0.
5. An OR gate corresponds to a .... switching circuit.
6. An XOR gate has an output 1 when its outputs are ....
7. An XOR gate has an output 0 when all its inputs are ....
8. The output of an AND gate is FALSE if any of its input is ....
9. .... gate is also called all-or-nothing gate.
10. In logic multiplication, $A \times A =$ ....
11. The only function of a NOT gate is to .... the input.
12. When $A = 0$, $B = 1$ and $C = 1$, the value of Boolean expression $(AB + C)$ is.......... .
13. The value of $\bar{1} + \bar{1} =$ ....
14. $C = \bar{0}.0 =$ ....
15. The value of $\overline{A + A}$ is ....
16. A......... gate has an output of 1 only when all its inputs are 0.
17. Both NOR and .... gates are called universal gates.
18. An XNOR gate gives an output of 1 if its both inputs are at the .... logic level.
19. A full adder has .... inputs and two outputs.
20. For adding 4-bit numbers, we need .... full adders connected in parallel.
21. The largest decimal numbers which a parallel adder consisting of four full adders can add is ....
22. A full subtractor contains two half subtractors and one .... gate.

B. Answer True or False

1. Logic algebra is essentially the same as numerical algebra.
2. Logic 1 represents positive voltage in positive logic and negative voltage in a negative logic.
3. In a digital system, it hardly matters whether we use positive logic or negative logic.

4. All variables in Boolean algebra can only assume either the value 0 or 1.
5. The OR gate has an output 1 when either A or B is 1.
6. All inputs of an OR gate must be OFF to turn the output OFF.
7. An exclusive OR gate has the output ON when its inputs are either all ON or all OFF.
8. In logic addition, $A + A = 2A$.
9. Logic addition can be represented by a parallel switching circuit.
10. An AND gate performs logic addition.
11. In an AND gate, all inputs must be ON to have an output.
12. All that a NOT gate does is to invert the input signal.
13. Double complementation gives the original value.
14. The logic NOT function can have more than one input.
15. In logic multiplication, $A.A = A^2$.
16. In logic algebra, ORing means adding and ANDing means multiplying.
17. An OR gate in a positive logic system becomes an AND gate in the negative logic system.
18. The truth table of a 3-input logic gate contains six rows.
19. In a bubbled AND gate, inversion takes place before ANDing.
20. A bubbled OR gate is equivalent to a NAND gate.
21. A NOR gate has an output 1 only when all its inputs are 1.
22. The output of a NOR gate is TRUE only when all its inputs are FALSE.
23. A NAND gate must have both inputs ON to have its output OFF.
24. An XNOR gate produces no output if its two inputs are at opposite logic levels.
25. A full-adder consists of two half-adders only.
26. For adding 6-bit numbers, we need a parallel adder consisting of 6 full adders.

C. Multiple Choice Items

1. A logic gate is an electronic circuit which
   (*a*) makes logic decisions
   (*b*) allows electron flow only in one direction
   (*c*) works on binary algebra
   (*d*) alternates between 0 and 1 values
2. In positive logic, logic state 1 corresponds to
   (*a*) positive voltage
   (*b*) higher voltage level
   (*c*) zero voltage level
   (*d*) lower voltage level
3. In negative logic, the logic state 1 corresponds to
   (*a*) negative voltage
   (*b*) zero voltage
   (*c*) more negative voltage
   (*d*) lower voltage level
4. The output of a 2-input OR gate is zero only when its
   (*a*) both inputs are 0
   (*b*) either input is 1
   (*c*) both inputs are 1
   (*d*) either input is 0
5. An XOR gate produces an output only when its two inputs are
   (*a*) high (*b*) low
   (*c*) different (*d*) same
6. An AND gate
   (*a*) implements logic addition
   (*b*) is equivalent to a series switching circuit
   (*c*) is an any-or-all gate
   (*d*) is equivalent to a parallel switching circuit
7. When an input electrical signal A = 10100 is applied to a NOT gate, its output signal is
   (*a*) 01011 (*b*) 10101
   (*c*) 10100 (*d*) 00101
8. The only function of a NOT gate is to
   (*a*) stop a signal
   (*b*) recomplement a signal
   (*c*) invert an input signal
   (*d*) act as a universal gate
9. A NOR gate is ON only when all its inputs are
   (*a*) ON (*b*) positive
   (*c*) high (*d*) OFF
10. The truth table shown in Fig. 33.66 is for a/an .... gate.
   (*a*) XNOR
   (*b*) OR
   (*c*) AND
   (*d*) NAND
   (*e*) XOR

A	B	C
0	0	0
0	1	1
1	0	1
1	1	1

Fig. 33.66

11. For getting an output from an XNOR gate, its both inputs must be
   (*a*) high
   (*b*) low
   (*c*) at the same logic level
   (*d*) at the opposite logic levels

12. The arrangement shown in Fig. 33.67 performs the logic function of a/an........ gate.
   (*a*) AND (*b*) NAND
   (*c*) OR (*d*) XOR

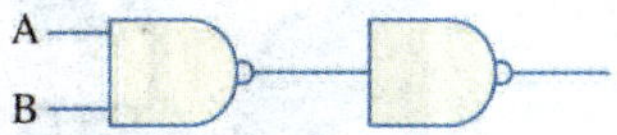

Fig. 33.67

13. In a certain 2-input logic gate, when A = 0, B = 0, then C = 1 and when A = 0, B = 1, then again C = 1. It must be .... gate.
   (*a*) XOR (*b*) AND
   (*c*) NAND (*d*) NOR

14. The logic symbol shown in Fig. 33.68 represents
   (*a*) single-output AND gate
   (*b*) NAND gate
   (*c*) NAND gate used as NOT gate
   (*d*) NOR gate

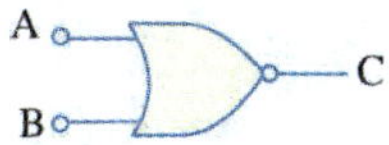

Fig. 33.68

15. The output from the logic gate shown in Fig. 33.69 will be available when inputs ...... are present.
   (*a*) A and C
   (*b*) B and C
   (*c*) A, B and C
   (*d*) A and B

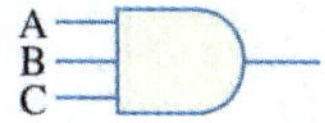

Fig. 33.69

16. To get an output 1 from circuit of Fig. 33.70, the input must be A B C
   (*a*) 0 1 0 (*b*) 1 0 0
   (*c*) 1 0 1 (*d*) 1 1 0

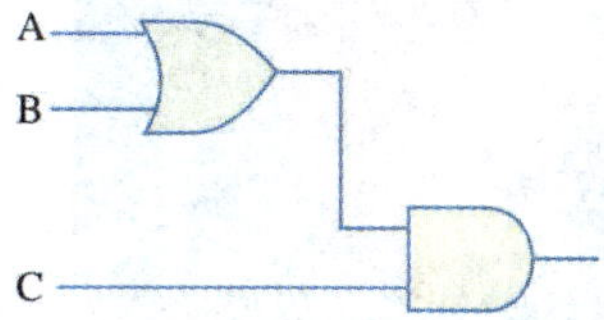

Fig. 33.70

ANSWERS

A. Fill in the blanks

1. logic	**2.** 0	**3.** OR	**4.** both
5. parallel,	**6.** different	**7.** 1	**8.** FALSE
9. AND	**10.** A	**11.** invert	**12.** 1
13. 0	**14.** 1	**15.** $\overline{A}$	**16.** NOR
17. NAND	**18.** same	**19.** 3	**20.** 4
21. 15	**22.** OR		

B. True or False

1. F	**2.** F	**3.** T	**4.** T	**5.** F	**6.** T
7. F	**8.** F	**9.** T	**10.** F	**11.** T	**12.** T
13. T	**14.** F	**15.** F	**16.** T	**17.** T	**18.** F
19. T	**20.** T	**21.** F	**22.** T	**23.** T	**24.** T
25. F	**26.** T				

C. Multiple Choice Items

1. *a*	**2.** *b*	**3.** *d*	**4.** *a*	**5.** *c*	**6.** *b*
7. *a*	**8.** *c*	**9.** *d*	**10.** *a*	**11.** *c*	**12.** *a*
13. *c*	**14.** *d*	**15.** *c*	**16.** *c*		

CHAPTER 34

Boolean Algebra

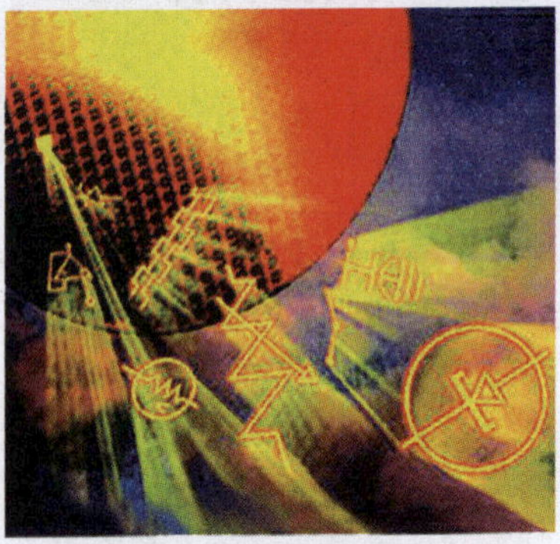

1. Introduction
2. Unique Feature of Boolean Algebra
3. Laws of Boolean Algebra
4. Equivalent Switching Circuits
5. De Morgan's Theorems
6. Duals

34.1. Introduction

Boolean algebra, named after its pioneer George Boole (1815-1864) is the algebra of **logic** presently applied to the operation of computer devices. The rules of this algebra are based on human reasoning. It originated from the study of how we reason, what lines of reasoning are valid and what constitutes proof etc.

Starting with his investigation of the laws of thought, Boole developed in 1854 a mathematical system of logic in which he expressed truth functions *as symbols* and then manipulated these symbols to arrive at a conclusion. His new system was not the **ordinary numerical algebra** we know from our high school days but a totally new system called **logic algebra.** For example, in Boolean algebra $A + A = A$ and not $2A$ as is the case in ordinary algebra.

George Boole (1815-1864).

Boolean algebra remained in the realm of philosophy till 1938 when Claude E. Shannon used it to solve relay logic problems. As we know, all thinking and logic is concerned with finding answers to binary or two-valued questions like: is it good or bad, right or wrong, true or false etc. This binary nature of logic is exactly like the binary working of relay and switching circuits where relay is either energised or not, light is ON or OFF or pulse is present or not. Because of its very logical nature, Boolean algebra is ideal for the design and analysis of logic circuits used in computers. Moreover, it provides an economical and straightforward way of describing computer circuitry and complicated switching circuits. As compared to other mathematical tools of analysis and design, Boolean algebra has the advantages of simplicity, speed and accuracy.

34.2. Unique Feature of Boolean Algebra

As we know, the different variables used in ordinary algebra can have *any value* including plus and minus values. There is no restriction on the value they can assume. For example, in the equation $2x + 3y = z$, the variables x, y and z can take on any value available in the entire field of real numbers.

However, the variables used in Boolean algebra have a unique property *i.e.*, they can assume *only one of the two possible values of 0 and 1*. Each of the variable used in a logical or Boolean equation can assume only the value 0 or 1. For example, in the logical equation $A + B = C$, **each** of the three variables A, B and C can have only the values of either 0 or 1. *This point must be clearly taken note of by the reader for easy understanding of the laws of Boolean algebra.*

34.3. Laws of Boolean Algebra

As stated earlier, Boolean algebra is a system of Mathematics based on *logic*. It has its own set of fundamental laws which are necessary for manipulating different Boolean expressions.

1. OR Laws

These four laws have already been discussed in the previous chapter. These are

Law 1. $A + 0 = A$ Law 2. $A + 1 = 1$

Law 3. $A + A = A$ Law 4. $A + \overline{A} = 1$

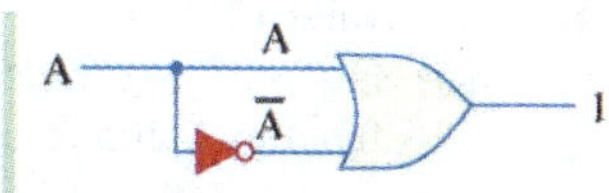

Fig. 34.1

The expression given in Law 4 can be understood with the help of Fig. 34.1. Consider the following two possibilities :

(*i*) When $A = 0, \overline{A} = 1$ $\therefore A + \overline{A} = 0 + 1 = 1$

(*ii*) When $A = 1, \overline{A} = 0$ $\therefore A + \overline{A} = 1 + 0 = 1$

2. AND Laws

Law 5. $A.0 = 0$ Law 6. $A.1 = A$

Law 7. $A.A = A$ Law 8. $A.\overline{A} = 0$

The expression for Law 8 can be easily understood with the help of the logic circuit of Fig. 34.2. Consider the following two possibilities :

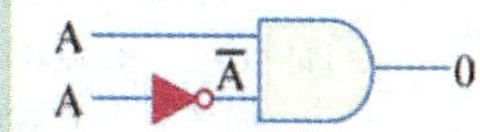

Fig. 34.2

(*i*) When $A = 0, \overline{A} = 1$ $\therefore A.\overline{A} = 0.1 = 0$

(*ii*) When $A = 1, \overline{A} = 0$ $\therefore A.\overline{A} = 1.0 = 0$

3. Laws of Complementation

Law 9. $\overline{0} = 1$ Law 10. $\overline{1} = 0$

Law 11. if $A = 0$, then $\overline{A} = 1$ Law 12. if $A = 1$, then, $\overline{A} = 0$

Law 13. $\overline{\overline{A}} = A$

4. Commutative Laws

These laws allow *change in the position* of variables in OR and AND expressions.

Law 14 $A + B = B + A$ Law 15. $A.B = B.A$

These two laws express the fact that the order in which a combination of terms is performed does not affect the final result of the combination.

5. Associative Laws

These laws allow removal of brackets from logical expression and regrouping of variables.

Law 16. $A + (B + C) = (A + B) + C$

Law 17. $(A + B) + (C + D) = A + B + C + D$ Law 18. $A.(B.C) = (A.B).C$

6. Distributive Laws

These laws permit factoring or multiplying out of an expression.

Law 19. $A(B + C) = AB + AC$ Law 20. $A + BC = (A + B)(A + C)$

Law 21. $A + \overline{A}.B = A + B$

7. Absorptive Laws

Law 22. $A + AB = A$ Law 23. $A.(A + B) = A$

Law 24. $A.(\overline{A} + B) = AB$

The above laws can be used to prove any given Boolean identity and also for simplifying complicated expressions.

34.4. Equivalent Switching Circuits

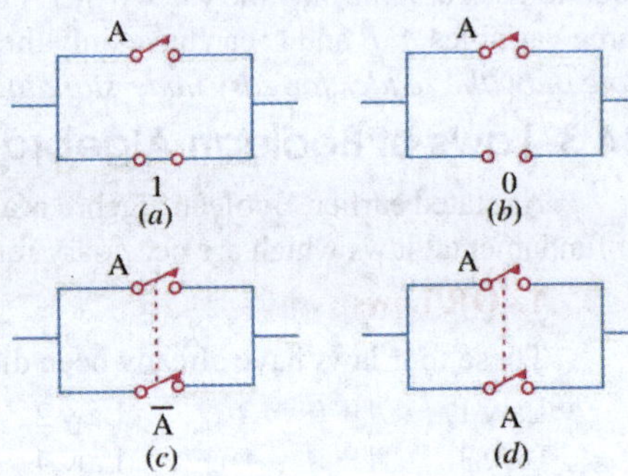

Fig. 34.3

The equivalent circuits to illustrate some of the OR and AND laws given above are shown in Fig. 34.3.

(*i*) Fig. 34.3 (*a*) illustrates $A + 1 = 1$. Here, lower switch is permanently closed representing 1. Hence, value of the OR function is 1 (*i.e.*, it is ON) whatever the value of A.

(*ii*) Fig. 34.3 (*b*) represents $A + 0 = A$. Here, function value is determined by A alone.

(*iii*) In Fig. 34.3 (*c*) when A opens, $\overline{A}$ closes and *vice versa*. Obviously, whatever the position of A, the circuit would always be ON proving that $A + \overline{A} = 1$.

(*iv*) Fig. 34.3 (*d*) proves $A + A = A$. It shows that final result depends on the value of A alone. If $A = 0$, the two switches are open, hence circuit is OFF. If $A = 1$, both switches are closed. Hence, circuit is ON.

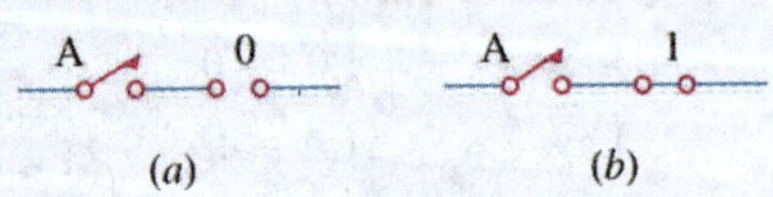

Fig. 34.4

(*v*) In Fig. 34.4 (*a*), the circuit is permanently OFF irrespective of the value of A. It is due to *permanent* open (0) in the circuit. Hence, it proves $A.0 = 0$.

(*vi*) Fig. 34.4 (*b*) shows that circuit conditions will depend solely on the position of the switch. If $A = 1$, circuit is ON (1) and when $A = 0$, circuit is OFF (0). It is all due to the presence of a *permanent* short (1) in the series circuit. Hence, everything depends on A.

Example 34.1. *Prove the following Boolean identity :*

$$AC + ABC = AC$$

Solution. Taking the left-hand side expression as y, we get

$$y = AC + ABC = AC(1 + B)$$

Now, $1 + B = 1$ — Law 2

$\therefore$ $y = AC.1 = AC$ — Law 6

$\therefore$ $AC + ABC = AC$

Example 34.2. *Prove the following Boolean identity:*

$$(A + B)(A + C) = A + BC$$

Solution. Putting the left-hand side expression equal to *y*, we get

$$\begin{aligned} y &= (A + B)(A + C) \\ &= AA + AC + AB + BC && \text{— Law 19} \\ &= A + AC + AB + BC && \text{— Law 7} \\ &= A + AB + AC + BC \\ &= A(1 + B) + AC + BC && \text{— Law 19} \\ &= A + AC + BC && \text{— Law 2} \\ &= A(1 + C) + BC && \text{— Law 19} \\ &= A + BC && \text{— Law 2} \end{aligned}$$

$$\therefore \quad (A + B)(A + C) = A + BC$$

Example 34.3. *Prove the following Boolean identity:*

$$A + \bar{A}B = A + B$$

Solution. Let the left-hand side expression be put equal to *y*.

$$\begin{aligned} \therefore \quad y &= A + \bar{A}B = A.1 + \bar{A}B && \text{— Law 6} \\ &= A(1 + B) + \bar{A}B && \text{— Law 2 and 14} \\ &= A.1 + AB + \bar{A}B && \text{— Law 19} \\ &= A + BA + B\bar{A} && \text{— Law 6 and 15} \\ &= A + B(A + \bar{A}) && \text{— Law 19} \\ &= A + B.1 && \text{— Law 4} \\ &= A + B && \text{— Law 6} \end{aligned}$$

$$\therefore \quad A + \bar{A}B = A + B$$

Example 34.4. *Prove the following Boolean identity :*

$$(A + B)(A + \bar{B})(\bar{A} + C) = AC$$

Solution. Let the left-hand side expression be represented by *y*.

$$\begin{aligned} \therefore \quad y &= (A + B)(A + \bar{B})(\bar{A} + C) = (AA + A\bar{B} + BA + B\bar{B})(\bar{A} + C) \\ &= (A + AB + A\bar{B})(\bar{A} + C) \\ &= [A(1 + B) + A\bar{B}](\bar{A} + C) && \because \; B\bar{B} = 0 \\ &= (A + A\bar{B})(\bar{A} + C) = A(1 + \bar{B})(\bar{A} + C) \\ &= (A.1)(\bar{A} + C) = A(\bar{A} + C) = A\bar{A} + AC = AC && \because A\bar{A} = 0 \end{aligned}$$

Example 34.5. *Prove the following Boolean identity :*

$$ABC + A\bar{B}C + AB\bar{C} = A(B + C)$$

Solution. Equating the left-hand side expression to *y*, we have

$$\begin{aligned} y &= ABC + A\bar{B}C + AB\bar{C} = AC(B + \bar{B}) + AB\bar{C} \\ &= AC + AB\bar{C} && \text{— Law 4} \\ &= A(C + B\bar{C}) \\ &= A(C + B) && \text{— Law 21} \\ &= A(B + C) && \text{— Law 14} \end{aligned}$$

Hence, it proves the given identity.

Example 34.6. *Simplify the following Boolean expression :*

$$AB\bar{C} + A\bar{B}C + \bar{A}BC + ABC + A\bar{B}\bar{C}$$

Solution. Bringing together those terms which have two common letters, we get

$$y = AB\bar{C} + ABC + A\bar{B}\bar{C} + A\bar{B}C + \bar{A}BC$$

$$= AB(\bar{C} + C) + A\bar{B}(\bar{C} + C) + \bar{A}BC$$

$$= AB + A\bar{B} + \bar{A}BC \qquad \text{— Law 4}$$

$$= A(B + \bar{B}) + \bar{A}BC \qquad \text{— Law 4}$$

$$= A + \bar{A}BC = A + BC \qquad \text{— Law 21}$$

Example 34.7. *Using truth table, prove that $A + \bar{A}B = A + B$ and illustrate the equivalence with the help of a switching circuit.*

Solution. Since there are only two variables A and B, their number of possible combination is $2^2 = 4$ in terms of 0 and 1.

Table 34.1

A	B	$\bar{A}$	$\bar{A}B$	$A + \bar{A}B$	$A + B$
0	0	1	0	0	0
0	1	1	1	1	1
1	0	0	0	1	1
1	1	0	0	1	1

As seen, $\bar{A}$ is the negative of A. In the fourth column of Table 34.1, $\bar{A}$ has been ANDed with B. In the fifth column, A has been ORed with $\bar{A}B$. The values in last column have been obtained by ORing A with B. By comparing results of column 5 and 6, the equivalence between the two statements can be proved.

Switching circuit of Fig. 34.5 (*a*) represents $(A + \bar{A}B)$. In this circuit, when A is open, $\bar{A}$ is closed and *vice versa*. It can be shown that this circuit becomes closed when either A or B is closed.

(*i*) when A is closed, then $\bar{A}$ opens. Circuit is completed *via* the upper branch;

(*ii*) keeping A open, when we close B, the circuit again becomes closed *via* lower branch because $\bar{A}$ is already closed (due to A being open).

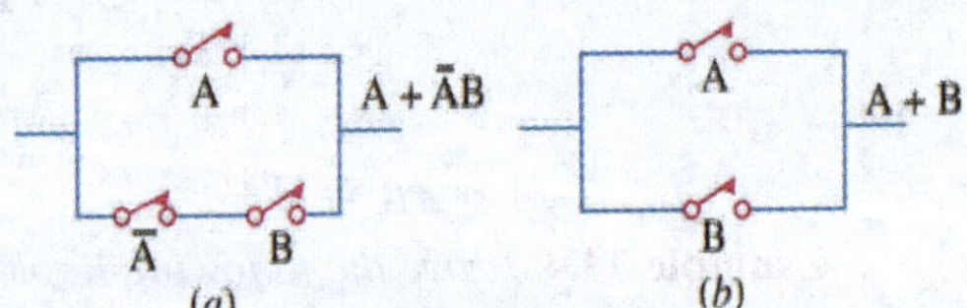

Fig. 34.5

Hence, all that we have to do for closing the circuit of Fig. 34.5 (*a*) is to close either switch A or B. It is exactly what circuit of Fig. 34.5 (*b*) does.

34.5. De Morgan's Theorems

These two theorems (or rules) are a great help in *simplifying complicated logical expressions.* The theorems can be stated as under :

De Morgan (1806-1881).

Law 25. $\overline{A + B} = \bar{A} \cdot \bar{B}$ Law 26. $\overline{A.B} = \bar{A} + \bar{B}$

The first statement says that *the complement of a sum equals the product of complements*. The second statement says that *the complement of a product equals the sum of the complements.* In fact, it allows transformation from a sum-of-products form to a product-of-sum form.

As seen from the above two laws, the procedure required for taking out an expression from under a NOT sign is as follows :

1. *complement the given expression i.e., remove the overall* NOT *sign*
2. *change all the ANDs to ORs and all the ORs to ANDs.;*
3. *complement or negate all individual variables.*

As an illustration, take the following example

$$\overline{A + BC} = A + BC \qquad \text{— step 1}$$
$$= A\,(B + C) \qquad \text{— step 2}$$
$$= \bar{A}\,(\bar{B} + \bar{C}) \qquad \text{— step 3}$$

Next, consider this example,

$$(A + B + \bar{C})(\bar{A} + B + C) = (\bar{A} + \mathrm{B} + \bar{C})(\bar{A} + \mathrm{B} + \mathrm{C}) \qquad \text{— step 1}$$
$$= \bar{A}B\bar{C} + \bar{A}BC \qquad \text{— step 2}$$
$$= \bar{\bar{A}}\bar{B}\bar{\bar{C}} + \bar{\bar{A}}\bar{B}\bar{C} \qquad \text{— step 3}$$
$$= A\bar{B}C + A\bar{B}\bar{C}$$

This process is called **demorganization.** It should, however, be noted that opposite procedure would be followed in bringing an expression under the NOT sign.

Let us bring the expression $\bar{A} + \bar{B} + \bar{C}$ under the NOT sign.

$$\bar{A} + \bar{B} + \bar{C} = \bar{\bar{A}} + \bar{\bar{B}} + \bar{\bar{C}} \qquad \text{— step 3}$$
$$= A + B + C$$
$$= ABC \qquad \text{— step 2}$$
$$= \overline{ABC} \qquad \text{— step 1}$$

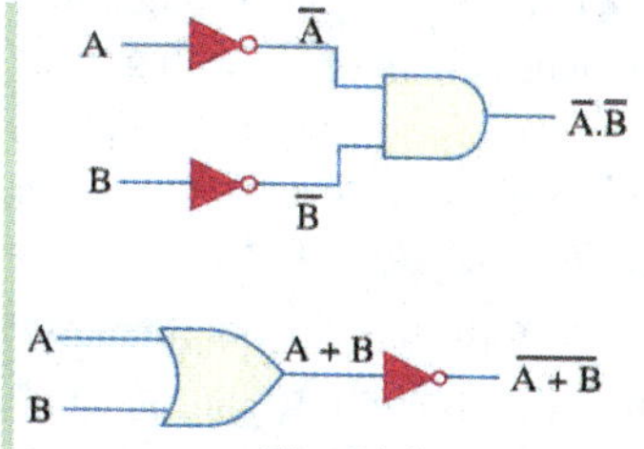

Fig. 34.6

Fig. 34.6 shows the circuits to illustrate De Morgan's theorems.

As seen, basic logic function can be either an OR gate or an AND gate.

Example 34.8. *Demorganize the expression*

$$\overline{(A + B)(C + D)}$$

Solution. The procedure is as explained above.

$$\overline{(A + B)(C + D)} = (A + B)\,(C + D) \qquad \text{— step 1}$$
$$= (AB) + (CD) \qquad \text{— step 2}$$
$$= \bar{A}\,\bar{B} + \bar{C}\,\bar{D} \qquad \text{— step 3}$$

Example 34.9. *Simplify the expression:* $(AB + C)\,(AB + D)$

Solution. Let

$$y = (AB + C)\,(AB + D)$$
$$= ABAB + ABD + ABC + CD \qquad \text{— Law 19}$$
$$= AABB + ABD + ABC + CD$$
$$= AB + ABD + ABC + CD \qquad \text{— Law 7}$$
$$= AB\,(1 + D) + ABC + CD = AB + ABC + CD \qquad \text{— Law 2}$$
$$= AB\,(1 + C) + CD = AB + CD$$

$\therefore \quad (AB + C)\,(AB + D) = AB + CD$

Example 34.10. *Find switching circuits for the following logic expressions :*

(*i*) $A.\,(B + C)$ (*ii*) $A\bar{B} + CD$ (*iii*) $(\bar{A}B + AC)\,\bar{C}$

Solution. (*i*) **A. (B + C)**

Here, switches *B* and *C* have been ORed *i.e.*, connected in parallel. This parallel circuit is connected in series with switch *A* because (*B* + *C*) has been ANDed with *A*. Hence, the circuit becomes as shown in Fig. 34.7 (*a*). As seen, it is a series-parallel circuit.

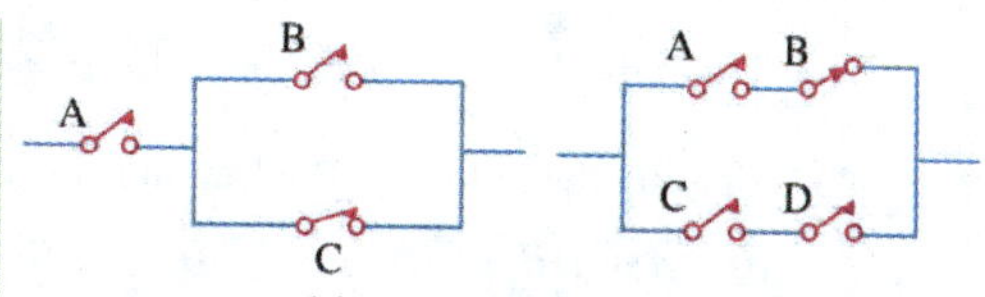

(*a*) Fig. 34.7 (*b*)

(*ii*) $\mathbf{A\overline{B} + CD}$

Here, $A\,\overline{B}$ has been ORed with CD. We can easily make out that a 2-branched circuit is needed for this logic expression. One branch contains switch A in series with $\overline{B}$ (with one contact point shown raised to indicate negation) and the other branch contains two series-connected switches as shown in Fig. 34.7 (*b*).

(*iii*) $\mathbf{(\overline{A}B + AC)\overline{C}}$

From the look of it, we can make out that it consists of a series-parallel circuit as shown in Fig. 34.8. $\overline{A}B$ has been ORed with AC *i.e.*, $\overline{A}B$ and AC are in parallel. Of course, $\overline{A}$ and B are in series in one branch where as A and C are in series in the other branch. Both these parallel branches are in series with $\overline{C}$.

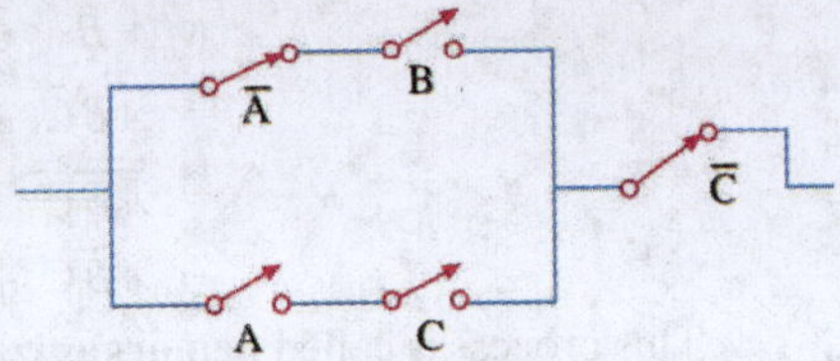

Fig. 34.8

Example 34.11. *Prove that 3-input NAND gate of Fig. 34.9 (a) is equivalent to the bubbled AND gate of Fig. 34.9 (b).*

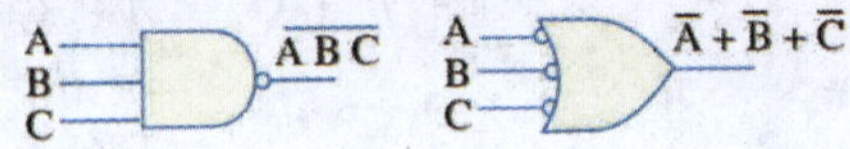

(*a*) (*b*)

Fig. 34.9

Solution. The output of NAND gate is $\overline{ABC}$ and that of bubbled OR gate is $\overline{A} + \overline{B} + \overline{C}$. We have to show that the above two expressions are equivalent.

De Morgan's theorem can be used to prove the above equivalence,

$$\overline{ABC} = ABC \quad \text{— step 1}$$
$$= A + B + C \quad \text{— step 2}$$
$$= \overline{A} + \overline{B} + \overline{C} \quad \text{— step 3}$$

34.6. Duals

Basic duality underlies all Boolean algebra. *Each expression has its dual which is as true as the original expression.* For getting the dual of a given Boolean expression, the procedure is to convert

1. all 1s to 0s and all 0s to 1s.
2. all ANDs to ORs and all ORs to ANDs.

The dual so obtained *is also found to be true.*

Some of the Boolean relations and their duals are given in Table 34.2.

Table 34.2

Relation	*Dual Relation*
$A \cdot 0 = 0$	$A + 1 = 0$
$A \cdot A = 1$	$A + A = A$
$A \cdot \overline{A} = 0$	$A + \overline{A} = 1$
$A \cdot 1 = A$	$A + 0 = A$
$A \cdot (A + B) = A$	$A + AB = 1$
$A(\overline{A} + B) = AB$	$A + \overline{A}B = A + B$

CONVENTIONAL PROBLEMS

1. Prove the following identities :

   (*i*) $AB + A\overline{B} = A$ (*ii*) $A + \overline{A}B = A + B$ (*iii*) $(A + B)(A + \overline{B}) = A$

2. Simplify the following Boolean expressions :

   (*i*) $A\overline{A}C$ (*ii*) $ABCD - ABD$ (*iii*) $ABCD + A\overline{B}CD$

   [(i) 0 (ii) ABD (iii) ACD]

3. Simplify the following Boolean functions :

   (*i*) $A + \overline{A}B + AB$ (*ii*) $A\overline{B} + \overline{A}B + AB + \overline{A}\overline{B}$ (*iii*) $(AB + C)(AB + D)$

   (*iv*) $A - \overline{B}C(A + \overline{B}C)$ (*v*) $A[B + C\overline{(AB + AC)}]$

   [(i) A + B (ii) 1 (iii) AB + CD (iv) A (v) AB]

4. Simplify the following Boolean expressions :

(i) $A\overline{B}C + AB\overline{C} - ABC$ (ii) $\overline{A}\overline{C} + B\overline{C} + \overline{A}BC + ABC$

(iii) $AD + A\overline{B}\overline{C} + \overline{B}C\overline{D} + \overline{A}CD + \overline{A}BC\overline{D} \supset$

(iv) $A\overline{B}\overline{D} + AB\overline{C}\overline{D} + ABC\overline{D} + \overline{A}\overline{B}\overline{D} + \overline{A}BC\overline{D}$

[(i) A(B + C) (ii) B + $\overline{A}\overline{C}$ (iii) AD + A$\overline{B}$ + $\overline{A}$C (iv) $\overline{B}\overline{D}$ + A$\overline{D}$ + C$\overline{D}$]

5. Use truth tables to verify the following identities :

(i) $\overline{A} + B = \overline{A}.B$ (ii) $A + BC = (A + B)(A - C)$ (iii) $A(\overline{A} + B) = AB$

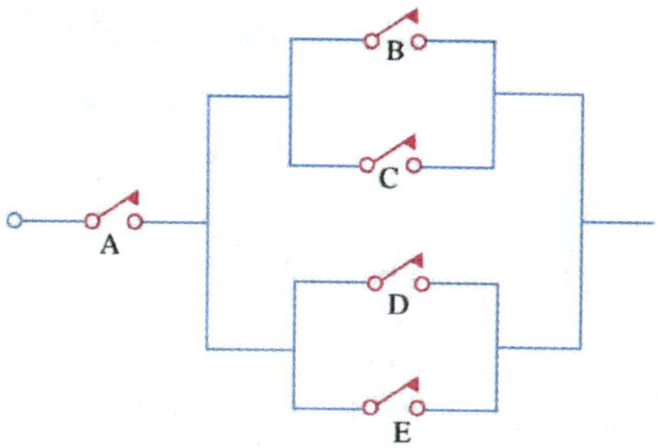

Fig. 34.10

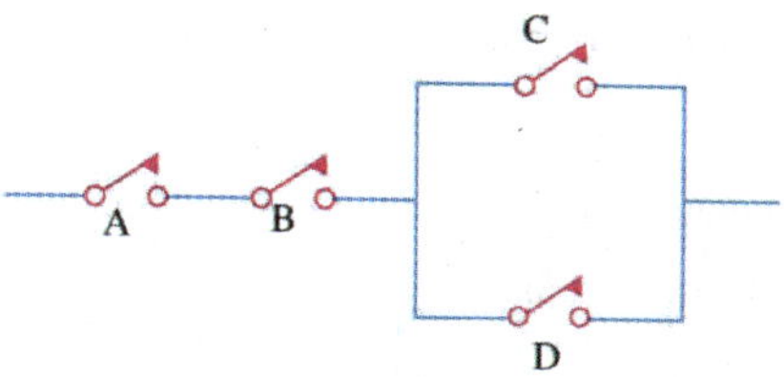

Fig. 34.11

6. Write the logic equation for the switching circuit of Fig. 34.10 **[A(B + C + D + E]**

7. Write down the logical expression which describes the working of the circuit of Fig. 34.11.
[A . B(C + D)]

SELF EXAMINATION QUESTIONS

A. Fill in the blanks with most appropriate word(s) or numerical value(s).

1. Boolean algebra is the algebra of ....
2. All variables in Boolean algebra can have only one of the two values: 0 and ....
3. According to OR laws, ... + $\overline{A}$ = 1.
4. The AND law gives $A.\overline{A}$ = ...
5. According to the laws of complementation, A = ...
6. According to the Absorptive Laws of Boolean algebra, $A + AB$ = ...
7. According to De Morgan's theorem, $A + \overline{B}$ = ...
8. Each Boolean relation has its ....
9. When simplified, $A\overline{B} + \overline{AB}$ = ... + ...

B. Answer True or False

1. Boolean algebra is based on human reasoning.
2. As in ordinary algebra Boolean variables can also take on any value including 0.
3. In the logical algebra, $A + A = 2A$.
4. Distributive laws apply equally to the Boolean algebra.
5. According to the Absorptive Laws of Boolean algebra.

$$A(\overline{A} + B) = A$$

6. De Morgan's theorems allow the taking out of an expression from under a NOT sign.
7. Every Boolean equation has two duals.
8. The dual of $A.1 = A$ is $\overline{A}.\overline{A} = 0$

C. Multiple Choice Items

1. Boolean algebra is essentially based on
   (a) symbols (b) logic
   (c) truth (d) numbers
2. The first person who used Boolean algebra for the design of relay switching circuits was
   (a) Aristotle (b) Boole
   (c) Shannon (d) Ramanujam
3. Different variables used in Boolean algebra can have values of
   (a) 0 or 1 (b) low or high
   (c) true or false (d) ON or OFF

4. According to the algebra of logic, $(A + \overline{A})$ equals
   (a) A (b) 1
   (c) 0 (d) $\overline{AA}$
5. According to the Absorptive Laws of Boolean algebra, expression $(A + AB)$ equals
   (a) A (b) B
   (c) AB (d) $\overline{A}$
6. When we demorganize $\overline{AB}$, we get
   (a) $\overline{AB}$ (b) $\overline{A} + \overline{B}$
   (c) $\overline{A + B}$ (d) $\overline{A}\overline{B}$
7. The dual of the statement $(A + 1) = 1$ is
   (a) $A.1 = A$ (b) $A.0 = 0$
   (c) $A + A = A$ (d) $A.A = 1$
8. The expression $\overline{ABC}$ can be simplified to
   (a) $\overline{A}.\overline{B}.\overline{C}$ (b) $AB + BC + CA$
   (c) $AB + \overline{C}$ (d) $\overline{A} + \overline{B} + \overline{C}$

ANSWERS

A. Fill in the blanks

1. logic	2. 1	3. A	4. 0	5. $\overline{A}$
6. A	7. $\overline{A}.\overline{B}$	8. dual	9. $\overline{A} + \overline{B}$	

B. True or False

1. T	2. F	3. F	4. T	5. F
6. T	7. F	8. F		

C. Multiple Choice Items

1. *b*	2. *c*	3. *a*	4. *b*	5. *a*
6. *c*	7. *b*	8. *d*		

CHAPTER 35

Logic Families

1. Main Logic Families
2. Saturated and Non-saturated Logic Circuits
3. Characteristics of Logic Families
4. RTL Circuit
5. DTL Circuit
6. TTL Circuit
7. TTL Subfamilies
8. ECL Circuit
9. I^2L Circuit
10. MOS Family
11. PMOS Circuit
12. NMOS Circuit
13. CMOS Circuit

35.1. Main Logic Families

Most digital systems are designed by combining various logic functions discussed in Chapter 33. All these logic circuits are available in IC modules and are divided into many 'families'. Each family is classified by abbreviations which indicate the type of logic circuit used. For example, RTL means resistor-transistor logic. We will discuss the following seven transistor logic families although the first two are, at present, of historic interest only.

1. **resistance-transistor logic (RTL)** —it was the first family group of logic circuits to be developed and packaged in IC form in *early* 1960s;
2. **diode-transistor logic (DLT)**—it followed RTL in *late* 1960s;
3. **transistor-transistor logic (TTL) or (T^2L)** —was introduced in the *early* 1970s;
4. **Schottky TTL**—was introduced to improve the speed of TTL;
5. **emitter-coupled logic (ECL)**—it is the fastest logic line currently available;
6. **integrated-injection logic (I^2L)** —it is one of the latest of the bipolar types of logic;
7. **complementary metal-oxide semiconductor (CMOS)**—it has the lowest power dissipation of the currently-available logic circuits.

The various logic families discussed above possess different characteristics as detailed below.

35.2. Saturated and Non-saturated Logic Circuits

Those logic circuits in which transistors are driven into saturation are called **saturated** logic circuits or simply **saturated logic.** Those circuits which avoid saturation of their transistors are designated **non-saturated logic.**

The disadvantage of saturated logic is the delay that occurs when the transistor is brought out of saturation. When a transistor is saturated, its base is flooded with carriers. Even when base voltage is switched off, the base remains flooded for some time till all carriers leave it. The time required by the carriers to leave the base is called **saturation delay time** (t_s). Obviously, saturated logic circuits have low switching speeds whereas non-saturated type are much faster. TTL is the example of a saturated logic whereas ECL represents a non-saturated logic.

35.3. Characteristics of Logic Families

The logic families possess different characteristics because of which one may prove to be the best suited for a particular job. For example, in a certain system, speed may be the main requirement, whereas in another system minimum power dissipation might be the overriding consideration. The different characteristics of a logic family are as under:

1. speed 2. fan-in 3. fan-out 4. noise immunity 5. power dissipation

Speed of a logic gate is the time that elapses between the application of a signal to an input terminal and the resulting change in logical state at the output terminal. It takes into account the **transition times** (rise and fall times of a pulse) and **propagation delays.** Both these times increase as the loading is increased. The more inputs are attached to the output of a logic gate, the more load has to be handled by that output.

The fan-in of a logic gate is the **number of inputs** (coming from similar circuits) the logic gate can handle properly.

The fan-out of a logic gate is the maximum number of similar circuits that this gate **can drive** reliably. For example, if a gate has a fan-out of 8, this means that it can drive 8 units connected to its output without its output voltage falling outside the limits at which the logic levels 1 and 0 are specified.

Noise immunity is represented by the maximum induced noise voltage a logic circuit can withstand without a false change in its output state. It is also called **noise margin.** Too much noise voltage induced by stray electric and magnetic fields can lead to false triggering of logic levels in the circuit. Higher the noise margin, better the logic circuit.

Power is dissipated in a circuit as it switches from one state to another and also within all current-carrying resistors. Power dissipation of a gate should be as small as possible.

35.4. RTL Circuit

It is a *saturated* logic. It uses only transistors and resistors as circuit elements and also resistors in the input to each base. This family is based on the NOR circuit shown in Fig. 35.1. All other members of the family are made up of NOR cells or variations on them.

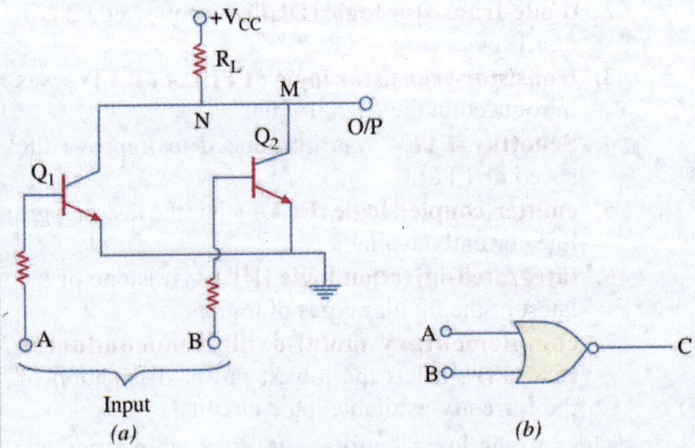

Fig. 35.1

Circuit Operation

We will assume ideal transistors. When both inputs A and B are 0 V (or logic 0) both transistors are turned OFF, hence point *M* goes to + V_{CC} so that output is logic 1.

If either or both input terminals are at + V_{CC} *i.e.,* are high (or logic 1), one or both transistors would be fully turned ON (*i.e.,* saturate) thereby reducing the voltage of point *N* to almost 0 V. Hence, output would be at logic 0.

It is seen that the output is at logic 1 only when *both inputs are at logic* 0 —the NOR logic function as shown in Fig. 35.1 (*b*).

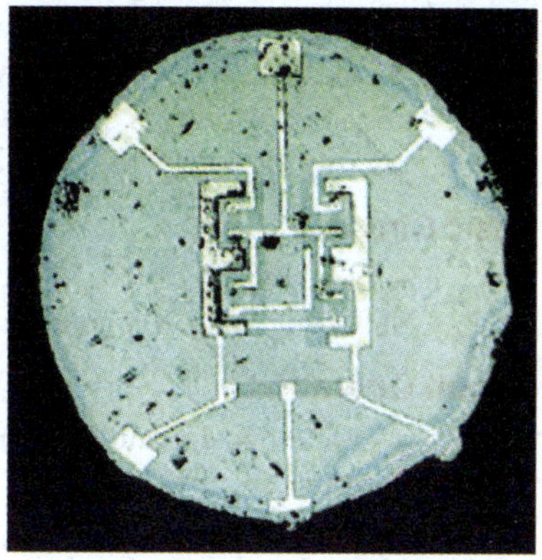

RTL integrated circuit.

The RTL family has the following characteristics:

1. relatively slow speed,
2. low fan-out of 6 and a fan-in of 4,
3. poor noise immunity,
4. expensive since resistors are required to be fabricated,
5. cannot operate at speeds above 4 MHz.

35.5. DTL Circuit

It is a *saturated* logic because transistors operate between cut-off and saturation. It was the next family to be introduced after RTL. It consists of diodes, resistors and transistors. The basic gate of this family performs NAND function. As shown in Fig. 35.2 (*a*), the circuit basically consists of a diode AND gate followed by a transistor inverter which leads to a NAND gate.

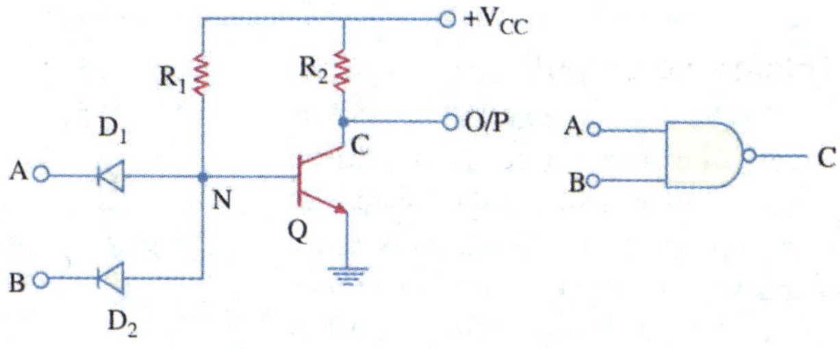

Fig. 35.2

Circuit Operation

1. When both D_1 and D_2 have positive voltage applied to them (logic 1), neither conducts and *Q* is turned ON by the current provided by V_{CC} through R_1. Since *Q* becomes saturated, point *C* is brought to 0 V (logic 0). Hence, output goes logic 0.
2. If either or both inputs are at 0 V (logic 0), the associated diode will conduct driving point *N* to ground *i.e.,* 0 V. Since there is no base voltage for *Q,* it will be cut OFF thereby driving point *C* and hence output to V_{CC} *i.e.,* logic 1.

It is seen that output is low (a logic 0) only when all inputs are high—the condition for a NAND gate.

The DTL family is characterised by

1. relatively lower speed,
2. comparatively better noise immunity,
3. transmission delay of 30 ns,
4. a fan-in of 8,
5. a fan-out of 5.

35.6. TTL Circuit

It is a *saturated* logic. It is the most widely-used circuit line since early 1970s because of its speed, good fan-in and fan-out and easy interface with other digital circuitry. The unique feature of this circuit is that it uses *multiple-emitter transistor* at the input which replaces the input diodes of the DTL. The number of emitters is equal to the desired fan-in of the circuit. Since a multi-emitter transistor is smaller in area than the diodes it replaces, the yield from a wafer is increased. Moreover, smaller area results in a *lower capacitance* to the substrate, thereby reducing circuit rise and fall times and hence *increasing its speed.* The family contains a very wide selection of circuit modules ranging from simple gates and flip-flops in SSI circuit series through various 'registers' in computers in MSI circuit series to micro-processor bit-slice chips in the LSI series.

Basic Circuit

The basic circuit of the TTL family is the NAND gate cell shown in Fig. 35.3. However, at present, NOR, OR and AND gate configurations have also been added to the series.

Circuit Operation

1. If both inputs *A* and *B* are high (logic 1), E/B junction of Q_1 is *reverse-biased* so that it has no emitter current. Hence, Q_1 is OFF. However, its C/B junction is *forward-biased* supplying base current to Q_2 from V_{CC} *via* R_1. As a result, transistor Q_2 is turned fully ON (*i.e.,* it becomes saturated) driving point *N* to 0 V. Hence, output is at logic 0.
2. When either or both inputs are at 0 V (logic 0), the associated E/B junction becomes forward-biased. The value of R_1 is so selected as to ensure that Q_1 is turned fully ON. The voltage at point *M* falls to 0 V with the result that base current for Q_2 is reduced to zero. Hence, Q_2 is cut OFF driving point *N* and the output to logic 1.

Totem Pole Output

The basic circuit of Fig. 35.3 is never used in practice. Its modified version with an added output stage is in common use (Fig. 35.4). This extra output stage is often known as *totem-pole* stage because the three output components Q_3, Q_4 and *D* are stacked, one on top of the other in the manner of a totem-pole. The circuit action is as follows:

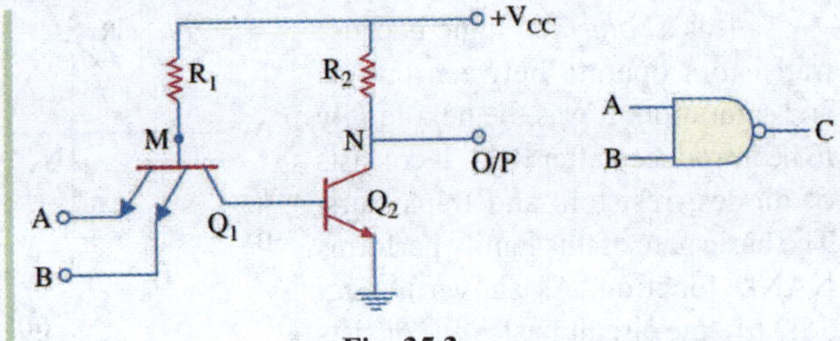

Fig. 35.3

1. When Input is High

In this case, the two input terminals have *positive* voltage (logic 1). The E/B junction is reverse-biased because of which there is no emitter current. Hence, Q_1 is OFF. Since C/B junction of Q_1 is forward-biased, base current of Q_2 flows from V_{CC} through R_1. Hence, Q_2 is turned ON. As a result, potential of point *N* falls so much that Q_3 is turned OFF. At the same time, Q_4 is turned ON by the voltage drop across R_3. Now, when Q_4 is ON, its collector potential (*i.e.,* potential of point *C*) is nearly that of its emitter. Hence, output is low *i.e.,* at logic 0.

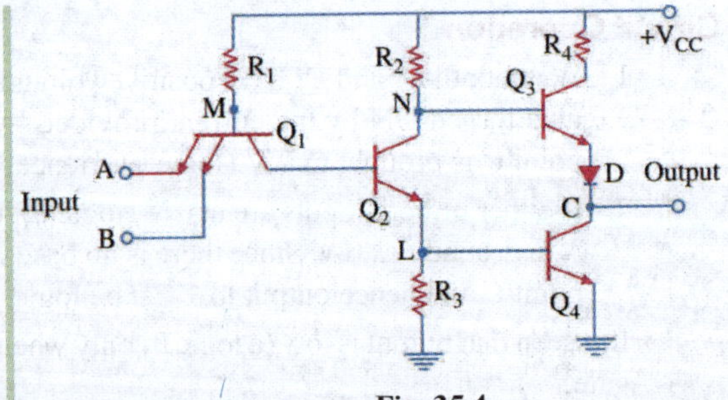

Fig. 35.4

In short, when inputs are at logic 1, Q_1 is OFF, Q_2 is ON, Q_3 is OFF and Q_4 is ON because of which output becomes logic 0.

2. When Input is Low

If any of the two inputs or both are low (logic 0), Q_1 turns ON and potential of its collector (point *M*) falls. Hence, Q_2 is turned OFF, grounding its emitter and the base of Q_4 so that Q_4 is also turned OFF.

Since *N* is at V_{CC}, it turns Q_3 ON. The potential of point *C* is V_{CC} minus drop in R_4, Q_3 and *D*. Since these drops do not amount too much, output is at logic 1.

It may be noted that when *even-numbered transistors are ON, the odd-numbered ones are OFF and vice-versa.*

The function of diode *D* in Fig. 35.4 is to prevent both Q_3 and Q_4 from being turned ON simultaneously. If both were to be ON at the same time, they would offer low impedance to the supply which will draw excessive current and produce large noise 'spikes' in the output. It may also be noted that the addition of a pair of totem pole transistor increases the operating speed and output current capability of this circuit. The standard TTL family has

1. greater speed than DTL,
2. less noise immunity (0.4 V),
3. average propagation delay per gate of 9 ns,
4. average power dissipation of 10 mW,
5. a fan-out of 10 meaning one output can drive 10 other TTL inputs,
6. a fan-in of 6.

Fig. 35.5 shows a pin-out of a BEL 7400 IC referred to as 'quad (quadruple) 2-input NAND gate chip'. As seen, there are four separate 2-input NAND gates of which any or all may be used at any point in time. It is manufactured by Bharat Electronics Ltd., (BEL) Bangalore, India.

35.7. TTL Subfamilies

TTL has several *subfamilies* having different speed and power dissipation characteristics as detailed below:

1. **74L00** series—the letter *L* standing for low power consumption. It has an average power dissipation of 1 mW per gate but an average propagation delay of 33 ns.
2. **74H00** series—the letter *H* standing for higher speed. It has a propagation delay of 6 ns but average power dissipation of 23 mW/gate.
3. **74S00**—the letter *S* representing Schottky. It has the highest speed because its average propagation delay is just 3 ns per gate. However, its average power dissipation is 23 mW/gate.
4. **74LS00**—it is the most recent variant of TTL family. It is called low-power Schottky TTL. It has an average propagation delay of 9.5 ns and an average power dissipation of 2 mW.

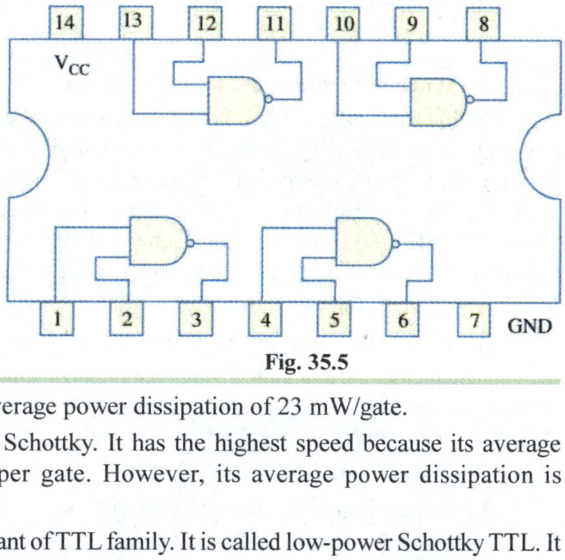

Fig. 35.5

35.8. ECL Circuit

It has the **highest speed** of any of the currently-available logic circuits. It is primarily due to the fact that transistors never operate fully saturated or cut-off. That is why ECL is known as *non-saturated* logic. It has propagation delay time varying from 0.75 ns to 2 ns. However, power dissipation is increased since one transistor is always in the active region.

Another feature of ECL is that it provides two outputs which are always complement of each other (Fig. 35.6). It is so because the circuit operation is based on a differential amplifier.

This family is particularly suited to monolithic fabrication techniques because logic levels are function of resistor ratios.

Circuit Operation

The basic circuit shown in Fig. 35.6 is a combined OR/NOR circuit and is operated from a

V_{EE} = –5.2 V supply. A built-in constant-current source provides current to the emitters. Strictly speaking, logic 1 is represented by –0.9 V (less negative) and logic 0 by –1.75 V (more negative). Please note that it is a *positive* logic. In negative logic, the functions would be AND/NAND. A reference voltage of –1.29 V is applied to the base of Q_3 from a built-in temperature-compensated reference voltage source (RVS).

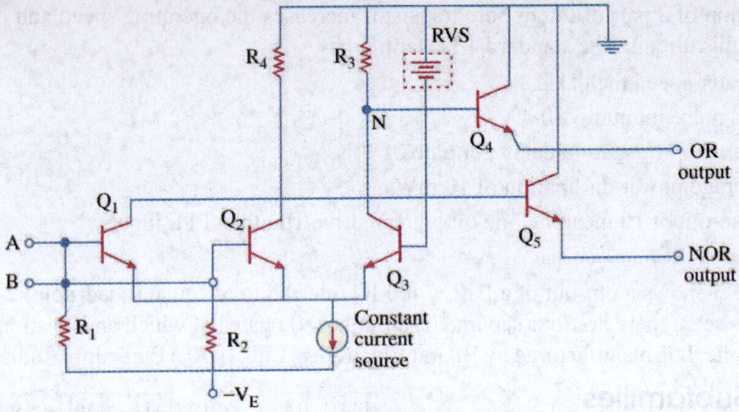

Fig. 35.6

1. When both inputs are logical 0 *i.e.*, –1.75 V

In this case, base potential of Q_3 is less negative (or more positive) than the base potential of either Q_1 or Q_2. Hence, Q_3 conducts whilst Q_1 and Q_2 do not. Only enough base current is drawn by Q_3 from RVS so as to remain out of saturation. The collector current of Q_3 develops a voltage of – 1.0 V across R_3 which makes Q_4 to conduct. Transistor Q_4 gives an output voltage at the emitter of about –1.0 – 0.7 = –1.7 V which represents logic 0. Since collector potentials of Q_1 and Q_2 are nearly zero (because they are cut-off), the output voltage at the emitter of Q_5 is 0 – 0.7 = – 0.7 V which is a logic 1. Obviously, the two outputs are complements of each other.

2. When either input A or B is at logical 1 *i.e.*, – 0.9 V

In that case, the associated transistor (either Q_1 or Q_2) is turned ON while Q_3 is turned OFF. The collector potentials of Q_1/Q_2 and Q_3 are opposite of the previous case. Hence, now Q_5 output is at logical 1 and Q_4 output is at logical 0.

Typical characteristics of an ECL family are :

1. propagation delay time per gate of 1 ns (meaning extremely fast speed),
2. power dissipation of 30 mW,
3. fan-out of 50,
4. fan-in of 5,
5. noise immunity 0.4 V.

There is another point worth noting about ECL family. Since *all its logic levels are negative, it is not directly compatible with other logic series*. For interconnecting an ECL circuit and another circuit of a different family, a special level-shifting circuit has to be provided between the two. Such level-shifting circuits are available in IC packages and are usually referred to as **interface circuits.**

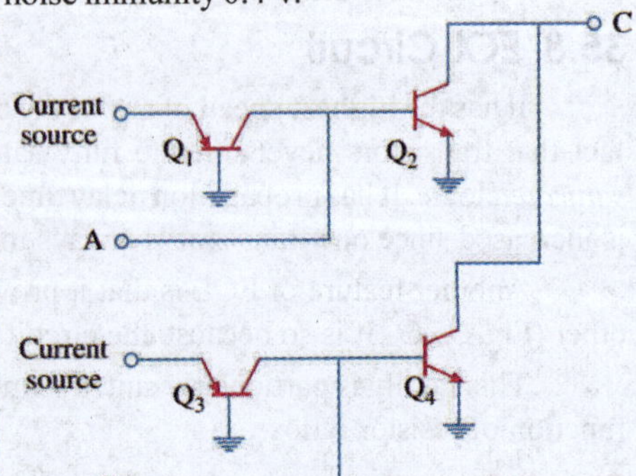

Fig. 35.7

35.9. I²L Circuit

It is the latest entry into the bipolar *saturated* logic field. It uses no biasing and loading resistors at all! Resistors require lot of power and space on an

IC chip. Hence, their elimination results in *higher density circuits operating at much reduce power.* Because of its **high speed** and **less power dissipation,** it is used in large computers. Such circuits are also used where high packing density is of prime consideration as in digital wrist watches. I^2L chips are capable of microwatt power dissipation yet can provide high currents when necessary to drive LED displays.

Another feature of I^2L (integrated injection logic) is that it is easy to fabricate.

Circuit Operation

I^2L NOR logic circuit is shown in Fig. 35.7. Here, transistors Q_1 and Q_2 act as current sources to the bases of Q_3 and Q_4 respectively.

If A input goes low, the current to the base of Q_2 will be shorted to ground which will result in Q_2 being turned OFF. The input B controls Q_4 in a similar way.

If A is high, base current flowing to Q_2 will turn it ON, making output C low. Same would be the case when B is high.

It is obvious that output would be high only when *both inputs A and B are low.* The output would be low when either A or B or both are high *i.e.,* NOR logic function.

35.10. MOS Family

It does not use bipolar transistors but just Enhancement-only MOSFETs (Art. 26.1). There are two kinds of digital MOS circuits.

(*a*) one which uses MOSFETs of one polarity either all of *N*-type called NMOS or all of *P*-type called PMOS but not both on the same chip,

(*b*) the other which employs both *N*-type and *P*-type MOSFETs on the same chip. It is called complementary MOS (CMOS).

The MOS family may be subdivided as under:

MOS circuit
- Non-complementary MOS
  - PMOS
  - NMOS
- Complementary MOS or CMOS

Since a FET requires small area, it is posssible to fabricate a large number of MOS circuits on a single small chip. Gating arrays with thousands of gates and flip-flops are manufactured in standard containers and are often used in IC memories and microprocessors.

35.11. PMOS Circuit

A PMOS NAND gate is shown in Fig. 35.8 (*a*). As seen, there are no resistors in the circuit. The gate consists of two E-only MOSFETs Q_1 and Q_2 as logic elements and the third one Q_3 as load resistor. When $-V_{DD}$ (say – 12 V) is applied, MOSFETs will be turned ON and when 0 V is applied they would be turned OFF. Hence, with positive logic, 0 V would be 1 and – 12 V would be 0 since 1 is assigned to the most positive voltage.

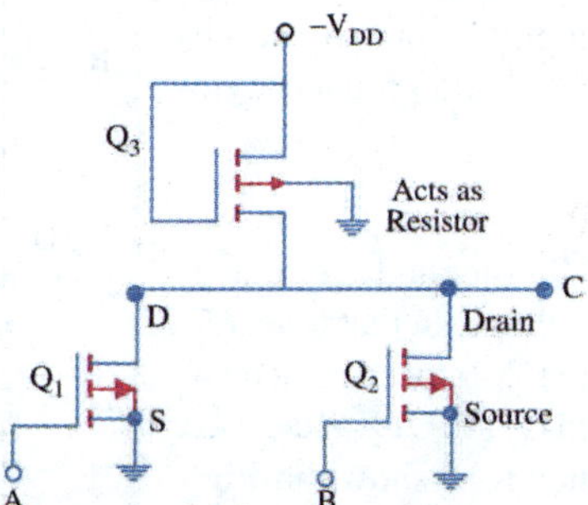

(*a*)

A	B	C
0	0	1
0	1	1
1	0	1
1	1	0

(*b*)

Fig. 35.8

Circuit Operation

1. If any of the two inputs A or B is at logic 0 (*i.e.,* – 12 V), the concerned MOSFET would turn ON thereby offering a low resistance from drain to source thus causing the output to be nearly 0 V *i.e.,* a logic 1.
2. The output can be at – 12 V *i.e.,* logic 0 only when both inputs A and B are at 0 V *i.e.,* logic1.

This is the NAND function as shown by the truth table of Fig. 35.8 (*b*).

Incidentally, the positive logic NAND gate of Fig. 35.8 (*a*) would be the *negative logic* NOR gate since the two are identical.

35.12. NMOS Circuit

In Fig. 35.9 (*a*) is shown a two-input NOR gate circuit consisting of two MOSFETs Q_1 and Q_2 acting as logic elements and Q_3 as a load resistor. For positive logic, 0 V will be logic 0 and positive voltage ($+V_{DD}$) will be logic 1.

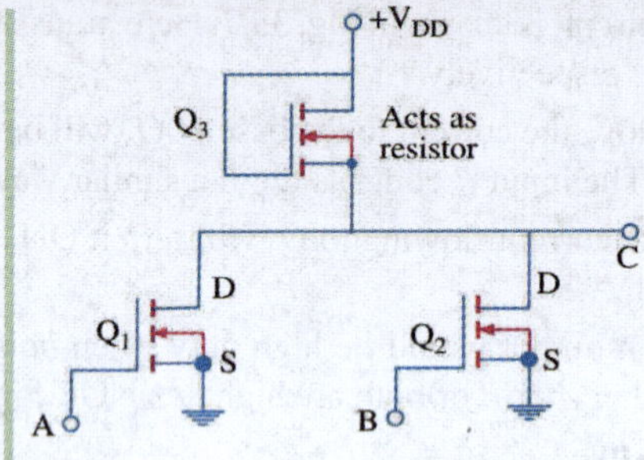

(*a*)

A	B	C
0	0	1
0	1	0
1	0	0
1	1	0

(*b*)

Fig. 35.9

Circuit Operation

1. If any of the inputs A or B is at logic 1, the corresponding MOSFET will conduct causing the output to go low *i.e.,* logic 0.
2. If both inputs A and B are at logic 0, then both MOSFETs will be OFF driving the output to logic 1.

This is NOR function as shown by the truth table of Fig. 35.9 (*b*).

35.13. CMOS Circuit

CMOS logic circuits use both PMOS and NMOS devices in the same circuit. It gives the advantage of *drastic decrease in power dissipation* (12 nW per gate) and *increase in speed of operation.* In fact, it has the **lowest power dissipation** amongst different logic families. It has very high packing density *i.e.,* larger number of circuits can be placed on a single chip. As a result, it is extensively used in VLSI circuits such as one-chip computers and memory systems. The newest silicon-on-suphire MOS (SOSMOS) is 2 to 4 times faster than the standard CMOS. Hence, they are being widely used for everything from electronic watches and calculators to micro-processors.

Fig. 35.10 (*a*) shows a CMOS NOR circuit which has two *N*-channel MOSFETs Q_1 and Q_2 and two *P*-channel MOSFETs Q_3 and Q_4. The two inputs A and B switch between $+V_{DD}$ (logic 1) and ground (logic 0).

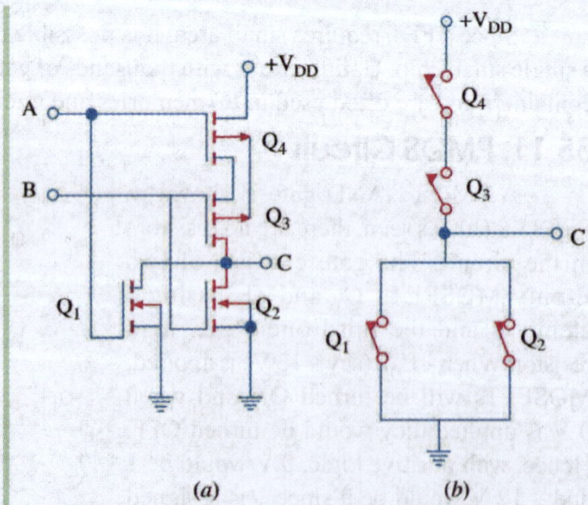

(*a*) (*b*)

Fig. 35.10

Circuit Operation

1. Let $A = B$ = logic 1 *i.e.,* have positive voltage. In that case, Q_1 and Q_2 are ON (closed switches) but Q_3 and Q_4 are OFF and act as open switches as shown in Fig. 35.10 (*b*). Hence, output C is logic 0.

2. If either A or B is at logic 1, then the associated N-channel MOSFET (Q_1 or Q_2) is turned ON but the associated P-channel MOSFET (Q_3 or Q_4) is turned OFF. Since either Q_3 or Q_4 would be OFF [Fig. 35.10 (*b*)], output C would be at logic 0.
3. When both A and B are at logic 0, Q_3 and Q_4 would be ON but Q_1 and Q_2 would be OFF—just the opposite of that in Fig. 35.10 (*b*). Hence, output C would be at logic 1 (remember, voltage across an open equals the supply voltage (Art. 2.9).

A	B	C
0	0	1
0	1	0
1	0	0
1	1	0

Fig. 35.11

The above logic represents a NOR function as shown in the truth table of Fig. 35.11. It would be observed from Fig. 35.10 that in each combination of A and B, there is at least *one open switch* between $+V_{DD}$ and ground. Hence, the gate draws *only leakage current from supply for any static state.* However, when the gate switches from one level to another, some power is consumed because two MOSFETs are *partly* ON at the same time. Because of this reason, power dissipated by CMOS circuit is a *function of input signal frequencey.* Higher the frequency, greater the power dissipation.

SELF EXAMINATION QUESTIONS

A. Fill in the blanks by most appropriate word (s) or numerical value (s).

1. The fan-in of a logic gate is given by the number of .... it can handle properly.
2. Speed of a logic gate depends on transition time and propagation ....
3. Noise immunity is generally expressed in ....
4. RTL employs only .... and resistors.
5. TTL uses multiple .... transistor.
6. As compared to DTL, TTL has .... speed.
7. ECL is .... logic because its transistors never go into saturation.
8. ECL is the .... currently-available logic circuit.
9. ECL family is used when maximum possible .... is the prime consideration.
10. All logic levels of an ECL are ....
11. We need an .... circuit for interconnecting an ECL circuit to another circuit of a different logic family.
12. I^2L circuit uses no biasing and loading ....
13. I^2L circuits are used when high .... density is of prime consideration.
14. MOS logic circuits use no ....

B. Answer True or False

1. Non-saturated logic is slower than the saturated logic.
2. Speed of a logic circuit decreases as its loading increases.
3. Higher the noise margin, better the logic circuit.
4. TTL family uses multiple-emitter transistor in the output circuit.
5. TTL is slower than DTL.
6. Basic gate of a standard TTL family is the NAND gate.
7. TTL circuit can be easily interfaced with other digital circuitry.
8. ECL represents non-saturated logic.
9. I^2L family uses no biasing and loading resistors.
10. I^2L is a bipolar non-saturated logic.
11. I^2L logic is built around NOR function.
12. NMOS circuit is faster than PMOS circuit because it uses electrons as carriers.
13. CMOS logic has the lowest static-state power dissipation.
14. CMOS logic is widely used in one-chip computers and memory systems.
15. CMOS family uses only fast NMOS circuits.

C. Multiple Choice Items

1. In a saturated bipolar logic circuit, transistors operate
   (*a*) in deep cut-off
   (*b*) over active region
   (*c*) in saturation
   (*d*) just short of saturation
2. Noise margin is expressed in
   (*a*) decibel (*b*) watt
   (*c*) volt (*d*) phon

3. DTL family employs
   (*a*) resistors and transistors
   (*b*) diode and resistor
   (*c*) diode and transistors
   (*d*) diodes, resistors and transistors
4. The chief advantage of Schottky TTL logic family is its least
   (*a*) power dissipation
   (*b*) propagation delay
   (*c*) fan-in
   (*d*) noise immunity
5. The main advantage claimed for ECL family of logic gates is its
   (*a*) very large fan-in
   (*b*) use of negative power supply voltage
   (*c*) extremely low propagation time
   (*d*) least power dissipation
6. Special feature of an I^2L logic circuit is that it
   (*a*) uses only high-value resistors
   (*b*) dissipates negligible power
   (*c*) is a bipolar saturated logic
   (*d*) is easy to fabricate
   (*e*) uses no biasing and loading resistors
7. A unique advantageous feature of CMOS family is its
   (*a*) use of NMOS circuits
   (*b*) power dissipation in nanowatt range
   (*c*) speed
   (*d*) dependence on frequency for power dissipation
8. CMOS circuits are extensively used for one-chip computers mainly because of their extremely
   (*a*) low power dissipation
   (*b*) large packing density
   (*c*) high noise immunity
   (*d*) low cost
9. CMOS family uses only
   (*a*) MOSFETs and resistors
   (*b*) NMOS circuits
   (*c*) MOSFETs
   (*d*) bipolar transistors
10. Power is drawn by a CMOS circuit only when
   (*a*) its output is high
   (*b*) its output is low
   (*c*) it switches logic levels
   (*d*) in static state

ANSWERS

A. Fill in the blanks

1. inputs	**2.** delays	**3.** volts	**4.** transistors	**5.** emitter
6. greater	**7.** non-saturated	**8.** fastest	**9.** speed	**10.** negative
11. interface	**12.** resistors	**13.** packing	**14.** resistor	

B. True or False

1. F	**2.** T	**3.** T	**4.** F	**5.** F	**6.** T	**7.** T	**8.** T
9. T	**10.** F	**11.** T	**12.** T	**13.** T	**14.** T	**15.** F	

C. Multiple Choice Items

1. *c*	**2.** *c*	**3.** *d*	**4.** *b*	**5.** *c*	**6.** *e*	**7.** *b*	**8.** *b*	**9.** *c*	**10.** *c*

CHAPTER 36

Transducers

1. Classification of Transducers
2. Resistive Position Transducer
3. Resistive Pressure Transducer
4. Inductive Pressure Transducer
5. Capacitive Pressure Transducer
6. Self-generating Inductive Transducers
7. Linear Variable Differential Transformer (LVDT)
8. Piezoelectric Transducer
9. Strain Gauge
10. Temperature Transducers
11. Resistance Temperature Detectors
12. Thermistor

36.1. What is a Transducer ?

In general terms, a transducer may be defined as any device that converts energy in one form to energy in another form.

Transducer

Most of the transducers either convert electrical energy into mechanical displacement and/or convert some non-electrical physical quantity (such as temperature, light, force, sound etc.) to an electrical signal.

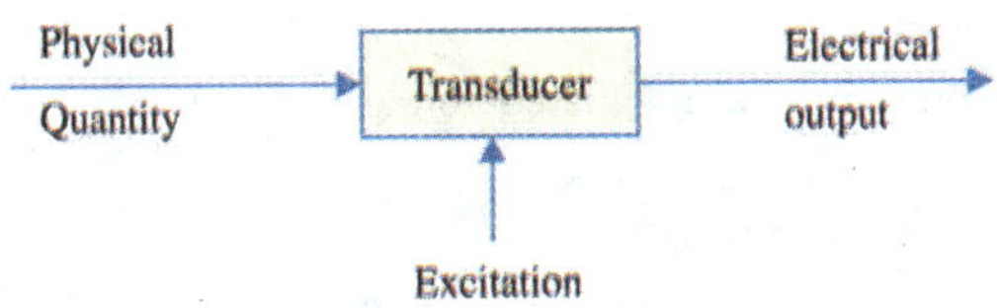

Fig. 36.1

In an electronic instrumentation system, the functions of a transducer (being the input device) are two-fold:

(*i*) to detect or sense the presence, magnitude and changes in the physical quantity being measured, and

(*ii*) to provide a proportional electrical output signal as shown in Fig. 36.1.

36.2. Classification of Transducers

Transducers may be classified according to their applications, method of energy conversion, electrical measuring principle involved, nature of output signal and so on.

Broadly speaking, transducers may be classified into the following groups:

1. passive transducers – they require an external power supply and their output is a measure of some variation in a physical parameter such as pressure, resistance and capacitance etc.;
2. self-generating transducers – they do not require an external power source. However, they provide an electrical output when stimulated by some physical form of energy.

A very useful listing of different transducers is given below:

1. Mechanical – strain gauge type for measuring force, torque and pressure etc.
2. Thermal – thermistors, thermocouples
3. Optical – photoconductive, photoemissive and photovoltaic (solar) cells
4. Acoustical – microphone
5. Magnetic – LVDT, permeameters
6. Chemical – pH and conductivity cells
7. Nuclear – Geiger-Muller tube, ionization chamber
8. Biological – electrocardiograph (ECG) and electroencephalograph (EEG)

36.3. Classification based on Electrical Principle Involved

A better way of classifying transducers is to arrange them according to the basic electrical measuring principle involved in converting the physical quantity (or its variations) into corresponding electrical quantity (or variations). Such an arrangement is given below:

(*a*) Variable-resistance Type

1. strain and pressure gauges, 2. thermistors, resistance thermometers,
3. photoconductive cell, 4. chemical conductive meter,
5. contact thickness gauge etc.

(*b*) Variable-inductance Type

1. linear variable differential transformer (LVDT),
2. reluctance pick-up 3. eddy current gauge.

(*c*) Variable-capacitance Type

1. capacitor microphone, 2. dielectric gauge, 3. pressure gauge.

(*d*) voltage-divider Type

1. potentiometer position sensor,2. pressure-actuated voltage divider.

(*e*) Voltage-generating Type

1. piezoelectric pick-up, 2. thermocouple,
3. photovoltaic (solar) cell, 4. rotational motion tachometer.

36.4. Resistive Position Transducer

It is used to sense the position of an object or the distance it has moved. The physical variable being measured is made to cause a resistance change in the sensing element.

Fig. 36.2 (*a*) shows one type of displacement transducer used in industrial measurement and control work. It consists of a resistance element *AB* with a sliding contact or wiper *W* linked with the object being monitored. As seen, the resistance between *W* and *B* *i.e.*, R_{BD} depends on the position of the object.

It is seen from Fig. 36.2 (*b*) that output voltage V_o depends on the slider position and is, therefore, a function of the shaft position.

The voltage V_o is given by

$$V_o = V\frac{R_2}{R_1 + R_2}$$

Obviously, V_o is proportional to R_2 *i.e.*, position of the wiper, *W*.

This voltage may be applied to a voltmeter calibrated in millimeter (mm) for visual display.

The above-described potentiometric principle is widely-used despite its limitations. Some of its advantages are as under:

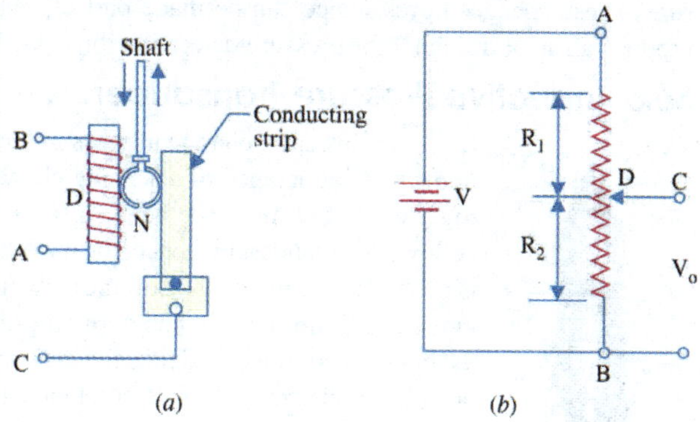

Fig. 36.2

1. its electrical efficiency is high,
2. it provides sufficient output to permit control operations without any further amplification,
3. the device may be excited by either dc or ac supply.

However, because of mechanical friction between the wiper and the resistance element *AB*, its life is limited. Moreover, it becomes noisy as the element wears out.

Example 36.1. *A displacement transducer with a shaft stroke of 75 mm is applied to the circuit of Fig. 36.2 (b). The total resistance of the potentiometer is 6 K and V = 4.5 V. When the wiper is 25 mm from end B, what is the value of* V_o*?*

Solution. Since full displacement of 75 mm corresponds to 6 K, a displacement of 25 mm corresponds to a resistance

$$R_2 = \frac{25}{75} \times 6,000 = 2,000\ \Omega$$

$$\therefore \quad V_o = V\frac{R_2}{R_1 + R_2} = 4.5 \times \frac{2000}{6000} = \mathbf{1.5\ V}$$

36.5. Resistive Pressure Transducer

It is based on an arrangement in which a change in pressure results in the change of resistance of a sensing element.

Fig. 36.3 shows two ways of mechanical arrangement for measuring pressure.

Fig. 36.3 (*a*) shows the bellows type whereas Fig. 36.3 (*b*) shows the diaphragm type. In Fig. 36.3 (*a*) when applied pressure changes, the spring-supported bellows *B* extends or contracts thereby moving the resistance slider up and down. Consequently, changes in the resistance of sensing resistive element become proportional to the changes in pres-

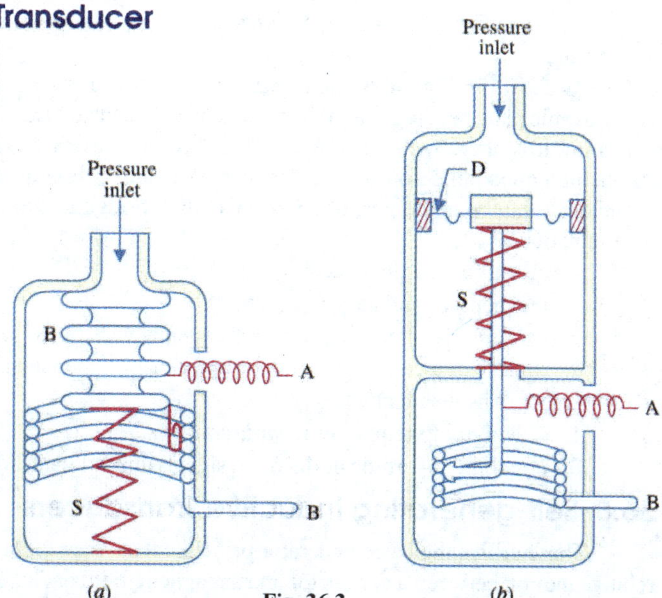

Fig. 36.3

sure. These changes in resistance can be made part of a bridge circuit and produce an ac or dc signal for pressure determination (Fig. 36.4).

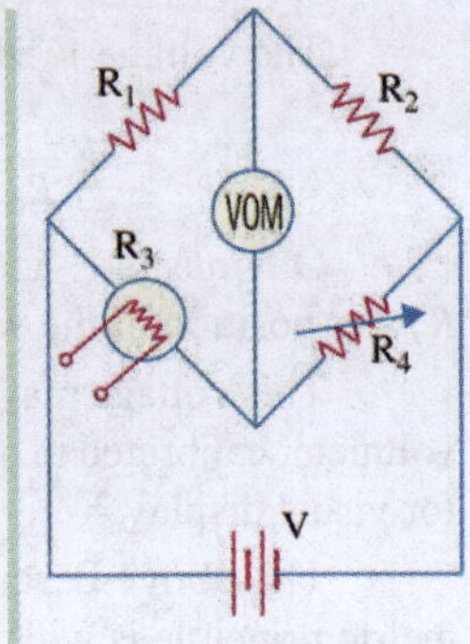

Fig. 36.4

36.6. Inductive Pressure Transducer

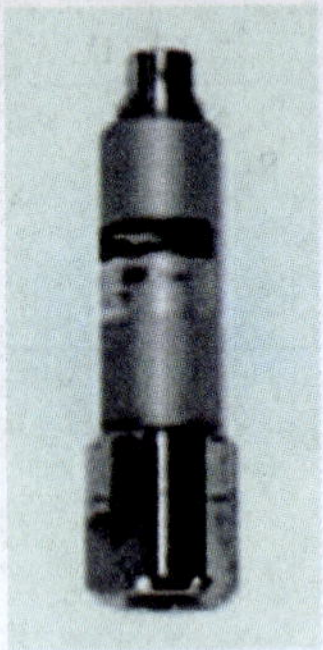
Pressure transducer.

In this case, change in pressure produces a change in the self-inductance of a sensing element. As shown in Fig. 36.5, when pressure being measured increases, the bellows *B* expands and, hence, the movable magnetic core is pushed more inside the coil thereby increasing its self-inductance. Similarly, when applied pressure decreases, the bellows contracts and thus the magnetic core is pulled out of the coil and, as a result, coil inductance is decreased.

The changes in self-inductance can be made a part of an ac inductance bridge and converted into an electrical signal for pressure measurement.

As compared to the previous resistive arrangement (Art. 36.5), the advantages of this inductive type are as under:

1. no moving parts are used thereby avoiding wear and tear,
2. no extra friction load is imposed on the measuring system.

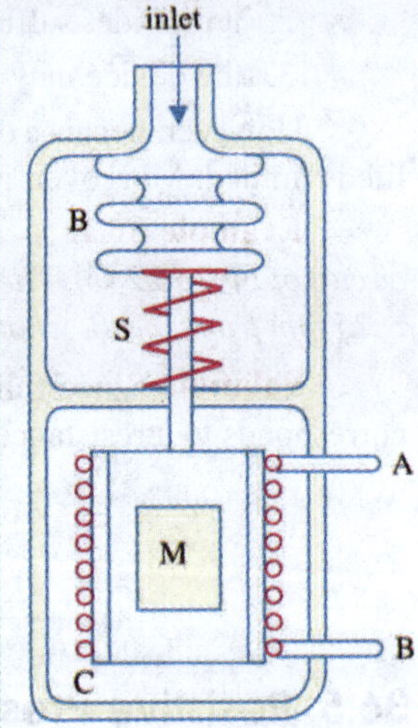

Fig. 36.5

36.7. Capacitive Pressure Transducer

In this arrangement, changes in pressure are detected by the variations in the capacitance of a parallel-plate capacitor whose one plate is free to move with change in pressure.

The capacitance of the capacitor shown in Fig. 36.6 is given by

$$C = \frac{\varepsilon_0 \varepsilon_r A}{d} \quad \text{where } A = \text{plate area,} \quad d = \text{plate separation}$$

As seen, $C \propto 1/d$ *i.e.,* capacitance varies inversely with plate separation.

As shown in Fig. 36.6, the capacitor consists of a static plate *N* and movable plate *M* (diaphragm) separated by a dielectric (usually, air). When pressure increases, *M* bends inwards, thereby decreasing *d* and, hence, increasing *C*. However, when pressure decreases, the diaphragm is pulled outwards, *d* is increased and, consequently, *C* is decreased. The changes in capacitance can be made part of an ac capacitive bridge to furnish an electrical signal for pressure determination.

A capacitive transducer is

1. inexpensive and easy to construct,
2. particularly effective for high-frequency variations,
3. capable of measuring both static and dynamic phenomena.

Amongst its disadvantages are

1. sensitivity to temperature variations,
2. possibility of erratic or distorted signals due to long leads.

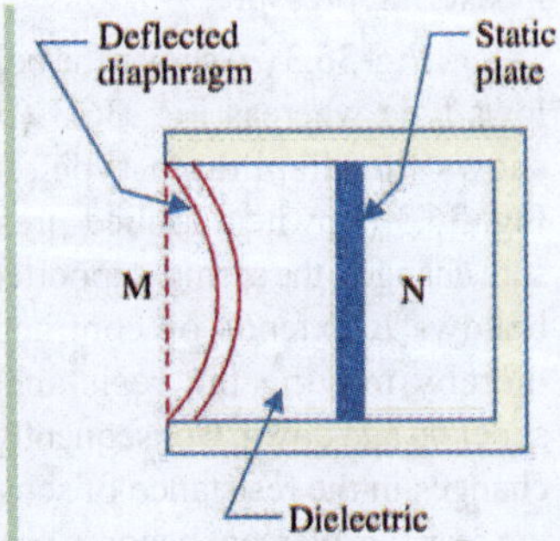

Fig. 36.6

36.8. Self-generating Inductive Transducers

These utilize the basic generator principle that when there is relative motion between a conductor and a magnetic field or flux an emf is induced in the conductor.

This relative motion between the conductor and the flux is provided by the physical quantity under measurement.

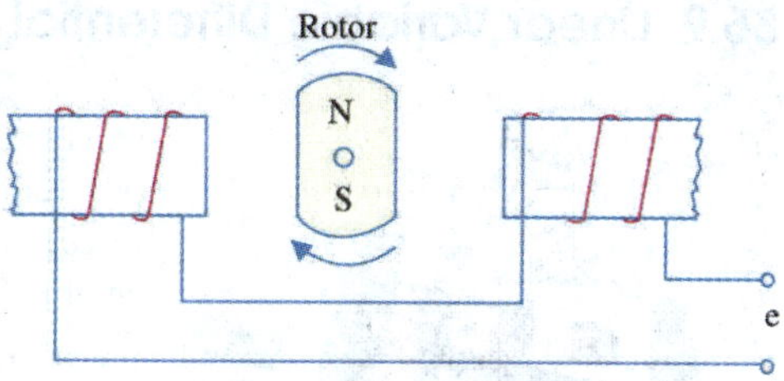

Fig. 36.7

(*a*) Tachometer

It is an inductive transducer that directly converts speed into an electrical signal as shown in Fig. 36.7. The object whose angular velocity is to be measured is directly coupled to the rotor which is made of permanent magnet having *N*- and *S*-poles. As the rotor rotates, its alternating flux cuts the stationary coils thereby inducing an alternating voltage '*e*' in them. The magnitude of '*e*' depends directly on the angular velocity of the rotor. This emf can be rectified and further amplified for being fed to a dc voltmeter, which is directly calibrated in revolutions per minute (rpm) instead of in volts.

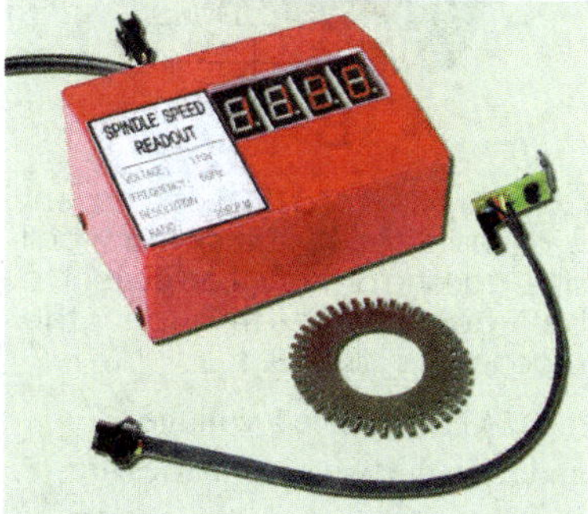

This spindle tachometer mounts on the top of the mini lathe headstock and displays the speed, to the rearest 20 rpm, of the spindle.

(*b*) Electromagnetic Flowmeter

It is also a self-generating inductive transducer and is particularly useful where the pipes cannot be opened to atmosphere *i.e.*, for flow measurement of liquids which contain suspended solids such as sewage or feed to paper mills etc. However, it should be remembered that it can measure the flow of only those materials which are electrically conductive.

As shown in Fig. 36.8, the meter is a small section of a pipe that is internally lined with an insulating material and has two diametrically opposite metallic electrodes with their ends in flush with the insulation.

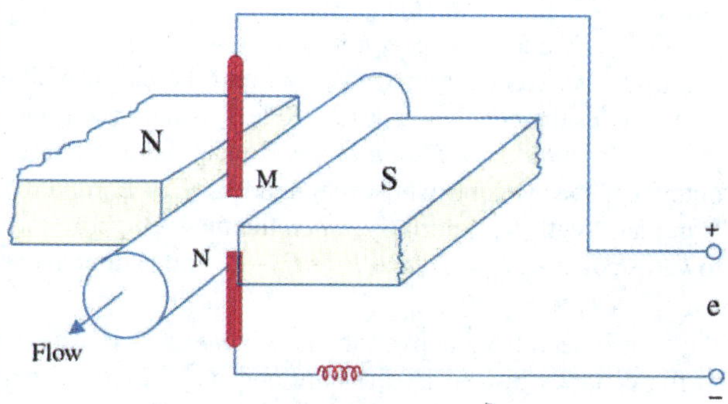

Fig. 36.8

As the conducting fluid passes through the magnetic field of the two permanent magnets, it cuts the flux and hence an emf '*e*' is induced in it whose magnitude is given by

$$e = Blv$$

Its direction can be found by using Fleming's Right-hand rule.

Here, l is the length of the conductor* *i.e.*, diameter *MN* of the pipe, B is the magnetic flux density and v is the velocity of the fluid. Since B and l are constant, $e \propto v$. Obviously, faster the rate of liquid flow, greater the induced emf '*e*'. This voltage can be fed to a voltmeter which is calibrated in velocity.

The electromagnetic flowmeter has the advantage that its readings are unaffected by variations in viscosity, density, temperature, pressure and conductivity etc., of the fluid. Moreover, it causes no drop in the fluid pressure. However, it requires certain minimum flow velocity for correct readings. Also, the electrodes must be always kept clean which can sometime present a serious problem as in the case of sewage flow.

* The fluid lying between electrodes *M* and *N* acts as a conductor which cuts the magnetic flux at right angles.

36.9. Linear Variable Differential Transformer (LVDT)

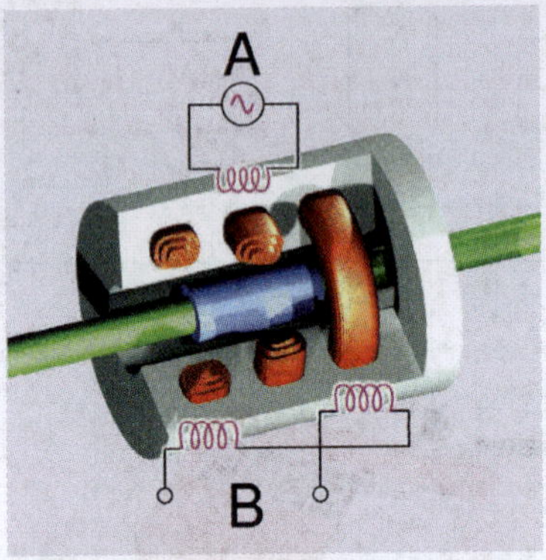

Cutaway view of a LVDT. The current is driven through the primary coil at *A*, causing an induction current to be generated through the secondary coils at *B*.

It is a passive inductive transducer (or magnetic displacement transducer). It is commonly employed to measure force (or weight, pressure and acceleration etc., which depend on force) in terms of the amount and direction of displacement of an object.

(i) Construction

As shown in Fig. 36.9, it consists of one primary winding and two secondary windings which are placed on either side of the primary mounted on the same magnetic core. The magnetic core is free to move axially inside the coil assembly and the motion being measured is mechanically coupled to it. The two secondaries *S*-1 and *S*-2 have *equal* number of turns but are connected in series-opposition so that emf 's induced in them are 180° out of phase with each other and, hence, cancel each other out. The primary is energised from a suitable ac source.

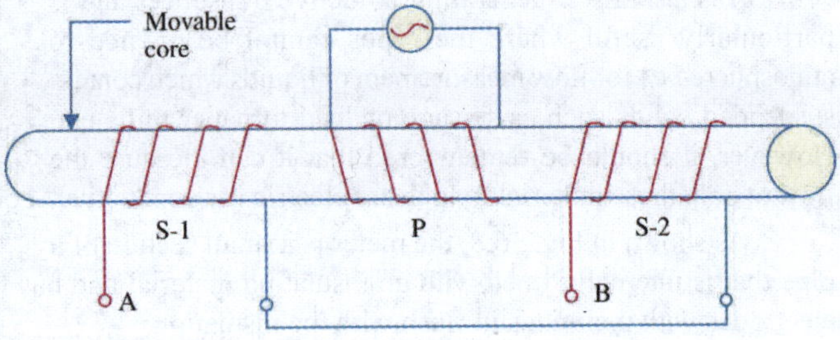

Fig. 36.9

The induced voltages E_1 and E_2 in the two secondaries are equal but *antiphase* as shown separately in Fig. 36.10. Hence, net secondary voltage across terminals *A* and *B* is zero.

(ii) Working

When the core is in the centre (called reference position), E_1 and E_2 are *equal and opposite*. Hence, they cancel out and the output voltage V_o is zero. When the external applied force moves the core towards coil *S*-2, E_2 is increased but E_1 is decreased in magnitude though they *are still antiphase with each other.* Hence, there is some net output voltage available whose magnitude is $(E_2 - E_1)$ and which is in phase with E_2. Similarly, when the magnetic core moves towards coil *S*-1, $E_1 > E_2$ and $V_o = E_1 - E_2$ and is in phase with E_1.

It is seen from above the *magnitude* of V_o is a function of the distance moved by the core and its *polarity* or *phase* indicates as to in which direction it has moved. If core is attached to a moving object, the magnitude of V_o gives the position of that object.

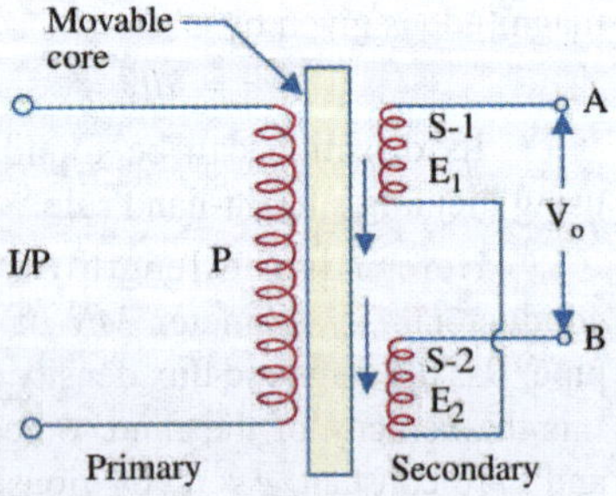

Fig. 36.10

(iii) Advantages

Some of the advantages of an LVDT are as under:

1. it gives higher output voltages for small changes in core position. It means that its sensitivity (V/m) is high. Sensitivity ranges from 50 mV/mm to 300 mV/mm;
2. it can operate over a temperature range from—265°C to 600°C;
3. it is available in radiation-resistant design for operation in nuclear reactors.

(iv) Applications

1. in accelerometers,
2. jet engine controls in close proximity to exhaust gases,

3. measurement of roll position,
4. measurement of material thickness in hot strip or slab steel mills.

36.10. Piezoelectric Transducer

Production of electric potential by means of mechanical pressure is called piezoelectric phenomenon. It is exhibited by crystals of quartz, tourmaline, Rochelle salt and ceramics such as barium titanate (Art. 36.23). That is why they are called piezoelectric transducers. Crystal and ceramic microphones and phonograph pick-up cartridges are common examples of devices using piezoelectric effect. Obviously they are of self-generating type.

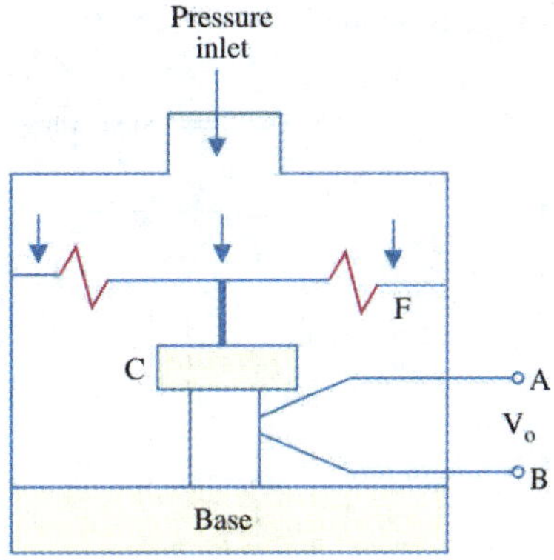

Fig. 36.11

In Fig. 36.11 is shown such a crystal placed between a solid base and the force-summing member *F**. When pressure is increased on the top surface of the crystals, a potential difference is developed across its opposite faces. The voltage V_o is found to be *directly proportional to the magnitude of the pressure.* However, if applied pressure is decreased, even then potential difference is developed though with opposite polarity.

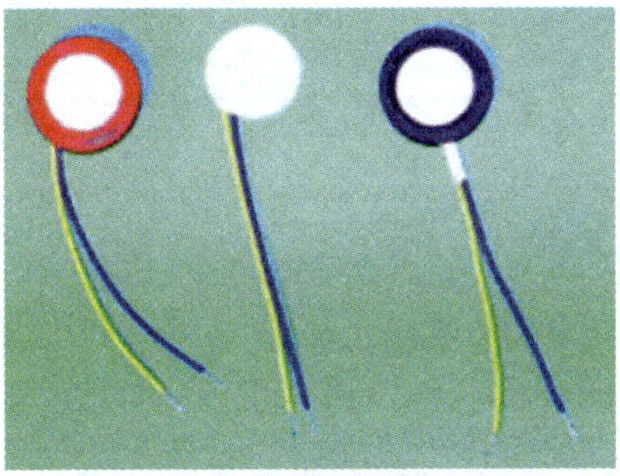

An ultrasonic Piezoelectric transducer is a disc shaped artificial crystal when it receives a suitable electric charge it immediately starts to vibrate and to generate ultrasonic waves of a predetermined frequency.

Conversely, if an ac voltage of suitable frequency is applied across the two opposite faces of such a crystal, it expands and contracts and starts vibrating with its natural frequency which is very stable. Because of this property, piezoelectric crystals are used in high-frequency accelerometers.

The principal disadvantage of this transducer is that it cannot measure static condition i.e., it produces voltage only when the applied pressure is changing.

36.11. Strain Gauge

It is a passive transducer which converts a given strain to a resistance change. In other words, it utilizes resistance variations of a wire to sense the strain produced by the force acting on the wire. It is extensively utilized for measuring weight, pressure, mechanical force or displacement etc.

(*i*) Construction

Fig. 36.12 (*a*) shows the construction of a bonded strain gauge. It consists of a few loops of fine constantan wire mounted on a flexible plate which is bonded or cemented to the member undergoing stress [Fig. 36.12 (*b*)]. The hairpin looping of the wire increases its length in the direction of the applied stress and, hence, increases the sensitivity of the gauge.

(*ii*) Working Principle

When a tensile stress is applied to the member (top of a jar, for example), it will elongate the wire thereby increasing its length but decreasing its cross-sectional area. Both these factors increase wire resistance because

$$R = \rho \frac{l}{A}$$

* The mechanical elements that are used to convert the applied force into a displacement are called force-summing devices.

The increase in wire resistance is proportional to the force causing the elongation. Hence, resistance of a strain gauge is indicative of the pressure or force applied to the surface to which the gauge is bonded.

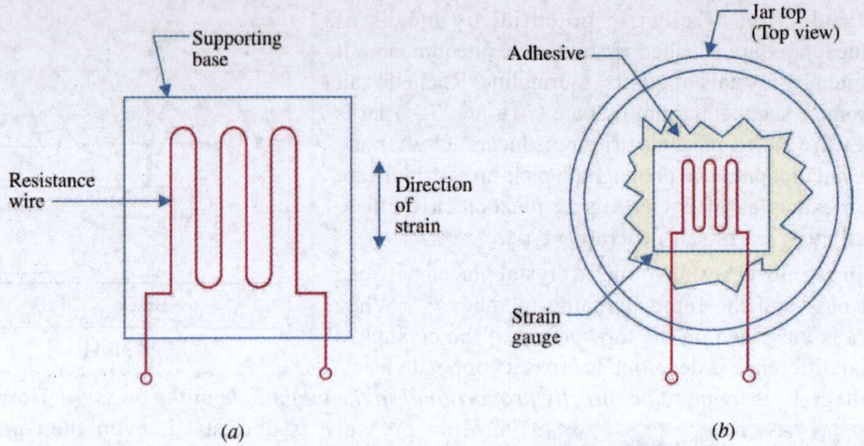

Fig. 36.12

(*iii*) Gauge Factor

The two physical quantities of interest are the fractional (*i*) change in resistance and (*ii*)change in length. The ratio of the two gives the gauge factor, *K*

$$\therefore \qquad K = \frac{\Delta R / R}{\Delta l / l} = \frac{\Delta R / R}{\text{strain}, \sigma}$$

The value of *K* for metallic gauges varies from – 12 (for nickel) to +6 (for alloys) whereas its value for semiconductor gauges varies from 50 to 200. Use of silicon strain gauge is spreading because it can measure even 'microstrain' due to its high gauge factor. Higher the gauge factor of a gauge, greater its sensitivity. For a given strain, greater resistance change means higher value of *K*.

It is seen from above that

$$\text{strain} = \frac{\Delta R / R}{K}$$

For a given wire and gauge of known value

$$\text{strain} \propto \Delta R$$

The changes in resistance can be measured with the help of the bridge shown in Fig. 36.13. The null indicator is calibrated in newton/m^2 to give the value of stress directly.

As seen from Fig. 36.13, a dummy gauge has been used alongwith the active one in order to compensate for temperature variations.

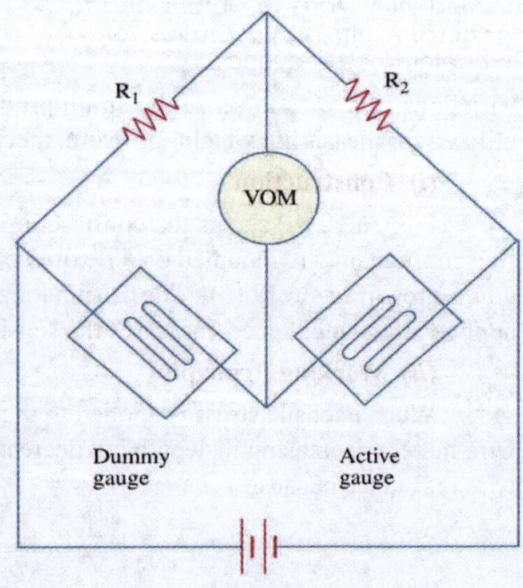

Fig. 36.13

Fig. 36.13

(*iv*) Advantages

The wire strain gauge has the following advantages:

1. it is rugged, easy to install and use,
2. in ac applications, it is capable of measuring variations upto 100 kHz,

3. it has excellent linearity and hysteresis characteristics,
4. it is comparatively less sensitive to shock and undesired vibrations.
5. it is available in multielement or rosette gauge form for simultaneous measurement of strain in more than one direction.

Example 36.2. *A wire strain gauge with a gauge factor K = 2 is bonded to a steel member which is subjected to a strain of 10^{-6}. If original no-strain resistance of the gauge is 120 Ω, calculate the change in gauge resistance.*

Solution. As seen from Art. 36.11,

$$\therefore \qquad K = \frac{\Delta R / R}{\sigma}$$

$$\Delta R = \sigma K R = 10^{-6} \times 2 \times 120 = \mathbf{240\ \mu\Omega}$$

Example 36.3. *A metallic strain gauge with K = 2 is bonded to a steel member which is subjected to a stress of 10.5 × 10^9 N/m^2. If modulus of elasticity for steel is 21 × 10^{12} N/m^2, calculate the fractional change in the resistance of the gauge due to the applied stress.*

Solution. The modulus of elasticity is given by

$$E = \frac{\text{stress}}{\text{strain}}$$

$$\text{strain, } \sigma = \frac{10.5 \times 10^9}{21 \times 10^{12}} = 5 \times 10^{-4}$$

Now,

$$K = \frac{\Delta R/R}{\Delta l/l} = \frac{\Delta R/R}{\text{strain}, \rho}$$

$$\therefore \qquad \Delta R/R = \sigma K$$

$$= 5 \times 10^{-4} \times 2 = 10^{-3} \text{ or } \mathbf{0.1\%}$$

36.12. Temperature Transducers

The control and measurement of temperature is of considerable importance in industry. Of all the major variables which need accurate measurement and control, well over half are related to the measurement of temperature. Without accurate information regarding the temperature of the materials used in various manufacturing processes, it would be well nigh impossible to produce thousands of the products available to the consumer to-day. Temperature transducers may be divided into the following four main categories:

1. **resistance temperature detectors (TRDs)** — these use platinum, nickel and tungsten elements etc., which have positive temperature coefficient;
2. **thermistors**—these are semiconductors having negative temperature coefficient;
3. **thermocouples**—these consist of a pair of dissimilar metal wires whose one end forms the temperature-sensing end.

36.13. Resistance Temperature Detectors

RTDs are made of a suitable length of platinum, nickel or tungssten wire whose resistance increases with increase in temperature *i.e.,* which has positive temperature coefficient. The relationship between temperature and resistance of the metallic wires is given by

$$R_t = R_0 (1 + \alpha t)$$

where R_t = conductor resistance at t °C

R_0 = conductor resistance at 0 °C

α = temperature coefficient of resistance

The α represents the percentage change in resistance per °C change in temperature. For example, $\alpha = 0.00385$ for platinum. It means that a platinum wire with $R_0 = 100\ \Omega$ when kept in an environment at 100°C will have its resistance increased to $100\ (1+100 \times 0.00385) = 138.5\ \Omega$.

Eq. (*i*) above may be written as

$$R_t - R_o = \alpha\, R_o . \Delta t \quad \text{or} \quad \Delta R = \alpha R_o . \Delta t$$

For given values of α and R_o, $\Delta R \ \alpha\ \Delta t$. Hence, by measuring ΔR (with the help of Wheatstone bridge circuit shown in Fig. 36.14) change in temperature can be found out. Alternatively, the null detector can be calibrated to read temperatures directly.

RTDs are commonly used for temperature measurement in the range – 180°C to 650°C. They are generally of the probe type for immersion in the medium whose temperature is to be measured (Fig. 36.15). For industrial purposes, the sensing element *i.e.*, resistance wire is mounted inside a protecting tube for guarding against any mechanical damage and contamination.

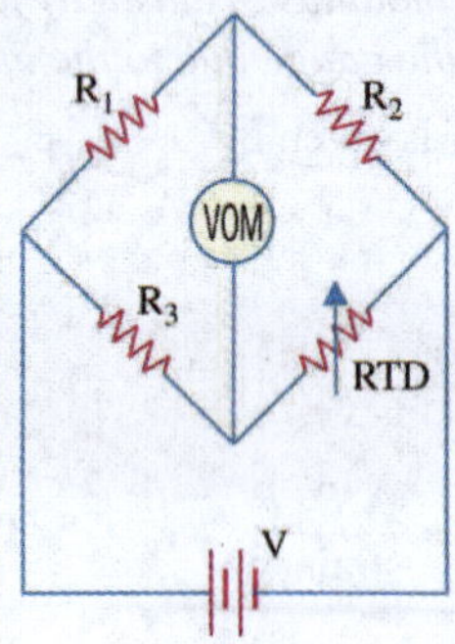

Fig. 36.14

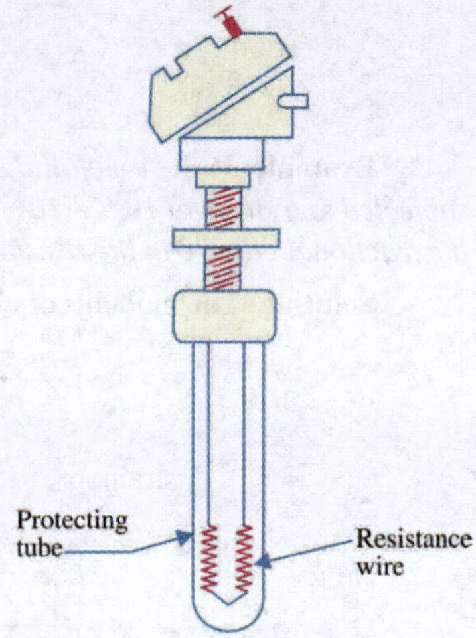

Fig. 36.15

Thermistor.

36.14. Thermistor

It is a temperature-sensitive semiconductor with a high negative temperature coefficient of resistance. In other words, resistance of a thermistor decreases considerably when its temperature increases. Likewise, its resistance increases when its temperature decreases. This high sensitivity to temperature changes makes the thermistor extremely well-suited for precise temperature measurement, control and compensation. Hence, they are widely used for such purposes particularly in the lower temperature range of –100°C to + 350°C.

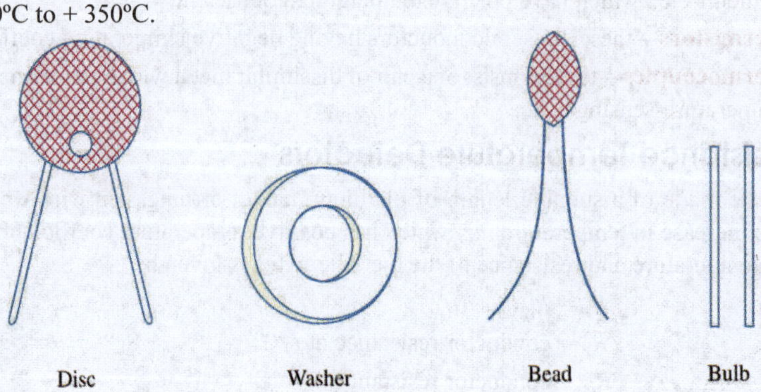

Fig. 36.16

(i) Construction

Thermistors are made by sintering mixtures of metallic oxides such as manganese, nickel, cobalt and copper etc. Their resistance varies from 0.5 Ω to 100 MΩ.

Thermistors are formed into rods, discs, washers and beads for special applications. Beads are as small as 0.15 mm in diameter and are either glass-coated or sealed in the tips of solid glass probes. These probes are used for measuring the temperatures of the liquids.

(ii) Response Time

It is the time required by the thermistor to heat up and change its resistance. It is expressed in terms of time constant which is defined *as the time required by the thermistor to reach 63.2 per cent of the total 'step' change in temperature.* Suppose, a thermistor is taken from a liquid at 100°C to another at 600°C. The step change in temperature is (600 – 100) = 500°C. Its 63.2 per cent = 500 × 63.2/100 = 316°C. The time required by the thermistor to increase its temperature by 316°C is called its time-constant. In this interval of time the thermistor temperature will increase from 100°C to (100 + 316) = 416°C. Usually, 5 time constants are required by the thermistor to achieve the new temperature.

The time constant is primarily determined by the type of metallic oxides used in the manufacture of the thermistor.

(iii) Advantages

Some of its advantages as a temperature measuring device are as under:

1. it is chemically stable,
2. it can be used in nuclear environments,
3. it can be connected in series-parallel arrangements for increased power-handling capacity,
4. it is used for measuring temperature even at places inaccessible to other thermometers.

(iv) Applications

Alongwith a Wheatstone bridge circuit, thermistors are used for measuring

1. temperature,
2. thermal conductivity,
3. attitude and wind velocity etc.

Thermistors are also used in limiting and regulating circuits

1. as time relays,
2. for integration of power pulses, and
3. as memory chips.

36.15. Thermocouples

This method of temperature measurement depends on the thermoelectric or Seebeck effect discovered by German physicist Thomas Seebeck.

(i) Construction and Working Principle

It consists of two *dissimilar* metal wires joined at one end as shown in Fig. 36.17. This end *A* is known as *hot or sensing* end (or junction). The other end is known as *cold or reference* end because it is kept at a lower but *constant* temperature (reference temperature). When end *A* is heated, an emf (called thermoelectric emf) in mV is produced between it and the cool ends *B* and *C*. The magnitude of this emf depends on (*i*) the *temperature difference* between the hot and cold junctions and (*ii*) the type of wire materials used. Since coil ends are kept at constant temperature (generally, room temperature), the millivolt output is indicative of the temperature of the hot junction *A* which is kept in touch with the object whose temperature is being measured. In fact, the millivoltmeter is calibrated directly in temperature rather than in mV.

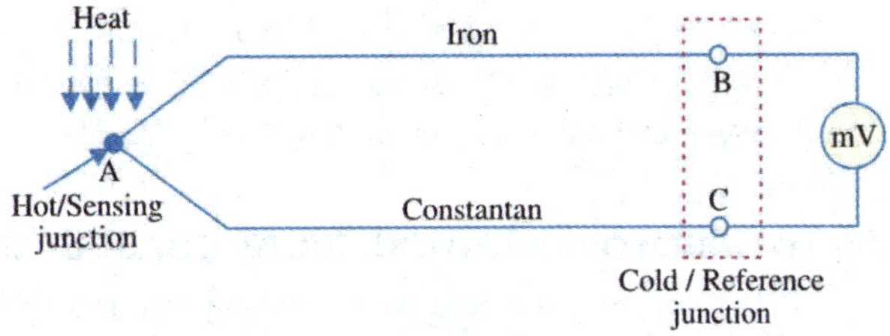

Fig. 36.17

Thermocouples are made from a number of different metals covering a wide range of temperature from –270ºC to + 2700ºC. They are available in simple uninsulated wire form, in insulated wire form or contained inside protecting probes or sheaths (of as small diameter as 0.25 mm). Some of the standard metal combinations commonly used in industry are given below alongwith the value of emf produced per 100ºC temperature difference between the hot and cold junctions.

1. chromel versus constantan — 7.5 mV
2. iron versus constantan — 5.5 mV
3. chromel versus alumel — 4 mV
4. nickel versus nickel-iron — 2 mV

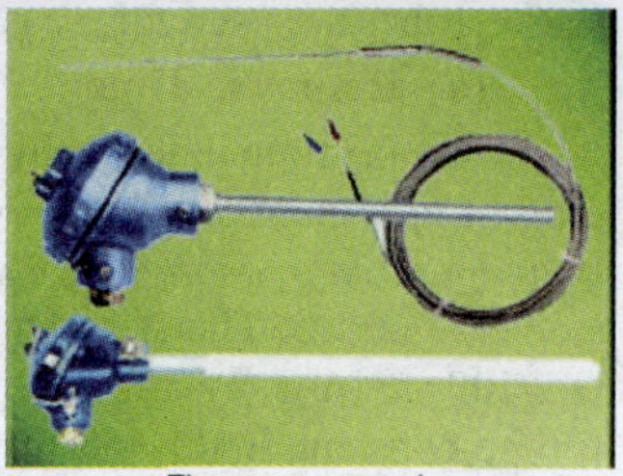
Thermocouple.

Thermocouples when connected in series or parallel as groups are called thermopiles.

(*ii*) Response Time

It is measured in terms of time constant which is defined as the time required by the temperature of the thermocouple to reach 63.2 per cent of the total step change in temperature. Usually, five time constants are needed to reach the final steady value of the temperature being measured.

(*iii*) Applications

1. in industry, thermocouples are used for measuring temperatures of industrial furnaces,
2. they are suitable for very low temperature measurements *i.e.,* in the cryogenic range required in industry, scientific research and medical instrumentation,
3. for monitoring the temperature of liquids and gases in storage and while flowing in pipes and ducts,
4. in medical work, extremely small-size thermocouple probes are used for measuring internal body temperature.

Example 36.4. *The hot junction of a thermocouple is shifted from an environment of 200ºC to an environment of 700ºC. If the time constant of thermocouple is one second, find the hot junction temperature after a time lapse of*

(*i*) *1 second* (*ii*) *2 seconds* (*iii*) *5 seconds*

Solution. (*i*) Step change in temperature = 700 – 200 = 500ºC. Its 63.2 per cent = 0.632 × 500 = 316ºC. Hence, temperature of the hot junction after 1 seconds *i.e.,* one time-constant is

= 200 + 316 = **516ºC**

(*ii*) Step change in temperature for next time-constant = 700 – 516 = 184ºC. Its 63.2 per cent = 0.632 × 184 = 116.3ºC. Hence, hot junction temperature after 2 second is

= 516 + 116.3 = **632.3ºC**

(*iii*) Time interval of 5 seconds corresponds to five time-constants during which the thermocouple would have attained the new temperature of 700ºC. Hence, hot-junction temperature after 5 seconds is 700ºC.

36.16. Ultrasonic Temperature Transducers

They utilize ultrasonic waves having frequencies exceeding 20 KHz. Their wide field of application includes the following:

1. for measuring temperature extremes from cryogenic to plasma level,
2. for measuring rapid temperature fluctuations,
3. suitable for nuclear or other corrosive environments,
4. for measuring temperature distribution inside solid bodies ,
5. for remote sensing.

36.17. Photoelectric Transducers

These have been already discussed as detailed below

1. Photoconductive cells or photocells — Art. 16.7.
2. Photovoltaic cell — Art. 16.6.
3. Solar cell — Art. 16.6.
4. Photodiode — Art. 16.8.
5. Photomultiplier tube — Art. 16.5.
6. Photoemissive cell — Art. 16.4.

36.18. Various Types of Microphones

Microphone is an acoustic transducer which converts a sound wave (acoustic energy) into a varying electric current (electrical energy) that can be further amplified and transmitted to distant places by telephone lines or radio. At the receiving end, the electric currents are converted back into sound waves by loudspeakers or headphones. Hence, a microphone plays an extremely important role in broadcasting, recording and telephony etc.

Microphones may be of the following seven types:

1. carbon microphone
2. ribbon microphone
3. moving-coil microphone
4. crystal microphone
5. ceramic microphone
6. capacitor microphone
7. electret microphone etc.

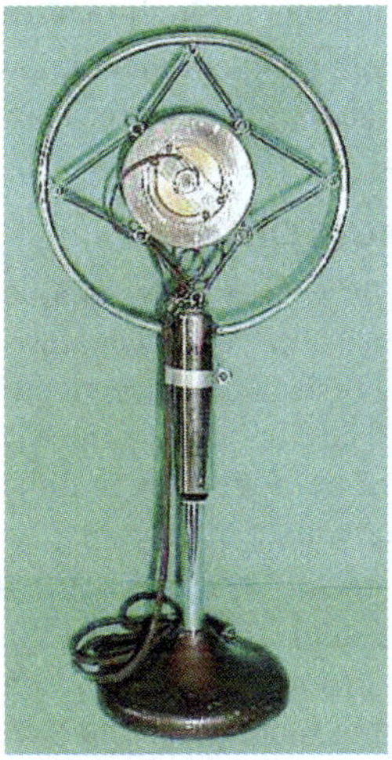
Double button carbon microphone.

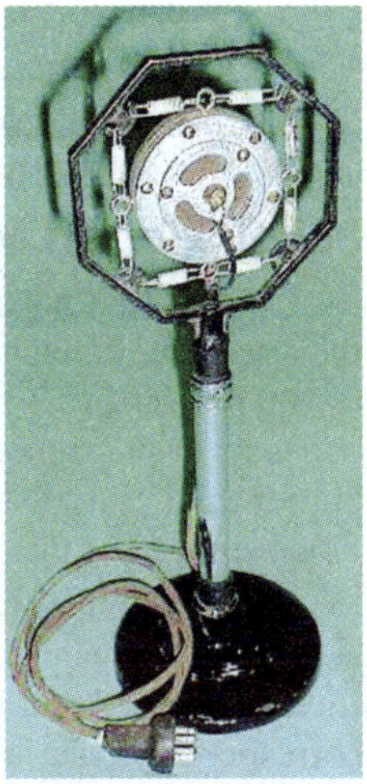
Single button carbon microphone.

36.19. Carbon Microphone

Carbon when in solid state is a fairly good conductor of electricity. But when packed in the form of powder or fine granules, it offers appreciable resistance which depends upon the closeness of the granules. If pressure is increased on loosely-packed carbon granules, then due to the increase in the contact area of the granules, the resistance is decreased. On the other hand, the resistance would increase if the carbon granules were to move apart due to decrease in pressure. The actual construction of such a microphone is shown in Fig. 36.18. The soundwaves falling on the very thin but highly flexible metal diaphragm set it into vibrations.

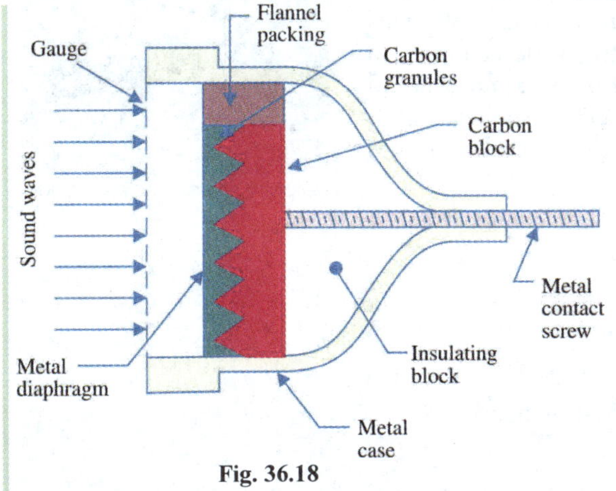

Fig. 36.18

Due to these vibrations, a varying pressure is applied on the carbon granules packed between the carbon block and the diaphragm.

This varies resistance of the microphone in step with the variations in the incident sound waves. If the microphone is connected in a circuit

(Fig. 36.19) containing a battery and a coil, then the current in the circuit varies according to the change in the speech. These variations in the circuit current set up an induced e.m.f. in the coil L_1 which may be further stepped up with the help of a transformer.

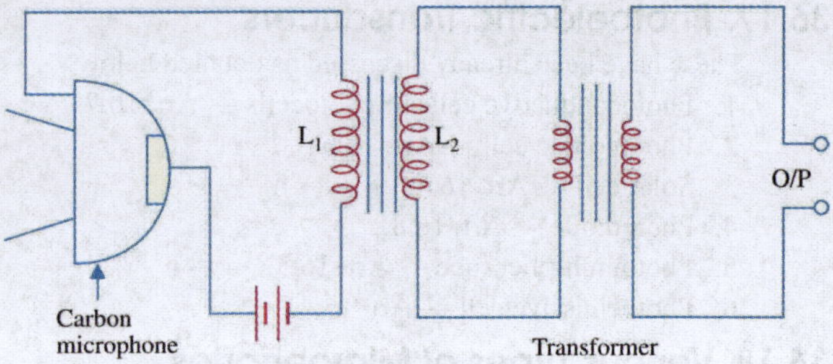

Fig. 36.19

It may be noted that the metal diaphragm has to be suitably damped by cotton wool in order to avoid prominent resonance effects. Such microphones are non-directional in nature, have good frequency response and give considerable output. However, they have the following serious drawbacks.

1. they are 'noisy' *i.e.,* they generate an objectionable background *hiss* (similar to the frying noise),
2. they are not suitable for high-quality reproduction,
3. they suffer from another defect known as 'packing' *i.e.,* clinging together of the granules due to the presence of moisture. This reduces the resistance of the microphone and renders it insensitive.

36.20. Ribbon Microphone

It is also known as *velocity* microphone and consists of a thin, stretched duralumin ribbon approximately 6 mm wide and 50 to 100 mm long (of 1 Ω resistance) suspended between the poles of a permanent magnet. The ribbon is clamped only at its ends and, but for these ends, is free to move back and forth in the magnetic field provided by the permanent magnet.

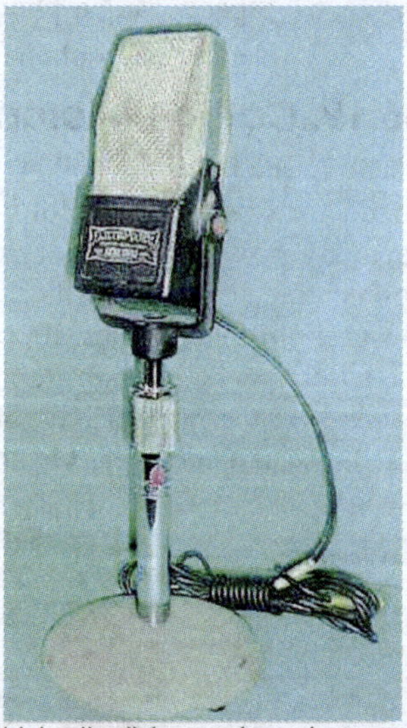

Velocity ribbon microphone.

When sound waves fall on the metallic ribbon, it moves in and out of the magnetic field. Due to the cutting of the magnetic flux, an induced emf is produced in the ribbon whose frequency corresponds to that of the incident sound. In this way, acoustic energy is converted into electrical energy.

Because of the low microphone resistance (about 1 Ω), a step-up transformer is mounted in the mike case itself. This transformer serves two purposes (*i*) firstly, while acting as an impedance matching transformer, it increases mike impedance to about 200 Ω thus enabling its direct connection to a preamplifier which has nearly the same input impedance (*ii*) secondly, it steps up the output voltage of the mike.

A ribbon mike is sensitive only to sounds coming from the front or back but not from sides.

Such microphones have

1. excellent high-frequency response,
2. flat response over audio range,
3. a smoother response than moving-coil mikes,
4. gentle trebble roll-off, and
5. no 'popping' problem.

36.21. Moving-Coil (MC) Microphone

Though both ribbon microphone and moving-coil microphone belong to the category of dynamic microphones, usually, the word 'dynamic' and 'moving coil' have become synonymous.

It consists of a light diaphragm shaped like a hollow cone which is clamped to a case around its outer rim as shown in Fig. 36.20 (*a*).

The diaphragm is made either of stiff cardboard or fibre (or more recently, of plastic materials) and has circular corrugations pressed into it. This makes it very springy and although it is clamped at the edges, yet even a slight pressure applied at its centre can make it move in and out. When sound waves fall on it, then due to variations in the pressure of the waves, the diaphragm moves in and out. Since the diaphragm is light and highly flexible, its vibratory motion is more or less faithful reproduction of the sound waveform.

To the rear or back side of the diaphragm is attached a very light cylindrical former of an insulating material on whch is wound the coil *C* having many turns of fine wire. The former, alongwith coil, fits into the gap of an annular magnet [Fig. 36.20 (*b*)].

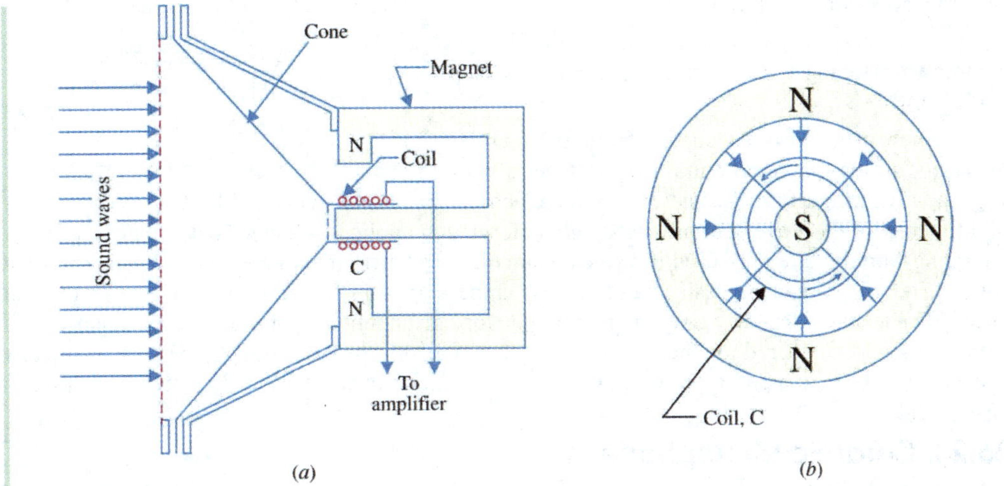

Fig. 36.20

When the sound waves fall on the diaphragm, it vibrates in and out. The coil attached to its rear also moves in and out in the strong magnetic field thereby cutting the magnetic flux. Hence, an oscillatory e.m.f. is induced in the coil *C* whose magnitude depends on the quickness of the movement of the coil. This oscillatory e.m.f. which represents the sound waves, can be amplfied further for driving a loudspeaker etc.

Moving-coil microphones are widely used because they

1. are dependable, rugged and have most reliable design for both indoor and outdoor work,
2. have the smoothest and most extended frequency response as compared to carbon, crystal and ceramic mikes,
3. provide much higher output signal voltage for a given sound pressure input than the ribbon mikes,
4. are quite suitable for high-quality reproduction.

36.22. Crystal Microphone

The action of such microphones depends on the piezoelectric effect exhibited by crystals of quartz, tourmaline and Rochelle salt. Such crystals develop a potential difference across their opposite faces when subjected to mechanical deformations such as squeezing, twisting or bending. The magnitude of the voltage developed is proportional to the extent of the deformation produced and its polarity is reversed when the direction of the deforming force is reversed. For example, if the voltage produced by the crystal is positive when twisted clockwise, then it would be negative when twisted anticlockwise.

Rochelle salt crystal is used for audio-frequencies whereas quartz crystal is used for high frequencies up to many megahertz (MHz).

As shown in Fig. 36.21, the centre of the diaphragm is attached to a drive pin whose other end contacts a sandwich of metal plates which contain the crystal. The lower end of the metal sandwich is fixed to the microphone housing but the upper end is free to move. When sound waves fall on the diaphragm, it vibrates to and fro thereby subjecting the crystal to mechanical stress. Hence, an alternating voltage is developed between the opposite faces of the crystal. This voltage is picked up by the metal plate electrodes (marked E) and then taken out for further necessary amplification.

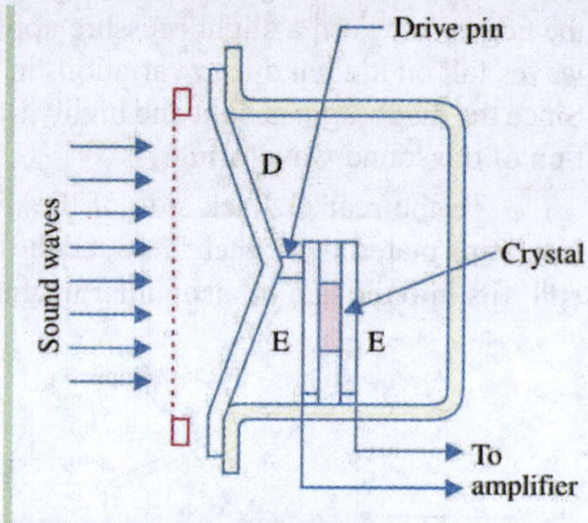

Fig. 36.21

In some types of microphones, the crystal slab acts as its own diaphragm but more slabs are necessary in that case.

Such microphones suffer from mechanical resonance of the diaphragm and their voltage falls off very rapidly at higher frequencies. Also, the electrical waveform generated by them is not a very faithful copy of the waveform of mechanical deformation applied to them. They are not useful for recording work but are only used in voice communication where main objective is intelligibility of sound. The crystal can be easily damaged by temperature and humidity. However, one of their outstanding feature is that they give very large output for a given sound input and so needless subsequent amplification as compared to other microphones. Since their impedance is high, they can be directly connected to the high input impedance of an amplifier. They can be made small enough to be used in hearing aids.

36.23. Ceramic Microphone

In this microphone, the transducer element is barium titanate (a ceramic) which has piezo-electric properties. Hence, it is just like a crystal microphone except that it has

1. better tolerance as regards heat and humidity,
2. slightly higher signal output for the same sound input.

36.24. Capacitor Microphone

The transducer used in mike is a parallel-plate capacitor whose one plate is variable (Art. 36.7). The capacitor is kept charged to a certain potential difference by a dc voltage called polarizing voltage. When sound falls on such a capacitor, its movable plate vibrates to and fro. As it moves inwards, capacitance C increases and hence increases its charging ability. Opposite condition prevails when the plate moves outwards. It is found that voltage variations of this capacitor are in step with the pressure variations of the incident sound. These voltage variations can be amplified to a level high enough to supply sufficient current for recording on magnetic tape.

Such microphones have

1. extremely high impedance (10 MΩ) and have built-in amplifier to avoid the use of connecting cable,
2. excellent frequency response,

3. low distortion,
4. excellent transient response due to the extremely small mass of their moving plate (i.e., diaphragm).

36.25. The Electret Microphone

It is similar to a capacitor microphone except that it does not require a polarizing voltage for its capacitor. Its transducer element is a self-polarised capacitor called electret capacitor. It is a specially-designed capacitor which once charged during its manufacture, holds its charge indefinitely thereby eliminating the need for a polarizing voltage source.

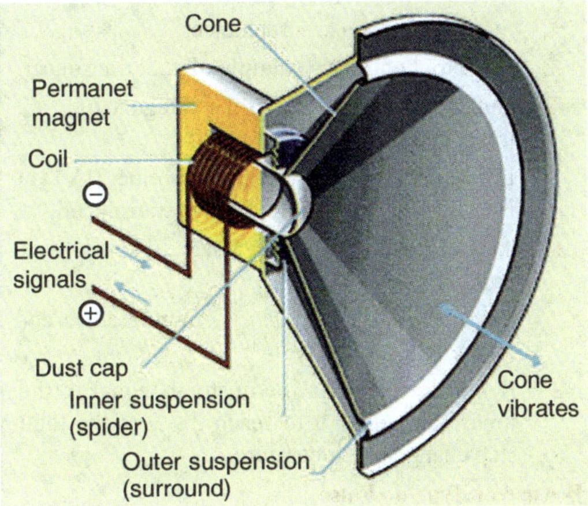

36.26. The Loudspeaker

A loudspeaker does just opposite of what a microphone does. It is a transducer which converts varying current back into sound waves identical with those which were collected by the microphone and converted into an alternating current. In essential details, a moving-coil loudspeaker resembles a moving-coil microphone except that its diaphragm is much larger.

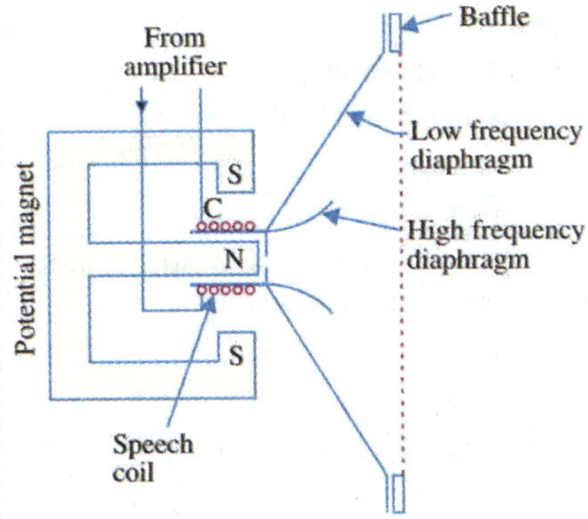

Fig. 36.22

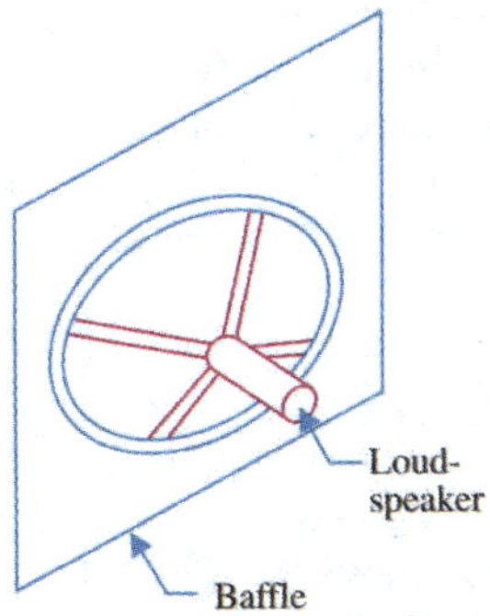

Fig. 36.23

As shown in Fig. 36.22, the moving-coil loudspeaker consists of fairly large diaphragm made of stiff paper or an aluminium alloy and is mounted on a large size baffle board. The front part of the conical diaphragm emits compression waves at the same time as the rear emits rarefactions. These two should not be allowed to mix up otherwise they will cancel each other out. Hence, a baffle board is very necessary to prevent a compression from the front uniting with a rarefaction in the rear (Fig. 36.23).

The moving coil (also called the 'speech coil') is wound on a light cylindrical former of carboard which is firmly attached to the diaphragm. This coil can move in and out parallel to its axis in the strong radial field of the electromagnet. When the amplified output from the microphone is passed through the coil, is acted upon by a mechanical force (Fleming's Left-hand rule) which varies according to the strength of the varying current carried by it. Hence, the speech coil performs an oscillatory motion and since it is firmly fixed to the diaphragm, the latter also starts vibrating thereby setting the air in front of it in vibratory motion. In this way, the sound is reproduced from the loudspeaker.

Since a single diaphragm cannot give satisfactory response for both low and high frequencies, hence all better class of loudspeakers are fitted with two diaphragms loosely coupled to each other by a flexible collar. One (called woofer) reproduces low frequencies whereas the other (called tweeter) looks after the higher ones.

SELF EXAMINATION QUESTIONS

A. Fill in the blanks with most appropriate word(s) or numerical value(s).

1. A transducer is device that .... energy in one form to energy in another form.
2. Strain gauge is a...... transducer.
3. A microphone is an example of .... transducer.
4. Quartz crystals belong to the category of .... transducers.
5. Linear variable differential transformer (LVDT) is a............................inductive transducer.
6. An LVDT has one primary but ...... identical secondaries.
7. Thermistors have .... temperature coefficient of resistance.
8. Response time of a thermistor is equal to the time required by it to reach .... of the total step change in temperature.

B Answer True or False

1. A transducer is an energy-conversion device.
2. A passive transducer is one which needs an external power supply.
3. Strain gauge is an electrical transducer.
4. A thermistor is variable-resistance transducer.
5. Electromagnetic flowmeter is an example of self-generating transducer.
6. All the three windings of an LVDT have equal number of turns.
7. A tourmaline crystal represents an acoustical transducer.
8. Higher the gauge factor, more sensitive a strain gauge.
9. Strain gauge converts strain to resistance change.
10. RTDs have negative temperature coefficient of resistance.
11. Thermistors are made of semiconductor materials.
12. Thermistor resistance can be as high as 100 MΩ.
13. Thermoelectric voltage produced by a thermocouple depends on the length of its two wires.
14. The voltage provided by a thermocouple is a function only of its hot junction.

C. Multiple Choice Items

1. A transducer is any device which
   (*a*) is used in electronic instruments
   (*b*) is self-generating
   (*c*) converts energy in one form into another form
   (*d*) only detects and senses the presence of a physical quantity
2. ..... represents magnetic transducer.
   (*a*) LVDT (*b*) Strain gauge
   (*c*) Thermocouple
   (*d*) Photoconductive cell
3. A microphone is classified as a .... transducer.
   (*a*) thermal (*b*) optical
   (*c*) magnetic (*d*) acoustical
4. A thermocouple is a .... type transducer.
   (*a*) variable-resistance
   (*b*) voltage generating
   (*c*) variable-inductance
   (*d*) voltage-divider
5. A linear variable differential transformer (LVDT) has
   (*a*) two primary windings
   (*b*) two series-aiding secondaries
   (*c*) movable magnetic core
   (*d*) an otuput voltage of zero
6. The principal disadvantage of a piezoelectric transducer is that it
   (*a*) can measure only force
   (*b*) cannot measure static conditions
   (*c*) is too small to handle
   (*d*) produces only dc voltage
7. Which of the following is *not* used in making resistance-temperature detectors?
   (*a*) copper (*b*) platinum
   (*c*) tungsten (*d*) nickel
8. RTDs can be used for measuring a temperature of ...... ºC.
   (*a*) –250 (*b*) 500
   (*c*) 1000 (*d*) 1500
9. A thermistor
   (*a*) is made of a semiconductor like silicon
   (*b*) has positive temperature-coefficient
   (*c*) cannot be used for measuring temperature below 0ºC
   (*d*) has negative temperature coefficient
10. A thermistor with a time constant of one second is shifted from a liquid at 100ºC to another at 200ºC. Its temperature after one second would be............... ºC.
   (*a*) 200 (*b*) 136.8
   (*c*) 163.2 (*d*) 150

ANSWERS

A. Fill in the blanks

1. converts	**2.** mechanical	**3.** acoustical	**4.** self-generating
5. passive	**6.** two	**7.** negative	**8.** 63.2

B. True or False

1. T	**2.** T	**3.** F	**4.** T	**5.** T	**6.** F	**7.** F
8. T	**9.** T	**10.** F	**11.** T	**12.** T	**13.** T	**14.** F

C. Multiple Choice Items

1. *c*	**2.** *a*	**3.** *d*	**4.** *b*	**5.** *c*	**6.** *b*
7. *a*	**8.** *b*	**9.** *d*	**10.** *c*		

37 CHAPTER

Electronic Instruments

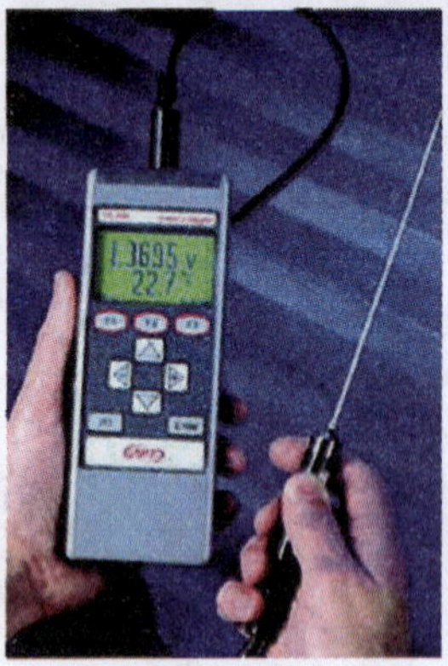

1. Analog and Digital Instruments
2. The Basic Meter Movement
3. Characteristics of Moving Coil Meter Movement
4. Variations of Basic Meter Movement
5. Converting Basic Meter to DC Ammeter
6. Multirange DC Voltmeter
7. Loading Effect of a Voltmeter
8. Ohmmeter
9. The Multimeter
10. Rectifier Type AC Meter
11. Electronic Voltmeters
12. The Direct Current VTVM
13. Comparison of VOM and VTVM
14. Direct Current FET VM
15. Electronic Voltmeter for Alternating Currents

37.1. Introduction

The scientific and technological progress of any nation depends on its ability to measure, calculate and finally, estimate the unknown. Also, the success of an engineer or technician is judged by his ability to measure precisely and to correctly interpret the circuit performance. There are three ways of making such measurements :

(*a*) **by mechanical means**–like measuring gas pressure by Bourdon pressure gauge.

(*b*) **by electrical means**–like measuring potential difference with an electrical voltmeter.

(*c*) **by electronic means**–which is a very sensitive way of detecting the measured quantity because of amplification provided by the active electron device.

The electronic instruments generally have higher sensitivity, faster response and greater flexibility than mechanical or electrical instruments in indicating, recording and, where required, in controlling the measured quantity.

37.2. Analog and Digital Instruments

The deflection type instruments with a scale and movable pointer are called *analog* instruments. The deflection of the pointer is a function of (and, hence, analogous to) the value of the electrical quantity being measured.

Digital instruments are those which use logic circuits

and techniques to obtain a measurement and then display it in numerical-reading (digital) form. The digital readouts employ either LED displays or liquid crystal displays (LCD).

Some of the advantages of digital instruments over analog instruments are as under:

1. easy readability,
2. greater accuracy,
3. better resolution,
4. automatic polarity and zeroing.

37.3. Functions of Instruments

Functionally, different instruments may be divided into the following three categories:

1. Indicating instruments

These are the instruments which indicate the instantaneous value of the quantity being measured at the time it is being measured. The indication is in the form of pointer deflection (analog instruments) or digital readout (digital instruments). Ammeters and voltmeters are examples of such instruments.

2. Recording instruments

Such instruments provide a graphic record of the variations in the quantity being measured over a selected period of time. Many of these instruments are electromechanical devices which use paper charts and mechanical writing instruments such as an inked pen or stylus.

Electronic recording instruments are of two types :

(a) ***null type***—which operate on a comparison basis.

(b) ***galvanometer type***—which operate on deflection type.

3. Controlling instruments

These are widely used in industrial processes. Their function is to control the quantity being measured with the help of information fed back to them by monitoring devices. This class forms the basis of automatic control systems (automation) which are extensively employed in science and industry.

37.4. Electronic versus Electrical Instruments

Both electrical and electronic instruments measure electrical quantities like voltage and current etc. Purely electrical instruments do not have any built-in amplifying device to increase the amplitude of the quantity being measured. The common dc voltmeter based on moving-coil meter movement is clearly an electrical instrument.

The electronic instruments always include in their make-up some active electron device such as vacuum tube, semiconductor diode or an integrated circuit etc.

As seen, the main distinguishing factor between the two types of instruments is the presence of an electron device in the electronic instruments. Of course, movement of electrons is common to both types, their main difference being that control of electron movement is more effective in electronic instruments than in electrical instruments.

Although electronic instruments are usually more expensive than their electrical counterparts, they offer following advantages for measurement purposes:

1. since electronic instruments can amplify the input signal, they possess very high sensitivity *i.e.*, they are capable of measuring extremely small (low-amplitude) signals,
2. because of high sensitivity, their input impedance is increased which means less loading effect when making measurements,
3. they have greater speed *i.e.*, faster response and flexibility,
4. they can monitor remote signals.

37.5. Essentials of an Electronic Instrument

As shown in Fig. 37.1, an electronic instrument is made up of the following three elements:

1. Transducer

It is the first sensing element and is required only when measuring a non-electrical quantity, say, temperature or pressure. Its function is to convert the non-electrical physical quantity into an electrical signal.

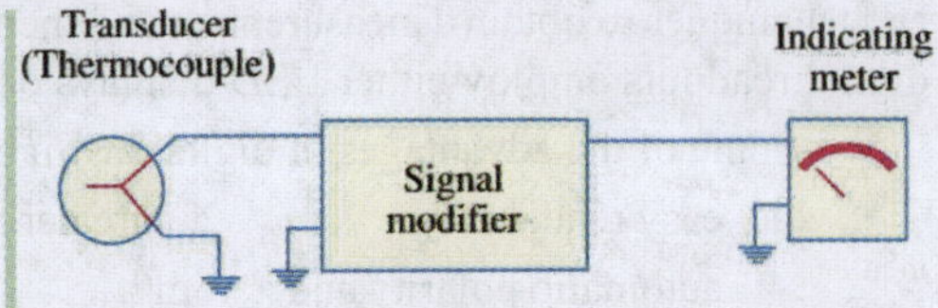

Fig. 37.1

Of course, a transducer is not required if the quantity being measured is already in the electrical form.

2. Signal Modifier

It is the second element and its function is to make the incoming signal suitable for application to the indicating device.

For example, the signal may need amplification before it can be properly displayed. Other types of signal modifiers are: voltage dividers for reducing the amount of signal applied to the indicating device or waveshaping circuits such as filters, rectifiers or choppers etc.

3. Indicating Device

For general purpose instruments like voltmeters, ammeters or ohmmeters, the indicating device is usually a deflection type meter as shown in Fig. 37.1. In digital readout instruments, the indicating device is of digital design.

37.6. Measurement Standards

All instruments, whether electrical or electronic, are calibrated at the time of manufacture against a measurement standard.

1. International Standards

These are defined by international agreement and are maintained at the international Bureau of Weights and Measurements in Paris.

2. Primary Standards

These are maintained at national standards laboratories in each country. They are not available for use outside these laboratories. Their principal function is to calibrate and verify the secondary standards used in industry.

3. Secondary Standards

These are the basic reference standards used by industrial laboratories and are maintained by the particular industry to which they belong. They are periodically sent to national laboratory for calibration and verification against primary standards.

4. Working Standards

These are the main tools of a measurement laboratory and are used to check and calibrate the instrument used in the laboratory.

37.7. The Basic Meter Movement

It is also called D' Arsonval meter movement or a permanent-magnet moving-coil (PMMC) meter movement. Since it is widely used in electronic instruments, it is worthwhile to discuss its construction and principle of operation.

1. Construction

As shown in Fig. 37.2, it consists of a permanent horse-shoe magnet with soft iron pole pieces attached to it. Between the two pole-pieces is situated a cylinder-shaped soft iron core around which moves a coil of fine wire wound on a light metal frame. The metal frame is mounted in jewel bearings so that it can rotate freely. A light pointer attached to the moving coil moves up-scale as the coil rotates when current is passed through it. The rotating coil is prevented from continuous rotation by a spring which provides restoring torque.

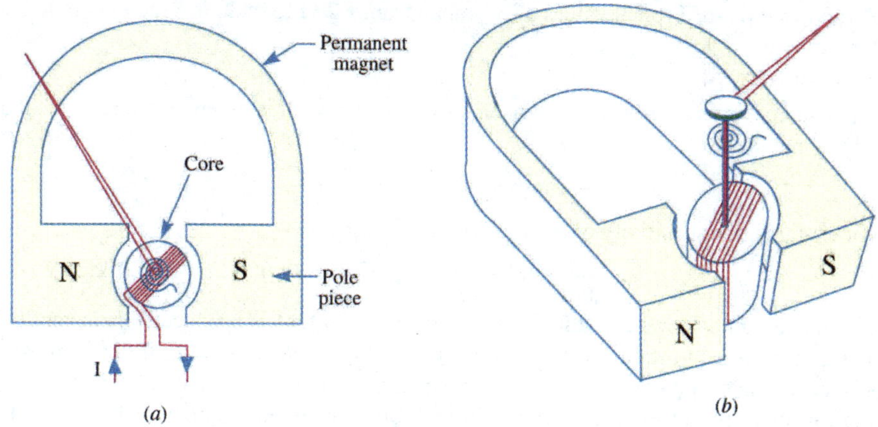

Fig. 37.2

The moving coil movement described above is being increasingly replaced by taut-band movement in which the moving coil and the pointer are suspended between bands of spring metal so that the restoring force is torsional. The bands perform two functions (*i*) they support the coil and (*ii*) they provide restoring torque thereby eliminating the pivots and jewels used with coil spring movement.

As compared to pivoted movement, the taut-band has the advantages of

1. greater sensitivity *i.e.*, smaller full-scale deflection current,
2. ruggedness,
3. minimal friction,
4. easy to manufacture.

2. Principle of Operation

This meter movement works on the *motor* principle and is a current-responding device. The deflection of the pointer is directly proportional to the amount of current passing through the coil.

When direct current flows through the coil, the magnetic field so produced reacts with the field of the permanent magnet. The resultant force turns the coil along with its pointer. The amount of deflection is directly proportional to the amount of current in the coil. Hence, their scale is linear. With correct polarity, the pointer reads up-scale to the right whereas incorrect polarity forces the pointer off-scale to the left.

37.8. Characteristics of Moving Coil Meter Movement

We will discuss the following three characteristics:

(*i*) full-scale deflection current (I_m), (*ii*) internal resistance of the coil (R_m), (*iii*) sensitivity (S).

1. Full-scale Deflection Current (I_m)

It is the current needed to deflect the pointer all the way to the right to the last mark on the calibrated scale. Typical values of I_m for D'Arsonval movement vary from 2 μA to 30 mA.

It should be noted that for smaller currents, the number of turns in the moving coil has to be *more* so that the magnetic field produced by the coil is strong enough to react with the field of the permanent magnet for producing reasonable deflection of the pointer. Fine wire has to be used for reducing the weight of the moving coil, but it increases its resistance. Heavy currents need thick wire but lesser number of turns so that resistance of the moving coil is comparatively less. The schematic symbol is shown in Fig. 37.3.

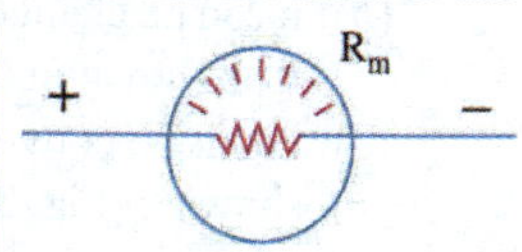

Fig. 37.3

2. Internal Resistance (R_m)

It is the dc ohmic resistance of the wire of the moving coil. A movement with smaller I_m has higher R_m and *vice versa*. Typical values of R_m range from 1.2 Ω for a 30 mA movement to 2 kΩ for a 50 μA movement.

3. Sensitivity (*S*)

It is also known as *current sensitivity* or *sensitivity factor*. It is given by the reciprocal of full-scale deflection current I_m.

$$\therefore \qquad S = \frac{1}{I_m} \text{ ohm/volt}$$

For example, the sensitivity of a 50 μA meter movement is

$$S = \frac{1}{50\ \mu\text{A}} = \frac{1}{50 \times 10^{-6}}\ \Omega/\text{V} = 20{,}000\ \Omega/\text{V} = 20\ \text{k}\Omega/\text{V}$$

The above figure shows that a full-scale deflection of 50 μA is produced whenever 20,000 Ω of resistance is present in the meter circuit *for each volt of applied voltage*. It also represents the ohms-per-volt rating of the meter. The sensitivity of a meter movement depends on the strength of the permanent magnet and number of turns in the coil. Larger the number of turns, smaller the amount of current required to produce full-scale deflection and, hence, higher the sensitivity. A high current sensitivity means a high quality meter movement. It also determines the lowest range that can be covered when the meter movement is modified as an ammeter (Art. 37.10) or voltmeter (Art. 37.12).

37.9. Variations of Basic Meter Movement

The basic moving-coil system discussed in Art. 37.7 can be converted into an instrument to measure dc as well as ac quantities like current, voltage and resistance etc. Without any modification, it can carry a maximum current of I_m and can withstand a maximum dc voltage $v = I_m R_m$.

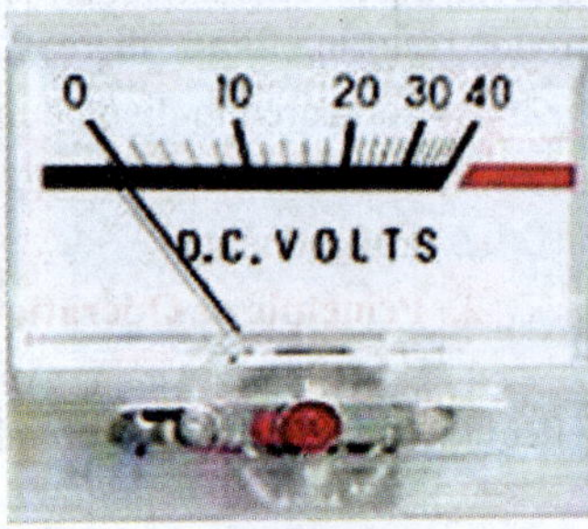

DC-voltmeter.

1. DC instruments

(*a*) it can be made into a dc ammeter, milliammeter or microammeter by adding a suitable shunt resistor R_{sh} in parallel with it as shown in Fig. 37.4 (*a*),

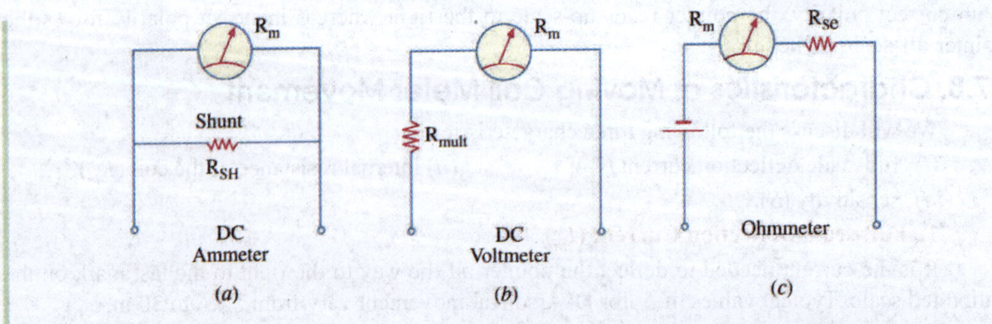

Fig. 37.4

(*b*) it can be changed into a dc voltmeter by connecting a multiplier resistor R_{mult} in series with it as shown in Fig. 37.4 (*b*),

(*c*) it can be converted into an ohmmeter with the help of a battery and series resistor R_{se} as shown in Fig. 37.4 (*c*).

2. AC Instruments

(*a*) it can be changed into an ac audio-frequency ammeter or voltmeter by simply adding an extra rectifier as shown in Fig. 37.5 (*a*), (*b*) it can be converted into a radio-frequency ammeter or voltmeter by adding a thermocouple as shown in Fig. 37.5 (*b*).

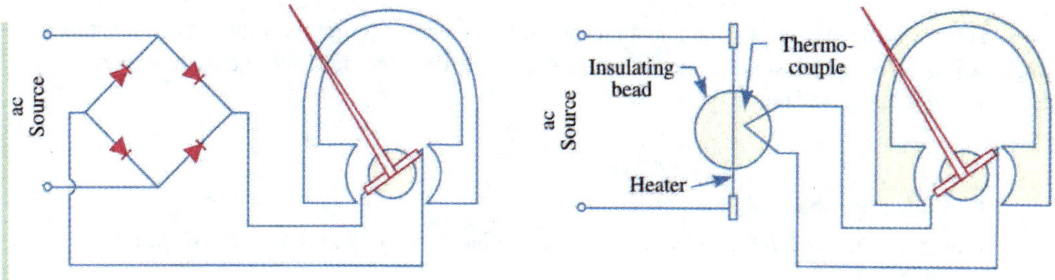

(*a*) (*b*)

Fig. 37.5

The above modifications of the basic meter movement have been tabulated below:

Basic Meter Movement

- DC Instruments
  - dc ammeter by using a shunt resistor
  - dc voltmeter by using series multiplier resistor
  - ohmmeter by using battery and series resistor
- AC Instruments
  - audio-frequency ac ammeter or voltmeter by using a rectifier
  - radio-frequency ammeter or voltmeter by using a thermocouple

37.10. Converting Basic Meter to DC Ammeter

As stated earlier and again shown in Fig. 37.6 (*a*), the basic meter movement can carry a maximum current of I_m *i.e.*, its full-scale deflection current. However, its current range can be increased (*i.e.*, multiplied) to any value by connecting a low resistance (called shunt resistance R_{sh}) in parallel with it as shown in Fig. 37.6 (*b*). The shunted meter works as an ammeter *with an extended range*.

Suppose, we want to measure a line current of I with the help of this meter movement. Obviously, the value of R_{sh} should be such as to shunt or bypass a current of $(I - I_m)$. As seen, range extension is from I_m to I. The ratio $I/I_m = n$ is known as the *multiplying power* or *multiplying factor* of the shunt. It means that a shunt allows the meter to measure current I which is n times larger than I_m.

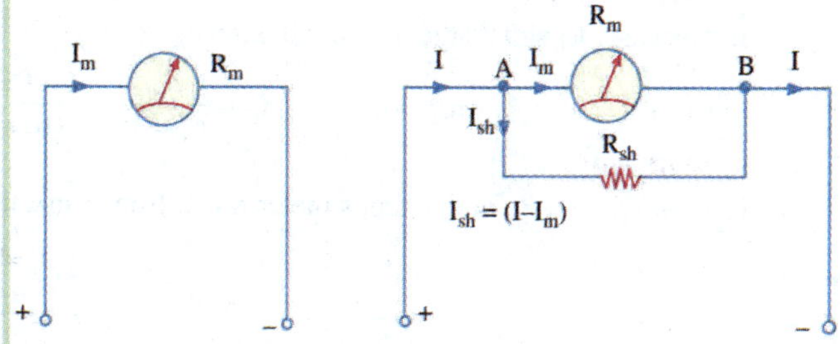

(*a*) (*b*)

Fig. 37.6

Value of R_{sh}

In Fig. 37.6 (*b*), voltage across the meter and the shunt is the same because they are joined in parallel.

$$\therefore \qquad I_m R_m = I_{sh} \cdot R_{sh} = (I - I_m) R_{sh}$$

$$\therefore \quad R_{sh} = \frac{I_m}{(I - I_m)} \cdot R_m = \frac{1}{(I/I_m - 1)} \cdot R_m \quad \therefore \quad R_{sh} = \frac{R_m}{(n-1)}$$

Here, n is the multiplying factor of the shunt. It is seen that larger the value of n *i.e.*, greater the range extension required, smaller the shunt resistance needed. Incidentally, it may be noted that the resistance of the *shunted* meter is

$$= R_m \parallel R_{sh} = \frac{R_m R_{sh}}{R_m + R_{sh}}$$

It is much less than either R_m or R_{sh}.

Example 37.1. *It is required to convert a 5 mA meter with 20 Ω internal resistance into a 5 A ammeter. Calculate*

(a) the value of shunt resistance required, (b) multiplying factor of the shunt.

Solution. Here, $I = 5$ A, $I_{sh} = 5$ mA = 0.005 A, $R_m = 20\ \Omega$

$$(a) \qquad R_{sh} = \frac{I_m}{(I - I_{sh})} \cdot R_m = \frac{0.005}{(5 - 0.005)} \times 20 = \mathbf{0.02\ \Omega\ (approx)}$$

$$(b) \qquad n = \frac{I}{I_{sh}} = \frac{5}{0.005} = \mathbf{1000}$$

Note. $R_{sh} = \dfrac{R_m}{(n-1)} = \dfrac{20}{(1000-1)} = \dfrac{20}{999} = 0.02\ \Omega$ —as found above

Fig. 37.7 shows such an ammeter connected in a load circuit.

37.11. Multirange Meter

The shunt resistance discussed above gives only a single range ammeter. By using universal shunt (also called Ayrton shunt), we can obtain a multirange ammeter as shown in Fig. 37.8.

It is seen that by changing the switch position from *A* to *B* to *C* and finally to *D*, the current range can be extended as desired.

1. Switch at A

Here, the meter is unshunted and so can read up to its full-scale deflection current of 1 mA only.

2. Switch at B

In this case, R_1 shunts the meter and extends its range to 10 mA *i.e.*, increases it ten times.

Since $\qquad n = 10 \quad \therefore \quad R_1 = \dfrac{R_m}{(n-1)} \cong \dfrac{100\Omega}{(10-1)} = 11.11\ \Omega$

3. Switch at C

Here, R_2 shunts R_m and extends meter range from 1 mA to 0.1 A *i.e.*, to 100 mA. Obviously,

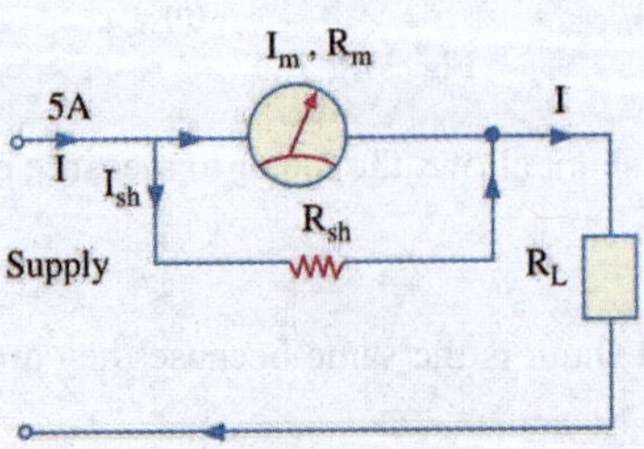

Fig. 37.7

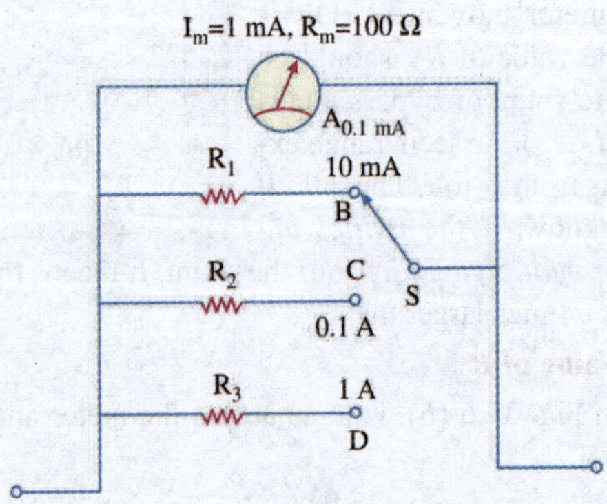

Fig. 37.8

$$n = \frac{100}{1} = 100 \quad \therefore \quad R_2 = \frac{100\ \Omega}{(100-1)} = 1.01\ \Omega$$

4. Switch at D

In this case, R_3 shunts R_m and extends the current range of the meter from 1 mA to 1.0 A *i.e.*, to 1000 mA. Hence, $n = 1000/1 = 1000$.

$$\therefore \qquad R_3 = \frac{100\ \Omega}{(1000-1)} = 0.1001\ \Omega$$

Incidentally, it may be noted that greater the range extension, smaller the shunt resistance.

Alternative Method

An alternative circuit for range extension is shown in Fig. 37.9. It is called 'add on' method of shunting the meter because resistances can be added one after another for changing the range. Unlike in Fig. 37.8, there is no possibility of the meter being in the circuit without any shunt.

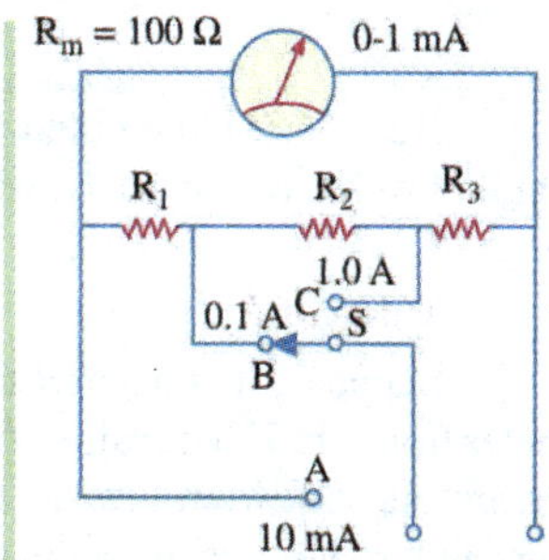

Fig. 37.9

As seen, the universal shunt consists of three resistances R_1, R_2 and R_3. How they are connected as a shunt is determined by the switch position. When S is at position A, the combination $(R_1 + R_2 + R_3)$ becomes connected across R_m. When S is at position B, $(R_2 + R_3)$ become connected in parallel across $(R_1 + R_m)$ and so on.

1. Switch at A

In this case, multiplying factor $n = 10\ \text{mA}/1\ \text{mA} = 10$

$$\therefore \qquad (R_1 + R_2 + R_3) = \frac{R_m}{(n-1)} = \frac{100}{9}\ \Omega \qquad ...(i)$$

2. Switch at B

Here, $(R_2 + R_3)$ is in parallel with $(R_1 + R_m)$ or $(R_1 + 100)$. Also, $n = 100/1 = 100$

$$\therefore \qquad R_2 + R_3 = \frac{R_1 + 100}{99} \qquad ...(ii)$$

3. Switch at C

In this position, R_3 is in parallel with $(R_1 + R_2 + 100)$ and $n = 1000/1 = 1000$

$$\therefore \qquad R_3 = \frac{R_1 + R_2 + 100}{999} \qquad ...(iii)$$

Solving for R_1, R_2 and R_3 from Eqs. (*i*), (*ii*) and (*iii*) we have

$$R_1 = 10\ \Omega,\ R_2 = 1\ \Omega \text{ and } R_3 = 1/9\ \Omega$$

37.12. Measurement of Current

While measuring current flowing in a circuit, following two points must be kept in mind :

1. The current meter must be connected in series with the circuit where current is to be measured (Fig. 37.7). The full circuit current cannot flow through the meter unless it is made a series component.
2. The dc meter must be connected with the correct polarity for the pointer to read up-scale to the right. Reversed polarity deflects the pointer down-scale to the left forcing it against the stop which can sometime bend the pointer.

37.13. Converting Basic Meter to DC Voltmeter

The basic meter movement can measure a maximum voltage of $I_m R_m$ which is very small [Fig. 37.10 (*a*)]. However, its voltage range can be extended to any value by connecting a large resistance in series with it as shown in Fig. 37.10 (*b*). The series resistance is also called *multiplier resistance* because it multiplies the voltage reading capability of the meter many times. It is usually connected inside the voltmeter case.

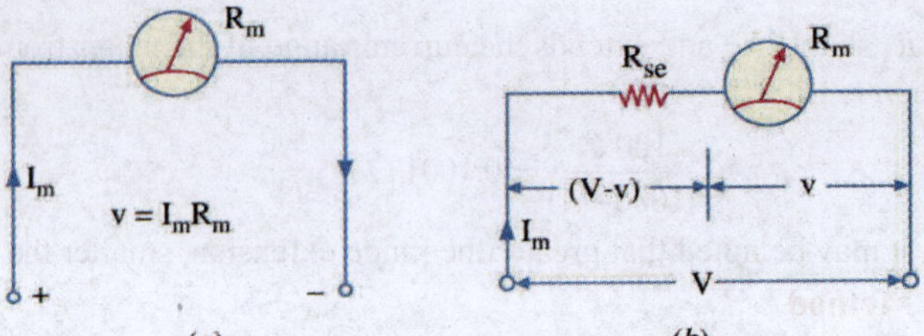

Fig. 37.10

But it should be noted that the voltmeter is connected in parallel with the load across which the voltage is to be measured (Fig. 37.11).

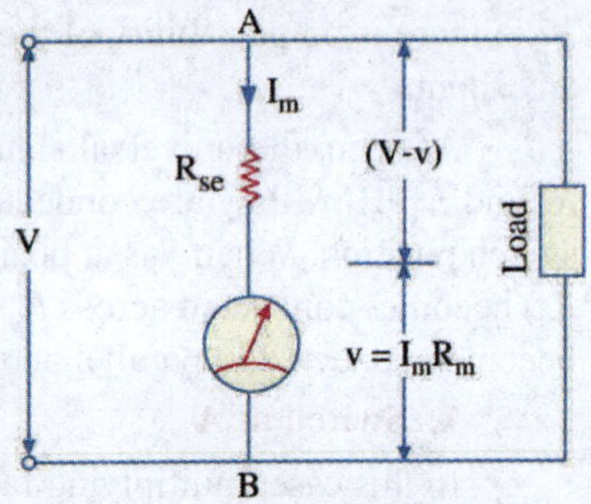

Fig. 37.11

Value of R_{sc}

Suppose, it is desired to extend the voltage range of the meter from v to V. The ratio V/v is known as the *voltage multiplication*. As seen from Fig. 37.11, drop across R_{se} is $(V - v)$ and current through it is the same as meter current *i.e.*, I_m

$$\therefore \quad I_m R_{se} = (V - v) \qquad \ldots(i)$$

$$\therefore \quad R_{se} = \frac{V - v}{I_m} = \frac{V - I_m R_m}{I_m} = \frac{V}{I_m} - R_m$$

The voltage multiplication (m) can be found from Eq. (i) above.

Dividing both sides by v, we get

$$\frac{I_m R_{se}}{v} = \frac{V}{v} - 1$$

$$\therefore \quad \frac{V}{v} = 1 + \frac{I_m R_{sc}}{v} = 1 + \frac{I_m R_{se}}{I_m R_m}$$

$$\therefore \quad \frac{V}{v} = \left(1 + \frac{R_{se}}{R_m}\right) \quad \therefore \; m = \left(1 + \frac{R_{se}}{R_m}\right)$$

It is seen that for a given meter, higher the series resistance, greater the voltage range extension.

Example 37.2. *A 50 mA meter movement with an internal resistance of 1 kΩ is to be used as dc voltmeter of range 50 V. Calculate the*

(a) *multiplier resistance required and* (b) *voltage multiplication.*

Solution. (a) $\quad R_{se} = \frac{V}{I_m} - R_m$

$$= \frac{50}{50 \times 10^{-6}} - 1000$$

$$= 10^6 - 1000 = 999{,}000\ \Omega = \mathbf{999\ k\Omega}$$

37.14. Multirange DC Voltmeter

A multirange voltmeter with ranges of 0 –5 V, 0 –25 V and 0 –50 V is shown in Fig. 37.12. Different values of resistors R_1, R_2 and R_3 can be found in the same way as in Art. 37.11. It would be found that for the meter movement shown in the figure.

$$R_1 = 99\text{ K}, R_2 = 499\text{ K}, R_3 = 999\text{ K}$$

It is seen that higher the voltage range, greater the multiplier resistance required (in almost the same proportion as the ranges).

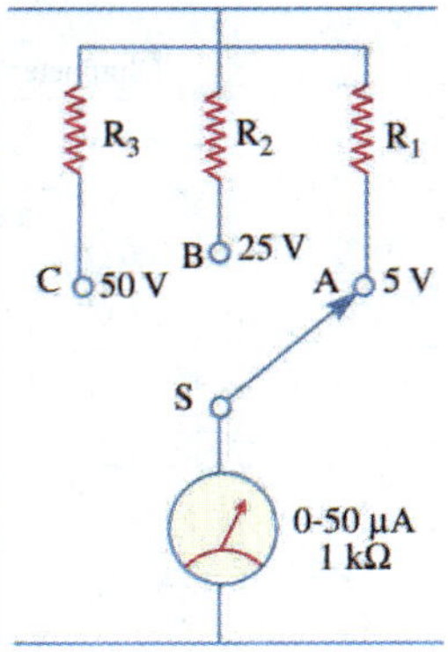

Fig. 37.12

37.15. Loading Effect of a Voltmeter

When the voltmeter resistance is not high as compared to the resistance of the circuit across which it is connected, the measured voltage becomes less. The decrease in voltage may be negligible or it may be appreciable depending on the sensitivity (ohms-per-volt rating) and input resistance of the voltmeter. It is called voltmeter *loading effect* because the voltmeter loads down the circuit across which it is connected. Since input resistance of electronic voltmeter is very high (10 MΩ or more), loading is not a problem in their case.

Consider the circuit shown in Fig. 37.13 in which two 15 K resistors are connected in series across a 100 V dc source. The drop across each is 50 V. Now, suppose, that a 30 K voltmeter is connected across R_2 to measure voltage drop across it. Due to loading effect of the voltmeter, the reading is reduced from 50V to 40 V as explained below. As seen from Fig. 37.13 (*b*), combined resistance of R_2 and voltmeter is 15 K || 30 K = 10 K.

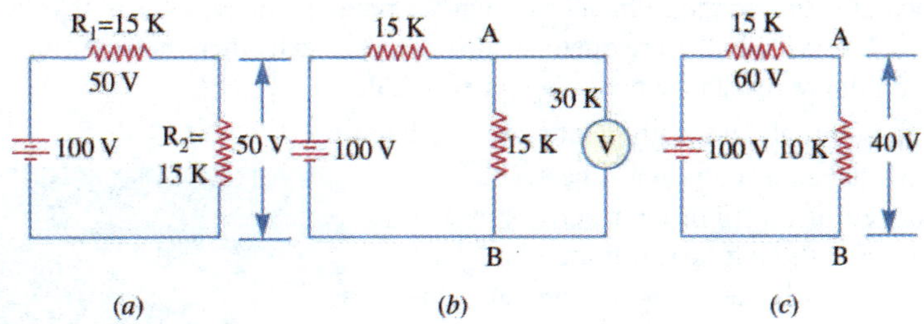

Fig. 37.13

$$\text{drop across 10 K} = \frac{10}{10+15} \times 100 = 40\text{ V}$$

Loading effect can be minimized by using a voltmeter whose resistance is as high as possible as compared to that of the circuit across which it is connected.

Correction Formula

The loading effect can be neutralized by using the following formula :

$$V_{corr} = V_{meas} + \frac{R_1 R_2}{R_V (R_1 + R_2)} . V_{meas}$$

where
V_{corr} = corrected voltage reading
V_{meas} = measured voltage reading
R_V = voltmeter resistance
R_1, R_2 = voltage dividing resistances in the circuit

In the above case,

$$V_{corr} = 40 + \frac{15 \times 15}{30\,(15+15)} \times 40 = 40 + 10 = \mathbf{50\ V}$$

37.16. Ohmmeter

The basic meter movement can be used to measure resistance if it is combined with a battery and a current-limiting resistance as shown in Fig. 37.14 (*a*). In that case, it is known as an ohmmeter.

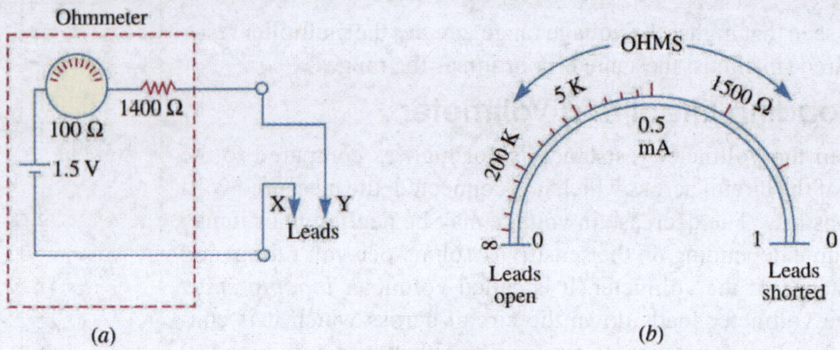

Fig. 37.14

For measuring resistance, the ohmmeter leads *X-Y* are connected across the unknown resistance after switching off the power in the circuit under test. Only in that case, the ohmmeter battery can provide current for the meter movement. Since the amount of current depends on the amount of external resistance, the meter scale can be calibrated in ohms (instead of mA).

When the leads *X-Y* are shorted, meter current is 1.5 V/(100 + 1400) Ω = 1 mA. The meter shows full-scale deflection to the right. The ohmmeter reading corresponds to 0 Ω because external resistance is zero. When leads *X-Y* are open *i.e.*, do not touch each other, meter current is zero. Hence, it corresponds to infinite resistance on the ohmmeter scale.

Following points about the ohmmeter are worth noting:

1. the resistance scale is non-linear *i.e.*, it is expanded at the right near zero ohm and crowded at the left near infinite ohm. This non-linearity is due to the reciprocal function $I = V/R$;
2. the ohmmeter reads up-scale regardless of the polarity of the leads because direction of current is determined by the internal battery;
3. at half-scale deflection, external resistance equals the internal resistance of the ohmmeter;
4. the test leads should be shorted and 'ZERO OHMS' control adjusted to bring the pointer to zero on each range.

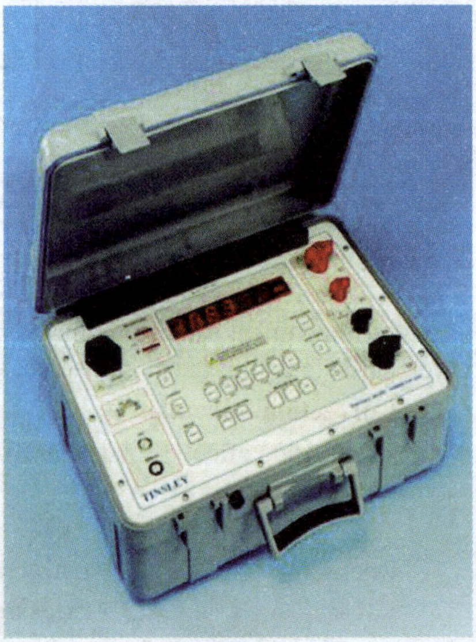

Fig. 37.15. Digital micro-ohmmeter.

Fig. 37.15 shows a digital micro-ohmmeter having a range of 1 μΩ – 2 kΩ with 3½ digit, 7-segment LED display. It has a basic accuracy of ± 0.2% ± 1 digit and is based on latest design using MOS LSI ICs and glass epoxy PCB.

Fig. 37.16 shows a battery-operated portable digital milli-ohmmeter having a measurement range of 200 mΩ–2 kΩ with an accuracy of ± 0.5% ± 1 digit. It has a 3½ digit 7-segment LED display.

37.17. The Multimeter

It is extensively used in cable industry, motor industry, transformer and switchgear industry. It is also called volt-ohm-milliammeter (VOM). It is a general purpose instrument having the necessary circuitry and switching arrangement for measuring ac/dc voltage or ac/dc current or resistance. It is simple, compact and portable because the only power it uses is the battery for the ohmmeter.

Multimeters may be of analog type (Fig. 37.17) or digital type (Fig. 37.18). The analog type is of the pointer and scale type *i.e.*, it uses the basic D'Arsonval meter movement. However, digital multimeters (DMMs) are becoming increasingly popular because of their easy readability, numerical display and improved accuracy.

Fig. 37.17 shows the photograph of an analog multimeter designed primarily for electrical, electronic, radio and TV engineers and technicians. It sells under the brand name of Motwane Multimeter 8 X Mark–III.

It is a 5-function, 30-range meter which measures high ac/dc voltages from 0 to 2.5 kV and ac/dc currents from 0 to 10 A. Its three resistance ranges cover from 0 to 20 MΩ. It is reputed for its excellent reliability, operational simplicity and easy portability.

Fig. 37.18 depicts a digital multimeter which can measure dc voltages upto 1000 V, ac voltages upto 1000 V, ac/dc currents from 200 μA to 1 A and resistances from 200 Ω to 20 MΩ.

It has 3½ digit readout with automatic decimal point and automatic polarity indication.

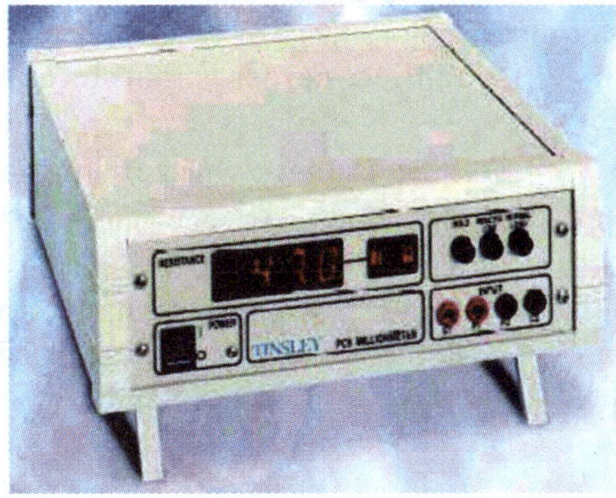

Fig. 37.16. Digital milli-ohmmeter

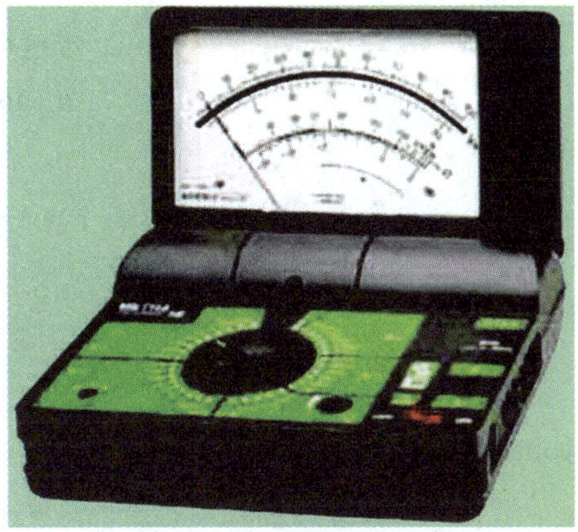

Fig. 37.17. Analog Multimeter

Fig. 37.18.

In Fig. 37.19 is shown a hand-held autoranging, digital multimeter (DMM) having high-contrast, 3½ digit LCD readout. Based on LSI ICs, it has been designed for speed, accuracy and reliability.

37.18. Rectifier Type AC Meter

The D'Arsonval meter movement can be used for measuring alternating quantities provided a rectifier is added to the measuring circuit. A similar rectifier arrangement is found as part of AC VOLTS function in multimeters (Art. 37.17). Such ac meters are more widely used than either (costly but more accurate) dynamometer type or more delicate thermal and hotwire type.

(*a*) With Half-wave Rectifier

The circuit of an ac voltmeter using half-wave diode rectifier is shown in Fig. 37.20. Here, a half-wave rectifier has been combined in series with a dc meter movement.

When used as a dc voltmeter (*i.e.*, without rectifier) it would have (say, for example) a range of 10 V. However, if an ac voltage of rms value 10 V is applied across input terminals *AB*, it would read 4.5 V.

It is so because an ac voltage of rms value 10 V has a peak value of $10 \times \sqrt{2} = 14$ V and an average value of $0.636 \times 14 = 9$ V. Since in the half-wave rectified output, one half-cycle is absent, the average for the full cycle is 9/2 – 4.5 V. The meter movement will, therefore, read 4.5 V *i.e.*, 45% of the dc value. It may also be noted that ac sensitivity of a half-wave ac meter is only 45 per cent of the dc sensitivity.

Fig. 37.19

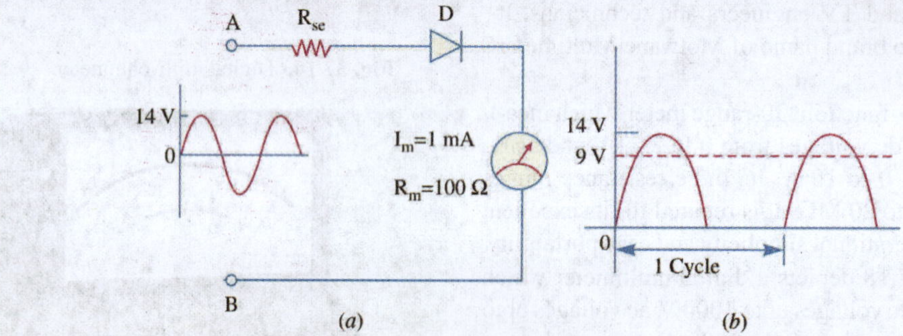

Fig. 37.20

(*b*) With Full-wave Rectifier

The circuit is shown in Fig. 37.21. In this case, the meter reading would be 90% of rms input voltage *i.e.*, 90% of the dc value.

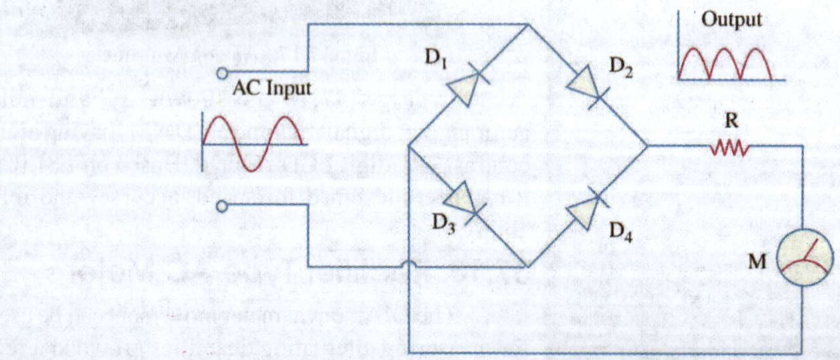

Fig. 37.21

37.19. Electronic Voltmeters

A VOM can be used to measure voltages but it lacks both sensitivity and high input resistance. Moreover, its input resistance is different for each range. The electronic voltmeter (EVM), on the other hand, has input resistance ranging from 10 M Ω to 100 M Ω, thus producing less loading of the circuit under test than the VOM. Another advantage of EVM is that its input resistance remains constant over all ranges.

Two types of voltmeters are in use to-day (*i*) analog and (*ii*) digital. However, a distinction must be made between a digital instrument and an instrument with digital readout. A digital instrument is one which uses internal circuitry of digital design. A digital readout instrument is one whose measuring circuitry is of analog design but the indicating device is of digital design.

The electronic voltmeters go by a variety of names reflecting the technology used.

(*i*) vacuum-tube voltmeter (VTVM) – it uses vacuum tubes with deflection meter movement, (Fig. 37.22).

(*ii*) digital voltmeters like transistor voltmeter (TVM) and FET voltmeter (FET VM) .

37.20. The Direct Current VTVM

It is used for measuring dc voltages but can measure ac voltage (by using a suitable rectifier).

In the instrument, input voltages are measured in terms of corresponding plate current readings.

1. Single Tube VTVM

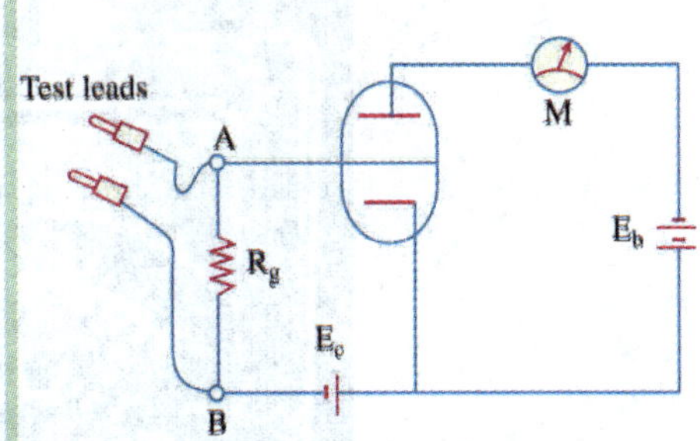

Fig. 37.22

As shown in Fig. 37.22 it consists of one triode and the dc meter movement M which is calibrated in volts rather than in terms of plate current. R_g is the usual grid-leak resistance. The voltage to be measured is applied across the grid and cathode with the help of two leads *A* and *B*. The voltage should be so applied that the grid remains negative but must not be driven towards cut-off.

The quiescent plate current value is marked as 0 V on the meter scale which is linear. As different voltages are applied to the grid, plate current changes according to the magnitude of these voltages.

The principal disadvantage of this single-tube voltmeter is that if E_b or E_c changes, I_p changes, thereby disturbing the scale calibration. Hence, the meter gives wrong voltage readings.

Another drawback is that meter (quiescent) current flows and gives deflection even when input voltage is zero.

(*ii*) Two tube VTVM

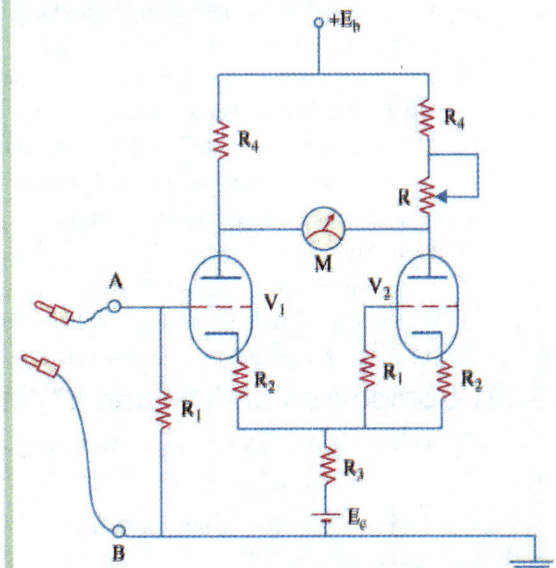

Fig. 37.23

It uses two similar triodes and is also known as balanced bridge type voltmeter. The meter movement is connected across the plates of the triodes V_1 and V_2 and indicates the voltage being measured. The variable resistance *R* in the plate circuit of V_2 is for zero adjustment of the meter. As shown in Fig. 37.23, the voltage to be measured is applied across the grid cathode circuit of V_1 whereas V_2 is used as the neutralizing tube.

Operation

Under no-input voltage condition, the two plate currents are equal so that both plates are at the same potential. Hence, meter *M* reads zero. In practice, the two plate currents are never exactly equal so that *M* always shows a very small reading. However, meter needle can be adjusted to read zero under this condition by adjusting *R*.

The voltage to be measured is applied across the grid of V_1. Consequently, the plate current of V_1 changes and is no longer equal to that of V_2. As a result, plate potentials of V_1 and V_2 become unequal and a current passes through *M* which is proportional to the voltage being measured.

Fig. 37.24 shows the photograph of a VTVM which can measure full-scale ac/dc voltage of 1500 V and has an input resistance of 11 MΩ.

Fig. 37.24

It has a frequency range from 15 Hz to 5 MHz for ac measurements. It can also be used to measure resistances upto 1000 MΩ.

Advantages

1. It has negligible 'shunting' or loading effect on the circuit under test. It is due to its very high input resistance (typically 11 MΩ).
2. It virtually requires no current from the measured circuit.
3. It has wide frequency response.
4. Meter accuracy is not affected by temperature variations.
5. Meter cannot burn out accidentally due to wrong connections.
6. It can be used for measuring resistances.

Disadvantages

1. It requires warm-up period.
2. It cannot measure current directly due to its extremely high internal resistance.
3. AC current measurements in rms values are accurate only for sine waves.

37.21 Comparison of VOM and VTVM

The points of similarity and dissimilarity between the two are given below :

VOM	*VTVM*
1. DC voltmeter resistance changes with range	it is the same (11 MΩ or so) for all ranges
2. measures ac volts	measures ac volts
3. measures current directly	cannot measure current directly because of its high resistance
4. ohm ranges are usually upto $R \times 10{,}000\ \Omega$	these are upto $R \times 1\ \text{M}\Omega$
5. zero-ohm adjustment has to be changed for each range	zero-ohm adjustment is the same for all ranges
6. is portable	is usually plugged into power line

37.22. Direct Current FET VM

The schematic diagram of a FET VM using difference amplifier is shown in Fig. 37.25. The two FETs are identical so that increase in the source current of one FET is offset by corresponding decrease in the source current of the other. The two FETs form the lower arms of the balanced bridge circuit, whereas the two drain resistors R_D form the upper arms. The meter movement is connected across the drain terminals of the FETs.

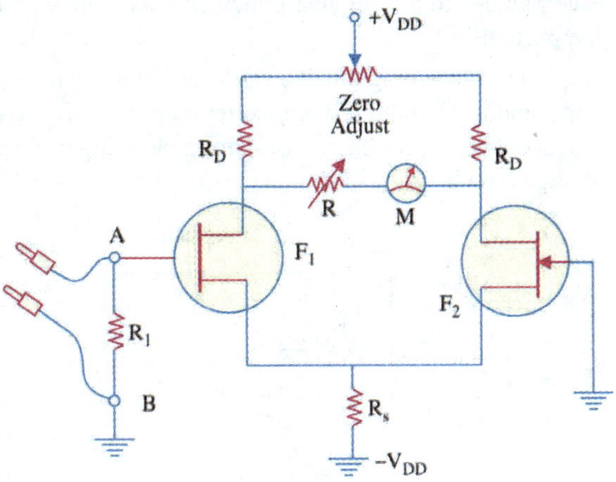

Fig. 37.25

The circuit is balanced under zero-input-voltage condition provided the two FETs are identical. In that case, there would be no current through M. Zero-Adjust potentiometer is used to get null deflection in case there is a small current through M under zero-signal condition.

Full-scale calibration is adjusted with the help of variable resistor R.

When positive voltage is applied to the gate of F_1, some current flows through M. The magnitude of this current is found to be proportional to the voltage being measured. Hence, meter is calibrated in volts to indicate input voltage.

37.23. Electronic Voltmeter for Alternating Currents

A VTVM provides high input resistance (11 MΩ or so) for dc measurements but much lower resistance (1 MΩ or so) for ac measurements. High input resistance for ac work can, however, be provided by placing a high-impedance amplifier between the input and the rectifier. Unfortunately, it is not possible in the case of VTVMs because vacuum tubes are bulky and expensive. With the availability of solid-state devices it has become possible to provide very high input impedance even for ac measurements.

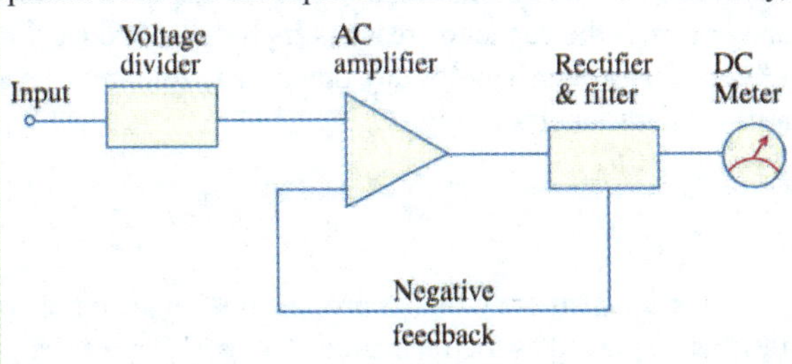

Fig. 37.26

The block diagram of such an EVM for ac measurements is shown in Fig. 37.26 where voltage divider allows range selection. The amplifier provides the necessary gain to establish voltmeter sensitivity as well as high input impedance. The negative feedback ensures stability and accurate overall gain.

37.24. The Digital Voltmeter (DVM)

Such a voltmeter displays measurements of dc or ac voltages as discrete numerals instead of pointer deflections on a continuous scale as in analog instruments. As compared to other voltmeters, a DVM offers the advantages of:

1. greater speed,
2. higher accuracy and resolution,
3. no parallax,
4. reduced human error,
5. compatibility with other digital equipment for further processing and recording.

With the development and perfection of IC modules, the size and power requirement of DVMs have reduced to a level where they can compete with conventional analog instrument both in price and portability.

The block diagram of a DVM based on dual-slope technique is shown in Fig. 37.27. The dual-slope analog-digital (A-D) converter consists of five basic blocks : an Op-Amp used as an integrator, a level comparator, a basic clock (for generating timing pulses), a set of decimal counters and a block of logic circuitry.

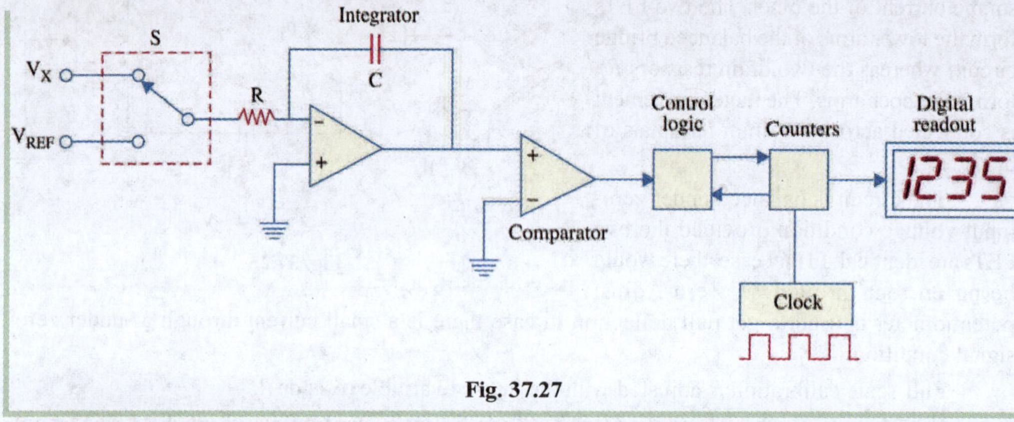

Fig. 37.27

The unknown voltage V_x is applied through switch S to the integrator for a known period of time T as shown in Fig. 37.28. This period is determined by counting the clock frequency in decimal counters. During time period T, C is charged at a rate proportional to V_x.

At the end of time interval T, S is shifted to the reference voltage V_{ref} of *opposite* polarity. The capacitor charge begins to decrease with time and results in a downward linear ramp voltage. During the second period a known voltage (*i.e.*, V_{ref}) is observed for an unknown time (t). This unknown time t is determined by counting timing pulses from the clock until the voltage across the capacitor reaches its basic reference value (reference may be ground or any other basic reference level). From similar triangles of Fig. 37.28.

$$\frac{V_x}{T} = \frac{V_{ref}}{t} \qquad \therefore \quad V_x = \frac{T}{t} \times V_{ref}$$

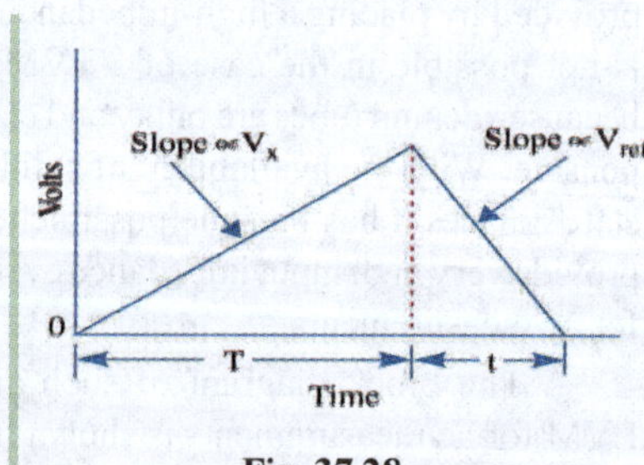

Fig. 37.28

The count after t which is proportional to the input voltage V_x is displayed as the measured voltage.

By using appropriate signal conditioners, currents, resistances and ac voltages can be measured by the same instrument.

DVMs are often used in data processing systems or data logging systems. In such systems, a number of analog input signals are scanned sequentially by an electronic system and then each signal is converted into an equivalent digital value by the A/D converter in the DVM. The digital value is then transmitted to a printer along with the information about the input line from which the signal has been derived. The whole data is then printed out. In this way, a large number of input signals can be automatically scanned or processed and their values either printed or logged.

Fig. 37.29 shows a portable digital dc micro-voltmeter (Agronic-112). It has a measurement range of 1 μV–1000 V with an accuracy of ± 0.2% ± 1 digit. It uses latest MOS LSI ICs and glass epoxy PCB. It has 3½ digit, 7-segment LED display and is widely-used by the testing and servicing departments of industries, research laboratories, educational institutions and service centres.

37.25. Cathode Ray Oscilloscope (CRO)

It is generally referred to as oscilloscope or scope and is the basic tool of an electronic engineer and technician as voltmeter, ammeter and wattmeter are those of an electrical engineer or electrician. The CRO provides a two-dimensional visual display of the signal waveshape on a screen thereby allowing an electronic engineer to 'see' the signal in various parts of the circuit. It, in effect, gives the electronic engineer an eye to 'see' what is happening inside the circuit itself. It is only by 'seeing' the signal waveforms that he can correct errors, understand mistakes in the circuit design and thus make suitable adjustments.

Fig. 37.29. Digital voltmeter.

An oscilloscope can display and also measure many electrical quantities like ac/dc voltages, time, phase relationships, frequency and a wide range of waveform characteristics like rise-time, fall-time and overshoot etc. Non-electrical quantities like pressure, strain, temperature and acceleration etc., can also be measured by using different transducers to first convert them into an equivalent voltage.

As seen from the block diagram of an oscilloscope (Fig. 37.30), it consists of the following major sub-systems :

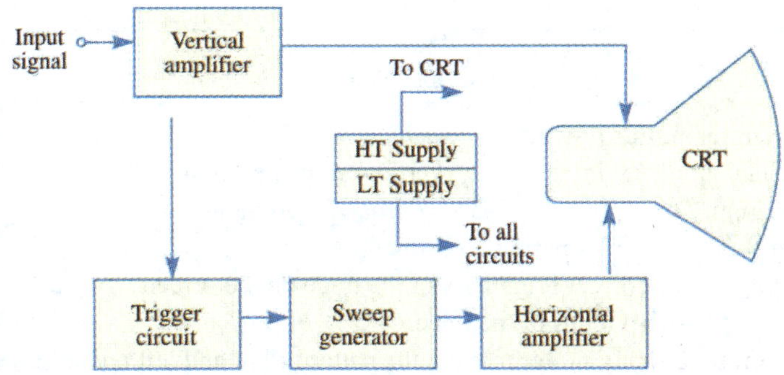

Fig. 37.30

1. **Cathode Ray Tube (CRT)**—it displays the quantity being measured.
2. **Vertical amplifier**—it amplifies the signal waveform to be viewed.
3. **Horizontal amplifier**—it is fed with a sawtooth voltage which is then applied to the *X*-plates.
4. **Sweep generator**—produces saw- tooth voltage waveform used for horizontal deflection of the electron beam.
5. **Trigger circuit** —produces trigger pulses to start horizontal sweep.
6. **High and low-voltage power supply.**

The operating controls of a basic oscilloscope are shown in Fig. 37.31.

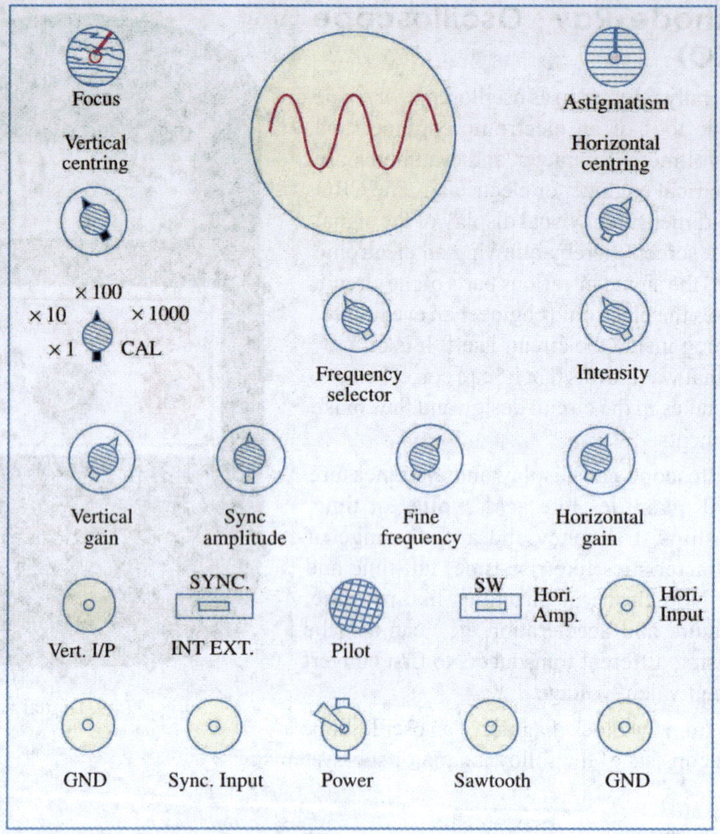

Fig. 37.31

The different **terminals** provide

1. horizontal amplifier input,
2. vertical amplifier input ,
3. sync. input,
4. Z-axis input ,
5. external sweep input.

As seen, different **controls** permit adjustment of

1. **Intensity** — for correct brightness of the trace on the screen.
2. **Focus** — for sharp focus of the trace.
3. **Horizontal centring** — for moving the pattern right and left on the screen.
4. **Vertical centring** — for moving the pattern up and down on the screen.
5. **Horizontal gain (also Time/div or Time/cm)** — for adjusting pattern width.
6. **Vertical gain (also volt/div or volt/cm)** — for adjusting pattern height
7. **Sweep frequency** — for selecting number of cycles in the pattern.
8. **Sync. voltage amplitude** — for locking the pattern

The different **switches** permit selection of:

1. sweep type,
2. sweep range,
3. sync. type.

A CRO can operate upto 50 MHz, can allow viewing of signals within a time span of a few nanoseconds and can provide a number of waveform displays simultaneously on the screen. It also has the ability to hold the displays for a short or long time (of many hours) so that the original signal may be compared with one coming on later.

37.26. Cathode Ray Tube (CRT)

It is the 'heart' of an oscilloscope and is very similar to the picture tube in a television set.

Construction

The cross-sectional view of a general-purpose electrostatic deflection CRT is shown in Fig. 37.32. Its four major components are:

1. an electron gun – for producing a stream of electrons,
2. focussing and accelerating anodes—for producing a narrow and sharply-focussed beam of electrons,
3. horizontal and vertical deflecting plates–for controlling the path of the beam,
4. an evacuated glass envelope with a phosphorescent screen which produces bright spot when struck by a high velocity electron beam.

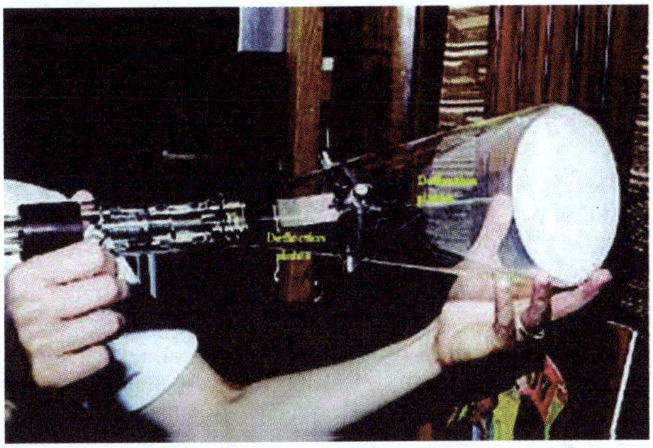

A photography of cathode ray tube.

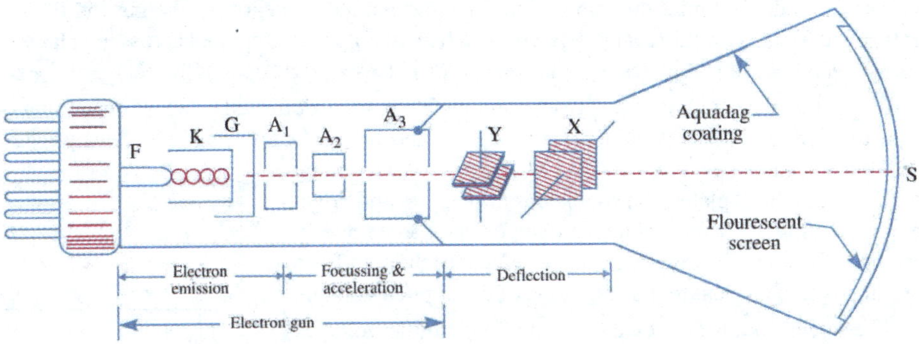

Fig. 37.32

As shown, a CRT is a self-contained unit like any electron tube with a base through which leads are brought out for different pins.

1. Electron Gun Assembly

The electron gun assembly consists of an indirectly-heated cathode K, a control grid G, a pre-accelerator anode A_1, focussing anode A_2 and an accelerating anode A_3. The sole function of the electron gun assembly is to provide a focussed beam of electrons which is accelerated towards the flourescent screen. The electrons are given off by thermionic emission from the cathode. The control grid is a metallic cylinder with a small aperture in line with the cathode and kept at a negative potential with respect to K. The number of electrons allowed to pass through the grid aperture (and, hence, the beam current) depends on the amount of the control grid bias. Since the intensity (or brightness) of the spot S on the screen depends on the strength of beam current, the knob controlling the grid bias is called the *intensity control*.

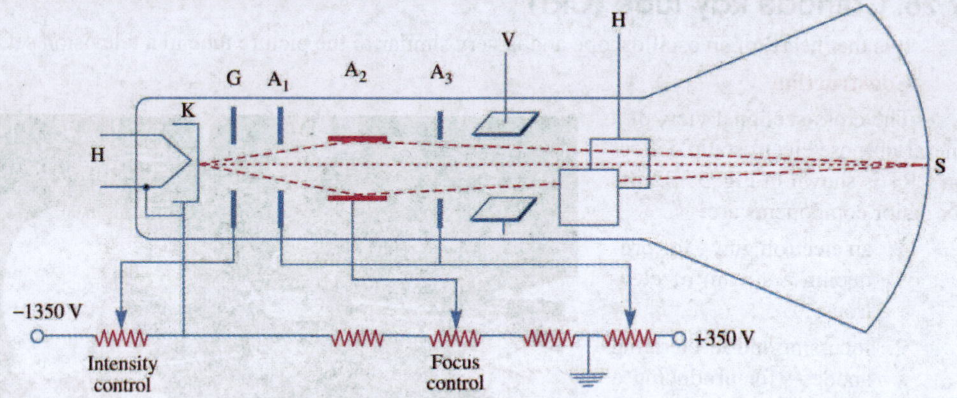

Fig. 37.33

The anodes A_1 and A_3, which are both at positive potential with respect to K, operate to accelerate the electron beam (Fig. 37.33). The cylindrical focussing anode A_2, being at negative potential, repels electrons from all sides and compresses them into a fine beam. The knob controlling the potential of A_2 provides the *focus control.*

2. Deflecting Plates

Two sets of deflecting plates are used for deflecting the thin pencil-like electronic beam both in the vertical and horizontal directions. The first set marked Y (nearer to the gun) is for vertical deflection and X-set is for horizontal deflection. When no potential is applied across the plates, beam passes between both sets of plates undeflected and produces a bright spot at the centre of the screen.

If upper Y-plate is given a positive potential, the beam is deflected upwards depending on the value of the applied potential. Similarly, the beam (and hence the spot) deflects downwards when lower Y-plate is made positive. However, if an *alternating* voltage is applied across the Y-plates, the spot keeps moving up and down thereby producing a vertical luminous trace on the screen due to persistence of vision. The maximum displacement of the spot from its central position is equal to the amplitude of the applied voltage.

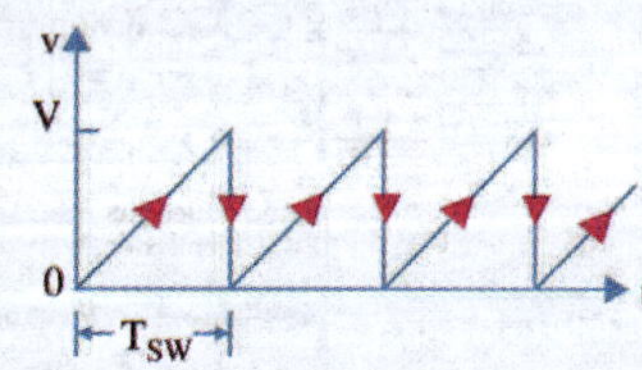

Fig. 37.34

The screen spot is deflected horizontally if similar voltages are applied to the X-plates. The dc potentials on the Y- and X-plates are adjustable by means of *centring controls.*

It must be remembered that the signal to be displayed on the screen is always applied across the Y-plates. The voltage applied across X-plates is a ramp voltage *i.e.*, a voltage which increases linearly with time. It has a sawtooth waveform as shown in Fig. 37.34. It is also called horizontal time-base or sweep voltage. It has a sweep time of T_{sw}.

3. Glass Envelope

It is funnel-shaped having a phosphor-coated screen at its flared end. It is highly-evacuated in order to permit the electron beam to traverse the tube easily. The inside of the flared part of the tube is coated with a conducting graphite layer called Aquadag which is maintained at the same potential as A_3. This layer performs two functions (*i*) it accelerates the electron beam after it passes between the deflecting plates and (*ii*) collects the electrons produces by secondary emission when electron beam strikes the screen. Hence, it prevents the formation of negative charge on the screen.

The screen itself is coated with a thin layer of a flourescent material called phosphor. When struck by high-energy electrons, it glows. In other words, it absorbs the kinetic energy of the electrons and converts it into light — the process being known as *flourescence.* That is why the screen is called *flourescent screen.* The colour of the emitted light depends on the type of phosphor used.

37.27. Deflection Sensitivity of a CRT

Fig. 37.35 shows the upward deflection of an electron beam when it passes between the vertical or *Y*-plates of a CRT. The beam deflects upwards because the upper *Y*-plate has been made positive with respect to the lower plate. Reversing the polarity of the applied voltage would, obviously, cause the beam to deflect downwards.

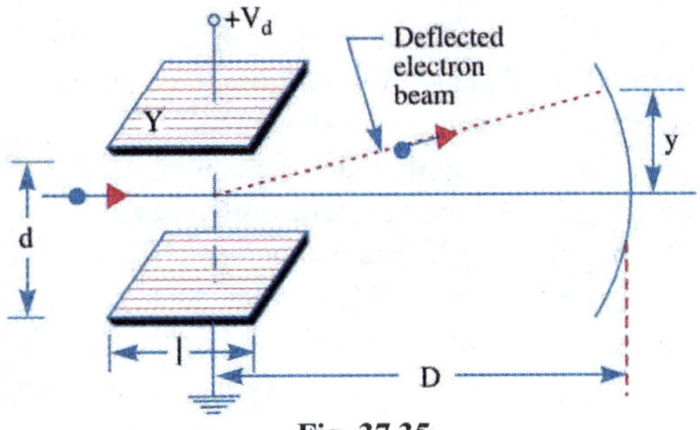

Fig. 37.35

The vertical deflection of the beam is

$$y = \frac{1}{2}.D.\frac{l}{d}.\frac{V_d}{V_A}$$

where V_A is the accelerating voltage applied to the electrons which make up the electron beam.

The deflection sensitivity of a CRT is defined as the vertical deflection of the beam on the screen per unit deflecting voltage.

$$\therefore \qquad S = \frac{y}{V_d}$$

Using the above equation, we get $S = \dfrac{l\,D}{2\,d\,V_A}$

The deflection factor which is defined as the reciprocal of deflection sensitivity is given by

$$G = 1/S$$

Substituting the value of *S* from above

$$G = 2.\frac{d}{l}.\frac{V_A}{D} \text{ volt/metre}$$

37.28. Normal Operation of a CRO

The signal to be viewed or displayed on the screen is applied across the *Y*-plates of a CRT. But to see its waveform or pattern, it is essential to spread it out horizontally from left to right. It is achieved by applying a sawtooth voltage wave (produced by a time base generator) to *X*-plates. Under these conditions, the electron beam would move uniformly from left to right thereby graphing vertical variations of the input signal versus time. Due to repetitive tracing of the viewed waveform, we get a continuous display because of persistence of vision. However, for getting a stable stationary display on the screen, it is essential to synchronize the horizontal sweeping of the beam (sync) with the input signal across *Y*-plates. The signal will be properly synced only when its frequency equals the sweep-generator frequency.

In general, for proper synchronization of time-base with the signal, the condition is

$$T_{sw} = n\,T_s$$

where T_s is the time-period of the signal and *n* is an integer.

If $n = 1$, then $T_{sw} = T_s$ *i.e.*, time-periods of the sweep voltage and input signal voltage are equal, then one cycle of the signal would be displayed as shown in Fig. 37.36 (*a*).

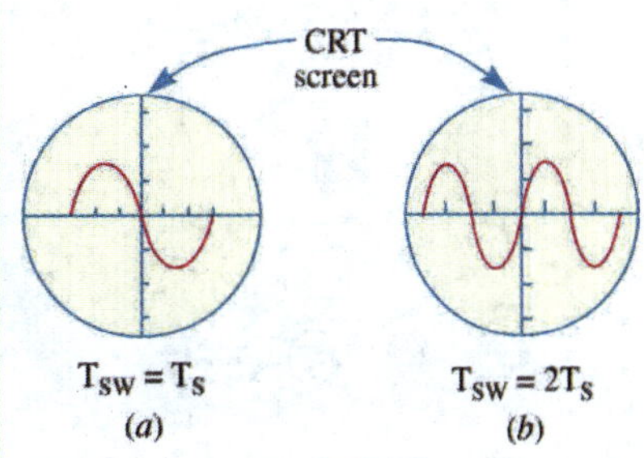

Fig. 37.36

On the other hand, if T_{sw} is twice T_s, then two cycles of the signal voltage would be displayed as shown in Fig. 37.36 (*b*). Obviously, three full cycles of the input voltage would be spread out on the screen when

$$T_{sw} = 3\,T_s.$$

Internal Synchronization

The periodic sawtooth voltage which is applied to *X*-plates for horizontal sweep (or scan) of the beam across the screen is usually provided by the unijunction relaxation oscillator (Art. 29.5). When the sawtooth voltage falls abruptly to zero, the beam experiences no horizontal deflection and hence flies back almost instantly to the original (central) position.

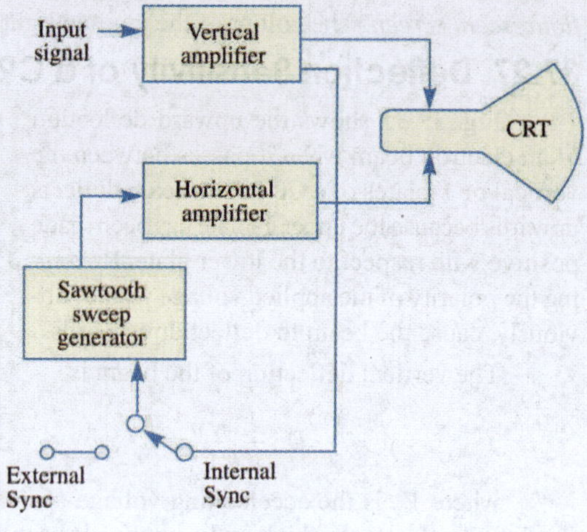

Fig. 37.37

The usual method of synchronizing the input signal is to use a portion of the input signal to trigger the sweep generator so that the frequency of the sweep signal is locked or synchronized to the input signal. It is called internal sync because the synchronization is obtained by internal wiring connection as shown in the block diagram of Fig. 37.37.

37.29. Triggered and Non-triggered Scopes

Oscilloscopes may be classified into two basic types:

1. triggered sweep type, 2. recurrent sweep (free-running) type.

Triggered oscilloscopes, being more sophisticated, are generally used in industrial laboratories and plants, in engineering and technical school laboratories and in all those applications which require study of low- and high-frequency waveforms, for accurate measurement of time and timing relationships etc.

A non-triggered oscilloscope is generally used in servicing work where a certain amount of waveform error can be tolerated and bandwidth requirements are limited to a few MHz.

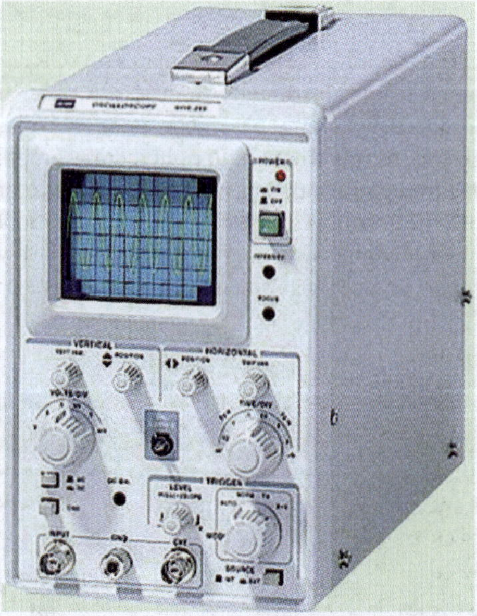

Fig. 37.38. A non-triggered oscilloscope.

The sweep (or ramp) generator which produces sawtooth voltage for *X*-deflection plates is present in both types of scopes. In non-triggered oscilloscopes, this generator runs continuously (recurrent sweep) and the control and calibration of the sweep is based on the repetition frequency of the sweep. For producing a stable stationary display, the sweep frequency has to be forced into synchronization with the input signal on the *Y*-plates. This is done by manually adjusting the free-running sweep frequency to a value very close to signal frequency (or some submultiple of it) and then depending on the internal sync signal (derived from the input) for locking the sweep generator into exact step. Unfortunately, this method is limited to the display of signals which have constant frequency and amplitude. Hence, voice or music signals from a microphone cannot be displayed on this scope because it has to be readjusted for each new change in frequency. Moreover, a free-running or recurrent time base cannot display less than one complete cycle of the input signal on the scope screen. On the other hand, triggered time base can be adjusted to pick out a small part of a waveform which can then be expanded horizontally for evaluation of waveform details.

The triggered oscilloscope is provided with a triggered (or driven) sweep. Here the input signal is caused to generate pulses that trigger the sweep thereby ensuring that the sweep is necessarily in step with the trigger that drives it. Hence, screen display remains stable inspite of variations in the frequency or amplitude of the input signal. It means that there is automatic mode of triggering in such scopes. Consequently, input signals of very short duration can be displayed for the simple reason that sweep is initiated by a trigger pulse derived from the waveform under observation.

Fig. 37.38 illustrates a triggered oscilloscope (Agronic-14) having a bandwidth 0 – 6 MHz, vertical sensitivity of 10 mV/div and horizontal sweep rate varying from 0.2 μs/div to 0.1 s/div.

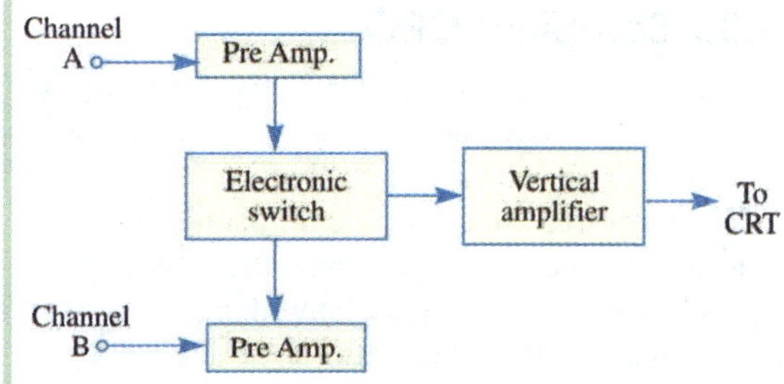

Fig. 37.39

37.30. Dual Trace CRO

Such oscilloscopes are used extensively by industrial firms and research laboratories. They produce a dual-trace display by means of electronic switching of two separate input signals. As shown in the block diagram of Fig. 37.39, there are two vertical input circuits marked channel *A* and *B* with identical preamplifiers. The outputs of these preamplifiers are fed to an electronic switch which alternately connects them to the main vertical amplifier of the oscilloscope. In this type of scope, there is only one electron beam. The electron switch is also capable of selecting a variety of display modes.

In Fig. 37.40 is shown a dual trace oscilloscope (type VOS-26) which has a bandwidth of 5–15 MHz, vertical sensitivity of 10 mV/cm to 30 V/cm and calibrated sweep speed from 0.3 μs/cm to 10 ms/cm. It is a very sensitive yet simple oscilloscope which assures long life and easy maintenance.

37.31. Dual Beam CRO

Such a CRO has two sets of vertical deflection plates and has two electronic beams which produce two separate traces on the scope screen by using the same set of horizontal deflection plates. This scope makes it possible to observe two time-related waveforms at different points *i.e.*, the electronic circuit.

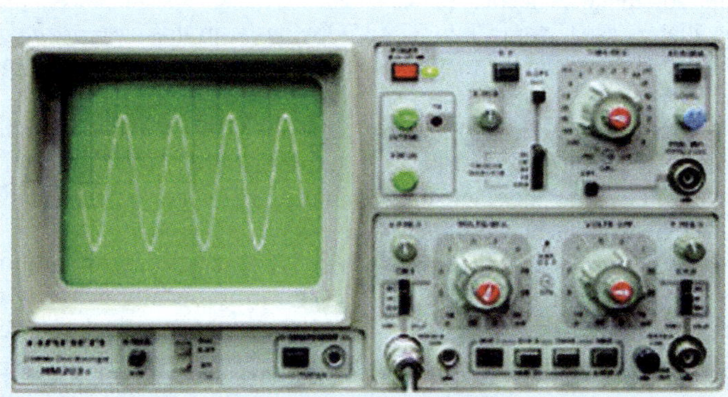

Fig. 37.40. Dual Beam CRO.

Such a scope does not have the same number of display modes as the dual-trace scope yet it is ideally suited for different input signals.

37.32. Storage Oscilloscope

It can retain a CRT display for 10 to 150 hours after the pattern is first produced on the screen. It uses the phenomenon of secondary electron emission to build up and store electrostatic charges on the surface of an insulated target. Such oscilloscopes are especially useful.

1. For real-time observation of events that occur only once.
2. For displaying the waveform of a very low-frequency signal.

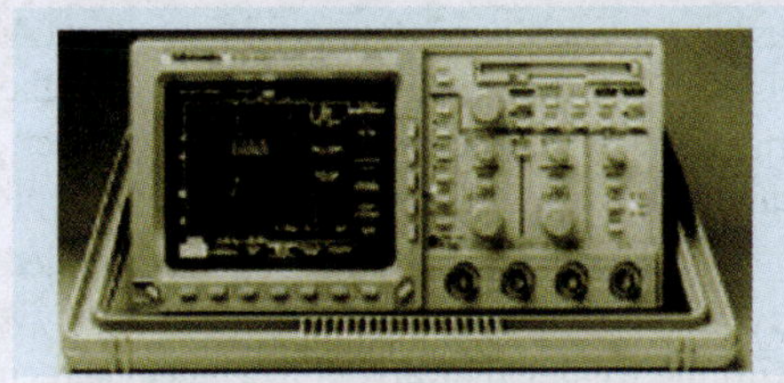

Fig. 37.41. 4 channel storage oscilloscope.

Fig. 37.41 shows a storage oscilloscope having variable storage time of from 30 seconds to 1 hour, a bandwidth of 50 MHz, a sensitivity of 5 mV/cm to 20 V/cm and 21 calibrated speeds from 0.1 μs/cm.

37.33. Sampling CRO

It is specifically meant to observe very high frequency repetitive electric signals by using the sampling technique. Such high-frequency signals cannot be viewed by conventional oscilloscopes because its frequency range is limited by the gain-bandwidth product of its vertical amplifier. The sampling technique 'slows down' the signal frequency many thousands of times thereby making it easier to view it on the screen.

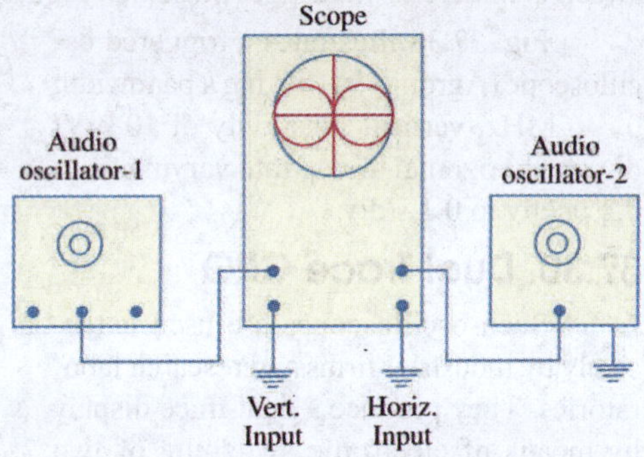

Fig. 37.42

37.34. Digital Readout CRO

It provides digital readout of the signal information such as voltage or time etc., in addition to the conventional CRT display. It consists of a high-speed laboratory CRO and an electronic counter, both contained in one cabinet.

37.35. Lissajous Figures

Lissajous figures (or patterns) are named in honour of the French scientist who first obtained them geometrically and optically. They illustrate one of the earliest uses to which the CRO was put.

Lissajous patterns are formed when two sine waves are applied simultaneously to the vertical and horizontal deflecting plates of a CRO. The two sine waves may be obtained from two audio oscillators as shown in Fig. 37.42. Obviously, in this case, *a sine wave sweeps a sinewave input signal.* The shape of the Lissajous pattern depends on the frequency and phase relationship of the two sine waves.

Two sine waves of the *same* frequency and amplitude may produce a straight line, an ellipse or a circle depending on their phase difference (Fig. 37.43).

In general, the shape of Lissajous figures depends on (*i*) amplitude, (*ii*) phase difference and (*iii*) ratio of frequency of the two waves.

Lissajous figures are used for (*i*) determining an unknown frequency by comparing it with a known frequency, (*ii*) calibrating an audio oscillator with a known-frequency signal and (*iii*)checking audio amplifiers and feedback networks for phase shift.

37.36. Frequency Determination with Lissajous Figures

The unknown signal is applied across one set of deflecting plates and a known signal across the other. By studying the resultant Lissajous pattern, unknown frequency can be found.

Depending on the frequency ratio, the various patterns obtained are shown in Fig. 37.44. The ratio of the two frequencies is given by

$$\frac{f_H}{f_v} = \frac{\text{No. of points of horizontal tangncy}}{\text{No. of points of vertical tangency}}$$

In Fig. 37.44 (*a*), there is one point of tangency along the horizontal as well as vertical axis. Hence, $f_H = f_v$ *i.e.*, the two signals have the same frequency. In Fig. 37.44 (*e*) $f_H / f_v = 3/2$. In other words $f_H = 1.5 f_v$.

It should be noted in passing that this method of frequency determination has limitations and is being discarded gradually because low-cost digital frequency counters are becoming increasingly available in the market (Fig. 37.45). The two main limitations of this method are as under:

(*i*) the numerator and denominator of the frequency ratio must be whole numbers,

(*ii*) the maximum ratio of frequencies that can be used is 10:1. Beyond that, the Lissajous patterns become too complex to analyse.

Fig. 37.45 shows an 8-digit digital frequency meter (Agronic-82) which can measure frequencies from 1 Hz to 10 MHz. It is extensively used in R and D laboratories, testing, servicing and development departments of industries and educational institutions etc.

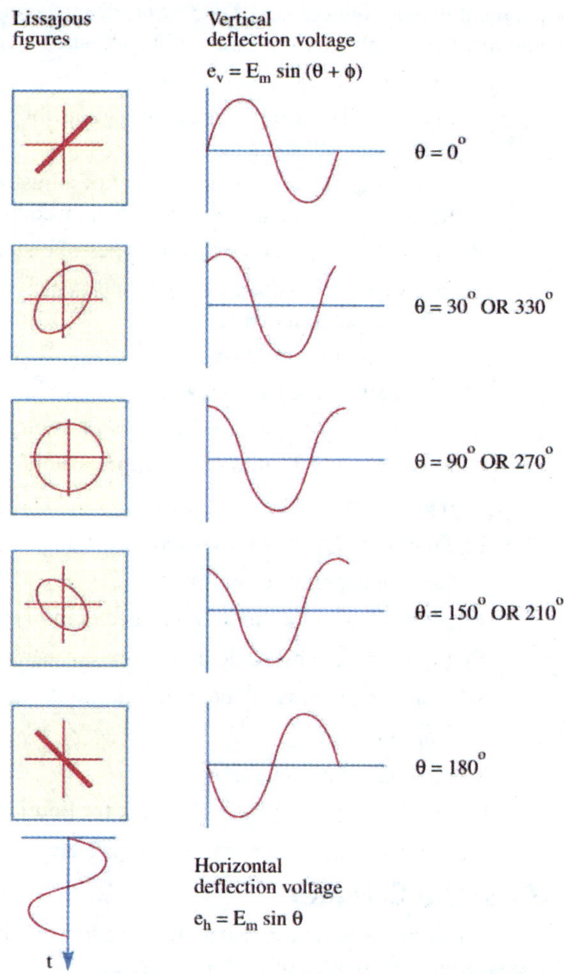

Fig. 37.43

37.37. Applications of a CRO

As stated earlier, no other instrument in electronic industry is as versatile as a CRO. In fact, a modern oscilloscope is the most useful single piece of electronic

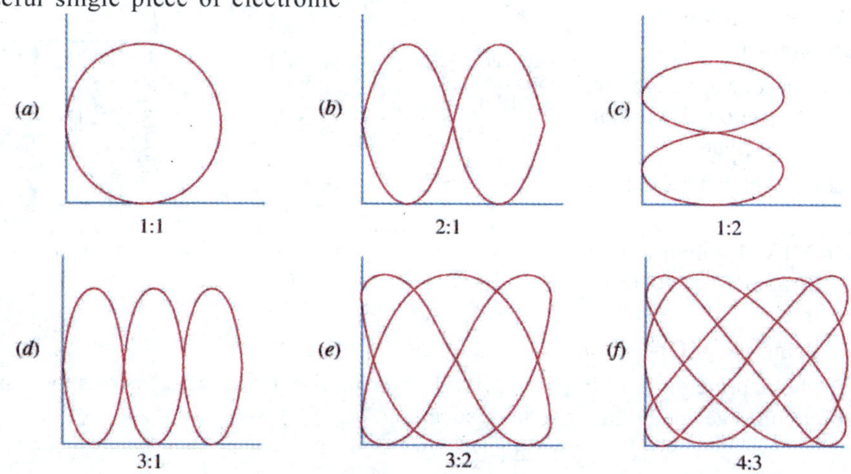

Fig. 37.44

equipment that not only removes guesswork from technical troubleshooting but makes it possible to determine the trouble quickly. Some of its uses are as under:

(*a*) In Radio Work

1. to trace and measure a signal throughout the RF, IF and AF channels of radio and television receivers;
2. it provides the only effective way of adjusting FM receivers, broadband high-frequency RF amplifiers and automatic frequency control circuits;
3. to test AF circuits for different types of distortions and other spurious oscillations;
4. to give visual display of waveshapes such as sine waves, square waves and their many different combinations;
5. to trace transistor curves;
6. to visually show the composite synchronized TV signal;
7. to display the response of tuned circuits etc.

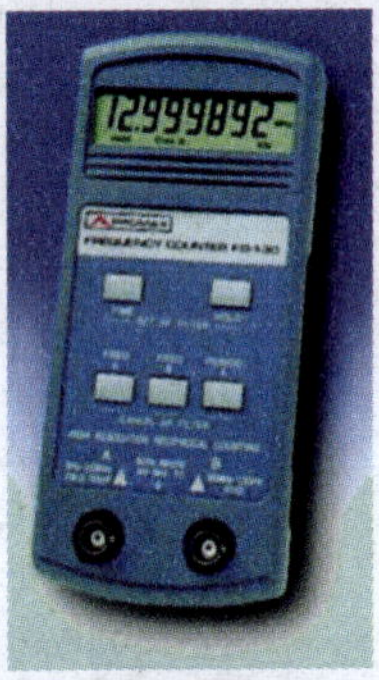

Fig. 37.45

(*b*) Scientific and Engineering Applications

1. measurement of ac/dc voltages,
2. finding B/H curves for hysteresis loop,
3. for engine pressure analysis,
4. study of stress, strain, torque, acceleration etc.,
5. frequency and phase determination by using Lissajous figures,
6. radiation patterns of antenna,
7. amplifier gain,
8. modulation percentage,
9. complex waveform as a short-cut for Fourier analysis,
10. standing waves in transmission lines etc.

37.38. The Q Meter

This instrument is designed to measure some of the electrical properties of coils and capacitors by measuring the *Q*-value of an *R-L-C* circuit.

1. Construction

As shown in Fig. 37.46, it essentially consists of

(*i*) a frequency-calibrated continuously-variable RF oscillator,

(*ii*) a calibrated variable capacitor *C*,

(*iii*) VTVM which is calibrated to read *Q* directly.

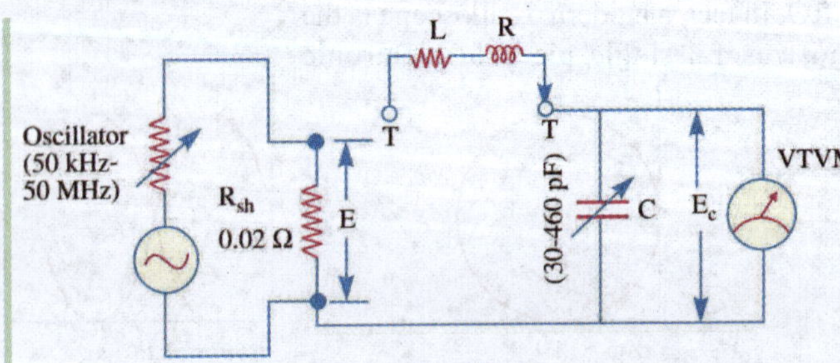

Fig. 37.46

2. Principle of Operation

The basic principle used in *Q*-meter is the resonant rise of the voltage across the capacitor in an *R-L-C* circuit. The condition for series resonance (Art. 19.7) is

$$X_L = X_C \quad \text{and} \quad E = IR$$

The value of circuit Q is

$$Q = \frac{x_L}{R} = \frac{X_C}{R} = \frac{E_C}{E}$$

If the applied voltage E is constant, then $Q \propto E_C$. Hence, by measuring voltage drop across C under resonant conditions, Q can be found. Alternatively, VTVM can be calibrated in terms of Q (rather than voltage).

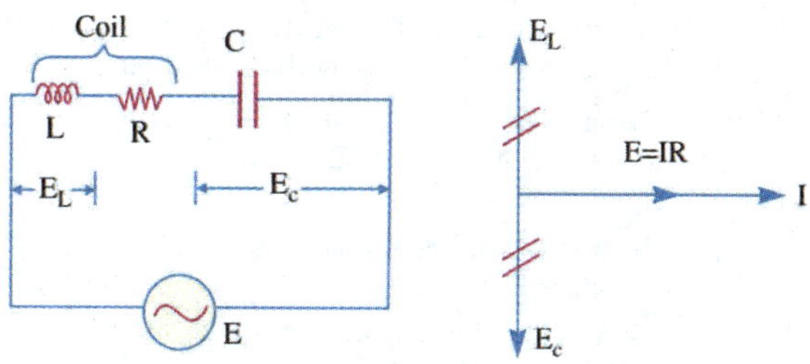

Fig. 37.47

As seen from Fig. 37.46, the oscillator delivers current to a very small (0.02 Ω) shunt resistance R_{sh} thereby developing a voltage E across it. It becomes the applied voltage for the RLC circuit and corresponds to the source voltage E of Fig. 37.47.

It is measured by a thermocouple meter marked 'Multiply Q by'. The voltage E_C across variable C is measured by the VTVM. The value of Q-factor is given by $Q = E_C / E$.

For making measurement, the unknown coil is connected to the test terminals TT of the instrument and the circuit is tuned to resonance

(*i*) either by setting the oscillator to a given frequency and varying C,

(*ii*) or by keeping value of C constant and adjusting the oscillator frequency.

The reading on the VTVM must be multiplied by the index setting of the 'Multiply Q by' meter in order to obtain the actual value of Q.

3. Applications

Some of the specialized uses of this instrument are to measure

1. Q of a coil,
2. inductance and capacitance,
3. distributed capacitance of a coil,
4. Q and p.f. of a dielectric material,
5. mutual inductance of coupled circuits,
6. coefficient of coupling,
7. critical coupling,
8. reactance and effective resistance of an inductor at operating frequency,
9. bandwidth of a tuned circuit etc.

The above list does not exhaust the number of its possible applications. It has been very truly said that if ever an RF problem exists, a Q meter can always provide the answer.

SELF EXAMINATION QUESTIONS

A. Fill in the blanks with most appropriate word(s) or numerical value(s).

1. The deflection type instruments with a scale and a pointer are called .... instruments.
2. Instruments serve three basic functions of indicating, recording and ....
3. Digital instruments use .... circuits and techniques to obtain measurements.
4. As compared to analog instruments, digital instruments have easy readability and higher ....
5. Input impedance of an electronic instrument is very ....
6. The three essential elements of an electronic instrument are : a .... , signal modifier and an indicating device.

7. Higher the full-scale deflection current of a meter movement, ... its internal resistance.
8. The sensitivity of a meter movement is given by the .... of its full-scale deflection current.
9. A meter movement with full-scale deflection current of 50 μ A has a sensitivity of .... kΩ/V.
10. A high input-resistance voltmeter has .... loading effect.
11. In general, the voltage scales on a VTVM are linear whereas ohms scale is ....
12. When resistance is to be measured, all power in the external circuit must be turned ....
13. A VOM is used for measuring voltage, current and ....
14. An ac instrument using a half-wave rectifier reads .... per cent of the dc value.
15. The input resistance of a VOM is .... on each range.
16. A VTVM has .... loading effect on the circuit under test.
17. The CRO provides a .... dimensional visual display of the signal waveshape on its screen.
18. The heart of an oscilloscope is ....
19. In a 'scope', the signal to be observed is applied across .... deflecting plates.
20. The voltage applied across the X-plates of a 'scope' is a .... voltage.
21. The deflection sensitivity of a CRT is defined as the vertical deflection of the beam per unit .... voltage.
22. For viewing two cycles of the input signal, sweep-generator frequency in a 'scope' must be .... the signal frequency.
23. In internal sync of a 'scope', a portion of the input signal is used to trigger the .... generator.
24. In a triggered type 'scope', sweep generator runs only when input signal is .... across the *Y*-plates.

B. Answer True or False

1. Instruments with numerical readout are called digital instruments.
2. Electronic instruments can amplify the input signal whereas electrical instruments cannot.
3. Digital meters are used where speed and ease of reading are of prime consideration.
4. Current sensitivity of a meter depends on the current passing through it.
5. A 50 μA meter movement has a sensitivity of 20 kΩ/V.
6. The loading effect of a voltmeter is determined by its ohms-per-volt rating.
7. Larger the sensitivity of the voltmeter, lesser the loading effect on the circuit under test.
8. Electronic voltmeters have sensitivities in the megohm range.
9. All non-digital ohmmeters have to be zeroed before each measurement.
10. A linear scale is one which is represented by a straight line.
11. Zeroing a voltmeter is not necessary because it will measure whatever is applied across its input terminals.
12. The OHMS ADJUST control of a VOM must be checked each time the resistance range is changed.
13. Polarity is not important when measuring resistance.
14. A VTVM has negligible loading effect on the circuit under test because its input resistance is very high.
15. The input resistance of a FET VM can be as high as 100 MΩ.
16. A digital voltmeter has higher resolution than an analog one.
17. A CRO permits an electronic engineer to 'see' what is happening inside an electronic circuit.
18. A CRO presents a 3-dimensional visual display at the signal waveshape.
19. CRT is the heart of a CRO.
20. The input signal to be viewed is applied across the X-plates of a CRT.
21. The deflection sensitivity of a CRT is expressed in volt/meter.
22. In a scope, horizontal sweeping of the beam is synced with the input signal.
23. When sweep voltage frequency is twice the input signal frequency, we will see two complete signal cycles on the screen.
24. In a non-triggered oscilloscope, the time-base generator runs continuously.

C. Multiple Choice Items

1. The main difference between the electronic and electrical instruments is that an electronic instrument contains
   (*a*) an electronic device
   (*b*) a transducer

(*c*) a digital readout
(*d*) electrons

2. The essential elements of an electronic instrument are
(*a*) transducer
(*b*) signal conditioner
(*c*) indicating device
(*d*) all of the above

3. The current sensitivity of a meter is expressed in
(*a*) ampere (*b*) ohm/ampere
(*c*) ohm/volt (*d*) ampere/division

4. The basic meter movement can be converted into an ohmmeter by connecting a ...... with it.
(*a*) high resistance in series
(*b*) low resistance in parallel
(*c*) battery in series
(*d*) battery and a variable resistance in series

5. The D'Arsonval meter movement can be converted into an audio-frequency ac ammeter by adding a ...... to it.
(*a*) thermocouple (*b*) rectifier
(*c*) chopper (*d*) transducer

6. In a linear meter, half-scale deflection occurs when there is ...... per cent of the rated current through its coil
(*a*) 100 (*b*) 25
(*c*) 50 (*d*) 75

7. A 0–1 mA meter has a sensitivity of
(*a*) 1 kΩ/V (*b*) 1 mA
(*c*) 1 kΩ (*d*) 1000 A

8. Loading effect is principally caused by ...... instruments.
(*a*) high resistance
(*b*) low-sensitivity
(*c*) high-sensitivity
(*d*) high-range

9. A multimeter is used to measure
(*a*) resistance (*b*) current
(*c*) voltage (*d*) all of the above

10. A sinusoidal voltage of rms value 10 V is applied to a D'Arsonval movement connected in series with a half-wave rectifier. It will show a reading of ...... volt
(*a*) 9 (*b*) 4.5
(*c*) 10 (*d*) 7.7

11. A VTVM produces negligible loading effect on a circuit under test primarily because
(*a*) it virtually draws no current from the circuit
(*b*) of its very high internal resistance
(*c*) it uses high vacuum tubes
(*d*) it is a null deflection instrument

12. The signal to be observed on the screen of an oscilloscope is applied
(*a*) across its *X*-plates
(*b*) across its *Y*-plates
(*c*) to the horizontal amplifier
(*d*) to the trigger circuit

13. Production of a steady stationary display of a signal waveform on the scope screen is due to
(*a*) persistence of vision
(*b*) flourescent material of the screen
(*c*) proper sync between the signal and the sweep generator
(*d*) electrostatic focussing of the electron beam.

14. Two complete signal cycles would be displayed on the screen of a scope when time-period of the sweep generator is ...... the signal time period.
(*a*) half (*b*) twice
(*c*) equal (*d*) thrice

15. A non-triggered oscilloscope is one which
(*a*) has no sweep generator
(*b*) cannot produce a stable stationary screen display
(*c*) has a continuously running time-base generator
(*d*) can display a portion of the input signal waveform

16. The operation of a *Q*-meter is based on
(*a*) self-induction
(*b*) series resonance
(*c*) mutual induction
(*d*) eddy currents

ANSWERS

A. Fill in the blanks

1. analog	2. controlling	3. logic	4. accuracy	5. high
6. transducer	7. lower	8. reciprocal	9. 20	10. less
11. non-linear	12. OFF	13. resistance	14. 45	15. different
16. negligible	17. two	18. CRT	19. Y	20. ramp
21. deflecting	22. half	23. sweep	24. present	

B. True or False

1. F	2. T	3. T	4. F	5. T	6. T
7. T	8. T	9. T	10. F	11. F	12. T
13. T	14. T	15. T	16. T	17. T	18. F
19. T	20. F	21. F	22. T	23. F	24. T

C. Multiple Choice Items

1. *a*	2. *d*	3. *c*	4. *d*	5. *b*	6. c
7. *a*	8. *b*	9. *d*	10. *b*	11. *b*	12. *b*
13. *c*	14. *b*	15. *c*	16. *b*		

CHAPTER 38

Fibre Optics

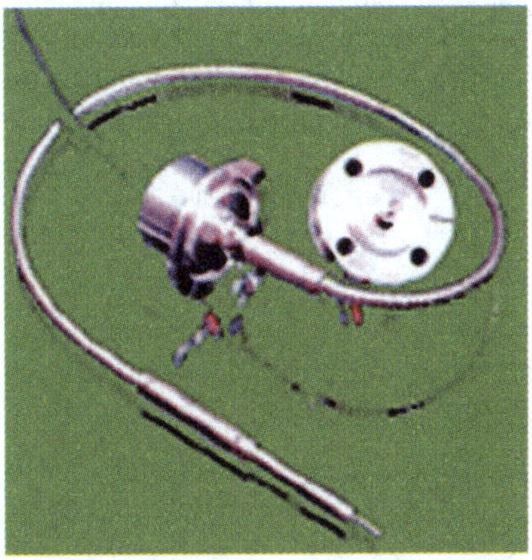

1. Structure of Optical Fibres
2. Classification of Optical Fibres
3. Plastic Fibres
4. Refraction and Snell's Law
5. Total Internal Reflection
6. Light Propagation through an Optical Fibre
7. Acceptance Angle and Numerical Aperture
8. Intermodal Dispersion
9. Intramodal Dispersion
10. Calculation of Losses
11. Optical Fibre Cable
12. Multifibre Cable
13. Splicing and Connectors
14. Fusion Splices
15. Mechanical Splices
16. Connectors
17. Connection Losses
18. Fabrication of Optical Fibres

38.1. Fibre Optics

About one and a half decade ago, most electronic com munication was carried out with the help of copper cables which were in the form of twisted pairs, co-axial cables or copper waveguides. Communication was achieved by sending electrical signals through the copper cable. In recent years, a new medium called optical fibres OF has been introduced. In OF communication, light signals are used instead of electrical signals.

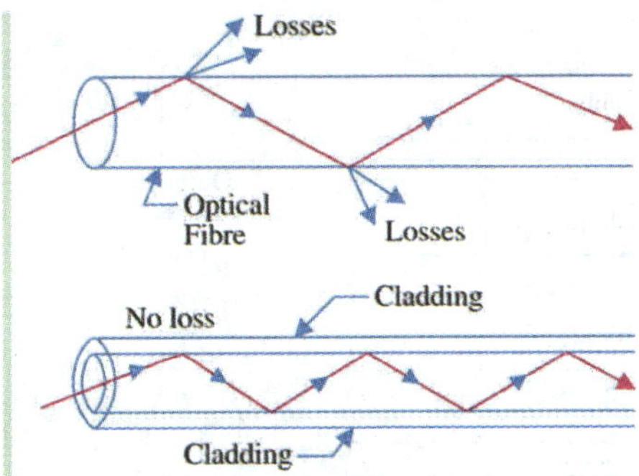

Fig. 38.1

An OF is a transparent rod usually made of glass or clear plastic through which light can propagate. The light signal travels through the rod from the transmitter to the receiver and can be easily detected at the receiving end of the rod. A modern fibre consist of an optical rod core coated with a cladding. The refractive index of the rod is higher than the refractive index of the cladding meterial in order to utilize the phenomenon of total internal reflection for the propagation of the light through the rod. The light used is coherent light and not the ordinary composite light.

Fig. 38.1 (*a*) shows an uncladded glass fibre and a claded fibre through which the light travels. As seen, there is loss of light to the surroundings in the uncladed fibre whereas there is no such loss in the claded fibre of Fig. 38.1 (*b*).

38.2. Structure of Optical Fibres

An optical fibre essentially consists of a very thin fibre at its centre surrounded by cladding which is made of a material less denses than the fibre material as shown in Fig. 38.2 (*a*).

The fibres are drawn from a furnace containing molten silica (SiO_2) with small amounts of additives such as GeO_2 to permit the control of refractive index. Most of the cable bulk is made up of strengthening and buffering materials for mechanical, moisture and chemical protection. Such a fibre is shown in Fig. 38.2 (*b*).

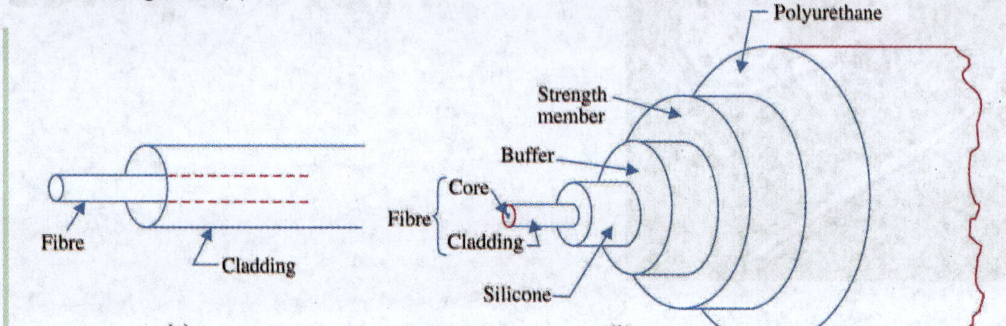

Fig. 38.2

38.3. Classification of Optical Fibres

Optical fibres can be classified into three groups according to the way light propagates down the fibre core.

1. Multimode step-index fibre,
2. Single-mode step-index fibre, and
3. Multimode graded-index fibre.

The mode refers to the path the light will travel through the fibres. One of the basic differences among different types of fibres is their refractive index profile.

The three structures are illustrated in Fig. 38.3 (*a*). For overall performance in terms

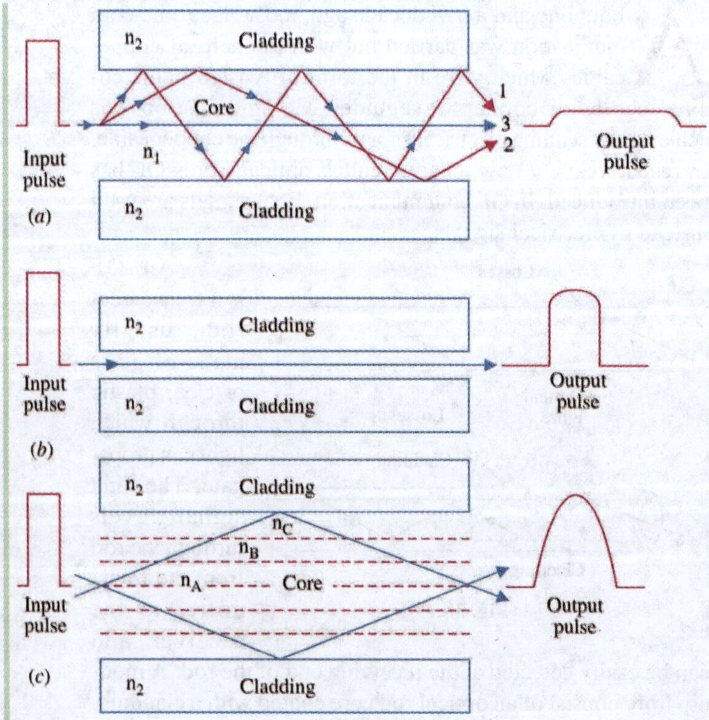

Fig. 38.3

of data rate and attenuation, the single-mode fibre has the best characteristic *i.e.,* it has highest data rate and least attenuation. However, it is the most costly and the most difficult to work with because of its extremely small size (the core is only 5.10 μm in diametre). In addition, the small size makes it difficult to inject light energy into the cable which makes it essential to use laser transmitter devices. The multimode fibre has the worst characteristics.

Fig. 38.3 (*a*) shows a multimode step-index fibre. The core is made of glass or plastic of a single index value and the cladding material also has a single index value. There is a single step between the two index values *i.e.,* between the refractive indices of the core and the cladding material and this difference in index value is relatively large. It means that the critical angle is small; hence there will be many paths available for light to travel down the fibre. In Fig. 38.3 (*a*), three light rays have been shown travelling along three paths or modes. Since they travel in the fibre core of constant index, they travel with the same velocity. But due to the difference in their path lengths, they arrive at the output at different times. Ray 3 travels the shortest distance and arrives at the output first. On the other hand, ray 1 travels the farthest and is the last to arrive at the output. This spreading of arrival times at the output causes a distortion known as modal dispersion. The effect of modal dispersion is shown in Fig. 38.3 (*a*). As seen, a narrow pulse of light energy at the input produces an attenuated output pulse dispersed over a wide period time. This modal dispersion limits the data rate that the optic fibre cable can handle. As the data rate goes up, the input pulses become narrower and closer together. Because of modal dispersion, the output pulses start to run together at higher data rates.

Modal dispersion can be removed by using either single-mode step-index fibres or multi-mode graded-index fibres. The single-mode step-index fibre uses a different index with two changes. First, the difference in the indexes between the core and the cladding is small. This causes a large critical angle (almost 90°) which means that the light rays must travel almost parallels with the junction of the core and the cladding. Second, the size of the core, is greatly reduced so that distance from side to side is small and all light paths through the core are practically of the same length. The nett effect of these two changes is that all light rays travelling through the core travel the same distance thereby reducing modal dispersion to a negligible amount as shown in Fig. 38.3 (*b*). Obviously such fibres can handle much higher data rates as compared to multimode step-index fibre.

Fig. 38.3 (*c*) shows a multimode graded-index fibre, whose core is made with its centre having an index of refraction which decreases with distance from the centre. In Fig. 38.3 (*c*), centre of the fibre core has the highest index of n_A such that $n_A > n_B > n_C$. As we know, higher the index of material, slower the velocity of light travelling through it. The light rays travelling through the middle part will travels the slowest and those near the edge will travel the fastest. Hence, all light rays travelling through different parts of the fibre reach the output at the same time. Since light rays enter together and also exit together, the modal dispersion is prevented.

The performance of multimode graded-index fibres lies between that of multimode step-index fibres and single-mode step-index fibres. They are also in the middle when it comes to price and workability.

38.4. Plastic Fibres

We must distinguish between plastic-clad and all-plastic fibres. In the former, only the cladding is plastic, while in the latter, both the core and cladding are plastic. The characteristics of a plastic-clad fibre are similar to those of an all-glass fibre except that the plastic-clad fibre is much more sensitive to abrasive damage. Moreover, an all-plastic fibre is stronger mechanically but has poorer transmission characteristics. Its loss is much higher and its bandwidth is lower than a glass fibre. Hence, it is rarely used in commercial communication systems. Since it is easier to handle and does not break as easily, it is often used for short-distance low BW applications such as closed circuits television (CCT) and demonstration systems etc.

38.5. Propagation of Light

Light is an electromagnetic wave which travels in vacuum $c = 3\times10^8$ m/s. This velocity is independent of frequency. However, the velocity of light propagation in a medium depends on the particular material and is always less than c. Most light consist of a group of frequencies rather than a wave with a signal frequency. For example, the visible light of a typical incandescent lamp contains blue light (0.6×10^{15} Hz) to red light (0.4×10^{15} Hz) and lower.

If we call the group velocity of light waves through a certain medium as v, then $v/c = n$ where n is called the refractive index of that material. When a light wave travels through a medium, its wavelength changes but not its frequency. If λ is the light wavelength in vacuum, (or air), then the wavelength λ in a material having refractive index of n can be found as under:

In vacuum, $\lambda = c/f$, in the medium, $\lambda' = v/f$

$$\therefore \qquad \lambda'/\lambda = v/c \text{ or } \lambda' = \lambda \times v/c = n_1\lambda$$

It may be noted that the refractive index of a material varies with the wavelength of light passing through it. For example, flint glass has a refractive index of 1.66 at

$$\lambda = 400 \text{ nm and } 1.62 \text{ at } \lambda = 700 \text{ nm}$$

Example 38.1. *Compute the velocity and wavelength of light of frequency $f = 0.5 \times 10^{15}$ Hz when travelling through glass having a refractive index of $n_1 = 1.5$.*

Solution. $v = c/n_1 = 3 \times 10^8 / 1.5 = \mathbf{2 \times 10^8}$ **m/s**

$\lambda = v/f = 2 \times 10^8 / 0.5 \times 10^{15} = \mathbf{400}$ **nm**

38.6. Refraction and Snell's Law

According to the theory of geometrical optics, light can be considered as consisting of narrow rays. When such a light ray travels from one medium to another having different refractive indexes of n_1 and n_2 both reflection and refraction take place as shown in Fig. 38.4.

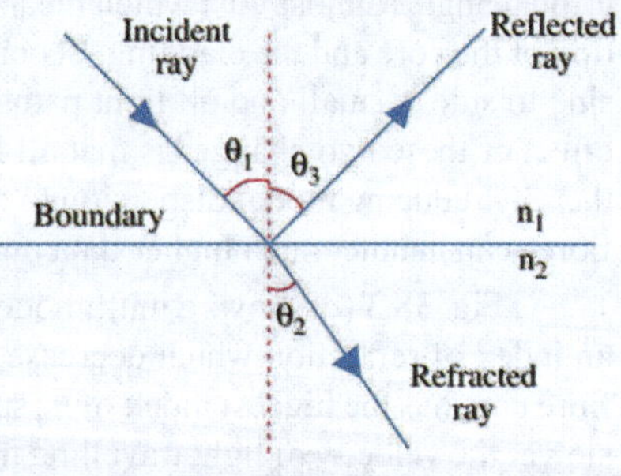

Fig. 38.4

In the case of reflection $\theta_1 = \theta_3$ but in the case of refraction, θ_2 is less than θ_1 because light is going from a rare to denser medium *i.e.*, n_2, n_1. The relation between θ_1 and θ_2 is given by Snell's law of refraction, according to which

$$n_1 \sin\theta_1 = n_2 \sin\theta_2 \text{ or } \sin\theta_1 / \sin\theta_2 = n_2 / n_1$$
$$= v_1 / v_2$$

The amount of light reflected depends on the angle of incidence and refractive indices. For incident angles smaller than 30º, the reflected light is less than 5% of the incident light. Approximate fraction of light that reflected for an incident angle 0º is given by

$$\text{loss} = [(n_1\ n_2)/(n_1 + n_2)]^2$$

Example 38.2. *Find the per cent of power loss when light travels from air (n=1) and water (n = 1.33). Assume a 0º incident angle.*

Solution. Loss $= [(n_1 - n_2)/(n_1 + n_2)]^2 = [(1.33 - 1)/(1.33 + 1)]^2 = 0.02 =$ **2%**

38.7. Total Internal Reflection

A ray of light travelling in a medium of higher refractive index (n_1) and directed towards one of lower refractive index (n_2) passes into the medium only if the angle of incidence is not too large. If the angle of incidence is increased, a limit is reached when the ray does not enter the second medium at all but is totally reflected at the interface.

Consider a light source O lying in a medium of higher refractive index (n_1) and emitting light rays in all directions (Fig. 38.5). First consider ray OA, striking the interface at an angle of incidence θ_1. It refracts into (optically rarer medium) with an angle of refraction n_2. Obviously, $r_1 > \theta_1$.

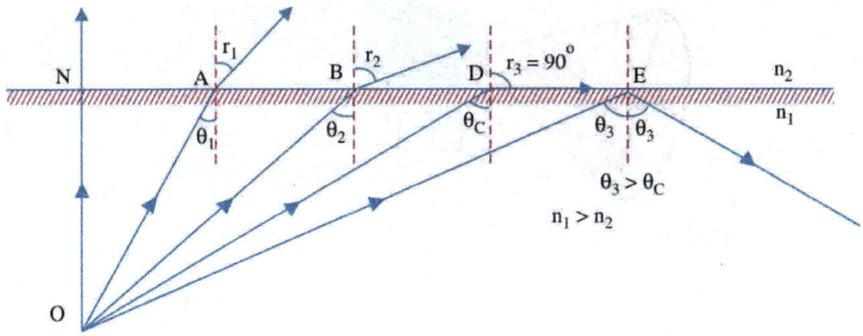

Fig. 38.5

The next ray OB is incident at a greater angle and refracts at still greater angle n_2. It will be seen that as the angle of incidence is increased the angle of refraction is also increased. The ray OD is incident at such an angle that the corresponding angle of refraction has become 90°. Hence, the refracted ray DE grazes along the boundary between the two media.

This angle of incidence for which the corresponding angle of refraction is 90° is called the angle of critical incidence for the two media and is denoted by θ_c in the figure.

If any other ray, say, ray OE is now incident at an angle θ_3 greater than the angle of critical incidence θ_c, then its angle of refraction will be greater than 90° hence, it will turn back into the first medium. This ray is said to suffer total internal reflection (not refraction). It is found that this totally internally reflected ray obeys the ordinary laws of reflection.

In the case of ray OD

angle of incidence $= \theta_c$, angle of refraction = 90°

$$\therefore \quad \frac{\sin \theta_C}{\sin 90°} = \text{ref. index of second medium w.r.t. to the first medium}$$

$$= \frac{n_c}{n_1} \quad \text{or} \quad \sin \theta_c = n_2 / n_1; \;\; \theta_c = \sin^{-1}(n_2/n_1)$$

Summing up the conditions for total internal reflection, we have

(*i*) the ray of the light must travel from a medium of higher refractive index to a medium of lower refractive index

(*ii*) the angle of incidence at the interface must be greater than the angle of critical incidence for the two media concerned.

38.8. Light Propagation through an Optical Fibre

Let us consider a fibre having a core of uniform refractive index n_1 surrounded by a cladding of uniform refractive index n_2 [Fig. 38.6 (*a*)]. The fibre core carries the light signal which is confined within the core by the total internal reflection from the cladding. As seen, the incident ray in the core makes an angle θ_1 with the normal to the axis while the refracted ray in the cladding makes an angle θ_2 with it. According to Snell's law, $n_1 \sin \theta_1 = n_2 \sin \theta_2$. If the incident ray is to be reflected back into the core (rather than be refracted into the cladding), angle of refraction θ_2 must be equal to or greater than 90° as shown in Fig. 38.6 (*b*). Substituting $\sin \theta_2 = 1$ in the above equation, we get $n_1 \sin \theta_c \geq n_2$ or $\theta_c \geq \sin^{-1}(n_2/n_1)$. For total internal reflection to take place, cladding has a lower refractive index than the core. If, for example, $n_1 = 1.54$ and $n_2 = 1.51$, $\theta_c = \sin(1.51/1.54) = 79°$.

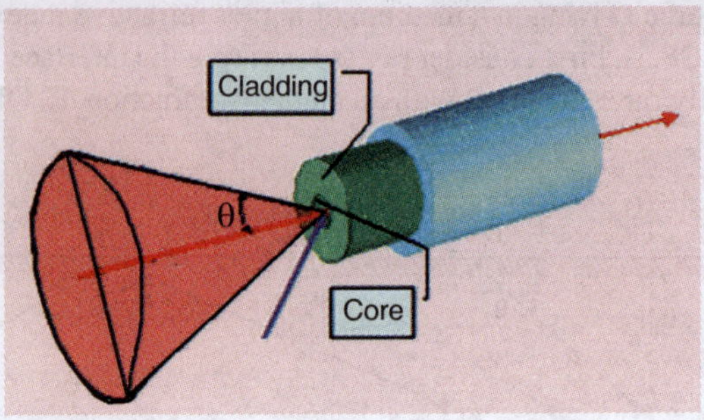

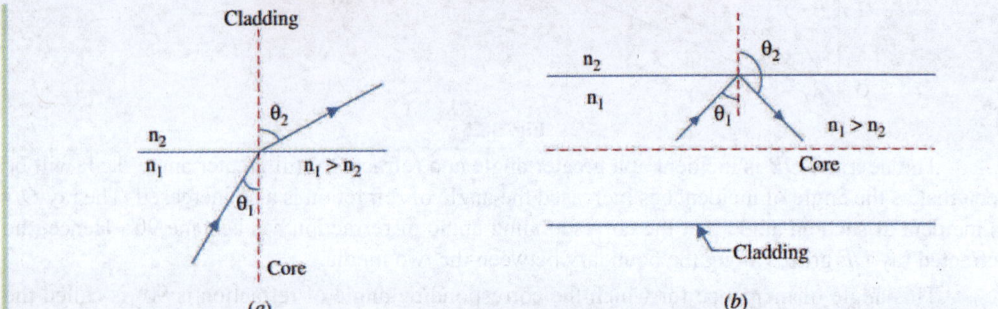

Fig. 38.6

38.9. Acceptance Angle and Numerical Aperture

It can be shown that a light ray can enter an optic fibre only if it lies within a cone having a semiangle less than a certain maximum value known as half acceptance angle α_{max}. As shown in Fig. 38.7, a light ray is entering a fibre from air ($n = 1$) at a angle of α to the axis. On entering the fibre at point A, the ray is refracted as per Snell's law according to which

$$1 \times \sin\alpha = n_1 \sin(90^\circ \times \theta_1)$$

or, $$\sin\alpha = n_1 \cos\theta_1$$

where θ_1 is the angle of incidence at the cladding. Since ray has suffered total internal reflection, $\theta_1 > \theta_c$. According to the laws of total internal reflection, the light ray would be reflected from the cladding back into the core only when $\theta_1 \geq \theta_c$. Let the maximum value of α which makes $\theta_1 = \theta_c$ be designed as α_{max}. Obviously, minimum value of θ_1 is θ_c and this angle gives the corresponding maximum value of α.

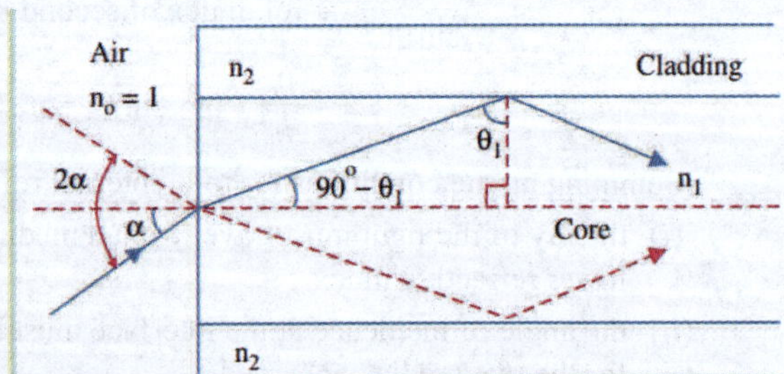

Fig. 38.7

In that case, $\sin\alpha_{max} = n_1 \cos\theta_c$

Substituting the value of $\cos\theta_c = \sqrt{1 - \sin^2\theta_c}$,

we get $$\sin\alpha_{max} = n_1\sqrt{1 - \sin^2\theta_c} = \sqrt{n_1^2 - n\sin_1^2\theta_c}$$

$$= \sqrt{n_1^2 - n_2^2}.$$

The value of $\sin\alpha_{max}$ is called the numerical aperture (NA). Hence, NA = $\sin\alpha_{max} =. \sqrt{n_1^2 - n_2^2}$. Because $n_1 \cong n_2$, the acceptance angle is small. For example, if $n_1 = 1.54$ and $n_2 = 1.51$, $\alpha_{max} = 18^\circ$.

Because of this small value, accurate alignment of lasers and fibres is needed to make sure that most of the light enters the fibre.

The value of the NA can be expressed in terms of the relative refractive index difference Δ which is defined as

$$\Delta = \frac{n_1^2 - n_2^2}{2 \times n_1^2} = \frac{(NA)^2}{2\, n_1^2}$$

$$\therefore \quad (\text{NA})^2 = n_1^2 - n_2^2 = 2n_1^2 \times \Delta \quad \text{or} \quad \text{NA} = n_1 \times \sqrt{2\,\Delta}$$

An approximate expression of Δ can be obtained as under:

$$\Delta = \left(n_1^2 - n_2^2\right)/\left(2 \times n_1^2\right)$$

$$= [(n_1 + n_2) \times (n_1 - n_2)] / (2 \times n_1^2)$$

If n_1 is approximately equal to n_2, then

$$= 2 \times n_1 \times (n_1 - n_2) / (2 \times n_1^2)$$

$$= (n_1 - n_2) / n_1 = (n_1 - n_2) / n_2$$

Example 38.3. *An optic fibre has the following characteristics:*
Fibre index, $n_1 = 1.36$ *and* $\Delta = 0.025$
Calculate the value of the numerical aperture and the acceptance angle.

Solution. $\text{NA} = n_1 \times \sqrt{2\,\Delta} = 1.36 \times \sqrt{(2 \times 0.025)} = \mathbf{0.3}$

$$\alpha = \sin^{-1}(\text{NA}) = \sin^{-1}(0.3) = 17.7^\circ$$

Since half-acceptance angle is 17.7°, the full acceptance angle is $2\alpha = 2 \times 17.7^\circ = \mathbf{35.4^\circ}$

Example 38.4. *In a step-index fibre, core index* $n_1 = 1.45$ *and cladding index* $n_2 = 1.41$*. Find (i) NA (ii) Δ and (iii) full acceptance angle.*

Solution. (*i*) $\text{NA} = \sqrt{1.45^2 - 1.41^2} = \sqrt{2.1025 - 1.9881}$

$$= \sqrt{0.1144} = \mathbf{0.338}$$

(*ii*) $\Delta = (\text{NA})^2 / 2\, n_1^2 = 0.1144 / 2 \times 1.45^2 = 0.039 = \mathbf{3.9\%}$

(*iii*) $\alpha = \sin^{-1}(0.338) = \mathbf{19.75^\circ}$

38.10. Dispersion

The term dispersion describes the pulse broadening effect in fibres. The dispersion is produced by the properties of the core material and the line width of the light signal passing through the fibre. As shown in Fig. 38.8 the pulse that appears at the output of a fibre is wider than the input pulse. Dispersion may, therefore, be defined as the output light pulse width produced by an input pulse of zero line width *i.e.,* a monochromatic light which has single wavelength. In that case, pulse width available at the output would be totally a result of fibre dispersion. Let us take a practical case of an input pulse of width W_1, and an output pulse of width W_2 with $W_2 > W_1$. The fibre dispersion dT is defined as

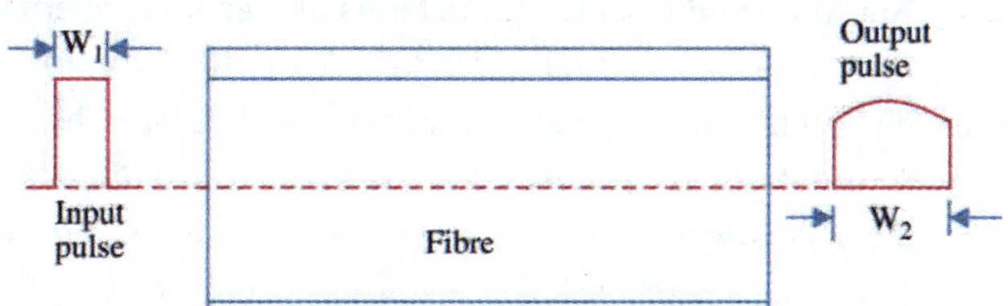

Fig. 38.8

$$\text{dT} = \sqrt{(W_2^2 - W_1^2)}$$

Dispersion is measured in units of time either in nanoseconds (10^{-9} s) or picoseconds (10^{-12}s).

The total dispersion produced by a fibre depends directly on its length. Hence, manufacturers give the dispersion per unit length of the fibre either in nanoseconds per kilometre (ns/km) or picoseconds per kilometre (ps/km). This permits the users to calculate the expected total dispersion for a given length of fibre. If L is the length of the fibre in kilometre, then the total fibre dispersion is given by

$$dT = L \times (\text{dispersion / km})$$

The dispersion can be divided into two general categories (*i*) intermodal dispersion and (*ii*)intramodal dispersion.

38.11. Intermodal Dispersion

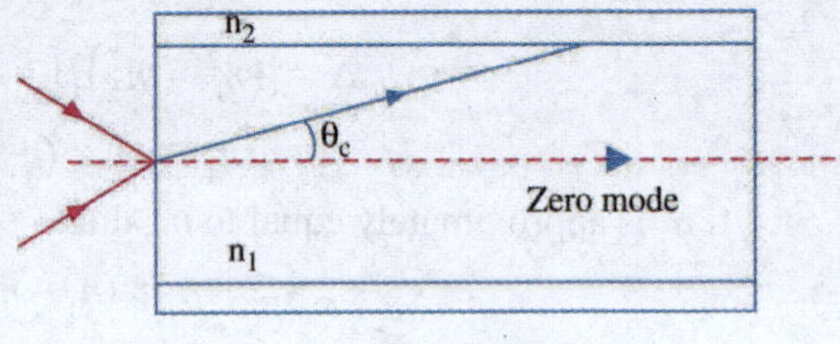

Fig. 38.9

This type of dispersion results from the fact that the light wave propagates through a fibre in different modes as explained in Art. 38.3. There is a difference in the propagation times for the different modes which produces dispersion between these different modes. For full understanding of this dispersion, consider two extreme modes as shown in Fig. 38.9 the critical mode propagating at angle θ_c and another mode where the propagation angle is zero *i.e.,* zero mode. A pulse of light launched into the fibre will propagate along the fibre in both modes.

For zero mode, travel time will be minimum and is given by

$$T_{do} = L / v_1 = L / (c / n_1) = L \times n_1 / c$$

where v_1 is the velocity of light through the core material, *c* is the velocity of light in vacuum and L is the length of the fibre. For the light rays travelling at the critical angle θ_c, the delay would be maximum and is given by

$$I_{dc} = (L / \cos\theta_c) / v_1 = (L / \cos\theta_c) / (c / n_1) = (L \times n_1) / (\cos\theta_c \times c)$$

the time difference $(T_{dc} - T_{do})$ represents the pulse width dT at the output

$$\therefore \quad dT = T_{dc} - T_{do} = (L \times n_1) / (\cos\theta_c \times c) - L \times n_1/c$$

Using the relation $\cos\theta_c = n_2 / n_1$, we get

$$dT = [(L \times n_1)/c)] \times \{(n_1 \times n_2)/n_2]$$

where n_1 and n_2 are the refractive indices of the core and cladding respectively.

As given earlier in Art. 38.9, the expression $(n_1 - n_2) / n_2 = \Delta$. Hence, the pulse width can be expressed as

$$dT = (L \times n_1/c) \times \Delta$$

Substituting the value of Δ in terms of numerical aperture (Art. 38.9), we get

$$dT = [L \times (N.A.)^2] / (2 \times n_1 \times c) \quad \text{or} \quad dT / L = (N.A.)^2 / (2 \times n_1 \times c)$$

where dT is the amount of pulse broadening and dT/L is the broadening per unit length of the fibre.

Example 38.5. *A train of light pulses is transmitted through a 500 m fibre with $n_1 = 1.4$ and $n_2 = 1.35$. Find the total dispersion assuming input pulse of zero line width.*

Solution.
$$\Delta = (n_1^2 - n_2^2) / (2 \times n_1^2)$$
$$= (1.4^2 - 1.35^2) / (2 \times 1.4^2)$$
$$= (1.96 - 1.82) / (2 \times 1.96) = 0.0357$$

The total dispersion for 500 m length of the fibre is

$$dT = (L \times n_1/c) \times \Delta$$
$$= [(500 \times 1.4) / (3 \times 10^8)] \times 0.0357 = \mathbf{83.3\ ns}$$

38.12. Intramodal Dispersion

It is due to the fact that the light signal propagating in fibre does not consist of a single frequency but a group of frequencies. It is also called chromatic dispersion. It is related to the line width of the light pulse. It is often expressed in terms of picoseconds per kilometre per nanometre of line width.

38.13. Fibre Characteristics

The characteristics of an optical fibre depend on the type of material used and its physical shape and size. Things such as glass composition, diameter of the fibre, the variations of the refractive index within the fibre, imperfections and small bends (called microbends), directly affect its performance. However, while discussing fibre characteristics, we will pay attention only to different types of fibre losses.

38.14. Fibre Losses

The three major causes of light losses in a fibre are as under :

***(i)* Material Loss.** It is due to the absorption of light by the fibre material. It includes absorption due to light interacting with the molecular structure of the material as well as loss due to material impurities. The loss due to the atomic structure of the material itself is relatively small. Losses due to impurities can be reduced by better manufacturing.

***(ii)* Light Scattering.** Light is scattered by the molecules of the material due to structural imperfections and impurities. The scattered light does not propagate down the fibre; it is lost.

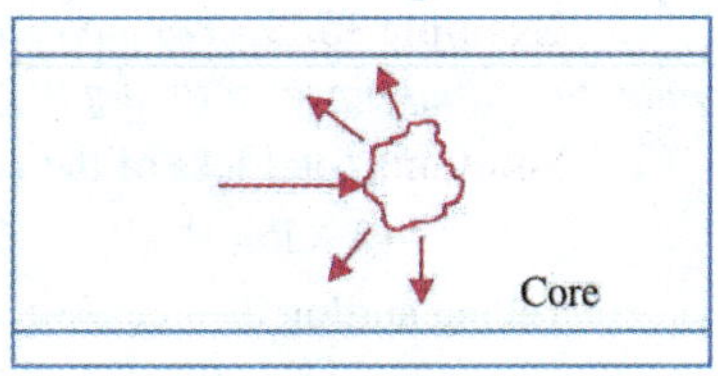

Fig. 38.10

As shown in Fig. 38.10, when light is scattered by an obstruction, it produces power loss. In optical fibres, obstruction refers to density variations in the material which produce changes in the refractive index. The index variations behave like point sources scattering light in all directions. Scattering can also be due to macrobends *i.e.,* fibre deformation. This type of loss has been made negligible with the help of improved manufacturing techniques.

***(iii)* Waveguide and Bend Losses.** These are caused by imperfections and deformations of the fibre structure which cause radiation of light away from the fibre. Fig. 38.11 (*a*) shows radiation loss caused by a change in diameter. As seen in the absence of deformation, the light ray would have been confined inside the fibre. Very small changes in the core size (*i.e.,* microbends) also cause radiation of light as well as backscatter as shown in Fig. 38.11 (*b*). Obviously, in both instances, the loss results from scattering of light.

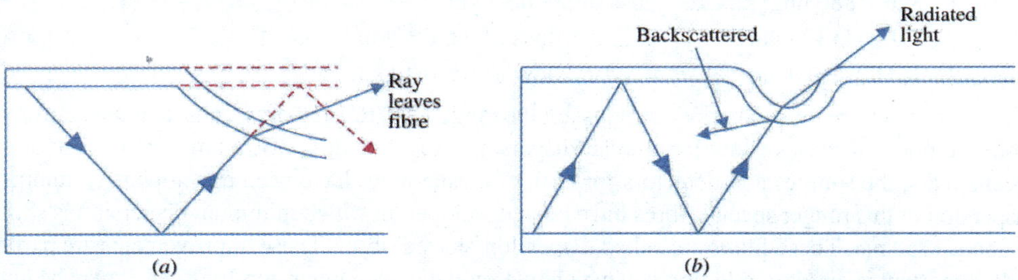

Fig. 38.11

All the above losses are dependent on the wavelength of light used. By carefully choosing the operating wavelength, the above losses can be minimized.

38.15. Calculation of Losses

As shown in the Fig. 38.12, if P_i is the power input to the fibre and P_o the power output, the fibre loss is given by the ratio P_o / P_i. The decibel loss is given by

loss = 10 log P_o / P_i decibel

Since the loss increases with fibre length, the loss in usually given in decibels per kilometre (dB/km).

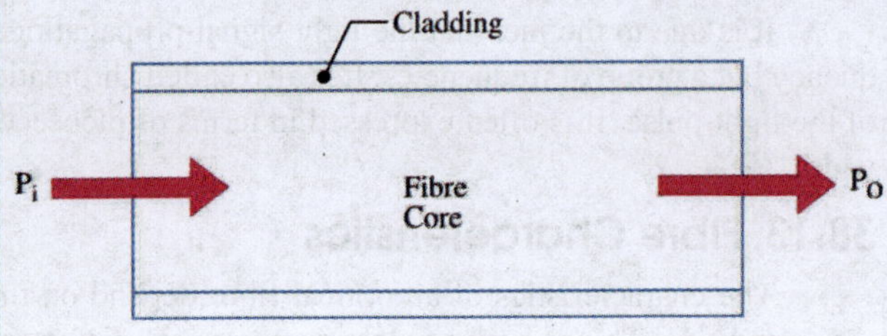

Fig. 38.12

Example 38.6. *An optical fibre of length 150 m has input power of 10 μW and output power of 9 μW. Compute the loss in dB/km.*

Solution. Decibel loss = 10 × log P_o / P_i

$$= 10 \times \log 9 / 10 = -0.458 \text{ dB}$$

Since this loss is for 150 m = 0.15 km, hence loss for 1 km is

$$= -0.458 \times (1/0.15) = \mathbf{-3.053 \text{ dB/km}}$$

The negetive sign denotes a loss.

Example 38.7. *A communication system uses a 10 km fibre having a loss of 2.5 dB/km. Compute the output power if the input power is 500 μW.*

Solution. Total loss of the fibre = 10 × (–2.5) = –25 dB

$$\therefore \quad 10 \times \log P_o / P_i = -25 \text{ or } \log P_o / P_i = -2.5$$

Taking antilog of – 2.5, we get

$$P_o / P_i = 0.00316 \text{ or } P_o / 500 = 0.00316$$

$$\therefore \quad P_o = 500 \times 0.00316 = \mathbf{1.58\ \mu W}$$

38.16. Choice of Wavelength

In the early stages of fibre optic technology, the wavelength of 820 nm was used mostly because both sources and detectors were easily available for this particular wavelength. For better performance of an optic fibre, that wavelength should be chosen which produces minimum losses and minimum dispersion.

Now-a-days, three low-loss **windows** have been found where losses from various causes are relatively low. The losses for high-quality glass fibres for these wavelengths are given below:

1. 820 – 880 nm .... loss of 2.2 dB / km
2. 1200 – 1300 nm .... loss of 0.6 dB/ km
3. 1550 – 1610 nm .... loss of 0.2 dB / km

As seen from the point of view of losses, the range 1550 to 1610 nm is most suitable. However, from the point of view of low intramodal dispersion, wavelength of 1300 nm is most suitable. In recent years, the sources and detectors for both of these ranges have been developed. To minimize dispersion in this range, special fibres have been developed in which minimum dispersion is shifted to about 1550 nm. These fibres are called dispersion-shifted fibres. These improvements are particularly important in single-mode fibres in which intermodal dispersion is non-existent. It may be noted that intermodal dispersion is usually much larger than intramodal dispersion.

38.17. Optical Fibre Cable

A typical glass fibre is about 100 – 250 μm in diameter. It is brittle and highly susceptible to damage such as scratches. Since the size of the fibre is small, this damage increases losses. Hence, it is not advisable to use a bare optical fibre. It has to be protected. The amount of protection a fibre needs varies from one application to another. In simple applications, a fibre protected by a thin buffer coating would be alright while a transoceanic fibre would need considerable protection during transportation, installation and operation.

The outside protection provided to an optical fibre has the following features:

1. it gives mechanical strength to the fibre structure,
2. it protects that fibre against breakage and damage,
3. it protects the fibre from excessive bending,
4. it permits easy field installation and maintenance,
5. it provides abrasion protection,
6. it protects the fibre from environmental conditions *i.e.,* it protects it from moisture and chemical contaminants.

Optical fibre cable.

Cabling improves the mechanical characteristics of a fibre without causing deterioration of its optical properties. We have single-fibre cables and multifibre cables. A multifibre cable makes better use of space than does a single-fibre cable because the fibres share common protecting and strengthening members. As the number of fibres in a cable increases, the cost per fibre decreases. Multifibre cables are ideal for trunk transmission links in which many messages travel the same route.

Fig. 38.13 show two single-fibre cables constructed in two basic ways: loose buffer and tight buffer. In a loose buffer cable [Fig. 38.13 (*a*)], the fibre lies loosely inside a surrounding plastic tube and is free of stress. The fibre can adjust itself within the tube when the cable is distorted. In this way, microbending is almost completely eliminated.

Moisture protection can be added by placing foam or jelly more inside the tube. External strengthening members have been added to the fibre cable for withstanding pulling, shearing and bending.

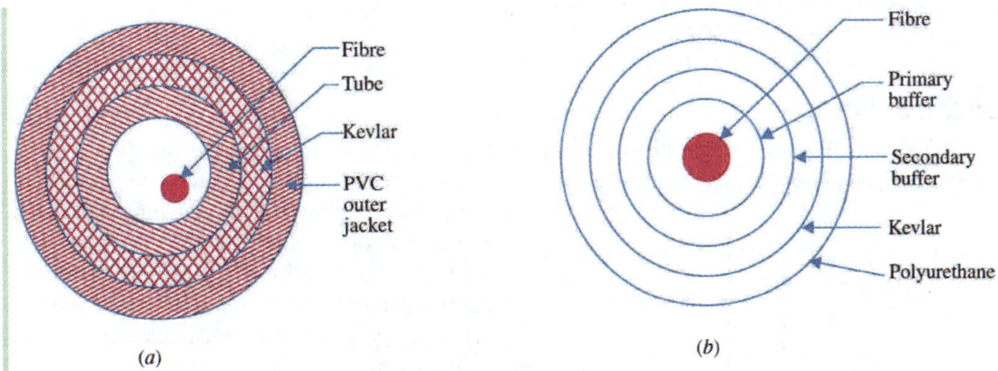

Fig. 38.13

Fig. 38.13 (*b*) presents a typical tight-fibre cable in which the protective jacket is tightly mounted around the buffer and fibre. The jacket provides cut and abrasion protection and is made of materials like polyurethane, polyethylene and polyvinyl chloride (PVC).

38.18. Multifibre Cable

Such a cable contains many fibres much like the conventional multiwire copper cable. Fig. 38.14, shows a number of loose buffer (or loose tube) cables combined into a multifibre cable. Each time, is mechanically isolated by spiralling loosely inside an oversized buffer jacket. This permits considerable expansion and contraction of the cable due to temperature changes and minimizes chance of failure from stress fatigue. Fig. 38.15 shows a multifibre cable centralized strengthening members and an armoured sheath. The fibres are individually buffered and strengthened. A corrugated aluminum sheath provides resistance to crushing forces and to water seepage.

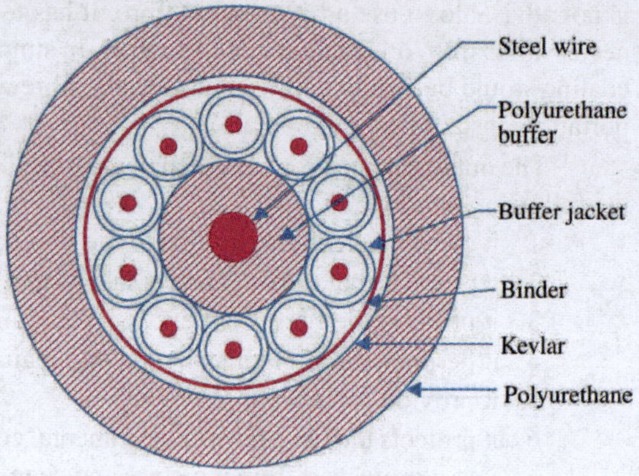

Fig. 38.14

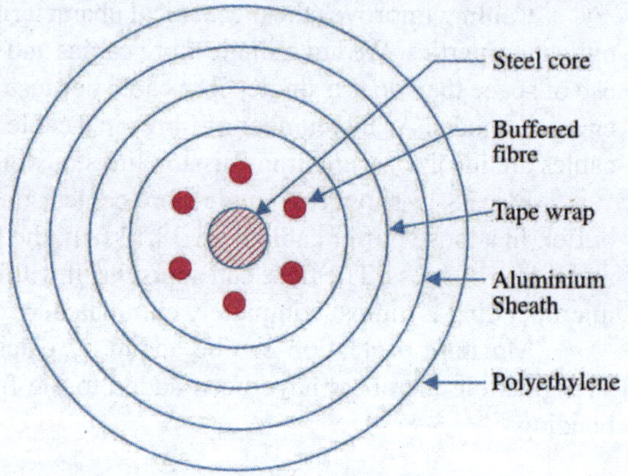

Fig. 38.15

38.19. Splicing and Connectors

Splicing means making permanent fibre joints either for repairing a broken connection or extending an optical link. In this way, the two fibre ends are attached together permanently. When a metallic wire breaks, the two ends of the broken wire can be easily spliced either by twisting the two broken edges or by melting the solder. The losses in solder joints are negligibly small. However, splicing of optic fibres has to be done very carefully otherwise in case of misalignment, light will escape at the junction and result in power loss.

Connectors provide easy connect-disconnected of two fibre ends which must come together as close as possible without damaging the fibres. The mounting of the connectors into the fibre requires special care and precision. The connectors can be mated and unmated repeatedly and rather easily.

38.20. Splicing

The two fibre edges can be connected by fusion splice which fuses the two fibre ends by melting the glass or by mechanical splice which mechanically lines up the two ends and permanently positions them there. Index-matching gel is added at the junction to eliminate any variations in refractive index from fibre to air to fibre. The gel is transparent and has the same refractive index as the fibre core. Its presence decreases fibre-separation losses.

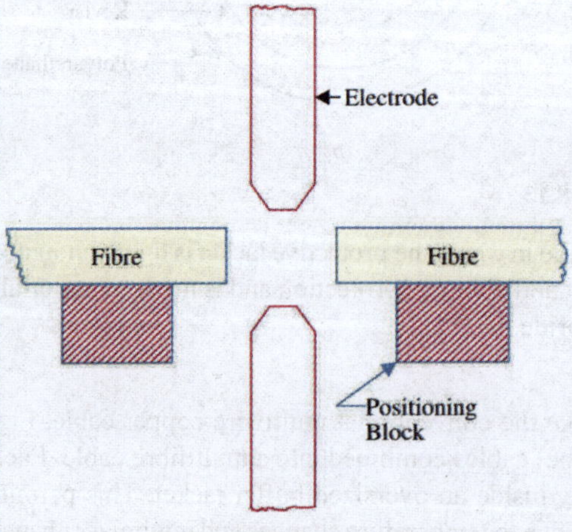

Fig. 38.16

38.21. Fusion Splices

Fusion splices are produced by welding two glass fibres as shown in Fig. 38.16. First, the edges are aligned and, after that, appropriate butting pressure is applied. Then, heat is applied usually with the help of an electric arc. Once the fibres are fused, they become 'one'. The splice is usually covered by a plastic jacket in order to protect it from damage in handling. The loss introduced by a fusion splice is about 0.1 dB.

38.22. Mechanical Splices

Mechanical splicing consists of aligning the two fibre edges and locking them in place with the help of various positioning devices and optical cement (Fig. 38.17). A number of mechanical splice configurations are shown in Fig. 38.18. Their purpose is to align the two fibre edges for splicing.

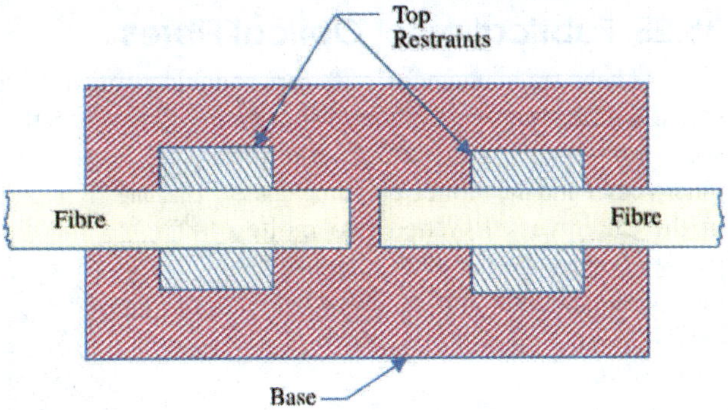

Fig. 38.17

The V-groove is probably the simplest mechanical splice. The bared fibres to be joined are placed in the groove with proper angular alignment. The two fibres can slide in the groove until they touch each other. They are then epoxied permanently into position. If the epoxy is index-matched to the fibre, even small gaps can be tolerated with little loss. The precision sleeve has a central hole just large enough for insertion of the cladded fibre. An index-matching epoxy or gel is applied to the fibre ends before insertion into the sleeve.

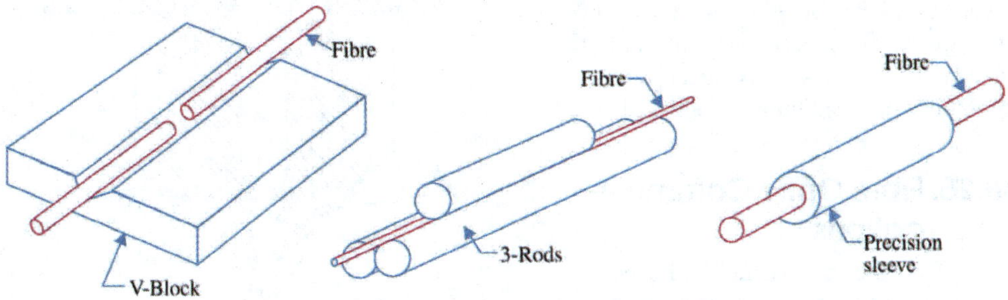

Fig. 38.18

Three precision glass or metal rods can be positioned to align the two fibres. Rod diameters are chosen so that the hole formed at their junction is just large enough to accept the fibre. Index-matching epoxy is applied to the fibres which are then inserted into the hole until they touch. A heat-shrinkable sleeve is placed over the assembly. When heat is applied, it secures the rods and squeezes against the fibre.

38.23. Connectors

A large variety of single-fibre connectors are available ranging from plastic snap-on to precision steel screw-on connectors. Similarly, there is a large variety of multifibre connectors. The single-mode fibre connectors require a much higher degree of precision than do multimode connectors because, in their case, even a very small misalignment can cause substantial loss.

38.24. Connection Losses

The optical losses in a fibre-to-fibre connection splice or by connector are of two kinds : misalignment losses and losses resulting from mismatches in fibre characteristics such as core area, numerical aperture and core profile. Single-mode fibres are much more sensitive to misalignment and fibres mismatches than multimode fibres.

Most connectors are designed to produce a butt joint / placing the fibre ends as close together as possible. Butt designs include the straight-sleeve, tapered-sleeve and overlap connectors.

38.25. Fabrication of Optical Fibres

There are many techniques for producing optical fibres but we will describe only one method called double crucible method for producing fibres directly. This method is shown in Fig. 38.19. As seen, molten core glass is placed in the inner vessel and the molten cladding-glass in the outer vessel. When two molten glasses come together at the base of the outer container, they form a glass-cladded core. This molten mixture is then pulled into a fibre.

At first glance, it might appear that this technique can produce only step-index fibres. But it is not true. Graded fibres can be produced by allowing the core and cladding glasses to interdiffuse after they come together. This diffusion causes a gradual change of refractive index between that of the core glass and the cladding glass. Glass can be continually added to the crucibles making it possible to obtain long continuous lengths of fibre.

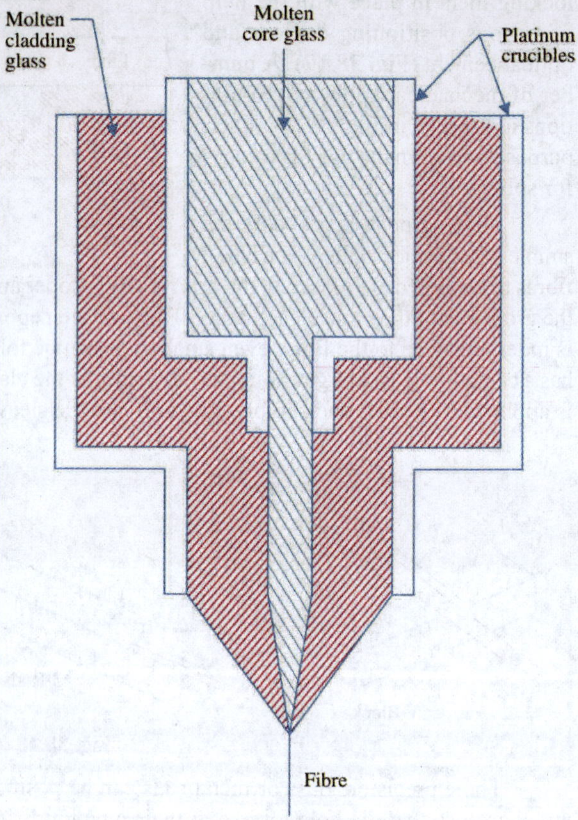

Fig. 38.19

38.26. Fibre Optic Communications

The fibre optic technology has been developed mostly for use in communication. A communication system depends on two technologies (*i*) fibre optic technology which involves the light source and detectors and the fibre itself and (*ii*) the communication technology which relates to the processing of data, modulation and demodulation etc.

In most of the fibre optic communication systems, information is carried as intensity modulated light. Fibre optic communication, particularly in voice-telephone application, is mostly digital and the transmitted light intensity is either high (ON) or low (OFF) because only two signal levels are recognized. Noise immunity is much higher for digital transmission than for analog transmission because the detecting circuit needs only to recognize two levels that are usually well separated. Also, the system does not have to be fast enough or to have short response time.

Data rate (bits per second) is the rate of information transfer and not necessarily the number of pulses or transitions per second. The number of signal transitions per second is called the baud rate which is characteristic of a communication system. Telephone line systems operate at 9600 Bd or lower whereas very high-speed systems operate at 2 GBd or more. For binary systems, baud and bit rates are often interchanged.

38.27. Advantages of Optic Fibres

The optic fibres are made of either glass or transparent plastic. The basic material for glass fibre is silicon dioxide (SiO_2) which is cheap and plentiful. Transparent plastic is also cheaply and readily available. The optic fibres have greater information carrying capacities than metallic conductors. In the case of telephone links, the cost comparison between optic fibres and metallic wires is made on the basis on cost per metre per telephone channel rather than just cost per metre. Economic comparison should also include the cost of installation, operation and maintenance. Since fibres are lighter and smaller, hence, for long paths, fibre cables are cheaper to transport and easier to install than metal cables. For example, take the case of a fibre cable which has fibre of diameter 125 μm and closed in a plastic sheath with an outer diameter of 2.5mm. Its weight is 6 kg/km and the loss is 5 dB/km. A comparable metallic coaxial cable has an outer diameter of 28.5 mm and a weight of 1100 kg/km. Its attenuation is 22.5 dB/km for a signal of 100 MHz.

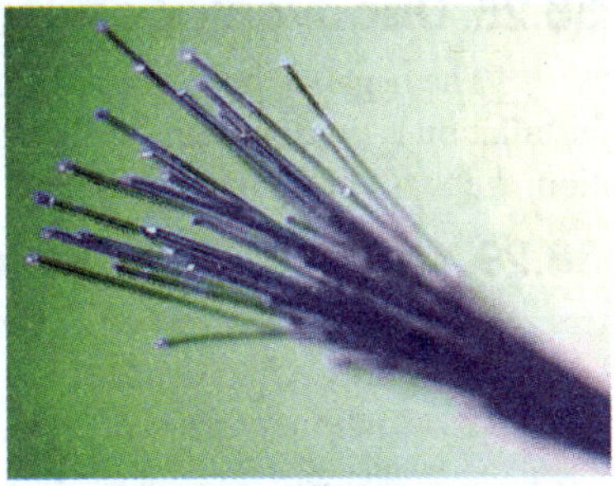

Fibre Optic communication needs fast devices with good llight conversion efficiencies

Fibres and fibre cables are very strong and flexible. Some fibres are so slender that they do not break when wrapped around curves of only a few centimeter radius.

The addition of a plastic sheath increases the tensile strength of a fibre transmission line.

One of the most important advantages of fibres is that they can carry large amount of information in either digital or analog form.

An optic fibre is well protected from external interference and coupling with other communication channels whether they are electric or optical. It is due to the fact that an optical fibre made of either glass or plastic is an insulator. No electric current can flow through it either due to the transmitted signal or due to the external radiations striking it. Moreover, the light wave within the optic fibre is trapped due to internal reflection so that none can leak out during transmission to interfere with signals in other fibres. Similarly, no light signal can couple into the fibre from outside. Hence, fibres have excellent rejection of radio frequency interference and of electromagnetic interference. Hence, radio and television station, radar and other signals cannot introduce any interference in the fibre. Similarly, electromagnetic interference caused by lightning and sparking etc., does not effect the fibres. Hence, optic fibres excel at rejecting externally caused background noise. Since a fibre can isolate itself from its environments, a large number of fibres can be packed together in a cable to transmit many channels of information along a single patch. No crosstalk occurs.

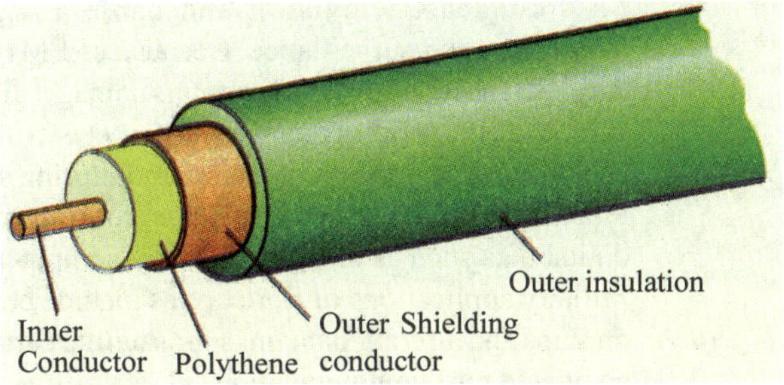

The structure of a coaxial cable which carries radio waves.

As compared to copper, corrosion due to water or chemicals is less for glass. Glass fibre themselves can stand high temperature before deteriorating.

Fibre offers a degree of security and privacy. Because fibres do not radiate the energy within them, it is difficult for an intruder to detect the signal being transmitted.

Fibres are available in long lengths, reducing the need for numerous splices of one kilometre is a common length but continuous strands of several kilometres have been produced.

38.28. Disadvantages

The negative point of fibres is the optic connector, its cost is high, its loss is also high and its installation is time consuming. The reasons for this situation are well understood. For a good connection, the two fibres must be aligned very precisely.

38.29. Application of Fibre Optic Communication

1. The small size and large information-carrying capacity of optic fibres is much high as compared to copper twisted-pair cables in telephone systems.
2. Continuous passive links (no repeaters) more than 100 km long have been produced. With repeaters, messages over thousands of kilometres of fibre can be sent. Because of low losses, separation between repeaters in a fibre system is greater than in a coaxial link.
3. Because repeaters can be spaced very far apart in fibre systems, underwater fibre links have been designed to span the oceans. The low weight of fibre cables as compared to coaxial lines gives them distinct advantages for submerged cable applications because of the relative ease of transporting and laying the fibres.
4. Optic fibre communications are compatible with electrified railways because they are not effected by EMI. Similarly fibres can be placed near high-voltage power lines without adverse effects whereas wire systems would be noisy. Optic cables can be suspended directly from power line towers or poles if clearance space permits and if the load can be tolerated. Alternatively, the fibre can be embedded in a wire cable.
5. Fibres are used in applications that are primarily video *i.e.*, broadcast television and cable T.V., remote monitoring and surveillance. Because of EMI rejection and low susceptibility to lightning damage, fibre optic video transmission successfully competes with coaxial cable for surveillance and remote monitoring systems.
6. Fibre systems are particularly suited for transmission of digital data such as that generated by computers.
7. Military applications of fibre optics include communications, command and control links on ships and aircraft, data links for satellite earth stations and transmission lines for tactical command post communication.

The satellite receiving dish at Telecom's station where international telephone calls are relayed.

Summarizing the above, we can state that important fibre characteristics are low weight, small size, EMI rejection and no signal radiation. On aircrafts and ships, the reduced shock, fire and spark hazards are significant assets.

SELF EXAMINATION QUESTIONS

A. Fill in the blanks with most appropriate word(s) or numerical value(s)

1. An optical fibre is a .... rod usually made of glass or clear plastic.
2. A modern optic fibre consists of an optical rod core coated with a ....
3. In an optical fibre, the refractive index of cladding material is .... than that of the fibre core material.
4. Propagation of light through a fibre core depends on the phenomenon of .... internal reflection.
5. The numerical aperture of an optical fibre depends on the core refractive index and sine of the .... angle.

6. Usually, the losses in an optical fibre are expressed in .... per kilometre.
7. Microbends in an optical fibre lead to .... loss.
8. For overall performance in terms of data rate and attenuation, the single-mode step-index fibre has the .... characteristics.
9. As compared to an all glass fibre, an all-plastic fibre has .... transmission characteristics.
10. Fibre optic communication particularly in voice-telephone applications is largely.................... .
11. A typical glass fibre is about 100-250 .... in diameter.
12. On has to .... the two fibre edges to repair a broken connection.
13. The losses introduced by a fusion splice are about .... dB.
14. Optical fibres either of glass or plastic are ....
15. Fibre systems are particularly suited for transmission of .... data such as that generated by computers.
16. Splices are generally .... fibre joints.
17. Fibre connectors are designed to produce a .... joint.

B. Answer True or False

1. The core of an optical fibre and its cladding have the same optical characteristics.
2. Cladding of a glass fibre reduces energy loss from it.
3. The fibre is lighter and less bulky than an equivalent copper cable.
4. In a fibre optic communication system, transmission is more secure and private.
5. Total internal reflection takes place when light travels from a medium of lower refractive index to a medium of higher refractive index.
6. In a step-index fibre, the refractive index changes in step fashion from the centre of the fibre to its cladding.
7. As compared to single-mode fibres, multimode fibres have the worst performance characteristics.
8. All-plastic fibres have lower bandwidth than all-glass fibres.
9. A typical glass-fibre is about 100-250 μm in diameter.
10. Basic splicing techniques include fusing the two fibres or bonding them together in an alignment structure.
11. Splices are generally permanent fibre joints whereas connectors can be mated and unmated repeatedly.
12. The basic material for glass fibres is silicon dioxide.
13. Optical fibres are good conductors of electricity.
14. Fibre optic transmission is immune to electromagnetic interference.
15. Optical fibres do not radiate energy carried by them.
16. Fibre optic systems are particularly suited for transmission of analog data.
17. Optical fibres are coated with a buffer immediately after being drawn.
18. Multifibre cables are not suited for trunk transmission links in which many messages travel the same route.
19. The performance of an optical fibre is measured in terms of its attenuation, pulse spreading and numerical aperture.
20. The wavelength range 1550-1610 nm produces minimum losses in an optical fibre.

C. Multiple Choice Items

1. Till recently, electronic communication was accomplished by sending electrical signals through copper cables. However, optical fibre communication uses .... signals instead of electrical signals.
   (*a*) light (*b*) analog
   (*c*) digital (*d*) unmodulated
2. An optical fibre is a transparent rod usually made of
   (*a*) clear plastic (*b*) glass
   (*c*) copper (*d*) either (*a*) or (*b*)
3. Optical fibres, whether made of glass or plastic, are
   (*a*) fragile (b) brittle
   (*c*) insulators (d) leaky
4. An optical fibre excels at rejecting
   (*a*) electromagnetic interference
   (*b*) radio-frequency interference
   (*c*) crosstalk
   (*d*) all of the above
5. Fibre systems are particularly suited for transmission of .... data such as that generated by computers.
   (*a*) digital (*b*) analog
   (*c*) binary (*d*) modulated
6. The velocity of light is independent of
   (*a*) wavelength

(*b*) medium refractive index
(*c*) frequency
(*d*) all of the above

7. Total internal reflection can take place when light travels from
(*a*) air to glass (*b*) glass to air
(*c*) rarer to a denser medium
(*d*) denser to a rarer medium

8. In an optic fibre, the propagation angle of light must be equal to or less than the .... angle.
(*a*) acceptance (*b*) incident
(*c*) critical (*d*) refraction

9. The numerical aperture of an optical fibre depends on
(*a*) core refractive index
(*b*) critical angle
(*c*) both of the above
(*d*) none of the above

10. The main energy losses in an optical fibre consist of
(*a*) material loss
(*b*) loss due to light scattering
(*c*) bend loss
(*d*) all of the above

11. Optical fibres are coated with .... immediately after being drawn.
(*a*) buffer (*b*) plastic
(*c*) kevlar (*d*) polyurethene

12. The structure of an optical fibre consists of
(*a*) multimode step-index
(*b*) graded index
(*c*) single-mode step-index
(*d*) all of the above

13. Very high-speed telephone line systems operate at .... gigabaud (GBd) or more.
(*a*) 2 (*b*) 4
(*c*) 6 (*d*) 8

14. Amongst different types of fibres, single-mode step-index fibre has
(*a*) highest data rate
(*b*) lowest attenuation
(*c*) highest thickness
(*d*) both (*a*) and (*b*)

15. A plastic-clad fibre has similar characteristics as an all-glass fibre except that it is more sensitive to .... damage.
(*a*) corrosion (*b*) abrasive
(*c*) chemical (*d*) moisture

16. An all-plastic fibre is strong mechanically as compared to an all-glass fibre but has
(*a*) higher losses
(*b*) lower bandwidth
(*c*) lower tensile strength
(*d*) both (*a*) and (*b*)

17. High-quality glass fibres have least losses in the light wavelength range of .... nanometre (nm)
(*a*) 820-880 (*b*) 1200-1320
(*c*) 1550-1610 (*d*) 1620 -1710

18. The core diameter of a single-mode step-index fibre lies in the range of .... micrometre (μm).
(*a*) 5-10 (*b*) 10-20
(*c*) 20-30 (*d*) 30-50

19. Fibre-to-fibre connections are made by
(*a*) splicing (*b*) connectors
(*c*) twisting (*d*) either (*a*) or (*b*)

20. Splices are generally .... fibre joints.
(*a*) temporary (*b*) butt
(*c*) permanent (*d*) welded

21. In fibres, dispersion means pulse ....
(*a*) narrowing (*b*) broadening
(*c*) distortion (*d*) rise-time

22. Fibre dispersion is usually defined in terms of time measured in
(*a*) nanoseconds (*b*) picoseconds
(*c*) both of the above
(*d*) none of the above

23. Intermodal dispersion results from the fact that the light waves propagate through a fibre
(*a*) in different modes
(*b*) along different routes
(*c*) as a group of frequencies
(*d*) with same velocity

24. Intermodal dispersion in a fibre depends on the refractive index of
(*a*) core (*b*) cladding
(*c*) surrounding medium
(*d*) both (*a*) and (*b*)
(*e*) none of the above

ANSWERS

A. Fill in the blanks

1. transparent	2. cladding	3. less	4. total	5. critical
6. decibel	7. energy	8. best	9. poorer	10. digital
11. micrometre	12. splice	13. 0.1	14. insulators	15. digital
16. permanent	17. butt			

B. True or False

1. F	2. T	3. T	4. T	5. F	6. T
7. T	8. T	9. F	10. T	11. T	12. T
13. F	14. T	15. T	16. F	17. T	18. F
19. T	20. T				

C. Multiple Choice Items

1. *a*	2. *d*	3. *c*	4. *d*	5. *a*	6. *c*
7. *d*	8. *c*	9. *c*	10. *d*	11. *a*	12. *d*
13. *a*	14. *d*	15. *b*	16. *d*	17. *c*	18. *a*
19. *d*	20. *c*	21. *b*	22. *c*	23. *a*	24. *d*

Index

A

Absolute permeability, 88
— permittivity, 62
Absorptive laws, 618
Accumulator, 76
AC generator, 100
— meter, 665, 669
Active region, 316
Adder, 566, 608
— block diagram, 609
— full, 609
— half, 609
— OP- AMP, 566
— parallel binary, 610
Addition, binary, 578
— logic, 597
Adjacent channel, 535
Advantages of optic fibres, 698
Alloy junction, 201
Alpha, cut-off frequency, 438
— transistor, 280
Alternating current, 100
— average value, 103
— crest factor, 99
— effective value, 102
— form factor, 103
— frequency, 101
— generator, 100
— nonsinusoidal, 110
— peak-to-peak value, 102
— peak value, 102
— phase, 104
— rms value, 102
— wave form, 99
Alternating current circuits, 114
— impedance, 115
— parallel type, 143
— receiver, 543
— transmitter, 531
Ammeter, ac, 659
— dc, 659
Amplifiers, 383
— cascaded, 402, 386
— CB, 372
— CC, 376
— CE, 373
— classes of, 379
— differential, 416, 425
— distortion, 396
— feedback, 442
— FET, 462
— gain, 372, 374, 377
— impedance-coupled, 409
— coupling, 404
— cut-off frequencies, 438
— multistage, 402
— operational, 562
— push-pull, 391
— RC coupled, 404
— single stage, 370
— small signal, 353
— transformer-coupled, 411
Amplitude, 101
— distortion, 398
— modulated wave, 527-528
— modulation, 522-523, 530, 531
Associative law, 549
Astable multivibrator, 509
Atomic mass number, 155
— number, 156
— structure, 156
Attenuation, 147
Autotransformer, 94
Avalanche breakdown, 183

— photodiode, 240
Average value, 103
Ayrton shunt, 660

B

Balanced amplifiers, 391
Bands in solids, 162
Band pass filter, 149
— stop filter, 150
Bandwidth, 120, 446
Barkhausen criterion, 490
Barrier potential, 179
Base bias, 320
— voltage divider, 323
— with collector feedback, 322
Battery, lead-acid, 80
— nickel cadmium, 78
— primary, 76
— rating, 80
— secondary, 76
— silver-oxide, 80
Beta, cut-off frequency, 329
— rule, 299
— sensitivity, 320
— transistor, 280
Biasing, transistor, 275
— FET, 461
— rules, 276
Binary numbers, 573
— addition, 578
— complement, 580
— decimal conversion, 574
— division, 584
— hex conversion, 589
— multiplication, 583
— octal conversion, 587
— subtraction, 579
Bipolar transistor, 274
Bistable multivibrator, 509
Bleeder resistor, 256
Bohr atom, 155
Bonds in solids, 163
Boolean, algebra, 617
— laws, 617
— identities, 600
Blocking oscillator, 515
Bridge rectifier, 249
Breakdown, avalanche, 183
— voltage, 190
— Zener, 183
Bubbled gates, 603
Buffer amplifier, 565
Bulk resistance, 190

C

Capacitor, 60
— ceramic, 65
— charging of, 72
— diode, 201
— discharging, 131
— electrolytic, 65
— energy stored, 69
— leakage resistance, 68
— mica, 65
— microphone, 649
— paper, 64
— parallel, 69
— polyester, 65
— series, 68
— time constant, 72
— variable, 66
— voltage rating, 67
Capacitance, 61
— distributed, 650
— farad, unit, 62
— stray, 68
Capacitive coupling, 413
— reactance, 73
Carbon microphone, 650
— resistors, 49
Carrier, charge, 171
— power, 528
— swing, 534
— wave, 520
Carry end-around, 581
Cascaded amplifiers, 402, 438
Cathode-ray oscilloscope, 671
— applications, 679
— digital read-out, 678
— dual trace, 677

— internal synchronization, 676
— non triggered type, 676
— sampling type, 678
— storage type, 677
— triggered type, 676
CB amplifier, 337
— circuit, 279, 335, 353, 360
— formulas, 295
CC amplifier, 350
— circuit, 282, 354
— formulas, 298
CE amplifier, 344
— circuit, 281, 335, 354, 360
— formulas, 296
Cell, alkaline, 77, 78
— carbon-zinc, 77, 78
— manganese alkaline, 78
— nickle-cadmium, 78
— mercury, 79
— photovoltaic, 81
— primary, 76
— secondary, 76
— silver oxide, 80
— solar, 81
Ceramic capacitors, 65
— microphone, 650
Channel, FET, 453
— width, 526
Chassis, ground, 2
Chip, 555
Clampers, 202, 205
Classes of amplifiers, 379
— A, 379, 381
— AB, 380
— B, 379
— C, 379
Classification of optic fibres, 686
Clippers, 202
CMOS logic circuit, 632
Coefficient, coupling, 56, 146
— termperature resistance, 644
Colpitts oscillator, 494
Common-mode rejection ratio, 424
Commutative law, 617
Complement, binary, 580
Complementary symmetry, 417
Complex wave, 110
Conduction band, 164
Constant-k filter, 148
Constant-voltage source, 37
Controlled rectification, 263
Converter, 542
CMOS IC chip, 631
Coupled circuits, 145
Coupling, 145, 403
— capacitor, 117
— coefficient, 56
— direct, 404
— impedance, 404
— indirect, 146
— RC, 404
— transformer, 404
Covalent bonds, 163
Critical Angle, 689
Cross-over distortion, 393
CRT, 673
Crystal microphone, 649
— oscillator, 497, 498
Current, ac, 102
— hole, 173
— regulation, 259
Cut-off,
— clipping, 318
— frequency, 147
— point, 302
— region, 291

D

D' Arsonval movement, 656
Darlington pair, 419
— and emitter follower, 350
Dead zone, 393
Decibel, system, 430, 433
— unit, 434
Decimal, binary conversion, 576
— octal conversion, 586
Deflection, CRT, 675
Degaussing, 87
Demodulation, 538
De Morgan's theorem, 620
Demorganisation, 621
Depletion region, 178
— MOSFET, 464, 465

Detection, 538
Detector, 539
— AM, 538
— amplifier, 540
— FM, 540
Deviation ratio, 535
Diac, 482
Differential amplifier, 422, 425
Differentiator, 569
Diffused junction, 201
Diffusion capacitance, 184
— AND gate, 598, 599
— binary numbers, 595
— ICs, 554
— instruments, 654
— multimeter, 664
— NAND gate, 606
— OR gate, 595-598
— voltmeter, 669
— XNOR gate, 606
— XOR gate, 606
Digital meters, 669
Diode, detector, 539
— fabrication, 200
— rectifier, 244
— testing, 193
Diodes, avalanche, 240
— Esaki, 219
— light emitting, 243
— photo, 239
— PIN, 223
— Schottky, 225
— semiconductor, 188
— tunnel, 219
— varactor, 222
— Zener, 212
Direct-coupled amplifier, 404
Direct-current circuits, 11
— Kirchhoff's laws, 30
— meters, 659
— networks, 35
— Ohm's law, 3
— parallel, 18
Dispersion, 691
Distortion, amplitude, 397
— cross-over, 393
— frequency, 398
— harmonic, 397
— intermodulation, 398
— phase, 398
Distributive law, 618
Donor elements, 169
Doping, semiconductor, 168
Double dadd method, 575
— side-band(DSB), 530
Drift current, 171
DSB—SC, 530
DTL—logic family, 627
Duals, 622

E

Eccles-Jordan MV, 509
ECL logic family, 629
Effective value, 102
Efficiency, 246
Electret microphone, 651
Electrolytic capacitors, 65
Electromagnetism, 86
Electron, distribution, 157
— orbits, 156
— suborbits, 157
Emmitter follower, 350
End-around carry, 581
Energy band diagram, 169
— levels, 161
— orbital, 159
Energy, in capacitor, 69
— in inductor, 59,
Envelope detection, 539
Epitaxial growth, 556
Equivalent circuit, 198, 199, 355, 544, 350
— switching circuits, 618
Esaki diode, 219
Exclusive OR gate, 598
Extrinsic semiconductor, 168

F

Fabrication, diode, 200, 558
— optic fibres, 698
Fan-in, 626
— out, 626
Farad, unit, 62

Feedback, amplifiers, 442
— current-series, 448
— current-shunt, 448
— negative, 442, 443, 447
— oscillator, 489
— positive, 442, 443
— stability, 443
— voltage-series, 447
— voltage-shunt, 448
Fermi level, 170
Ferrite beads, 86
— core, 86
Ferromagnetic materials, 86
FET, 453, 462, 469, 492, 494, 496
— oscillator, 496
— voltmeter, 669
Fibre characteristics
—losses, 693
—optics, 685-703
Field effect transistors, 453
— characteristics, 457
— transconductance, 460
Filters, rectifier, 251
Filters, band-pass, 149
— band-stop, 150
— costant-k, 148
— high pass, 149
— low pass, 148
— L-type, 148
— π-type, 150, 255
— T-type, 150
Flasher, 133
Flat package, 557
Flip-flop, 513
Flowmeter, 639
FM, defined, 522
— detection, 540
— receiver, 544
Form factor, 103, 244
Forward, current, 191
— voltage, 191
Floating zono technique, 200
Flux density, 87
— magnetic, 87
FM, 533
— comparison with AM, 544
— deviation ratio, 535
— detection, 540
— frequency deviation, 534
— modulation index, 535, 537
— per cent modulation, 535
— receiver, 544
— sidebands, 536
Free-running multivibrator, 509
Frequency, audio, 102
— conversion, 542
— cut-off, 148
— defined, 101
— deviation, 534
— distortion, 398
— harmonic, 398
— modulation, 533
— radio, 520
— response, 413
— resting, 534
— stability, 489
— spectrum, 520
Full adder, 609
— subtractor, 611
— wave rectifier, 247
Fusible resistors, 52

G

Gain, amplifier, 371, 374, 377
— bandwidth product, 439
— stability, 445
Gate, logic, 594-612
— AND, 598, 599
— bubbled, 603
— exclusive OR, 598
— NAND, 606
— NOR, 604
— NOT, 601
— OR, 595-598
— XNOR, 606
— XOR, 606
Gauge factor, 642
Ground chassis, 2
Grown junction, 201

H

Half adder, 609
— power frequencies, 120
— subtractor, 611
— wave rectifier, 244

Harmonics, 112
Hartley oscillator, 493
— FET, 494
Heterodyne principle, 543
Hexadecimal number system, 589
— binary conversion, 590
High pass filter, 149
Hole flow, 165, 173
— formation, 165
Hot—carrier diode, 225
h—parameters, 357
Hybrid equivalent circuit, 361
— parameters, 357
Hysteresis loss, 89

I

ICs, 549 - 570
Ideal diode, 193
I2L logic family, 630
Impedance, coil, 59
— matching, 95
— stepping up, 352
Inclusive OR gate, 598
Inductance, 55
— energy stored in, 59
— mutual, 56, 58
— series combination, 58
— stray, 58
— time constant, 128, 130
— troubles, 59
Inductive kick, 130
— reactance, 59
Inductor, 55
Instantaneous value, 102
Instruments, electronic, 654-681
— analog and digital, 654
— cathode-ray oscilloscope, 671
— digital voltmeters, 669
— essentials of, 655
— functions of, 655
— moving coil, 657
— multimeter, 664
— ohmmeter, 663
— rectifier type, 665
— versus electrical, 655
— voltmeters, 666
— VOM, 666
— VTVM, 667
Insulators, 166
Integrated circuits, 549-570
— applications, 560
— fabrication, 558
— classification of, 553
— digital, 554
— hybrid, 552
— linear, 554
— monolithic, 551, 555, 559
— terminology, 554
— thick film, 552
— thin film, 552
Integrator, OP-AMP, 567
Inter modal Dispersion, 693
Intermodulation distortion, 398
Interstage coupling, 403
Interinsic, conduction, 173
— semiconductor, 167
— stand—off ratio, 473
Ionic bonds, 163

J

JFET, 453, 457, 462
junction, PN, 177
— capacitance, 184
— resistance, 190
— voltage, 180

K

Krichhoff's laws, 30
Knee voltage, 180

L

Large signal amplifiers, 370
Latching devices, 471, 476
Lattice, crystal, 154
L-type filter, 148
Leakage current, 190
— resistance, 68
LED display, 236
Level detector, 515
Limiter, 544
Linear amplifier, 564
— distortion, 397
— IC, 554
— detector, 539

Lissajous figures, 678
Load line, 315, 327
Loading effect, voltmeter, 663
Logic, negative, 595
— families, 625-631
— positve, 595
Logic gates, 594-612
— AND, 598
— NAND, 605
— NOR, 604
— NOT, 601
— OR, 596-598
— XNOR, 606
— XOR, 598
Lower sideband, 525
— frequency, 525
Low pass filters, 148
L/R time constant, 128
LSI chip, 551
LVDT, 640

M

Magnetising force, 88
Majority carriers, 170
Mark-to-space ratio, 505
Masks, diffusion, 555
Maximum power transfer theorem, 44
MC instruments, 657
Measurement standards, 656
Mercury cell, 79
Metallic bonds, 164
Metallization, 557
Meters, analog, 654
— dc, 658
— digital, 669
— FET, 595
— VTVM, 595
Microphones, capacitor, 650
— carbon, 647
— ceramic, 650
— crystal, 649
— electret, 651
— moving coil, 649
— ribbon, 648
Miller effect, 437
Minority carriers, 170
mobile charge carriers, 171
Mobility, 172
Modulation, 521, 522
— amplifier, 532
— amplitude, 522, 530, 531
— frequency, 522, 533
— index, 523, 535
— percent, 523, 535
— phase, 522
— oscillator, 531
Monolithic ICs, 551, 555
Monostable multivibrator, 509
MOS IC, 560
— logic family, 631
MOSFET, 464, 465, 466, 467
— logic family, 631
Moving-coil microphone, 649
— meter, 657
Multifibre cable, 695
Multirange meter, 660
Multimode G.I., 687
— S.I. fibre, 687
Multistage amplifiers, 402
Multivibrators, 508
— astable, 509
— bistable, 513
— monostable, 511
— one shot, 509
Mutual inductance, 56

N

NAND gate, 605
Negative, logic, 595
— resistance, 220
— scaler, 564
Network theorems, 35
Nickel-cadmium cell, 78
NMOS circuit, 632
Noise, 399
— factor, 399
— margin, 626
Non-linear distrotion, 397
— inverting amplifier, 564
— resistor, 5
— saturated logic, 626
Non-sinusoidal oscillators, 504-515
— weveform, 504
Norton's theorem, 41

NOT gate, 601
— function, 601
N-type semiconductor, 168
Number systems, 572-591
— binary, 573
— decimal, 573
— hexadecimal, 589
— octal, 585
Numerical Aperture, 690

O

Octal number system, 585
— advantages of, 588
— binary conversion, 588
— decimal conversion, 586
Ohm's law, ac circuit, 115
— dc circuits, 3
— magnetic circuits, 89
Ohmmeter, 663
One's complement, 581
OP-AMP, 561, 562, 564
Operation amplifier, 562
— adder, 566
— application, 564
— comparator, 570
— differentiator, 569
— gain, 434
— ideal, 562
— integrator, 567
— subtractor, 566
— summer, 566
— virtual ground, 563
Optical fibre cable, 694
Orbital electrons, 156
— energy, 159
OR gate, 595, 596, 597
— exclusive, 598
— inclusive, 596
Oscillators, beat frequency, 543
— Clapp, 496
— Colpitts, 494
— crystal, 497, 499
— Hartely, 493
— Pierce, 499, 499
— unijunction, 473
— Wien bridge, 501
Oscilloscope, 671
— free running, 676
— sampling, 678
— strorage, 677
— triggered, 676

P

Parallel, binary adder, 610
Parameters, hybrid, 357
Pass-band, 147
Peak inverse voltage, 192
Peak point voltage, 474
Permeability absolute, 88
— relative, 89
Permeance. 89
Permittivity, 62
Phase angle, 104
— control, 478
— distortion, 398
— modulation, 522
— shift oscillator, 500
Photoconductive cell, 237
Photolithographic process, 556
Photodiode, 239
— avalanche, 240
— PIN, 240
— P-N, 240
Photoelectric devices, 81
— transducers, 647
Photoemissive cell, 236
Phototube, 236
Photovoltaic cell, 81
Pezoelectric effect, 497
— transducer, 641
PIN diode, 223
Pinch-off bias, 456
Pitch, 520
PIV, 192
Plastic Fibres, 687
PMOS logic circuit, 631
P—N junction, 177
— equivalent circuit, fabrication, 185
Polarity, IR drops, 12
Positive logic, 595
—scaler, 564
Potentiometers, 50
Power, ac circuits,
— efficiency, 383
— rectangle, 382
Pre-ohmic etch., 557

Pulse, duration
— repetition frequency, 505
— waveform, 505
Push pull amplifier, 391, 394, 395
P-type semiconductor, 168

Q

Quadrature detector, 541
Q—factor, 60, 116, 141
— meter, 680
— point, 318

R

Radio frequencies, 520
Radix, 573, 585
RC coupling, 404
— oscillator, 500
Receiver, FM, 544
— superhet, 543
Rectifier circuits, 244
— bridge, 249
— full-wave, 247
— half-wave, 244
Regenerative feedback, 442
Regulated power supply, 242
Regulation, 256
Regulators, current, 259
— transistor, 259
— Zener, 213, 257
Relaxation oscillators, 496, 505, 509
Reluctance, 90
Resistances, parallel, 18
— series, 11
Resistive circuits, 11
Resistivity, 173
Resistors, carbon, 48
— cermet, 49
— colour code, 52
— fusible, 52
— metal—film, 49
— power rating, 49
— tolerance, 50
— troubles, 53
— variable, 50, 47
— wirewound, 47
Resonance, curve, 119
— parallel, 122
— series, 119, 140
— sharpness, 120
Resonant frequency, 120, 140
Resting frequency, 533
Reverse saturation current, 182, 192
— voltage, 192
Ribbon microphone, 648
Ripple, dc, 243
— factor, 254
RMS value, 102
Roll-off frequencies, 435
RTD, 643
RTL logic family, 626

S

Saturation current, 182, 191
— point, 302
Saturated logic, 625
Sawtooth waveform, 505
Scale of integration, 551
Schmitt trigger, 514
Schottky barrier diode 225
SCR, 268, 476
Scribing, 557
Secondary cells, 76
Selectivity, 120, 140
Semiconductors, 166
— covalent bonds, 163
— doping, 168
— extrinsic, 168
— intrinsic, 168
— PN junction, 77
Series ac circuits, 114
— capacitances, 68
— cells, 7
— resistances,
— resonance, 12
— voltage dividers, 15
Series—parallel circuits, 21, 22, 23
Sharpness of resonance, 120
SHF band, 520
Shielding, Magnetic, 87, 94
Shockley's equation, 458
Shunts, meter, 659
Side bands, 536
— frequencies, 525
Signal-to-noise ratio, 399

Silicon controlled rectifier (SCR), 286, 476
Singal-Model, S.I. Fibre, 687
Single shot MV, 509
Sinusoidal oscillatiors, 485-503
Skin effect, 485-503
Snell's Law, 688
Solar cell, 81
Splicing, 696, 697
Squarer, 515
Square wave, 505
SSB—SC, 530
— TC, 530
SSB generation, 530
Stability factor, 320
Stabilization, bias, 320
Stop-band filter, 147
Storage capacitance, 184
Strain gauge, 641
Stray capacitance, 68
Substrate, 552
Subtractor, 566, 608
— OP-AMP, 566
Subtraction, binary, 579
Summer, OP-AMP, 566
Superheterodyne receiver, 543
Superposition theorem, 35
Switching circuits, logic, 618
Synchroniziation, 676

T

T-type filters, 150
Temperature coefficient of resistance, 227, 644
T-equivalent circuit, 353
Theft alarm, 479
Theorem, De Morgan's, 620
— maximum power transfer, 44
— Norton's, 41
— superposition, 35
— Thevenin's, 38, 324
Thermal runaway, 285
Thermistor, 227, 544
Thevenin's theorem, 38, 324
Thin—film IC, 552
Thyristor, 472
Time base, 675
— constant, 127, 128, 135
— period, 101
T-model, 353
Total Internal Refection, 688
Totem pole, 628
Transconductance, FET, 458
Transducers, 635-651, 656
— classification of, 635
Transformer, 90
— audio, 91
— coupling, 403
— impedance, 93
— impedance matching, 95
— radio frequency, 91
— power, 92
— turn ratio, 92
— tuned, 142
— working, 92
Transistors, ac/dc analysis, 320
— alpha, 279
— amplifiers, 370
— cut-off frequencies, 291
— beta, 281
— biasing, 275, 319
— CB configuration, 279, 283, 295
— CC configuration, 282, 298
— CE configuration, 281, 284, 291
— characteristics, 281
— configuration, comparison, 378
— detector, 540
— currents, 277
— fault location, 303
— FET, 453
— h-parameters, 359, 318, 362
— leakage currents, 283
— saturation, T-model, 353
— unijunction, 472
Transition capacitance, 184
Triac, 480
Trigger, Schmitt., 514
Triggering CRO, 675
Truth tables, 595, 598, 599, 605, 609, 611
TTL logic family, 628
— subfamilies, 629
Tuning, 120
— circuit, 139
— dial, 121
— ratio, 121
Tunnel diode, 219

Tweeter, 651
Two's complement, 580, 582

U

Ultrasonic transducer, 646
Unijunction oscillator, 475
— transistor, 472
Unity follower, 565
Universal gate, 537
— stabilization circuit, 305
Univibrator, 609
Unregulated power supply, 242
Upper side frequency, 525
— band, 526

V

Variable, inductor, 57
— regulation, 215
— resistor, 50
Valence band, 164
— electrons, 159
Varactors, 222
Varicap diode, 222
VCE, importance of, 300
VHF band, 520
Voltage dividers, 243, 260
— division, 24
— doubles, 261, 262
— multipliers, 260
— quadruplers, 262
— regulations, 215, 256
— regulators, 215, 257
Volt—ampere characteristics, 219
— Zener diode, 213
Voltmeter, analog, 670
— electronic, 669
— digital, 670
— loading effect, 663
— ohm-per-volt, 663
VOM, 664
VTVM, 667

W

Wafer, 556
Wire-wound resistors, 48
Wave trap, 144
Work, 6

X

XNOR gate, 606
XOR gate, 606

Y

Y—deflecting plates, 674

Z

Z-impedance, 114
Zener biasing, 213
— breakdown, 183
— current, 213
— diode, 212
Zero IR drop, 12
Zero—decibel reference, 434
— ohm adjust, 664
— reference level, 1